Tyler Cutforth
3629 Wellington Ave.
Vancouver, B. C.
Canada V5R 4Z4

434-7061

VOLUME I   GENERAL PRINCIPLES

MOLECULAR BIOLOGY OF THE GENE

FOURTH EDITION

James D. Watson   COLD SPRING HARBOR LABORATORY

Nancy H. Hopkins   MASSACHUSETTS INSTITUTE OF TECHNOLOGY

Jeffrey W. Roberts   CORNELL UNIVERSITY

Joan Argetsinger Steitz   YALE UNIVERSITY

Alan M. Weiner   YALE UNIVERSITY

The Benjamin/Cummings Publishing Company, Inc.

Menlo Park, California • Reading, Massachusetts • Don Mills, Ontario
Wokingham, U.K. • Amsterdam • Sydney • Singapore
Tokyo • Madrid • Bogota • Santiago • San Juan

Cover art is a computer-generated image of DNA interacting with the Cro repressor protein of bacteriophage λ. The image was prepared by the Graphic Systems Research Group at the IBM U.K. Scientific Centre.

Editor: Jane Reece Gillen
Production Supervisor: Karen K. Gulliver
Editorial Production Supervisor: Betsy Dilernia
Cover and Interior Designer: Gary A. Head
Contributing Designers: Detta Penna, Michael Rogondino
Copy Editor: Janet Greenblatt
Art Coordinator: Pat Waldo
Art Director and Principal Artist: Georg Klatt
Contributing Artists: Joan Carol, Cyndie Clark-Huegel, Barbara Cousins, Cecile Duray-Bito, Jack Tandy, Carol Verbeek, John and Judy Waller

Library of Congress Cataloging-in-Publication Data
Molecular biology of the gene.

  Rev. ed. of: Molecular biology of the gene / James D. Watson. 3rd ed. c1976.
  Bibliography
  Includes index.
  Contents: v. 1. General principles.
  1. Molecular biology. 2. Molecular genetics.
I. Watson, James D., 1928–   . [DNLM: 1. Cytogenetics.
2. Molecular Biology. QH 506 M7191]
QH506.M6627 1987   574.87'328   86-24500
ISBN 0-8053-9612-8

ABCDEFGHIJ-MU-89876

The Benjamin/Cummings Publishing Company, Inc.
2727 Sand Hill Road
Menlo Park, California 94025

About the Authors

James D. Watson is the Director of the Cold Spring Harbor Laboratory. He spent his undergraduate years at the University of Chicago and received his Ph.D. in 1950 from Indiana University. Between 1950 and 1953 he did postdoctoral research in Copenhagen and Cambridge, England. While at Cambridge, he began the collaboration that resulted in 1953 in the elucidation of the double-helical structure of DNA. (For this discovery, Watson, Francis Crick, and Maurice Wilkins were awarded the Nobel Prize in 1962.) Later in 1953 he went to the California Institute of Technology. He moved to Harvard in 1955, where he taught and did research on RNA synthesis and protein synthesis until 1976. While at Harvard he also wrote the first, second, and third editions of *Molecular Biology of the Gene*, which were published in 1965, 1970, and 1976, respectively. He has been at Cold Spring Harbor since 1968, where his major interest has been the induction of cancer by viruses.

Nancy H. Hopkins is a Professor of Biology at the Massachusetts Institute of Technology. She graduated from Radcliffe College in 1964 and did graduate work at Yale and Harvard, receiving her Ph.D. in Molecular Biology and Biochemistry from Harvard in 1971. After postdoctoral work at the Cold Spring Harbor Laboratory, she joined the faculty at M.I.T., where she teaches and does research on the molecular biology of retroviruses. She is the primary author of Chapters 23 through 27 in Volume II of this edition of *Molecular Biology of the Gene*.

Jeffrey W. Roberts is a Professor of Biochemistry at Cornell University. He received a B.A. in Physics and Liberal Arts from the University of Texas in 1964 and a Ph.D. in Biophysics from Harvard in 1970. He was a postdoctoral fellow at Harvard and also did research at the MRC Laboratory of Molecular Biology in Cambridge, England, before going to Cornell in 1974. His current research interests are genetic regulation in bacteria and phages, in particular the regulation of transcription and the control of DNA repair functions. He is the primary author of Chapters 11, 12, 13, 16, and 17 of this text.

Joan Argetsinger Steitz is a Professor of Molecular Biophysics and Biochemistry at Yale University. She graduated from Antioch College in 1963 and received a Ph.D. from Harvard in 1967. She did postdoctoral work at the MRC Laboratory of Molecular Biology before joining the Yale faculty in 1970. Her research interests have always focused on the structures and functions of RNA molecules; her current research is on gene expression in mammalian cells, with an emphasis on the roles of small RNA-protein complexes. A member of the National Academy of Sciences, she is a recipient of the National Medal of Science, among other awards. She is the primary author of Chapters 14, 15, 20, and 21 of this text.

Alan M. Weiner is a Professor of Molecular Biophysics and Biochemistry at Yale University. He graduated from Yale College in 1968 and received his Ph.D. from Harvard in 1973. After postdoctoral work at Stanford University and M.I.T., he returned to Yale as a faculty member in 1976. His current research concentrates on the structure, function, and evolution of mammalian genes for small nuclear RNA species. He is the primary author of Chapters 22 and 28 in Volume II of this text.

Preface

Today no molecular biologist knows all the important facts about the gene. This was not the case in 1965 when the first edition of *Molecular Biology of the Gene* appeared. Then there were few practicing molecular biologists and not too many facts to learn. So what we knew about DNA and RNA could easily be explained to beginning college students. That year the final codons of the genetic code were being assigned, and everyone at the forefront of research could regularly assemble in the modest lecture hall at Cold Spring Harbor. Five years later, when the second edition appeared, our numbers were rising rapidly. Yet, despite the emerging popularity of molecular biology, it was still quite uncertain if the future would be as intellectually meaningful as the years just after the discovery of the double helix. The isolation of the first repressors and the demonstration that they bind specifically to control sequences in DNA seemed to some pioneers in DNA research to mark the end of the years of germinal discovery. With no means to isolate the genes of any higher organism, much less any way to know their nucleotide sequences, any pathway to understanding how genes guide the differentiation events that give rise to multicellular organisms seemed impossibly remote.

Happily, these worries did not last long. By the time the third edition of *Molecular Biology of the Gene* was published (1976), recombinant DNA procedures had given us the power to clone genes. Moreover, there was reason to believe that highly reliable methods to rapidly sequence long stretches of DNA would soon be available. As this new era of molecular biology began, however, there initially was widely voiced concern that recombinant DNA procedures might generate dangerous and pathogenic new organisms. It was not until after much deliberation that in 1977 the cloning of the genes of higher organisms began in earnest. The third edition could barely mention the potential of recombinant DNA, and of necessity its brief discussions of how genes function in eucaryotic organisms were tentative, and sometimes quite speculative.

It is only in this fourth edition that we see the extraordinary fruits of the recombinant DNA revolution. Hardly any contemporary experiment on gene structure or function is done today without recourse to ever more powerful methods for cloning and sequencing genes. As a result, we are barraged daily by arresting new facts of such importance that we seldom can relax long enough to take comfort in the accomplishments of the immediate past. The science described in this edition is by any measure an extraordinary example of human achievement.

Because of the immense breadth of today's research on the gene, none of us can speak with real authority except in those areas where our own research efforts are concentrated. Thus it was clear from the first discussions about the fourth edition that writing it would be beyond the capability of any one scientist who also had other major responsibilities. So the task of preparing this edition has required several authors. We also realized that it would be a formidable undertaking to keep the book within a manageable length; even by adopting a larger page format, we saw no way not to exceed a thousand pages. DNA can no longer be portrayed with the grandeur it deserves in a handy volume that would be pleasant to carry across a campus. Although this edition could have been shortened by eliminating the introductory material found in the first eight chapters, we never seriously considered this alternative. To do so would remove the background material that so many readers of previous editions have found valuable, and which has let many novices in molecular biology use this book as their first real introduction to gene structure and function.

Now that we are at last finished, we find that the book is even longer than we had planned. In part this happened because we are two years behind schedule, and 150 additional pages were needed to accommodate the immediate past. We also seriously underestimated how many words and illustrations would be required to describe the extraordinary variety of gene structures and functions that underlie the complexity of eucaryotic cells. We therefore have made the decision to split the fourth edition into two volumes. In the first volume we cover the general principles that govern the structure and function of both procaryotic and eucaryotic genes. It can be used as the sole text for a one-term course in molecular biology at the undergraduate level. The second volume concentrates on those specialized aspects of the gene that underlie multicellular existence, and it concludes with a chapter on the evolution of DNA. In this edition the second volume is appreciably smaller than the first. This will not be true of subsequent editions. Now that it is at last possible to study differentiation at the DNA level, we can easily foresee the time when, in fact, more than one volume will be required for even an introductory description of how genes are organized and expressed in the specialized cells of multicellular organisms.

We hope that this new edition, like its predecessors, will be found to be a highly suitable text for teaching at the undergraduate level, and that it also will provide all molecular biologists with an easy reference to the basic facts about genes. We have shown sections of the manuscripts to a variety of colleagues who are listed as reviewers. Their comments have been taken seriously, and we hope that the final manuscript faithfully reflects their expertise. Any mistakes that remain are, of course, our responsibility. Those who have made major contributions by writing or rewriting large sections of the text are Thomas Steitz (Chapter 6), Ira Herskowitz (Chapters 18 and 19), John Coffin (Chapter 24), and Brent Cochran (Chapter 25). Their generous contributions of specialized knowledge has vastly upgraded those portions of the book. In addition, John Coffin, Scott Powers, Haruo Saito, Lisa Steiner, and Parmjit Jat helped with the references for various chapters in Volume II. The excellent index was prepared by Maija Hinkle.

Equally important have been the efforts at Cold Spring Harbor of Andrea Stephenson, whose competent secretarial assistance helped coordinate our diverse labors, and Susan Scheib, whose intelligent attention to detail kept the manuscript and the galleys moving on a forward course. We also wish to acknowledge the pleasure of working with the staff of The Benjamin/Cummings Publishing Company, including Editor-in-Chief Jim Behnke and Production Supervisors Karen Gulliver and Betsy Dilernia. In particular we wish to thank Jane Gillen, who has functioned as the responsible editor during the entire writing and production of the book. An especially satisfying aspect of the process has been seeing rough drawings come alive through the efforts of the talented illustrator Georg Klatt, who has been responsible for the vast majority of the hundreds of new drawings prepared for this edition, and whose commitment and interest have greatly improved the book. And finally we gratefully acknowledge the strong support of our families throughout this endeavor, which was of course far more difficult and protracted than we ever foresaw.

James D. Watson

Nancy H. Hopkins

Jeffrey W. Roberts

Joan Argetsinger Steitz

Alan M. Weiner

Reviewers

John Abelson, California Institute of Technology
Bruce Alberts, University of California-San Francisco
Manny Ares, Yale University
Spyros Artanvanis-Tsakonas, Yale University
Piet Borst, The Netherlands Cancer Institute
Andy Brenda, Cold Spring Harbor Laboratory
Clifford Brunk, University of California-Los Angeles
Tom Cech, University of Colorado
Brent Cochran, Massachusetts Institute of Technology
John Coffin, Tufts University
Nick Cozzarelli, University of California-Berkeley
Don Crothers, Yale University
Steve Dellaporta, Cold Spring Harbor Laboratory
Ashley Dunn, Ludwig Institute for Cancer Research,
  Melbourne
Gary Felsenfeld, National Institutes of Health
John Fessler, University of California-Los Angeles
Wally Gilbert, Harvard University
Nigel Godson, New York University
Howard Green, Harvard Medical School
Nigel Grindley, Yale University
Carol Gross, University of Wisconsin
Ronnie Guggengheim, Cold Spring Harbor Laboratory
Gary Gussin, University of Iowa
David Helfman, Cold Spring Harbor Laboratory
Roger Hendrix, University of Pittsburgh
Ira Herskowitz, University of California-San Francisco
Bob Horvitz, Massachusetts Institute of Technology
David Housman, Massachusetts Institute of Technology
Martha Howe, University of Wisconsin
Richard Hynes, Massachusetts Institute of Technology
Amar Klar, Cold Spring Harbor Laboratory
Nancy Kleckner, Harvard University
Larry Klobutcher, University of Connecticut
Chuck Kurland, Uppsala University
Peter Laird, The Netherlands Cancer Institute
Richard Losick, Harvard University
Tony Mahowald, Case Western Reserve University
Will McClure, Carnegie Mellon University
Bill McGinnis, Yale University
Jeffrey Miller, University of California-Los Angeles
Paul Modrich, Duke University
Peter Moore, Yale University

Steve Mount, University of California-Berkeley
Kim Mowry, Yale University
Tim Nelson, Yale University
Harry Noller, University of California-Santa Cruz
Bill Nunn, University of California-Irvine
Paul Nurse, Imperial Cancer Research Fund
Kim Nysmath, Laboratory of Molecular Biology,
  Cambridge
Norm Pace, Indiana University
Carl Parker, California Institute of Technology
Sheldon Penman, Massachusetts Institute of
  Technology
Scott Powers, Cold Spring Harbor Laboratory
Mark Ptashne, Harvard University
Charles Radding, Yale University
David Raulet, Massachusetts Institute of Technology
Phil Robbins, Massachusetts Institute of Technology
Hugh Robertson, Rockefeller University
Lucia Rothman-Denes, University of Chicago
Earl Ruley, Massachusetts Institute of Technology
Haruo Saito, Harvard University
Robert Schleif, Brandeis University
Thomas Shenk, Princeton University
Gerry Smith, Hutchinson Cancer Center, University of
  Washington
Deborah Steege, Duke University
Lisa Steiner, Massachusetts Institute of Technology
Tom Steitz, Yale University
Bruce Stillman, Cold Spring Harbor Laboratory
Bill Studier, Brookhaven National Laboratory
Bob Symons, University of Adelaide
Susumu Tonegawa, Massachusetts Institute of
  Technology
Olke Ulhenbeck, University of Colorado
Axel Ullrich, Genentech
Chris Walsh, Massachusetts Institute of Technology
Jim Wang, Harvard University
Bob Webster, Duke University
Allan Wilson, University of California-Berkeley
Barbara Wold, California Institute of Technology
Sandra Wolin, University of California-San Francisco
Keith Yamamoto, University of California-San Francisco
Jorge Yunis, University of Minnesota

Brief Contents

Detailed Contents

VOLUME I   GENERAL PRINCIPLES

CHAPTER 5
THE IMPORTANCE OF WEAK CHEMICAL INTERACTIONS   126

Part VI
Regulation of Gene Function in Bacterial Cells   463

CHAPTER 16
REGULATION OF PROTEIN SYNTHESIS AND FUNCTION IN BACTERIA   465

Part VII
Facing Up to Eucaryotic Cells 549

CHAPTER 19
RECOMBINANT DNA AT WORK   595

Part VIII
The Functioning of Eucaryotic Chromosomes   619

Detailed Contents

VOLUME II   SPECIALIZED ASPECTS

CHAPTER 23
THE GENERATION OF IMMUNOLOGICAL SPECIFICITY 832

CHAPTER 24
THE EXTRAORDINARY DIVERSITY OF EUCARYOTIC VIRUSES   898

Note to the Reader

The following features are intended to add to the usefulness of this book as both a text and a reference.

- **Pagination of Volumes I and II** of this text is consecutive, with the first chapter of Volume II (Chapter 22) beginning on page 745. (Cross-references refer to chapter or page numbers only.)

- **The index** at the end of Volume I covers that volume only; the index at the end of Volume II covers both volumes.

- **Key terms** within the text are highlighted by boldface type at the point in a chapter where the first full definition and major discussion of each term occur. Boldface type is also used in the index to identify the page where the full definition appears.

- **The concept headings,** which originated in the first edition of this text, have been retained. In addition, the longer chapters of Volume II have been subdivided into several major sections set off by briefer headings, to help organize the material for the reader. A complete list of all the headings in both volumes may be found in the Detailed Contents beginning on page x.

- **Summaries** follow the main text for each chapter.

- **Bibliographies** at the ends of the chapters provide a bridge to the scientific literature. Included are a relatively short list of recommended **General References,** which are mainly books and review articles, and a longer list of **Cited References.** The Cited References include the original papers in which important discoveries were first reported as well as a selection of more recent papers. The citations of these references within the text appear as **superscript numbers** that accompany text headings. Thus the Cited References provide a convenient way of finding more detailed information on specific topics.

- **Color plates** showing computer graphics are included in both volumes. Volume I (following p. 64) includes Computer Images of DNA (Plate 1), Protein Structure at Increasing Levels of Complexity (Plate 2–3), Interaction of Repressors with DNA (Plate 4–5), and The Complete Structure of Multiprotein Complex (Plate 6). Volume II includes Immunoglobulin and Viral Hemagglutinin (Plate 7) and Capsid Structure of Icosahedral Viruses (Plate 8).

I

HISTORICAL BACKGROUND

CHAPTER 1

The Mendelian View of the World

It is easy to consider human beings unique among living organisms. We alone have developed complicated languages that allow meaningful and complex interplay of ideas and emotions. Moreover, great civilizations have developed and changed our world's environment in ways inconceivable for any other form of life. Hence, there has always been a tendency to think that something special differentiates us from every other species. This belief has found expression in the many religions through which we have sought out the reason for our existence and, in doing so, have tried to establish workable rules for conducting our lives. Until a little more than a century ago, it seemed natural to think that just as every human life begins and ends at a fixed time, the human species must also have been created at a fixed moment, with the same series of events leading to the creation of all other forms of life.

This belief was first seriously questioned 125 years ago, when Charles Darwin and Alfred R. Wallace proposed their theories of evolution, based on the selection of the most fit. They stated that the various forms of life are not constant, but are continually giving rise to slightly different animals and plants, some of which are adapted to survive and multiply more effectively. At the time of this theory, they did not know the origin of this continuous variation, but they did correctly realize that these new characteristics must persist in the progeny if such variations are to form the basis of evolution.

At first, there was a great furor against Darwin, most of it coming from people who did not like to believe that humans and the rather obscene-looking apes could have a common ancestor, even if this ancestor had lived some 10 to 20 million years ago. There was also initial opposition from many biologists who failed to find Darwin's evidence convincing. Among these was the famous Swiss-born naturalist Jean L. Agassiz, then at Harvard, who spent many years writing against Darwin and Darwin's champion, Thomas H. Huxley, the most successful of the popularizers of evolution. But by the end of the nineteenth century, the scientific argument was almost complete; both the current geographic distribution of plants and animals and their selective occurrence in the fossil records of the geologic past were explicable only by postulating that continuously evolving groups of organisms had descended from a common ancestor. Today, evolution is an accepted fact for everyone but a fundamentalist minority, whose objections are based not on reasoning but on doctrinaire adherence to religious principles.

Figure 1-1
Electron micrograph of a thin section from a cell of the African violet. The thin primary cellulose cell wall and the nucleus, containing a prominent nucleolus, are clearly visible. The cytoplasmic ground substance is heavily laden with ribosomes, visible as small black dots. The profiles of a network of hollow membranes, the endoplasmic reticulum, can be seen scattered throughout the cell. (Courtesy of K. R. Porter and M. C. Ledbetter.)

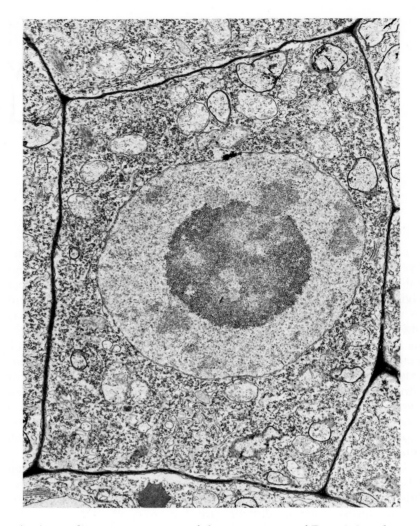

An immediate consequence of the acceptance of Darwinian theory is the realization that life first existed on our Earth some four billion years ago in a simple form, possibly resembling the bacteria—the simplest variety of life known today. Of course, the very existence of such small bacteria tells us that the essence of the living state is found in very small organisms. Evolutionary theory further suggests that the same basic principles of life apply to all living forms.

The Cell Theory[1]

The same conclusion was independently stated by the second great principle of nineteenth-century biology, the **cell theory.** This theory, first put forward convincingly in 1839 by the German microscopists Matthias Schleiden and Theodor Schwann, proposed that all plants and animals are constructed from small fundamental units called cells, and that all cells arise from other cells. It was later discovered that all cells are surrounded by a membrane and usually contain an inner body, the nucleus, which is also surrounded by a membrane, called the nuclear membrane (Figures 1-1 and 1-2). Most importantly, cells arise from other cells by the process of cell division. Most cells are capable of growing and of splitting roughly equally to give two daughter cells. At the same time, the nucleus divides so that each daughter cell can receive a nucleus.

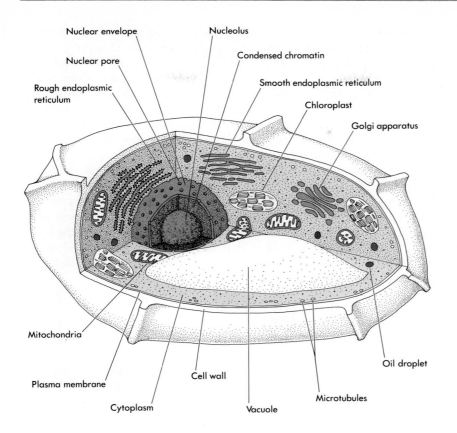

Nuclear envelope
Nucleolus
Nuclear pore
Condensed chromatin
Rough endoplasmic reticulum
Smooth endoplasmic reticulum
Chloroplast
Golgi apparatus
Mitochondria
Oil droplet
Plasma membrane
Cell wall
Microtubules
Cytoplasm
Vacuole

Figure 1-2
A schematic view of the plant cell shown in Figure 1-1. The various components are not drawn to scale.

Mitosis Maintains the Parental Chromosome Number

Each nucleus encloses a fixed number of linear bodies called **chromosomes** (Figure 1-3). Before cell division, each chromosome duplicates to form two identical chromosomes. This process, first accurately observed by Walter Flemming in 1879, doubles the number of nuclear chromosomes. During nuclear division, one of each pair of daughter chromosomes moves into each daughter nucleus (Figure 1-4). As a result of these events (now collectively termed **mitosis**), the chromosomal complement of daughter cells is usually identical to that of the parental cell.

During most of a cell's life, its chromosomes exist in a highly extended linear form. Prior to cell division, however, they condense into much more compact bodies. The duplication of chromosomes occurs chiefly when they are in the extended state characteristic of interphase (the various stages of cell division are defined in Figure 1-4). One part of the chromosome, however, always duplicates during the contracted metaphase state; this is the **centromere,** a body that controls the movement of the chromosome during cell division. The centromere always has a fixed location specific to a given chromosome; in some chromosomes it is near one end, and in others it occupies an intermediate region.

When a chromosome is completely duplicated except for the centromere, it is said to consist of two **chromatids.** A chromatid becomes a chromosome as soon as its centromere has divided and is no longer shared with another chromatid. As soon as one centromere becomes two, the two daughter chromosomes begin to move away from each other.

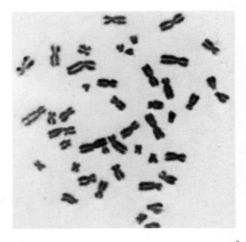

Figure 1-3
Chromosomes of a human male. In this micrograph, the chromosomes have already duplicated in preparation for cell division. (Courtesy of M. Siniscalco, M.D., and B. Alhadeff, Memorial Sloan-Kettering Cancer Center, New York.)

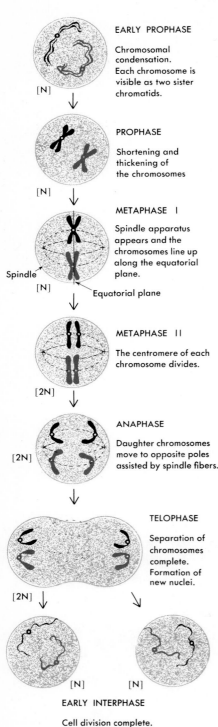

EARLY PROPHASE

Chromosomal condensation. Each chromosome is visible as two sister chromatids.

[N]

PROPHASE

Shortening and thickening of the chromosomes

[N]

METAPHASE I

Spindle apparatus appears and the chromosomes line up along the equatorial plane.

Spindle

[N]

Equatorial plane

METAPHASE II

The centromere of each chromosome divides.

[2N]

ANAPHASE

Daughter chromosomes move to opposite poles assisted by spindle fibers.

[2N]

TELOPHASE

Separation of chromosomes complete. Formation of new nuclei.

[2N]

[N]   [N]

EARLY INTERPHASE

Cell division complete. Chromosomes elongate.

Figure 1-4
Diagram of mitosis in the nucleus of a haploid cell containing two (nonhomologous) chromosomes. *N* refers to the number of chromosomes in the haploid cell.

The regular lining up of chromosomes during the metaphase stage is accompanied by the appearance of the **spindle**. This is a cellular region, shaped like a spindle, through which the chromosomes of higher organisms move apart during the anaphase stage. Much of the spindle region is filled with long, thin fibers, called microtubules. These fibers are largely responsible for chromosome movements on the spindle. Microtubules attach to the chromosomes at the centromeres, and as the daughter chromosomes move toward the spindle poles during anaphase, the centromeres are in the lead.

Objects called **nucleoli** are also present in the nucleus of almost every plant and animal cell. There is often one nucleolus per haploid set of chromosomes, and in some cells the nucleolus is connected to a specific chromosome. The functional role of the nucleolus was once thought by some biologists to be related to the formation of the spindle. Now, however, it is clear that the nucleolus is involved in the synthesis of ribosomes, small particles within the cell on which all proteins are synthesized.

Meiosis Reduces the Parental Chromosome Number

One important exception was found to the mitotic process. After conclusion of the two cell divisions that form the sex cells (the gametes: the sperm and the egg), the number of chromosomes is reduced to one-half of its previous number (Figure 1-5). This set of processes is called **meiosis**. In higher plants and animals, each specific type of chromosome is normally present as two copies of homologous chromosomes (the **diploid** state). In gamete formation, the resulting sperm or egg usually encloses only one of each chromosome type (the **haploid** state). Union of sperm and egg during fertilization results in a **zygote** containing one homologous chromosome from the male parent and another from the female parent. Thus, the normal diploid chromosome constitution is restored.

Although most cells are diploid in higher plants and animals, the haploid state is the most frequent condition in lower plants and bacteria, the diploid number existing only briefly following sex-cell fusion. In these organisms, meiosis usually occurs almost immediately after fertilization to produce haploid cells. Figure 1-6 shows a generalized life cycle that includes individual organisms in both haploid and diploid states.

The cell theory thus tells us that all cells come from preexisting cells. All the cells in adult plants and animals are derived from the division and growth of a zygote, itself formed by the union of two other cells, the sperm and the egg. All growing cells contain chromosomes, usually two of each type, and here again, new chromosomes always arise through division of previously existing bodies.

The Cell Theory Is Universally Applicable

Although the cell theory developed from observations about higher organisms, it holds with equal force for the more simple forms of life, such as protozoa and bacteria. Each bacterium or protozoan is a single cell, whose division ordinarily produces a new cell identical to its parent, from which it soon separates. In the higher organisms, on the

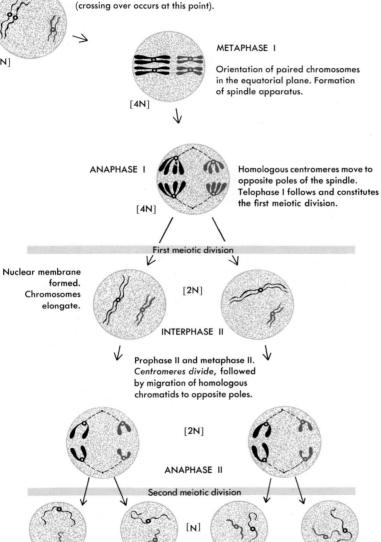

PROPHASE I

Two pairs of homologous chromosomes are shown in this imaginary diploid cell. Chromosomes become visible as single strands.

[2N]

Homologous chromosomes undergo pairing. Later, each chromosome becomes visible as two chromatids (crossing over occurs at this point).

[2N]

METAPHASE I

Orientation of paired chromosomes in the equatorial plane. Formation of spindle apparatus.

[4N]

ANAPHASE I

Homologous centromeres move to opposite poles of the spindle. Telophase I follows and constitutes the first meiotic division.

[4N]

First meiotic division

Nuclear membrane formed. Chromosomes elongate.

[2N]

INTERPHASE II

Prophase II and metaphase II. *Centromeres divide,* followed by migration of homologous chromatids to opposite poles.

[2N]

ANAPHASE II

Second meiotic division

[N]

Final result is four haploid cells.

Figure 1-5
Diagram of meiosis in an organism containing two pairs of homologous chromosomes in each diploid cell.

other hand, not only do the daughter cells often remain together, but they also frequently differentiate into radically different cell types (such as nerve or muscle cells) while maintaining the chromosome complement of the zygote. Here, new organisms arise from the highly differentiated sperm and egg, whose union initiates a new cycle of division and differentiation.

Thus, although a complicated organism like *Homo sapiens* contains a large number of cells (up to 5×10^{12}), all these cells arise initially from a single cell. The zygote contains all the information necessary for the growth and development of an adult plant or animal. Again, the living state per se does not demand the complicated interactions that occur in complex organisms; its essential properties can be found in single growing cells.

Mendelian Laws[2]

The most striking attribute of a living cell is its ability to transmit hereditary properties from one cell generation to another. The products of heredity must have been noticed by early humans, who witnessed the passing of characteristics, like eye or hair color, from parents to offspring. Its physical basis, however, was not understood until the first years of the twentieth century, when, during a remarkable period of creative activity, the chromosomal theory of heredity was established.

Hereditary transmission through the sperm and egg became known by 1860, and in 1868 Ernst Haeckel, noting that sperm consists largely of nuclear material, postulated that the nucleus is responsible for heredity. Almost 20 years passed before the chromosomes were singled out as the active factors, because the details of mitosis, meiosis, and fertilization had to be worked out first.

When this was accomplished, it could be seen that, unlike other cellular constituents, the chromosomes are equally divided between daughter cells. Moreover, the complicated chromosomal changes that reduce the sperm and egg chromosome number to the haploid number during meiosis became understandable as necessary for keeping the chromosome number constant. These facts, however, merely suggested that chromosomes carry heredity.

Proof came at the turn of the century with the discovery of the basic rules of heredity. These rules, named after their original discoverer, Gregor Mendel, had in fact been first proposed in 1865, but the climate of scientific opinion had not been ripe for their acceptance. They were completely ignored until 1900, despite some early efforts on Mendel's part to interest the prominent biologists of his time. Then Hugo De Vries, Karl Correns, and Erich Tschermak, all working independently, realized the great importance of Mendel's forgotten work. All three were plant breeders doing experiments related to Mendel's, and each reached similar conclusions before they knew of Mendel's work.

Principle of Independent Segregation

Mendel's experiments traced the results of breeding experiments (genetic crosses) between strains of peas differing in well-defined characteristics, like seed shape (round or wrinkled), seed color (yellow or green), pod shape (inflated or wrinkled), and stem length (long or short). His concentration on well-defined differences was of great importance; many breeders had previously tried to follow the inheritance of more gross qualities, like body weight, and were unable to discover any simple rules about their transmission from parents to offspring. After ascertaining that each type of parental strain bred true (that is, produced progeny with particular qualities identical to those of the parents), Mendel made a number of crosses between parents (P) differing in single characteristics (such as seed shape *or*

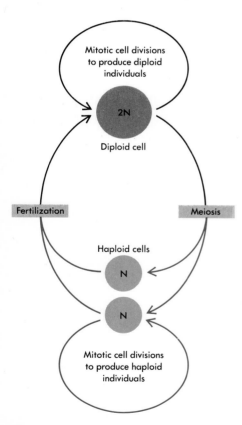

Figure 1-6
The alternation of haploid and diploid states that constitute the sexual life cycle. The haploid and diploid individuals may be unicellular or multicellular, depending on the organism.

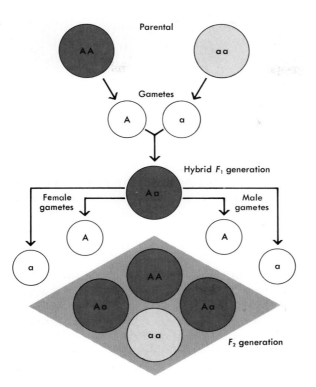

Figure 1-7
How Mendel's first law (independent segregation) explains the 3:1 ratio of dominant to recessive phenotypes among the F₂ progeny. *A* represents the dominant gene and *a* the recessive gene. The colored circles represent the dominant phenotype, the gray circles the recessive phenotype.

seed color). All the progeny (F₁ = first filial generation) had the appearance of *one* parent. For example, in a cross between peas having yellow seeds and peas having green seeds, all the progeny had yellow seeds. The trait that appears in the F₁ progeny is called **dominant,** whereas the trait that does not appear in F₁ is called **recessive.**

The meaning of these results became clear when Mendel made genetic crosses between F₁ offspring. These crosses gave the important result that the recessive trait reappeared in approximately 25 percent of the F₂ progeny, whereas the dominant trait appeared in 75 percent of these offspring. For each of the seven traits he followed, the ratio in F₂ of dominant to recessive traits was always approximately 3:1. When these experiments were carried to a third (F₃) progeny generation, all the F₂ peas with recessive traits bred true (produced progeny with the recessive traits). Those with dominant traits fell into two groups: one-third bred true (produced only progeny with the dominant trait); the remaining two-thirds again produced mixed progeny in a 3:1 ratio of dominant to recessive.

Mendel correctly interpreted his results as follows (Figure 1-7): The various traits are controlled by pairs of factors (which we now call **genes**), one factor derived from the male parent, the other from the female. For example, pure-breeding strains of round peas contain two genes for roundness (*RR*), whereas pure-breeding wrinkled strains have two genes for wrinkledness (*rr*).* The round-strain gametes each have one gene for roundness (*R*); the wrinkled-strain gametes each have one gene for wrinkledness (*r*). In a cross between *RR* and *rr*, fertilization produces an F₁ plant with both genes (*Rr*). The seeds look round because *R* is dominant over *r*. We refer to the appearance

---

*One or several letters may be used to represent a particular gene. The dominant variety of the gene (i.e., the dominant allele) may be indicated by a capital letter (e.g., *R*), a superscript + (e.g., *r⁺*), or a + standing alone.

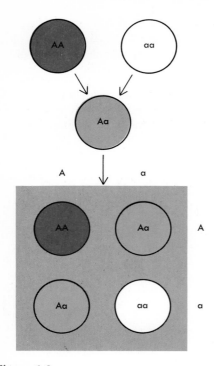

Figure 1-8
The inheritance of flower color in the snapdragon. One parent is homozygous for red flowers (*AA*) and the other homozygous for white flowers (*aa*). No dominance is present, and the heterozygous F₁ flowers are pink. The 1:2:1 ratio of red, pink, and white flowers in the F₂ progeny is shown by appropriate coloring.

(physical structure) of an individual as its **phenotype,** and to its genetic composition as its **genotype.** Individuals with identical phenotypes may possess different genotypes; thus, to determine the genotype of an organism, it is frequently necessary to perform genetic crosses for several generations. The term **homozygous** refers to a gene pair in which both the maternal and paternal genes are identical (e.g., *RR* or *rr*). In contrast, those gene pairs in which paternal and maternal genes are different (e.g., *Rr*) are called **heterozygous.**

It is important to notice that a given gamete contains only one of the two genes present in the organism it comes from (e.g., either *R* or *r*, but never both) and that the two types of gametes are produced in equal numbers. Thus, there is a fifty-fifty chance that a given gamete from an F₁ pea will contain a particular gene (*R* or *r*). This choice is purely random. We do not expect to find *exact* 3:1 ratios when we examine a limited number of F₂ progeny. The ratio will sometimes be slightly higher and other times slightly lower. But as we look at increasingly larger samples, we expect that the ratio of peas with the dominant trait to peas with the recessive trait will approximate the 3:1 ratio more and more closely.

The reappearance of the recessive characteristic in the F₂ generation indicates that recessive genes are neither modified nor lost in the **hybrid** (*Rr*) generation, but that the dominant and recessive genes are independently transmitted and so are able to segregate independently during the formation of sex cells. This **principle of independent segregation** is frequently referred to as Mendel's first law.

Some Genes Are Neither Dominant Nor Recessive

In the crosses reported by Mendel, one of each gene pair was clearly dominant, and the other recessive. Such behavior, however, is not universal. Sometimes the heterozygous phenotype is intermediate between the two homozygous phenotypes. For example, the cross between a pure-breeding red snapdragon (*Antirrhinum*) and a pure-breeding white variety gives F₁ progeny of the intermediate pink color. If these F₁ progeny are crossed among themselves, the resulting F₂ progeny contain red, pink, and white flowers in the proportion of 1:2:1 (Figure 1-8). Thus, it is possible here to distinguish heterozygotes from homozygotes by their phenotype. We also see that Mendel's laws do not depend on whether one *gene* of a gene pair is dominant over the other.

Principle of Independent Assortment

Mendel extended his breeding experiments to peas differing by more than one characteristic. As before, he started with two strains of peas, each of which bred pure when mated with itself. One of the strains had round yellow seeds; the other, wrinkled green seeds. Since round and yellow are dominant over wrinkled and green, the entire F₁ generation produced round yellow seeds. The F₁ generation was then crossed within itself to produce a number of F₂ progeny, which were examined for seed appearance (phenotype). In addition to the two original phenotypes (round yellow; wrinkled green), two new types (**recombinants**) emerged: wrinkled yellow and round green.

Again Mendel found he could interpret the results by the postulate of genes, if he assumed that each gene pair was independently trans-

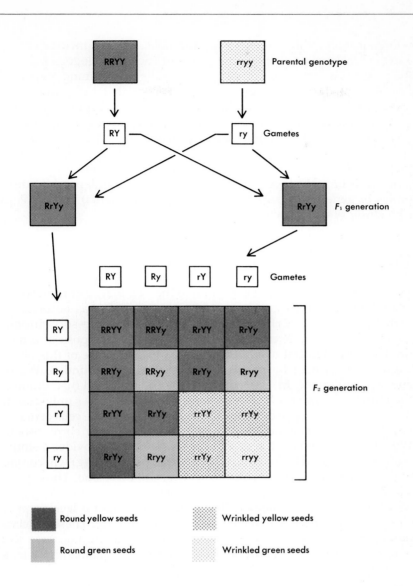

Figure 1-9
How Mendel's second law (independent assortment) operates. In this example, the inheritance of yellow (Y) and green (y) seed color is followed together with the inheritance of round (R) and wrinkled (r) seed shapes. The R and Y alleles are dominant over r and y. The genotypes of the various parents and progeny are indicated by letter combinations, and four different phenotypes are distinguished by appropriate shading.

mitted to the gamete during sex-cell formation. This interpretation is shown in Figure 1-9. Any one gamete contains only one type of inherited factor from each gene pair. Thus, the gametes produced by an F_1 (*RrYy*) will have the composition *RY*, *Ry*, *rY*, or *ry*, but never *Rr*, *Yy*, *YY*, or *RR*. Furthermore, in this example, all four possible gametes are produced with equal frequency. There is no tendency of genes arising from one parent to stay together. As a result, the F_2 progeny phenotypes appear in the ratio 9 round yellow, 3 round green, 3 wrinkled yellow, and 1 wrinkled green. This **principle of independent assortment** is frequently called Mendel's second law.

Chromosomal Theory of Heredity[3]

A principal reason for the original failure to appreciate Mendel's discovery was the absence of firm facts about the behavior of chromosomes during meiosis and mitosis. This knowledge was available, however, when Mendel's laws were reannounced in 1900, and was seized upon in 1903 by the American Walter S. Sutton. In his classic paper, "The Chromosomes in Heredity," Sutton emphasized the importance of the fact that the diploid chromosome group consists of two morphologically similar sets and that, during meiosis, every gamete receives only one chromosome of each homologous pair. He then used this fact to explain Mendel's results by assuming that genes are parts of the chromosome. He postulated that the yellow- and green-seed genes are carried on a certain pair of chromosomes and that the round- and wrinkled-seed genes are carried on a different pair. This hypothesis immediately explains the experimentally observed 9:3:3:1 segregation ratios. Although Sutton's paper did not prove the chromosomal theory of heredity, it was immensely important, for it brought together for the first time the independent disciplines of genetics (the study of breeding experiments) and cytology (the study of cell structure).

Chromosomal Determination of Sex

There exists one important exception to the rule that all chromosomes of diploid organisms are present in two copies. It was observed as early as 1890 that one chromosome (then called an accessory chromosome and now the X chromosome) does not always possess a morphologically identical mate. The biological significance of this observation was clarified by the American cytologist Edmund B. Wilson and his student, N. M. Stevens, in 1905. They showed that although the female carries two X chromosomes, the male carries only one. In addition, in some species (including humans), the male cells contain a unique chromosome, not found in females, called the Y chromosome. Wilson and Stevens pointed out how this situation provides a simple method of sex determination. Whereas every ovum (egg) will contain one X chromosome, only half the sperms will carry one. Thus, fertilization of an ovum by an X-bearing sperm leads to an XX zygote, which becomes a female; and fertilization by a sperm lacking an X chromosome gives rise to male offspring (Figure 1-10). These observations provided the first clear linking of a definite chromosome to a hereditary property. In addition, they elegantly explained how male and female zygotes are created in equal numbers.

The Importance of *Drosophila*[4, 5]

Initially, all breeding experiments used genetic differences already existing in nature. For example, Mendel used seeds obtained from seed dealers, who must have obtained them from farmers. The existence of alternative forms of the same gene (**alleles**) raises the question of how they arose. One obvious hypothesis states that genes can change (mutate) to give rise to new genes (mutant genes). This hypothesis was first seriously tested, beginning in 1908, by the great American biologist Thomas H. Morgan and his young collaborators, the geneticists Calvin B. Bridges, Hermann J. Muller, and Alfred H. Sturtevant. They worked with the tiny fly *Drosophila melanogaster*. This fly, which normally lives on fruit, was found to be easily maintained under laboratory conditions, where a new generation can be produced every 14 days. Thus, by using *Drosophila* instead of more slowly multiplying organisms like peas, it was possible to work at least 25 times faster and much more economically. The first mutant found was a male with white eyes instead of the normal red eyes. It spontaneously appeared in a culture bottle of red-eyed flies. Because essentially all *Drosophila* found in nature have red eyes, the gene lead-

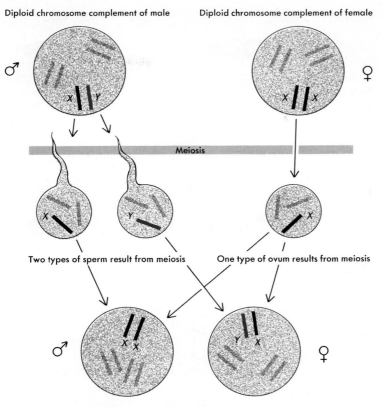

Diploid chromosome complement of male   Diploid chromosome complement of female

♂   ♀

X   Y   X   X

Meiosis

X   Y   X

Two types of sperm result from meiosis   One type of ovum results from meiosis

♂   X   X   ♀   Y   X

Sex determined by the type of sperm entering the ovum

Figure 1-10
How sex chromosomes operate. Here is a case in which males carry one *X* and one *Y* chromosome, and females carry two *X* chromosomes. This is the situation in both humans and *Drosophila*. In certain species there is no *Y* chromosome, so that diploid male cells contain one less chromosome than diploid female cells.

ing to red eyes was referred to as the **wild-type gene;** the gene leading to white eyes was called a **mutant gene** (allele).

The white-eye mutant gene was immediately used in breeding experiments (Figure 1-11), with the striking result that the behavior of the allele completely paralleled the distribution of an *X* chromosome (i.e., was sex-linked). This immediately suggested that this gene might be located on the *X* chromosome, together with those genes controlling sex. This hypothesis was quickly confirmed by additional genetic crosses using newly isolated mutant genes. Many of these additional mutant genes also were sex-linked.

Gene Linkage and Crossing Over[6, 7]

Mendel's principle of independent assortment is based on the fact that genes located on different chromosomes behave independently during meiosis. Often, however, two genes do not assort independently, because they are located on the same chromosome (**linked genes**). Numerous examples of nonrandom assortment were found as soon as a large number of mutant genes became available for breeding analysis. In every well-studied case, the number of linked groups was identical with the haploid chromosome number. For example, there are four groups of linked genes in *Drosophila* and four morphologically distinct chromosomes in a haploid cell.

Linkage, however, is in effect never complete. The probability that two genes on the same chromosome will remain together during meiosis ranges from just less than 100 percent to about 50 percent.

This means that there must be a mechanism for exchanging genes on homologous chromosomes. This mechanism is called **crossing**

Figure 1-11
The inheritance of a sex-linked gene in *Drosophila*. Genes located on sex chromosomes can express themselves differentially in male and female progeny, because if there is only one *X* chromosome present, recessive genes on this chromosome are always expressed. Here are two crosses, both involving a recessive gene (*w*, for white eye) located on the *X* chromosome. In (a), the male parent is a white-eyed (*wY*) fly, and the female is homozygous for red eye (*WW*). In (b), the male has red eyes (*WY*) and the female white eyes (*ww*). The letter *Y* stands here not for an allele, but for the *Y* chromosome, present in male *Drosophila* in place of a homologous *X* chromosome. There is no gene on the *Y* chromosome corresponding to the *w* or *W* gene on the *X* chromosome.

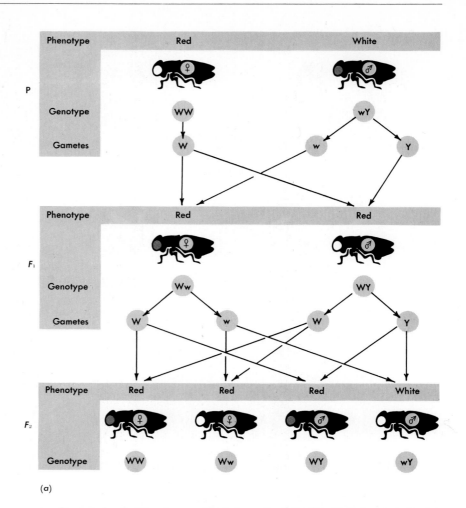

(a)

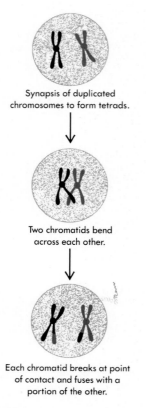

Synapsis of duplicated chromosomes to form tetrads.

Two chromatids bend across each other.

Each chromatid breaks at point of contact and fuses with a portion of the other.

Figure 1-12
Janssens' hypothesis of crossing over.

over. Its cytological basis was first described by the Belgian cytologist F. A. Janssens. At the start of meiosis, through the process of **synapsis,** the homologous chromosomes form pairs with their long axes parallel. At this stage, each chromosome has duplicated to form two chromatids. Thus, synapsis brings together four chromatids (a tetrad), which coil about one another. Janssens postulated that, possibly because of tension resulting from this coiling, two of the chromatids might sometimes break at a corresponding place on each. This could create four broken ends, which might rejoin crossways, so that a section of each of the two chromatids would be joined to a section of the other (Figure 1-12). Thus, recombinant chromatids might be produced that contain a segment derived from each of the original homologous chromosomes. Formal proof of Janssens' hypothesis that chromosomes physically interchange material during synapsis, however, had to await the passage of more than 20 years. Then, in 1931, Barbara McClintock and Harriet Creighton, working at Cornell University with the corn plant *Zea mays*, devised an elegant cytological demonstration of chromosome breakage and rejoining (Figure 1-13).

Chromosome Mapping[8]

Morgan and his students, however, did not await formal cytological proof of crossing over before exploiting the implication of Janssens' hypothesis. They reasoned that genes located close together on a

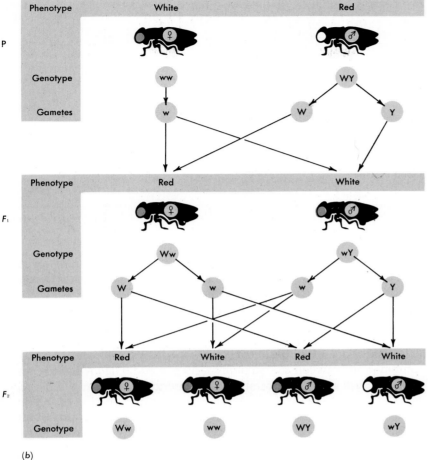

Figure 1-11
(*Continued*)

(b)

chromosome would assort with one another much more regularly (close linkage) than genes located far apart on a chromosome. They immediately saw this as a way to locate (map) the relative positions of genes on chromosomes and thus to produce a **genetic map.** The way they used the frequencies of the various recombinant classes is very straightforward. Consider the segregation of three genes all located on the same chromosome. The arrangement of the genes can be determined by means of three crosses, in each of which two genes are followed (two-factor crosses). A cross between *AB* and *ab* yields four progeny types: the two parental genotypes (*AB* and *ab*) and two recombinant genotypes (*Ab* and *aB*). A cross between *AC* and *ac* similarly gives two parental combinations as well as the *Ac* and *aC* recombinants, whereas a cross between *BC* and *bc* produces the parental types and the recombinants *Bc* and *cB*. Each cross will produce a specific ratio of parental to recombinant progeny. Consider, for example, the fact that the first cross gives 30 percent recombinants, the second cross 10 percent, and the third cross 25 percent. This tells us that genes *a* and *c* are closer together than *a* and *b* or *b* and *c* and that the genetic distances between *a* and *b* and *b* and *c* are more similar. The gene arrangement that best fits these data is *a-c-b* (Figure 1-14).

The correctness of gene order suggested by crosses of two gene factors can usually be unambiguously confirmed by three-factor crosses. When the three genes used in the preceding example are followed in the cross *ABC* × *abc*, six recombinant genotypes are found (Figure 1-15). They fall into three groups of reciprocal pairs. The rar-

Figure 1-13

Demonstration of physical exchanges between homologous chromosomes. In most organisms, pairs of homologous chromosomes have identical shapes. Occasionally, however, the two members of a pair are not identical, one being marked by the presence of extra-chromosomal material or compacted regions that reproducibly form knoblike structures. McClintock and Creighton, having found one such pair, used them to show that crossing over involves actual physical exchanges between the paired chromosomes. In the experiment shown here, the homozygous *c, wx* progeny had to arise by crossing over between the *C* and *wx* loci. When such *c, wx* offspring were cytologically examined, no knob chromosomes were seen, showing that the knobbed *C* region had been physically replaced by a knobless *c* section. The colored box in the figure identifies the chromosomes of the homozygous *c, wx* offspring.

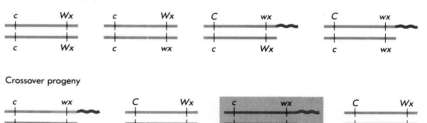

est of these groups arises from a double crossover. By looking for the least frequent class, it is often possible to instantly confirm (or deny) a postulated arrangement. The results in Figure 1-15 immediately confirm the order hinted at by the two-factor crosses. Only if the order is *a-c-b* does the fact that the rare recombinants are *AcB* and *aCb* make sense.

The existence of multiple crossovers means that the amount of recombination between the outside markers *a* and *b* (*ab*) is usually less than the sum of the recombination frequencies between *a* and *c* (*ac*) and *c* and *b* (*cb*). To obtain a more accurate approximation of the distance between the outside markers, we calculate the probability (*ac* × *cb*) that when a crossover occurs between *c* and *b*, a crossover also occurs between *a* and *c*, and vice versa (*cb* × *ac*). This probability subtracted from the sum of the frequencies expresses more accurately the amount of recombination. The simple formula

$$ab = ac + cb - 2(ac)(cb)$$

is applicable in all cases where the occurrence of one crossover does not affect the probability of another crossover. Unfortunately, accurate mapping is often disturbed by *interference* phenomena, which can either increase or decrease the probability of correlated crossovers.

Using such reasoning, the Columbia University group headed by Morgan had by 1915 assigned locations to more than 85 mutant genes in *Drosophila* (Table 1-1), placing each of them at distinct spots on one of the four linkage groups, or chromosomes. Most importantly, all the genes on a given chromosome were located on a line. The gene arrangement was strictly linear and never branched. Figure 1-16 shows the genetic map of one of the chromosomes of *Drosophila*. Dis-

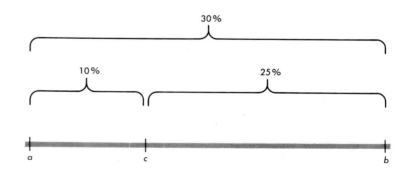

Figure 1-14

Assignment of the tentative order of three genes on the basis of 3 two-factor crosses.

tances between genes on such a map are measured in **map units,** which are related to the frequency of recombination between the genes. Thus, if the frequency of recombination between two genes is found to be 5 percent, the genes are said to be separated by five map units. Because of the high probability of double crossovers between widely spaced genes, such assignments of map units can be considered accurate only if recombination between closely spaced genes is followed.

Even when two genes are at the far ends of a very long chromosome, they will show not less than 50 percent linkage (assort together at least 50 percent of the time) because of multiple crossovers. The two genes will be separated if an odd number of crossovers occur between them, but they will end up together if an even number occur between them. Thus, in the beginning of the genetic analysis of *Drosophila*, it was often impossible to determine whether two genes were on different chromosomes or at the opposite ends of one long chromosome. Only after large numbers of genes had been mapped was it possible to convincingly demonstrate that the number of linkage groups equalled the number of cytologically visible chromosomes. It was then possible for Morgan, Sturdevant, Muller, and Bridges to publish their definitive book *The Mechanism of Mendelian Heredity,* which first announced the general validity of the chromosomal basis of heredity. We now rank this concept, along with the theories of evolution and the cell, as a major achievement in our quest to understand the nature of the living world.

Many Genes Control the Red Eye

Inspection of the list of mutant genes in Table 1-1 reveals an important fact. Many different genes act to influence a single characteristic. For example, 13 of the genes discovered by 1915 affect eye color. When a fly is homozygous for a mutant form of any one of these genes, the eye color is not red, but a different color, distinct for the mutant gene (e.g., carnation, vermilion). Thus, there is no one-to-one correspondence between genes and complex characteristics like eye color or wing shape. Instead, the development of each characteristic is controlled by a series of events, each of which is controlled by a gene. We might make a useful analogy with the functioning of a complex machine like the automobile. There are a number of separate parts—such as the motor, the brakes, the radiator, and the fuel tank—all of which are essential for proper operation of the vehicle. Although a fault in any one part may cause the car to stop functioning properly, there is no reason to believe that the presence of that component alone is sufficient for proper functioning.

The Origin of Genetic Variability Through Mutations[9-15]

It now became possible to understand the hereditary variation that is found throughout the biological world and that forms the basis of the theory of evolution. Genes are normally copied exactly during chromosome duplication. Rarely, however, changes (**mutations**) occur in genes to give rise to altered forms, most, *but not all,* of which function less well than the wild-type alleles. This process is necessarily rare;

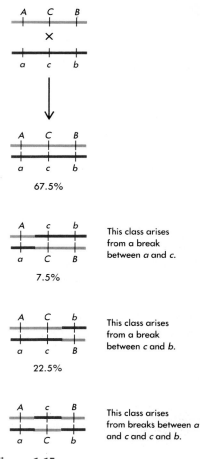

Figure 1-15
The use of three-factor crosses to assign gene order. The least frequent pair of reciprocal recombinants must arise from a double crossover. The percentages listed for the various classes are the theoretical values expected for an infinitely large sample. When finite numbers of progeny are recorded, the exact values will be subject to random statistical fluctuations.

Table 1-1 The 85 Mutant Genes Reported in *Drosophila melanogaster* in 1915*

Name	Region Affected	Name	Region Affected
Group 1			
Abnormal	Abdomen	Lethal, 13	Body, death
Bar	Eye	Miniature	Wing
Bifid	Venation	Notch	Venation
Bow	Wing	Reduplicated	Eye color
Cherry	Eye color	Ruby	Leg
Chrome	Body color	Rudimentary	Wing
Cleft	Venation	Sable	Body color
Club	Wing	Shifted	Venation
Depressed	Wing	Short	Wing
Dotted	Thorax	Skee	Wing
Eosin	Eye color	Spoon	Wing
Facet	Ommatidia	Spot	Body color
Forked	Spine	Tan	Antenna
Furrowed	Eye	Truncate	Wing
Fused	Venation	Vermilion	Eye color
Green	Body color	White	Eye color
Jaunty	Wing	Yellow	Body color
Lemon	Body color		
Group 2			
Antlered	Wing	Jaunty	Wing
Apterous	Wing	Limited	Abdominal band
Arc	Wing	Little crossover	Chromosome 2
Balloon	Venation	Morula	Ommatidia
Black	Body color	Olive	Body color
Blistered	Wing	Plexus	Venation
Comma	Thorax mark	Purple	Eye color
Confluent	Venation	Speck	Thorax mark
Cream II	Eye color	Strap	Wing
Curved	Wing	Streak	Pattern
Dachs	Leg	Trefoil	Pattern
Extra vein	Venation	Truncate	Wing
Fringed	Wing	Vestigial	Wing
Group 3			
Band	Pattern	Pink	Eye color
Beaded	Wing	Rough	Eye
Cream III	Eye color	Safranin	Eye color
Deformed	Eye	Sepia	Eye color
Dwarf	Size of body	Sooty	Body color
Ebony	Body color	Spineless	Spine
Giant	Size of body	Spread	Wing
Kidney	Eye	Trident	Pattern
Low crossing over	Chromosome 3	Truncate intensf.	Wing
Maroon	Eye color	Whitehead	Pattern
Peach	Eye color	White ocelli	Simple eye
Group 4			
Bent	Wing	Eyeless	Eye

*The mutations fall into four linkage groups. Since four chromosomes were cytologically observed, this indicated that the genes are situated on the chromosomes. Notice that mutations in various genes can act to alter a single character, such as body color, in different ways.

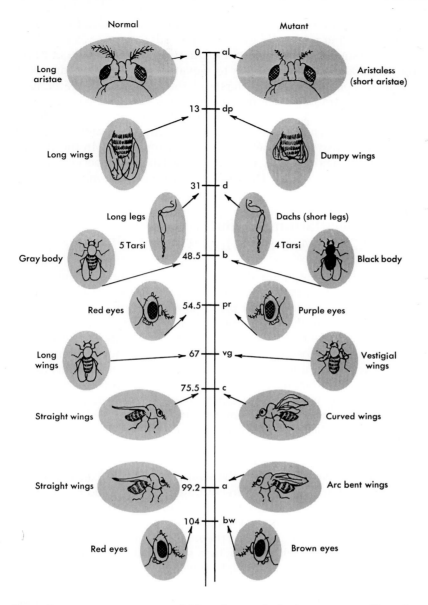

Normal Mutant

Long aristae — 0 — al — Aristaless (short aristae)

Long wings — 13 — dp — Dumpy wings

Long legs — 31 — d — Dachs (short legs)

Gray body   5 Tarsi   4 Tarsi   Black body

Red eyes — 48.5 — b

— 54.5 — pr — Purple eyes

Long wings — 67 — vg — Vestigial wings

Straight wings — 75.5 — c — Curved wings

Straight wings — 99.2 — a — Arc bent wings

Red eyes — 104 — bw — Brown eyes

Figure 1-16
The genetic map of chromosome 2 of *Drosophila melanogaster.*

otherwise, many genes would be changed during every cell cycle, and offspring would not ordinarily resemble their parents. There is, instead, a strong advantage in there being a small but finite mutation rate; it provides a constant source of new variability, necessary to allow plants and animals to adapt to a constantly changing physical and biological environment.

Surprisingly, however, the results of the Mendelian geneticists were not avidly seized upon by the classical biologists, then the authorities on the evolutionary relations between the various forms of life. Doubts were raised about whether genetic changes of the type studied by Morgan and his students were sufficient to permit the evolution of radically new structures, like wings or eyes. Instead, these biologists believed that there must also occur more powerful "macromutations," and that it was these that allowed great evolutionary advances.

Gradually, however, doubts vanished, largely as a result of the efforts of the mathematical geneticists Sewall Wright, Ronald A. Fisher, and J. B. S. Haldane. They showed that considering the great age of the earth, the relatively low mutation rates found for *Drosophila* genes, together with only mild selective advantages, would be sufficient to allow the gradual accumulation of new favorable attributes. By the 1930s, biologists began to reevaluate their knowledge on the origin of species and to understand the work of the mathematical geneticists. Among these new Darwinians were the biologist Julian Huxley (a grandson of Darwin's original publicist, Thomas Huxley), the geneticist Theodosius Dobzhansky, the paleontologist George Gaylord Simpson, and the ornithologist Ernst Mayr. In the 1940s, all four wrote major works, each showing from his special viewpoint how Mendelianism and Darwinism were indeed compatible.

Early Speculations About What Genes Are and How They Act[16–19]

Almost immediately after the rediscovery of Mendel's laws, geneticists began to speculate about both the chemical structure of the gene and the way it acts. No real progress could be made, however, because the chemical identity of the genetic material remained unknown. Even the realization that both nucleic acids and proteins are present in chromosomes did not really help, since the structure of neither was at all understood. The most fruitful speculations focused attention on the fact that genes must be, in some sense, self-duplicating. Their structure must be exactly copied every time one chromosome becomes two. This fact immediately raised the profound chemical question of how a complicated molecule could be precisely copied to yield exact replicas.

Some physicists also became intrigued with the gene, and when quantum mechanics burst on the scene in the late 1920s, the possibility arose that in order to understand the gene, it would first be necessary to master the subtleties of the most advanced theoretical physics. Such thoughts, however, never really took root, since it was obvious that even the best physicists or theoretical chemists would not worry about a substance whose structure still awaited elucidation. There was only one fact that they might ponder: Hermann Muller and L. J. Stadler's independent 1927 discoveries that X-rays induce mutations. Since there is a greater possibility that an X-ray will hit a larger gene than a smaller gene, the frequency of mutations induced in a given gene by a given X-ray dose yields an estimate of the size of this gene. But even here, so many special assumptions had to be made that virtually no one, not even the estimators themselves, took the estimates very seriously.

Bands (Genes?) Along the Giant Salivary Chromosomes of *Drosophila*[20, 21]

Unfortunately, no information on what genes are could be obtained by directly looking at ordinary chromosomes with the light microscope. Most chromosomes are far too small for such analysis, usually being just large enough to be clearly delineated from one another. Moreover, when they are visible during mitosis, they are highly supercoiled into arrangements about which there was much disagreement among the leading cytologists of that period. Much fresh hope was thus generated when in 1934 the chromosomes from the salivary glands of the fly *Chironomus tentans,* and soon thereafter of *Drosophila,* were found to be some thousand times larger than their equivalents in all other cells. Moreover, the individual chromosomes were not only highly extended, but marked by hundreds of discrete, dark-staining bands (Figure 1-17). Most importantly, the locations and sizes of the bands were highly reproducible, allowing very precise cytological maps to be made. These could be used to delineate the exact points where chromosomal segments of *Drosophila* switched partners during the rare chromosomal exchanges (translocations) that occur following accidental chromosomal breaks. Careful examination of the salivary chromosomes also showed that some mutations in fact represented discrete deletions or insertions of one or more bands.

By 1941, some 5059 different bands had been mapped, a number only severalfold larger than the number of genes that had been revealed by genetic analysis. This approximate similarity prompted the speculation that the bands in fact represented the genes, with the less darkly staining interbands representing the nonspecific chromosomal material. There was no way at that time, however, to follow up on this hypothesis, much less rule out the alternative possibility that the darkly stained masses were genetically inert, with the true genes being the less compacted interband material.

Preliminary Attempts to Find a Gene-Protein Relationship[22–25]

The most fruitful early endeavors to find a relationship between genes and proteins examined the ways in which gene changes affect which proteins are present in the cell. At first this study was difficult, since no one knew anything about the proteins that were present in structures such as the eye or the wing. It soon became clear that genes with simple metabolic functions would be easier to study than genes affecting gross structures. One of the first useful examples came from a study of a hereditary disease affecting amino acid metabolism. Spontaneous mutations occur in humans affecting the ability to metabolize the amino acid phenylalanine. When individuals homozygous for the mutant trait eat food containing phenylalanine, their inability to convert the amino acid to tyrosine causes a toxic level of phenylpyruvic acid to build up in the bloodstream. Such diseases, examples of "inborn errors of metabolism," suggested as early as 1909 to the English physician A. E. Garrod that the wild-type gene is responsible for the presence of a particular enzyme, and that in a homozygous mutant, the enzyme is congenitally absent.

Garrod's general hypothesis of a gene-enzyme relationship was extended in the 1930s by work on flower pigments by Haldane and Rose Scott-Moncrieff in England, studies on the hair pigment of the guinea pig by Wright in the U.S., and research on the pigments of insect eyes by A. Kuhn in Germany and by Boris Ephrussi and George W. Beadle, working first in France and then in California. In all cases, evidence revealed that a particular gene affected a particular step in the formation of the respective pigment whose absence changed, say, the color of a fly's eyes from red to ruby. However, the lack of fundamental knowledge about the structures of the relevant enzymes ruled out deeper examination of the gene-enzyme relationship, and no assurance could be given either that most genes control the synthesis of proteins (by then it was suspected that all enzymes were proteins) or that all proteins are under gene control.

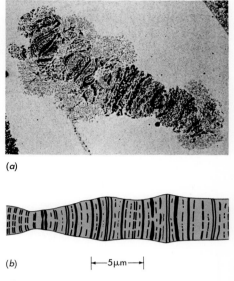

(a)

(b) |——5μm——|

Figure 1-17
(a) Electron micrograph of chromosome 4 from a salivary gland of *Chironomus tentans.* [Reproduced with permission from B. Daneholt, *Cell* 4 (1975):1.] (b) Diagram of a portion of a salivary gland chromosome (the right arm of chromosome 3) of *Drosophila melanogaster.* [After P N. Bridges, *J. Heredity* 32 (1941).]

As early as 1936, it became apparent to the Mendelian geneticists that future experiments of the sort successful in elucidating the basic features of Mendelian genetics were unlikely to yield productive evidence about how genes act. Instead, it would be necessary to find biological objects more suitable for chemical analysis. They were aware, moreover, that contemporary knowledge of nucleic acid and protein chemistry was completely inadequate for a fundamental chemical attack on even the most suitable biological systems. Fortunately, however, the limitations in chemistry did not deter them from learning how to do genetic experiments with chemically simple molds, bacteria, and viruses. As we shall see, the necessary chemical facts became available almost as soon as the geneticists were ready to use them.

Uneasiness About Cytoplasmic Inheritance and Movable Genetic Elements[26–33]

The neat picture of the Mendelian world with its discrete nuclear chromosomes along which genes were sited at fixed locations was not without its exceptions. Some genetic markers, already well studied by the mid-1930s, did not map at all to chromosomes but rather showed "maternal inheritance," being transmitted only through the cytoplasm-rich egg—never through sperm. These misbehaving genetic traits must somehow reside in the cytoplasm, raising the question of whether such examples negate the general validity of the chromosomal theory of heredity. Since many of these maternally transmitted traits involve losses of photosynthetic pigments, speculations arose that the chloroplasts, the organelle carriers of these pigments, might in fact be tiny bacterium-like cells endowed with their own genetic material (chromosomes?). But there existed neither the means to purify away chloroplasts from other cellular constituents nor any way to show that one of their molecular constituents was a carrier of genetic information.

Equally puzzling were reports first emanating from research on *Drosophila* of extremely unstable genes that showed very high mutation rates. How such hypervariable genetic elements originated first began to be understood from Barbara McClintock's genetic analysis of the corn plant. This seminal research began soon after her arrival at Cold Spring Harbor in 1942 and culminated with her 1951 announcement that several key genetic elements, which she named "controlling elements," did not have fixed locations. Instead, they could apparently move from one chromosomal site to another, without in any way disturbing the general architecture of the chromosomes involved. When these controlling elements jumped, they frequently affected the functioning of the genes into which they became inserted; hence, their designation as control elements. Why only certain genes so moved, and in fact whether such movements played any role in the normal development of the corn plant, could not then be ascertained. So most geneticists saw no alternative but to sweep control elements under the rug (together with cytoplasmic inheritance) until the nature of more conventional chromosomal genes could be better explained.

Summary

The study of living organisms at the biological level has led to three great generalizations: (1) Darwin's and Wallace's theory of evolution by natural selection, which tells us that today's complex plants and animals are derived by a continuous evolutionary progression from the first primitive organisms; (2) the cell theory, the realization that all organisms are composed of cells; and (3) the chromosomal theory of heredity, the understanding that the function of chromosomes is what controls heredity.

All cells contain chromosomes, normally duplicated prior to a cell division process (mitosis) that produces two daughter cells, each with a chromosomal complement identical to that of the parental cell. In haploid cells, there is just one copy of each type of chromosome; in diploid cells, there are usually two copies (pairs of homologous chromosomes). A diploid cell arises by fusion of a male and a female haploid cell (fertilization), whereas haploid cells are formed from a diploid cell by a distinctive form of cell division (meiosis) that reduces the chromosome number to one-half its previous number.

Chromosomes control heredity because they are the cellular locations of genes. Hereditary factors were first discovered by Mendel in 1865, but their importance was not realized until the start of the twentieth century. Each gene can exist in a variety of different forms called alleles. Mendel proposed that a hereditary factor (now known to be a gene) for each hereditary trait is given by each parent to each of its offspring. The physical basis for this behavior is the distribution of homologous chromosomes during meiosis: One (randomly chosen) of each pair of homologous chromosomes is distributed to each haploid cell. When two genes are on the same chromosome, they tend to be inherited together (linked). Genes affecting different characteristics are sometimes inherited independently of each other; this is because they are located on different chromosomes. In any case, linkage is seldom complete, because homologous chromosomes attach to each other during meiosis and often break at identical spots and rejoin crossways (crossing over). Crossing over attaches genes initially found on a paternally derived chromosome to gene groups originating from the maternal parent.

Different alleles from the same gene arise by inheritable changes (mutations) in the gene itself. Normally, genes are extremely stable and are exactly copied during chromosome duplication; mutation occurs only rarely and usually has harmful consequences. It does, however, play a positive role, since the accumulation of the rare favorable mutations provides the basis for the genetic variability that the theory of evolution presupposes.

The most suitable chromosomes for assigning genes to specific chromosomal locations are the giant chromosomes of the salivary glands of insects. Such chromosomes are highly extended and are marked by hundreds (or thousands, depending on the species) of discrete, dark-staining bands. Through studies of the segregation patterns following chromosomal rearrangement, genes can be assigned locations in or adjacent to specific bands. Such maps, however, do not reveal whether the dense bands or the interband regions are the genetically active factors.

For many years, the structure of genes and the chemical way in which they control cellular characteristics were a mystery. As soon as large numbers of spontaneous mutations had been described, it became obvious that a one gene–one characteristic relationship does not exist and that all complex characteristics are under the control of many genes. The most sensible idea, postulated by Garrod in 1909, was that genes affect the synthesis of enzymes. However, the tools of Mendelian geneticists—organisms such as the corn plant, the mouse, and even the fruit fly *Drosophila*—were not suitable for detailed chemical investigations of gene-protein relations. For this type of analysis, work with much simpler organisms became indispensable.

The neat picture of discrete genes located linearly at fixed positions along a chromosome does not always hold true, since some genetic markers were found to reside in the cytoplasm; some organelles (e.g., chloroplasts) might contain their own genetic material. Moreover, genes with extraordinarily high mutation rates were found in rare instances. In maize (corn), these highly mutable elements were found to jump from one chromosome site to another, affecting the functioning of the genes into which they became inserted. However, most geneticists saw no alternative but to ignore these exceptions until the nature of conventional chromosomal genes was clarified.

Bibliography

General References

Alberts, B., D. Bray, J. Lewis, M. Raff, K. Roberts, and J. D. Watson. 1983. *Molecular Biology of the Cell.* New York: Garland.

Ayala, F. J., and J. A. Kiger, Jr. 1984. *Modern Genetics.* 2nd ed. Menlo Park, Calif.: Benjamin/Cummings.

Carlson, E. J. 1966. *The Gene Theory: A Critical History.* Philadelphia: Saunders. A survey of the development of genetic ideas beginning with Mendel.

Mayr, E. 1982. *The Growth of Biological Thought: Diversity, Evolution, and Inheritance.* Cambridge, Mass., and London: Harvard University Press. Traces the development of the earliest problems in biology, with historical and modern references to evolutionary and genetic ideas.

Moore, J. 1972. *Heredity and Development.* 2nd ed. Oxford, Eng., and New York: Oxford University Press. An elegant introduction to genetics and embryology, with emphasis on the historical approach.

Moore, J. 1972. *Readings in Heredity and Development.* Oxford, Eng.: Oxford University Press. Reproduces many of the key papers in the development of the chromosomal theory of heredity, including those of Sutton.

Peters, J. A. 1959. *Classic Papers in Genetics.* Englewood Cliffs, N.J.: Prentice-Hall. A collection of reprints of the most significant papers in the history of genetics, up to Benzer's fine-structure analysis of the gene.

Cited References

1. Wilson, E. B. 1925. *The Cell in Development and Heredity*. 3d ed. New York: Macmillan.
2. Olby, R. C. 1966. *Origins of Mendelism*. London: Constable and Company Ltd. The growth of ideas about inheritance and variation that eventually led to the Mendelian solution.
3. Sutton, W. S. 1903. "The Chromosome in Heredity." *Biol. Bull.* 4:231–251.
4. Sturtevant, A. H., and G. W. Beadle. 1962. *An Introduction to Genetics*. New York: Dover. Now available in paperback form; this book, originally published in 1939, remains a classic statement of the results of *Drosophila* genetics.
5. Morgan, T. H. 1910. "Sex-Linked Inheritance in *Drosophila*." *Science* 32:120–122.
6. McClintock, B., and H. B. Creighton. 1931. "A Correlation of Cytological and Genetical Crossing Over in *Zea Mays*." *Proc. Nat. Acad. Sci.* 17:492–497.
7. Sturtevant, A. H. 1913. "The Linear Arrangement of Six Sex-Linked Factors in *Drosophila* as Shown by Mode of Association." *J. Exp. Zool.* 14:39–45.
8. Morgan, T. H., A. H. Sturtevant, H. J. Muller, and C. B. Bridges. 1915. *The Mechanism of Mendelian Heredity*. New York: Holt, Rinehart & Winston.
9. Dobzhansky, T. 1941. *Genetics and the Origin of Species*. 2nd ed. New York: Columbia University Press.
10. Huxley, J. 1943. *Evolution: The Modern Synthesis*. New York: Harper & Row. The grandson of T. H. Huxley explores Mendelism, evolutionary trends, and genetic systems.
11. Mayr, E. 1942. *Systematics and the Origin of Species*. New York: Columbia University Press. A correlation of the evidence and points of view of systematics with those of other biological disciplines, particularly genetics and ecology.
12. Simpson, G. G. 1944. *Tempo and Mode in Evolution*. New York: Columbia University Press. A synthesis of paleontology and genetics.
13. Haldane, J. B. S. 1932. *The Courses of Evolution*. New York: Harper & Row.
14. Fisher, R. A. 1930. *The Genetical Theory of Natural Selection*. Oxford, Eng.: Clarendon Press.
15. Wright, S. 1931. "Evolution in Mendelian Populations." *Genetics* 16:97–159.
16. Muller, H. J. 1927. "Artificial Transmutation of the Gene." *Science* 46:84–87.
17. Carlson, E. 1981. *Genes, Radiation, and Society: The Life and Work of J. H. Muller*. Ithaca, N.Y.: Cornell University Press.
18. Stadler, L. J. 1928. "Mutations in Barley Induced by X-Rays and Radium." *Science* 110:543–548.
19. Lea, D. E. 1947. *Actions of Radiations on Living Cells*. New York: Macmillan.
20. Bridges, P. N. 1941. "A Revision of the Salivary Gland 3R-Chromosome Map of *Drosophila melanogaster*." *J. Heredity* 32:299–300.
21. Painter, T. S. 1934. "The Morphology of the X Chromosomes in the Salivary Glands of *D. melanogaster* and a New Type of Chromosome Map for this Element." *Genetics* 19:98–99.
22. Garrod, A. E. 1908. "Inborn Errors of Metabolism." *Lancet* 2:1–7, 73–79, 142–148, 214–220.
23. Beadle, G. W., and B. Ephrussi. 1937. "Development of Eye Color in *Drosophila*: Diffusible Substances and Their Interrelations." *Genetics* 22:76–86.
24. Scott-Moncrieff, R. 1936. "A Biochemical Survey of Some Mendelian Factors for Flower Color." *J. Genetics* 32:117–170.
25. Wright, S. 1941. "The Physiology of the Gene." *Physiol. Rev.* 21:487–527.
26. McClintock, B. 1951. "Chromosome Organization and Gene Expression." *Cold Spring Harbor Symp. Quant. Biol.* 16:13–57. First extensive report on movable elements in maize.
27. McClintock, B. 1984. "The Significance of Responses of Genome to Challenge." *Science* 226:792–800. This Nobel address elegantly traces and discusses how the concept of movable genes arose.
28. Caspari, E. 1948. "Cytoplasmic Inheritance." *Adv. Genetics* 2:1–66.
29. Correns, C. 1937. *Nicht Mendelnde Vererbung*. Edited by F. von Wettstein. Berlin: Borntraeger.
30. Rhoades, M. M. 1946. "Plastid Mutations." *Cold Spring Harbor Symp. Quant. Biol.* 11:202–207.
31. Sager, R. 1972. *Cytoplasmic Genes and Organelles*. New York: Academic Press.
32. Sonneborn, T. M. 1950. "The Cytoplasm in Heredity." *Heredity* 4:11–36.
33. Winge, O., and O. Laustsen. 1940. "On a Cytoplasmic Effect of Inbreeding in Homozygous Yeast." *Comptes Rend. Laboratoire de Carlsberg* 23:17–40.

CHAPTER 2

Cells Obey the Laws of Chemistry

In Darwin's time, chemists were already asking whether living cells work by the same chemical rules as nonliving systems. By then, cells had been found to contain no chemical elements unique to living material. Also recognized early was the predominant role of carbon, a major constituent of almost all types of biological molecules. The initial tendency to distinguish between carbon compounds, like those in living matter, and all other molecules is reflected in the division of modern chemistry into organic chemistry (the study of most compounds containing carbon atoms) and inorganic chemistry. We now know that this distinction is artificial and has no biological basis. There is no purely chemical way to decide whether a compound has been synthesized in a cell or in a chemist's laboratory.

Nonetheless, through the first quarter of this century, many scientists thought that some vital force outside the laws of chemistry differentiate the animate from the inanimate. Part of the reason for the persistence of this "vitalism" was that the success of the biologically oriented chemists (now usually called biochemists) was limited. Although the techniques of the organic chemists were sufficient to work out the structures of relatively small molecules like glucose (Table 2-1), there was increasing awareness that many of the most important molecules in the cell—the so-called macromolecules—were far too large to be pursued by even the best of organic chemists.

For many years, the most important group of macromolecules was believed to be the proteins because of the growing evidence that all enzymes are proteins. Initially, there was controversy as to whether enzymes are small molecules or macromolecules. It was not until 1926 that the enzymatic nature of a crystalline protein was demonstrated by the American biochemist James Sumner.[1] The controversy was then practically settled. But even this important discovery did not dispel the general aura of mystery about proteins. In those days, the complex structures of proteins were undecipherable by available chemical tools, so as late as 1940, many scientists still believed that these molecules would eventually be shown to have features unique to living systems.

There was also a general belief that genes, like enzymes, might be proteins. There was no direct evidence, but the high degree of specificity of genes suggested to most people who speculated on their nature that they could only be proteins, by then known to occur in the chromosomes. Another class of molecules, the nucleic acids, were also found to be a common chromosomal component, but these were

Table 2-1 Some Important Classes of Small Biological Molecules

Class	Characteristics	Example
Aliphatic hydrocarbon	Linear or branched molecules containing only carbon and hydrogen	Ethane
Aromatic hydrocarbon	Ring-shaped hydrocarbons containing alternating single and double bonds	Benzene
Pyrimidine	An aromatic compound of the formula $C_4H_4N_2$ (or a derivative thereof)	Uracil
Purine	An aromatic compound of the formula $C_5H_4N_4$ (or a derivative)	Guanine
Alcohol	A hydrocarbon skeleton substituted with one to several OH groups	Ethanol
Phosphate ester	Molecule formed from alcohols and phosphoric acid with the elimination of H_2O	Glucose-1-Ⓟ*
Nucleoside	Contains a pentose sugar linked to either a purine or a pyrimidine base, through a C–N bond	Adenosine

*In this book, Ⓟ is frequently used as an abbreviation for phosphoric acid (H_3PO_4), phosphate ion (PO_4^{3-}), or a phosphate group ($-PO_3^{2-}$). See Table 2-2.

Table 2-1 *(Continued)*

Class	Characteristics	Example
Nucleotide	Phosphate ester of a nucleotide	Adenosine-5-Ⓟ or adenylic acid
Carboxylic acid	Hydrocarbon skeleton with one to several COO^- groups	Acetic acid
Hydroxy acid	Substituted hydrocarbon containing both an OH and a COOH group	Lactic acid
Keto acid	Substituted hydrocarbon containing both a keto group and a COOH group	Pyruvic acid
Amino acid	Substituted hydrocarbon containing both NH_2 and COOH groups; both groups are usually charged. The general formula is where R represents a group that varies from amino acid to amino acid. There are 20 amino acids, each with a distinctive R group.	Glycine
Sugar (monosaccharide)	Polyhydroxyl molecuole containing a C=O group (either an aldehyde or a ketone). The most common sugars have 3 (triose), 4 (tetrose), 5 (pentose), or 6 (hexose) carbon atoms.	Ribose

initially thought to be relatively small molecules and incapable of carrying sufficient information to function as genes.

Besides general ignorance of the structures of the large molecules in the cell, the feeling was often expressed that there is something unique about the three-dimensional organization of the cell that gives it its living feature. This argument was sometimes phrased in terms of the impossibility of ever understanding all the exact chemical interactions of the cell. More frequently, however, it took the form of the prediction that some new natural laws, as important as the cell theory or the theory of evolution, would have to be discovered before the essence of life could be understood. But these almost mystical ideas never led to meaningful experiments and, in their vague form, could never be tested. Progress was made instead only by biologically oriented chemists and physicists patiently attempting to devise new ways of solving the structures of more and more complex biological molecules. But for many years, there were no triumphs to shout. The chemists and biologists usually moved in different and sometimes hostile worlds, the biologist often denying that the chemist would ever provide the real answers to the important riddles of biology. Always not too far back in some biologists' minds was the feeling, if not the hope, that something more basic than mere complexity and size separated biology from the bleak, inanimate world of a chemical laboratory.

The Concept of Intermediary Metabolism

As soon as the organic chemists began to identify some of the various cellular molecules, it became clear that food molecules are extensively transformed after they enter an organism. In no case does a food source contain all the different molecules present in a cell. On the contrary, in some cases practically all the organic molecules within an organism are synthesized inside it. This point is easily seen by observing cellular growth on well-defined food sources, for example, the growth of yeast cells using the simple sugar glucose as the sole source of carbon. Soon after its entry into the cells, glucose is chemically transformed into a large variety of molecules necessary for the building of new cellular components. These chemical transformations rarely occur in one step; instead, intermediate compounds are produced. These intermediate compounds often have no cellular function besides forming part of a pathway leading to the synthesis of a necessary structural component, such as an amino acid.

The sum total of all the chemical reactions occurring in a cell is referred to as the **metabolism** of the cell. Correspondingly, the molecules involved in these transformations are often called **metabolites.** **Intermediary metabolism** is the term used to describe the chemical reactions involved in the transformation of food molecules into essential cellular building blocks.

Cellular Energy Is Generated by Oxidation-Reduction Reactions

By the middle of the nineteenth century, it was known that the food (initially of plant origin) eaten by animals and bacteria is only partly transformed into new cellular building blocks, some of it being burned by combustion with oxygen (O_2) to yield carbon dioxide (CO_2), water (H_2O), and energy. At the same time, it was becoming clear that in green plants, the reverse process also operates. These two processes, called **respiration** and **photosynthesis,** respectively, can be summarized by the following equations:

Respiration:

$$C_6H_{12}O_6 \text{ (glucose)} + 6O_2 \longrightarrow$$
$$6CO_2 + 6H_2O + \text{energy (in the form of heat)} \quad \textit{(2-1)}$$
Occurs in both plants and animals.

Photosynthesis:

$$6CO_2 + 6H_2O + \text{energy (from the sun)} \longrightarrow$$
$$C_6H_{12}O_6 \text{ (glucose)} + 6O_2 \quad \textit{(2-2)}$$
Occurs only in plants, algae, and certain bacteria.

Both these equations can be thought of as the sum total of a lengthy series of oxidation-reduction reactions.

In respiration, organic molecules such as glucose are oxidized by molecular oxygen to form C=O bonds (Table 2-2), which contain *less* usable energy (energy that can do work) than the starting C–H, C–OH, and C–C bonds. Energy is given off in respiration, just as it is when any organic compound burns at high temperatures outside the cell, to produce carbon dioxide, water, and energy in the form of heat. In contrast, during photosynthesis, the energy from the light quanta of the sun is used to reduce carbon dioxide to molecules that contain *more* usable energy.

When these relationships were first worked out, no one knew how the energy obtained during respiration is put to advantage. It was clear that somehow a useful form of energy has to be available to enable living organisms to carry out a variety of forms of work, such as muscular contraction and selective transport of molecules across cell membranes. Even then, it seemed unlikely that the energy obtained from food is first released as heat, because at the temperature at which life exists, heat energy cannot be effectively used to synthesize new chemical bonds. Thus, since the awakening of an interest in the chemistry of life, a prime challenge has been to understand the generation of energy in a useful form.

Most Biological Oxidations Occur Without Direct Participation of Oxygen[2, 3]

Because oxygen is so completely necessary for the functioning of animals, it was natural to guess that oxygen would participate directly in all oxidations of carbon compounds. Actually, most biological oxidations occur in the absence of oxygen. This is possible because, as first proposed around 1912 by the German biochemist H. O. Wieland, most biological oxidations are actually dehydrogenations. A compound is *oxidized* when we remove a pair of hydrogen atoms from it (Figure 2-1). However, hydrogen atoms cannot merely be removed; they must be transferred to another molecule, which is then said to be *reduced* (Figure 2-2). In these reactions, as in all other **oxidation-reduction reactions,** every time one molecule is oxidized, another must be reduced. There are several different molecules whose role is to receive hydrogen atoms. All are medium-size (MW about 500) organic

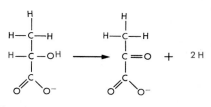

Figure 2-1
Oxidation of an organic molecule by removal of a pair of hydrogen atoms. This figure shows the oxidation of lactate to pyruvate.

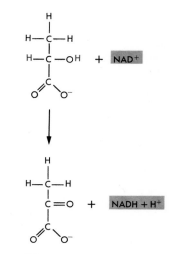

Figure 2-2
The participation of nicotinamide adenine dinucleotide (NAD^+) in the oxidation of lactate to pyruvate. The oxidizing agent here is NAD^+, and the hydrogen donor is lactate; lactate is oxidized and NAD^+ is reduced.

Table 2-2 Important Functional Groups in Biological Molecules

Group	Molecular Example	Biological Significance

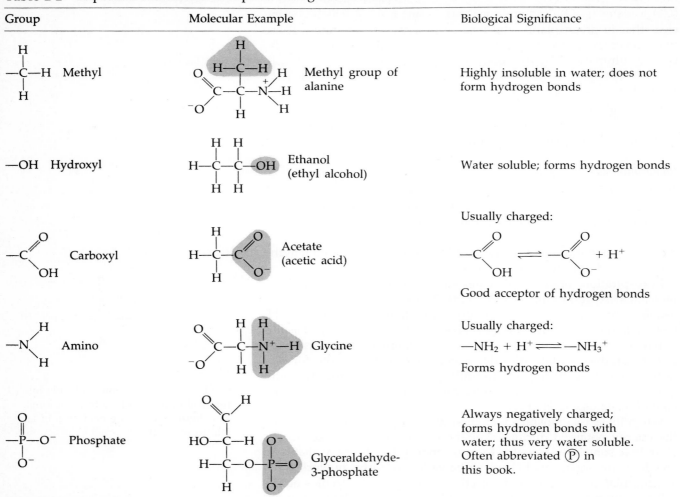

molecules that associate with specific proteins to form active enzymes. The protein components alone (**apoenzymes**) have no enzymatic activity. Only when the smaller partner is present will activity be present. Hence, these smaller molecules are named **coenzymes.** (We should note that not all coenzymes participate in oxidation-reduction reactions; some coenzymes function in other types of metabolic reactions.)

Although the involvement of coenzymes in oxidative reactions was hinted at by 1910, it was not until the early 1930s that their cardinal significance was appreciated. Then the work of the great German biochemist Otto Warburg and the Swedish chemists Ulf S. von Euler and Theodor Theorell established the structure and action of several of the most important coenzymes: nicotinamide adenine dinucleotide (NAD+, earlier called diphosphopyridine nucleotide or DPN, Figure 2-3), flavin mononucleotide (FMN), and flavin adenine dinucleotide (FAD).

Coenzymes, like enzymes, function over and over and are not used up in the course of a reaction. This is because the hydrogen atoms transferred to them do not remain permanently attached, but are transferred by a second oxidation-reduction reaction, usually to another enzyme, and sometimes to oxygen itself (Figure 2-4). Coenzymes are thus continually oxidized and reduced. Furthermore, although oxygen may not be directly necessary for a given reaction, we

Table 2-2 (*Continued*)

Group	Molecular Example	Biological Significance

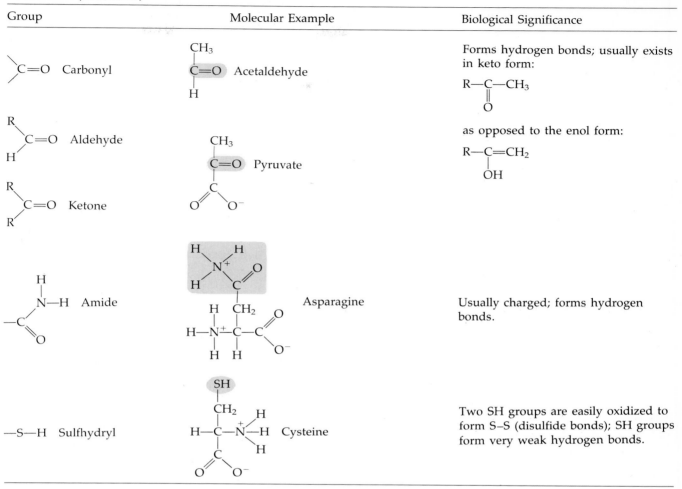

For the Carbonyl / Acetaldehyde row:
Forms hydrogen bonds; usually exists in keto form:

$$R-\underset{\underset{O}{\|}}{C}-CH_3$$

as opposed to the enol form:

$$R-\underset{\underset{OH}{|}}{C}=CH_2$$

For the Amide / Asparagine row:
Usually charged; forms hydrogen bonds.

For the Sulfhydryl / Cysteine row:
Two SH groups are easily oxidized to form S–S (disulfide bonds); SH groups form very weak hydrogen bonds.

see that it is often necessary indirectly, since it must be available to oxidize the coenzyme molecules to make them available for accepting additional pairs of hydrogen atoms (or electrons).

The Breakdown of Glucose

Much of the early work in intermediary metabolism dealt with the transformation of glucose into other molecules. Glucose was emphasized not only because it plays a central role in the metabolic economy of cells, but also for a practical reason; the alcohol (ethanol) produced when wine is made from grapes is derived from the breakdown of glucose. In 1810, the chemist J. Gay-Lussac demonstrated the production of ethanol by this process, and by 1837, the essential role of yeast was established. The production of the alcohol in wine is not a spontaneous process but normally requires the presence of living yeast cells.

Also important in the initial work with glucose was the French microbiologist Louis Pasteur, who discovered that the breakdown of glucose to alcohol does not require air; to distinguish it from reactions requiring oxygen, he used the term **fermentation.** He also showed that ethanol is not the only product of glucose fermentation but that there are other products, such as lactic acid and glycerol.

Figure 2-3
Nicotinamide adenine dinucleotide, a very important coenzyme. Shown here is the conversion between the oxidized form (abbreviated NAD$^+$), an acceptor of hydrogen atoms, and the reduced form (abbreviated NADH), a donor of hydrogen atoms. The release of hydrogen atoms decreases the free energy of a molecule; the acceptance of them increases the free energy.

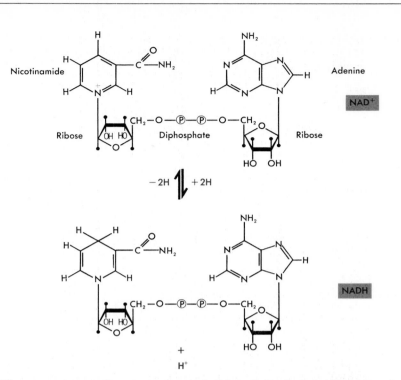

The next great advance came with Eduard Buchner's discovery in 1897 that the living cell per se is not necessary for fermentation and that a cell-free extract from yeast can, by itself, transform glucose into ethanol. Not only was this step conceptually important, but it also provided a much more practical system for studying the chemical steps of fermentation. When working with cell-free systems, it is relatively easy to add or subtract components thought to be involved in the reaction; when living cells are being used, it is often very difficult, and sometimes impossible, to transfer specific compounds in an unmodified form across the cell membrane.

Over the next 40 years, cell-free extracts were used by a large number of distinguished biochemists, including the Englishmen Arthur Harden and William Young and the Germans Gustav Embden and Otto Meyerhof, to work out the exact chemical pathways of glucose degradation (Figure 2-5). The important generalization emerged that the reactions involved in the breakdown of glucose to pyruvate,* originally known as the *Embden-Meyerhof pathway* and now called **glycolysis,** are not peculiar to alcoholic fermentation in yeast but occur in many other cases of glucose utilization as well. Perhaps the most significant discovery was made by Meyerhof. He showed that when muscles contract in the absence of oxygen, the carbohydrate food reserve of glycogen is broken down via glucose to lactic acid (anaerobic glycolysis). Thus, it became clear that not only can microorgan-

*The terms *pyruvic acid* and *pyruvate* are used interchangeably. Technically, pyruvate refers to the negatively charged ion. Likewise, lactic acid is often called lactate; glutamic acid, glutamate; citric acid, citrate; and so on.

Figure 2-4
The transfer of a pair of hydrogen atoms from one coenzyme to another. In this series of reactions, the final hydrogen acceptor is molecular oxygen (O$_2$). FAD usually exists in combination with a specific protein, forming a flavoprotein.

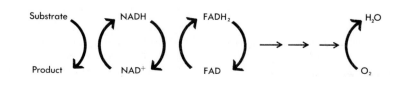

isms obtain their energy and carbon via the Embden-Meyerhof pathway, but energy involved in muscle contraction is also generated by the same pathway.

The ubiquity of the Embden-Meyerhof pathway supported a growing feeling about "the unity of biochemistry." That is, the basic biochemical reactions upon which cell growth and division depend are the same, or very similar, in all cells—those of microorganisms as well as those of higher plants and animals. This unity was not surprising to the more astute biologists, many of whom were largely preoccupied with the consequences of evolutionary theory. Given that humans and fish are descended from a common ancestor, it should not be surprising that many of their cell constituents are similar.

Metabolic Importance of Phosphorus and the Generation of ATP[4]

As early as 1905, the phosphorus atom was implicated in a vital role in metabolism. Harden and Young found that alcoholic fermentation occurs only when inorganic phosphate (PO_4^{3-}) is present. This discovery was followed over the next two decades by the isolation of a large number of intermediary metabolites containing phosphate (Ⓟ) groups attached to carbon atoms by phosphate ester linkages,

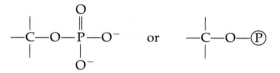

The significance of such phosphorylated intermediates was unclear for 25 years. Then, in the 1930s, Meyerhof and Fritz Lipmann had the crucial insight that phosphate esters enable cells to trap much of the energy of the chemical bonds present in their food molecules. During fermentation, several intermediates are created (e.g., D-1,3-diphosphoglyceric acid, Figure 2-6) that contain what are popularly known as "high-energy" phosphate ester bonds (see Chapter 6 for more details). These high-energy phosphate groups are usually transferred to acceptor molecules, where they can serve as sources of chemical energy for vital cellular processes, such as motion, generation of light, and (as we shall see in Chapter 6) the efficient biosynthesis of necessary cellular molecules. The most important of the acceptor molecules is *adenosine diphosphate* (ADP, Figure 2-7). Addition of a high-energy phosphate group to ADP forms **adenosine triphosphate (ATP)**.

$$\text{ADP} + Ⓟ \rightleftharpoons \text{ATP}$$

The discovery of the role of ADP as an acceptor molecule and that of ATP as a donor of high-energy phosphate groups was one of the most important landmarks of modern biology. Until the roles of these molecules were known, there was complete mystery about how cells obtained energy. There was constant speculation about how cellular existence was incompatible with the second law of thermodynamics, which states that in a closed system, the amount of disorder (entropy) invariably increases. The apparent paradox disappeared, however, as soon as it was seen how animal cells could trap and utilize the energy in food molecules. At that time, the mechanism by which the sun's

Figure 2-5
The stepwise degradation of glucose to pyruvate. This collection of consecutive reactions is called glycolysis. The name of the enzyme that catalyzes each step is shown in a colored rectangle. Additional enzymatic steps can convert pyruvate to other products, such as lactate or ethanol.

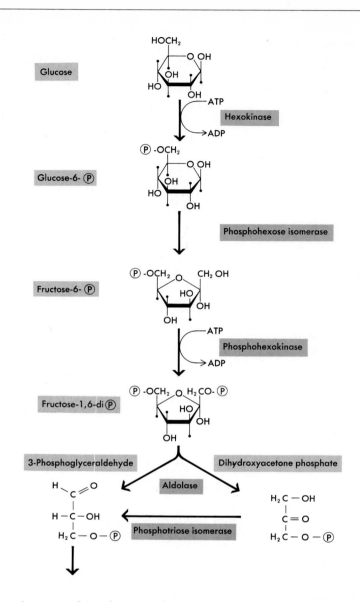

energy is trapped in photosynthesis was not known. Here again, the primary action of the sun's energy is now known to be the generation of ATP.

Most Specific Cellular Reactions Require a Specific Enzyme

As the glycolytic pathway was being worked out, it became clear that each step requires a separate enzyme (see Figure 2-5). Each of the enzymes acts by combining with the molecules involved in the particular reaction (the **substrates** of the enzymes). For example, glucose and ATP are substrates for the enzyme hexokinase. When glucose and ATP molecules interact on the surface of hexokinase, in what is called the **active site** of the enzyme, the terminal phosphate of ATP is transferred to a glucose molecule to form glucose-6-phosphate (Figure 2-8).

The essence of an enzyme is its ability to speed up, or **catalyze,** a reaction involving the making or breaking of a specific covalent bond. In the absence of enzymes, most of the covalent bonds of biological

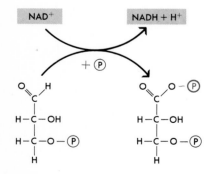

Figure 2-6
The formation of an energy-rich phosphate ester bond coupled with the oxidation of 3-phosphoglyceraldehyde by NAD to form 1,3-diphosphoglyceric acid. The symbol ~ signifies that the bond is of the high-energy variety.

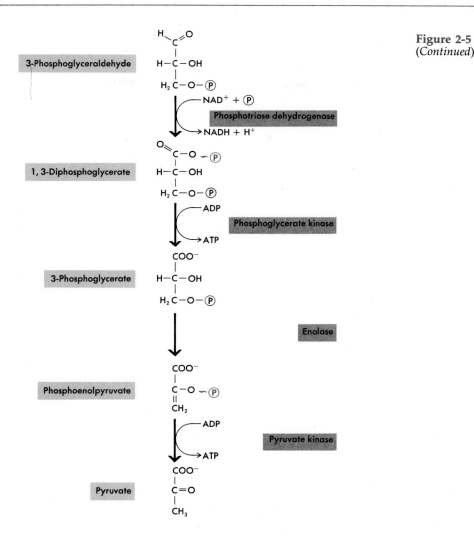

Figure 2-5
(Continued)

molecules are very stable and decompose only under high, nonphysiological temperatures; for example, only at several hundred degrees Celsius is glucose appreciably oxidized by oxygen in the absence of enzymes. Enzymes must therefore act by somehow lowering the temperature at which a given bond becomes unstable. A physical chemist would say that an enzyme lowers the "activation energy" of the reaction. How this is done is finally being understood at the molecular level. The three-dimensional structure of many enzymes is now known, and there are plausible chemical theories about how enzymes work (see Figure 2-25). Thus, there is no reason to suspect that still-undiscovered laws of chemistry underlie enzyme action. Numerous examples already exist in inorganic chemistry where well-defined molecules catalyze reactions between other molecules.

A very important characteristic of enzymes is that they are never consumed in the course of a reaction; once a reaction is complete, they are free to adsorb new molecules and function again. On a biological time scale (seconds to years), enzymes can work very fast, some being able to catalyze as many as 10^6 reactions per minute; often, no successful collision of substrates will occur for years when enzymes are absent.

The degree of specificity for particular substrates varies from enzyme to enzyme. Not all enzymatic reactions are highly specific. For example, a number of enzymes can break down a variety of different

Figure 2-7
The formation of ATP (adenosine-5'-triphosphate) from ADP and an energy-rich phosphate group. Here the donor of the high-energy phosphate is phosphoenolpyruvate.

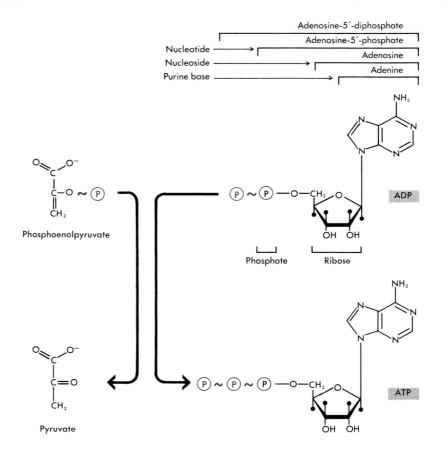

proteins to their component amino acids. They are specific only in the sense that they catalyze the breakdown of a specific type of covalent bond, the peptide bond, and will not, for example, degrade the phosphodiester linkages of the nucleic acids.

The Key Role of Pyruvate: Its Utilization Via the Krebs Cycle[5, 6]

Attempts to understand the generation of ATP in the presence of oxygen occurred parallel with the study of fermentation and glycolysis. It was immediately obvious from the amounts of ATP generated by fermentation and glycolysis that these processes could account for only a small fraction of total ATP production in the presence of oxygen. This means that ATP production in the presence of oxygen does not cease once glucose has been degraded as far as pyruvate, but that pyruvate itself must be further transformed via energy-yielding reactions requiring the presence of oxygen.

The first real breakthrough in understanding how this happens came with discoveries made by the biochemists Albert Szent-Györgyi, Carl Martius and Franz Knoop, and Hans A. Krebs. Their work revealed the existence of a cyclic series of reactions (now usually called the **Krebs cycle,** citric acid cycle, or tricarboxylic acid cycle) by which pyruvate is oxidatively broken down to yield carbon dioxide and pairs of hydrogen atoms that attach to oxidized coenzyme molecules. Before pyruvate enters the Krebs cycle, it is transformed into a key molecule called acetyl coenzyme A or **acetyl-CoA,** known before its chemical identification as "active acetate" (Figure 2-9). This important intermediate, discovered in 1949 by Lipmann working in Boston, then combines with oxaloacetate to yield citrate. A series of at least nine additional steps then occur to yield four pairs of hydrogen atoms and two molecules of carbon dioxide (Figure 2-10). The pairs of hydrogen atoms never exist free, but are transferred to specific coenzyme molecules.

The Krebs cycle should be viewed as a mechanism for breaking down acetyl-CoA to two types of products: the completely oxidized carbon dioxide molecules, which cannot be used as an energy source, and the reduced coenzymes, whose further oxidation yields most of the energy used by organisms growing in the presence of oxygen.

Oxidation of Reduced Coenzymes by Respiratory Enzymes[7]

During the Krebs cycle, there is no direct involvement of molecular oxygen. Oxygen is involved only after the hydrogen atoms (or their electrons) have been transferred through an additional series of oxidation-reduction reactions that involve a series of closely linked enzymes, most of which contain iron atoms. These enzymes are often collectively called the respiratory enzymes. Their existence was hinted at late in the nineteenth century, but it was not until the period from 1925 to 1940 that their significance was appreciated, largely as a result of the work of Warburg and the Polish-born David Keilin, who spent most of his scientific life in Cambridge, England. The exact number of enzymes involved differs from organism to organism. Nevertheless, the general picture is similar in all cases. Figure 2-11 shows the basic features of the eucaryotic respiratory chain.

The chain operates by a series of coupled oxidation-reduction reactions, each of which releases energy. Cytochromes (iron-containing proteins) function as electron carriers in the respiratory chain. Thus, the energy present in the reduced coenzymes is released not all at once but in a series of small packets. If NADH were instead directly oxidized by molecular oxygen, a great amount of energy would be released, which would be impossible to couple efficiently with the formation of the high-energy bonds of ATP.

Oxidative Phosphorylation Generates Much More ATP Than Fermentation[8, 9, 10]

From 1925 to 1940, most biochemists concentrated on following the path of hydrogen atoms (or electrons) through the linked, energy-yielding oxidation-reduction reactions. Until the end of this period, only slight attention was given to how the energy is released in a useful form. Then H. M. Kalckar of Denmark and V. A. Belitzer of the Soviet Union observed ATP formation coupled with oxidation-reduction reactions in cell-free systems (1938–1940).

Further understanding did not come quickly, since most of the enzymes involved could not be obtained in pure soluble form. These troubles were not resolved until it was realized that the normal sites of **oxidative phosphorylation** (the synthesis of ATP in the presence of oxygen) in plant and animal cells are large, highly organized subcellular organelles, the **mitochondria** (singular, **mitochondrion**). Using

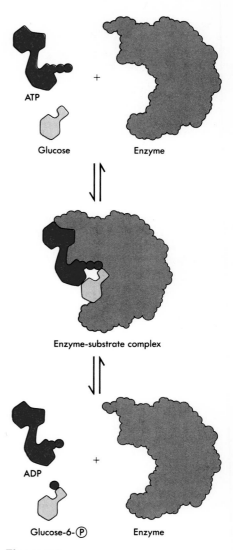

Figure 2-8
Simple diagram of enzyme action. The substrates glucose and ATP bind to the enzyme at its active site to form an enzyme-substrate complex. The substrates react to form products, which are released from the enzyme. In this example, the enzyme is hexokinase.

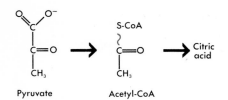

Figure 2-9
The transformation of pyruvate to acetyl-CoA. The transformation is written in a greatly simplified form; actually several steps are required, in which the coenzymes thiamine pyrophosphate and lipoic acid are involved. Acetyl-CoA is an extremely important intermediate, for it is formed not only from glucose via pyruvate, but also by the degradation of fatty acids.

intact mitochondria, it is easy to observe the oxidative generation of ATP; this was first demonstrated in 1949 by the Americans Albert L. Lehninger and Eugene Kennedy. Now we know that three ATP molecules are generated for each pair of hydrogen atoms that pass through the respiratory chain. It is believed, however, that at least six separate oxidation-reduction steps occur in the chain.

Roughly 20 times more energy is released by the respiratory chain than by the initial breakdown of glucose to pyruvate. This explains why the growth of cells under aerobic conditions is so much more efficient than growth without air: If oxygen is not present, pyruvate cannot accumulate, since its formation (see Figure 2-5) demands a supply of unreduced NAD^+. The amount of NAD^+ within cells, like that of other coenzymes, is, however, very small. In the absence of oxygen, it is rapidly converted to NADH as glucose is oxidized by the Embden-Meyerhof pathway. Thus, for the continued generation of ATP without oxygen (fermentation), a device other than the respiratory chain must be used to oxidize the reduced NADH. Often, pyruvate itself serves as the hydrogen acceptor. A typical reduced end product is lactic acid, which is formed during the anaerobic contraction of muscles as well as during the anaerobic growth of many bacteria (Figure 2-12).

ATP Is Generated During Photosynthesis (Photophosphorylation)[11, 12]

The ultimate source of the various food molecules used by most microorganisms and animals is photosynthesis, during which energy from sunlight is converted into energy present in covalent bonds. How this happens chemically did not become clear until the basic energy relations within animals and bacteria were understood. Largely as a result of the work of the American biochemist Daniel Arnon, it was discovered in 1959 that the primary action of the sun's light quanta is to phosphorylate ADP to ATP.

This photophosphorylation takes place in the **chloroplasts,** which are complicated, chlorophyll-containing organelles found in most cells capable of trapping the energy of the sun. Thus, the controlled release of energy in plants and photosynthetic bacteria depends on the same energy carrier important in nonphotosynthetic bacterial and animal cells—ATP. The first step is the capture of the light quanta by the green pigment molecule chlorophyll to excite one of its electrons to a high-energy state.

Then, this energy is released gradually by a series of coupled oxidation-reduction reactions mediated by specific electron carrier molecules (Figure 2-13). Oxidative phosphorylation and photophosphorylation thus turn out to be surprisingly similar processes, essentially differing only in the way energy is initially introduced.

Photosynthesis, however, is unique in an important way. It is the only significant cellular event that utilizes any energy source other than the covalent bond. All other important cellular reactions are accompanied by a decrease in the energy contained in covalent bonds. We might be tempted to guess that the ability to photosynthesize must have been a primary feature of the first forms of life. It is, however, difficult to imagine that very early forms possessed the complicated chloroplast structures necessary for photosynthesis. Instead, there is good reason to believe that early in the history of the Earth, a

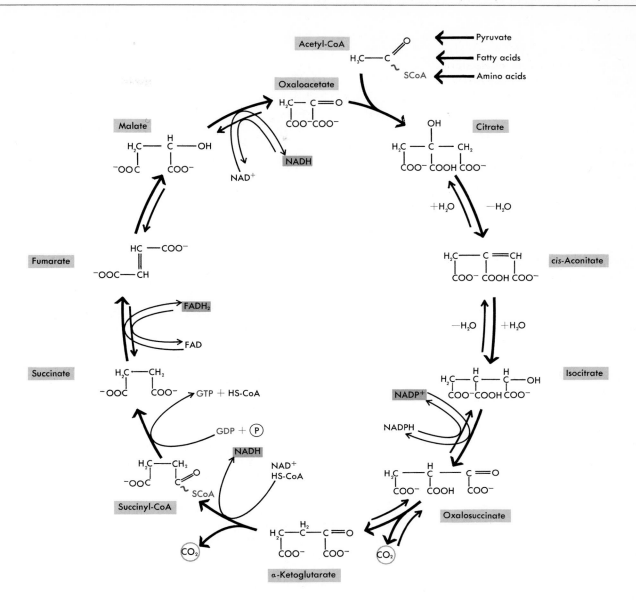

Figure 2-10
The Krebs cycle, often called the citric acid cycle or tricarboxylic acid (TCA) cycle. Colored rectangles highlight the reduced coenzymes that carry pairs of hydrogen atoms stripped off the cycle intermediates.

unique chemical environment allowed the spontaneous creation of a large number of carbon-containing compounds. Energy derived from these organic molecules then served as the food energy supply for the first forms of life. As these original organic molecules were depleted and living matter increased, a strong selective advantage developed for those cells that evolved a photosynthetic structure to provide a means of increasing the amount of organic molecules. Today, essentially all glucose molecules are formed using chemical energy originating in photosynthesis.

The Chemiosmotic Generation of ATP from ADP and Phosphate[13]

At first it seemed obvious that the energy released by the cytochrome-mediated oxidation-reduction reactions must momentarily be stored in high-energy compounds capable of phosphorylating ADP to ATP. But some 30 years of searching failed to yield convincing evidence for

Figure 2-11
The respiratory chain. The heavy black arrows represent the flow of electrons through the three major respiratory enzyme complexes. These oxidation-reduction reactions release, in small packets, the energy present in NADH molecules. Two electrons are transferred from NADH to oxygen. Ubiquinone and cytochrome *c* serve as carriers between the complexes.

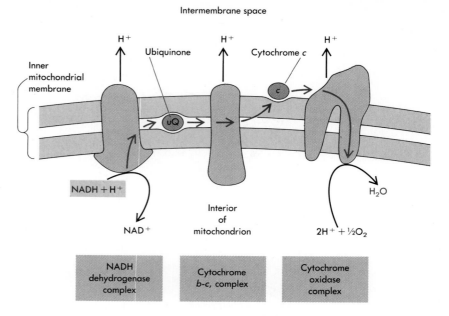

even one such intermediate. Equally troublesome was the observation that only intact mitochondria or chloroplasts make ATP. Unbroken surrounding membranes are required for the coupling of the oxidation-reduction reactions and ATP generation. Resolution of these unexpected complications came in the early 1960s from the work of the English biochemist Peter Mitchell. He proposed that the location of the cytochromes and other electron carriers within membranes allows the generation of gradients in which more hydrogen ions are present outside the mitochondrion or chloroplast than within. Such gradients form because the respiratory enzyme complexes are so placed that three hydrogen ion pairs are displaced outwardly as each pair of electrons goes down the respiratory chain from NAD to oxygen. These hydrogen ion gradients provide the energy for the union of ADP and phosphate to form ATP, a reaction catalyzed by the membrane protein ATP synthetase (Figure 2-14). This explanation, known as the **chemiosmotic theory,** was originally greeted with much skepticism by a biochemical world molded by a succession of new high-energy intermediates. Now, however, Mitchell's idea is recognized as one of the major conceptual advances in the development of modern biochemistry.

Figure 2-12
The fermentation of glucose to yield lactate. In the absence of oxygen, the NADH produced during pyruvate formation is oxidized to reduce pyruvate to lactate. When oxygen is present, the NADH is oxidized through the respiratory chain, and no lactate is produced.

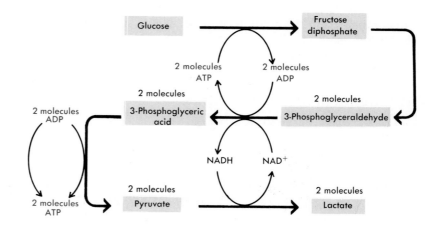

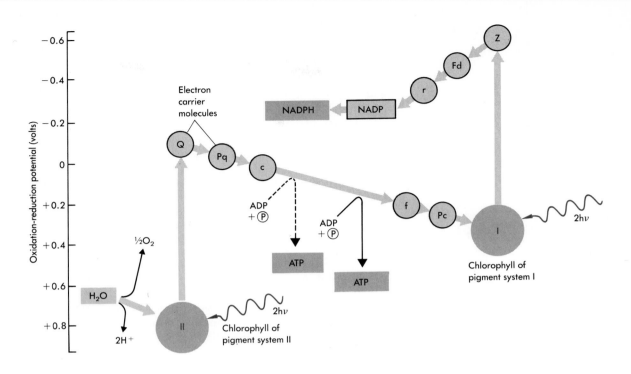

Vitamins and Other Dietary Requirements

Although some microorganisms, such as the bacteria *Escherichia coli*, can use glucose as their sole carbon and energy source, not all bacteria and none of the higher animals can use glucose to synthesize all the metabolites necessary for life. For example, rats are unable to synthesize 11 of the 20 amino acids present in their proteins; thus, their food supply must contain substantial amounts of these molecules.

In addition to dietary requirements for compounds with important structural roles, there is often a need for very small amounts of specific organic molecules. These molecules, needed in trace amounts only, are called **vitamins**. For many years, these molecules seemed quite mysterious. Now we realize that the vitamins are closely related to the coenzymes. Some are precursors of coenzymes and some are coenzymes themselves. For example, the vitamin niacin is used in the synthesis of NAD^+. The fact that coenzymes, like enzymes, are able to function over and over explains why vitamins are needed only in trace amounts.

Thus, there is nothing unusual about the fact that a particular molecule is sometimes required for growth; such requirements are fully explicable in chemical terms. It seems likely that the genes necessary for the synthesis of certain molecules were lost during evolution, resulting in the need for specific growth factors. There would be no selective advantage to an organism's retaining a specific gene if the corresponding metabolite were always available in its food supply.

Figure 2-13
A simplified scheme to show the passage of a pair of electrons (gray arrows) from water to NADP during photosynthesis, coupled with the formation of one or possibly two molecules of ATP. (NADP is a coenzyme similar to NAD; it has an additional phosphate group.) The NADPH and ATP formed are then used to incorporate carbon dioxide into carbohydrates. An electron, excited by the absorption of a photon by the pigment system II, is passed via a series of carrier molecules, by oxidation-reduction reactions, to the chlorophyll of pigment system I. This replaces an electron, which is emitted following the absorption of another photon, and which is passed by further carriers to reduce NADP. ATP is generated somewhere between the two carriers *c* and *f*. The electron emitted from system II is replaced by one taken from water, coupled with the release of oxygen. Q and Z are the primary electron acceptors of systems II and I, respectively. (Pq = plastoquinone, f = cytochrome *f*, Pc = plastocyanin, Fd = ferrodoxin, r = reductase, c = cytochrome b_{563}.)

The Polymeric Nature of Cellular Macromolecules

In striking contrast with the success of biochemists in understanding the behavior of small molecules like the amino acids and nucleotides, scientists interested in large molecules had arrived at only partial an-

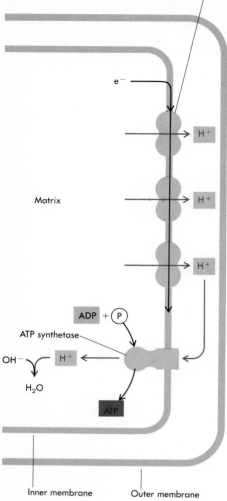

Respiratory enzyme complex

e⁻

Matrix

ADP + P

ATP synthetase

OH^- H^+

H_2O

ATP

Inner membrane   Outer membrane

Figure 2-14
Simplified scheme showing the chemiosmotic theory of the generation of ATP from ADP and phosphate. Electrons pass down a chain of membrane-bound, cytochrome-containing enzyme complexes and generate a hydrogen ion gradient across the membrane. This gradient then drives the synthesis of ATP, coupled with the formation of water (see Figure 2-11).

Figure 2-15
(a) General formula for an amino acid, where R is a side group that is different for each amino acid. Under physiological conditions, amino acids are usually in the ionized form (zwitterion). (b) Portion of a protein molecule. Three amino acids are shown linked together by peptide bonds (color). The chain of Cs and Ns forms the "backbone" of the protein.

swers before 1950 (see Table 2-3). Their only real success with macromolecules involved the polysaccharide glycogen, which is formed by the regular polymerization of large numbers of glucose molecules. Glycogen is enzymatically broken down when needed into its energy-yielding glucose monomers. The understanding of glycogen's structure, however, had little biological impact; as a highly regular polymer, it was structurally monotonous. The molecules that biologists really wanted to unravel were the proteins (because many of them are enzymes) and the nucleic acids (because they are chromosomally located).

Both proteins and nucleic acids are inherently much more complex than polysaccharides. Early in the twentieth century, the great German chemist Emil Fischer had established that proteins are polymers composed of large numbers of nitrogen-containing organic molecules called **amino acids,** linearly linked together by **peptide bonds** (Figure 2-15). (Hence chains of amino acids are called **polypeptides.**) Twenty different amino acids are found in proteins, each with a specific side group (Figure 2-16). The two nucleic acids, DNA and RNA, are also polymeric molecules, each consisting of a chain made up of four different nucleotides. Each nucleotide building block of DNA or RNA has a common sugar-phosphate group attached to a specific side group that is either a purine or a pyrimidine (see Table 2-3).

A feature common to all these biological polymers is that the individual subunits in the polymer chain contain two hydrogen atoms and one oxygen atom less than the simple monomers from which they are synthesized. Synthesis thus involves the release of water. When the polymers are degraded to yield smaller molecules, one water molecule is incorporated for each bond broken (Figure 2-17). Degradative reactions in which water uptake is required are known as hydrolytic reactions; the process is called **hydrolysis.** There are many types of hydrolytic reactions, since many small molecules can be broken down to still smaller products by the addition of water. Under normal physiological cell conditions, spontaneous hydrolysis of any of the important polymers or small molecules is *very rare*. Hydrolysis, however, is vastly speeded up by the presence of specific enzymes. For example, the enzymes pepsin and trypsin specifically catalyze the hydrolytic breakdown of proteins.

Distinction Between Regular and Irregular Polymers

Table 2-3 reveals an important difference between the polysaccharides, like glycogen, and the proteins and nucleic acids. Polysaccharides are invariably constructed by regular (or semiregular) joining of

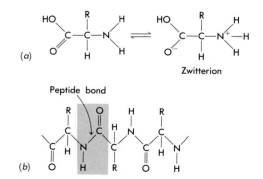

(a)

Zwitterion

Peptide bond

(b)

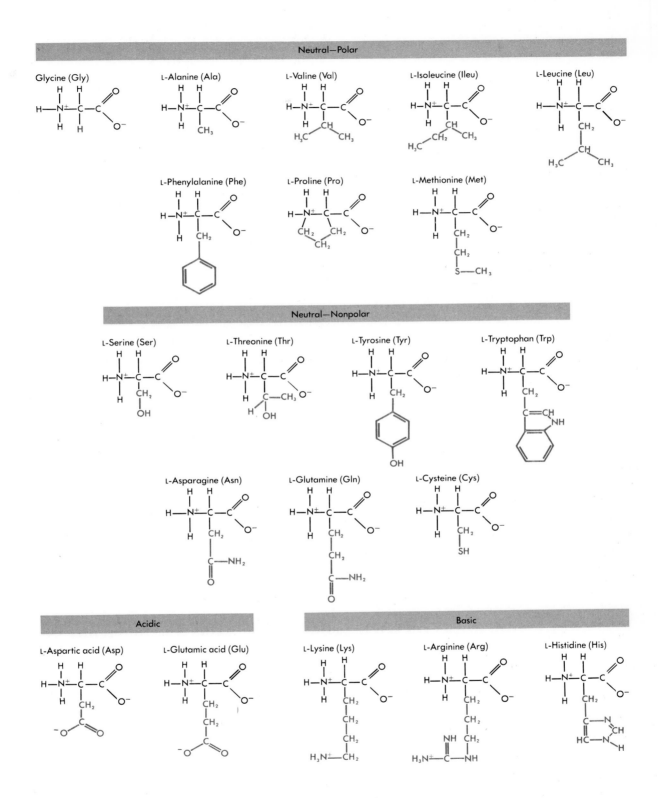

Figure 2-16

The 20 common amino acids found in proteins. The amino acids are grouped by the chemical characteristics of their side groups.

Figure 2-17
Hydrolysis of a tripeptide to form three amino acids. Breaking the two peptide bonds (shown in color) requires the addition of two water molecules.

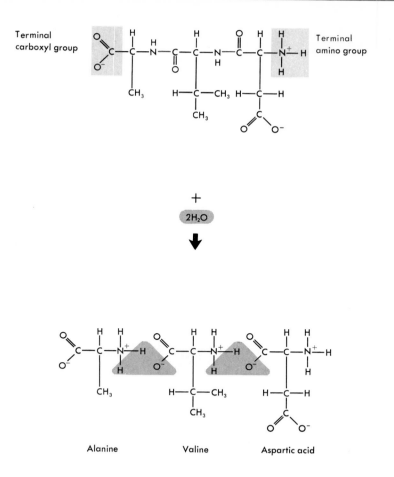

Alanine   Valine   Aspartic acid

one or two different kinds of monosaccharide building blocks. In contrast, proteins may contain up to 20 different amino acids, and nucleic acids contain 4 different nucleotides. Moreover, in the nucleic acids and proteins, the order of subunits is highly irregular and varies greatly from one specific molecule to another. While polysaccharide synthesis from monosaccharides need only involve the making of one or two types of backbone bonds, the synthesis of nucleic acids and proteins demands, in addition, a highly efficient mechanism for choosing and ordering the correct subunits.

Proteins and Nucleic Acids Are Inherently Labile[14]

A major complication to working with proteins and nucleic acids was that they appeared to be much more labile (less stable) than most small molecules. Extremes of temperature and pH (acidity or alkalinity) cause them to lose their natural shapes (**denaturation**) and sometimes to precipitate irreversibly out of solution in an inactive form. Thus, great care had to be taken in isolating them, and sometimes it was necessary to perform the entire isolation process at temperatures near 0°C. At first it was thought that only proteins were subject to denaturation, but now it is clear that nucleic acid molecules can also denature during isolation if proper precautions are not taken.

Until 1946 (just after World War II ended), the techniques of organic chemistry were the main tools for studying most small molecules. Thus, the success of a biochemist working on intermediary metabo-

Table 2-3 Structural Organization of Several Important Biological Polymeric Macromolecules

Macromolecule	Monomeric Units	Number of Different Monomers	General Monomer Formula	Fixed or Irregular Chain Length	Linkage Between Monomers
Glycogen (a polysaccharide)	Glucose	1		Indefinite: may be > 1000	1–4-Glycosidic linkage $-C_1-O-C_4-$
DNA (deoxyribonucleic acid)	Deoxyribonucleotides	4: deoxyadenylate deoxyguanylate deoxythymidylate deoxycytidylate	Purine-dexoyribose-$\circled{P}$ or pyrimidine-deoxyribose-$\circled{P}$	Genetically fixed: may be > 10^7	3′–5′-Phosphodiester linkage
RNA (ribonucleic acid)	Ribonucleotides	4: adenylate guanylate uridylate cytidylate	Purine-ribose-$\circled{P}$ or pyrimidine-ribose-$\circled{P}$	Genetically fixed: often > 3000	3′–5′-Phosphodiester linkage
Protein	L-Amino acids	20: glycine, alanine, serine, etc.	side group	Genetically fixed: usually varies between 100 and 1000	Peptide linkage

lism often depended on his or her ability as an organic chemist. In work with proteins and nucleic acids, however, most of the initial stages of research did not rely on the analytical techniques of the organic chemist. Instead, the protein chemist, before even starting to worry about the detailed structure of a protein, needed to work very hard to be sure that the protein was both chemically pure and biologically active. Gentle isolation techniques had to be devised that avoided the usual strong acids and alkalies of organic analysis. Then additional techniques were needed to reveal whether the product was homogeneous and to provide data on molecular size. For these sorts of questions, the help of physical chemists was indispensable, and there developed a new line of research investigating the physical-chemical properties of macromolecules in solution. It concerned itself with topics like the osmotic properties of macromolecular solutions and the movement of macromolecules under electrical and centrifugal forces.

Perhaps the most striking contribution of physical chemistry to the study of biological macromolecules was the development, in the 1920s, of centrifuges that could rotate at very high speeds (**ultracentrifuges**) and cause the rapid sedimentation of proteins and nucleic acids. The initial development of the ultracentrifuge was the work of The Svedberg, of Sweden, after whom the unit of sedimentation (S for Svedberg) was named. Ultracentrifuges equipped with optical devices to observe exactly how fast the molecules sedimented

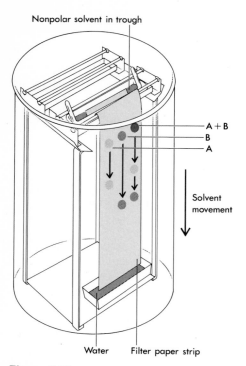

Nonpolar solvent in trough

A + B
B
A

Solvent movement

Water   Filter paper strip

Figure 2-18

Partition chromatography. Molecules in a mixture can be separated on the basis of their different solubilities in polar versus nonpolar solvents. Here we show one of many methods developed for paper chromatography of amino acids or peptides. A drop of the mixture is placed on a strip of filter paper, which is then hung in a closed vessel containing some water (the polar solvent). The water vapor moistens the paper. One end of the paper dips into a trough filled with the nonpolar solvent. As the nonpolar solvent moves along the moist surface of the paper, it carries solutes with it. Each substance in the original mixture moves a distance proportional to its relative solubilities in the two solvents. The finished chromatogram is dried and then stained with ninhydrin, which forms colored reaction products with the colorless amino acids.

were extremely valuable in obtaining data on the molecular weights of proteins and in establishing the concept that proteins, like smaller biological molecules, are of discrete molecular weights and shapes. This work revealed that the sizes of proteins vary greatly, with a continuous range in weights between the extremes of approximately 10,000 and 1,000,000.

Implications of Chromatography[15]

The means to determine the exact order in which amino acids are linked together to form a particular protein remained a great puzzle until 1951. This was partly because there are 20 different amino acids (see Figure 2-16) and their proportions vary from one type of protein to another. Until about 1942, the methodological problems involved in amino acid separation and identification were formidable, and most organic chemists chose to work with simpler molecules.

This state of affairs changed completely in 1942, when the Englishmen A. J. P. Martin and R. L. M. Synge developed separation methods that depended on the relative solubilities of the several amino acids in two different solvents (**partition chromatography**). Particularly useful were methods by which amino acids were separated on strips of paper (Figure 2-18). With these new techniques, it became a routine matter to separate quantitatively the 20 amino acids found in proteins.

Amino Acids Within Polypeptide Chains Are Linearly Linked Together in Unique Sequences[16, 17]

By breaking down polypeptide chains randomly into many smaller fragments and using the newly developed chromatographic methods to reveal their amino acid compositions, Frederick Sanger, working at Cambridge University in the early 1950s, established the first definitive order of amino acids within a protein. The molecule he examined, the hormone insulin, contains two separate polypeptide chains, A and B, held together by pairs of disulfide bonds (S—S) between sulfur atoms on the side groups of opposing cysteine residues. Sanger's work was a milestone in the study of proteins, for it was the first demonstration of the now accepted principle that a given type of protein contains a unique sequence of amino acids.

Synthesis of the S—S bonds holding the A and B insulin chains together occurs after the respective cysteine residues have been incorporated into their polypeptide chains. This was first suggested by experiments in which the separate chains were mixed together and then observed often to form the correct S—S bonds between appropriate partner cysteines. Many years later, D. F. Steiner at the University of Chicago showed that the A and B insulin chains are originally part of a single nonfunctional polypeptide chain (proinsulin) (Figure 2-19). As the chain is synthesized, it spontaneously adopts a three-dimensional configuration that brings close together the cysteine residues destined to form S—S bonds to each other. After these bonds form, the intact polypeptide chain is cleaved twice by a highly specific polypeptide-clipping enzyme, thereby releasing an inactive polypeptide from the now two-chained, functional insulin molecule.

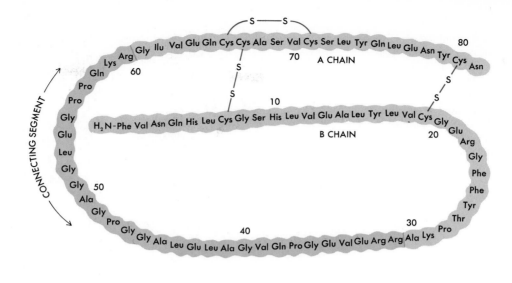

Figure 2-19
Structure of bovine proinsulin. The amino acids removed during its conversion to insulin are shown in color.

The 25-Year Loneliness of the Protein Crystallographers[18, 19, 20]

An equally significant step in understanding macromolecules was the effective extension of X-ray crystallographic techniques to their study. **X-ray crystallography** utilizes the diffraction of X-rays by crystals to give precise data about the three-dimensional arrangement of the atoms in the crystals. The first successful use of X-ray diffraction was in 1912, when the Englishman Lawrence Bragg solved the structure of sodium chloride (NaCl). This success immediately initiated research on the structures of molecules of increasing complexity.

The technique used in the first X-ray diffraction studies of small molecules consisted of guessing the structure, calculating the theoretical diffraction pattern predicted by this structure, and then comparing the calculated pattern with the observed pattern. This method was practical for studying relatively simple structures, but was not often useful in the study of larger structures. It took much insight on the part of Bragg and the great American chemist Linus Pauling, in the 1920s, to solve the structures of some complicated inorganic silicate molecules. Clearly, however, proteins were too complicated for even the best chemist to guess their three-dimensional structures. Thus, the early protein crystallographers knew that until new methods for structural determination were found, they would have no results to present to the impatient biochemists, who were increasingly eager to know what proteins actually looked like.

The first serious X-ray diffraction studies on proteins began in the mid-1930s in J. D. Bernal's laboratory in Cambridge, England. Here it was found that although dry protein crystals gave very poor X-ray patterns, wet crystals often gave beautiful pictures. Unfortunately, however, there was no logical method available for their interpretation. Nonetheless, Bernal's student Max Perutz, an Austrian then in England, slowly increased the pace of his work (begun in 1937) with the oxygen-carrying blood protein **hemoglobin.** He had chosen hemoglobin for several reasons: Not only is it one of the most important of all animal proteins, but it is also easy to obtain, and it forms crystals that lend themselves well to crystallographic analysis. For many years, however, no very significant results emerged either from

Figure 2-20
The three-dimensional structure of myoglobin as determined by X-ray crystallography. Each numbered circle represents the central carbon of one amino acid; other atoms are not shown. Note the eight segments of α helices, labeled A through H. Also, NA is the amino terminal, and HC is the carboxyl terminal. (Copyright Irving Geis.)

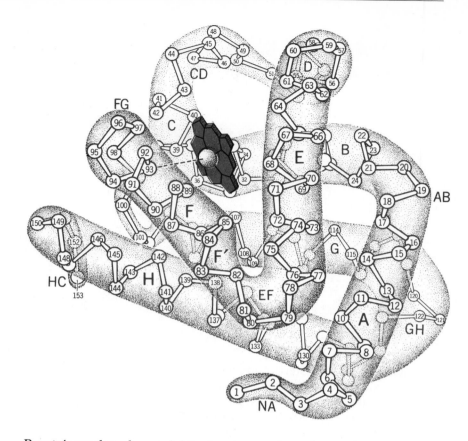

Perutz's work on hemoglobin structure or from the work begun in the same laboratory in 1947 by John Kendrew on the structure of the muscle protein myoglobin. This protein, which, like hemoglobin, combines with oxygen, had the added advantage of being only a fourth the size of hemoglobin (MW = 17,000).

During that lonely period of no real results, there was only one triumph. Pauling correctly guessed from stereochemical considerations that amino acids linked together by peptide bonds would sometimes tend to assume helical configurations, and he proposed in 1951 that a helical configuration, which he called the alpha (α) helix, would be an important element in protein structure. (The α helix will be discussed later in this chapter and in Chapter 5.) Support for Pauling's α-helix theory came soon after its announcement when Perutz demonstrated that several synthetic polypeptide chains containing only one type of amino acid exist as α helices.

It was not until 1959 that Perutz and Kendrew got their answers. An essential breakthrough, which occurred in 1953, showed how the attachment of heavy atoms to protein molecules could logically lead from the diffraction data to the correct structures. For the next several years, these heavy-atom methods were exploited at a pace undreamed of 20 years before, largely as a result of the availability of high-speed electronic computers. Then, to everyone's delight, the X-ray diffraction measurements could at last be translated into the enormously complicated arrangement of atoms in myoglobin (Figure 2-20) and hemoglobin (see Chapter 3). Most importantly, their molecular configurations were found to obey in every respect the chemical laws that govern the shape of smaller molecules. Absolutely no new laws of nature are involved in the construction of proteins. This was no surprise to the biochemists.

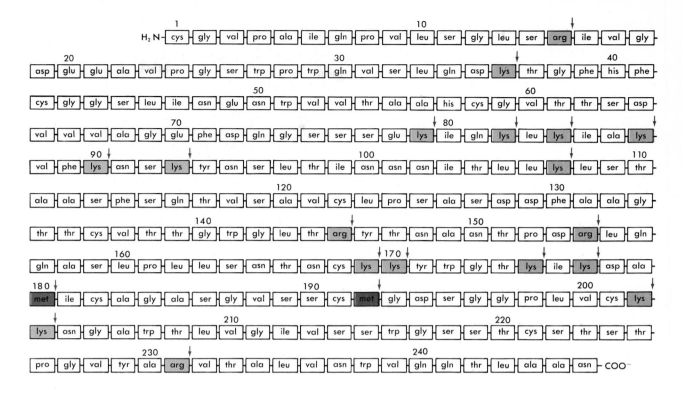

Figure 2-21

The amino acid sequence of the protein chymotrypsinogen. Key to the establishment of such protein sequences is the isolation of smaller, well-defined polypeptide fragments. The enzyme trypsin, for example, cleaves on the carboxyl side of arginine and lysine residues (light color). The so-called tryptic fragments have proved invaluable for working out almost all protein sequences. Larger fragments are generally made by treatment with the reagent cyanogen bromide, which cleaves the peptide bond on the carboxyl side of the much less common amino acid methionine (dark color).

The Primary Structure of Proteins[21]

With the establishment of those first three-dimensional structures, it became possible to analyze the arrangements of atoms within proteins at three levels of order. The first level, the **primary structure,** is the description of the covalent bonds within a protein. It tells us the number of polypeptide chains and the sequence of amino acids within them. For example, the primary structure of myoglobin is a single polypeptide chain of 153 amino acids. Many other proteins also contain only a single type of chain. One such protein is chymotrypsinogen, with 246 amino acids. After the conversion of this protein to its active form, chymotrypsin, it catalyzes the hydrolysis of several highly specific peptide bonds. The determination of the sequence of chymotrypsinogen in the early 1960s required some 15 years of collective work by several talented chemists (Figure 2-21). Today, new experimental techniques allow such sequences to be established much more quickly. With moderate luck, less than a year may be sufficient for an expert protein chemist to work out a sequence 200 to 300 amino acids long.

Many other proteins are more complicated in that they possess several types of polypeptide chains. For example, hemoglobin contains both α and β chains, each present in two copies ($\alpha_2\beta_2$). The primary structure also specifies which cysteine residues are held together by disulfide bonds (S—S). Disulfide bonds are important in helping a protein maintain a particular shape by linking together either two parts of a single polypeptide chain or two different chains. As was shown in Figure 2-19, they are what hold together the two chains of the insulin molecule. In chymotrypsinogen, there are five disulfide bridges, each linking specific cysteine residues in the single chain. In chymotrypsin, which is derived from chymotrypsinogen by polypeptide chain cleavage, two of these disulfide bonds connect separate chains. In other cases (e.g., hemoglobin), the several chains are held together by weaker, noncovalent bonds (see next section).

In addition, a number of proteins have attached to them nonprotein components called **prosthetic groups,** which play a vital role in their function. In contrast to coenzymes, prosthetic groups are bound very tightly to their protein, sometimes by covalent bonds. They are often metal-organic compounds. Both myoglobin and hemoglobin contain the prosthetic group *heme,* a metal-organic compound closely related to the porphyrin component of chlorophyll. Heme combines with oxygen and enables hemoglobin and myoglobin to bind oxygen.

A characteristic feature of prosthetic groups is that they possess very little functional activity unless they are attached to a polypeptide partner. Heme by itself, for example, combines with oxygen in an effectively irreversible fashion. Only when heme is attached to either myoglobin or hemoglobin does it possess the quality of reversibly binding oxygen. Then it can release bound oxygen when it is needed under conditions of oxygen scarcity.

Secondary Structures of Proteins May Be Sheets or Helices

Individual extended polypeptide chains are conformationally unstable and frequently either assume contracted helical configurations or aggregate side by side to form sheetlike structures. The driving force that leads to either of these two alternatives is the strong tendency of the carbonyl (CO) and imino (NH) groups that flank the peptide bonds to form hydrogen bonds, the most specific of the "weak" (noncovalent) secondary bonds (see Chapter 5 for an extensive discussion of their actions). The carbonyl group next to one peptide bond can hydrogen-bond either to an imino group flanking a peptide bond several residues removed from it on the same chain, or to an imino group flanking a peptide bond on a different polypeptide chain. The resulting hydrogen-bonding patterns are highly repetitive and lead to regular polypeptide conformations that are known as the **secondary structures** of proteins.

The stereochemically most satisfactory hydrogen-bonded helical conformation of the polypeptide backbone is the **α helix,** in which 3.6 amino acids are found for every turn of the helix (Figure 2-22). Significant sections of the polypeptide chains of myoglobin and hemoglobin were found to be so arranged, and now the α helix is recognized as a major secondary structural feature of many, if not most, proteins. Equally important are regular, hydrogen-bonded, lateral associations of extended polypeptide chains into sheets, called **β sheets** (Figure 2-23). β sheets are especially favored by the presence of large numbers of glycine residues. Their existence was first suggested by X-ray crystallographic analysis of the fibrous protein of silk, carried out in the 1930s by W. T. Astbury and his collaborators in Leeds, England. More recently, β sheets have also been found within a very large number of globular (nonfibrous) proteins.

Tertiary Structures of Proteins Are Exceedingly Irregular

The **tertiary structure** of a protein is its three-dimensional form. It is, in many cases, very irregular. Few proteins are built up from α helices or β sheets alone, or even from a simple combination of the two.

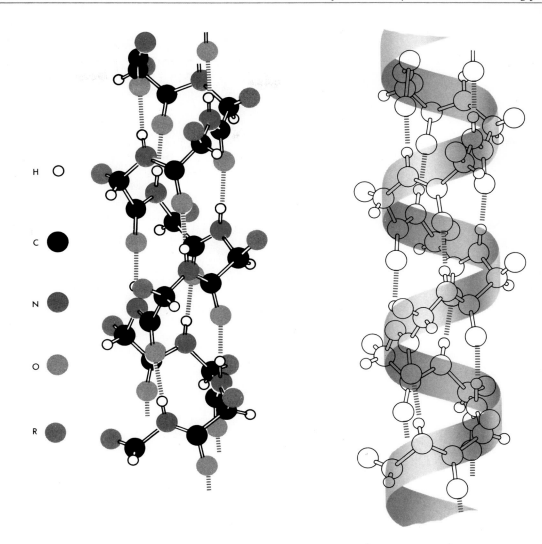

Instead, most proteins contain both regular and irregular regions. Some proteins, in fact, seem to have almost no regular structure. There are a number of stereochemical reasons why α helices, β sheets, and other regular secondary arrangements are not found more extensively. One reason is that the amino acid proline does not contain an amino group, and so wherever it occurs, the regular hydrogen bonding must be interrupted.

The most important reason for irregularity in protein structures, however, arises from the diverse chemical nature of the amino acid side groups (see Figure 2-16). Each of these side groups will tend to make the energetically most favorable secondary interactions with other chemical groups. As an example, the free hydroxyl group on tyrosine will tend to assume a position where it can form a hydrogen bond. The considerable energy of the bond would be lost if, for example, the hydroxyl group were next to an isoleucine side group, which is incapable of forming a hydrogen bond. Furthermore, the side groups of several amino acids, like valine and leucine, are insoluble in water, whereas others, like those of glutamic acid and lysine, are highly soluble. It thus makes chemical sense that the water-insoluble side groups are found stacked next to one another in the interior of a protein, whereas the external surface contains groups that mix easily with water.

Figure 2-22

A polypeptide chain folded into a helical configuration called the α helix. All the backbone atoms have identical orientations within the molecule. It may be looked at as a spiral staircase in which the steps are formed by amino acids. There is an amino acid every 1.5 Å along the helical axis. The distance along the axis required for one turn is 5.3 Å, giving 3.6 amino acids per turn. The helix is held together by hydrogen bonds between the carbonyl (CO) group of one residue and the imino (NH) group of the fourth residue down along the chain. [Molecular structure after L. Pauling, *The Nature of the Chemical Bond*, 3rd ed. (Ithaca, N.Y.: Cornell University Press, 1960), p. 500, with permission.]

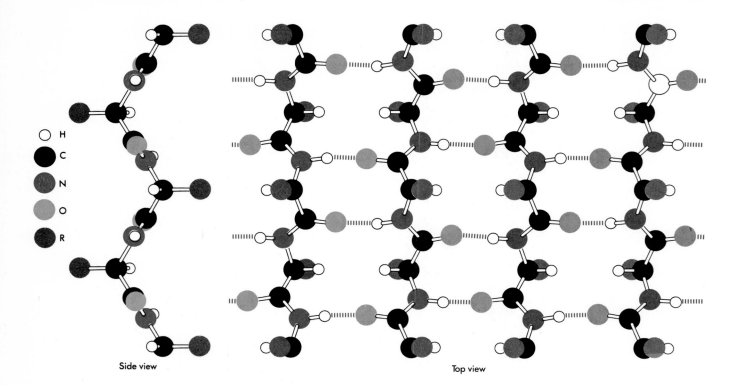

H
C
N
O
R

Side view

Top view

Figure 2-23

Several extended polypeptide chains held together by hydrogen bonds in a configuration called a β sheet. [Molecular structure after L. Pauling, *The Nature of the Chemical Bond,* 3rd ed. (Ithaca, N.Y.: Cornell University Press, 1960), p. 501, with permission.]

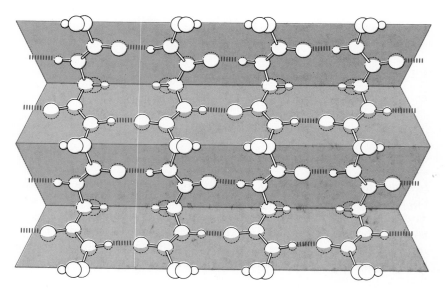

The tertiary configuration thus represents the energetically most favorable arrangement of the polypeptide chain. Each specific sequence of amino acids takes up the particular "native" arrangement that makes possible a maximum number of favorable atomic contacts between it and its normal environment. This concept is strongly supported by very striking experiments in which high temperature or some other unnatural condition denatures the protein to give randomly oriented, biologically inactive polypeptide chains. When the denatured chains are gently returned to their normal environment, some of them are found to **renature**—to resume their native conformation with full biological activity.

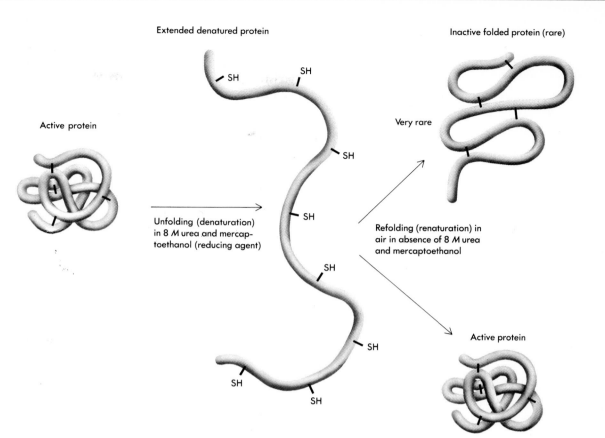

Extended denatured protein

Inactive folded protein (rare)

Active protein

Very rare

Unfolding (denaturation)
in 8 *M* urea and mercap-
toethanol (reducing agent)

Refolding (renaturation) in
air in absence of 8 *M* urea
and mercaptoethanol

Active protein

S–S Bonds Form Spontaneously Between Correct Amino Acid Residues[22]

In many cases, renaturation of a disordered protein to an active form involves not only the formation of thermodynamically favorable weak bonds, but also the making of specific disulfide (S–S) bridges. This was first shown by experiments with the enzyme ribonuclease, a protein consisting of one polypeptide chain of 124 amino acids, cross-linked by four specific S–S bonds. The native, active configuration of the enzyme can be destroyed by reducing the S–S groups to sulfhydryl (SH) groups in the presence of the denaturing agent 8 *M* urea and the reducing agent mercaptoethanol (Figure 2-24). When the urea is removed, oxygen in the air reoxidizes the SH groups to yield S–S bonds identical to those found in the original molecule. A given SH group reassociates, not randomly with any of the other seven SH groups in the molecule, but rather with a specific SH group brought into close contact with it by the folding of the polypeptide chain. Thus, S–S bridges are not a primary reason for the peculiar folding of the chain. They might be better viewed as a device for increasing the stability of an already stable configuration.

Visualization of the Active Sites of Enzymes[23, 24]

Although for decades organic chemists had speculated about how specific amino acid side groups might speed up certain chemical reactions, convincing explanations had to await detailed three-dimensional structural analysis. Such an analysis was first accom-

Figure 2-24
The fate of S–S bonds during protein denaturation and renaturation. When the denaturing agents are removed, most of the polypeptide chains resume the native configuration with the original S–S bonds. Only a few polypeptide chains fold up in an inactive form characterized by a different set of S–S bonds than those found in the native molecules.

plished in London in 1967 when D. C. Phillips and his colleagues at the Royal Institution worked out the structure of the enzyme lysozyme. Knowing its conformation enabled them to propose a precise chemical mechanism for the way lysozyme breaks down specific polysaccharide chains into their component sugar groups. This work, together with that of other scientists who have since established the conformations of over 35 additional enzymes, indicates that the chemical mechanisms of enzyme-catalyzed reactions can be as well understood as the best-known chemical reactions of pure organic chemistry.

Particularly pleasing has been the clear establishment of the way a number of proteolytic enzymes catalyze specific breaks in polypeptide chains. The working out of the three-dimensional structures of trypsin and chymotrypsin reveals very similar molecular shapes, with their active sites sharing many common amino acids. Three of these amino acids—serine, histidine, and aspartic acid—are directly involved in the catalytic process, with each residue playing identical roles in both enzymes (Figure 2-25). We now see that amino acid sequence data by themselves seldom can reveal profound information on how proteins function.

Lipid Bilayers Give Cellular Membranes Their Sheetlike Properties[25]

The cytoplasm of a cell is surrounded by a lipid-containing semipermeable membrane that prevents escape of cellular constituents to the outside environment as well as facilitates the selective entry into the cell of food molecules and inorganic ions needed for growth and cellular maintenance. Because these membranes are very thin and highly insoluble in aqueous environments, their underlying structural organization was long a mystery. Key to the working out of the so-called unit membrane structure was the early study of the red blood cells (erythrocytes) of mammals. These cells lack a chromosome-containing nucleus as well as any other internal membranous organelles (such as mitochondria); the only membranes they contain are their external plasma membranes. Most of their lipids, as well as those of all other membranes, are **phospholipids**; they contain phosphate "head" groups attached to long hydrocarbon tails. The four major phospholipids found in all higher cells are phosphatidylserine,

Figure 2-25 (Opposite)

Involvement of specific amino acid residues in the catalytic breakdown of peptide bonds by trypsin. The catalytic mechanism of trypsin accelerates the hydrolysis of peptide bonds by providing intermediate states for the reaction and by smoothing the transition from one intermediate to the next. Enzyme and substrate must first come together (1) to form a precisely oriented complex (2). The oxygen atom of serine 195 then bonds covalently to the substrate carbon, forming a tetrahedral intermediate (3); the proton, or hydrogen ion, from serine 195 is transferred to the substrate nitrogen. The proton transfer breaks the peptide bond, and the first product is liberated (4). The remaining complex is called an acyl enzyme; it breaks down to regenerate free enzyme in steps that are symmetrical with those of the first half of the process. A water molecule enters the reaction (5), its hydroxyl group forming with the substrate another tetrahedral intermediate (6). The hydrogen ion from the water molecule is transferred to serine 195, breaking the covalent bond between enzyme and substrate (7). The second product is now freed (8); its departure is hastened by repulsion between negatively charged carboxyl groups of product and aspartic acid 102. Histidine 57 and aspartic acid 102 participate in the proton transfers. [Redrawn from "A Family of Protein-Cutting Proteins," by R. Stroud. *Sci. Amer.* 231 (1974):76. Copyright © 1974 by Scientific American, Inc. All rights reserved.]

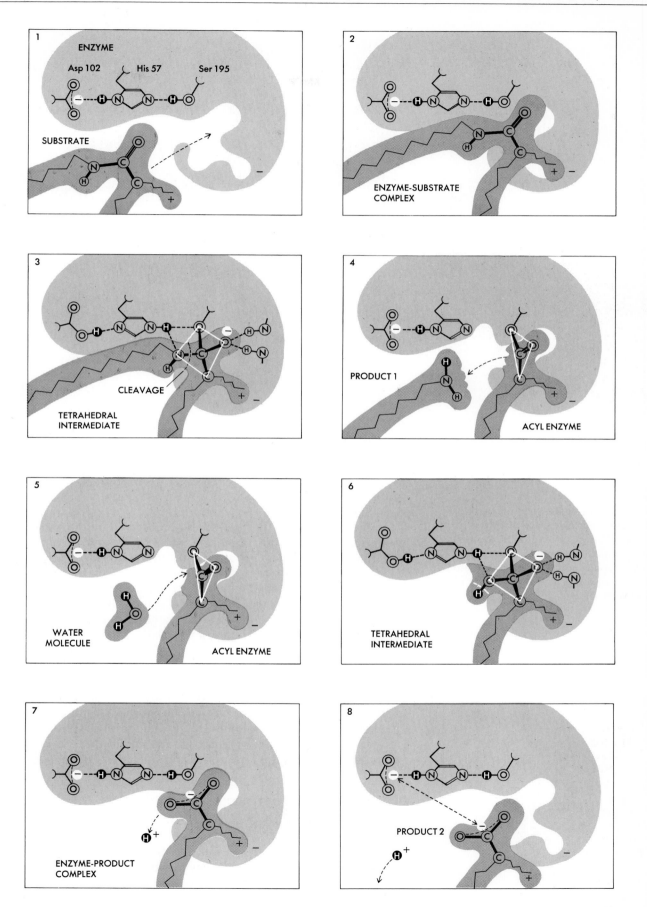

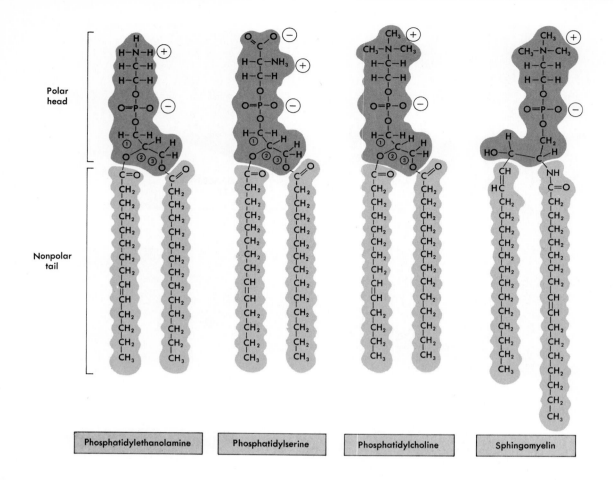

| Phosphatidylethanolamine | Phosphatidylserine | Phosphatidylcholine | Sphingomyelin |

Figure 2-26
The structure of the four major phospholipids found in eucaryotic cell membranes. The exact lengths of the fatty acid nonpolar tails are variable. ①, ②, and ③ indicate the three carbon atoms of glycerol, on which three of the phospholipids are based.

phosphatidylethanolamine, phosphatidylcholine, and sphingomyelin (Figure 2-26).

In the late 1930s, J. F. Danielli and H. Davson first postulated that lipid building blocks are organized into a 50 Å thick bilayer in which the charged phosphate groups are all on the outside. Such a bilayer arrangement neatly places all the water-insoluble hydrocarbon tails in contact with one another, leaving the highly charged phosphate head regions free to interact with the polar water environment. Probably from the very start of life on Earth, membranes possessed the same bilayer arrangement as that present today.

Most cell membranes also contain several types of **glycolipids,** in which one to several water-soluble sugar groups are attached to a hydrophobic tail made up of a fatty acid and the amino alcohol sphingosine (Figure 2-27). **Steroids** are also found within most bilayers, with cholesterol present in large amounts in virtually all vertebrate membranes. Like phospholipids, glycolipids and steroids are arranged so that their hydrophobic groups are in the center of the bilayer with their charged groups facing outward (Figure 2-28).

Insertion of Membrane Proteins into Lipid Bilayers[26]

Most membranes contain a variety of different protein molecules directly inserted into the lipid bilayer. Some pass right through the bilayer, having groups that protrude from both its sides. Those amino

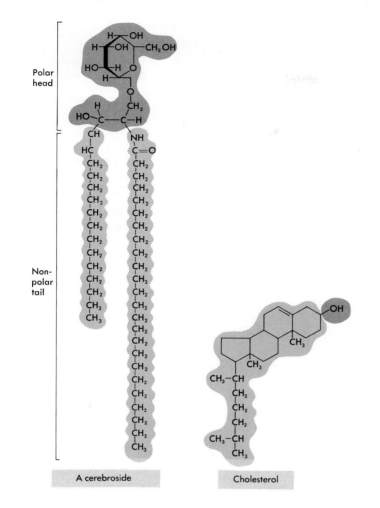

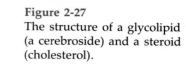

Figure 2-27
The structure of a glycolipid (a cerebroside) and a steroid (cholesterol).

A cerebroside

Cholesterol

acid side chains that lie within the bilayer are largely hydrophobic. In contrast, hydrophilic amino acids predominate in the regions that face out onto the surfaces (Figure 2-29). Depending on their function, membranes have very specific protein components. Some of the most highly specialized membranes contain only one major protein component (e.g., the protein of the retinal membrane is almost exclusively composed of the visual pigment rhodopsin).

Although most electron micrographs suggest that membranes are symmetrical, the two sides of a given membrane have fundamentally different chemical compositions. Phosphatidylserine and phosphatidylethanolamine face toward the cytoplasmic (inner) side, while phosphatidylcholine and sphingomyelin are present largely on the exterior surface. Those proteins that are inserted into membranes have specific regions that normally face exclusively inward or outward. The regions that face outward may have sugar groups covalently attached to them, making these proteins **glycoproteins.** The location of sugars only on the outer-facing side of membrane proteins suggests that the glycosylating enzymes that transfer sugars to membrane proteins are found exclusively on the outer membrane surface.

Until very recently, it has not been possible to crystallize membrane proteins, and conventional X-ray crystallographic techniques could not be used to work out their three-dimensional structures. Crystalline two-dimensional sheets of the purple membrane protein of the

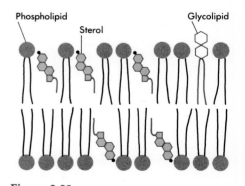

Figure 2-28
How various lipids are arranged in the cell membrane. Hydrocarbon chains are in the center of the bilayer, and the polar heads face outward.

Figure 2-29
How glycophorin, a large glycoprotein, may be inserted in the red blood cell membrane.

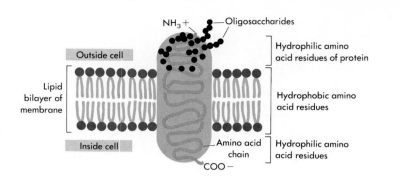

photosynthetic bacterium *Halobacterium halobium*, however, have been analyzed using electron micrograph image reconstruction techniques. In 1975, R. Henderson and P. N. T. Unwin in Cambridge, England, showed that the purple membrane protein contains seven closely packed α helices lying nearly perpendicular to the plane of the membrane (Figure 2-30).

Semifluidity of Cell Membranes[27, 28]

There is now very good evidence that the individual phospholipid and protein molecules do not maintain fixed orientations but are in constant movement within the two-dimensional fabric of a given membrane. Molecules that at one moment may be in touch, can move to opposite sides of a 10 μm diameter cell within an hour or so. Membranes thus can be thought of as two-dimensional liquids, virtually all of whose components are in constant diffusional motion. They have the essential quality of a self-sealing rubber tire, and holes quickly disappear owing to diffusion of nearby hydrocarbon molecules. An equally important property is the potential ability of adjacent membranous sacs to fuse on contact. Membranous sacs also have the capacity to divide when seemingly random motion creates unstable pinched-off regions. Probably most of these fusion and division events are not random occurrences but responses to well-defined stimuli. Conceivably, they are mediated by forces exerted on the lipid bilayers that momentarily create minute tears in the two-dimensional surface. If so, fusion and division events may be the result of rearrangements of such torn surfaces, whereby inner surfaces fuse with other inner surfaces while outer surfaces match up with outer surfaces. Figure 2-31 shows the probable mechanism by which membrane fusion and division events lead to **pinocytosis** (the uptake of macromolecular particles from the solution surrounding a cell) and the opposite process, **exocytosis.**

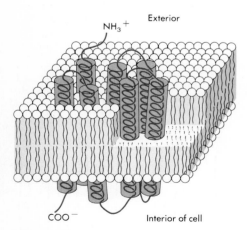

Figure 2-30
Purple membrane protein of *Halobacterium halobium*. This protein has seven α helices perpendicular to the membrane plane. Here the helices are represented by colored cylinders. [After R. Henderson, *Ann. Rev. Biophys. Bioeng.* 6 (1977):87.]

Membranes Contain Pumps and Gates[29, 30, 31]

The phospholipid component by itself provides a firm barrier to the transport across the membrane of most cellular molecules, the vast majority of which contain one to several polar groups that effectively cannot enter the hydrophobic hydrocarbon environment that characterizes the interior of a lipid bilayer. The functioning of membranes as agents that selectively transport food molecules, inorganic ions, and intracellular metabolites from one side of a membrane to the other depends on many of the hundreds of different proteins that characterize the membranes of most cells. Many of these proteins act as active pumps, moving molecules against concentration gradients. To do so, they couple the breakdown of intracellular ATP with the inward or outward movement of desired food molecules or inorganic ions. Exactly how these directed movements occur cannot be determined until the exact conformations of the respective membrane proteins have been worked out. Among the best-known cellular pumps are the several polypeptide chains that aggregate together to form the **Na⁺-K⁺pump** that maintains the high potassium and low sodium intracellular levels relative to the extracellular medium. The movement of these two ions in opposite directions is linked to ATP breakdown, most likely through cyclical conformational changes in the respective polypeptide chains induced by phosphorylation of key amino acid residues. Such pumps play essential roles in the cellular economy of many resting cells; about one-third of the total cellular ATP is used to maintain intracellular sodium and potassium levels.

In addition to such active-transport processes, other membrane proteins function as passive gates through which metabolites and ions can flow from regions of higher concentration to regions of lower concentration. In vertebrates, the best-known gates are the **gap junctions,** composed of six identical polypeptides surrounding a 15–20 Å diameter channel through which most small molecules (e.g., sugars, amino acids, and nucleotides) can flow virtually unimpeded from one cell to another (Figure 2-32). Gap junctions do not necessarily remain open. When the normally very low (10^{-8} *M*) intracellular calcium levels rise, the gap-junction proteins undergo conformation changes that close the gap.

Hormones and Their Receptors[32, 33]

By the late nineteenth century, it was realized that the cells of multicellular organisms are in part controlled by molecules emanating from far distant cells. These chemical agents, given the name **hormones** by the English physiologists W. M. Bayliss and E. H. Starling in 1909, fall into three main chemical groups: amino acid derivatives, polypeptides, and steroids (derived from cholesterol). Only over the past twenty years have their modes of action begun to be established at the molecular level. In all cases, they act by regulating preexisting cellular reactions, not by themselves acting as enzymes or coenzymes. To do so, they must first bind to specific receptors on their effector cells, unions which then lead to chains of events that dramatically alter preexisting patterns of enzyme synthesis or function.

The cellular receptors for steroid hormones (e.g., the female estrogenic hormones) are specific proteins located largely in the nucleus (Figure 2-33). As many as 30,000 such receptors exist in appropriate target cells. No specific receptors need exist on the cell's surface, since the highly lipid-soluble steroids passively penetrate lipid bilayers, pass through the cytoplasm, and enter the nucleus. After steroid hormones bind to their receptors, the hormone-receptor complexes affect gene function through their binding to specific chromosomal segments. In contrast, the receptors for hormones derived from polypeptides and amino acids are specific membrane proteins. The insulin receptor, for example, is a membrane-sited polypeptide with an approximate molecular weight of 135,000 daltons. In target fat cells, it is

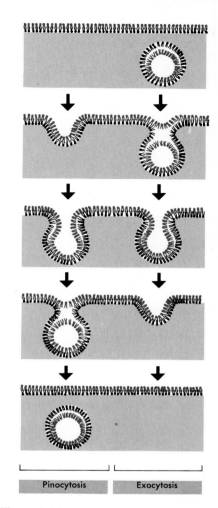

Pinocytosis   **Exocytosis**

Figure 2-31
Pinocytosis (left) and exocytosis (right) probably occur by symmetrical mechanisms of membrane fusion and division. (The outer half of the membrane lipid bilayer appears in color; the inner, cytoplasmic half, appears in black.)

Figure 2-32
A gap junction. Each gap-junction pore consists of an aggregation of identical polypeptides that form a channel through which small molecules can flow between two cells. [After L. Makowski et al., *J. Cell Biol.* 74 (1977):629.]

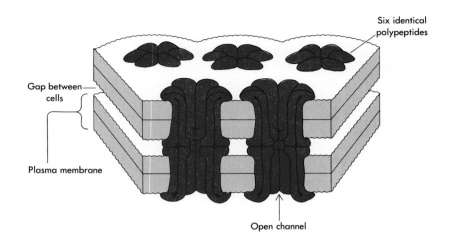

present in some 10^4 copies per cell and represents slightly less than 1 percent of the membrane protein.

cAMP as the Key Intermediate in the Action of Many Polypeptide Hormones[34, 35, 36]

Many surface-binding proteins act in part through raising or lowering the amount of the key cellular mediator **cyclic AMP (cAMP)**. Its action as a "second messenger" was elucidated in the early 1960s by the classic investigations of the American biochemist E. W. Sutherland at Vanderbilt University in Nashville, Tennessee. Cyclic AMP is formed from ATP by the enzyme adenylate cyclase, which is bound to the cytoplasmic side of plasma membranes. So located, the activity of adenylate cyclase is influenced by the binding of hormones to their specific surface receptors (Figure 2-34). Such influences, however, are not direct, with the hormone-receptor complexes inducing changes in the conformations of a still incompletely understood class of proteins known as the G proteins (because of their ability to bind the ATP-like nucleotide GTP). In turn, the conformationally altered G proteins directly regulate adenylate cyclase activity.

After its synthesis, cAMP works by stimulating the activity of protein kinases, enzymes that transfer terminal phosphate groups from ATP to specific amino acids on target proteins. Phosphorylation alters the enzymatic activities of these proteins, either raising or lowering them. The skeletal muscle enzyme glycogen synthetase, for example, is inactivated after phosphorylation by a cAMP-activated protein kinase. In contrast, phosphorylation of glycogen phosphorylase makes it more active. Through these dual actions, cAMP simultaneously stimulates glycogen breakdown and inhibits its synthesis.

The use of second messengers like cAMP greatly amplifies the action of single-hormone molecules, since for every polypeptide- or amino acid-derived hormone bound by its receptor, hundreds of cAMP molecules are made. In turn, each cAMP molecule, by binding to an appropriate protein kinase, activates or inactivates many hundreds of specific target proteins. Of equal biological importance is the amplification of the signals carried by individual steroid hormone molecules. By making a specific gene work much faster or slower, the synthesis of its protein product is drastically increased or decreased.

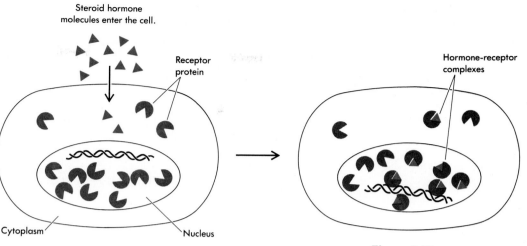

Figure 2-33
Steroid hormones diffuse across the plasma membrane and bind to specific receptors within the cell. The hormone-receptor complex affects gene function by binding to chromatin in the nucleus.

The Classical Period of Biochemistry Is Drawing to a Close

Until recently, the study of the chemical reactions underlying the living processes has been dominated by the need to define the cast of key characters (molecules) and the enzymatic pathways by which they become transformed into one another. These pursuits have been successful beyond any reasonable anticipation. Now all major features of living cells—the generation of useful energy from food molecules and the sun, the ways energy-rich molecules promote the making of desired chemical bonds, and, as we shall relate in the next chapter, the functioning of genes—can be understood in terms of well-defined chemical principles.

Biochemistry has become an enormous discipline whose basic metabolic and biosynthetic principles will probably never be seriously modified. So over the next several decades, we shall be slowing down our search for still undiscovered key enzymes. Instead, more and more future effort will go to the working out of the mechanisms of enzymatic reactions at the molecular and atomic levels. For example, how light quanta are collected by the photosynthetic apparatus, or how a membrane protein pumps ions, increasingly have become the important biochemical questions that only chemically oriented scientists (as opposed to biologically oriented investigators) will be able to answer.

These changes should not, however, in any way distress those of us who have become interested in biochemistry through our curiosity about genes. As we shall see in the later chapters of this book, it is only because the basic facts of biochemistry are now so firm that we shall soon be able to move on from the way genes store and express hereditary information to attack the still unsolved major biological problem of how fertilized eggs develop into the marvelously complex forms of higher plants and animals. Only by feeling that the essence of the living single cell is now well within our grasp, can we so optimistically proclaim that the essence of multicellular existence may now be an achievable objective.

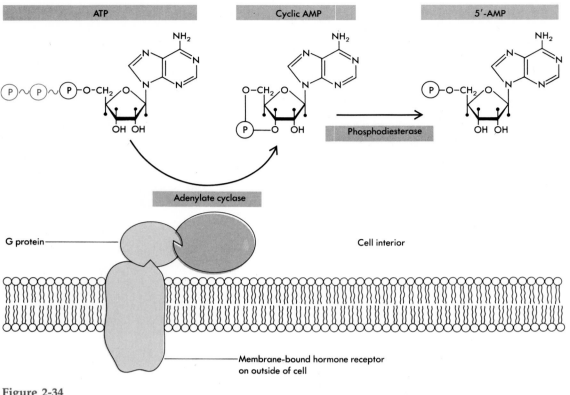

Figure 2-34
Cyclic AMP synthesis and breakdown. Cyclic AMP (cAMP) is synthesized
by the membrane-bound enzyme adenylate cyclase. The synthesis of
cAMP may be stimulated, via membrane-bound receptors and G proteins,
by polypeptide- or amino acid-derived hormones outside the cell. Cyclic
AMP is broken down inside the cell by a phosphodiesterase.

Summary

The growth and division of cells are based on the same laws of chemistry that control the behavior of molecules outside of cells. Cells contain no atoms unique to the living state; they can synthesize no molecules that the chemist, with inspired hard work, cannot someday make. Thus, there is no special chemistry of living cells. A biochemist is not someone who studies unique types of chemical laws, but a chemist interested in learning about the behavior of molecules found within cells (biological molecules).

The growth and division of cells depend on the availability of a usable form of chemical energy. The energy initially comes from the energy of the sun's light quanta, which is converted by photosynthetic plants into cellular molecules, some of which are used as food sources by various microorganisms and animals. The most striking initial triumphs of the biochemists told us how food molecules are transformed into other cellular molecules and into useful forms of chemical energy. The energy within food molecules largely resides within the covalent bonds of reduced carbon compounds; it is released when these molecules are transformed by oxidation-reduction reactions to carbon compounds of a higher degree of oxidation. For most forms of life, the ultimate oxidizing agent is molecular oxygen. The products of the complete oxidation of organic molecules like glucose are carbon dioxide and water.

Most organic molecules, however, are not oxidized directly by oxygen, but by diverse organic molecules, often coenzymes, such as the coenzyme NAD^+. The reduced coenzyme (e.g., NADH) is itself oxidized by another molecule (such as FAD) to yield a new reduced coenzyme ($FADH_2$) and the original coenzyme in the oxidized form (NAD^+). After several such cycles, molecular oxygen directly participates to end the oxidation-reduction chain, giving off water.

The energy released during the oxidation-reduction cycles is not released entirely as heat; in fact, more than half the energy is converted into new chemical bonds. Phosphorus atoms play a key role in the transformation. Phosphate esters are formed that have a higher usable energy content than most covalent bonds. These phosphate groups are transferred in a high-energy form to acceptor molecules such as ADP. A phosphate group is added to

ADP to form ATP—a primary step in photosynthesis called photophosphorylation. The ADP → ATP transformation is at the heart of energy relations in all cells.

Until a few decades ago, chemists' understanding of the cell's very large molecules, proteins and nucleic acids, was much less firm than it is today. Proteins and nucleic acids are complex structures, and only recently have the physical and chemical techniques needed for their study been developed. Among the most important techniques have been partition chromatography, analytical ultracentrifugation, and X-ray crystallography as extended to the study of large molecules. Now we know practically all the important features of many large proteins like myoglobin, hemoglobin, and trypsin.

The primary structure of proteins consists of amino acids linked together in a unique sequence by covalent bonds. These chains then acquire secondary structure by assuming helical configurations (α helix) or aggregating side by side to form β sheets. Both are held together by hydrogen bonds between adjacent carbonyl and imino groups. The tertiary structure is very irregular owing largely to the diverse side groups of the amino acid striving to form the most energetically stable configuration. Intra- and interchain cysteine (S–S) bonds also contribute to the stability of many proteins. The three-dimensional folding (tertiary structure) of a protein brings these cysteine residues into close contact, thereby allowing S–S bonds to form between correct partners.

All cellular membranes contain phospholipid bilayers arranged so that the hydrophilic phosphate groups are on the outside of the membrane and the hydrophobic hydrocarbon tails are on the inside. Proteins, many of which function as pumps to control the movement of molecules into and out of the cells, are dispersed throughout the membrane, usually extending through the membrane in a semifluid form in which they do not necessarily remain in fixed positions. Sodium and potassium concentrations are maintained against the concentration gradient via a Na^+-K^+ pump. In resting cells, such pumps collectively utilize approximately one-third the ATP generated. Small molecules can flow virtually unimpeded across the membranes of closely opposed cells via channels called gap junctions, which open and close, depending on physiological need. Hormones are molecules that function in cells at locations distant from their site of synthesis. They may be amino acid derivatives, polypeptides, or steroids. In acting, they bind to specific receptors on or in their target or effector cells and then function to regulate preexisting cellular reactions such as enzyme synthesis or function. Steroid hormone-receptor complexes affect cellular functions by binding intracellular receptors, which, in turn, bind to chromosomal segments and affect gene function. In contrast, polypeptide hormones, after binding to their membrane-located receptors, induce changes in the still little understood membrane-bound G proteins, which, in turn, regulate the activity of adenylate cyclase, the enzyme that forms cAMP from ATP. Each cAMP molecule can subsequently bind to an appropriate protein kinase, which, in turn, through its phosphate-adding capacity, activates or inactivates many specific target proteins.

Bibliography

General References

Fruton, J. S. 1972. *Molecules and Life*. New York: Wiley. A detailed guide to the history of biochemistry.
Lehninger, A. L. 1982. *Principles of Biochemistry*. New York: Worth.
Stryer, L. 1981. *Biochemistry*. 2nd ed. San Francisco: Freeman.
Zubay, G. 1983. *Biochemistry*. Reading, Mass.: Addison-Wesley.

Cited References

1. Sumner, J. 1926. "The Isolation and Crystallization of the Enzyme Urease." *J. Biol. Chem.* 69:435–439.
2. Krebs, H. 1981. *Otto Warburg: Biochemist and Eccentric*. Oxford, Eng., and New York: Oxford University Press.
3. Baldwin, E. 1959. *Dynamic Aspects of Biochemistry*. 3rd ed. New York: Cambridge University Press. This is one of the few texts in biochemistry deserving to be called a classic. Now it is best read for the way in which coenzymes participate in enzyme transfers.
4. Lipmann, F. 1971. *Wanderings of a Biochemist*. New York: Wiley. A look back at the author's key role in the discovery of high-energy bonds.
5. Krebs, H. A. 1970. "The History of the TCA Cycle." *Persp. Biol. Med.* 14:154–170.
6. Lipmann, F., and D. Novelli. 1950. "Coenzyme A in Citric Acid Synthesis." *J. Biol. Chem.* 182:213–228.
7. Keilin, D. 1966. *The History of Cell Respiration and Cytochromes*. New York: Cambridge University Press.
8. Kalckar, H. M. 1969. *Biological Phosphorylations*. Englewood Cliffs, N.J.: Prentice-Hall. Contains a collection of many of the classic papers that led to the understanding of the manyfold importance of high-energy bonds.
9. Kennedy, E. P., and A. L. Lehninger. 1949. "Oxidation of Fatty Acids and Tricarboxylic Acid Cycle Intermediates by Isolated Rat Liver Mitochondria." *J. Biol. Chem.* 179:957–972.
10. Lipmann, F. 1941. "Metabolic Generation and Utilization of Phosphate Bond Energy." *Adv. Enzymol.* 1:99–162.
11. Arnon, D. I. 1959. "Conversion of Light into Chemical Energy." *Nature* 184:10–21. Reviews the effect of sunlight in the conversion of ADP to ATP and traces the evolution of photochemical events.
12. Arnon, D. I. 1984. "The Discovery of Photosynthetic Phosphorylation." *Trends in Biochem. Sci.* 9:258–262.
13. Mitchell, P. 1979. "Keilin's Respiratory Chain Concept and Its Chemiosmotic Consequences." *Science* 206:1148–1159. The evolution of the chemiosmotic principle.
14. Svedberg, T., and K. O. Peterson. 1940. *The Ultracentrifuge*. Oxford, Eng., and New York: Oxford University Press.
15. Martin. A. J. P., and R. L. M. Syne. 1941. "A New Form of Chromotography Employing Two Liquid Phases." *Biochem. J.* 35:1358–1368. Original presentation of the theory of partition chromatography.
16. Sanger, F., and H. Tuppy. 1951. "The Amino Acid Sequence in the Phenylalanyl Chain of Insulin." *Biochem. J.* 49:463–490.
17. Steiner, D. F., O. Hallund, A. Rubenstein, S. Cho, and C. Bayliss. 1968. "Isolation and Properties of Proinsulin, Intermediate Forms, and Other Minor Components from Crystalline Bovine Insulin." *Diabetes* 17:725–736.
18. Pauling, L., R. B. Corey, and H. R. Branson. 1951. "The Structure of Proteins: Two Hydrogen Bonded Helical Configurations of the Polypeptide Chain." *Proc. Nat. Acad. Sci.* 37:205–211.
19. Perutz, M. F. 1962. *Proteins and Nucleic Acids*. Amsterdam, London, and New York: Elsevier.

20. Kendrew, J. 1966. *The Thread of Life: An Introduction to Molecular Biology.* Cambridge, Mass.: Harvard University Press.
21. Dickerson, R. E., and I. Geis. 1969. *The Structure and Action of Proteins.* New York: Harper & Row. A well-illustrated introduction to the ways polypeptide chains fold up.
22. Anfinsen, C. B., E. Haber, M. Sela, and F. H. White, Jr. 1961. "The Kinetics of Formation of Native Ribonucleases During Oxidation of the Reduced Polypeptide Chain." *Proc. Nat. Acad. Sci.* 47:1309–1314.
23. Phillips, D. 1968. "The Three Dimensional Structure of an Enzyme Molecule." In *The Molecular Basis of Life,* ed. R. H. Haynes and P. C. Hanawalt. San Francisco: Freeman. pp. 52–64. A description of the work that solved the structure of lysozyme, the first enzyme for which it was possible to correlate structure and function.
24. Stroud, R. M. 1974. "A Family of Protein-Cutting Proteins." *Sci. Amer.* 231:74–88. Describes the structure and functioning of trypsin and chymotrypsin-like enzymes.
25. Davson, H., and J. F. Danielli. 1952. *Permeability of Natural Membranes.* New York: Cambridge University Press.
26. Henderson, R., and P. N. T. Unwin. 1975. "Three Dimensional Model of Purple Membrane Obtained by Electron Microscopy." *Nature* 257:28–32.
27. Alberts, B., D. Bray, J. Lewis, M. Raff, K. Roberts, and J. D. Watson. 1983. *Molecular Biology of the Cell.* New York: Garland. See Chapter 6, "The Plasma Membrane," pp. 255–318.
28. Singer, S. and G. Nicholson. 1972. "The Fluid Mosaic Model of the Structure of Cell Membranes." *Science* 175:720–731.
29. Skou, J. C., and J. G. Norby, eds. 1979. *Na+-K+ ATPase: Structure and Kinetics.* New York: Academic Press.
30. Unwin, P. N. T., and P. D. Ennis. 1984. "Two Configurations of a Channel-Forming Membrane Protein." *Nature* 307:609–613. The protein oligimer forming the gap-junction channel is analyzed in two Ca^{++}-sensitive states.
31. Bennett, M. V. L., and D. C. Spray, eds. 1985. *Gap Junctions.* Cold Spring Harbor, N.Y.: Cold Spring Harbor Laboratory.
32. Jensen, E. V., G. L. Greene, L. E. Closs, E. R. DeSombre, and M. Nadji. 1982. "Receptors Reconsidered—A 20 Year Perspective." *Rec. Prog. Horm. Res.* 38:1–40.
33. Yamamoto, K. R., and B. M. Alberts. 1976. "Steroid Receptors: Elements for Modulation of Eukaryotic Transcription." *Ann. Rev. Biochem.* 45:721–746.
34. Sutherland, E. W., T. W. Rall, and T. Menon. 1962. "Adenyl Cyclase. I. Distribution, Preparation, and Properties." *J. Biol. Chem.* 237:1220–1227.
35. Sutherland, E. W. 1972. "Studies on the Mechanism of Hormone Action." *Science* 177:401–407. Nobel laureate speech on the discovery of cAMP as second messenger.
36. Gilman, A. G. 1984. "G Proteins and Dual Control of Adenylate Cyclase." *Cell* 36:577–579.

Plate 1

Computer Images of DNA

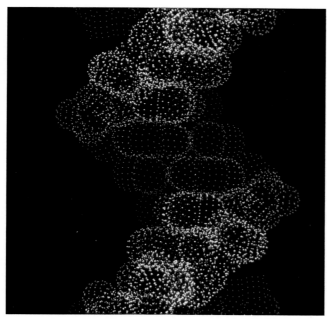

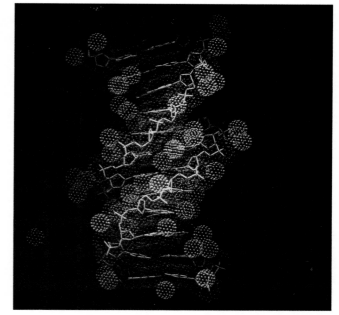

A. Side view of B-DNA with the individual atoms represented as dot-covered spheres. Carbon is green, oxygen is red, nitrogen is blue, and phosphorus is yellow. The sugar-phosphate backbone on the far side of the structure is not shown. Image courtesy of the Computer Graphics Laboratory of the University of California at San Francisco.

B. Model of B-DNA (skeleton in yellow and surface indicated with red dots) showing locations of water molecules (green spheres) bound to a crystallographically determined DNA dodecamer. Image courtesy of the Institut de Biologie Moleculaire, Strasbourg.

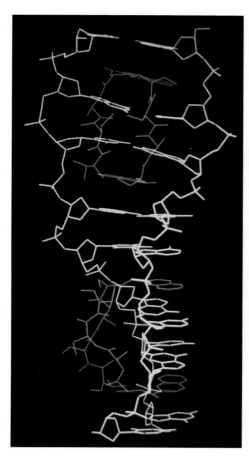

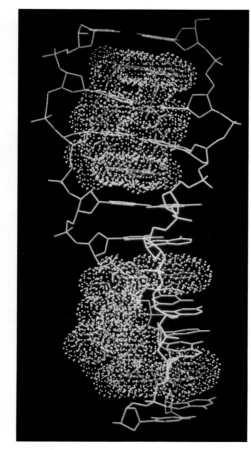

C. Skeletal structure of a DNA octamer (pink) cocrystallized with the intercalator antibiotic triostin A (blue). Binding of the triostin causes the right-handed DNA double helix to partially unwind. In the right-hand image, aqua dots indicate the surfaces of the two triostin molecules. Images courtesy of A. Rich, Department of Biology, M.I.T.

Plate 2

Protein Structure at Increasing Levels of Complexity

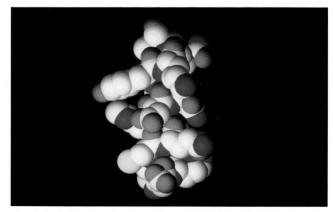

A. Polypeptide backbone atoms of a portion of α helix, with the atoms shown as space-filling. Carbon is white, oxygen is red, and nitrogen is blue. Note how the oxygen and nitrogen atoms are placed for hydrogen bonding.

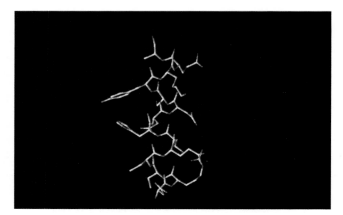

B. As in A, but showing the amino acid side chains pointing outwards. Hydrogen atoms are not shown. The yellow atoms are sulfur.

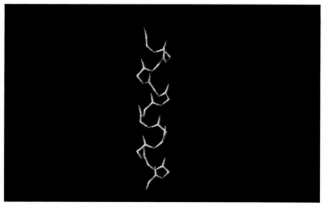

C. Skeletal model of the α helix backbone.

D. Skeletal model of the α helix with side chains.

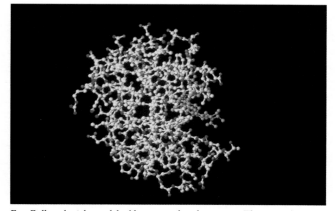

E. Ball-and-stick model of hen egg-white lysozyme. The trisaccharide substrate NAM-NAG-NAM is shown in purple.

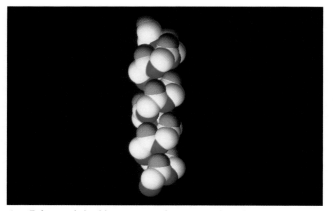

F. α-carbon skeleton of lysozyme, tracing the course of the polypeptide backbone.

All images on plates 2 and 3 courtesy of the Graphic Systems Research Group, IBM U.K. Scientific Centre.

Plate 3

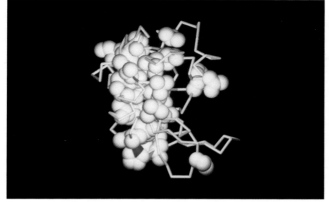

G. Lysozyme backbone with hydrophobic side chains (Val, Ile, Leu, Phe, Pro, Met) shown as space-filling.

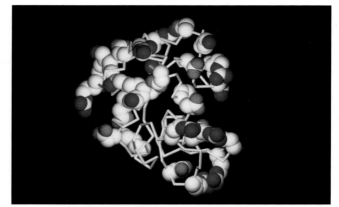

H. Lysozyme backbone with hydrophilic side chains (Asp, Glu, Lys, Arg) shown as space-filling.

I. Lysozyme backbone with "hydro-indifferent" side chains (Ser, Thr, Gln, Asn, Cys, Gly, Ala, Trp, Tyr, His) shown as space-filling.

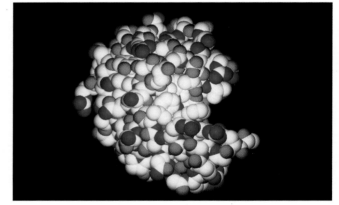

J. Lysozyme with all atoms (except hydrogens) shown as space-filling.

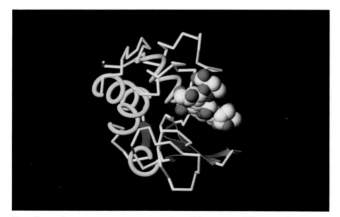

K. Secondary structure representation of lysozyme. The α-carbon—α-carbon "bonds" are white; α helices are shown as yellow spirals; two strands making up one β sheet are shown as blue arrows; three strands forming another β sheet are shown as red arrows. (The arrows point toward the carboxyl end of the polypeptide chain.) The substrate trisaccharide is shown as space-filling.

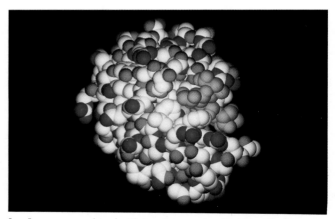

L. Lysozyme combined with its substrate trisaccharide, with all atoms shown as space-filling. The substrate atoms are in purple.

Plate 4

Interaction of Repressors with DNA

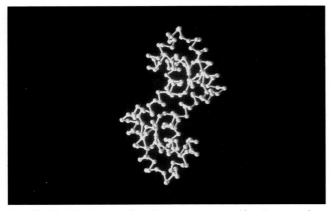

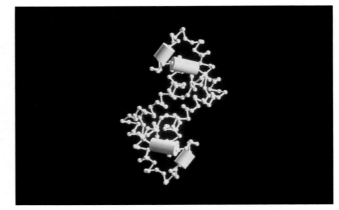

A. Model of the bacteriophage λ repressor protein (the *cI* gene product) viewed facing the side that binds to DNA. The white lines trace the α-carbon backbones of the two identical polypeptides that make up this dimeric molecule.

B. As in A, but highlighting the helix 2-turn-helix 3 motif (see page 473). The helices are represented by solid cylinders.

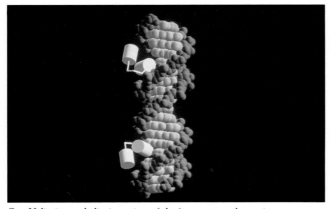

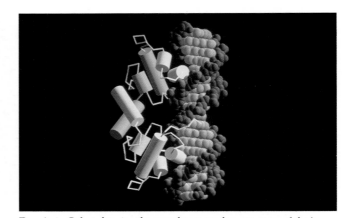

C. Helix 2-turn-helix 3 portion of the λ repressor, shown in a secondary-structure representation (white), bound to a space-filling model of DNA (sugar-phosphate backbone red, bases blue).

D. As in C, but showing the complete secondary structure of the λ repressor. The α helices are shown as cylinders and the rest of each polypeptide chain as an α-carbon skeleton. Domains of the repressor are differentiated with green and yellow color; helix 2-turn-helix 3 portions are highlighted in white. Images A–D courtesy of the Graphic Systems Research Group, IBM U.K. Scientific Centre.

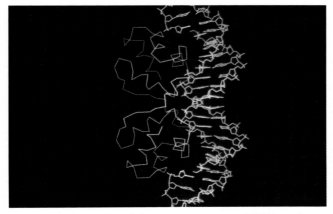

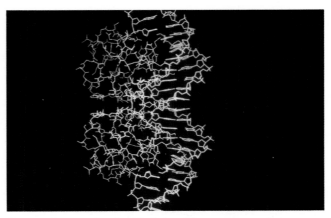

E. The repressor protein of phage 434 interacting with DNA, as determined by direct crystallographic analysis. The orange region of the DNA ⸱e 434 operator. Only the α-carbon skeleton of the repressor is shown.

F. As in E, except that the side chains of the repressor are shown. Images E and F courtesy of J.E. Anderson and S.C. Harrison, Department of Biochemistry, Harvard University.

Plate 5

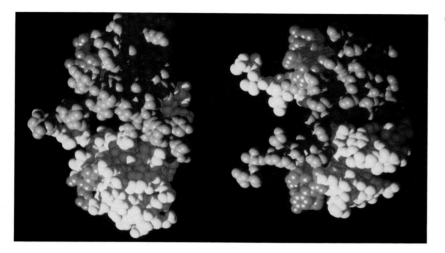

G. Space-filling models of the CI repressor (left) and the Cro repressor (right) of λ. The two identical subunits of each dimeric molecule are shown in different shades of blue. Helix 3 is shown in red.

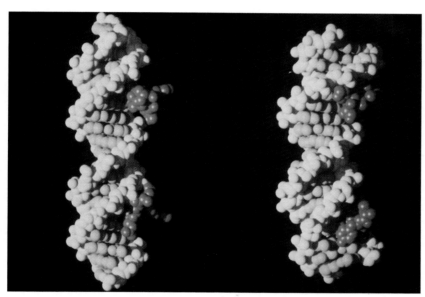

H. Helix 3 region (red) of CI and Cro repressors shown bound to their respective operator sequences. The phosphates which when ethylated reduce binding are colored yellow.

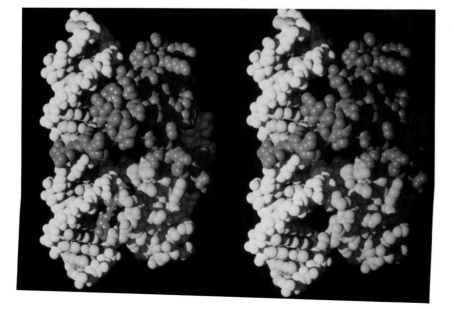

I. CI repressor bound to its operator with and without its helix 3 region (red). Images G–I courtesy of W.F. Anderson, Department of Biochemistry, University of Alberta.

Plate 6

The Complete Structure of a Multiprotein Complex

This spectacular computer image shows the membrane-spanning photosynthetic reaction center from the purple photosynthetic bacterium *Rhodopseudomonas viridis*. The four proteins that make up the complex are shown in a ribbon representation. The cytochrome subunit (Cyt) is shown in green, with its four covalently attached heme groups in brown. The light subunit (L) is red; the medium subunit (M) is blue; and the heavy subunit (H) is violet. The L and M subunits surround the pigments (4 bacteriochlorophyll *b* molecules, 2 bacteriopheophytin *b* molecules, 1 menaquinone molecule—all in yellow). The reaction center has a total of eleven membrane-spanning, hydrophobic α helices: 5 each from the L and M subunits and 1 from the H subunit. Image courtesy of H. Michel and J. Deisenhofer, Max-Planck-Institut für Biochemie, Munich.

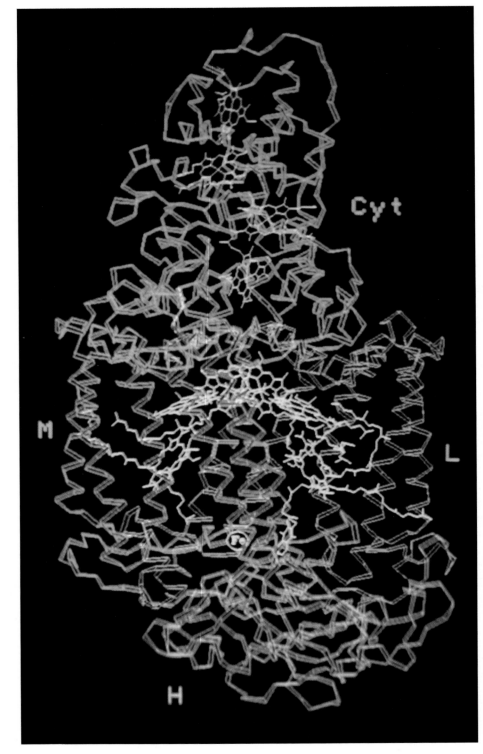

CHAPTER 3

Nucleic Acids Convey Genetic Information

That special molecules might carry genetic information was appreciated by geneticists long before the problem claimed the attention of chemists. By the 1930s, geneticists began speculating as to what sort of molecules could have the kind of stability that the gene demanded yet be capable of permanent, sudden change to the mutant states that must provide the basis of evolution. But until the mid-1940s, there appeared to be no direct way to attack the chemical essence of the gene. Though it was known that chromosomes possessed a unique molecular constituent, deoxyribonucleic acid (DNA), there was no way to show that this constituent carried genetic information, as opposed to serving merely as a molecular scaffold for a still undiscovered class of proteins especially tailored to carry genetic information.

So it made sense to approach the nature of the gene by asking how genes function within cells. Beginning in 1941, research on the mold *Neurospora* had begun to generate increasingly strong evidence supporting the then 30-year-old hypothesis of Garrod that genes work by controlling the synthesis of specific enzymes (the one gene–one enzyme hypothesis). Thus, given that all known enzymes had, by this time, been shown to be protein, the key problem was the way genes participate in the synthesis of proteins. From the very start of serious speculation, the simplest hypothesis was that genetic information within genes determines the order of the 20 different amino acids within the polypeptide chains of proteins.

In attempting to test this proposal, intuition was of little help to even the best biochemists, since there is no logical way to use enzymes as ordering tools. Such schemes would require, for the synthesis of a single type of protein, as many ordering enzymes as there are amino acids in the respective protein. But since all known enzymes are themselves proteins, still additional ordering enzymes would be necessary to synthesize the ordering enzymes. And so on. This is clearly a paradox, unless we assume a fantastically interrelated series of syntheses in which a given protein has many different enzymatic specificities. With such an assumption, it might be possible (and then only with great difficulty) to visualize a workable cell. It did not seem likely, however, that most proteins would be found to carry out multiple tasks. In fact, all our knowledge pointed to the opposite conclusion of one protein, one function.

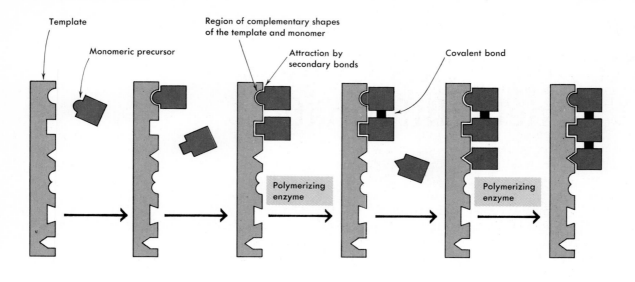

Figure 3-1
Formation of a specific polymeric molecule on a template surface.

The Concept of Template Surfaces[1-4]

It was therefore necessary to throw out the idea of ordering proteins with enzymes and to predict instead the existence of a specific surface, the **template,** that attracts the amino acids (or their activated derivatives) and lines them up in the correct order (Figure 3-1). Then a specific enzyme common to all protein synthesis could make the peptide bonds. Furthermore it was necessary to assume that the templates also have the capacity of serving either directly or indirectly as templates for themselves (self-duplication). That is, in some way their specific surfaces must be exactly copied to give new templates. Again, scientists could not invoke the help of specific enzymes, for this immediately led back to the "enzyme cannot make enzyme" paradox.

Templates Must Be in the Size Range of Their Products

The finding that proteins possess unique amino acid sequences demanded the existence of specific templates on which their amino acid building blocks are laid down. These templates themselves had to be macromolecules, at least as large as their polypeptide products. This becomes clear when we examine the nature of the chemical forces that attract small molecules to their templates. It was expected that these forces would be the same as those that attract substrates to their enzymes. These are ionic bonds; hydrogen bonds in which an electropositive hydrogen atom is attracted to an electronegative atom, such as oxygen or nitrogen; and van der Waals forces (see Chapter 5 for a complete discussion of these forces).

Since all these forces operate only over very small distances (<5 Å), templates can order small molecules only when they are in close contact on the atomic level. Thus, it was correctly assumed that the specific attracting regions of the templates would be in the same size range as the amino acid side groups in the protein products.

Attraction of Opposites Versus Self-Attraction

Can a polypeptide chain serve as a template for its own synthesis? If it could, it would greatly reduce the chemical prerequisites for life. The problems of protein synthesis and template replication would then be the same, and the additional biochemical complexity required to maintain a special class of template molecules would be unnecessary. This conceptual possibility found no support, however, from close inspection of the amino acid side groups. There was no chemical reason why, for example, the occurrence of valine on a template should preferentially attract the specific side group of another valine molecule. In fact, none of the amino acid side groups have specific affinities for themselves. Instead, it was much easier to imagine molecules with opposite or complementary features attracting each other. Negative charges obviously attract positive groups, and hydrogen atoms can form hydrogen bonds only to electronegative atoms like oxygen or nitrogen. Similarly, molecules can be specifically attracted by van der Waals forces only when they possess complementary shapes, so that a cavity in one molecule can be filled with a protruding group of another molecule.

There is no way, however, to group the 20 amino acids into 10 pairs whose side groups have mutually self-complementary shapes. A formal scheme remained, however, that could conceivably allow the existence of protein templates. We could imagine the existence of 20 different specific molecules, which we call connectors. Each connector might possess two identical surfaces complementary in charge and/or shape to a given amino acid. The intervention of these connector molecules could then make possible the lining up of amino acids in a sequence identical to that of the template polypeptide chain. However, there was no evidence for such molecules. Instead, as we shall soon see, a specific template class (the nucleic acids) does in fact exist.

A Chemical Argument Against the Existence of Protein Templates[5]

The failure of proteins to evolve a template role may have originated in the composition of the amino acid side groups. The argument can be made that no template whose specificity depends on the side groups of closely related amino acids, like valine or alanine, can ever be copied with the accuracy demanded for efficient cellular existence. This follows from the fact that some amino acids are chemically very similar. For example, valine and isoleucine differ only by the presence of an additional methyl group in isoleucine. Likewise, glycine and alanine also differ by only one methyl group (see Figure 2-16). This close chemical similarity immediately posed the question of whether any copying process could be sufficiently accurate to distinguish between such closely related molecules. A good speculative guess was that each amino acid in a hereditary molecule would have to be copied with an accuracy of not more than one error in 10^8. On the other hand, a semirigorous chemical argument could be made that no chemical reaction could distinguish between molecules differing by only one methyl group with an accuracy of less than one error in 10^6. When the proper experiments were later carried out, amino acids like leucine were inserted into polypeptide chains with no greater than a 99.9 percent accuracy (one error in 10^3). Thus, proteins did not have the "smell" of a hereditary molecule.

Establishing the Chemical Constituents of DNA and RNA[6, 7]

If proteins were not the genetic molecules, then the prime candidates had to be the nucleic acids. Though prominent features of most cells, they had yet to be ascribed any biological function. Dealing with them chemically, however, was initially difficult, since, like proteins, they are macromolecules, constructed from large numbers of their unique building blocks, the nucleotides. Moreover, in addition to DNA, which by the late nineteenth century had already been recognized as a major constituent of chromosomes, there also existed a second form of nucleic acid, called ribonucleic acid (RNA). Early on, it had been found to be a major constituent of yeasts, and for many years there was the erroneous belief that RNA was the plant nucleic acid while DNA was the nucleic acid of animals. However, with the development of better ways to distinguish DNA from RNA, all cells, including those of bacteria, were found to contain both DNA and RNA.

Both DNA and RNA are constructed from four main nucleotide building blocks. Each of these nucleotides contains a phosphate group linked to a five-carbon-atom sugar group, which, in turn, is joined to a flat aromatic molecule that can be either a double-ringed purine or a single-ringed pyrimidine (Figure 3-2). Since they contain the sugar deoxyribose, the nucleotides of DNA are called **deoxyribonucleotides,** while those of RNA, which contain the sugar ribose, are known as **ribonucleotides.** DNA and RNA are both built up from two purine-containing nucleotides and two pyrimidine-containing molecules. The purines of both DNA and RNA are the same, adenine and guanine. The pyrimidine cytosine is found in both nucleic acids, while the pyrimidine thymine is limited to DNA, being replaced in RNA by the very similar pyrimidine uracil.

In both DNA and RNA, the nucleotides are joined together by covalent bonds linking the phosphate group of one nucleotide to a hydroxyl group on the sugar of the adjacent nucleotide. Such bonds, called **phosphodiester linkages,** were known to exist, but the details, such as the exact atoms that linked together the nucleic acids, were long to remain unknown. Also initially confusing were the true sizes of DNA and RNA within cells. The original separation techniques used to isolate them seriously degraded them, and only beginning in the mid-1930s was it shown that both could be present in very large forms. Especially crucial were the experiments of chemists R. Signer, E. Hammarsten, and T. Caspersson, who found that gently prepared DNA had molecular weights even larger than most proteins (greater than 500,000). They also deduced that DNA was a very elongated molecule, a conjecture fully confirmed by 1950 with the advent of easy-to-use electron microscopes, which revealed DNA to be an extraordinarily thin molecule with a diameter of only 20 Å. Individual DNA molecules had lengths of many thousands of angstroms, telling us that they were built up from thousands of nucleotide building blocks (Figure 3-3).

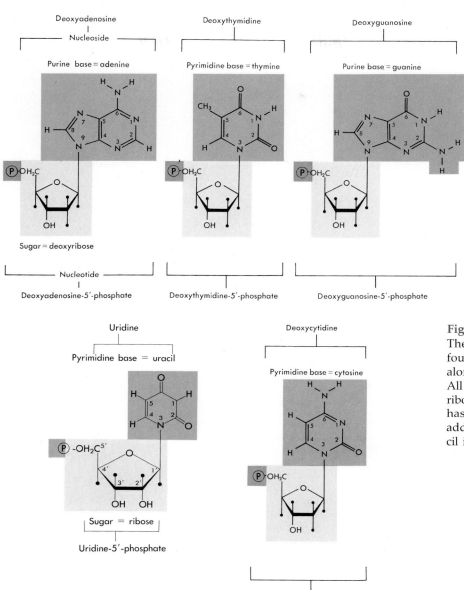

Figure 3-2
The nucleotides of DNA and RNA. The four nucleotides of DNA are shown, along with one of the RNA nucleotides. All RNA nucleotides have the sugar ribose (instead of deoxyribose), which has a hydroxyl group on carbon 2. In addition, RNA has the pyrimidine uracil instead of thymine.

Avery's Bombshell:
DNA Can Carry Genetic Specificity[8, 9]

That DNA might be the key genetic molecule emerged most unexpectedly from studies on pneumonia-causing bacteria. In 1928, the English microbiologist Frederick Griffith had made the startling observation that nonvirulent strains of the bacteria became virulent when mixed with their heat-killed pathogenic counterparts. That such transformations from nonvirulence to virulence represented hereditary changes was shown by using descendants of the newly pathogenic strains to transform still other nonpathogenic bacteria. This raised the possibility that when pathogenic cells are killed by heat, their genetic components remain undamaged and, liberated somehow from the heat-killed cells, can pass through the cell wall of the living recipient cells and subsequently undergo genetic recombi-

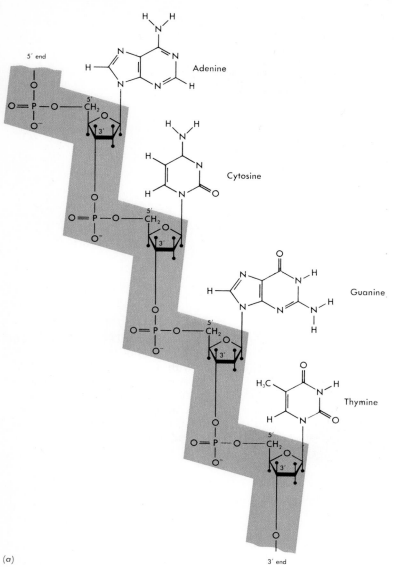

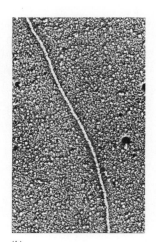

(b)

(a)

Figure 3-3
(a) A portion of a DNA polynucleotide chain, showing the 3'-5' phosphodiester linkages that connect the nucleotides. Phosphate groups connect the 3' carbon of one nucleotide with the 5' carbon of the next. (b) Electron micrograph of a portion of a DNA molecule.

nation with the recipient's genetic apparatus (Figure 3-4). Subsequent research has confirmed this genetic interpretation. Pathogenicity reflects the action of the *S* (smooth) gene, which codes for a key enzyme involved in the synthesis of the carbohydrate-containing capsule that surrounds most pneumonia-causing bacteria. When the *R* (rough) allele of this gene is present, no capsule is formed and the respective cells are not pathogenic.

Within several years after Griffith's original observation, extracts of the killed bacteria were found capable of inducing hereditary transformations, and a search could begin for the chemical identity of the transforming agent. At that time, the vast majority of biochemists still believed that genes were proteins. It therefore came as a great surprise when in 1944, after some ten years of research, American microbiologist Oswald T. Avery and his colleagues Colin M. MacLeod and Maclyn McCarty, at the Rockefeller Institute in New York, made the momentous announcement that the active genetic principle was DNA (Figure 3-5). Supporting their conclusion were key experiments showing that the transforming activity of their highly purified active fractions was destroyed by pancreatic deoxyribonuclease, a recently purified enzyme that specifically degrades DNA molecules to their nucleotide building blocks and has no effect on the integrity of protein molecules or RNA. The addition of either pancreatic ribonuclease (which degrades RNA) or various proteolytic (protein-destroying) enzymes had no influence on transforming activity.

The Amount of Chromosomal DNA Is Constant[10, 11]

Even though the transformation results were clear-cut, there was initially great skepticism about their general applicability; many, if not the majority of, interested scientists doubted that they would be found relevant to anything but certain strains of bacteria. Thus, the momentous nature of Avery's discovery was only gradually appreciated.

One important confirmation came from studies on the chemical nature of chromosomes. DNA was found to be located almost exclusively in the nucleus and almost never where detectable chromosomes were absent. Moreover, the amount of DNA per diploid set of chromosomes was shown, through the cytochemical analysis of A. E. Mirsky and H. Ris of the Rockefeller Institute and by the Vendrelys in France, to be constant for a given organism and equal to twice the amount present in the haploid sperm cells (Table 3-1). Another type of evidence that favored DNA as the genetic molecule was the observation that it is metabolically stable. It is not rapidly made and broken down like many other cellular molecules; once atoms are incorporated into DNA, they do not leave it so long as healthy cell growth is maintained.

Viral Genes Are Also Nucleic Acids[12]

Even more important confirmatory evidence came from chemical studies with viruses and virus-infected cells. It was possible by 1950 to obtain a number of essentially pure viruses and to determine which types of molecules were present in them. This work led to the very important generalization that all viruses contain nucleic acid. Since there was at that time a growing realization that viruses contain genetic material, the question immediately arose as to whether the nucleic acid component was the viral chromosome. The first crucial test of the question came from isotopic study of the multiplication of T2, a bacterial virus (**bacteriophage,** or **phage**) containing a DNA core and a protective shell built up by the aggregation of a number of different protein molecules. In these experiments, performed in 1952 by A. D. Hershey and M. Chase working at Cold Spring Harbor on Long Island, the protein coat was labeled with the radioactive isotope ^{35}S, and the DNA with the radioactive isotope ^{32}P. The labeled virus was then used to follow the fates of the phage protein and nucleic acid as phage multiplication proceeded, particularly to see which labeled atoms from the parental virus entered the host cell and later appeared in the progeny phage.

Clear-cut results emerged from these experiments; much of the parental nucleic acid and none of the parental protein was detected in

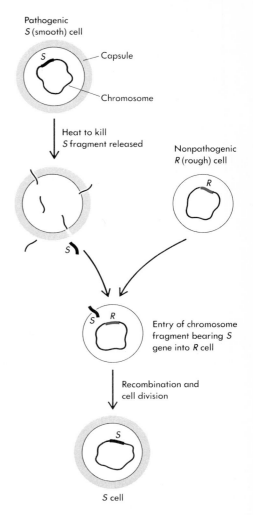

Figure 3-4
Transformation of a genetic characteristic of a bacterial cell (*Streptococcus pneumoniae*) by addition of heat-killed cells of a genetically different strain. Here we show an *R* cell receiving a chromosomal fragment containing the *S* gene. Since most *R* cells receive other chromosomal fragments, the efficiency of transformation for a given gene is usually less than 1 percent.

Figure 3-5
The chemical method used in the original isolation of a chemically pure transforming agent. [After F. W. Stahl, *The Mechanics of Inheritance* (Englewood Cliffs, N.J.: Prentice-Hall, 1964), Fig. 2.3, with permission.]

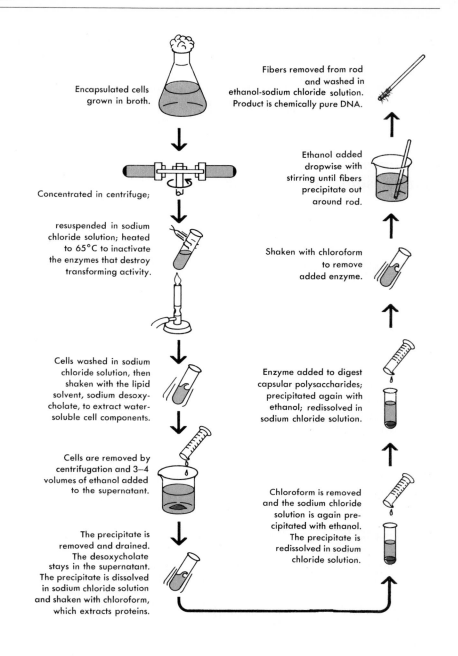

Encapsulated cells grown in broth.

Concentrated in centrifuge;

resuspended in sodium chloride solution; heated to 65°C to inactivate the enzymes that destroy transforming activity.

Cells washed in sodium chloride solution, then shaken with the lipid solvent, sodium desoxycholate, to extract water-soluble cell components.

Cells are removed by centrifugation and 3–4 volumes of ethanol added to the supernatant.

The precipitate is removed and drained. The desoxycholate stays in the supernatant. The precipitate is dissolved in sodium chloride solution and shaken with chloroform, which extracts proteins.

Chloroform is removed and the sodium chloride solution is again precipitated with ethanol. The precipitate is redissolved in sodium chloride solution.

Enzyme added to digest capsular polysaccharides; precipitated again with ethanol; redissolved in sodium chloride solution.

Shaken with chloroform to remove added enzyme.

Ethanol added dropwise with stirring until fibers precipitate out around rod.

Fibers removed from rod and washed in ethanol-sodium chloride solution. Product is chemically pure DNA.

Table 3-1 Amount of DNA per Nucleus in Beef Tissues

Organ	DNA (pg)*	Probable Chromosome Number
Thymus	6.6	Diploid
Liver	6.4	Diploid
Pancreas	6.9	Diploid
Kidney	5.9	Diploid
Sperm	3.3	Haploid

*1 picogram (pg) = 10^{-12} gram.
SOURCE: Data from R. Vendrely, "The DNA Content of the Nucleus," in E. Chargaff and J. N. Davidson, eds., *The Nucleic Acids*, vol. 2. New York: Academic Press, 1955.

the progeny phage (Figure 3-6). Moreover, it was possible to show that little of the parental protein ever enters the bacteria; instead, it stays attached to the outside of the bacterial cell, performing no function after the DNA component has passed in. This point was neatly shown by violently agitating infected bacteria after the entrance of the DNA; the protein coats were shaken off without affecting the ability of the bacteria to form new phage particles.

With some viruses it is now possible to do an even more convincing experiment. For example, purified DNA from the mouse virus polyoma can enter mouse cells and initiate a cycle of viral multiplication producing many thousands of new polyoma particles. The primary function of viral protein is thus to protect its genetic nucleic acid component in its movement from one cell to another. Thus no reason exists for the assignment of any genetic role to protein molecules.

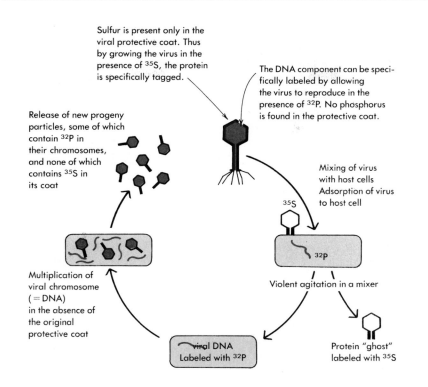

Sulfur is present only in the viral protective coat. Thus by growing the virus in the presence of ^{35}S, the protein is specifically tagged.

The DNA component can be specifically labeled by allowing the virus to reproduce in the presence of ^{32}P. No phosphorus is found in the protective coat.

Release of new progeny particles, some of which contain ^{32}P in their chromosomes, and none of which contains ^{35}S in its coat

Mixing of virus with host cells
Adsorption of virus to host cell

^{35}S

^{32}P

Multiplication of viral chromosome (=DNA) in the absence of the original protective coat

Violent agitation in a mixer

viral DNA Labeled with ^{32}P

Protein "ghost" labeled with ^{35}S

Figure 3-6
Demonstration that only the DNA component of T2 carries the genetic information and that the protein coat serves only as a protective shell.

Chargaff's Rules[13, 14]

The paper chromatographic techniques that had first proved so useful in revealing the chemical makeup of proteins allowed the biochemist Erwin Chargaff, then working in New York at the College of Physicians and Surgeons of Columbia University, to analyze the nucleotide composition of DNA. By 1949 his data showed not only that the four different nucleotides are not present in equal amounts but also that the exact ratios of the four nucleotides vary from one species to another (Table 3-2). These findings opened up the possibility that it is

Table 3-2 Data Leading to the Formulation of Chargaff's Rules

Source	Adenine to Guanine	Thymine to Cytosine	Adenine to Thymine	Guanine to Cytosine	Purines to Pyrimidines
Ox	1.29	1.43	1.04	1.00	1.1
Human	1.56	1.75	1.00	1.00	1.0
Hen	1.45	1.29	1.06	0.91	0.99
Salmon	1.43	1.43	1.02	1.02	1.02
Wheat	1.22	1.18	1.00	0.97	0.99
Yeast	1.67	1.92	1.03	1.20	1.0
Hemophilus influenzae	1.74	1.54	1.07	0.91	1.0
E-coli K2	1.05	0.95	1.09	0.99	1.0
Avian tubercle bacillus	0.4	0.4	1.09	1.08	1.1
Serratia marcescens	0.7	0.7	0.95	0.86	0.9
Bacillus schatz	0.7	0.6	1.12	0.89	1.0

SOURCE: After E. Chargaff et al., *J. Biol. Chem.* 177 (1949).

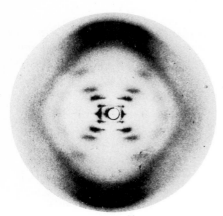

Figure 3-7
The key X-ray photograph involved in the elucidation of the DNA structure. This photograph, taken by Rosalind Franklin at Kings College, London, in the winter of 1952–1953, confirmed the guess that DNA was helical. The helical form is indicated by the crossways pattern of X-ray reflections (photographically measured by darkening of the X-ray film) in the center of the photograph. The very heavy black regions at the top and bottom tell that the 3.4 Å thick purine and pyrimidine bases are regularly stacked next to each other, perpendicular to the helical axis. [Reproduced from R. E. Franklin and R. Gosling, *Nature* 171 (1953):740, with permission.]

the precise arrangement of nucleotides within a DNA molecule that confers its genetic specificity.

Chargaff's experiments also showed that the relative ratios of the four bases were not random. The number of adenine (A) residues in all DNA samples was equal to the number of thymine (T) residues, while the number of guanine (G) residues equaled the number of cytosine (C) residues. The fundamental significance of the A = T and G = C relationships (Chargaff's rules) could not emerge, however, until serious attention was given to the three-dimensional structure of DNA.

The Double Helix[15-23]

While work was proceeding on the X-ray analysis of protein structure, a smaller number of scientists were trying to solve the X-ray diffraction pattern of DNA. The first diffraction patterns were taken in 1938 by the Englishman Astbury and used DNA supplied from Sweden by Hammarsten and Caspersson. It was not until after the war (1950–1952), however, that high-quality photographs were taken by M. H. F. Wilkins and Rosalind Franklin, working in London at King's College (Figure 3-7). These photographs suggested not only that the underlying DNA structure was helical but that it was composed of more than one polynucleotide chain—either two or three. Only then, however, were all the covalent bonds of DNA being unambiguously established. In 1952, a group of organic chemists working in the Cambridge, England, laboratory of Alexander Todd, showed that 3'-5' phosphodiester bonds regularly link together the nucleotides of DNA (see Figure 3-3).

Because of interest in Linus Pauling's α helix, in 1951 an elegant theory of diffraction of helical molecules was developed in England by W. Cochran, Francis Crick, and V. Vand. This theory made it easy to test possible DNA structures on a trial-and-error basis. The correct solution, a complementary double helix (see Chapter 9 for details), was found in 1953 by Crick and James D. Watson, then working in England in the laboratory of Perutz and Kendrew. Their arrival at the correct answer was in large part dependent on finding the stereochemically most favorable configuration compatible with the X-ray diffraction data of the King's College group.

In the double helix, the two DNA chains are held together by hydrogen bonds (a weak noncovalent chemical bond; see Chapter 5 for details) between pairs of bases on the opposing strands (Figure 3-8). This base pairing is very specific: The purine adenine only base-pairs to the pyrimidine thymine, while the purine guanine only base-pairs to the pyrimidine cytosine. In double-helical DNA, the number of A residues must thus be equal to the number of T residues, while the number of G and C residues must likewise be equal. Hence, Chargaff's rules. As a result, the sequence of the bases of the two chains of a given double helix have a complementary relationship with the sequence on one strand exactly defining that of its partner strand.

The establishment of the double helix immediately began a profound revolution in the way many geneticists analyzed their data. The gene was no longer a mysterious entity whose behavior could be investigated only by breeding experiments. Instead, it quickly became a real molecular object about which chemists could think objectively, as they did about smaller molecules such as pyruvate and

NAD. Most of the excitement, however, came not merely from the fact that the structure was solved, but also from the nature of the structure. Before the answer was known, there had always been the worry that it would turn out to be dull, revealing nothing about how genes replicate and function. Fortunately, however, the answer was immensely exciting. The two intertwined strands of complementary structures suggested that one strand serves as the specific surface (template) upon which the other strand is made (Figure 3-8). If this hypothesis were true, then the fundamental problem of gene replication, about which geneticists had puzzled for so many years, was, in fact, conceptually solved.

Finding the Polymerases That Make DNA[24]

Rigorous proof that a single DNA chain is the template upon which a complementary DNA chain is laid down had to await the development of test-tube (in vitro) systems for DNA synthesis. These came much faster than anticipated by the molecular geneticists, whose world until then had been far removed from that of the biochemist well versed in the procedures needed for enzyme isolation. Leading this biochemical assault on DNA replication was (and still is) the American biochemist Arthur Kornberg, who by 1956 had demonstrated DNA synthesis in cell-free extracts of bacteria. Over the next several years, Kornberg went on to show that a specific polymerizing enzyme was needed to catalyze the linking together of the building block precursors of DNA. The nucleotide building blocks utilized (dATP, dGTP, dCTP, and dTTP) had earlier been revealed by him to be energy rich and ultimately dependent on the generation of ATP for their synthesis. The enzyme involved, DNA polymerase I, was the first enzyme found to be involved in the synthesis of polydeoxynucleotides. It links the nucleotide precursors by 3'-5' phosphodiester bonds (Figure 3-9). Furthermore, it works only in the presence of DNA, which is needed to order the four nucleotides in the polynucleotide product.

DNA polymerase I recognizes only the regular sugar-phosphate portion of the nucleotide precursors and so cannot determine sequence specificity. This is neatly demonstrated by allowing the enzyme to work in the presence of DNA molecules that contain varying amounts of A-T and G-C base pairs. In every case, the enzymatically synthesized product has the base ratios of the template DNA (Table 3-3). During this cell-free synthesis, no synthesis of proteins or any other molecular class occurs, unambiguously eliminating any non-DNA compounds as intermediate carriers of genetic specificity. There is thus no doubt that DNA is the direct template for its own formation. As we might expect, the enzymatic product, like the primer, has a double-helical structure.

The ability of DNA polymerase I to make complementary copies of DNA chains immediately suggested that this might be the major enzyme in *E. coli* involved in linking up nucleotides during DNA replication. But as we shall see later, DNA replication is a complex process involving several different polymerizing enzymes. Discovery of the first DNA polymerase, however, was of immense importance in showing that the assembly of complementary nucleotide sequences was amenable to relatively straightforward test-tube analysis.

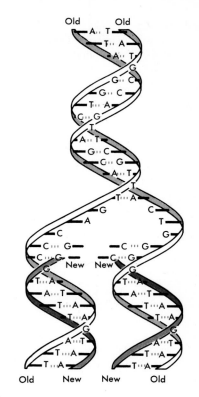

Figure 3-8
The replication of DNA. The newly synthesized strands are shown in color.

Figure 3-9
Enzymatic synthesis of a DNA chain catalyzed by DNA polmerase I.

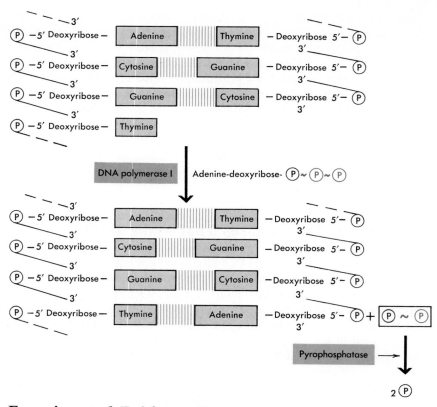

Experimental Evidence Favors Strand Separation During DNA Replication[25]

Simultaneously with Kornberg's research, in 1958 M. Meselson and F. W. Stahl, then at the California Institute of Technology, carried out an elegant experiment in which they separated parental from daughter DNA molecules and in so doing showed that the two strands of the double helix permanently separate from each other during DNA replication (Figure 3-10). Their success was due in part to the use of the heavy isotopes ^{15}N and ^{13}C as tags to differentially label the parental and daughter DNA strands. Bacteria grown in a medium containing the heavy isotope ^{15}N have denser DNA than bacteria grown

Table 3-3 A Comparison of the Base Composition of Enzymatically Synthesized DNA and Their DNA Templates

Source of DNA Template	Base Composition of the Enzymatic Product				$\dfrac{A + T}{G + C}$ in Product	$\dfrac{A + T}{G + C}$ in Template
	Ade-nine	Thy-mine	Gua-nine	Cyto-sine		
Micrococcus lysodeikticus (a bacterium)	0.15	0.15	0.35	0.35	0.41	0.39
Aerobacter aerogenes (a bacterium)	0.22	0.22	0.28	0.28	0.80	0.82
Escherichia coli	0.25	0.25	0.25	0.25	1.00	0.97
Calf thymus	0.29	0.28	0.21	0.22	1.32	1.35
Phage T2	0.32	0.32	0.18	0.18	1.78	1.84

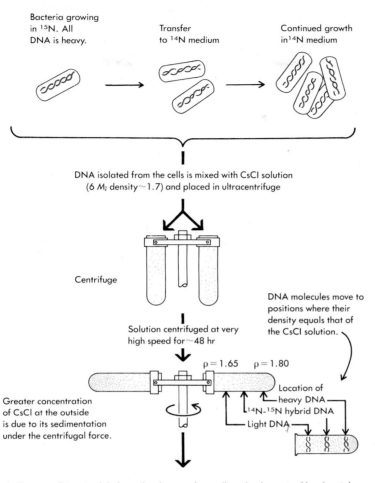

Bacteria growing in ¹⁵N. All DNA is heavy.

Transfer to ¹⁴N medium

Continued growth in ¹⁴N medium

DNA isolated from the cells is mixed with CsCl solution
(6 *M*; density~1.7) and placed in ultracentrifuge

Centrifuge

Solution centrifuged at very high speed for~48 hr

DNA molecules move to positions where their density equals that of the CsCl solution.

$\rho = 1.65$ $\rho = 1.80$

Location of heavy DNA
¹⁴N-¹⁵N hybrid DNA
Light DNA

Greater concentration of CsCl at the outside is due to its sedimentation under the centrifugal force.

The location of DNA molelcules within the centrifuge cell can be determined by ultraviolet optics. DNA solutions absorb strongly at 260nm.

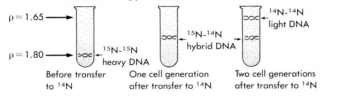

$\rho = 1.65$

$\rho = 1.80$

¹⁵N-¹⁵N heavy DNA

Before transfer to ¹⁴N

¹⁵N-¹⁴N hybrid DNA

One cell generation after transfer to ¹⁴N

¹⁴N-¹⁴N light DNA

Two cell generations after transfer to ¹⁴N

Figure 3-10
Use of a cesium chloride (CsCl) density gradient to demonstrate the separation of complementary strands during DNA replication.

under normal conditions with ¹⁴N. Also contributing to the success of the experiment was the development of procedures for separating heavy from light DNA in density gradients of heavy salts like cesium chloride. When high centrifugal forces are applied, the solution becomes more dense at the outside of the centrifuge cell. If the correct initial solution density is chosen, the individual DNA molecules will move to the central region of the centrifuge cell, where their density equals that of the salt solution. There the heavy molecules will band at a higher density than the light molecules. If bacteria containing heavy DNA are transferred to a light medium (containing ¹⁴N) and allowed to grow, the precursor nucleotides available for use in DNA synthesis will be light; hence, DNA synthesized after transfer will be distinguishable from DNA made before transfer.

If DNA replication involves strand separation, definite predictions can be made about the density of the DNA molecules found after various growth intervals in a light medium. After one generation of growth, all the DNA molecules should contain one heavy strand and one light strand and thus be of intermediate hybrid density. This result is exactly what Meselson and Stahl observed. Likewise, after two generations of growth, half the DNA molecules were light and half hybrid, just as strand separation predicts.

DNA was thus proved to be a "semiconservative" process in which the single strands of the double helix remain intact (are conserved) during a replication process that distributes one parental strand into each of the two daughter molecules. Ruled out by these experiments were any "conservative" replication schemes in which the two parental strands only temporarily separate when they function as templates for daughter DNA strands. While conservative DNA replication always seemed stereochemically implausible, only through the Meselson-Stahl experiments was the semiconservative replication of DNA transformed from a very attractive hypothesis into a solid fact on which to base further experimentation.

The Genetic Information Within DNA Is Conveyed by the Sequence of Its Four Nucleotide Building Blocks

The finding of the double helix had effectively ended any controversy about whether DNA was the primary genetic substance. Even before strand separation during DNA replication was experimentally verified, the main concern of molecular geneticists had turned to how the genetic information of DNA functions to order amino acids during protein synthesis. With all DNA chains capable of forming double helices, the essence of their genetic specificity had to reside in the linear sequences of their four nucleotide building blocks. Thus, as information-containing entities, DNA molecules were by then properly regarded as very long words (as we shall see later, they are now best considered very long sentences) built up from a four-letter alphabet (A, G, C, and T). Even with only four letters, the number of potential DNA sequences (4^N, where N is the number of letters in the sequence) is very very large for even the smallest of DNA molecules; a virtually infinite number of different genetic messages can exist. Later it will be demonstrated that many bacterial genes contain some 1500 base pairs. The number of potential genes of this size is 4^{1500}, a value much larger than the number of different genes that could have existed in all the chromosomes present since the origin of life.

Proof That Genes Control Amino Acid Sequences in Proteins[26]

The first experimental demonstration that genes (DNA) control amino acid sequences arose from the study of the hemoglobin present in humans suffering from the genetic disease sickle-cell anemia. If the sickle (*s*) gene is present in both homologous chromosomes, a severe anemia results characterized by the red blood cells having a sickle shape. If only one *s* gene is present and the allele in the homologous chromosome is normal (+), then the anemia is less severe and the red blood cells appear almost normal in shape. The type of hemoglobin in red blood cells is likewise correlated with the genetic pattern. In the *ss* case, the hemoglobin is abnormal, characterized by a solubility different from that of normal hemoglobin, whereas in the *+s* condition, half the hemoglobin is normal and half sickle.

Wild-type hemoglobin molecules are constructed from two kinds of polypeptide chains: α chains and β chains (see Figure 5-32). Each chain has a molecular weight of about 16,100 daltons. Two α chains and two β chains are present in each molecule, giving hemoglobin a molecular weight of about 64,650 daltons. The α and β chains are controlled by distinct genes so that a single mutation will affect either the α or the β chain but not both. In 1957, V. M. Ingram at Cambridge University showed that sickle hemoglobin differs from normal hemoglobin by the change of one amino acid in the β chain; at position 6, the glutamic acid found in wild-type hemoglobin is replaced by valine. Except for this one change, the entire amino acid sequence is identical in normal and mutant hemoglobin. This result reveals that the sickle gene mutation involves a very specific change in the template for hemoglobin and strongly hints that all the information required to order hemoglobin (if not all other protein) sequences is present in DNA. Subsequent studies of amino acid sequences in hemoglobin isolated from other forms of anemia completely supported this proposal; here sequence analysis showed that each specific anemia is characterized by a single amino acid replacement at a unique site along the polypeptide chain (Figure 3-11).

DNA Cannot Be the Template That Directly Orders Amino Acids During Protein Synthesis[27, 28, 29]

Although DNA must carry the information for ordering amino acids, it was quite clear that the double helix itself could not be the template. Ruling out a direct role for DNA were experiments showing that protein synthesis occurs at sites where DNA is absent. The location of protein synthesis in all eucaryotic cells appeared to be the cytoplasm, which is separated by the nuclear membrane from the chromosomally located DNA.

A second information-containing molecule therefore had to exist that obtains its genetic specificity from DNA, after which it moves to the cytoplasm to function as the template for protein synthesis. Attention from the start focused on the still functionally obscure second class of nucleic acids, RNA. Sweden's Caspersson and Belgium's J. Brachet had found RNA to reside largely in the cytoplasm; and it was easy to imagine single DNA strands, when not serving as templates for complementary DNA strands, acting as templates for complementary RNA chains.

RNA Is Chemically Very Similar to DNA

Mere inspection of RNA structure shows how it can be exactly synthesized on a DNA template. Chemically, it is very similar to DNA. It, too, is a long, unbranched molecule containing four types of nucleotides linked together by 3'-5' phosphodiester bonds (Figure 3-12). Two differences in its chemical groups distinguish RNA from DNA. The first is a minor modification of the sugar component (see Figure

Figure 3-11
A summary of some established amino acid substitutions in human hemoglobin variants.

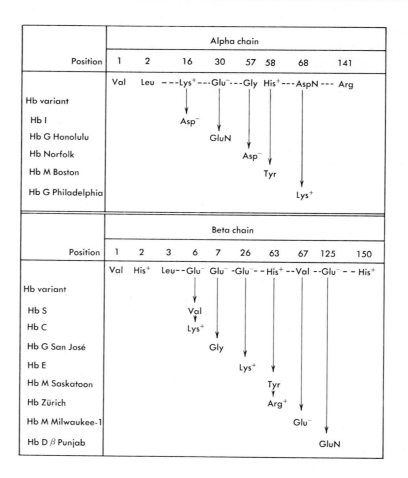

3-2). The sugar of DNA is deoxyribose, whereas RNA contains ribose, identical to deoxyribose except for the presence of an additional OH (hydroxyl) group. The second difference is that RNA contains no thymine, but instead contains the closely related pyrimidine uracil. Despite these differences, however, polyribonucleotides have the potential for forming complementary helices of the DNA type. Neither the additional hydroxyl group nor the absence of the methyl group found in thymine affects RNA's ability to form double-helical structures held together by hydrogen-bonded base pairs.

RNA Is Usually Single-Stranded

RNA molecules do not usually have complementary base ratios (Table 3-4). The amount of adenine does not often equal the amount of uracil, and the amounts of guanine and cytosine also usually differ from each other. This tells us that most RNA does not possess a regular hydrogen-bonded structure but, unlike double-helical DNA, exists as single polyribonucleotide strands. Because of the absence of regular hydrogen bonding, these single-stranded molecules do not have a simple, regular structure like DNA. This structural uncertainty initially caused much pessimism, for it was generally believed that we would have to know the template's structure before we could attack the problem of how the template selects amino acids during protein synthesis. Fortunately, as we will see later, this hunch was wrong.

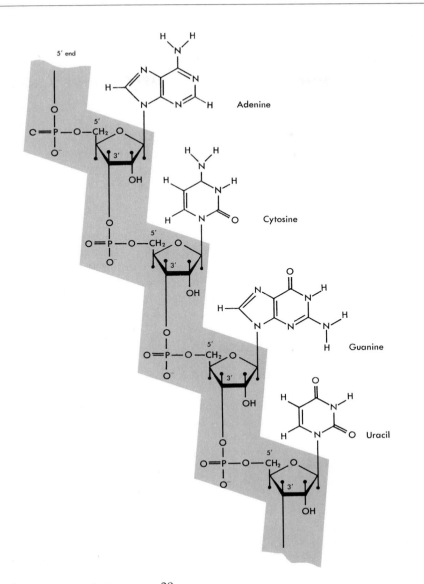

Figure 3-12
A portion of a polyribonucleotide (RNA) chain.

The Central Dogma[30]

By the fall of 1953, the working hypothesis was adopted that chromosomal DNA functions as the template for RNA molecules, which subsequently move to the cytoplasm, where they determine the arrangement of amino acids within proteins. In 1956, Francis Crick referred to this pathway for the flow of genetic information as the **central dogma.**

$$\text{Duplication} \left(\overset{\curvearrowright}{\text{DNA}} \xrightarrow{\text{Transcription}} \text{RNA} \xrightarrow{\text{Translation}} \text{Protein} \right.$$

Here the arrows indicate the directions proposed for the transfer of genetic information. The arrow encircling DNA signifies that DNA is the template for its self-replication. The arrow between DNA and RNA indicates that all cellular RNA molecules are made on ("transcribed off") DNA templates. Correspondingly, all proteins are determined by ("translated on") RNA templates. Most importantly, the last two arrows were presented as unidirectional; that is, RNA sequences are never determined by protein templates, nor was DNA then imagined ever to be made on RNA templates. That proteins

Table 3-4   The Base Composition of RNA from Various Sources

RNA Source	Proportion of the Four Main Bases			
	Adenine	Uracil	Guanine	Cytosine
E. coli	0.24	0.22	0.32	0.22
Proteus vulgaris (a bacterium)	0.26	0.19	0.31	0.24
Euglena (an alga)	0.22	0.21	0.30	0.27
Turnip yellow mosaic virus	0.23	0.22	0.17	0.38
Poliomyelitis virus	0.30	0.25	0.25	0.20
Rat kidney	0.19	0.20	0.30	0.31

never serve as templates for RNA has stood the test of time. However, as related later in this chapter, RNA chains sometimes do act as templates for DNA chains of complementary sequence. But such reversals of the normal flow of information are very rare events compared with the enormous number of RNA molecules made on DNA templates. Thus, the central dogma as originally proclaimed some 30 years ago still remains essentially valid.

The Adaptor Hypothesis of Crick[31]

At first it seemed simplest to believe that the RNA templates for protein synthesis were folded up to create cavities on their outer surfaces specific for the 20 different amino acids. The cavities would be so shaped that only one given amino acid would fit, and in this way RNA would provide the information to order amino acids during protein synthesis. By 1955, however, Crick became disenchanted with this conventional wisdom, arguing that it would never work. In the first place, the specific groups on the four bases of RNA (A, U, G, and C) should mostly interact with water-soluble groups. Yet, the specific side groups of many amino acids (e.g., leucine, valine, and phenylalanine) strongly prefer interceptions with water-insoluble (hydrophobic) groups. In the second place, even if somehow RNA could be folded so as to display some hydrophobic surfaces, there was no way for an RNA template to discriminate accurately between chemically very similar amino acids like glycine and alanine or valine and isoleucine, both of which differ only by the presence of single methyl (CH_3) groups. Crick thus proposed that prior to incorporation into proteins, amino acids are first attached to specific adaptor molecules, which in turn possess unique surfaces that can bind specifically to bases on the RNA templates.

The Test-Tube Synthesis of Proteins[32, 33, 34]

The discovery of how proteins are synthesized required the development of cell-free extracts capable of carrying on the essential synthetic steps. These were first effectively developed beginning in 1953 at the Massachusetts General Hospital in Boston by P. C. Zamecnik and his collaborators. Key to their success were the recently available radioactively tagged amino acids, which they used to mark the trace amounts of newly made proteins, as well as high-quality, easy-to-use, preparative ultracentrifuges for fractionation of their cellular extracts. Early on, the cellular site of protein synthesis was pinpointed to be the

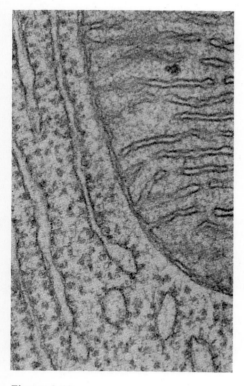

Figure 3-13
Electron micrograph (105,000×) of a portion of a cell in the pancreas of a bat, showing a portion of a mitochondrion (upper right) and large numbers of ribosomes. Some ribosomes exist free; others are attached to a membranous component, the endoplasmic reticulum. (Courtesy of K. R. Porter.)

ribosomes, small RNA-containing particles in the cytoplasm of all cells actively engaged in protein synthesis (Figure 3-13).

Also to emerge from these first experiments was the realization that ATP is required as the energy source for peptide bond formation. Several years later, Zamecnik, by then collaborating with M. B. Hoagland, went on to make the seminal discovery that prior to their incorporation into proteins, amino acids are first attached to what we now call **transfer RNA** (**tRNA**) molecules by a class of enzymes called aminoacyl synthetases. Transfer RNA accounts for some 10 percent of all cellular RNA (Figure 3-14).

This was a totally unexpected discovery except to Crick, who had previously speculated that his proposed "adaptors" might be short RNA chains, since their bases would be able to base-pair with appropriate groups on the RNA molecules that served as the templates for protein synthesis. As we shall relate later in greater detail (Chapter 14), the transfer RNA molecules of Zamecnik and Hoagland are in fact Crick's adaptors. They each contain several adjacent bases (the anticodon) that bind specifically during protein synthesis to successive groups of several bases (codons) along the RNA templates.

The Paradox of the Unspecific-Appearing Ribosomes[35, 36]

About 85 percent of cellular RNA is found in ribosomes, and since its absolute amount is greatly increased in cells engaged in large-scale protein synthesis (e.g., pancreas and liver cells and rapidly growing bacteria), **ribosomal RNA** (**rRNA**) was initially thought to be the template for ordering amino acids. But as soon as the ribosomes of *E. coli* began to be carefully analyzed, several disquieting features emerged. First, all *E. coli* ribosomes, as well as those from all other organisms, are composed of two unequally sized subunits, each containing RNA, that reversibly either stick together or fall apart, depending on the surrounding ion concentration. Second, all the rRNA chains within the small subunits are of similar chain lengths (about 1500 bases in *E. coli*), as are the rRNA chains of the large subunits (about 3000 bases). Third, the base composition of both the small and large rRNA chains is approximately the same (high in G and C) in all known bacteria, plants, and animals despite wide variations in the AT/GC ratios of the respective DNAs. This was not to be expected if the rRNA chains were in fact a large collection of different RNA templates made of a large number of different genes. If that were the case, the basic composition of the rRNA chains should be strongly correlated with that of their respective genomes. Moreover, we might expect the chain lengths of the various RNA templates to be dissimilar and correlated with the sizes of their polypeptide products. Thus, neither the small nor large class of rRNA had the feel of template RNA.

Discovery of Messenger RNA (mRNA)[37–40]

Cells infected with phage T4 provided the ideal system to find the true template. Following infection by this virus, cell synthesis of host *E. coli* RNA stops, and the only RNA synthesized is transcribed off the infecting T4 DNA. Most strikingly, not only does T4 RNA have a base composition very similar to T4 DNA, but it does not become assembled into a ribosomal component by binding to the ribosomal proteins that normally associate with rRNA to form ribosomes.

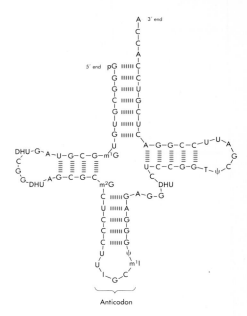

Figure 3-14
Yeast alanine tRNA structure, as determined by R. W. Holley and his associates. The anticodon in this tRNA recognizes the codon for alanine in the mRNA. Several modified nucleosides exist in the structure: ψ = pseudouridine, T = ribothymidine, DHU = 5,6-dihydrouridine, I = inosine, m^1G = 1-methylguanosine, m^1I = 1-methylinosine, and m^2G = N,N-dimethylguanosine.

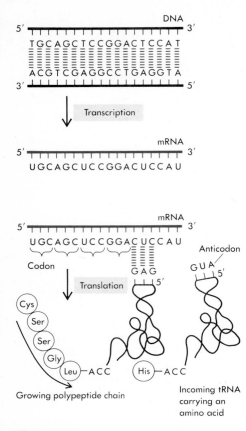

Figure 3-15

Transcription and translation. The nucleotides of mRNA are assembled to form a complementary copy of one strand of DNA. Each group of three is a codon that is complementary to a group of three nucleotides in the anticodon region of a specific tRNA molecule. When base pairing occurs, an amino acid carried at the other end of the tRNA molecule is added to the growing protein chain.

Instead, T4 RNA, after first attaching to previously existing ribosomes, moves across their surface to bring its bases into positions where they can bind to the appropriate tRNA–amino acid precursors for protein synthesis (Figure 3-15). In so acting, T4 RNA orders the amino acids and is thus the long-sought-for RNA template for protein synthesis. Because it carries the information from DNA to the ribosomal sites of protein synthesis, it is called **messenger RNA (mRNA)**. The observation of T4 RNA binding to *E. coli* ribosomes, first made in the spring of 1960, was soon followed with evidence for a separate messenger class of RNA within uninfected *E. coli* cells, thereby definitively ruling out a template role for any rRNA. Instead, in ways that we shall discuss more extensively in Chapter 14, the rRNA components of ribosomes, together with some 50 different ribosomal proteins that bind to them, serve as the factories for protein synthesis, functioning to bring together the tRNA–amino acid precursors into positions where they can read off the information provided by the messenger RNA templates.

Only some 4 percent of total cellular RNA is mRNA, whose sizes show the expected large variations in length, depending on the proteins for which they code. Hence, it is easy to understand why mRNA was first overlooked. Because only a small segment of mRNA is attached at a given moment to a ribosome, which is moving across it, a single mRNA molecule can simultaneously be read by several ribosomes. Most ribosomes are found as parts of **polyribosomes,** ranging in size from 6 to more than 50 members (Figure 3-16).

Messenger RNA molecules are always translated in the same direction, starting near the 5′ end and finishing near the 3′ end. It is thus the 5′ end of an mRNA molecule that first attaches to a ribosome.

Enzymatic Synthesis of RNA Upon DNA Templates[41, 42, 43]

As messenger RNA was being discovered, the first of the enzymes that transcribe RNA off DNA templates was being independently isolated in the labs of American biochemists J. Hurwitz, A. Stevens, and

Figure 3-16

Diagram of a polyribosome. Each ribosome attaches at a start signal at the 5′ end of an mRNA chain and synthesizes a polypeptide as it proceeds along the molecule. Several ribosomes may be attached to one mRNA molecule at one time; the entire assembly is called a polyribosome.

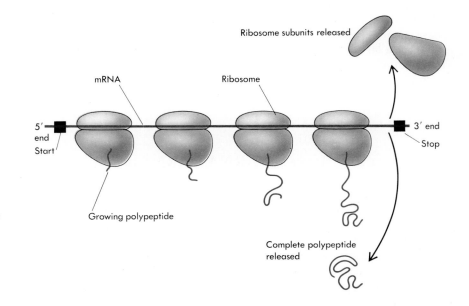

Figure 3-17
Enzymatic synthesis of RNA upon a
DNA template, catalyzed by RNA
polymerase. As the RNA is synthesized, pyrophosphate is split off each
incoming nucleotide; the enzyme pyrophosphatase catalyzes the breakdown
of pyrophosphate to phosphate ions.

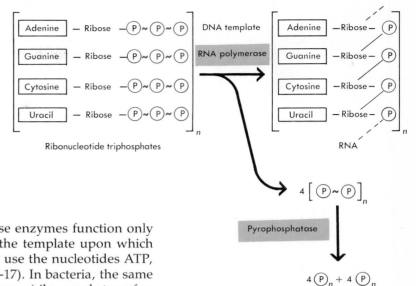

S. B. Weiss. Called **RNA polymerases,** these enzymes function only
in the presence of DNA, which serves as the template upon which
single-stranded RNA chains are made, and use the nucleotides ATP,
GTP, CTP, and UTP as precursors (Figure 3-17). In bacteria, the same
enzyme makes each of the major RNA classes (ribosomal, transfer,
and messenger), using appropriate segments of chromosomal DNA
as their templates. Direct evidence that DNA lines up the correct ribonucleotide precursors came from seeing how the RNA base composition varied with the addition of DNA molecules of different AT/GC
ratios. In every enzymatic synthesis, the RNA AU/GC ratio was
roughly similar to the DNA AT/GC ratio (Table 3-5).

Table 3-5   Comparison of the Base Composition of Enzymatically
Synthesized RNAs with the Base Composition of Their Double-
Helical DNA Templates

Source of DNA Template	Composition of the RNA Bases				$\dfrac{A + U}{G + C}$ Observed	$\dfrac{A + T}{G + C}$ in DNA
	Adenine	Uracil	Guanine	Cytosine		
T2	0.31	0.34	0.18	0.17	1.86	1.84
Calf thymus	0.31	0.29	0.19	0.21	1.50	1.35
E. coli	0.24	0.24	0.26	0.26	0.92	0.97
Micrococcus lysodeikticus (a bacterium)	0.17	0.16	0.33	0.34	0.49	0.39

During transcription, only one of the two strands of DNA becomes
translated into RNA. This makes sense, because the messages carried
by the two strands, being complementary but not identical, might be
expected to code for completely different polypeptides. The synthesis
of RNA always proceeds in a fixed direction, beginning at the 5′ end
and concluding with the 3′-ended nucleotide (see Figure 3-16). Transcription and translation thus occur in the same direction.

By this time, there was firm evidence for the postulated movement
of RNA from the DNA-containing nucleus to the ribosome-containing
cytoplasm. By briefly exposing cells to radioactively labeled RNA precursors, the site of RNA synthesis was shown to be the nucleus.
Within an hour, most of this RNA had left the nucleus to be observed
in the cytoplasm (Figure 3-18).

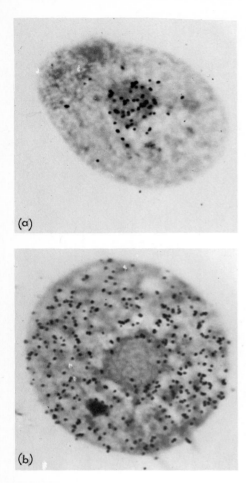

Figure 3-18
Demonstration that RNA is synthesized in the nucleus and moves to the cytoplasm. (a) Autoradiograph of a cell (*Tetrahymena*) exposed to radioactive cytidine for 15 minutes. Superimposed on a photograph of a thin section of the cell is a photograph of an exposed silver emulsion. Each dark spot represents the path of an electron emitted from a ^{3}H (tritium) atom that has been incorporated into RNA. Almost all the newly made RNA is found within the nucleus. (b) Autoradiograph of a similar cell exposed to radioactive cytidine for 12 minutes and then allowed to grow for 88 minutes in the presence of nonradioactive cytidine. Practically all the label incorporated into RNA in the first 12 minutes has left the nucleus and moved into the cytoplasm. [Courtesy of D. M. Prescott, University of Colorado Medical School; reproduced from *Progr. Nucleic Acid Res.* III, 35 (1964), with permission.]

Repressors and Activators of Gene Expression[44–49]

How cells control the exact amounts of their proteins became open for detailed analysis as the general pathway for the flow of genetic information became clear. Particularly intriguing were those proteins whose amounts changed greatly, depending on their need. Long serving as the key model system was the bacterial enzyme β-galactosidase, which breaks the sugar lactose down into the smaller sugars glucose and galactose. In the absence of lactose, the bacterium *E. coli* makes an exceedingly small number of molecules of this enzyme, while in its presence, each *E. coli* cell contains thousands of such molecules. Leading this research were the French microbiologists J. Monod and F. Jacob, working at the Institut Pasteur in Paris. In the late 1950s, they observed that the functioning of the gene that codes for the order of amino acids in β-galactosidase is itself controlled by a separate regulatory gene. Monod and Jacob postulated its gene product to be a **repressor** that in the absence of lactose specifically turns off the functioning of the structural gene. Only very small numbers of this repressor are normally present in an *E. coli* cell, and it was not until 1966 that it was first isolated at Harvard by B. Mueller-Hill and W. Gilbert. Soon afterward, they demonstrated that it bound just before the beginning of the β-galactosidase structural gene and prevented its transcription into RNA.

Meanwhile, M. Ptashne and N. Hopkins, also at Harvard, showed that a key regulatory gene of the bacterial virus λ also coded for a protein repressor that acted by binding to specific DNA segments and preventing their functioning. Not all DNA-binding regulatory proteins act negatively. Some turn up, or **activate,** DNA expression by binding to other DNA sequences. After binding to cyclic AMP, the catabolic activator protein (CAP) can stimulate the functioning of several sugar-utilizing genes (e.g., arabinose, lactose, and galactose) by binding to appropriate segments of DNA.

Establishing the Genetic Code[50–55]

Given the existence of 20 amino acids but only four bases, groups of several nucleotides must somehow specify a given amino acid. Groups of two, however, would specify only 16 (4 × 4) amino acids. So from 1954, the start of serious thinking about what the genetic code might be like, most attention was given to how triplets (groups of three) might work, even though they obviously would provide more permutations (4 × 4 × 4) than needed if each amino acid was specified by only a single triplet. The assumption of colinearity was then very important. It held that successive groups of nucleotides along a DNA chain code for successive amino acids along a given polypeptide chain. That colinearity does in fact exist was shown by elegant mutational analysis on bacterial proteins, carried out in the early 1960s at Stanford University by C. Yanofsky and at Cambridge, England, by A. Brenner. Equally important were the genetic analyses by Brenner and Crick, which in 1961 first established that groups of three nucleotides are used to specify individual amino acids.

But which specific groups of three bases (codons) determine which specific amino acids could only be learned by biochemical analysis. Here the major breakthrough came when M. W. Nirenberg and J. H. Matthaei, then both working together at the National Institute of

Table 3-6   The Genetic Code

Second Position

		U	C	A	G	
First Position	**U**	UUU ⎤ Phe UUC ⎦ UUA ⎤ Leu UUG ⎦	UCU ⎤ UCC ⎥ Ser UCA ⎥ UCG ⎦	UAU ⎤ Tyr UAC ⎦ UAA *Stop* UAG *Stop*	UGU ⎤ Cys UGC ⎦ UGA *Stop* UGG Trp	U C A G
	C	CUU ⎤ CUC ⎥ Leu CUA ⎥ CUG ⎦	CCU ⎤ CCC ⎥ Pro CCA ⎥ CCG ⎦	CAU ⎤ His CAC ⎦ CAA ⎤ Gln CAG ⎦	CGU ⎤ CGC ⎥ Arg CGA ⎥ CGG ⎦	U C A G
	A	AUU ⎤ AUC ⎥ Ile AUA ⎦ AUG Met	ACU ⎤ ACC ⎥ Thr ACA ⎥ ACG ⎦	AAU ⎤ Asn AAC ⎦ AAA ⎤ Lys AAG ⎦	AGU ⎤ Ser AGC ⎦ AGA ⎤ Arg AGG ⎦	U C A G
	G	GUU ⎤ GUC ⎥ Val GUA ⎥ GUG ⎦	GCU ⎤ GCC ⎥ Ala GCA ⎥ GCG ⎦	GAU ⎤ Asp GAC ⎦ GAA ⎤ Glu GAG ⎦	GGU ⎤ GGC ⎥ Gly GGA ⎥ GGG ⎦	U C A G

Third Position (right margin label)

Health in Bethesda, Maryland, observed, also in 1961, that the addition of the synthetic polynucleotide poly U (UUUUU . . .) to a cell-free system capable of making proteins leads to the synthesis of polypeptide chains containing only the amino acid phenylalanine. The nucleotide groups UUU thus must specify phenylalanine. Use of increasingly more complex, defined polynucleotides as synthetic messenger RNAs rapidly led to the identification of more and more codons. Particularly important in completing the code was the use of polynucleotides like AGUAGU, put together by the Indian organic chemist H. G. Khorana, then working in Madison, Wisconsin. Completion of the code in 1966 revealed that 61 out of the 64 possible permuted groups corresponded to amino acids, with most amino acids being coded by more than one nucleotide triplet (Table 3-6).

Start and Stop Signals Are Also Encoded Within DNA[56-59]

Initially, it was guessed that translation of an mRNA molecule would commence at one end and finish when the entire mRNA message had been read into amino acid sequences. But, in fact, translation both starts and stops at internal positions. Thus, signals must be present within DNA (and its mRNA products) to initiate and terminate translation. First to be worked out were the stop signals. Three separate codons (UAA, UAG, and UGA), first known as **nonsense codons,** do not correspond to any amino acids but instead serve as chain-termi-

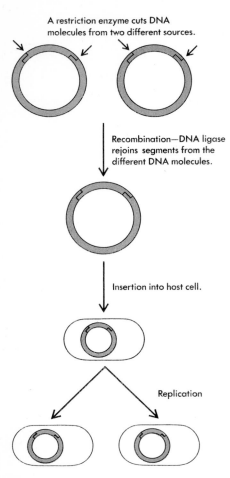

A restriction enzyme cuts DNA molecules from two different sources.

Recombination—DNA ligase rejoins segments from the different DNA molecules.

Insertion into host cell.

Replication

Figure 3-19
The joining of DNA from two different sources via recombinant DNA techniques. DNA ligase joins DNA segments previously cut with restriction enzymes.

nating (stop) signals. More complicated is the way translational start signals are encoded. The amino acid methionine starts all polypeptide chains, but the triplet (AUG) that codes for these initiating methionines also codes for methionine residues that have internal locations. The AUG codons at which polypeptide chains start are preceded by specific purine-rich blocks of nucleotides that serve to attach mRNA to ribosomes (see Chapters 14 and 15 for details).

The Extreme Utility of Enzymes That Cut DNA into Well-Defined Fragments[60, 61, 62]

Signals to start and stop the synthesis of RNA chains also are encoded within DNA. These are more complicated than the signals involved in translational starts and stops and could only be revealed when techniques became available for DNA sequencing. For all practical purposes, we could not observe it in action, as even the smallest of natural DNA molecules were much too big to be individually looked at by the methods of chemistry. This seemingly hopeless picture fortunately did not last long. By 1970, Hamilton Smith at Johns Hopkins Medical School had isolated the first enzyme that cuts DNA at a very specific nucleotide sequence. Over the next few years, a number of different such enzymes were isolated, each from a distinct type of bacterium, and there are now over a hundred different specificities.

In general, each such cutting enzyme (**restriction enzyme**) recognizes a unique sequence of four or six nucleotides (Table 3-7) and so cuts a given DNA molecule about once every several hundred or thousand bases. The availability of these enzymes quickly changed the course of DNA research, enabling the biochemist for the first time to experiment on well-defined DNA fragments of the size that, for example, has permitted the development of high-powered methodologies for sequencing DNA molecules. These cutting tools were first most effectively used in the isolation of distinct DNA fragments from viral DNAs ranging from several thousand to 50,000 nucleotides in length. Such relatively small DNAs generally yield fragments of all different lengths, which can be easily separated from each other by newly developed gel separation techniques.

Recombinant DNA[63–66]

The most important early consequence of restriction enzyme availability was the impetus it gave to the development of what are now called **recombinant DNA** procedures. Prior research on the enzymology of DNA replication had resulted in the discovery of the enzyme **DNA ligase,** which joins together (ligates) DNA chains. Using this ligase, the "restriction" DNA fragments generated by restriction enzyme cutting of a preexisting DNA molecule can be randomly linked to each other.

The possibility was thus conceived of joining together pieces arising from different parental molecules, and the first such attempts were performed in the laboratory of P. Berg at Stanford University in 1972. The first effective insertion of recombinant DNA into cells occurred the following year, when S. Cohen of Stanford and H. Boyer at San Francisco used a tiny bacterial chromosome (plasmid) as one of the DNA partners to be joined together (Figure 3-19). Following entry into a suitable bacterium, the tiny recombinant chromosome multi-

Table 3-7 Recognition Sequences of Several Restriction Enzymes*

Microbial Origin	Enzyme	Recognition Site
Escherichia coli KY13	*Eco* RI	Cut bond ↓ ⁵′GAA | TTC / CTT | AAG ↑ (Axis of symmetry)
Hemophilus influenzae Rd	*Hin* dII	⁵′GTPy | PuAC / CAPu | PyTG ↑
	Hin dIII	↓ ⁵′AAG | CTT / TTC | GAA ↑
Hemophilus parainfluenzae	*Hpa* I	⁵′GTT | AAC / CAA | TTG ↑
	Hpa II	↓ ⁵′CC | GG / GG | CC ↑
Hemophilus aegyptius	*Hae* III	⁵′GG | CC / CC | GG ↑

*Note that each sequence has twofold symmetry.

plied, and a clone of the recombinant DNA piece was generated as its host bacterium itself multiplied.

It is impossible to overstate the effect of recombinant DNA procedures, not only in their promotion of ongoing DNA research itself but also through their generation of new fields of research (such as cellular immunology at the molecular level).

Extraordinary New Techniques Aid in Sequencing DNA[67, 68, 69]

Only a decade ago, there was no simple way to determine the sequences of short DNA chains containing only a few nucleotides. Today, the sequences of restriction-enzyme-generated DNA fragments several hundred nucleotides in length can be routinely accomplished in two to three days. In fact, the complete sequences of the bacterial viruses T7 (39,936 base pairs) and λ (48,513 base pairs) are already worked out. And even longer viral genomes (e.g., herpesvirus, about 1.5×10^5 base pairs) can be established now within a one- to two-year interval. Two powerful new techniques for DNA sequencing, one developed at Harvard by A. M. Maxam and Gilbert, the other worked out in Cambridge, England, by Sanger, underlie these spectacular advances (see Chapter 9 for details). Both these procedures depended on the availability of restriction enzymes. In obtaining the necessary amounts of needed DNA fragments, recombinant DNA procedures are now routinely employed.

Many DNA sequences are already known that code from proteins whose amino acid sequences had previously been worked out. They not only completely confirm the codon assignments worked out using synthetic polynucleotides but also afford us the opportunity to determine the actual codons used when a given amino acid is specified by more than one codon. By now, DNA sequencing is a much more routine task than protein sequencing, and the sequences of many proteins are in fact first being known through the sequences of their corresponding genes.

RNA Can Also Function as a Genetic Molecule[70–73]

In addition to its role as the template for protein synthesis, RNA has the capacity to act, like DNA, as a genetic molecule. That RNA can store genetic information was first hinted in the late 1930s, when several plant viruses were found to contain RNA but no DNA. Proof that this RNA component was indeed the viral genetic material did not emerge, however, until 1956, when Alfred Gierer, working in Tübingen, Germany, showed that highly purified RNA from tobacco mosaic virus was infectious in the absence of any viral protein (Figure 3-20). Since then, a variety of other viruses (including influenza and polio) have been determined also to have RNA as their genetic molecule.

Because the structure of RNA is very similar to that of DNA, it seemed logical to assume that RNA replication followed the same principles established for DNA. Proof that RNA can be so replicated came first from experiments with a class of small viruses that multiply in bacteria and which contain single-stranded RNA molecules as their genetic component. A key breakthrough came from the 1965 isolation at the University of Illinois by I. Haruna and S. Spiegelman of a virally coded *replicase*, an enzyme that catalyzes the formation of new infectious viral RNA molecules in the presence of the appropriate precursors. How these new molecules are made was at first debatable, but by 1968, C. Weissmann and J. August, independently working in different New York laboratories, conclusively demonstrated the involvement of a double-stranded RNA intermediate. The self-replication of RNA as well as DNA thus involves base pairing between chains of complementary sequences.

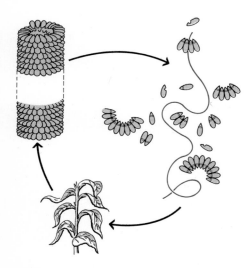

Figure 3-20
Proof that RNA is the genetic component of TMV. The rod-shaped TMV particles can easily be separated into their protein and RNA components, which can then be separately tested for their ability to initiate virus infection. Only the RNA molecules have this ability. The virus particles produced by infecting with pure RNA are identical to those resulting from infection with intact virus.

Reverse Transcription of RNA Yields DNA[74, 75]

All RNA viruses do not replicate using double-stranded RNA intermediates. The infecting RNA of **retroviruses** (earlier known as RNA tumor viruses; Chapter 24) serves as a template for the synthesis of a single-stranded DNA complement that in turn functions as a template for a complementary DNA chain. The resulting double-stranded DNA, generally after insertion in a host cell chromosome, then acts as a template for the synthesis of an RNA chain identical to that which initiated the viral life cycle (Figure 3-21). The flow of genetic information back from RNA to DNA does not present any conceptual problems, since RNA-DNA hybrid double helices also exist, though only fleetingly, when RNA is transcribed off single-stranded DNA templates. The direction of information flow may be determined by the particular enzyme available. Catalyzing the making of DNA on RNA during retroviral infection is the viral-specific enzyme **reverse transcriptase.** It was its discovery in 1970, by S. Mizutani and H. M. Temin at Madison and by D. Baltimore at MIT, that first led to general acceptance of the idea that information can flow back from RNA to DNA. In the absence of retrovirus infection, very little, if any, reverse transcriptase activity is present in normal cells; thus, under normal conditions, effectively very little DNA is made on RNA templates.

Interrupted Genes Within Eucaryotic Chromosomes[76, 77]

Whereas sequencing of the first bacterial DNA fragment confirmed previous structural speculations about bacterial genes, a completely different picture emerged when the first eucaryotic genes were cloned and sequenced. Suprisingly, it was found that the nucleotide sequences that specify amino acids for hemoglobin and ovalbumin are frequently interrupted by blocks of nucleotides that have no coding information. Those nucleotide segments that code for amino acids are called **exons,** while those segments that are not translated into amino acids are called **introns.** Most genes coding for mRNA products in higher eucaryotic cells are split. In the cell, they are transcribed into very large precursor RNA molecules, which subsequently have their introns removed by specific enzymes within the nucleus. In the process, the remaining exons are spliced together to form smaller RNAs, which then move into the cytoplasm where they function as mRNA. Later it was found that many tRNAs and some tRNA genes are also split, with their introns also being removed by RNA splicing events. Why so many genes of higher cells are split while their bacterial equivalents are not is not yet understood. In any case, an inevitable consequence is that eucaryotic genes are frequently very large, with many such genes having much more of their DNA existing as introns than as exons.

Unlimited Vistas for Genetics at the Molecular Level

Until 1953, genes were perceived to be at the heart of the existence of life and yet very mysterious. Once the double helix was revealed, however, biologists had the reference point for further research. In particular, biochemists could now effectively focus their highly developed skills for working out enzymatically catalyzed chemical reactions that underlie genetic phenomena. The discovery of the genetic code by 1966 was an enormous scientific achievement, and the rapidity with which molecular genetics has advanced over the past three decades has surprised and pleased virtually everyone. Now, with powerful new recombinant DNA techniques and DNA-sequencing procedures in hand, we stand in marvel at the genetic knowledge revealed in just the last decade, and we predict with awe where future DNA research will take us.

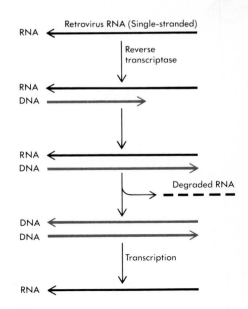

Figure 3-21
Reverse transcription of RNA to yield DNA. The DNA-RNA hybrid strand is highly unstable and exists only fleetingly.

The long-mysterious genetic signals that govern eucaryotic cells and that are now being discovered will open up new, direct avenues into the nature of diseases like cancer, as well as begin to reveal the essence of cell differentiation. Moreover, we can soon expect to uncover those molecular tricks that underlie the organization of nerve cells into the intricate cellular networks that in turn underlie perception, memory, and thinking. The gene will thus remain the key to future efforts to probe the nature of life.

Summary

Specific classes of molecules within cells carry the genetic information that determines the order of the 20 different amino acids in proteins. Such genetic molecules function by presenting specific surfaces, or templates, to which the correct amino acids (activated derivatives) are sequentially attracted and ordered prior to their polymerization. A specific template exists for each specific protein. When a cell grows and divides, the number of protein templates must also double. Each template must thus itself be formed on the surface of another template.

The templates that carry the genetic information of cellular chromosomes are always DNA molecules. All DNA molecules are very large in comparison to all other cellular molecules, being very long linear polymers built from large numbers of 4 different nucleotide building blocks connected by phosphodiester linkages (polynucleotide molecules). Nucleotide sequence along a given chain is very irregular, with a specific DNA sequence conveying each genetic message. Two of the nucleotides contain purine bases (adenine and guanine), while the other two nucleotides contain pyrimidine bases (thymine and cytosine). In virtually all DNA molecules, two polynucleotide chains are twisted around each other to form a regular double helix. The two chains within the double helix are held together by hydrogen bonds between pairs of bases. Adenine is always joined to thymine, and guanine is always bonded to cytosine. The existence of the base pairs means that the sequence of nucleotides along the two chains are not identical, but complementary. If the sequence of one chain is known, then the sequence of its partner is automatically known. The finding of this relationship suggested a mechanism for the replication of DNA in which each strand serves as a template for its complement. Proof for this hypothesis came from (a) the observation that the two strands of each double helix do separate during each round of DNA replication in cells and (b) the discovery of a specific enzyme (DNA polymerase I) that uses single-stranded segments of DNA as templates for the polymerization of the immediate precursors of DNA (the deoxynucleotide triphosphates dATP, dGTP, dTTP, and dCTP) into chains of complementary sequence.

The sequence of base pairs does not serve as the direct template for protein synthesis. Instead, it is used as a template to make RNA the other form of nucleic acid, which is primarily found in the cytoplasm. In turn, many of these RNA products function as templates for protein synthesis. RNA is chemically very similar to DNA, also being composed of four different nucleotide building blocks, two of which contain the purines adenine and guanine and two of

which contain the pyrimidines uracil and cytosine. The synthesis of an RNA chain upon a DNA chain utilizes the same base-pairing rules involved in DNA self-duplication (except that uracil base-pairs like thymine). And as with the synthesis of DNA, a specific enzyme, in this case an RNA polymerase, is needed to catalyze the formation of covalent bonds between the immediate precursors, the ribonucleotide triphosphates ATP, GTP, UTP, and CTP.

During the synthesis of RNA upon DNA, a process called transcription, only one of the two strands of a given double helix functions as a template. The RNA products of transcription are thus single-stranded. Prior to amino acid polymerization into proteins, the 20 different amino acids are first transformed into energy-rich precursors in which a given amino acid is attached to a small transfer RNA (tRNA) molecule. Amino acids do not have specific regions that bind to RNA templates. Instead, they are recognized by the specific enzymes (the aminoacyl synthetases) that attach them to specific tRNA molecules. Each tRNA molecule contains a group of three nucleotides (the anticodon) that uses base pairing to seek out a group of three adjacent nucleotides (a codon) on a messenger RNA molecule, the RNA class that carries the ordering information. Successful codon-anticodon interactions only occur on the surfaces of ribosomes, the factories for protein synthesis. Ribosomes themselves consist of large amounts of ribosomal RNA (rRNA) as well as many different protein components. During protein synthesis, mRNA molecules move across the active surfaces of ribosomes, bringing successive codons into position to determine successive amino acids along polypeptide chains (the process of translation). Out of the 64 ($4 \times 4 \times 4$) potential codons, 61 are used to specify amino acids, while 3 are used to provide chain-terminating signals. Many amino acids are thus determined by more than one codon. The amino acids specified by the various codons (the genetic code) are essentially the same for all forms of life.

Many factors determine the rate at which specific segments of DNA function. Much control is effected at the level of transcription by specific DNA-binding proteins, some of which repress RNA synthesis (repressors, or negative control proteins), while others are necessary for its initiation (activators, or positive control proteins).

Over one hundred different enzymes are now known that recognize and cut DNA at very specific nucleotide sequences. Almost all such "restriction enzymes" recognize unique sequences of four to six nucleotides. Their availability allowed the isolation of DNA fragments of well-defined length and hastened the development of high-powered methods for precisely sequencing DNA. The availability of restriction enzymes also quickly led to the generation of "recombinant DNA" molecules and the means to clone specific DNA fragments (i.e., isolate specific genes). Sequencing of genes led to the realization that many eucaryotic genes are split. The nucleotide segments (exons) that code for amino acids are often interrupted by long DNA segments (introns) that do not code for amino acids. Both exons and introns are transcribed into long RNA precursors from which the introns are spliced out to yield functional RNA products.

RNA molecules are the genetic components of many small viruses as well as the templates for their viral-specific proteins. Replication of many RNA chromosomes involves the formation of complementary RNA chains. Other RNA viruses (retroviruses) replicate via complementary DNA intermediates using the enzyme reverse transcriptase. The initially single-stranded DNA products are then converted into double-stranded DNA molecules; then this DNA, generally after insertion into a host chromosome, acts as a template for the synthesis of RNA chains identical to those that started the viral life cycle.

Bibliography

General References

Cairns, J., G. S. Stent, and J. D. Watson, eds. 1966. *Phage and the Origins of Molecular Biology.* Cold Spring Harbor, N.Y.: Cold Spring Harbor Laboratory.

Judson, H. F. 1979. *The Eighth Day of Creation.* New York: Simon & Schuster.

Kornberg, A. 1980. *DNA Replication.* San Francisco: Freeman.

McCarty, M. 1985. *The Transforming Principle: Discovering that Genes Are Made of DNA.* New York: Norton.

Olby, R. 1975. *The Path to the Double Helix.* Seattle: University of Washington Press.

Portugal, F. H., and J. S. Cohen. 1980. *A Century of DNA: A History of the Discovery of the Structure and Function of the Genetic Substance.* Cambridge, Mass.: MIT Press.

Stent, G. S., and R. Calendar. 1978. *Molecular Genetics: An Introductory Narrative.* 2nd ed. San Francisco: Freeman.

Watson, J. D. 1968. *The Double Helix.* New York: Atheneum (text and paperback editions); New York: New American Library, 1969 (paperback).

Watson, J. D. 1980. *The Double Helix: A Norton Critical Edition.* Ed. G. S. Stent. New York: Norton.

Cited References

1. Friedrich-Freksa, H. 1940. "Bei der Chromosomen Konjuation Wirksame Krafte und ihre Bedeutung für die Identische Verdopplung von Nucleoproteinen." *Naturwissenshaffen* 28:376–379.
2. Pauling, L., and M. Delbruck. 1940. "The Nature of the Intermolecular Forces Operative in Biological Processes." *Science* 92:77–79.
3. Muller, H. J. 1947. "The Gene." *Proc. Roy. Soc. London* (B) 134:1–37. A lecture given in 1945 in which a distinguished geneticist traces the history of the gene concept and speculates about how it might function as a template.
4. Dounce, A. L. 1952. "Duplicating Mechanisms for Peptide Chain and Nucleic Acid Synthesis." *Enzymologia* 15:251–258.
5. Loftfield, R. B. 1963. "The Frequency of Errors in Protein Biosynthesis." *Biochem. J.* 89:82–92.
6. Signer, R., T. Caspersson, and E. Hammarsten. 1938. "Molecular Shape and Size of Thymonucleic Acid." *Nature* 141:122.
7. Chargaff, E., and J. N. Davidson, eds. 1955. *The Nucleic Acids.* Vols 1 and 2. New York: Academic Press.
8. Griffith, F. 1928. "The Significance of Pneumonococcal Types." *J. Hyg.* 27:113–159.
9. Avery, O. T., C. M. MacLeod, and M. McCarty. 1944. "Studies on the Chemical Nature of the Substance Inducing Transformation of Pneumococcus Types." *J. Exp. Med.* 79:137–158.
10. Mirsky, A. E., and H. Ris. 1949. "Variable and Constant Components of Chromosomes." *Nature* 163:666–667.
11. Boivin, A., R. Vendrely, and C. Vendrely. 1948. *Compt. Rend. Acad. Sci.* 226:1061.
12. Hershey, A. D., and M. Chase. 1952. "Independent Function of Viral Protein and Nucleic Acid on Growth of Bacteriophage." *J. Gen. Physiol.* 36:39–56.
13. Chargaff, E., E. Vischer, R. Doniger, C. Green, and F. Misani. 1949. "The Composition of the Desoxypentose Nucleic Acids of Thymus and Spleen." *J. Biol. Chem.* 177:405–416.
14. Chargaff, E. 1951. "Structure and Function of Nucleic Acids as Cell Constituents." *Fed. Proc.* 10:654–659.
15. Astbury, W. T. 1951. "X-Ray Studies of Nucleic Acids." *Symp. Soc. Exp. Biol. I. Nucleic Acid* 1:66–76.
16. Brown, D. M., and A. R. Todd. 1952. "Nucleotides Part X: Some Observations on Structure and Chemical Behavior of the Nucleic Acids." *J. Chem. Soc.* Pt. 1: 52–58.
17. Dekker, C. A., A. M. Michaelson, and A. R. Todd. 1953. "Nucleotides Part XIX. Pyrimidine Deoxyribonucleoside Diphosphates." *J. Chem. Soc.* Pt. 1: 947–951.
18. Cochran, W., F. H. C. Crick, and V. Vand. 1952. "The Structure of Synthetic Polypeptides I. The Transform of Atoms on a Helix." *Acta Cryst.* 5:581–586.
19. Franklin, R. E., and R. G. Gosling. 1953. "Molecular Configuration in Sodium Thymonuclease." *Nature* 171:740–741.
20. Watson, J. D., and F. H. C. Crick. 1953. "Molecular Structure of Nucleic Acids: A Structure for Deoxyribose Nucleic Acid." *Nature* 171:737–738.
21. Watson, J. D., and F. H. C. Crick. 1953. "Genetical Implications of the Structure of Deoxyribonucleic Acid." *Nature* 171:964–967.
22. Wilkins, M. H. F., A. R. Stokes, and H. R. Wilson. 1953. "Molecular Structure of Deoxypentose Nucleic Acid." *Nature* 171:738–740.
23. Crick, F. H. C., and J. D. Watson. 1954. "The Complementary Structure of Deoxyribonucleic Acid." *Proc. Roy. Soc.* (A) 223:80–96.
24. Kornberg, A. 1960. "Biological Synthesis of Deoxyribonucleic Acid." *Science* 131:1503–1508.
25. Meselson, M., and F. W. Stahl. 1958. "The Replication of DNA in *Escherichia coli.*" *Proc. Nat. Acad. Sci.* 44:671–682.
26. Ingram, V. M. 1957. "Gene Mutations in Human Hemoglobin: The Chemical Difference Between Normal and Sickle Cell Hemoglobin." *Nature* 180:326–328.
27. Brachet, J. 1947. *Embryologia Chimique.* Paris, France: Masson.
28. Brachet, J. 1957. *Biochemical Embryology.* New York: Academic Press.
29. Caspersson, T. 1950. *Cell Growth and Cell Function.* New York: Norton.
30. Crick, F. H. C. 1958. "On Protein Synthesis." *Symp. Soc. Exp. Biol.* 12:548–555.
31. Crick, F. H. C. 1955. "On Degenerate Template and the Adaptor Hypothesis." A note for the RNA Tie Club, unpublished. Mentioned in Crick's 1957 discussion, pp. 25–26, in "The Structure of Nucleic Acids and Their Role in Protein Synthesis." *Biochem. Soc. Symp.* no. 14, Cambridge University Press.
32. Hoagland, M. B., E. B. Keller, and P. C. Zamecnik. 1956. "Enzymatic Carboxyl Activation of Amino Acids." *J. Biol. Chem.* 218:345–358.
33. Hoagland, M. B., M. L. Stephenson, J. F. Scott, L. I. Hecht, and P. C. Zamecnik. 1958. "A Soluble Ribonucleic Acid Intermediate in Protein Synthesis." *J. Biol. Chem.* 231:241–257.
34. Holley, R. W., J. Apgar, G. A. Everett, J. T. Madison, M. Mar-

quisse, S. H. Merrill, J. R. Penswick, and A. Zamir. 1965. "Structure of a Ribonucleic Acid." *Science* 147:1462–1465.

35. Tisséires, A., J. D. Watson, D. Schlesinger, and B. R. Hollingworth. 1959. "Ribonucleoprotein Particles from *Escherichia coli.*" *J. Mol. Biol.* 1:221.

36. Watson, J. D. 1963. "Involvement of RNA in Synthesis of Proteins." *Science* 140:17–26.

37. Volkin, E., and L. Astrachan. 1956. "Phosphorus Incorporation in *E. coli* Ribonucleic Acid After Infection with Bacteriophage T2." *Virology* 2:146–161.

38. Brenner, S., F. Jacob, and M. Meselson. 1961. "An Unstable Intermediate Carrying Information from Genes to Ribosomes for Protein Synthesis." *Nature* 190:576–581.

39. Gros, F., H. Hiatt, W. Gilbert, C. G. Kurland, R. W. Risebrough, and J. D. Watson. 1961. "Unstable Ribonucleic Acid Revealed by Pulse Labelling of *Escherichia coli.*" *Nature* 190:581–585.

40. Hall, B. D., and S. Spiegelman. 1961. "Sequence Complementarity of T2 DNA and T2-Specific RNA." *Proc. Nat. Acad. Sci.* 47:137–146.

41. Hurwitz, J., A. Bresler, and R. Diringer. 1960. "The Enzymatic Incorporation of Ribonucleotides into Polyribonucleotides and the Effect of DNA." *Biochem. Biophys. Res. Comm.* 3:15–19.

42. Weiss, S. B. 1960. "Enzymatic incorporation of ribonucleotide triphosphates into the interpolynucleotide linkages of ribonucleic acid." *Proc. Nat. Acad. Sci.* 46:1020–1030.

43. Stevens, A. 1960. "Incorporation of the Adenine Ribonucleotide into RNA by Cell Fractions from *E. coli* B." *Biochem. Biophys. Res. Comm.* 3:92–96.

44. Jacob, F., and J. Monod. 1961. "Genetic Regulatory Mechanisms in the Synthesis of Proteins." *J. Mol. Biol.* 3:318–356.

45. Gilbert, W., and B. Muller-Hill. 1966. "Isolation of the Lac Repressor." *Proc. Nat. Acad. Sci.* 56:1891–1898.

46. Ptashne, M. 1967. "Isolation of the λ Phage Repressor." *Proc. Nat. Acad. Sci.* 57:306–313.

47. Ptashne, M., and N. Hopkins. 1968. "The Operators Controlled by the Phage Repressor." *Proc. Nat. Acad. Sci.* 60:1282–1287.

48. Zubay, G., D. Schwartz, and J. Beckwith. 1970. "Mechanism of Activation of Catabolite-Sensitive Genes: A Positive Control System." *Proc. Nat. Acad. Sci.* 66:104–110.

49. Beckwith, J. R., and D. Zipser, eds. 1970. *The Lactose Operon.* Cold Spring Harbor, N.Y.: Cold Spring Harbor Laboratory.

50. Nirenberg, M. W., and J. H. Matthaei. 1961. "The Dependence of Cell-Free Protein Synthesis in *E. coli* upon Naturally Occurring or Synthetic Polyribonucleotides." *Proc. Nat. Acad. Sci.* 47:1588–1602.

51. Crick, F. H. C. 1963. "The Recent Excitement in the Coding Problem." *Prog. Nucleic Acid Res.* 1:164–217. A superb analysis of the state of the coding problem as of 1962.

52. Sarabhai, A. S., A. O. W. Stretton, S. Brenner, and A. Bolte. 1964. "Co-Linearity of the Gene with the Polypeptide Chain." *Nature* 201:13–17.

53. Yanofsky, C., B. C. Carlton, J. R. Guest, D. R. Helinski, and U. Henning. 1964. "On the Colinearity of Gene Structure and Protein Structure." *Proc. Nat. Acad. Sci.* 51:266–272.

54. Nishimura, S., D. S. Jones, and H. G. Khorana. 1965. "The In-Vitro Synthesis of a Copolypeptide Containing Two Amino Acids in Alternating Sequence Dependent upon a DNA-Like Polymer Containing Two Nucleotides in Alternating Sequence." *J. Mol. Biol.* 13:302–324. Demonstration of the fact that each codon contains three nucleotides.

55. 1966. "The Genetic Code." *Cold Spring Harbor Symposia on Quantitative Biology,* Vol. 31 (1966), Cold Spring Harbor publ., Cold Spring Harbor, N.Y.

56. Adams, J. M., and M. R. Capecchi. 1966. "N-Formyl-methionyl-sRNA as the Initiator of Protein Synthesis." *Proc. Nat. Acad. Sci.* 55(1):147–155.

57. Brenner, A., A. O. W. Stretton, and S. Kaplan. 1965. "Genetic Code: The Nonsense Triplets for Chain Termination and Their Suppression." *Nature* 206:994–998.

58. Zipser, D. 1967. "UGA: A Third Class of Suppressible Polar Mutants." *J. Mol. Biol.* 29:441–445.

59. Sambrook, J. F., D. P. Fan, and S. Brenner. 1967. "A Strong Suppressor Specific for UGA." *Nature* 214:452–453.

60. Linn, S., and W. Arber. 1968. "Host Specificity of DNA Produced by *E. coli*. X. In Vitro Restriction of Phage fd Replicative Form." *Proc. Nat. Acad. Sci.* 59:1300–1306.

61. Smith, H. O., and K. W. Wilcox. 1970. "A Restriction Enzyme from *Hemophilus Influenzae*. I. Purification and General Properties." *J. Mol. Biol.* 51:379–391.

62. Roberts, R. J. 1983. "Restriction and Modification Enzymes and Their Recognition Sequences." *Nucleic Acid Res.* 11:r135–r167.

63. Mertz, J. E., and R. W. Davis. 1972. "Cleavage of DNA by RI Restriction Endonuclease Generates Cohesive Ends." *Proc. Nat. Acad. Sci.* 69:3370–3374.

64. Jackson, D., R. Symons, and P. Berg. 1972. "Biochemical Method for Inserting New Genetic Information into DNA of Simian Virus 40: Circular SV40 DNA Molecules Containing Lambda Phage Genes and the Galactose Operon of *E. coli*." *Proc. Nat. Acad. Sci.* 69:2904–2909.

65. Lobban, P., and A. D. Kaiser. 1973. "Enzymatic End-to-End Joining of DNA Molecules." *J. Mol. Biol.* 79:453–471.

66. Cohen, S., A. Chang, H. Boyer, and R. Helling. 1973. "Construction of Biologically Functional Bacterial Plasmids in Vitro." *Proc. Nat. Acad. Sci.* 70:3240–3244.

67. Sanger, F., and A. R. Coulson. 1975. "A Rapid Method for Determining Sequences in DNA by Primed Synthesis with DNA Polymerase." *J. Mol. Biol.* 94:444–448.

68. Maxam, A. M., and W. Gilbert. 1977. "A New Method of Sequencing DNA." *Proc. Nat. Acad. Sci.* 74:560–564.

69. Sanger, F., S. Nicklen, and A. R. Coulson. 1977. "DNA Sequencing with Chain-Terminating Inhibitors." *Proc. Nat. Acad. Sci.* 74:5463–5467.

70. Spiegelman, S., I. Haruna, I. B. Holland, G. Beaudreau, and D. Mills. 1965. "The Synthesis of a Self-Propagating and Infectious Nucleic Acid with a Purified Enzyme." *Proc. Nat. Acad. Sci.* 54:919–927. The first report of the in vitro replication of infectious viral (Qβ) RNA.

71. August, J. T., A. K. Banerjee, L. Eoyang, M. T. F. de Fernandez, K. Hori, C. H. Kuo, U. Rensing, and L. Shapiro. 1968. "Synthesis of Bacteriophage Qβ RNA." *Cold Spring Harbor Symp. Quant. Biol.* 33:73–83.

72. Weissmann, C., G. Felix, and H. Slor. 1968. "In Vitro Synthesis of Phage RNA: The Nature of the Intermediates." *Cold Spring Harbor Symp. Quant. Biol.* 33:83–101.

73. Zinder, N., ed. 1975. *RNA Phages.* Cold Spring Harbor, N.Y.: Cold Spring Harbor Laboratory. A collection of articles bringing together new facts about these fascinating viruses.

74. Temin, H. M., and S. Mizutani. 1970. "Viral RNA-Dependent DNA Polymerase." *Nature* 226:1211–1213.

75. Baltimore, D. 1970. "Viral RNA-Dependent DNA Polymerase." *Nature* 226:1209–1211.

76. Breathnach, R., J. L. Mandel, and P. Chambon. 1977. "Ovalbumin Gene Is Split in Chicken DNA." *Nature* 270:314–319.

77. Jeffreys, A. J., and R. A. Flavell. 1977. "The Rabbit Beta-Globin Gene Contains a Large Insert in the Coding Sequence." *Cell* 12:1097–1108.

II

CHEMICAL FACTS
AND PRINCIPLES

CHAPTER 4

A Chemist's Look at the Bacterial Cell

The smallest self-contained living entities governed by the genetic information of DNA are the bacteria. Because they are visible only when viewed under the microscope and do not contain morphologically distinct chromosomes that separate on mitotic spindles, their true nature was at first obscure. The question long persisted as to whether bacteria, like all other cells, have a conventional hereditary apparatus based on genes or whether they represent some radically different form of life governed by interrelated metabolic pathways that utilize the laws of chemical kinetics to generate the cellular molecules needed for cell growth and division.

Today, the latter speculation appears at best a silly aberration by biologically naive chemical kineticists. However, only in 1943 did convincing evidence begin to appear supporting the existence in bacteria of discrete genes capable of spontaneous mutation.[1] Soon afterward, DNA was pinpointed as the carrier of genetic information and then localized in bacteria within distinct "nucleoid" bodies by the methods of cytochemistry. When the 1946 finding of genetic recombination in the bacterium *E. coli* opened up the possibility of systematic genetic crosses, research on bacteria acquired a momentum that only recently has begun to be challenged by efforts on other forms of life.[2]

Bacterial Cells Do Not Have Nuclei[3, 4]

Although the cells of most bacteria are smaller than other types of cells, there is considerable size variation between the smallest and largest bacteria, whose cells approach the size of simple unicellular fungi (e.g., yeast). It is thus not size per se that makes a cell a bacterium. Instead, it is the absence of a discrete nucleus that defines bacteria and their very close relatives, the blue-green algae (now often called cyanobacteria), as evolutionarily distinct forms of life. No nuclear membrane separates the DNA (chromosomes) of bacteria and blue-green algae from the cytoplasm in which protein synthesis occurs.

Organisms lacking nuclei are called **procaryotes,** while those organisms whose cells contain nuclei are called **eucaryotes** (*caryon* means "nucleus" in Greek). Even the smallest of eucaryotic cells differ quite fundamentally from procaryotic cells in the way their genetic information is organized as well as in their patterns of RNA and protein synthesis (see Table 4-1). Bacteria accomplish these ends in more straightforward ways than eucaryotes, and it is unlikely that this sim-

Table 4-1 Differences Between Procaryotic and Eucaryotic Cells

Feature	Procaryotes	Eucaryotes
Genetic Organization		
Nuclear membrane	Absent	Present
Number of different chromosomes	1	>1
Chromosomes with histones	Absent	Present
Nucleolus	Absent	Present
Genetic exchange	Plasmid-mediated, unidirectional	By gamete fusion
Cell Structures		
Endoplasmic reticulum	Absent	Present
Golgi apparatus	Absent	Present
Lysosomes	Absent	Present
Mitochondria	Absent	Present
Chloroplasts	Absent	Present in plants
Ribosome size	70 S	80 S
Microtubules	Absent	Present
Cell wall with peptidoglycan	Present, except mycoplasma and archaebacteria	Absent
Some Functional Attributes		
Phagocytosis	Absent	Sometimes present
Pinocytosis	Absent	Sometimes present
Site of electron transport	Cell membrane	Organelle membranes
Cytoplasmic streaming	Absent	Present

SOURCE: After Stanier et al., *Introduction to the Microbiol World* (Englewood Cliffs, N.J.: Prentice-Hall, 1976), pp. 86–87.

plicity is a matter of chance. While evolving to multiply more rapidly than any other cells, bacteria have become as simple as their nutritional requirements allow. It is this economy in their molecular components that made them the obvious organisms with which to initially establish the detailed chemical pathways through which genes control the life of the cell.

Bacteria Grow Under Simple, Well-Defined Conditions[5, 6, 7]

There exists an enormous number of different types of bacteria, and they vary not only in size and shape but also in the nutritional conditions best suited for their growth and survival. Some bacteria, for example, are aerobes and grow only in the presence of oxygen; others are anaerobes and multiply only in the absence of oxygen; and still others, facultative anaerobes, can change their exact mixture of enzymes to allow growth in both environments. While most bacteria derive their energy from breaking down externally derived food molecules, others have evolved photosynthetic pigments to let them use sunlight to make ATP. Independent of their nutritional specialization, most bacteria grow free as single cells, separating from each other as soon as cell division occurs. In general, it has proved easy to grow them in the laboratory once their nutritional requirements have been worked out. Most importantly, in contrast to mammalian cells, which require a large variety of growth factors, many bacteria will grow well on a simple, well-defined diet, or medium. For example, the bacteria *Escherichia coli (E. coli)* will grow in an aqueous solution containing just glucose and several inorganic ions (Table 4-2).

Table 4-2 A Simple Synthetic Growth Medium for *E. coli**

NH_4Cl	1.0 g
$MgSO_4$	0.13 g
KH_2PO_4	3.0 g
Na_2HPO_4	6.0 g
Glucose	4.0 g
Water	1000 ml

*Traces of other ions (e.g., Fe^{2+}) are also required for growth. Usually, these are not added separately, since they are normally present as contaminants in either the added inorganic salts or the water itself.

Figure 4-1
Electron micrograph of a group of intact
E. coli cells.

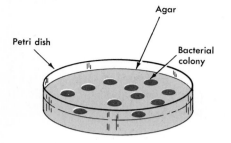

Figure 4-2
The multiplication of single bacterial
cells to form colonies. *E. coli* cells are
usually not motile. Thus, when a cell
divides on a solid surface, the two
daughter cells and all their descendants
tend to remain next to one another.
After 24 hours at 37°C, each initial living cell gives rise to a solid mass of
cells.

The growth of a specific bacterium is usually not dependent on the
availability of a specific carbon source. Most bacteria are highly adapt-
able as to which organic molecules they can use as their carbon and
energy sources. Glucose can be replaced by a number of other organic
molecules, and the greater the variety of food molecules supplied, the
faster a bacterium generally grows. For example, if *E. coli* grows only
on glucose, about 60 minutes are required at 37°C to double the cell
mass. But if glucose is supplemented by the various amino acids and
purine and pyrimidine bases, then only 20 minutes are necessary for
the doubling of cell mass. This shortening of **generation time** is due to
the direct incorporation of the dietary components into proteins and
nucleic acids, sparing the cell the task of carrying out the synthesis of
their building blocks. There is a lower limit, however, to the time
necessary for a cell generation; no matter how favorable the growth
conditions, an *E. coli* cell is unable to divide more than once every 20
minutes.

E. coli Is the Best-Understood Organism at the Molecular Level[8, 9]

The most intensively investigated and correspondingly best-known
bacterium at both the biochemical and genetic levels is *E. coli*. Its
prominence as the premier organism used to probe the essential features of life started in the early 1940s through its adoption by a group
of young physicists, chemists, and geneticists. These scientists believed that only by working on the simplest of biological systems
would they ever be able to come to grips with the gene at its most
fundamental level. So they initiated research on *E. coli*, a common
inhabitant of the human intestine. This organism possesses the very
desirable properties of small size, lack of pathogenicity to most other
organisms, and extreme ease of cultivation under laboratory conditions. First extensively used to study the multiplication of its viruses,
the bacteriophages (phages), *E. coli* became most interesting in its
own right when it became the first bacterium shown to have discrete
sexes that genetically recombine.

 E. coli quickly became the obvious bacterium in which to search for
mutations that blocked the synthesis of essential metabolites. Such
mutations were relatively easy to find and map, and they soon
proved indispensable in the working out of the enzymatic steps
through which bacterial molecules are either made or broken down.
Today, virtually a thousand different *E. coli* genes have already been
assigned chromosomal locations, and it would make no sense to start
serious work on another bacterium if *E. coli* can be used to solve a
particular problem. As we shall see in subsequent chapters, the combined methods of genetics and biochemistry are so powerful that it is
undesirable to perform biochemical studies on an organism with
which genetic analysis is not possible.

The Growth Parameters of *E. coli*[10–14]

The average *E. coli* cell is rod shaped and about 2 μm in length and
1 μm in diameter (Figure 4-1). It grows by increasing in length, followed by a fission process that generates two cells of equal length.

Growth occurs best at temperatures around 37°C, suiting it for existence in the intestines of higher mammals, where it is frequently found as a harmless parasite. It will, however, regularly grow and divide at temperatures as low as 20°C. Cell growth proceeds much more slowly at low temperatures; the generation time under otherwise optimal conditions is about 120 minutes at 20°C.

Cell number and size are often measured by observing the bacteria under the light microscope (and occasionally the electron microscope). Viewing a cell in this way, however, does not tell us whether the cell is alive or dead. We can determine whether a bacterium is alive or dead only by seeing whether or not it forms daughter cells. This observation is usually made by spreading a small number of cells on top of a solid agar surface, which has been supplemented with the nutrients necessary for cell growth (Figure 4-2). If a cell is alive, it will grow to form two daughter cells, which in turn give rise to subsequent generations of daughter cells. The net result after 12 to 24 hours of incubation at 37°C is discrete masses, or **colonies,** of bacterial cells. Provided that the colonies do not overlap, each must have arisen from a single bacterial cell.

The growth of bacteria may also be followed in liquid nutrient solutions. If a nutrient medium is inoculated with a small number of rapidly dividing bacteria from a similar medium, the bacteria will continue dividing with a constant division time, doubling the number of bacteria each generation time. Thus, the number of bacteria increases in an exponential (logarithmic) fashion (Figure 4-3). *Exponential growth* continues until there are so many cells that the initial optimal nutritional conditions no longer exist. One of the first factors to limit growth is usually the supply of oxygen. When the number of cells is low, the oxygen available by diffusion from the liquid surface is sufficient; but as the number of cells increases, additional oxygen is needed. It is often supplied by bubbling oxygen through the solution or shaking the solution rapidly. But even with violent aeration, growth rates begin to slow down after the cell density reaches about 10^9 cells per milliliter, and a tendency develops for the cells being produced to be shorter. Finally, at cell densities of about 5×10^9 cells per milliliter, cell growth is discontinued for nutritionally related reasons that are not yet clear. The term **growth curve** is frequently used to describe the increase of cell numbers as a function of time.

In most growing bacterial cultures, the exact division time of the cell varies, so that even if a culture has started from a single cell, after a few generations, cells can be found at various stages of the division cycle at any given moment. Such growth is frequently called *unsynchronized growth.* Over the past ten years, tricks have been developed to isolate bacterial cells at the same stage of the cell cycle. These tricks can be used to obtain several generations of **synchronized cell growth** (Figure 4-4). Then, because of slightly unequal division times, the growth curve gradually acquires an unsynchronized appearance.

Even Small Cells Are Very Complex

Even cells as small as those of *E. coli* present great difficulties when we study them at the molecular level. At first sight, the problem of understanding the essential features of *E. coli* seems insurmountable, for on a chemical scale, even the smallest cells are fantastically large.

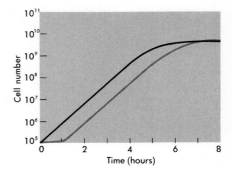

Figure 4-3
Growth curve of *E. coli* cells at 37°C. The black line shows the increase in cell number following the inoculation of a sterile, nutrient-rich solution (glucose, salts, amino acids, purines, pyrimidines) with 10^5 cells from an *E. coli* culture in an exponential phase of growth. If this growth curve had been started instead from cells in a slow-multiplying, nearly saturated culture, the growth would not have begun immediately; rather, a lag period of approximately 1 hour would have preceded exponential growth (colored line).

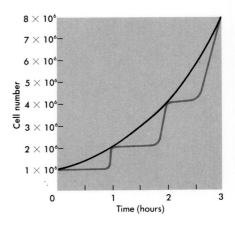

Figure 4-4
The growth curve of a synchronized *E. coli* culture growing on glucose as the sole carbon source is given by the colored line. The black line shows the increase in cell number of an unsynchronized culture. In this example, the degree of synchronization lessens noticeably in the second and third cycles of growth.

Table 4-3   The Main Classes of Biological Molecules in Bacteria

	General Description	Functions	Building Blocks
Proteins	Molecules containing C, H, O, N, and sometimes S, which are built up from amino acids.	Most proteins are enzymes. Some, usually present in very large numbers per cell, are used to build up essential structures, such as the cell wall, the cell membrane, the ribosomes, muscle fiber, and nerves.	Amino acid
Lipids	Molecules insoluble in water. They include triglycerides, which are built up by the combination of glycerol and three long-chain fatty acids, and lecithins, in which one fatty acid is replaced by choline. Sometimes, glycerol is replaced by sphingosine. Phosphorus is present in many of the lipids (phospholipids).	Triglycerides are a main storehouse of energy-rich food. They degrade to give acetyl-CoA. Phospholipids are an essential component of all membranes. Their insolubility in water is related to their control of permeability.	(Fatty acids; n usually 16 or 18) — Glycerol — Sphingosine — Choline
Carbohydrates	Molecules containing C, H, and O, usually in ratios near 1:2:1. Polysaccharides are built up from simple sugars (monosaccharides) such as glucose and galactose. In some cases, the sugars contain amino groups (e.g., glucosamine).	Some, such as cellulose and pectin, are used to construct strong, protective cell walls; others, such as glycogen, provide a form of storing glucose.	β-D-Glucose — β-D-Galactose — β-D-Glucosamine
Nucleic Acids	Long, linear molecules containing P, as well as C, H, O, and N, which are built up from nucleotides	There are two main classes of nucleic acids in cells: DNA is the primary genetic component of all cells; RNA usually functions in the synthesis of proteins. In some viruses, RNA is the genetic material.	Bases: Purine, Pyrimidine + Pentose sugar: Ribose, Deoxyribose + Phosphate → Nucleotide

Although an *E. coli* cell is about five hundred times smaller than an average cell in a higher plant or animal (which has a diameter of approximately 10 μm), it nonetheless has a wet weight of approximately 10^{-12} gram (5×10^{11} daltons, where a dalton is the weight of one hydrogen atom). This number, which initially may seem very small, is immense on a chemist's scale, since it is 3×10^{10} times greater than the weight of a water molecule (MW = 18 daltons). Furthermore, this mass reflects the highly complex arrangement of a large number of different carbon-containing molecules.

There is also seemingly infinite variety in the chemical nature of the molecules contained in an *E. coli* cell. Fortunately, it is possible to distribute most of the larger molecules into several well-defined classes possessing common arrangements of atoms. These classes are the proteins, lipids, carbohydrates, and nucleic acids (Table 4-3). Many molecules possess chemical groups common to several of these categories, so the classification of such molecules is necessarily arbitrary. Also in the cell are many smaller organic molecules (such as amino acids, purine and pyrimidine nucleotides, and various coenzymes), very small inorganic molecules (e.g., O_2 and CO_2), and numerous electrically charged inorganic ions (Table 4-4). Among the very small molecules is water (H_2O), the most common molecule in all cells and a solvent for most biological molecules, through which diffusion from one cellular location to another can occur quickly.

At present, we can make only an approximate guess of the number of chemically different molecules within a single *E. coli* cell. Each year, many new molecules are discovered, so the known molecules clearly represent only a fraction of those we shall eventually know. Already over 800 small molecules are implicated in its metabolism (Table 4-5), and with a slightly larger number of genes (and their proteins) already identified, we can say with confidence that *E. coli* must easily possess over 2000 different molecules. This is surely an underestimate, and later we shall give reasons for suspecting that as many as 4500 different molecules might in fact be found within the average growing cell. Thus, we must immediately admit that the structure of a cell will never be understood in the same way that we understand water or glucose molecules. Not only will the three-dimensional structures of most cellular proteins remain unsolved, but their location within cells often remain imprecisely defined.

It is thus not surprising that many chemists, after periods of enthusiasm for studying "life," silently return to the world of pure chemistry. Increasingly, however, others are becoming lifelong converts to the fascination of biology once they realize (1) that all macromolecules are polymeric molecules built up from smaller monomers, (2) that there exist well-defined chains of successive chemical reactions in cells (metabolic pathways), and (3) that there is a limit to the number of proteins that can exist in a cell owing to the fact that each cell contains a finite amount of DNA. This number, in turn, leads to an estimate of the number of enzymes and hence small molecules that a cell can possess. Most likely, at least half the proteins in a cell are enzymes, each of which catalyzes a specific metabolic reaction. The approximate number of different types of small molecules could be estimated if we knew, on the average, how many specific enzymes are needed for the metabolism of the average small molecule. At present, it seems a good guess that the number lies between one and two.

Table 4-4 Inorganic Ions Found in a Bacterial Cell After Growth in a Glucose-Minimal Medium

Ion	Function
K^+	Principal cation, cofactor for certain enzymes
NH_4^+	Principal form of inorganic nitrogen for assimilation
Mg^{2+}	Cofactor for a large number of enzymes
Ca^{2+}	Cofactor for certain enzymes
Fe^{2+}	Present in cytochromes and other enzymes
Mn^{2+}	Cofactor for several enzymes
Mo^{2+}	Present in several enzymes
Co^{2+}	Present in vitamin B_{12} and its coenzyme derivatives
Cu^{2+}	Present in several enzymes
Zn^{2+}	Present in several enzymes
Cl^-	Not required for many bacteria
SO_4^{2-}	Main source of sulfur in most media
PO_4^{3-}	Participant in many metabolic reactions

SOURCE: Reproduced from J. L. Ingraham et al., *The Growth of the Bacterial Cell* (Sunderland, Mass.: Sinauer Associates, 1983), p. 20, with permission.

Table 4-5 The Various Molecules in a Bacterium Growing in Glucose-Minimal Medium

Molecule	Approximate Number of Kinds
Amino acids and their precursors and derivatives	120
Nucleotides and their precursors and derivatives	100
Fatty acids and their precursors	50
Sugars, carbohydrates, and their precursors	250
Quinones, poly-isoprenoids, porphyrins, vitamins, other coenzymes and prosthetic groups, and their precursors	300

SOURCE: Reproduced from J. L. Ingraham et al., *The Growth of the Bacterial Cell* (Sunderland, Mass.: Sinauer Associates, 1983), p. 19, with permission.

Cellular Proteins Can Be Displayed on Two-Dimensional Gels[15, 16]

The amount of DNA in the *E. coli* chromosome is about 2.5×10^9 daltons, representing some 4×10^6 base pairs. If all this DNA codes for amino acids (see Chapter 15), then a total of $(4 \times 10^6) \div 3 = 1.33 \times 10^6$ amino acids are coded for by the nucleotide sequences of *E. coli* DNA. As the average *E. coli* protein (MW 4×10^4) contains some 360 amino acids, $(1.33 \times 10^6) \div (3.6 \times 10^2) = {\sim}3600$ represents the maximum number of proteins that are ever likely to be described within *E. coli*.

This number can now be compared with the actual number of different *E. coli* proteins revealed by newly developed procedures for two-dimensional gel electrophoresis. (The basic methods of gel electrophoresis are described in Figure 4-5.) The new procedures make possible the display of the total cell protein as discrete spots on a thin rectangular layer of polyacrylamide gel. First, the proteins are separated on the basis of their net charge by isoelectric focusing in one dimension; then they are separated on the basis of their apparent size as revealed by their rates of movement in the presence of SDS (sodium dodecyl sulfate, an anionic detergent) when the current is applied in the second dimension. Figure 4-6 shows a gel in which the acidic proteins of a growing *E. coli* cell are separated into some 1400 spots. With few exceptions, each spot represents a distinct polypeptide chain coded by a separate gene. The basic proteins were not separated in that experiment, but these can be visualized by using different gel conditions to reveal the existence of still another 300 different proteins. In these gel systems, any *E. coli* protein present in at least 30 molecules per cell can be seen. By measuring and calibrating the intensity of individual spots, the number of copies of individual proteins can be seen to vary over a 10^5-fold range, with some proteins present in over 100,000 copies per cell.

Some 200 different spots have already been identified as known enzymes and structural (e.g., ribosomal) proteins, and no serious obstacle (except for tedium) prevents the identification of still many others. The exact pattern shown by a given two-dimensional gel varies with the exact nutritional conditions. *E. coli* cells growing under nutritionally rich conditions, for example, contain many more copies of the various proteins used to construct ribosomes than when they subsist on a simple, minimal medium.

The number of spots (proteins) observed to date (about 1700) is much less than the number predicted (about 3500) if all the *E. coli* DNA codes for protein products. Only time will tell whether there exist large numbers of proteins still undetected because they exist in only a few copies or are only present under nutritional conditions not normally encountered.

The Anatomy of *E. coli* as Revealed by the Electron Microscope[17, 18, 19]

A direct way to appreciate the morphological simplicity of *E. coli* is through electron microscope examination of very thin sections cut from a rapidly growing cell. These sections reveal *E. coli* to have a simple, saclike structure in which an external envelope surrounds a membrane-free, dense granular cytoplasm (called the cytosol), which

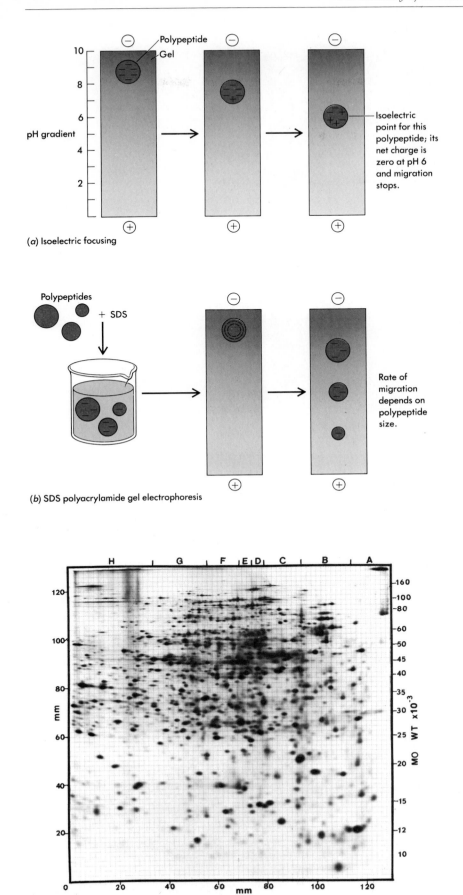

(a) Isoelectric focusing

(b) SDS polyacrylamide gel electrophoresis

Figure 4-5
Gel electrophoresis separates macromolecules on the basis of their migration through a gel under the influence of an electric field. This figure illustrates the principles underlying two electrophoretic methods used to separate polypeptides. In both cases, proteins are first treated to denature them and to separate their subunits, and the electrophoresis is carried out in a gel of polyacrylamide. (a) Isoelectric focusing depends on the fact that the net charge on a polypeptide changes with pH; the electrophoresis is performed in a gel in which a pH gradient has been established. At a pH characteristic for each polypeptide (its isoelectric point), the net charge on the molecule is zero, and it ceases to migrate. (b) Electrophoresis in the presence of the negatively charged detergent sodium dodecyl sulfate (SDS) separates polypeptides by size. The SDS coats each polypeptide, giving it a roughly uniform negative charge. The smaller the polypeptide, the more readily it moves through the pores in the gel toward the positive pole. (Nucleic acid molecules, which intrinsically have uniform negative charge, can also be separated by size with gel electrophoresis; an agarose gel is usually used.) The locations and amounts of the macromolecules on the completed gel may be determined by using special dyes or autoradiography.

Figure 4-6
Two-dimensional gel of *E. coli* K-12, strain W3110, cell extract showing the acidic proteins separated into approximately 1400 spots. The grid overlay provides coordinates for individual spots. Letters A through H denote zones of increasing isoelectric points. [Courtesy of Dr. Frederick C. Neidhardt, *Microbiol. Revs.* 47 (1983):232.]

Figure 4-7
(a) Electron micrograph of an *E. coli* B cell prepared by freeze substitution with no pretreatment. Bar = 0.5 μm. [Courtesy of J. A. Hobot et al., *J. Bacteriol.* 162 (1985):964.] (b) Schematic diagram of such a cell, highlighting the nucleoid regions (color) and showing various other cellular components (not to scale). Most of the DNA resides in the relatively ribosome-free nucleoid region, which is in close association with the ribosomes, theoretically to provide for optimal efficiency in translation.

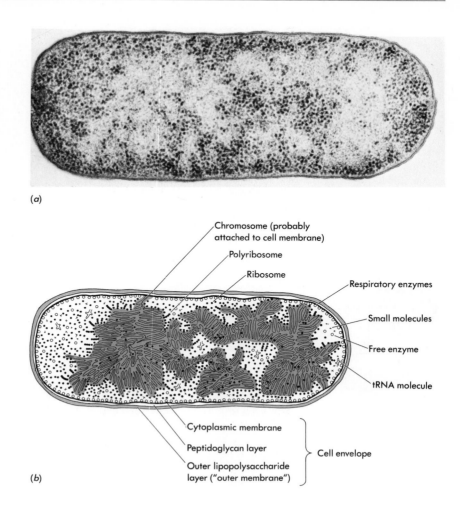

(a)

(b)

contains an apparently less dense fibrous DNA component (Figure 4-7). The granular appearance of the cytosol reflects the very large number of ribosomes found in all growing bacterial cells. More ribosomes per unit dry mass are found in bacteria than in any other kind of cell. This makes possible the high rate of protein synthesis needed to allow bacteria to double their number as rapidly as once every 20 minutes at 37°C. Very high concentrations of free enzymes also exist in bacteria, and the dry/wet ratio of bacteria can be as high as 30 percent, much greater than the 10 percent ratio characteristic of typical higher eucaryotic cells. The metabolic rate of bacteria is correspondingly much higher, with the amount of oxygen and substrates consumed per unit of time frequently 10 to 100 times greater than that consumed by eucaryotic cells.

Bacteria Display Their Individuality Through Their Envelopes[20–25]

The envelopes of bacteria are extremely complex compared to their tightly packed, relatively undifferentiated cytosol (Figure 4-8). The innermost layer of the envelope is always a phospholipid bilayer into which many functionally distinct proteins are inserted. Surrounding this "cytoplasmic facing membrane" is a rigid shell of covalently

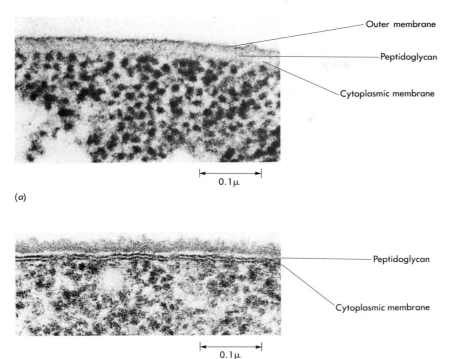

(a)

Outer membrane

Peptidoglycan

Cytoplasmic membrane

0.1μ

(b)

Peptidoglycan

Cytoplasmic membrane

0.1μ

Figure 4-8
Comparison of Gram-negative and Gram-positive cell envelopes. (a) Electron micrograph of a cross-section of a Gram-negative *E. coli* cell envelope with outer membrane (lipopolysaccharide) and cytoplasmic membrane. [Courtesy of J. A. Hobot et al., *J. Bacteriol.* 160 (1985):143–152.] (b) Electron micrograph of the peptidoglycan layer and cytoplasmic membrane of Gram-positive *Bacillus cereus*. (Courtesy of M. T. Silva.) Bars = 0.1 μm.

linked carbohydrates and amino acids that has been given the name **peptidoglycan** (Figure 4-9). It is through their peptidoglycan component that bacteria obtain their structural integrity. If it is enzymatically removed (e.g., by lysozyme), the effectively naked bacteria assume spherical shapes. Depending on the specific bacterium, the peptidoglycan shell can be relatively thin (as in *E. coli*) or composed of many effective layers (as in bacteria belonging to the *Bacillus* groups).

Those bacteria that have thick peptidoglycan shells are very readily stained by the dyes crystal violet and iodine, first used by the Danish microbiologist Christian Gram. Such peptidoglycan-rich bacteria are known as Gram-positive bacteria, while those bacteria possessing only thin layers of rigid peptidoglycan are collectively known as Gram-negative bacteria. To compensate for their thin peptidoglycan shells, Gram-negative cells possess a second phospholipid-containing membrane that surrounds the peptidoglycan shell. This **outer membrane** is characterized by unique lipopolysaccharides that have their highly specific (immunogenic) polysaccharide side groups projecting outward, as well as by the presence of a very abundant lipoprotein that anchors the outer membrane to the peptidoglycan shell. Inserted into the outer membrane also are the matrix proteins, or **porins,** which form specific pores through which extracellular molecules must pass to reach the inner cytoplasmic membrane (Figure 4-10). Usually, such passages are passive processes governed by the laws of diffusion. In other cases, movement through these pores is facilitated by at least 30 different **transport proteins,** or receptors, that make possible the entry of molecules too large for simple passage through the more common porin channels. How such transport proteins work remains to be established, with the possibility still open that they form specialized pores.

Figure 4-9
(a) The chemical structure of a peptidoglycan unit from *E. coli*. Within the peptidoglycan layer, the two sugars alternate with each other in chains. A tetrapeptide is attached to each N-acetylmuramic acid residue. Adjacent tetrapeptides may be directly bonded to each other or linked by an oligopeptide bridge (not shown) to form a crosslink between two chains. These crosslinks make the peptidoglycan layer in effect one giant molecule. Other amino acids are occasionally found in some bacteria. (b) The likely arrangement of peptidoglycan within the *E. coli* cell envelope. The peptidoglycan is thought to form a periplasmic gel that is more crosslinked and thus denser toward the outer membrane. (After a drawing by J. A. Hobot.)

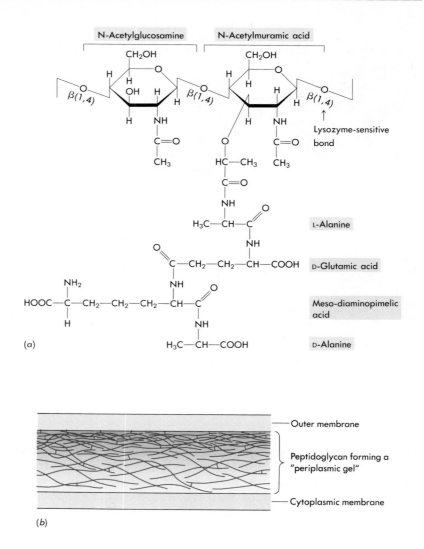

(a)

(b)

Until recently, little was known about the structure of the "periplasmic space" between the inner and outer lipid-containing membranes. This region was known to contain three important classes of vital proteins: (1) hydrolytic enzymes, which initiate the degradation of food molecules; (2) specific binding proteins that help initiate the transport of certain food molecules across the inner cytoplasmic membrane; and (3) specific chemoreceptors used to measure the concentration of nutrients (or poisons) in the environment so that their respective cells can move toward (or away) from them. Also, the periplasmic region was known to contain peptidoglycan, which was thought to exist in a thin, discrete layer near the outer membrane. Now, however, we believe that peptidoglycan actually fills the periplasmic space, forming a periplasmic gel that is highly fluid near the inner membrane and more compact near the outer membrane (see Figure 4-9b). The gel is heavily hydrated, and periplasmic proteins can freely diffuse within it.

Although it is generally thought that Gram-positive cells do not have an outer lipopolysaccharide layer, regular-appearing patterns suggestive of proteins have recently been seen on their outer surfaces. Whether these patterns are, in fact, proteins in a thin lipopoly-

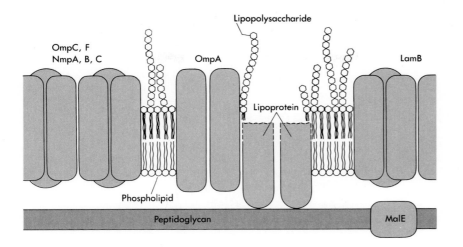

Figure 4-10
The major outer-membrane proteins of Gram-negative bacteria. Notice that OmpA and the porins OmpC, OmpF, NmpA, NmpB, NmpC, and LamB extend through the membrane. LamB is an outer-membrane protein that functions as a porin for transportation of maltose and maltodextrins (it is also a receptor for bacteriophage λ). MalE is a periplasmic maltose-binding protein. [After M. J. Osborn and H. C. P. Wu, *Ann. Rev. Microbiol.* 34 (1980).]

saccharide bilayer remains to be worked out. It is speculated that such a layer might not have been discovered earlier because of its relative sparcity compared to the thick peptidoglycan layer found in Gram-positive cells. If this speculation proves correct, the difference between Gram-negative and Gram-positive cells will simply be a quantitative difference in the amount of peptidoglycan in each.

The Inner Cytoplasmic Membrane Carries Out the Chemiosmotic Generation of ATP[26, 27]

Unlike eucaryotic cells, where the aerobic generation of ATP is made possible by the passage of protons through the membranes of mitochondria, the site of oxidative phosphorylation in bacteria is the inner, phospholipid-containing cytoplasmic membrane. Inserted within it are all the major proteins involved in the translocation of hydrogen ions, including the flavoproteins, quinones, iron-sulfur ("FeS") proteins, and cytochromes. Their combined action leads to lower hydrogen ion concentrations in the cytosol compared to the exterior medium as well as differences in electrical charge across the inner membrane. Together they create the proton motive force, which provides the energy for a number of vital cellular processes that occur on the cytoplasmic membrane, including the active transport of food molecules and certain inorganic ions into the cytosol, the turning of flagella, and the generation of ATP from ADP by cytoplasmic membrane-bound ATP synthetase (Figure 4-11). This latter enzyme is a complex molecular aggregate composed of at least ten different polypeptide chains, six of which combine to form knobs that protrude into the interior of the cell. The action of this membrane-based complex is reversible. The complex can either generate electrical charge differences across the cytoplasmic membrane (membrane potential) by hydrolysis of ATP to ADP (hence its alternative name, ATPase) or generate ATP from ADP and inorganic phosphate by the proton motive force.

We know very little about the relative locations of the various proteins of the cytoplasmic membrane. Studies using two-dimensional gel electrophoresis indicate that at least 200 different proteins are so located in the membrane; but this number is probably an underestimation of the true complexity. Clearly, we are just at the beginning of

Figure 4-11
Vital cellular processes that occur on the cytoplasmic membrane. Proton transfer and the generation of ATP provide the energy for activities such as active transport of food molecules and the turning of flagella. [After "How Cells Make ATP," by P. C. Hinkle and R. E. McCarthy. *Sci. Amer.* 238 (1978). Copyright © 1978 by Scientific American, Inc. All rights reserved.]

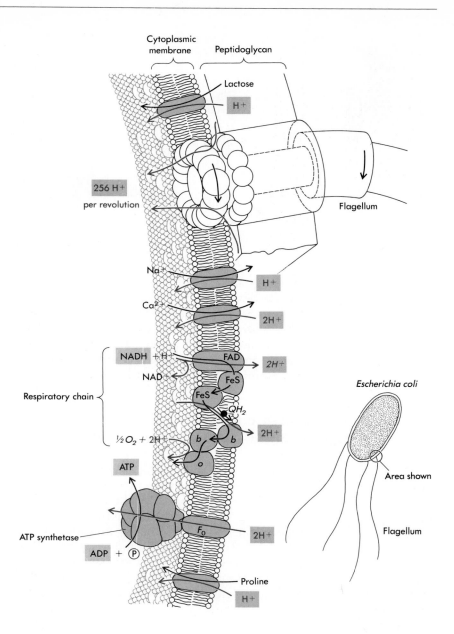

a real understanding of how membranes function at the molecular level. Fortunately, for the first time, true crystals suitable for X-ray crystallographic examination have recently been obtained for several integral membrane proteins. Through the elucidation of their three-dimensional conformations, we may know within the next decade how molecules selectively move through membranes.

Flagella and Pili Extend Outward from *E. coli* Surfaces[28-32]

E. coli, like many other bacteria, possesses two forms of extended surface appendages, flagella and pili. Though they are both long, linear bodies that arise from the cytoplasmic membrane, they are neither structurally nor functionally related. The bacterial **flagellum** (plural, **flagella**) is a very long, thin (200 Å diameter) organelle of locomotion, which is formed mainly by the helical aggregation of

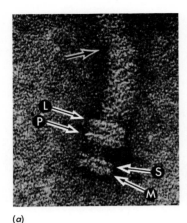

(a)

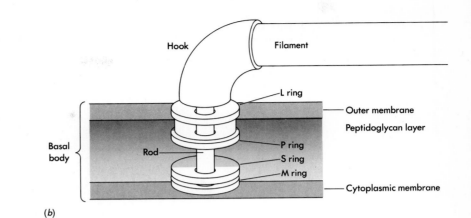

(b)

large numbers of a protein building block called flagellin. In reality, it is structurally quite complex. It is composed of three parts: (1) the long filament made up of flagellin, which can extend 15 to 20 μm into the medium, (2) the basal structure that anchors the flagellum to the cell envelope, and (3) the hook, a short, curved structure that connects the filament to the basal structure (Figure 4-12). Flagella serve as locomotory organelles by rotating the basal structures in either a clockwise or counterclockwise direction. In *E. coli*, counterclockwise rotation leads to the formation of a stable bundle of the several (about five) rotating flagella that propels the cell smoothly forward. In contrast, clockwise rotation disperses the bundle and leads to a chaotic tumbling of the cell in many directions (see Figure 4-23).

While the rotating filaments clearly function as propellers, the exact function of the hook is less clear. Like the filament, it is built up by the aggregation of a number of identical protein subunits. Much more complicated is the basal structure, which consists of at least ten different types of protein subunits that come together to form rings surrounding a thinner rod that is inserted into the cytoplasmic membrane. Exactly how the basal structure rotates remains to be worked out, though it is clear that the movement of hydrogen atoms across the cell membrane (the proton motive force) is involved. How specific attractants or repellents set into motion the reversals between clockwise and counterclockwise rotations has yet to be discovered.

Much simpler both in structure and in function are the **pili** (singular, **pilus**) (Figure 4-13). Composed of only single protein building blocks called pilin, they are adhesive organs that can extend 10 μm or more into the medium. They function by attaching bacteria to other surfaces, often the glycoprotein components on the exterior of higher cells (such as the epithelial cells that line our intestines). In addition to the more common type of pili present in large numbers on every bacterial cell, most male bacteria contain one to several morphologically similar pili called **sex pili,** which function to initiate contacts with female cells.

Figure 4-12
(a) The basal end of a flagellum from *E. coli.* The top arrow marks the junction between the hook and the filament. The L, P, S, and M rings (see part b) are also indicated (500,000×). (Courtesy of M. L. DePamphlis and J. Adler.) (b) Diagram of the hook–basal body complex.

Condensation of Bacterial DNA into Nucleoid Bodies[33]

All the essential genes of *E. coli* are present on a single chromosome, a single DNA molecule, which genetic experiments first revealed to be circular. Single cells, growing exponentially under favorable nutritional conditions, generally contain two to four such molecules in the

Figure 4-13
Escherichia coli possessing type 1 pili, as seen with the electron microscope. (Courtesy of Steven Clegg.)

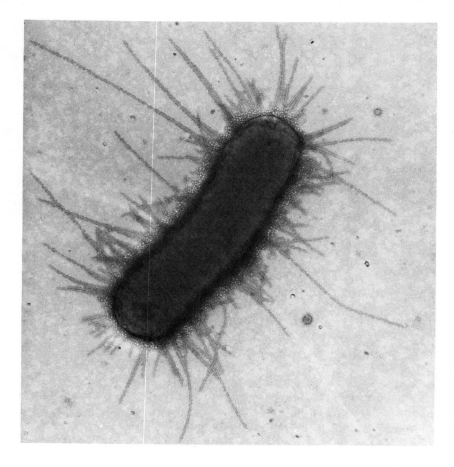

act of duplicating themselves, beginning at origins of replication present at one unique site on each chromosome. In such multiplying cells, the cell division process lags behind the DNA replication process, and only when cell growth has virtually ceased do many *E. coli* cells contain only one chromosome. The reason why rapidly growing cells contain multiple copies of the same chromosome only recently became clear. It relates to the fact that there is a fundamental upper limit to the rate of DNA replication. For reasons not yet understood, only 750 base pairs and no more can be polymerized per second at a given replicating fork (see Chapter 10). Even given the bidirectional pattern of the DNA replication, some 40 minutes are needed to complete the replication of the *E. coli* chromosome (4×10^6 base pairs). We might thus predict that an *E. coli* cell can never divide faster than once every 40 minutes. But in fact, as we mentioned earlier, optimally fed cells multiply once every 20 minutes. Resolution of this paradox came with the realization that a new round of DNA replication can be initiated before the previous round is completed. The replication of a given *E. coli* chromosome can thus occur during two successive cell cycles.

The length of a fully extended *E. coli* chromosome would be $4 \times 10^6 \times 3.4$ Å, or approximately 1 mm. However, it never even exists partially extended; electron microscopic examination reveals its compaction into irregularly shaped bodies, the **nucleoids** (see Figure 4-7). Despite their similarity in name, nucleoids bear no relation to conventional nuclei, since they are not surrounded by any form of phospholipid-containing membrane. Also distinguishing all procaryotic cells from their eucaryotic equivalents is the absence of conventional *his-*

tones, the highly conserved class of basic proteins around which eucaryotic DNA coils to form the nucleosome particles that give to chromatin (the underlying DNA-protein complex of chromosomes) its granular (beaded) appearance. Despite their lack of typical histones, however, gently prepared *E. coli* chromosomes initially display sections of beaded appearance that quickly transform into thinner fibers resembling pure DNA. So *E. coli* DNA must also, in part, be complexed with proteins that lead to its compaction. In fact, two small histonelike basic proteins have recently been found in *E. coli,* semitightly bound to the DNA. Their amounts, however, are insufficient to compact all the *E. coli* DNA into nucleosome-like particles. Instead, the most important factor that leads to the formation of nucleoid bodies may be enzymatically induced supercoiling (see Chapter 9), which of necessity leads to the compaction of the circular double helical DNA.

Also still a mystery is how progeny bacterial chromosomes correctly partition themselves into their respective daughter cells. Evidence suggests that regions of each bacterial chromosome are associated with the inner, or cytoplasmic, membrane of the bacterial envelope, with the occasional claim that the attached regions include the unique points where replication initiates (origins of replication). But even today, we have no good techniques to cleanly separate the cytoplasmic membrane from other cellular components.

Transcription and Translation Occur in the Same Cellular Compartment[34]

At any given moment, the *E. coli* chromosome is being transcribed at the rate of 60 nucleotides per second (at 37°C) into some 400 to 800 unique mRNA chains, some 100 different tRNA molecules, and some 700 precursor molecules for rRNA. Each of these transcription events requires the participation of a single RNA polymerase molecule. So at a given moment, some 1600 RNA polymerase molecules, representing about 1 percent of the total bacterial protein, are at work on a single *E. coli* chromosome. Even though only some 4 percent of the total bacterial RNA is mRNA, its average half-life of perhaps no more than 1½ minutes requires the participation of approximately one-half of a bacterial cell's RNA polymerase for its synthesis.

Very soon after an mRNA chain begins to be made, its leading (5') end becomes attached to a free ribosome, which in turn commences to translate its message and in so doing to move toward the 3' end still in the process of extending itself along its DNA template. By the time a given mRNA chain is completely transcribed, it usually has attached to it a string of ribosomes successively carrying polypeptide chains of ever-increasing length (Figure 4-14). With transcription occurring at approximately 60 nucleotides per second, the time required to make an average-size mRNA chain containing 2000 to 3000 nucleotides approaches 1 minute, an interval not much shorter than the average time (1.5 minutes) a given nucleotide remains incorporated within a given mRNA molecule. Thus, by the time the synthesis of many large mRNA molecules is completed, they may already have begun to be broken down. A large fraction, if not a majority, of the ribosomes collected together on a single mRNA molecule (the assembly is called a *polyribosome* or *polysome*) may thus be physically attached to the bacterial chromosomes. Such linkages, however, do not

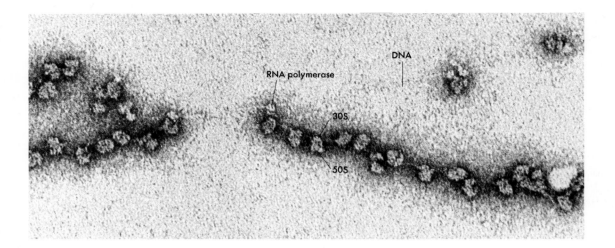

Figure 4-14
Groups of ribosomes (polysomes) moving across mRNA chains being formed on *E. coli* DNA. (Courtesy of Barbara Hamkalo.)

trap significant numbers of ribosomes within the DNA-containing nucleoid regions, since the extended lengths of average-size polysomes (10 to 15 ribosomes) approach the diameter of the *E. coli* cell itself.

The dilemma remains, however, as to how the nascent mRNA molecules with their attached strings of ribosomes separate from their DNA templates. In principle, this could occur either through rotation of the DNA or by the untwisting of RNA around DNA. Arguing against massive rotatory movements by the mRNA is the fact that many polyribosomes become securely fastened to the surrounding inner membrane so that their protein products can be built into the cell envelope. But it is equally unlikely for RNA to separate away from DNA by a coupling of transcription to the rapid rotation events required for DNA replication. Not only does DNA replication occur at a rate some ten times faster than RNA synthesis, but also various genes along the *E. coli* chromosome are transcribed in different directions. Perhaps those DNA segments being transcribed contain temporary single-stranded cuts along their backbone, which give them the capacity for the unimpeded rotation needed to separate away from their RNA products.

Membrane-Bound Polyribosomes Make the Protein Components of the Bacterial Envelope[35]

Slightly over half the dry mass of *E. coli* is protein. Of this, some 20 percent is used to form the ribosomes, about 50 percent consists of the thousand or so enzymes found floating in the cytosol, and the remaining 30 percent is used to construct the cell envelope. Both the cytosolic proteins and the ribosomal proteins are made on free polyribosomes. In contrast, the envelope proteins are made on polyribosomes attached to the inner (cytoplasmic) membrane of the envelope. Their peptide products are thus directly extruded into the surrounding envelope (Figure 4-15). Some of the completed protein ends up in the inner membrane and some in the periplasmic space. In addition, some of the protein first transferred to the inner membrane moves to the outer membrane, perhaps through tubular connections between the two membrane systems. Not only proteins, but all newly made phospholipids, appear first in the inner membrane before diffusing to the outer membrane.

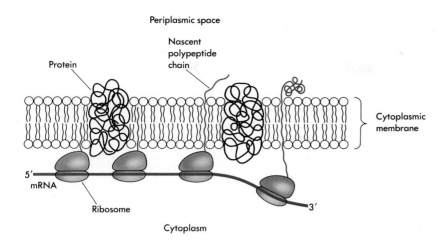

Periplasmic space

Nascent
polypeptide
chain

Protein

Cytoplasmic
membrane

5'
mRNA

Ribosome

3'

Cytoplasm

Figure 4-15
Synthesis of envelope proteins on polyribosomes attached to the cytoplasmic
membrane. As translation proceeds, the
nascent chain is transferred vectorially
across the membrane bilayer. Its final
destination may be the cytoplasmic
membrane, the periplasmic space, or
the outer membrane.

Metabolic Pathways Interconnect a Thousand Different Small Molecules[36, 37]

We can see that all the molecules in a bacterial cell arise from cellular transformations of food molecules when we grow *E. coli* on a simple, well-defined medium containing glucose. Under these conditions, glucose is the only organic source of carbon, and all the cellular carbon compounds (except for a few derived from carbon dioxide) result from enzyme-mediated chemical transformations that commence as the glucose molecules are broken down to smaller fragments. These smaller molecules, in turn, are used to form the many diverse building blocks whose polymerization creates the macromolecules from which major cellular organelles are assembled. The exact way in which all these transformations occur (collectively known as **metabolism**) is enormously complex, and most biochemists concern themselves with studying only a small fraction of the total interactions.

Fortunately, most of the major metabolic features of *E. coli* are common not only to all bacteria but also to all of life; therefore, what can be learned by focusing on *E. coli* is widely applicable. Figure 4-16 shows some of the more important types of chemical reactions that interconnect the various small molecules of *E. coli*, starting with its intake of glucose. Much information about this "intermediary metabolism" comes from experiments utilizing radioactively labeled food molecules. For example, if we expose *E. coli* for several seconds (a "pulse") to ^{14}C-labeled glucose, the radioactive atoms can be detected almost immediately in molecules chemically similar to glucose, such as glucose-6-phosphate. Only later do labeled atoms find their way into amino acids and nucleotides. The amount of time before radioactivity appears in the various compounds corresponds roughly to the number of enzymatic reactions separating glucose from the various intermediate metabolites.

Such experiments are relatively easy to interpret because most of the small molecules within cells are present in relatively small amounts. Only about 4 percent of the mass of *E. coli* is occupied by its collection of a thousand or so intermediary metabolites. Individual molecules thus have only fleeting existence before transformation into other small molecules or polymerization into macromolecules.

Figure 4-16 shows that various key intermediates in glucose degradation have several possible fates. They may be completely degraded

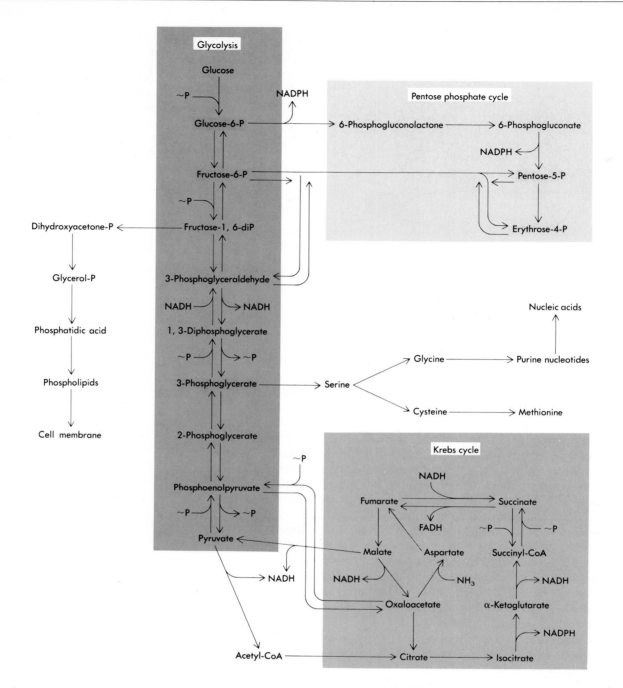

Figure 4-16
Some of the main metabolic pathways of *E. coli*. The 12 key precursor metabolites are outlined.

via the Embden-Meyerhof pathway (glycolysis), the Krebs cycle, and the respiratory chain to yield carbon dioxide and water. During this process, ADP is converted to ATP. Alternatively, intermediates may be used to initiate a series of chemical reactions that end with the biosynthesis of vital building-block molecules such as amino acids or nucleotides. For example, dihydroxyacetone phosphate is used as a precursor for the lipid constituent glycerol, whereas 3-phosphoglyceric acid is the beginning metabolite in a series of reactions that lead to the formation of the amino acids serine, glycine, and cysteine.

Connected groups of biosynthetic or degradative reactions are referred to as **metabolic pathways.** Once a molecule has started on a pathway, it often has no choice but to undergo a series of successive transformations. Not all pathways, however, are necessarily linear. Some are branched, with the intermediates at branch points being transformed into one of two possible compounds. Sometimes, the intermediates that have several alternative fates are very important in cellular metabolism. Perhaps the most important such compound is acetyl-CoA; not only is it the main precursor of lipids, but it also supplies the acetate residues consumed by the Krebs cycle.

The metabolic fates of the majority of molecules, however, are much more limited. An average molecule can be either broken down to compound x or used as an intermediate in the biosynthesis of compound y; thus, each such metabolite is able to combine specifically with only two different enzymes.

Biosynthetic Reactions Commence with Twelve Key Precursor Metabolites

All the biosynthetic pathways begin with one or another of a small group of molecules called *key precursor metabolites* (Table 4-6). There exist just 12 of these precursor metabolites, from which some 75 building blocks (e.g., amino acids, purines and pyrimidines, and fatty acids) and coenzyme products are derived by specific sets of enzymatic reactions. These pathways, however, are not universally present in all bacteria. Many bacteria lack one or more specific biosynthetic routes and correspondingly require the respective end product(s) to be externally supplied for growth. Not surprisingly, it is those bacteria normally found in environments rich in organic material that have lost one or more specific biosynthetic pathways. When such a pathway is present, however, it is almost always composed of the same set of enzymatic reactions, which either begin directly with one of the 12 key precursor metabolites or branch off from an intermediate or end product of another pathway.

Degradative Pathways Are Distinct from Biosynthetic Pathways

When *E. coli* is growing with glucose as its *sole* carbon source, all its amino acids must be synthesized from metabolites derived from glucose. There is a distinct biosynthetic pathway for each of the 20 amino acids (one example is shown in Figure 4-17). But *E. coli* can also grow in the absence of sugar, using any of the 20 amino acids as a sole carbon source. This means that there must also exist 20 pathways of amino acid degradation by which the carbon and nitrogen atoms of the amino acids are usefully freed to form key metabolite compounds such as α-ketoglutarate (Figure 4-18). These compounds can then be used in the synthesis of other amino acids. Degradative pathways also exist for the various lipids, the purine and pyrimidine nucleotides, many pentose and hexose sugars, and so on.

That distinct degradative pathways are almost never simply the reverse of biosynthetic pathways is to be expected. As we shall see in Chapter 6, most biosynthetic reactions require energy and often involve the breakdown of ATP, whereas degradative reactions by their very function must eventually generate ATP in addition to supplying carbon and nitrogen skeletons.

The Extraordinary Diversity of Fueling Reactions

The degradative reactions that produce the 12 key precursor metabolites, ATP, and the needed reducing power (NADH) are called *fueling reactions*. While many fueling reactions are common to all bacteria,

Table 4-6   The Twelve Key Precursor Metabolites

Glucose-6-phosphate
Fructose-6-phosphate
3-Phosphoglyceraldehyde
3-Phosphoglycerate
Phosphoenolpyruvate
Pyruvate
Acetyl-CoA
α-Ketoglutarate
Succinyl-CoA
Oxaloacetate
Pentose-5-phosphate
Erythrose-4-phosphate

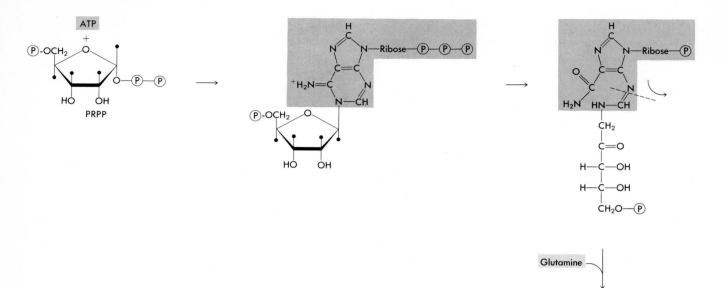

Figure 4-17
The biosynthesis of histidine from
5-phosphoribosyl-1-pyrophosphate
(PRPP). The resulting histidine contains
the five carbons of PRPP, the amide-
nitrogen of glutamine, and the one
nitrogen and one carbon of the
pyrimidine ring of ATP.

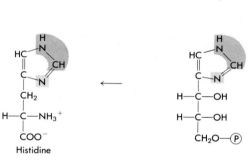

others are unique to particular species. Those fueling reactions common to all cells, called the *central pathways,* are involved in the deriva-
tion of the key metabolic precursors from each other. They can function linearly, as in glycolysis, to replace key metabolites drained off
for biosynthesis, or they can function cyclically, as in the Krebs cycle,
to produce carbon dioxide and water as well as to generate ATP and
NADH. Thus, there must be very great flexibility in the way the central pathways operate.

The real diversity of the fueling reactions is expressed by the peripheral pathways used when bacteria grow on compounds that are
not intermediates in the central pathways. The enormous variety of
potentially valuable sources of carbon, nitrogen, energy, and reducing power has led to the evolution of the approximately 4000 known
forms of bacteria that can utilize almost every carbon compound naturally existing on Earth with the exception of diamonds and coal. Even
within a single type of bacterium, the number of alternative fueling
reactions that can be called into action, depending on the nutritional
environment, is enormous. *E. coli* probably possesses at least 75 different peripheral pathways made up of at least 200 to 300 different
enzymatic reactions. These numbers must be even larger in many
other bacteria, particularly the pseudomonads, which can grow on a
much larger, more varied collection of inorganic molecules than can
E. coli.

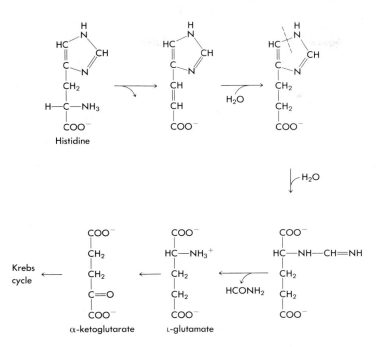

Figure 4-18
Degradative pathway of histidine.

Peripheral Pathways Are Turned On or Off as Units[38]

At any given time under constant nutritional conditions, a bacterial cell may possess the enzymes for only a single peripheral pathway. Even when several different sugars are available as food sources, an *E. coli* cell tends to have only the enzymes for the pathway that generates the most energy. For example, in the presence of both glucose and galactose, *E. coli* only possesses the enzymes needed to break down glucose. Only when the glucose supply is exhausted does it make the enzymes needed to metabolize galactose.

The set of enzymes that forms a peripheral pathway is thus present or absent as a unit. This makes obvious biological sense, for it would be wasteful to possess only half a pathway and thereby produce intermediate compounds that cannot be used. Likewise, it would make no sense to possess the enzymes needed to metabolize food molecules that are not present in the environment. The genes coding for enzymes of peripheral pathways constitute a significant fraction of the *E. coli* genome, and if they functioned in the absence of need, a significant fraction of the cell's protein-synthesizing capacity would be tied up to no avail.

Rapid Turnover of Bacterial mRNA Molecules Allows Speedy Changes in Gene Expression

Given both their potentially short generation times as well as the rapid speed at which their nutritional environment can change, bacteria need to quickly call into action new peripheral pathways as well as to stop the synthesis of enzymes no longer needed. They accomplish

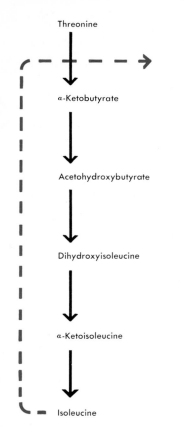

Figure 4-19
The pathway of isoleucine biosynthesis starting from threonine. The dotted colored line shows that isoleucine inhibits the enzyme (threonine deaminase) that transforms threonine into α-ketobutyrate.

this by having the synthesis of many of their mRNA molecules dependent on the receipt of specific signals indicating that given nutrients are available to be broken down (this is control at the level of transcription). Moreover, bacterial mRNA molecules, unlike those of higher eucaryotic cells, are unstable, with average half-lives of only 1 to 2 minutes. When a given type of mRNA is no longer needed and its synthesis stops, its inherent instability leads quickly to the disappearance of the preexisting molecules (see Chapter 16 for more details). Even finer adaptation to need is achieved by a second set of control signals that operates at the level of translation. A number of proteins (e.g., certain ribosomal proteins), if present in excess, block the functioning of their mRNA templates by binding to them and thereby preventing further ribosomal attachment and commencement of translation.

Regulation of Protein Function by Feedback Inhibition[39, 40, 41]

The catalytic activity of many proteins is affected by their ability to bind to specific small molecules. Because of this property, the activity of enzymes may be blocked when they are not needed. Consider, for example, what happens when an *E. coli* cell growing on minimal glucose medium is suddenly supplied with the amino acid isoleucine. Immediately, the synthesis of the mRNA molecules that code for the specific enzymes utilized in isoleucine biosynthesis ceases. Without a further control mechanism, preexisting enzymes will cause continued isoleucine production, now unnecessary because of the extracellular supply. Wasteful synthesis, however, almost never occurs, because high levels of isoleucine block the activity of the enzyme involved in the first step of its biosynthesis from threonine (Figure 4-19). This inhibition is due to the binding of isoleucine to the enzyme threonine deaminase. Thus bound, the enzyme is unable to convert threonine to α-ketobutyrate. Because the association between the enzyme and isoleucine is weak and reversible, relatively high isoleucine concentrations must exist before most of the enzyme molecules are inactivated. This very specific inhibition is called **feedback inhibition** or end-product inhibition because accumulation of a product prevents its further formation. Usually, only the first step in a metabolic chain is blocked. With the first reaction blocked, there is no accumulation of unwanted intermediates; so inhibition of the remaining enzymes would serve no purpose.

The final enzymatic step in the synthesis of a feedback inhibitor is often separated by several intermediate metabolic steps from the substrate (or product) of the enzyme involved in the first step of its biosynthesis. Figure 4-20 gives another example of this situation. The structure of the inhibitor may thus only loosely resemble that of the substrate of the inhibited enzyme, so that we would not expect an end-product inhibitor to combine with the active site (region that binds the substrate) of the enzyme it inactivates. Instead, the end product reversibly binds to a second site on the enzyme, yet nevertheless causes the enzyme activity to be blocked. The inhibitor acts by causing a change in the precise enzyme shape (an **allosteric transformation**), which prevents the enzyme from combining with its substrate (Figure 4-21). Such proteins whose shapes are changed by the

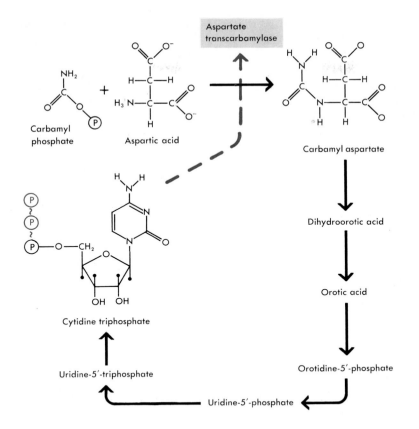

Cytidine triphosphate

Uridine-5'-triphosphate

Uridine-5'-phosphate

Carbamyl phosphate

Aspartic acid

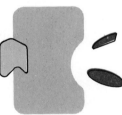
Aspartate transcarbamylase

Carbamyl aspartate

Dihydroorotic acid

Orotic acid

Orotidine-5'-phosphate

Figure 4-20
Diagram showing how feedback inhibition controls the biosynthesis of pyrimidines in *E. coli*. Note that the end product, cytidine triphosphate (CTP), has no structural similarity to either the substrate or the product of the first reaction in the pathway.

Enzyme

Substrates

Enzyme-substrate complex

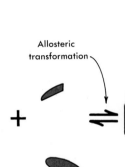

Allosteric transformation

End-product inhibitor

Enzyme

Substrates

Inactive enzyme

Figure 4-21
Schematic view of how the binding of an end-product inhibitor inhibits an enzyme by causing an allosteric transformation.

binding of specific small molecules at sites other than the active site are called *allosteric proteins;* correspondingly, the small molecules that bring about allosteric transformations are called *allosteric effectors.* The chemical forces binding specific feedback inhibitors to proteins are weak secondary forces such as hydrogen bonds, salt linkages, and van der Waals forces and do not involve covalent bonds. Hence, feedback inhibition can be quickly reversed once the end-product concentration is again reduced to a low level.

Protein Modifications That Lead to Modulations in Enzymatic Activity[42]

A second way enzyme activity can be regulated is by reversible covalent modification. Here the enzyme is not directly sensitive to the concentration of a small molecule, but instead to the action of another enzyme that adds or removes a modifying group. Phosphate, adenyl, uridyl, and methyl groups all are known to modify different proteins and thus to affect their function. In *E. coli*, the activity of the major enzyme for ammonia (and thus nitrogen) uptake, glutamine synthetase, is blocked if the enzyme is adenylated. Its activity decreases progressively as more of its 12 identical subunits acquire an adenyl group on a particular tyrosine side chain. As is true for feedback-inhibited enzymes, the activity of glutamine synthetase ultimately depends on the cellular concentration of its product, glutamine, although very indirectly. The enzyme that adds and removes adenyl groups from glutamine synthetase (adenyltransferase) is controlled by a protein (P_{II}) that in turn is controlled (through addition and removal of uridyl groups) by another enzyme, uridyltransferase, which finally is directly responsive to glutamine concentration (Figure 4-22). Why glutamine synthetase is controlled by this succession of steps rather than directly by glutamine we do not know, but it is a reasonable guess that the regulatory pathway branches in the cell; thus, adenyltransferase, P_{II}, and uridyltransferase also might affect other enzymes responsive to glutamine concentration. Also, other conditions besides the glutamine concentration might affect the

Figure 4-22
Modulation of glutamine synthetase activity. (a) The synthetic pathway from α-ketoglutarate to glutamine. (b) When ammonia is scarce, the concentration of α-ketoglutarate is high and that of glutamine low. The concentrations of glutamine and α-ketoglutarate indirectly affect glutamine synthetase activity by influencing the activity of the enzyme that controls the uridylation of a protein called P_{II}. (c) In turn, the state of the P_{II} protein influences the activity of the enzyme that controls the conversion between the active and inactive (adenylated) states of glutamine synthetase.

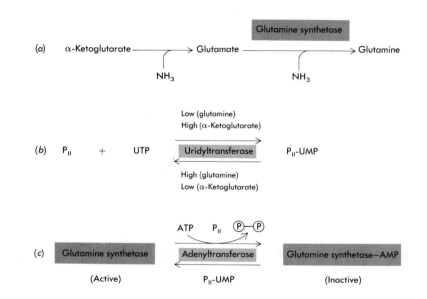

adenyltransferase and thus change the activity of glutamine synthetase.

Methylation of Transducer Proteins in Chemotaxis[43-46]

It is through the addition and removal of methyl groups that chemical attractants or repellents control the functioning of bacterial flagella. This movement toward or away from certain chemicals is called **chemotaxis.** *E. coli* cells are propelled in one direction when a set of flagella rotating together counterclockwise collects in a bundle that acts as a propeller. If the concentration of an attractant increases, this rotation is prolonged, so that a cell tends to "run" ("swim") toward the attractant. If a repellent is encountered, on the other hand, the direction of rotation is reversed to clockwise, destroying the efficiency of the propeller and causing the cell to "tumble" for a while (Figure 4-23). When tumbling stops, the cell is likely to move off in a different direction and so escape the repellent. When no chemical is present, the cell is in a neutral state in which it alternately runs and tumbles.

For many attractants and repellents, the signal to swim or to tumble is transmitted from the outside into the cell by **transducer proteins** that span the cell's cytoplasmic membrane. The protein's NH_2-terminal portion extending outside the cell either binds the attractant (or repellent) directly or is bound by other periplasmic proteins that detect the chemical. In either case, when the transducer protein is contacted from the outside, it undergoes some change in its intracellular (COOH-terminal) portion that ultimately signals the flagellar motor to go either counterclockwise (in response to an attractant) or clockwise (in response to a repellent). What this change might be is completely unknown; one guess, however, is that an attractant or repellent stimulates an enzymatic activity in the intracellular domain of the transducer to synthesize a signal molecule that diffuses to the flagellar motor and switches it from one direction to the other. How the cell actually behaves in a complex environment depends on the combined

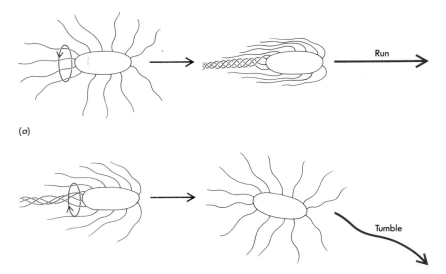

(a)

(b)

Run

Tumble

Figure 4-23
Mechanism of movement during chemotaxis in *E. coli*. (a) An attractant causes a counterclockwise rotation of the flagella, which form a bundle to propel the cell forward. (b) A repellent causes the bundle to rotate clockwise, the bundle falls apart, and the bacterium tumbles.

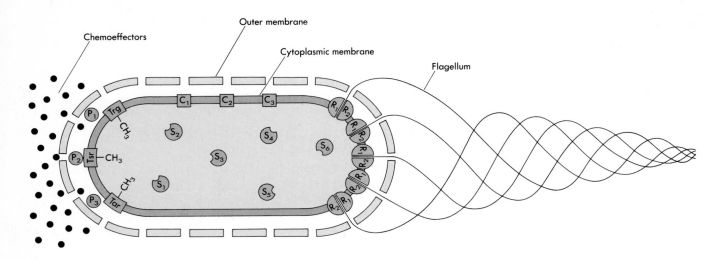

Figure 4-24
The mechanism of chemotaxis in *E. coli.* Small molecules such as glucose, ribose, galactose, and oxygen may serve as chemoeffectors that pass through outer-membrane pores and interact with receptor proteins in the periplasmic space (P_1, P_2, etc.) and in the cytoplasmic membrane (C_1, C_2, etc.). Messages generated by these interactions are focused through the Tar, Trg, and Tsr proteins (methyl-accepting transducer proteins), which become methylated to a degree determined by the amount of chemoeffector present. These signals are then processed by cytoplasmic proteins (S_1, S_2, etc.) to alter the level of certain response regulators, which in turn interact with proteins R_1 and R_2. Changes in these proteins, which surround the bases of the flagella, somehow regulate the flagellar rotation. [After D. E. Koshland, Jr., *Bacterial Chemotaxis as a Model Behavioral System* (New York: Raven Press, 1980.)]

effect of all signals sent to its motor by all transducing proteins (and certain other signaling systems) that may be sensing a variety of chemicals (Figure 4-24).

A further property of the transducer protein is **adaptation.** If the new concentration of attractant or repellent persists, the transducer soon adapts to this new condition and no longer transmits a signal. This means that the transducer responds to *change* and not to a given concentration of an attractant or repellent. Adaptation is caused by modification of a transducer through methylation and demethylation of about four glutamyl groups on its cytoplasmic side. Methylation favors the state in which a transducer signals clockwise rotation (tumbling), whereas demethylation promotes counterclockwise rotation (running). The modification that occurs opposes the last chemotactic signal, but takes about a minute to occur. Thus, exposure to an attractant eventually leads to greater methylation, allowing more tumbling again. And exposure to a repellent leads to demethylation, so that tumbling once again alternates with periods of swimming. Probably the same conformational change of a transducer protein that produces a signal to the flagellar motor also allows it to bind either the methylation or demethylation enzyme, so that the initial effect of the attractant or repellent is inevitably neutralized.

The Bacterial Cell Is a Precisionally Fine-Tuned Machine

The day has long passed when the question should be asked whether there is more than the laws of chemistry behind the functioning of the bacterial cell. We now see the bacterium as an extraordinarily sophisticated set of interrelated molecules that harmoniously work together in highly predictable ways to ensure the growth and selective survival of more of its kind. At the heart of these remarkable, almost clockwork-type, machines are the DNA molecules that encode, with total precision, sets of commands that bring into action molecules needed to cope with ever-changing nutritional potentials. What is equally important is that DNA has the capacity to incorporate within its structure new changes that will permit further evolution of the cell into forms needed to prosper successfully upon the Earth's continuously changing face.

Mycoplasmas Are the Smallest Living Organisms[47, 48]

Existing with genomes some four times smaller than those of *E. coli* are the smallest of procaryotic cells, the mycoplasmas (Figure 4-25). They were at one time referred to as pleuropneumonia-like organisms owing to their initial discovery as disease-causing agents responsible for many of the pneumonia-like respiratory illnesses that affect higher vertebrates, including humans. These tiny, irregularly shaped, flexible organisms essentially have no cell walls and usually live in close contact with cells of higher eucaryotic organisms, on whose surfaces they survive as parasites, deriving many of their essential metabolites from body fluids like those found in our nasal passages. Because they live only in such nutritionally rich environments, these organisms have lost many of the key biosynthetic pathways common to most cells and, in so doing, have lost much of the DNA needed by the more freely living microbes. The smallest of mycoplasmas have chromosomes containing no more than 8 to 10×10^5 base pairs, limiting them to at most 800 to 1000 different proteins. We still do not know how these proteins function (enzymatically versus structurally). Because of their complex nutritional requirements, mycoplasmas are not easy to grow in culture. So despite the obvious attractiveness of their small genomes, they are still studied largely for their disease-causing properties and not because they are probably the smallest organisms on Earth today.

The present restriction of mycoplasmas to very specific ecological niches no doubt reflects the fact that they have too little DNA to compete on equal footing with the true bacteria, most of which possess roughly the same amount of DNA as *E. coli*. Bacterial genomes, as they now exist, represent trade-offs between the advantages of a low DNA content (with the corresponding relatively short time needed for chromosome replication) and the possession of sufficient metabolic pathways to ensure survival in nutritionally unpredictable environments.

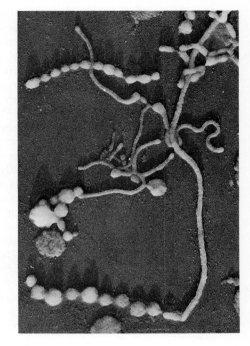

Figure 4-25
Electron micrograph of *Mycoplasma mycoides*. (Courtesy of Alan Rodwell.)

Summary

The simplest cells to use in studying the essential chemical features of cell growth and division are the bacteria. They are generally 500 times smaller than the average cell of a higher plant or animal. By far the best-understood bacterium is *Escherichia coli (E. coli)*, which weighs about 2×10^{-12} gram (10^{12} daltons), of which some 70 percent is water. Even with this small size, an *E. coli* cell is quite complicated chemically, containing at least 3000, and more likely 6000, different molecules. At least 1000 of these components are "small" molecules, and the remainder are macromolecules. The vast majority are either nucleic acids or proteins.

All the essential genes of *E. coli*, if not of all bacteria, are present on a single circular DNA molecule whose replication starts at a fixed point and proceeds bidirectionally. Each chromosome is greatly compacted by supercoiling into irregularly arranged nucleoid bodies that occupy about 10 percent of the cytoplasmic volume. No membranes separate the chromosomal DNA from the remainder of the cytoplasm, with the result that in all bacteria, the processes of DNA replication, DNA transcription into RNA, and protein synthesis all occur in the same cellular compartment. Ribosomes attach to newly made mRNA molecules soon after their synthesis, and most mRNA molecules are covered by ribosomes carrying nascent polypeptide chains of ever-increasing length. How given RNA molecules become released from their constantly rotating DNA templates is still very much a mystery, as is the way progeny DNA molecules become partitioned into daughter cells.

The cytoplasm of *E. coli*, like that of all cells, is surrounded by a phospholipid-containing cytoplasmic membrane into which are inserted hundreds of different proteins. These proteins become part of the membrane through a complex set of events, in which ribosomes position themselves on the cytoplasm-facing membrane surface so that their polypeptide products grow into and then through it. Among the cytoplasmic membrane proteins are the vital collections of proteins involved in generating ATP through the proton motive force created by movements of electrons through the respiratory chain. The immediate space outside the cytoplasmic ("inner") membrane is occupied by a peptidoglycan gel, which, in turn, is attached to a second "outer" phospholipid membrane. Major components of this outer membrane include highly immunogenic lipopolysaccharide-containing proteins as well as matrix proteins (porins) that form specific holes through which extracellular molecules must pass to reach the inner cytoplasmic membrane.

The holes within the peptidoglycan gel constitute the periplasmic space. In it are found hydrolytic enzymes that initiate the degradation of food molecules, binding proteins involved in initiating transport across the cytoplasmic membrane, and chemoreceptors that sense attractants and repellents.

Many bacteria, including *E. coli*, contain two forms of linear surface appendages that arise from the cytoplasmic membrane, penetrate the periplasmic space, and extend out through the external surface. The most conspicuous are the flagella, organelles of locomotion. They generate movement by using the energy produced by the proton motive force to rotate. Rotation in a counterclockwise direction leads to formation of bundles of flagella that propel cells smoothly forward; clockwise rotation, on the other hand, disperses the bundle and causes the bacteria to move erratically (tumble). A flagellum is an inherently complex structure containing a basal structure and hook as well as an extended filament made up of very large numbers of the basic protein building block flagellin. Much less complex are the even thinner pili, filaments made from single protein building blocks called pilin. Pili often function to attach bacteria to other surfaces, often glycoproteins on the surfaces of higher cells.

All bacterial molecules arise from metabolic transformation, usually starting from simple food molecules like glucose. Individual small molecules usually exist only fleetingly before transformation into other small molecules or polymerization into macromolecules. Most compounds cannot be directly transformed into a large number of other compounds. Instead, each compound is a step in a series of reactions leading either to the degradation of a food molecule or to the biosynthesis of a necessary cellular molecule such as an amino acid or fatty acid. Cellular metabolism is the sum total of a large number of such pathways connected in such a way that products of degradative pathways can be used to initiate specific biosynthetic pathways.

Connecting the biosynthetic and degradation pathways are the so-called central pathways that lead to the production of the 12 key precursor metabolites that initiate individual biosynthetic pathways. While central pathways operate virtually all the time, peripheral biosynthetic or degradative pathways are turned on or off as units, depending on need. Such decisions are often made for biosynthetic reactions by feedback (end-product) inhibition of enzyme function. An end-product metabolite can reversibly combine with the first enzyme involved in its specific biosynthetic pathway. This combination transforms the enzyme into an inactive form. The end-product inhibitor does not combine with the enzymatically active site but binds instead to a second site on the enzyme, causing a change in the enzyme shape. Proteins whose shapes and activities are changed by combination with other molecules are called allosteric proteins. Modulation in enzyme activity can also be effected by reversible covalent modifications, which are affected by still other enzymes. Additions of adenyl, phosphate, and uridyl groups can drastically influence the functioning of given enzymes.

The complexity of the metabolic map of an organism is related to the amount of genetic information (DNA) in the organism. The amount of DNA in a cell places an upper limit on the different number of enzymes the cell can produce. *E. coli* possesses sufficient DNA to code for the amino acid sequences of 3000 to 4500 different proteins. About 1700 different small molecules have now been detected in *E. coli*. The metabolic reactions that we already know of in *E. coli* must account for at least one-fifth of its total metabolism.

Bibliography

General References

Brock, T. D. 1984. *Biology of Microorganisms.* 4th ed. Englewood Cliffs, N.J.: Prentice-Hall. A good basic introduction to general microbiology.

Davis, B. D., R. Dulbecco, H. N. Eisen, and H. S. Ginsberg. 1980. *Microbiology.* 3rd ed. New York: Harper & Row. An excellent text that is medically oriented but also contains good general information.

Ingraham, J. L., O. Maaløe, and F. C. Neidhardt. 1983. *The Growth of the Bacterial Cell.* Sunderland, Mass.: Sinauer Associates. An excellent, more advanced text.

Joklick, W. K., H. P. Willet, and D. B. Amos, eds. 1984. *Zinsser Microbiology.* New York: Appleton-Century-Crofts. Text emphasizes medical microbiology but also contains good general information.

Losick, R., and L. Shapiro. 1984. *Microbial Development.* Cold Spring Harbor, N.Y.: Cold Spring Harbor Laboratory.

Stanier, R. Y., J. L. Ingraham, M. L. Wheelis, and P. R. Painter. 1986. *The Microbial World.* 5th ed. Englewood Cliffs, N.J.: Prentice-Hall. A superb treatment of microbiology that is somewhat more detailed than the text by Brock.

Cited References

1. Luria, S. E., and M. Delbrück. 1943. "Mutations in Bacteria from Virus Sensitivity to Virus Resistance." *Genetics* 28:491–511.
2. Lederberg, J., and E. L. Tatum. 1946. "Novel Genotypes in Mixed Cultures of Biochemical Mutants of Bacteria." *Cold Spring Harbor Symp. Quant. Biol.* 11:113–114.
3. Deley, J., and K. Kersters. 1975. "Biochemical Evolution in Bacteria." *Comp. Biochem.* 29B:1.
4. Margulis, L. 1970. *Origin of Eukaryotic Cells.* New Haven, Conn.: Yale University Press.
5. Bauchop, T., and S. R. Elsden. 1960. "The Growth of Microorganisms in Relation to Their Energy Supply." *J. Gen. Microbiol.* 23:457–469.
6. Marr, A. G., E. H. Nilson, and D. J. Clark. 1963. "The Mainte-

nance Requirements of *Escherichia coli.*" *Ann. N.Y. Acad. Sci.* 102:536–548.

7. Van Niel, C. B. 1971. "Techniques for the Enrichment, Isolation and Maintenance of the Photosynthetic Bacteria." *Methods in Enzymol.* 23:3–28. Includes a broad overview of the ecology of photosynthetic bacteria.

8. Delbrück, M. 1946. "Experiments with Bacterial Viruses (Bacteriophages)." *Harvey Lecture Series* 41:161–187.

9. Delbrück, M. 1966. "A Physicist Looks at Biology." In *Phage and the Origins of Molecular Biology,* ed. J. Cairns, G. S. Stent, and J. D. Watson. Cold Spring Harbor, N.Y.: Cold Spring Harbor Laboratory, pp. 9–22.

10. Helmstetter, C. E., and J. D. Cummings. 1963. "Bacterial Synchronization by Selection of Cells at Division." *Proc. Nat. Acad. Sci.* 50:767–774.

11. Herendeen, S. L., R. A. van Bogelen, and F. C. Neidhart. 1979. "Levels of Major Proteins of *E. coli* During Growth at Different Temperatures." *J. Bacteriol.* 139:185–194.

12. Anderson, K. B., and K. von Meyenburg. 1980. "Are Growth Rates of *E. coli* in Batch Culture Limited by Respiration?" *J. Bacteriol.* 144:114–123.

13. Mendelson, N. H. 1982. "Bacterial Growth and Division: Genes, Structures, Forces and Clocks." *Microbiol. Revs.* 46:341–375.

14. Ratkowsky, D. A., J. Olley, T. A. Meekin, and A. Ball. 1982. "Relationship Between Temperature and Growth Rate of Bacterial Cultures." *J. Bacteriol.* 149:1–5.

15. O'Farrell, P. H. 1975. "High Resolution Two-Dimensional Electrophoresis of Proteins." *J. Biol. Chem.* 250:4007–4021.

16. Phillips, T. A., P. L. Bloch, and F. C. Neidhart. 1980. "Protein Identifications on O'Farrell's 2-Dimensional Gels: Locations of 55 Additional *E. coli* Proteins." *J. Bacteriol.* 144:1024–1033.

17. Kellenberger, E. 1960. "The Physical State of the Bacterial Nucleus." In *Microbial Genetics,* ed. W. Hayes and R. C. Clowes. New York: Cambridge University Press, pp. 39–60.

18. Costerton, J. W. 1979. "The Role of Electron Microscopy in the Elucidation of Bacterial Structure and Function." *Ann. Rev. Microbiol.* 33:459–479.

19. Dubochet, J., A. W. McDowall, B. Menge, E. N. Schmid, and K. G. Lickfeld. 1983. "Electron Microscopy of Frozen-Hydrated Bacteria." *J. Bacteriol.* 155:381–390.

20. Cronan, J. 1978. "Molecular Biology of Bacterial Membrane Lipids." *Ann. Rev. Biochem.* 47:163–189.

21. Osborn, M. J., and H. C. P. Wu. 1980. "Proteins of the Outer Membrane of Gram-Negative Bacteria." *Ann. Rev. Microbiol.* 34:369–422.

22. Shockman, G. D., and J. F. Barrett. 1983. "Structure, Function and Assembly of Cell Walls of Gram-Positive Bacteria." *Ann. Rev. Microbiol.* 37:501–527.

23. Hobot, J. A., E. Carlemalm, W. Villiger, and E. Kellenberger. 1985. "Periplasmic Gel: New Concept Resulting from the Reinvestigation of Bacterial Cell Envelope Ultrastructure by New Methods." *J. Bacteriol.* 160:143–152.

24. Nogami, T., and S. Mizushima. 1983. "Outer Membrane Porins Are Important in the Maintenance of the Surface Structure of *E. coli* Cells." *J. Bacteriol.* 156:402–408.

25. Sutcliffe, J., R. Blumenthal, A. Walter, and J. Foulds. 1983. "*E. coli* Outer Membrane Protein K Is a Porin." *J. Bacteriol.* 156:867–872.

26. Hinkle, P. C., and R. E. McCarthy. 1978. "How Cells Make ATP." *Sci. Amer.* 238(3):104–123.

27. Davis, B. D., and P. C. Tai. 1980. "The Mechanism of Protein Secretion Across Membranes." *Nature* 283:433–438.

28. Berg, H. C., and R. A. Anderson. 1973. "Bacteria Swim by Rotating Their Flagellar Filaments." *Nature* 245:380–382.

29. Silverman, M., and M. I. Simon. 1977. "Bacterial Flagella." *Ann. Rev. Microbiol.* 31:397–419.

30. Achtman, M., G. Morelli, and S. Schwuchow. 1978. "Cell-Cell Interactions in Conjugating *E. coli:* Role of F Pili and Fate of Mating Aggregates." *J. Bacteriol.* 135:1053–1061.

31. McMichael, M. C., and J. T. Ou. 1979. "Structure of Common Pili from *E. coli.*" *J. Bacteriol.* 138:969–975.

32. Mett, H., L. Kloetzlen, and K. Vosbeck. 1983. "Properties of Pili from *E. coli* SS142 that Mediate Mannose-Resistant Adhesion to Mammalian Cells." *J. Bacteriol.* 153:1038–1044.

33. Hobot, J. A., W. Villiger, J. Escaig, M. Maeder, A. Ryder, and E. Kellenberger. 1985. "Shape and Fine Structure of Nucleoids Observed on Sections of Ultrarapidly Frozen and Cryosubstituted Bacteria." *J. Bacteriol.* 162:960–971.

34. Millar, O. L., B. R. Beatty, B. A. Hamkalo, and C. A. Thomas, Jr. 1970. "Electron Microscopic Visualization of Transcription." *Cold Spring Harbor Symp. Quant. Biol.* 35:505–512.

35. Silhavy, T. J., S. A. Benson, and S. D. Emr. 1983. "Mechanisms of Protein Localization." *Microbiol. Revs.* 47(3):313–344.

36. Atkinson, D. E. 1977. *Cellular Energy Metabolism and Its Regulation.* New York: Academic Press.

37. Gottschalk, G. 1979. *Bacterial Metabolism.* New York: Springer-Verlag.

38. Beckwith, J. R., and D. Zipser, eds. 1970. *The Lactose Operon.* Cold Spring Harbor, N.Y.: Cold Spring Harbor Laboratory.

39. Umbarger, H. E. 1956. "Evidence for a Negative Feedback Mechanism in the Biosynthesis of Isoleucine." *Science* 123:848.

40. Yates, R. A., and A. B. Pardee. 1956. "Control of Pyrimidine Biosynthesis in *E. coli* by a Feedback Mechanism." *J. Biol. Chem.* 221:757–770.

41. Umbarger, H. E. 1978. "Amino Acid Synthesis and Its Regulation." *Ann. Rev. Biochem.* 47:533–606.

42. Ginsberg, A., and E. R. Stadtman. 1973. "Regulation of Glutamine Synthetase in *E. coli.*" In *The Enzymes of Glutamine Metabolism,* ed. S. Prusiner and E. R. Stadtman. New York: Academic Press, pp. 9–43.

43. Alder, J. 1969. "Chemoreceptors in Bacteria." *Science* 166:1588–1597.

44. DeFranco, A. L., and D. E. Koshland, Jr. 1980. "Multiple Methylation in Processing of Sensory Signals During Bacterial Chemotaxis." *Proc. Nat. Acad. Sci.* 77(5):2429–2433.

45. Koshland, E. E., Jr. 1981. "Biochemistry of Sensing and Adaptation in a Simple Bacterial System." *Ann. Rev. Biochem.* 50:765–770.

46. Boyd, A., and M. I. Simon. 1982. "Bacterial Chemotaxis." *Ann. Rev. Physiol.* 44:501–517.

47. Smith, P. F. 1971. *The Biology of Mycroplasmas.* New York: Academic Press.

48. Kawauchis, Y., A. Muto, and S. Osawa. 1982. "The Protein Composition of *Mycoplasma capriola.*" *Mol. Gen. Genetics* 188:7–11.

CHAPTER 5

The Importance of Weak Chemical Interactions

Until now, we have focused our attention on discrete organic molecules and, following classical organic chemistry, have emphasized the covalent bonds that hold them together. It takes little insight, however, to realize that this type of analysis is inadequate for describing a cell and that we must also concern ourselves with the exact shapes of molecules and with the forces that bind them together in an organized fashion. The distribution of molecules within cells is not random, and we must ask ourselves what chemical laws determine this distribution. Clearly, covalent bonding cannot be involved; by definition, atoms united by covalent bonds belong to the same molecule.

The arrangement of distinct molecules within cells is controlled by chemical bonds much weaker than covalent bonds. Atoms united by covalent bonds are capable of weak interactions with nearby atoms. These interactions, sometimes called "secondary bonds," occur not only between atoms of different molecules, but also between atoms within the same molecule. Weak bonds are important in determining which molecules lie next to each other and in giving shape to flexible molecules such as the polypeptides and polynucleotides. It is therefore useful to understand the nature of weak chemical interactions and to understand how their "weak" character makes them indispensable to cellular existence. The most important such bonds include van der Waals bonds, hydrophobic bonds, hydrogen bonds, and ionic bonds.

Definition and Some Characteristics of Chemical Bonds[1]

A **chemical bond** is an attractive force that holds atoms together. Aggregates of finite size are called molecules. Originally, it was thought that only covalent bonds hold atoms together in molecules; but now, weaker attractive forces are known to be important in holding together many macromolecules. For example, the four polypeptide chains of hemoglobin are held together by the combined action of several weak bonds. It is thus now customary also to call weak positive interactions chemical bonds, even though they are not strong enough, when present singly, to effectively bind two atoms together.

Chemical bonds are characterized in several ways. An obvious characteristic of a bond is its strength. Strong bonds almost never fall

126

apart at physiological temperatures. This is why atoms united by covalent bonds always belong to the same molecule. Weak bonds are easily broken, and when they exist singly, they exist fleetingly. Only when present in ordered groups do weak bonds last a long time. The strength of a bond is correlated with its length, so that two atoms connected by a strong bond are always closer together than the same two atoms held together by a weak bond. For example, two hydrogen atoms bound covalently to form a hydrogen molecule (H:H) are 0.74 Å apart, whereas the same two atoms held together by van der Waals forces are 1.2 Å apart.

Another important characteristic is the maximum number of bonds that a given atom can make. The number of covalent bonds that an atom can form is called its **valence.** Oxygen, for example, has a valence of two: It can never form more than two covalent bonds. There is more variability in the case of van der Waals bonds, where the limiting factor is purely steric. The number of possible bonds is limited only by the number of atoms that can simultaneously touch each other. The formation of hydrogen bonds is subject to more restrictions. A covalently bonded hydrogen atom usually participates in only one hydrogen bond, whereas an oxygen atom seldom participates in more than two hydrogen bonds.

The angle between two bonds originating from a single atom is called the **bond angle.** The angle between two specific covalent bonds is always approximately the same. For example, when a carbon atom has four single covalent bonds, they are directed tetrahedrally (bond angle = 109°). In contrast, the angles between weak bonds are much more variable.

Bonds differ also in the **freedom of rotation** they allow. Single covalent bonds permit free rotation of bound atoms (Figure 5-1), whereas double and triple bonds are quite rigid. Bonds with partial double-bond character, such as the peptide bond, are also quite rigid. For that reason, the carbonyl (C–O) and imino (N–H) groups bound together by the peptide bond must lie in the same plane (Figure 5-2). Much weaker, ionic bonds, on the other hand, impose no restrictions on the relative orientations of bonded atoms.

Chemical Bonds Are Explainable in Quantum-Mechanical Terms

The nature of the forces, both strong and weak, that give rise to chemical bonds remained a mystery to chemists until the quantum theory of the atom (quantum mechanics) was developed in the 1920s. Then, for the first time, the various empirical laws about how chemical bonds are formed were put on a firm theoretical basis. It was realized that all chemical bonds, weak as well as strong, are based on electrostatic forces. Quantum mechanics provided explanations for covalent bonding by the sharing of electrons and also for the formation of weaker bonds.

Chemical-Bond Formation Involves a Change in the Form of Energy

The spontaneous formation of a bond between two atoms always involves the release of some of the internal energy of the unbonded atoms and its conversion to another energy form. The stronger the

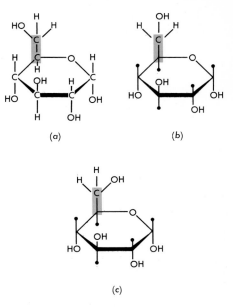

Figure 5-1
Rotation about the C_5–C_6 bond in glucose. This carbon-carbon bond is a single bond, and so any of the three configurations, (a), (b), or (c), may occur.

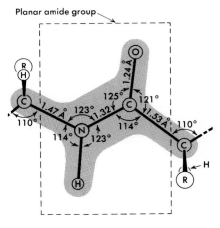

Figure 5-2
The planar shape of the peptide bond. Shown here is a portion of an extended polypeptide chain. Almost no rotation is possible about the peptide bond because of its partial double-bond character:

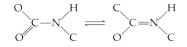

All the atoms in the gray area must lie in the same plane. Rotation is possible, however, around the remaining two bonds, which make up the polypeptide configurations. [After L. Pauling, *The Nature of the Chemical Bond*, 3rd ed. (Ithaca, N.Y.: Cornell University Press, 1960), p. 498, with permission.]

bond, the greater the amount of energy released upon its formation. The bonding reaction between two atoms A and B is thus described by

$$A + B \longrightarrow AB + energy \qquad (5\text{-}1)$$

where AB represents the bonded aggregate. The rate of the reaction is directly proportional to the frequency of collision between A and B. The unit most often used to measure energy is the calorie, the amount of energy required to raise the temperature of 1 gram of water from 14.5°C to 15.5°C. Since thousands of calories are usually involved in the breaking of a mole of chemical bonds, most energy changes within chemical reactions are expressed in kilocalories per mole.

However, atoms joined by chemical bonds do not remain together forever, since there also exist forces that break chemical bonds. By far the most important of these forces arises from heat energy. Collisions with fast-moving molecules or atoms can break chemical bonds. During a collision, some of the kinetic energy of a moving molecule is given up as it pushes apart two bonded atoms. The faster a molecule is moving (the higher the temperature), the greater the probability that, upon collision, it will break a bond. Hence, as the temperature of a collection of molecules is increased, the stability of their bonds decreases. The breaking of a bond is thus always indicated by the formula

$$AB + energy \longrightarrow A + B \qquad (5\text{-}2)$$

The amount of energy that must be added to break a bond is exactly equal to the amount that was released upon formation of the bond. This equivalence follows from the first law of thermodynamics, which states that energy (except as it is interconvertible with mass) can be neither made nor destroyed.

Equilibrium Between Bond Making and Breaking

Every bond is thus a result of the combined actions of bond-making and bond-breaking forces. When an equilibrium is reached in a closed system, the number of bonds forming per unit time will equal the number of bonds breaking. Then the proportion of bonded atoms is described by the following mass action formula:

$$K_{eq} = \frac{conc^{AB}}{conc^A \times conc^B} \qquad (5\text{-}3)$$

where K_{eq} is the **equilibrium constant,** and $conc^A$, $conc^B$, and $conc^{AB}$ are the concentrations of A, B, and AB, respectively, in moles per liter. Whether we start with only free A and B, with only the molecule AB, or with a combination of AB and free A and B, at equilibrium the proportions of A, B, and AB will reach the concentrations given by K_{eq}.

The Concept of Free Energy[2, 3]

There is always a change in the form of energy as the proportion of bonded atoms moves toward the equilibrium concentration. Biologically, the most useful way to express this energy change is through the physical chemist's concept of **free energy,** denoted by the symbol *G*.* We shall not give a rigorous description of free energy in this text nor show how it differs from the other forms of energy. For this, the reader must refer to a chemistry text that discusses the second law of thermodynamics. It must suffice to say here that *free energy is energy that has the ability to do work.*

The second law of thermodynamics tells us that a decrease in free energy (ΔG is negative) always occurs in spontaneous reactions. When equilibrium is reached, however, there is no further change in the amount of free energy ($\Delta G = 0$). The equilibrium state for a closed collection of atoms is thus the state that contains the least amount of free energy.

The free energy lost as equilibrium is approached is either transformed into heat or used to increase the amount of entropy. We shall not attempt to define entropy here except to say that the amount of entropy is a measure of the amount of disorder. The greater the disorder, the greater the amount of entropy. The existence of entropy means that many spontaneous chemical reactions (those with a net decrease in free energy) need not proceed with an evolution of heat. For example, when sodium chloride (NaCl) is dissolved in water, heat is absorbed rather than released. There is, nonetheless, a net decrease in free energy because of the increase in disorder of the sodium and chlorine ions as they move from a solid to a liquid phase.

K_{eq} Is Exponentially Related to ΔG

Clearly, the stronger the bond, and hence the greater the change in free energy (ΔG) that accompanies its formation, the greater the proportion of atoms that must exist in the bonded form. This common-sense idea is quantitatively expressed by the physical-chemical formula

$$\Delta G = -RT \ln K_{eq} \quad \text{or} \quad K_{eq} = e^{-\Delta G/RT} \tag{5-4}$$

where R is the universal gas constant, T is the absolute temperature, ln is the logarithm (of K_{eq}) to the base e, K_{eq} is the equilibrium constant, and $e = 2.718$.

Insertion of the appropriate values of R (1.987 cal/deg-mol) and T (298 at 25°C) tells us that ΔG values as low as 2 kcal/mol can drive a bond-forming reaction to virtual completion if all reactants are present at molar concentrations (Table 5-1).

Covalent Bonds Are Very Strong

The ΔG values accompanying the formation of covalent bonds from free atoms such as hydrogen or oxygen are very large and negative in sign, usually −50 to −110 kcal/mol. Equation 5-4 tells us that K_{eq} of the bonding reaction will be correspondingly large, and so the concentration of hydrogen or oxygen atoms existing unbound will be very small. For example, with a ΔG value of −100 kcal/mol, if we start with 1 mol/L of the reacting atoms, only one in 10^{40} atoms will remain unbound when equilibrium is reached.

---

*It was formerly the custom in the United States to refer to free energy by the symbol *F*. Now, however, most texts use the international symbol *G*, which honors the great nineteenth-century physicist Josiah Gibbs.

Table 5-1 The Numerical Relationship Between the Equilibrium Constant and ΔG at 25°C

K_{eq}	ΔG, kcal/mol
0.001	4.089
0.01	2.726
0.1	1.363
1.0	0
10.0	−1.363
100.0	−2.726
1000.0	−4.089

Weak Bonds Have Energies Between 1 and 7 kcal/mol

The main types of weak bonds important in biological systems are the van der Waals bonds, hydrophobic bonds, hydrogen bonds, and ionic bonds. Sometimes, as we shall soon see, the distinction between a hydrogen bond and an ionic bond is arbitrary. The weakest bonds are the van der Waals bonds. These have energies (1 to 2 kcal/mol) only slightly greater than the kinetic energy of heat motion. The energies of hydrogen and ionic bonds range between 3 and 7 kcal/mol.

In liquid solutions, almost all molecules form a number of weak bonds to nearby atoms. All molecules are able to form van der Waals bonds, whereas hydrogen and ionic bonds can form only between molecules that have a net charge (ions) or in which the charge is unequally distributed. Some molecules thus have the capacity to form several types of weak bonds. Energy considerations, however, tell us that molecules always have a greater tendency to form the stronger bond.

Weak Bonds Are Constantly Made and Broken at Physiological Temperatures

The energy of the strongest weak bond is only about ten times larger than the average energy of kinetic motion (heat) at 25°C (0.6 kcal/mol). Since there is a significant spread in the energies of kinetic motion, many molecules with sufficient kinetic energy to break the strongest weak bond always exist at physiological temperatures.

Enzymes Are Not Involved in Making or Breaking Weak Bonds

The average lifetime of a single weak bond is only a fraction of a second. Thus, cells do not need a special mechanism to speed up the rate at which weak bonds are made and broken. Not surprisingly, therefore, enzymes never participate in reactions involving only weak bonds.

The Distinction Between Polar and Nonpolar Molecules

All forms of weak interactions are based on attractions between electric charges. The separation of electric charges can be permanent or temporary, depending on the atoms involved. For example, the oxygen molecule (O:O) has a symmetric distribution of electrons between its two oxygen atoms, so each of its two atoms is uncharged. In contrast, there is a nonuniform distribution of charge in water (H:O:H), in which the bond electrons are unevenly shared (Figure 5-3). They are held more strongly by the oxygen atom, which thus carries a considerable negative charge, whereas the two hydrogen atoms together have an equal amount of positive charge. The center of the positive charge is on one side of the center of the negative charge. A combination of separated positive and negative charges is

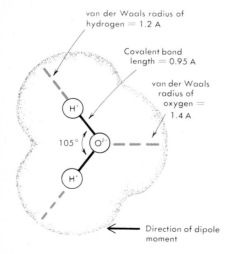

van der Waals radius of hydrogen = 1.2 A

Covalent bond length = 0.95 A

van der Waals radius of oxygen = 1.4 A

105°

Direction of dipole moment

Figure 5-3
The structure of a water molecule.

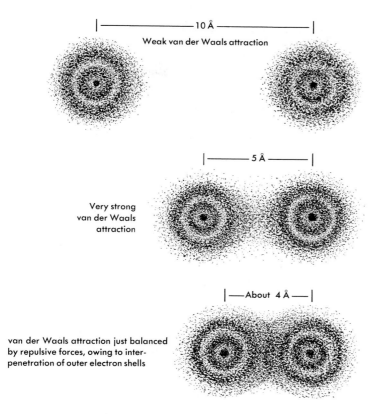

|———— 10 Å ————|
Weak van der Waals attraction

|———— 5 Å ————|

Very strong
van der Waals
attraction

|—About 4 Å —|

van der Waals attraction just balanced
by repulsive forces, owing to inter-
penetration of outer electron shells

Figure 5-4
Variation of van der Waals forces with distance. The atoms shown in this diagram are atoms of the inert rare gas argon. (Redrawn from *General Chemistry*, 2nd ed., by Linus Pauling. W. H. Freeman and Company, copyright © 1953, p. 322, with permission.)

called an electric **dipole moment.** Unequal electron sharing reflects dissimilar affinities of the bonding atoms for electrons. Atoms that have a tendency to gain electrons are called **electronegative** atoms. **Electropositive** atoms have a tendency to give up electrons.

Molecules (such as H_2O) that have a dipole moment are called **polar molecules. Nonpolar molecules** are those with no effective dipole moments. In methane (CH_4), for example, the carbon and hydrogen atoms have similar affinities for their shared electron pairs, so neither the carbon nor the hydrogen atom is noticeably charged.

The distribution of charge in a molecule can also be affected by the presence of nearby molecules, particularly if the affected molecule is polar. The effect may cause a nonpolar molecule to acquire a slightly polar character. If the second molecule is not polar, its presence will still alter the nonpolar molecule, establishing a fluctuating charge distribution. Such induced effects, however, give rise to a much smaller separation of charge than is found in polar molecules, resulting in smaller interaction energies and correspondingly weaker chemical bonds.

Van der Waals Forces[4]

Van der Waals bonding arises from a nonspecific attractive force originating when two atoms come close to each other. It is based not on the existence of permanent charge separations, but rather on the induced fluctuating charges caused by the nearness of molecules. It therefore operates between all types of molecules, nonpolar as well as polar. It depends heavily on the distance between the interacting groups, since the bond energy is inversely proportional to the sixth power of distance (Figure 5-4).

Table 5-2 Van der Waals Radii of the Atoms in Biological Molecules

Atom	van der Waals radius (Å)
H	1.2
N	1.5
O	1.4
P	1.9
S	1.85
CH_3 group	2.0
Half thickness of aromatic molecule	1.7

There also exists a more powerful van der Waals repulsive force, which comes into play at even shorter distances. This repulsion is caused by the overlapping of the outer electron shells of the atoms involved. The van der Waals attractive and repulsive forces balance at a certain distance specific for each type of atom. This distance is the so-called **van der Waals radius** (Table 5-2 and Figure 5-5). The van der Waals bonding energy between two atoms separated by the sum of their van der Waals radii increases with the size of the respective atoms. For two average atoms, it is only about 1 kcal/mol, which is just slightly more than the average thermal energy of molecules at room temperature (0.6 kcal/mol).

This means that van der Waals forces are an effective binding force at physiological temperatures only when several atoms in a given molecule are bound to several atoms in another molecule. Then the energy of interaction is much greater than the dissociating tendency resulting from random thermal movements. For several atoms to interact effectively, the molecular fit must be precise, since the distance separating any two interacting atoms must not be much greater than the sum of their van der Waals radii (Figure 5-6). The strength of interaction rapidly approaches zero when this distance is only slightly exceeded. Thus, the strongest type of van der Waals contact arises when a molecule contains a cavity exactly complementary in shape to a protruding group of another molecule. This is the type of situation thought to exist between an antigen and its specific antibody (Figure 5-7). In this instance, the binding energies sometimes can be as large as 20 to 30 kcal/mol, so that antigen-antibody complexes seldom fall apart. The bonding pattern of polar molecules is rarely dominated by van der Waals interactions, since such molecules can acquire a lower energy state (lose more free energy) by forming other types of bonds.

Hydrogen Bonds[5, 6]

A hydrogen bond is formed between a covalently bound donor hydrogen atom with some positive charge and a negatively charged, covalently bound acceptor atom (Figure 5-8). For example, the hydrogen atoms of the imino (NH) group are attracted by the negatively charged keto (CO) oxygen atoms. Sometimes, the hydrogen-bonded atoms belong to groups with a unit of charge (e.g., NH_3^+ or COO^-). In other cases, both the donor hydrogen atoms and the negative acceptor atoms have less than a unit of charge.

The biologically most important hydrogen bonds involve hydrogen atoms covalently bound to oxygen atoms (O–H) or nitrogen atoms (N–H). Likewise, the negative acceptor atoms are usually nitrogen or oxygen. Table 5-3 lists some of the most important hydrogen bonds. In the absence of surrounding water molecules, bond energies range between 3 and 7 kcal/mol, the stronger bonds involving the greater charge differences between donor and acceptor atoms. Hydrogen bonds are thus weaker than covalent bonds, yet considerably stronger than van der Waals bonds. A hydrogen bond, therefore, will hold two atoms closer together than the sum of their van der Waals radii, but not so close together as a covalent bond would hold them.

Hydrogen bonds, unlike van der Waals bonds, are highly directional. In the strongest hydrogen bonds, the hydrogen atom points directly at the acceptor atom (Figure 5-9). If it points more than 30° away, the bond energy is much less. Hydrogen bonds are also much

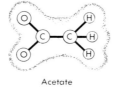

Acetate

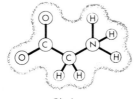

Glycine

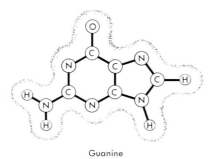

Guanine

Figure 5-5
Drawings of several molecules with the van der Waals radii of the atoms shown in color.

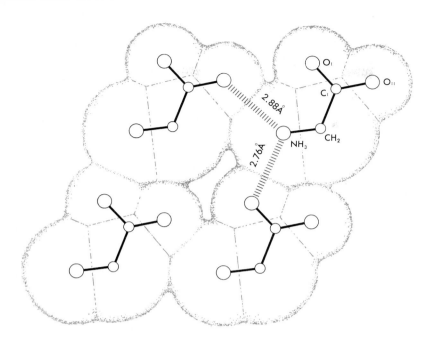

Figure 5-6
The arrangement of molecules in a layer of a crystal formed by the amino acid glycine. The packing of the molecules is determined by the van der Waals radii of the groups, except for the N–H⁅⁆O contacts, which are shortened by the formation of hydrogen bonds. [After L. Pauling, *The Nature of the Chemical Bond,* 3rd ed. (Ithaca, N.Y.: Cornell University Press, 1960), p. 262, with permission.]

more specific than van der Waals bonds, since they demand the existence of molecules with complementary donor hydrogen and acceptor groups.

Some Ionic Bonds Are Hydrogen Bonds

Many organic molecules possess ionic groups that contain one or more units of net positive or negative charge. The negatively charged mononucleotides, for example, contain phosphate groups (PO_3^{3-}) with three units of negative charge, whereas each amino acid (except proline) has a negative carboxyl group (COO^-) and a positive amino group (NH_3^+), both of which carry a unit of charge. These charged groups are usually neutralized by nearby, oppositely charged groups. The electrostatic forces acting between the oppositely charged groups are called ionic bonds. Their average bond energy in an aqueous solution is about 5 kcal/mol.

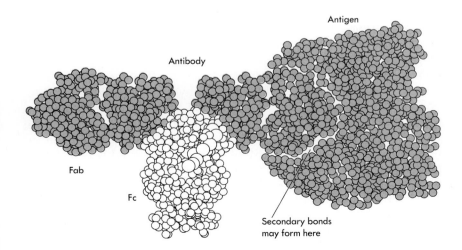

Figure 5-7
Antibody-antigen interaction. Complementary structure allows van der Waals contact between antibody and antigen. Note that the antigen may totally enclose the Fab portion (colored) of the antibody. The other, unoccupied Fab arm may interact with another antigen of similar type.

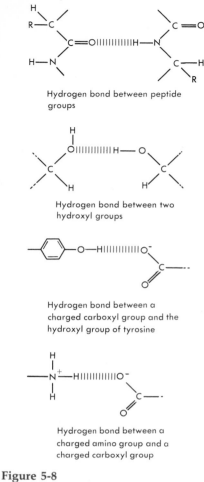

Hydrogen bond between peptide groups

Hydrogen bond between two hydroxyl groups

Hydrogen bond between a charged carboxyl group and the hydroxyl group of tyrosine

Hydrogen bond between a charged amino group and a charged carboxyl group

Figure 5-8
Example of hydrogen bonds in biological molecules.

Table 5-3 Approximate Bond Lengths of Biologically Important Hydrogen Bonds

Bond	Approximate H bond length. (Å)
O–H⫶⫶⫶⫶O	2.70 ± 0.10
O–H⫶⫶⫶⫶O$^-$	2.63 ± 0.10
O–H⫶⫶⫶⫶N	2.88 ± 0.13
N–H⫶⫶⫶⫶O	3.04 ± 0.13
N$^\pm$H⫶⫶⫶⫶O	2.93 ± 0.10
N–H⫶⫶⫶⫶N	3.10 ± 0.13

In many cases, either an inorganic cation like Na$^+$, K$^+$, or Mg^{2+} or an inorganic anion like Cl$^-$ or SO$_4^{2-}$ neutralizes the charge of ionized organic molecules. When this happens in aqueous solution, the neutralizing cations and anions do not carry fixed positions because inorganic ions are usually surrounded by shells of water molecules and so do not directly bind to oppositely charged groups. Thus, it is now believed that in water solutions, electrostatic bonds to surrounding inorganic cations or anions are usually not of primary importance in determining the molecular shapes of organic molecules.

On the other hand, highly directional bonds result if the oppositely charged groups can form hydrogen bonds to each other. For example, COO$^-$ and NH$_3^+$ groups are often held together by hydrogen bonds. Since these bonds are stronger than those that involve groups with less than a unit of charge, they are correspondingly shorter. A strong hydrogen bond can also form between a group with a unit charge and a group having less than a unit charge. For example, a hydrogen atom belonging to an amino group (NH$_2$) bonds strongly to an oxygen atom of a carboxyl group (COO$^-$).

Weak Interactions Demand Complementary Molecular Surfaces[7]

Weak binding forces are effective only when the interacting surfaces are close. This proximity is possible only when the molecular surfaces have **complementary structures,** so that a protruding group (or positive charge) on one surface is matched by a cavity (or negative charge) on another. That is, the interacting molecules must have a lock-and-key relationship. In cells, this requirement often means that some molecules hardly ever bond to other molecules of the same kind, because such molecules do not have the properties of symmetry necessary for self-interaction. For example, some polar molecules contain donor hydrogen atoms and no suitable acceptor atoms, whereas other molecules can accept hydrogen bonds but have no hydrogen atoms to donate. On the other hand, there are many molecules with the necessary symmetry to permit strong self-interaction in cells, water being the most important example.

Water Molecules Form Hydrogen Bonds

Under physiological conditions, water molecules rarely ionize to form H$^+$ and OH$^-$ ions. Instead, they exist as polar H–O–H molecules with both the hydrogen and oxygen atoms forming strong hydrogen bonds. In each water molecule, the oxygen atom can bind to two external hydrogen atoms, whereas each hydrogen atom can bind to one adjacent oxygen atom. These bonds are directed tetrahedrally (Figure 5-10), so in its solid and liquid forms, each water molecule tends to have four nearest neighbors, one in each of the four directions of a tetrahedron. In ice, the bonds to these neighbors are very rigid and the arrangement of molecules fixed. Above the melting temperature (0°C), the energy of thermal motion is sufficient to break the hydrogen bonds and to allow the water molecules to change their nearest neighbors continually. Even in the liquid form, however, at any given instant most water molecules are bound by four strong hydrogen bonds.

Weak Bonds Between Molecules in Aqueous Solutions

The average energy of a secondary bond, though small compared to that of a covalent bond, is nonetheless strong enough compared to heat energy to ensure that most molecules in aqueous solution will form secondary bonds to other molecules. The proportion of bonded to nonbonded arrangements is given by Equation 5-4, corrected to take into account the high concentration of molecules in a liquid. It tells us that interaction energies as low as 2 to 3 kcal/mol are sufficient at physiological temperatures to force most molecules to form the maximum number of strong secondary bonds.

The specific structure of a solution at a given instant is markedly influenced by which solute molecules are present, not only because molecules have specific shapes, but also because molecules differ in which types of secondary bonds they can form. Thus, a molecule will tend to move until it is next to a molecule with which it can form the strongest possible bond.

Solutions, of course, are not static. Because of the disruptive influence of heat, the specific configuration of a solution is constantly changing from one arrangement to another of approximately the same energy content. Equally important in biological systems is the fact that metabolism is continually transforming one molecule into another and so automatically changing the nature of the secondary bonds that can be formed. The solution structure of cells is thus constantly disrupted not only by heat motion, but also by the metabolic transformations of the cell's solute molecules.

Organic Molecules That Tend to Form Hydrogen Bonds Are Water Soluble

The energy of hydrogen bonds per atomic group is much greater than that of van der Waals contacts; thus, molecules will form hydrogen bonds in preference to van der Waals contacts. For example, if we try to mix water with a compound that cannot form hydrogen bonds, such as benzene, the water and benzene molecules rapidly separate from each other, the water molecules forming hydrogen bonds among themselves while the benzene molecules attach to one another by van der Waals bonds. It is therefore impossible to insert a non-hydrogen-bonding organic molecule into water.

On the other hand, polar molecules such as glucose and pyruvate, which contain a large number of groups that form excellent hydrogen bonds (e.g., =O or OH), are soluble in water (i.e., they are hydrophilic as opposed to hydrophobic). While the insertion of such groups into a water lattice breaks water-water hydrogen bonds, it results simultaneously in the formation of hydrogen bonds between the polar organic molecule and water. These alternative arrangements, however, are not usually as energetically satisfactory as the water-water arrangements, so that even the most polar molecules ordinarily have only limited solubility.

Thus, almost all the molecules that cells acquire, either through food intake or through biosynthesis, are somewhat insoluble in water. These molecules, by their thermal movements, randomly collide with other molecules until they find complementary molecular surfaces on which to attach and thereby release water molecules for water-water interactions.

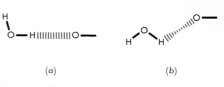

(a)   (b)

Figure 5-9
Directional properties of hydrogen bonds. (a) The vector along the covalent O–H bond points directly at the acceptor oxygen, thereby forming a strong bond. (b) The vector points away from the oxygen atom, resulting in a much weaker bond.

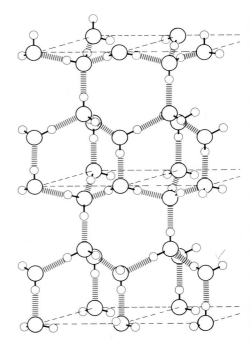

Figure 5-10
Diagram of a lattice formed by water molecules. The energy gained by forming specific hydrogen bonds between water molecules favors the arrangement of the molecules in adjacent tetrahedrons. Oxygen atoms are indicated by large circles, hydrogen atoms by small circles. Although the rigidity of the arrangement depends on the temperature of the molecules, the pictured structure is nevertheless predominant in water as well as in ice. [After L. Pauling, *The Nature of the Chemical Bond*, 3rd ed. (Ithaca, N.Y.: Cornell University Press, 1960), p. 465, with permission.]

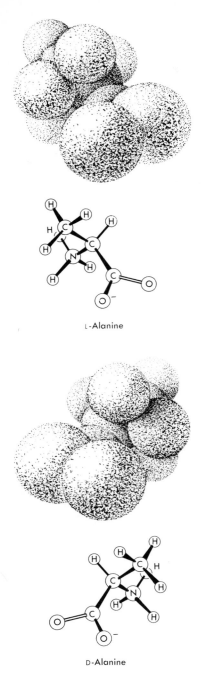

Figure 5-11
The two stereoisomers of the amino acid alanine. (Redrawn from *General Chemistry*, 2nd ed., by Linus Pauling. W. H. Freeman and Company, copyright © 1953, p. 598, with permission.)

The Uniqueness of Molecular Shapes and the Concept of Selective Stickiness

Even though most cellular molecules are built up from only a small number of chemical groups, such as OH, NH_2, and CH_3, there is great specificity as to which molecules tend to lie next to each other. This is because each molecule has unique bonding properties. One very clear demonstration comes from the specificity of stereoisomers. For example, proteins are always constructed from L-amino acids, never from their mirror images, the D-amino acids (Figure 5-11). Although the D- and L-amino acids have identical covalent bonds, their binding properties to asymmetric molecules are often very different. Thus, most enzymes are specific for L-amino acids. If an L-amino acid is able to attach to a specific enzyme, the D-amino acid is unable to bind.

Most molecules in cells can make good "weak" bonds with only a small number of other molecules, partly because all molecules in biological systems exist in an aqueous environment. The formation of a bond in a cell depends not only on whether two molecules bind well to each other, but also on whether the bond will permit their water solvent to form the maximum number of hydrogen bonds.

Hydrophobic "Bonds" Stabilize Macromolecules[8, 9]

The strong tendency of water to exclude nonpolar groups is frequently referred to as **hydrophobic bonding.** Some chemists like to call all the bonds between nonpolar groups *in a water solution* hydrophobic bonds (Figure 5-12). In a sense this term is a misnomer, for the phenomenon that it seeks to emphasize is the absence, not the presence, of bonds. (The bonds that tend to form between the nonpolar groups are due to van der Waals attractive forces.) On the other hand, the term *hydrophobic bond* is often useful, since it emphasizes the fact that nonpolar groups will try to arrange themselves so that they are not in contact with water molecules. Hydrophobic bonds are important both in the stabilization of proteins and complexes of proteins with other molecules and in the partitioning of proteins into membranes. They may account for as much as one-half the total free energy of protein folding.

Consider, for example, the different amounts of energy generated when the amino acids alanine and glycine are bound, in water, to a third molecule that has a surface complementary to alanine. A methyl group is present in alanine but not in glycine. When alanine is bound to the third molecule, the van der Waals contacts around the methyl group yield 1 kcal/mol of energy, which is not released when glycine is bound instead. From Equation 5-4, this small energy difference alone would give only a factor of 6 between the binding of alanine and glycine. However, this calculation does not take into consideration the fact that water is trying to exclude alanine much more than glycine. The presence of alanine's CH_3 group upsets the water lattice much more seriously than does the hydrogen atom side group of glycine. At present, it is still difficult to predict how large a correction factor must be introduced for this disruption of the water lattice by the hydrophobic side groups. A current guess is that the water tends to exclude alanine, thrusting it toward a third molecule, with a hydrophobic force 2 to 3 kcal/mol larger than the forces excluding glycine.

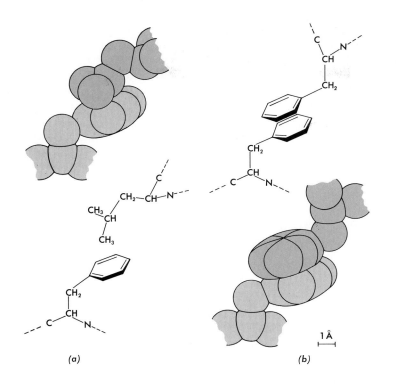

Figure 5-12
Examples of van der Waals (hydrophobic) bonds between the nonpolar side groups of amino acids. The hydrogens are not indicated individually. For the sake of clarity, the van der Waals radii are reduced by 20 percent. The structural formulas adjacent to each space-filling drawing indicate the arrangement of the atoms. (a) Phenylalanine-leucine bond. (b) Phenylalanine-phenylalanine bond. [After H. A. Scheraga, in *The Proteins*, 2nd ed., vol. 1, ed. H. Neurath (New York: Academic Press, 1963), p. 527, with permission.]

(a)   (b)

We thus arrive at the important conclusion that the energy difference between the binding of even the most similar molecules to a third molecule (when the difference between the similar molecules involves a nonpolar group) is at least 2 to 3 kcal/mol greater in the aqueous interior of cells than under nonaqueous conditions. Frequently, the energy difference is 3 to 4 kcal/mol, since the molecules involved often contain polar groups that can form hydrogen bonds.

The Advantage of ΔG Between 2 and 5 kcal/mol

We have seen that the energy of just one secondary bond (2 to 5 kcal/mol) is often sufficient to ensure that a molecule preferentially binds to a selected group of molecules. Moreover, these energy differences are not so large that rigid lattice arrangements develop within a cell; that is, the interior of a cell never crystallizes, as it would if the energy of secondary bonds were several times greater. Larger energy differences would mean that the secondary bonds seldom break, resulting in low diffusion rates incompatible with cellular existence.

Weak Bonds Attach Enzymes to Substrates

Secondary forces are necessarily the basis by which enzymes and their substrates initially combine with each other. Enzymes do not indiscriminately bind all molecules, having noticeable affinity only for their own substrates.

Since enzymes catalyze both directions of a chemical reaction, they must have specific affinities for both sets of reacting molecules. In some cases, it is possible to measure an equilibrium constant for the binding of an enzyme to one of its substrates (Equation 5-4), which consequently enables us to calculate the ΔG upon binding. This calculation in turn hints at which types of bonds may be involved. For ΔG values between 5 and 10 kcal/mol, several strong secondary bonds are the basis of specific enzyme-substrate interactions. Also worth noting is that the ΔG of binding is never exceptionally high; thus, enzyme-substrate complexes can be both made and broken apart rapidly as a result of random thermal movement. This explains why enzymes can function quickly, sometimes as often as 10^6 times per second. If enzymes were bound to their substrates, or more importantly to their products, by more powerful bonds, they would act much more slowly.

Most Molecular Shapes Are Determined by Weak Bonds

The shapes of numerous molecules are automatically given by the distribution of covalent bonds. This inflexibility occurs when groups are attached by covalent bonds about which free rotation is impossible. Rotation is possible only when atoms are attached by single bonds. (For example, the methyl groups of ethane, H_3C-CH_3, rotate about the carbon-carbon bond.) When more than one electron pair is involved in a bond, rotation does not occur; in fact, the atoms involved must lie in the same plane. Thus, the aromatic purine and pyrimidine rings are planar molecules 3.4 Å thick. There is no uncertainty about the shape of any aromatic molecules. They are almost always flat, independent of their surrounding environment.

On the other hand, for molecules containing single bonds, the possibility of rotation around the bond suggests that a covalently bonded molecule exists in a large variety of shapes. This theoretical possibility, however, is seldom realized, because the various possible three-dimensional configurations differ in the number of weak bonds that can be formed. Generally, there is one configuration that has significantly less free energy than any of the other geometric arrangements.

It is no simple matter to guess the correct shape of a large molecule from its covalent-bond structure; although one configuration for a protein or a nucleic acid may be energetically much more suitable than any other, we cannot yet derive it from our knowledge of bond energies. The number of possible configurations of a nucleic acid or even a small protein molecule is immense. Given present techniques for building molecular models, a single person (or even a small group of people) armed with the best computers cannot rapidly calculate the sum of the energies of the weak bonds for each possible configuration. The reason for our present inability to derive protein and nucleic acid structures is that our knowledge about the nature of weak bonds is still very incomplete. In many cases, we are not sure about either the exact bond energies or the possible angles the bonds can form to each other.

Thus, protein and nucleic acid shapes still can be revealed only by X-ray diffraction analysis. Fortunately, these experimental structure determinations are beginning to yield some general rules that tell us which weak chemical interactions are most important in governing the molecular shapes of large molecules. In particular, these rules emphasize the vital importance of interactions of the macromolecules with water, by far the most common molecule in all cells.

Helical Arrangements of Polypeptide Backbones[10–17]

In Chapter 2 we saw that polypeptide chains have regular linear backbones in which specific groups (e.g., –CO–NH–) repeat over and over along the molecule. Often, these regular groups are arranged in helical configurations held together by secondary bonds. The helix is a natural conformation for a regular linear polymer like the polypeptide backbone, since it places each monomer group in an identical orientation within the molecule. Each monomer thereby forms the same group of secondary bonds as every other monomer. It is important to emphasize that helical symmetry does not evolve from the particular shape of the monomer, but is instead the natural consequence of a unique monomer arrangement that is significantly more stable than all other arrangements.

The most stable arrangement of a polypeptide backbone is the α helix. It is a right-handed helix, repeating every 5.4 Å along the helical axis, containing a regular pattern of hydrogen bonding between carbonyl and imino groups on the same chain (see Figure 2-23). Not only are its hydrogen bonds so oriented for maximum stability, but its atoms pack together to form a stereochemically perfect central core lacking energetically unfavorable holes (Figure 5-13). The nonintegral number of amino acids per turn (3.6) is a consequence of the precise geometry of the polypeptide chain. If, for example, four amino acids were used per turn, the hydrogen bonds would not be so neatly formed, nor would the individual backbone atoms fit together so well.

There are two energetically favorable forms of maximally extended (β) polypeptide backbones. In both forms, adjacent amino acids are related by 180° rotations, thereby projecting their respective side groups in opposite directions (see Figure 2-24). With such extended chains, there is no possibility of hydrogen bonds forming between carbonyl and imino groups relatively close to each other on the same chain. Instead, hydrogen bonding is restricted to groups belonging to different chains or to very distant parts of the same chain, which because of their twofold symmetry can only come together as sheets.

The two different forms of β sheets differ in the relative orientations of their chains (Figure 5-14). In one, the adjacent chains run in the same amino to carboxyl direction to produce a parallel β sheet. In the other, the adjacent chains run in opposite directions to yield an antiparallel β sheet. In both the parallel and antiparallel β sheets, all

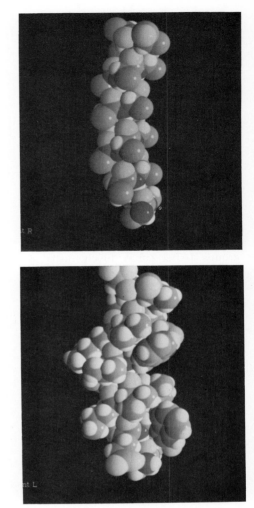

Figure 5-13
Space-filling model of an α helix constituting part of the structure of myoglobin. The model on the top shows the α helix without its amino acid side chains. (Courtesy of Richard J. Feldmann.)

Figure 5-14
Two different forms of β sheets. (a) An antiparallel β sheet, in which the adjacent chains run in opposite directions. (b) A parallel β sheet, in which the adjacent chains run in the same direction. The peptide groups (several are shaded with color) lie in the plane of the sheet, while the R groups (not shown here for clarity) extend perpendicularly to the plane of the sheet. Actual β sheets in globular proteins are usually less regular than that shown here, as many sheets are slightly twisted (see Figure 5-20).

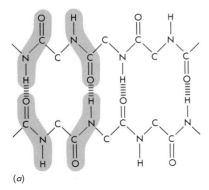

(a)

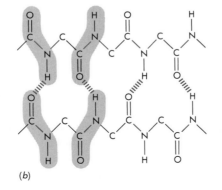

(b)

Figure 5-15
Structural representation of hen egg-white lysozyme by a ribbonlike model that emphasizes secondary structure—α helices and β sheets. The arrows on the ribbons representing the β chains point toward the carboxyl end of the polypeptide, and they make it immediately apparent that the β chains form an antiparallel β sheet. In this molecule, we see the tendency of amino acid side groups to twist the backbone α helices and β sheets into much less regular configurations. These changes in the structure of the regular backbone serve to maximize the strength of the secondary bonds formed by the side groups. [After Zubay, *Biochemistry*, p. 95.]

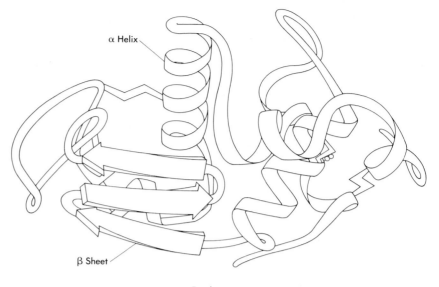

Egg lysozyme

the peptide groups lie approximately in the plane of the sheet. Therefore, the interchain bonds that link them together can be energetically very favorable and of equal strength to those that hold together α helices.

In no case, however, are polypeptide chains of proteins regular polymers containing identical monomers. Instead, they invariably have irregular side groups attached to the regular backbone. Thus, we need not necessarily expect either a helical structure or a β sheet. A three-dimensional arrangement that is energetically very satisfactory for the backbone groups often produces very unsatisfactory bonding of the side groups. The three-dimensional structures of the polypeptide chains of proteins are thus compromises between the tendency of regular backbones to form either α helices or β sheets and the tendency of the side groups to twist the backbone into much less regular configurations that maximize the strength of the secondary bonds formed by the side groups (Figure 5-15).

This argument leads to the interesting prediction that in aqueous solutions, proteins containing very large numbers of nonpolar side groups will tend to be more stable than proteins containing mostly polar groups. If we disrupt a polar molecule held together by a large number of internal hydrogen bonds, the decrease in free energy is often small, since the polar groups can then hydrogen-bond to water. On the other hand, when we disrupt molecules having many nonpolar groups, there is usually a much greater loss in free energy because the disruption necessarily inserts nonpolar groups into water.

Interactions Between Nonpolar Side Groups Stabilize Proteins[18]

Almost one-half of the amino acid residues in proteins have nonpolar side groups that interact with each other rather than lying next to either polar side groups or the polar water molecules. Polypeptide chains within the aqueous portions of cells (as opposed to within the lipid environments of membranes) will thus be energetically driven to assume configurations that effectively isolate their nonpolar side groups within internal regions into which water molecules cannot penetrate. The free energy that would be lost by failure to so internalize the many hydrophobic groups of a given protein is very large and provides by far the major force that drives disordered, newly synthesized polypeptide chains into their "native," usually highly compacted, states. In contrast, very little real energy is gained by water-soluble proteins when they make their correct hydrogen bonds, since their polar backbone atoms and polar side groups can make nearly as favorably hydrogen bonds to water in the disordered (denatured) state as they can to themselves in the native state.

The Specific Conformation of a Protein Results from Its Pattern of Hydrogen Bonds

While the energy stabilizing a protein is provided by hydrophobic interactions, the specific conformation of a protein structure is provided by hydrogen bonds. The energy of hydrophobic "bonds" has no directional component, whereas hydrogen bonds require proper distances and angles. In general, all hydrogen-bond donors and acceptors within a protein's interior have suitable mates. Presumably, failure to make a hydrogen bond in a protein interior is energetically costly, at the rate of a few kilocalories per hydrogen bond. The vitally important role of hydrogen bonds in proteins is thus to destabilize the incorrect structures rather than to stabilize the correct one.

The necessity of satisfying all the hydrogen-bond donors and acceptors on the polypeptide backbone (two per residue) provides the energetic driving force that leads to the creation of the large sections of α helices and β sheets found in most proteins. The only way that a polypeptide can traverse the nonaqueous interior of a protein, as it must, and satisfy the hydrogen-bonding necessity is through formation of regular secondary structures. Side chains do not have enough donors and acceptors to do the job. Thus, all large proteins contain significant regions of β sheets, α helices, or both. Of course, some polypeptide sections must be less regular to allow their chains to turn at the ends of α helices and β chains.

Proteins embedded in the nonaqueous interior of membranes are stabilized in the opposite way. Hydrophobic interactions partition these proteins into membranes, but do not stabilize their structures within the lipid bilayer. Hydrogen bonds, on the other hand, must assume a very important role in stabilizing the correct structures of membrane proteins (Figure 5-16).

α Helices Come Together at Specific Angles[19, 20, 21]

A number of long, thin fibrous proteins (e.g., keratins and myosin tails) are constructed by the supercoiling of α helices about each other with their amino acid sequences arranged so that the nonpolar side groups are internally located. The twisting of the chains around each other reflects the nonintegral (3.6 residues per turn) nature of the α helix, which allows the side groups to neatly pack together only when the α helices interact at an angle of 18° from parallel. If the α helices remained perfectly rigid, they could stay in contact for only a few residues. But by supercoiling in a left-handed direction, neatly

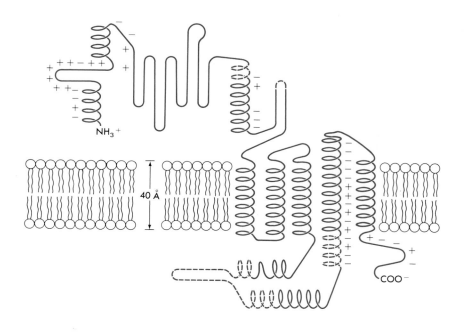

Figure 5-16
An acetylcholine receptor subunit. Each polypeptide chain of this membrane protein traverses the membrane bilayer five times. The charges shown represent the average charges on the side chains. Outside the membrane the protein is predominantly hydrophilic; inside the membrane, it is mostly hydrophobic.

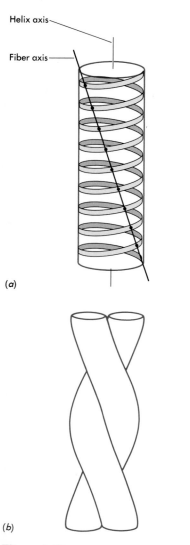

Figure 5-17
The coiled coil of α helices in α keratin. (a) One α helix and the positions within the helix of the fiber and helix axes when the two α helices are coiled about each other. (b) In the coiled coil, the side groups of each helix pack neatly together when the helices interact at an angle of 18°.

packed, highly stable, coiled coils are created (Figure 5-17). Fibrous molecules containing such coiled coils are frequently elastic, with their α helices becoming slightly uncoiled when subjected to tension.

Much shorter sections of interlocked α helices exist in a number of globular proteins. The signal that frequently leads the α helices to bend back on themselves is the presence of a proline residue, whose ringed structure prevents it from forming an intramolecular hydrogen bond along the α helix. Its occasional occurrence in the middle of a helix causes the helix to bend or kink. As a result of packing considerations, interacting α helices in globular proteins are related by one of three angles between their helix axes. Ridges formed by the side chains of one helix will pack into the grooves between side chains of another helix if the helix axes are inclined at 18°, 60°, or 90° to each other (Figure 5-18).

β Sheets Twist in a Right-Handed Fashion[22, 23, 24]

At one time, β sheets were thought to be flat. However, when they became visualized within known proteins by X-ray crystallography, they invariably appeared to have right-handed twists. The origin of this twisting lies in the slightly twisted nature of fully extended (β) polypeptide chains. They do not have the perfect twofold rotational symmetry initially postulated, but have slight intrinsic right-handed twists (Figure 5-19). To form an antiparallel β sheet, the direction of an extended β chain must suddenly reverse, and as with the α helix, proline residues are often used to effect such directional switches. The formation of parallel β sheets is more complicated, involving crossover sequences that usually assume α-helical conformations (Figure 5-20).

Twisted parallel β sheets are the dominant structural motif of many proteins. In some proteins, they are organized into saddlelike arrangements, while in other proteins they come together to form barrel-shaped structures (Figure 5-21). No protein is stable as a single-layered twisted β sheet or α helix; at least two layers (ββ or αα) are

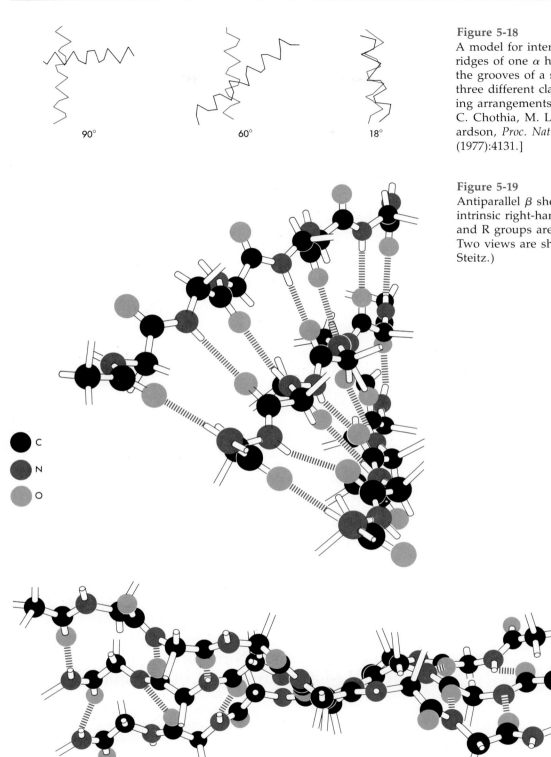

Figure 5-18
A model for interlocked α helices. The ridges of one α helix pack neatly into the grooves of a second α helix. The three different classes of possible packing arrangements are shown. [After C. Chothia, M. Levitt, and D. Richardson, *Proc. Nat. Acad. Sci.* 74 (1977):4131.]

Figure 5-19
Antiparallel β sheet showing the slight intrinsic right-handed twist. Hydrogens and R groups are not shown as balls. Two views are shown. (Courtesy of T. Steitz.)

required to isolate the hydrophobic residues that must come together in a central core. Although pairs of twisted antiparallel β sheets form the essential structure of many proteins (Figure 5-22), a twisted parallel sheet needs to be surrounded by additional sequences on both its sides. These sequences principally form α helices, which pack neatly with each other and with the two surfaces of the β sheet.

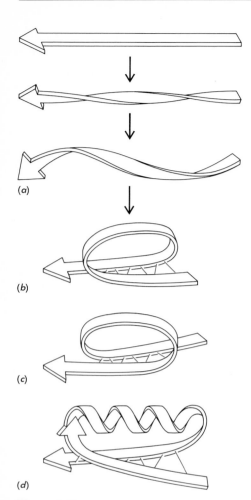

Figure 5-20

The formation of parallel β sheets.
(a) The tendency of β sheets to twist right-handedly (b) produces sheets with right-handed crossover connections, although (c) left-handed connections also exist. (d) Frequently, crossover sequences assume helical conformation. [After G. Zubay, *Biochemistry* (Reading, Mass.: Addison-Wesley, 1983), p. 95].

Supersecondary Structures That Have Functional Implications

Although an oversimplification, a protein's structure can often be described as repeating patterns of supersecondary structural motifs. For example, myohemerythrin, cytochrome c', and tobacco mosaic virus (TMV) coat protein each contain *four antiparallel α helices* (Figure 5-23), and these can be constructed from two αα units. Some proteins that are virtually all antiparallel β sheets, such as rubredoxin and immunoglobulin, consist of three (or more) ββ motifs. And all parallel β sheets are built up from βαβ units.

Some supersecondary structures have very specific functions and are therefore found in only certain kinds of proteins. For example, a supersecondary structure (helix-turn-helix) with two α helices at nearly right angles to each other was initially found in the DNA binding region of both the catabolite gene activator protein (CAP) of *E. coli* (see Figure 5-26) and the Cro repressor of phage λ (see Figure 5-30) and may well be a structural motif for most, if not all, gene regulatory proteins.

Most Proteins Are Modular, Containing Two or Three Domains[25, 26]

The subunits of soluble proteins vary in size from less than 100 amino acids for some peptide hormones to as large as 2000 amino acid residues for myosin. The smallest enzyme polypeptides have molecular weights of about 13,000 daltons (120 residues), but most are between 20,000 and 70,000 daltons for a single subunit. Protein subunits larger than about 20,000 daltons are usually formed from two or more domains (Figure 5-24). The term **domain** is used to describe a compact, folded part of the structure that appears separate from the rest, as if it would be stable in solution on its own (which is often demonstrated to be the case). There is a colinear correspondence between the structural domains and the amino acid sequence of a protein. That is, a single domain is formed from a continuous amino acid sequence and not portions of sequence that are scattered throughout the polypeptide. This is an important point when considering possible genetic mechanisms by which multidomain proteins have evolved.

A Surprisingly Small Number of Structural Motifs Are Found in Domains[27, 28]

Determination of the first half dozen protein structures showed a bewildering variety of protein folding motifs, implying the existence of an infinite number of protein structures. However, now that we know the three-dimensional structures of more than 200 proteins, it appears that a relatively small number of domain structural motifs account for most of the large variety of protein activities. Although an accurate estimate is not possible, the number of truly unique domain motifs might lie somewhere between 50 and 100 rather than in the thousands.

Specific kinds of domain motifs are often associated with particular kinds of activities. One frequently observed motif has been termed the *dinucleotide fold* because it is always found in enzymes, such as lactate dehydrogenase, that bind the coenzyme NAD$^+$ (Figure 5-25).

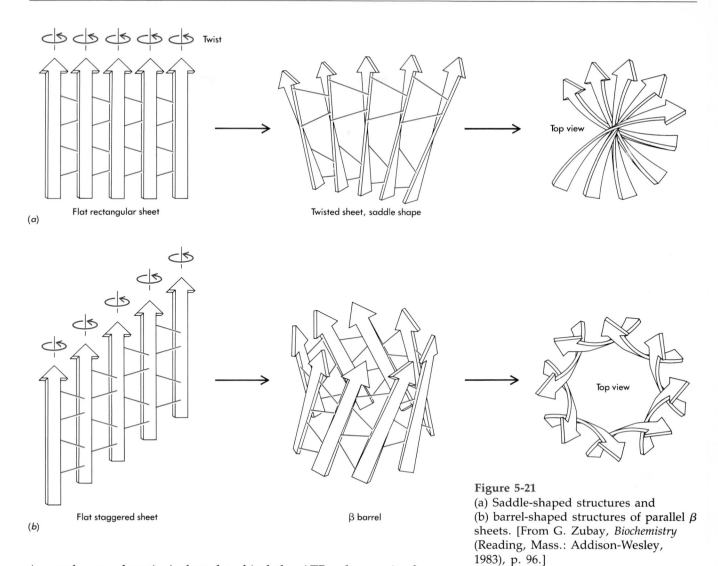

(a) Flat rectangular sheet → Twisted sheet, saddle shape → Top view

Twist

(b) Flat staggered sheet → β barrel → Top view

Figure 5-21
(a) Saddle-shaped structures and (b) barrel-shaped structures of parallel β sheets. [From G. Zubay, *Biochemistry* (Reading, Mass.: Addison-Wesley, 1983), p. 96.]

An analogous domain is found to bind the ATP substrate in those kinases whose structure is known, and indeed, it seems to be observed in all enzymes that use ATP as a substrate. What is common to the domains that bind ATP or NAD^+ is a central, parallel β sheet with α helices on both sides and the nucleotide binding site on the carboxyl end of the β strands. What varies is number and detailed arrangement of the α helices and, to a far lesser extent, the order of the β strands. These observed similarities and differences raise interesting questions about the evolution of domains like the dinucleotide fold. Did all ATP and NAD^+ binding domains evolve from a common ancestral precursor, or is this dinucleotide fold the only kind of protein fold that will function correctly?

Different Protein Functions Arise from Various Domain Combinations

The various functional properties of proteins appear to arise from their modular construction in much the same way as computers with different specifications can be assembled from the appropriate modular components. Numerous examples can be given. Dehydrogenases consist of two domains. One is a common dinucleotide binding do-

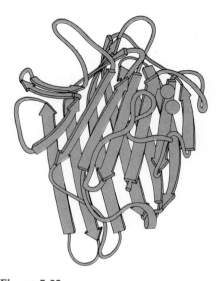

Figure 5-22
Concanavalin A, a protein formed from twisted antiparallel β sheets. (Courtesy of Jane Richardson.)

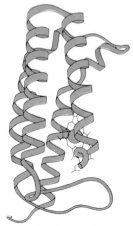

Cytochrome c'

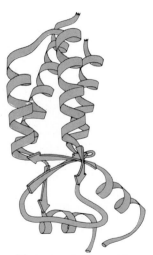

Tobacco mosaic virus protein

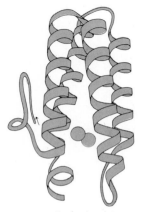

Myohemerythrin

Figure 5-23
Proteins with the similar structural
motif of four antiparallel α helices.
[After G. Zubay, *Biochemistry* (Reading,
Mass.: Addison-Wesley, 1983), p. 101.]

main that binds the coenzyme NAD$^+$. The other domain binds substrate and has the catalytic site; its structure varies among dehydrogenases.

The gene regulatory repressor and activator proteins provide another example of modular construction. The *lac* repressor and the catabolite gene activator protein (CAP) of *E. coli* both have subunits containing multiple domains. The crystal structure of CAP shows two domains: A larger domain is seen to bind a molecule of cyclic AMP in its interior, while the smaller domain appears to be involved in the recognition and binding of CAP to specific DNA sequences (Figure 5-26). There are very significant amino acid sequence similarities between the cAMP-binding domain of CAP and the regulatory subunit of cAMP-dependent protein kinase, suggesting that the cAMP-binding domain of both proteins evolved from the same precursor. In CAP, this cAMP-binding domain is attached to the DNA-binding domain, so that changes in cAMP levels control transcription levels. In the kinase, the cAMP-binding domain regulates the activity of the first enzyme in a cascade of enzymes that result in the breakdown of stored glycogen. The DNA-binding domain of CAP, on the other hand, shows structural and amino acid sequence similarities to many other gene regulatory proteins, suggesting a common ancestral precursor to the DNA binding domains of regulatory proteins.

The *lac* repressor has a similar modular construction. It has a small domain whose sequence is related to that portion of CAP that binds to specific DNA sequences, and it has a large domain that displays amino acid sequence similarity to the galactose binding protein. Once again, domains with different functions (galactose binding and specific-DNA binding) are combined into a protein with another function: control of gene expression.

We Can Now Determine Most Protein Structures by Crystallography[29, 30]

It took several research workers 7 to 15 years to solve the three-dimensional structures of the first proteins. Thanks to advances in crystallization techniques, computers and programming techniques, and methods of X-ray diffraction data measurement, it is now possible for one or two people to solve an average protein structure in one or two years. A great deal less time is needed to determine the structure of a modification of a known protein, such as a mutant protein. Therefore, ascertaining precisely how small molecule substrates or inhibitors bind to proteins of known structure can usually be accomplished in a few weeks if the small molecule diffuses into the crystals and binds without disturbing the crystal.

As a result of gene-cloning technology and improved methods of crystallization, almost all globular proteins, however rare normally, can now be studied by crystallographic techniques. For example, the crystal structure of the antiviral agent interferon is currently being determined, when only a short time ago only the minutest quantities were available for clinical and biochemical tests. The same is true for an entire host of proteins and nucleic acids of particular interest to molecular biologists—proteins involved in replication, recombination, transcription, and gene regulation, to name a few.

Even membrane proteins are now amenable to three-dimensional structural analysis. Using appropriate nonionic detergents like octal-glucoside to solubilize membrane proteins, researchers have already

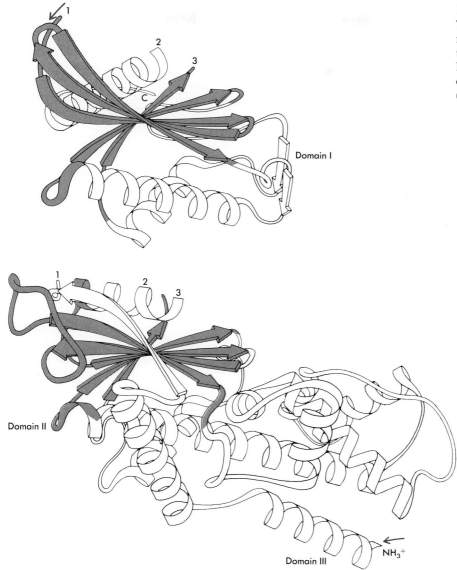

Figure 5-24
The three domains of hexokinase. Domains I and II are related by a 180° rotation, which aligns points 1, 2, and 3. Domains II and III are connected by a continuous piece of α-helical backbone. (Courtesy of Jane Richardson.)

crystallized the pore-forming protein (porin); the proton pump bacteriorhodopsin; and a major component in photosynthesis, the photosynthetic reaction center. The last, a complex of four different proteins, has been solved at high resolution, giving us our first detailed view of the molecular architecture of an integral transmembrane protein complex (see color section).

In theory, there is no limit to the size of the macromolecular assembly whose detailed molecular structure can be determined by diffraction methods. Already, the structures of several plant viruses with molecular weights of millions of daltons have been solved, and even the much more complex ribosomes have been crystallized.

Protein Structure Is the Same in Crystals and Solution

Is the structure of a protein the same in the crystal as it is in solution under similar conditions, or does the process of crystallization significantly alter the structure? Macromolecule crystals, unlike crystals of

Figure 5-25
Schematic drawing of an idealized dinucleotide binding domain. Each β chain and α helix is assigned a letter.

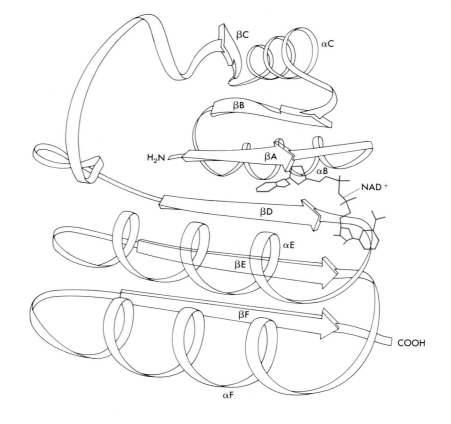

small molecules, are very soft and actually consist of 40 to 60 percent solvent. The aqueous channels in crystals are so large that in many cases substrate molecules can diffuse into crystals and bind to enzyme active sites. The enzymatic activities of ribonuclease and carboxypeptidase A crystals have been measured and are only a few times smaller than their activities in solution. Thus, these enzymes must have very similar active-site structures in both the solution and crystalline states. Further, many protein structures have been solved in several, very different crystal forms, sometimes using crystals ob-

Figure 5-26
CAP complex with cAMP interacting with a bent DNA. The DNA binding domain is colored and the cAMP binding domain is gray. Each vertex on the polypeptide chain represents the α carbon of an amino acid residue. (Courtesy of T. Steitz.)

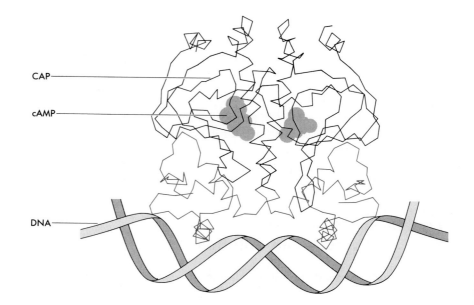

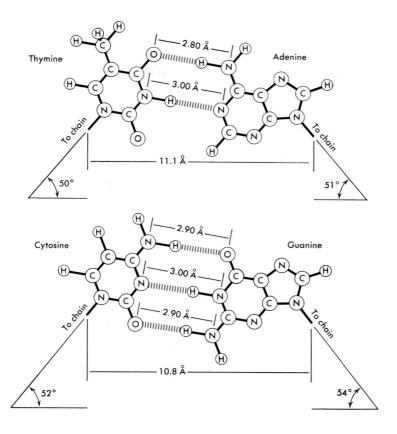

Figure 5-27
The hydrogen-bonded base pairs of DNA. Adenine is always attached to thymine by two hydrogen bonds, whereas guanine always bonds to cytosine by three hydrogen bonds. The obligatory pairing of the smaller pyrimidine with the larger purine allows the two sugar-phosphate backbones to have identical helical configurations. All the hydrogen bonds in both base pairs are strong, since each hydrogen atom points directly at its acceptor atom (nitrogen or oxygen).

tained from solvents as different as ammonium sulfate and alcohol. The molecular structures are found to be identical within experimental error.

DNA Can Form a Regular Helix

At first glance, DNA looks even more unlikely to form a regular helix than does an irregular polypeptide chain. Not only does DNA have an irregular sequence of side groups, but in addition, all its side groups are hydrophobic. Both the purines (adenine and guanine) and the pyrimidines (thymine and cytosine), even though they contain polar C=O and NH_2 groups, are quite insoluble in water because their flat sides are completely hydrophobic.

Nonetheless, DNA molecules usually have regular helical configurations. This is because most DNA molecules contain two polynucleotide strands that have complementary structures (see Chapter 9 for more details). Both internal and external secondary bonds stabilize the structure. The two strands are held together by hydrogen bonds between pairs of complementary purines and pyrimidines (Figure 5-27). Adenine is always hydrogen-bonded to thymine, whereas guanine is hydrogen-bonded to cytosine. In addition, virtually all the surface atoms in the sugar and phosphate groups form bonds to water molecules.

The purine-pyrimidine base pairs are found in the center of the DNA molecule. This arrangement allows their flat surfaces to stack on top of each other, thereby limiting their contact with water. This stacking arrangement would be much less satisfactory if only one polynucleotide chain were present. A single chain could not have a regular backbone because the pyrimidines are smaller than the purines, and so the angle of helical rotation would have to vary with the sequence of bases. The presence of complementary base pairs in double-helical DNA makes a regular structure possible, since each base pair is of the same size.

DNA Molecules Are Stable at Physiological Temperatures

The double-helical DNA molecule is very stable at physiological temperatures for two reasons. First, disruption of the double helix would bring the hydrophobic purines and pyrimidines into greater contact with water, which is very unfavorable. Second, individual DNA molecules have a *very large number of weak bonds,* arranged so that most of them cannot break without the simultaneous breaking of many others. Even though thermal motion is constantly breaking apart the terminal purine-pyrimidine pairs at the ends of each molecule, the two chains do not usually fall apart because the hydrogen bonds in the middle are still intact (Figure 5-28). Once a break occurs, the most likely next event is the reforming of the same hydrogen bonds to restore the original molecular configuration. Sometimes, of course, the first breakage is followed by a second one, and so forth. Such multiple breaks, however, are quite rare, so that double helices held together by more than ten nucleotide pairs are very stable at room temperature. The same principle also governs the stability of most protein molecules. Stable protein shapes are never due to the presence of just one or two weak bonds, but always represent the cooperative result of a number of weak bonds.

Ordered collections of hydrogen bonds become less and less stable as their temperature is raised above physiological temperatures. At physiologically abnormally high temperatures, the simultaneous breakage of several weak bonds is more frequent. After a significant number have broken, a molecule usually loses its original form (the process of denaturation) and assumes an inactive (denatured) configuration.

Weak Bonds Correctly Position Proteins Along DNA Molecules[31]

DNA never exists completely naked within cells, but at all times has a significant fraction of its surface covered with a large variety of protein molecules. The bonds that hold these proteins onto DNA are the same collection of weak bonds that give to DNA and the various proteins their own specific three-dimensional configurations. The most common DNA-binding proteins play a structural role by packaging and compacting the huge amount of DNA that must be fitted into the cell. For example, the nucleus of a human cell is only 10 μm (10^{-5} meter) across but contains roughly 3 meters of double-stranded DNA.

Best known of the chromosomal proteins are the **histones,** a group of small, positively charged proteins found in association with the DNA of all eucaryotic cells. The histones bind to DNA largely through ionic bonds between the negatively charged phosphate

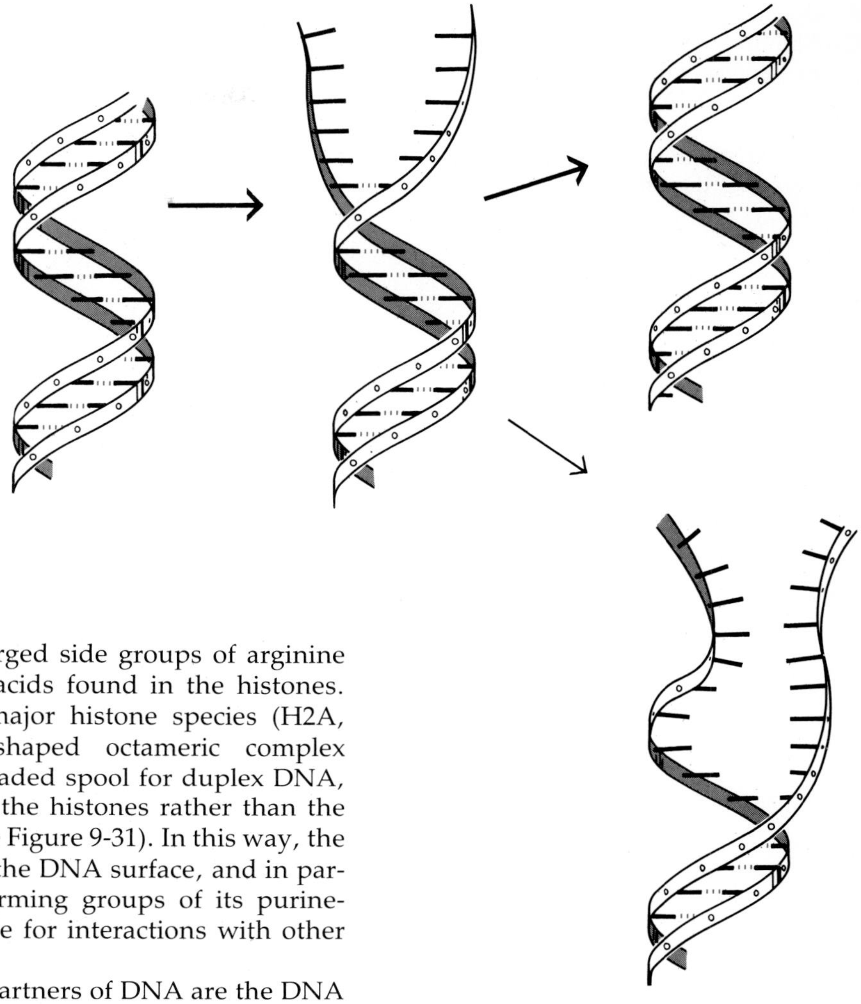

Figure 5-28
The breaking of terminal hydrogen bonds in DNA by random thermal motion. Because the internal hydrogen bonds continue to hold the two chains together, the immediate reforming of the broken bonds is highly probable. Also shown is the very rare alternative: the breaking of additional hydrogen bonds and the consequent disentangling of the chains.

groups of DNA and the positively charged side groups of arginine and lysine, the most common amino acids found in the histones. More importantly, because the four major histone species (H2A, H2B, H3, and H4) form a disk-shaped octameric complex ($H2A_2H2B_2H3_2H4_2$) that acts like a threaded spool for duplex DNA, the DNA is able to wrap itself around the histones rather than the histones wrapping around the DNA (see Figure 9-31). In this way, the DNA can be compacted while much of the DNA surface, and in particular many of the hydrogen-bond-forming groups of its purine-pyrimidine base pairs, remain accessible for interactions with other DNA binding proteins.

Prominent among these nonhistone partners of DNA are the DNA polymerases needed for its replication, the RNA polymerases used in its transcription, and a large variety of regulatory proteins that help ensure that individual genes function (are transcribed) when needed. All these proteins recognize the specific DNA segments on which they are to work through bonding to specific groups of base pairs. In contrast to the histones, most regulatory proteins that bind to DNA have more negatively than positively charged amino acids on their surfaces. Such electrostatic charge interaction is exemplified by the interaction between DNA and its enzyme CAP (Figure 5-29). Like CAP, the vast majority of cytoplasmic proteins have net negative charges; they may have evolved to prevent unwanted nonspecific aggregation between positively and negatively charged molecules.

Sequence Recognition Results from Protein-DNA Complementarity[32, 33]

Sequence-specific protein-DNA interactions arise in large part from a complementarity in shape that permits weak interactions between protein side chains and the amino and keto groups of the bases exposed in the major groove (see Chapter 9). For a regulatory protein to make a specific interaction, the protein must move closer to the DNA than in its unspecific complex, pushing a polypeptide α helix into the

Figure 5-29
CAP binding to DNA, showing the electrostatic charge interaction between enzyme and DNA. The negatively charged DNA interacts with the positively charged amino acid residues (colored area) of the protein. (Courtesy of T. Steitz.)

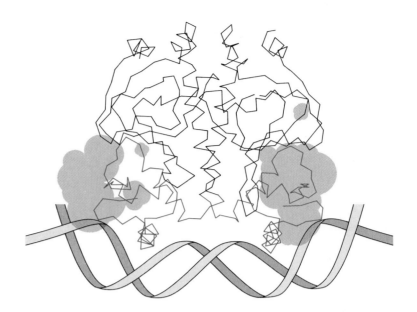

major groove of the DNA. Side chains emanating from this helix can make hydrogen bonds with the edges of the bases. If the two macromolecules are complementary in hydrogen-bond donors and acceptors, a stable, specific complex is made; if not, the protein diffuses away.

There are at least two ways in which a protein might interact with duplex DNA. The structures of several such proteins (λ phage repressor, Cro repressor, and CAP) complexed with DNA suggest that protruding α helices fit into the DNA major groove. (In Figure 5-30, we see protrusions on the Cro protein fitting into grooves in the DNA.) In contrast, the *E. coli* DNA polymerase I has a handlike structure with a deep cleft into which duplex DNA can bind (Figure 5-31). In this case, it appears that the protein may partially or totally surround the DNA. Likewise, some repressors may have flexible "arms" that may be able to embrace the DNA.

Figure 5-30
Interaction of DNA with Cro repressor, a dimer. (Courtesy of Dr. Jane M. Burridge.)

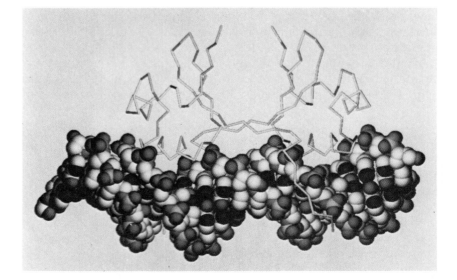

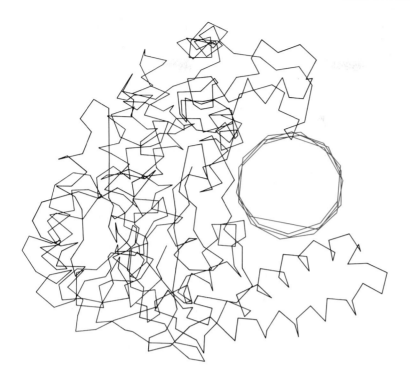

Figure 5-31
The α-carbon backbone of a fragment of
DNA polymerase I (black) complexed
with DNA (color). The protein can
nearly surround the DNA, which is
shown looking down its axis. (Courtesy
of T. Steitz.)

Sliding Diffusion of Proteins Along DNA Molecules[34]

Many DNA binding proteins can bind nonspecifically to the DNA
backbone as well as specifically to the base pairs. Mediating their
nonspecific attachments are patches of positively charged amino
acids located at sites very close to those that bind to the base pairs. In
these nonspecific bindings, the proteins usually lie some 10 Å to 15 Å
farther from the DNA than in specific binding, ruling out even the
occasional hydrogen-bond interaction while allowing significant electrostatic attraction. Though such nonspecific binding may be 10^6 to
10^7 times less tight, there is always a much larger number of nonspecific than specific attachment sites, particularly for those proteins that
recognize only a single DNA site. At a given moment, therefore,
many more copies of a given binding protein are likely to be attached
nonspecifically than are found at their biologically important destination. Now it is strongly suspected that these nonspecific unions are
not unavoidable nuisances but a device to speed up the rate at which
a given regulatory protein reaches its appropriate target. This would
be the case if such nonspecifically bound proteins are constrained by
their charge interaction to diffuse linearly along DNA and do not
randomly pop on and off DNA in their "search" for their specific
binding sites. By being so restricted to linear movements, they will
reach their targets much faster than if they had the freedom for three-
dimensional diffusion.

Not all the movements of proteins bound to DNA occur by purely
thermal events. Some DNA binding proteins involved in DNA replication (see Chapter 10) utilize the breakdown of ATP to produce cycles of reversible conformational changes that propel them in fixed
directions until they meet their targets.

Modifying Proteins with Lipid Side Groups Promotes Their Association to Membranes[35, 36]

Transmembrane proteins possessing separate domains protruding onto the inner and outer sides of lipid bilayers are known as **integral membrane proteins.** They become so sited by being synthesized on membrane-bound polyribosomes that can directly place them into appropriate membrane sections. Other "membrane proteins" are much less tightly associated and may have none of their amino acids directly inserted into the hydrophobic lipid interior. Instead, they bind to the cytoplasmic surface of the membrane. Such "nonintegral" membrane proteins frequently contain single lipid molecules covalently attached to an amino acid on their surface. Such lipid groups, being totally insoluble in water, automatically seek out other lipid groups and do so most often by directly inserting themselves into the inner side of the lipid bilayer. In some proteins (e.g., some protein kinases), lipids are attached via amide bonds to amino terminal glycine groups. In other proteins, lipids are attached by ester linkages to the hydroxyl groups of amino acids like serine and threonine.

Most Medium-Size and Almost All Large Protein Molecules Are Aggregates of Smaller Polypeptide Chains

Earlier we pointed out how the realization that macromolecules are all polymers constructed from small, regular monomers, such as the amino acids, greatly simplified the problem of solving macromolecular structure. It has recently become clear that most of the very large proteins are regular aggregates of polypeptide chains, containing up to 400 amino acids apiece. (The relationship among the polypeptide chains making up such a protein is termed its **quaternary structure.**) For example, the protein ferritin, which functions in mammals to store iron atoms, has a molecular weight of about 480,000. It contains, however, not just one long polypeptide chain of 4000 amino acids, but 20 identical smaller polypeptide chains of about 200 amino acids each. Similarly, the protein component of tobacco mosaic virus (TMV) was originally thought to have the horrendous molecular weight of 36 million. Most fortunately, it was subsequently discovered that each TMV particle contains 2150 identical smaller protein molecules, each containing 158 amino acids. Even much smaller protein molecules are frequently constructed from a number of polypeptide chains. Hemoglobin, which has a molecular weight of only 64,500, contains four polypeptide chains, two α chains and two β chains, each of which has a molecular weight of about 16,000 (Figure 5-32).

In all three examples, as with most other protein aggregates, the polypeptide subunits are held together by secondary bonds. This fact is known because the polypeptides can be dispersed by the addition of reagents (e.g., urea) that tend to break secondary bonds but not covalent bonds. Moreover, weak bonds are not the only force holding macromolecular units together. In some proteins, such as insulin, disulfide bonds (S–S) between cysteine residues are also a binding force.

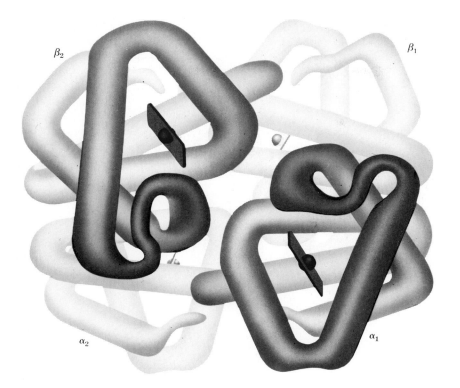

β_2 β_1

α_2 α_1

Figure 5-32
The four polypeptide chains of hemo-globin. (Copyright Irving Geis.)

Subunits Are Economical

Both the construction of polymers from monomers and the use of polymeric molecules themselves as subunits to build still larger molecules reflect a general building principle applicable to all complex structures, nonliving as well as living. This principle states that it is much easier to reduce the impact of construction mistakes if we can discard the faulty subunits before they are incorporated into the final product. For example, let us consider two alternative ways of constructing a molecule with a million atoms. In scheme 1, we build the structure atom by atom; in scheme 2, we first build a thousand smaller units, each with a thousand atoms, but subsequently put the subunits together into the million-atom product. Now consider that our building process randomly makes mistakes, inserting the wrong atom with a frequency of 10^{-5}. Let us assume that each mistake results in a nonfunctional product.

Under scheme 1, each molecule will contain, on the average, ten wrong atoms, and so almost no good products will be synthesized. Under scheme 2, however, mistakes will occur in only 1 percent of the subunits. If there is a device to reject the bad subunits, then good products can be easily made, and the cell will hardly be bothered by the presence of the 1 percent of nonfunctional subunits. This concept is the basis of the assembly line, in which complicated industrial products, such as radios and automobiles, are constructed. At each stage of assembly, there are devices to throw away bad subunits. In industrial assembly lines, mistakes were initially removed by human hands; now, automation often replaces manual control. In cells, mistakes are sometimes controlled by the specificity of enzymes. If a monomeric subunit is wrongly put together, it usually will not be

Figure 5-33
Ligand-induced conformational
changes. The ligand is represented by a
color circle.

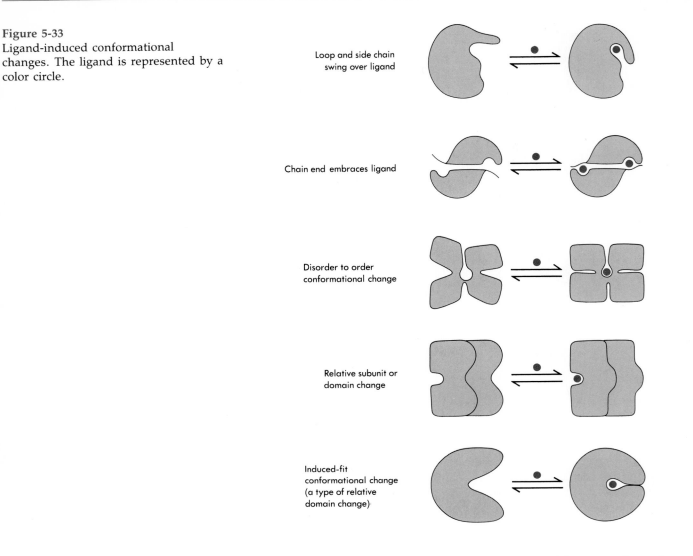

Loop and side chain
swing over ligand

Chain end embraces ligand

Disorder to order
conformational change

Relative subunit or
domain change

Induced-fit
conformational change
(a type of relative
domain change)

recognized by the polymer-making enzyme and hence will not be incorporated into a macromolecule. In other cases, faulty substances are rejected because they are unable to spontaneously become part of stable molecular aggregates.

Allostery: Alternative Shapes for Single Proteins[37, 38, 39]

The binding of either small or large molecules (*ligands*) to proteins often results in a substantial change in the conformation of the protein. These ligand-induced conformational changes can alter a protein's activity, enabling some proteins to function like an on-off light switch. Control of metabolism and gene expression is often modulated by such molecular protein switches (see Figure 4-20).

At least four different kinds of ligand-induced protein conformational changes have been observed (Figure 5-33). The simplest involve alterations around the binding site of the ligand itself, either a loop and side chain swinging over the ligand, as occurs when substrates bind to carboxypeptidase, or the polypeptide chain end moving to interact with the ligand. The latter change occurs upon the

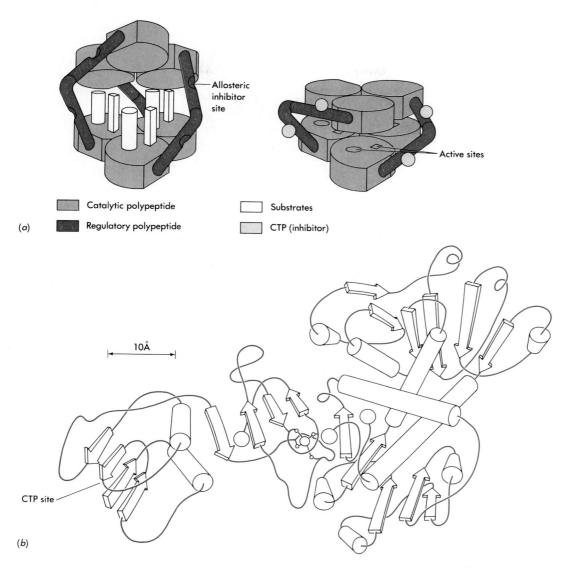

(a)

Catalytic polypeptide

Regulatory polypeptide

Substrates

CTP (inhibitor)

Allosteric inhibitor site

Active sites

10Å

CTP site

(b)

binding of several repressors to DNA. These changes at the ligand binding site allow the protein to surround or embrace the ligand in a way that would not be possible if the protein were rigid. A third effect of ligand binding is often to stabilize a fixed protein conformation in a small part of the protein that is very flexible or disordered in its absence. When RNA binds to TMV coat protein to form the intact virus particle, such a disorder-to-order transition is induced at the RNA binding site. Once again, flexibility in the protein structure allows the RNA to bind before a binding site that surrounds the RNA is completed.

Perhaps the largest category of protein conformational change and the one that is most significant for regulation is the relative subunit or domain change. In this case, the binding of a ligand to one subunit results in a change in the relative orientation of subunits—a change in the quaternary structure of an oligomeric protein. Such changes in quaternary structure can have profound effects on the properties of binding sites that are very distant from the site to which the first ligand binds. The allosteric modification of the enzyme aspartate transcarbamoylase by its ligand, CTP, provides a good example of such an interaction (Figure 5-34). Also, ligand binding can alter the

Figure 5-34
The allosteric modification of aspartate transcarbamoylase (ATCase) by CTP. ATCase is a large molecule consisting of six catalytic subunits that bind its substrates and six regulatory subunits that are devoid of catalytic activity but bind CTP. (a) When CTP is bound to a regulatory subunit, small changes occur that compact the polypeptides, thereby closing the active sites to substrates.
(b) Shown here are the structure of one of the regulatory chains and the site where CTP is thought to bind. The cylinders represent α helices. [Part (b) after Monaco, H. L., J. L. Crawford, and W. N. Lipscomb. *Proc. Nat. Acad. Sci.* 75 (1978): 5278.]

Figure 5-35
Glucose binding to hexokinase A from yeast. (a) Free hexokinase and the glucose substrate (color). (b) Glucose-hexokinase complex. Note the change in enzyme structure after the glucose is bound. (Courtesy of T. Steitz.)

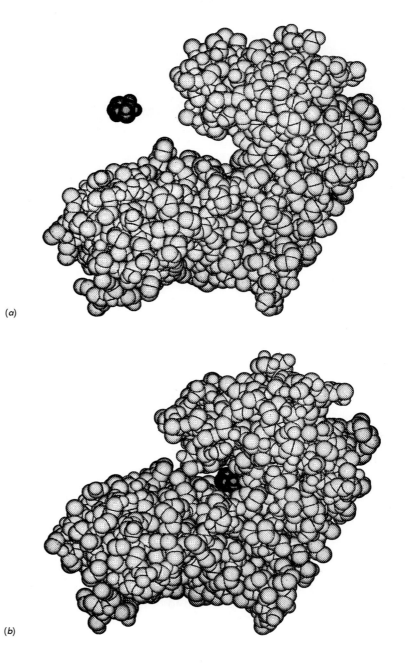

(a)

(b)

relative orientation of two domains in some multidomain proteins. For example, binding of the substrate glucose to hexokinase A results in a movement of two domains relative to each other, closing off the cleft in which the glucose binds and burying the substrate (Figure 5-35). Such a conformational change brought about by a substrate is called an **induced fit.** In this case, the correct binding site for ATP does not exist until glucose, the acceptor of the high-energy phosphate of ATP, is already bound. In this way, the shape change prevents the kinase from hydrolyzing the high-energy ATP molecule.

Ligand-induced quaternary structure changes often allow the binding of a ligand of one kind to alter the affinity or activity of a protein for a ligand of a different kind. The "interaction" between binding sites for ligands of different kinds is called **allosteric interaction.** When the binding of one ligand (e.g., cAMP to CAP) increases the protein's affinity for another ligand (e.g., DNA), the allosteric interaction is termed **positive cooperativity.** Alternatively, **negative cooperativity** describes the allosteric effect of one ligand reducing the protein's affinity for another ligand. An example of negative cooperativity is the binding of inducer to the *lac* repressor protein, which reduces the protein's affinity for operator DNA. Because of these allosteric proteins, the binding of a protein to DNA and thus its ability to alter gene expression can be switched on or off, depending on the presence or absence of a small-molecule metabolite.

The Principle of Self-Assembly[40, 41, 42]

With ΔG values of 1 to 5 kcal/mol, not only are single weak bonds spontaneously made, but also structures held together by several weak bonds are spontaneously formed. For example, an unstable, folded polypeptide chain tends to assume a large number of random configurations as a result of thermal movements. Most of these conformations are thermodynamically unstable. Inevitably, however, thermal movements bring together groups that can form weak bonds. These groups tend to stay together because more free energy is lost when they form than can be regained by their breakage. Thus, by a random series of movements, the polypeptide chain gradually assumes a configuration in which most, if not all, the atoms have fixed positions within the molecule.

Aggregation of separate molecules also occurs spontaneously. The protein hemoglobin furnishes a clear example (Figure 5-36). It can be broken apart by the addition of reagents such as urea, which break secondary bonds to yield half-molecules with molecular weights of 32,000. If, however, the urea is removed, the half-molecules quickly aggregate to form functional hemoglobin molecules. The surface structure of the half-molecules is very specific; they bind only to each other and not with any other cellular molecules.

This same principle of self-assembly operates to build even larger and more complicated structures, such as the cell membrane and the cell wall. Both are mosaic surfaces containing large numbers of various molecules, some large, like proteins, and others much smaller, like lipids. At present, practically nothing is known about the precise arrangement of the molecules in these very large, complex structures. Nonetheless, there is every reason to believe that the constituent molecules form stable contacts only with other molecules in the cell membrane (or wall). This situation is easy to visualize in the case of lipids, which are extremely insoluble in water because of their long, nonpolar hydrocarbon chains. Newly synthesized lipids have a stronger tendency to attach to other lipids in the cell membrane or cell wall by van der Waals forces than to enter some other, polar area, such as the aqueous (polar) interior of the cell.

The nucleic acid and protein components of viruses also come together spontaneously in the correct shape during the formation of new (progeny) viral particles. Enzymes often play no role in the aggregation of these components, since with many viruses, the formation of new covalent bonds is not needed either to build a stable protein coat or to inject its nucleic acid core. Only weak secondary bonds are needed. This last point was first shown for tobacco mosaic virus (TMV). Here the rod-shaped particles can be gently broken down and their free RNA and coat protein components separated. When they are again mixed together, new infectious particles, identical to the original rods, quickly form (Figure 5-37).

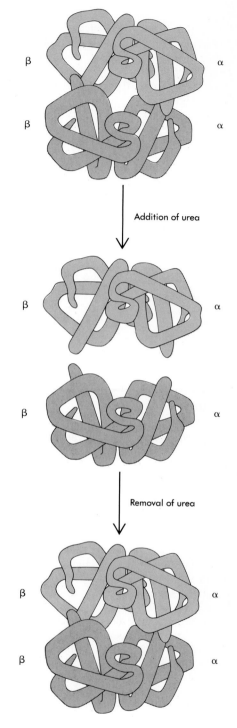

Addition of urea

Removal of urea

Figure 5-36
Formation of an active hemoglobin molecule from two half-molecules. Each hemoglobin molecule contains two α chains and two β chains. When placed in urea (a reagent that destabilizes weak bonds), the native molecule falls apart to two halves, each containing one α chain and one β chain. Upon removal of the urea, the halves reassociate to form the complete molecule.

Figure 5-37
Formation of a TMV particle from its protein subunits and its RNA molecule. [After H. Fraenkel-Conrat, *Design and Function at the Threshold of Life: The Viruses* (New York: Academic Press, 1962), Figure 18.]

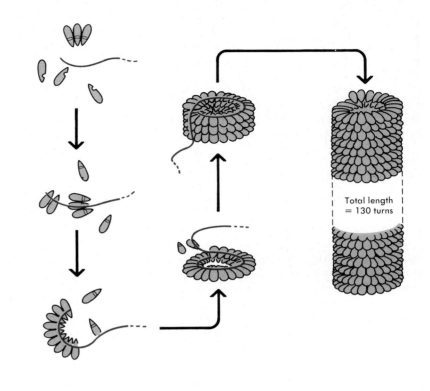

Total length = 130 turns

Summary

Many important chemical events in cells do not involve the making or breaking of covalent bonds. The cellular location of most molecules depends on weak, or secondary, attractive or repulsive forces. In addition, weak bonds are important in determining the shape of many molecules, especially very large ones. The most important of these weak forces are hydrogen bonds, van der Waals interactions, hydrophobic bonds, and ionic bonds. Even though these forces are relatively weak, they are still large enough to ensure that the right molecules (or atomic groups) interact with each other. For example, the surface of an enzyme is uniquely shaped to allow specific attraction of its substrates.

The formation of all chemical bonds, weak interactions as well as strong covalent bonds, proceeds according to the laws of thermodynamics. A bond tends to form when the result would be a release of free energy (negative ΔG). For the bond to be broken, this same amount of free energy must be supplied. Because the formation of covalent bonds between atoms usually involves a very large negative ΔG, covalently bound atoms almost never separate spontaneously. In contrast, the ΔG values accompanying the formation of weak bonds are only several times larger than the average thermal energy of molecules at physiological temperatures. Single weak bonds are thus frequently being made and broken in living cells.

Molecules having polar (charged) groups interact quite differently from nonpolar molecules (in which the charge is symmetrically distributed). Polar molecules can form good hydrogen bonds, whereas nonpolar molecules can form only van der Waals bonds. The most important polar molecule is water. Each water molecule can form four hydrogen bonds to other water molecules. Although polar molecules tend to be soluble in water (to various degrees), nonpolar molecules are insoluble because they cannot form hydrogen bonds with water molecules.

Every distinct molecule has a unique molecular shape that restricts the number of molecules with which it can form strong secondary bonds. Strong secondary interactions demand both a complementary (lock-and-key) relationship between the two bonding surfaces and the involvement of many atoms. Although molecules bound together by only one or two secondary bonds frequently fall apart, a collection of these weak bonds can result in a stable aggregate. The fact that double-helical DNA never falls apart spontaneously demonstrates the extreme stability possible in such an aggregate. The formation of such aggregates can proceed spontaneously, with the correct bonds forming in a step-by-step fashion (the principle of self-assembly).

The shape of polymeric molecules is determined by secondary bonds. All biological polymers contain single bonds about which free rotation is possible. They do not, however, exist in a variety of shapes, as might be expected, because the formation of one of the possible configurations generally involves a maximum decrease in free energy. This energetically preferred configuration is thus exclusively formed. Some polymeric molecules have regular helical backbones held in shape by sets of regular internal secondary bonds between backbone groups. Regular helical structures cannot be formed, however, if the specific side groups occur in positions in which they cannot form favorable weak bonds. This situation occurs in many proteins where an irregular distribution of nonpolar side groups forces the backbone into a highly irregular conformation, permitting the nonpolar groups to form van der Waals bonds with each other. Irregularly distributed side groups do not always lead, however, to nonhelical molecules. In the DNA molecule, for example, the specific pairing of purines with pyrimidines in a double-stranded helix allows the nonpolar aromatic groups to stack on top of each other in the center of the molecule.

Polypeptide chains in the aqueous environments of cells are energetically driven to assume unique configurations that isolate their nonpolar side groups into internal regions of their proteins into which water cannot penetrate. Such hydrophobic bonding provides the major force that converts newly made polypeptide chains into "native," usually much more compacted, shapes. The exact configuration adopted results from hydrogen-bond formation between suitable donor and acceptor groups within a protein's interior. The need for these internal groups to be properly hydrogen-bonded creates the large sections of α helices and β sheets found in most proteins.

Many native proteins contain several discrete folded sections (domains) that are stable by themselves and which arise from a continuous amino acid sequence. Combinations of such domains account for a large variety of all known proteins. The number of truly unique domains probably falls between 50 and 100. Each domain is often associated with a specific functional activity (e.g., DNA binding).

Binding of specific proteins to specific sequences along DNA molecules also involves the formation of weak bonds, usually hydrogen bonds between groups on DNA bases and appropriate acceptor (or donor) groups on proteins. Many important regulatory proteins have a protruding two α-helix domain that fits into the major groove of DNA. Other DNA binding proteins contain deep clefts that can effectively surround sections of a double helix. Individual DNA binding proteins also contain regions that allow nonspecific bonding to the DNA backbone. This permits their linear diffusion along DNA, allowing them to reach their specific target sequences more quickly.

Bibliography

General References

Blum, H. F. 1962. *Time's Arrow and Evolution*. New York: Harper & Row. Chapter 3 provides a clear introduction to the thermodynamics applicable to biological systems.

Creighten, T. E. 1983. *Proteins*. San Francisco, Calif.: Freeman.

Dickerson, R. E., and I. Geis. 1969. *The Structure and Action of Proteins*. Menlo Park, Calif.: Benjamin/Cummings.

———. 1983. *Hemoglobin: Structure, Function, Evolution and Pathology*. Menlo Park, Calif.: Benjamin/Cummings.

Fersht, A. R. 1983. *Enzyme Structure and Mechanism*, 3rd ed. San Francisco, Calif.: Freeman.

Fletterick, R. J., T. Schroer, and R. J. Matela. 1985. *Molecular Structure; Macromolecules in Three Dimensions*. Palo Alto, Calif.: Blackwell Scientific Publishers. Includes basic principles and a guide to model building.

Lehninger, A. L. 1971. *Bioenergetics*. 3rd ed. Menlo Park, Calif.: Benjamin/Cummings. The laws of thermodynamics are neatly outlined for the beginning biology student in the first several chapters.

Pauling, L. 1960. *The Nature of the Chemical Bond*. 3rd ed. Ithaca, N.Y.: Cornell University Press. A great classic of chemical literature emphasizing the hydrogen bond.

Richardson, J. S. 1981. "The Anatomy and Taxonomy of Protein Structure." *Adv. Prot. Chem.* 34:168–364.

Schultz, G., and R. H. Schirmer. 1979. *Principles of Protein Structure*. New York: Springer-Verlag.

Zubay, G. 1983. *Biochemistry*. Reading, Mass.: Addison-Wesley. Includes an excellent treatment of the principles of protein structure.

Cited References

1. Gray, H. B. 1964. *Electrons and Chemical Bonding*. Menlo Park, Calif.: Benjamin/Cummings.
2. Klotz, I. M. 1967. *Energy Changes in Biochemical Reactions*. New York: Academic Press. Easy-to-read work on energetics, including the concepts of entropy, free energy, and the dependence of chemical potential on concentration of reactants and products.
3. Morowitz, H. J. 1970. *Entropy for Biologists*. New York: Academic Press.
4. Kabat, E. A. 1978. "The Structural Basis of Antibody Complimentarity." *Adv. Prot. Chem.* 32:1–76.
5. Donohue, J. 1968. "Selected Topics in Hydrogen Bonding." In *Structural Chemistry and Molecular Biology*, ed. A. Rich and N. Davidson. San Francisco, Calif.: Freeman.
6. Marsh, R. E. 1968. "Some Comments on Hydrogen Bonding in Purine and Pyrimidine Bases." In *Structural Chemistry and Molecular Biology*, ed. A. Rich and N. Davidson. San Francisco, Calif.: Freeman.
7. Pauling, L. 1944. Appendix. In *The Specificity of Serological Reactions* by K. Landsteiner. Cambridge, Mass.: Harvard University Press.
8. Hildebrand, J. H. 1979. "Is There a Hydrophobic Effect?" *Proc. Nat. Acad. Sci.* 76:194.
9. Hagler, A. T., and J. Moult. 1978. "Computer Simulation of Solvent Structure Around Biological Macromolecules." *Nature* 272:222–226.
10. Chothia, C. 1985. "Principles of Proteins That Determine the Three-Dimensional Structure." *Ann. Rev. Biochem.* In press.
11. Ramachandran, G. N., and V. Sasisekharan. 1968. "Conformations of Polypeptides and Proteins." *Adv. Prot. Chem.* 28:283–437. Describes the conformational space of polypeptide backbones.
12. Anfinsen, C. B., and H. A. Scheraga. 1975. "Experimental and Theoretical Aspects of Protein Folding." *Adv. Prot. Chem.* 29:205–299.
13. Schultz, G., and R. H. Shirmer. 1979. *Principles of Protein Structure*. New York: Springer-Verlag, pp. 67–78.
14. Lesk, A. M. O., and C. Chothia. 1980. "How Different Amino Acid Sequences Determine Similar Protein Structures. I. The Structure and Evolutionary Dynamics of the Globins." *J. Mol. Biol.* 136:225–270.
15. Hol, W. G., L. M. Halie, and C. Sander. 1982. "Dipoles of the Alpha Helix and Beta Sheet: Their Role in Protein Folding." *Nature* 294:532–536.
16. Lesk, A. M., and K. D. Hardman. 1982. "Computer-Generated Schematic Diagrams of Protein Structures." *Science* 216:539–540.
17. Novotny, J., R. Bruccoleri, and M. Karplus. 1984. "An Analysis of Incorrectly Folded Protein Models: Implications for Structure Predictions." *J. Mol. Biol.* 177:787–818.

18. Privilov, P. L. 1979. "Stability of Proteins." *Adv. Prot. Chem.* 33:167–236.
19. Crick, F. H. C. 1952. "Is α-Keratin a Coiled Coil?" *Nature* 170:882–883.
20. Pauling, L. 1953. "Compound Helical Configurations of Polypeptide Chains: Structure of Proteins of the α-Keratin Type." *Nature* 171:59–61.
21. Levitt, M., and C. Chothia. 1976. "Structural Patterns in Globular Proteins." *Nature* 261:552–557.
22. Chothia, C., M. Levitt, and D. Richardson. 1977. "Structure of Proteins: Packing of α-Helices and β Pleated Sheets." *Proc. Nat. Acad. Sci.* 74:4130–4134.
23. Sternberg, M. J. E., and J. M. Thornton. 1977. "On the Conformation of Proteins: The Handedness of the Connections Between Parallel β Strands." *J. Mol. Biol.* 110:269–283.
24. Cohen, F., and M. Sternberg. 1980. "Packing of α-Helices onto β Pleated Sheets and the Anatomy of αβ Proteins." *J. Mol. Biol.* 143:95–128.
25. Rose, G. E. 1979. "Hierarchic Organization of Domains in Globular Proteins." *J. Mol. Biol.* 134:447–470.
26. Novokhatny, V. V., S. A. Kudinov, and P. L. Privalov. 1984. "Domains in Human Plasminogen." *J. Mol. Biol.* 179:215–232.
27. Adams, M. J., G. C. Ford, R. Koekoek, P. J. Lentz, Jr., A. McPherson, Jr., M. R. Rossmann, I. E. Smiley, R. W. Schevitz, and A. J. Wonacott. 1970. "Structure of Lactate Dehydrogenase at 2.8 Å Resolution." *Nature* 227:1098–1103.
28. Steitz, T. A., I. T. Weber, and J. B. Matthew. 1982. "Catabolite Gene Activator Protein: Structure, Homology with Other Proteins, and Cyclic AMP and DNA Binding." *Cold Spring Harbor Symp. Quant. Biol.* 47:419–426.
29. Sherwood, D. 1976. *Crystals, X-rays, and Proteins.* New York: Wiley.
30. Blundell, T. L., and L. N. Johnson. 1976. *Protein Crystallography.* New York: Academic Press. An advanced text.
31. Sperling, R., and E. J. Wachtel. 1981. "The Histones." *Adv. Prot. Chem.* 34:1–52.
32. Ohlendorf, D. H., W. F. Anderson, R. G. Fisher, Y. Takeda, and B. W. Matthews. 1982. "The Molecular Basis of DNA-Protein Recognition Inferred from the Structure of the Cro Repressor." *Nature* 298:718–723.

33. Frederick, C. A., J. Grable, M. Melia, C. Samudzi, L. Jen-Jacobson, B.-C. Wang, P. Greene, H. W. Boyer, and J. M. Rosenberg. 1984. "Kinked DNA in Crystalline Complex with EcoRl Endonuclease." *Nature* 309:327–331.
34. von Hippel, P. H., A. Revzin, C. A. Gross, and A. C. Wang. 1974. "Non-Specific DNA Binding of Genome Regulating Protein as a Biological Control Mechanism. I. The *lac* Operon. Equilibrium Aspects." *Proc. Nat. Acad. Sci.* 71:4808–4812.
35. Carr, S. A., K. Biemann, S. Shoji, D. C. Parmelei, and K. Titani. 1982. "n-Tetradecanoyl Is the NH2-Terminal Blocking Group of the Catalytic Subunit of Cyclic AMP-Dependent Protein Kinase from Bovine Cardiac Muscle." *Proc. Nat. Acad. Sci.* 79:6128–6131.
36. Marchildon, G. A., J. E. Casnellie, K. A. Walsh, and E. G. Krebs. 1984. "Covalently Bound Myristate in a Lymphoma Tyrosine Protein Kinase." *Proc. Nat. Acad. Sci.* 81:7679–7682.
37. Gerhart, J. C., and A. B. Pardee. 1963. "The Effect of the Feedback Inhibitor CTP on Subunit Interactions in Aspartate Transcarbamylase." *Cold Spring Harbor Symp. Quant. Biol.* 28:491–496. Study of an enzyme system that was important in developing the concept of allostery.
38. Monod, J., J.-P. Changeux, and F. Jacob. 1963. "Allosteric Proteins and Cellular Control Systems." *J. Mol. Biol.* 6:306–329.
39. Monaco, H. L., J. L. Crawford, and W. N. Lipscomb. 1978. "Three-Dimensional Structures of Aspartate Carbamoyltransferase from *E. coli* and of Its Complex with Cytidine Triphosphate." *Proc. Nat. Acad. Sci.* 75:5276–5280.
40. Fraenkel-Conrat, H., and R. C. Williams. 1955. "Reconstitution of Active Tobacco Mosaic Virus from Its Inactive Protein and Nucleic Acid Components." *Proc. Nat. Acad. Sci.* 41:690–698.
41. Crick, F. H. C., and J. D. Watson. 1957. "Virus Structure: General Principles." In Ciba Foundation Symposium on *The Nature of Viruses*, ed. G. E. W. Wolstenholme and E. C. P. Millar. London: Churchill, pp. 5–13.
42. Caspar, D. L. D., and A. Klug. 1962. "Physical Principles in the Construction of Regular Viruses." *Cold Spring Harbor Symp. Quant. Biol.* 27:1–24.

CHAPTER 6

Coupled Reactions
and Group Transfers

In the previous chapter we looked at the formation of weak bonds from the thermodynamic viewpoint. Each time a potential weak bond was considered, the question was posed, Does its formation involve a gain or a loss of free energy? because only when ΔG is negative does the thermodynamic equilibrium favor a reaction. This same approach is equally valid for covalent bonds. The fact that enzymes are usually involved in the making or breaking of a covalent bond does not in any sense alter the requirement of a negative ΔG.

On superficial examination, however, many of the important covalent bonds in cells appear to be formed in violation of the laws of thermodynamics, particularly those bonds joining small molecules together to form large polymeric molecules. The formation of such bonds involves an increase in free energy. Originally, this fact suggested to some people that cells had the unique ability to work in violation of thermodynamics and that this property was, in fact, the real "secret of life."

Now, however, it is clear that these biosynthetic processes do not violate thermodynamics but rather are based on different reactions from those originally postulated. Nucleic acids, for example, do not form by the condensation of nucleoside phosphates; glycogen is not formed directly from glucose residues; proteins are not formed by the union of amino acids. Instead, the monomeric precursors, using energy present in ATP, are first converted to high-energy "activated" precursors, which then spontaneously (with the help of specific enzymes) unite to form larger molecules. In this chapter, we shall illustrate these ideas by concentrating on the thermodynamics of peptide (protein) and phosphodiester (nucleic acid) bonds. First, however, we must briefly look at some general thermodynamic properties of covalent bonds.

Food Molecules Are Thermodynamically Unstable

There is great variation in the amount of free energy possessed by specific molecules. This is because covalent bonds do not all have the same bond energy. As an example, the covalent bond between oxygen and hydrogen is considerably stronger than the bond between hydrogen and hydrogen or oxygen and oxygen. The formation of an O–H bond at the expense of O–O or H–H will thus release energy. Energy considerations therefore tell us that a sufficiently concentrated mixture of oxygen and hydrogen will be transformed into water.

163

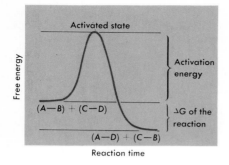

Figure 6-1
The energy of activation of a chemical reaction:
(A–B) + (C–D) → (A–D) + (C–B).

A molecule thus possesses a larger amount of free energy if linked together by weak covalent bonds than if it is linked together by strong bonds. This idea seems almost paradoxical at first glance, since it means that the stronger the bond, the less energy it can give off. But the notion automatically makes sense when we realize that an atom that has formed a very strong bond has already lost a large amount of free energy in this process. Therefore, the best food molecules (molecules that donate energy) are those molecules that contain weak covalent bonds and are therefore thermodynamically unstable.

For example, glucose is an excellent food molecule, since there is a great decrease in free energy when it is oxidized by oxygen to yield carbon dioxide and water. On the other hand, carbon dioxide is not a food molecule in animals, since in the absence of the energy donor ATP, it cannot spontaneously be transformed to more complex organic molecules, even with the help of specific enzymes. Carbon dioxide can be used as a primary source of carbon in plants only because the energy supplied by light quanta during photosynthesis results in the formation of ATP.

The Distinction Between Direction and Rate of a Reaction

The chemical reactions by which molecules are transformed into other molecules that contain less free energy do not occur at significant rates at physiological temperatures in the absence of a catalyst. This is because even a weak covalent bond is, in reality, very strong and is only rarely broken by thermal motion within a cell. For a covalent bond to be broken in the absence of a catalyst, energy must be supplied to push apart the bonded atoms. When the atoms are partially apart, they can recombine with new partners to form stronger bonds. In the process of recombination, the energy released is the sum of the free energy supplied to break the old bond plus the difference in free energy between the old and the new bond (Figure 6-1).

The energy that must be supplied to break the old covalent bond in a molecular transformation is called the **activation energy.** The activation energy is usually less than the energy of the original bond because molecular rearrangements generally do not involve the production of completely free atoms. Instead, a collision between the two reacting molecules is required, followed by the temporary formation of a molecular complex (the **activated state**). In the activated state, the close proximity of the two molecules makes each other's bonds more labile, so that less energy is needed to break a bond than when the bond is present in a free molecule.

Most reactions of covalent bonds in cells are therefore described by

$$(A\text{–}B) + (C\text{–}D) \longrightarrow (A\text{–}D) + (C\text{–}B) - \Delta G \qquad (6\text{-}1)$$

The mass action expression for such reaction is

$$K_{eq} = \frac{conc^{A\text{-}D} \times conc^{C\text{-}B}}{conc^{A\text{-}B} \times conc^{C\text{-}D}} \qquad (6\text{-}2)$$

where $conc^{A\text{-}B}$, $conc^{C\text{-}D}$, and so on, are the concentrations of the several reactants in moles per liter. Here, also, the value of K_{eq} is related to ΔG by Equation 5-4 (Table 6-1).

Since energies of activation are generally between 20 and 30 kcal/mol, activated states practically never occur at physiological temperatures. High activation energies should thus be considered barriers preventing spontaneous rearrangements of cellular covalent bonds.

These barriers are enormously important. Life would be impossible if they did not exist, for all atoms would be in the state of least possible energy. There would be no way to temporarily store energy for future work. On the other hand, life would also be impossible if means were not found to selectively lower the activation energies of certain reactions. This also must happen if cell growth is to occur at a rate sufficiently fast so as not to be seriously impeded by random destructive forces, such as ionizing or ultraviolet radiation.

Enzymes Lower Activation Energies

Enzymes are absolutely necessary for life. The function of enzymes is to speed up the rate of the chemical reactions requisite to cellular existence by lowering the activation energies of molecular rearrangements to values that can be supplied by the heat of motion (Figure 6-2). When a specific enzyme is present, there is no longer an effective barrier preventing the rapid formation of the reactants possessing the lowest amounts of free energy. Enzymes never affect the nature of an equilibrium: They merely speed up the rate at which it is reached. Thus, if the thermodynamic equilibrium is unfavorable for the formation of a molecule, the presence of an enzyme can in no way bring about the molecule's accumulation.

Since enzymes must catalyze essentially every cellular molecular rearrangement, knowing the free energy of various molecules cannot by itself tell us whether an energetically feasible rearrangement will, in fact, occur. The rate of the reactions must always be considered. Only if a cell possesses a suitable enzyme will the reaction be important.

A Metabolic Pathway Is Characterized by a Decrease in Free Energy

Thermodynamics tells us that all biochemical pathways must be characterized by a decrease in free energy. This is clearly the case for degradative pathways, in which thermodynamically unstable food molecules are converted to more stable compounds, such as carbon dioxide and water, with the evolution of heat. All degradative pathways have two primary purposes: (1) to produce the small organic fragments necessary as building blocks for larger organic molecules and (2) to conserve a significant fraction of the free energy of the original food molecule in a form that can do work, by coupling some of the steps in degradative pathways with the simultaneous formation of high-energy molecules such as ATP, which can store free energy.

Not all the free energy of a food molecule is converted into the free energy of high-energy molecules. If this were the case, a degradative pathway would not be characterized by a decrease in free energy, and there would be no driving force to favor the breakdown of food molecules. Instead, we find that all degradative pathways are characterized by a conversion of at least one-half the free energy of the food

Table 6-1 The Relationship Between K_{eq} and ΔG ($\Delta G = -RT \ln K_{eq}$)

K_{eq}	ΔG (kcal/mol)
10^{-6}	8.2
10^{-5}	6.8
10^{-4}	5.1
10^{-3}	4.1
10^{-2}	2.7
10^{-1}	1.4
10^{0}	0.0
10^{1}	-1.4
10^{2}	-2.7
10^{3}	-4.1

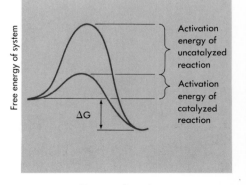

Progress of reaction

Figure 6-2
Enzymes (color curve) lower activation energies and thus speed up the rate of the reaction. Note that ΔG remains the same, since the equilibrium position remains unaltered.

molecule into heat or entropy. For example, it is now believed that in cells, approximately 40 percent of the free energy of glucose is used to make new high-energy compounds, the remainder being dissipated into heat energy and entropy.

High-Energy Bonds
Hydrolyze with Large Negative ΔG

A high-energy molecule contains one or more bonds whose breakdown by water (*hydrolysis*) is accompanied by a large decrease in free energy (5 kcal/mol or more). The specific bonds whose hydrolysis yields these large negative ΔG values are called **high-energy bonds,** a somewhat misleading term, since it is not the bond energy but the free energy of hydrolysis that is high. Nonetheless, the term *high-energy bond* is generally employed, and for convenience, we shall continue this usage by marking high-energy bonds with the symbol $\sim$.

The energy of hydrolysis of the average high-energy bond (7 kcal/mol) is very much smaller than the amount of energy that would be released if a glucose molecule were to be completely degraded in one step (688 kcal/mol). A one-step breakdown of glucose would be inefficient in making high-energy bonds. This is undoubtedly the reason why biological glucose degradation requires so many steps. In this way, the amount of energy released per degradative step is of the same order of magnitude as the free energy of hydrolysis of a high-energy bond.

The most important high-energy compound is ATP. It is formed from inorganic phosphate (P) and ADP, using energy obtained either from degradative reactions or from the sun (photosynthesis). There are, however, many other important high-energy compounds. Some are directly formed during degradative reactions; others are formed using some of the free energy of ATP. Table 6-2 lists the most important types of high-energy bonds. All involve either phosphate or sulfur atoms. The high-energy pyrophosphate bonds of ATP arise from the union of phosphate groups. The pyrophosphate linkage ((P)$\sim$(P)) is not, however, the only kind of high-energy phosphate bond: The attachment of a phosphate group to the oxygen atom of a carboxyl group creates a high-energy acyl bond. It is now clear that high-energy bonds involving sulfur atoms play almost as important a role in energy metabolism as those involving phosphorus. The most important molecule containing a high-energy sulfur bond is acetyl-CoA. This bond is the main source of energy for fatty acid biosynthesis.

The wide range of ΔG values of high-energy bonds (see Table 6-2) means that calling a bond "high-energy" is sometimes arbitrary. The usual criterion is whether its hydrolysis can be coupled with another reaction to effect an important biosynthesis. For example, the negative ΔG accompanying the hydrolysis of glucose-6-phosphate is 3 to 4 kcal/mol. But this ΔG is not sufficient for efficient synthesis of peptide bonds, so this phosphate ester bond is not included among high-energy bonds.

High-Energy Bonds
Necessary for Biosynthetic Reactions

The construction of a large molecule from smaller building blocks often requires the input of free energy. Yet, a biosynthetic pathway, like a degradative pathway, would not exist if it were not character-

Table 6-2 Important Classes of High-Energy Bonds

Class	Molecular Example	ΔG of Reaction, kcal/mole
Pyrophosphate	(P) ~ (P) pyrophosphate	(P) ~ (P) $\rightleftharpoons$ (P) + (P) $\Delta G = -6$
Nucleoside diphosphates	Adenosine—(P) ~ (P) (ADP)	ADP $\rightleftharpoons$ AMP + (P) $\Delta G = -6$
Nucleoside triphosphates	Adenosine—(P) ~ (P) ~ (P)(ATP)	ATP $\rightleftharpoons$ ADP + (P) $\Delta G = -7$ ATP $\rightleftharpoons$ AMP + (P) ~ (P) $\Delta G = -8$
Enol phosphates	Phosphoenolpyruvate (PEP)	PEP $\rightleftharpoons$ pyruvate + (P) $\Delta G = -12$
Aminoacyl adenylates		AMP ~ AA $\rightleftharpoons$ AMP + AA $\Delta G = -7$
Guanidinium phosphates	Creatine phosphate	Creatine ~ P $\rightleftharpoons$ Creatine + P $\Delta G = -8$
Thioesters	Acetyl-CoA	Acetyl CoA $\rightleftharpoons$ CoA-SH + acetate $\Delta G = -8$

ized by a net decrease in free energy. This means that many biosynthetic pathways demand an external source of free energy. These free-energy sources are the high-energy compounds. The making of many biosynthetic bonds is coupled with the breakdown of a high-energy bond, so that the net change of free energy is always negative. Thus, high-energy bonds in cells generally have a very short life. Almost as soon as they are formed during a degradative reaction, they are enzymatically broken down to yield the energy needed to drive another reaction to completion.

Not all the steps in a biosynthetic pathway require the breakdown of a high-energy bond. Often, only one or two steps involve such a bond. Sometimes this is because the ΔG, even in the absence of an externally added high-energy bond, favors the biosynthetic direction. In other cases, ΔG is effectively zero or may even be slightly positive. These small positive ΔG values, however, are not significant so long as they are followed by a reaction characterized by the hydrolysis of a

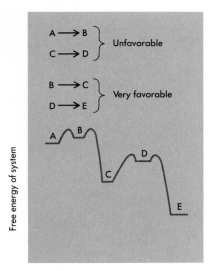

Free energy of system

Progress of reaction

Figure 6-3
Free-energy changes in a multistep metabolic pathway, A → B → C → D → E. Two steps (A → B and C → D) do not favor the A → E direction of the reaction, since they have small positive ΔG values. However, they are insignificant owing to the very large negative ΔG values provided in steps B → C and D → E. Therefore, the overall reaction favors the A → E conversion.

high-energy bond. Rather, it is the *sum* of all the free-energy changes in a pathway that is significant (Figure 6-3). It does not really matter that the K_{eq} of a specific biosynthetic step is slightly (80:20) in favor of degradation if the K_{eq} of the succeeding step is 100:1 in favor of the forward biosynthetic direction.

Likewise, not all the steps in a degradative pathway generate high-energy bonds. For example, only two steps in the lengthy glycolytic (Embden-Meyerhof) breakdown of glucose generate ATP. Moreover, there are many degradative pathways that have one or more steps requiring the breakdown of a high-energy bond. The glycolytic breakdown of glucose is again an example. It uses up two molecules of ATP for every four that it generates. Here, of course, as in every energy-yielding degradative process, more high-energy bonds must be made than consumed.

Peptide Bonds Hydrolyze Spontaneously

The formation of a dipeptide and a water molecule from two amino acids requires a ΔG of 1 to 4 kcal/mol, depending on which amino acids are being bound. This ΔG value decreases progressively if amino acids are added to longer polypeptide chains; for an infinitely long chain, the ΔG is reduced to about 0.5 kcal/mol. This decrease reflects the fact that the free, charged NH_3^+ and COO^- groups at the chain ends favor the hydrolysis (breakdown accompanied by the uptake of a water molecule) of nearby peptide bonds.

These positive ΔG values by themselves tell us that polypeptide chains cannot form from free amino acids. In addition, we must take into account the fact that water molecules have a much, much higher concentration than any other cellular molecules (generally over 100 times higher). All equilibrium reactions in which water participates are thus strongly pushed in the direction that consumes water molecules. This is easily seen in the definition of equilibrium constants. For example, the reaction forming a dipeptide,

$$\text{amino acid(A)} + \text{amino acid(B)} \longrightarrow \text{dipeptide(A–B)} + H_2O \quad \text{(6-3)}$$

has the following equilibrium constant:

$$K_{eq} = \frac{\text{conc}^A \times \text{conc}^B}{\text{conc}^{A-B} \times \text{conc}^{H_2O}} \quad \text{(6-4)}$$

where concentrations are given in moles per liter. Thus, for a given K_{eq} value (related to ΔG by the formula $\Delta G = -RT \ln K_{eq}$), a much greater concentration of water means a correspondingly smaller concentration of the dipeptide. The relative concentrations are therefore very important. In fact, a simple calculation shows that hydrolysis may often proceed spontaneously even when the ΔG for the nonhydrolytic reaction is -3 kcal/mol.

Thus, in theory, proteins are unstable and, given sufficient time, will spontaneously degrade to free amino acids. On the other hand, in the absence of specific enzymes, these spontaneous rates are too slow to have a significant effect on cellular metabolism. That is, once a protein is made, it remains stable unless its degradation is catalyzed by a specific enzyme.

Coupling of Negative with Positive ΔG

Free energy must be added to amino acids before they can be united to form proteins. How this happens became clear with the discovery of the fundamental role of ATP as an energy donor. ATP contains three phosphate groups attached to an adenosine molecule (adenosine $-O-\textcircled{P}\sim\textcircled{P}\sim\textcircled{P}$). When one or two of the terminal $\sim\textcircled{P}$ groups are broken off by hydrolysis, there is a significant decrease of free energy.

Adenosine$-O-\textcircled{P}\sim\textcircled{P}\sim\textcircled{P}$ + H$_2$O $\longrightarrow$

$\qquad$ adenosine$-O-\textcircled{P}\sim\textcircled{P}$ + $\textcircled{P}$ $\qquad$ ($\Delta G = -7$ kcal/mol) $\quad$ *(6-5)*

Adenosine$-O-\textcircled{P}\sim\textcircled{P}\sim\textcircled{P}$ + H$_2$O $\longrightarrow$

$\qquad$ adenosine$-O-\textcircled{P}$ + $\textcircled{P}\sim\textcircled{P}$ $\qquad$ ($\Delta G = -8$ kcal/mol) $\quad$ *(6-6)*

Adenosine$-O-\textcircled{P}\sim\textcircled{P}$ + H$_2$O $\longrightarrow$

$\qquad$ adenosine$-O-\textcircled{P}$ + $\textcircled{P}$ $\qquad$ ($\Delta G = -6$ kcal/mol) $\quad$ *(6-7)*

All these breakdown reactions have negative ΔG values considerably greater in absolute value (numerical value without regard to sign) than the positive ΔG values accompanying the formation of polymeric molecules from their monomeric building blocks. The essential trick underlying these biosynthetic reactions, which by themselves have a positive ΔG, is that they are coupled with breakdown reactions characterized by negative ΔG of greater absolute value. Thus, during protein synthesis, the formation of each peptide bond ($\Delta G = +0.5$ kcal/mol) is coupled with the breakdown of ATP to AMP and pyrophosphate, which has a ΔG of -8 kcal/mol. This results in a net ΔG of -7.5 kcal/mol, more than sufficient to ensure that the equilibrium favors protein synthesis rather than breakdown.

Activation Through Group Transfer

When ATP is hydrolyzed to ADP and phosphate, most of the free energy is liberated as heat. Since heat energy cannot be used to make covalent bonds, a coupled reaction cannot be the result of two completely separate reactions, one with a positive ΔG, the other with a negative ΔG. Instead, a coupled reaction is achieved by two or more successive reactions. These are always *group-transfer* reactions: reactions, not involving oxidations or reductions, in which molecules exchange functional groups. The enzymes that catalyze these reactions are called transferases. Consider the reaction

$$(A-X) + (B-Y) \longrightarrow (A-B) + (X-Y) \qquad (6-8)$$

In this example, group X is exchanged with component B. Group-transfer reactions are arbitrarily defined to exclude water as a participant. When water is involved,

$$(A-B) + (H-OH) \longrightarrow (A-OH) + (BH) \qquad (6-9)$$

This reaction is called a hydrolysis, and the enzymes involved are called hydrolases.

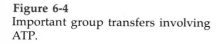

Figure 6-4
Important group transfers involving ATP.

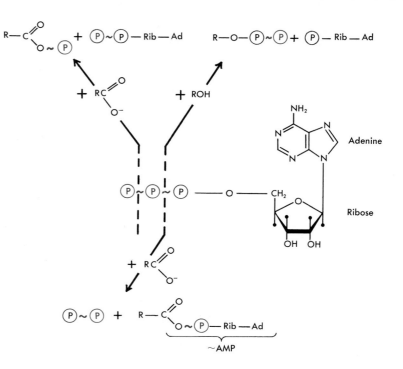

The group-transfer reactions that interest us here are those involving groups attached by high-energy bonds. When such a high-energy group is transferred to an appropriate acceptor molecule, it becomes attached to the acceptor by a high-energy bond. Group transfer thus allows the transfer of high-energy bonds from one molecule to another. For example, Equations 6-10 and 6-11 show how energy present in ATP is transferred to form GTP, one of the precursors used in RNA synthesis:

Adenosine—Ⓟ~Ⓟ~Ⓟ + guanosine—Ⓟ ⟶

   adenosine—Ⓟ~Ⓟ + guanosine—Ⓟ~Ⓟ   *(6-10)*

Adenosine—Ⓟ~Ⓟ~Ⓟ + guanosine—Ⓟ~Ⓟ ⟶

   adenosine—Ⓟ~Ⓟ + guanosine—Ⓟ~Ⓟ~Ⓟ   *(6-11)*

The high-energy Ⓟ~Ⓟ group on GTP allows it to unite spontaneously with another molecule. GTP is thus an example of what is called an **activated molecule;** correspondingly, the process of transferring a high-energy group is called **group activation.**

ATP Versatility in Group Transfer

In Chapter 2, we emphasized the key role of ATP synthesis in the controlled trapping of the energy of food molecules. In both oxidative and photosynthetic phosphorylations, energy is used to synthesize ATP from ADP and phosphate:

Adenosine—Ⓟ~Ⓟ + Ⓟ + energy ⟶

   adenosine—Ⓟ~Ⓟ~Ⓟ   *(6-12)*

Since ATP is thus the original biological recipient of high-energy groups, it must be the starting point of a variety of reactions in which high-energy groups are transferred to low-energy molecules to give them the potential to react spontaneously. ATP's central role utilizes the fact that it contains two high-energy bonds whose splitting releases specific groups. This is seen in Figure 6-4, which shows three important groups arising from ATP: Ⓟ~Ⓟ, a pyrophosphate group; ~AMP, an adenosyl monophosphate group; and ~Ⓟ, a phosphate group. It is important to notice that these high-energy groups retain their high-energy quality only when transferred to an appropriate acceptor molecule. For example, although the transfer of a ~Ⓟ group to a COO⁻ group yields a high-energy COO~Ⓟ acyl-phosphate group, the transfer of the same group to a sugar hydroxyl group (–C–OH), as in the formation of glucose-6-phosphate, gives rise to a low-energy bond (less than 5 kcal/mol decrease in ΔG upon hydrolysis).

Activation of Amino Acids by Attachment of AMP

The activation of an amino acid is achieved by transfer of an AMP group from ATP to the COO⁻ group of the amino acid:

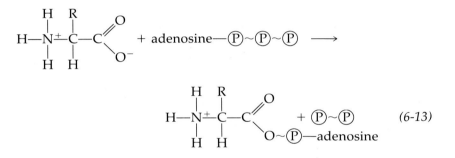

(R represents the specific side group of the amino acid.) The enzymes that catalyze this type of reaction are called aminoacyl synthetases. Upon activation, an amino acid (AA) is thermodynamically capable of being efficiently used for protein synthesis. Nonetheless, the AA~AMP complexes are not the direct precursors of proteins. Instead, for a reason we shall explain in Chapter 14, a second group transfer must occur to transfer the amino acid, still activated at its carboxyl group, to the end of a tRNA molecule:

$$\text{AA\sim AMP} + \text{tRNA} \longrightarrow \text{AA\sim tRNA} + \text{AMP} \qquad (6\text{-}14)$$

A peptide bond then forms by the condensation of the AA~tRNA molecule onto the end of a growing polypeptide chain:

AA~tRNA + growing polypeptide chain (of n amino acids) $\longrightarrow$

tRNA + growing polypeptide chain (of n + 1 amino acids) (6-15)

Thus, the final step of this "coupled reaction," like that of all other coupled reactions, necessarily involves the removal of the activating group and the conversion of a high-energy bond into one with a lower free energy of hydrolysis. This is the source of the negative ΔG that drives the reaction in the direction of protein synthesis.

Nucleic Acid Precursors Activated by the Presence of (P)~(P)

Both types of nucleic acid, DNA and RNA, are built up from mononucleotide monomers (nucleoside phosphate). Mononucleotides, however, are thermodynamically even less likely to combine than amino acids. This is because the phosphodiester bonds that link the former together release considerable free energy upon hydrolysis (-6 kcal/mol). This means that nucleic acids will spontaneously hydrolyze, at a slow rate, to mononucleotides. Thus, it is even more important that activated precursors be used in the synthesis of nucleic acids than in the synthesis of proteins.

Recently, it has been found that the immediate precursors for both DNA and RNA are the nucleoside-5'-triphosphates. For DNA, these are dATP, dGTP, dCTP, and dTTP (d stands for deoxy); for RNA, the precursors are ATP, GTP, CTP, and UTP. ATP thus not only serves as the main source of high-energy groups in group-transfer reactions, but is itself a direct precursor for RNA. The other three RNA precursors all arise by group-transfer reactions like those described in Equations 6-10 and 6-11. The deoxytriphosphates are formed in basically the same way: After the deoxymononucleotides have been synthesized, they are transformed to the triphosphate form by group transfer from ATP:

$$\text{Deoxynucleoside—(P) + ATP} \longrightarrow$$
$$\text{deoxynucleoside—(P)~(P) + ADP} \quad (6\text{-}16)$$

$$\text{Deoxynucleoside—(P)~(P) + ATP} \longrightarrow$$
$$\text{deoxynucleoside—(P)~(P)~(P) + ADP} \quad (6\text{-}17)$$

These triphosphates can then unite to form polynucleotides held together by phosphodiester bonds. In this group-transfer reaction, a pyrophosphate bond is broken and a pyrophosphate group released:

$$\text{Deoxynucleoside—(P)~(P)~(P)}$$
$$\text{+ growing polynucleotide chain (of } n \text{ nucleotides)} \longrightarrow$$
$$\text{(P)~(P) + growing polynucleotide chain (} n + 1 \text{ nucleotides)} \quad (6\text{-}18)$$

This reaction, unlike that which forms peptide bonds, does not have a negative ΔG. In fact, the ΔG is slightly positive (about 0.5 kcal/mol). This immediately poses the question, since polynucleotides obviously form: What is the source of the necessary free energy?

The Value of (P)~(P) Release in Nucleic Acid Synthesis

The needed free energy comes from the splitting of the high-energy pyrophosphate group that is formed simultaneously with the high-energy phosphodiester bond. All cells contain a powerful enzyme, pyrophosphatase, which breaks down pyrophosphate molecules almost as soon as they are formed:

$$\text{Ⓟ~Ⓟ} \longrightarrow 2 \text{ Ⓟ} \qquad (\Delta G = -7 \text{ kcal/mol}) \qquad (6\text{-}19)$$

The large negative ΔG means that the reaction is effectively irreversible. This means that once Ⓟ~Ⓟ is broken down, it never reforms.

The union of the nucleoside monophosphate group (Equation 6-16), coupled with the splitting of the pyrophosphate groups (Equation 6-19), has an equilibrium constant determined by the combined ΔG values of the two reactions: (0.5 kcal/mol) + (-7 kcal/mol). The resulting value ($\Delta G = -6.5$ kcal/mol) tells us that nucleic acids almost never break down to reform their nucleoside triphosphate precursors.

Here we see a powerful example of the fact that often it is the free-energy change accompanying a *group of reactions* that determines whether a reaction in the group will take place. Reactions with small, positive ΔG values, which by themselves would never take place, are often part of important metabolic pathways in which they are followed by reactions with large negative ΔG values. At all times we must remember that a single reaction (or even a single pathway) never occurs in isolation; rather, the nature of the equilibrium is constantly being changed through the addition and removal of metabolites.

Ⓟ~Ⓟ Splits Characterize Most Biosynthetic Reactions

The synthesis of nucleic acids is not the only reaction where direction is determined by the release and splitting of Ⓟ~Ⓟ. In fact, the generalization is emerging that essentially all biosynthetic reactions are characterized by one or more steps that release pyrophosphate groups. Consider, for example, the activation of an amino acid by the attachment of AMP. By itself, the transfer of a high-energy bond from ATP to the AA~AMP complex has a slightly positive ΔG. Therefore, it is the release and splitting of ATP's terminal pyrophosphate group that provides the negative ΔG that is necessary to drive the reaction.

The great utility of the pyrophosphate split is neatly demonstrated when we consider the problems that would arise if a cell attempted to synthesize nucleic acid from nucleoside diphosphates rather than triphosphates. Phosphate, rather than pyrophosphate, would be liberated as the backbone phosphodiester linkages were made. The phosphodiester linkages, however, are not stable in the presence of significant quantities of phosphate, since they are formed without a significant release of free energy. Thus, the biosynthetic reaction would be easily reversible; as soon as phosphate began to accumulate, the reaction would begin to move in the direction of nucleic acid breakdown (mass-action law). Moreover, it is not possible for a cell to remove the phosphate groups as soon as they are generated (thus preventing this reverse reaction), since all cells need a significant internal level of phosphate to grow. Thus, the use of nucleoside triphosphates as precursors of nucleic acids is not a matter of chance.

This same type of argument tells us why ATP, and not ADP, is the key donor of high-energy groups in all cells. At first this preference seemed arbitrary to biochemists. Now, however, we see that many reactions using ADP as an energy donor would occur equally well in both directions.

Summary

The biosynthesis of many molecules appears, at a superficial glance, to violate the thermodynamic law that spontaneous reactions always involve a decrease in free energy (ΔG is negative). For example, the formation of proteins from amino acids has a positive ΔG. This paradox is removed when we realize that the biosynthetic reactions do not proceed as initially postulated. Proteins, for example, are not formed from free amino acids. Instead, the precursors are first enzymatically converted to high-energy activated molecules, which, in the presence of a specific enzyme, spontaneously unite to form the desired biosynthetic product.

Many biosynthetic processes are thus the result of "coupled" reactions, the first of which supplies the energy that allows the spontaneous occurrence of the second reaction. The primary energy source in cells is ATP. It is formed from ADP and inorganic phosphate, either during degradative reactions (e.g., fermentation or respiration) or during photosynthesis. ATP contains several high-energy bonds whose hydrolysis has a large negative ΔG. Groups linked by high-energy bonds are called high-energy groups. High-energy groups can be transferred to other molecules by group-transfer reactions, thereby creating new high-energy compounds. These derivative high-energy molecules are then the immediate precursors for many biosynthetic steps.

Amino acids are activated by the addition of an AMP group, originating from ATP, to form an AA~AMP molecule. The energy of the high-energy bond in the AA~AMP molecule is similar to that of a high-energy bond of ATP. Nonetheless, the group-transfer reaction proceeds to completion because the high-energy Ⓟ~Ⓟ molecule, created when the AA~AMP molecule is formed, is broken down by the enzyme pyrophosphatase to low-energy groups. Thus, the reverse reaction, Ⓟ~Ⓟ + AA~AMP → ATP + AA, cannot occur.

Almost all biosynthetic reactions result in the release of Ⓟ~Ⓟ. Almost as soon as it is made, it is enzymatically broken down to two phosphate molecules, thereby making impossible a reversal of the biosynthetic reaction. The great utility of the Ⓟ~Ⓟ split provides an explanation for why ATP, not ADP, is the primary energy donor. ADP cannot transfer a high-energy group and at the same time produce Ⓟ~Ⓟ groups as a by-product.

Bibliography

General References

Dugas, H., and C. Penney. 1981. *Biorganic Chemistry: A Chemical Approach to Enzyme Action.* New York: Springer-Verlag.

Jencks, W. P. 1969. *Catalysis in Chemistry and Enzymology.* New York: McGraw-Hill.

Kornberg, A. 1962. "On the Metabolic Significance of Phosphorolytic and Pyrophosphorolytic Reactions." In *Horizons in Biochemistry,* ed. M. Kasha and B. Pullman. New York: Academic Press, pp. 251–264.

Krebs, H. A., and H. L. Kornberg. 1957. "A Survey of the Energy Transformation in Living Material." *Ergeb. Physiol. Biol. Chem.*

Exp. Pharmakol. 49:212. A classical comprehensive review which, although many of its facts have been modified by subsequent research, can still be used with great profit.

Lehninger, A. L. 1982. *Principles of Biochemistry.* New York: Worth.

Smith, E. L., and R. L. Hill. 1983. *Principles of Biochemistry.* 7th ed. New York: McGraw-Hill.

Stryer, L. 1981. *Biochemistry.* 2nd ed. San Francisco: Freeman.

Walsh, C. 1979. *Enzymatic Reaction Mechanisms.* San Francisco: Freeman.

Wood, W. B., J. H. Wilson, R. M. Benbow, and L. Hood. 1981. *Biochemistry: A Problems Approach.* 2nd ed. Menlo Park, Calif.: Benjamin/Cummings.

III

BACTERIAL GENETICS

CHAPTER 7

The Genetic Systems Provided by *E. coli* and Its Viruses

The genes of *E. coli* and its viruses, the bacteriophages (phages), are now understood better than those of any other form of life. They provide the simplest examples of what genes are and how the genetic code operates. Moreover, skillful exploitation of the *E. coli* genetic properties revealed that the information needed to properly regulate the rate at which specific genes work is also encoded within the nucleotide sequences of DNA. The more we learn about *E. coli,* the more we understand the extraordinary plasticity of its genetic systems and the more we appreciate how superbly programmed this organism is to face the highly changeable environments that all bacteria encounter daily.

Key to much, if not most, of our knowledge about the genetic systems of *E. coli* has been the use of mutants. By noting a specific aberration in a particular mutant, we can often reveal the function of the normal cell. For example, if a mutant *E. coli* unable to synthesize lactose is found to lack a specific gene group, we can surmise that that gene group is responsible for lactose synthesis. Without mutants, we would still be in the dark as to how gene expression is regulated, and we would still be unaware of many essential features of DNA replication.

The Intrinsic Advantages of Using Microorganisms for Genetic Research[1, 2, 3]

That microorganisms were the obvious route to discovering the nature of the gene became clear as soon as geneticists began to focus on the gene-protein relationship. This happened in the 1930s, when biosynthetic processes were discovered to be series ("pathways") of enzymatically catalyzed steps resembling the multistep procedures that skilled chemists invariably needed to make sophisticated organic molecules. Directly pursuing gene-enzyme connections, however, was virtually impossible. No way was known to select for mutations in genes that controlled already known biochemical pathways. Even when apparently suitable mutations were found in the fruit fly *Dro-*

sophila, it was not clear how to proceed. Direct comparison of the chemical compositions of mutant and wild-type flies was impractical because of the small size of the flies and the difficulties of breeding enough of them for chemical analysis (even though their generation time is relatively short—about 14 days). Furthermore, since fruit flies are diploid organisms, most mutations are not detected; this is because most mutations are recessive, and recessive mutations are not usually detected when the dominant wild-type allele is also present.

In contrast, single-celled eucaryotic microorganisms, such as the green alga *Chlamydomonas,* the yeast *Saccharomyces cerevisiae,* and the mold *Neurospora crassa,* which were among the first subjects of genetic analysis, were more attractive experimental organisms. First, they are usually haploid, so that the ease of genetic analysis does not depend on whether a mutation is dominant or recessive. In work with diploids, several generations of genetic crosses must often be carried out to detect the presence of a particular mutant gene; in a haploid organism, the mutant gene can express itself almost immediately, since there is no masking dominant allele present. Second, microorganisms can multiply very rapidly. There are enormous advantages to working with organisms that can produce a new cell generation every 2 hours rather than working with a plant like corn, which under the most favorable conditions can only produce two generations per year for study. When necessary, grams (and even kilograms) of desired mutant microorganisms can be grown under simple laboratory conditions.

At first, the fact that many microorganisms do not have easily recognizable features except for the morphology of their colonies on solid media made genetic analysis difficult. It was hard to know when the microorganisms contained mutations. But as soon as they could be characterized by their metabolic defects, the study of their genetics could blossom.

The first microorganism successfully used in studying gene-enzyme relationships was the mold *Neurospora crassa,* which, like *E. coli,* grows on a simple minimal glucose-containing medium. The essential trick was to seek out genes that controlled the biosynthetic pathways for vital metabolites that could either be ingested directly from the surrounding environment or made internally starting from one of the key precursor metabolites. Such genes were found by subjecting a population of *Neurospora* cells to ultraviolet (UV) light to increase the number of mutant cells. Searches were then made for descendants of treated cells that could only multiply in minimal medium when it was supplemented with one of a variety of known metabolites (e.g., amino acids, purine and pyrimidine bases, and vitamins). Such nutritionally restricted cells might, if all went well, lack the ability to synthesize one of the supplemented compounds. They were therefore tested to see if their growth on minimal medium could be restarted by the addition of a *single* known metabolite. The first gene to be so identified controlled the synthesis of one of the enzymes involved in the biosynthesis of the vitamin niacin. Over the next several years, many other *Neurospora* **growth factor** mutants were found that blocked the synthesis of many additional metabolites. In some cases, the specific enzyme activity lacking in the mutant cell could be identified, thereby directly demonstrating that a given gene controls the synthesis of a specific enzyme (the **one gene–one enzyme hypothesis**).

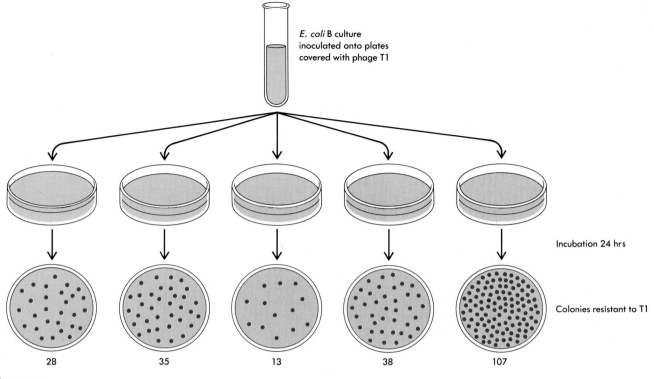

E. coli B culture
inoculated onto plates
covered with phage T1

Incubation 24 hrs

Colonies resistant to T1

| 28 | 35 | 13 | 38 | 107 |

Figure 7-1
Fluctuation analysis of bacterial resistance to phage. A total of 0.05 cc of the original culture was plated on each of five nutrient agar plates previously plated with an excess of bacteriophage T1. Microscopic analysis revealed that lysis of *E. coli* B cells followed rapidly. However, upon further incubation, colonies resulting from division of surviving resistant bacteria were formed. If the resistant bacteria represented adaptive responses to actual physical contact with T1, the same number of surviving colonies should be seen on each plate. The large fluctuation of the numbers of resulting colonies, however, proves that the mutants arose spontaneously prior to exposure to the phage.

Bacteria Have Genes That Mutate Spontaneously[4]

The first "biochemical" mutant in *Neurospora* was found in 1941, several years before convincing evidence appeared of a conventional hereditary mechanism within bacteria. Before that time, many variants of bacterial colony morphology were already known, and there was preliminary evidence for the existence of rare stable strains resistant to such antibacterial agents as antibiotics, viruses, and poisons. Nevertheless, there was much speculation that these variants were not the result of spontaneous genetic alterations but were direct adaptive (Lamarkian) responses to the environment. Today, such Lamarkian explanations appear to be totally without merit. However, in the absence of any evidence for chromosomes within bacteria, a novel form of proof had to be devised showing that bacterial variants occur spontaneously in the absence of discrete signals from the environment.

Such proof first arose out of experiments carried out in 1943 that measured the frequencies of newly occurring *E. coli* variants resistant to attack by the bacteriophage T1. Such variants arise with a frequency of approximately 1 in 10 million (10^{-7}). If they represented adaptive responses to actual physical contact with T1, then roughly the same number of phage-resistant cells should be detected in duplicate cultures of bacteria containing some 10^8 cells, each such culture arising from the multiplication of a single cell introduced into the pure medium. If, on the other hand, they represented spontaneous mutations, then an occasional culture would contain large jackpots of mutants, because they would result from mutations that occurred early in the growth of the culture. Moreover, there would also occur other cultures totally lacking any phage-resistant mutants. The re-

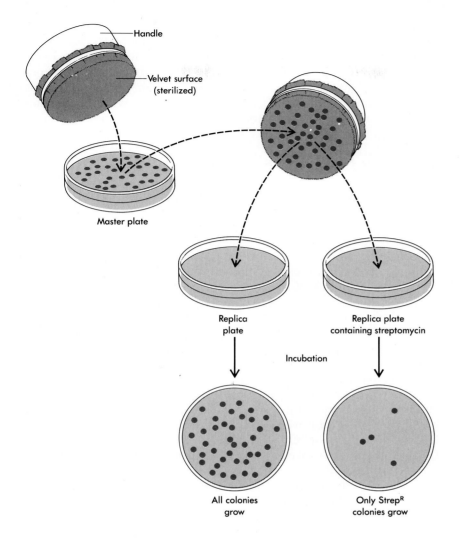

Handle

Velvet surface
(sterilized)

Master plate

Replica
plate

Replica plate
containing streptomycin

Incubation

All colonies
grow

Only Strep^R
colonies grow

Figure 7-2
Replica plating. A piece of sterile velvet cloth is pressed down on an agar plate containing many colonies, picking up cells in the process. The velvet is then pressed against another agar plate to transfer the cells, which then grow into colonies oriented exactly like those of the original (master) plate. Antibiotic-resistant colonies, such as streptomycin-resistant (Strep^R) colonies, can be identified by adding streptomycin to a second replica plate. Only Strep^R cells will grow on the replica plate with streptomycin.

sults of such "fluctuation analysis" unequivocally showed that resistance to phages as well as to antibiotics originated before the bacteria had contact with phages or antibacterial agents (Figure 7-1). Through these elegant experiments, adaptive (Lamarkian) explanations for bacterial variability lost all credibility. It became more logical to believe that all stable hereditary variations in bacteria represented changes in genetic material similar to that possessed by eucaryotic organisms.

The Great Power of Replica Plating[5]

An even simpler demonstration of the spontaneous nature of bacterial variants came a decade later through the development of **replica plating** (Figure 7-2). In this procedure, a flat, sterile piece of velvet cloth is placed face down on top of an agar plate containing a hundred to a thousand tiny bacterial colonies. Through this process, some living bacteria from each colony are transferred to the velvet. Subsequent application of the velvet to a sterile fresh agar surface allows each parent colony to give rise to a new colony positioned identically to those found on the original plate. Many identical replicas of an original plate of colonies can be so obtained and each tested for their possession of cells resistant to the antibiotic streptomycin, for example. The same sets of colonies display streptomycin resistance, showing that they existed on the original agar plate long before any contact with the streptomycin. Since its original use, replica plating has advanced bacterial (*E. coli*) genetics by allowing single colonies to be rapidly screened for large numbers of genetic markers.

The Use of Mutagens

Most mutant genes studied by the early Mendelian geneticists arose spontaneously. Now there is increasing use of mutations induced by external agents such as ionizing radiation, ultraviolet light, and certain chemicals. These agents, collectively called **mutagens,** greatly increase the rate at which mutant genes arise. Mutagens act quite indiscriminately; no presently known physical or chemical mutagen increases the possibility of mutating a given gene without also increasing the possibility of mutating all other genes.

For many years, the various forms of radiation were the most powerful known mutagens. Now chemical mutagens are usually used because they require no special radiation sources and because some of them produce a much higher percentage of mutated genes. Mutagenic chemicals fall into three classes (see Chapter 11 for more details): (1) nucleotide-base analogues, which because of their close structural similarity to the normal four bases of DNA can be incorporated into DNA (afterward their different base-pairing possibilities lead to mistakes in DNA replication); (2) chemicals that react with DNA to change the bases into forms that base-pair abnormally; and (3) DNA intercalating agents, which by inserting themselves between the base pairs also lead to frequent mistakes in DNA replication. The most powerful known mutagens are certain alkylating agents like ethyl methane sulfonate, which add methyl or ethyl groups to the rings of the DNA bases. One of these agents, nitrosoguanidine, acts preferentially at the DNA replicating forks (see Chapter 10), frequently producing clusters of closely linked changes in DNA base sequences.

Mutations That Lead to *E. coli* Growth Factor Requirements[6]

A further key step in the development of *E. coli* as a system for the study of genetics came in 1944 with the realization that mutations similar to those earlier obtained in *Neurospora* could be obtained that affected its ability to synthesize essential metabolites. For example, *E. coli* ordinarily grows well with only glucose as a carbon source; but as a result of specific mutations, there are mutant *E. coli* strains that will grow only when their normal medium is supplemented with specific metabolites.

Such **growth factor** mutations are very easy to work with: To test for their presence, we only need to grow a suspected mutant both in the presence and in the absence of a metabolite, for example, the amino acid arginine (Figure 7-3). If a mutation inhibiting arginine biosynthesis has occurred, the bacteria will grow only in the presence of arginine as a supplement. The use of this approach quickly led

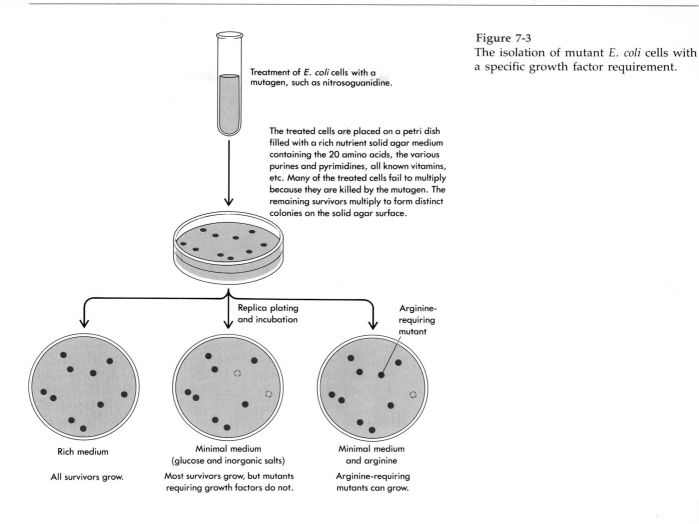

Treatment of *E. coli* cells with a mutagen, such as nitrosoguanidine.

The treated cells are placed on a petri dish filled with a rich nutrient solid agar medium containing the 20 amino acids, the various purines and pyrimidines, all known vitamins, etc. Many of the treated cells fail to multiply because they are killed by the mutagen. The remaining survivors multiply to form distinct colonies on the solid agar surface.

Replica plating and incubation

Arginine-requiring mutant

Rich medium

All survivors grow.

Minimal medium (glucose and inorganic salts)

Most survivors grow, but mutants requiring growth factors do not.

Minimal medium and arginine

Arginine-requiring mutants can grow.

Figure 7-3
The isolation of mutant *E. coli* cells with a specific growth factor requirement.

(with the help of mutagens) to the isolation of a large number of different growth factor mutations. Their existence has vastly speeded up elucidation of the enzymatic steps involved in the biosynthesis of many important intermediary metabolites.

The Highly Diverse Phenotypes of Bacterial Mutants[7]

Mutant cells that have lost their ability to synthesize essential metabolites are called **auxotrophs** ("increased growth requirements"), while wild-type cells lacking such requirements are called **prototrophs.** For example, there exists a class of *E. coli* auxotrophs unable to metabolize a diverse collection of sugars (for example, lactose, glucose, maltose) as well as other auxotrophs unable to metabolize one of these food molecules.

Some cells contain mutations that are only expressed under certain conditions (**conditional mutants**), as in the case of cells that can grow at 30°C but are unable to grow at 37°C. Most temperature-sensitive mutants synthesize enzyme products that are unstable at the restrictive temperature (37°C in our example), yet sufficiently active at the permissive temperature (30°C) to yield a wild-type phenotype. Conditional mutations can occur in virtually all genes, including those that control the steps in macromolecular synthesis, modification, and assembly into supermolecular structures.

Many valuable mutations involve the resistance to poisonous compounds like antibiotics. For example, most *E. coli* cells are rapidly killed by small amounts of streptomycin. Very rarely (one in 10^9 cell generations) there occur mutations (Strep^R) that make the cells resistant to certain amounts of this compound. Also easily observed are mutations that cause bacteria to be resistant to their viruses, the phages. One such mutation in *E. coli* strain B confers resistance to phage T1; these mutant cells are designated B/1. Correspondingly, *E. coli* strain B cells resistant to T2 and T4 are called B/2 and B/4, respectively.

Mutant Phenotypes Need Not Be Immediately Expressed[8]

Although they are genetically haploid, bacterial cells tend to contain more than one chromosome (two to four identical chromosomes); therefore, the time taken for a mutant gene to be expressed depends on whether the mutant allele is dominant or recessive. If it is dominant, the cellular phenotype becomes mutant as soon as sufficient protein products become synthesized. This interval may be no more than several minutes. However, most mutant alleles are recessive, and here the time taken for expression can be considerable. To start with, at least two cell divisions must occur before cells containing four chromosomes become genetically homogeneous. Even then, the resulting cells are likely to contain significant amounts of the wild-type protein product in the cytosol. Consider, for example, what happens when an *E. coli* cell becomes resistant to phage T4. Attachment of T4 to the surface of *E. coli* occurs by its binding to a specific outer membrane protein, a porin called OmpC, which normally is present in some 10^5 copies per cell. Resistance develops when the gene coding for OmpC is inactivated. As only one such porin receptor is necessary for T4 attachment, many cell divisions (up to 11) may have to occur before the mutant cell becomes totally resistant to T4.

Enriching Mutants Through Direct Selection, Counterselection, or Physical Selection[9, 10, 11]

Advancement of the genetic analysis of *E. coli* has depended on three different enrichment techniques for increasing the percentage of mutant cells in comparison to their wild-type equivalents: (1) **direct selection,** which utilizes growth conditions that greatly favor the growth of the desired mutant; (2) **counterselection,** which makes use of conditions that kill the parent cells; and (3) **physical selection,** which involves unique properties of mutant cells that enable them to be physically separated from their parent cells.

Direct selection is usually employed to isolate mutant cells relatively resistant to well-defined chemical (e.g., antibiotic), physical (e.g., ultraviolet radiation), or biological (e.g., phage) agents. Since most such agents kill the parent cells, at least several cycles of cell division must occur following exposure to a mutagen in order to increase the numbers of mutants before the direct selective agent is applied (Figure 7-4). Direct selection can also be used to isolate rever-

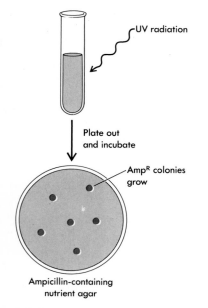

UV radiation

Plate out and incubate

Amp^R colonies grow

Ampicillin-containing nutrient agar

Figure 7-4

An example of direct selection. Bacterial cells are selected for ampicillin resistance after exposure to a mutagen.

tants, cells that have regained a function lacked by either auxotrophic parents or by carbon or nitrogen source mutants.

Counterselection usually involves application of agents that selectively kill *growing* cells (Figure 7-5). The first such method ever used utilized the fact that penicillin kills only growing cells. In the presence of penicillin, the crosslinks that tie together newly made peptidoglycan precursors cannot be made. This leads to weakened cell walls that burst from exposure to the cell's internal osmotic pressure. Thus, if mutagenized cells are placed in a minimal medium and then exposed to penicillin, the only cells not killed by the penicillin should be new auxotrophic mutants that are unable to synthesize essential metabolites. Another method for killing growing cells involves exposure to the thymine analogue 5-bromouracil. When incorporated into bacterial DNA, it makes the cells particularly sensitive to killing by ultraviolet light. Still another procedure utilizes parent cells that are unable to synthesize thymine. In the absence of thymine, these cells die once they begin to grow (thymineless death). If, however, they have another defect that prevents them from initiating growth, they remain alive. Selection for specific amino acid auxotrophs can thus be achieved by suspending mutagenized thymine-requiring cells in an enriched medium lacking thymine and the respective amino acid.

Physical selection often distinguishes mutant cells from normal cells on the basis of size. For example, cells unable to divide at high temperatures because of a temperature conditional mutation tend to be smaller at high temperatures. These cells can then be detected by cell-sorting systems that discriminate size differences on the basis of the amount of light they intercept. Another physical separation technique selectively divides a group of cells into different locations within a medium. For example, nonmotile mutants unable to respond to chemical attractants cannot swim away from nutrient-depleted regions in which their relative concentration becomes much enriched over their normal equivalents (Figure 7-6).

Colored Dyes Are Used to Detect Desired Metabolic Mutants[12, 13]

Carbon source (e.g., lactose) mutants of *E. coli* can be easily detected using identification media that dye the medium surrounding bacteria differentially, depending on whether the sugar in question can be metabolized. These methods utilize the fact that when the sugar is fermented sufficient acid is produced to lower the pH of the surrounding medium. The lower pH then triggers a change in the color of the dye (Figure 7-7). Such media can support equally well the growth of cells that can or cannot ferment the sugar. Colonies unable to utilize lactose, for example, can be selected for by looking for colonies growing in the presence of lactose that do not acidify the medium and thus do not change the color of the medium.

Brute Force Isolation[14, 15]

In many other cases, detection of desired mutants is much more laborious and depends on direct measurements of the amount of desired enzymes or cellular components. Such brute force approaches work only following application of highly efficient mutagens. Their use

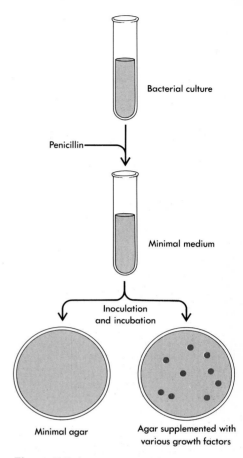

Figure 7-5
An example of counterselection. Selection is for a bacterial mutant unable to grow in the absence of a specific growth factor (e.g., arginine). Cells are added to minimal medium (lacking growth factors), and penicillin is added to kill all actively growing cells. Mutants that were not growing owing to a lack of growth factors are not killed by the penicillin and thus are selected for. They can then be seen by plating them on a supplemented agar plate.

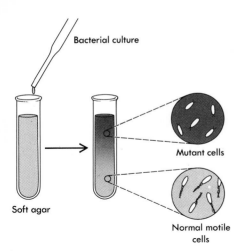

Figure 7-6
An example of physical selection. Cells with mutant flagella rendering them nonmotile can be selected for by inoculating a tube of soft agar with a bacterial culture previously exposed to a mutagen. Normal cells will move toward the bottom of the tube as nutrients in the upper portions of the tube are used up. The mutant nonmotile cells remain in the upper portion. An indicator dye can be used to detect the areas of nutrient depletion (colored area).

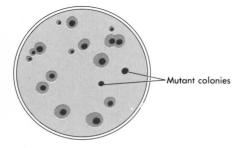

Figure 7-7
The use of pH-sensitive dyes to detect metabolic mutants. Mutants unable to ferment lactose in the medium do not release acid and thus do not change the color of the medium surrounding them.

now allows the isolation of virtually any "loss-of-function" mutation by screening some 10,000 treated colonies. An *E. coli* strain lacking the enzyme ribonuclease I was the first mutant isolated by massive brute force screening, and soon after a much sought after strain lacking virtually all of its DNA polymerase I was so obtained.

Proof of the Existence of Genetic Recombination Within *E. coli*[16]

The isolation of mutations involving growth factors, antibiotic resistance, and phage resistance was quickly followed by experiments demonstrating the occurrence of genetic recombination (and hence a sexual process) in *E. coli.* Twofold use was made of the mutant genes. First, they were used as conventional genetic markers since their segregation patterns revealed the chromosomal arrangement of genes. Second, they provided a method of detecting a genetic recombination process occurring in only a small fraction of the population during a given time. In *E. coli,* for example, simple morphological examination of bacterial cells gave no clues that cell fusion and genetic recombination existed. To detect recombination, it was necessary to devise an experiment in which only the recombinant cells would be able to multiply. This was done by using parent strains with specific growth requirements such that they could not multiply in minimal media.

In these experiments, conducted in 1946, two strains of bacteria, each possessing different growth requirements, were mixed together (Figure 7-8). Neither strain alone was able to grow in the absence of its specific growth factors; the amino acids threonine and leucine were required by one strain, the vitamin biotin and the amino acid methionine by the other. After the two strains were mixed together, a small number of cells were able to grow without any growth factors added to the medium. This result showed that the growing cells had somehow acquired wild-type copies of each of their mutant genes, strongly suggesting that *E. coli* has a sexual phase that can bring together the chromosomes of two different cells. Crossing over could then take place between two chromosomes carrying mutations in different genes and lead, in some cases, to a chromosome with wild-type alleles for all necessary genes. Genetic analysis confirmed this hypothesis, and through the genetic analysis done over the last three decades, *E. coli* has become one of the best understood organisms.

Bacterial Viruses (Phages) Provide Simple, Easy-to-Study Chromosomes[17, 18]

Our knowledge of the genetics of several bacterial viruses (usually called bacteriophages or phages) has made enormous strides along with the expansion of our knowledge of bacteria. Their discovery in 1914 produced great excitement, for it was hoped that they might afford an effective and simple way to combat bacterial diseases in the days before the discovery of antibiotics. Phages, however, were never medically useful because their bacterial hosts readily mutate into forms resistant to viral infection. Thus, almost everybody lost interest in the phages, and few studied their replication until the late 1930s. Then, a small group of biologists and physicists (who came to be known as "the phage group"), intrigued by the problem of gene rep-

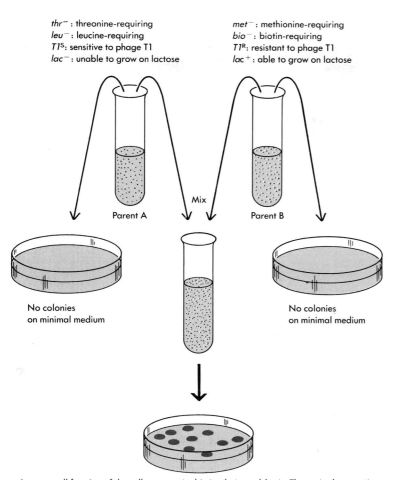

thr⁻ : threonine-requiring
leu⁻ : leucine-requiring
T1ˢ: sensitive to phage T1
lac⁻ : unable to grow on lactose

met⁻ : methionine-requiring
bio⁻ : biotin-requiring
T1ᴿ: resistant to phage T1
lac⁺ : able to grow on lactose

Parent A

Mix

Parent B

No colonies
on minimal medium

No colonies
on minimal medium

A very small fraction of the cells are *met⁺*, *bio⁺*, *thr⁺*, and *leu⁺*. They arise by genetic recombination, as shown by examination of the *lac* and *T1* markers. In addition to the parent *lac⁻ T1ˢ* and *lac⁺ T1ᴿ* genotypes, there are found *lac⁻ T1ᴿ* and *lac⁺ T1ˢ* cells.

Figure 7-8
The use of growth factor requirements to demonstrate sexuality in *E. coli*. Parent A, parent B, and a mixture of the two are plated out on minimal medium. Only the mixture gives rise to colonies.

lication, began to investigate the reproduction of several phages that multiplied in *E. coli*. They chose to work with phages rather than with the equally tiny plant or animal viruses because under laboratory conditions, it is far easier to grow bacterial cells than plant or animal cells.

Before the phage group came into being, the biological significance of phages, as well as that of animal and plant viruses, was obscure, and the question most often asked was, Are phages (and other viruses) living? Now we realize that all viruses are small pieces of genetic material (DNA or RNA), each enclosed within a protective protein-rich coat that facilitates its transportation from one cell to another. Progeny viruses closely resemble their parents because they contain identical chromosomes. Viruses are no more alive than isolated chromosomes; both the chromosomes of cells and those of viruses can duplicate only in the complex environment of a living cell. Nonetheless, viruses have been of immense value in our understanding of how cells live. For more than 30 years (1940–1973), they afforded the only convenient systems for studying the consequences of the sudden introduction of new genetic material into a cell.

Most work with phages has been—and still is—concentrated on a small collection that multiplies in *E. coli* or closely related bacteria. Arbitrarily, they have been given names like T1, T2, P1, F1, or λ. Now

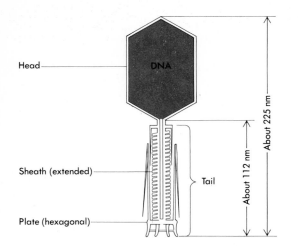

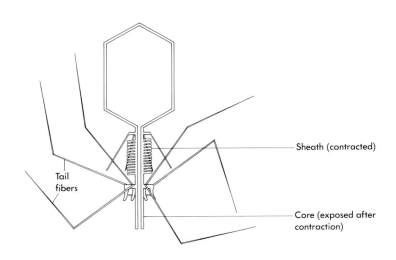

Intact phage

Phage after DNA expulsion

Figure 7-9
Electron micrograph and diagrammatic representation of bacteriophage T4.

among the best known of the DNA-containing phages are T4 (Figure 7-9), T7, and ϕX174. Also well characterized are the group of small spherical *E. coli* phages that use RNA as their genetic material. Among these are the phages f2, R17, and MS2.

Phages Do Not Grow by Gradually Increasing in Size[19]

The life cycle of an average cell involves a gradual increase in size to approximately twice its initial size, followed by a division process (mitosis) producing two identical daughter cells. Phages, however, do not multiply in this fashion. They are not produced by the fission of a large, preexisting particle; on the contrary, all the virus particles of a given variety have approximately the same (and in some cases identical) masses.

Upon infection of a bacterial cell, there is a temporary disappearance of the original parent virus particle. During this phase, which is called the **eclipse period,** the parent virus breaks down and releases its chromosome into the bacterial cytosol, leaving the empty protective protein coat, or **capsid,** outside the bacterial cell. Then the free viral chromosome serves as a template to direct the synthesis of new viral components. This breakdown process is an obligatory feature of viral multiplication. As long as the phage chromosome is tightly enclosed within the capsid, it cannot provide the information to direct the synthesis of either new chromosomes or new coat protein molecules.

The growth cycles of *E. coli* phage start when a phage particle collides with a sensitive bacterium and attaches to a specific receptor on the surface of its outer membrane (Figure 7-10). Different phages attach to different receptors. Phage λ, for example, attaches to a protein involved in the transport of maltose into *E. coli*, while T4 attaches to one of the porin proteins (OmpC) used to construct narrow channels spanning the outer membrane. How the viral chromosome passes through either the outer or inner cell membrane remains to be worked out. Many phage particles possess enzymes capable of breaking down components of bacterial envelopes. It is probable that they have specifically evolved to make the necessary holes.

Following its entry into the cytosol, the phage chromosome and the subsequent daughter chromosomes continue to duplicate to eventually form 100 to 10,000 new chromosomes. These, in turn, become encapsulated into newly synthesized capsids, forming a large number of new bacteriophage particles. The protein coats are encoded by phage genes that instruct the protein-synthesizing machinery of the host cell to construct the new capsids. The growth cycle is complete when the bacterial cell wall breaks open (lyses) and releases the progeny particles into the surrounding medium.

Phages (Viruses) Are Parasites at the Genetic Level

Viruses are of necessity parasites. It is impossible for them to reproduce without host cells, on which they are totally dependent for metabolic precursors and for ribosomes, the cellular organelles necessary for making proteins. It is not their small size per se that makes them viruses but the fact that they do not possess the capacity to synthesize their components independently. To replicate, they must always insert their nucleic acid into a functional cell. This feature enables us to distinguish the larger viruses, such as that causing smallpox, from very small cellular organisms, including the mycoplasmas and such intracellular disease-causing parasites as the rickettsias. The latter two organisms are distinct from viruses, since they contain both DNA and RNA and grow by increasing in size and then splitting into two smaller cells. At no time do their cell membranes break down, and all their proteins are synthesized upon their own protein-synthesizing machinery.

Phages Form Plaques

The presence of viable phage particles in a solution can be quickly demonstrated by adding some of the solution to the surface of a nutrient agar plate on which bacteria susceptible to the phage are rapidly multiplying. If no phage particles are present, the rapidly dividing bacteria will form a uniform surface layer of bacteria. But if even one phage particle is present, it will attach to a bacterium and multiply to form several hundred new progeny phage particles, which are then released by dissolution (lysis) of the cell wall some 15 to 60 minutes after the start of phage infection. Each of these several hundred progeny particles can then attach to a new bacterium and multiply. After several such cycles of attachment, multiplication, and release, all the bacteria in the immediate region of the original phage particles are killed. These regions of killed bacteria appear as circular holes, or **plaques,** in the lawn of healthy bacteria (Figure 7-11).

Phages Also Mutate[20, 21, 22]

The plaques formed by a given type of phage are quite characteristic and can often be distinguished from those of genetically distinct phages. For example, the plaques of phage T2 can easily be distin-

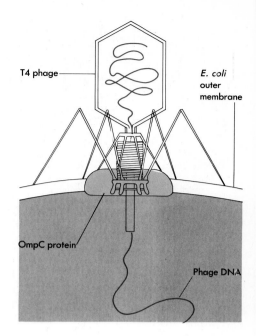

T4 phage — E. coli outer membrane

OmpC protein

Phage DNA

Figure 7-10
Injection of bacteriophage T4 DNA into an *E. coli* cell. Phage T4 attaches to the outer membrane porin OmpC.

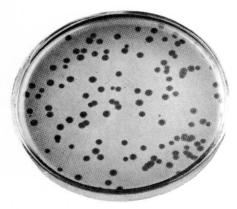

Figure 7-11
Photograph of phage T2 plaques on a lawn of *E. coli* bacteria growing in a Petri plate. (From *Molecular Biology of Bacterial Viruses* by Gunther S. Stent. W. H. Freeman and Company, copyright © 1963, p. 41, with permission.)

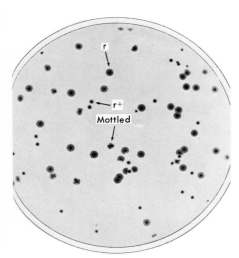

Figure 7-12
Photograph of mutant phage plaques, showing a mixture of T2 r^+ (wild-type) plaques and T2 r (rapid-lysis mutant) plaques. The mottled plaques result from the simultaneous growth of both r and r^+ phages in the same plaque. (From *Molecular Biology of Bacterial Viruses* by Gunther S. Stent. W. H. Freeman and Company, copyright © 1963, p. 177, with permission.)

guished from those made by phage λ or by phage f2. More significantly, mutations occur that change the morphology of phage plaques. We do not usually know the biochemical basis of these plaque differences, but this does not matter. The important fact is that these differences are usually reproducible and simple to score. It is easy to examine the morphology of thousands of plaques to see if any look different from the plaques made by the wild-type phage. In this way, a large number of different plaque-type mutations have been found (Figure 7-12).

Another class of mutations changes the ability of the phage to adsorb to bacteria. For example, wild-type T2 cannot multiply on *E. coli* strain B/2, because the B/2 mutation changes the cell surface, thereby preventing the attachment of T2. Mutant T2 particles can, however, multiply on B/2: They are called T2h (h for host), and they are able to adsorb because they possess altered tail fibers.

Another very large and important class of phage mutants consists of phages that can multiply at 25°C but not at 42°C (i.e., they are temperature-sensitive conditional lethals). Although we still cannot always pinpoint the exact reason why these mutants do not multiply at the higher temperature, in many cases the high temperature destroys the three-dimensional structure of a protein necessary for their reproduction. This type of mutation has proven very useful because a large number of viral genes can mutate to a temperature-sensitive form.

The existence of conditional lethal mutations has allowed phage geneticists to find mutations in essentially all the genes of any phage they wish to investigate. Although genetic recombination between mutant phages was not discovered until 1945, phages are among the most completely characterized genetic objects yet studied.

Phage Crosses[23, 24]

More than one phage particle at a time can grow in a single bacterium. If several particles adsorb at once, the chromosome from each enters the cell and duplicates to form a large number of new copies. So long as the chromosomes exist free (unenclosed by a protective coat), they can cross over (recombine) with similar chromosomes. This is shown by infecting cells with two or more genetically distinct phage particles and finding recombinant genetic types among the progeny particles. For example, it is easy to obtain mutant T2 particles differing from the wild type by two mutations, one in an *h* gene, which allows the particles to grow on *E. coli* strain B/4 (a strain resistant to wild-type T2 particles), and the other in an *r* gene, which causes the particles to form larger and clearer plaques than wild-type T2 particles. These double mutant phages are designated T2 *hr* and the wild type is called T4 h^+r^+. When an *E. coli* cell is infected simultaneously with a T2 h^+r^+ and a T2 *hr* phage, four types of progeny particles are found: the parent genotypes h^+r^+ and *hr* and the recombinant genotypes hr^+ and h^+r (Figure 7-13).

The frequency with which recombinants appear depends on the particular mutants used in the cross. Crosses between some pairs of markers give almost 50 percent recombinant phages; crosses between others give somewhat lower recombinant values, and sometimes almost no recombinants are found. This immediately suggests that viruses also have unbranched genetic maps, a suggestion confirmed by

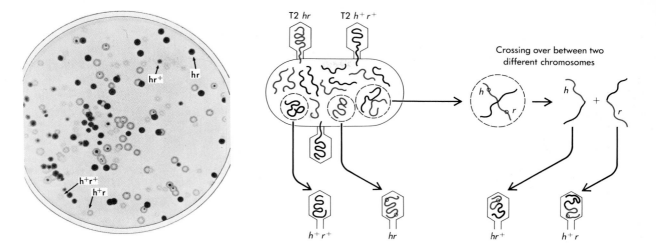

Figure 7-13
Phage recombination in cells infected
with two different strains of phage T2.
Within the cell, crossing over occurs
between the free phage chromosomes,
both identical and different. Recombinant and parent phages are identified
by plating and observing the plaques.
The technique used to see all four prog-
eny types (hr, hr^+, h^+r, h^+r^+) is to look
for plaques on a mixture of strain B
and B/2 cells. Only phages possessing the h
gene can kill both B and B/2 cells.
Phages with the h^+ gene kill only B
cells, and their plaques look turbid
owing to the presence of live B/2 cells.
The r gene causes the formation of
larger, clearer plaques than those
formed by the wild type. (From *Molecular Biology of Bacterial Viruses* by Gunther S. Stent. W. H. Freeman and Company, copyright © 1963, p. 185, with
permission.)

intense analysis of a large number of independently isolated mutations. The best-known phage genetic maps are those of T4 (Figure
7-14), T7, φX174, and λ. As we shall see in Chapter 17, the locations of
viral genes are far from random, with genes grouped together according to when they function during their respective phage life cycles.

Phage Crosses Involve Multiple Pairings

The genetic structure of phage chromosomes is quite similar to that of
cellular chromosomes. Nonetheless, it is important to note a distinct
feature of phage crosses. In conventional meiosis, each chromosome
pairs just once, whereas in a viral cross, pairing and crossing over
may occur repeatedly throughout the period when free chromosomes
are present in the infected cell. Thus, a given phage chromosome may
participate in several pairing and crossing-over events. We can demonstrate this fact by infecting a cell with three phage particles, each
with a distinct genetic marker, and finding single progeny particles
that have derived chromosomal regions from all three original phage
particles. Thus, the products of a phage cross are essentially the products of a large number of distinct pairings and crossings over. Moreover, there is great variation among viruses in the amount of crossing
over; for example, a chromosome of phage T4 crosses over an average
of 5 to 10 times during each growth cycle, while a phage λ chromosome crosses over only 0.5 time. The occurrence of many distinct
crossovers does not, however, restrict our ability to map the genetic
markers, since the general rule still holds that genes located close
together seldom recombine.

Another way in which a phage cross differs from recombination in
conventional meiosis is that phages are not separated into male and
female particles. Differentiation into sexes can be considered a device
to bring about cell fusion between genetically distinct organisms.
Phage particles, however, do not have to fuse, since genetic recombination occurs when two phage chromosomes of different genotypes
enter a single host bacterial cell. Moreover, in cells infected with several genetically distinct phages, crossing over can occur between both
genetically identical and genetically different chromosomes; thus,
only a fraction of the crossovers in a cell infected with different
phages result in recombinant chromosomes of a new genotype. There

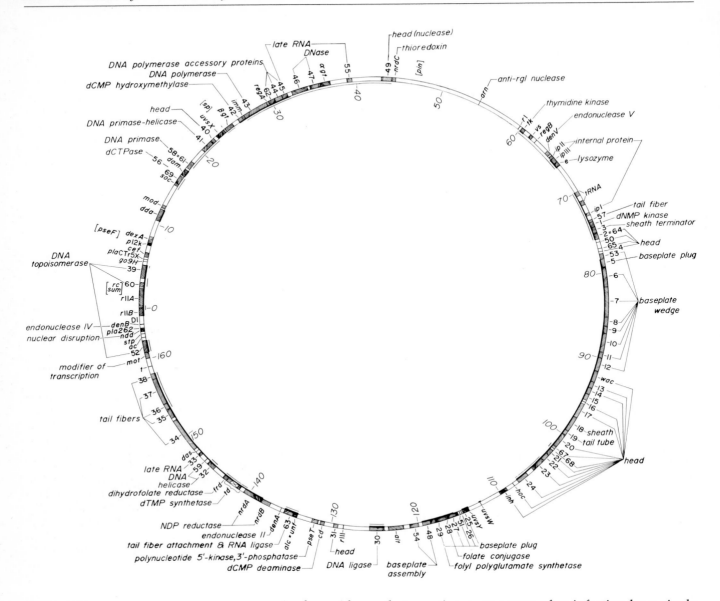

Figure 7-14
The genetic map of bacteriophage T4.
The units (inner circle) are kilodaltons.
(Courtesy of Dr. Burton S. Guttman
and Dr. Elizabeth M. Kutter.)

is also evidence that crossing over occurs after infection by a single phage particle. However, this phenomenon cannot be revealed by genetic analysis, since all the progeny chromosomes are genetically identical; thus, experiments with isotopically labeled phage chromosomes are needed for its demonstration.

Two Distinct Sexes Are Found in *E. coli*[25–33]

When sexuality in *E. coli* was first discovered, it seemed simplest to believe that a conventional cycle of cell fusion occurred followed by meiotic segregation. Now we know, however, that the *E. coli* sexual cycle has distinctive features complicating conventional analysis. As in higher organisms, there exist male and female cells, but these do not fuse completely, allowing their two sets of chromosomes to intermix and form two complete diploid genomes. Instead, the transfer is always unidirectional, with male chromosomal material moving into female cells; the converse movement of female genes into male cells never occurs.

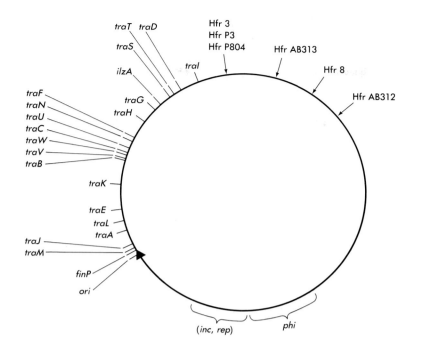

Figure 7-15
The genetic map of the F factor of *E. coli*. Note that about a third of the F⁺ DNA consists of genes involved in the transfer of the F factor to female cells (*tra* genes). Also present are genes associated with fertility inhibition (*fin*), phage inhibition (*phi*), incompatibility (*inc*), replication (*rep*), and immunity to lethal zygosis (*ilz*). The origin of transfer replication (*ori*) and the positions where insertion elements on F recombine with the bacterial chromosome to create Hfrs are also indicated.

Male and female *E. coli* cells are distinguished by the presence of a distinct supernumerary sex chromosome called the **F (fertility) factor.** When it is present as a discrete body, the cells are male (F⁺) and capable of transferring genes into female cells. In its absence, *E. coli* cells are female (F⁻) and act as recipients for gene transfer from male cells. Like the main *E. coli* chromosome, the F factor is a circular double-helical DNA molecule present in one copy per cell. It only carries approximately 94×10^3 base pairs—¹⁄₄₀ the amount of genetic information contained in the main chromosome. About one-third of the F⁺ DNA consists of 19 transfer (*tra*) genes specifically involved in the transfer of male genetic material into female cells (Figure 7-15). Among these genes are those responsible for synthesis of the sex-specific **F pili.**

When F⁺ cells are mixed with F⁻ cells, conjugal pairs form by the attachment of a male sex (F) pilus to the surface of a female cell (Figure 7-16). Attachment then triggers a set of still not understood events that initiates DNA replication of the F⁺ chromosome and leads to the transfer of a progeny F⁺ chromosome to the F⁻ cell (see Chapter 10 for details). It is important to note that the only chromosomal material transferred by F⁺ cells is the F⁺ factor itself (Figure 7-17). None of the genes present on the main chromosome are transferred.

Pair formation and F⁺ transfer occur very efficiently between F⁺ and F⁻ cells. As a result, all the cells in mixed cultures rapidly become male (F⁺) donor cells.

Integration of F (Fertility) Factors into Main Bacterial Chromosomes Creates Hfr Strains[34]

Transfer of most bacterial genes from male to female cells only occurs when, through a crossing-over process, the F⁺ factor becomes integrated into the cell's single main chromosome (Figure 7-18). Such **Hfr**

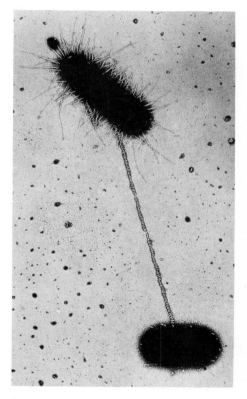

Figure 7-16
The attachment of a male F pilus to the surface of a female cell. This is followed by the replication of the F chromosome and its transfer to the female cell, which then becomes F⁺. (Courtesy of C. C. Brinton, Jr., and Judith Carnahan.)

Figure 7-17
The transfer of F⁺ DNA to an F⁻ cell. The size of the F DNA is exaggerated for clarity. One strand of the F factor is cleaved, and replication commences by a rolling circle mechanism (see Chapter 11). The newly replicated F factor passes into the F⁻ cell at a point of direct contact between the two cells after they are linked by the F pilus. Note that only the F DNA is transferred.

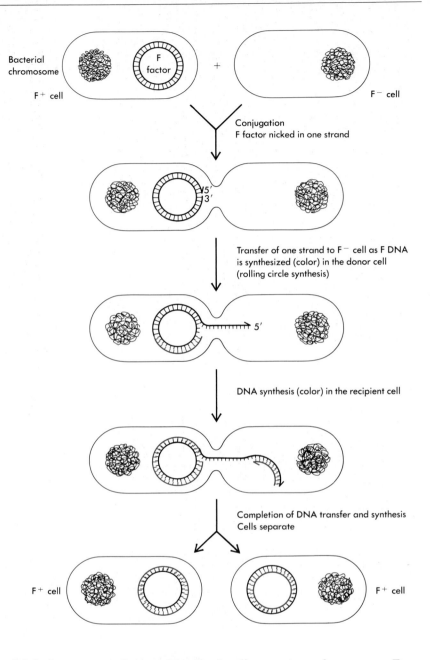

Bacterial chromosome

F factor

F⁺ cell

F⁻ cell

Conjugation
F factor nicked in one strand

Transfer of one strand to F⁻ cell as F DNA is synthesized (color) in the donor cell (rolling circle synthesis)

DNA synthesis (color) in the recipient cell

Completion of DNA transfer and synthesis
Cells separate

F⁺ cell

F⁺ cell

(high frequency of recombination) cells remain male, possess F sex pili, and fuse with F⁻ cells as frequently as F⁺ cells do. When an Hfr cell joins to an F⁻ cell, conjugation-induced replication of F DNA begins, and because the leading edge of the F⁺ factor is now attached to the main chromosome, transfer of the main chromosome follows (Figure 7-19). Because DNA replication is involved, transfer is inherently a lengthy process, requiring some 100 minutes at 37°C to bring a complete male genome into a female cell. Complete transfer, however, rarely occurs. The conjugal unions between male and female cells are very fragile, usually breaking and terminating transfer before it has finished. Since the main chromosomal genes are inserted into the F⁺ factor, only a portion of the F⁺ genome is usually transferred into recipient F⁻ cells (Figure 7-20). Consequently, the recipient cells usually remain female, in contrast to F⁺ × F⁻ matings, where the recipient cell invariably becomes converted to a male (F⁺) cell.

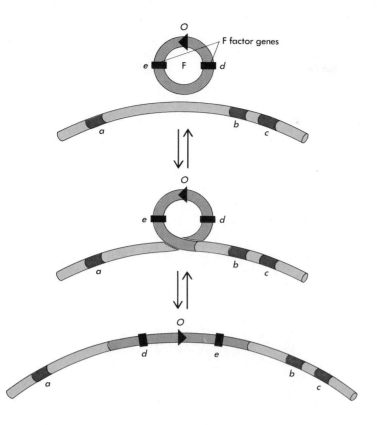

Figure 7-18
Integration of the F factor into the main chromosome, leading to the creation of an Hfr state. The F factor gene *O* is the origin of oriented chromosome transfer from the male cell to the recipient cell; *d* and *e* stand for two other F factor genes, and *a*, *b*, and *c* stand for genes on the main, circular chromosome. A single crossing-over event, as shown, leads to the integration of F between genes *a* and *b*. The resulting Hfr chromosome transfers *a* as one of its first genes and *b* as one of its last when conjugating with an F⁻ cell. In this diagram and the next two, double-stranded DNA is represented by a single line.

At first, the fact that only part of the male main chromosome enters the female cell made genetic analysis more difficult. Then it was realized that a fixed end of the male chromosome, specific for a given Hfr strain, always enters the female cell first and that the relative frequency with which male genes are incorporated into the recombinant chromosome is a measure of how close they are to the entering end. Moreover, it is possible to break apart the male and female cells artificially by violent agitation (this was first done in a mixing machine, the Waring Blender); matings can be made and the couples violently agitated at fixed times during the process. (These experiments were known as "the blender experiments"). If the pairs are disrupted soon after mating, only the genes very close to the forward end will have entered the cell (Figure 7-21). It is thus possible to roughly map *E. coli* gene positions merely by observing the time intervals required for various alleles carried by the male cells to enter female cells.

Bacterial Chromosomes Are Circular[35]

The genetic map obtained through analysis of interrupted matings is the same as that arrived at by analysis of frequencies of various recombinant classes. As in the chromosomal maps of higher organisms, the bacterial genes are arranged on an unbranched line. However, there is one important distinction: The genetic map of *E. coli* is a circle.

Within a given Hfr strain, the male chromosome always initiates DNA duplication prior to DNA transfer at the point of F factor insertion. The place where replication initiates, however, is not the same

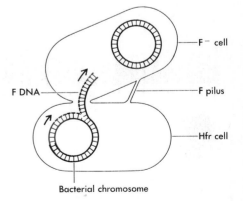

Figure 7-19
Movement of the bacterial chromosome from an Hfr cell into an F⁻ cell. The chromosome is led by a segment of the F factor DNA. The F DNA is actually only a tiny fraction of the total cell DNA.

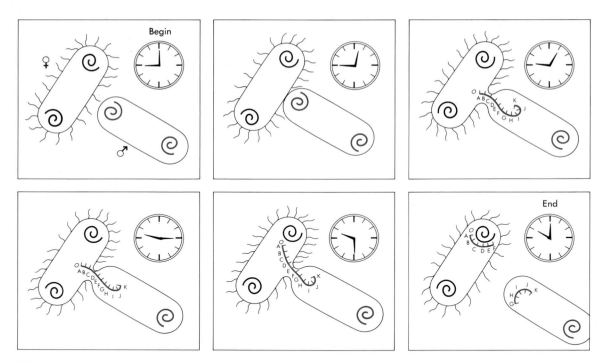

Figure 7-20
Conjugation between F⁻ and Hfr bacteria, as shown in a classic diagram, drawn before the bacterial chromosome was known to be circular. Since the transfer is a lengthy process, the fragile conjugal unions are usually broken before transfer is complete. Genes on the donor chromosome are denoted by capital letters. [After F. Jacob and E. L. Wollman, *Sexuality and the Genetics of Bacteria* (New York: Academic Press, 1961).]

in all strains, since the F factor can become integrated at any one of several preferred sites. There is one Hfr strain, for example, in which some of the genes involved in the synthesis of threonine and leucine are transferred soon after mating; in another strain, however, a gene involved in methionine synthesis is among the first to enter the female cell (Table 7-1). The existence of many different Hfr strains, each with its own specific sites of integration, has been very important in assigning *E. coli* gene locations. If there was only one entering point, it would be nearly impossible to assign even a rough order to those genes that are transferred last, since they would enter the female cell only very rarely.

The biological significance of circular chromosomes is still somewhat of a mystery. However, circular chromosomes are also found in many viruses, so the *E. coli* form should not be viewed as a strange

Figure 7-21
The frequency of donor Hfr marker genes (*azi, T1, lac, gal*) appearing among the resulting recombinants after an interrupted mating experiment. This procedure makes it possible to roughly map *E. coli* gene positions. Here the relative gene order would be *azi-T1-lac-gal*. [After F. Jacob and E. L. Wollman, *Sexuality and the Genetics of Bacteria* (New York: Academic Press, 1961).]

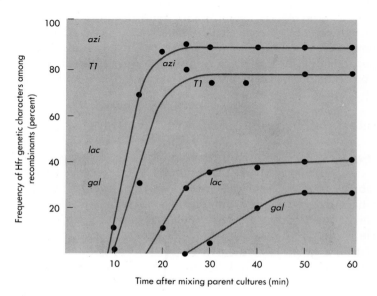

Table 7-1 Order of Genes in Conjugal Transfer in Different Hfr Strains

Hfr Strain	Order of Gene Transfer
Hayes	O-thr-leu-azi-ton-pro-lac-pur-gal-trp-his-gly-str-mal-xyl-mtl-ile-met-thi
Hfr 1	O-leu-thr-thi-met-ile-mtl-xyl-mal-str-gly-his-trp-gal-pur-lac-pro-ton-azi
Hfr 2	O-pro-ton-azi-leu-thr-thi-met-ile-mtl-xyl-mal-str-gly-his-trp-gal-pur-lac
Hfr 3	O-pur-lac-pro-ton-azi-leu-thr-thi-met-ile-mtl-xyl-mal-str-gly-his-trp-gal
Hfr 4	O-thi-met-ile-mtl-xyl-mal-str-gly-his-trp-gal-pur-lac-pro-ton-azi-leu-thr
Hfr 5	O-met-thi-thr-leu-azi-ton-pro-lac-pur-gal-trp-his-gly-str-mal-xyl-mtl-ile
Hfr 6	O-ile-met-thi-thr-leu-azi-ton-pro-lac-pur-gal-trp-his-gly-str-mal-xyl-mtl
Hfr 7	O-ton-azi-leu-thr-thi-met-ile-mtl-xyl-mal-str-gly-his-trp-gal-pur-lac-pro
AB311	O-his-trp-gal-pur-lac-pro-ton-azi-leu-thr-thi-met-ile-mtl-xyl-mal-str-gly
AB312	O-str-mal-xyl-mtl-ile-met-thi-thr-leu-azi-ton-pro-lac-pur-gal-trp-his-gly
AB313	O-mtl-xyl-mal-str-gly-his-trp-gal-pur-lac-pro-ton-azi-leu-thr-thi-met-ile

SOURCE: From F. Jacob and E. L. Wollman, *Sexuality and the Genetics of Bacteria* (New York: Academic Press, 1961).

exception. We shall return to the topic of circles when we look at the precise chemistry of the chromosome.

Over 900 Genes Have Already Been Mapped on the *E. coli* Chromosome[36, 37, 38]

By now, some 900 different genes have been located on the *E. coli* chromosome (Figure 7-22). Early on in gene mapping, it became clear that many genes with related functions (e.g., components of a biosynthetic or degradative pathway) were closely grouped on the chromosome. Such related gene clusters, called **operons,** are transcribed into single messenger RNA molecules, thereby ensuring that if one enzyme member of a metabolic pathway is present, its partner enzyme will also be present (see Chapter 16). Already, some 260 of the *E. coli* genes have been assigned to some 75 different operons, and the figure will probably go much higher, since less than half the *E. coli* genes have been identified so far. Moreover, we still remain ignorant of much of *E. coli*'s metabolism and so cannot yet spot the related functions of many adjacent genes.

Even without such definitive knowledge, most, if not all, the genes coding for certain key metabolic functions have been found (Table 7-2). Well characterized are the genes coding for the enzymes needed to synthesize amino acids, purines and pyrimidines, fatty acids, and vitamins, as well as the genes controlling the catabolism of most known carbon and nitrogen sources. The genes that code for more than 50 different protein components of ribosomes are also fully characterized.

Except for their grouping into operons, there appears no obvious reason for the relative position of most genes on the *E. coli* map. Genes controlling the biosynthesis of small molecules, fueling reactions, and macromolecular synthesis and assembly are found in all sections of the map. Moreover, while the genetic maps of *E. coli* and close enteric (intestinal) relatives (e.g., *Shigella, Salmonella, Enterobacter*) are quite similar, inversions of gene order occur. The map of *Salmonella typhimurium* is virtually identical to that of *E. coli* except that a segment containing some 10 percent of the genome is inverted with

Table 7-2 The Genes for Proteins Involved in Glycolysis

Gene Protein	Map Position
aceE pyruvate dehydrogenase (decarboxylase component)	3
aceF pyruvate dehydrogenase (dihydrolipoyltransferase component)	3
cxm methylglyoxal synthesis	6
eno enolase	59
fda fructose diphosphate aldolase	63
fpk fructose-1-phosphate kinase	46
fdp fructose diphosphatase	95
gap glyceraldehyde-3-phosphate dehydrogenase	39
glk glucokinase	51
kba ketose-*bis*-phosphate aldolase	69
pgi glucose phosphate isomerase	91
pfkA 6-phosphofructokinase I	88
pfkB suppressor of *pfkA*	38
pfkC modifier of 6-phosphofructokinase activity	58
pfl pyruvate-formate lyase	20
pgk phosphoglycerate kinase	63
pgm phosphoglucomutase	15
pps phosphoenolpyruvate synthase	37
tpiA triosephosphate isomerase	88

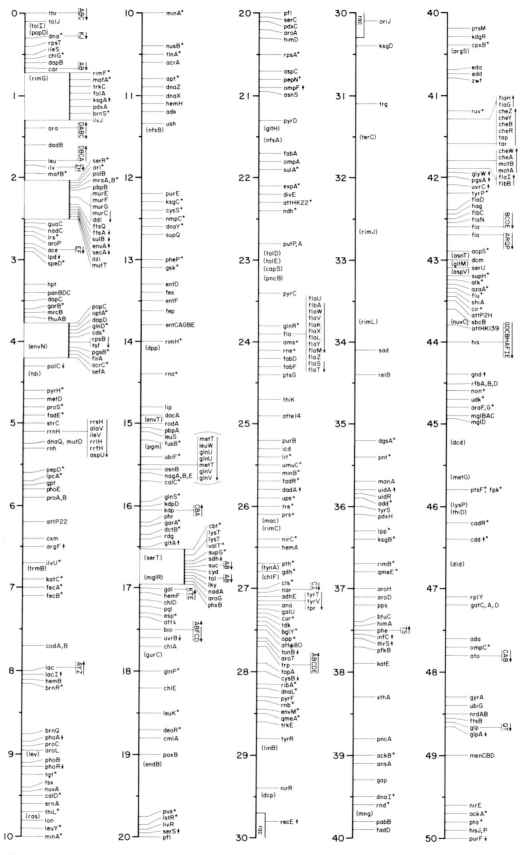

Figure 7-22
The complete genetic map of *E. coli*. [Courtesy of Barbara J. Bachman, *Microbiol. Rev.* 47 (1983):180]

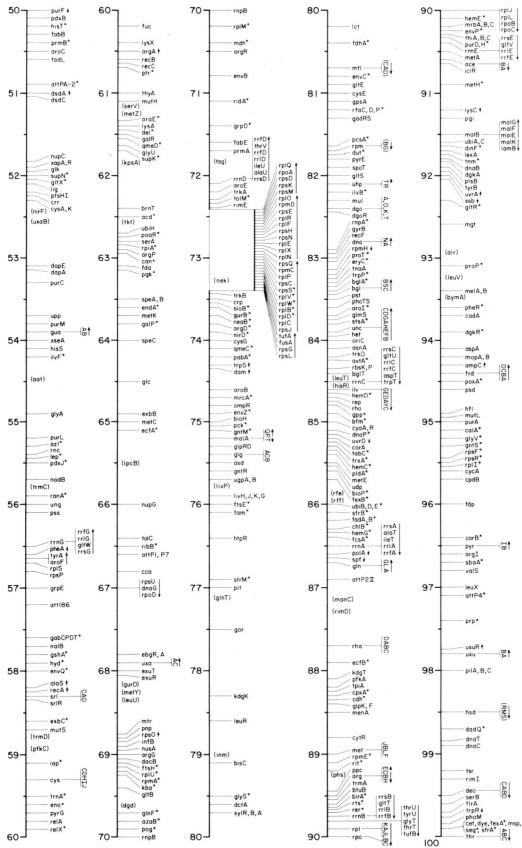

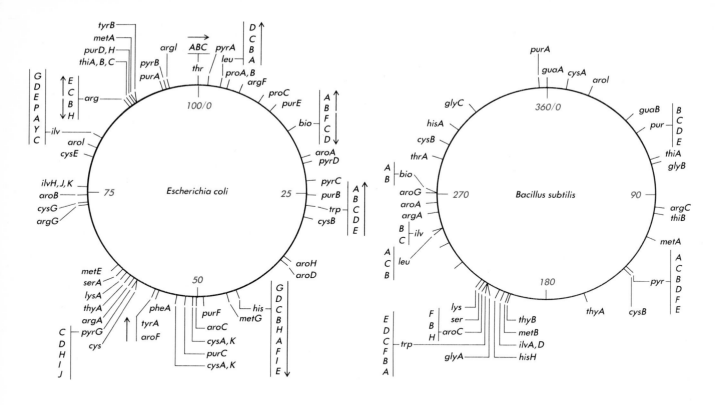

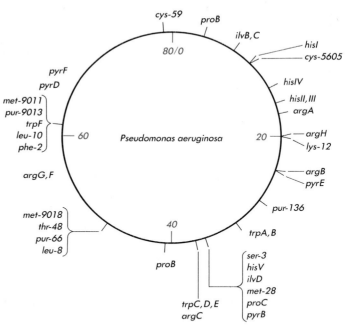

Figure 7-23
The genetic maps of *E. coli*, *B. subtilis* and *P. aeruginosa*, showing some of the comparable genes for these three bacteria. Note the different gene orders. (Data from J. L. Ingraham et al., *The Growth of the Bacterial Cell* pp. 178, 179, with permission.)

no obvious physiological consequence. Bacteria much more distantly related to *E. coli*, such as *Bacillus subtilus* and *Pseudomonas aeruginosa*, have completely different gene orders (Figure 7-23).

The orientation of a given operon along the chromosome also appears to be inconsequential. Because only one strand of a given operon's DNA is transcribed, it is either transcribed in the clockwise or counterclockwise direction. Of the 50 operons or genes whose transcription direction is known, 27 are transcribed in the clockwise direction and 23 in the counterclockwise direction.

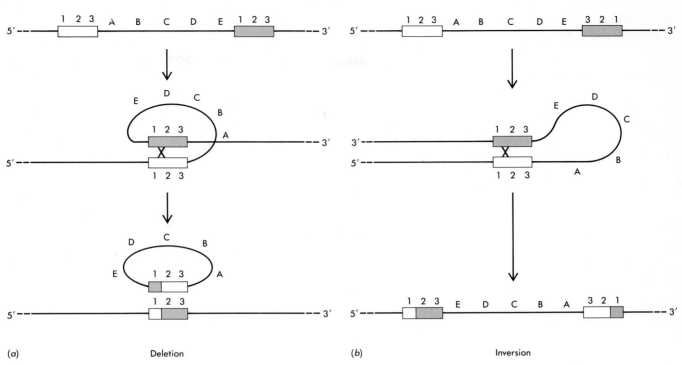

(a)   Deletion   (b)   Inversion

Figure 7-24
Crossing over between two identical genes on the same chromosome. (a) When two genes oriented in the same direction recombine, one gene and the DNA segment between them are deleted. (b) However, when the genes are oriented in opposite directions recombine, there is no deletion, and the segment between them becomes inverted.

Bacterial Genes Are Usually Present as Single Copies[39]

Virtually all *E. coli* genes are present in only one copy per genome. Multiple copies on the same chromosome are inherently unstable, since crossing over can occur between them, leading to either deletions or inversions of the chromosomal segments lying between. Deletion results when crossing over occurs between two genes oriented in the same direction, while inversions result when two genes oriented in opposite directions recombine (Figure 7-24). Multiple copies not only of the same gene but also of two different genes with highly similar structures will tend to recombine. A gene's inherent stability thus demands that it not be too similar to any other genes. Duplications of minuscule chromosomal regions, however, are not strongly selected against, since the chance of crossing over occurring between them is almost negligible.

Despite their inherent long-term instability, duplications of certain genes arise with frequencies as high as one in 10^4 cells. Thus, a significant fraction of the cells in any culture can be expected to carry duplicated genetic material somewhere. Such duplications are usually generated by illegitimate crossing-over events that involve short, partially homologous sections of DNA on dividing daughter chromosomes (Figure 7-25). Such duplication events, by creating new stretches of homologous DNA on the same chromosome, automatically tend to be followed by further gene amplification or elimination events.

The fact that so little duplicated genetic material exists, despite its continual new generation, tells us that it seldom conveys any selective advantage to those bacteria possessing it. Imagine, for example, a duplication that adds 10 percent more DNA to the *E. coli* genome. Because of this added DNA, bacteria carrying it will have their mini-

Figure 7-25
A deletion and a duplication generated by an illegitimate crossing-over event between homologous sections of DNA on daughter chromosomes.

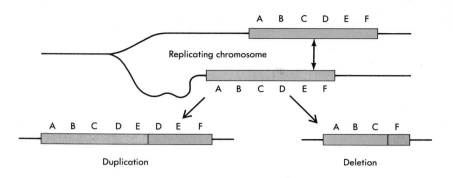

mum division time raised by 10 percent and so will be at a great disadvantage unless this additional DNA codes for particularly needed gene products. Duplications of the lactose operon, for example, have a positive selective advantage in cells growing solely on very limited amounts of lactose or on lactose derivatives that are very inefficiently metabolized. However, the moment such selective pressures are removed, the disadvantages of an unnecessary, if not actually harmful, genetic duplication manifest themselves. Then new variants are selected that eliminate the duplicated DNA through new crossing-over events.

There Are Multiple Copies of Genes Coding for Ribosomal RNA[40]

Most *E. coli* strains contain seven sets of the genes coding for the ribosomal RNA precursor that becomes enzymatically cleaved to 16S, 23S, and 5S mature ribosomal RNAs (Figure 7-26). Only through their possession of multiple rRNA genes can *E. coli* cells make enough rRNA chains to assemble the very large number of ribosomes needed to double their protein content every 20 minutes. It is important to note a very basic difference between those relatively few genes that code for rRNA and tRNA and the many more numerous genes that code for the many kinds of mRNA. The number of rRNA and tRNA molecules in a cell is inherently limited by the rate at which these genes are transcribed. In contrast, the number of products of those genes that code for proteins is potentially much higher. This is because each mRNA molecule is subsequently translated many times, yielding at least 50 to 100 identical protein products. There is thus no need for multiple copies of genes for proteins, even in the case of genes that code for ribosomal proteins, each of which accounts for about 0.5 percent of the total cellular protein.

Because the rRNA genes (rDNA) of necessity have virtually identical sequences coding for their functional rRNA products, unequal crossing over between two daughter *E. coli* chromosomes will occasionally occur. The result will either be a contraction or expansion of the number of rRNA genes. The deletion events that lead to contraction, however, will simultaneously delete essential genes lying between and so lead to the death of a small fraction of daughter *E. coli* cells each cell generation. Correspondingly, there will be produced a small number of cells with a duplicated DNA section containing one or more rRNA genes. At high growth rates, these tend to survive, but soon lose the duplicated DNA section when their growth rate is slowed down.

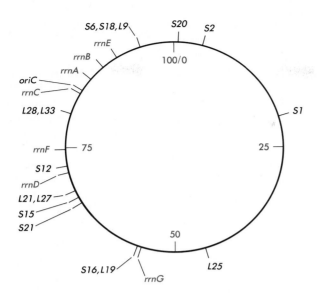

Figure 7-26
The genetic map of the *E. coli* K12 chromosome, showing the positions of the seven sets of ribosomal RNA genes (*rrnA* to *rrnG*), as well as the positions of some of the ribosomal protein genes (S and L refer to the small and large ribosomal subunits). The gene *oriC* is the origin of replication. (Data from J. L. Ingraham et al., *The Growth of the Bacterial Cell* p. 76, with permission.)

Most rRNA Genes Are Located Near the Origin of Replication

In *E. coli* cells duplicating their DNA, there is a greater concentration of genes near the point at which the bidirectional replication of DNA starts (map position 83) than near the terminus of replication (map position 33). Within cells dividing every 40 minutes, where the time required for DNA replication equals that required for a cell division, the copy number of genes near the origin of replication is almost twice that of those near the terminus. In very rapidly growing cells, where DNA replication reinitiates at the origin before the previous round of replication has been completed, genes near the origin will have copy numbers nearly four times those at the terminus. It is thus not surprising that most of the rRNA genes cluster around the origin of replication. If, on the other hand, they were located near the termination point, they would have to be present in even higher numbers to maintain an equivalent rate of protein synthesis.

The various control circuits that determine the level at which a given gene functions are thus assumed to have evolved with some regard for the gene's relative position on the chromosome. This may be the reason why gene location is so conserved among related enteric bacteria, despite the obvious ease with which new aberrantly rearranged chromosomes appear.

Transposable Genetic Elements Appear in *E. coli*[41, 42, 43]

Long sacred to genetic thinking was the conviction that genes have fixed chromosomal locations that change only as a consequence of chromosomal rearrangements resulting from illegitimate crossing over between incompletely homologous short sections of DNA. Then, in the late 1940s, genetic experiments with maize (corn) showed that certain genetic elements regularly "jump" to new locations. In so jumping, or **transposing,** they often inactivate the genes into which they become inserted. Thus, the existence of movable elements is often first revealed by their capacity to disturb the functioning of other previously characterized genes.

For almost 20 years, the significance of the jumping genes of maize was unclear. Then a number of apparently random mutations that blocked gene function in *E. coli* were found to arise by the insertion of large blocks of genetic material (insertion sequences). Soon after, several of these sequences were found to have the jumping property of the maize elements. Then, through the use of the newly developed methods of recombinant DNA and DNA sequencing (discussed in Chapter 19), the nature of the *E. coli* transposing elements quickly became established at the molecular level. They were found to be DNA sequences that code for enzymes that bring about the insertion of an identical copy of themselves into a new DNA site. Transposition events involve both recombination and replication processes that frequently generate two daughter copies of the original transposable element; one remains at the parental site, while the other appears at the target site.

The insertion of transposable elements invariably disrupts the integrity of their target genes. Since transposable elements carry signals for the initiation of RNA synthesis, they sometimes activate previously dormant genes. Insertion of these elements also may cause deletions, inversions, and chromosomal fusions. By possessing these elements, the *E. coli* genome has thus acquired the potential for much genetic plasticity.

Two main classes of transposable elements have been identified in bacteria, insertion sequences and transposons (Figure 7-27). **Simple transposons** (also called **insertion sequences,** or **IS,**) are transposable elements that carry no genetic information except that necessary for transposition. Key components include specific terminal DNA sequences as well as the coding region for an enzyme (transposase) that recognizes the ends. **Complex transposons** are transposable elements in which the terminal sequences enclose additional genetic material unconnected with transposition. In some complex transposons, two complete IS elements enclose one to several functional bacterial genes. By being contained within a transposon, a specific gene, such as one that confers resistance to an antibiotic, acquires a greatly heightened capacity to move with the transposon from one DNA molecule to another, if not from one bacterium to another. Antibiotic resistance genes that have been observed to spread quickly among a population of previously susceptible bacteria are often found to be inserted within transposons. For further discussion of transposons, see Chapter 11.

Plasmids

DNA molecules that possess their own origins of replication and so are capable of self-duplication are called **replicons.** The circular main bacterial chromosome is a replicon, as are the much smaller autonomously replicating circular DNA molecules that we call **plasmids** (Figure 7-28). Plasmids are not merely tiny chromosomes that in every other way resemble the much larger DNA molecules we designate as chromosomes. Plasmid DNAs usually encode no essential functions, and bacteria lacking them usually multiply normally. Rather, plasmids act as dispensable accessory sources of DNA that provide unique functions to bacteria. The F factors, for example, are plasmids that encode the proteins that allow bacteria to conjugate and donate

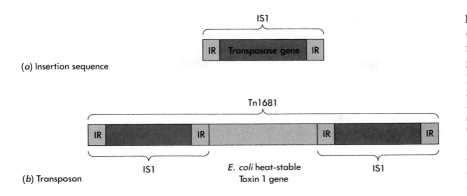

(a) Insertion sequence

IS1

IR | Transposase gene | IR

Tn1681

IR | | IR | | IR | | IR

IS1

E. coli heat-stable
Toxin 1 gene

IS1

(b) Transposon

Figure 7-27
(a) A simple transposon (also called an insertion sequence) carries only the genetic information necessary for transposition (known as the transposase gene) and is often enclosed by short regions of inverted repeat (IR) sequences. (b) The complex transposon Tn1681 is composed of two insertion sequences and contains genetic information in addition to that needed for transposition. In this case, the transposon carries the gene for a heat-stable toxin.

genetic material. The resistance transfer factors (RTF) carry the several genes that confer simultaneous multiple resistances to many different antibiotics (e.g., ampicillin, streptomycin, and tetracycline). While some plasmids, such as the F factor, are present in 1:1 correspondence with the bacterial chromosomes, other plasmids, particularly those very small ones that carry only one to several genes, can be present in many more copies (up to 50) per bacterium and replicate independently of the main chromosome.

Small plasmids are present in many different bacteria, since they exist in multiple copies per cell and thus allow gene amplification. Any gene that they carry is present in correspondingly large numbers, thereby opening up the possibility that very large numbers of their protein products are made. Such plasmids thus provide the ideal vehicles to carry the many multiple genes needed for high-level antibiotic resistance, where detoxifying enzymes are required in massive amounts. Likewise, they are ideal vehicles for producing many copies of enzymes needed to break down inefficiently metabolized food sources.

It is important to note that a given gene is present only once on a given plasmid. Unlike the multiple copies of ribosomal RNA genes present on the main *E. coli* chromosome, the copy number of a gene present on a plasmid cannot be expanded or contracted by crossing over, since it is present in only one copy (i.e., it is haploid). It is possible, however, for a *whole plasmid*, even when present in multiple copies, to be lost by chance segregation events during cell division. In the absence of positive selective pressure, small numbers of plasmid-free cells are constantly generated. Accessory genes thus tend to survive best when they are included within a transposon that gives them the potential to move onto other plasmids, if not into other cells by first jumping onto F factors.

Genetic Elements That Invert[44]

Long puzzling to scientists were certain bacterial traits that alternate between one phenotype and another at frequencies far higher than would be expected if they arose through mutational mechanisms (Table 7-3). The best-studied case involved the bacterium *Salmonella* (a close relative of *E. coli*), whose flagella are constructed either from the protein designated H1 or the protein designated H2, but never both. Changes from H1 to H2 (or vice versa) occur as frequently as once in every 1000 cell divisions, much too frequently to be caused by sponta-

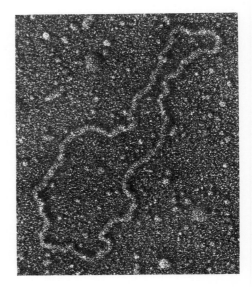

Figure 7-28
Electron micrograph of plasmid pSC101. (Courtesy of Stanley N. Cohen, Stanford University.)

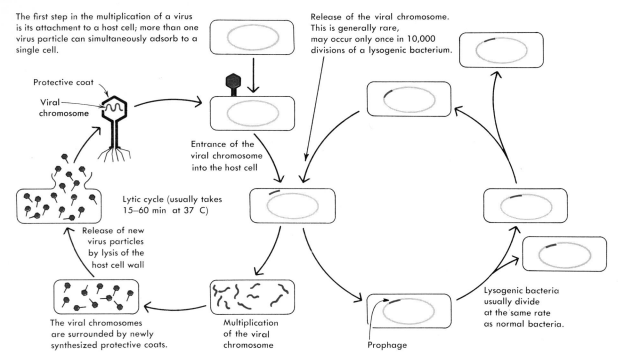

The first step in the multiplication of a virus is its attachment to a host cell; more than one virus particle can simultaneously adsorb to a single cell.

Release of the viral chromosome. This is generally rare, may occur only once in 10,000 divisions of a lysogenic bacterium.

Protective coat

Viral chromosome

Entrance of the viral chromosome into the host cell

Lytic cycle (usually takes 15–60 min at 37 C)

Release of new virus particles by lysis of the host cell wall

The viral chromosomes are surrounded by newly synthesized protective coats.

Multiplication of the viral chromosome

Prophage

Lysogenic bacteria usually divide at the same rate as normal bacteria.

Figure 7-29
The life cycle of a lysogenic bacterial virus. After its chromosome enters a host cell, it sometimes immediately multiplies like a lytic virus and at other times becomes transformed into a prophage. The lytic phase of its life cycle is identical to the complete life cycle of a lytic (nonlysogenic) virus. Lytic bacterial viruses are so called because their multiplication results in the rupture (lysis) of the bacteria.

Table 7-3   Examples of Bacterial Traits That Frequently Alternate Between Two Different Phenotypes

Phenotype	Organism
Colony morphology	*Escherichia coli*
Pigmentation	*Streptomyces reticuli*
Pigmentation	*Myxococcus xanthus*
Pilin	*Escherichia coli*
Surface antigen	*Borrelia hermsii*
Vi antigen	*Citrobacter freundii*
H and O antigens	*Salmonella*
Bioluminescence	*Vibrio harveyi*
Host range	*Bacteriophage Mu*

neous mutations. Only within the last five years has it become clear that chromosomal flip-flops (inversions) lead to the alternative expression of the H1 or H2 gene. When the controlling DNA segment is oriented in one direction, synthesis of the H1 protein occurs and H2 synthesis is repressed. After the controlling DNA segment inverts, H2 synthesis starts up while H1 synthesis slows down (see Chapter 11 for details).

Insertion of Phage Chromosomes into the Chromosomes of *E. coli*[45, 46, 47]

Some phages (e.g., phage λ) do not always multiply upon entering a host cell. Instead, their DNA becomes inserted into a host bacterial chromosome. Then the phage chromosome is, for all practical purposes, an integral part of its host chromosome and is duplicated, like the bacterial chromosome, just once every cell generation (Figure 7-29). A phage chromosome that is integrated into a host chromosome is called a **prophage.** Those bacteria containing prophages are called lysogenic bacteria, and those phages whose chromosomes can become prophages are known as **lysogenic phages.**

In contrast, some phages (e.g., T2, T7) always multiply when they enter a host cell and thus often lyse the cell because of the large numbers of phage particles created. These are called **lytic phages.** It is often difficult to know when a bacterium is lysogenic. We can only be sure when the phage chromosome is released from the host chromosome and the multiplication process that forms new progeny particles commences. Why only certain phages have evolved to form lysogenic associations and what advantages result from such associations are still open questions.

How phage chromosomes are inserted into bacterial chromosomes was at first very mysterious. Now we realize that such insertions do not always follow the same pathways and that two very different routes are used. The best-understood route is that used by phage λ. Integration of λ is achieved by crossing over between the *E. coli* chromosome and a circular form of the λ chromosome. Prior to integration, the originally linear λ chromosome circularizes and attaches to a very homologous short section of the *E. coli* chromosome. Both the host chromosome and the phage λ chromosome then break and rejoin in such a way that the broken ends of the λ chromosome join to the broken ends of the bacterial chromosome, instead of to each other, thereby inserting the λ chromosome into the *E. coli* chromosome (Figure 7-30). The prophage form of λ detaches from the host chromosome by the reverse process. The prophage-host chromosome junctions pair prior to a crossover event that ejects the λ chromosome. The now free λ genome can then begin to multiply as if it were the chromosome of a lytic virus.

The sections of genetic similarity between the λ chromosome and the *E. coli* chromosome are minuscule compared to the length of the *E. coli* chromosome. So the probability that ordinary crossing-over events occur between them is very small. Frequent crossing over between them is made possible by specific phage proteins that not only bring together the homologous sections, but also make the necessary cuts that allow the respective DNAs to exchange partners (see Chapter 11).

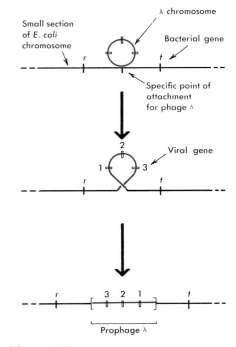

Figure 7-30
Insertion of the chromosome of phage λ into the *E. coli* chromosome by crossing over.

Phage Mu Is a Transposable Genetic Element[48]

At first, phage Mu (mutator) was thought to behave in a completely unique fashion. It never multiplies in a simple lytic way, but always commences its life cycle by inserting itself into the *E. coli* chromosome at totally random locations. The resulting prophagelike insertions often inactivate (mutate) the genes into which they become inserted (Figure 7-31). Thus, Mu was originally known for its mutagenic properties. Later in the life cycle, upon receipt of an appropriate signal, the Mu prophage somehow generates many new copies of itself, which then reinsert themselves at random into the *E. coli* chromosome. Subsequent DNA-cutting events release the new copies, which are then encapsulated by protein, thereby forming large numbers of progeny Mu particles. The nature of phage Mu became clear only when transposable elements were discovered and compared in their behavior to phage Mu. Then phage Mu was ascertained to be an especially large transposon that encodes not only the enzyme (transposase) that inserts it into the *E. coli* chromosome but also the large number of proteins necessary to construct the outer protective shell that surrounds the Mu chromosome. For further discussion of phage Mu, see Chapter 17.

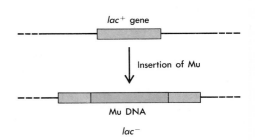
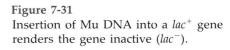

Figure 7-31
Insertion of Mu DNA into a *lac*+ gene renders the gene inactive (*lac*−).

Phages Occasionally Carry Bacterial Genes[49, 50, 51]

Not only can bacterial genes be transferred through mating or transposition, but they can also be passively carried from one bacterium to

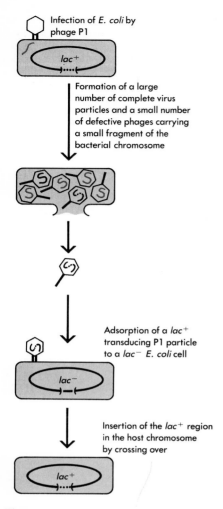

Infection of *E. coli* by phage P1

Formation of a large number of complete virus particles and a small number of defective phages carrying a small fragment of the bacterial chromosome

Adsorption of a *lac⁺* transducing P1 particle to a *lac⁻ E. coli* cell

Insertion of the *lac⁺* region in the host chromosome by crossing over

Figure 7-32
Transduction, the passive transfer of genetic material from one bacterium to another by means of carrier phage particles.

another by phage particles, a process called **transduction.** This happens when a virus particle (a transducing phage) accidentally encapsulates sections of a host chromosome. When this virus particle (usually biologically inactive because its own viral chromosome is incomplete or totally missing) attaches to a host cell, the fragment of bacterial chromosome is injected into the cell. It then can engage in crossing over with the host chromosome. If the transducing phage has been grown on a bacterial strain genetically different from the strain subsequently infected with the phage, a genetically altered bacterium may be produced (Figure 7-32). For example, a suspension of P1 phage particles grown on a strain of *E. coli* that is able to metabolize lactose will contain a small number of particles carrying the gene (*lac⁺*) required to utilize lactose. Addition of such phages to an *E. coli* strain unable to use lactose for growth (*lac⁻*) will transform a small number of the *lac⁻* bacteria to the *lac⁺* form by means of genetic recombination.

Because transduction is very rare, one might think that it would not be a useful tool for probing chromosomal structure. On the contrary, it has been most helpful in telling us whether two genes are located close to each other and what their exact order is. This is because the segment of the bacterial chromosome carried by a single transducing particle is relatively short, so that only genes located very close to each other will be enclosed in the same transducing particle. Thus, by determining the frequencies with which groups of genes can be transduced by the same phage particle, we can very accurately establish their relative locations.

In addition to transducing phages like P1, which randomly carry small sections of the *E. coli* chromosome, there are also specialized transducing phages that carry specific bacterial genes. Certain defective λ phage particles possess hybrid chromosomes in which a section of the λ chromosome is linked to a fragment of *E. coli* chromosome. Such hybrid chromosomes result when the λ prophage improperly recombines out of the *E. coli* chromosome to generate a λ DNA fragment still attached to bacterial genes that lie immediately adjacent to the site into which the λ DNA inserts on the *E. coli* chromosome (Figure 7-33). Purification of these specialized transducing phages therefore affords a way to obtain large numbers of certain *E. coli* genes, and it was through such a procedure that the genetically well-studied gene involved in the utilization of the sugar lactose was first molecularly investigated. Unfortunately, most prophages preferentially insert themselves at only a few chromosomal sites, and most genes cannot easily be isolated within specialized transducing particles.

Transfer of Genetic Traits Via Incorporation of Purified DNA[52, 53, 54]

Bacterial genes can also move from one cell to another in the form of pure DNA. The process by which a bacterium absorbs pure DNA from its surroundings and functionally integrates the exogenous DNA into its own genetic material is called **transformation.** As mentioned in Chapter 3, it was through the discovery of this phenomenon that DNA was pinpointed as the key genetic molecule. First observed in the Gram-positive bacterium *Streptococcus pneumoniae*, transforma-

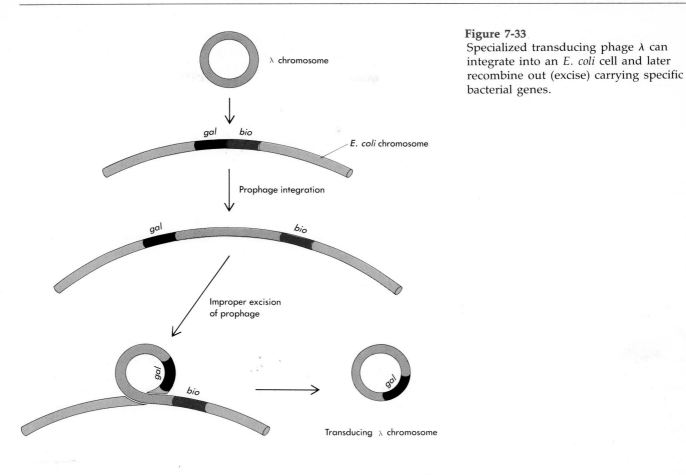

λ chromosome

gal bio

E. coli chromosome

Prophage integration

gal bio

Improper excision
of prophage

gal

bio

gal

Transducing λ chromosome

Figure 7-33
Specialized transducing phage λ can
integrate into an *E. coli* cell and later
recombine out (excise) carrying specific
bacterial genes.

tion was found to occur naturally in many Gram-positive and Gram-negative bacteria. While initially the argument could be made that transformation is accidental and plays no significant role in nature, this is no longer a tenable belief. Indeed, some bacteria (e.g., *Hemophilus* and *Neisseria*) possess special proteins that mediate the entry of DNA into cells and that only recognize and take up DNA from related species (Figure 7-34). *Hemophilus* cells do this by binding specific short DNA sequences (e.g., 5'-AAGTGGGTCA-3') that are characteristic of the *Hemophilus* genome and occur some 600 times on its chromosome (or once per 4000 base pairs). Only double-stranded DNA is so absorbed, after which it somehow enters into the host bacterial chromosome as single-stranded segments that base-pair to appropriate single-stranded regions of the host DNA, perhaps at the replicating forks.

Many other bacteria, including *E. coli*, do not seem to have evolved or retained ways of taking up functionally active DNA. A powerful trick, however, is now available for "artificially" introducing functional DNA. In 1970, it was discovered that DNA added in the presence of high amounts of Ca^{2+} can transform. This Ca^{2+}-induced artificial transformation works in *E. coli* only when the added DNA is a complete replicon capable of self-duplication. Plasmid DNAs and intact viral chromosomes transform with high efficiencies, while linear fragments of bacterial DNA transform at much lower levels. The failure of linear DNAs to transform well is thought to result from their

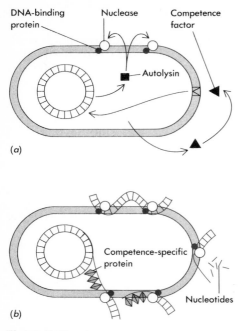

(a)

(b)

Figure 7-34

Transformation, in which a bacterial cell takes up free DNA. (a) Development of competence. Competence factor (▲) interacts with a cell-surface receptor (M) and causes certain competence-specific proteins to be expressed. One of these is an autolysin (■), which exposes a membrane-associated DNA-binding protein (●) and a nuclease (○). (b) Transformation. A long fragment of dsDNA is bound at several sites to the cell surface. It is progressively nicked and cut. One strand of the cut fragment is degraded by the nuclease, a process that releases nucleotides (−) into the medium. The other strand becomes associated with a competence-specific protein (◆). In this form it enters the cell and recombines by a process that involves replacement of one strand of the endogenote by the donor. (After Smith, 1981.)

digestion by periplasmic DNases before they can be brought into the cytosol. *E. coli* strains deficient in such nucleases are much more effectively transformed by exogenous DNA fragments.

The realization that Ca^{2+} treatment renders virtually all bacteria susceptible to transformation has had enormous consequences. It provides a way to functionally introduce any piece of DNA, after its insertion into a suitable plasmid, into any desired bacterial genome. Thus, bacterial transformation, which first told the world of science that DNA was the genetic molecule, now gives us a most versatile alternative to viral infection as the means to monitor the consequences of the sudden entry of specific DNA into cells. As we shall see in the next section, because it allows the introduction of test-tube-made recombinant DNA into cells, transformation provides a means of genetically engineering cells as well as making available highly powerful techniques for the isolation of desired genes.

Recombinant DNA Techniques Allow the Isolation of Single Genes[55, 56]

Until a few years ago, genetic crosses in their many manifestations were the key tools for revealing what genes are and how they function. Through genetic manipulations, not only could the location of genes on their respective chromosomes be found, but in a few cases, desired genes could be recombined on the phage λ chromosome. There they could be used both for molecular analysis and to search for postulated regulatory molecules (e.g., repressors) whose binding to specific genetic elements was believed to control the functioning of adjacent genes. Whether a desired gene could be so isolated on a transducing phage, however, was not only very chancy, but could easily consume many years of hard work.

The scope of genetic research, even with the simplest bacterial and viral systems, has thus dramatically improved as a result of the development of recombinant DNA procedures. These procedures make possible the isolation and **cloning** (duplication) of any gene essential for cellular growth and for which mutations exist that block its proper functioning. With genetically well-characterized microorganisms, the isolation of a given desired gene is no longer a major challenge.

Particularly straightforward are the techniques for the isolation of many *E. coli* genes (Figure 7-35). They start with the fragmentation of the *E. coli* chromosome (4×10^6 base pairs) into some thousand different discrete pieces by using an appropriate DNA **restriction** (cutting) enzyme that recognizes a specific nucleotide sequence that occurs on the average every several thousand base pairs along the *E. coli* chromosome. These chromosomal fragments are then mixed with an excess of DNA from an *E. coli* plasmid that has been converted into a linear form by the same sequence-specific restriction enzyme. Addition of the DNA-joining enzyme, DNA ligase, randomly links the two sets of fragments together, leading to the generation of many recombinant plasmid DNA circles into which have been inserted discrete fragments of the *E. coli* chromosome. Found within this recombinant DNA "library" are some thousand different recombinant plasmids, collectively containing all the DNA of the *E. coli* chromosome.

To use such an *E. coli* library to isolate a desired gene, we must first mix the library in the presence of Ca^{2+} with the respective mutant

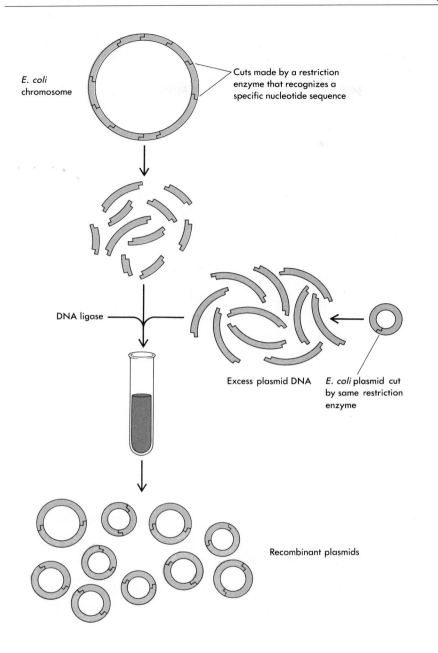

Figure 7-35
Formation of a recombinant DNA "library."

E. coli chromosome

Cuts made by a restriction enzyme that recognizes a specific nucleotide sequence

DNA ligase

Excess plasmid DNA

E. coli plasmid cut by same restriction enzyme

Recombinant plasmids

bacteria (e.g., a strain that requires a specific amino acid for growth) under conditions that favor the uptake of the plasmid DNA. The bacteria are then placed under the appropriate selective conditions for growth (absence of the needed amino acid). The only cells that can multiply to form bacterial colonies will be those that have taken up specific recombinant plasmids that carry the normal gene needed for the biosynthesis of the amino acid. These colonies can in turn be used to grow a much larger number of genetically identical bacteria, from which the small plasmid DNA circles can be easily separated from the much larger main bacterial chromosomes. By cleaving these recombinant plasmids with the right restriction enzyme, pure copies of the desired *E. coli* gene can be separated away from the plasmid DNA.

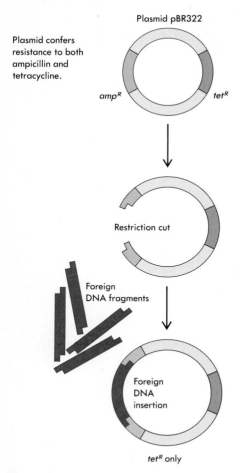

Plasmid pBR322

Plasmid confers resistance to both ampicillin and tetracycline.

amp^R tet^R

Restriction cut

Foreign DNA fragments

Foreign DNA insertion

tet^R only

Figure 7-36
Plasmid pBR322 contains restriction sites into which foreign DNA can be inserted. A number of these sites are in genes conferring antibiotic resistance on the bacterium containing them. When foreign DNA is inserted into one of these genes, the gene is usually inactivated. Here, foreign DNA is shown inserting in amp^R, the gene for ampicillin resistance. The resulting loss of ampicillin resistance is a preliminary indicator that foreign DNA has been inserted.

The Development of Increasingly Sophisticated Cloning Vectors[57–61]

Vectors are autonomously replicating DNA units into which DNA fragments are inserted for gene cloning. Genes taken up by these vectors are multiplied (cloned) as the vector replicates, to yield numbers suitable for molecular analysis. Now the best-known plasmid vector is **pBR322**, which has genetically tailored cutting sites into which DNA fragments can be inserted without affecting plasmid self-replication (Figure 7-36). DNA fragments of up to 5 to 10 kilobase pairs (kbp) are very efficiently cloned in this vector. Larger fragments, however, often tend to be unstable and lose most of their DNA. The multiplication rates of plasmids slow down as they get larger, with those that have lost large chunks of their foreign DNA eventually predominating. When larger DNA pieces must be cloned, they often are placed in specially constituted λ vectors into which foreign fragments as large as 15 kbp are easily inserted without disturbing the normal packaging of the λ chromosome into functional virus particles (Figure 7-37). Still larger DNA pieces of up to 50 kbp can now be cloned in the so-called **cosmid vectors,** constructed from the two ends of the λ chromosome. Thus, depending on the size of the DNA needed to be cloned, DNA fragments are inserted into plasmid, λ phage, or cosmid vectors.

The initial plasmid vectors multiplied only in their natural host cells. DNA sequences around the origins of replication control this specificity. More recently, newly designed vectors have been made using recombinant DNA procedures that contain two distinct origins of replication derived from, for example, *E. coli* and yeast or *E. coli* and human. These vectors offer extended host specificity and are called **shuttle vectors.** They are now extensively used to transfer genes between very different cell types, and in so doing are helping to more clearly define the rules governing gene expression.

In Vitro Mutagenesis[62, 63]

Until recently, genetic analysis started with the finding of mutant organisms followed by a series of genetic crosses to locate the genes in which the mutations had occurred. Now an alternative approach is possible. First the desired gene (e.g., that for a repressor molecule) is isolated and then sequenced, after which mutations are created within it. Then the phenotypic consequences of these in vitro-made mutations are observed by reinserting the mutant gene back into suitable cells. Underlying this "reverse genetics" is a set of recently developed procedures for in vitro mutagenesis. Both small deletions and nucleotide substitutions can now be easily made around the sites of restriction enzyme cuts. Becoming even more powerful is a technique that uses well-defined oligonucleotides as primers onto which DNA chains can be enzymatically extended. By synthesizing primers containing specific base changes from the normal sequence, mutant genes can be created bearing these same changes. Through this technique, a much deeper analysis of how specific proteins are constructed and function will soon be possible. Equally important is the possibility of further pinpointing the DNA nucleotide sequences that control transcription and translation.

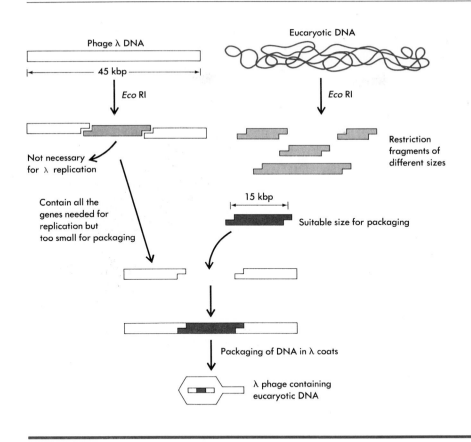

Figure 7-37
Phage λ DNA can act as a cloning vector for eucaryotic genes or other foreign DNA fragments of approximately 15 kbp in length.

Phage λ DNA

|←————— 45 kbp —————→|

Eco RI

Eucaryotic DNA

Eco RI

Not necessary for λ replication

Contain all the genes needed for replication but too small for packaging

Restriction fragments of different sizes

|← 15 kbp →|

Suitable size for packaging

Packaging of DNA in λ coats

λ phage containing eucaryotic DNA

Summary

The genes whose structure and functioning we now understand the best are those of bacteria, in particular *E. coli* and its viruses (phages). Like other haploid organisms, *E. coli* has the advantages of a very short life cycle and ease of manipulation under controllable laboratory conditions. Vital for the accumulation of this precise genetic knowledge has been the use of mutants. Through them we have been able to reveal sexual phenomena in bacterial cells, once thought too simple to make use of genetic recombination, and also to establish genetic maps revealing the nature of their chromosomes. Bacterial mutations involving growth factors and resistance to specific antibiotics and viruses are particularly useful, since they allow easy separation of wild-type and mutant organisms. There are numerous ways to increase the number of newly arising mutants through the application of increasingly sophisticated mutagenic agents. Many powerful techniques also exist to select desired rare mutants from very large populations of wild-type organisms. Most useful are agents like penicillin, which selectively kills growing wild-type cells, thereby enriching the population of mutant cells unable to multiply under certain environmental conditions (e.g., in the absence of a growth factor).

Bacterial conjugation involves the attachment of male and female cells followed by the movement of male chromosomal material into recipient female cells. The essence of a male cell is the presence of a specific DNA segment, which can exist free as a tiny supernumerary chromosome (the F factor) or as a segment integrated into the main and much larger bacterial chromosome. When the F factor exists free (as in F⁺ cells), only F factor DNA is transferred into female (F⁻) cells following mating between a male (F⁺) and a female (F⁻) cell. If, however, the F factor has become part of the *E. coli* chromosome (as in Hfr—high frequency of recombination—cells), then the entire male chromosome has the potential to move into the female cell, with such movement commencing within a section of F DNA. The movement of a male chromosome into a female cell often aborts in mid-passage, leading to the insertion of incomplete male chromosomal fragments into female cells. Thus, conjugation between Hfr cells and F⁻ cells rarely leads to conversion of the female cell into a male cell.

Bacterial viruses (phages), like all other viruses, contain single tiny chromosomes surrounded by protein-containing protective coats that allow the chromosomes to move from one cell to another. Cycles of phage replication always start with the injection of phage chromosomes into the host cells. Viruses can multiply only in cells and are best characterized as parasites at the genetic level. The growth of many phages leads to the obligatory death of their host cells (lytic phages). Other phages may have lysogenic life cycles in which their chromosomes become inserted into their host bacterial chromosomes in apparently passive (prophage) form. For the replication of such lysogenic phages, the prophage forms must detach from the respective bacterial chromosomes and commence cycles of phage chromosome multiplication. Occasionally, phage particles become assembled containing bacterial DNA instead of phage DNA. Such transducing phages, following infection of a suitable bacterial host, provide an alternate way to transfer genes from one bacterium to another. Phage chromosomes, like bacterial chromosomes, are capable of genetic exchange (crossing over) when present in the same bacterial cell. There is no differentiation into male and female phages, with genetic recombination being generally assured by simultaneous infection of a given cell by more than one genetically distinct phage.

Genetic recombination can also be initiated by the introduction of free DNA. How such free DNA enters recipient cells is still a mystery. Ca^{2+} somehow increases efficiency of such transformation events. In some bacteria, special proteins and DNA sequences are necessary for such DNA transfers, suggesting that movement of free DNA is frequently an important means of achieving genetic recombination.

Small, autonomously replicating DNA molecules like the F factor are present in many bacteria. Now known as plasmids, these relatively tiny DNA molecules do not segregate regularly at cell division and are invariably lost from a small fraction of all daughter cells. Thus, they usually are not the sole bearers of essential genes required for growth and multiplication. Because, however, they frequently are present in high copy number, they provide a means to amplify the number of any essential single-copy gene that becomes inserted into their genome. Insertion on a plasmid, for example, can lead to rapid amplification of genes needed to achieve high-level resistance to antibiotics.

The location of genes on a given bacterial chromosome is relatively fixed, with those genes whose products are needed in abnormally large amounts (e.g., ATP synthetase) frequently being located at the origin of replication. However, specialized genetic segments (transposons) exist that have the capacity to insert copies of themselves at other chromosomal sites. Other genetic segments have the capacity to flip-flop their relative orientation along a chromosome and in so doing to alter their functioning. Such flip-flop events demand the occurrence of inverted duplicated sequences at the termini of such segments. Their high frequencies reflect the existence of site-specific recombination, which promotes enzymes that recognize the inverted terminal sequences.

Traditional genetic procedures are increasingly being supplemented by recombinant DNA techniques. In these analyses, new in vitro procedures for site-specific mutagenesis are allowing mutant DNAs to be made for a "reverse" genetic approach in which mutations become produced before their phenotypic consequences are known.

Bibliography

General References

Adelberg, E. A., ed. 1966. *Papers on Bacterial Genetics.* 2nd ed. Boston: Little, Brown.

Davis, R. W., D. Botstein, J. R. Roth. 1980. *Advanced Bacterial Genetics.* Cold Spring Harbor, N.Y.: Cold Spring Harbor Laboratory. A manual for genetic engineering.

Hayes, W. 1968. *The Genetics of Bacteria and Their Viruses.* 2nd ed. New York: Wiley.

Ingraham, J. L., O. Maaloe, and F. C. Neidhardt. 1983. *Growth of the Bacterial Cell.* Sunderland, Mass.: Sinauer.

Miller, J. H. 1972. *Experiments in Molecular Genetics.* Cold Spring Harbor, N.Y.: Cold Spring Harbor Laboratory. Concentrates on the molecular genetics of *E. coli.*

O'Brien, S. J., ed. 1984. *Genetic Maps 1984.* Vol. 3. Cold Spring Harbor, N.Y.: Cold Spring Harbor Laboratory. A compilation of linkage and restriction maps of genetically studied organisms.

Silhavy, T. J., M. L. Berman, and L. W. Enquist. 1984. *Experiments with Gene Fusions.* Cold Spring Harbor, N.Y.: Cold Spring Harbor Laboratory. Demonstrates the use of gene fusions, transposable elements, and methods of recombinant DNA for genetic analysis in *E. coli.*

Stent, G. S., ed. 1965. *Papers on Bacterial Viruses.* 2nd ed. Boston: Little, Brown.

Stent, G. S., and R. Calendar. 1978. *Molecular Genetics: An Introductory Narrative.* 2nd ed. San Francisco: Freeman.

Timmis, K. N., and A. Putiler, eds. 1979. *Plasmids of Medical, Environmental, and Commercial Importance.* Amsterdam, London, and New York: Elsevier.

Cited References

1. "Heredity and Variation in Microorganisms." 1946. *Cold Spring Harbor Symp. Quant. Biol.* Vol. 11. Cold Spring Harbor, N.Y.: Cold Spring Harbor Laboratory.

2. Beadle, G. W., and E. L. Tatum. 1941. "Genetic Control of Biochemical Reactions in *Neurospora.*" *Proc. Nat. Acad. Sci.* 27:499–506.

3. Beadle, G. W., and E. L. Tatum. 1945. "Neurospora. II. Methods of Producing and Detecting Mutations Concerned with Nutritional Requirements." *Amer. J. Bot.* 32:678–686.

4. Luria, S. E., and M. Delbruck. 1943. "Mutations of Bacteria from Virus Sensitivity to Virus Resistance." *Genetics* 28:491–511.

5. Lederberg, J., and E. M. Lederberg. 1952. "Replica Plating and Indirect Selection of Bacterial Mutants." *J. Bacteriol.* 63:399–406.

6. Gray, C. H., and E. L. Tatum. 1944. "X-Ray Induced Growth Factor Requirements in Bacteria." *Proc. Nat. Acad. Sci.* 30:404–410.

7. Newcombe, H. B. 1948. "Delayed Phenotypic Expression of Spontaneous Mutations in *E. coli.*" *Genetics* 33:447–476.

8. Hayes, W. 1957. "The Phenotypic Expression of Genes Determining Various Types of Drug Resistance Following Their Inheritance by Sensitive Bacteria." In *Drug Resistance in Microorganisms,* ed. G. E. W. Wolstenhlome and C. M. O'Connor. Ciba Foundation Symposium. London: Churchill, p. 197.

9. Davis, B. D. 1948. "Isolation of Biochemically Deficient Mutants of Bacteria by Penicillin." *Amer. Chem. Soc.* 70:4267.

10. Lederberg, J. 1950. "Isolation and Characterization of Biochemical Mutants of Bacteria." In *Methods in Medical Research,* vol. 3, ed. J. H. Comrie, Jr. Chicago: Year Book, p. 5.

11. Bonhoeffer, F., and H. Schaller. 1965. "A Method for Selective Enrichment of Mutants Based on the High Ultraviolet Sensitivity of DNA Containing 5-bromouracil." *Biochem. Biophys. Res. Comm.* 20:93–97.

12. Lederberg, J. 1948. "Detection of Fermentation Mutants with Tetrazolium." *J. Bacteriol.* 56:695.

13. Lin, E. C. C., S. A. Lerner, and S. E. Jorgensen. 1962. "A Method for Isolating Constitutive Mutants for Carbohydrate-Containing Enzymes." *Biochem. Biophys. Acta* 60:422.

14. Gesteland, R. F. 1966. "Isolation and Characterization of Ribonuclease I Mutants of *E. coli.*" *J. Mol. Biol.* 16:67–84.
15. De Lucia, P., and J. Cairns. 1969. "Isolation of an *E. coli* Strain with a Mutation Affecting DNA Polymerase." *Nature* 224:1164–1166.
16. Lederberg, J., and E. L. Tatum. 1946. "Novel Genotypes in Mixed Cultures of Biochemical Mutants of Bacteria." *Cold Spring Harbor Symp. Quant. Biol.* 11:113–114.
17. Stent, G. S. 1963. *Molecular Biology of Bacterial Viruses.* San Francisco: Freeman.
18. Mathews, C. K., E. M. Kutter, G. Mosig, and P. B. Berget. 1983. *Bacteriophage T4.* Washington, D.C.: Amer. Soc. for Microbiol.
19. Doermann, A. 1953. "The Vegetative State in the Life Cycle of Bacteriophage: Evidence for Its Occurrence, and Its Genetic Characterization." *Cold Spring Harbor Symp. Quant. Biol.* 18:3–11.
20. Luria, S. E. 1945. "Mutations of Bacterial Viruses Affecting Their Host Range." *Genetics* 30:84–99.
21. Hershey, A. D. 1946. "Mutations of Bacteriophage with Respect to Types of Plaque." *Genetics* 31:620–640.
22. Epstein, R. H., A. Bolle, C. M. Steinberg, E. Kellenberg, E. Boy de la Tour, R. Chevalley, R. S. Edgar, M. Susman, G. H. Denhardt, and A. Lielausis. 1963. "Physiological Studies on Conditional Lethal Mutants of the Bacteriophage T4D." *Cold Spring Harbor Symp. Quant. Biol.* 28:375–394.
23. Hershey, A. D., and R. Rotman. 1949. "Genetic Recombination Between Host Range and Plaque-Type Mutants of Bacteriophage in Single Bacterial Cells." *Genetics* 34:44–71.
24. Hershey, A. D. 1958. "The Production of Recombinants in Phage Crosses." *Cold Spring Harbor Symp. Quant. Biol.* 23:19–46.
25. Hayes, W. 1952. "Recombination in *Bact. coli* K-12: Unidirectional Transfer of Genetic Material." *Nature* 169:118–119.
26. Jacob, F., and E. L. Wollman. 1961. *Sexuality and the Genetics of Bacteria.* N.Y.: Academic Press.
27. Brinton, C. C., P. Gemski and J. Carnahan. 1964. "A New Type of Bacterial Pilus Genetically Controlled by the Fertility Factor of *E. coli* K12." *Proc. Nat. Acad. Sci.* 52:776–783.
28. Crawford, E. M., and R. A. Gesteland. 1964. "The Adsorption of Bacteriophage R-17." *Virology* 22:165–167.
29. Broda, P., J. R. Beckwith, and J. Scaife. 1965. "The Characterization of a New Type of F-Prime Factor in *E. coli* K-12." *Genet. Res.* 5:489–493.
30. Ou, J. T., and T. F. Anderson. 1970. "Role of Pili in Bacterial Conjugation." *J. Bacteriol.* 102:648–654.
31. Manning, P., and M. Achtman. 1979. "Cell-to-Cell Interactions in Conjugating *Escherichia coli:* The Involvement of the Cell Envelope." In *Bacterial Outer Membranes,* ed. M. Inouye. New York: Wiley, pp. 409–447.
32. Minkley, E. G., Jr., S. Polen, C. C. Brinton, Jr., and K. Ippen-Ihler. 1976. "Identification of the Structural Gene for F-Pilin." *J. Mol. Biol.* 108:111–121.
33. Achtman, M., G. Morelli, and S. Schwuchow. 1978. "Cell-Cell Interactions in Conjugating *E. coli:* Role of F Pili and Fate of Mating Aggregates." *J. Bacteriol.* 135:1053–1061.
34. Wollman, E. L., F. Jacob, and W. Hayes. 1956. "Conjugation and Genetic Recombination in *E. coli.*" *Cold Spring Harbor Symp. Quant. Biol.* 21:141–162.
35. Cairns, J. 1963. "The Bacterial Chromosome and Its Manner of Replication as Seen by Autoradiography." *J. Mol. Biol.* 6:208–213.
36. Henner, D. J., and J. A. Hoch. 1980. "The *Bacillus subtilis* Chromosome." *Microbiol. Revs.* 44:57–82.
37. Royle, P. L., H. Matsumoto, and B. W. Holloway. 1981. "Genetic Circularity of the *Pseudomonas aeruginosa* PAO Chromosome." *J. Bacteriol.* 145:145–155.
38. Bachman, B. J. 1983. "Linkage Map of *E. coli* K-12, ed. 7." *Microbiol. Revs.* 47:180–230.
39. Anderson, R. P., and J. R. Roth. 1977. "Tandem Genetic Duplications in Phage and Bacteria." *Ann. Rev. Microbiol.* 31:473–506.
40. Lindahl, L., and J. M. Zengel. 1982. "Expression of Ribosomal Genes in Bacteria." *Adv. Genetics* 21:53–121.
41. Bukhari, A. I., J. A. Shapiro, and S. L. Adhya. 1977. *DNA Insertion Elements, Plasmids and Episomes.* Cold Spring Harbor, N.Y.: Cold Spring Harbor Laboratory.
42. Shapiro, J. A. 1983. *Mobile Genetic Elements.* New York: Academic Press.
43. "Moveable Genetic Elements." 1980. *Cold Spring Harbor Symp. Quant. Biol.* Vol. 45. Cold Spring Harbor, N.Y.: Cold Spring Harbor Laboratory.
44. Simon, M. I., and M. Silverman. 1983. "Recombinational Gene Expression in Bacteria." In *Gene Function in Procaryotes,* ed. J. Beckwith, J. Davies, and J. A. Gallant. Cold Spring Harbor, N.Y.: Cold Spring Harbor Laboratory, pp. 211–228.
45. Lwoff, A. 1953. "Lysogeny." *Bacteriol. Revs.* 17:269–337.
46. Campbell, A. 1962. "Episomes." *Adv. Genetics* 11:101–145.
47. Herskowitz, I., and D. Hagen. 1980. "The Lysis-Lysogeny Decision of Phage λ: Explicit Programming and Responsiveness." *Ann. Rev. Genetics* 14:399–446.
48. Bukhari, A. I. 1976. "Bacteriophage Mu as a Transposable Element." *Ann. Rev. Genetics* 10:389–412.
49. Zinder, N. D., and J. Lederberg. 1952. "Genetic Exchange in *Salmonella.*" *J. Bacteriol.* 64:679–699.
50. Campbell, A., and E. Balbinder. 1959. "Transduction of the Galactose Region of *E. coli* K-12 by the Phages of λ and λ-434 Hybrid." *Genetics* 44:309–319.
51. Shapiro, J., L. MacHattie, L. Eron, G. Ihler, K. Ippen, and J. Beckwith. 1969. "Isolation of Pure lac Operon DNA." *Nature* 224:768–774.
52. Smith, H. O., D. B. Danner, and R. A. Deich. 1981. "Genetic Transformation." *Ann. Rev. Biochem.* 50:41–68.
53. Raiband, O., M. Mock, and M. Schwartz. 1984. "A Technique for Integrating Any DNA Fragment into the Chromosome of *E. coli.*" *Gene* 29:231–241.
54. Mandel, M., and A. Higa. 1970. "Calcium-Dependent Bacteriophage DNA Infection." *J. Mol. Biol.* 53:159–162.
55. Old, R. W., and S. B. Primrose. 1981. *Principles of Gene Manipulation.* 2nd ed. Berkeley and Los Angeles: University of California Press.
56. Watson, J. D., J. Tooze, and D. T. Kurtz. 1983. *Recombinant DNA: A Short Course.* San Francisco: Freeman.
57. Bolivar, F., R. L. Rodrigues, P. J. Greene, M. C. Betlach, H. L. Hayneker, H. W. Boyer, J. Crosa, and S. Falkow. 1977. "Construction and Characterization of New Cloning Vehicles. II. A Multipurpose Cloning System." *Gene* 2:95–113.
58. Collins, J., and B. Hohn. 1978. "Cosmids: A Type of Plasmid Gene Cloning Vector That Is Packageable in Vitro in Bacteriophage Heads." *Proc. Nat. Acad. Sci.* 75:4242–4246.
59. Hohn, B., and J. Collins. 1980. "A Small Cosmid for Efficient Cloning of Large DNA Fragments." *Gene* 11:241–298.
60. Hicks, J. B., J. N. Strathern, A. J. S. Klar, and S. L. Dellaporta. 1982. "Cloning by Complementation in Yeast: The Mating Type Genes." In *Genetic Engineering—Principles and Methods,* vol. 4, ed. J. K. Stelow and A. Hollaender. New York: Plenum, pp. 219–248.
61. Sullivan, M. A., R. E. Yasbin, and F. E. Young. 1984. "New Shuttle Vectors for *Bacillus subtilis* and *Escherichia coli* Which Allow Rapid Detection of Inserted Fragments." *Gene* 29:21–26.
62. Watson, J. D., J. Toose, and D. T. Kurtz. 1983. "In Vitro Mutagenesis." In *Recombinant DNA: A Short Course.* San Francisco: Freeman, Chap. 8.
63. Shortle, D., D. DiMaio, and D. Nathans. 1981. "Directed Mutagenesis." *Ann. Rev. Genetics* 15:265–294.

CHAPTER 8

The Fine Structure of Bacterial and Phage Genes

Until the advent of recombinant DNA procedures, genetic crosses constituted the only real tool we had to investigate genes. Not only did they provide the means of ordering genes along chromosomes, but they also gave us our first detailed knowledge of the topology (fine structure) of the gene itself. This elucidation of the structure of the gene was possible because crossing over occurs not only between genes, but also *within a single gene*. In fact, in bacteria and their phages, virtually all crossing over occurs within, rather than between, the genes. This fact was recognized in the early 1950s, just after the double helix was discovered. Until then, the chromosome was generally viewed as a linear collection of genes held together by some nongenetic material, somewhat like a string of pearls. We now know that all chromosomes are likely to contain only single, usually immensely large, DNA molecules, each carrying the information needed to order the amino acids in many different protein molecules (see Chapter 9). All crossing over must thus occur by the breakage and reunion of DNA itself.

The probability of any crossing-over event occurring within a given gene is much lower than the probability of crossing over occurring somewhere within an entire chromosome. Its detection therefore demands that very large numbers of progeny (greater than 100,000) be looked at. Even with the relative ease of manipulating the fruit fly *Drosophila,* intragenic recombination long remained a controversial issue. It is very impractical and tedious to look at the more than 10,000 progeny arising from a single *Drosophila* cross. Only when microorganisms were adopted as the preferred tools for molecular genetics did intragenic crossing over become established as a powerful means for revealing important details about gene structure and function.

Recombination Within Genes Allows Construction of a Gene Map[1, 2]

The first incisive results on the genetic structure of the gene came from work with the *rIIA* and *rIIB* genes of the bacterial virus T4. These are two adjacent genes that influence the length of the T4 life cycle; mutations in either therefore affect the size of plaques produced in a

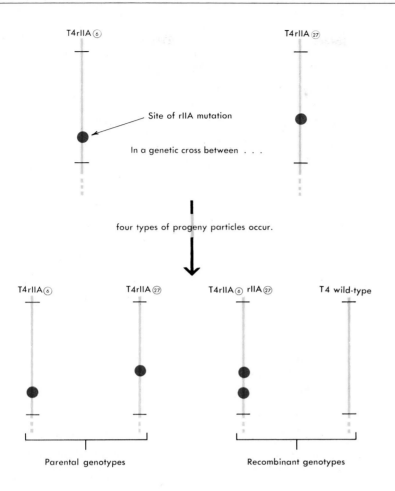

Site of rIIA mutation

In a genetic cross between . . .

four types of progeny particles occur.

Parental genotypes

Recombinant genotypes

Figure 8-1

The use of T4 *rII* mutations in the demonstration of crossing over within the gene. Equal numbers of wild-type and double-*r* recombinants occur. The wild-type recombinants are easily found, because they are the only progeny genotype that will form plaques on K12(λ). It is much harder to identify the double *rIIA₆rIIA₇* recombinants, inasmuch as their plaques are indistinguishable from single-*r* plaques. To detect them, it is necessary to isolate a large number of progeny *r* viruses with the *r* phenotype and use them for new genetic crosses with both parent *r* strains. The double-*r* mutants will not produce wild-type recombinants with either of the single-*r* mutants.

bacterial layer growing on an agar plate. The presence of either an *rIIA* or *rIIB* mutation causes a shorter life cycle of T4 phage within an *E. coli* cell. T4-infected cells on an agar plate normally do not break open and release new progeny phage until several hours after they have been infected. Cells infected with *rII* mutants, however, always break open more rapidly. Thus, *rII* mutants produce larger plaques than wild-type phage. The *rII* mutants were chosen to work with because of the possibility of detecting a very small number of wild-type particles among a very large number of mutants. Although the wild type and the *rII* mutants grow equally well on *E. coli* strain B, there is another strain, *E. coli* K12(λ), on which only the wild type can multiply. Thus, when the progeny of a genetic cross between two different *rII* mutants are added to K12(λ), only the wild-type recombinants form plaques. Even as few as one wild-type recombinant per 10^6 progeny is easily detected.

Over 2000 independent mutations in the *rIIA* and *rIIB* genes were isolated and used in breeding experiments. In a typical cross, *E. coli* strain B bacteria were infected with two phage particles, each bearing an independently isolated *rIIA* or *rIIB* mutation. As the virus particles multiplied, genetic recombination occurred. The progeny were then grown on *E. coli* K12(λ) to detect wild-type recombinants. Normal particles were found in a very large fraction of the crosses, indicating recombination within the *rIIA* or *rIIB* gene (Figure 8-1). If recombination occurred only between genes, it would be impossible to produce wild-type recombinants by crossing two phage particles with muta-

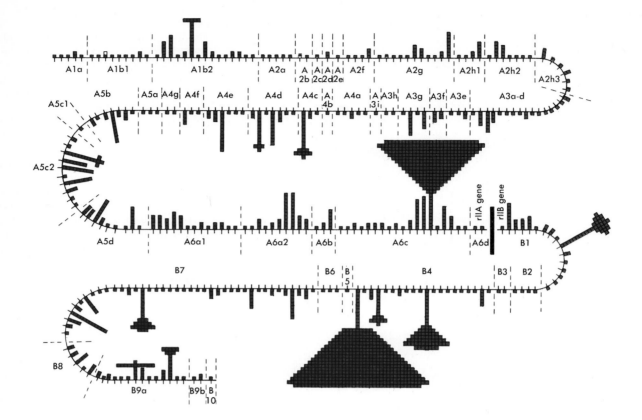

Figure 8-2
The genetic map of the *rIIA* and *rIIB* genes of phage T4. This map shows the assignment of a large number of different mutations in specific regions (A1a, A1b₁, etc.) of the chromosome. In most cases, the order of the mutations within a given region has not yet been determined. Each square corresponds to an independent occurrence of a mutation at a given site on the chromosome. Thus, some regions ("hot spots") mutate much more frequently than others. [From S. Benzer, *Proc. Nat. Acad. Sci.* 47, 410 (1961), with permission.]

tions in the same gene. A large spectrum of recombination values was found, just as in crosses between mutants in separate genes. This spread of values indicates that some mutant genes lie closer together than others and allows for the construction of genetic maps for the two *rII* genes. Examination of these genetic maps yields the following striking conclusions:

- A large number of different sites of mutation (mutable sites) occur within a gene. This number exceeds some 500 collectively for the *rIIA* and *rIIB* genes (Figure 8-2).
- The *rIIA* and *rIIB* genetic maps are unambiguously linear, strongly hinting that the gene itself has a linear construction.
- Most mutations are changes at only one mutable site. Many genes containing such mutations are able to be restored to the original wild-type gene structure by the process of undergoing a second (reverse) mutation at the same site as the first mutation.
- Other mutations can cause the deletion of one or more mutable sites. These reflect a physical loss of part of the *rII* gene (Figure 8-3). Mutations deleting more than one mutable site are highly unlikely to mutate back to the original gene form.

The genetic fine structure of a number of other viral and bacterial genes has since been extensively mapped. The lengths of these maps vary from gene to gene, suggesting that some gene products are larger than others. Although no other study has been as extensive as the *rII* work, each points to the same conclusion: that all genes have a very large number of sites at which mutation can occur and that these mutable sites are arranged in a strictly linear order. The geneticist's

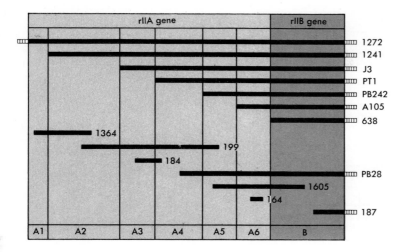

Figure 8-3
Deletion mutations within the *rII* region of T4. About 10 percent of the spontaneous *rII* mutations do not map at a distinct point. They are due to deletions of a large number of adjacent mutable sites. Some deletions, for example 1272, involve both the *rIIA* and *rIIB* genes. The small rectangles indicate that the deletion most likely extends into the adjacent gene. The existence of deletion mutations has considerably facilitated genetic mapping. By crossing a newly isolated *rII* mutant with a number of deletion mutations covering increasingly larger regions, it is possible to assign an approximate map location to the new mutant. In this map, the size of the *rIIB* gene has been arbitrarily reduced.

view of a gene is thus a discrete chromosomal region that is responsible for a specific cellular product and that consists of a linear collection of potentially mutable units or sites, each of which can exist in several alternative forms and between which crossing over can occur.

The Complementation Test Determines If Two Mutations Are in the Same Gene

Since mutations in both the *rIIA* and *rIIB* genes result in a larger plaque, it is natural to ask why they are considered two genes. Would it not be simpler to consider them parts of the same gene? The answer is straightforward. If *E. coli* strain K12(λ) is infected simultaneously with T4 *rIIA* and T4 *rIIB* mutants, the chromosomes of both viruses multiply, and progeny virus is produced (Figure 8-4). In contrast, simultaneous infection of the K12(λ) with either two different T4 *rIIA* or two different T4 *rIIB* mutants results in no viral multiplication. This demonstrates that *rIIA* and *rIIB* genes carry out two different functions, each necessary for T4 multiplication on K12(λ).

In infection with a single T4 *rIIA* mutant, the *rIIB* gene functions normally, but no active form of the *rIIA* product is available, so that no viral multiplication is possible. Likewise, in infection with a single *rIIB* mutant, only an active *rIIA* product arises. Viral multiplication occurs only when we simultaneously infect a bacterium with an *rIIA* phage mutant and an *rIIB* phage mutant, because each mutant chromosome is able to produce the gene product that its infecting partner is unable to make. Two chromosomes can thus complement each other when the mutations are present in distinct genes, thereby giving a positive **complementation test** result. (This experiment is not affected by the possibility of intragenic crossing over between mutants to produce wild-type phage because the number of such recombinants is very small; in contrast, complementation between different mutant genes yields a normal number of progeny particles.)

It is easy to perform complementation tests with phage mutants. All that is necessary is to infect a bacterium with two different mutants at once. This automatically creates a cell containing one copy of each mutant chromosome. These tests were often useful in telling us

Figure 8-4
The demonstration that the *rII* region consists of two distinct genes that can complement each other during simultaneous infection.

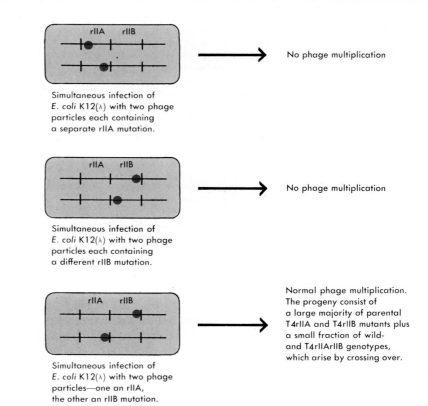

rIIA   rIIB

No phage multiplication

Simultaneous infection of
E. coli K12(λ) with two phage
particles each containing
a separate rIIA mutation.

rIIA   rIIB

No phage multiplication

Simultaneous infection of
E. coli K12(λ) with two phage
particles each containing
a different rIIB mutation.

rIIA   rIIB

Normal phage multiplication.
The progeny consist of
a large majority of parental
T4rIIA and T4rIIB mutants plus
a small fraction of wild-
and T4rIIArIIB genotypes,
which arise by crossing over.

Simultaneous infection of
E. coli K12(λ) with two phage
particles—one an rIIA,
the other an rIIB mutation.

that a chromosomal region thought to contain only one gene actually produced several gene products and so must contain a corresponding number of genes. Until recently, it was more difficult to carry out complementation tests with normally haploid cells like *E. coli*. Laborious genetic tricks, too complicated to be described here, had to be employed to construct special strains with some chromosome sections present twice (partially diploid strains).

Now, recombinant DNA techniques greatly facilitate the use of complementation tests. The essential trick is to place one of the two mutant genes on a plasmid that can be introduced through the Ca^{2+} technique into a recipient strain carrying the second mutant gene (see Chapter 7). Since it is arranged that the plasmid also carry an antibiotic resistance gene, cells that have taken up DNA can be selected for by growing them in the presence of that antibiotic to ensure that the recipient cell carries both mutant genes. If the wild-type phenotype is restored, then the two mutations must have occurred in two different genes, each producing different polypeptide products (Figure 8-5).

Genetic Control of Protein Function[3, 4, 5]

It was not until the 1940s, when a variety of growth factor mutants became available, first in *Neurospora* and then in *E. coli*, that the idea that genes control the synthesis of single proteins (enzymes) became generally adopted. One of the first proofs involved the biosynthesis of the amino acid arginine. Its pathway starts with glutamic acid and proceeds via eight chemical reactions, each catalyzed by a distinct

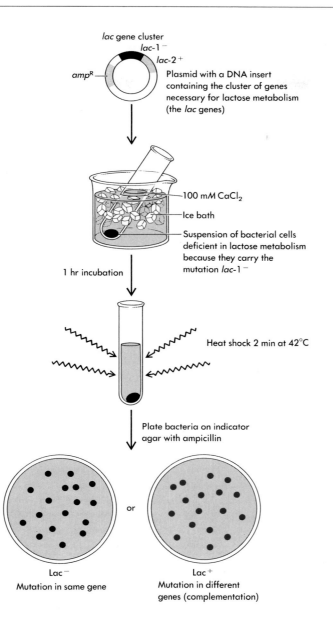

Figure 8-5
A complementation test using recombinant DNA techniques. A plasmid carrying a mutant gene affecting lactose utilization (*lac⁻*) and an ampicillin resistance gene is introduced via the calcium technique into a bacterium also carrying a *lac⁻* mutation. The bacterium is then plated on an agar plate containing an indicator dye to distinguish between the two cell phenotypes Lac⁻ and Lac⁺ (see Figure 7-8). Since the medium also contains ampicillin, only cells that have taken up the plasmid DNA grow. If the wild-type phenotype (Lac⁺) is restored, then the plasmid and chromosomal mutations must be located in two different *lac* genes.

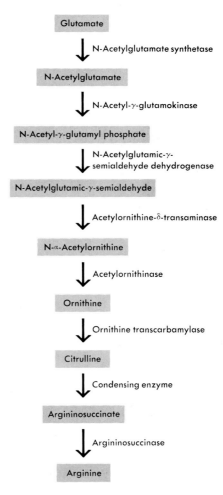

Figure 8-6
Pathway of arginine biosynthesis in *E. coli.*

enzyme (Figure 8-6). For each of these eight steps, mutants have been isolated that fail to carry out that specific enzymatic reaction. This suggests that a separate gene controls the presence of each enzyme, a hypothesis confirmed by the absence of the respective enzyme activity in cell extracts prepared from such mutant cells.

The biosynthesis of histidine provides another beautiful example of the direct relationship between genes and enzymes (Figure 8-7). Specific bacterial mutations result in the absence of each of the ten enzymes necessary for histidine biosynthesis. In this case, all the genes are tightly clustered together in one operon (set of functionally related genes; Figure 8-8), which is transcribed into one very long mRNA molecule. At first it was thought that the order of the genes in such operons correspond to the order in which their respective enzymes act in biosynthesis. However, the first two enzymes in histidine biosynthesis were found to be located at opposite ends of the cluster. When we discuss operons in detail in Chapter 16, we shall learn that those enzymes at the beginning of operons often tend to be

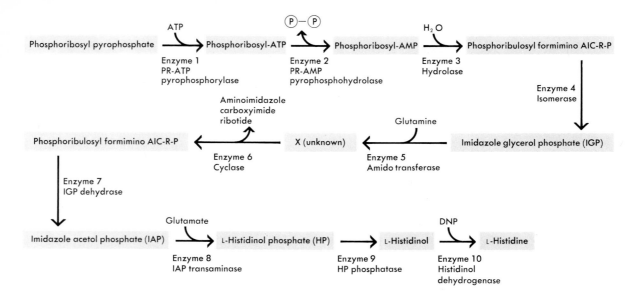

Figure 8-7
The pathway for histidine biosynthesis in *Salmonella typhimurium*. This bacterium, which is closely related to *E. coli*, appears to have a chromosome with a similar gene arrangement.

made in larger amounts than those present at the termini. Thus, the order of genes may have evolved to reflect the relative activities (turnover rates) of the enzymes.

One Gene–One Polypeptide Chain

Initially, the preceding ideas on gene-protein relationships were summarized by the slogan one gene–one protein (enzyme). Now we realize that this concept is more correctly stated as **one gene–one polypeptide.** A number of important proteins are composed of several polypeptide chains, each of which is the product of a separate gene. As later discussed, the *E. coli* enzyme tryptophan synthetase is composed of two distinct chains, each made from a separate gene. Likewise, *E. coli* RNA polymerase has separate genes for each of the four main polypeptides (α, β, β', and ω) used to construct it. However, the existence of several distinct polypeptide chains within functional proteins does not necessarily imply that separate genes code for each of their chains. Many such multichained proteins arise through proteolytic cleavage of single polypeptide chains. The A and B chains of insulin, for example, arise through two cuts in their single-chained proinsulin precursor (see Figure 2-19). Therefore, we now see that the complementation test tells us not whether two genes code for different proteins, but whether they control two different polypeptide chains.

Identifying the Protein Products of Genes[6]

Even now, it is sometimes difficult to find the protein product of a given gene. Just a few years ago, this was often a particularly de-

Figure 8-8
Clustering of the genes involved in the biosynthesis of histidine by the bacterium *Salmonella typhimurium*. The gene order was determined by transduction experiments. Each gene is responsible for the synthesis of one of the ten enzymes needed to transform phosphoribosyl pyrophosphate into histidine. Here the genes are designated by numbers that correspond to the enzyme numbers of Figure 8-7. Enzyme 1 is responsible for catalyzing the first reaction in the biosynthesis, enzyme 2 for the second step, and so on. The names of these enzymes are given in Figure 8-7.

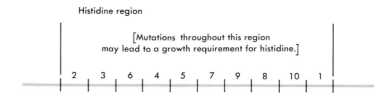

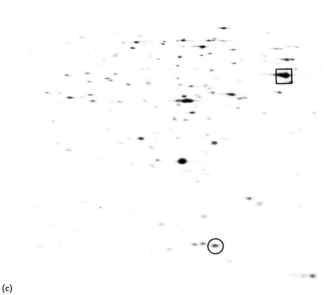

(a)

(b)

(c)

Figure 8-9
Almost all the polypeptides of *E. coli* K-12 can be resolved as radioactive spots on two-dimensional O'Farrell focusing gels. (a) The gel from an extract of wild-type cells. The box encloses the *groEL* product; the circle encloses the *groES* gene (very faint). These gene products are essential for the morphogenesis of coliphages, and for the growth of the uninfected cell. (b) Gel from a mixture of wild-type extract and extract from cells with a missense mutation in *groEL*. The arrows point to a double spot; the left, lowermost being the product of the normal gene, and the one displaced upward and to the right being the product of the mutated gene. (c) Gel from wild-type cells harboring a multicopy plasmid (a derivative of pBR322) carrying the *groESL* operon. Both the *groEL* (box) and the *groES* (circle) protein products are overproduced as a result of the high gene dosage. By such means, the products of even essential genes (for which nonsense or deletion mutations are lethal) can be tentatively identified. (After Dr. Frederick C. Neidhardt.)

manding objective if the protein played only a minor structural role or possessed an enzymatic activity that had not yet been discovered. Now, with two-dimensional gel electrophoretic separation techniques, spotting the right protein is much easier. This method works particularly well for those genes whose mutated forms either fail to produce any product, yield a truncated protein product, or code for amino acid differences that change the electric charge of the protein product, because these mutants form spots that differ in location or are entirely absent from the gel, and thus the change is immediately apparent (Figure 8-9). Fortunately, a large fraction of the mutant genes that we isolate produce radically altered products that can easily be spotted on gels. This is not a matter of chance, since the best mutants to analyze are usually those with totally nonfunctional gene products. Those mutant proteins that differ from their wild-type equivalents only by a single amino acid replacement often retain par-

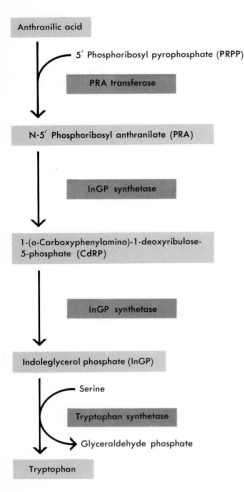

Figure 8-10
Last steps in the pathway of tryptophan biosynthesis.

tial biological activity (they are said to be leaky) and thus are not as easy to work with as those that produce either no product or a drastically rearranged product.

For a long time, the identity of the two proteins coded by the *rIIA* and *rIIB* genes of phage T4 remained unknown. Now they are known to be membrane proteins of molecular weights 86,000 (*rIIA*) and 30,000 (*rIIB*), both present in minor amounts. Unfortunately, still unclear are their exact metabolic functions within T4-infected cells. Thus, for many years there has been a strong tendency to restrict intensive genetic analysis to those genes whose protein products are easy to isolate and whose metabolic or structural roles are well established.

Recessive Genes Frequently Do Not Produce Functional Products

Most mutant genes are recessive with respect to wild-type genes. This fact, puzzling to early geneticists, is now partially understood in terms of the gene-protein relationship. The recessive phenotype often results from the failure of mutant genes to produce any functional protein (enzyme). In heterozygotes, however, there is often present one "good" gene and, correspondingly, a number of "good" gene products. Because the wild-type gene is present in only one copy in heterozygotes, it is possible that there are always fewer good copies of the relevant protein in heterozygotes than in individuals with two wild-type genes. If this were the case, we might guess that the heterozygous phenotype would tend to be intermediate between the two homozygous phenotypes. Usually, however, this does not happen for one of two reasons. Either there are still enough good enzyme molecules to catalyze the metabolic reaction of concern, even though the total number of molecules is reduced, or the recessive gene is not noticeable because control mechanisms cause the single wild-type gene in a heterozygote to produce more gene product than does each wild-type gene in a homozygote. In Chapter 16, we shall discuss how the rates at which bacterial genes act are controlled.

Colinearity of the Gene and Its Polypeptide Product[7]

The best-understood example of the relationship between the order of the mutable sites in a gene and the order of their corresponding amino acid replacements involves the *E. coli* enzyme tryptophan synthetase, one of the several enzymes involved in tryptophan synthesis (Figure 8-10). This enzyme consists of two easily separated polypeptide chains, A and B, neither of which is enzymatically active by itself. A large number of mutants unable to synthesize tryptophan lack a functional A chain in their tryptophan synthetase molecules. When these mutants were genetically analyzed, it was found that changes at a large number of different mutable sites could give rise to inactive A chains. Accurate mapping of these mutants revealed that they all could be unambiguously located on the linear genetic map shown in Figure 8-11. It was possible to isolate the inactive A chains from many of these mutants and to begin to compare their amino acid sequences with the sequence of the wild-type A chain, which contains 267

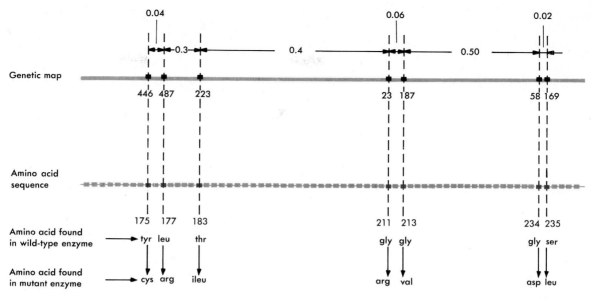

amino acids. This sequence allows us to see how the location of a mutation within a gene is correlated with the location of the replaced amino acid in its polypeptide chain product. Since both genes and polypeptide chains are linear, the simplest hypothesis is that amino acid replacements are in the same relative order as the mutationally altered sites in the corresponding mutant genes. This was most pleasingly demonstrated in 1964. The location of each specific amino acid replacement is exactly correlated with its location along the genetic map, a property called **colinearity.** Thus, successive amino acids in a polypeptide chain are controlled, or coded, by successive regions of a gene.

Mutable Sites Are the Base Pairs Along the Double Helix

In all bacterial genes extensively mapped, the large number of linearly arranged mutable sites that have been found in each gene, and between which genetic recombination (crossing over) is possible, leaves us no choice but to conclude that these sites are the specific base pairs along the DNA of the respective gene (Figure 8-12). A given mutable site can thus exist in any of four different states, AT, TA, GC, or CG. Many mutations are therefore likely to represent simple switches from one state to another. The genetic data that reveal deletions and insertions of genetic material must now be thought of in terms of the addition or deletion of discrete blocks of one to very many base pairs. The three classes of mutations resulting from changes in the sequence of nucleotide bases are illustrated in Figure 8-13.

By carefully studying the fine details of genetic maps, we should be able to obtain important information about the corresponding DNA. However, not every change in base sequence leads to easily observed changes in the corresponding protein. In the genetic code, many amino acids are specified by more than one **codon** (set of three adja-

Figure 8-11
Colinearity of the gene and its protein product. Here is the genetic map for one-fourth of the gene coding for the amino acid sequences in the *E. coli* protein tryptophan synthetase A. The designation 0.04, for example, refers to map distances (frequencies of recombination) between tryptophan synthetase mutations A446 and A487. The numbers in the amino acid sequence refer to their position in the 267 residues of the A protein. Following convention, the amino terminal end of the segment is on the left.

Figure 8-12
The relationship of mutations in the *rII* region of the phage T4 chromosome to the structure of DNA.

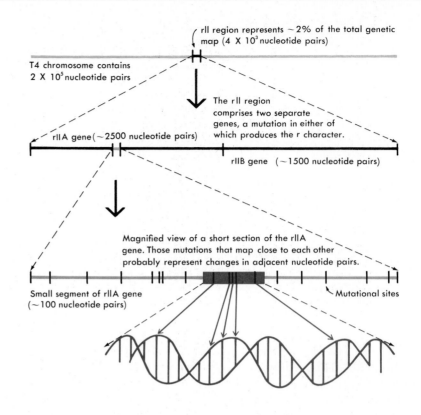

rII region represents ~2% of the total genetic map (4 X 10³ nucleotide pairs)

T4 chromosome contains 2 X 10⁵ nucleotide pairs

The rII region comprises two separate genes, a mutation in either of which produces the r character.

rIIA gene (~2500 nucleotide pairs)

rIIB gene (~1500 nucleotide pairs)

Magnified view of a short section of the rIIA gene. Those mutations that map close to each other probably represent changes in adjacent nucleotide pairs.

Small segment of rIIA gene (~100 nucleotide pairs)

Mutational sites

cent bases), which means that in many cases, base-pair substitutions will not lead to any amino acid replacements. Moreover, as we document later, many of the amino acids in proteins are not essential, and when they are replaced by somewhat similar amino acids, the proteins often retain full activity. The number of observed mutable sites therefore seriously underrepresents the number of base pairs within the corresponding gene.

There Are Four Alternative Structures for Each Mutable Site[8, 9]

As anticipated, enzymatically inactive tryptophan synthetase molecules resulting from independent mutations at the same mutable site (as shown by failure to give wild-type recombinants) do not always contain the same amino acid replacement. For example, changes in a single mutable site that specifies the amino acid at position 213 results in the replacement of glycine by either glutamic acid or valine. Inspection of the genetic code (see Chapter 15) indicates that in the wild-type strain, this glycine must be specified by either GGA or GGG codons and that the mutable site under study specifies the G in the middle position of this codon. When this G is replaced by U, valine (GUA or GUG) becomes inserted into the glycine site while its replacement by A generates the glutamic acid (GAA or GAG) substitution. Further study of this particular mutable site might eventually turn up the anticipated third replacement in which a G to C switch leads to the appearance of alanine (GCA or GCG).

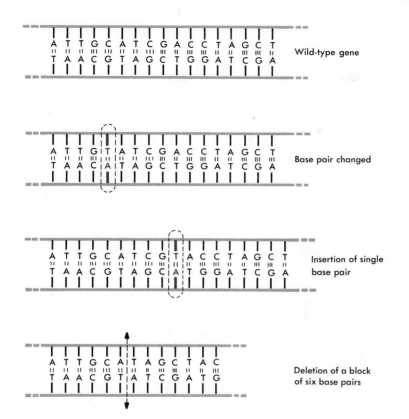

Figure 8-13
Three classes of mutations result from introducing defects in the sequence of bases (A, T, G, C) attached to the backbone of the DNA molecule. In one class, a base pair is simply changed from one into another (i.e., GC to AT). In the second class, a base pair is inserted (or deleted). In the third class, a block of base pairs is deleted (or inserted).

Single Amino Acids Are Specified by Several Adjacent Nucleotide Bases

We expected to find that given amino acids within a particular protein are specified by adjacent mutable sites. This point was first demonstrated in the tryptophan synthetase A gene, where the relevant evidence came from study of the tryptophan synthetase fragment illustrated in Figure 8-14. Treatment of the wild-type strain with a mutagen had given rise to mutant A23, in which arginine replaces glycine (this time at position 212), and mutant A46, in which glutamic acid replaces glycine at the same position. The difference between A23 and A46 does not represent changes to alternative forms of the same mutable site, since a genetic cross between A23 and A46 yields a number of wild-type recombinants (glycine in position 212). If these changes were at the same mutable site, no wild-type recombinants would be produced. Moreover, the very low observed frequency of the wild-type recombinants is compatible with the prediction from the genetic code that these mutable sites are adjacent to each other.

Additional genetic evidence that confirms the separate locations of the A23 and A46 mutable sites comes from observing how A23 and A46 themselves mutate upon treatment with mutagens. After exposure to a mutagen, both strains give rise to new strains, some of which contain active tryptophan synthetase A chains with glycine in position 212. These reverse mutations most likely involve changing the altered mutable sites back to the original wild-type configuration. However, strains containing active tryptophan synthetase also arise

Figure 8-14
Demonstration that a single amino acid is specified by more than one mutable site. We now know that the mutable sites are DNA bases and the codons are actually bases complementary to these in mRNA. (After Emanual J. Murgola.)

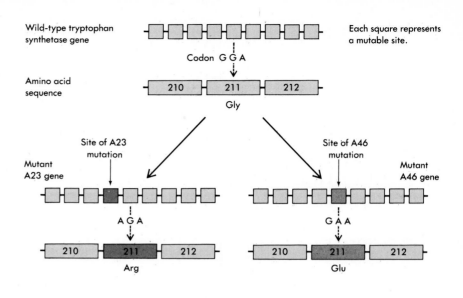

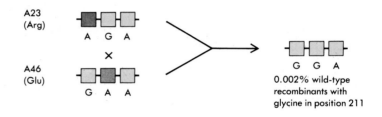

Genetic cross between mutants A23 and A46:

in which the amino acid in position 212 is replaced by another amino acid. Most significantly, the type of replacement differs for strains A23 and A46. Besides back-mutating to glycine, strain A23 mutates to threonine and serine, whereas A46 mutates to alanine and valine in addition to glycine. The failure of A23 ever to give rise to alanine or valine and the failure of A46 ever to mutate to threonine or serine is very difficult to explain if their differences from wild type are based on alternative configurations of the same mutable site. But these mutational patterns make perfect sense if glycine at the 212 position is coded by GGA with the A23 mutation to arginine representing a G to A change at the first position of the codon to give rise to AGA and the A46 mutation to glutamic acid occurring at the middle (second) position to give rise to GAA. Their divergent subsequent mutations to serine and threonine and to alanine and valine, respectively, can also be understood by inspecting the genetic code (Figure 8-15).

Single Amino Acid Substitutions Usually Do Not Alter Enzyme Activity

The ability of a polypeptide chain to be enzymatically active does not require an exactly specified amino acid sequence. This is shown by examination of the new mutant strains obtained by treating strains A23 and A46 with mutagens. The possession of either glycine or serine in position 212 yields a fully active enzyme, whereas threonine in

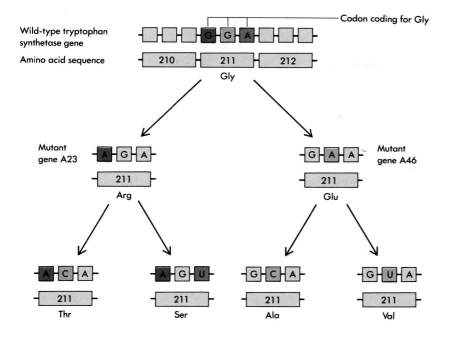

Figure 8-15
Formation of mutants A23 and A46 and their subsequent mutations. Notice that Thr and Ser cannot result from a single base change to the codon for Glu; likewise, Ala and Val cannot result from only one base change to the codon for Arg. Therefore, the A23 and A46 mutants must occur from mutations at two different mutable sites, as shown in Figure 8-14.

the same position yields an enzyme with reduced activity, demonstrating that the activity of an enzyme does not demand a perfectly unique amino acid sequence (Figure 8-16). In fact, evidence now indicates that amino acid replacements in many parts of a polypeptide chain can occur without seriously modifying catalytic activity. However, one sequence may often be best suited to a cell's particular needs, and it is this sequence that is encoded by the wild-type allele. Even though other sequences are almost as good, they will tend to be selected against in evolution.

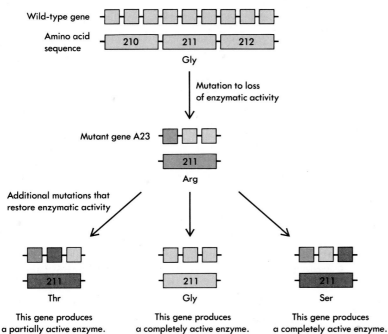

Figure 8-16
Evidence that many amino acid replacements do not result in loss of enzymatic activity.

A Second Amino Acid Replacement May Cancel Out the Effect of the First[10]

The conclusion that minor changes to amino acid sequence do not significantly alter enzyme activity is extended by the finding that some mutations that convert inactive mutant enzymes to active forms may work by causing a second amino acid replacement in the mutant enzyme. Consider mutant A46, which produces inactive tryptophan synthetase because of the substitution of glutamic acid for glycine at ~~protein~~ 212. *amino acid* In this case, distant second-site mutations that result in the active enzyme occasionally emerge. For example, the second-site mutation A446 is located one-tenth of a gene length away from the first mutation. The double mutant A46A446 produces active enzyme molecules containing two amino acid replacements: the original glycine-to-glutamic acid shift and a tyrosine-to-cysteine shift located 36 amino acids away (Figure 8-17).

The second shift can be studied independently of the first by obtaining recombinant cells with only the A446 mutation. Most interestingly the A446 change, when present alone, also results in an inactive enzyme. We thus see that a combination of two wrong amino acids can produce an enzyme with an active three-dimensional configuration. However, only occasionally do two wrong amino acids cancel out each other's faults. For example, double mutants containing A446 and A23, or A446 and A187, do not produce active enzyme. At this time, it does not seem wise to speculate on how the various amino acid residues are folded together in the three-dimensional configuration and why only some combinations are enzymatically active. This kind of analysis must await the establishment of the three-dimensional structure of tryptophan synthetase.

The Very Drastic Consequences of the Insertion or Deletion of Single Base Pairs[11, 12]

Early on in the analysis of mutant proteins, it became clear that the vast majority of mutants being isolated did not yield the minimally altered proteins, bearing single amino acid replacements, that would arise through the change of one type of base pair into one of its three alternatives. Instead, most mutants represented changes that led to drastically altered gene products, often containing many fewer amino acids and with many of their amino acid sequences bearing no relationship to the wild-type polypeptide products. The nature of these mutants first became apparent through the proposal that such mutations usually represented either insertions or deletions of single nucleotide pairs. The drastic effect of these insertion or deletion events is a consequence of the fact that mRNA molecules are read in successive blocks of three nucleotides, called codons. AUG codons, which code for the methionine residues found at the amino terminal ends of newly synthesized polypeptide chains, are the signal for ribosomes to begin reading the mRNA molecule about to be translated into a protein. Since reading always begins at the appropriate AUG condon, the mRNA molecules are aligned on the ribosomes so that their messages are read in the correct **reading frame.**

If, however, a single base pair is inserted or deleted in a coding sequence, the triplets that designate amino acids become completely changed beginning at the site of insertion or deletion (Figure 8-18).

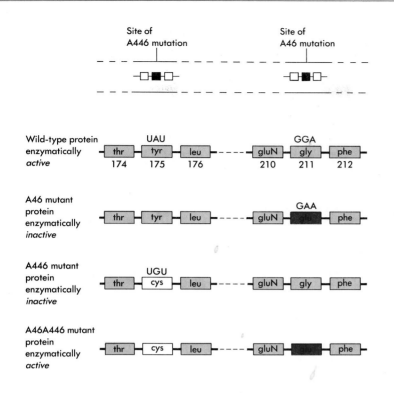

For example, if normally the gene sequence ATTAGACAC . . . is read as (ATT)(AGA)(CAC) . . . , then the insertion of a new nucleotide C in the fourth position of that sequence creates ATTCAGACAC, which is read as (ATT)(CAG)(ACA)(C . . .). These new triplets may code for entirely different amino acids. A similar consequence follows from a deletion. Moreover, the crossing of two deletion or two insertion mutants yields double mutants in which the reading frame is still misplaced.

Reversion of Insertion or Deletion Mutants

Active (or partially active) genes are regenerated by crossing over between an insertion and a nearby deletion. Such events restore the correct reading frame except in the short region between the mutations (see Figure 8-18). If the affected gene region is nonessential (e.g., the early section of the T4 *rIIB* gene), then the resulting protein product is fully functional. In other cases, the short segments of inappropriate amino acids are only mildly disadvantageous, and partial activity results. No activity, however, will usually be found if the inappropriate codons include any of the three that signify chain termination (UAA, UAG, or UGA). Their presence inevitably results in incomplete fragments of the wild-type polypeptide.

It is also sometimes possible to obtain functional genes by producing recombinants containing three closely spaced insertions or deletions (Figure 8-19). In contrast, recombinants containing four nearby insertions or deletions produce only nonfunctional polypeptides. These later experiments were performed in 1961, before the basic outlines of the genetic code were known. They in fact provided the first good evidence that the genetic code was likely to be read in groups of three as opposed to groups of two or four.

Figure 8-18
Mutations that add or remove a base shift the reading frame of the genetic message.

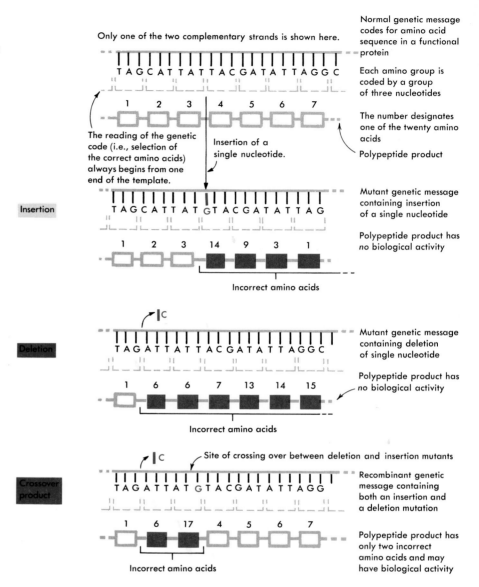

Only one of the two complementary strands is shown here.

Normal genetic message codes for amino acid sequence in a functional protein

T A G C A T T A T T A C G A T A T T A G G C

Each amino group is coded by a group of three nucleotides

1 2 3 4 5 6 7

The number designates one of the twenty amino acids

Polypeptide product

The reading of the genetic code (i.e., selection of the correct amino acids) always begins from one end of the template.

Insertion of a single nucleotide.

Insertion

T A G C A T T A T GT A C G A T A T T A G

Mutant genetic message containing insertion of a single nucleotide

1 2 3 14 9 3 1

Polypeptide product has *no* biological activity

Incorrect amino acids

Deletion

T A G A T T A T T A C G A T A T T A G G C

Mutant genetic message containing deletion of single nucleotide

1 6 6 7 13 14 15

Polypeptide product has *no* biological activity

Incorrect amino acids

Site of crossing over between deletion and insertion mutants

Crossover product

T A G A T T A T G T A C G A T A T T A G G

Recombinant genetic message containing both an insertion and a deletion mutation

1 6 17 4 5 6 7

Polypeptide product has only two incorrect amino acids and may have biological activity

Incorrect amino acids

Cloned Genes Can Be Sequenced[13–17]

Virtually all the essential features of the genetic code were deduced by 1966 from the coding properties of either enzymatically or chemically synthesized mRNA molecules and from the accumulated knowledge of genetic fine structure that we have just detailed. No real genes were directly analyzed, however, since at that time there were no procedures either to sequence DNA or to isolate desired genes. But with the arrival of recombinant DNA and of powerful methods for DNA sequencing, the nature of genetic research has dramatically changed. No longer are genetic crosses the prime vehicle for probing genes. The quickest and most direct way to proceed is now the cloning and sequencing of relevant genetic material. As indicated in the previous chapter, it is now a relatively straightforward matter to isolate any *E. coli* gene that codes for a function that can be selected for by one of the many enrichment procedures.

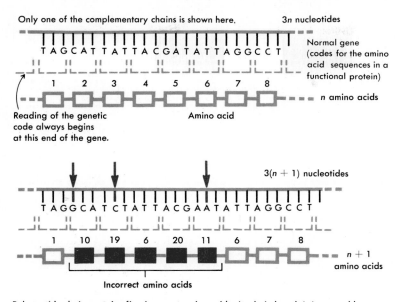

Only one of the complementary chains is shown here.

3n nucleotides

T A G C A T T A T T A C G A T A T T A G G C C T

Normal gene (codes for the amino acid sequences in a functional protein)

1 2 3 4 5 6 7 8

n amino acids

Reading of the genetic code always begins at this end of the gene.

Amino acid

3(n + 1) nucleotides

T A G G C A T C T A T T A C G A A T A T T A G G C C T

1 10 19 6 20 11 6 7 8

n + 1 amino acids

Incorrect amino acids

Polypeptide chain contains five incorrect amino acids; its chain length is increased by one amino acid. It may have some biological activity depending upon how the five wrong amino acids influence its 3-D structure.

Figure 8-19
When three nucleotides are added close together, the genetic message is scrambled only over a short region. The same type of result is achieved by the deletion of three nearby nucleotides.

Already, a large number of *E. coli* genes have been completely or partially sequenced. In all cases, the codons found to specify given amino acids are those predicted by the genetic code (Figure 8-20). This agreement between prediction and result, though inherently very satisfying, surprised no one, since the experimental evidence used to deduce the genetic code was effectively unassailable (see Chapter 15). Also as predicted, the coding segments of virtually all mRNAs start with the AUG codon and always conclude with a chain-terminating codon (UAA, UAG, or UGA).

Untranslated Sequences at the Beginnings and Ends of mRNA Molecules[18–23]

When mRNA was first discovered, it seemed simplest to assume that the translation events would begin at one end of the molecule and then move along in steps of three nucleotides until the other end was reached. This was a very naive view, adopted before the discoveries that methionine initiates all polypeptide chains and that specific codons specify chain termination. Now we realize that untranslated sequences exist at both the 5' end of the mRNA, near which translation begins, and at the 3' end, near which translation stops (Figure 8-21). Hence, there must be internal signals in mRNA that mark the starting and stopping sites for translation. With the exception of a small purine-rich block of nucleotides that functions to position ribosomes at the correct AUG start codon, the untranslated regions probably play no role in translation and are of variable lengths, ranging from 20 to more than 100 nucleotides, depending on the particular mRNA species.

These seemingly unnecessary extra sequences only make sense

Figure 8-20
The nucleotide sequence and corresponding amino acids of the *trpC* gene of the tryptophan operon of *E. coli*. Note that the codons found to specify given amino acids are those predicted by the genetic code. [Courtesy of Dr. Charles Yanofsky, *Nucleic Acid Res.* 9 (1981):6657.]

```
TRP C
       3330        3340        3350        3360        3370        3380        3390        3400        3410
ATG ATG CAA ACC GTT TTA GCG AAA ATC GTC GCA GAC AAG GCG ATT TGG GTA GAA GCC CGC AAA CAG CAA CCG CTG GCC AGT TTT CAG
MET GLN THR VAL LEU ALA LYS ILE VAL ALA ASP LYS ALA ILE TRP VAL GLU ALA ARG LYS GLN GLN GLN PRO LEU ALA SER PHE GLN
  1                          10                         20
       3420        3430        3440        3450        3460        3470        3480        3490        3500
GAG GTT CAG CCG AGC ACG CGA CAT TTT TAT GAT GCG TCA CAG GGT GCG CGC ACG GTG TTT ATT CTG GAG TGC AAG AAA GCG TCG CCG TCA
GLU VAL GLN PRO SER THR ARG HIS PHE TYR ASP ALA LEU GLN GLY ALA ARG THR ALA PHE ILE LEU GLU CYS LYS LYS ALA SER PRO SER
 30                         40                          50
       3510        3520        3530        3540        3550        3560        3570        3580        3590
AAA GGC GTG ATC CGT GAT GAT TTC GAT CCA GCA CGC ATT GCC GCC ATT TAT AAA CAT TAC GCT TCG GCA ATT TCG GTG CTG ACT GAT GAG
LYS GLY VAL ILE ARG ASP ASP PHE ASP PRO ALA ARG ILE ALA ALA ILE TYR LYS HIS TYR ALA SER ALA ILE SER VAL LEU THR ASP GLU
 60                         70                          80
       3600        3610        3620        3630        3640        3650        3660        3670        3680
AAA TAT TTC AGG GGT AGC TTT AAT TTC CTC CCC ATC GTC AGC CAA ATC GCC CCG CAG CCG ATT TTA TGT AAA GAC TTC ATT ATC GAC CCT
LYS TYR PHE ARG GLY SER PHE ASN PHE LEU PRO ILE VAL SER GLN ILE ALA PRO GLN PRO ILE LEU CYS LYS ASP PHE ILE ILE ASP PRO
 90                         100                         110
       3690        3700        3710        3720        3730        3740        3750        3760        3770
TAC CAG ATC TAT CTG GCG CGC TAT TAC CAG GCC GAT GCC TGC TTA TTA ATG CTT TCA GTA CTG GAT GAC GAC CAA TAT CGC CAG CTT GCC
TYR GLN ILE TYR LEU ALA ARG TYR TYR GLN ALA ASP ALA CYS LEU LEU MET LEU SER VAL LEU ASP ASP ASP GLN TYR ARG GLN LEU ALA
120                         130                         140
       3780        3790        3800        3810        3820        3830        3840        3850        3860
GCC GTC GCT CAC AGT CTG GAG ATG GGG GTG CTG ACC GAA GTC AGT AAC GAG GAG GAG CAG GAG CGC GCC ATT GCA TTG GGA GCA AAG GTC
ALA VAL ALA HIS SER LEU GLU MET GLY VAL LEU THR GLU VAL SER ASN GLU GLU GLU GLN GLU ARG ALA ILE ALA LEU GLY ALA LYS VAL
150                         160                         170
       3870        3880        3890        3900        3910        3920        3930        3940        3950
GTT GGC ATC AAC AAC CGC GAT CTG CGT GAT TTG TCG ATT GAT CTC AAC CGT ACC CGC GAG CTT GCG GCG AAA CTG GGG CAC AAC GTG ACG
VAL GLY ILE ASN ASN ARG ASP LEU ARG ASP LEU SER ILE ASP LEU ASN ARG THR ARG GLU LEU ALA ALA LYS LEU GLY HIS ASN VAL THR
180                         190                         200
       3960        3970        3980        3990        4000        4010        4020        4030        4040
GTA ATC AGC GAA TCC GGC ATC AAT ACT TAC GCT CAG GTG CGC GAG TTA AGC CAC TTC GCT AAC GGT TTT CTG ATT GGT TCG GCG TTG ATG
VAL ILE SER GLU SER GLY ILE ASN THR TYR ALA GLN VAL ARG GLU LEU SER HIS PHE ALA ASN GLY PHE LEU ILE GLY SER ALA LEU MET
210                         220                         230
       4050        4060        4070        4080        4090        4100        4110        4120        4130
GCC CAT GAC GAT TTG CAC GCC GCC GTG CGC CGG GTG TTG CTG GGT GAG AAT AAA GTA TGT GGC CTG ACG CGT GGG CAA GAT GCT AAA GCA
ALA HIS ASP ASP LEU HIS ALA ALA VAL ARG ARG VAL LEU LEU GLY GLU ASN LYS VAL CYS GLY LEU THR ARG GLY GLN ASP ALA LYS ALA
240                         250                         260
       4140        4150        4160        4170        4180        4190        4200        4210        4220
GCT TAT GAC GCG GGC GCG ATT TAC GGT GGG TTG ATT TTT GCG ACA TCA CCG CGT TGC GTC AAC GAT GAA CAG GCG CAG GAA GTG ATG GCT
ALA TYR ASP ALA GLY ALA ILE TYR GLY GLY LEU ILE PHE ALA THR SER PRO ARG CYS VAL ASN ASP GLU GLN ALA GLN GLU VAL MET ALA
270                         280                         290
       4230        4240        4250        4260        4270        4280        4290        4300        4310
GCG GCA CCG TTG CAG TAT GTT GGT GTG TTC CGC AAT CAC GAT ATT GCC GAT GTG GTG GAC AAA GCT AAG GTG TTA TCG CTG GTG GCA GTG
ALA ALA PRO LEU GLN TYR VAL GLY VAL PHE ARG ASN HIS ASP ILE ALA ASP VAL VAL ASP LYS ALA LYS VAL LEU SER LEU VAL ALA VAL
300                         310                         320
       4320        4330        4340        4350        4360        4370        4380        4390        4400
CAA CTG CAT GGT AAT GAA GAA CTG CTG TAT ATC GAT ACG CTG CGT GAA GCT CTG CCA GCA CAT GTT GCC ATC TGG AAA GCA TTA AGC GTC
GLN LEU HIS GLY ASN GLU GLU LEU LEU TYR ILE ASP THR LEU ARG GLU ALA LEU PRO ALA HIS VAL ALA ILE TRP LYS ALA LEU SER VAL
330                         340                         350
       4410        4420        4430        4440        4450        4460        4470        4480        4490
GGT GAA ACC CTG CCC GCC CGC GAG TTT CAG CAC GTT GAT AAA TAT GTT TTA GAC AAC GGC CAG GGT GGA AGC GGG CAA CGT TTT GAC TGG
GLY GLU THR LEU PRO ALA ARG GLU PHE GLN HIS VAL ASP LYS TYR VAL LEU ASP ASN GLY GLN GLY GLY SER GLY GLN ARG PHE ASP TRP
360                         370                         380
       4500        4510        4520        4530        4540        4550        4560        4570        4580
TCA CTA TTA AAT GGT CAA ACG CTT GGC AAC GTT CTG CTG GCG GGG GGC TTA GGC GCA GAT AAC TGC GTG GAA GCG CAA ACC GGC TGC
SER LEU LEU ASN GLY GLN THR LEU GLY ASN VAL LEU LEU ALA GLY GLY LEU GLY ALA ASP ASN CYS VAL GLU ALA ALA GLN THR GLY CYS
390                         400                         410
       4590        4600        4610        4620        4630        4640        4650        4660        4670
GCC GGA CTT GAT TTT AAT TCT GCT GTA GAG TCG CAA CCG GGC ATC AAA GAC GCA CGT CTT TTG GCC TCG GTT TTC CAG ACG CTG CGC GCA
ALA GLY LEU ASP PHE ASN SER ALA VAL GLU SER GLN PRO GLY ILE LYS ASP ALA ARG LEU LEU ALA SER VAL PHE GLN THR LEU ARG ALA
420                         430                         440
TAT TAA
TYR END
450
```

Figure 8-21
Untranslated sequences at the beginning (a) and the end (b) of the tryptophan operon. (a) The untranslated sequence (in color) at the beginning of the *trpE* gene of the operon showing the codon for methionine (boxed) signalling the beginning of translation. (b) The untranslated sequences (in color) at the end of the *trpA* gene at the end of the operon. The stop codon is boxed and the termination region is marked by an arrow. Other evidence indicates that termination may sometimes occur farther down the molecule. (After Terry Platt, *Nucl. Acids Res.* 9 (1981):6651–6653.

```
          1               20              40
(a)   AAGTTCACGTAAAAAGGGTATCGACA ATG AAAGCAATTTT

              6690        6710                     ↓
(b)   CGTTGCGCT TAA TCCCACAGCCGCCAGTTCCGCTGGCGGCATTTTAA
```

when we consider how mRNA molecules are synthesized upon their DNA templates. Signals must be encoded into DNA that not only say to start RNA synthesis or to stop RNA synthesis but also that control the rates at which these processes occur. Those signal sequences that start RNA synthesis are called **promoters.** Their existence was first predicted through the discovery of mutants blocked in the initiation of transcription. Only a decade later, through DNA sequencing, was their fine structure revealed. In general, *E. coli* promoters are positioned just before the starting sites for transcription (at the 5' end) and so do not become transcribed into mRNA. Often situated between promoters and amino acid coding sequences are regulatory sequences called **operators,** to which specific DNA-binding proteins known as repressors can bind and thereby prevent RNA polymerase from attaching and initiating transcription (Figure 8-22). Operator sequences usually straddle RNA start sites and so when present have their 3'-terminal sections transcribed to form the beginning segments of the 5' untranslated "leader" segments. See Chapter 16 for details.

Equally complicated are the signals used to stop RNA synthesis, a process whose frequency also can be regulated. These stop signals

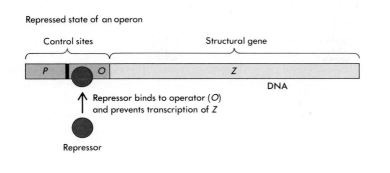

Repressed state of an operon

Figure 8-22
Repressors bind to operators to prevent transcription. When repressor molecules are not present, RNA polymerase can bind to DNA at the promoter site and initiate transcription. Shown here is an operon showing the relative locations of a promoter (*P*), operator (*O*), and structural gene (*Z*).

generally are located before the actual termination sites and when so sited are found as components of the 3' untranslated regions (see Chapter 13). The 3' untranslated regions may also be important in conferring stability on mRNAs. (The structure of mRNA is discussed in more detail in later chapters.)

Transcriptional Units Are the Fundamental Segments of Chromosomal Activity

Not all the DNA in bacteria (or in any other cells) codes for polypeptide chains. Some 1 percent of the *E. coli* chromosome codes for RNA chains that never become translated (e.g., tRNA and rRNA). The discrete DNA segments that code for these untranslatable RNAs are also regarded as genes, since they mutate and recombine like all other DNA regions. The designations rDNA and tDNA refer to those genes coding respectively for ribosomal and transfer RNA.

DNA molecules (chromosomes) should thus be functionally regarded as linear collections of discrete transcriptional units, each designed for the synthesis of a specific RNA molecule. Whether such "transcriptional units" should now be redefined as genes, or whether the term *gene* should be restricted to the smaller segments that directly code for individual mature rRNA or tRNA molecules or for individual polypeptide chains, is now an open question. This dilemma would not exist if all transcriptional units coded for either a single polypeptide chain or one mature RNA chain. But many, if not most, mRNA molecules in bacteria code for a number of functionally interrelated proteins (e.g., the several enzymes that make up a biosynthetic pathway). Likewise, most mature rRNA and tRNA molecules are synthesized as parts of larger RNA molecules that later become enzymatically cleaved into smaller functional units (see Chapter 14). Yet, the literature of genetics that has developed over the past three decades has defined genes in terms of their final polypeptide (structural RNA) products. Therefore, replacing the term *operon* with *gene* would probably cause more confusion than enlightenment at this time. Instead, it may actually make more sense to restrict the term *gene* to those DNA sequences that code for amino acids (and polypeptide chain termination) or that code for functional RNA chains, treating all transcriptional control regions as extragenic elements. DNA chromosomes, so defined, would be linear collections of genes interspersed with promoters, operators, and RNA chain-termination signals.

trpL	AAAGGGTATCGACAATGAAA
trpE	ATTAGAGAATAACAATGCAA
trpD	CAGGAGACTTTCTGATGGCT
trpC	CACGAGGGTAAATGATGCAA
trpB	TAAGGAAAGGAACAATGACA
trpA	CGAGGGGAAATCTGATGGAA

Figure 8-23
Short nucleotide segment between adjacent genes in the *trp* operon. Initiator and stop codons are boxed. [Courtesy of Dr. Terry Platt, *Nucl. Acids Res.* 9 (1981):6659]

This nontraditional definition is now the preferred way of viewing the bacterial chromosome, and we shall adopt this definition in our subsequent discussion of bacterial and phage genes. However, when the genes of eucaryotic organisms are later described, we shall see that they, too, present serious obstacles to being defined in terms of single gene products. Here again, logic may tell us that if we wish to continue to designate the gene as the unit of chromosomal function, then it should refer to those DNA segments that become transcribed into discrete RNA chains rather than to those DNA segments that provide the genetic code for given polypeptide chains (or are directly transcribed into the sequences of mature rRNA or tRNA molecules).

Gaps Between Genes Can Be Very Short[24, 25]

Usually, there is much less crossing over between mutations located at adjacent ends of two contiguous genes within a given bacterial operon (transcriptional unit) than between two mutations at extreme ends of even relatively short genes. This suggested that adjacent genes within operons are often separated by only a few base pairs (Figure 8-23). Confirmation of this point has come with the advent of DNA-sequencing procedures. They reveal, for example, that in the tryptophan operon, gaps of at most several nucleotides exist between adjacent genes. Also being accumulated are data on the distances in base pairs between different transcriptional units, which also are not large in bacteria. This makes biological sense, since such spacer DNA would be carrying out no function and would increase the time needed to replicate the bacterial chromosome containing it.

There Is Agreement Between the Genetic Map and the Corresponding Distance Along a DNA Molecule[26]

Ever since crossing-over analysis began, geneticists have wondered how closely their genetic maps correspond to actual physical lengths along chromosomes. The visualization, in the 1930s, of *Drosophila*'s highly extended salivary chromosomes first permitted the physical mapping of the sites of mutations. Good correlations between the physical and genetic distances along some chromosomal stretches were shown. But these views of chromosomes in the light microscope told us nothing about what was happening within the constituent DNA molecules. Firm answers at the molecular level came only when it became possible to locate the sites of specific mutations along genetically well-defined DNA molecules.

The first high-resolution data emerged through the mapping of well-defined deletions of large numbers of nucleotides from phage λ DNA. This initially was achieved using the technique of DNA-DNA hybridization. In this procedure, the two strands of the λ DNA double helix are separated by heating the DNA to near 100°C. At that temperature, the DNA will denature; virtually all the hydrogen bonds holding the double helix together break, and the resulting free single strands unwind from each other. Because the + and − strands have sufficiently different base compositions, they can be separated by density differences in a cesium chloride (CsCl) solution. Reformation

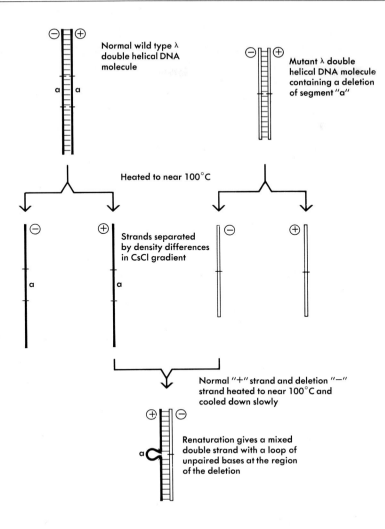

Figure 8-24
Mapping deletions by electron microscope visualization of renatured DNA molecules.

of the double helices from separated single strands (renaturation) occurs if the single strands are mixed together, heated to near 100°C, and then gradually cooled.

If, however, renaturation occurs between a normal + strand and a − strand containing a deletion, the section of the + strand complementary to the deleted − section cannot form hydrogen bonds. It exists as a single-stranded loop extended out from the predominantly double-stranded molecule (Figure 8-24). These loops are easily detected in the electron microscope (Figure 8-25), allowing different deletions to be localized precisely along the λ DNA molecule. Such data, taken together with the extensively studied λ genetic map, give the picture shown in Figure 8-26. It reveals a much closer correspondence between the genetic map and the actual physical structure of the DNA than had been anticipated. Thus, at medium resolution, the probability of crossing over is approximately equal throughout the DNA molecule. Exceptions to the correspondence between the genetic and physical maps are obvious in Figure 8-26, and caution must be taken not to overinterpret genetic linkage data.

Figure 8-25
(a) An electron micrograph (left) of a heteroduplex formed between one strand of wild-type λ and the complementary strand of λ*b2b5,* and (b) an interpretive drawing. λ*b2b5* contains a deletion (*b2*) and a nonhomologous section (*b5*) unable to pair with its λ counterpart. [Reproduced with permission from B. C. Westmoreland, W. Szybalski, and H. Ris, *Science,* 163, 1343 (1969).]

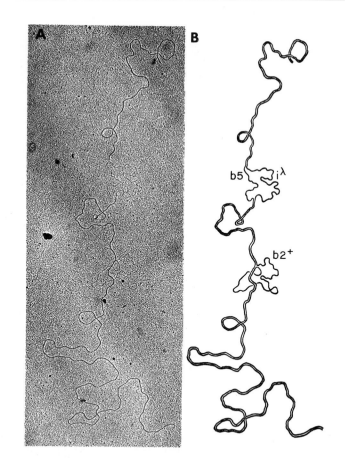

The Eventual Sequencing of the Entire *E. coli* Chromosome[27]

Major new insights into the genetic organization of *E. coli* and of bacteria in general may now have to await the sequencing of sufficiently long chromosomal segments that will reveal the actual density of coding sequences along the DNA. Preferably, this will be initiated on segments containing many genes whose polypeptide products are known and whose amino acid sequences can be compared with DNA sequences so obtained. It should not, however, be necessary to have prior knowledge of a gene to spot its presence along a DNA sequence. Its existence should be revealed by the occurrence of an extensive open reading frame uninterrupted by the presence of chain-terminating codons. The three chain-terminating codons (UAA, UAG, and UGA) occur by chance once in approximately every 20 codons, and inspection of known genes invariably reveals that the two unused reading frames are filled with chain-terminating codons (Figure 8-27). Computers are now routinely employed to search out long stretches of open reading frame that start with AUG codons preceded by ribosome binding site sequences. These long, open reading frames (ORFs) very likely are coding sequences for still undiscovered polypeptides.

At this time, we cannot predict when the complete *E. coli* sequence of some 4×10^6 base pairs will be determined. Since this task is

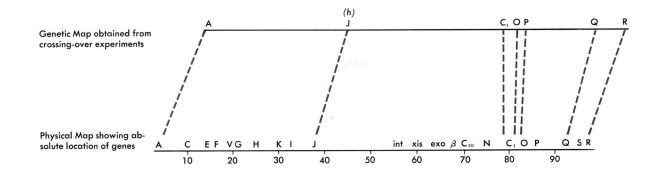

(h)

Genetic Map obtained from
crossing-over experiments

Physical Map showing ab-
solute location of genes

A   C   E F VG H   K I   J   int xis exo β C₁₁₁   N   C₁ O P   Q S R
    10   20   30   40   50   60   70   80   90

Figure 8-26
Comparison of the λ (recombination) map with the true physical map. [Recombination map after P. Amati and M. Meselson, *Genetics* 51 (1965):369–379; physical map after a map prepared by W. Szybalski, University of Wisconsin.]

Figure 8-27
Unused reading frames of the *trpC* gene in the *trp* operon of *E. coli* showing many chain terminating codons (boxed). (After Dr. Terry Platt.)

clearly too big for any average-size laboratory, the *E. coli* sequence is emerging in small bits of ever-increasing size. Still unforeseen improvements in cloning and sequencing procedures may have to occur before one or more individuals decide that the rewards of obtaining and collecting such massive sequence data are commensurate with the tedium that would have to be endured with such an undertaking. Hopefully, this moment will not be too far in the future. The way we view *E. coli* is undoubtedly affected by our knowledge of the number of polypeptides encoded by its DNA. If we were to find that it has the potential to make 5,000 different polypeptides, it is bound to appear an infinitely harder objective to characterize completely than if it could synthesize only some 2,000 different polypeptides. We already know the approximate functions carried out by some 1000 different *E. coli* genes.

TRP C

	3330		3340				3350				3360				3370				3380				3390				3400			3410

ATG ATG CAA ACC GTT TTA GCG AAA ATC GTC GCA GAC AAG GCG ATT TGG GTA GAA GCC CGC AAA CAG CAG CAA CCG CTG GCC AGT TTT CAG
MET GLN THR VAL LEU ALA LYS ILE VAL ALA ASP LYS ALA ILE TRP VAL GLU ALA ARG LYS GLN GLN GLN PRO LEU ALA SER PHE GLN
1   10   20

GAG GTT CAG CCG AGC ACG CGA CAT TTT TAT GAT GCG CTA CAG GGT GCG CGC ACG GCG TTT ATT CTG GAG TGC AAG AAA GCG TCG CCG TCA
GLU VAL GLN PRO SER THR ARG HIS PHE TYR ASP ALA LEU GLN GLY ALA ARG THR ALA PHE ILE LEU GLU CYS LYS LYS ALA SER PRO SER
30   40   50

AAA GGC GTG ATC CGT GAT GAT TTC GAT CCA GCA CGC ATT GCC GCC ATT TAT AAA CAT TAC GCT TCG GCA ATT TCG GTG CTG ACT GAT GAG
LYS GLY VAL ILE ARG ASP ASP PHE ASP PRO ALA ARG ILE ALA ALA ILE TYR LYS HIS TYR ALA SER ALA ILE SER VAL LEU THR ASP GLU
60   70   80

Summary

Crossing over occurs within the gene, thereby making possible the mapping of mutations within it. Like the order of genes on a chromosome, the arrangement of mutable sites in a gene is strictly linear. Within average-size bacterial genes, there are many hundreds of sites in which changes lead to detectable mutations. Sometimes, it is initially difficult to determine whether a chromosomal region contains more than one gene. This uncertainty can usually be resolved by introducing into a given cell more than one chromosome (or chromosomal fragment) containing mutations within that region. If the two mutants can complement each other to restore function, their mutations must be in different genes (the complementation test).

Almost all bacterial genes (the exception being those DNA segments that code for rRNA and tRNA) are responsible for the synthesis of a specific polypeptide chain and contain the information that specifies the sequence of its amino acids. Genes and their polypeptide products are colinear, with successive regions along a gene controlling the successive amino acids along its polypeptide chain product. The discovery of the double helix led to the postulation that mutable sites are the base pairs along DNA molecules. There are four alternative conformations (AT, TA, GC, and CG) for each mutable site, with genetic analysis of well-defined polypeptides showing that each amino acid is coded by adjacent groups of base pairs. Genetic experiments exploring the consequence of mutations that insert or delete individual base pairs found that groups of three base pairs code for each amino acid. Insertions or deletions of base pairs often have more drastic consequences than simple base-pair replacement; they frequently lead to the utilization of the wrong group of codons (reading frame) on a given mRNA template.

Such genetic inferences have now been completely confirmed by the sequence analysis of genes whose polypeptide products already have been sequenced. In all cases, the amino acid coding segments inferred from knowledge of the genetic code were completely confirmed. DNA sequencing provided important new insights by revealing the existence of untranslated base pairs at the beginnings and ends of all mRNA molecules. Gaps between bacterial genes are often very short. The vast majority of the bacterial chromosomes are now thought to be composed of base pairs that code for amino acids. Exactly how many genes are present in the *E. coli* chromosome awaits complete sequencing (4×10 base pairs). Until recently, this task seemed beyond the capability of molecular biology. Now there is reason to believe that this task will be completed by the mid-1990s.

Bibliography

General References

Glass, R. E. 1982. *Gene Function: E. coli and Its Heritable Elements.* Berkeley: University of California Press.

Hayes, W. 1968. *The Genetics of Bacteria and Their Viruses.* 2nd ed. New York: Wiley.

Inouye, M. 1983. *Experimental Manipulation of Gene Expression.* New York: Academic Press.

Lewin, B. M. 1974. *Gene Expression I.* New York: Wiley.

Lewin, B. M. 1985. *Genes.* 2nd ed. New York: Wiley.

Pontecorvo, G. 1958. *Trends in Genetic Analysis.* New York: Columbia University Press. A classical essay on the fine-structure analysis of the gene.

Stent, G. S., and R. Calendar. 1978. *Molecular Genetics: An Introductory Narrative.* 2nd ed. San Francisco: Freeman.

Cited References

1. Benzer, S. 1955. "Fine Structure of a Genetic Region in Bacteriophage." *Proc. Nat. Acad. Sci.* 41:344–354.
2. Singer, B. S., S. T. Shinedling, and L. Gold. 1983. "The rII Genes: A History and a Prospectus." In *Bacteriophage T4,* ed. C. K. Mathews, E. M. Kutter, G. Mosig, and P. B. Berget. Washington, D.C.: Amer. Soc. for Microbiol., pp. 327–333.
3. Ames, B. N., B. Garry, and L. A. Herzenberg. 1960. "The Genetic Control of the Enzymes of Histidine Biosynthesis in *Salmonella typhimurium*." *J. Gen. Microbiol.* 22:369–378.
4. Maas, W. K. 1961. "Studies on Repression of Arginine Biosynthesis in *E. coli*." *Cold Spring Harbor Symp. Quant. Biol.* 31:151–162.
5. Gorini, L., W. Gundersen, and M. Burger. 1961. "Genetics of Regulation of Enzyme Synthesis in the Arginine Biosynthetic Pathway of *E. coli*." *Cold Spring Harbor Symp. Quant. Biol.* 26:173–182.
6. Pribnow, D., D. C. Sigurdson, L. Gold, B. S. Singer, C. Napoli, J. Brosius, T. J. Dull, and H. F. Noller. 1981. "rII Cistrons of Bacteriophage T4. DNA Sequence Around the Intercistronic Divide and Positions of Genetic Landmarks." *J. Mol. Biol.* 149:337–376.
7. Yanofsky, C., B. C. Carlton, J. R. Guest, D. R. Helinski, and U. Henning. 1964. "On the Colinearity of Gene Structure and Protein Structure." *Proc. Nat. Acad. Sci.* 51:226–272.
8. Yanofsky, C., J. Ito, and V. Horn. 1966. "Amino Acid Replacements and the Genetic Code." *Cold Spring Harbor Symp. Quant. Biol.* 31:151–162.
9. Helinski, D. R., and C. Yanofsky. 1962. "The Correspondence Between Genetic Data and the Position of Amino Acid Alteration in A Protein." *Proc. Nat. Acad. Sci.* 48:173–183.
10. Helinski, D. R., and C. Yanofsky. 1963. "A Genetic and Biochemical Analysis of Second Site Reversion." *J. Biol. Chem.* 238:1043–1048.
11. Crick, F. H. C., L. Barnett, S. Brenner, and J. Watts-Tobin. 1961. "General Nature of the Genetic Code for Proteins." *Nature* 192:1227–1232.
12. Streisinger, G., Y. Okada, J. Emrich, J. Newton, A. Tsugita, E. Terzaghi, and M. Inouye. 1966. "Frameshift Mutations and the Genetic Code." *Cold Spring Harbor Symp. Quant. Biol.* 31:77–84.
13. Yanofsky, C., T. Platt, I. P. Crawford, B. P. Nichols, G. E. Christie, M. VanCleemput, and A. M. Wu. 1981. "The Complete Nucleotide Sequence of the Tryptophan Operon of *E. coli*." *Nucleic Acid Res.* 9:6647–6668.
14. Farabaugh, P. J. 1978. "Sequence of the *lac I* Gene." *Nature* 274:765–771.
15. Buchel, D. E., B. Gronneborne, and B. Muller-Hill. 1980. "Sequence of the Lactose Permease Gene." *Nature* 283:541–545.
16. Wallace, R. G., N. Lee, and A. V. Fowler. 1980. "The *araC* gene of *Escherichia coli*: Transcriptional and Translational Start Points and Complete Nucleotide Sequence." *Gene* 12:179–190.
17. Wurtzel, E. T., M.-Y. Chou, and M. Inouye. 1982. "Osmoregulation of Gene Expression. I. DNA Sequences of the *ompR* Gene of the *ompB* Operon of *E. coli* and Characterization of Its Gene Product." *J. Biol. Chem.* 257:13685–13691.
18. Bronson, M. J., C. Squires, and C. Yanofsky. 1973. "Nucleotide Sequences from Tryptophan Messenger RNA of *Escherichia coli*. The Sequences Corresponding to the Amino-Terminal Region of the First Polypeptide Specified by the Operon." *Proc. Nat. Acad. Sci.* 70:2335–2339.
19. Platt, T. 1978. "Regulation of Gene Expression in the Tryptophan Operon of *Escherichia coli*." In *The Operon,* ed. J. H. Miller and W. S. Reznikoff. Cold Spring Harbor, N.Y.: Cold Spring Harbor Laboratory, pp. 263–302.
20. Guarente, L. P., D. H. Mitchell, and J. Beckwith. 1977. "Transcription Termination at the End of the Tryptophan Operon of *E. coli*." *J. Mol. Biol.* 112:423–436.
21. Mitchell, D. H., W. S. Reznikoff, and J. Beckwith. 1976. "Genetic Transfusions That Help to Define a Transcription Termination Region in *E. coli*." *J. Mol. Biol.* 101:441–457.
22. Beckwith, J. R., and D. Zipser. 1970. *The Lactose Operon.* Cold Spring Harbor, N.Y.: Cold Spring Harbor Laboratory.
23. Miller, J. G., K. Ippen, J. G. Scaife, and J. R. Beckwith. 1968. "The Promoter and Operator Region of the *lac* Operon of *E. coli*." *J. Mol. Biol.* 38:413–420.
24. Platt, T., and C. Yanofsky. 1975. "An Intercistronic Region and Ribosome Binding Site in Bacterial Messenger RNA." *Proc. Nat. Acad. Sci.* 72:2399–2403.
25. Steitz, J. A. 1978. "Genetic Signals and Nucleotide Sequences in Messenger RNA." In *Biological Regulation and Control,* ed. R. Goldberger. New York: Plenum.
26. Westmoreland, B. C., W. Szybalski, and H. Ris. 1969. "Mapping of Deletions and Substitutions in Heteroduplex DNA Molecules of Bacteriophage Lambda by Electron Microscopy." *Science* 163:1343–1348.
27. Staden, R., and A. D. McLaughlin. 1982. "Codon Preference and Its Use in Identifying Protein Coding Regions in Long DNA Sequences." *Nucleic Acid Res.* 10:141–156.

IV

DNA IN DETAIL

CHAPTER 9

The Structures of DNA

The realization that DNA is the prime genetic molecule, carrying all the hereditary information within chromosomes, immediately focused attention on its structure. Only by understanding how its atoms are linked together by covalent bonds and then establishing their three-dimensional arrangements in space would we be able to learn how DNA carries the genetic messages that are replicated when chromosomes divide to produce two identical copies of themselves. Not surprisingly, there initially were fears that DNA might have very complicated and perhaps bizarre structures that differ radically from one gene to another. Great relief, if not general elation, was thus generated when the fundamental DNA structure was found to be the double helix. It told us that all genes have roughly the same three-dimensional form and that the differences between two genes reside in the order and number of their four nucleotide building blocks along their complementary strands.

Now, some 30 years after the discovery of the double helix, we realize that the DNA structure is not as simple as first thought. For example, the chromosomes of some small viruses have single-stranded, not double-helical, molecules. Moreover, some DNA sequences permit the double helix to twist in the left-handed sense, as opposed to the right-handed sense originally formulated for DNA's general structure. And while some DNA molecules are linear, others are circular. Still additional complexity comes from the supercoiling (further twisting) of the double helix, often around cores of DNA-binding proteins. The supercoiling can so distort the basic double helix that sections of it may untwist to form sections of single-stranded DNA.

These complexities were often first perceived as annoying deviations from perfection. By now, however, they have been so firmly documented that we realize there is no one structure of DNA and will henceforth refer to the "structures" of DNA. Clearly, it will only be through understanding why these "variant" DNAs exist that we will finally comprehend the myriad devices used by DNA to govern the life of the cell.

DNA Is Usually a Double Helix[1-5]

The most important feature of DNA is that it usually consists of two complementary polymeric chains twisted about each other in the form of a regular right-handed double helix (see Figure 3-7). Each

240

chain is a polynucleotide, a polymeric collection of nucleotides in which the sugar of each nucleotide is linked by a phosphate group to the sugar of the adjacent nucleotide. There are four main nucleotides, each containing a deoxyribose residue, a phosphate group, and a purine or pyrimidine base (see Figure 3-2). There are two single-ringed pyrimidines, thymine (T) and cytosine (C), and two double-ringed purines, adenine (A) and guanine (G). In the polynucleotide chain, the sugar and phosphate groups are always linked together by the same chemical bonds (3′–5′ phosphodiester linkages). Hence, this part of the molecule, called the sugar-phosphate backbone, is very regular. In contrast, the order of the purine and pyrimidine residues along a chain is highly irregular, varying from one molecule to another. Both the purine and pyrimidine bases are flat, relatively water-insoluble molecules that tend to stack above each other roughly perpendicular to the direction of the helical axis.

The Two Chains of the Double Helix Have Complementary Sequences

The two chains are joined together by hydrogen bonds between pairs of bases. Adenine is always paired with thymine and guanine with cytosine (see Figure 5-28). Only these arrangements are possible, for two purines would occupy too much space to allow a regular helix, and two pyrimidines would occupy too little space. The strictness of these pairing rules results in a complementary relation between the sequences of bases on the two intertwined chains. For example, if we have a sequence 5′-ATGTC-3′ on one chain, the opposite chain must have the sequence 3′-TACAG-5′. A stereochemical consequence of the way the AT and GC base pairs come together is that the phosphodiester linkages of the two polynucleotide chains are oriented in opposite directions. Thus, if the double helix is inverted by 180°, it superficially looks the same (Figure 9-1).

The two glycosidic bonds that attach a base pair to its sugar rings do not lie directly opposite each other. As a result, the two sugar-phosphate backbones of the double helix are not equally spaced along the helical axis, and the grooves that form between the backbones are not of equal size (Figure 9-2). The floor of the major groove is filled with nitrogen and oxygen atoms that belong to the "top" of the base pairs (as drawn in the figure) and that project inward from their respective sugar-phosphate backbones. In contrast, the floor of the minor groove is filled with base-pair nitrogens and oxygens that project outward toward their backbones. The hydrogen-bonding potential of the major groove shows much greater dependence on base sequence than that of the minor groove. This finding led to the speculation that proteins attach to sequence-specific DNA segments by forming hydrogen bonds predominantly to specific groups positioned along the major groove.

Each Base Has Its Preferred Tautomeric Form[6, 7]

An important chemical feature of DNA is the position of the hydrogen atoms in the purine and pyrimidine bases. Before 1953, many chemists thought that some of these hydrogen atoms randomly moved from one ring nitrogen or oxygen atom to another and so

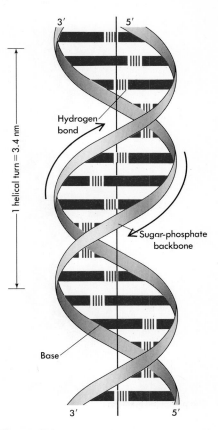

Figure 9-1
If inverted 180°, the double helix looks superficially the same, owing to the complementary nature of its component DNA strands. Turn the book upside down to demonstrate this.

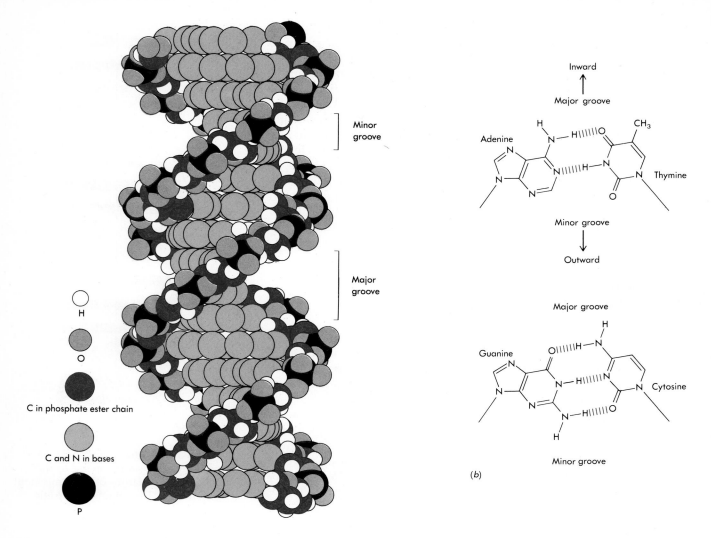

(a)

(b)

Figure 9-2
(a) The positions of the major and minor grooves of the double helix and (b) the relative directions in which the base-pair nitrogens and oxygens project in each groove. The nitrogens and oxygens of the floor of the major groove project toward the center of the DNA molecule, while those of the minor groove project toward the outer edge of the DNA molecule. (The structure in this figure is the form of DNA called B-DNA, discussed later.)

could not be assigned a fixed location. Now we realize that although such movements, called **tautomeric shifts,** do occur, these hydrogens have preferred atomic locations. The nitrogen atoms attached to the purine and pyrimidine rings are usually in the *amino* (NH_2) form and only rarely assume the *imino* (NH) configuration (Figure 9-3a). Likewise, the oxygen atoms attached to the C6 atoms of guanine and thymine normally have the *keto* (C—O) form and only rarely take up the *enol* (COH) configuration (Figure 9-3b).

It is essential to the biological functioning of DNA that the hydrogen atoms have relatively stable locations. If the hydrogen atoms had no fixed positions, adenine could often pair with cytosine and guanine with thymine, and the sequence of the bases on the two intertwined chains would not necessarily be complementary. If that were the case, then DNA could not function as a genetic molecule, for it is the complementary relationship between the opposing chain sequences that gives DNA its capacity for self-replication. As we shall discuss in great detail in the next chapter, replication of the double helix involves strand separation followed by formation of complementary DNA chains using the now free single strands as templates to attract the appropriate partner bases dictated by AT and GC base-pairing rules. The imino forms of adenine and cytosine and the enol forms of guanine and thymine must indeed occur only very rarely

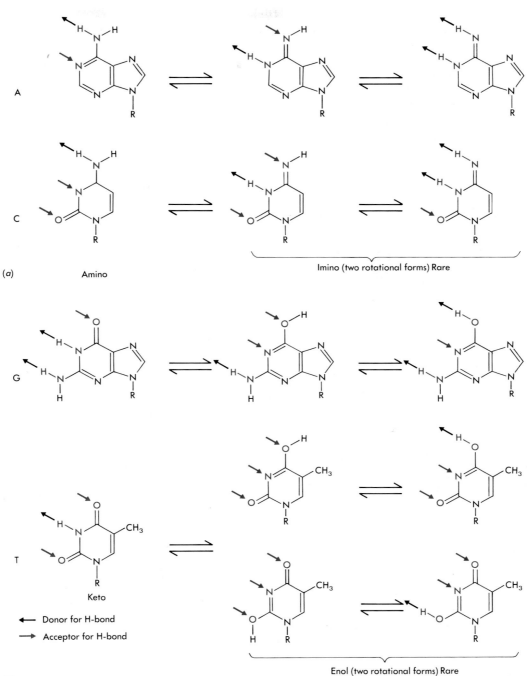

(a) Amino

Imino (two rotational forms) Rare

← Donor for H-bond

→ Acceptor for H-bond

Keto

Enol (two rotational forms) Rare

(b)

compared to their amino and keto alternatives. Otherwise, the many errors (mutations) incurred during DNA replication would be incompatible with orderly cell growth and division.

DNA Renatures as Well as Denatures[8–11]

DNA molecules with a high GC content are more resistant to thermal melting than AT-rich molecules. When double-helical DNA molecules are heated above physiological temperatures (to near 100°C), their hydrogen bonds break and the complementary strands often

Figure 9-3

Amino ⇌ imino and keto ⇌ enol tautomerism. (a) The A and C nucleotide bases are usually in the amino form but rarely form the imino configuration. (b) The bases G and T are usually in the keto form but are rarely found in the enol configuration. Atoms shown in color may participate in hydrogen-bond formation.

Figure 9-4
Dependence of DNA denaturation on G + C content. The greater the G + C content, the higher the temperature must be to denature the DNA strand. DNA from different sources was dissolved in two different solutions at pH 7.0. The points represent the temperature at which the DNA denatured, graphed against the G + C content. [Data from J. Marmur and P. Doty, *J. Mol. Biol.* 5 (1962):120.]

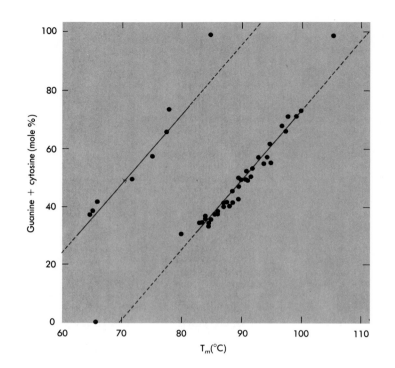

separate from each other, a process called DNA denaturation. Because each GC base pair is held together by three hydrogen bonds rather than the two hydrogen bonds holding together each AT base pair, higher temperatures are necessary to separate GC-rich strands than to break apart AT-rich molecules (Figure 9-4). At intermediate temperatures or in the presence of intermediate amounts of destabilizing agents such as alkali or formamide, DNA molecules are partially denatured. In such molecules, those regions of the double helix that are relatively rich in AT base pairs melt apart, while regions rich in GC base pairs retain their double-helical character (see Figure 9-23).

Even complete denaturation is not necessarily an irreversible phenomenon. When heated solutions of totally denatured DNA are slowly cooled, single strands often meet their complementary strands and reform regular double helices. This ability to renature DNA molecules permits artificial hybrid DNA molecules to be formed by slowly cooling mixtures of denatured DNA from two different species. For example, hybrid molecules can be formed containing one strand from a man and one from a mouse. Only a fraction (25 percent) of the DNA from a man can form hybrids with mouse DNA. This is not surprising, since it merely means that some genes of man are very similar to those of a mouse, whereas others have quite different nucleotide sequences. Before the development of rapid DNA-sequencing methods, this molecular technique was widely employed in establishing the genetic similarity of the various taxonomic groups.

Many Very Small Viruses Have Single-Stranded DNA Chromosomes[12]

At first it was thought that all DNA molecules are double-stranded except during replication, when a small region around the replicating fork (see Figure 10-10) is temporarily in a non-hydrogen-bonded, single-stranded form. It therefore came as quite a surprise when experiments revealed that the DNA of several groups of small bacterial viruses exists as single-stranded molecules in which the amount of A is not equal to the amount of T and the amount of G does not equal the amount of C. Among these single-stranded phages are the spherically shaped φX174 and S13 viruses and the rod-shaped f1 and M13 viruses. The chromosomes of the parvoviruses, minute viruses that infect many vertebrate organisms, are also single-stranded DNA molecules.

When these single-stranded viral chromosomes enter their host cells, they serve as templates for the formation of complementary strands. The resulting double helices in turn serve as templates for new single strands that then become incorporated into new virus particles (Figure 9-5). Thus, the fundamental mechanism for ordering nucleotides during the synthesis of single-stranded DNA is basically the same as that used for double-helical DNA. Nucleotide selection always occurs by attraction of the complementary base. Single-stranded DNA replication usually differs from double-helical replication in that it uses only one of the two complementary strands (the − strand) as a template for the progeny (+) strand.

Although single-stranded DNA can serve as the chromosome of a small virus, it would not be very effective in storing genetic information within the chromosomes of cells. In bacteria and in many eucaryotic cells (e.g., the haploid phase of yeast cells), there usually is only one copy of a given gene. If interphase chromosomes contained single-stranded DNA, with the double helix existing only briefly during mitosis, then the daughter cells would contain two completely different sets of genetic information. This follows from the fact that the two complementary chains do not have identical base sequences and hence would code for entirely different amino acid sequences. The double-strand ensures that each daughter cell will contain the same genetic information. The double-stranded form also permits effective DNA repair mechanisms to exist. For example, when one strand is damaged by exposure to X-rays, the remaining strand can provide the genetic information to rebuild the damaged section. It is probably no accident that only very small viral chromosomes are single-stranded. If they were to become, say, the size of T2, they would present too large a target for chain-damaging events.

Single-Stranded DNA Has a Compact Structure[13, 14]

Single-stranded DNA molecules have a strong tendency to fold back on themselves to form irregular double-helical hairpin loops whenever their sequences permit significant numbers of nucleotides to base-pair (Figure 9-6). Imperfect as these loops are, they nonetheless are energetically favored under physiological salt conditions, since they allow much more effective stacking of the flat hydrophobic surfaces of the bases than is possible in any fully extended single-stranded structure. Often, as many as half the bases of single-stranded DNA are so hydrogen-bonded, and their molecules are highly compacted. If, however, there are only minimal numbers of neutralizing cations (e.g., less than or equal to $0.01\ M$ Na$^+$), then the electrostatic repulsion between the partially unneutralized phosphates will keep the single-stranded chains highly extended, and there will be fewer hairpin loops.

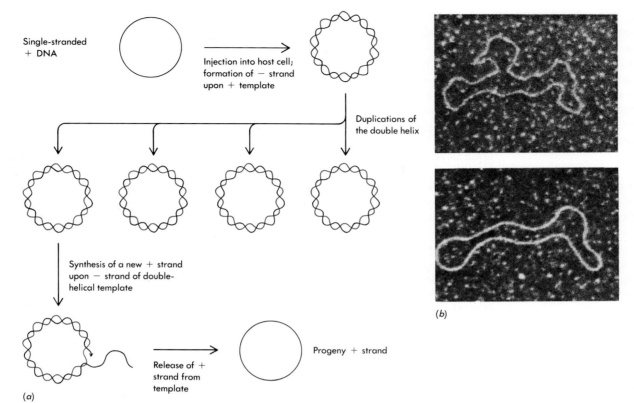

Figure 9-5

(a) The replication of a single-stranded DNA molecule. Each double helix generally serves as the template for the formation of a large number of new + strands. (b) Electron micrograph of a single-stranded (upper) and double stranded (lower) circular DNA molecule from phage φX174. (Courtesy of David Dressler and Kirston Koths, Harvard Medical School.)

Single-stranded DNA is always denser than double-helical DNA. Much more complete hydrogen bonding to water is possible for the atoms of the highly regular double helix than for the atoms of the inherently irregularly shaped single-stranded DNA. In the absence of base-pairing hydrogen bonds, van der Waal's interactions take over and bring the atoms of DNA closer together than would be the case if they formed regular hydrogen bonds with each other. An analogy can be made with the behavior of water, which becomes less dense as it forms the regular, hydrogen-bonded lattice of crystalline ice.

Rigorous Crystallographic Proof of the Double Helix[15, 16, 17]

X-ray diffraction patterns from parallel-oriented fibers of DNA provided the first clues that the polynucleotide chains of DNA have helical conformations. They also told us that DNA was a multichained molecule composed of two or three polynucleotide chains. However, the double-helical nature of DNA did not directly emerge from X-ray diffraction analysis alone. Equally important was the use of model building in which the question was asked, Given the chemical features of a single DNA chain, into what three-dimensional helical conformations could it fold? The simplest models were those in which the sugar-phosphate backbones were on the outside with the bases stacking in the center. The double helix emerged when it was realized that thymine and guanine had keto, not enol, configurations and that by base-pairing adenine to thymine and guanine to cytosine, a stereochemically pleasing, regular helical molecule resulted (i.e., the AT base pairs had the same "shape" as the GC base pairs). Immediately, a structural explanation was provided for the equivalence of the num-

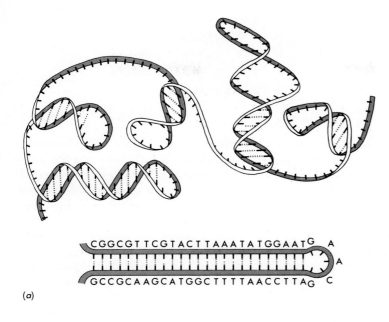

CGGCGTTCGTACTTAAATATGGAAT G A
GCCGCAAGCATGGCTTTTAACCTTA G C

(a)

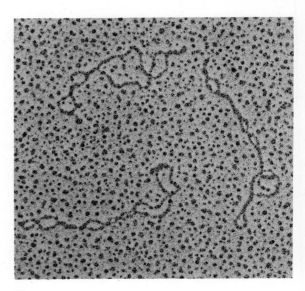

(b)

bers of adenine and thymine residues and of guanine and cytosine residues in most DNA molecules. Since the complementary relationship between the sequences of bases on the two opposing chains also suggested an obvious replication mechanism for DNA, there seemed every reason to assume that the double helix was the correct proposal.

Rigorous proof, however, could only come using X-ray diffraction procedures, and yet for many years it was only possible to say that the experimental data were compatible with the base-paired right-handed double helix derived by model building. The diffraction patterns obtainable from imprecisely oriented DNA fibers formed from DNA molecules of highly heterogeneous sequences could not provide sufficient information to locate individual atoms and thus establish definitive molecular structures like those obtained for protein crystals (e.g., hemoglobin). So while no molecular biologist seriously doubted that DNA was a base-paired double helix, for a long time there was no way to confirm that it was right-handed or that the base pairs were in fact held together by the originally proposed hydrogen bonds.

Solid proof came more than 25 years later, when organic chemists finally devised methods for the synthesis of short DNA chains called oligonucleotides. By mixing synthetic oligonucleotides with other oligonucleotides of complementary sequence, identical, well-defined double helices could be generated and crystallized. X-ray analysis of such complementary oligonucleotides containing all four bases

for example, $\frac{GGTATACC}{CCATATGG}$

revealed right-handed double helices held together by the long-postulated base pairs (see the A-DNA in Figure 9-9).

In the right-handed helices, the two members of each base pair frequently do not lie exactly in the same plane but display a "propeller twist" arrangement that stabilizes their respective helices by in-

Figure 9-6
(a) Schematic folding of a single-stranded DNA molecule to form irregular double-helical hairpin loops in regions of sequence homology. (b) Electron micrograph of single-stranded bacteriophage fd DNA showing regions of double-helical hairpin loops. [Courtesy of John E. Hearst, see C.-K. J. Shen, A. Ikoku, and J. E. Hearst. *J. Mol. Biol.* 127 (1979):167.]

Figure 9-7
The propeller twist between the purine and pyrimidine base pairs of a right-handed helix. The C1' atoms of deoxyribose are shifted up and down by the twist and may serve to stabilize the helix by increasing the stacking overlaps between bases. The angle of the twist varies and may be as large as 45° in some cases.

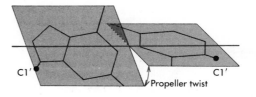

creasing the stacking overlaps between bases (Figure 9-7). However, simple propeller twists generate too-close contacts between the individual purines, since, as the larger members, they extend out beyond the centers of their respective base pairs. Such unacceptable purine overlaps are avoided either by an alteration in the rotation angle between the closely opposing base pairs or by a sliding of one of the base pairs over its neighbor so that purines are pushed out of the main stack of bases. How these factors compensate for each other depends on the exact sequence of bases. As a result, individual DNA molecules are never perfectly regular double helices. Instead, their exact conformations depend on which base pair (AT, TA, GC, or CG) is present at each position along the double helix (Figure 9-8).

Alternative Forms of Right-Handed DNA[18–22]

There are two well-characterized forms of right-handed DNA, referred to as A and B. They differ in the distance required to make a complete helical turn (the **pitch**), in the way their deoxyribose groups are bent, or puckered, and in the angle of the tilt that the base pairs make with the helical axis (Table 9-1). The B conformation is the longer, thinner form with its 3.4 Å thick purine-pyrimidine base pairs oriented approximately perpendicular to the helical axis. Separating its sugar-phosphate backbones are two very noticeable grooves (the major groove and the minor groove) on its external surface. B-DNA converts to A-DNA by tilting the base pairs some 30° so that successive base pairs occur every 2.7 Å along the helical axis. A-DNA molecules are thus shorter and squatter than their B equivalents, with their base pairs pushed away from the helical axis and their sugar-phosphate backbones separated by a much deeper and thinner minor groove than found in B-DNA (Figure 9-9).

In solution, DNA usually assumes the B form. Ordinarily the A conformation is only found when relatively little water is available to interact with the double helix. Relatively unhydrated DNA fibers tend to assume the A form, as do most crystals formed from complementary oligonucleotides, because the chemical reagents used in crystallization effectively dehydrate the DNA by binding water molecules. It is still unclear whether any significant sections of pure A-form DNA exists within cells. However, the DNA-RNA hybrid double helices, which exist only fleetingly when RNA is synthesized on single-stranded DNA chains, probably have A-like configurations; RNA chains cannot fold up into B-form double helices because they possess an extra hydroxyl group on each of their sugar residues. Instead, all double-helical RNA regions so far examined (such as the double-stranded RNA molecules from certain viruses or the double-helical sections of tRNA molecules) take up an A-like form in which the base pairs are strongly tilted away from the perpendicular position.

Polypurine-Polypyrimidine Double Helices Have Mixed A and B Properties[23]

Sections of double helices in which one strand is exclusively polypurine (e.g., poly dA) and its complementary strand is exclusively polypyrimidine (e.g., poly dT) show mixed A and B features. While the purine and pyrimidine bases have essentially the same tilt, their sugar-phosphate backbones have very different conformations. The sugar-phosphate backbone of the polypurine strand is A-like, while that of the polypyrimidine strand is B-like, marked by different puckering of the deoxyribose groups. Such "heteronomous" (obeying different rules) DNA sections appear to be unable to bind normally with histones to form nucleosomes (see Chapter 18). If so, the appearance of polypurine or polypyrimidine sequences along DNA molecules may be signals that mark nucleosome-free regions of DNA.

Alternating Anti and Syn Conformations Allow Transition into Left-Handed Helices[24-30]

DNA containing alternating purine and pyrimidine residues can fold up into left-handed as well as right-handed helices. In the left-handed helix, the fundamental repeating unit usually is a purine-pyrimidine dinucleotide, with the glycosyl bond connecting the base to the deoxyribose group oriented in the *anti* form at pyrimidine residues and in the *syn* configuration at purine residues. In right-handed DNA, the glycosyl bond is always oriented anti; so it is the different configuration at the purine nucleotides that usually generates left-handed helices (Figure 9-10). The alternating anti-syn configurations give the backbone of left-handed DNA a zigzag look (hence its designation of Z-DNA), which distinguishes it from the right-handed A and B forms (see Figure 9-9).

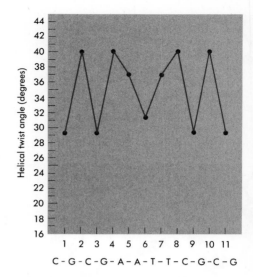

Figure 9-8
A given DNA molecule is never perfectly regular; the helical twist angle depends on the specific nucleotide bases present, reflecting the different stacking of each nucleotide base. [Data from Richard E. Dickerson, *Cold Spring Harbor Symp. Quant. Biol.* 47 (1982), p. 15.]

Table 9-1   A Comparison of the Structural Properties of A-, B-, and Z-DNAs as Derived from Single-Crystal X-Ray Analysis

	Helix Type		
	A	B	Z
Overall proportions	Short and broad	Longer and thinner	Elongated and slim
Rise per base pair	2.3 Å	3.32 Å (0.19 Å)*	3.8 Å
Helix-packing diameter	25.5 Å	23.7 Å	18.4 Å
Helix rotation sense	Right-handed	Right-handed	Left-handed
Base pairs per helix repeat	1	1	2
Base pairs per turn of helix	~11	~10	12
Mean rotation per base pair	33.6°	35.9° (4.2°)*	−60°/2
Pitch per turn of helix	24.6 Å	33.2 Å	45.6 Å
Tilt of base normals to helix axis	+19°	−1.2° (4.1°)*	−9°
Base-pair mean propeller twist	+18°	+16° (7°)*	~0°
Helix axis location	Major groove	Through base pairs	Minor groove
Major-groove proportions	Extremely narrow but very deep	Wide and of intermediate depth	Flattened out on helix surface
Minor-groove proportions	Very broad but shallow	Narrow and of intermediate depth	Extremely narrow but very deep
Glycosyl-bond conformation	anti	anti	anti at C, syn at G

SOURCE: R. E. Dickerson et al., *Cold Spring Harbor Symp. Quant. Biol.* 47 (1982):14. Reproduced by permission.

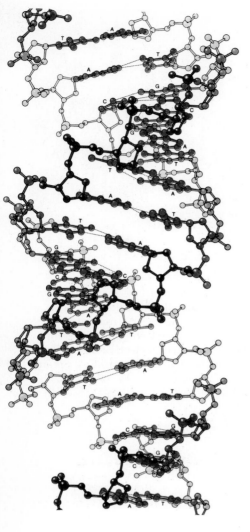
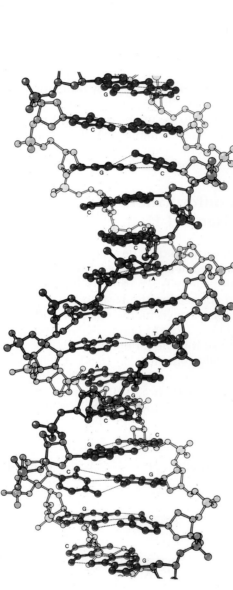
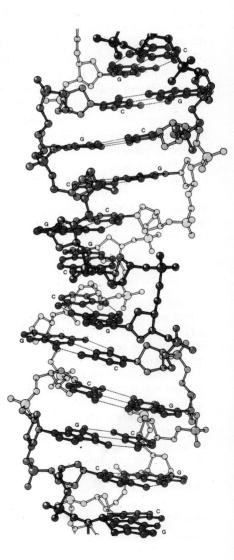

Figure 9-9
Comparison of A, B, and Z DNA. The bases are colored and the backbones are in grey. Note the zigzag course of the sugar-phosphate backbone chain in the Z-DNA. [Drawings by Irving Geis, reproduced from R. E. Dickerson's ''The DNA Helix and How It Is Read,'' *Sci. Am.* 249 (1983):100–104, by permission.]

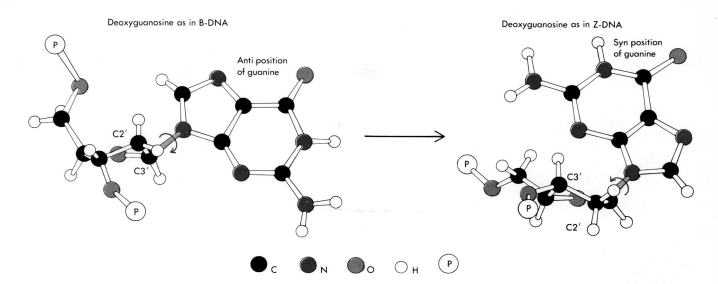

Deoxyguanosine as in B-DNA

Anti position of guanine

Deoxyguanosine as in Z-DNA

Syn position of guanine

C2′ C3′ P P

C3′ P P C2′

C N O H P

Overall, Z-DNA is more elongated and slimmer than B-DNA, with the base pairs of Z-DNA tilted in such a way that their successive distances along the helical axis are increased to 3.8 Å (see Figure 9-10). Twelve base pairs are found within each helical turn of 45.6 Å. Only one deep groove, corresponding to the minor groove of B-DNA, characterizes the external surface of Z-DNA. The relative narrowness of the groove places the base pairs as much on the outside of the helix as the sugar-phosphate backbone. The sugar-phosphate backbone of Z-DNA is not as fully extended as that of A- or B-DNA. In solution, the alternating purine-pyrimidine residues assume the left-handed configuration only in the presence of high concentrations of positively charged ions (e.g., Na⁺) that neutralize the negatively charged phosphate groups. At lower salt concentrations, they form typical B configurations.

Because GC base pairs are much more stable than AT base pairs, it has proved much easier to obtain crystals of short, alternating dG-dC double helices than of corresponding short dA-dT repeats. Left-handed DNA was thus first crystallized using alternating dG-dC oligonucleotides. More complex sequences such as

CGCATGCG
GCGTACGC

were subsequently shown to form perfect crystals of Z-DNA, and now it is believed that any alternating purine-pyrimidine sequence of at least six base pairs under suitable ionic conditions can assume the Z form.

At first it was thought that only sections of DNA containing regularly repeating purine-pyrimidine sequences could form Z-DNA. Then, in 1985, the sequence

ᵐ⁵C GATᵐ⁵C G
Gᵐ⁵CTA Gᵐ⁵C (m = methylated)

was found by X-ray methods to crystallize as Z-DNA. This means that alternating purine-pyrimidine sequences are not necessary to make Z-DNA if the cytosine residues are methylated at the 5′ position (see following section for discussion of cytosine methylation). Thus, it is

Figure 9-10
In right-handed B-DNA, the glycosyl bond (color) connecting the base to the deoxyribose group is always in the anti position, while in left-handed Z-DNA it rotates in the direction of the arrow, forming the syn conformation at the purine (here guanine) residues but remains in the regular anti position (no rotation) in the pyrimidine residues. It is this change to the syn position in the purine residues that generates the zig-zag (anti to syn) conformation and thus allows the transition from B-DNA to Z-DNA. The rotation causing the change from anti to syn also causes the ribose group to undergo a change in its pucker. Note that C3′ and C2′ switch locations. [After A. H.-J. Wang, et al. *Cold Spring Harbor Symp. Quant. Biol.* (1982), p. 41.]

Z-DNA

B-DNA

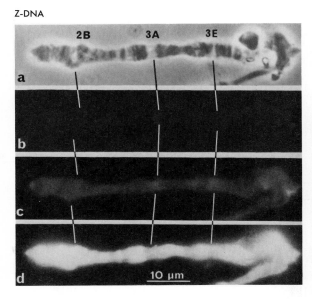

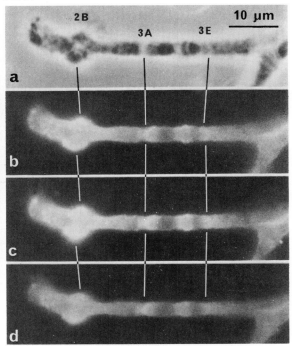

Figure 9-11
Demonstration of antibody binding in Z-DNA and B-DNA in the X polytene chromosome of *Drosophila melanogaster*. In the Z-DNA (left), (b) addition of anti-Z antibodies results in no detectable binding, but when first treated with 45 percent acetic acid for (c) 5 seconds and (d) 30 seconds, some anti-Z binding may be seen to occur. The acid removes histones, exposing naked supercoiled DNA. In contrast, B-DNA (right) antibodies (a) bind readily even without acid treatment, and binding may be even slightly decreased in long (30-second) exposure to acid treatment. [Courtesy of R. J. Hill and B. D. Stollar, *Nature* 305 (1983):338–340.]

not yet possible to pinpoint potential Z-DNA segments simply by scanning nucleotide sequences.

The amount of Z-DNA within cells remains unknown, with the best guess now that it accounts for, at most, only tiny segments of a given cell's DNA. That some cellular DNA may in fact adopt this conformation is suggested by experiments that purport to identify proteins that bind specifically to Z-DNA and not to B-DNA. However, only one such protein has yet been even semirigorously characterized. Conceivably, by so binding, such proteins help stabilize in the Z form DNA that would otherwise assume the B form.

Perhaps because of its rarity, Z-DNA is much more immunogenic than B-DNA, against which antibodies are only infrequently produced. Addition of anti-Z antibodies to native *Drosophila* chromosomes results in no detectable binding (Figure 9-11). However, after weak-acid treatment, which removes the histone components from the chromosomes and exposes naked, supercoiled DNA, some anti-Z antibody binding occurs. (No binding occurs to DNA whose supercoiling has been removed, suggesting that DNA must be under torsional strain to go left-handed.) In contrast, anti-B antibodies readily attach to many regions of native chromosomes, confirming early evidence, obtained by X-ray diffraction techniques, that the vast majority of chromosomal DNA is in the B form.

Methylation of Specific Cytosine and Adenine Residues After Their Incorporation into DNA

A small fraction of the cytosine residues in the DNA of most chromosomes contain methyl groups attached to the C5 atoms (5-methyl cytosine, or m^5C). These methyl groups are added by specific enzymes, the **DNA methylases,** after cytosine residues are incorporated into DNA chains (Figure 9-12).

Methyl groups also can be added to adenine groups to form

6-methyladenine. Such groups are added after DNA synthesis by a DNA methylase distinct from that which modifies cytosine residues. In procaryotic cells, methylated adenine is more common than methylated cytosine, while in eucaryotes, almost all methyl groups are added only to cytosine residues.

In eucaryotic cells, DNA methylases modify predominantly those cytosines linked on their 3' sides to guanine ($^{5'}CG^{3'}$), a dinucleotide combination that is present much less often in eucaryotic genomes than would be expected by chance alone. Not all $^{5'}CG^{3'}$ groups become so methylated, and vast differences in the extent of methylation of given genes have been detected among the different cell types (tissues) of multicellular organisms. Now it is suspected that undermethylation of certain CG sequences in key control regions often promotes gene expression. There are, however, important exceptions to this rule: Heavily methylated genes may be expressed, unexpressed genes may be unmethylated, and the methylation level may gradually decrease over a period of hours after transcription of a gene has begun. Moreover, methylated DNA has not been found in yeast or in *Drosophila*. Thus, demethylation may be a consequence of gene activation rather than its cause or prerequisite (see Chapter 21).

DNA Methylation Favors the B to Z Transition[31-34]

Methylation of key cytosine residues may influence gene functioning through its potentiation of the conversion of B-DNA to Z-DNA. At the cation levels found in most cells, alternating CG segments are most likely to exist as B-DNA. However, after the cytosines have been methylated, the equilibrium switches to favor the Z form. The reason for this switch is that in B-DNA, the methyl groups protrude into the aqueous environment of the major groove, a destabilizing feature; but in the Z form, the same methyl groups form a stabilizing hydrophobic patch that otherwise must be filled with water (Figure 9-13).

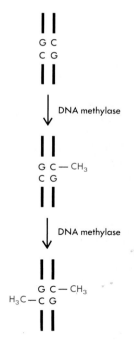

Figure 9-12
Action of the eucaryotic DNA methylase which modifies cytosine.

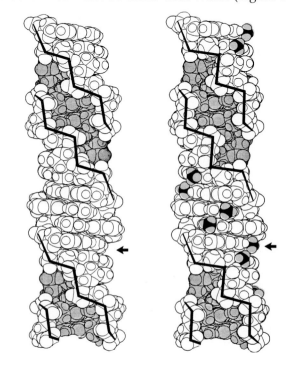

Figure 9-13
Crystal structures of (dC–dG)$_3$ and (m^{5}dC–dG)$_3$ showing the structure of Z-DNA in both its unmethylated (left) and methylated (right) forms, respectively. The black line traces the sugar-phosphate backbone, and the carbons of the methyl groups are shown in black. The arrow marks a depression in the unmethylated form, which is filled with the hydrophobic methyl group in the methylated form, which thus stabilizes the structure. [Courtesy of Dr. A. H.-J. Wang, reproduced from *Cold Spring Harbor Symp. Quant. Biol.* 47 (1982):41.]

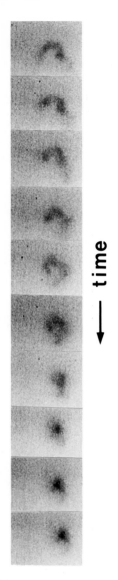

time

Figure 9-14
Spontaneous shape fluctuation and deformation of the double helix in solution. Shown here is a single folded T4 DNA molecule. Micrographs were taken at 0.1-second intervals. Bar = 10 μm. [Courtesy of Dr. M. Yanagida, reproduced from *Cold Spring Harbor Symp. Quant. Biol.* 47 (1982):179, Section 4.]

Spontaneous Deformations of the Double Helix in Solution[35–38]

In solution, B-DNA is quite a plastic structure. It is constantly subjected to localized thermally induced fluctuations in the arrangements of its atoms, which cause individual molecules to bend, twist, and stretch. Such distortions are not generally the consequence of momentary losses of hydrogen bonds between base pairs. Instead, they often represent changes in the rotational angles of the covalent bonds along the polynucleotide backbones that arise from sequence-dependent variations in the twist per base pair, reflecting the different ways the four base pairs optimally stack on each other. These can lead to gentle bends of the helical axis that partially close up the major groove and generate minor kinking of the stacked base pairs, which pushes them into the minor groove. As a result, the direction in which a given double helix points can change greatly over distances as short as several hundred angstroms (100 base pairs or 10 turns of the double helix). Thus, when individual DNA molecules are visualized in solution through the binding of a suitable fluorescent dye, they do not have long, linear shapes. Instead, they assume roughly spherical shapes, as expected for randomly coiled semirigid rods (Figure 9-14). Most dramatic are the occasional, drastic changes in the overall shape of DNA, which suggest abrupt kinking. In any case the fact that bends and kinks arise spontaneously indicates that relatively little energy is needed to deform the double helix. If the double helix can more effectively bind DNA binding proteins by bending, we can assume it will do so.

Sequence-Specific Bending and Kinking of DNA[39, 40]

The ability of all DNA to bend gently or to kink is accentuated by the presence of specific nucleotide sequences. Examination of a DNA molecule that moves anomalously fast through gels (and which thus has a more collapsed or kinked form than most DNA) reveals a set of four CAAAAAT (or CAAAAAAT) segments. These were found to be separated by single turns (ten nucleotides) of the double helix (Figure 9-15). Conceivably, the CAAAAAT sequence has adopted an unconventional (perhaps "heteronomous") conformation whose borders with conventional B-DNA are regions of predictable kinking.

Bending can also be the result of binding to specific proteins. The binding of the regulatory protein CAP to its normal binding site near the promoter of the *lac* operon leads to a more collapsed DNA conformation in that region. Even more direct evidence for kinking now comes from X-ray crystallographic analysis of DNA complexes with histones in nucleosomes and with the restriction enzyme *Eco* RI (see Figure 9-37).

Unwinding of the Double Helix by the Insertion of Flat, Ringed Molecules[41–44]

A number of flat, ring-shaped molecules bind to DNA by inserting themselves (**intercalating**) between the base pairs of the double helix. In this manner, they often act as inhibitors of DNA or RNA synthesis and thus act like drugs in living systems, killing cells or inhibiting

...GAATTCC CAAAAAT GT CAAAAAAT AGG CAAAAAAT GC CAAAAAT CCCAAAC...

their growth. They were originally discovered through their antimicrobial and anticancer attributes. Each time these intercalating agents become inserted into a DNA molecule, they necessarily *increase* the spacing of successive base pairs along the helical axes to roughly 7 Å, almost the distance between phosphate atoms in a fully extended polynucleotide chain (Figure 9-16). In this situation, very little rotation is possible around the helical axis, and so intercalating agents not only extend double helices, but extensively unwind them. For example, when a molecule of either ethidium bromide (an inhibitor of DNA synthesis) or actinomycin D (a powerful inhibitor of RNA synthesis) intercalates, the rotation angle between the two adjacent base pairs is reduced from 36° to 10°. The fact that intercalation occurs so readily indicates that it must be energetically favored, with the van der Waals bonds holding the inserted molecules to the base pairs being stronger than those found between conventionally stacked base pairs. Intercalation is additional evidence for the *metastability* of the double-helical structure—its ability to temporarily assume many inherently unstable configurations that normally quickly revert back to the standard B conformation. One such metastable variant must be short stretches of extended chains with gaps separating adjacent base pairs, into which an intercalating agent may bind.

The Chromosomes of Viruses, *E. coli,* and Yeast Are Single DNA Molecules[45–48]

The first estimates of the average molecular weights of DNA centered at about a million, the size needed to encompass an average-size bacterial gene coding for some 300 to 400 amino acids. Therefore, it

Figure 9-15
Segment of a kinetoplast DNA molecule found to move anomalously fast through gels. Note that the four $CA_{5-6}T$ sequences start about every ten bases—approximately a single turn of the double helix. These sequences allow the DNA to kink and thus become more compact. [Data from H.-M. Wu and D. M. Crothers, *Nature* 308 (1984):509.]

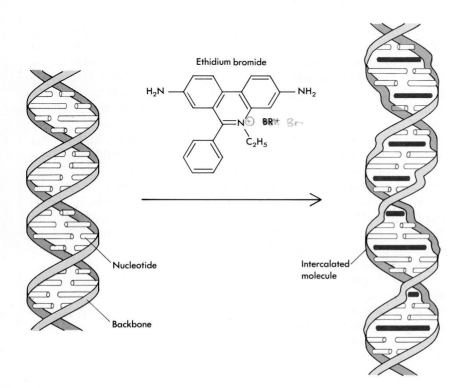

Ethidium bromide

Nucleotide

Backbone

Intercalated molecule

Figure 9-16
The intercalation of ethidium bromide into a DNA molecule. Note that the ethidium bromide increases the spacing of successive base pairs, distorts the regular sugar-phosphate backbone, and decreases the pitch of the helix.

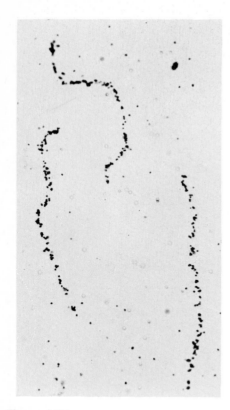

Figure 9-17
Autoradiograph of several T2 chromosomes. The total length of each is 52 μm. (Courtesy of J. Cairns.)

seemed natural to equate single DNA molecules with single genes. However, these early reports were inaccurate owing to DNA breakage during its isolation and study. Now it is clear that virtually all undegraded DNA molecules contain the information of at least several genes. The most certain molecular weight values come from DNA-containing viruses. Regardless of whether the DNA content is relatively small or large, each virus particle contains a single DNA molecule. For example, all the DNA of the small monkey (simian) virus SV40 is present within a single molecule of molecular weight ~3×10^6 daltons (5×10^3 base pairs, or 5 kbp), while the DNA molecule of the large bacterial virus T2 has a molecular weight of 1.2×10^8 daltons (2×10^2 kbp) (Figure 9-17). In these cases, the entire viral chromosome, rather than separate genes, corresponds to a single DNA molecule. Likewise, the chromosome of an *E. coli* cell is a single DNA molecule whose molecular weight is about 2.5×10^9 daltons (about 4×10^3 kbp) and whose extended length is roughly 1 mm.

DNA molecules of the yeast *Saccharomyces cerevisiae* range in size between the T2 and *E. coli* DNAs, with their average size being that expected if each of the yeast's 17 chromosomes contain one DNA molecule. Thus, the centromeres of yeast chromosomes (if not those of all eucaryotic chromosomes) do not represent discontinuities between separate DNA molecules, but instead are specialized regions of DNA evolved to interact with the spindle bodies upon which chromosomal segregation occurs during cell division. As of yet, there is no direct proof that the very much larger chromosomes of higher plants and animals (often 50 times larger than the *E. coli* chromosomes) also contain only one DNA molecule. However, we now expect that the one chromosome—one DNA molecule rule will hold for the chromosomes of all organisms. In any case, some DNA molecules are larger than any other biological molecules by several powers of ten.

Circular Versus Linear DNA Molecules[49, 50, 51]

Our evidence, which came largely from autoradiography and electron microscopy, initially suggested that all DNA molecules are linear and have two free ends. But when it became possible to take a better look at undegraded DNA, many DNA molecules were found to be circular (see Figure 7-29). For instance, the small monkey DNA virus SV40 has a 5000-base-pair circular double-helical chromosome; the similarly short chromosomes of single-stranded phages are likewise circular, as are almost all autonomously replicating plasmid DNAs. Most, if not all, bacterial chromosomes are also circular. We suspect that the DNA found in the rare circular chromosomes of higher cells are likewise circular molecules. Circular shapes were initially puzzling, since they seemed to present obstacles to the untwisting of double helices during DNA replication. Only when specific enzymes that first snip and then join DNA chains were discovered did the untwisting dilemma disappear. (These enzymes, the topoisomerases, are discussed later in this chapter.)

Equally important is the realization that linear DNA molecules do not have the uncomplicated structures first envisioned for them. As we see in the next chapter, replicating the ends of DNA molecules is not a straightforward process. Thus, all linear DNA molecules have evolved special tricks to replicate their ends, as well as to prevent their free ends from either being nibbled back by cellular enzymes or being enzymatically ligated together to form even larger chromosomes. For example, the ends of the linear adenovirus chromosomes have specialized proteins covalently attached to the 5' ends of strands; and poxvirus chromosomes have closed hairpin loops at their ends, with their basic structures being that of largely self-complementary circular single strands. Still other linear viral chromosomes (e.g., T2 and T7) have the same sequences at both ends, allowing the ends of the chromosome to recombine together during DNA replication to form very long, end-to-end aggregates (concatamers) (Figure 9-18).

Moreover, some DNA molecules that are linear when isolated from a virus particle (e.g., phage λ) are found as circles inside the host cell. This tells us that the linear and circular forms of such DNA molecules are interconvertible. A molecular mechanism that converts a DNA circle to the linear form (or vice versa) is shown in Figure 9-19. Starting with a closed circle, a specific enzyme introduces two breaks, one in the + strand and one in the − strand. As these breaks are very close to each other, the intervening hydrogen bonds occasionally are broken by thermal agitation, and then the circle unfolds. Most importantly, the resulting linear form contains single-stranded ends that have complementary nucleotide sequences (*"sticky ends"*). Thus, at a later time, the linear form can base-pair and resume a circular configuration. If the missing phosphodiester bonds are then reformed, the covalent circle is regenerated.

There is solid evidence that phage λ DNA interconverts between its circular and linear forms using precisely this mechanism. Later we shall see that the duplication of such "sticky" DNAs involves circular replicative intermediates. Thus, the linear form is most likely an adaptation for injection of the viral chromosome through the narrow phage tail.

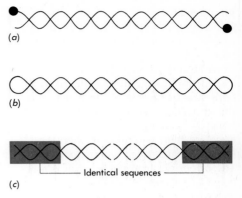

(a)

(b)

(c) —— Identical sequences ——

Figure 9-18
Structures of some viral DNAs. (a) Adenovirus DNA is linear and double-stranded and has specialized proteins covalently attached to its 5' ends. (b) Poxvirus DNA has closed hairpin loops at its ends. (c) The linear DNA of certain phages has several hundred base pairs of identical sequences at the two ends, allowing the formation of long, end-to-end concatamers. All three molecules are actually very much longer than shown in this figure.

Supercoiling of Circular DNA Molecules[52–56]

As DNA is a very flexible structure, its exact molecular parameters are a function of both the surrounding ionic environment and the nature of the DNA-binding proteins with which it is complexed. Because

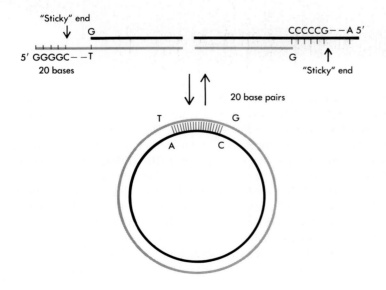

Figure 9-19
Interconversion of the linear and circular forms of λ DNA.

Figure 9-20
Generation of supercoiled DNA by cutting and ligating. In a circular DNA molecule, the number of times the DNA strands twist about each other (called the linkage number) is fixed. If both strands are cut (1) and rotated in the opposite direction from that of the twist of the helix (2), and if the cut ends are then rejoined (3), then the DNA molecule negatively supercoils (4). The upper electron micrograph is a relaxed (nonsupercoiled) DNA molecule of bacteriophage PM2. The lower electron micrograph shows the phage in its supertwisted form. [Electron micrographs courtesy of J. C. Wang, *Sci. Am.* 247 (1982):97.]

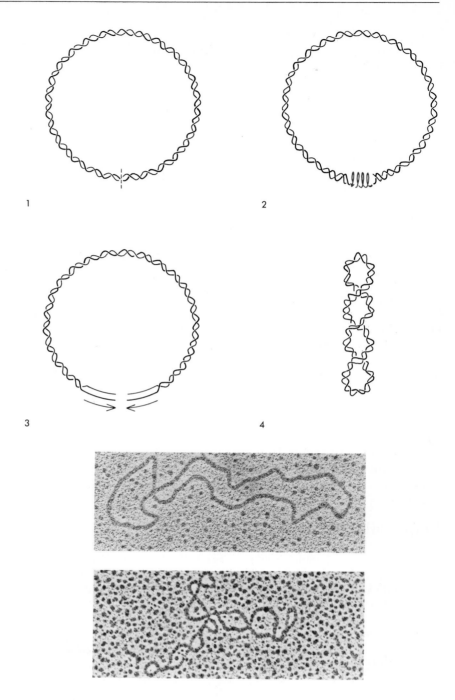

their ends are free, linear DNA molecules can freely rotate to accommodate changes in the number of times the two chains of the double helix twist about each other. But once the two ends become covalently linked to form circular DNA molecules, the absolute number of times the chains twist about each other (the **linkage number**) cannot change. For the most part, changes in the average number of base pairs per turn of the double helix will necessarily be accommodated by the formation of the appropriate number of supercoils in the opposite direction. **Supercoiling** refers to the further twisting of double-helical DNA molecules. Untwisting of the double helix usually leads to supercoiling in the negative (left-handed) direction (Figure 9-20), while overtwisting leads to positive (right-handed) supercoiling.

The supercoiled state is inherently less stable than uncoiled DNA, and the cutting, or **nicking,** of a single strand instantly converts a supercoiled molecule into its simple (**relaxed**) circular state (Figure 9-21). Because circular DNA molecules become increasingly compacted as they become more supercoiled, relaxed DNA is easily distinguished from supercoiled DNA by its slower sedimentation in a centrifugal field and by the longer time it takes to move through an agarose gel. At the present time, gel electrophoresis affords the most direct way to measure supercoiling, since it can separate molecules that differ by only single turns (Figure 9-22).

Localized Denaturation Within Supercoiled DNA[57, 58, 59]

The stress present within negatively supercoiled DNA molecules sometimes leads to localized denaturation, in which the complementary strands come apart. When this happens, the free energy lost by breaking the hydrogen bonds between base pairs is less than that gained by the unbending of the double helix. Those regions that denature tend to be rich in the relatively weak AT base pairs, while segments rich in the much stronger GC base pairs are usually the last to denature (Figure 9-23).

Localized denaturation should be especially favored for DNA segments containing self-complementary regions capable of forming hydrogen-bonded hairpin structures called **cruciform loops** (Figure 9-24). In such cases, relatively few net hydrogen bonds are lost and considerable energy is gained from the loss of one or more supercoils dominating the final equilibrium. There is considerable debate, however, about whether this phenomenon takes place in living cells. For example, in recent experiments, when a supercoiled plasmid DNA containing a perfect 68-base-pair inverted palindromic sequence was isolated very gently from bacteria, none of it was found as stable cruciforms. Rigorous proof for the absence of this cruciform could be obtained because a restriction enzyme cutting site is located at the exact center of the inverted palindrome: If a cruciform were present, this *Eco* RI recognition site (GAATTC) would be located within the terminal single-stranded hairpin loop and thus be resistant to digestion with *Eco* RI endonuclease, which can only cut double-stranded DNA (Figure 9-25).

The absence of any DNA uncut by *Eco* RI proved that none of the supercoiled plasmid DNA had rearranged into stable cruciforms. Apparently, within the bacterial cell, neither thermal energy alone nor the combined action of various enzymes (such as topoisomerases and proteins that bind to single-stranded DNA) is able to denature a sufficient number of base pairs in the supercoiled plasmid molecule so that the cruciform structure can begin to form. Thus, while cruciforms may be the most *energetically* stable configuration for certain negatively supercoiled DNAs, such structures appear to be *kinetically* forbidden in at least some situations.

Negative supercoils can also be lost through the conversion of any alternating purine-pyrimidine residues within the DNA molecules from the B form to the Z form of the DNA double helix. When supercoils are present, conversion into the Z form occurs at much lower salt concentrations and is much less dependent on methylation of key cytosine residues than when supercoils are absent.

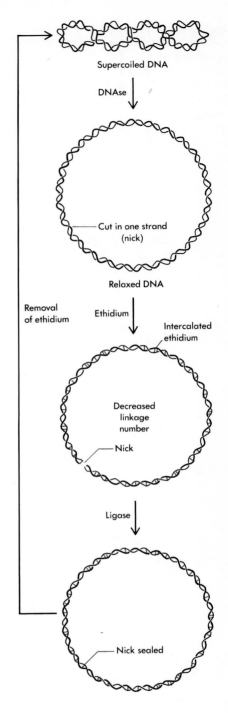

Figure 9-21
Use of nicking and intercalation to interconvert supercoiled and relaxed DNA. When one strand is nicked, a supercoiled circular DNA molecule relaxes. Ethidium intercalated into the DNA molecule increases the spacing of base pairs, unwind the double helix by 26°, and thus decrease the linkage number. DNA ligase can rejoin the free ends. When ethidium is removed, the DNA molecule supercoils negatively.

Figure 9-22
Gel electrophoresis separation of re-laxed and supercoiled DNA. The speed with which the DNA molecules migrate increases as the number of superhelical turns increases. (Courtesy of J. C. Wang.)

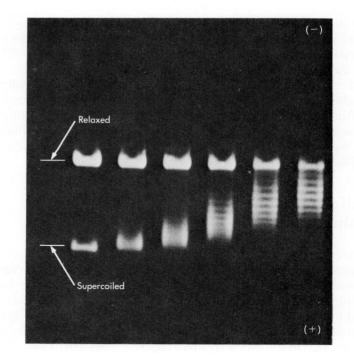

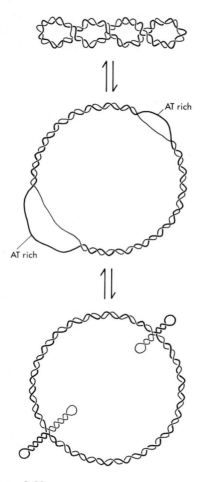

Figure 9-23
Upon partial denaturation, supercoiled DNA may convert to its unsupercoiled relaxed form. Areas of the DNA mole-cules relatively rich in the AT base pair tend to denature first, since it is held together by two hydrogen bonds rather than the three hydrogen bonds holding together the GC base pair. The single-stranded areas may then convert to more stable cruciform structures if there are inverted homologous sequences in the denatured regions.

Most Cellular DNA Exists as Protein-Containing Supercoils[60–63]

Just because isolated DNA molecules would energetically prefer to be in the relaxed state does not mean that the "natural state" of DNA is relaxed. In fact, virtually all DNA within both procaryotic and eucary-otic cells exists in the negative supercoiled state. As such it seldom, if ever, occurs as free DNA, but is complexed with specific DNA-bind-ing proteins to form compacted molecules called **chromatin.** How this compaction occurs is best understood for eucaryotic chromatin, whose most prominent DNA-binding components are the histones. *Histones* are relatively small, positively charged, arginine- and lysine-rich proteins that aggregate together to form discrete ellipsoid-shaped packets (histone cores) around which the DNA supercoils. The result-ing **nucleosomes** (see Chapter 20) give to chromatin its beaded ap-pearance (Figure 9-26).

Since the same collection of histones binds to nearly all sections of DNA, histones are thought to play essentially structural roles, as op-posed to enzymatic or regulatory roles. Only about half the mass of chromatin proteins is histone. The remaining proteins are not nearly as well characterized, with their exact functions yet to be determined.

Viral DNA chromosomes are also complexed with proteins, both when multiplying within cells and when packaged into virus parti-cles. The SV40 viral chromosome is at all times compacted into 24 histone-containing nucleosomes identical in structure to those exist-ing in the chromosomes of their host cells (Figure 9-27). When freed from its protein components, SV40 DNA is found as a negative super-coil. In contrast, no histones are complexed with the DNA found in adenovirus particles. These vertebrate viruses encode a unique DNA binding protein to help package their DNA within their protein cap-sules. When, however, an adenovirus DNA functions within a cell, its own DNA-binding protein is released and replaced with the same histone components that bind to all other cellular DNA.

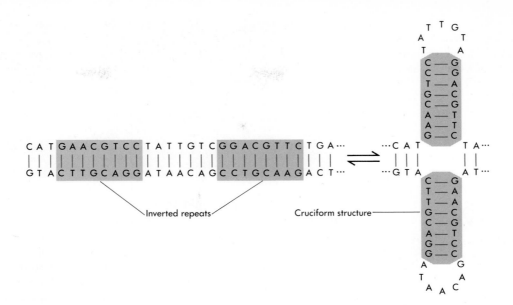

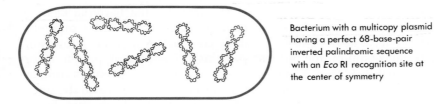

Inverted repeats   Cruciform structure

Figure 9-24
DNA segments containing self-complementary regions, such as the inverted repeat sequences shown here, can form hydrogen-bonded cruciform loops.

DNA Supercoils Twice Around Each Nucleosome[64, 65]

In forming the nucleosomes of eucaryotic cells, 200-base-pair-long segments of DNA negatively (left-handedly) supercoil twice around the histone cores (Figure 9-28). The energy lost by supercoiling may

Bacterium with a multicopy plasmid having a perfect 68-base-pair inverted palindromic sequence with an *Eco* RI recognition site at the center of symmetry

Gentle lysis to prevent thermal rearrangement of supercoiled plasmid DNA

GAATTC
CTTAAG

"Native" supercoiled plasmid with duplex *Eco* RI site that is susceptible to *Eco* RI cutting

Native structure rearranges into cruciform after partial denaturation by heating or treatment with phenol.

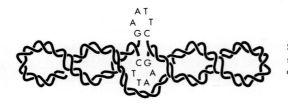

Single-stranded *Eco* RI recognition site is resistant to *Eco* RI digestion in cruciform structure

Figure 9-25
Formation of stable cruciform structures from inverted palindromic sequences. It is now questionable whether these do indeed form within cells.

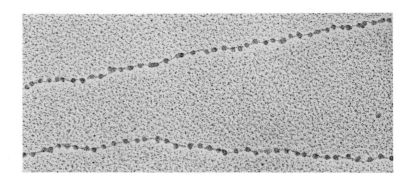

Figure 9-26
Nucleosomes give chromatin strands a beadlike appearance. Magnification 100,000. (Courtesy of Victoria Foe.)

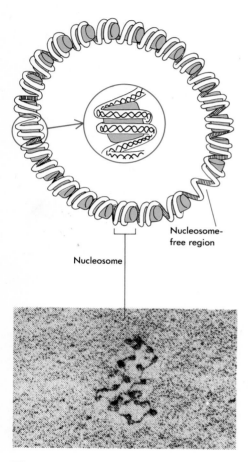

Figure 9-27
SV40 chromosome showing its 24 histone-containing nucleosomes. Often such minichromosomes contain a nucleosome-free region located near the origin of replication (see Chapter 24).

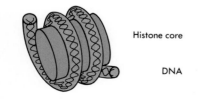

Figure 9-28
DNA coiling about a histone core.

be partially compensated by the energy gained from the ionic and hydrogen bonds formed between the histones and the DNA. The number of supercoils in solution need not be the same as found in chromatin. In chromatin, the double helix is slightly more twisted than in solution, with 10 base pairs per helical turn in chromatin versus 10.5 base pairs per helical turn in naked DNA in solution. As a result, there are fewer superhelices found in solution than found in chromatin.

Procaryotic Cells Contain Histonelike DNA-Binding Proteins[66, 67]

For many decades, it was believed that bacterial DNA, unlike eucaryotic DNA, occurs naked and does not have a compacted chromatinlike arrangement. But recently, using more gentle preparative procedures, condensed *E. coli* chromosomal sections have been seen that are organized into beadlike packets from which small, basic proteins analogous to the histones can be extracted. That bacterial DNA is complexed with proteins that have histonelike properties undoubtedly reflects its need to be regularly compacted in order to function. Crystallization of one of these proteins, DNA-binding protein II, reveals that its 9500-dalton chains associate in pairs as dimers containing extended arginine-containing arms able to interact with the phosphates of one turn of the DNA backbone. So it is likely that dimers sited adjacently along procaryotic DNA are oriented into a helical arrangement with the DNA bound on the outside (Figure 9-29). Interestingly, the average *degree of supercoiling* (i.e., the number of superhelical twists per ten base pairs) is about 0.05 for all naturally occurring DNA supercoils, regardless of their source. This may in part be because the underlying chromatin structure is so similar within both procaryotic and eucaryotic cells.

Topoisomerases Change the Linkage Numbers of Supercoiled DNAs[68–75]

That supercoiling not only occurs but is biologically very important is affirmed by the existence of the enzyme topoisomerase II, which specifically generates the negative supercoils that characterize so much of chromosomal DNA. **Topoisomerases** (Table 9-2) are a group of enzymes that convert (isomerize) one topological version of DNA into another. They do so by changing the linkage number, which is the number of times two DNA chains twist around each other. Procary-

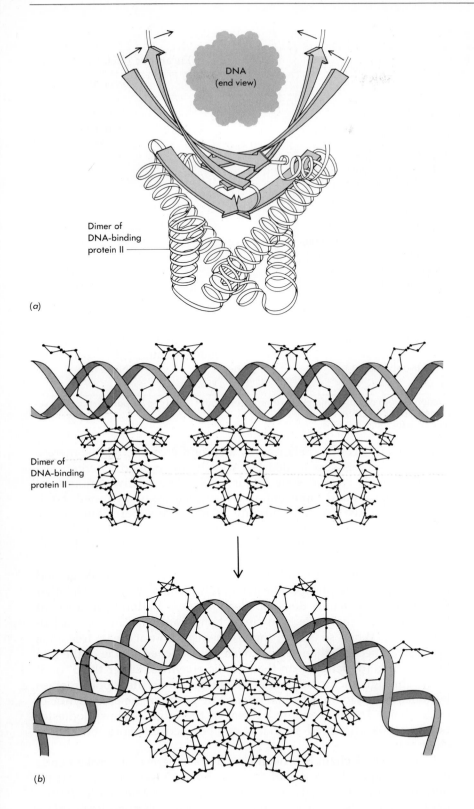

(a)

(b)

Figure 9-29
The histonelike DNA binding protein II of *E. coli.* (a) The structure of the dimer of DNA binding protein II bound to DNA viewed end-on. (b) How DNA binding protein II might induce supercoiling of DNA to form nucleosomes. The arms of the binding proteins are thought to interact with the DNA and each other, while electrostatic interactions at the other end of the proteins draw them together and in so doing bend the DNA. It is thought that 8 to 10 protein dimers and 80 to 100 nucleotide bases are needed to generate one turn of the supercoil. Here the DNA is not shown supercoiled, to prevent a confusing superimposition of the protein dimers.

otic topoisomerase II (sometimes called **gyrase**) uses the energy of ATP to generate negative supercoils by untwisting DNA in the left-handed direction. Its eucaryotic counterpart, however, may be capable of using ATP to generate negative supercoils only in the presence of other, yet to be described proteins. The supercoiling action of topo-

Table 9-2 DNA Topoisomerases

	Procaryotic (*E. coli*)		Eucaryotic (Hela)	
	Type I	Type II	Type I	Type II
Approximate MW	100,000	400,000	100,000	309,000
Subunits	Monomer	Tetramer (A_2B_2)	?	Dimer (subunit MW = 172,000)
Gene	*topA*	*gyrA* Polypeptide MW 105,000; activity inhibited by nalidixic acid and oxolinic acid *gyrB* Polypeptide MW 95,000; activity inhibited by coumermycin and novobiocin	?	?
Covalent intermediate with DNA	At 5′ Ⓟ	Subunit A Tyrosine covalent bond to 5′ Ⓟ	At 3′ Ⓟ	At 5′ Ⓟ
Relaxation*	Mg^{2+} required; ATP-independent	ATP-independent	No Mg^{2+} required; ATP-independent	ATP-dependent
Negative supercoiling	None	ATP-dependent	None	None†

*Eucaryotic type I topoisomerases usually catalyze relaxation of both negative and positive supercoils, but *E. coli* type I topoisomerase relaxes positive supercoils only when it is presented with a single-stranded DNA region.
†It may participate in supercoiling only in the presence of additional proteins.

isomerase II is counterbalanced by a second enzyme, topoisomerase I, which converts supercoiled DNA to the unstrained, energetically more favorable relaxed state (Figure 9-30). The relative amounts of topoisomerase I and topoisomerase II in cells are finely tuned, tending to create just the right amount of negative supercoiling. Thus, mutations that lower the number of topoisomerase I molecules are viable only if the number of topoisomerase II molecules also decreases.

Both topoisomerase I and II work by catalyzing the breakage and rejoining of DNA phosphodiester bonds. Unlike virtually all other enzymes, their action does not lead to new patterns of covalent bonds. Rather, their role is to create temporary gaps in polynucleotide chains. When so acting, topoisomerases do not create free ends but instead become themselves covalently attached to one of the two broken ends (depending on the specific enzymes, either the 3′ or 5′ end). Through such catalysis, a tyrosine group on the topoisomerase becomes linked to the terminal phosphate group of the cut polynucleotide chain. The free energy present in the original phosphodiester bond is thus preserved so that it can be used to rejoin the broken chain.

When a DNA chain is broken by a topoisomerase, its broken ends do not fall apart but are held together by the topoisomerase. At no time do the broken ends have the capacity to rotate freely. If that happened, topoisomerases would be limited to only completely relaxing double helices. Instead, individual topoisomerases function to make discrete changes of either one positive turn (topoisomerase I) or two negative turns (topoisomerase II) in the linkage number. Such discrete steps result from DNA chains passing through either transient single-stranded breaks (topoisomerase I) or transient double-stranded breaks (topoisomerase II) (Figures 9-31 and 9-32).

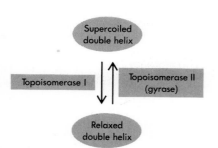

Figure 9-30
The counterbalancing actions of topoisomerase I and topoisomerase II (gyrase).

Linkage number = n   Linkage number = n + 1

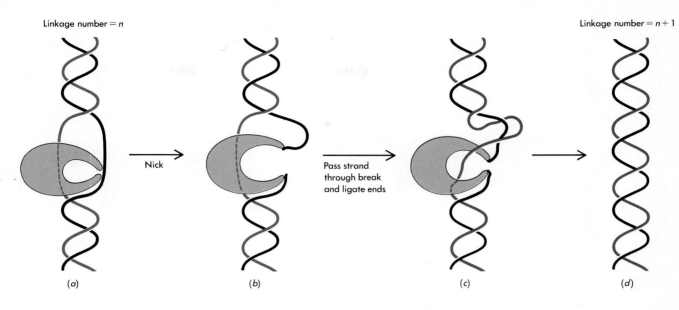

(a)   (b)   (c)   (d)

Topoisomerases are constructed in such a way that they can undergo reversible conformational changes that create cavities through which DNA chains can pass. The ATP requirement for bacterial topoisomerase II (gyrase)-induced supercoiling most likely reflects the use of ATP in mediating the necessary conformational changes that lead to chain passage (Figure 9-33). Topoisomerase I, however, relaxes supercoiled DNA without requiring a cyclical input of energy. Complete understanding of how topoisomerase I and II work can occur only when their precise structures are determined through X-ray crystallographic procedures. Even now, however, the prediction can be made that double helices wrap themselves around topoisomerase II in positive supercoils prior to the start of the cutting-rejoining cycle. Without this restraint on the orientation of the DNA, topoisomerase II would break as well as make negative supercoils, and the net result of its action would be the eventual relaxation of the double-helical substrate to the uncoiled state.

Figure 9-31
The mechanism of action of procaryotic topoisomerase I. (a) The enzyme binds to DNA, unwinding the double helix. (b) It then nicks one strand and prevents the free rotation of the helix by remaining bound to each broken end. (c) The enzyme passes the other strand through the break and ligates the cut ends, thereby increasing the linkage number of the DNA by 1. (d) The enzyme falls away and the strands renature, leaving a DNA with one less negative supercoil. [Redrawn from F. Dean, et al. *Cold Spring Harbor Symp. Quant. Biol.* 47 (1982):773.]

Long, Linear DNA Molecules May Be Divided into Looped, Supercoiled Domains[76–80]

The fact that linear DNA molecules have free ends would seem to preclude their possession of supercoiled segments. Yet, examination of gently isolated linear eucaryotic chromosomes shows that their chromatin is organized into a large number of successive looped domains organized along a scaffold containing two major DNA-binding proteins. Somehow, attachment to the scaffold prevents the rotation of one domain from being transmitted to adjacent sections, since the DNA within each of these loops independently supercoils and may be under different torsional strain. Now there is much evidence both from *Drosophila's* giant salivary chromosomes and from the looped lampbrush chromosomes of the salamander that the individual loops represent functional units of chromatin, with given loops usually being transcribed into one or more very long RNA molecules. How the protein scaffold might maintain independently supercoiled loops was a complete mystery until recently. Now it is less so, following the discovery that a major component of the scaffold is none other than topoisomerase II.

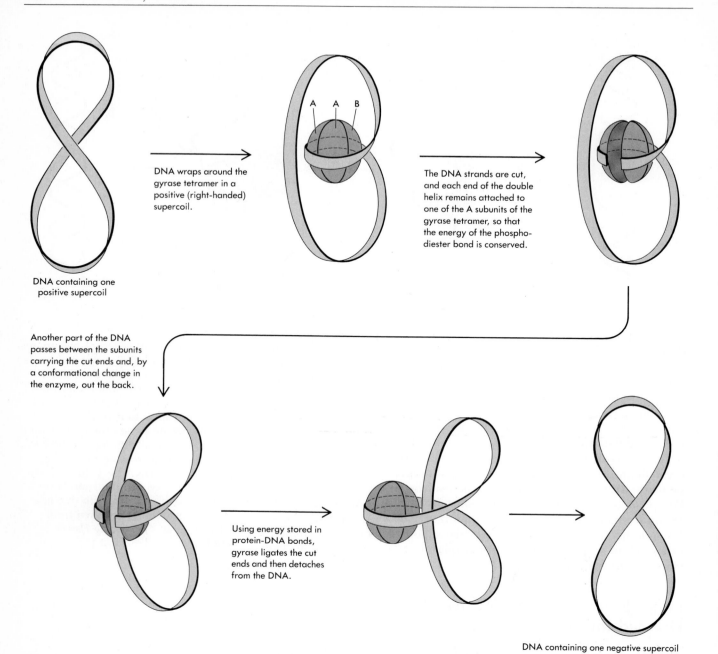

DNA containing one positive supercoil

DNA wraps around the gyrase tetramer in a positive (right-handed) supercoil.

The DNA strands are cut, and each end of the double helix remains attached to one of the A subunits of the gyrase tetramer, so that the energy of the phosphodiester bond is conserved.

Another part of the DNA passes between the subunits carrying the cut ends and, by a conformational change in the enzyme, out the back.

Using energy stored in protein-DNA bonds, gyrase ligates the cut ends and then detaches from the DNA.

DNA containing one negative supercoil

Figure 9-32
A possible mechanism of action for procaryotic topoisomerase II (gyrase). For simplicity, the starting DNA is shown to have one positive supercoil, which then wraps around the enzyme. In fact, the starting DNA might originally have numerous positive or negative supercoils. The action of topoisomerase II is to decrease the linking number by two, a change that, in this case, is represented by the difference between one positive and one negative supercoil. (Redrawn from A. J. H. Wang et al. *Sci. Amer.* 247:97.)

Not only are the long, linear chromosomes of higher organisms divided into looped, functional domains, but the circular *E. coli* DNA chromosome also is somehow divided into some 50 looped domains (each about 20 μm long) whose supercoiling is independently controlled. No scaffold has yet been isolated from any bacterial chromosome. Indeed, the best way now to search for it may be to look for topoisomerase II activity in very gently ruptured cells.

Generation of Unique DNA Fragments by Restriction Enzymes[81, 82]

Enzymes that cut the phosphodiester bonds of polynucleotide chains are called **nucleases.** Those nucleases that preferentially break internal bonds are known as **endonucleases,** while those that clip off terminal nucleotides are referred to as **exonucleases** (Figure 9-34).

For a long time, it was thought that all endonucleases were relatively unspecific, cutting polynucleotide bonds regardless of the surrounding nucleotide sequence. Over the past decade, however, a new class of endonucleases that cleaves only specific nucleotide sequences has been discovered in procaryotic cells. These specific nucleases, which collectively are referred to as *restriction enzymes*, generally make breaks only within "recognition" sequences that exhibit twofold symmetry around a given point:

$$\text{for example, } \frac{\text{AAG CTT}}{\text{TTC GAA}}.$$

Because the same sequence (running in opposite directions) is found on both strands, such enzymes always create double-stranded breaks (Table 9-3). These breaks can occur almost simultaneously, since many, if not most, restriction enzymes bind as dimers to both component DNA strands at their recognition sites.

Restriction enzymes were initially discovered through their ability to break down ("restrict") foreign, as opposed to native, DNA. A given restriction enzyme can distinguish between DNA native to the cells in which it is synthesized and DNA from another species. Within cells, the main (and possibly only) function of restriction enzymes is to destroy foreign DNA segments before they can recombine into host chromosomal DNA. Their discovery was totally unanticipated, since there was no reason to believe that the introduction of foreign DNA was either a common event or so damaging that specific enzymes would have evolved to nullify their genetic potential.

By now, over 100 different endonuclease specificities are known from examination of restriction enzymes isolated from more than 300 different procaryotic organisms. While restriction enzymes are found in most procaryotic cells, they are absent from eucaryotic cells. What accounts for this difference remains a mystery. Most restriction enzymes recognize groups of four to six specific nucleotides, although a few recognize groups of eight. The inverted identical sequences on the two DNA strands are not always continuous; one or more unspecified base pairs often separate them. The sites of cleavage usually lie within the recognition sequences. Sometimes, blunt-ended fragments are produced, but more often, the cleavage sites are positioned off center, generating short, single-stranded tails (Figure 9-35). Quite frequently, two different enzymes that bind to the same recognition site cut between different pairs of nucleotides. So far, a preponderance of G and C bases have been found at recognition sites, and only recently has a restriction enzyme (*Aha* III) been found that recognizes a sequence composed entirely of A and T residues (TTTAAA). Whether this bias reflects the stronger base pairing between guanine and cytosine residues is not known.

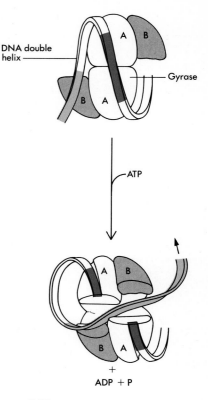

Figure 9-33

Conformational changes in procaryotic topoisomerase II (gyrase) allow the DNA duplex to pass through itself. A number of different models for gyrase action are consistent with current data. Figure 9-32 shows a model in which the DNA duplex passes through the front of the enzyme and out the back, thus requiring two "gates" for DNA. A simpler model (whose first two steps are shown here) requires only one "gate" since the DNA duplex passes through the front face of the enzyme into an internal cavity, and then out again through the same face. The A subunits first cut the DNA duplex and separate to allow another region of the same duplex to pass between the cut ends into the internal cavity. Later the A subunits reclose, the cut ends are ligated, and finally the A subunits separate again to let the intact DNA duplex back out of the internal cavity. The B subunits serve as a protein hinge for opening the DNA "gate." Both the A and B subunits are shown schematically, and are known to have a much more complex structure. [Adapted from A. Morrison and N. R. Cozzarelli, *Proc. Natl. Acad. Sci.* 78 (1981):1420.]

Figure 9-34
The action of endonucleases and exonucleases.

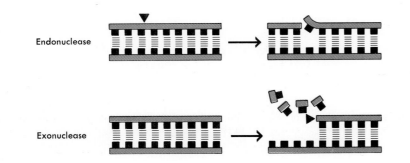

Table 9-3 The Recognition Sequences of Various Restriction Enzymes

Enzyme	Recognition Site		Enzyme	Recognition Site	
	Axis			**Axis of symmetry**	
	Cut bond ↓			Cut bond ↓	
Eco RI	5'-GAA	TTC	*Hin* dII	5'-GTPy↓	PuAC
	CTT	AAG ↑		CAPu	↑PyTG
Hin dIII	A↓AG	CTT	*Hpa* I	GTT↓	AAC
	TTC	GAA ↑		CAA	↑TTG
Hpa II	C↓C	GG	*Hae* III	GG↓	CC
	GG	CC ↑		CC	↑GG
Acy I	GPu↓C	GPyC	*Afl* III	↓ACPu	PyGT
	CPyG	CPuG ↑		TGPy	PuCA ↑
Aha II	GPu↓C	GPyC	*Ava* I	↓CPyC	GPuG
	CPyG	CPuG ↑		GPuG	CPyC↑
Cfr I	Py↓GG	CCPu	*Gdi* II	Py↓GG	CCG
	PuCC	GGPy ↑		PuCC	GGC ↑
Hae II	PuGC	GC↓Py	*Hgi* CI	↓GGPy	PuCC
	PyCG↑	CGPu		CCPu	PyGG ↑
Hgi JII	GPuG	CPy↓C	*Nsp* CI	PuCA	TGPy
	CPyC↑	GPuG		PyGT↑	ACPu
Alu I	AG↓	CT	*Asu* II	TT↓C	GAA
	TC	↗GA		AAG	CTT↑
Cla I	AT↓C	GAT	*Bse* PI	↓GCG	CGC
	TAG	CTA ↑		CGC	GCG ↑

Arrows denote the site of cleavage; Pu = purine; Py = pyrimidine. All base sequences are shown with the 5' end of the upper strand at the left.

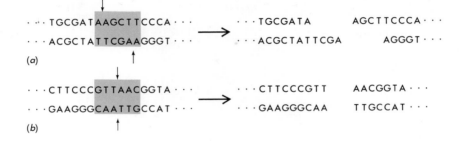

A given combination of four bases will occur by chance only once every several hundred base pairs, while specific groups of six will randomly occur only once every several thousand base pairs. Thus, restriction enzyme cutting generally leads to fragments of significant length. For example, the six-base-pair recognition sequence of the enzyme *Eco* RI occurs only once in the SV40 DNA (5243 base pairs), while the enzyme *Hin* dIII breaks SV40 at five different sites. Moreover, sequences present in one DNA are frequently absent in other DNAs. For instance, phage T7 DNA (39,936 base pairs) does not contain even one example of the six-base-pair sequences recognized by *Eco* RI.

Kinking in *Eco* RI-DNA Recognition Site Complexes[83]

How a restriction enzyme binds to DNA was first revealed by X-ray analysis of the complex formed between *Eco* RI and the DNA segment

TCGCGAATTCGCG
GCGCTTAAGCGCT

containing within it the recognition sequence GAATTC. Such static complexes are made in the absence of Mg^{2+}, whose presence is necessary for *Eco* RI to hydrolyze its substrate. The two polypeptide subunits of *Eco* RI effectively wrap around the double helix, making tight complementary contacts with bases lining the major groove. No contacts are made with the minor groove. The DNA conformation differs from both the classical A and B forms; the distortions are concentrated in three abrupt kinks that separate four blocks of three base pairs each. Because of the symmetry of the DNA sequence, the first and third kinks are identical and differ from the central kink. The two identical terminal DNA segments have A-like conformations, while the central block resembles B-DNA. The terminal kinks reflect transition states separating A- and B-DNA. The central kink within the block of B-DNA contains a dramatic unwinding of the double helix by 25° and a 12° bend of the DNA that enlarges the major groove to permit access by *Eco* RI (Figure 9-36). This structure differs significantly from that found for the same sequence when not bound to a restriction enzyme. Especially unique to the complex is the central kink, which probably only exists very briefly in the absence of binding by *Eco* RI.

Figure 9-35
The action of restriction enzymes.
(a) Restriction enzymes often cut the DNA molecule off center from the central axis of the recognition site, resulting in short, single-stranded tails. Here a sequence contains the recognition site for *Hin* dIII, which cleaves the sequence at the points marked by the arrows.
(b) Other restriction enzymes, such as *Hpa* I shown here, cut at the central axis, resulting in blunt-ended DNA fragments. In both (a) and (b), the restriction enzyme recognition sequence is shaded.

Figure 9-36
Interaction of B-DNA with *Eco* RI. The arrow points to the central kink in the block of B-DNA interacting with the *Eco* RI protein. The central kink unwinds the double helix by 25° and bends the DNA by 12°, which consequently enlarges the major groove. (Courtesy of J. Rosenberg.)

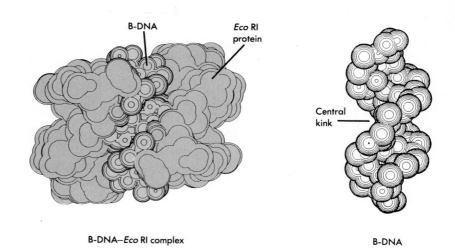

B-DNA—*Eco* RI complex

B-DNA

Methylated Recognition Sites Protect Cells from Their Own Restriction Enzymes[84, 85]

Restriction enzymes are prevented from acting on the DNA native to their organism of origin by companion enzymes that add methyl groups to specific adenine or cytosine residues within their recognition sites (Figure 9-37). These highly specific methylases, unrelated to the more general methylases responsible for creating the m^5C and the m^6A found within procaryotic cells, place methyl groups only on appropriate bases within their recognition sites rather than elsewhere on the DNA molecule. Thus, a given bacterial chromosome is protected only from cutting by its own restriction enzymes, not from cutting by enzymes that bind to different recognition sites. Only one of the two complementary strands of DNA at a recognition site need be methylated for protection. Such **hemimethylation** is necessary, since if hemimethylated DNA were susceptible to cleavage, newly replicated chromosomes might often be cleaved before the cell's methylase had acted to form fully methylated DNA.

Although a given restriction enzyme and its companion methylase both bind to the same specific groups of bases, there is no obvious amino acid homology between their polypeptides, and they are encoded by completely different genes. The functional forms of methylases are monomers, perhaps because their substrates are singlechain segments.

Separating DNA Fragments on Agarose Gels[86, 87]

The DNA fragments that result from restriction enzyme cutting are easily separated and displayed by electrophoresis through agarose gels. The rate at which DNA pieces move in such gels depends on their lengths; the distances traveled decrease as the fragment lengths increase. After DNA has moved through a gel, it is easily visualized by the addition of dyes like ethidium bromide, which bind to DNA and generate visible bands corresponding to discrete fragments whose lengths can be established through calibration with DNA molecules of known weight (Figure 9-38). DNA moves unharmed

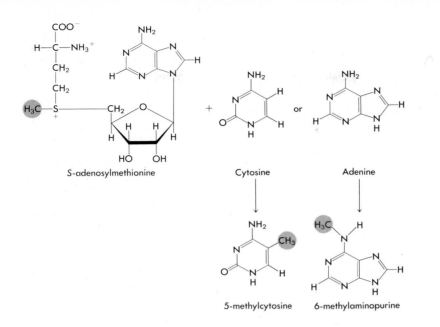

Figure 9-37
Specific adenine and cytosine residues in restriction enzyme recognition sites may become methylated to prevent the restriction enzymes from acting on DNA native to the cell. *S*-adenosylmethionine is the methyl donor.

through agarose gels and can be recovered as biologically functional double helices.

The ordering of DNA restriction fragments along a chromosome or chromosomal segment gives rise to a **restriction map** (Figure 9-39). The first such maps were obtained when viral chromosomes were digested with enzymes such as *Hin* dII, *Hin* dIII, and *Eco* RI. Ordering the fragments was accomplished by studying the temporary appearance of incompletely cut fragments as digestion went to completion. For example, SV40 is first converted into a linear molecule that is in turn progressively cut up into 11 smaller fragments. Maps produced by different enzymes are necessarily different; initially, the most useful are those made by enzymes whose recognition sequences are rare in DNA and which thereby generate small numbers of easily separable fragments.

Using a Methylase to Create Extended Restriction Enzyme Recognition Sequences[88, 89]

Through the clever employment of two highly specific methylases, extended recognition sequences for the restriction enzyme *Dpn* I can be constructed in vitro that contain eight and ten base pairs. Even the shortest of these sequences should occur by chance only once every 65,000 (4^8) base pairs in a given DNA molecule, thereby creating the opportunity to break even very large DNA molecules into a relatively small number of pieces. This ability to make abnormally long recognition sequences rests on a unique property of the restriction enzyme *Dpn* I to cut only DNA that has methylated adenine (^mA) within its restriction site:

$$G^mA \ TC$$
$$C \ T^mAG \cdot$$

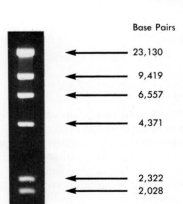

Lambda DNA
Hind III DIGEST

Base Pairs

23,130

9,419

6,557

4,371

2,322
2,028

Figure 9-38
Separation of DNA restriction fragments by electrophoresis through an agarose gel. The smaller fragments move more rapidly than the larger ones; thus, the sizes of the restriction fragments can be estimated by comparing the distances they travel through the gel to the distances traveled by DNA molecules of known size. (Courtesy of D. Helfman.)

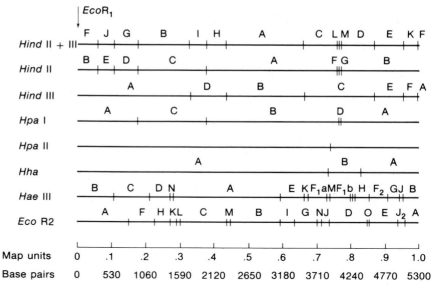

Figure 9-39

Cleavage map of the SV40 genome. The zero point of the map is the *Eco* R1 site. For clarity, the circular genome is shown opened at the R1 site, and the cleavage sites (and resulting fragments) for each restriction enzyme are indicated on a separate line.

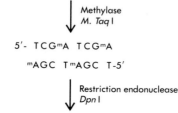

(a)

(b)

Figure 9-40

Creation of long restriction enzyme recognition sequences. (a) *M. Tag* I is a methylase that adds a methyl group to the adenine of its recognition sequence of repeated TCGA. This creates an eight-base-pair recognition sequence for the restriction enzyme *Dpn* I. (b) An even longer, ten-base-pair *Dpn* I recognition sequence is generated by methylase *M. Cla* I, which recognizes and methylates two partially overlapping ATCGA sequences. Such long restriction sequences occur very rarely in DNA and thus allow the cleavage of large DNA molecules into relatively few fragments. The recognition sequences of the methylases are shaded on one DNA strand. [After M. McClelland et al., *Proc. Nat. Acad. Sci.* 81 (1984):984]

Thus, *Dpn* I ordinarily lacks the ability to cut DNA from most organisms (e.g., vertebrate DNA is not broken by it). Such DNA, however, becomes susceptible to *Dpn* I following exposure to certain of those methylases whose recognition sequences overlap the *Dpn* I sequence and which add methyl groups to the appropriate adenine residues. For example, methylation by *M. Taq* I at a direct repeat of its TCGA methylation recognition sequence produces an eight-base-pair *Dpn* I restriction recognition sequence (Figure 9-40a). An even longer, ten-base-pair *Dpn* I recognition sequence is generated by methylation by *M. Cla* I at two partially overlapping *M. Cla* I methylation recognition sequences (Figure 9-40b).

Cleavage at these inherently rare recognition sequences should prove very useful in the generation of restriction maps of DNAs containing more than 10^5 base pairs. Until very recently, there was no way to physically separate the very large DNA molecules that can result from the construction of these sites. Now, however, through the use of pulse field electrophoresis (electrophoresis using short pulses of electricity in two different directions) through agarose gels, DNA molecules containing up to 10^6 base pairs can be routinely handled. Already, the *E. coli* chromosome (4×10^6 base pairs) has been cut into some 20 different fragments (Figure 9-41), and its restriction map should soon be available.

Ligating DNA Fragments to Create Recombinant DNA[90-91]

The complementary single-stranded tails produced by restriction enzymes that make staggered cuts tend to associate briefly with each other by base-pairing. Such unstable unions can be made permanent by adding the joining (ligating) enzyme DNA ligase (Figure 9-42). This enzyme plays a vital role in DNA replication and recombination within cells by catalyzing the formation of new phosphodiester bonds. Ligase acts only in the presence of ATP (or NAD⁺), which provides the energy needed to make the backbone bonds. At first, ligase was thought capable of linking together only those fragments bearing complementary tails. If, however, all reactants are present in high concentrations, then blunt-ended DNA fragments can also be joined together. The frequency of such random bonding, however, is very low, and frequently complementary tails are enzymatically added to blunt-ended fragments to promote their joining (Figure 9-43).

Since any two fragments bearing complementary tails will tend to base-pair with each other, the pieces that ligase permanently joins together need not have been previously linked. Ligase, together with restriction enzymes that make staggered cuts (such as *Eco* RI), thus provides a simple way to produce recombinant DNA, recombining in the test tube DNA pieces originating from different DNA molecules. The generation of recombinant DNA molecules by itself would have had little immediate impact if it had not soon been followed by the development of procedures that allow the selective amplification of such newly created DNA (a process known as *cloning*). The heart of gene-cloning procedures involves the insertion of desired DNA sequences into autonomously replicating DNA molecules (**cloning vectors**) that multiply to form many identical clones after their reintroduction back into suitable host cells (see Figure 7-29). Most frequently used as cloning vectors today are several autonomously replicating plasmid and viral DNAs.

Libraries of Cloned DNA Fragments[92]

Although electrophoresis through agarose gels now provides a practical way to isolate restriction fragments from viral chromosomes, bacterial chromosomes, or plasmids, it cannot be used to isolate single fragments from the much bigger chromosomes of higher cells. Even those enzymes that cut within relatively rare recognition sequences produce so many fragments from any higher cell DNA that the fragments move in gels as smears rather than as discrete bands. For example, the 8×10^9 base-pair human genome is cut into approximately 2 million different fragments by *Eco* R1.

Study of individual pieces of cellular DNA only became possible with the advent of recombinant DNA procedures. These allow the separation of even very similar-size DNA pieces by individually inserting them into cloning vectors. All the restriction fragments from genomes even as large as *Drosophila* can be inserted in only some 40,000 different *E. coli* plasmids. The subsequent multiplication of these individual vectors (known as a "gene library") can generate all the pure DNA ever needed for exhaustive molecular characterization. Now the making of a gene library containing all the DNA sequences from a given cell is a routine task. Often far from trivial, however, is the identification of the library member that interests us. Fortunately, there is a large number of sophisticated, non-brute-force methods for selecting the desired genes. Some of these techniques are described in Chapter 19. However, it is worth emphasizing here that only through the development of gene libraries have recombinant DNA procedures so drastically raised the sights of biological research.

Figure 9-41

Pulse field electrophoresis gel of a *Not1* restriction enzyme digest of *E. coli* DNA. The 20 fragments range in size from 25 kb to 450 kb, not including the largest fragment, whose size is unknown. [Courtesy of C. L. Smith.]

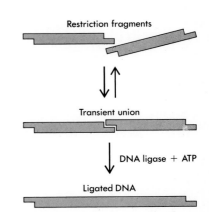

Figure 9-42

In the presence of ATP or NAD⁺, DNA ligase can covalently join DNA fragments that were previously in only a transitory, unstable union.

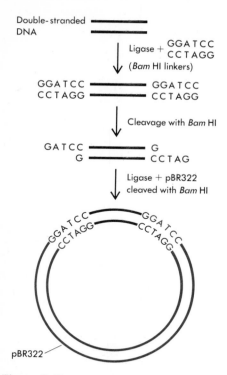

Figure 9-43
Double-stranded oligonucleotides called *Bam* HI linkers are used to add complementary ends to blunt-ended double-helical DNA. The linkers are first ligated to both ends of the given DNA. Then the product is treated with *Bam* HI, which generates "sticky ends" to promote joining to other DNA fragments, such as the plasmid pBR322 shown here.

Very Long DNA Segments Can Be Rapidly Sequenced[93-96]

Following close upon the discovery of restriction enzymes and development of recombinant DNA methodologies has been the formulation of two powerful methods for accurately establishing the order of the four bases (A, G, C, and T) along given DNA fragments. One method uses chemical reagents (e.g., dimethylsulfate) that react with specific bases to break DNA preferentially at given nucleotides. By correlating the appearance of fragments of specified length with the specific base destroyed by chemical attack, the exact order of bases along the original unbroken DNA strand can be determined (Figure 9-44). This method, as well as the enzymatic method described next, makes use of the fact that electrophoresis through polyacrylamide gels can resolve single chains of DNA differing in length by only a single nucleotide. On a given gel, over 300 bands can be resolved. These chemical procedures, first used to reveal the sequences of key regulatory regions in bacterial DNA (e.g., the *lac* operon), soon allowed the entire sequence of the plasmid cloning vector pBR322 (4362 base pairs) to be worked out by a single scientist in only one year.

The second widely used method for DNA sequencing uses an enzyme, DNA polymerase I, to produce different-size DNA fragments that when subsequently separated on gels allow, as in the first method, the direct reading of the nucleotide order. In this method, the single-stranded DNA being analyzed is used as a template for the synthesis of complementary DNA chains using as substrates chain-terminating 2'3'-dideoxynucleotides as well as the four normal nucleotide precursors of DNA. These nucleotide analogs become incorporated at the growing ends of DNA chains, but do not allow further elongation, since they cannot form phosphodiester bonds with incoming precursors. Inclusion of a small amount of a given dideoxynucleotide in a reaction mixture leads to its occasional random incorporation at all the sites usually occupied by its normal equivalent. A series of incompletely elongated DNA chains are thereby produced that are specifically terminated by the given dideoxy residue. To determine a sequence, four separate DNA synthesis reactions, each involving a different chain-terminating nucleotide, are run and their product chains electrophoresed side by side on a polyacrylamide gel to display bands of ever-increasing length (Figure 9-45). This method quickly revealed the complete sequence of the phage T4 DNA (5577 base pairs) and more recently has established that of phage λ DNA (48,513 base pairs).

Now the dideoxy method is increasingly used in preference to the chemical fission approach. Once dideoxynucleotides became commercially available, the ease with which this technique could be mastered became apparent, and it is likely to dominate sequencing efforts for the next decade.

The Significance of Many Palindromic Sequences Remains Unclear[97, 98]

An unexpected consequence of the first serious DNA sequencing was the discovery of many inverted repetitious sequences in which the same (or almost the same) sequence runs in opposite directions if one

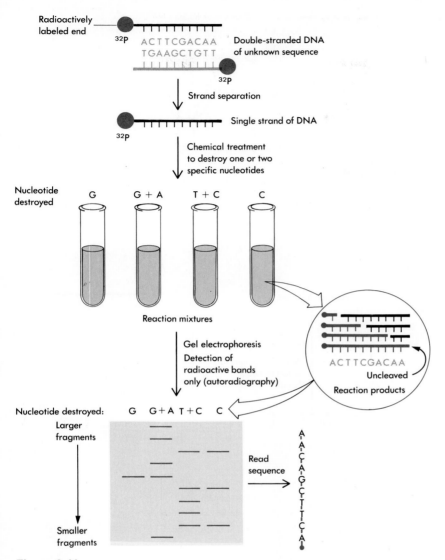

Figure 9-44
DNA-sequencing method developed by Maxam and Gilbert. This method
uses chemical reagents to destroy specific nucleotide bases and thus break
the DNA molecule at specific sites. First the strands of the DNA molecule
are labeled radioactively at one end (usually the 5' end), and the two
strands are separated (only one will be sequenced). Then aliquots of the
chosen strand are treated with four different chemical reagents that break
the strand at one or two specific nucleotides; the treatment is limited so
that at most a single residue of the susceptible base(s) in the molecule will
react. Thus, in each reaction mixture, a nested set of radioactive fragments is generated, as shown here for only the reaction mixture that destroys C residues. Finally, gel electrophoresis is used to separate the
products of each reaction by size. The pattern of radioactive bands seen
on X-ray film immediately reveals the sequence.

switches from one strand to its complement at the central axis of
symmetry, for example

GTATCC GGATAC
CATAGG CCTATG.

Figure 9-45

DNA-sequencing method developed by Sanger. A dideoxynucleotide is incorporated into a growing DNA strand, subsequently stopping chain growth, since it cannot form a phosphodiester bond with the next incoming nucleotide. Four different reactions are run, each with a different dideoxynucleotide. The products of each reaction are a series of incompletely elongated segments, which are separated by gel electrophoresis. As in the Maxam-Gilbert method, the sequence can be read from the bands produced in the gel.

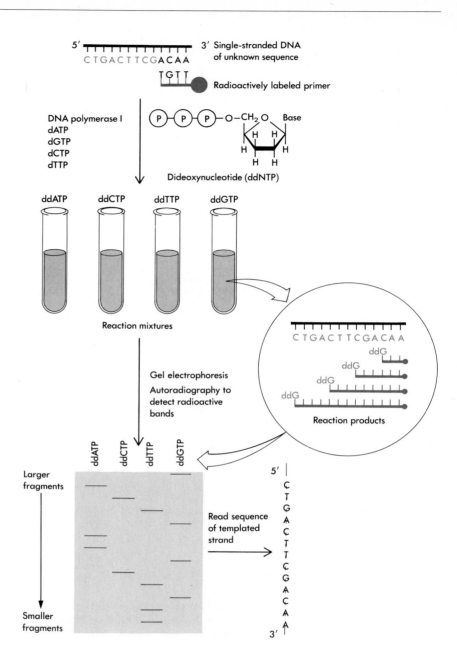

Some inverted palindromes (a palindrome is a word that reads the same backward and forward), like those recognized by certain restriction enzymes (see Table 9-3), are relatively short (three to ten bases in a given direction). Others are much longer, usually being somewhat imperfect as well as being interrupted by long stretches of unrelated sequences. Inverted sequences separated by hundreds to thousands of unrelated nucleotides mark the ends of certain movable genetic elements, while others permit site-specific crossing-over events that allow flip-flop genetic elements to function in alternative orientations.

Many of the medium-size, continuous, inverted palindromes probably serve as DNA recognition sites for **multimeric proteins** composed of identical subunits (e.g., repressors). The significance of the longer continuous palindromes is less obvious. As mentioned earlier in this chapter, when they are nearly perfect and present on super-coiled DNA, they have the potential to undergo semistable steric rearrangements in which the complementary sequences on each DNA strand base-pair to form cruciform (hairpin) loops extending out from the main helical axes. Such looping would remove some of the stress of supercoiling as well as generate structures that might be differentially recognized by specific DNA binding proteins.

In any case, those sections of RNA chains transcribed off palindromic regions will tend to form hydrogen-bonded hairpin loops, the stability of which will be directly related to how perfectly symmetrical the palindromic sequences are. Some palindromes may have evolved to endow structural compactness upon their single-stranded RNA products.

The Storing of DNA Sequence Information in Computers[99]

Computers are now routinely used to store DNA sequences. Such information is directly fed into the computer as the correct bases are read off gels, thereby avoiding errors that inevitably would occur if very long sequences were manually copied. Once in the computer, sequences can be scanned for restriction enzyme recognition sites, start and stop signals for RNA synthesis, the presence of inverted palindromes, preferred segments of potential Z-DNA (alternating purine and pyrimidine stretches), and homologies to other known DNA sequences. They can also be read to give amino acid sequences of proteins. High-powered software programs to carry out these searches are available from many sources, and the very large amounts of DNA sequence information already available requires the use of increasingly more powerful computers. The time has passed when a small, primitive desk-top computer can do the job, and much recombinant DNA research requires access to sophisticated minicomputers (e.g., VAX).

DNA sequencing itself can be speeded up through the use of a powerful computer. The working out of restriction maps for long sequences can consume much more time than DNA sequencing itself. It thus often makes sense to sequence large numbers of potentially overlapping DNA fragments and then use a computer to search for the sequence overlaps that directly order the fragments along their respective chromosomes. This approach only became practical when the M13 variation of the dideoxy sequencing method was developed, in which the DNA fragments to be analyzed are inserted at a precise spot in the single-stranded DNA chain of phage M13 (Figure 9-46). The subsequent DNA synthesis (sequencing) reactions all commence by elongating a single, short (15 to 17 bases) DNA **primer** that is complementary in sequence to the M13 region immediately adjacent to the foreign DNA insert. By using the same primer for all reactions, sequencing by the M13 method becomes almost child's play, allowing sequencing to be determined at rates never even imagined only a few years ago (about 2000 base pairs a week). Viral chromosomes containing more than 10^5 base pairs are now being sequenced, and the *E. coli* chromosome of about 4×10^6 base pairs is currently in the process of being worked out. Still seemingly too large are the chromosomes of higher vertebrates, but even here, selected key segments will eventually be sequenced.

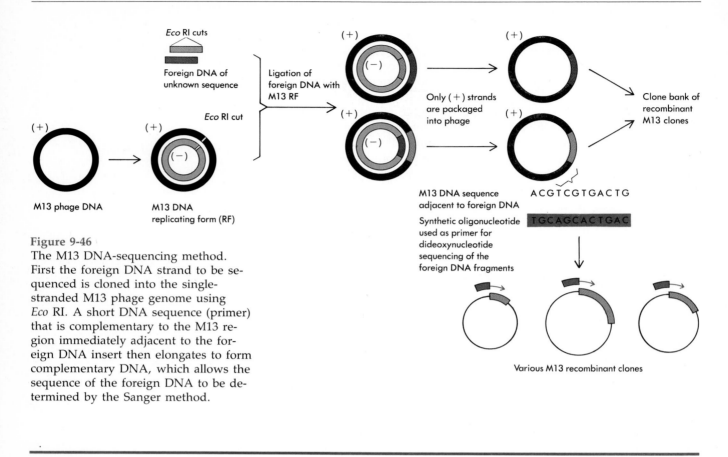

Figure 9-46
The M13 DNA-sequencing method. First the foreign DNA strand to be sequenced is cloned into the single-stranded M13 phage genome using *Eco* RI. A short DNA sequence (primer) that is complementary to the M13 region immediately adjacent to the foreign DNA insert then elongates to form complementary DNA, which allows the sequence of the foreign DNA to be determined by the Sanger method.

Summary

Most DNA occurs as right-handed double helices held together by hydrogen bonds between the bases adenine and thymine and between the bases guanine and cytosine. The strict base-pairing rules (AT and CG) underlying all double helices reflect the highly fixed locations of the hydrogen atoms in the purine and pyrimidine bases. Adenine and cytosine almost always exist in the amino as opposed to the imino tautomeric forms, while guanine and thymine almost always exist in the keto as opposed to enol forms. The biological functioning of DNA depends on these highly preferred tautomeric forms for the accurate synthesis of complementary DNA or RNA chains upon single-stranded DNA templates.

The two strands of the double helix fall apart (denature) upon exposure to high temperature, extremes of pH, or any agent that causes the breakage of the individual hydrogen bonds. Upon slow return to normal cellular conditions, the denatured single strands can specifically reassociate to biologically active double helices (renature).

While all cellular DNA occurs as double helices, the genetic component of certain small viruses is single-stranded DNA. Such single-stranded (+) DNA chromosomes can become converted into +/− double helices after entry into cells. They then may use their − strands as templates to make new single-stranded + progeny chains identical to those found in the parent viral chromosomes. Single-stranded DNA molecules are more compact than double helices and often form imperfectly base-paired hairpin loops that allow more effective stacking of the hydrophobic flat surfaces of the purine and pyrimidine bases. True crystals of DNA can be obtained from short, chemically synthesized polydeoxynucleotides. The precise determination of their structures is then possible using X-ray crystallographic procedures. The structures determined by this method confirm the essential features of DNA postulated in 1953 from model building. Right-handed DNA has two major forms: a compacted A form containing tilted base pairs, and a more elongated and wider B form, in which the almost flat base pairs lie approximately perpendicular to the helical axis. Both the A and B forms have major grooves that expose the insides and minor grooves that expose the outsides of the base pairs to the surrounding environment.

DNA segments containing alternating purine and pyrimidine bases can become converted to left-handed Z helices. In Z-DNA, the purines have syn conformations and the pyrimidines have the anti conformations characteristic of right-handed DNA. Transition into the Z form is favored by methylation of cytosine residues. Methylation also allows some nonalternating purine-pyrimidine sequences to become left-handed. Z-DNA is more elongated and slimmer than either A- or B-DNA and contains only one surface indentation (the minor groove). Almost all double helices in solution or in cells are in the B form and only very rarely are found to be in the Z form.

In solution, B-DNA constantly bends and twists. Abrupt kinks sometimes form, often favored by DNA sequences that affect the exact ways the four different base pairs can stack on each other. Some highly transient kinks allow the insertion (intercalation) of flat, ringed heterocyclic molecules into DNA. This insertion increases the translation of successive base pairs along the DNA axis and thereby extensively unwinds the DNA molecule.

Almost all cellular DNA molecules are extremely large, with only one DNA molecule now thought to reside within a given chromosome. The smallest DNA molecules are found within viruses; the tiniest contains only several thousand base pairs. Even the smallest cellular chromosomes (bacterial) contain at least a million base pairs. Some cellular DNA molecules within cells exist in linear forms, but these are usually of viral origin. The remainder are circular and often exist as supercoils; their two chains are slightly more (positive supercoils) or slightly less (negative supercoils) twisted around each other than in unconstrained linear double helices. The extent of DNA supercoiling within cells is in part controlled by enzymes called topoisomerases, which either add or remove supercoils. Within cells, most DNA is negatively supercoiled, partially owing to its attachment to specific DNA binding proteins. The best-understood of these proteins are the core histones that bind to DNA to form the nucleosomes that dominate chromatin structures. When bound to DNA, these core histones exist as ellipsoid-shaped octomers around which the DNA makes two superhelical turns, which slightly untwist the double helices. When separated from the histones, circular DNA becomes negatively supercoiled. It was originally assumed that linear DNA molecules would never appear supercoiled within cells. However, supercoiled regions of linear chromosomes exist, perhaps reflecting the binding of currently undetected specific protein and possibly RNA molecules.

Negative supercoiling of circular DNA molecules is sometimes partially relieved by the creation of localized denatured regions within DNA segments rich in the relatively weak AT base pairs. Localized denaturation is especially favored for segments containing self-complementary regions capable of forming hydrogen-bonded hairpin (cruciform) loops. Negative supercoils may also be lost through the conversion of sections of right-handed DNA into left-handed DNA.

Key tools for the analysis of DNA molecules are a group of bacterial enzymes that cut at specific nucleotide sequences. These enzymes, known as restriction enzymes, generally make breaks only within specific "recognition" sequences that exhibit twofold symmetry. As a result, they usually make cuts through both strands. Enzymes with over 100 different specificities have been isolated from various bacteria. Within a cell, a given restriction enzyme may function to degrade foreign DNA that may have entered the cell. The enzyme is prevented from breaking down the DNA native to the cell by very specific DNA methylases, which add methyl groups only to specific adenine or cytosine residues within the recognition sequences. All bacterial cells that contain a restriction enzyme thus also contain a corresponding methylase. Only one strand within a recognition sequence needs to be methylated to prevent its degradation. In this way, newly replicated hemimethylated DNA is protected from destruction. Restriction enzymes and their corresponding methylases have no common sequences and are coded by different genes.

The DNA fragments created by restriction enzyme cuts can be easily separated by electrophoresis on agarose gels, which separates fragments by their different rates of movement. The larger the fragment, the more slowly it moves. Movement through such gels does not harm DNA, which can be recovered in biologically active form. The original order of DNA restriction fragments along a chromosome (or chromosomal segment) is known as a restriction map. Maps produced by different enzymes are usually different; the most useful are those maps made by enzymes whose recognition sequences are rare in DNA.

Addition of the joining enzyme, DNA ligase, to mixtures of restriction fragments emanating from two genetically different DNA molecules creates recombinant DNA molecules. Such joining (ligation) occurs much more readily with those fragments containing single-stranded tails generated by cuts that occur outside of the center of the recognition sequence. However, blunt-ended fragments can also be joined together, although at a much slower rate. The most useful recombinant DNA molecules are those in which the desired DNA segments are inserted into autonomously replicating DNA molecules (cloning vectors). Collections of large numbers of different DNA segments from unique cellular or organismal sources that are each inserted into separate cloning vectors are referred to as "gene libraries."

Restriction fragments containing several hundred base pairs now can be rapidly sequenced by either of two very efficient DNA-sequencing methods. Continuous sequences of several thousand base pairs can now be determined within a week, and the sequences of viral chromosomes of up to several hundred thousand base pairs can be established in a year or two. Efforts are now being made to work out the sequence of the *E. coli* chromosome, a task that may be accomplished by 1990.

Bibliography

General References:

Kornberg, A. 1980. *DNA Replication*. San Francisco, Calif.: Freeman.

Kornberg, A. 1982. *1982 Supplement to DNA Replication*. San Francisco, Calif.: Freeman.

Linn, S. M., and R. J. Roberts, eds. 1982. *Nucleases*. Cold Spring Harbor, N.Y.: Cold Spring Harbor Laboratory.

Saenger, W. 1984. *Principles of Nucleic Acid Structure*. New York: Springer-Verlag.

Sarma, R. H., ed. 1981. *Bimolecular Stereodynamics*. Vols. 1 and 2. Guilderland, N.Y.: Adenine Press.

"Structures of DNA." 1982. *Cold Spring Harbor Symposium on Quantitative Biology*. Vol. 47. Cold Spring Harbor, N.Y.: Cold Spring Harbor Laboratory.

Cited References

1. Franklin, R. E., and R. G. Gosling. 1953. "Molecular Configuration in Sodium Thymonucleate." *Nature* 171:740–741.
2. Watson, J. D., and F. H. C. Crick. 1953. "Molecular Structure of Nucleic Acids: A Structure for Deoxynucleic Acids." *Nature* 171:737–738.
3. Wilkins, M. H. F., A. R. Stokes, and H. R. Wilson. 1953. "Molecular Structure of Deoxypentose Nucleic Acids." *Nature* 171:738–740.
4. Crick, F. H. C., and J. D. Watson. 1954. "The Complementary Structure of Deoxyribonucleic Acid." *Proc. Roy. Soc.* (A) 223:80–96.
5. Langridge, R., H. R. Wilson, C. W. Hooper, M. H. F. Wilkins, and L. D. Hamilton. 1960. "The Molecular Configuration of Deoxyribonucleic Acid. I. X-Ray Diffraction Study of a Crystalline Form of the Lithium Salt." *J. Mol. Biol.* 2:19–37.
6. Wolfenden, R. V. 1969. "Tautomeric Equilibria in Inosine and Adenosine." *J. Mol. Biol.* 40:307–310.
7. Pieber, M., P. A. Kroon, J. H. Prestegard, and S. I. Chan. 1973. "Erratum: Tautomerism of Nucleic Acid Bases." *J. Amer. Chem. Soc.* 95:3408.
8. Doty, P., J. Marmur, J. Eigner, and C. Schildkraut. 1960. "Strand Separation and Specific Recombination in Deoxyribonucleic Acids: Physical Chemical Studies." *Proc. Nat. Acad. Sci.* 46:461–476.
9. Marmur, J., and L. Lane. 1960. "Strand Separation and Specific Recombination in Deoxyribonucleic Acids: Biological Studies." *Proc. Nat. Acad. Sci.* 46:453–461.
10. Schildkraut, C. L., J. Marmur, and P. Doty. 1961. "The Formation of Hybrid DNA Molecules and Their Use in Studies of DNA Homologies." *J. Mol. Biol.* 3:595–617.
11. Marmur, J., R. Rownd, and C. L. Schildkraut. 1963. "Denaturation and Renaturation of Deoxyribonucleic Acid." *Prog. Nucleic Acid Res.* 1:231–300.
12. Denhardt, D. T., D. Dressler, and D. S. Ray. 1978. *The Single Stranded DNA Phages.* Cold Spring Harbor, N.Y.: Cold Spring Harbor Laboratory.
13. Schaller, H., H. Voss, and S. Gucker. 1969. "Structure of the DNA of Bacteriophage fd. II. Isolation and Characterization of a DNA Fraction with Double Strand-Like Properties." *J. Mol. Biol.* 44:445–458.
14. Shen, C.-K. J., A. Ikoku, and J. E. Hearst. 1979. "A Specific DNA Orientation in the Filamentous Bacteriophage fd as Probed by Psoralen Crosslinking and Electron Microscopy." *J. Mol. Biol.* 127:163–175.
15. Wing, R. M., H. R. Drew, T. Takano, C. Broka, S. Tanaka, K. Itakura, and R. E. Dickerson. 1980. "Crystal Structure Analysis of a Complete Turn of B-DNA." *Nature* 287:755–758.
16. Shakked, Z., D. Rabinovich, W. B. T. Cruse, E. Egert, O. Kennard, G. Sala, S. A. Salisbury, and M. A. Viswamitra. 1981. "Crystalline A-DNA: The X-Ray Analysis of the Fragment d(G-G-T-A-T-A-C-C)." *Proc. Roy. Soc. Lond.* (B) 213:479.
17. Dickerson, R. E. 1983. "The DNA Helix and How It Is Read." *Sci. Amer.* 249:94–111.
18. Arnott, S., F. Hutchinson, M. Spencer, M. H. F. Wilkins, W. Fuller, and R. Langridge. 1966. "X-Ray Diffraction Studies of Double Helical Ribonucleic Acid." *Nature* 211:227–232.
19. Wang, A. H.-J., S. Fujii, J. H. van Boom, and A. Rich. 1982. "Molecular Structure of the Octamer d(G-G-C-C-G-G-C-C): Modified A-DNA." *Proc. Nat. Acad. Sci.* 79:3968–3972.
20. Wang, A. H.-J., S. Fujii, J. H. van Boom, G. A. van der Marel, S. A. A. van Boeckel, and A. Rich. 1982. "Molecular Structure of r(GCG)d(TATACGC): A DNA-RNA Hybrid Helix Joined to Double Helical DNA." *Nature* 299:601–604.
21. Dickerson, R. E. 1983. "Base Sequence and Helix Structure Variation in B and A DNA." *J. Mol. Biol.* 166:419–441.
22. Conner, B. N., C. Yoon, J. L. Dickerson, and R. E. Dickerson. 1984. "Helix Geometry and Hydration in an A-DNA Tetramer: CCGG." *J. Mol. Biol.* 174:663–695.
23. Arnott, S., R. Chandrasekaran, I. H. Hall, and L. C. Puigjaner. 1983. "Heteronomous DNA." *Nucleic Acid Res.* 11:4141–4155.
24. Pohl, F. M., and T. M. Jovin. 1972. "Salt-Induced Co-Operative Conformational Change of a Synthetic DNA: Equilibrium and Kinetic Studies with Poly(dG-dC)." *J. Mol. Biol.* 67:375–396.
25. Wang, A. H.-J., G. J. Quigley, F. J. Kolpak, J. L. Crawford, J. H. van Boom, G. van der Marel, and A. Rich. 1979. "Molecular Structure of a Left-Handed DNA Fragment at Atomic Resolution." *Nature* 282:680–686.
26. Hill, R. J., and B. D. Stollar. 1983. "Dependence of Z DNA Antibody Binding to Polytene Chromosomes on Acid Fixation and DNA Torsional Strain." *Nature* 305:338–340.
27. Rich, A., A. Nordheim, and A. H.-J. Wang. 1984. "The Chemistry and Biology of Left-Handed Z-DNA." *Ann. Rev. Biochem.* 53:791–846.
28. Azorin, F. and A. Rich. 1985. "Isolation of Z-DNA Binding Proteins from SV40 Minichromosomes: Evidence for Binding to the Viral Control Region." *Cell* 41:365–374.
29. Kmiec, E. B., K. J. Angelides, and W. K. Holloman. 1985. "Left-handed DNA and the Synaptic Pairing Reaction Promoted by Ustilago Rec1 Protein." *Cell* 40:139–145.
30. Wang, A. H.-J., R. Gessner, G. A. van der Marel, J. H. van Boom, and A. Rich. 1986. "A Crystal Structure of Z-DNA Without an Alternating Purine-Pyrimidine Sequence." *Nature.* In press.
31. Razin, A., and A. D. Riggs. 1980. "DNA Methylation and Gene Function." *Science* 210:604–610.
32. Behe, M., and G. Felsenfeld. 1981. "Effects of Methylation on Synthetic Polynucleotides. The B-Z Transition in Poly(dG-m5dC) Poly(dG-m5dC)." *Proc. Nat. Acad. Sci.* 78:1619–1623.
33. Fujii, S., A. H.-J. Wang, G. van der Marel, J. H. van Boom, and A. Rich. 1982. "The Molecular Structure of (m5dC-dG)3: The Role of the Methyl Group on 5-Methyl Cytosine in Stabilizing Z-DNA." *Nucleic Acid Res.* 10:7879–7892.
34. Doerfler, W. 1983. "DNA Methylation and Gene Activity." *Ann. Rev. Biochem.* 52:93–124.
35. Sarma, R. H., ed. 1980. *Nucleic Acid Geometry and Dynamics.* New York: Pergamon Press.
36. Fratini, A. V., M. L. Kopka, H. R. Drew, and R. E. Dickerson. 1982. "Reversible Bending and Helix Geometry in a B-DNA Dodecamer: CGCGAATT^{Br}CGCG." *J. Biol. Chem.* 257:14686–14707.
37. Levitt, M. 1982. "Computer Simulation of DNA Double Helix Dynamics." *Cold Spring Harbor Symp. Quant. Biol.* 47:251–262.
38. Yanagida, M., Y. Hiraoka, and I. Katsura. 1982. "Dynamic Behaviors of DNA Molecules in Solution Studied by Fluorescence Microscopy." *Cold Spring Harbor Symp. Quant. Biol.* 47:177–187.
39. Marini, J. C., S. D. Levene, D. M. Crothers, and P. T. Englund. 1982. "Bent Helical Structure in Kinetoplast DNA." *Proc. Nat. Acad. Sci.* 79:7664–7668.
40. Wu, H.-M., and D. M. Crothers. 1984. "The Locus of Sequence-Directed and Protein-Induced DNA Bending." *Nature* 308:509–513.
41. Bauer, W., and J. Vinograd. 1968. "The Interaction of Closed Circular DNA with Intercalative Dyes." *J. Mol. Biol.* 33:141–171.
42. Wang, J. C. 1974. "The Degree of Unwinding of the DNA Helix by Ethidium. I. Titration of Twisted PM2 DNA Molecules in Alkaline Cesium Chloride Density Gradients." *J. Mol. Biol.* 89:783–801.
43. Wang, A. H.-J., J. Nathans, G. van der Marel, J. H. van Boom, and A. Rich. 1978. "Molecular Structure of a Double Helical DNA Fragment Intercalator Complex Between DeoxyCpG and a Terpyridine Platinum Compound." *Nature* 276:471–474.
44. Record, M. T., Jr., S. J. Mazur, P. Melancon, J.-H. Roe, S. L. Shaner, and L. Unger. 1981. "Double Helical DNA: Conformations, Physical Properties and Interactions with Ligands." *Ann. Rev. Biochem.* 50:997–1024.
45. Cairns, J. 1961. "An Estimate of the Length of the DNA Molecule of T2 Bacteriophage by Autoradiography." *J. Mol. Biol.* 3:756.
46. Cairns, J. 1962. "A Minimum Estimate for the Length of the DNA of *Escherichia coli* Obtained by Autoradiography." *J. Mol. Biol.* 4:407.
47. Fangman, W. L. 1978. "Separation of Very Large DNA Molecules by Gel Electrophoresis." *Nucleic Acid Res.* 5:653–665.
48. Carle, G. F., and M. V. Olson. 1984. "Separation of Chromosomal DNA Molecules from Yeast by Orthogonal-Fixed-Alternative Gel Electrophoresis." *Nucleic Acid Res.* 12:5647–5664.
49. Fiers, W., and R. L. Sinsheimer. 1962. "The Structure of the DNA of the Bacteriophage φX174. I. The Action of Exopolynucleotidases." *J. Mol. Biol.* 5:408–419.
50. Dulbecco, R., and M. Vogt. 1963. "Evidence for a Ring Struc-

ture of Polyoma Virus DNA." *Proc. Nat. Acad. Sci.* 50:236–243.

51. Weil, R., and J. Vinograd. 1963. "The Cyclic Helix and Cyclic Coil Forms of Polyoma Viral DNA." *Proc. Nat. Acad. Sci.* 50:730–738.

52. Vinograd, J., J. Lebowitz, R. Radloff, R. Watson, and P. Laipis. 1965. "The Twisted Circular Form of Polyoma Viral DNA." *Proc. Nat. Acad. Sci.* 53:1104–1111.

53. Vinograd, J., J. Lebowitz, and R. Watson. 1968. "Early and Late Helix-Coil Transitions in Closed Circular DNA: The Number of Superhelical Turns in Polyoma DNA." *J. Mol. Biol.* 33:173–197.

54. Keller, W. 1975. "Determination of the Number of Superhelical Turns in SV40 DNA by Gel Electrophoresis." *Proc. Nat. Acad. Sci.* 72:4876–4880.

55. Crick, F. H. C. 1976. "Linking Numbers and Nucleosomes." *Proc. Nat. Acad. Sci.* 73:2639–2643.

56. Bauer, W. R., F. H. C. Crick, and J. H. White. 1980. "Supercoiled DNA." *Sci. Amer.* 243:118–133.

57. Lilley, D. M. J. 1980. "The Inverted Repeat as a Recognizable Structural Feature in Supercoiled DNA Molecules." *Proc. Nat. Acad. Sci.* 77:6468–6472.

58. Courey, A. J., and J. C. Wang. 1983. "Cruciform Formation in a Negatively Supercoiled DNA May Be Kinetically Forbidden Under Physiological Conditions." *Cell* 33:817–829.

59. Wells, R. D., R. Brennan, K. A. Chapman, T. C. Goodman, P. A. Hart, W. Hillen, D. R. Kellogg, M. W. Kilpatrick, R. D. Klein, J. Klysik, P. F. Lambert, J. E. Larson, J. J. Miglietta, S. K. Neuendorf, T. R. O'Connor, C. K. Singleton, S. M. Stirdivant, C. M. Veneziale, R. M. Wartell, and W. Zacharias. 1982. "Left Handed DNA Helices, Supercoiling, and the B-Z Junction." *Cold Spring Harbor Symp. Quant. Biol.* 47:77–84.

60. McGhee, J. D., and G. Felsenfeld. 1980. "Nucleosome Structure." *Ann. Rev. Biochem.* 49:1115–1156.

61. Kornberg, R. D., and A. Klug. 1981. "The Nucleosome." *Sci. Amer.* 244:52–79.

62. Burlingame, R. W., W. E. Love, B.-C. Wang, R. Hamlin, N.-H. Xuong, and E. N. Moudrianakis. 1985. "Crystallographic Structure of the Octameric Histone Core of the Nucleosome at a Resolution of 3.3 Å." *Science* 228:546–553.

63. Morse, R. H. and C. C. Cantor. 1985. "Nucleosome Core Particles Suppress the Thermal Untwisting of Core DNA and Adjacent Linker DNA." *Proc. Nat. Acad. Sci.* 82:4653–4657.

64. Klug, A., L. C. Lutter, and D. Rhodes. 1982. "Helical Periodicity of DNA On and Off the Nucleosome as Probed by Nucleases." *Cold Spring Harbor Symp. Quant. Biol.* 47:285–292.

65. Richmond, T. J., J. T. Finch, B. Rushton, D. Rhodes, and A. Klug. 1984. "Structure of the Nucleosome Core Particle at 7 Å Resolution." *Nature* 311:532–537.

66. Tanaka, I., K. Appelt, J. Dijk, S. W. White, and K. S. Wilson. 1984. "3-Å Resolution Structure of a Protein with Histone-Like Properties in Procaryotes." *Nature* 310:376–381.

67. Rouviere-Yaniv, J., and M. Yaniv. 1979. " *E. coli* DNA Binding Protein HU Forms Nucleosome-Like Structures with Circular Double Stranded DNA." *Cell* 17:265–274.

68. Brown, P. O., and N. R. Cozzarelli. 1979. "A Sign Inversion Mechanism for Enzymatic Supercoiling of DNA." *Science* 206:1081–1084.

69. Gellert, M. 1981. "DNA Topoisomerases." *Ann. Rev. Biochem.* 50:879–910.

70. Dean, F., M. A. Krasnow, R. Otter, M. M. Matzuk, S. J. Spengler, and N. R. Cozzarelli. 1982. "*Escherichia coli* Type-1 Topoisomerases: Identification, Mechanism and Role in Recombination." *Cold Spring Harbor Symp. Quant. Biol.* 47:769–777.

71. DiNardo, S., K. A. Voelkel, R. Sternglanz, A. E. Reynolds, and A. Wright. 1982. "*Escherichia coli* DNA Topoisomerase I Mutants Have Compensatory Mutations at or Near DNA Gyrase Genes." *Cold Spring Harbor Symp. Quant. Biol.* 47:779–784.

72. Wang, J. C. 1982. "DNA Topoisomerases." *Sci. Am.* 247:94–109.

73. Liv, L. 1983. "DNA Topoisomerases—Enzymes That Catalyze the Breaking and Rejoining of DNA." *CRC Crit. Rev. Biochem.* 15:1–9.

74. Drilica, K. 1984. "Biology of Bacterial Deoxyribonucleic Acid Topoisomerases." *Microbiol Revs.* 48:273–302.

75. Kirchhausen, T., J. C. Wang, and S. C. Harrison. 1985. "DNA Gyrase and its Complexes with DNA: Direct Observation by Electron Microscopy." *Cell* 41:933–943.

76. Worcel, A., and E. Burgi. 1972. "On the Structure of the Folded Chromosome of *Escherichia coli*." *J. Mol. Biol.* 71:127–147.

77. Pettijohn, D. E., and R. Hecht. 1973. "RNA Molecules Bound to the Folded Bacterial Genome Stabilize DNA Folds and Segregate Domains of Supercoiling." *Cold Spring Harbor Symp. Quant. Biol.* 38:31–41.

78. Luchnik, A. N., V. V. Bakayev, and V. M. Glaser. 1982. "DNA Supercoiling: Changes During Cellular Differentiation and Activation of Chromatin Transcription." *Cold Spring Harbor Symp. Quant. Biol.* 47:793–801.

79. Earnshaw, W. C., B. Halligan, C. A. Cooke, M. M. S. Heck, and L. F. Liu. 1985. "Topoisomerase II Is a Structural Component of Mitotic Chromosome Scaffolds." *J. Cell Biol.* 100:1706–1715.

80. Earnshaw, W. C., and M. M. S. Heck. 1985. "Localization of Topoisomerase II in Mitotic Chromosomes." *J. Cell Biol.* 100:1716–1725.

81. Danna, K., and D. Nathans. 1971. "Sequence Specific Cleavage of Simian Virus 40 DNA by Restriction Endonucleases of *Hemophilus influenzae*." *Proc. Nat. Acad. Sci.* 68:2913–2917.

82. Smith, H. 1979. "Nucleotide Sequence Specificity of Restriction Endonucleases." *Science* 205:455–462.

83. Frederick, C. A., J. Grable, M. Melia, C. Samudzi, L. Jen-Jacobsen, B.-C. Wang, P. Greene, H. W. Boyer, and J. M. Rosenberg. 1984. "Kinked DNA in Crystalline Complex with EcoR1 Endonuclease." *Nature* 309:327–330.

84. Arber, W., and S. Linn. 1969. "DNA Modification and Restriction." *Ann. Rev. Biochem.* 38:467–500.

85. Smith, J. D., W. Arber, and U. Kuhnlein. 1972. "Host Specificity of DNA Produced by E. coli. XIV. The Role of Nucleotide Methylation in *in Vivo* B-Specific Modification." *J. Mol. Biol.* 63:1–8.

86. Sharp, P. A., B. Sugden, and J. Sambrook. 1973. "Detection of Two Restriction Endonuclease Activities in *Hemophilus Parainfluenzae* Using Analytical Agarose-Ethidium Bromide Electrophoresis." *Biochemistry* 12:3055–3062.

87. Aaij, C., and P. Borst. 1972. "The Gel Electrophoresis of DNA." *Biochem. Biophys. Acta* 269:192–200.

88. McClelland, M., L. G. Kessler, and M. Bittner. 1984. "Site-Specific Cleavage of DNA at 8- and 10-Base-Pair Sequences." *Proc. Nat. Acad. Sci.* 81:983–987.

89. Schwartz, D. C., and C. R. Cantor. 1984. "Separation of Yeast Chromosome-Sized DNAs by Pulsed Field Gradient Gel Electrophoresis." *Cell* 37:67–75.

90. Mertz, J. E., and R. W. Davis. 1972. "Cleavage of DNA by R1 Restriction Endonuclease Generates Cohesive Ends." *Proc. Nat. Acad. Sci.* 69:3370–3374.

91. Cohen, S., A. Chang, H. Boyer, and R. Helling. 1973. "Construction of Biologically Functional Bacterial Plasmids *in Vitro*." *Proc. Nat. Acad. Sci.* 70:3240–3244.

92. Maniatis, T., R. C. Hardison, E. Lacy, J. Lauer, C. O'Connell, D. Quon, G. K. Sim, and A. Efstratiadis. 1978. "The Isolation of Structural Genes from Libraries of Eucaryotic DNA." *Cell* 15:687–701.

93. Maxam, A. M., and W. Gilbert. 1977. "A New Method of Sequencing DNA." *Proc. Nat. Acad. Sci.* 74:560–564.

94. Sanger, F., G. M. Air, B. G. Barrell, N. L. Brown, A. R. Coulson, J. C. Fiddes, C. A. Hutchison III, P. M. Slocombe, and M. Smith. 1977. "Nucleotide Sequence of Bacteriophage ϕX174." *Nature* 265:687–695.

95. Sanger, F., S. Nicklen, and A. R. Coulson. 1977. "DNA Sequencing with Chain Terminating Inhibitors." *Proc. Nat. Acad. Sci.* 74:5463–5467.

96. Sutcliffe, G. 1979. "Complete Nucleotide Sequence of the *E. coli* Plasmid pBR322." *Cold Spring Harbor Symp. Quant. Biol.* 43:77–90.

97. Thomas, C. A., Jr., R. E. Pyeritz, D. A. Wilson, B. M. Dancis, C. S. Lee, M. D. Bick, H. L. Haung, and B. H. Zimm. 1973. "Cyclodromes and Palindromes in Chromosomes." *Cold Spring Harbor Symp. Quant. Biol.* 38:353–370.

98. Collins, J. 1981. "Instability of Palindromic DNA in *E. coli*." *Cold Spring Harbor Symp. Quant. Biol.* 45:409–416.

99. Soll, D., and R. J. Roberts, eds. 1984. *The Applications of Computers to Research Nucleic Acids II. Parts 1 and 2.* Oxford, Eng.: IRL Press.

CHAPTER 10

The Replication
of DNA

When the double helix was first discovered, the feature that most excited the world of genetics was the complementary relationship between the sequences of bases on its intertwined polynucleotide chains. It seemed unimaginable that such a complementary structure would not be utilized as the basis for DNA replication. In fact, it was the self-complementary nature of DNA that finally led most geneticists and biologists to accept Avery's conclusion that DNA, not some form of protein, was the carrier of genetic information.

Earlier in our discussion of how templates act, we emphasized that two identical surfaces will not attract each other and that it is much easier to visualize the attraction of groups with opposite shape or charge. Thus, without any detailed structural knowledge, we might guess that a molecule as complicated as the gene could not be directly copied. Instead, replication would involve the formation of a molecule complementary in shape, and this, in turn, would serve as a template to make a replica of the original molecule. So some geneticists, in the days before detailed knowledge of protein or nucleic acid structure, wondered whether DNA served as a template for a specific protein that in turn served as a template for a corresponding DNA molecule.

But as soon as the self-complementary nature of DNA became known, the idea that protein templates might play a role in DNA replication was discarded, for it was immensely simpler to postulate that each of the two strands of every parental DNA molecule served as a template for the formation of a daughter strand complement. Although from the start this hypothesis seemed too good not to be true, experimental support nonetheless had to be generated. Happily, within five years of the discovery of the double helix, there emerged both decisive evidence for the separation of the complementary strands during DNA replication and firm enzymological proof that DNA alone can function as the template for the synthesis of new DNA strands.

With these results, the problem of how genes replicate was in one sense solved. But in another and equally valid sense, the study of DNA replication had only begun. As this chapter will show, the replication of even the simplest DNA molecule is a complicated, multistep process, involving many more enzymes than was initially anticipated following the discovery of the first DNA polymerizing enzyme. Fortunately, it now seems likely that all the key enzymatic steps involved in the replication of both linear and circular DNA have been described and that the general picture of DNA replication presented here will not be seriously modified.

Strand Separation Requires Untwisting of the Double Helix[1,2]

In even the smallest DNA molecules, the two chains of the double helix are twisted around each other many times and must be untwisted an equal number of times for DNA replication to occur. This feature initially troubled many scientists who were otherwise enamored of the implications of DNA's double-helical structure. But untwisting linear DNA molecules need not lead to insuperable problems if the unreplicated sections of DNA can rotate around their axes, much like the rotating cables of speedometers (Figure 10-1). Since the DNA molecule is very thin, it should encounter virtually no frictional resistance, and thus the energy input needed for rotation should be negligible.

No conceptual difficulty arises from the need to break the hydrogen bonds joining the base pairs in the parental templates. Although these bonds are highly specific, they are also relatively weak, and no enzymes are required to make or break them. Hydrogen bonds thus provide highly suitable means of specifically holding together a template and its complementary replica.

Base Pairing Should Permit Very Accurate Replication[3]

In Chapter 3, we stated that the chemistry of the various amino acid side groups argued against their employment as templates. Just the opposite, however, is true of the purine and pyrimidine bases, because each of them can form several hydrogen bonds. These bonds are ideal template forces, for they are highly specific, in contrast to van der Waals interactions, whose attractive forces are both weaker and virtually independent of particular chemical groupings.

The average energy of a hydrogen bond is about 3 kcal/mol, or eight times the average energy of thermal motion of molecules at room temperature. This value gives us an estimate of how often a given group (e.g., an amino group) will be hydrogen-bonded. Under ordinary cellular conditions, the ratio of bonding to nonbonding is about $10^4:1$. This means that two molecules held together by several hydrogen bonds are almost never found in the nonbonded form. Because of the great specificity of the hydrogen bonds, the frequency with which adenine will attach to cytosine during DNA replication might be as low as 10^{-8} times the frequency of its bonding to thymine ($10^{-4} \times 10^{-4}$, since two hydrogen bonds are involved). The replication of the GC pair should be even more accurate, since it is held together by three hydrogen bonds rather than two. However, the actual selection process never reaches this high level of accuracy, because occasionally (perhaps about once in 10^4 times) the base selected will be in the "wrong" (imino or enol) tautomeric form, which must lead to mispairing. Now we realize that only through proofreading of the initially selected bases does the DNA replication process achieve its measured level of accuracy (error levels between 10^{-8} and 10^{-12}). (Mispairing and proofreading are discussed in detail later in the chapter.)

These arguments are not diminished by the fact that the bases can also form almost equally strong hydrogen bonds with the hydrogen and oxygen atoms of water. It does not matter if a potential thymine

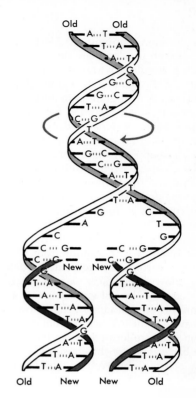

Figure 10-1
Untwisting of linear DNA strands during replication. The strands untwist by rotating about the axis of the unreplicated DNA double helix.

Figure 10-2

Denaturation mapping. (a) Electron micrograph of a partially denatured T7 DNA molecule. (b) Mapping of the location of the denatured regions shown in (a). (c) A histogram summarizing the location of single-stranded regions in a large number of denatured molecules. The arrow in (a) points to a single-stranded loop-out that is present in the DNA molecule even before partial denaturation. The loop-out occurs because this was not just an ordinary T7 DNA molecule but a "heteroduplex" molecule (see Chapter 9) derived by re-annealing separated strands from wild-type T7 DNA with strands from T7 DNA molecules that contained a deletion known by genetic mapping to be located near the left end. It is this additional physical landmark that allows us to say that the AT-rich, partial denaturation sites are located in the left half of the bacteriophage chromosome. (Courtesy of Drs. John Wolfson and David Dressler.)

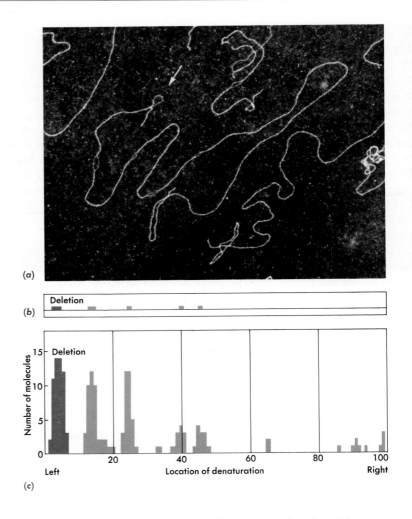

site is temporarily filled with several water molecules. These water molecules cannot be chemically joined to the growing polynucleotide chain, and they will soon diffuse away and be replaced by the suitable hydrogen-bonding nucleotide.

It is not yet possible to give a similar *quantitative* argument for why two purines or two pyrimidines never accidentally bond to each other. Both a purine-purine or a pyrimidine-pyrimidine pair will distort the sugar-phosphate backbone and thus may be energetically unfavorable. However, the large number of atoms involved makes the precise calculation of energy differences impossible. Part of the difficulty arises from the fact that the precise locations of the atoms in DNA are not known to an accuracy of 0.1 Å. Furthermore, it is unclear what effect such distortions of the double helix would have on the action of the enzyme that catalyzes the formation of the internucleotide covalent bonds.

Partial Denaturation Maps[4, 5]

Brief exposure of linear DNA molecules to denaturing conditions (e.g., high temperature, high or low pH, or hydrogen-bond breaking reagents) preferentially unwinds those sections rich in AT base pairs, leaving intact regions held together largely by the much stronger GC base pairs. When such partially denatured molecules are treated with

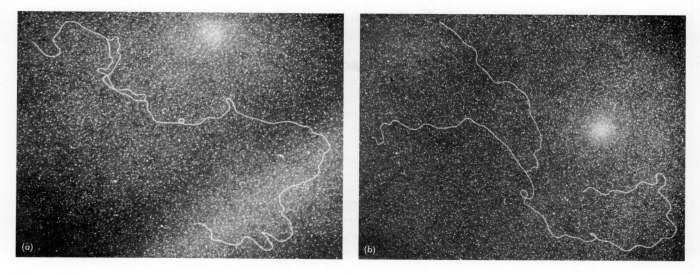

Figure 10-3
T7 DNA replication. (a) Electron micrograph of a T7 DNA molecule that
has just begun replication and has a small replicating bubble centered
about a point 17 percent from the left end of the genetic map. (b) A more
advanced replicating molecule whose Y shape is the result of the left replicating fork having run off its end. (Courtesy of J. Wolfson and D. Dressler.)

formaldehyde, the single-stranded regions become stabilized, since
formaldehyde combines with the free amino groups of adenine and
prevents formation of the AT base pairs. Visualization of such treated
molecules in the electron microscope reveals the melted regions as
single-stranded "bubbles" that occur reproducibly at fixed positions
along a given DNA molecule (Figure 10-2). Such "denaturation
maps" now provide an easy way to distinguish one end of a linear
DNA molecule from the other and so precisely define the locations
where DNA replication is initiated.

Visualization of the
Replication of a Linear DNA Molecule[6, 7]

It is easy to isolate DNA in the process of replication and visualize it
by electron microscopy. When this was first done for the linear T7
viral DNA, it was expected that replication would be seen to commence at either one or both ends, since it is easier to imagine strand
separation occurring at terminal, rather than internal, sections of a
double helix. Surprisingly, however, replication was found to commence internally, always at a point (called the **origin of replication**)
some 17 percent along the molecule from the left end (Figure 10-3).
Once DNA synthesis starts, it moves in both directions, generating
eye (-O-)-shaped intermediates that convert to Y-shaped molecules
when the left-hand replicating fork reaches the end of the DNA.
These structures clearly rule out a two-step replication process in
which the parental strands first completely unwind and only later act
as templates to form new double helices. In all such replication intermediates, only traces of unwound single-stranded regions can be
seen. Thus, daughter polynucleotide strands are synthesized almost
as soon as the parental strands separate.

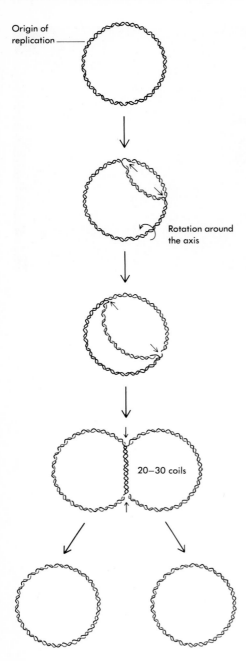

Origin of replication

Rotation around the axis

20–30 coils

Figure 10-4
Simplified version of the bidirectional replication of circular DNA molecules. Only a fraction of the replication intermediates observed with electron microscopy resemble these structures, since the molecules with intact parent strands take up supercoiled configurations.

θ-Shaped Intermediates Form During the Replication of Circular DNA[8, 9]

Duplication of circular DNA does not involve the temporary creation of linear DNA. Visualization of the replicating intermediates of many ring chromosomes reveals that the parental strands maintain a circular form throughout replication. They always possess a θ-like shape that comes into existence by the initiation of a replicating bubble at the origin of replication (Figure 10-4). Circular λ DNA, for example, has its replication origin within gene *O*, and synthesis proceeds outwardly in both the clockwise and counterclockwise directions.

From the moment of its discovery, circular DNA posed the question of how two covalently closed intertwined strands could unravel during replication. As long as each chain remains intact, even minor untwisting of one section of a circular double helix forces the creation of positive supercoils in the opposing direction. The semiconservative replication of such DNA therefore demands temporary cuts in one or both of its polynucleotide backbones. The problem thus becomes how and when the breaks are made. About this there could only be speculation until the discovery of the topoisomerases, DNA nicking/closing enzymes that can either further supercoil or relax DNA (see Chapter 9). The transient single-stranded cuts made by topoisomerase I, for example, can create a molecular swivel around which supercoiled parental DNA strands can twist.

Now it seems clear that both parental strands remain intact throughout most of the replication cycle (Figure 10-5). The unreplicated portion of the θ-shaped intermediate becomes periodically supercoiled in either of two possible ways: Negative superhelices (untwists) may be made by topoisomerase II (gyrase) in anticipation of subsequent chain elongation, and positive superhelices may be made by elongation of replicating bubbles in the absence of chain cuts. The latter, positive supertwisting is clearly limited in scope and should halt replication bubble growth owing to the steric impossibility of creating further positive supercoils. Such halted bubbles could resume growth if a topoisomerase type I molecule made the single-stranded cuts that convert the highly supercoiled unreplicated section into the relaxed configuration. Alternatively, the essential cutting agent needed for further movement may be topoisomerase II (gyrase), acting to untwist the positively supertwisted DNA.

Overall Chain Growth Occurs in Both 5′→3′ and 3′→5′ Directions

The opposing chain directions (5′→3′ and 3′→5′) of the two strands of a double helix mean that the two daughter strands being synthesized at each replicating fork must also run in opposite directions. Therefore, the overall direction of chain growth must be 5′→3′ for one daughter strand and 3′→5′ for the other daughter strand (Figure 10-6). However, all known forms of DNA polymerase, the only type of enzyme that can add nucleotide precursors to DNA, extend chains only in the 5′→3′ direction, since the chemical reaction catalyzed by these enzymes allows a nucleotide triphosphate to react only with the free 3′-OH end of a growing polynucleotide strand (Figure 10-7). The question thus arose whether other replicating enzymes had been missed, specifically, those that would add nucleotides onto free 5′

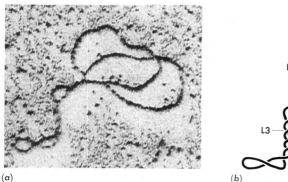

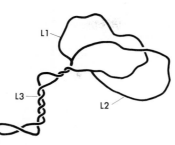

(a)   (b)

Figure 10-5
Twisted replicating molecules of circular SV40 DNA: (a) Electron micrograph of an early phase of replication. (b) Diagrammatic representation showing the unreplicated, supercoiled parent strands. L1 and L2 indicate portions already replicated, while L3 indicates the unreplicated portion, which takes up a supercoiled configuration. Note that through the replication cycle, the DNA remains intact. [From N. P. Salzman et al., *Cold Spring Harbor Symp. Quant. Biol.* 48 (1973):257.]

ends. Despite a very extensive search, no enzyme with such specificity has ever been found.

Small Fragments Are Precursors of Many Long Chains: The Leading Versus the Lagging Strand[10, 11]

Resolution of the apparently paradoxical 3'→5' direction of synthesis came with the discovery that deoxyribonucleotide substrates are not always added directly to the very long daughter polynucleotide. Instead, they are sometimes found as parts of much shorter DNA chains (100 to 1000 bases long) that later link up to the main daughter strands through the action of the polydeoxyribonucleotide-joining enzyme, DNA ligase. This finding opened up the possibility that a chain whose final direction of growth is 3'→5' might, in fact, be formed by the joining of smaller chains (often called **Okazaki fragments,** after their discoverer), each of which grows in the conventional 5'→3' manner (Figure 10-8). In this way, the same set of enzymes could be responsible for making all progeny strands. Proof that this indeed is the case came from direct studies of the ends of growing Okazaki fragments to which precursor nucleotides had recently been added. All such experiments show growth exclusively at the 3' end.

Now it is very clear that the Okazaki fragments are precursors only to half of the daughter chains—those that grow overall in the 3'→5' direction. Synthesis of the strand that grows overall in the 5'→3' direction for the most part precedes that of the strand appearing to grow 3'→5'. This difference in rates of growth between the leading (5'→3') and lagging (3'→5') strands produces a small section of single-stranded parental DNA at the replication fork (Figure 10-9). It becomes converted to double-stranded DNA upon the initiation of an Okazaki fragment whose synthesis continues until it reaches the terminus of the previously synthesized Okazaki fragment, to which it then becomes joined. Okazaki fragments only commence synthesis at

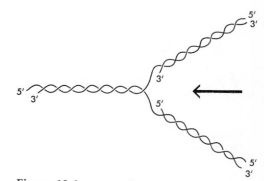

Figure 10-6
Overall direction of progeny chain growth at a replicating fork.

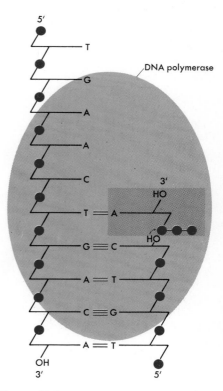

Figure 10-7
Reaction of a nucleoside triphosphate with the 3'-OH end of a growing DNA chain.

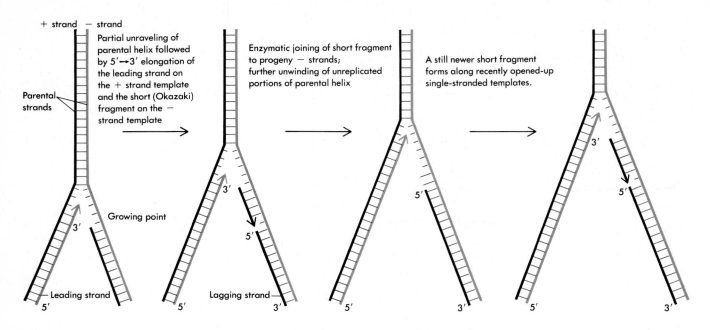

Figure 10-8
Hypothesis that replication of lagging progeny strands involves the joining together of short fragments, each synthesized in the 5'→3' direction.

precise points along single-stranded DNA templates, and it is the distance between these signals that determine the average length of Okazaki fragments formed during replication of any particular DNA.

Test-Tube Replication of Complete Double Helices[12–16]

While biochemists were initially very pleased to find any test-tube incorporation of nucleotides into DNA-like molecules, now their objectives are much more ambitious. They not only want systems where DNA synthesis occurs rapidly, but also aim for conditions that carefully reflect the process by which in vivo DNA synthesis occurs. One goal was the replication of complete DNA molecules rather than the mere addition of nucleotides to growing chains of an undefined structure. Now, after much intermittent progress, conditions have been found that clearly mimic those present in growing cells. DNA synthesis now occurs rapidly in cell-free extracts, and the products are often indistinguishable from intracellularly produced DNA molecules. Extracts made from *E. coli* cells, for example, not only can make linear T7-length molecules, but also can carry out the complete semiconservative replication of circular plasmid DNA. In the latter case, replicating bubbles start at the same sites in vitro that are used as origins in vivo, and the θ-shaped intermediates grow until replication is complete and the resulting circular double helices separate (see Figure 10-4). Most importantly, these daughter circles can serve as templates for still additional rounds of semiconservative replication. *E. coli* extracts are now being fractionated to reveal the nature and mode of action of each enzyme involved in the replication process. With these observations, our all-too-speculative ideas about certain key steps in DNA replication and its control may soon be replaced by a rigorous description of the proteins and reactions involved.

Precise conditions are also being worked out for the in vitro conversion of the single-stranded DNA of viruses like φX174 and M13 into their double-helical replicative intermediates. Surprisingly, the enzymology involved in converting single-stranded φX174 DNA into its duplex replicative form shows major differences from that involved in making an M13 DNA duplex. In particular, several bacterial DNA replication genes are required for φX174 replication but not for M13 replication. So we must face the complication that apparently similar DNA molecules may employ different modes of replication.

There Are Three Types of *E. coli* DNA Polymerase[17, 18, 19]

The exact ways in which enzymes are used to make complete DNA chains are much more complicated than initially suspected when the original DNA polymerase (now called DNA polymerase I) was first investigated. At least three different DNA polymerases have been identified within *E. coli*, and two appear to have distinct roles in DNA replication (Table 10-1). While DNA polymerase I was long thought to be the major enzyme that joins together deoxyribonucleotides, we now realize that a more recently discovered enzyme, DNA polymerase III, is the main polymerizing agent. In contrast, DNA polymerase I is largely used for filling in gaps between the small Okazaki fragment precursors that are formed during lagging-strand synthesis (see Figure 10-8). No specific task in DNA replication has been assigned to DNA polymerase II, and since mutant cells that have very few (if any) molecules of DNA polymerase II survive, this enzyme may be dispensable and its role tricky to define.

Correction of Mistakes by 3′→5′ Exonuclease Action[20, 21]

Extensive purification of all three types of DNA polymerase reveals that each enzyme has a 3′→5′ exonuclease activity as well as a polymerizing ability. Possession of this exonuclease specificity means that every DNA polymerase molecule has the ability to cut back nascent polynucleotides as well as to extend them. In the presence of even moderate levels of the deoxyribonucleotide triphosphate precursors, however, synthesis is overwhelmingly favored over degradation. The question thus arose why this specific degradative potential should be

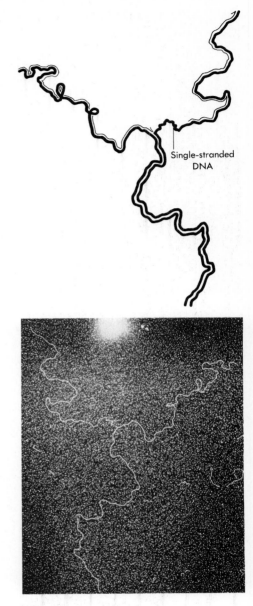

Single-stranded DNA

Figure 10-9
Different rates of movement of the leading and lagging strands lead to a small section of single-stranded parental DNA at the replication fork. (Courtesy of David Dressler and John Wolfson, Harvard Medical School.)

Table 10-1 Properties of Polymerases I, II, and III of *E. coli*

	Pol I	Pol II	Pol III
Functions			
Polymerization: 5′→3′	+	+	+
Exonuclease: 3′→5′	+	+	+
Exonuclease: 5′→3′	+	−	+
Polymer synthesis de novo	+	−	−
General			
Size, kdal	109	120	140
Molecules per cell, estimated	400		10–20
Turnover number, estimated*	(1)	0.05	15
Structural genes known	*polA*	*polB*	*dnaE, N, Z*
Conditional lethal mutants	Yes	No	Yes

*Nucleotides polymerized at 37°/min/molecule of enzyme, relative to Pol I, which is near 600.
+ and − represent the presence and absence, respectively, of the property listed.
▨ = distinction from Pol I
▨ = distinction between Pol II and Pol III
SOURCE: After A. Kornberg, *DNA Replication* (San Francisco: Freeman, 1980), p. 169.

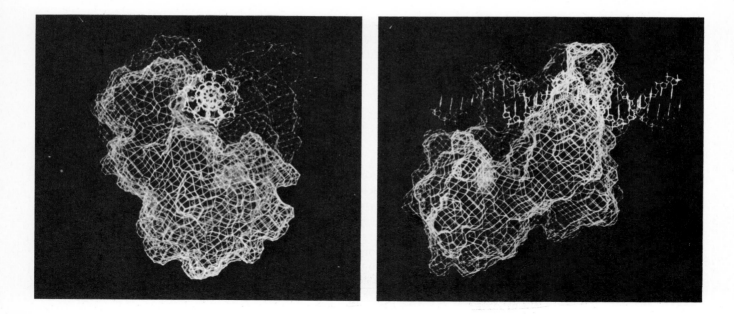

Figure 10-10
Binding of DNA to the Klenow fragment (a fragment of DNA polymerase I) as shown (left) from above and (right) from the side. [Courtesy of T. A. Steitz, D. L. Ollis, and P. Brick from *Nature* 313 (1985):765.]

coupled (within the same polypeptide chain) to polymerizing action. At first, possession of a specific degradative action made no sense. But the reason for its presence became obvious when it was shown that the exonuclease acted preferentially on incorrectly paired bases. If, by chance, the wrong base becomes added onto the 3' end of a growing chain, it has a very high probability of being removed before the next base is added. The 3'→5' exonuclease activity thus provides a proofreading ability that gives DNA replication much higher fidelity than it would have if selection were the result of only the initial base-pairing step.

A constant need for proofreading may be the reason that no enzyme has evolved that adds deoxyribonucleotides onto the 5' end of a DNA chain. If growth were to occur in this direction, the growing end would always be terminated by a triphosphate group, whose high-energy bonds are necessary for the addition of the next nucleotide. Removal of the most recently added deoxyribonucleotide from such a triphosphate-terminated molecule during proofreading would thus create a 5'-monophosphate-ended chain that would be energetically incapable of being extended by DNA polymerase. Some form of ligaselike reaction could conceivably do the job, but such a step would probably slow down DNA replication markedly. Thus, a polymerase that adds nucleotides to the 5' end of a growing DNA chain would never make sense.

DNA Polymerases Act Processively[22]

Once a DNA polymerase has begun adding nucleotides to growing DNA chains, it remains attached to the chain, continuing to add new nucleotides until a signal is reached that tells it to detach. How such processive action occurs remained a complete mystery until the elucidation of the first three-dimensional structure of a DNA polymerase. Such structural work requires large amounts of highly purified polymerases, a condition now possible through the advent of recombi-

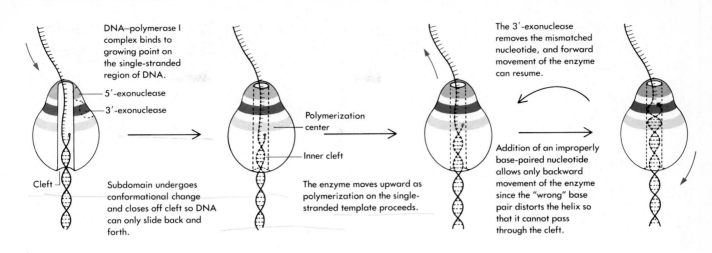

DNA–polymerase I complex binds to growing point on the single-stranded region of DNA.

5'-exonuclease

3'-exonuclease

Cleft

Subdomain undergoes conformational change and closes off cleft so DNA can only slide back and forth.

Polymerization center

Inner cleft

The enzyme moves upward as polymerization on the single-stranded template proceeds.

The 3'-exonuclease removes the mismatched nucleotide, and forward movement of the enzyme can resume.

Addition of an improperly base-paired nucleotide allows only backward movement of the enzyme since the "wrong" base pair distorts the helix so that it cannot pass through the cleft.

nant DNA procedures that attach desired genes to control regions that permit their high-level expression. The first polymerase so examined was a large carboxy-terminal fragment of DNA polymerase I called the **Klenow fragment,** which possesses polymerase activity as well as the 3'→5' exonuclease activity that edits out mismatched nucleotides. This fragment contains two distinct domains: The larger domain contains a deep, positively charged, 20 Å diameter cleft into which DNA binds (Figure 10-10), while the smaller domain may contain a nucleotide-binding region. A flexibly attached subdomain most likely has the ability to close off the cleft after DNA binds. When bound in such a manner, the DNA may only be able to slide either backward or forward through the cleft. Conceivably, DNA containing a misincorporated base that cannot properly base-pair has a distorted shape that will not allow it to slip easily through the semirigid 20 Å diameter cleft. When so blocked, such abnormally terminated DNA may only diffuse backward, coming into extended contact with the 3'→5' clipping (proofreading) exonuclease that will remove the misincorporated base (Figure 10-11).

Helicases Unwind the Double Helix in Advance of the Replication Fork[23-26]

The two strands of the double helix at the replicating fork do not spontaneously come apart frequently enough to permit the rapid rates of DNA polymerization observed in vivo. To facilitate strand separation, cells contain a class of specific proteins called **helicases,** which bind to ATP and to single-stranded sections of DNA. They use energy derived from the breakdown of their bound ATP to ADP to move progressively along and increase the rate of DNA strand separation. Two different helicases are often involved. One (helicase II or III) attaches to the template for the lagging strand and moves in the 5'→3' direction. The other helicase (Rep protein) binds to the strand directing synthesis of the leading strand and moves in the 3'→5' direction (Figure 10-12). Somehow, the conversion of the bound ATP to ADP must lead to conformational changes that propel the helicases unidirectionally along the DNA strands to which they have become attached.

Figure 10-11

Diagrammatic view of proofreading after incorporation of a mismatched nucleotide base into DNA.

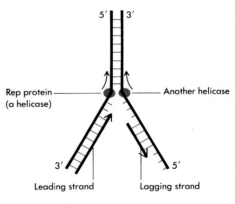

5' 3'

Rep protein (a helicase)

Another helicase

3'

5'

Leading strand

Lagging strand

Figure 10-12

Helicases unwind the DNA double helix in advance of a replicating fork. The rep protein (a helicase) moves on one strand in the 3'→5' direction, while another helicase (e.g., helicase II, or III) moves on the complementary strand in the 5'→3' direction. (From *1982 Supplement to DNA Replication* by Arthur Kornberg, W. H. Freeman and Company, copyright © 1982, with permission.)

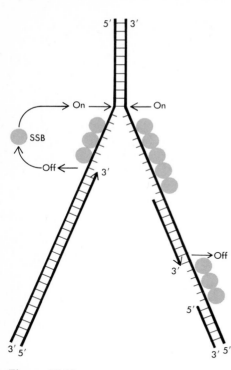

Figure 10-13
Hypothetical scheme for the action of single-stranded binding protein (SSB) at a replicating fork. The protein binds to single-stranded regions, facilitates replication, and is then recycled. (Courtesy of B. Alberts.)

Stabilization of Exposed Single-Stranded DNA by Single-Stranded DNA Binding Proteins[27]

The newly created single-stranded regions do not remain free for long after the helicases progress along a double helix. They quickly become covered by large amounts of specific single-stranded DNA binding proteins (designated **SSB**), which, unlike the helicases, do not bind ATP and have no known enzymatic function. The major SSB within *E. coli* is a polypeptide of 177 amino acids that normally exists as a tetramer that binds to adjacent groups of 32 nucleotides on a given DNA chain. The binding of one SSB tetramer promotes the binding of another to an adjacent section of a single-stranded DNA, a process called **cooperative binding** (Figure 10-13). Single-stranded DNA that is tightly covered by SSB has a rigid, semiextended form without bends or kinks. This rigid configuration is essential for the functioning of DNA as a template upon which complementary DNA chains can first be initiated and then elongated. Were the single-stranded DNA not held extended, its template capability would be repeatedly compromised by its tendency to fold back on itself to form imperfectly base-paired hairpin loops.

Initiation of Okazaki Fragments by RNA Primers: Primase and Primosomes[28–31]

As the reaction requirements for the several DNA polymerases became established, the unexpected fact emerged that each such enzyme could only add nucleotides onto preexisting polynucleotide chains. None of these polymerases could initiate new DNA chains, which raised the question of whether still undiscovered DNA polymerases might exist whose only role was to start new DNA chains. Now we suspect that such enzymes play only peripheral roles and that almost all DNA synthesis commences by the use of **RNA primers.** Starting points for DNA synthesis are thus recognized not by specific DNA polymerases but by enzymes that transcribe DNA to generate RNA chains that serve as primers. However, the classic RNA polymerases, the cellular enzymes normally used to make RNA off DNA templates, are involved in making only a small fraction of such primers. The primers for Okazaki fragments are made by a special class of enzymes called **primases,** which recognize specific DNA sequences along single-stranded DNA chains. The *E. coli* primase is a single polypeptide of 60 kilodaltons (kdal), present in 50 to 100 copies per cell. By itself, it is relatively inactive and only functions effectively when complexed with six or seven other polypeptides to form an aggregate called a **primosome.**

This multipolypeptide complex moves processively 5'→3' on the lagging-strand template to prime the repeated initiation of Okazaki fragments needed to put together the lagging strands. Exactly what functions of the primosome are carried out by its various components largely remains unknown. These components somehow must work cooperatively to ensure the movement of the primosome along the DNA, the displacement of SSBs, the recognition of appropriate start sites, and the polymerization of ribonucleotides into RNA.

Thus, the primosome moves along the DNA template strand in the direction opposite to the direction in which it must move when cata-

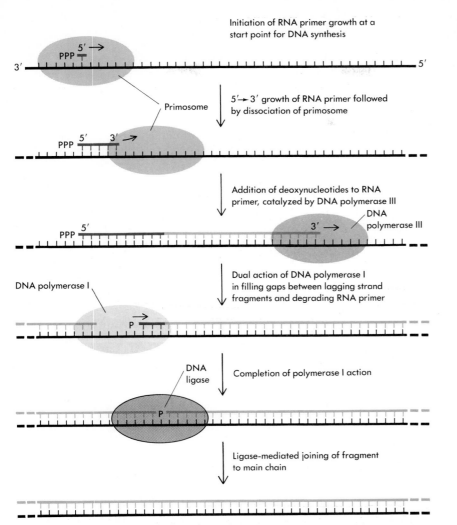

Initiation of RNA primer growth at a
start point for DNA synthesis

Primosome

$5' \rightarrow 3'$ growth of RNA primer followed
by dissociation of primosome

Addition of deoxynucleotides to RNA
primer, catalyzed by DNA polymerase III

DNA
polymerase III

DNA polymerase I

Dual action of DNA polymerase I
in filling gaps between lagging strand
fragments and degrading RNA primer

DNA
ligase

Completion of polymerase I action

Ligase-mediated joining of fragment
to main chain

Figure 10-14
Use of an RNA primer in the initiation
of lagging strand fragments.

lyzing synthesis of its RNA chain products. This may explain why
only very short RNA primers, usually containing only three to five
bases, are normally made (Table 10-2). In a given organism (such as a
virus), all the RNA primers have very similar sequences, reflecting
the fact that primases synthesize RNA at specific sequences along
their lagging-strand DNA templates.

Removal of the RNA primer occurs later through the action of other
enzymes. One enzyme that has this capacity is DNA polymerase I,
whose primary task may be to fill in the gaps between RNA-primed
DNA fragments (Figure 10-14). In doing so, it adds nucleotides to the
3' end of the fragment at its back and simultaneously cuts away the
RNA primer that lies ahead. The snipping off of ribonucleotides is
catalyzed by a completely different active site from that involved in its
$3' \rightarrow 5'$ proofreading capacity (see Figure 10-14). This $5' \rightarrow 3'$ exonuclease activity can remove deoxyribonucleotides as well as ribonucleo-
tides, and so its use is not restricted to primer removal. As we shall
see later in Chapter 12, DNA polymerase I is also employed to remove sections of DNA damaged by radiation attack.

Why RNA primers are used to start virtually all DNA chains probably has its origin in the need to prevent a lethal number of mistakes
near the starting points of DNA chains. The laying down of the first
few nucleotides along a DNA template may be inherently much less

Table 10-2 Primer RNAs for Discontinuous DNA Replication

Genome	Chain Length of RNA linked to DNA Fragments	Primary Products
T7 phage	1–5	pppApCpC/A(pN)$_{1-2}$* (N: A and C rich)
T4 phage	1–5	pppApC(pN)$_3$
E. coli	1–3	Not determined
φX174 phage	1–5†	‡
Sea urchin	1–~8	(p)ppA/G(pN)$_7$
Polyoma, SV40 (isolated nuclei)	~10	pppA/G(pN)$_{\sim9}$
Animal cells	~9	

*The complementary template sequence that signals the primer is: 3'-pCpTpGpG/Tp–; G/T signifies that either G or T is found in this position.
†Of primers linked to the complementary strand in the SS→RF conversion in vitro, 39% were 1 residue long, 18% 2 residues, and 83% between 1 and 5 residues; comparable values for the reaction in vivo were: 1 residue 48%, 2 residues 23%, and 1 to 5 residues 92%.
‡The 5'-terminal residue is A in about 80% of the primers in vitro.
SOURCE: Data from T. Okazaki in A. Kornberg, *1982 Supplement to DNA Replication* (San Francisco: Freeman, 1982).

accurate than when nucleotides add onto long sections of an already formed double helix. So there may be great selective advantages for an initiation mechanism that allows easy discrimination and subsequent elimination of the starting nucleotide sequences.

DNA Polymerase III Is an Aggregate of Seven Different Polypeptides[32–36]

The active form of DNA polymerase III (pol III) is an assembly, or **holoenzyme,** of seven different polypeptides, all of which must be present for correct biological functioning (Table 10-3). One of these polypeptides, α, carries the 5'→3' exonuclease activities as well as the polymerizing function, while the ε polypeptide possesses the 3'→5' exonuclease activities. One or more of the remaining polypeptides bind an ATP molecule needed for DNA Pol III to commence synthesis at the end of an RNA primer. The functions of the remaining polypeptides are not known. The number of DNA Pol III holoenzyme molecules in an *E. coli* cell is small, and given the slow rate at which they initiate synthesis, it appears that once one such molecule has commenced polymerization, it generally continues processively until the end of its template is reached.

Two copies of the DNA pol III holoenzyme may be present at the replicating fork, giving rise to the possibility that one such dimeric complex can simultaneously carry out the synthesis of both the leading and lagging progeny strands. This could happen if the lagging-strand template is looped backward (perhaps around the polymerase) so that it can associate with a polymerizing component in the same orientation as the leading-strand template (Figure 10-15). According to this scheme, the primer made by the primosome would be extended by DNA pol III as the lagging-strand template is drawn through the holoenzyme complex. Then, when the elongating chain reaches the previously synthesized small Okazaki fragments, the loop would be relaxed. By this time, leading-strand synthesis would make available a new stretch of unpaired lagging-strand template that

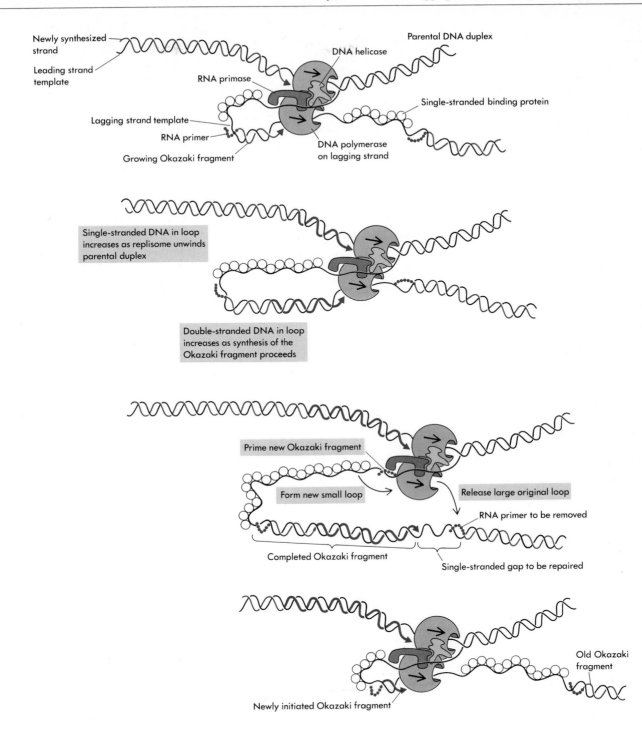

Figure 10-15
The simultaneous synthesis of both leading and lagging progeny strands by a dimeric complex of the DNA pol III holoenzyme in association with the primosome and helicases (the collection of which is known as a replisome.) [After Keith Roberts.]

could loop backward and participate in the formation of the next small lagging-strand Okazaki fragment. By such a mechanism, leading-strand synthesis would never rush too far ahead of lagging-strand synthesis. Moreover, the primosome would be automatically translocated along the replicating DNA with the same speed as the DNA polymerizing components of DNA pol III.

A complex of DNA replication proteins (see Table 10-3)—the dimeric polymerase, a primosome, and a helicase—is of ribosome size. The name **replisome** has been proposed for this collective entity, which carries out the above-mentioned replication process (see Figure 10-15).

Table 10-3   Replication Proteins of *E. coli*

Protein	Native Mass, kdal	Subunits	Function
SSB	74	4	Single-strand binding
Protein i	66	3 ⎫	
Protein n	28	2 ⎪	Primosome assembly
Protein n′	76	1 ⎬ c	and function
Protein n″	17	1 ⎪	
DnaC	29	1 ⎪	
DnaB	300	6 ⎭	
Primase	60	1 ⎱	Primer synthesis
Pol III Holoenzyme	(760)	(2)	
α	140	1 ⎫	
ε	25	1 ⎪	Processive chain
θ	10	1 ⎪	elongation
β	37	1 ⎬ × 2	
γ	52	1 ⎪	
δ	32	1 ⎪	
τ	83	1 ⎭	
Pol I	102	1	Gap filling, primer excision
Ligase	74	1	Ligation
Topoisomerase II (gyrase)	400	4	Supercoiling
GyrA	210	2	
GyrB	190	2	
Rep	65	1	Helicase
Helicase II	75	1	Helicase
DnaA	48		Origin of replication
Topoisomerase I	100	4	Relaxing negative supercoils

SOURCE: After A. Kornberg, *1982 Supplement to DNA Replication* (San Francisco: Freeman, 1982), Table S11-2.

Mutations That Block DNA Synthesis[37–40]

Increasingly important in finding out how enzymes like DNA pol III work is the discovery of mutations that block their functioning (Table 10-4). A large number of genes that can contain mutations that block DNA synthesis have already been located on the *E. coli* chromosome (Figure 10-16). Some, such as *dnaA*, control the formation of new replicating forks but have no effect on the movement of preexisting forks. Other genes participate in the movement of preexisting forks either by coding for components of the polymerization process (*dnaE* codes for the α polypeptide of DNA pol III) or by coding for components of the primosome (*dnaG* codes for primase). Mutations have also been found in the gene *polA*, which codes for DNA polymerase I, and in *lig*, the gene coding for DNA ligase. These mutations do not stop most DNA synthesis, but lead to the accumulation of unusually large numbers of unsealed Okazaki fragments. By now, a number of the genes involved in DNA synthesis have been cloned, allowing their polypeptide products to be expressed in abnormally large amounts.

Table 10-4   Replication Genes of *E. coli*[a]

Gene	Map Location, minutes	In Vivo Phenotype of Mutant	Protein and in Vitro Function
dnaA	82	Slow stop; defective origin initiation	Unknown
dnaB	91	Quick stop	DnaB protein; prepriming
dnaC	99	Slow or quick stop (depending on the mutation)	DnaC protein complexes with DnaB protein
dnaE (polC)	4	Quick stop	Subunit α of Pol III holoenzyme
dnaG	66	Quick stop, defective initiation of fragments	Primase
dnaI	39	Slow stop	Unknown
dnaJ	0.5	Slow stop	Unknown
dnaK	0.5	Slow stop	Unknown
dnaL	28	Quick stop	Unknown
dnaP	84	Phenethyl alcohol resistance; slow stop	Unknown
dnaT	95–99	Regulates termination	Unknown
dnaZ	10	Quick stop	Subunit γ of Pol III holoenzyme; chain growth
cou	82	Coumermycin (and novobiocin) resistance	DNA topoisomerase II subunit β; nicking/closing
lig	51	Accumulation of replication fragments	DNA ligase; covalently seals DNA nicks
nalA	48	Nalidixate (and oxolinate) resistance	DNA topoisomerase II subunit α; ATPase
polA	85	Defective in DNA repair	DNA polymerase I; gap filling, RNA excision
rep	83	Slowed fork movement	ATP-using helicase
ssb	91	Quick stop	SS DNA binding protein

SOURCE: After A. Kornberg, *DNA Replication* (San Francisco: Freeman, 1980).

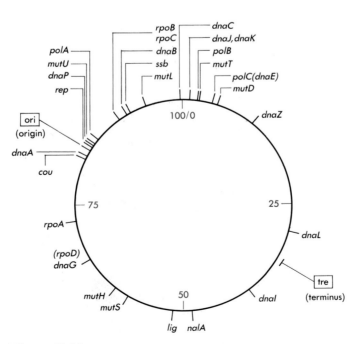

Figure 10-16
Genetic map of *E. coli* K12 showing loci involved in DNA replication.

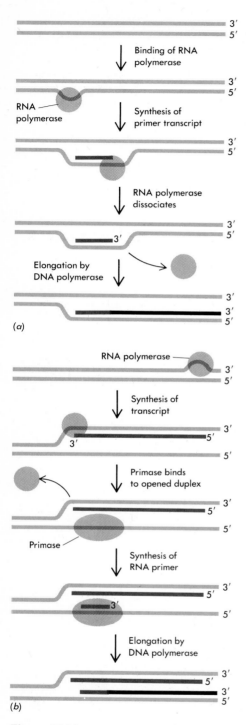

(a)

Binding of RNA polymerase

RNA polymerase

Synthesis of primer transcript

RNA polymerase dissociates

Elongation by DNA polymerase

RNA polymerase

Synthesis of transcript

Primase binds to opened duplex

Primase

Synthesis of RNA primer

Elongation by DNA polymerase

(b)

Figure 10-18
Alternative schemes for how RNA polymerase, at the origin of DNA replication, acts to make RNA primers. (a) RNA polymerase makes RNA, which, after processing by RNase H (not shown), acts as primer. (b) RNA made by RNA polymerase opens the double helix to allow binding of a primosome, which will make the actual primer.

Figure 10-17
Consensus nucleotide sequence of the minimal origin of bacterial chromosomes. The minimal origin of *E. coli* is enclosed within the box. A large capital letter means that the same nucleotide base is found in all six origins; the small capital letter is used when the same nucleotide is present in five of the six sequences; the lowercase letter is used when the same nucleotide is found in three or four of the six origins but only two different nucleotides are found at the site; the letter "n" is used when three or four of the four possible nucleotides, or two different nucleotides plus a deletion, are found at the site. A dot (·) means that the nucleotide in the bacterial sequence is the same as that in the consensus sequence. Bold capital letters locate positions where single-base substitutions produce an *oriC⁻* phenotype in *E. coli*. Restriction enzyme recognition sites are indicated, and arrows denote the A-protein binding sites, R1, R2, R3, R4. [Courtesy of J. Zyskind and D. Smith, *Proc. Nat. Acad. Sci.* 80 (1983):1164–1168.]

Initiating Replicating Forks at Origins of Replication[41–50]

Origins of replication are by definition distinct from the many sites at which Okazaki fragments are initiated. Bacterial and viral DNA molecules usually contain just one such origin. The length of DNA comprising an origin can now be sharply delineated using cloned F plasmid DNA, since its size is known and it inserts at replication start sites. In this way, the unique *E. coli* origin has been narrowed to a region of 245 base pairs (Figure 10-17) lying between the gene for asparagine synthetase and the operon controlling ATP synthetase (see Figure 7-23). In vitro mutagenesis has shown that only a fraction of the bases within this origin of replication are essential for its functioning. Perhaps it is these bases that are recognized by the protein(s) specifically involved in the initiation of the replicating fork.

The key event in initiation of a replicating fork is the commencement of leading-strand synthesis. Once the polymerization of the leading strand has started, polymerization of the lagging strand soon follows, using the mechanism described earlier. An obligatory step in the starting of all leading strands is the transcription by RNA polymerase molecules of origin DNA sequences into short RNA chains. For some origins (e.g., that of the plasmid Col E) these RNA chains may function as the primers for leading strands after unneeded regions are removed by digestion with the enzyme **RNase H** (H stands for "hybrid"), which specifically digests the RNA strand of DNA-RNA hybrid molecules. However, most of these short origin RNA transcripts probably have no primer role. Instead, their only function seems to be to separate the two DNA strands of the origin so as to expose those specific sequences around which primosomes assemble prior to the synthesis of primers (Figure 10-18). This latter process is called **transcriptional activation.**

No matter how they are made, the presence of an RNA primer does not automatically mean that a DNA chain will begin growth upon it, since RNA chains usually dissociate immediately from their DNA templates. Normally favoring the release of most newly made RNA chains is the strong tendency of the separated DNA strand to reassociate to form double-helical DNA. Therefore, there must be molecules that specifically prevent such DNA-DNA reassociations, perhaps by binding to origins of replication.

Initiation of leading strands in many (but not all!) cases requires the presence of one or more additional proteins that function only during initiation. Among these are the dna A protein of *E. coli* and the O protein of phage λ, both of which bind to four highly conserved nine-base-pair-long recognition sequences. How either of these DNA-binding proteins functions is unknown. It is possible that they act by providing a site favoring the assembly of the seven different DNA pol III components into their functional holoenzyme form.

Even less clear are the tricks used to ensure bidirectional replication once DNA synthesis has commenced. We might naively predict that the elongation at the 3' end of the first lagging-strand (Okazaki) fragment would routinely proceed past the origin of replication, becoming in effect the leading strand for a new replicating fork moving in the opposite direction. If this were true, replication would be bidirectional for virtually all replication starts. In fact, however, in a minority of replication starts, synthesis does not proceed bidirectionally. These unidirectional starts may reflect the proposal that the same DNA pol III complex (replisome) normally replicates both the leading and lagging strands simultaneously. For replication to proceed in the opposite direction, a second DNA pol III holoenzyme complex may need to be assembled.

Completing the Ends of Linear DNA Molecules[51, 52, 53]

The realization that RNA primers for DNA synthesis are ordinarily excised raises the question of how the extreme ends of linear DNA can be completed. Not only must a small RNA fragment initiate at the terminal 3' nucleotide of the template strand, but then some way must be found to fill in the sequence after ribonucleotide removal. A resolution of this dilemma was first understood for T7 DNA, all of whose terminal deoxyribonucleotides conventionally base-pair, but which has the surprising property of having redundant ends; the sequence (about 160 base pairs) found at the left end is exactly repeated at the right end of the molecule. Moreover, when such linear T7 molecules replicate, they do not generate unit-length progeny molecules, but instead generate very long **concatamers** (units linked end to end) containing the complete genome sequence repeated over and over (Figure 10-19).

Why linear DNA might pass through a concatamer stage can be seen when we consider the incompleteness of the two daughter mole-

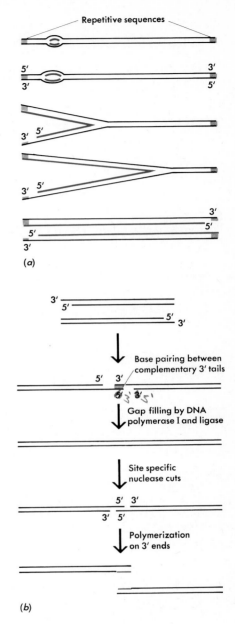

Figure 10-19
Replication of T7 DNA molecules.
(a) The 3' ends of the linear DNA are first incompletely replicated. (b) The formation of concatamers then makes possible the completion of replication by DNA polymerase working 5'→3'.

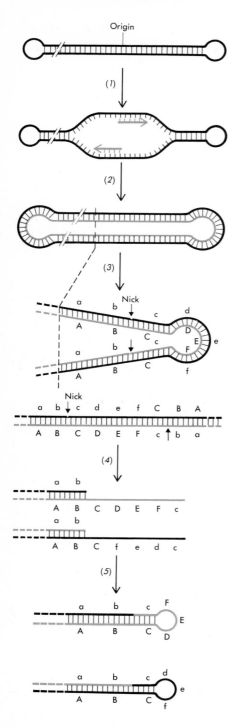

cules produced by the initial replication of a linear double helix. One end of each daughter molecule will be completely replicated, while the other end will have a 3'-ended single-stranded tail. If there are redundant terminal sequences, then the two tails will be complementary and can base-pair to form a two-unit length concatamer (see Figure 10-19). Following gap filling and ligase action, this two-unit DNA can be replicated to again produce unfinished progeny that will tend to stick together to produce a four-unit molecule; and so forth. Unit-length molecules can be produced from such concatamers by the action of specific endonucleases that create staggered nicks. These can easily be filled in by a DNA polymerase synthesizing in the 5'→3' direction to yield double helices identical to those present in the original parental phage particle.

A second suggested way the ends of linear viral chromosomes might be preserved is illustrated by the poxviruses, which have hairpin loops joining together their DNA strands at both their termini. Replication of such viral DNAs might begin at single internal initiation sites with subsequent bidirectional replication converting the single-stranded hairpins into double-helical equivalents. How such double-stranded hairpins become resolved into daughter single-stranded hairpins is unclear. One proposal starts with two staggered cuts displaced about the center of the hairpin, followed by the denaturation of the central DNA that leads to the separation of the two ends. The resulting single-stranded tails snap back into floppy loops, which DNA ligase can act upon to create new, totally closed hairpins (Figure 10-20).

Still much less clear is how the linear chromosomes that characterize all eucaryotic organisms replicate their ends. Each such chromosome is now thought to contain a single DNA molecule; yet there is no morphological evidence suggesting end-to-end linkage during DNA replication. At first it seemed logical to believe that their ends (also called telomeres) would be found to have closed hairpin loops of fixed length like the pox viruses. However, recent evidence favors the alternative hypothesis that the ends of eucaryotic chromosomes are maintained by enzymes that add new terminal DNA sequences to replace those that otherwise would be lost during each replication cycle. Such a terminal repair process was first suggested by experiments with the small, gene-sized DNA molecules present in the macronucleus of the ciliated protozoan *Tetrahymena* (Chapter 22). The ends of all these linear macronuclear DNA molecules have tandemly repeated blocks of the sequence

$$3'\text{-AACCCC-}5'$$
$$5'\text{-TTGGGG-}3'$$

Figure 10-20
Replication of linear viral DNA molecules whose ends are joined by hairpin loops. Bidirectional replication beginning at the origin (1) converts the single-stranded structure to a double-helical equivalent (2). Two staggered cuts are made (3), and the molecules are rearranged (4) and ligated (5) to complete the replication cycle. In (3) through (5), half molecules are shown.

that vary in number from 30 to 70. Variation in the number of these repetitive units prompted the search for—and recent discovery of—an enzyme that adds 5'-TTGGGG-3' to the 3' end of any preexisting single-stranded 5'-TTGGGG-3' sequence. The function of this completely new type of enzymatic activity is surely to prevent the ends of chromosomes from growing progressively shorter because of failure to completely replicate their 5'-terminal ends. Once the TTGGGG-adding enzyme has extended the 3' end of the telomere, synthesis of the complementary 3'-AACCCC-5' repeats could be primed by a conventional primase, by looping back of the 3' end of the 5'-TTGGGG-3' strand, or by a specialized priming protein (see Figure 10-26).

Complete Separation of Two Parental Circular Strands Requires the Participation of an Enzyme Like Topoisomerase II[54, 55]

There is no end-completing dilemma to contend with for the replication of circular DNA; there the gaps created by removal of the RNA primers each eventually become filled in by extension of the 3' termini growing around the circle. The occurrence of so many DNAs in circular form may therefore be explained, at least in part, by the ease with which circles ensure their own complete replication.

Bidirectional replication normally proceeds around circular DNA molecules until the two opposing replicating forks meet each other, usually at some site approximately 180° across from the origin of replication. No special termination signals are thought to exist, since deletions of DNA that remove the regions where the replicating forks usually meet do not prevent union of the daughter strands and their subsequent separation from each other. The final separation of the daughter double helices, however, is not an automatic affair. The completion of replication of the individual strands usually leaves the daughter double helices intertwined by 20 to 30 coils about each other (see Figure 10-4). The only way to separate such intertwined circles is the creation of double-stranded cuts that allow the double helices to pass through each other. In vitro, topoisomerase II catalyzes the separation of such intertwined double helices, and it would be most surprising if this capability were not used by cells to complete their replication.

Rolling Circles Are Alternative Means for Replicating Circular DNA[56, 57, 58]

An alternative way to replicate circular DNA is the **rolling circle mechanism.** This replication scheme underlies the multiplication of many viral DNAs, of bacterial F (fertility) factors during mating, and of the DNA in certain cases of gene amplification (see Chapter 22). In this mechanism, DNA synthesis starts with a specific cut at the origin of replication in one specific strand (+) of the parental duplex circle. The 5' end then becomes displaced from the duplex, thereby permitting DNA Pol III to add deoxyribonucleotides to the free 3' OH. As replication proceeds, the 5' end of the cut strand is rolled out as a free tail of increasing length (Figure 10-21), which soon becomes covered by the single-stranded DNA-binding proteins. Such a replicating structure is called a rolling circle because the unraveling of the SSB-bound single-stranded DNA is accompanied by rotation of the double-helical template about its axis. Such unraveling does not create any topological problem, since the circle is held together by only one intact DNA strand, which will freely rotate about many of the covalent bonds contained in its backbone.

When the rolling circle mechanism is used to replicate double-stranded DNA, the 5'-ended tails quickly become converted to double-stranded DNA through the formation of Okazaki fragments that grow from RNA primers started by primase molecules contained in the locomotory primosomes. The active force that then continues to unwind the 5' tail from the rolling circle is probably not the polymerization reactions occurring at the 3' end but the movement of the Pol III–primosome complex (replisome) propelled by its helicase com-

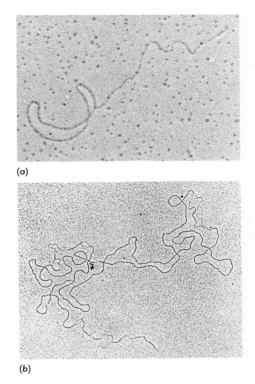

(a)

(b)

Figure 10-21
Electron micrographs of the rolling circle intermediates in (a) single-stranded φX174 DNA and (b) the replication of double-stranded phage P2 DNA. In (b), partial denaturing conditions were used to show sequence homologies between circle and tail sections. [Part (a) courtesy of D. Dressler; part (b) courtesy of R. B. Inman.]

ponents (Figure 10-22). Elongation sometimes goes on to produce tails many times the complete length of the original circle. This happens, for example, in the later stages of phage λ DNA replication. How these long tails become converted into progeny-length DNA molecules is not yet established. The best guess is that the tails are cut by specific endonucleases like those that convert λ circles into λ linear forms (see Figure 9-21). The resulting unit-length progeny linear forms could remain linear and become encapsulated into mature virus particles. Alternatively, they could use their single-stranded complementary ends to form new circular double helices capable of becoming new circles.

Conversion of Circular Single-Stranded Viral DNA Chromosomes into Double-Helical Replicative Forms[59, 60]

RNA primers are needed to convert the circular single-stranded (+) DNA chromosomes of certain small viruses (the phages φX174, G4, and M13) into their double-helical (+/−) replicating forms (RFs). Replication of these viral DNAs commences after the infecting + strands enter suitable host cells and become coated with single-stranded DNA-binding proteins (SSBs). Subsequently, the small RNA chains that prime synthesis of the respective − strands are made. No one mechanism is used to produce these primers; the route followed depends on the specific + strand template (Figure 10-23). Full primosomes assemble on φX174 DNA, while primase alone appears necessary to synthesize phage G4 RNA primers. In phage M13 DNA, host *E. coli* RNA polymerase recognizes a small hairpin loop to which SSBs cannot adhere and uses this site to make a small primer. The RNA primers made in each of these processes are removed after the − strand synthesis catalyzed by DNA Pol III goes almost to completion around the circular template. The small gap remaining is filled in by DNA polymerase I, whose 5'→3' exonuclease also removes the RNA nucleotides lying immediately ahead.

Synthesis of Circular + Strands upon Circular Phage Replicating Forms[61]

All the RFs of single-stranded DNA phages use rolling circles both to produce more copies of themselves and to produce the + strand progeny that becomes encapsulated in new phage particles. How this happens is best understood for phage φX174 whose nicking enzyme, the **A protein,** is necessary to start the rolling circle and is encoded by the phage chromosome. This enzyme recognizes a single origin site on the covalently closed RF, and in making its cut, becomes covalently bound at the 5' end of the cut DNA. As the replicating fork proceeds around the + strand, the 5' tail of the displaced + strand does not hang loose, but instead remains complexed through its attached nicking enzyme to the DNA Pol III holoenzyme (perhaps to a replisome) acting at the replicating fork (Figure 10-24). Movement of the replicating fork one complete turn around the rolling circle brings the attached nicking enzyme into position to make the next specific cut at the origin of replication. In making such a cut and generating a new round of rolling circle synthesis, the enzyme becomes detached

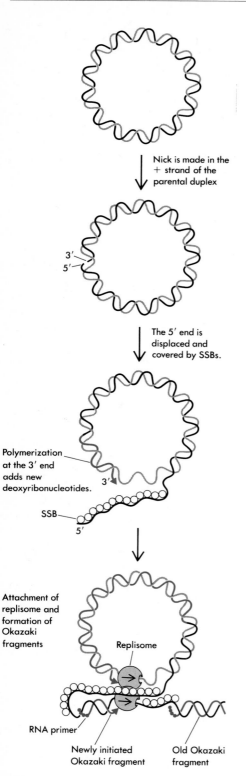

Nick is made in the + strand of the parental duplex

3'
5'

The 5' end is displaced and covered by SSBs.

Polymerization at the 3' end adds new deoxyribonucleotides.

3'

SSB

5'

Attachment of replisome and formation of Okazaki fragments

Replisome

RNA primer

Newly initiated Okazaki fragment

Old Okazaki fragment

Figure 10-22
Replication of double-stranded circular DNA molecules through the rolling circle mechanism. The active force that unwinds the 5' tail is the movement of the replisome propelled by its helicase components.

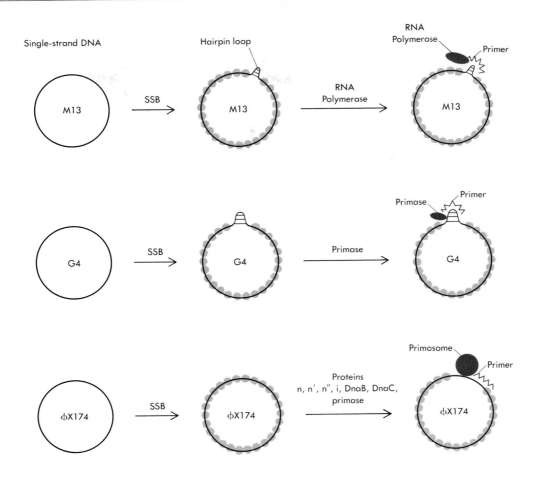

from its previously bound 5′ tail. The newly free 5′ end in turn links onto the newly made 3′ end, thereby yielding a circular unit-length + progeny strand. At no time in these nicking/closing cycles is any ATP employed, so the nicking (initiation) enzyme bears a resemblance to classical topoisomerase I, except that much longer intervals pass between the cutting and rejoining events in the case of the nicking enzyme.

Initially, all the + strand progeny produced during phage infection become converted into daughter RFs through the use of RNA primers. Later, as phage infection proceeds, the daughter circular + strands become covered with phage-specific proteins, which not only prevents the formation of RNA primers but serve as nucleation sites for assembly of new progeny phage particles.

Figure 10-23
Three mechanisms by which RNA primers are synthesized in single-stranded viral DNA molecules. (After Arthur Kornberg, *1982 Supplement to DNA Replication* p. S109, W. H. Free- and Company, copyright © 1982, with permission.)

Synthesis and Transfer of Single-Stranded DNA During Bacterial Mating[62]

When bacteria mate, the DNA that the male transfers into the female is single-stranded, not double-stranded. It is synthesized by the rolling circle mechanism started with the nicking of one specific strand of the male chromosome. This strand is then elongated at its 3′ end, leading to the displacement of growing 5′-ended single-stranded DNA. Most likely, the enzyme that makes the specific cut, like those that start off other rolling circles, remains attached to the 5′ end and

Figure 10-24

ϕX174 DNA replication involves two replicating forms (RFs). RF I, a double-stranded closed circle, results from the synthesis of a − strand complementary to the phage particle's + strand. RF II, a double-stranded open circle, is the rolling circle intermediate for the synthesis of new + strands. The phage's A protein plays a major role in the latter process.

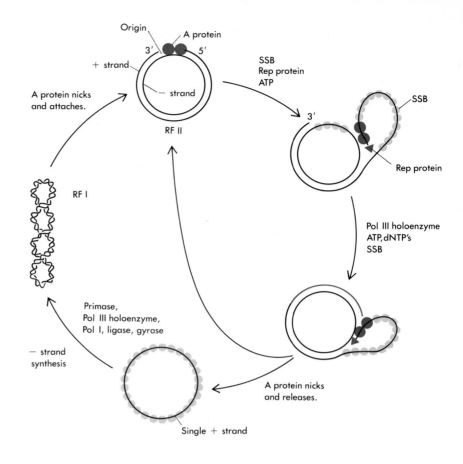

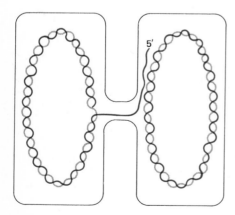

Figure 10-25

Transfer of single-stranded DNA from a male bacterium into a female via the rolling circle mechanism.

somehow serves to guide the single-stranded DNA into the female cell (Figure 10-25).

This transfer mechanism couples DNA replication and mating in such a way that genetically deficient male parents are never produced. What happens once the male DNA is inside the recipient cells remains unclear. Some of it may be converted into a normal double-helical molecule, or, as single-stranded DNA, it may be exchanged for a portion of the original female chromosome.

Linear Viral DNAs That Replicate Without Using RNA Primers[63-71]

A number of viruses, some of which multiply in procaryotic cells and others in eucaryotic cells, possess linear DNA molecules whose synthesis does not utilize RNA primers. Best known are the linear DNAs found in *Bacillus* phage ϕ29 as well as in the adenoviruses, a group of viruses that infect vertebrate cells. Unlike the superficially similar-looking T7 DNA, ϕ29 and adenovirus DNAs do not replicate using bidirectional replicating forks starting from an internal origin. Instead, replication always begins directly at the ends of the DNA, with equal probability of starting at either end. Although each adenoviral and ϕ29 DNA molecule possesses identical sequences at its two ends, these redundant sections are oriented in opposite directions, so their DNA replication cannot generate concatamers during replication. Nor do the two complementary DNA strands both serve as templates when replication proceeds from a given end. Instead, only one strand at each end acts as a template. This form of DNA synthesis involves a displacement process that leads to the generation of single-stranded tails representing the released 5'→3' oriented parental strands. Following their complete displacement and hydrogen bonding of their complementary 5'- and 3'-ended sequences, they in turn can serve as templates for new 5'→3' progeny strands starting at their 3' termini (Figure 10-26).

The first real clue to how such DNA synthesis can start de novo was the finding that the 5' ends of all growing adenoviral chains contain a specific protein attached to the terminal cytosine nucleotide. Before DNA replication starts, terminal protein becomes covalently attached to the 5' phosphate of deoxycytidine monophosphate, soon to be the 5'-terminal nucleotide in the newly synthesized DNA. The deoxycytidine base-pairs with the opposing 3'-end G residue on the template strand. This process requires a unique, virus-coded DNA polymerase that utilizes the terminal protein–dCTP complex as a primer and then synthesizes DNA in a processive fashion down the remaining 36,000 nucleotides needed to complete the chain. Complete replication requires the action of a cellular type I topoisomerase to remove torsional strain induced by the displacement synthesis. This is a case where even a linear DNA, presumably not attached at its ends, cannot replicate without the action of a topoisomerase. Still remaining to be worked out is the way the 5'→3' oriented parental strand is displaced. Conceivably, the adenovirus-coded 72 kdal single-stranded DNA binding protein that is abundantly made during adenovirus multiplication has some helicase-like function.

Regulation of DNA Synthesis[47, 72, 73]

How DNA synthesis is regulated in various cells is almost as mysterious today as it was when the first DNA polymerases were isolated. Until the general principles by which DNA synthesis is initiated became established, there was no way to probe this problem at the level of defined molecules. Now, however, it looks as if the way the leading DNA chains are started may soon be worked out at the molecular level. Then we may be able to search for the tricks that restrict DNA replication to limited periods within the cell cycle and somehow ensure that a given origin is used only once during a given cycle of chromosome replication. Perhaps key initiating proteins, like the dnaA protein of *E. coli*, somehow modify the DNA on which they act so that the origins cannot be used again as starting sites for DNA synthesis until they become modified back to their original condition (e.g., a methylation-demethylation cycle). But so far, there is no evidence of clear-cut chemical modifications occurring specifically at replication origins.

Incisive answers may come from analysis of plasmid DNAs. The replication of the *E. coli* plasmid ColE1, for example, is regulated at least in part by the synthesis of RNA molecules, transcribed in the opposite direction from the same DNA sequences that give rise to the RNA primers for leading strand synthesis. Such "anti-sense" RNA somehow blocks primer initiation of DNA synthesis, conceivably by annealing to the primers to form RNA-RNA double helices. The signals used to control when this anti-sense RNA is made are yet to be discovered.

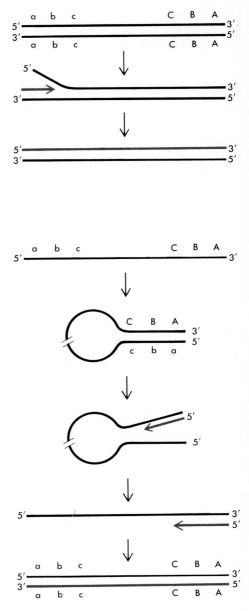

Figure 10-26
Replication of linear DNA (such as adenovirus and phage φ29) that does not use RNA primers. Replication begins at either end of the DNA, and only one strand serves as a template. The displaced single-stranded tail separates from the parental DNA molecule and forms a hairpin loop, a result of complementary sequences oriented in the opposite direction at either end. This tail then serves as a template for a new progeny strand starting at the 3' terminus.

Equally valuable information may come from study of eucaryotic plasmid DNAs that replicate only once per cell cycle and thus may be subject to the same general control mechanisms that prevent origins of replication from functioning more than once per cell cycle. The study of bovine papilloma virus (BPV) DNA should be particularly useful as it normally multiplies as a plasmid whose copy number is tightly regulated along with the copy number of the host cell's chromosomes.

Are Membranes Involved in DNA Replication and Segregation?[74]

The question of how DNA molecules (chromosomes) are correctly partitioned during cell division (mitosis) so that each daughter cell receives one of each parental chromosome still has no clear answer. It has always been natural to postulate a connection between the cell (or nuclear) membrane and specific regions of chromosomes, but despite many claims, no firm evidence yet exists for well-defined molecules that bind DNA to surrounding membranes. On the contrary, it is now clear that none of the stages in DNA replication requires the presence of any membrane component. This fact, however, does not mean that membranes and DNA do not need to specifically bind to each other. But as long as it remains so difficult to purify cellular membranes away from other cellular fractions, this area of investigation may continue to fail to yield convincing answers.

There Are Several Functionally Distinct DNA Polymerases Within Every Eucaryotic Cell[75]

Only recently has a major effort been made to understand DNA replication at the enzymatic level within eucaryotic cells. Three distinct forms of DNA polymerase (α, β, and γ) have been characterized in all well-studied cell systems (Table 10-5). The α DNA polymerase, the form responsible for most DNA synthesis, carries out the role played by DNA polymerase III in bacteria. Like pol III, it is a holoenzyme with a molecular weight of 220 kdal made up of several polypeptides, the largest of which (about 150 kdal) carries the polymerizing activity. So far, no form functionally equivalent to pol I has been identified, although a newly discovered δ form found in the calf thymus may fit the bill. Still functionally obscure is the much smaller β DNA polymerase, although there are hints that it may be involved in DNA repair. Already much better understood is the γ DNA polymerase, which is exclusively located within mitochondria (and perhaps chloroplasts) where it functions to replicate the mitochondrial DNA molecules. It catalyzes a displacement-type synthesis (Figure 10-27) in which only

Table 10-5 Eucaryotic DNA Polymerases α, β, and γ

	α	β	γ
Location	Nucleus	Nucleus	Mitochondria
Function	DNA replication	Repair	mtDNA replication
Mass, kdal	120–220	30–50	150–300

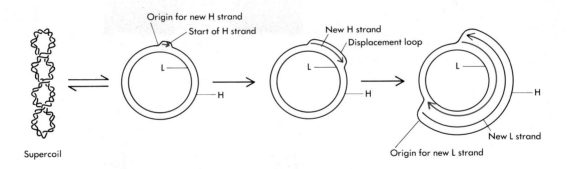

one strand (H) is initially copied. Not until the replicating fork moves halfway around the genome does synthesis commence on the L-strand template. As expected, topoisomerase molecules are found within mitochondria. They can catalyze the nicking and closing events that make possible the untwisting of the parental strands and subsequent separation of the daughter double helices.

The Existence of a Cell-Free System for Replicating SV40 DNA[76, 77]

Working out the exact enzymology of eucaryotic DNA synthesis will require the development of well-defined in vitro systems similar to those now being used to reveal how the *E. coli* chromosome is replicated. In the case of *E. coli*, real progress was first made using the well-defined DNA's of *E. coli*'s viruses and plasmids. In the same way, real progress on the eucaryotic front has come through studying eucaryotic systems that replicate adenovirus DNA, even though adenovirus DNA replicates by a strand displacement mechanism rather than through the growth of replicating bubbles, as is typical for eucaryotic chromosomes. Much excitement has thus been generated by the recent working out of conditions for the in vitro replication of biologically active SV40 chromosomes, whose synthesis is initiated at a single site along its circular DNA molecule. Most importantly, this system works only in the presence of the SV40 T-antigen (see Chapter 24), a virally coded protein that binds to several sites surrounding the SV40 origin of replication and which from in vivo experiments is known to play an essential role in the initiation of SV40 DNA replication. Further fractionation of this system should soon reveal important facts regarding the exact composition of the α DNA polymerase holoenzyme, which is responsible for most of the synthesis. Also available in the near future will be data on how the RNA primers are made for both leading- and lagging-strand synthesis as well as the roles played by the several cellular topoisomerases.

From such analysis we should at last begin to understand what role, if any, the several major histones play during DNA replication. In the absence of hard facts, it had seemed most plausible that nucleosomes would dissociate from DNA in the course of its replication. This is not the case, however. As related later, nucleosomes are conserved intact during DNA replication, with the surprising result that they are not equally partitioned to the two daughter double helices, but remain attached to only one of the two strands along a given stretch of a double helix.

Finally, in understanding how T-antigen promotes SV40 synthesis, we should acquire clues that will let us look at how the multiple

Figure 10-27
Replication of mitochondrial DNA molecules. Replication of the H strand commences first. When the expanding displacement loop has traveled halfway around the circle, the replication of the L strand begins. (After Arthur Kornberg, *DNA Replication* p. 605, W. H. Freeman and Company, copyright © 1980, with permission.)

origins of replication along a given eucaryotic chromosome (see Chapter 21) are signaled to commence DNA replication. Although this may not be an easy goal to achieve, its accomplishment should provide many important dividends in understanding how the growth of cells is controlled.

Summary

Replication of DNA double helices to yield two identical daughter molecules, besides requiring the breakage of the hydrogen bonds holding base pairs together, necessarily involves untwisting of the intertwined parental chains. For this to happen, unreplicated sections must rotate around their axes, a process that need not require significant energy because the double helix is very thin. Strand separation and untwisting go hand in hand with synthesis of new strands to generate moving replicating forks that move away from the starting point (origins) for DNA replication. Origins of replication on a given DNA molecule are very precisely defined. On linear viral DNA molecules, they usually occur internally to generate bidirectionally growing replicating bubbles (-O-shape molecules), which become Y-shaped molecules, until replication is completed. Circular DNA molecules usually have θ-shaped replication intermediates containing two replicating forks that move away from the origin until they meet on the opposite side of the DNA circle.

Within each cell, there are several types of DNA polymerase, which add on deoxyribonucleotides to preexisting DNA chains in a $5' \rightarrow 3'$ direction. In *E. coli*, three different DNA polymerases have been identified. Besides their polymerizing activity, DNA polymerases usually contain a $5' \rightarrow 3'$ exonuclease on one side of the molecule, and on the other side a $3' \rightarrow 5'$ exonuclease. The $3' \rightarrow 5'$ exonucleases are capable of removing misincorporated nucleotides and thus have a vital proofreading role, which reduces the error level of DNA replication. After proofreading, the average probability of an error in the insertion of a new nucleotide may be as low as 10^{-8} to 10^{-12}. DNA polymerase I (and probably all other polymerases) is constructed with a 20 Å diameter cleft through which a DNA double helix moves. Conceivably, the incorporation of a wrong base prevents this movement and thereby allows the $3' \rightarrow 5'$ exonuclease component the opportunity to remove the latest incorporated base.

The opposing chain directions of the two strands of the double helix ($5' \rightarrow 3'$ and $3' \rightarrow 5'$) requires the two daughter strands also to run in opposite directions. All known DNA polymerases, however, extend individual chains only in the $5' \rightarrow 3'$ direction. Resolution of this paradox came from the realization that while the $5' \rightarrow 3'$ strand of a replicating fork grows continuously, the $3' \rightarrow 5'$ strand grows discontinuously, being assembled from a group of DNA fragments each individually made in the $5' \rightarrow 3'$ direction. The strand growing continuously $5' \rightarrow 3'$ (the leading strand) moves slightly ahead of the strand that is assembled discontinuously $3' \rightarrow 5'$ (the lagging strand).

Separation of the two strands at the growing fork is speeded up by DNA-binding proteins called helicases, which use the energy of ATP to move along the just-separated DNA chains at growing forks. Soon after the two parental chains have separated through the helicase action, they become covered by specific proteins (single-stranded DNA-binding proteins) that keep the individual chains in semiextended, conceivably helical configurations. New deoxyribonucleotides are added onto leading daughter strands by the DNA polymerase III holoenzyme, a complex assembly of seven different polypeptides, all of which must be present for rapid processive growth. Although the DNA polymerase III holoenzyme also makes most of each lagging-strand fragment, it does not initiate the growth of these fragments. Each lagging-strand fragment is initiated by an RNA primer made by a second complex assembly of polypeptides called the primosome. Primosomes start RNA synthesis at very specific recognition sequences. The distance between the recognition sequences regulates the size of the lagging-strand fragments. DNA polymerase I fills in the remaining gaps between fragments and catalyzes, through its $5' \rightarrow 3'$ exonuclease component, removal of the RNA primer at the 5' end of the adjacent fragment.

Still unclear is whether any universal mechanism exists for the RNA priming of the leading strands. To start a replicating bubble, RNA polymerase–directed RNA synthesis must occur at the origin region. This raises the possibility that such RNA chains (after processing?) become the primers for leading strands. The alternative possibility, however, is that the function of this synthesis is to temporarily separate the two DNA chains at the origin so that a primosome can recognize an appropriate start signal (the transcriptional activation hypothesis). The starting of a leading strand also frequently demands the presence of specific proteins that bind sequences very close to the starting point for DNA synthesis.

The replication of the DNA of many single-stranded phages occurs in two main steps: (1) conversion of a circular single-stranded + template, using a primosome and the DNA polymerase III holoenzyme, into a double-helical +/− replicative form, and (2) the synthesis off an RF template of a + single-stranded circle using a rolling circle mechanism. Rolling circles are initiated by the cutting at the origin of replication of the + strand by a specific nicking enzyme. This cut is followed by a continuous displacement of the 5'-ended strand accompanied by DNA growth onto the 3'-ended fragment. DNA synthesis progresses completely around the circle to the origin. There a nicking enzyme makes a new cut that releases a complete + single strand, which immediately adopts a circular configuration. Circular double-helical DNA can also replicate by the rolling circle mechanism. Double-stranded tails are generated, which then can be enzymatically cut into unit-length linear fragments with sticky ends that make possible their subsequent circularization.

Initiation of replicating bubbles, however, does not ensure the successful synthesis of linear DNA molecules. Unless some special trick is used, each cycle of replication will produce shorter and shorter 5'-ended chains. Phage T7 DNA has identical sequences at its ends, which allow end-to-end union of progeny molecules to form a very long molecule, which subsequently can be cut up into single genomes with complete ends. A completely different device is used by the pox viruses, which have closed hairpin loops whose cutting provide the needed 3' primers. How the linear DNAs of cellular chromosomes maintain their ends remains unclear. One possibility is that their integrity is maintained by specific enzymes, which during every cycle of chromosome duplication add specific repetitive sequences (e.g., GGGGTT) to the 3' ends. If such sequences are recognized by appropriate primosomes, then these chromosomes will stay approximately constant in length and never become nonfunctional.

Certain linear viruses (e.g., adenoviruses and φ29 phage) that have proteins covalently attached to their 5' ends do not start DNA synthesis with replicating bubbles. Instead, they employ a displacement process in which the 5' end of each strand is displaced by growth of a new 5'-end strand starting at the 3' end of the parental strand. RNA primers are not used; instead, 5'-ended growth is initiated by the nucleotide covalently bound to a terminal protein molecule.

Continued progress in understanding the essential mechanisms of DNA replication demands the development of cell-free systems for making biologically active DNA molecules. While systems that replicate procaryotic DNAs have been in use for over a decade, only during the past several years has it been possible to generate biologically active eucaryotic DNA in vitro. Particularly important is the recent development (1984) of cell-free systems in which circular SV40 DNA is replicated. This synthesis requires the presence of the SV40 T-antigen, a protein previously known to be essential for in vivo SV40 synthesis and which binds to the origin of replication. Hopefully, the study of DNA synthesis in these systems will begin to provide insights into how eucaryotic chromosomal DNA is initiated, elongated, and terminated.

Bibliography

General References

Boyer, P. D., ed. 1981. "Nucleic Acids." *The Enzymes*. Vol. 14(A). New York: Academic Press.

Denhardt, D. T., D. Dressler, and D. S. Ray. 1978. *The Single-Stranded DNA Phages*. Cold Spring Harbor, N.Y.: Cold Spring Harbor Laboratory.

DePamphilis, M. L. and P. M. Wassarman. 1982. "Organization and Replication of Papovavirus DNA." In *Organization and Replication of Viral DNA*, ed. A. S. Kaplan, Boca Raton, FL: CRC Press, pp. 38–113.

Kornberg, A. 1980. *DNA Replication*. San Francisco: Freeman.

Kornberg, A. 1982. *1982 Supplement to DNA Replication*. San Francisco: Freeman.

"DNA: Replication and Recombination." 1978. *Cold Spring Harbor Symposium on Quantitative Biology*. Vol. 43. Cold Spring Harbor, N.Y.: Cold Spring Harbor Laboratory.

"Structures of DNA." 1982. *Cold Spring Harbor Symposium on Quantitative Biology*. Vol. 47. Cold Spring Harbor, N.Y.: Cold Spring Harbor Laboratory.

Cited References

1. Delbrück, M. 1954. "On the Replication of Deoxyribonucleic Acid (DNA)." *Proc. Nat. Acad. Sci.* 40:783–788.
2. Delbrück, M., and G. Stent. 1957. "On the Mechanism of DNA Replication." In *The Chemical Basis of Heredity*, ed. W. D. McElroy and B. Glass. Baltimore: Johns Hopkins Press, pp. 699–736.
3. Brown, T., O. Kennard, G. Kneale, and D. Rabinovich. 1985. "High-Resolution Structure of a DNA Helix Containing Mismatched Base-Pairs." *Nature* 315:604–606. Gives an opposing view as to the extent of helix deformation due to mismatched base pairs.
4. Inman, R. B. 1966. "A Denaturation Map of the λ Phage DNA Molecule Determined by Electron Microscopy." *J. Mol. Biol.* 18:464–476.
5. Mulder, C., and H. Delius. 1972. "Specificity of the Break Produced by Restricting Endonuclease R1 in Simian Virus 40 DNA, as Revealed by Partial Denaturation Mapping." *Proc. Nat. Acad. Sci.* 69:3215–3219.
6. Dressler, D., J. Wolfson, and M. Magazin. 1972. "Initiation and Reinitiation of DNA Synthesis During Replication of Bacteriophage T7." *Proc. Nat. Acad. Sci.* 69:998–1002.
7. Wolfson, J., D. Dressler, and M. Magazin. 1972. "Bacteriophage T7 DNA Replication: A Linear Replicating Intermediate." *Proc. Nat. Acad. Sci.* 69:499–504.
8. Cairns, J. 1966. "The Bacterial Chromosome." *Sci. Amer.* 214:36–44.
9. Salzman, N. P., G. C. Fareed, E. D. Sebring, and M. M. Thoren. 1973. "The Mechanism of SV40 DNA Replication." *Cold Spring Harbor Symp. Quant. Biol.* 38:257–265.
10. Okazaki, T., and R. Okazaki. 1969. "Mechanism of DNA Chain Growth. IV. Direction of Synthesis of T4 Short DNA Chains as Revealed by Exonucleolytic Degradation." *Proc. Nat. Acad. Sci.* 64:1242–1248.
11. Wolson, J., and D. Dressler. 1972. "Regions of Single-Stranded DNA in the Growing Points of Replicating Bacteriophage T7 Chromosomes." *Proc. Nat. Acad. Sci.* 69:2682–2686.
12. Goulian, M., A. Kornberg, and R. L. Sinsheimer. 1967. "Enzymatic Synthesis of DNA, XXIV. Synthesis of Infectious Phage φX174 DNA." *Proc. Nat. Acad. Sci.* 58:2321–2328.
13. Wickner, R. B., M. Wright, S. Wickner, and J. Hurwitz. 1972. "Conversion of φX174 and fd Single-Stranded DNA to Replicative Forms in Extracts of *Escherichia coli*." *Proc. Nat. Acad. Sci.* 69:3233–3237.
14. Sakakibara, Y., and J.-I. Tomizawa. 1974. "Replication of Colicin El Plasmid DNA in Cell Extracts." *Proc. Nat. Acad. Sci.* 71:802–806.
15. Masker, W. E., and C. C. Richardson. 1976. "Bacteriophage T7 Deoxyribonucleic Acid Replication *in Vitro*. VI. Synthesis of Biologically Active T7 DNA." *J. Mol. Biol.* 100:557–567.
16. Tsurimoto, T., and K. Matsubara. 1982. "Replication of Bacteriophage λ DNA." *Cold Spring Harbor Symp. Quant. Biol.* 47:681–691.
17. Knippers, R. 1970. "DNA Polymerase II." *Nature* 228:1050–1053.
18. Kornberg, T., and M. L. Gefter. 1971. "Purification and DNA Synthesis in Cell-Free Extracts: Properties of DNA Polymerase II." *Proc. Nat. Acad. Sci.* 68:761–764.

19. Kornberg, A. 1978. "Aspects of DNA Replication." *Cold Spring Harbor Symp. Quant. Biol.* 43:1–9.

20. Brutlag, D., and A. Kornberg. 1972. "Enzymatic Synthesis of Deoxyribonucleic Acid. XXXVI. A Proofreading Function for the 3'→5' Exonuclease Activity in Deoxyribonucleic Acid Polymerases." *J. Biol. Chem.* 247:241–248.

21. Livingston, D. M., and C. C. Richardson. 1975. "Deoxyribonucleic Acid Polymerase III of *Escherichia coli*: Characterization of Associated Exonuclease Activities." *J. Biol. Chem.* 250:470–478.

22. Ollis, D. L., P. Brick, R. Hamlin, N. G. Xuong, and T. A. Steitz. 1985. "Structure of Large Fragment of *Escherichia coli* DNA Polymerase I Complexed with dTMP." *Nature* 313:762–765.

23. Scott, J., and A. Kornberg. 1978. "Purification of the Rep Protein of *E. coli*. An Enzyme Which Separates Duplex Strands in Adenovirus Replication." *J. Biol. Chem.* 253:3292–3304.

24. Yarranton, G. T., R. H. Das, and M. L. Gefter. 1979. "Enzyme-Catalyzed DNA Unwinding." *J. Biol. Chem.* 254:11997–12001.

25. Geider, K., and H. Hoffmann-Berling. 1981. "Proteins Controlling the Helical Structure of DNA." *Ann. Rev. Biochem.* 50:233–260.

26. Arai, N., K. Arai, and A. Kornberg. 1981. "Complexes of Rep Protein with ATP and DNA as a Basis for Helicase Action." *J. Biol. Chem.* 256:5287–5293.

27. Sancar, A., K. R. Williams, J. W. Chase, and W. D. Rupp. 1981. "Sequences of the *ssb* Gene and Protein." *Proc. Nat. Acad. Sci.* 78:4274–4278.

28. Brutlag, D., R. Schekman, and A. Kornberg. 1971. "A Possible Role for RNA Polymerase in the Initiation of M13 DNA Synthesis." *Proc. Nat. Acad. Sci.* 68:2826–2829.

29. McMacken, R., K. Ueda, and A. Kornberg. 1977. "Migration of *Escherichia coli dnaB* Protein on the Template DNA Strand as a Mechanism in Initiating DNA Replication." *Proc. Nat. Acad. Sci.* 74:4190–4194.

30. Arai, K., and A. Kornberg. 1981. "Unique Primed Start of Phage φX174 DNA Replication and Mobility of the Primosome in a Direction Opposite Chain Synthesis." *Proc. Nat. Acad. Sci.* 78:69–73.

31. Ogawa, T., T. A. Baker, A. van der Ende, and A. Kornberg. 1985. "Initiation of Enzymatic Replication at the Origin of the *Escherichia coli* Chromosome: Contributions of RNA Polymerase and Primase." *Proc. Nat. Acad. Sci.* 82:3562–3566.

32. Wickner, S. H. 1978. "DNA Replication Proteins of *Escherichia coli*." *Ann. Rev. Biochem.* 47:1163–1191.

33. McHenry, C., and A. Kornberg. 1977. "DNA Polymerase III Holoenzyme of *Escherichia coli*: Purification and Resolution into Subunits." *J. Biol. Chem.* 252:6478–6484.

34. Alberts, B. M., B. P. Bedinger, T. Formosa, C. V. Jongeneel, and K. N. Kreuzer. 1982. "Studies on DNA Replication in the Bacteriophage T4 in Vitro Systems." *Cold Spring Harbor Symp. Quant. Biol.* 47:655–668.

35. Kornberg, A. 1982. *1982 Supplement to DNA Replication.* San Francisco: Freeman, pp. 124–127.

36. Scheuermann, R. H. and H. Echols. 1984. "A Separate Editing Exonuclease for DNA Replication: The ε Subunit of *Escherichia coli* DNA Polymerase III Holoenzyme." *Proc. Nat. Acad. Sci.* 81:7747–7751.

37. DeLucia, P., and J. Cairns. 1969. "Isolation of an *E. coli* Strain with a Mutation Affecting DNA Polymerase." *Nature* 224:1164–1166.

38. Wechsler, J. A., and J. D. Gross. 1971. *Escherichia coli* Mutants Temperature-Sensitive for DNA Synthesis." *Mol. Gen. Genetics* 113:273–284.

39. Dumas, L. B. 1978. "Requirements for Host Gene Products in Replication of Single-Stranded Phage DNA in Vivo." In *The Single-Stranded DNA Phages,* ed. D. T. Denhardt, D. Dressler, and D. S. Ray. Cold Spring Harbor, N.Y.: Cold Spring Harbor Laboratory, pp. 341–359.

40. Wechsler, J. A. 1978. "The Genetics of *E. coli* Replication." In *DNA Synthesis: Present and Future,* ed. M. Kohiyama and I. Molineux. New York: Plenum Press, p. 49.

41. Hirota, Y., M. Yamada, A. Nishimura, A. Oka, K. Sugimoto, K. Asada, and M. Takanami. 1981. "The DNA Replication Origin (*ori*) of *Escherichia coli*: Structure and Function of the *ori*-Containing DNA Fragment." *Prog. Nucleic Acid Res.* 26:33–48.

42. Von Meyenburg, K., and F. G. Hansen. 1980. "The Origin of Replication, *ori* C, of the *Escherichia coli* Chromosome: Genes Near *ori* C and Construction of *ori* C Deletion Mutations." *ICN-UCLA Symp.* 19:137–159.

43. Kaguni, M., and A. Kornberg. 1984. "Replication Initiated at the Origin (*oriC*) of the *E. coli* Chromosome Reconstituted with Purified Enzymes." *Cell* 38:183–190.

44. Marians, K. J. 1984. "Enzymology of DNA in Replication in Prokaryotes." *CRC Crit. Rev. Biochem.* 17:153–215.

45. Nossal, N. G. 1983. "Prokaryotic DNA Replication Systems." *Ann. Rev. Biochem.* 53:581–615.

46. Grosschedl, R., and G. Hobom. 1979. "DNA Sequences and Structural Homologies of the Replication Origins of Lambdoid Bacteriophages." *Nature* 277:621–627.

47. Tomizawa, J. 1984. "Control of ColE1 Plasmid Replication: The Process of Binding of RNAI to the Primer Transcript." *Cell* 38:861–870.

48. Tsurimoto, T., and K. Matsubara. 1984. "Multiple Initiation Sites of DNA Replication Flanking the Origin Region of λdv Genome." *Proc. Nat. Acad. Sci.* 81:7402–7406.

49. LeBowitz, J. H., M. Zylicz, C. Georgopoulos, and R. McMacken. 1985. "Initiation of DNA Replication on Single-Stranded DNA Templates Catalyzed by Purified Replication Proteins of Bacteriophage λ and *Escherichia coli*." *Proc. Nat. Acad. Sci.* 82:3988–3992.

50. Van der Ende, A., T. A. Baker, T. Ogawa, and A. Kornberg. 1985. "Initiation of Enzymatic Replication at the Origin of the *Escherichia coli* Chromosome: Primase as the Sole Priming Enzyme." *Proc. Nat. Acad. Sci.* 82:3954–3958.

51. Watson, J. D. 1972. "Origin of Concatemeric T7 DNA." *Nature New Biol.* 239:197–201.

52. Baroudy, B. M., S. Venkatesan, and B. Moss. 1982. "Structure and Replication of Vaccinia Virus Telomeres." *Cold Spring Harbor Symp. Quant. Biol.* 47:723–729.

53. Shampay, J., J. W. Szostak, and E. H. Blackburn. 1984. "DNA Sequences of Telomeres Maintained in Yeast." *Nature* 310:154–157.

54. Sundin, O., and A. Varshavsky. 1981. "Arrest of Segregation Leads to Accumulation of Highly Intertwined Catenated Dimers: Dissection of the Final Stages of SV40 DNA Replication." *Cell* 25:659–669.

55. DiNardo, S., K. Voelkel, and R. Sternglanz. 1984. "DNA Topoisomerase II Mutant of *Saccharomyces cerevisiae*: Topoisomerase II Is Required for Segregation of Daughter Molecules at the Termination of DNA Replication." *Proc. Nat. Acad. Sci.* 81:2616–2620.

56. Gilbert, W., and D. Dressler. 1968. "DNA Replication: The Rolling Circle Model." *Cold Spring Harbor Symp. Quant. Biol.* 33:473–484.

57. Dressler, D., D. Hourcade, K. Koths, and J. Sims. 1978. "The DNA Replication Cycles of the Isometric Phages." In *The Single-Stranded DNA Phages,* ed. D. T. Denhardt, D. Dressler, and D. S. Ray. Cold Spring Harbor, N.Y.: Cold Spring Harbor Laboratory, pp. 187–214.

58. Bastia, D., N. Sueoka, and E. Cox. 1975. "Studies in the Late Replication of Phage Lambda: Rolling Circle Replication of the Wild Type and a Partially Suppressed Strain Oam29Pam80." *J. Mol. Biol.* 98:305–320.

59. McMacken, R., L. Rowen, K. Ueda, and A. Kornberg. 1978. "Priming of DNA Synthesis on Viral Single-Stranded DNA in Vitro." In *The Single-Stranded DNA Phages,* ed. D. T. Denhardt, D. Dressler, and D. S. Ray. Cold Spring Harbor, N.Y.: Cold Spring Harbor Laboratory, pp. 273–285.

60. Wickner, S. 1978. "Conversion of Phage Single-Stranded DNA to Duplex DNA in Vitro." In *The Single-Stranded DNA Phages,* ed. D. T. Denhardt, D. Dressler, and D. S. Ray. Cold Spring Harbor, N.Y.: Cold Spring Harbor Laboratory, pp. 255–271.

61. Eisenberg, S., J. F. Scott, and A. Kornberg. 1978. "Enzymatic Replication of φX174 Duplex Circles: Continuous Synthesis." *Cold Spring Harbor Symp. Quant. Biol.* 43:295–302.

62. Warren, G. J., A. J. Twigg, and D. J. Sherratt. 1978. "ColEl Plasmid Mobility and Relaxation Complex." *Nature* 274:259–261.

63. Sussenbach, J. S., P. C. van der Vliet, D. J. Ellens, and H. S.

Jansz. 1972. "Linear Intermediates in the Replication of Adenovirus DNA." *Nature New Biol.* 239:47–49.

64. Rekosh, D. M. K., W. C. Russell, A. J. D. Bellett, and A. J. Robinson. 1977. "Identification of a Protein Linked to the Ends of Adenovirus DNA." *Cell* 11:283–295.

65. Kelly, T. J., Jr., and R. L. Lechner. 1978. "The Structure of Replicating Adenovirus DNA Molecules: Characterization of DNA-Protein Complexes from Infected Cells." *Cold Spring Harbor Symp. Quant. Biol.* 43:721–728.

66. Challberg, M. D., and T. J. Kelly, Jr. 1979. "Adenovirus DNA Replication in Vitro." *Proc. Nat. Acad. Sci.* 76:655–659.

67. Lichy, J. H., K. Nagata, B. R. Friefeld, T. Enomoto, J. Field, R. A. Guggenheimer, J.-E. Ikeda, M. S. Horwitz, and J. Hurwitz. 1982. "Isolation of Proteins Involved in the Replication of Adenoviral DNA in Vitro." *Cold Spring Harbor Symp. Quant. Biol.* 47:731–740.

68. Sogo, J. M., J. A. García, M. A. Peñalva, and M. Salas. 1982. "Structure of Protein-Containing Replicative Intermediates of *Bacillus subtilis* Phage ϕ29 DNA." *Virology* 116:1–18.

69. Stillman, B. W., and F. Tamanoi. 1982. "Adenoviral DNA Replication: DNA Sequences and Enzymes Required for Initiation in Vitro." *Cold Spring Harbor Symp. Quant. Biol.* 47:741–750.

70. Blanco, L., and M. Salas. 1984. "Characterization and Purification of a Phage ϕ29-encoded DNA Polymerase Required for the Initiation of Replication." *Proc. Nat. Acad. Sci.* 81:5325–5329.

71. Watabe, K., M. Leusch, and J. Ito. 1984. "Replication of Bacteriophage ϕ29 DNA *in Vitro:* The Roles of Terminal Protein and DNA Polymerase." *Proc. Nat. Acad. Sci.* 81:5374–5378.

72. Grummt, F. 1978. "Diadenosine 5′,5‴-P^i,P^4-Tetraphosphate Triggers Initiation of in Vitro DNA Replication in Baby Hamster Kidney Cells." *Proc. Nat. Acad. Sci.* 75:371–375.

73. Kornberg, A. 1980. "Regulation of Replication." In *DNA Replication.* San Francisco: Freeman, Chap. 13.

74. Kornberg, A. 1982. *1982 Supplement to DNA Replication.* San Francisco: Freeman, pp. S144–S146.

75. Kornberg, A. 1982. "Eukaryotic DNA Polymerase." In *1982 Supplement to DNA Replication.* San Francisco: Freeman, Chap. S6.

76. Li, J. J., and T. J. Kelly. 1984. "Simian Virus 40 DNA Replication in Vitro." *Proc. Nat. Acad. Sci.* 81:6973–6977.

77. Stillman, B. W., and Y. Gluzman. 1985. "Replication and Supercoiling of SV40 DNA in Cell-Free Extracts from Human Cells." *Mol. Cell. Biol.* 5:2051–2060.

CHAPTER 11

Recombination at the Molecular Level

All DNA is recombinant DNA. In engineering relatively simple changes in gene expression by combining and modifying DNA segments in the test tube, we are only simulating natural processes of recombination and mutation that have acted throughout evolution to construct complex chromosomes. Genetic exchange works constantly to blend and rearrange chromosomes, most obviously during meiosis, when homologous chromosomes cross over in every generation. Genetic maps derived from early measurements of the frequency of crossing over between different genes gave the first real information about chromosome structure: that genes are arranged in a fixed, linear order. A most important recent discovery is that gene order does change, although rarely: Movable DNA segments called transposons occasionally jump around chromosomes, thus fundamentally altering chromosomal structure. In addition to neatly moving genes, transposons also scramble DNA, making deletions, inversions, and other rearrangements. It is becoming clear that such changes are a critical feature of chromosome evolution, particularly in eucaryotic cells.

We now appreciate that recombination is not accidental, but is instead an essential cellular process catalyzed by enzymes that cells encode and regulate for the purpose. Besides providing genetic variation, recombination enzymes allow cells to retrieve sequences lost when DNA is damaged by radiation or chemical accidents, by replacing the damaged section with an undamaged DNA strand from a homologous chromosome. Furthermore, special types of recombination regulate gene expression. By switching specific segments within chromosomes, cells put dormant genes into sites where they can be expressed, even creating new protein-coding regions. Moreover, recent discoveries, in revealing the molecular mechanisms of recombination, have provided new tools for the deliberate manipulation of genes.

Crossing Over Is Due to Breakage and Rejoining of Intact DNA Molecules[1]

Until recently, there was not even a superficial understanding of the molecular basis of crossing over. The classical picture of crossing over, developed in the 1930s from cytological observations, hypothesized that during meiosis, the paired, coiled chromosomes were sometimes physically broken at the chromatid level as a result of ten-

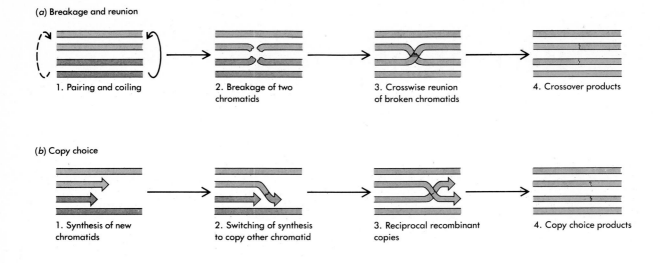

Figure 11-1
Representation of two possible mechanisms originally considered for crossing over between chromosomes. Each line represents a separate chromatid. According to the current idea that each chromatid contains a single double-helical DNA molecule, and our understanding of semiconservative replication of the double helix, a simple copy-choice mechanism with no strand breakage would not be possible.

sion created by their contraction. The broken ends could then relieve the tension by crossways reunion, creating two reciprocally recombinant chromatids (Figure 11-1a). According to this model, recombination occurs after chromosome duplication is complete—that is, when there are four chromatids in each pair of synapsed chromosomes. This hypothesis fell into disfavor about 1955, when geneticists found that crossing over occurs within the gene, by then realized to be part of a DNA molecule. A seemingly inconvenient consequence was that to produce exactly reciprocal recombinants, the breakage points must necessarily lie between the same nucleotides in the two homologous chromatids. Otherwise, recombination would generate new DNA molecules differing in length from the parental molecules.

As a result of this dilemma, enthusiasm developed for a hypothesis relating recombination to chromosome duplication. This alternative hypothesis proposes that during replication of the paired chromosomes, the new DNA strand being formed along the paternal chromosome (for example) switches to the maternal one that it thereafter copies. If the complementary replica of the maternal strand also switches templates when it reaches the same point, two reciprocally recombinant strands are formed. This hypothetical process is called **copy choice.** A fundamental distinction between the two hypotheses lies in their prediction of the physical origin of recombinant chromosomes: Either the recombinant chromosomes inherit physical material from the two parental chromosomes following breakage and reunion, or the recombinant chromosomes produced by copy choice are synthesized from new material.

These two hypotheses were tested by experiments using isotopically heavy (^{13}C, ^{15}N) parental phage λ particles (Figures 11-2 and 11-3). Here, again, the heavy isotopes were used so that a cesium chloride (CsCl) gradient could distinguish between parental and daughter DNA strands. Genetic crosses between heavy (or between heavy and light) phage particles were made in *E. coli* cells growing in a light (^{12}C, ^{14}N) medium. Under these conditions, all of the newly synthesized viral DNA molecules are derived from light precursors. Thus, if copy choice is the correct mechanism, all the recombinant particles should be light. On the other hand, if recombinants are derived by breakage and reunion, some of the recombinant phage particles will contain heavy atoms derived from the parental chromo-

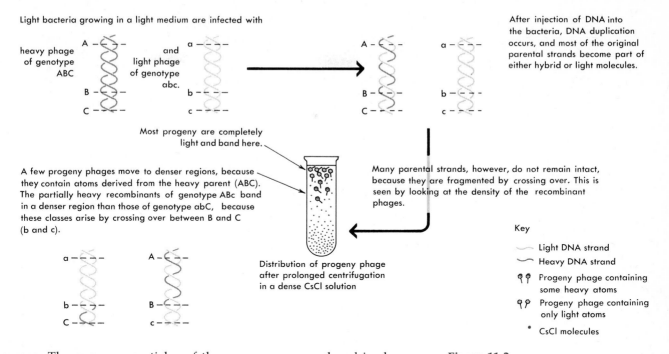

Light bacteria growing in a light medium are infected with

heavy phage of genotype ABC

and light phage of genotype abc.

After injection of DNA into the bacteria, DNA duplication occurs, and most of the original parental strands become part of either hybrid or light molecules.

Most progeny are completely light and band here.

A few progeny phages move to denser regions, because they contain atoms derived from the heavy parent (ABC). The partially heavy recombinants of genotype ABc band in a denser region than those of genotype abC, because these classes arise by crossing over between B and C (b and c).

Many parental strands, however, do not remain intact, because they are fragmented by crossing over. This is seen by looking at the density of the recombinant phages.

Distribution of progeny phage after prolonged centrifugation in a dense CsCl solution

Key

⁓ Light DNA strand

⁓ Heavy DNA strand

φ φ Progeny phage containing some heavy atoms

φ φ Progeny phage containing only light atoms

• CsCl molecules

somes. The progeny particles of these crosses were placed in dense CsCl solutions and rapidly centrifuged to separate particles of different densities. Phage particles of various densities were then collected and genetically tested to see which were recombinants. The results were clear-cut, and to the surprise of most molecular biologists, showed that some recombinant particles contained heavy atoms (see Figure 11-2). Moreover, further experiments revealed that recombination can occur between nonreplicating DNA molecules (see Figure

Figure 11-2
The employment of heavy isotopes to study the mechanism of crossing over.

Figure 11-3
The demonstration that crossing over and DNA duplication are independent phenomena.

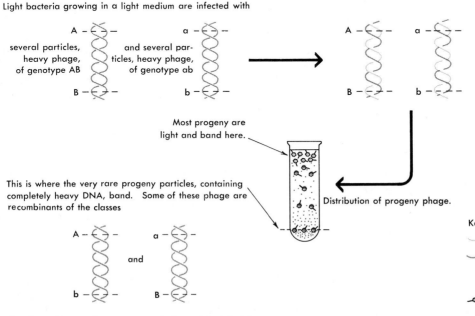

Light bacteria growing in a light medium are infected with

several particles, heavy phage, of genotype AB

and several particles, heavy phage, of genotype ab

After injection of the DNA molecules into the bacteria, most heavy strands become part of hybrid molecules. Rarely, however, an infecting molecule fails to duplicate and, when the new progeny particles are formed, these unreplicated molecules, in which the parental strands have never separated, become enclosed in new protein shells. This phenomenon enables us to ask whether the rare, completely heavy, DNA molecules are ever recombinants. Again use is made of a CsCl gradient to separate progeny particles of different density.

Most progeny are light and band here.

This is where the very rare progeny particles, containing completely heavy DNA, band. Some of these phage are recombinants of the classes

and

Distribution of progeny phage.

This shows that crossing over occurs between intact double helices and that extensive DNA synthesis is not involved in crossing over.

Key

⁓ Light DNA strand

⁓ Heavy DNA strand

• CsCl molecule

φ φ Progeny phage containing only light atoms

φ φ Progeny phage containing some heavy atoms

11-3). Breakage and reunion of intact double helices must therefore be the primary mechanism of crossing over in bacteriophage. In fact, the same is true in bacteria and higher organisms. Physical exchange of DNA between sister chromatids of mammalian cells can be visualized dramatically in tissue culture cells by means of a staining procedure that allows the chromatids to be distinguished (Figure 11-4).

Homologous Recombination Is Guided by Base Pairing

In retrospect it is obvious what mechanism most precisely aligns DNA molecules in crossing over, because we can hardly imagine any other: Complementary base-pairing between strands unwound from two different chromosomes puts the chromosomes in exact register. Crossing over thus generates **homologous recombination;** that is, it occurs between two regions of DNA containing identical or nearly identical sequences. Usually, these are sequences on two equivalent regions of homologous chromosomes, but crossing over also can be detected between homologous segments in nonequivalent regions, as long as the recombinant survives. Such unequal crossing over gives rise to duplications and deletions in chromosomes (Figure 7-25). In the laboratory, recombination between homologous segments on different DNAs is now used to construct new genetic variants. Thus, when a yeast DNA fragment cloned in a circular bacterial plasmid is introduced into a yeast cell, it finds and crosses over with its homologous sequence, thereby inserting the plasmid into a yeast chromosome (see Figure 19-3). At least in bacteria and yeast, any two homologous DNA segments will recombine in the cell, given one condition: One of them must have a break or a region of single-stranded DNA.

Recombination Is Initiated from Breaks or Gaps in DNA[2, 3]

A strong hint about the way crossing over begins is that breaks in DNA greatly stimulate it. Thus, ultraviolet irradiation, X-irradiation, and reactive chemicals that create double-strand breaks or gaps (single-stranded regions in otherwise double-stranded DNA) all increase crossing over. Mutations in bacterial genes for DNA ligase and DNA polymerase I that give rise to breaks and gaps by preventing proper sealing of DNA during replication or repair have the same effect. More direct proof is that a plasmid bearing a yeast DNA segment can be made to recombine with a yeast chromosome (as shown in Figure 19-3) up to a thousand times more efficiently by cutting the yeast segment with a restriction enzyme before the plasmid is introduced into the cell. Now we suspect that in a similar way, cellular enzymes actively induce recombination in meiosis by making occasional breaks along a pair of synapsed chromosomes. In fact, it is likely that all homologous recombination is initiated from discontinuities in DNA, whether their source is apparent or not. The usual semi-random nature of crossing over that allows a genetic map to be constructed probably results from these initiating sites occurring nearly at random.

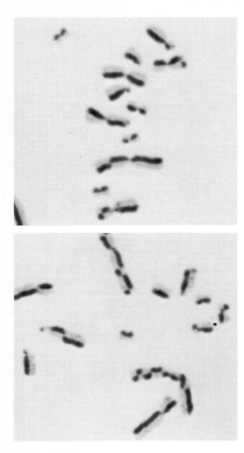

Figure 11-4
Reciprocal crossing over directly visualized in hamster cells in tissue culture. Chromosomes whose DNA contains bromodeoxyuridine in place of thymidine in both strands appear light after treatment with Giemsa stain, whereas those containing DNA substituted in only one strand appear dark. After two generations of growth in bromodeoxyuridine, one newly replicated chromatid has only one of its strands substituted, whereas its sister has both substituted. Thus, sister chromatids can be distinguished by staining. Then crossovers, called sister chromatid exchanges, are easily detected as alternating lengths of light and dark (top). Such exchanges increase dramatically when cells are treated by agents that damage DNA, like the drug adriamycin (bottom). (This is not *meiotic* crossing over between homologous chromosomes, which serves to create genetic diversity, but is an equivalent process occurring between *mitotic* sister chromosomes before they separate, which serves to repair DNA damage.) (Courtesy of Sheldon Wolff and Judy Bodycote.)

RecA Protein Matches Single-Stranded DNA to a Complementary Sequence in a Target Chromosome[4, 5, 6]

From what we know about the physical properties of DNA, there is no reason to expect that two molecules will break and rejoin simply because one is already gapped or broken. Indeed, genetic analysis shows that phage, bacteria, and yeast have enzymes to catalyze the biochemical steps in recombination, a fact that is no doubt true of all organisms. Our most detailed understanding of the mechanism of recombination comes from studies of the *recA* gene product of *E. coli*, an enzyme that carries out the fundamental reaction in recombination—that of making the base-paired hybrid that joins two DNA molecules.

The enzymatic properties of **RecA protein** show why gapped or broken DNA is required to initiate recombination. No reaction occurs if two homologous but completely base-paired helices are mixed with RecA protein. However, RecA protein specifically recognizes single-stranded DNA and anneals it to a complementary sequence in a homologous duplex, simultaneously displacing the resident strand (Figures 11-5 and 11-6). RecA protein first binds single-stranded DNA in a fixed ratio of about one polypeptide per five nucleotides, thus forming a DNA-protein filament; if ATP is present, RecA protein in the filament may then unwind successively different parts of the duplex, attempting to anneal the bound single strand to unwound regions until a complementary sequence is found. Once a small portion is

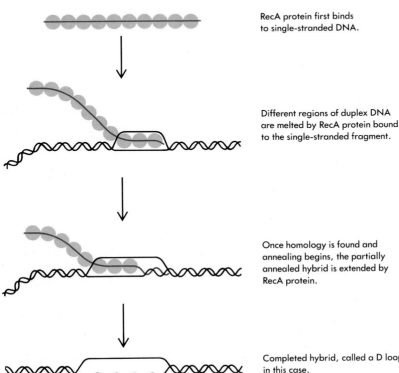

RecA protein first binds to single-stranded DNA.

Different regions of duplex DNA are melted by RecA protein bound to the single-stranded fragment.

Once homology is found and annealing begins, the partially annealed hybrid is extended by RecA protein.

Completed hybrid, called a D loop in this case.

Figure 11-5
Activity of RecA protein as demonstrated in vitro in a model reaction using DNA fragments.

Figure 11-6
Electron micrograph of RecA protein exchanging DNA strands in vitro. In this case, the reacting DNAs are a circular single strand and a linear double-stranded fragment that is homologous to a portion of the single-stranded circle. In the reaction, one strand of the double-stranded fragment is displaced by its equivalent segment in the circle. The final products are this displaced linear single strand, probably bound to the excess RecA protein that was present in the reaction mixture, and the mostly single-stranded circle that now contains a double-stranded region. Preparation of the sample for microscopy could have disrupted part of the structure, for example, the hypothetical complex of RecA protein and three DNA strands shown in Figure 11-7. The accompanying diagram identifies the structures seen in the electron micrograph. [Adapted from C. Radding, J. Flory, A. Wu, R. Kahn, C. Das Gupta, D. Gorda, M. Bianchi, and S. Tsang, *Cold Spring Harbor Symp.* 47 (1983):824, with permission.]

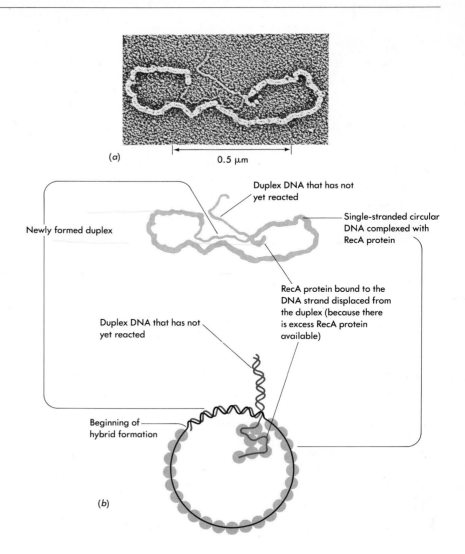

(a)

0.5 μm

Newly formed duplex

Duplex DNA that has not yet reacted

Single-stranded circular DNA complexed with RecA protein

RecA protein bound to the DNA strand displaced from the duplex (because there is excess RecA protein available)

Duplex DNA that has not yet reacted

Beginning of hybrid formation

(b)

properly annealed, the energy of ATP drives the pairing reaction to completion, moving in the 5′→3′ direction relative to the single strand. RecA protein is displaced as new hybrid DNA is formed. We describe below how this same reaction can form a bridge between two separate duplex DNAs, if one of them contains a single-stranded region. Breaks as well as gaps in DNA initiate recombination because they provide sites for nucleases to degrade one strand of the duplex, or for unwinding enzymes to enter, exposing single-stranded DNA that reacts with RecA protein.

Phages such as T7 and λ (lambda) encode simpler enzymatic devices to promote recombination: nucleases to expose single-stranded DNA and DNA-binding proteins to extend single-stranded DNA and facilitate annealing by random collisions. Perhaps these devices succeed because the concentration of phage sequences is high in infected cells and there is a good chance that complementary single-stranded regions will exist. Higher organisms, on the other hand, are expected to require sophisticated enzymes like RecA protein to search for homology because the abundance of a given sequence is very low. Indeed, there is already evidence that the fungus *Ustilago* has such an enzyme.

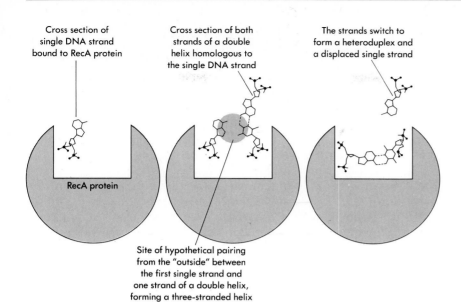

Cross section of single DNA strand bound to RecA protein

RecA protein

Cross section of both strands of a double helix homologous to the single DNA strand

The strands switch to form a heteroduplex and a displaced single strand

Site of hypothetical pairing from the "outside" between the first single strand and one strand of a double helix, forming a three-stranded helix

Figure 11-7
How a three-stranded helix might be involved in DNA strand exchange mediated by RecA protein. [Adapted from Howard-Flanders, P., S. West, and A. Stasiak, *Nature* 309 (1984):215–220, with permission.]

The Possibility of Three-Stranded and Four-Stranded DNA Helices in Recombination[7]

Initially it seemed likely that the specific pairing reaction that RecA protein promotes between single-stranded DNA and its complementary sequence in one strand of a double helix could occur only if the double helix first unwound so that new base pairing could occur. However, there is another, rather intriguing possibility: The single-stranded DNA might align in specific register with the duplex by wrapping around it. Model building shows that base-pair-specific contacts can be made between two separate duplexes, entwined in a four-stranded structure that is aligned by chemical groups exposed in the grooves of the two helices. Similarly, a single DNA strand bound to RecA protein might make these same specific contacts in a groove of the target duplex DNA and thus recognize its complementary sequence from the "outside" (Figure 11-7). Such a three-stranded helix would represent the first base-specific recognition and would be followed by unwinding of the duplex and formation of normal base pairing between the invading single strand and its complement. Clearly, initial pairing in a three-stranded helix should speed the search for homology, since each segment examined need not be first unwound.

Recombination Involves a Crossed-Strand Intermediate[8, 9]

The cross over intermediate that joins two DNA duplexes in recombination contains four DNA strands that switch pairing at the joining point to form a crossed-strand junction called the **Holliday structure,** after R. Holliday, who first proposed it (Figure 11-8). The existence of the Holliday structure was originally suggested by genetic experiments that detected heteroduplex regions in both recombining DNAs. **Heteroduplexes** are regions on recombinant DNA molecules where the two strands are not exactly complementary. Model building

(a)

Exchange site

(b)

Figure 11-8
The Holliday structure. (a) Photograph of a molecular model of sections of two double helices, showing that crossed-strand connections can be made without disruption of the individual DNA molecules. [Courtesy of N. Sigel and B. Alberts, *J. Mol. Biol.* 71 (1972):789.] (b) The mutual interchange of parallel strands between double helices (branch migration) as a consequence of right-handed axial rotation. [After T. Broker, *J. Mol. Biol.* 81 (1973):1.]

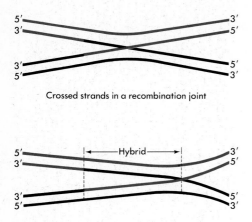

Crossed strands in a recombination joint

Diffusion of branch rightward ("branch migration"),
creating extensive hybrid regions in both duplexes

Figure 11-9
Diffusion of a recombination joint.

shows that free rotation of the backbone permits a strand to move from one double helix to another in the Holliday joint without loss of any potential base pairs or the creation of strained chemical bonds. Once two helices are so joined, the cross connection can easily diffuse by a zipperlike action in which equivalent bases on the two original molecules exchange places, a process called **branch migration** (see Figure 11-9). In this way, extensive sections of a given chain can move from one helix to another, frequently generating long sections of hybrid DNA that may contain heteroduplex regions.

Thus, strands can be exchanged between two DNA duplexes for some distance from the initiating break. How fully recombinant duplexes form can be understood from a second physical property of the structure: its ability to undergo isomerization, in which a simple steric rearrangement converts the bridging strand pair into outside strands

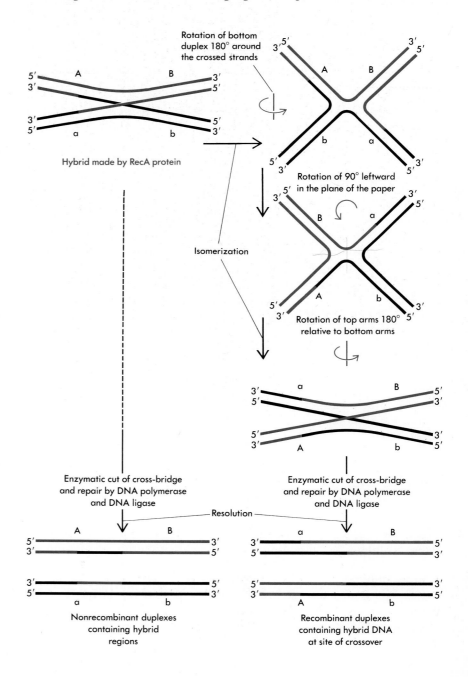

Figure 11-10
Isomerization of the recombination joint and resolution of two different isomers.

and vice versa (Figure 11-10). Such isomerizations can occur very frequently, so that a nuclease cut that completes crossing over (a process called resolution) can lead to a double exchange of two progeny strands or to single exchanges of all four strands, thus recombining the flanking arms of the chromosomes (Figure 11-10).

Direct proof that strand exchange occurs comes from electron microscopic visualization of phage DNA molecules isolated from infected cells, where recombination is frequent. The clearest results are obtained from phage λ, where denaturation mapping shows that cross-bridges always link together homologous regions of the pairing partners (Figure 11-11). Moreover, the individual DNA strands can be visualized in the bridging region, showing unambiguously their passage from one double helix to another.

Heteroduplexes

The frequent generation of heteroduplex segments during crossing over provides direct experimental support for the hypothesis that the fundamental recombination event involves pairing between regions of single-stranded DNA. Heteroduplexes arise when the primary pairing region, together with the adjacent region of branch migration (see above), encompasses sites of genetic differences between the two parental chromosomes. There is evidence of heteroduplex regions in all viruses whose genetics have been extensively analyzed. There is usually one such region, several thousand nucleotides long, on each T4 molecule. Heteroduplexes have also been well characterized in phage λ and again are much longer than might be expected if they reflected only the primary pairing event. So most of their length probably arises by branch migration transferring strands from one double helix to another.

Heteroduplex lifetime can be very short, because when the recombinant DNA molecules duplicate, the alternative alleles segregate out (Figure 11-12). This was first observed in phage systems, where mere

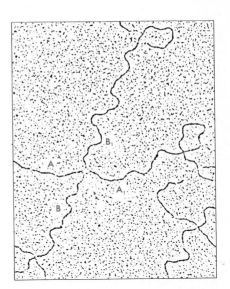

Figure 11-11
Electron micrograph visualization of crossover between two homologous regions of λ DNA. Here the Holliday joint has been partly unwound so that all four strands can be seen; this is the intermediate stage in isomerization (Figure 11-10). (Courtesy of R. Inman, University of Wisconsin.)

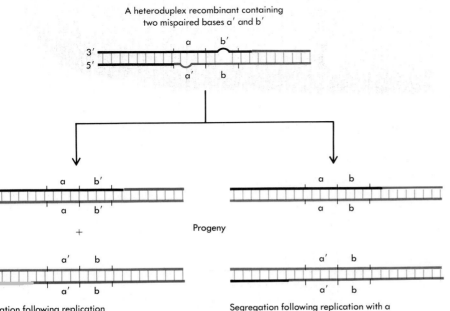

A heteroduplex recombinant containing two mispaired bases a' and b'

Progeny

+

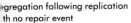
-gregation following replication
th no repair event

Segregation following replication with a
single repair event b' → b

Figure 11-12
Heteroduplex segregation with and without repair event.

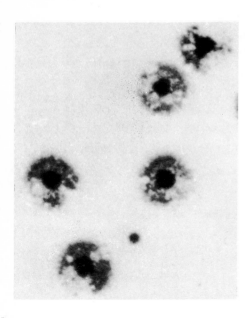

Figure 11-13
Photograph of several mottled plaques arising from segregation of a heteroduplex DNA molecule containing both *rII*+ and *rII* markers on opposing strands. (Courtesy of A. D. Hershey.)

inspection of recombinant plaques reveals mixtures of two genetically distinct types. When the DNA of a parental T4 phage particle contains a heteroduplex *rII* region, the resulting plaque has a mottled appearance, with some sections characteristic of the *rII* phenotype and others of the wild-type *rII*+ phenotype (Figure 11-13).

RecA Protein Makes Holliday Structures in Vitro[5]

Although RecA protein initially transfers only one strand into a DNA duplex, a second exchange follows naturally from the first. Figure 11-14 shows how RecA protein makes a complete recombination joint in vitro from two identical duplexes, one of which has a gap at one end. When annealing has been driven to the edge of the gap by RecA protein (see Figure 11-14c), the joint diffuses into the wholly duplex region as the top helix to the left of the gap unwinds, and simultaneously the second hybrid helix to the right of the crossover forms (see Figure 11-14d). It has been suggested that RecA protein may act in a second way at this stage, guiding the formation of a four-stranded helix composed of the two DNA duplexes in the region to the left of the exchange site, thereby accelerating the exchange of strands that allows the joint to move through the wholly duplex region (Figure 11-15).

To explain how recombination could be initiated in the cell when RecA protein binds to gaps in duplex DNA that are distant from an end, a refinement of the model reaction shown in Figure 11-14 is needed. A break must be made in the target DNA duplex, giving a free end from which the strands interwind to form a duplex. We sup-

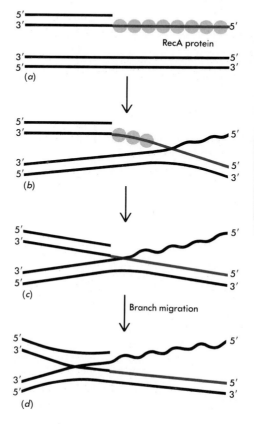

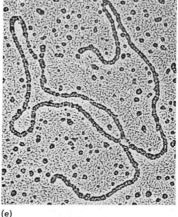

Figure 11-14
A model reaction by which RecA protein makes recombination joints in vitro. The two starting DNAs are identical, except that one is single-stranded at an end. The electron micrograph shows the product. (Courtesy of C. Radding.)

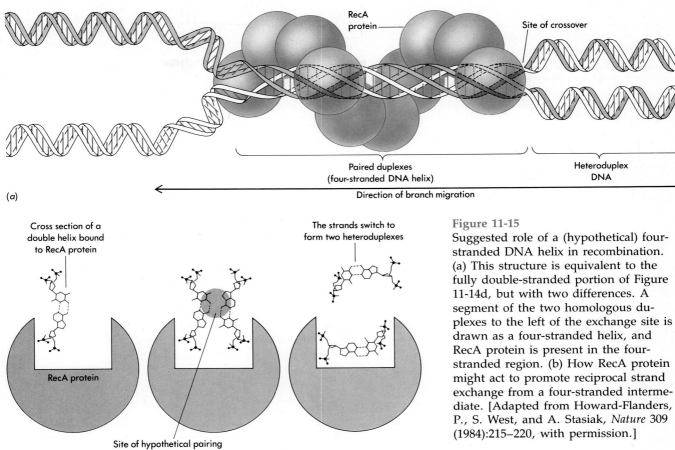

RecA protein

Site of crossover

Paired duplexes
(four-stranded DNA helix)

Heteroduplex
DNA

(a)

Direction of branch migration

Cross section of a
double helix bound
to RecA protein

The strands switch to
form two heteroduplexes

RecA protein

(b)

Site of hypothetical pairing
between two DNA duplexes
to form a four-stranded helix

Figure 11-15
Suggested role of a (hypothetical) four-stranded DNA helix in recombination. (a) This structure is equivalent to the fully double-stranded portion of Figure 11-14d, but with two differences. A segment of the two homologous duplexes to the left of the exchange site is drawn as a four-stranded helix, and RecA protein is present in the four-stranded region. (b) How RecA protein might act to promote reciprocal strand exchange from a four-stranded intermediate. [Adapted from Howard-Flanders, P., S. West, and A. Stasiak, *Nature* 309 (1984):215–220, with permission.]

pose that a nuclease recognizes an incipient hybrid and breaks the displaced strand, allowing the crossed-strand joint to form as it does in the original model reaction (Figure 11-16).

Alternatively, RecBC enzyme (described in the next section) may provide single-stranded DNA with a free end, which can be annealed to a homologous duplex as in the in vitro reaction illustrated in Figure 11-5. A subsequent cut in the DNA strand displaced from the duplex, followed by branch migration, would allow the Holliday joint to form.

RecBC Enzyme Initiates Recombination by Unwinding DNA and Nicking It at Specific Sites Named Chi[10, 11]

Although RecA protein carries out the fundamental pairing step of recombination, other proteins are involved in other steps of the process. Paramount among these is the RecBC enzyme, specified by the *E. coli* genes *recB* and *recC*, whose absence reduces the yield of recombinants after bacterial mating about a hundredfold. This complex enzyme with a molecular weight of about 300,000 daltons has both DNA-unwinding and nuclease activities that combine to expose single-stranded DNA with a free end, which in turn allows RecA protein to begin the pairing reaction. An important feature of the

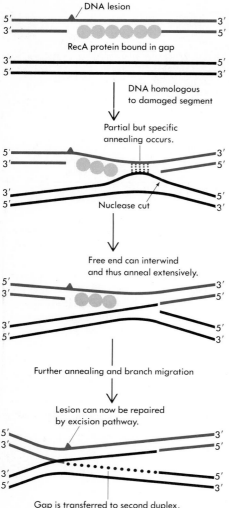

Figure 11-16

A likely pathway for initiation of recombination from a gap in vivo. Here the gap is shown to be adjacent to a **DNA lesion,** which is a chemical change in a DNA base that can destroy its ability to act as a template in replication and thus can interrupt synthesis of the complementary strand; the resulting gap is a region where the bare template DNA strand is exposed. DNA lesions and their repair are discussed in Chapter 12.

Figure 11-17

How the RecBC enzyme provides single-stranded DNA to initiate recombination.

RecBC enzyme is that it initiates unwinding only on DNA containing a free duplex end. First binding to the duplex end, it uses energy in ATP to travel along the duplex, unwinding and rewinding DNA as it goes. Since the RecBC enzyme unwinds DNA faster than it rewinds the strands, the unwound segment becomes progressively longer as the enzyme moves down the DNA. This unwound structure appears in electron micrographs as two single-stranded loops ("rabbit ears"), because the sites of unwinding and winding are both held by the enzyme as it moves (Figure 11-17).

A second activity of RecBC enzyme that acts while it is unwinding DNA is critical to its function in recombination. Genetic experiments show that the RecBC enzyme promotes recombination most frequently in DNA containing a site named Chi, which has the nucleotide sequence 5'-GCTGGTGG-3'. When Chi is present in DNA being unwound, a specific nuclease activity of RecBC enzyme cuts the exposed single strand near Chi, thus preventing its being rewound as the enzyme progresses (Figure 11-17). The consequence is to leave behind a free single-stranded tail and a gap, both of which are sites where RecA protein can bind to initiate DNA strand exchange with a homologous sequence.

E. coli DNA has about a thousand Chi sequences, or about one per five genes, and thus offers many opportunities for RecBC enzyme action. These sequences are not usually available to the enzyme in a normal cell, whose DNA has no free ends, but are particularly important in bacterial conjugation. Here a DNA end is introduced by the

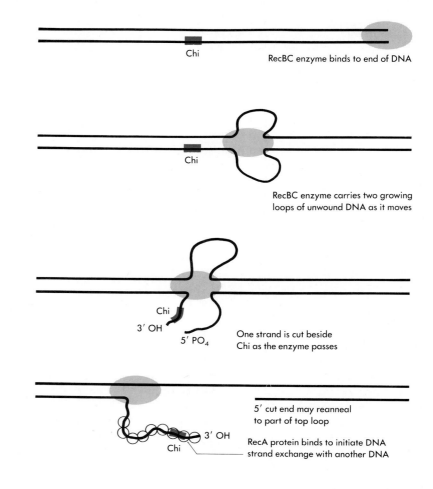

male bacterium, and it is thought that RecBC enzyme binds to the injected DNA at this end, moving along it and cutting at Chi sequences, thereby promoting recombination with the recipient cell's DNA. Ends also are introduced into a cell by phage P1 transduction and are present in phage λ–infected cells because free ends are made by the enzyme that cuts λ DNA for packaging into the phage particle; in both cases, the RecBC enzyme is required for efficient recombination. In fact, its function in phage λ recombination led to the discovery of Chi.

Other Proteins That Participate in Homologous Recombination[12]

Even in *E. coli*, we know only some of the genes and enzymes that contribute to recombination. Besides *recA*, *recB*, and *recC*, there are several other genes whose products are required for efficient recombination, although the specific functions of those products are unknown. One unidentified enzyme is the nuclease that cuts apart Holliday junctions and thus resolves recombinant DNA molecules. It is possible that the RecBC enzyme does this in addition to Chi-directed cutting, because it also has other, less specific nuclease activities.

Several proteins necessary for DNA synthesis also act in recombination. Single-stranded DNA-binding protein, which facilitates DNA synthesis by holding single-stranded DNA in an extended form accessible to replication enzymes (Chapter 10), also assists RecA protein in binding in an orderly way to single-stranded DNA. DNA polymerase fills gaps left when recombinant DNAs are cut apart, and finally, DNA ligase seals the nicks left by DNA polymerase.

Recombination Is Not Always Reciprocal at the Site of Crossing Over

Early investigation of crossing over between different genes revealed the seemingly obligatory occurrence of reciprocal recombinants (Figure 11-18). However, exceptions were discovered when the recombinants studied arose between nearby sites on the same gene. Then, nonreciprocal behavior was often observed. This phenomenon, called **gene conversion,** is best studied in organisms such as yeast or *Neurospora*, where all the products of a single meiotic event can be seen. Here, instead of always observing equal (2:2) segregation of the entering alleles, cases of 3:1 segregation are found. In these cases, one allele is lost, and an extra copy of the other appears.

The crossing-over hypothesis outlined earlier permits such exceptions. As shown in Figure 11-12, the final segregation pattern of genes localized around the crossover site may be affected by repair enzymes that recognize heteroduplex distortions in the double helix and randomly remove one of a mismatched base pair. Depending on the bases removed, either 2:2 or 3:1 ratios will be found. Seen from a broad perspective, however, the most striking thing about recombination, even at the molecular level, is still the prevalence of reciprocal recombinants. This is because heteroduplexes are short, relative to the distance between most recombining genes, so that neither of two genes under study is likely to be included in the heteroduplex that gives rise to a crossover between them.

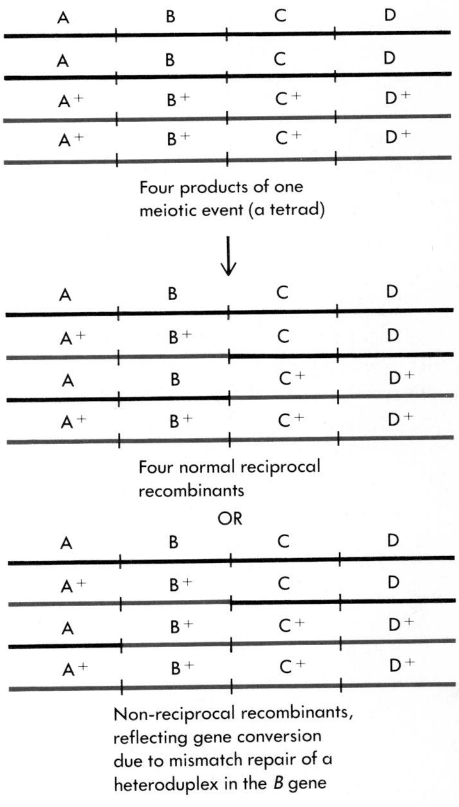

Four products of one meiotic event (a tetrad)

Four normal reciprocal recombinants
OR

Non-reciprocal recombinants, reflecting gene conversion due to mismatch repair of a heteroduplex in the *B* gene

Figure 11-18
Reciprocal and nonreciprocal recombination following crossover.

Insertions and Deletions
Arise from Errors in Crossing Over

The extreme accuracy of most crossing-over events depends on the correct juxtaposition of the complementary single-stranded regions during reformation of the hydrogen bonds. This correct alignment is assured by the uniqueness of most polynucleotide sequences. When a random chain of polynucleotides contains more than 10–12 nucleotides, it will almost never have the same sequence as another fragment of similar length. Thus, as long as the single-stranded segments are relatively long, it is very, very unlikely that a fragment of one gene will be mistakenly linked up to the wrong part of its own gene or to a different gene. In those rare cases, however, where two different chromosomal regions have considerable homology, we should expect occasional misjoinings, leading to the insertion or deletion of large blocks of nucleotides (Figure 11-19). Such accidental rearrangements are usually catalyzed by RecA protein, although other, unknown enzymes may sometimes be responsible.

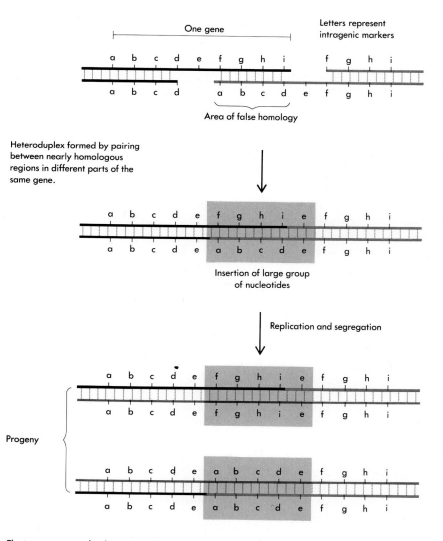

Figure 11-19

Origin of a multinucleotide insertion through pairing of nearly homologous sequences within the same gene.

Recombination Repairs Damaged DNA Molecules[3, 13]

We often think of crossing over as a mechanism that creates genetic diversity, but its most vital function is probably to repair DNA damage. This is revealed in studies of *recA⁻* bacteria and recombination-defective mutants of yeast, which are extremely sensitive to killing by radiation and by many chemicals that damage DNA. How recombinational repair works is implied by our description of how a gap in DNA leads to the recombination intermediate (see Figure 11-16). The gap is filled by DNA from a homologous duplex and is thus repaired. Repair is achieved whether or not the intermediate is cut so as to exchange flanking arms of the two helices; it only matters that the gap is filled with DNA of the correct sequence. Although a simple gap can be filled by DNA polymerase, a more serious problem is posed by gaps like the one shown in Figure 11-16 that contain DNA lesions such as thymine dimers that no longer can base-pair; these arise when replication necessarily stops at the damaged site and must reinitiate beyond it. The genetic information at the lesion is lost from both DNA strands and can be retrieved only by extracting it through recombination from a homologous duplex. The newly replicated sister DNA molecule is the most accessible source of the correct sequence, although a bacterial cell also usually has other copies of its chromosome.

A simple extension of this mechanism explains how RecA protein can repair the ultimate damage, a double-stranded break, which may be caused, for example, by X-irradiation. Degradation from each end of the break by an exonuclease leaves single-stranded ends, which then separately can invade a homologous sequence, forming two adjacent recombination joints (Figure 11-20). Alternatively, the RecBC enzyme might provide initiation sites by entering at each free end. When the two joints are cut, the parental helices are separated and

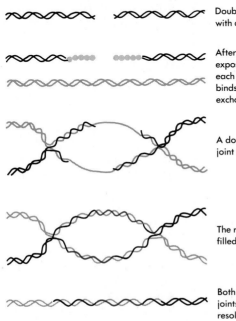

Double-strand break in DNA, with a missing segment

After unwinding, degradation by a nuclease exposes single-stranded DNA at each broken end; RecA protein binds and initiates strand exchange with homologous DNA

A double recombination joint is formed

The resulting gaps are filled by DNA polymerase

Both recombination joints are resolved, yielding two intact duplexes

Figure 11-20
Model for repair of a double-stranded DNA break by RecA protein. The fate of the single-stranded DNA that initiates the strand exchange is marked by colored shading.

the resulting gaps are filled by DNA polymerase and sealed by DNA ligase.

Site-Specific Recombination Gives Precise DNA Rearrangements[14–18]

By its very nature, crossing over generally preserves the order of DNA sequences in homologous chromosomes. However, in exceptional cases, cells also use a carefully regulated recombinational process that has the opposite consequence, rearranging sequences by recombination directed between special sites. DNA segments can be moved by **site-specific recombination,** often with the result that different genes or sets of genes are then expressed. The premier example, found in higher cells, is the construction of a great variety of specific antibody genes through site-specific DNA rearrangements of a single set of precursor sequences (Chapter 23).

Unlike crossing over, site-specific recombination is guided not by DNA sequence homology (although short homologies are involved), but by the location of binding sequences for specialized enzymes that catalyze breakage and rejoining. Site-specific recombination is initiated by regulatory processes that make available the correct enzymes, in contrast to the random breakage events that may expose any sequence to bind proteins like RecA and initiate crossing over.

Site-specific recombination was recognized first from genetic studies of bacteriophage λ, whose DNA integrates by recombination into a specific site on the *E. coli* chromosome and is carried thereafter in a passive state as a prophage (Chapter 17). There are two principal features of λ integration that are true of site-specific recombination in general: The exchange is reciprocal and preserves all preexisting DNA, and it occurs at a specific nucleotide within a short region of homology between the phage and bacterial DNA.

The Interacting Targets for Site-Specific Recombination by λ Integrase Share a Short Homologous Sequence But Differ Dramatically in Protein-Binding Sites[16, 17, 18]

Bacteriophage λ encodes an enzyme, **λ integrase,** that directs insertion of phage DNA into the *E. coli* chromosome by recombination across specific sites on both DNAs, generating one circular molecule out of two (Figure 11-21). Integrase is made abundantly in the early stage of phage infection, so that integration can occur in almost every infected cell. The reason λ integration is so well understood is that it occurs in vitro in a simple reaction mixture containing only four components: purified integrase protein, an *E. coli* accessory protein named **IHF** (for integration host factor), magnesium ions, and DNA containing the specific sites on phage and bacterial DNA (named *attP* and *attB*) across which recombination occurs. A convenient assay for integration uses a plasmid artificially constructed so that this single DNA contains both *attB* and *attP*, oriented such that the integration reaction recombines out the DNA between them, yielding two smaller DNA circles from the original circular plasmid (Figure 11-

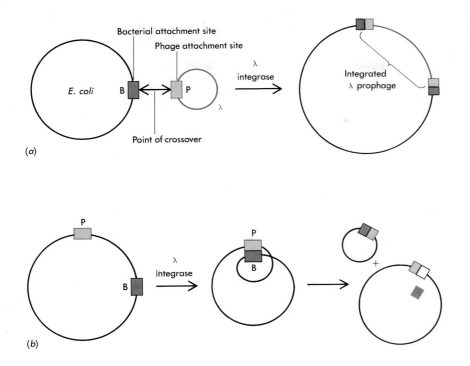

(a)

Figure 11-21
Integration of bacteriophage λ DNA into the *E. coli* chromosome by site-specific recombination. (a) The natural reaction, in which recombination joins two molecules. (b) An artificial substrate made by in vitro recombination to contain both phage and bacterial attachment sites. When integrase aligns and recombines P and B, the result is to produce two smaller DNA circles. Since these can be detected quickly by agarose gel electrophoresis, this reaction provides an easy assay for integrase activity.

(b)

21b). All steps of the integration reaction are coupled: Four strands are cut, exchanged, and religated without any stable intermediates appearing. This is much like a type II topoisomerase reaction (Chapter 9) in that both strands of each DNA duplex are cut and the phosphodiester bonds are remade without energy from ATP being required. In fact, integrase can act as a topoisomerase, relaxing supercoils that carry an *att* site or a similar sequence, in what is thought to be an aberrant variation of the normal integration reaction. As is true for topoisomerases, integrase makes a staggered cut, in this case with a seven-nucleotide overlap, in both the phage and bacterial sites (Figure 11-22).

Figure 11-22
Nucleotide sequence of both *attP* and *attB* core regions and of the hybrid recombinant sites. The core sequence is boxed, and vertical arrows mark the sites of exchange. The horizontal arrows between the strands are symmetrical integrase-binding sites that overlap the core sequence and serve to align integrase correctly at the sites of exchange.

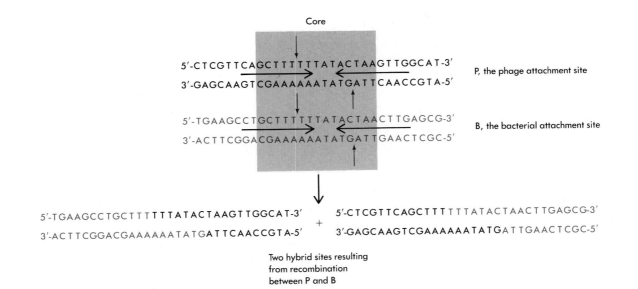

Two hybrid sites resulting from recombination between P and B

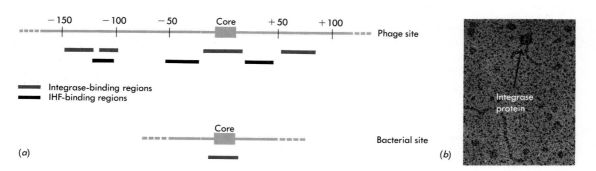

Figure 11-23
Binding of integrase and IHF to DNA.
(a) Structure of phage and bacterial attachment sites, showing where the integrase enzyme and the integration host factor bind to DNA. The scale is length in nucleotides. (b) Electron micrograph of integrase protein bound to a DNA restriction fragment containing *attP*. The condensation of DNA into the complex is evident from the greater apparent length of an identical DNA fragment (at the lower left) that is not bound by integrase. (Courtesy of Marc Better and Harrison Echols).

How does integrase choose these specific sites? Experiments to delimit the phage and bacterial sites by asking how much DNA adjacent to the actual exchange sites is required for integrase to bind and act show that *attP* encompasses a region of 250 base pairs, whereas *attB* consists of only about 20 base pairs. Both integrase and the host factor IHF bind at distinct places throughout *attP*, so that the integrase activity is fixed to phage DNA near the exchange site (Figure 11-23). The entire 250-nucleotide *attP* segment appears to be wound around integrase in a condensed nucleosome-like structure that may contain as many as eight integrase monomers of 40,000 daltons each (Figure 11-23b). The shorter bacterial sequence consists mostly of a 15-base-pair core that also is present in *attP*. Integrase also binds both core sequences, which are the actual crossover sites. If the core sequence is changed even slightly in *attP* or *attB* alone, the rate of recombination is greatly reduced; but if the core is changed identically in *attP* and *attB*, exchange is still efficient. Thus, integrase requires a sequence homology in the core as well as a particular sequence to which it binds. It is not known if integrase detects this homology between separate duplexes before it cuts and unwinds, requiring, for example, a specifically paired four-stranded helix of the type previously discussed, or if it cuts first but then tends to reverse the reaction unless it detects homology in the emerging seven-base-pair overlap.

When a λ prophage is induced to grow, integration is reversed (a process called **excision**), and both phage and bacterial DNAs are reconstituted intact. Prophage λ invokes excision by expressing a second protein, **excisionase,** which allows integrase now to catalyze recombination between the hybrid attachment sites of the prophage. A complex of integrase and excisionase binds tightly to a hybrid bacterial-prophage attachment site, unlike integrase alone, and this changed specificity undoubtedly underlies its ability to carry out the reverse reaction.

Site-Specific Recombination Regulates Gene Expression[19]

Recombination between two sites on a single DNA molecule has one of two consequences, depending on how the interacting sites are oriented. Recombination either removes the intervening segment or inverts it (Figure 11-24). Cells sometimes use recombinational inversion to choose between two alternate DNA arrangements that allow two different proteins or sets of proteins to be expressed. Curiously, this mechanism often regulates proteins that appear on the outside of the organism, for example, the tail protein of phage Mu (mutator), regulated by the invertible *gin* segment (Chapter 17), and the flagellar

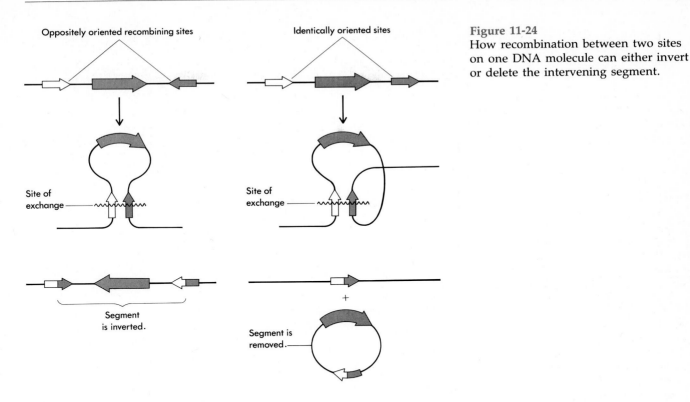

Figure 11-24
How recombination between two sites on one DNA molecule can either invert or delete the intervening segment.

antigen of the bacterium *Salmonella.* In a very complex case reminiscent of the construction of antibody genes, variation in the surface proteins of trypanosomes is achieved by a succession of DNA rearrangements. Perhaps recombination is used to control surface proteins because the switch can be absolute, allowing the organism to make one surface antigen exclusively and thus avoid even a slight reaction to antibodies directed by an infected host against other antigens.

Salmonella phase variation results from the alternate expression of its two flagellar proteins H1 and H2. At a given time, a cell expresses one of these, but never both together. The promoter for the *H2* gene is found on an adjacent invertible DNA segment 970 base pairs long, bounded by repeats (of 14 base pairs) oriented oppositely (Figure 11-25). This short homology appears to be used as the core sequence

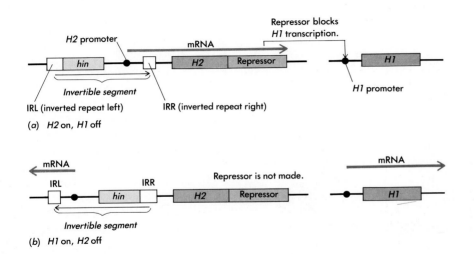

Figure 11-25
Control of alternate expression of *Salmonella* flagellar genes *H1* and *H2* by an invertible DNA segment. IRL and IRR are inverted repeats that make the inversion possible in the presence of the enzyme coded by *hin*.

homology is used during λ integration, to guide recombination. Because the homologous repeats point in opposite directions, the segment is inverted with each exchange. If the segment is pointed in one way, the promoter is beside gene *H2,* which can then be transcribed along with an adjacent gene encoding a repressor of the distant gene for flagellar protein H1; thus, H1 is repressed while H2 is transcribed. If the segment is inverted, H2 is no longer made because it has no promoter, but neither is the repressor of H1 made; thus, protein H1 is made. The invertible segment itself encodes the Hin enzyme that catalyzes inversion. Possibly the activity or expression of the Hin enzyme is increased when the cell's growth suffers, thus allowing a new surface protein to be tried.

A Remarkable Discovery: Transposons Move Genes to New and Unrelated Sites[20–27]

The existence of specialized mechanisms for site-specific recombination did not seriously affect our conviction that recombinational processes generally preserve the order of genes on chromosomes; site-specific rearrangements are limited and conservative and in higher organisms often are not inherited because they occur only in somatic cells. However, a further discovery showed the chromosomes do rearrange in a completely unexpected way. The most dramatic example is provided by genes specifying bacterial resistance to antibiotics such as tetracycline and penicillin, which can be carried on plasmids that have no homology with the bacterial chromosome. In a cell carrying such a plasmid, the resistance gene occasionally appears in the bacterial chromosome or in the progeny of an infecting phage, apparently by recombination in the absence of homology. Once resident in a new genome, it then can move again (e.g., from the bacterial DNA to another plasmid). These are rare events, typically occurring less than once per million cell divisions, but they are easily detectable because a genome carrying the resistance gene can be selected. Electron microscopy of heteroduplexes and restriction enzyme analysis both show that a new DNA sequence is inserted when the resistance gene is acquired. Such DNA segments that move without benefit of homology are called transposons. Known transposons range in length from 750 to 40,000 base pairs, about the same as small- to medium-size phages.

Two types of transposons have been found in bacteria: complex and simple, the latter also called **insertion sequences.** Complex transposons contain one to several genes besides those essential for the process of transposition; they are most easily recognized when they carry identifiable genetic markers such as genes that specify antibiotic resistance or toxin production. Insertion sequences carry only the genes necessary for their own transposition and can be detected in two ways. First, they interrupt and inactivate genes into which they insert. Second, they may contain promoters that allow RNA polymerase to transcribe and thus turn on adjacent genes. Although no specific function of insertion sequences in bacteria is known, they may act as natural agents of genetic change through their ability to revise genome structure and alter the expression of genes.

Complex transposons often have insertion sequences or remnants of insertion sequences at their ends. This suggests that complex transposons can be created when insertion sequences jump to both sides of a cellular gene, so that the entire assembly afterwards moves

Table 11-1 Some Insertion Sequences and Complex Transposons of *E. coli*

	Size (bp)	Target DNA Repeat (bp)	Known Functions or Proteins Encoded (Besides Transposase)
Insertion Sequences			
IS1	768	9	—
IS2	1,327	5	—
IS10-R	1,329	9	—
Complex Transposons			
Tn3	4,957	5	Ampicillin resistance
Tn5	5,700	9 (IS50 at end)	Kanamycin resistance
Tn10	9,300	9 (IS10 at end)	Tetracycline resistance
Tn681	2,100	9 (IS1 at end)	Heat-stable enterotoxin
Tn2571	23,000	9 (IS1 at end)	Resistance to chloramphenicol, fusidic acid, streptomycin, sulfonamides, and mercury

Adapted from N. Kleckner, *Ann. Rev. Genetics.* 15:354.

from the outer insertion sequence boundaries. The bounding insertion sequences may no longer be able to move independently, because they have changed in evolution to favor movement of the complex as a whole.

There is good evidence that transposons can be found in all organisms. Some transposons of *E. coli* are given in Table 11-1. Transposons of higher organisms include the Ty elements of yeast (Chapter 18), the copia elements of *Drosophila* (Chapter 20), and the retroviruses that inhabit cells of many species (Chapter 24).

Resistance Transfer Factors Are Constructed Mostly of Transposons and Can Specify Resistance to Several Drugs at Once

A serious medical problem arises when infecting bacteria become resistant to several antibiotics simultaneously by acquiring a plasmid carrying separate genes that confer resistance to each antibiotic. The evolution of multiple-resistance plasmids is made easy by the ability of transposons to recombine without homology and thus to gather together in a single plasmid, whose survival is assured by frequent use of several antibiotics together. The naturally occurring complex plasmid R1 evolved from transposons encoding resistance to chloramphenicol (Cm) and kanamycin (Km), and it also contains the intact transposon Tn4 carrying resistance to streptomycin (Sm), sulfonamide (Su), and ampicillin (Ap) (Figure 11-26). Tn4 itself includes a separate and independent transposon, Tn3, which is responsible only for the resistance to ampicillin. The whole assembly of resistance genes (an **r-determinant**) is attached to other segments containing genes for plasmid replication and for **transfer functions,** which promote conjugal contact of the host bacterial cell with other bacteria, thus allowing the plasmid to be transferred readily between cells. The very plastic nature of this modular assembly gives rise to many plasmid variants, both in the laboratory and in nature.

Figure 11-26
R1, a naturally occurring 94-kilobase complex plasmid that confers resistance to five antibiotics: chloramphenicol (Cm), streptomycin (Sm), sulfonamide (Su), ampicillin (Ap), and kanamycin (Km). The plasmid has two parts, which also can exist separately in other plasmids: the resistance determinant (r-determinant) and a resistance transfer factor (RTF), which encodes proteins that promote movement of plasmids between cells through a mating process. R1 was formed when an IS1 insertion sequence on one of these elements transposed to the other, fusing them into a stable cointegrate bearing a copy of IS1 at each of the junctions. (Adapted from S. N. Cohen in *DNA Insertion Elements, Plasmids, and Episomes,* Cold Spring Harbor Laboratory, 1977, p. 672, with permission.)

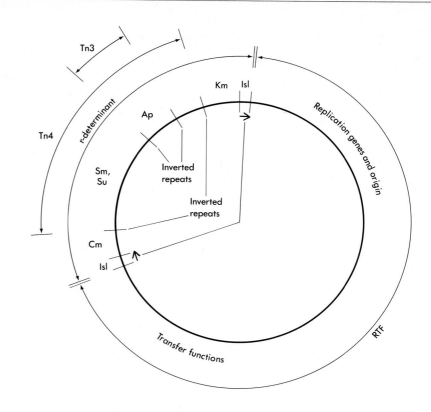

Transposons Encode Genes That Direct and Regulate Their Transposition

An antibiotic resistance gene is a passenger carried by transposon DNA that otherwise is devoted to directing and controlling the frequency of its own movement. Transposons have two features that are crucial to their behavior. First, 20 to 40 nucleotides of the sequence at one end is repeated nearly identically, but oriented oppositely, at the other end; these are called **inverted repeats.** Second, many (and probably all) transposons encode an enzyme called a **transposase,** which catalyzes their insertion into new sites.

The rarity of transposition results from the very small concentration of transposase in the cell. On average, less than one molecule of transposase is made by transposon Tn10 in each generation, an amount that allows Tn10 to transpose naturally only once in every 10^7 cell generations. However, if more transposase is provided by a plasmid engineered to contain the Tn10 transposase gene, then Tn10 transposes a thousand times more frequently. It seems likely that the natural transposition rate is also controlled by changes in transposase level.

One important determinant of transposase expression and function has been discovered recently. The transposase promoter of Tn10 and some other transposons contains the DNA sequence GATC, which is methylated in *E. coli* at the adenine nucleotide by the cellular *dam* methylase (Chapter 12). Since the sequence complementary to GATC is also GATC (when read in the 5'→3' direction), both DNA strands are methylated at this site. In this condition, the promoter is relatively inactive. However, after DNA replicates, the adenine in the newly made GATC sequence is not methylated by the *dam* methylase for a minute or so, and in this *hemimethylated* state, the promoter is much more active; thus, a small burst of transposase is made just after repli-

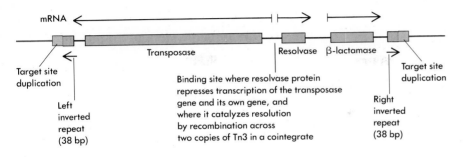

Figure 11-27
Structure of transposon Tn3. Tn3 encodes β-lactamase, an enzyme that destroys β-lactam antibiotics like penicillin and ampicillin, and the two enzymes transposase and resolvase; its length is 4957 nucleotides. The arrows above the genes show the directions of transcription. [Adapted from F. Heffron, B. J. McCarthy, H. Ohtsubo, and E. Ohtsubo, *Cell* 18 (1979):1153, with permission.]

cation. Similarly, the site in Tn10 DNA on which transposase acts contains GATC, and it is a much better substrate for transposase during the short time when only one strand is methylated. Therefore, Tn10 tends to transpose just after DNA replication. It is not yet known if other cellular conditions affect DNA methylation and in so doing change the transposition rate, or if transposition is also regulated in other ways.

Some transposons encode a second enzyme, a **resolvase,** that catalyzes a second stage of transposition (as we shall see in Figure 11-29). For Tn3, the resolvase protein also has a second and independent activity: It is a repressor of both its own gene and the transposase gene, serving to keep their normal levels of expression low (Figure 11-27). It represses both genes by binding a site between them in transposon DNA.

How Transposons Move[23, 27]

DNA sequence analysis of bacterial transposons like Tn3 and their junctions with target DNA provides strong clues to the mechanism by which they are transposed. First, we know that transposons move precisely, carrying all of the sequence inserted at the old site and none of the adjacent DNA. This could occur only if an enzyme, almost certainly transposase itself, recognizes both transposon ends; thus, the ends probably are identical because they are the common binding sites of transposase. Second, a precise duplication of 3 to 12 bases of the target DNA (depending on the transposon) is created at the site of insertion, one copy staying at each end of the transposon sequence (Figure 11-28). Some DNA synthesis must therefore occur during transposition to create the target site duplication, a clear difference from the site-specific recombination event of λ integration. Third, although most transposons move to almost any region of a new genome, they do not move completely at random, preferring instead to invade certain DNA sequences. Some attack specific four-

Figure 11-28
Nucleotide sequence of one target site of Tn3, with the crucial five-nucleotide segment highlighted. After Tn3 insertion, one such five-nucleotide sequence directly adjoins each of the identical (but oppositely directed) Tn3 ends.

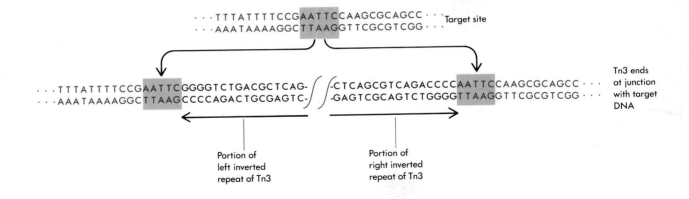

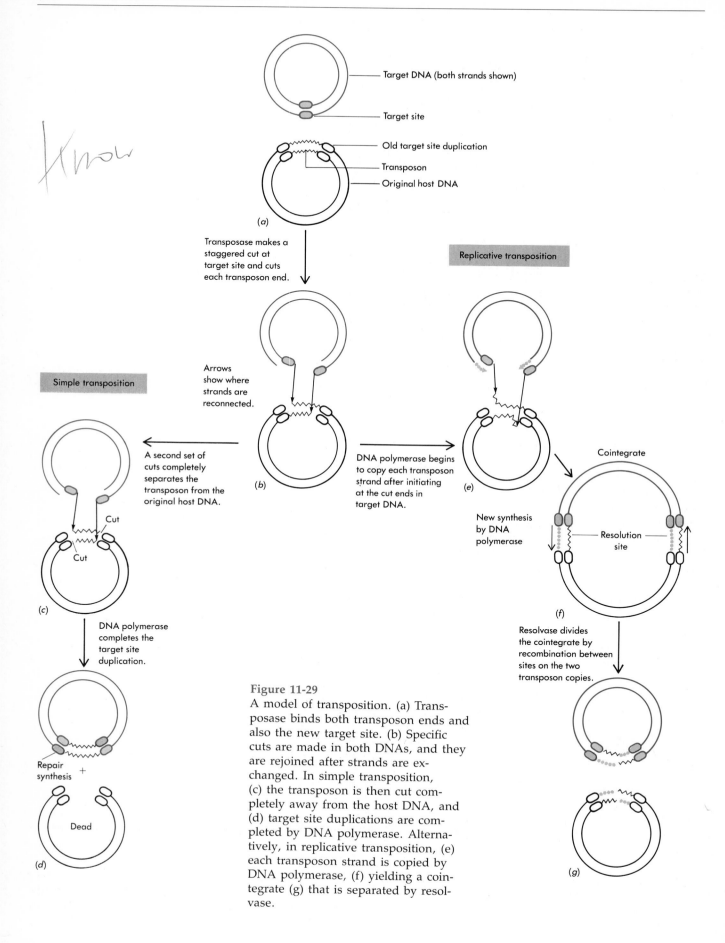

Figure 11-29
A model of transposition. (a) Transposase binds both transposon ends and also the new target site. (b) Specific cuts are made in both DNAs, and they are rejoined after strands are exchanged. In simple transposition, (c) the transposon is then cut completely away from the host DNA, and (d) target site duplications are completed by DNA polymerase. Alternatively, in replicative transposition, (e) each transposon strand is copied by DNA polymerase, (f) yielding a cointegrate (g) that is separated by resolvase.

or six-base-pair symmetrical sequences, just as restriction enzymes do.

Studies of enzymes that break and rejoin DNA strands, like λ integrase and DNA topoisomerases, suggest a model to explain how transposase may work (Figure 11-29). We suppose that transposase binds both ends of the transposon and also the target DNA sequence in which the transposon will insert. It then makes staggered cuts in the target like a restriction enzyme and simultaneously one cut at each transposon end, but on different strands. The freed transposon ends are ligated to the target site ends, thus connecting the two DNA molecules, which are bridged at each end by one strand of the original transposon (Figure 11-29). Here, there are two choices, depending on the transposon or its particular conditions of growth. In **simple transposition,** a second set of cuts is made at the other transposon ends, transferring the transposon entirely to the new DNA and leaving a lethal gap in the old host DNA (Figure 11-29c). Insertion of a few nucleotides by DNA polymerase completes the target site duplication (Figure 11-29d). Thus, simple transposition is a conservative process that moves the transposon to a new site. The second possibility is **replicative transposition,** in which the transposon is duplicated: One copy moves, but also one copy stays so that the original host DNA is not disrupted. In this case, no second cuts are made, but instead the second transposon strand remains to be copied by DNA polymerase, which initiates at the first cut end (Figure 11-29e). The resulting structure is called a **cointegrate,** and its existence is strong evidence for this pathway of transposition. A resolvase enzyme can then catalyze site-specific recombination across the two transposon copies in a reaction like λ integration, producing separate DNA molecules, each with one transposon copy.

It is clear how variations of these specific steps account for other activities of transposons. For example, if resolvase is absent, a cointegrate (like plasmid R1, Figure 11-27) is the final product. Or a cointegrate might be divided incorrectly, yielding two independent and rearranged DNAs. A close look at the model in Figure 11-29 reveals that transposition into a second site of the same DNA causes an inversion or a deletion, again radically changing the structure of a chromosome.

A thorough understanding of the mechanism of transposition will require that the individual steps be studied in vitro. Fortunately, transposition by bacteriophage Mu, which is both a phage and a transposon (Chapter 17), does occur in cell-free extracts, and we expect that this discovery will quickly lead to an understanding of exactly how transposons move.

Recent experiments show that some transposons of higher organisms, particularly those related to retroviruses, move in a very different way from bacterial transposons like Tn3. Thus, Ty elements of yeast (Chapter 18) are transcribed first into RNA, which is copied to DNA by a reverse transcriptase enzyme; this DNA copy is then recombined into a new site of the host chromosome by a mechanism that is not yet understood.

Summary

Crossing over, or homologous recombination, is the exchange of nearly identical segments between DNAs that share regions of sequence homology, for example, two homologous chromosomes. Crossing over creates genetic diversity by producing different combinations of particular genetic alleles and also allows repair of DNA damaged by radiation or chemicals through reconstruction of intact DNAs from segments. Recombinant chromosomes are made by joining separate preexisting pieces, rather than by template switching during a copying process. Combining segments are joined accurately at heteroduplex regions, which contain one strand derived from each parental DNA and which thus ensure correct alignment. Homologous recombination is initiated from gaps in DNA or from breaks that lead to single-stranded regions. One initiation pathway is provided in *E. coli* by the RecBC enzyme, which unwinds DNA and leaves single-stranded ends by cutting strands at sites named Chi. The *E. coli* recombination enzyme RecA protein binds single-stranded DNA and pairs it with a complementary strand in a separate DNA molecule, thus establishing a bridge between DNAs. At the completed bridge, called a Holliday joint, two DNAs exchange a homologous strand. The Holliday joint can move by branch migration, thereby extending the region of heteroduplex DNA. Recombinant DNAs are made when the Holliday joint is broken by nucleases.

Site-specific recombination is an exchange of particular segments of DNA that is guided not (primarily) by homology at the boundaries, but instead by binding sites for proteins that cut and rejoin segments at precise sites. Generally, site-specific recombination serves regulatory functions. Phage λ integrase inserts phage into bacterial DNA by site-specific recombination, creating the passive prophage state in which phage DNA is replicated by the host. Phase variation in *Salmonella* is the result of a site-specific recombination event that inverts a DNA segment and thereby determines the orientation of a promoter that controls genes for two different flagellar antigens.

Transposons are small, mobile DNA sequences that move around chromosomes with near total disregard for homology, producing insertions, deletions, and more complicated rearrangements. Often they carry genes that allow them to be selected, for example, genes specifying antibiotic resistance. Transposons also encode transposases, enzymes that carry out the DNA-cutting and ligation reactions by which transposons move.

Bibliography

General References

"Recombination at the DNA Level." 1984. *Cold Spring Harbor Symp. Quant. Biol.* 49. Recent research reports on most topics covered in this chapter.

Shapiro, J. A., ed. 1983. *Mobile Genetic Elements.* New York: Academic Press.

Stahl, F. 1979. *Genetic Recombination: Thinking About It in Phage and Fungi.* San Francisco: Freeman. An advanced and detailed but very readable text on classical genetic analysis, published just before RecA function was understood.

Cited References

1. Meselson, M., and J. J. Weigle. 1961. "Chromosome Breakage Accompanying Genetic Reconstruction in Bacteriophage." *Proc. Nat. Acad. Sci.* 47:857–868.
2. Orr-Weaver, T., J. Szostak, and R. Rothstein. 1981. "Yeast Transformation: A Model System for the Study of Recombination." *Proc. Nat. Acad. Sci.* 78:6354–6358.
3. Szostak, J., T. Orr-Weaver, R. Rothstein, and F. Stahl. 1983. "The Double-Strand Break Repair Model for Recombination." *Cell* 33:25–35.
4. Cox, M., and I. Lehman. 1981. "*RecA* Protein of *Escherichia coli* Promotes Branch Migration, a Kinetically Distinct Phase of DNA Strand Exchange." *Proc. Nat. Acad. Sci.* 78:3433–3437.
5. Das Gupta, C., A. Wu, R. Kahn, R. Cunningham, and C. Radding. 1981. "Concerted Strand Exchange and Formation of Holliday Structures by *E. coli* RecA Protein." *Cell* 25:507–516.
6. Radding, C. 1982. "Homologous Pairing and Strand Exchange in Genetic Recombination." *Ann. Rev. Genetics* 16:405–437.
7. Howard-Flanders, P., S. West, and A. Stasiak. 1984. "Role of RecA Protein Spiral Filaments in Genetic Recombination." *Nature* 309:215–220.
8. Holliday, R. 1964. "A Mechanism for Gene Conversion in Fungi." *Genet. Res.* 5:282–304.
9. Sigal, N., and B. Alberts. 1972. "Genetic Recombination: The Nature of a Crossed Stranded Exchange Between Two Homologous DNA Molecules." *J. Mol. Biol.* 71:789–793.
10. Kobayashi, I., M. Stahl, and F. Stahl. 1984. "The Mechanism of the Chi-*cos* Interaction in RecA-RecBC-Mediated Recombination in Phage λ." *Cold Spring Harbor Symp. Quant. Biol.* 49:497–506.
11. Ponticelli, A. S., D. W. Schultz, A. F. Taylor, and G. R. Smith. 1985. "Chi-Dependent DNA Strand Cleavage by RecBC Enzyme" *Cell* 41:145–151.
12. Clark, A. J. 1973. "Recombination Deficient Mutants of *E. coli* and Other Bacteria." *Ann. Rev. Genetics* 7:67–86.
13. Howard-Flanders, P. 1981. "Inducible Repair of DNA." *Sci. Amer.* 245:72–80.
14. Nash, H. 1975. "Integrative Recombination of Bacteriophage λ DNA *in Vitro*." *Proc. Nat. Acad. Sci.* 72:1072–1076.
15. Nash, H. 1981. "Integration and Excision of Bacteriophage λ: The Mechanism of Conservative Site Specific Recombination." *Ann. Rev. Genetics* 15:143–167.
16. Better, M., C. Lu, R. C. Williams, and H. Echols. 1982. "Site-Specific DNA Condensation and Pairing Mediated by the Int Protein of Bacteriophage λ." *Proc. Nat. Acad. Sci.* 79:5837–5841.
17. Ross, W., and A. Landy. 1983. "Patterns of λ Int Recognition in the Regions of Strand Exchange." *Cell* 33:261–272.
18. Weisberg, R., and A. Landy. 1983. "Site-Specific Recombination in Phage Lambda." In *Lambda II*, ed. R. Hendrix, J. Roberts, F. Stahl, and R. Weisberg. Cold Spring Harbor, N.Y.: Cold Spring Harbor Laboratory, pp. 211–250.
19. Simon, M., J. Zieg, M. Silverman, G. Mandel, and R. Doolittle. 1980. "Phase Variation in the Evolution of a Controlling Element." *Science* 209:1370–1374.
20. Heffron, F., B. J. McCarthy, H. Ohtsubo, and E. Ohtsubo. 1979. "DNA Sequence Analysis of the Transposon Tn3: Three Genes and Three Sites Involved in Transposition of Tn3." *Cell* 18:1153–1163.
21. Cohen, S., and J. Shapiro. 1980. "Transposable Genetic Elements." *Sci. Amer.* 242:40–50.
22. Kleckner, N. 1981. "Transposable Elements in Prokaryotes." *Ann. Rev. Genetics* 15:341–404.
23. Craigie, R., and K. Mizuuchi. 1985. "Mechanism of Transposition of Bacteriophage Mu: Structure of a Transposition Intermediate" *Cell* 41:867–876.
24. Morisato, D., J. Way, H. Kim, and N. Kleckner. 1983. "Tn10 Transposase Acts Preferentially on Nearby Transposon Ends *in Vivo*." *Cell* 32:799–807.
25. Harshey, R. M. 1984. "Transposition Without Duplication of Infecting Bacteriophage Mu DNA." *Nature* 311:580–581.
26. Roberts, D., B. C. Hoopes, W. R. McClure, and N. Kleckner. 1985. "IS10 Transposition Is Regulated by DNA Adenine Methylation" *Cell* 43:117–130.
27. Boeke, J. D., D. J. Garfinkel, C. A. Styles, and G. R. Fink. 1985. "Ty Elements Transpose Through an RNA Intermediate." *Cell* 40:491–500.

CHAPTER 12

The Mutability and Repair of DNA

Since the developmental potential of an organism is determined by its genes, DNA must necessarily change, or mutate, as organisms evolve. But evolutionary changes occur only rarely, and it is the stability of DNA, as it is maintained and replicated over generations, that is most impressive. Since living cells require the correct functioning of thousands of proteins, each of which could be damaged by a mutation at many different sites in its gene, it is clear that DNA sequences usually must be passed on unchanged if progeny are to have a good chance of survival. The challenge for the cell is twofold. First, the enzymatic machinery that replicates DNA must be inherently accurate. Second, the cell must repair accidental damage to DNA that would destroy its function. DNA is a complex organic molecule of finite chemical stability: Not only does it suffer spontaneous damage like loss of bases, but it also is assaulted by natural (and unnatural) chemicals and radiation that break its backbone and chemically alter the bases. Since mutations result when damage changes the coding properties of bases, an organism could not survive the natural rate of damage to its DNA without specific enzymatic mechanisms to repair damaged sites. In fact, DNA repair is so important that a bacterium may devote several percent of its genome to specifying and controlling the enzymes involved.

The Nature of Mutations

Naturally occurring mutations include almost all conceivable changes in DNA sequence. Mutations that have only a subtle effect on a gene product, such as temperature-sensitive mutations, are often simple switches of one base for another. However, many natural mutations destroy the function of a gene completely. These more drastic changes, called **null mutations,** include not only base switches and insertions or deletions of a base, but also more extensive insertions and deletions and even gross rearrangements of chromosome structure. Such changes might be caused, for example, by the insertion of a transposon, which typically places many thousands of bases of foreign DNA in the coding sequence of a gene (Chapter 11), or by aberrant actions of cellular recombination processes.

Mutation by Single-Base Change[1, 2]

There are a number of reasons to consider first the simplest type of mutation, the exchange of one base for another. First, base switches reflect the basic accuracy of DNA replication. Second, many important mutagens act by making single-base changes. Finally, single-base mutations are critical to evolution, because they change genes in ways that are subtle enough to yield useful variants.

What sorts of single-base changes occur naturally? To determine this in a practical way, we have to be able to pick out quickly those mutants that result from only a single change. This was first accomplished by examining large numbers of spontaneous mutations that create a translation stop codon in the *E. coli* lactose operon *I* gene, which codes for the lactose repressor. It can be verified that these mutations are single-base changes in some codon originally identical to the stop codon in two out of three positions, by showing that active repressor is once again made from the mutant gene in a cell containing a suppressor (which allows the stop codon to be translated as an amino acid; see Chapter 14). This analysis shows that all types of base switches occur: **transitions,** which are changes from one purine to the other or from one pyrimidine to the other, and **transversions,** which are changes from purine to pyrimidine or from pyrimidine to purine (Figure 12-1). A second important fact is that not all possible sites mutate with the same efficiency; some are "hot spots" (first discovered by S. Benzer in his classic genetic analysis of the phage T4 *rII* gene), at which mutations occur much more often than at most sites.

Single-base switches are usually reversible, and often the rate of "back" mutation to the normal nucleotide arrangement is similar in order of magnitude to the rate of change to the mutant arrangement. This fact represents an important way to distinguish base switches from more drastic alterations like large deletions, for which the reverse reaction (called **reversion**) is impossible. An intermediate case is single-base insertions or deletions; these may revert, but much less often than single-base switches. Most spontaneous single-base exchanges are simply rare failures in the replication process, arising when a nucleotide is added to the growing chain even though it does not pair normally with the template base.

Error Levels per Incorporated Nucleotide Range from 10^{-7} to 10^{-11}

Since the *lacI* system detects all instances of a particular base change at 80 distinct sites, it is simple to calculate how often each mutation appears when a cell's DNA replicates and thus the rate at which the replication system makes an error at each site. The average rate for all 80 sites is about 10^{-9} mistakes for each replication. However, hot spots produce mutants at 25 times the average rate, and other sites are at least 25 times *less* active than the average. Many other measurements of base substitutions at particular sites in other genes show that the rate varies from 10^{-7} to 10^{-11} mistakes per replication event.

Control of Mutation Levels by the Relative Efficiencies of Forward Polymerizing and Backward Nuclease Activities[3, 4]

At first it might appear that the error frequencies directly reflect the inherent accuracy of AT and GC base-pair formation. However, strong theoretical arguments contradict this idea: There is not enough difference between right and wrong base pairs for a polymerase to make as few as one error for every 10^9 nucleotides it incorporates.

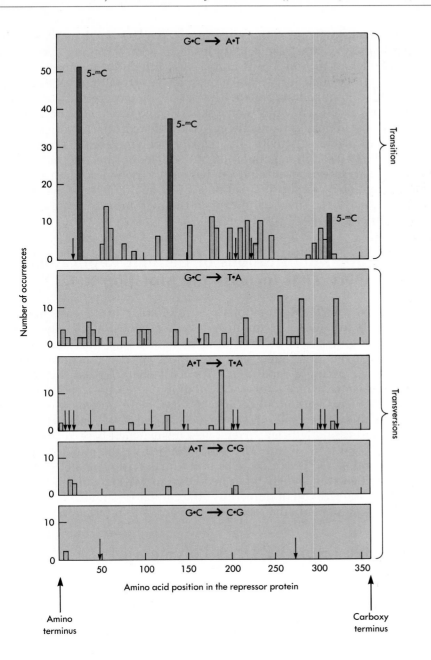

Figure 12-1
Nature and distribution of spontaneous mutations that create stop codons in the lactose repressor gene. This assay detects all four transversions and one of the two possible transitions. Other assays show that the other transition (AT→GC) is also found among spontaneous mutations. The vertical arrows indicate sites where mutations could have been detected but did not occur. The three cytosines methylated in the 5 position (5-me-C) are hot spots for mutation. [After J. H. Miller and K. B. Low, *Cell* 37 (1984):675–682.]

Furthermore, the lower limit of wrong base insertion should be set by the frequency that bases assume the "incorrect" tautomeric form (enol instead of keto, imino instead of amino), perhaps one in 10^4, much greater than the actual mutation frequency. This dilemma vanished with the discovery of the *proofreading* capability of DNA polymerases. Although the *initial* error frequencies during base-pair selection are consistent with measured tautomeric ratios, almost all the wrongly inserted bases are later removed by the 3'→5' exonuclease component of the respective DNA polymerase. At first, its association with every known bacterial DNA polymerase seemed most bizarre, but now we realize that the 3'→5' exonuclease activity is necessary for accurate DNA replication. In fact, studies with mutant DNA polymerase molecules show that mutation levels may be controlled by the ratio of backward (3'→5') nuclease activities to forward (5'→3') polymerizing capabilities. If the 3'→5' exonuclease is inefficient, then abnormally high mutation rates result. On the other hand, a very efficient 3′→5′ nuclease activity leads to a very low mutation rate.

The failure to find 3′→5′ exonuclease activity in eucaryotic DNA polymerases, which may make even fewer errors than bacterial enzymes, cannot be interpreted. The possibility that exonuclease activity of eucaryotic enzymes is on a separate and weakly bound subunit that is lost in purification is suggested by the recent discovery that the 3′→5′ exonuclease activity of *E. coli* DNA polymerase III is carried on a small, 25,000-dalton subunit (ε) of the holoenzyme, and not on the large, 140,000-dalton subunit (α) that has the polymerization activity. In exceptional cases, proofreading activity might not exist, a possibility that has been invoked to explain the high mutation rate of mitochondrial genomes.

Mutators: Modifications of Cellular Replication Machinery That Increase the Mutation Rate[5, 6]

There exist bacterial mutations whose effect is to increase the rate at which mutations accumulate in other genes. Generally, these **mutators** give rise to excess mutations in any gene that is examined and must therefore affect proteins whose normal functions are required for accurate replication. The best characterized mutator in *E. coli* is *mutD*, which alters the ε subunit of DNA polymerase III and thus interferes with proofreading. In vitro, DNA polymerase III containing this altered subunit shows a decreased 3′→5′ exonuclease activity and cannot efficiently remove a mismatched base at the growing end of a DNA chain. The genes defined by the mutators *mutH*, *mutL*, and *mutS* specify proteins that function in mismatch repair, a process that scans newly replicated DNA for wrongly inserted bases. The characterization of numerous other mutators should reveal other cellular components that contribute to accurate replication.

Slippage Errors by DNA Polymerase Cause Small Additions and Deletions[7]

Replication errors by DNA polymerase are not limited to single-base-pair changes. Deletions or additions of one or a few base pairs are thought to result from displacement ("looping out") of bases from either the template strand (giving a deletion) or the growing strand (giving an addition). These mistakes occur particularly at runs of identical bases in the DNA, where the looped-out structure can be stabilized by normal base pairing beyond the unpaired base (Figure 12-2). When the miscopied DNA strand acts as a template, the deletion or addition is usually copied accurately, fixing the mutation. An addition or deletion of one or a few base pairs that changes the translation reading frame within a protein coding sequence and thus completely disrupts its synthesis is often called a **frameshift mutation.**

Mutation by Large DNA Sequence Rearrangement[8, 9]

Many mutations are much less subtle than single-base changes, involving extensive rearrangements. These are often found among mutations selected to destroy the function of a gene. Large segments,

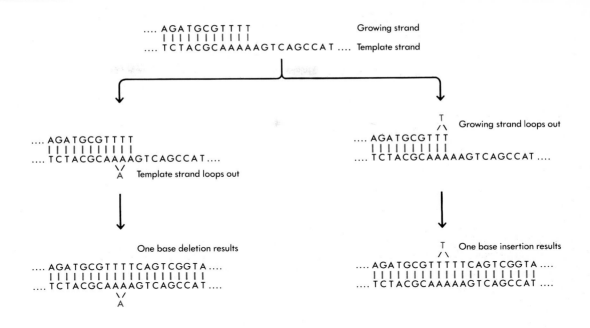

Figure 12-2
Small additions and deletions are caused by slippage at the replication growing point.

composed of hundreds or thousands of nucleotides, can be deleted by the aberrant operation of the cell's recombination enzymes (Chapter 11) or by errors in copying the template DNA strand. Similarly, segments can be inverted by recombination within a DNA molecule or exchanged between chromosomes of a higher cell, sometimes with the important consequence that genes near the site of exchange are no longer regulated correctly. Since chromosomes are generally stable, massive rearrangements must represent rare accidents that are preserved by selection or that happen to be harmless.

The Importance of Studying Chemical Mutagens[10, 11]

Since mutagenic chemicals greatly increase the probability that a mutation will occur, they must interfere in some important way with accurate replication of DNA. Mutagens are studied not only because they are useful for modifying genes and learning how accurate replication occurs, but also because naturally occurring mutagens affect all organisms on earth, including us. Bacteria and phages have allowed the mechanism of action of many mutagens to be worked out, and they also provide the most straightforward way to test if new chemicals may be dangerous mutagens that should be avoided.

Mutagens either act on DNA directly to change its template properties or in some way subvert replication so that a wrong base is inserted. We understand clearly how some mutagens work. **DNA-Reactive chemicals,** such as nitrous acid (HNO_2) or alkylating agents like methyl-nitrosoguanidine, act directly on DNA to change the bases into chemically distinct structures. These new bases often base-pair in a different way from the original bases, in which case the effect of the mutagen is to alter the genetic code directly (Figure 12-3). However, some modified bases cannot pair at all, and sometimes the damage completely removes the base from the DNA backbone; in these cases, mutations arise only through a special DNA repair process that introduces errors as part of the correction mechanism. (This process of error-prone repair is discussed later in this chapter.)

Figure 12-3
The oxidative deamination of DNA bases by nitrous acid, and its effects on subsequent base pairing. (a) Adenine is deaminated to hypoxanthine, which bonds to cytosine instead of to thymine. (b) Cytosine is deaminated to uracil, which bonds to adenine instead of to guanine. (c) Guanine is deaminated to xanthine, which continues to bond to cytosine, though with only two hydrogen bonds. Thymine and the uracil of RNA do not carry an amino group and so remain unaltered. [After W. Hayes, *The Genetics of Bacteria and Their Viruses* (Oxford, Eng.: Blackwell, 1964), p. 280, with permission.]

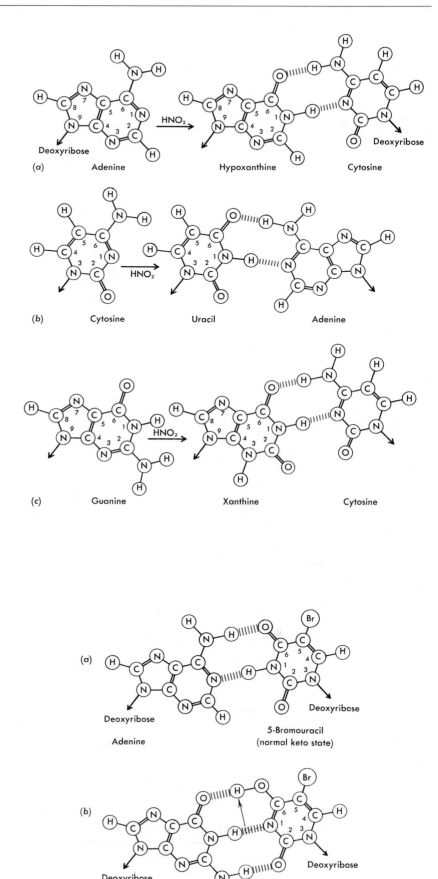

Figure 12-4
The base-pairing attributes of 5-bromouracil. (a) In the normal keto state, with a hydrogen atom in the N1 position, bromouracil bonds to adenine. (b) In the rare enol state, a tautomeric shift of this hydrogen atom determines specific pairing with guanine. [After W. Hayes, *The Genetics of Bacteria and Their Viruses* (Oxford, Eng.: Blackwell, 1964), p. 278, with permission.]

A second class of chemical mutagens is **base analogs,** which because of their similarity to the normal bases are incorporated (as nucleoside triphosphates) into DNA during normal replication. Their different structures, however, lead to less accurate base-pair formation than normal, causing frequent mistakes during the replication process. One of the most powerful base analog mutagens is 5-bromouracil, an analog of thymine. It is believed to cause mutations because its hydrogen atom at position 1 is not as firmly fixed as the corresponding hydrogen atom in thymine. Sometimes, this hydrogen atom is bonded to the oxygen atom attached to carbon atom 6 (Figure 12-4). When this happens, the 5-bromouracil can pair with guanine.

A third class of chemical mutagens consists of the **frameshift mutagens,** like proflavin, which cause the deletion or addition of a single base, or sometimes a few bases. Frameshift mutagens are flat molecules containing several connected rings (polycyclic) that bind to the equally flat purine or pyrimidine bases of DNA, just as the bases bind to or "stack" with each other in the double helix. These mutagens might act by binding to bases that loop out of either the template DNA strand or the growing strand during DNA synthesis (see Figure 12-2), thereby stabilizing the looped-out structure and greatly increasing the chance that it will lead to a mutation.

Cell Survival Depends on Continual Repair of DNA[12–15]

A mutant survives when its genetic change is not harmful, or, more rarely, is beneficial. But most mutations are damaging, and cells would not endure without enzymatic mechanisms to reverse the effects of natural mutagenic processes. As one example, the base cytosine suffers spontaneous deamination to the closely related base uracil, which pairs in DNA like thymine; thus, a GC base pair becomes an AT base pair when the DNA is replicated. Were this change not repaired, 1 in every 1000 cytosines in human DNA would become uracil during a lifetime, giving rise to an intolerable rate of mutation. Even when DNA repair is efficient, the rate of genetic change caused by natural damage may be much greater than that resulting from errors in DNA replication.

An even more immediate problem is posed by breaks in the DNA sugar-phosphate backbone and by bases so altered that they no longer even simulate natural bases and thus cannot pair at all. Clearly, such lesions prevent expression of any gene whose coding sequence they interrupt. Furthermore, they prevent DNA replication, because DNA polymerase almost always stops if it cannot make a correct base pair. Besides backbone breaks, which prevent a DNA strand from acting as a template, a typical lesion that prevents base pairing is the **pyrimidine dimer.** Ultraviolet radiation with a wavelength of about 260 nm is absorbed strongly by bases, and one consequence is the photochemical fusion of two adjacent pyrimidines into nonpairing structures like the cyclobutane-containing thymine dimer and the mutagenic "6-4" photoproduct (Figure 12-5). The DNA of a skin cell exposed to normal daylight, for example, would acquire thousands of dimers per day, were these not removed by repair enzymes. The human skin disease **xeroderma pigmentosum** is caused by a genetic defect in the enzymes that remove dimers and other ultraviolet-induced lesions. The UV component of sunlight results in abnormally high skin cell death and creates many skin cancer cells in people who have this disease.

Figure 12-5
Two important pyrimidine dimer photoproducts produced by ultraviolet light.

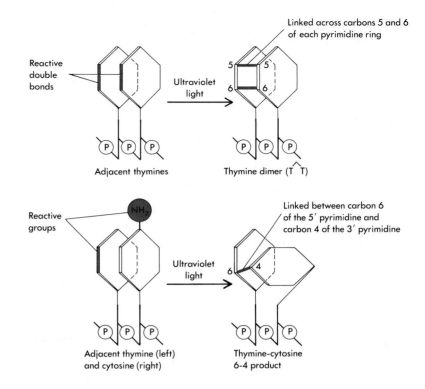

Some DNA Repair Enzymes Recognize and Reverse Specific Products of DNA Damage: Photolyase and O^6-Methylguanine Methyltransferase[14, 16–19]

Although most damage calls for more radical surgery, two common lesions in DNA can be repaired directly by enzymes that simply reverse the chemical change that engendered them. *Pyrimidine dimers* are the target of the universal enzyme **photolyase,** which binds a dimer and catalyzes a second photochemical reaction, this time using visible light, that changes the cyclobutane ring back into individual pyrimidine bases—a process called **photoreactivation.**

The second type of damage results from **alkylation,** the transfer of methyl or ethyl groups to reactive sites on the bases and to phosphates in the DNA backbone. Alkylating chemicals include nitrosamines and the very potent laboratory mutagen methyl-nitrosoguanidine (Figure 12-6). One of the most vulnerable sites of alkylation is the base guanine, which is particularly subject to methylation at the oxygen of carbon atom 6. The product O^6-methylguanine often mispairs with thymine, changing GC to AT when the DNA is replicated (Figure 12-7). This lesion is removed by the cellular enzyme O^6-**methylguanine methyltransferase,** encoded by the gene *ada,* which recognizes O^6-methylguanine in the DNA duplex and removes the methyl group by transferring it to an amino acid of the enzyme. The same enzyme also removes highly disruptive methyl groups from phosphates of the DNA backbone. A bizarre and extravagant feature of this reaction is that there is no means to regenerate the unmethylated enzyme: A new enzyme molecule is spent for each methyl group removed.

If growing *E. coli* are exposed to a low and relatively harmless concentration of nitrosoguanidine, they are afterward much more resis-

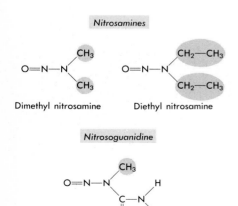

Figure 12-6
Some alkylating mutagens. The reactive alkyl groups are shaded.

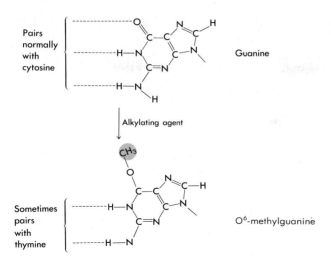

tant to the mutagenic and toxic effects of the chemical. This induced resistance, called the **adaptive response,** results from a hundredfold increase in the cell's content of O^6-methylguanine methyltransferase and a concurrent induction of the glycosylase that removes alkylated bases from the DNA backbone. The induction of both enzymes is blocked by mutations in the *ada* gene, whose product is therefore both an enzyme (the transferase) and a positive regulator of itself and the glycosylase. How the regulation works is unknown, but a reasonable guess is that the inducing signal is O^6-methylguanine itself and that the reaction of the transferase activity with methyl groups initiates the important regulatory event.

Most DNA Repair Depends on Information in the Complementary DNA Strand[14]

It is the double-stranded nature of DNA that allows almost any damage to be repaired, no matter what the chemical change may be. Because a double helix holds two interconvertible copies of the genetic message, the loss of one is not serious: The information necessary to renew a damaged segment is present in the complementary strand. Of course, this is true only if damage is limited to one of the two complementary bases; but generally, only one base or two adjacent bases on a strand are damaged, so the chance of losing both bases of a complementary pair is slight. It is difficult to imagine how stable complex genomes ever could exist without the inherent duplication of information the double helix provides. Genomes consisting of single-stranded DNA must survive without this advantage, and, in fact, the largest single-stranded DNA genomes we know are those of phages like ϕX174, whose 5000-nucleotide DNA is only one-thousandth the size of the *E. coli* chromosome.

There are several pathways of **excision repair,** in which a damaged base or backbone segment of DNA is removed. All lead to the same intermediate product: either a single-strand break or a gap in the DNA at the site of damage that provides a 3'-hydroxyl end from which DNA polymerase can initiate synthesis to replace the damaged segment. Cells mutant in the DNA polymerase I gene are very deficient in excision repair; only this enzyme is abundant and mobile enough to fill small gaps efficiently when heavy damage requires many sites to be repaired at once. In contrast, the replicative enzyme DNA polymerase III is a large and complex enzyme that remains engaged at the growing fork for long periods and is present in only a few copies per cell. The other common essential enzyme is DNA ligase, which seals the nick left after DNA polymerase I acts.

The UvrABC Endonuclease Removes a Twelve-Base Segment Surrounding a Site of Damage[20, 21]

As for other biochemical pathways, the isolation of mutants has been essential to understanding the mechanism of DNA repair. Besides those in the photolyase gene *(phr)*, mutations in *E. coli* sensitive to ultraviolet light (and thus defective in repairing this damage) occur mostly in three genes: *uvrA, uvrB,* and *uvrC.* Unlike photoreactivation, the repair system these genes encode is not specific for ultraviolet damage, but instead senses any serious damage-induced distortion of the DNA helix. The three genes encode separate subunits of a single enzyme that removes a damaged DNA segment by cutting the strand that contains it. For many years, it was thought that the enzyme specified by these genes made one cut in DNA and that the nick translation activity of DNA polymerase acting on the cut end removed the damaged section; some phage-encoded enzymes do in fact work this way. However, biochemical studies of the *E. coli* enzyme were frustrated until cloning of the three *uvr* genes allowed their polypeptide products to be made abundantly and purified separately. When they were mixed and assayed in vitro on DNA containing pyrimidine dimers, a more complex activity dependent on all three products was revealed: the UvrABC enzyme cuts on *both* sides of the dimer (lesion), freeing a 12-base oligonucleotide containing the lesion and leaving a 12-base gap that is filled by DNA polymerase and sealed by DNA ligase (Figure 12-8).

It is clear from its activity on different types of damage that the UvrABC nuclease does not detect any particular lesion directly, but instead recognizes the *absence* of a normal DNA shape. How could this work? The two cuts it makes are on the same side of the helix, a little more than one turn apart, whereas the lesion is more nearly on the other side. We imagine that the enzyme "feels" the DNA shape in the 12-base segment between the sites where its nuclease active centers are aligned, perhaps wrapping around the helix to touch the other side. If the enzyme cannot conform to the natural shape of DNA because a bulky lesion is in its way, the nuclease domains are activated and the healing cuts are made.

Damaged Bases Can Be Removed by Glycosylases[14, 22]

Cells have a second way to excise a damaged base, begun by a glycosylase enzyme that detects an individual unnatural base and removes it from the deoxyribose sugar, initially leaving the sugar-phosphate backbone intact. The "hole" that results is called an **AP site,** an abbreviation for *apurinic* (lacking A or G) or *apyrimidinic* (lacking C or T). These also may result from natural loss of bases through breakage of

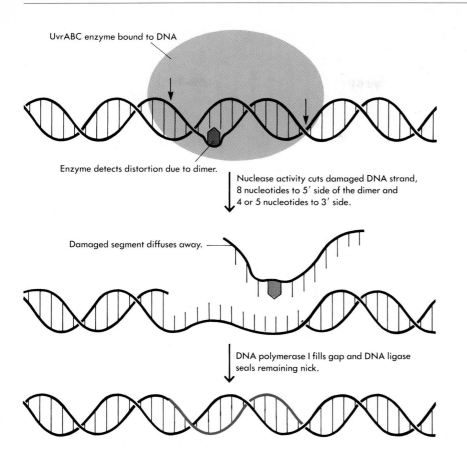

Enzyme detects distortion due to dimer.

Nuclease activity cuts damaged DNA strand, 8 nucleotides to 5′ side of the dimer and 4 or 5 nucleotides to 3′ side.

Damaged segment diffuses away.

DNA polymerase I fills gap and DNA ligase seals remaining nick.

UvrABC enzyme bound to DNA

Figure 12-8
Excision repair of dimer and other bulky lesion damage initiated by the UvrABC endonuclease.

the glycosidic bond. The "hole" is in turn recognized by an AP endonuclease that cuts the backbone, leaving a primer end from which DNA polymerase initiates synthesis to replace the missing nucleotide along with a few adjacent nucleotides (Figure 12-9).

There are many different glycosylases that remove specific lesions, in particular, aberrantly methylated bases. **Uracil-DNA glycosylase** corrects a problem caused by the natural instability of cytosine. Deamination of cytosine in DNA occurs spontaneously at a low rate to yield uracil. The glycosylase recognizes and removes uracil residues in DNA, allowing the C to be replaced by DNA polymerase. This repair system is foiled in a significant way if the cytosine is methylated on its 5 carbon, as occurs, for example, in sequences protected from restriction endonucleases. The 5-methylcytosine pairs normally in DNA and is not itself removed by a glycosylase. But deamination of 5-methylcytosine gives rise to thymine, not uracil. Thymine is a natural base and is not removed by uracil-DNA glycosylase or any other enzyme (Figure 12-10). Instead, it remains to pair with adenine in the next generation and to give rise to a mutation. Thus, 5-methylcytosines are hot spots for spontaneous mutation.

Mismatch Correction Removes Errors That Escape Proofreading[23, 24]

Some incorrectly paired bases escape even the proofreading activity of bacterial DNA polymerases. Like polymerization, proofreading has a substantial, but still finite, inherent accuracy. A good estimate is

Nucleotide containing damaged base
5′
3′
3′
5′
Glycosylase recognizes damaged base and cuts bond to sugar in backbone.
AP endonuclease cuts phosphodiester backbone beside missing base.
New DNA
DNA polymerase I initiates repair synthesis from free 3′-hydroxyl group, removing backbone fragment and translating nick several nucleotides into intact DNA.
DNA ligase seals the remaining nick.

Figure 12-9
Excision repair initiated by a glycosylase that recognizes and removes chemically damaged bases. As is shown, both the damaged nucleotide and a few adjacent nucleotides are replaced. When DNA polymerase 1 binds the free primer end, its 5′-3′ exonuclease activity (Chapter 10) cuts a few nucleotides ahead of the missing base, and its polymerization activity fills the entire gap of several nucleotides. This activity (in which DNA polymerase 1 simultaneously polymerizes new DNA and degrades DNA ahead of the growing site, a process that can continue some distance beyond the initial primer end) is called **nick translation.**

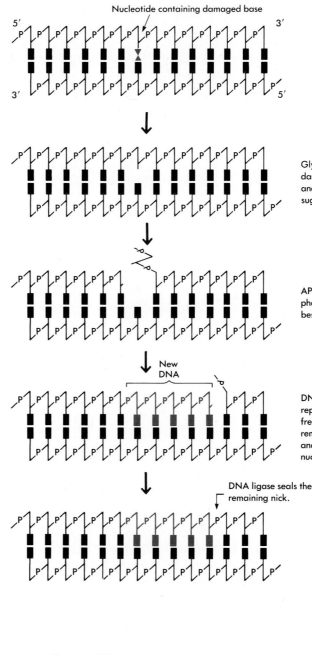

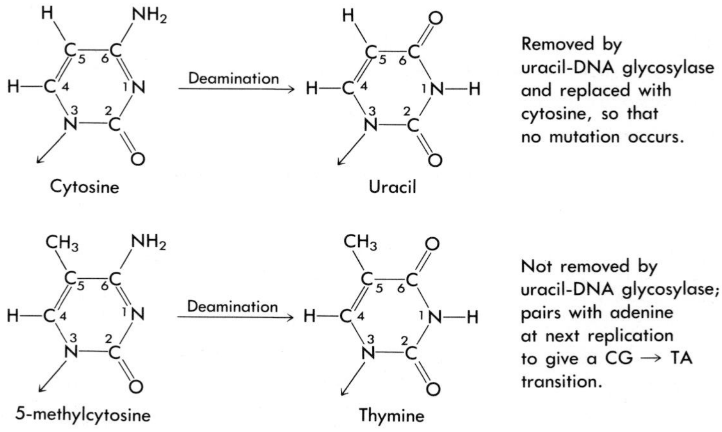

Deamination
Cytosine
Uracil
Removed by uracil-DNA glycosylase and replaced with cytosine, so that no mutation occurs.
Deamination
5-methylcytosine
Thymine
Not removed by uracil-DNA glycosylase; pairs with adenine at next replication to give a CG → TA transition.

Figure 12-10
Molecular basis of 5-methylcytosine hot spots: different deamination products of cytosine and 5-methylcytosine.

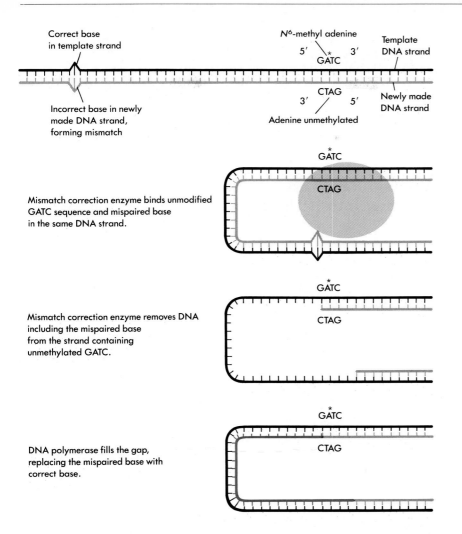

Figure 12-11
A model showing how the *E. coli* mismatch repair system could act to replace an incorrect and mispaired base in double-stranded DNA.

that DNA polymerase leaves about one mistake per 10^8 replicated base pairs. However, not even these usually appear as mutations, since the measured mutation rate can be as low as one mistake per 10^{10} or 10^{11} nucleotides. In *E. coli*, the final degree of accuracy is the responsibility of a **mismatch correction enzyme** encoded by genes *mutH, mutL,* and *mutS*. The enzyme scans newly replicated DNA for mismatched base pairs and removes a single-stranded segment containing the wrong nucleotide, thereby allowing a DNA polymerase to insert the correct base when it fills the resulting gap. The obvious problem that this entails is that of distinguishing which base of a mismatched pair is wrong, because both are natural components of DNA. If one of the pair were removed at random, the correct base would be taken about half of the time and a mutation would be ensured instead of prevented. Instead, a specific signal controlled by a timing device directs the mismatch excision system exclusively to the newly synthesized strand. The correction enzyme removes DNA containing a mismatched base only from a strand in which it also detects the DNA sequence GATC nearby (Figure 12-11). However, if a methylase encoded by the *dam* gene first modifies adenine in the GATC sequence to N^6-methyladenine, the correction enzyme cannot act and no excision occurs. Most GATC sequences in the cell become so modified after their synthesis, but only after a slight delay—

perhaps a few seconds or minutes. This is long enough for the mismatch correction enzyme to bind, so that the enzyme is directed to the DNA strand that has just been synthesized, rather than to the template strand. The correction enzyme detects not only singly mispaired bases but also small additions and deletions, so that it reduces the incidence of frameshift mutations as well as base substitutions.

Recombination Repairs DNA by Splicing into It Undamaged Strands Extracted from Duplex DNA[12, 25, 26]

Excision repair uses a complementary strand as a template to replace a damaged segment of DNA. But sometimes this template is not available, such as after a replication fork meets a lesion (e.g., a pyrimidine dimer) and separates the strands before excision repair can take place. Moreover, if both bases of a complementary pair are altered, perhaps crosslinked by a chemical like the carcinogen mitomycin C, then neither can act as a template for the other. Finally, a double helix that is broken straight across, particularly if a duplex segment is missing altogether, cannot be repaired directly with much chance of the connection being correctly remade. In all these cases, all information is lost at the site of damage, and it can be recovered only by taking a corresponding DNA segment from a separate but identical DNA molecule: This is called **recombinational repair.** An appropriate source of DNA almost always can be found in the cell. For example, when a lesion is left in a single-stranded segment by passage of the replication fork, the second daughter DNA carries the same sequence. Or if both strands are damaged in an unreplicated region, a vigorously growing bacterial cell often has one or more extra copies of its chromosome that can be used. Higher cells, of course, are generally diploid and carry the same, or nearly the same, sequence on each chromosome of a homologous pair. The key to recombinational repair is an enzyme that anneals the sequences on either side of a lesion to their complement in the undamaged DNA molecule, thus making the critical match that identifies the segment carrying the missing information. In *E. coli,* the RecA protein carries out this vital function (Chapter 11).

Inducibility of the SOS DNA Repair Genes[27]

Since a cell can often regulate the expression of genes according to a need for their products, it is not surprising that many DNA repair enzymes are induced by DNA damage. For example, methyltransferases are induced by abnormally alkylated DNA during the adaptive response (discussed earlier in this chapter). However, the most important and extensive group consists of the **SOS genes,** whose regulation is considered in detail in Chapter 16. The SOS genes are induced by damage that is severe enough to stop DNA synthesis altogether, rather than merely changing the pairing properties of bases. An example of an SOS-inducing lesion is a pyrimidine dimer, which cannot base-pair at all. When a replication fork meets a dimer,

Table 12-1   SOS Genes, Inducible by DNA Damage

Gene Name	Role in DNA Repair
Genes of known function	
uvrA *uvrB* *uvrC*	Encode excision endonuclease
umuD *umuC*	Encode proteins required for error-prone DNA repair and for most mutagenesis
sulA	Encodes protein that inhibits cell division, possibly to allow time for DNA repair
Genes involved in DNA metabolism whose specific role in DNA repair is unknown	
ssb	Encodes single-strand binding protein (SSB)
uvrD	Encodes helicase II (DNA unwinding protein)
himA	Encodes subunit of integration host factor, involved in site-specific recombination
recN	Involved in recombinational DNA repair
Genes of unknown function	
dinA *dinB* *dinD* *dinF*	

it stops and reinitiates some distance away, leaving a gap in the DNA to which the recombination enzyme RecA protein binds. Besides initiating DNA strand exchange (and thus recombinational repair), binding to single-stranded DNA activates RecA protein to carry out an enzymatic function entirely separate from recombination: to destroy by proteolytic cleavage the repressor of the SOS genes (the LexA repressor). In this way, RecA protein promotes DNA repair both by recombination and by mediating the induction of about 15 SOS genes (Table 12-1). Some of these genes have been discussed already: *uvrA*, *uvrB*, *uvrC*, and *recA* itself, all of which are expressed at a significant rate even without induction, but then much faster after the cell's DNA is damaged. Other SOS genes encode proteins involved in DNA synthesis and thus DNA repair as well. Meanwhile, the role of still other SOS genes is yet to be discovered.

Bases Too Damaged to Pair Give Rise to Mutations: Error-Prone Repair[28]

Although we know how base analogs and some modified bases cause mutations through mispairing, it is still unclear how mutations arise when bases are so damaged that they are unable to pair at all. Yet, this is the most significant process of all, since most natural mutagens and thus most **carcinogens** (cancer-causing agents; see below) act because cells can continue a DNA chain despite the apparently complete lack of template information at the site of a lesion. One important mutagenic structure of this sort is the ultraviolet-induced 6-4 photoproduct of two adjacent pyrimidine bases (see Figure 12-5). However, the extreme case is a DNA molecule lacking a base altogether (apurinic or apyrimidinic site). To show that such drastic damage can be repaired, we can use apurinic DNA made in vitro by brief treatment of single-stranded φX174 DNA with acid, which causes depurination of only one or a few nucleotides of the DNA. Since φX174 DNA is infectious to protoplasts (cells with part of their cell wall removed), it is possible to determine if the cell can repair the apurinic DNA molecule: Progeny phage are viable only if they have a nucleotide where the infecting DNA molecule had a hole. The cell does repair a fraction of the infecting apurinic DNA molecules, but it also usually makes a mutation in doing so, because there is no information by which it could choose the correct base to repair a single-stranded DNA molecule; the process is thus called **error-prone repair.** It seems likely that error-prone repair is an absolute last recourse for the cell, used only when there is no template available for accurate repair: It is better to incorporate a base that is probably wrong and let replication proceed than not to replicate at all. Naturally, the chance of mutation is very high, probably about three in four. Ironically, DNA repair is the *cause* of the mutation in this case.

Both UmuC and UmuD Proteins and RecA Protein Are Required for Error-Prone Repair[29, 30]

The major clue to the mechanism of error-prone repair is the involvement of two *E. coli* genes, *umuC* and *umuD* (for *U*V non*mu*table). These genes were originally identified as sites of mutations that prevent mutagenesis by ultraviolet irradiation. They are SOS genes, so that their products are abundant enough to act only after the cell is treated with an SOS inducer. (Error-prone repair is sometimes called **SOS repair**). Cells mutant in either gene are somewhat more sensitive than the wild type to killing by ultraviolet light, indicating that *umuC* and *umuD* promote survival by repairing damage. Because *umuC* and *umuD* are SOS genes, RecA protein is needed for error-prone repair, at least indirectly, to mediate induction of *umuC* and *umuD*. However, genetic experiments suggest that RecA protein also acts *directly* in error-prone repair.

How might error-prone repair work? We know that the mutations it results in are generally single-base changes at the site of the damage and not, for example, extensive replacements of DNA by some mechanism that splices in unrelated DNA sequences. Somehow, the UmuC and UmuD proteins cause a base to be inserted into the growing strand, even though no template base exists, a process that normally would be prohibited. This process has been named **bypass synthesis.** One model to explain bypass synthesis proposes that the UmuC and UmuD proteins might act to inhibit proofreading by the $3' \rightarrow 5'$ exonuclease activity of DNA polymerase. Another suggestion has been that UmuC and UmuD encode an entirely new DNA polymerase that can make mistakes. How RecA protein might be involved is also a mystery. One possibility follows from the fact that RecA protein attaches to DNA in the gap produced when a nonpairing lesion stops DNA synthesis: RecA protein might also bind the UmuC and UmuD proteins and thereby direct them to the site where error-prone DNA synthesis is required.

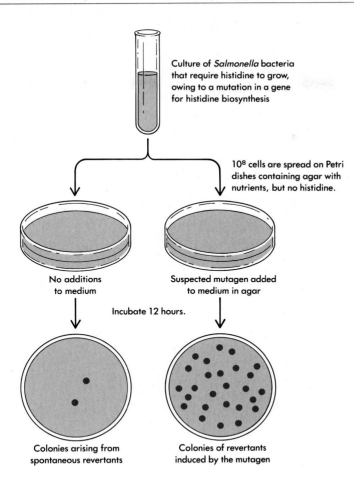

Culture of *Salmonella* bacteria that require histidine to grow, owing to a mutation in a gene for histidine biosynthesis

10^8 cells are spread on Petri dishes containing agar with nutrients, but no histidine.

No additions to medium

Suspected mutagen added to medium in agar

Incubate 12 hours.

Colonies arising from spontaneous revertants

Colonies of revertants induced by the mutagen

Figure 12-12
In the Ames test, the activity of a mutagen is detected as an increased incidence of wild-type revertants among auxotrophic bacteria grown in the presence of the chemical. Potent mutagens yield thousands of revertants from the 10^8 bacteria on a Petri dish.

Mutagens in Nature and Their Detection[31, 32]

It is important to identify mutagens and avoid human exposure to them for two reasons. First, random genetic change is more likely to be harmful than helpful in future generations. But even mutations that occur in somatic (nongerm) cells and thus are not passed to progeny threaten the organism. We now know, as was long suspected, that cancer is often the result of mutations in somatic cells that lead to uncontrolled cell growth. And a number of disorders associated with age, such as atherosclerosis, may be the expression of accumulated, unrepaired DNA damage. In fact, it has been suggested that the entire aging process may reflect irreversible damage to the DNA of somatic cells.

The clear demonstration that most mutagens are carcinogens (and vice versa) means that there is a simple way to determine if a substance is likely to be carcinogenic: We determine whether it is mutagenic to bacteria, whose short generation time allows a test to be made in a day or two (Figure 12-12). This method, now called the **Ames test,** was developed and refined by B. Ames, who used several tricks to make it very sensitive. Reversion to wild type of a mutation causing a growth defect, such as a requirement for the amino acid histidine, is measured, because even a few growing cells among many give rise to colonies that can easily be scored. The test bacteria bear mutations making their cell walls leaky, so that most compounds to be tested can permeate them, and they also carry a plasmid that expresses an efficient variant of the *umuC* and *umuD* genes, which are essential if most types of DNA damage are to be mutagenic.

A final requirement for carcinogens to be active as mutagens in the bacterial test reveals an important principle of mutagenesis in higher organisms. Most carcinogenic compounds, particularly natural products, are not mutagenic if they are applied directly to growing bacteria. First they must be *activated* by incubation with an extract of mammalian liver, a process that occurs naturally when foreign substances are metabolized in the liver. Activation converts carcinogens to electrophilic derivatives that react readily with the bases in DNA.

What are the important mutagens revealed by such tests as likely or certain carcinogens? It is proved from epidemiological studies that tobacco smoke is one of the most important carcinogens that humans encounter, and tobacco smoke is highly mutagenic in the Ames test. The potential danger of certain mutagenic industrial chemicals like ethylene dichloride is well known, and many other environmental contaminants may be significant carcinogens. But a more recent realization is that our greatest exposure to mutagens and carcinogens comes from eating—and not necessarily exotic foods, but even such staples as common vegetables. Plants make toxic chemicals as a defense against predators like insects, and many of these are metabolized in the liver into potent mutagens. The actual mutagens often are **oxidants** such as oxygen radicals that react with bases in DNA. Another dietary source of mutagens and a likely cause of cancer is lipids, which are metabolized to reactive oxidants by the liver. An optimistic note is that natural antioxidants like glutathione may counter many of the effects of these dietary hazards.

Summary

Organisms can survive only if their DNA is replicated faithfully and is protected from chemical and physical damage that would change its coding properties. The limits of accurate replication and repair of damage are revealed by the natural mutation rate. Thus, an average nucleotide is likely to be changed by mistake only about once every 10^9 times it is replicated, although error rates for individual bases vary by a hundredfold from this. Much of the accuracy of replication is inherent in the way DNA polymerase copies a template. The initial selection of the correct base is guided by complementary pairing. Accuracy is increased by proofreading, in which the $3' \rightarrow 5'$ exonuclease activity of DNA polymerase checks the last base inserted and removes a wrong one. Finally, in mismatch correction, the newly synthesized DNA strand is scanned by an enzyme that initiates replacement of DNA containing incorrectly paired bases. Cellular mutations called mutators that increase the natural mutation rate affect proteins involved in both proofreading and mismatch correction. Despite the safeguards, mistakes of all types occur: base substitutions, small and large additions and deletions, and gross rearrangements of DNA sequences.

Mutagens increase the natural rate of mutation. Base analogs simulate natural bases and are incorporated in their place, but pair less accurately and thus give rise to base switches. Reactive chemicals, including alkylating agents like nitrosamines and nitrosoguanidine, chemically change the bases so that pairing mistakes are made. Frameshift mutagens probably act by stabilizing looped-out regions in the template or growing strand during replication, ensuring that they are miscopied.

Cells have a large repertory of enzymes devoted to repairing DNA damage that would otherwise be lethal or that would modify DNA so as to engender damaging mutations. Photoreactivation is catalyzed by photolyase, an enzyme that reverses the photochemical fusion of adjacent pyrimidine bases into pyrimidine dimers caused by ultraviolet light. O^6-methylguanine methyltransferase enzymatically removes methyl groups from the dangerously reactive O^6 groups of guanine and probably from the phosphodiester backbone as well. The adaptive response, an induced resistance to alkylating agents that cause methylation, results from greatly increased synthesis of the transferase.

Although there are a few enzymes that simply reverse damage, a more universal process is excision repair, in which a damaged segment is removed and replaced through new DNA synthesis for which the undamaged strand serves as a template. In *E. coli*, excision repair is initiated in two ways. The UvrABC endonuclease cuts out a 12-base segment of a DNA strand that includes the lesion. Glycosylases remove only a damaged base, leaving a second step to an apurinic or apyrimidinic endonuclease, which cuts the backbone. Both pathways leave a primer end from which DNA polymerase can initiate nick translation to remove the damaged segment and renew the DNA. An alternative repair method, which is particularly important if excision repair cannot occur because no template for repair synthesis is available, is recombinational repair, in which an intact DNA strand is taken from a different but homologous duplex.

A central enzyme of DNA repair in *E. coli* is the RecA protein, which both catalyzes DNA strand exchange and thus promotes recombinational repair and also regulates a group of at least 15 DNA repair genes called the SOS genes. When cellular DNA is damaged, a protease function of RecA protein is activated to destroy the repressor (LexA repressor) of the SOS genes, leading to their induction. Two SOS genes, *umuC* and *umuD*, are required for error-prone repair, an important but still mysterious process by which mutations result from repair of DNA damage so severe that non-base-pairing lesions are created. The majority of all dangerous natural mutagens act through error-prone repair.

Mutagens are of concern to us because they permanently affect the genes that organisms inherit and because cancer is often caused by mutations in somatic cells that lead to uncontrolled growth. Thus, it is important to be able to identify and characterize mutagens to which organisms might be exposed. Bacterial tests allow this to be done conveniently because the short generation time of bacteria and our ability to handle many cells at once reveal the activity of a mutagen in a day. Natural mutagens are prevalent, but so are the biological defenses that organisms mount against them.

Bibliography

General References

Friedberg, E. C. 1985. *DNA Repair.* San Francisco: Freeman.
Kornberg, A. 1980. *DNA Replication.* San Francisco: Freeman. Also, *1982 Supplement to DNA Replication.* The essential textbook that describes the basic principles of replication and repair of DNA.

Cited References

1. Miller, J. H. 1983. "Mutational Specificity in Bacteria." *Ann. Rev. Genetics* 17:215–238.
2. Miller, J. H., and K. B. Low. 1984. "Specificity of Mutagenesis Resulting from the Induction of the SOS System in the Absence of Mutagenic Treatment." *Cell* 37:675–682.
3. Topal, M. D., and J. R. Fresco. 1976. "Complementary Base Pairing and the Origin of Substitution Mutations." *Nature* 263:285–293.
4. Loeb, L. A., and T. A. Kunkel. 1982. "Fidelity of DNA Synthesis." *Ann. Rev. Biochem.* 51:429–457.
5. Cox, E. C. 1976. "Bacterial Mutator Genes and the Control of Spontaneous Mutation." *Ann. Rev. Genetics* 10:135–156.
6. Echols, H., C. Lu, and P. M. J. Burgers. 1983. "Mutator Strains of *Escherichia coli, mutD* and *dnaQ*, with Defective Exonucleolytic Editing by DNA Polymerase III Holoenzyme." *Proc. Nat. Acad. Sci.* 80:2189–2192.
7. Ghosal, D., and H. Saedler. 1978. "DNA Sequence of the Mini-Insertion IS2-6 and Its Relation to the Sequence of IS2." *Nature* 275:611–617.
8. Albertini, A., M. Hofer, M. Calos, and J. Miller. 1982. "On the Formation of Spontaneous Deletions: The Importance of Short Sequence Homologies in the Generation of Large Deletions." *Cell* 29:319–328.
9. Ripley, L. S. 1982. "Model for the Participation of Quasi-Palindromic DNA Sequences in Frameshift Mutation." *Proc. Nat. Acad. Sci.* 79:4128–4132.
10. Roth, J. R. 1974. "Frameshift Mutations." *Ann. Rev. Genetics* 8:319–346.
11. Singer, B., and J. T. Kusmierek. 1982. "Chemical Mutagenesis." *Ann. Rev. Biochem.* 52:655–693.
12. Howard-Flanders, P. 1981. "Inducible Repair of DNA." *Sci. Amer.* 245:72–103.
13. Brash, D. E., and W. A. Haseltine. 1982. "UV-Induced Mutation Hotspots Occur at DNA Damage Hotspots." *Nature* 298:189–192.
14. Lindahl, T. 1982. "DNA Repair Enzymes." *Ann. Rev. Biochem.* 51:61–87.
15. Haseltine, W. 1983. "Ultraviolet Light Repair and Mutagenesis Revisited." *Cell* 33:13–17.
16. Samson, L., and J. Cairns. 1977. "A New Pathway for DNA Repair in *Escherichia coli.*" *Nature* 267:281–282.
17. Cairns, J., P. Robins, B. Sedgwick, and P. Talmud. 1981. "The Inducible Repair of Alkylated DNA." *Prog. Nucleic Acid Res. Mol. Biol.* 26:237–244.
18. Lindahl, T., B. Demple, and P. Robins. 1982. "Suicide Inactivation of the *E. coli* O^6-Methylguanine-DNA Methyltransferase." *EMBO J.* 1:1359–1363.
19. Teo, I., B. Sedgwick, B. Demple, B. Li, and T. Lindahl. 1984. "Induction of Resistance to Alkylating Agents in *E. coli:* The *ada* Gene Product Serves Both as a Regulatory Protein and as an Enzyme for Repair of Mutagenic Damage." *EMBO J.* 3:2151–2157.
20. Sancar, G. B., and W. D. Rupp, 1983. "A Novel Repair Enzyme: UVRABC Excision Nuclease of *Escherichia coli* Cuts a DNA Strand on Both Sides of the Damaged Region." *Cell* 33:249–260.
21. Yeung, A. T., W. B. Mattes, E. Y. Oh, and L. Grossman. 1983. "Enzymatic Properties of Purified *Escherichia coli* uvrABC Proteins." *Proc. Nat. Acad. Sci.* 80:6157–6161.
22. Coulondre, C., J. Miller, P. Farabaugh, and W. Gilbert. 1978. "Molecular Basis of Base Substitution Hotspots in *Escherichia coli.*" *Nature* 274:775–780.
23. Wagner, R., and M. Meselson. 1976. "Repair Tracts in Mismatched DNA Heteroduplexes." *Proc. Nat. Acad. Sci.* 73:4135–4139.
24. Lu, A.-L., S. Clark, and P. Modrich. 1983. "Methyl-Directed Repair of DNA Base-Pair Mismatches *in vitro.*" *Proc. Nat. Acad. Sci.* 80:4639–4643.
25. Radding, C. 1982. "Homologous Pairing and Strand Exchange in Genetic Recombination." *Ann. Rev. Genetics* 16:405–437.
26. Szostak, J., T. Orr-Weaver, R. Rothstein, and F. Stahl. 1983. "The Double-Strand Break Repair Model for Recombination." *Cell* 33:25–35.
27. Walker, G. C. 1984. "Mutagenesis and Inducible Responses to Deoxyribonucleic Acid Damage in *Escherichia coli.*" *Microbiol. Revs.* 48:60–93.
28. Schaaper, R. M., T. A. Kunkel, and L. A. Loeb. 1983. "Infidelity of DNA Synthesis Associated with Bypass of Apurinic Sites." *Proc. Nat. Acad. Sci.* 80:487–491.
29. Kato, T., and Y. Shinoura. 1977. "Isolation and Characterization of Mutants of *Escherichia coli* Deficient in Induction of Mutations by Ultraviolet Light." *Mol. Gen. Genetics* 156:121–131.
30. Elledge, S., and G. Walker. 1983. "Proteins Required for Ultraviolet Light and Chemical Mutagenesis: Identification of the Products of the *umuC* Locus of *Escherichia coli.*" *J. Mol. Biol.* 164:175–192.
31. Ames, B. 1979. "Identifying Environmental Chemicals Causing Mutations and Cancer." *Science* 204:587–593.
32. Ames, B. 1983. "Dietary Carcinogens and Anticarcinogens." *Science* 221:1256–1264.

V

THE STEPS IN
PROTEIN SYNTHESIS

CHAPTER 13

The Synthesis
of RNA upon
DNA Templates

Although the chemistry of DNA synthesis and RNA synthesis is very similar at the nucleotide level, the biological functions of the two processes are quite different. Whereas DNA synthesis must only replicate the genome accurately and is inherently unvarying, transcription of the genome into RNA reflects much of the complexity of genetic expression itself. Genes are usually regulated at the level of transcription, and all phases of RNA synthesis are subject to variation: where an RNA chain starts or stops on the DNA template, when it starts and stops, and how efficiently it does both. Much of the recent progress in understanding the molecular biology of gene expression has come from studying the DNA sequences and the proteins that control transcription, particularly the RNA polymerase enzymes themselves. The basic principles were first established for bacterial (mainly *E. coli*) RNA polymerase, but these principles are probably universal, despite a few important differences that eucaryotic enzymes reveal. In this chapter we will primarily consider transcription in bacteria and mention eucaryotic enzymes only briefly as a preview to Chapter 21.

The Structures of RNA

RNA is almost always single-stranded, in contrast to DNA, and is synthesized from the DNA template in a correspondingly different way. Before we consider their synthesis, it is important to point out that although they are single-stranded, RNA molecules nevertheless do have structures, some as definite and fixed as those of proteins and some much less specifically defined. Almost every RNA molecule has many short double-helical regions, which can form because the two sections of an RNA chain within a hairpin fold are in the correct antiparallel orientation to base-pair (Figure 13-1). In addition to the standard AU and GC base pairs, weaker GU base pairs (Chapter 14) also contribute to this secondary structure of single-stranded RNA. Even an RNA of random base sequence has much secondary structure, because short adjacent complementary sequences occur by chance. Furthermore, some RNAs (such as transfer RNA) have evolved to contain extensive internal complementary sequences that base-pair to give the whole molecule a specific shape, its tertiary structure. As we would expect, secondary structure also exists within

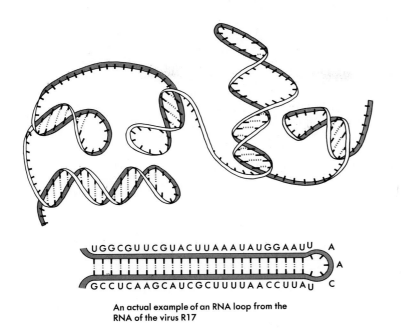

Figure 13-1
Schematic folding of an RNA chain, showing several double-helical regions held together by hydrogen bonds.

UGGCGUUCGUACUUAAAUAUGGAAUU A
A
GCCUCAAGCAUCGCUUUUAACCUUAU C

An actual example of an RNA loop from the
RNA of the virus R17

single chains of DNA when there is no complementary strand that allows a long duplex to form.

Two principles of chain folding that apply to polypeptides are probably equally true for RNA chains: The structure that each can form is encoded in the sequence of functional groups in the chain (whether amino acids or nucleotides), and this structure forms as the chain is being made. This is important to keep in mind, because as we shall see, the structure of a growing RNA chain can affect transcription of the remainder of the molecule.

Each cell contains many distinct RNA molecules, with lengths from less than 50 nucleotides to tens of thousands of nucleotides. These are almost always linear single strands, although a few circular RNAs are known to occur (Chapter 24).

Enzymatic Synthesis of RNA upon DNA Templates

The fact that RNA, like DNA, is a long, unbranched chain of four different nucleotides immediately suggests that the genetic information of DNA chains is transferred to a complementary sequence of RNA nucleotides. According to this hypothesis, the DNA strands at one or more stages in the cell cycle separate and function as templates onto which complementary ribonucleotides are attracted by DNA-like base pairing [adenine with thymine (uracil in RNA) and guanine with cytosine]. This hypothesis also implies that some control mechanism must dictate whether the separated DNA strands will function as templates for a complementary DNA strand or a complementary RNA strand.

Direct evidence for the hypothesis came from the discovery of the appropriate enzyme, RNA polymerase, which exists in virtually all cells; a cell of *E. coli* has about 3000 molecules of RNA polymerase. This enzyme links together ribonucleotides by catalyzing the forma-

Figure 13-2
Catalysis of 3′–5′ phosphodiester bond formation in RNA.

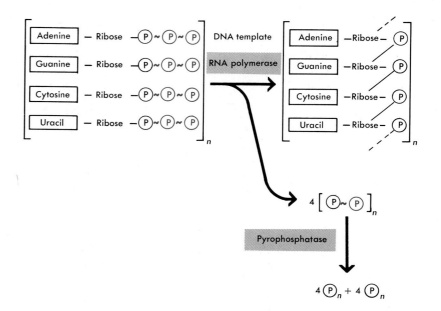

tion of the internucleotide 3′–5′ phosphodiester bonds that hold the RNA backbone together (Figure 13-2). It does so, however, only in the presence of DNA, a fact that suggests that DNA must line up the correct nucleotide precursors in order for RNA polymerase to work.

Although RNA polymerase normally uses double-stranded DNA in making RNA, the actual template is single-stranded DNA that the enzyme exposes by unwinding a bit of the helix as it passes. This is shown clearly in experiments with DNA from the virus φX174. This virus belongs to one of the viral classes that contain single-stranded DNA instead of the customary double-helical form. Only one of the two possible complementary DNA strands is present, and when it is used as the template for RNA polymerase, the enzymatic product has a complementary base sequence (Table 13-1). Moreover, in this special case, the RNA product remains attached to its DNA template, allowing the isolation of a hybrid DNA-RNA double helix. When the template is double-helical DNA, the RNA product quickly detaches from its template, and the two DNA strands again come together in specific register. This means that the front end of a messenger RNA becomes immediately available to bind to a ribosome (Chapter 14).

We thus see that the fundamental mechanism for the synthesis of RNA is very similar to that of DNA. In both cases, the immediate precursors are nucleoside triphosphates that use the energy in one of their pyrophosphate bonds to drive the reaction toward synthesis.

Table 13-1 Base Composition of Enzymatically Synthesized RNA Using Single-Stranded φX174 DNA as Template

	φX174 DNA	Observed Values of RNA Product	Predicted RNA Composition
Adenine	0.25	0.32	0.33
Uracil	0.33 (thymine)	0.25	0.25
Guanine	0.24	0.20	0.18
Cytosine	0.18	0.23	0.24
TOTAL	1.00	1.00	1.00

Also in both cases, a single enzyme works on all four possible nucleotides, whose correct selection is dictated by the obligatory need to base-pair with a polynucleotide template.

RNA synthesis is necessarily an extremely accurate process. There is no evidence, however, for any proofreading, and so the overall precision of transcription does not approach the almost perfect copying characteristic of DNA self-replication. But since cellular RNA is not self-replicating, the rare mistakes that do occur are not genetically perpetuated.

Along Each Gene Only One DNA Strand Acts as an RNA Template[1]

If each of the two DNA strands of a given gene served as an RNA template, each gene would produce two RNA products with complementary sequences. Since our genetic evidence tells us that each gene controls only one protein, we must assume that either only one of the two possible RNA strands is made or, if both are synthesized, only one is functional for some special reason. It appears that the former possibility is correct: In vivo, only one of the two possible RNAs is found. This can be seen by examining the RNA synthesized in vivo from the virus SP8, which multiplies in the bacterium *Bacillus subtilis*. The two complementary DNA strands of SP8, unlike those of most viruses, have quite different base compositions and can be relatively easily separated. It is thus possible to determine whether the RNA products have base sequences complementary to one or both of the DNA strands.

To make this determination, we use our ability to form artificial DNA-RNA hybrid molecules by mixing RNA molecules with single-stranded DNA molecules formed by heating double-helical DNA. Heating DNA molecules to temperatures just below 100°C breaks the hydrogen bonds holding the complementary strands together; they then quickly separate from each other (DNA denaturation). If the temperature is gradually lowered, the complementary strands again form the correct hydrogen bonds, and the double-helical form is regained (renaturation of DNA). If, however, this gentle cooling is done in the presence of single-stranded RNA that has been synthesized using the DNA as template, then DNA-RNA hybrids form as well as renatured DNA double helices (Fig. 13-3). These DNA-RNA hybrids are very specific and form only if stretches of the nucleotide sequence in the DNA are complementary to the RNA nucleotide sequences. This technique lets us see whether the in vivo RNA products will form hybrids with only one or with both of the complementary SP8 DNA strands. A clear result emerges: Only one DNA strand is copied.

The copying of only one DNA strand within a given gene explains why the base ratios of RNA need not be complementary, even though RNA is made on a DNA template. In a given DNA strand, there is no reason why the A should equal T, or the G equal C. These ratios must be 1:1 only when the corresponding bases on the two complementary chains are added together. Thus, we must assume that only in the exceptional DNA molecule will the single strands be found, on close inspection, to have even approximately complementary ratios. Correspondingly, only rarely will single-stranded RNA molecules be found in which A nearly equals U and G nearly equals C.

Figure 13-3
The use of DNA-RNA hybrids to show the complementarity in nucleotide sequence between an RNA molecule and one of the two strands of its DNA template. The left side of the diagram shows the formation of a hybrid molecule between an RNA molecule and one of the two strands of the template. The right side of the diagram shows why DNA-RNA hybrid formation allows complementarity to be detected: If the same RNA molecule is mixed with unrelated DNA, no hybrid molecules are formed.

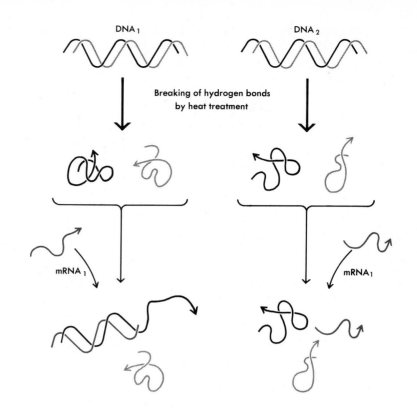

Differential copying of the two strands of a DNA molecule by RNA polymerase can also be observed in vitro. This result depends on the use of DNA templates that have not been severely damaged. If, for example, T7 DNA has been denatured to produce single-stranded regions that RNA polymerase can bind, then both strands are copied. However, on its natural template, intact double-stranded DNA, RNA polymerase binds only to specific sequences called promoters, which direct it to copy the same strand that is copied during in vivo viral reproduction (Figure 13-4).

The transcription patterns along other chromosomes are often not as clear-cut as with T7 or SP8. Both strands of the T4 and λ chromosomes are transcribed, one strand serving as the template for some genes (usually contiguous groups) while the remaining genes are transcribed along the other strand (see Figure 17-15). The same picture holds for the *E. coli* chromosome, which can be transcribed both clockwise and counterclockwise.

Synthesis of RNA Chains Occurs in a Fixed Direction

Each RNA chain, like each DNA chain, has a direction defined by the orientation of the sugar-phosphate backbone. The chain end terminated by the 5' carbon atom is called the 5' end, while the end containing the 3' carbon atom is called the 3' end (see Figure 3-13). As with DNA synthesis, it always seemed safe to assume that RNA chains grow sequentially in only one of the two possible polarities, 5'→3' or 3'→5'. If they grow 5'→3', then we expect the begin-

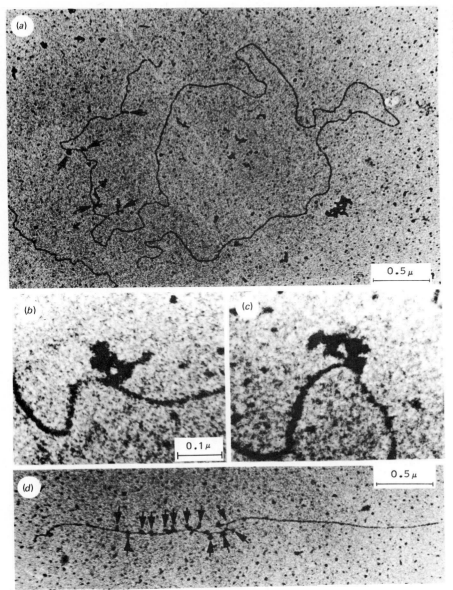

Figure 13-4
Electron micrographs of in vitro transcription of T7 DNA. (a) Several growing RNA chains can be seen attached to the DNA template. All of the RNA chains are transcribed off the same T7 strand, initiating their growth at three clustered promoters near the end of the molecule. Total synthesis time is 5 minutes at 37°C. (b, c) Views of two attached RNA chains are shown at higher magnification. (d) Synthesis was terminated after 2½ minutes at 37°C. Twelve nascent chains all at one end can be seen. RNA polymerase is present where the RNA chains are attached, but it is not visible. (Courtesy of Dr. R. Davis, Stanford University.)

ning nucleotide to possess a Ⓟ~Ⓟ~Ⓟ (triphosphate) group (Figure 13-5). But if the chain grows 3'→5', then the nucleotide at the growing end would contain the Ⓟ~Ⓟ~Ⓟ group. Firm evidence shows that the direction of growth is 5'→3': Newly inserted nucleotides are found at the 3' ends, while the Ⓟ~Ⓟ~Ⓟ groups are found attached to the nucleotides that commenced chain growth.

The 5'→3' direction is confirmed by work with the metabolic inhibitor 3'-deoxyadenosine, an analog of adenosine lacking the 3' oxygen that forms part of the internucleotide bond. When it is added to cells, 3'-deoxyadenosine is first phosphorylated to 3'-deoxyadenosine-Ⓟ~Ⓟ~Ⓟ and then joined to the 3' growing end. Because it contains no 3'-OH group, further nucleoside triphosphates cannot attach, and RNA synthesis stops. But this means that synthesis must be 5'→3', because otherwise the inhibitor could not be incorporated at all.

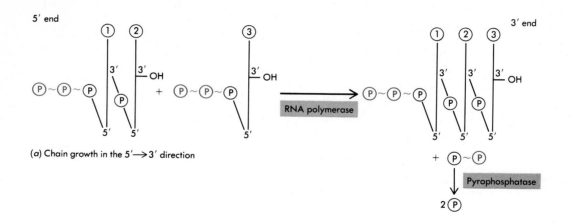

(a) Chain growth in the 5′→3′ direction

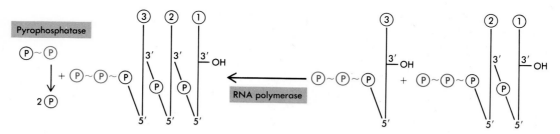

(b) Chain growth in the 3′→5′ direction

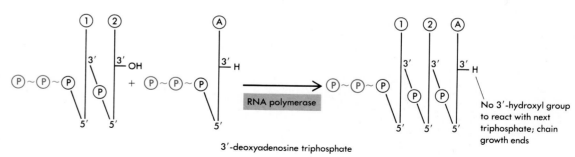

3′-deoxyadenosine triphosphate

(c) 3′-deoxyadenosine incorporation

Figure 13-5

Alternative directions for the synthesis of an RNA chain are shown in (a) and (b). Many types of experiments suggest that chains always grow in the 5′→3′ direction. Part (c) shows the structure of 3′-deoxyadenosine triphosphate and its incorporation at the 3′ end of an RNA chain growing 5′→3′, halting RNA synthesis.

Construction of Bacterial RNA Polymerase from Subunits

Like DNA polymerase III, bacterial RNA polymerase has a very complicated subunit structure. The active form, the holoenzyme, sediments at about 15S and contains five different polypeptide chains, β', β, σ, α, and ω, with respective molecular weights (for *E. coli*) of 155,000, 151,000, 70,000, 36,500, and about 11,000 daltons. No covalent bonds join the various chains; aggregation results from the formation of secondary bonds. Each specific chain, with the exception of α, which appears twice, is present once within the active molecule, giving the complete holoenzyme ($\beta'\beta\alpha_2\omega\sigma$) a molecular weight of about 450,000 (Figure 13-6 and Table 13-2). If it were a sphere, the diameter of RNA polymerase would be about 100 Å, the length of a 30-base-pair segment of duplex DNA; however, the fact that it binds over 60 nucleotides suggests instead an elongated shape.

It is tempting to relate the complexity of bacterial RNA polymerase to the many individual enzymatic steps that it carries out to achieve initiation, elongation, and termination of RNA chains. Yet, phage T7 encodes an RNA polymerase containing only a single chain with a molecular weight of 98,000 daltons that manages all of this as well (Chapter 17). Perhaps much greater complexity is necessary for the elaborate and varied transcriptional controls that the bacterial cell requires. It would be of great interest to know the detailed three-dimensional structure of RNA polymerase. However, its large size makes X-ray crystallography a difficult undertaking, and furthermore, the requisite crystals have not yet been produced.

The attachment of σ to the other chains is not very firm, so it is relatively easy to isolate a $\beta'\beta\alpha_2\omega$ aggregate. This specific grouping is known as the **core enzyme,** for it catalyzes the formation of the internucleotide phosphodiester bonds equally well in the presence or absence of σ. The catalytic center is thought to be in the β subunit of the core. Subunit β is the binding site of the important antibiotic **rifampicin** (and the related antibiotic rifamycin), a potent inhibitor of the enzyme that prevents initiation of an RNA chain but does not affect its subsequent elongation. Mutations that make the cell resistant to rifampicin map in *rpoB*, the gene for β, as do many nonlethal mutations that subtly affect RNA polymerase function. The antibiotic **streptolydigin** inhibits elongation of an RNA chain, and mutations conferring resistance to it also are in the gene encoding β.

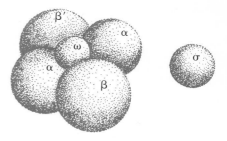

Figure 13-6
Schematic picture of RNA polymerase showing subunit construction.

Organization of RNA Polymerase Genes into Operons[2]

The genes for the *E. coli* RNA polymerase subunits (with the exception of ω, whose gene is unknown) are found in three operons that also contain essential genes for protein and DNA synthesis (Figure 13-7). Mostly, these genes encode some of the 50 or so ribosomal proteins (Chapter 14), although the σ operon also contains the gene for the replication protein DNA primase (Chapter 10). It is clear that

Table 13-2   Subunits of *E. coli* RNA Polymerase and Transcription Factors

	Gene Name	Map Position* (Minutes)	Polypeptide MW (Daltons)	Number in Enzyme	Function
RNA Polymerase Subunits					
β' (beta')	*rpoC*	89.5	155,000	1	DNA binding?
β (beta)	*rpoB*	89.5	151,000	1	Catalytic site for RNA polymerization
σ (sigma)	*rpoD*	66.5	70,000	1	Promoter recognition, initiation
α (alpha)	*rpoA*	72	36,500	2	?
ω (omega)	—	—	11,000	1	?
Transcription Factors					
ρ (rho)	*rho*	84.5	46,000	6	Termination
nusA	*nusA*	65	69,000	1	Elongation, termination

*The map position refers to the conventional (circular) *E. coli* genetic map (see Figure 7-22).
SOURCE: After G. Zubay, *Biochemistry* (Reading, Mass.: Addison-Wesley, 1983), p. 792. Used with permission.

Figure 13-7
The three operons that contain genes for the σ, β, β', and α subunits of RNA polymerase. The major promoters (P) are at the left, and RNA is synthesized rightward. Minor promoters (p), including one in the σ operon that is active in heat-shock conditions (p_{HS}), are found within the operons; they promote transcription only of genes to their right. The first two genes of the β operon, *L11* and *L1*, are sometimes considered to make up a separate operon, because there is a promoter before the *L10* gene (see Chapter 16). [After Z. F. Burton, C. Gross, K. Watanabe, and R. Burgess, *Cell* 32 (1983):335, with permission.]

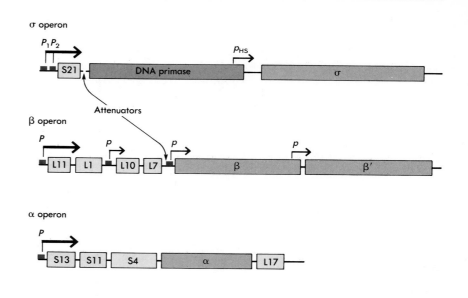

this congregation must reflect common regulation of proteins needed in amounts that are closely related to the growth rate of the cell. How these operons are regulated is unknown, although the σ operon has two tandem promoters, as do the ribosomal RNA operons, which are thought to be controlled by the molecule ppGpp* in response to growth rate (Chapter 16). The genes in these operons are not transcribed or expressed equally: Ribosomal protein genes at the beginning of the σ and β operons are followed by **attenuators,** partially effective termination sites that allow only a fraction of the RNA polymerase to transcribe the remainder of the operon. Corresponding to the unequal transcription within each operon, fewer RNA polymerase subunits (3000) than ribosomal proteins (20,000) are needed in each cell. However, this agreeable relation is confounded by the DNA primase gene, which precedes the σ gene, even though the cell has 60 times as many molecules of σ as primase (3000 versus 50); the explanation is that the mRNA segment encoding DNA primase is much less stable than the rest. The occurrence of several minor promoters within these operons, including one in the σ operon that is activated by heat shock (Chapter 16), bespeaks a much more complex regulation than we so far understand.

RNA Polymerase Recognizes Specific Starting Sequences in DNA: Promoters[3-8]

The σ subunit by itself has no catalytic function; its role is the recognition of start signals along the DNA molecule. In vitro experiments show that in the absence of σ, the core enzyme will occasionally initiate RNA synthesis, but this is the result of starting mistakes. RNA chains made by the core enzyme alone are started randomly along both strands of a given gene. When σ is present, however, the correct site is selected, indicating that the holoenzyme has the capacity to bind specifically to a "starting" sequence, the *promoter*.

---

*This is the common abbreviation for guanosine with a pyrophosphate on its 5' and 3' carbons. In our previous nomenclature it is Ⓟ~Ⓟ—G—Ⓟ~Ⓟ.

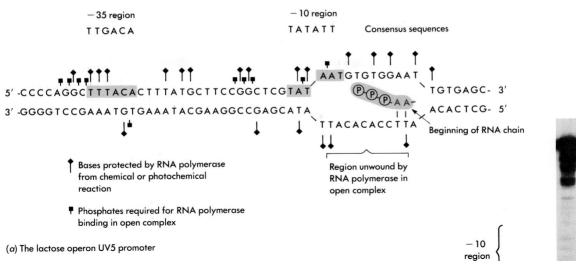

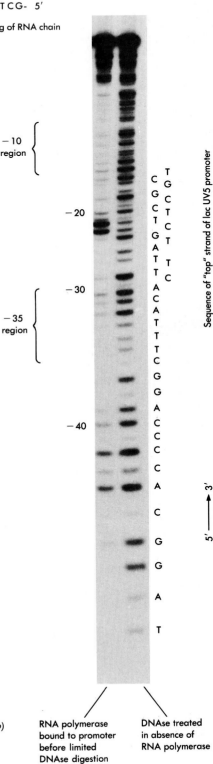

(a) The lactose operon UV5 promoter

Figure 13-8
Promoter consensus sequences and RNA polymerase contact sites.
(a) Below the −35 and −10 consensus sequences is the sequence of a particular promoter, named lactose UV5. RNA polymerase contacts bases (indicated by symbols above them) and phosphates in the helix backbone (indicated by symbols between the bases) throughout this promoter, but particularly near the −35 and −10 regions. In this particular experiment, it was found that at least the indicated 12-base segment is unwound, although a different measurement shows that a segment about 17 bases long is unwound. [After U. Siebenlist, R. Simpson, and W. Gilbert, *Cell* 20 (1980):270, with permission.] (b) "Footprint" analysis of the lactose UV5 promoter, revealing sequences protected by RNA polymerase from limited DNase digestion (see Chapter 16). This experiment shows that RNA polymerase covers 50 to 60 bases of DNA, blocking access by the relatively large nuclease to most phosphodiester bonds in this segment. [After A. Schmitz and D. Galas, *Nucleic Acid Res.* 6 (1979):124, with permission.]

What features of DNA structure direct bacterial RNA polymerase to bind only at promoters? Clearly, there must be nucleotide sequences giving DNA a specific local shape that fits RNA polymerase, just as any enzyme fits its substrate. A comparison of more than 100 promoters shows that two common sequences of six nucleotides occur about 10 base pairs and about 35 base pairs before the site where RNA synthesis starts. These are named the −35 (minus 35) and −10 (minus 10) sequences, according to the numbering scheme in which the DNA nucleotide encoding the beginning of the RNA chain is named +1. Few promoters have exactly the −10 and −35 **consensus sequences** shown in Figure 13-8, but most differ from them by only a few nucleotides. There is strong evidence that these sequences are the critical parts of the promoter. First, RNA polymerase contacts bases and phosphates of the DNA backbone in and alongside the −35 and −10 sequences. These contacts are revealed by "protection" experiments, in which the ability of RNA polymerase to shield regions of DNA from attack by chemicals and nucleases is measured, and by "chemical modification" experiments, in which chemical groups whose integrity is essential for RNA polymerase binding are identified (Chapter 16). Second, weaker promoters tend to differ in sequence somewhat from the consensus, whereas stronger promoters match it

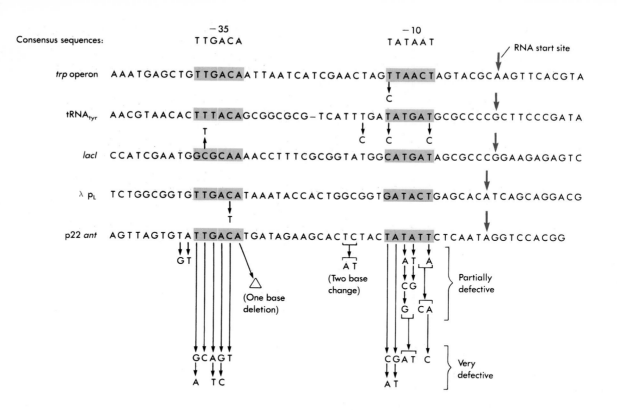

Figure 13-9
Mutations that decrease (downward arrows) and increase (upward arrow) initiation from five promoters. The strand identical in sequence to the RNA is shown. [After M. Rosenberg and D. Court, *Ann. Rev. Genetics* 13 (1979):319; and after P. Youderian, S. Bouvier, and M. Susskind, *Cell* 30 (1982): 843–853.]

more closely. Third, and most important, 75 percent of mutations that destroy promoter function change nucleotides in the common sequences (Figure 13-9); the other 25 percent are near the common sequences.

The two binding regions are separated by about 20 base pairs, or two complete turns of the double helix. Since both binding regions are therefore on one side of a DNA molecule, the enzyme must bind along one face of the helix, feeling in each region for the specific chemical shape made by base pairs in the bottoms of the grooves (Figure 13-10).

RNA Polymerase Unwinds DNA as It Binds the Promoter[9, 10]

E. coli RNA polymerase binds a promoter in two distinguishable steps (see Figure 13-10). First, it finds the promoter sequence and binds loosely in a "closed" complex. This initial recognition occurs mostly at the −35 region, since mutations that impair binding to the promoter change bases in the −35 region. Second, the closed complex is converted to an "open" complex, in which RNA polymerase is bound much more tightly. During this shift, about 17 base pairs beginning in the −10 region unwind, exposing the template strand against which RNA nucleotides are polymerized. It may be significant that the −10 region is rich in AT base pairs, because these come apart or "melt" much more easily than GC pairs. Not surprisingly, mutations that hinder formation of the open complex map in the −10 region. Many of these change the −10 sequence but preserve its AT content, showing that this region must have a specific shape for RNA polymerase to recognize, as well as being inherently susceptible to melting. Unwinding of the DNA as the open complex forms extends the stretch of

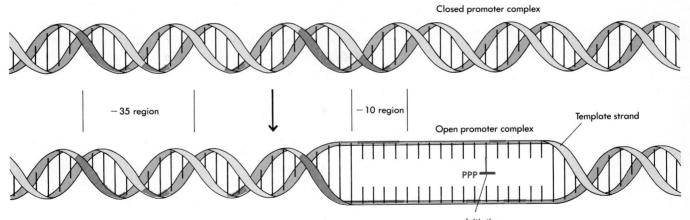

Closed promoter complex

| −35 region | | −10 region |

Open promoter complex

Template strand

PPP

Initiating
nucleotide

Figure 13-10
RNA polymerase binding to DNA in the closed and open promoter complexes. This interpretation, based on the experiment summarized in Figure 13-8, shows where RNA polymerase may touch promoter DNA during the initial binding (closed complex) and after the strands are separated to allow initiation of RNA chain growth (open complex). Sites of contact are shown in color. The ATP that initiates the RNA chain is shown, although its presence is not required for the open complex to form. [After U. Siebenlist, R. Simpson, and W. Gilbert, *Cell* 20 (1980):269–281.]

DNA touching RNA polymerase to include nucleotides beyond the −10 region that were initially on the other side of the DNA helix from the binding face (see Figure 13-10). Since these contacts can probably be made only if this DNA strand swings around as it unwinds, their formation may provide the energy for unwinding.

Different σ Factors Allow RNA Polymerase to Recognize Different Promoter Sequences[11–16]

The σ subunit must be present for RNA polymerase to engage a promoter, although the core enzyme alone can catalyze synthesis of an RNA chain. Can we distinguish whether it is σ or the core enzyme that recognizes the specific DNA sequences that constitute a promoter? A most important discovery shows that σ is responsible: there are other σ factors, distinct from the predominant σ that copurifies with core RNA polymerase (called σ^{70}), which direct RNA polymerase to initiate RNA synthesis at promoters whose sequences are quite different from the major bacterial promoters. These variant σ factors serve special functions in the cell, generally to redirect cellular RNA synthesis drastically when expression of a completely new set of genes is required. They were first discovered in the bacterium *Bacillus subtilis*, where their role is to begin expression of new proteins required for the process of sporulation (Chapter 22). In *E. coli*, a separate σ factor named σ^{32} promotes expression of the heat-shock genes, which probably serve special metabolic functions required when the temperature of cell growth is suddenly raised (Chapter 16). Still others are made by phages to invoke transcription of phage genes necessary at particular times of phage development (Chapter 17).

Since the promoter sequences that variant σ factors recognize are different from major bacterial promoters in both the −35 and −10 hexamers, it is possible either that σ binds both regions of DNA at once, even though it would have to be greatly elongated to span the distance required (about 70 Å), or that it binds the two regions in succession during the initiation process. However, a surprising recent discovery about the phage-encoded σ factor that directs synthesis of *E. coli* phage T4 "late" genes suggests that this σ factor may bind only to the −10 region: The −35 region of T4 "late" promoters can be changed (by genetic engineering) to other and unrelated sequences without harming promoter function, indicating that the −35 sequence is not involved in the specificity of RNA polymerase binding to T4 late promoters. Perhaps this means that promoter binding directed by the T4 σ is so strong that the initial binding site at -35 is no longer necessary; at least we can conclude that σ binds primarily at the -10 region in this case. Since crosslinking experiments show that σ binds nucleotides in the unwound region at the -10 hexamer of the standard promoter, one possibility is that it both recognizes the sequence at -10 and promotes the unwinding step that lets RNA synthesis begin.

Promoters Have Very Different Efficiencies[8, 17]

It is clear from our discussion of promoter sequences that some work better than others, meaning that they allow faster repeated initiation of RNA synthesis. Biochemical experiments show that both steps, initial binding and open complex formation, differ among promoters. In the cell, different promoters function at rates that vary by several orders of magnitude—from only one initiation in ten or more minutes to one initiation every second or two. This is a fundamental way in which the rate of gene expression is set.

Regulatory Proteins Affect Transcription Initiation by Binding Near or In Promoters

The rate of transcription of a particular gene is not necessarily fixed, but instead may change according to the needs of the cell in different conditions of growth; such a gene is said to be "regulated." In *E. coli*, regulation of transcription is often mediated by proteins that, by binding to DNA near or within the promoter, increase or decrease the rate at which RNA polymerase initiates RNA synthesis. These proteins include repressors, which block binding of RNA polymerase and thus inhibit transcription, and **activators,** which bind with RNA polymerase and stimulate its activity. Promoters that require activators in order to function efficiently often match the -35 consensus sequence poorly, suggesting that the activator substitutes for this part of the binding site. A protein that is a repressor of one promoter may be an activator of another, and vice versa, depending on the precise location of its binding site; the name of a regulatory protein usually reflects the function that was discovered first. Some regulatory proteins are presented in Table 13-3, and their mechanism of action is considered in detail in Chapters 16 and 17.

Upstream DNA Sequences Can Increase Promoter Activity[18]

Although the -10 and -35 hexamers of a promoter are the essential sequences where RNA polymerase both binds and acts, other nearby DNA sequences can affect promoter function as well. For example, the natural sequences between 50 and 150 nucleotides upstream of the start site for ribosomal RNA synthesis are essential for full promoter activity. If these are deleted and replaced by foreign DNA, such as DNA of a plasmid into which the ribosomal RNA genes are cloned, the rate of initiation decreases about tenfold. In other cases, similarly distant AT-rich DNA is thought to stimulate the rate of initiation. Although upstream sequences sometimes may be binding sites for unidentified activators that stimulate RNA polymerase directly, they also could act in other ways. For example, an upstream sequence might attract a topoisomerase that induces a favorable state of local

Table 13-3   Regulatory Proteins That Affect RNA Polymerase Function at Promoters

Protein	Effect on Initiation at Target Promoter	Function in Cell
Lactose repressor	Blocks	Prevents synthesis of enzymes that metabolize lactose when the sugar is absent.
Galactose repressor	Blocks	Prevents synthesis of enzymes that metabolize galactose when the sugar is absent.
LexA repressor	Blocks	Prevents synthesis of DNA repair enzymes until cellular DNA is damaged.
λ repressor	Blocks (promoters P_L and P_R)	Prevents synthesis of proteins required for lytic growth.
λ repressor	Activates (promoter P_{RM})	Increases synthesis of itself by stimulating its own promoter.
CAP protein	Activates	Increases synthesis of enzymes that metabolize other sugars when glucose is not available.
CAP protein	Blocks	Regulates its own gene and prevents synthesis of other proteins not needed in the absence of glucose.
AraC protein	Activates	Increases synthesis of proteins required to utilize arabinose when arabinose is available.
AraC protein	Blocks	Prevents synthesis of enzymes that metabolize arabinose when the sugar is not present.
λ cII protein	Activates	Stimulates synthesis of phage proteins required to establish lysogeny.

supercoiling. Or it could bring the DNA to a region where RNA polymerase has better access to it, perhaps by binding proteins fixed in the cell structure. Finally, it is possible that subtle differences in DNA structure imposed by specific sequences are transmitted a considerable distance along the helix, so that a particular upstream sequence might affect the detailed DNA structure in the -10 and -35 regions.

Superhelical Tension of the Template Affects Promoter Efficiency[19]

Since RNA polymerase must unwind DNA at the promoter before it initiates RNA synthesis, conditions that help or hinder unwinding should increase or decrease the rate of initiation. An important condition in the cell is the superhelical tension of the DNA template: The enzyme DNA gyrase keeps bacterial DNA negatively supertwisted, thus favoring unwinding. Simultaneously, negative supercoiling is kept in check by topoisomerase I, an enzyme that *relaxes* excess negative supercoils (see Chapter 9). Impairing the activity of these enzymes by mutation or by inhibitors generally affects promoter function in the expected way. Thus, the DNA gyrase inhibitor nalidixic acid inhibits expression of many genes. Mutations in the gene encoding topoisomerase I can have the opposite effect, increasing the expression of some genes. A mutation (*supX*) in the bacterium *Salmonella typhimurium* selected to reverse the effect of a defective leucine operon promoter turned out to inactivate topoisomerase I, providing direct evidence that the function of the leucine promoter is sensitive to supercoiling.

It is not always true that increased negative supercoiling stimulates promoter function; for some promoters, the opposite is true. In these cases, we guess that negative supercoiling changes the detailed structure of DNA at the promoter in a way that inhibits RNA polymerase function much more than easier melting of DNA stimulates it. The promoter from which mRNA for DNA gyrase is made is believed to have this property. Here there is a regulatory advantage: Synthesis of DNA gyrase decreases when there is enough in the cell to keep DNA negatively supercoiled to the proper extent.

RNA Chains Usually Initiate with pppA or pppG

Actual chain initiation occurs at one extreme of the region to which *E. coli* RNA polymerase binds, within the melted segment (Figures 13-8 and 13-10) and about 12 or 13 bases from the beginning of the −10 region. Thus, the catalytic center of the enzyme appears to be at its far end, toward the direction of RNA synthesis. The first nucleotide is usually pppA or pppG, but it is sometimes pppC and rarely pppU. Occasionally, initiation occurs at several adjacent nucleotides; thus, the enzyme is somewhat flexible, and it may scan part of the melted region to find a purine start, accepting a pyrimidine only as a second choice.

The nucleotide at the 5′ end of an RNA molecule isolated from a cell is not necessarily the one that initiated transcription. After synthesis, many RNA chains are cleaved by endonucleases to yield smaller fragments, many of which start with a pyrimidine. For example, although some tRNA chains start with U (Chapter 14), they never contain terminal triphosphate groups; instead, they are made by in vivo cleavage of longer chains. As virtually all cells contain a variety of nucleases, unambiguous data about initiating nucleotide sequences are likely to come from in vitro experiments, where, using highly purified enzymes, it is possible to exclude nuclease effects.

σ Dissociates After the First Nucleotides Are Joined[3, 20]

After chain elongation begins, σ dissociates from the core enzyme–DNA–nascent RNA complex and becomes free to attach to another core molecule (Figure 13-11). Conceivably, it is released merely because it no longer has any role to play. Alternatively, its continued presence might result in such tight binding to the promoter sequence

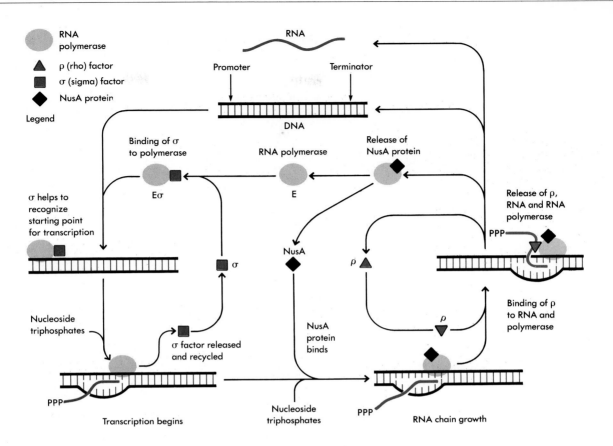

Figure 13-11
RNA synthesis. Here is a case where chain termination requires the ρ factor. In many other situations, RNA polymerase by itself can read the terminating signal.

that subsequent enzyme movement along the template would be effectively impossible. Chain elongation may best occur when the nascent chain–enzyme complex is held to the DNA template by the much weaker and sequence-unspecific interactions that occur between the core enzyme and DNA.

Although holoenzyme binds much more tightly to the promoter than core enzyme does to any DNA, the opposite behavior characterizes binding to non-promotor DNA. Thus, σ suppresses the nonspecific DNA binding activity of RNA polymerase, allowing the enzyme to travel lightly along the helix until it finds a promoter sequence. This weaker binding is important to the way holoenzyme finds a promoter quickly. By "diffusing" along DNA in one dimension, the holoenzyme encounters a promoter much faster than it could by diffusing around the cell in three dimensions.

Since transcribing RNA polymerase molecules lack σ, and perhaps half of the enzyme molecules are active in a cell in normal growth conditions, there is no reason that the number of σ and core subunits in the cell should be equal. Nonetheless, the numbers are not too different; probably a pool of free σ is maintained so that holoenzyme is quickly reconstituted when core enzyme is released at a termination site.

DNA Is Unwound and Rewound by RNA Polymerase as the Enzyme Advances[9, 10, 21, 22]

RNA polymerase unwinds DNA continuously as the enzyme extends the growing RNA chain, exposing the template strand against which RNA precursors are polymerized. The growing point of an RNA

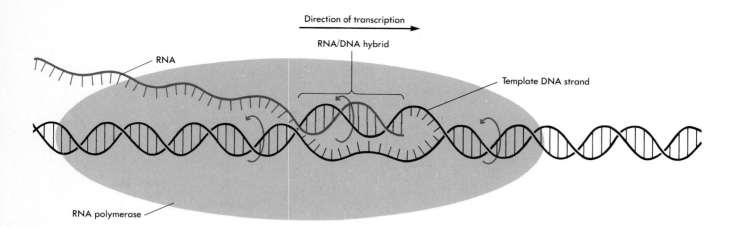

Figure 13-12
A model of the elongating complex of RNA polymerase, template DNA, and emerging RNA strand. Curved arrows indicate the rotation required for DNA to unwind as the template strand is exposed, for the RNA-DNA hybrid to form and unwind, and for the DNA strands to rewind again as the RNA chain is expelled. Although the enzyme is shown to be extended along nearly 60 nucleotides, as it is at the promoter, it is believed that only about 30 nucleotides of DNA actually are covered in the elongating complex. [After H. Gamper and J. Hearst, *Cell* 29 (1982):86, with permission.]

chain begins a region of RNA-DNA hybrid about 12 bases long, which then ends when the RNA chain leaves the template strand and the two DNA strands are rewound into double helix (Figure 13-12). The amount of DNA kept unwound by the elongating enzyme is a bit more than the length of the RNA-DNA hybrid, about 17 bases, the same amount that is unwound initially as the open complex forms at the promoter. DNA unwinding by RNA polymerase can be conveniently measured by allowing a topoisomerase to relax covalently closed circular DNA while it is being transcribed, then adding a detergent to remove the RNA polymerase and RNA, and finally measuring the number of supercoils in the DNA (Figure 13-13). One new negative supercoil appears for each 10.4 base pairs of DNA unwound at the time the topoisomerase acted; thus, by knowing the number of active enzymes, we can calculate the number of base pairs that each unwinds.

Although the elongating complex generally keeps about 17 bases of DNA unwound, this number can change dramatically when the RNA polymerase encounters special sequences. At terminators, for example, the DNA strands completely rewind and enzyme and RNA are released from the complex. The opposite occurs during transcription of the RNA primer for replication of plasmid ColE1, where rewinding of the DNA is inhibited at a specific site, and the remaining transcript remains as an RNA-DNA hybrid several hundred nucleotides long. It is not known why this occurs, although specific secondary structures of the preceding transcript seem to be involved.

The Transcription Factor NusA May Replace σ During Elongation and Termination of RNA Synthesis[23, 24, 25]

An RNA polymerase accessory factor named **NusA protein** was unrecognized for many years because the enzyme's basic activity does not require it and because NusA protein does not copurify with RNA polymerase. Its importance to RNA synthesis was realized when NusA protein was shown to bind tightly to the bacteriophage λ *N* gene protein, a regulatory factor whose function is to prevent termination of transcription during λ growth (Chapter 17). Like σ, NusA binds core RNA polymerase, but less tightly; thus, σ displaces NusA during purification. These properties suggest that in the cell, NusA

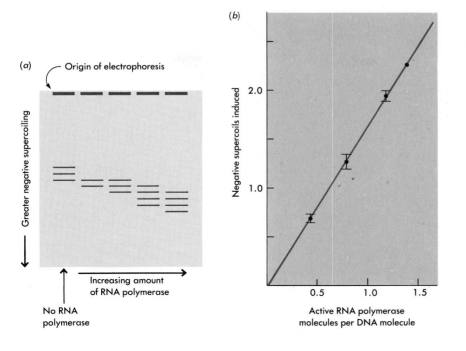

(a)

Origin of electrophoresis

Greater negative supercoiling

Increasing amount
of RNA polymerase

No RNA
polymerase

(b)

Negative supercoils induced

2.0

1.0

0.5   1.0   1.5

Active RNA polymerase
molecules per DNA molecule

Figure 13-13
RNA polymerase–induced unwinding
of DNA. (a) By allowing a topoisomerase to relax covalently closed, circular
DNA as it is transcribed, one can trap
the unwinding induced by an active
RNA polymerase molecule. When the
transcribing complex is disrupted by
denaturation of RNA polymerase, the
partially unwound DNA becomes negatively supercoiled and, being more
compact, migrates faster than relaxed
DNA in an agarose gel. (b) Graphing
the number of supercoils induced
against the number of active RNA polymerase molecules reveals the amount
of unwinding per enzyme: 1.6 supercoils per RNA polymerase, or an unwinding of 17 base pairs per enzyme
(1.6 turns of DNA multiplied by 10.5
base pairs per turn). [After H. Gamper
and J. Hearst, *Cell* 29 (1982):84, with
permission.]

binds to an RNA polymerase molecule that is making an RNA chain
as soon as σ is released from it. However, NusA is not fixed irreversibly to one RNA polymerase molecule elongating an RNA chain, because a single NusA molecule can act on several RNA polymerase
molecules during a transcription reaction. Therefore, even if there are
no more NusA molecules in the cell than are recovered during purification (a few hundred per cell), these can be expected to act on all
elongating RNA polymerase molecules. Presumably, σ again displaces NusA after RNA polymerase is released from the DNA
through its greater affinity for free core enzyme (see Figure 13-11).

It is difficult to assign a specific role to NusA protein in transcription, although genetic experiments show that it is an essential protein. In vitro, NusA has subtle but distinct effects on the rate of RNA
synthesis and on termination of RNA synthesis by RNA polymerase.
It clearly has an important function in mediating interactions with
regulatory factors such as the phage λ N gene protein.

Stop Signals Produce Chains of Finite Length[26-31]

There are signals in bacterial DNA called **terminators,** whose function
is to stop RNA synthesis at specific points along the DNA template.
At terminators, the enzyme stops polymerizing nucleotides into
RNA, releases the RNA chain, and leaves the DNA to eventually
reinitiate at another promoter. At some sites, termination requires an
accessory protein, **ρ factor** (rho factor), whereas at other sites, the
core enzyme alone carries out all these steps.

Much has been learned from sequence analysis of terminators in
bacteria (and phages), particularly those active without ρ factor. Two
distinct features of ρ-independent terminators are found: a dyad symmetry in the DNA, centered 15 to 20 nucleotides before the end of the

Figure 13-14
Sequence of a *ρ*-independent terminator and structure of the terminated RNA (the *trp* attenuator; Chapter 16). The mutations shown in gray partially or completely prevent termination. [After C. Yanofsky, *Nature* 289 (1981):751.]

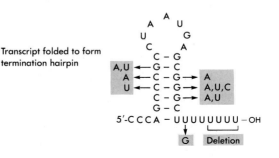

RNA, and a run of about six As in the DNA template strand, which are transcribed into Us at the end of the RNA (Figure 13-14). Although a dyad symmetry often marks a DNA-binding site for an oligomeric protein, its importance here is that its transcript can form a hairpin loop (Figure 13-15). Experiments in vitro show that termination is prevented by incorporation of altered bases that do not pair normally, arguing that formation of the hairpin is important for termination. Furthermore, most mutations selected to destroy the function of terminators are single nucleotide changes that disrupt the regular double-helical stem (see Figure 13-14). Analysis of terminator mutations also reveals the importance of the poly dA template sequence; interrupting this sequence by a base change or removing part of it by deletion inactivates the terminator.

How could these two structural features cause termination? One model suggests that the first half of the hairpin stem anneals to and thus prematurely extracts the second half from the RNA-DNA hybrid region at the growing site, leaving only the poly U sequence annealed; since an RNA-DNA hybrid consisting of polyribo U and polydeoxy A is very unstable, the RNA chain will be quickly expelled from the DNA duplex (see Figure 13-15).

The *ρ*-dependent terminators lack the poly A sequence characteristic of *ρ*-independent ones, and not all are able to form strong hairpins. We do not know how *ρ* substitutes for these sequence features, but there are several hints. First, *ρ* hydrolyzes ATP in the presence of RNA, suggesting that it binds nascent RNA and uses the energy of ATP in its activity, perhaps to extract the RNA from the enzyme and template. Second, *ρ* probably binds RNA polymerase as it acts, since certain *ρ* proteins altered by mutations in the *rho* gene are active in terminating RNA synthesis only with RNA polymerase modified by specific mutations in the gene encoding its *β* subunit. Third, we know that RNA polymerase itself recognizes the *ρ*-dependent termination sequence in DNA, and *ρ* acts afterward to release the RNA: The evidence is that RNA polymerase by itself pauses at terminators that require *ρ*, but then moves on unless *ρ* is present in the reaction. One model of *ρ* function suggests that even a weak hairpin causes polymerase to pause, allowing *ρ* to attach and to dissociate the RNA and polymerase enzymatically.

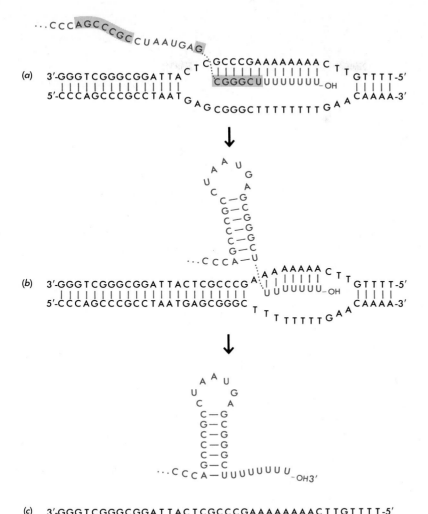

Figure 13-15
How a short duplex stem and a U-rich sequence in RNA might lead to termination and release. (a) The elongation complex has just finished synthesis of the U stretch. (b) RNA-RNA hybrid formation destroys part of the RNA-DNA elongation hybrid, leaving only the poly U stretch annealed. (c) The unstable poly U-poly dA hybrid dissociates, thus releasing the transcript. [After T. Platt, *Cell* 24 (1981):10.

There Are Three RNA Polymerases in Eucaryotic Cells

The synthesis of cellular RNA that is accomplished entirely by one RNA polymerase in bacteria is accomplished by three enzymes in all eucaryotic cells: RNA polymerases I, II, and III. Each has a specific role. RNA polymerase I (called pol I) makes ribosomal RNA; pol II makes mostly messenger RNA; and pol III makes transfer RNA, 5S ribosomal RNA, and a few other small RNA molecules. Like bacterial RNA polymerase, these are large and complex enzymes with a molecular weight of about a half million. They differ, however, in that they contain about ten, rather than five, subunits. There are some common subunits among the three enzymes, so it is likely that a more primitive enzyme became specialized in evolution for different functions in the complex eucaryotic cell. A further complexity of eucaryotic cells is that the genomes of mitochondria and chloroplasts are transcribed by their own organelle-specific RNA polymerases.

There exist for all three eucaryotic RNA polymerases, as for the bacterial enzyme, promoters and terminators that bound transcription units and regulate the rate at which they work. These signals are quite different for two, and probably for all three, of the enzymes. Promoters recognized by pol III have been well characterized, with the surprising result that the promoter is not in front of, but is instead within, the transcribed sequences (Chapter 21).

Since pol II makes messenger RNA and is thus the enzyme most directly involved in genetic regulation, there is great interest in understanding how it starts. Answers are just emerging, delayed by several difficulties. First, the enzyme as purified does not work on natural promoters; it shows only undirected transcription, reminiscent of bacterial core RNA polymerase unless other proteins, as yet not clearly defined, are added. These proteins might well include initiation factors like the bacterial σ, but the first indications are that not one, but several, different kinds of factors are missing. Second, analysis of mutants shows that the region of DNA that affects transcription is much longer than the bacterial promoter. Furthermore, certain DNA sequences important for the promoter's activity in the cell (named **enhancers**) are not active at all in the test tube. How enhancers work can only be guessed, but because they may be the key to understanding the basic features of regulation in eucaryotic cells (Chapter 21), enhancers are the subject of much current research.

Summary

RNA is made like DNA, in that an enzyme (RNA polymerase) links monomeric nucleoside triphosphate precursors into an RNA chain for which one DNA strand serves as a template. Although eucaryotic cells have three enzymes to make different types of RNA, bacterial cells have only one. *E. coli* RNA polymerase is a large enzyme of 450,000 molecular weight, constructed of five types of polypeptide chains: β, β', α, ω, and σ. The capacity to make RNA chains resides in the core enzyme, $\beta\beta'\alpha_2\omega$, whereas the role of the loosely bound σ subunit (σ factor) is to direct initiation to natural starting sites, called promoters. Besides the predominant σ factor, which is nearly as abundant as the core enzyme subunits, there are minor σ factors in *E. coli* and *B. subtilis* that direct initiation to distinct promoters.

Promoters have greatly varying efficiencies, corresponding to the different rates at which genes are expressed. Inspection of many promoters reveals a consensus sequence, a hypothetical ideal sequence that actual promoters approximate. The consensus sequence consists of two segments of six nucleotides each, located 35 and 10 bases before the RNA start site. Strong promoters match the consensus closely, whereas weak promoters match it less well. Most mutations affecting promoter function change bases in the consensus segments. Chemical and nuclease protection experiments show that RNA polymerase binds to both parts of the consensus sequence. Promoters recognized by variant σ factors have distinct consensus sequences, indicating that σ carries the specificity for promoter recognition. Promoters are the targets of many regulatory proteins, such as repressors, which prevent RNA polymerase binding, and activators, which stimulate RNA polymerase binding and thereby increase the rate of promoter function.

Initiation of RNA synthesis involves distinct steps. First, RNA polymerase binds the promoter in a closed complex. The closed complex is then converted to an open complex, in which a segment of DNA at the starting sequence is unwound. And finally, the first nucleotide bond is made. As RNA polymerase progresses along the DNA, a 17-base-pair region of DNA is kept unwound, continuously exposing the template DNA strand. The strands rewind as RNA polymerase passes, freeing the single-stranded RNA product. After initiation of RNA synthesis, σ is released from the enzyme and may be replaced by the transcription factor NusA protein during elongation of the RNA chain. NusA protein affects the behavior of RNA polymerase during elongation and termination of RNA synthesis, and it mediates the function of transcription antiterminators like the phage λ N gene protein.

Transcription ends at terminators, sequences that stop RNA polymerase and release both RNA and enzyme from the DNA. At many bacterial terminators, called ρ-independent terminators, RNA polymerase carries out these steps by itself. At other terminators, ρ-dependent terminators, RNA synthesis stops only if the termination factor ρ acts on the transcribing complex. The ρ-independent terminators consist of a symmetrical DNA sequence that forms a hairpin loop in the RNA, followed by an AT-rich segment encoding a stretch of uridines in the RNA, at the end of which the RNA is released. RNA polymerase also recognizes and pauses at ρ-dependent terminators, whereupon ρ binds the emerging RNA and catalyzes its release from the transcribing complex.

Bibliography

General References

Chamberlin, M. 1982. "Bacterial DNA-Dependent RNA Polymerases." In *The Enzymes*, vol. 15, part B, ed. P. Boyer. New York: Academic Press, p. 61.

Lewis, M. K., and R. R. Burgess. 1982. "Eukaryotic RNA Polymerases." In *The Enzymes*, vol. 15, ed. P. Boyer. New York: Academic Press, pp. 109–153.

Losick, R., and M. Chamberlin, eds. 1976. *RNA Polymerase*. Cold Spring Harbor, N.Y.: Cold Spring Harbor Laboratory. A collection of reviews and research articles about RNA polymerase and transcription, somewhat out of date but still very useful.

McClure, W. 1985. "Mechanism and Control of Transcription Initiation in Prokaryotes." *Ann. Rev. Biochem.* 54:171–204.

Platt, T. 1986. "Transcription Termination and the Regulation of Gene Expression." *Ann. Rev. Biochem.* 55, in press.

Von Hippel, P., D. Bear, W. Morgan, and J. McSwiggen. 1984. "Protein-Nucleic Acid Interactions in Transcription: A Molecular Analysis." *Ann. Rev. Biochem.* 53:389–446. An advanced review of the physical chemistry of transcription.

Cited References

1. Taylor, K., Z. Hradecna, and W. Szybalski. 1967. "Asymmetric Distribution of the Transcribing Regions on the Complementary Strands of Coliphage λ DNA." *Proc. Nat. Acad. Sci.* 57:1618–1625.
2. Burton, Z. F., C. Gross, K. Watanabe, and R. Burgess. 1983. "The Operon That Encodes the Sigma Subunit of RNA Polymerase Also Encodes Ribosomal Protein S21 and DNA Primase in *E. coli* K12." *Cell* 32:335–349.
3. Burgess, R., A. A. Travers, J. J. Dunn, and E. K. F. Bautz. 1969. "Factor Stimulating Transcription by RNA Polymerase." *Nature* 222:537–540.
4. Schmitz, A., and D. Galas. 1979. "The Interaction of RNA Polymerase and lac Repressor with the lac Control Region." *Nucleic Acid Res.* 6:111–137.
5. Siebenlist, U., R. Simpson, and W. Gilbert. 1980. "*E. coli* RNA Polymerase Interacts Homologously with Two Different Promoters." *Cell* 20:269–281.
6. Rodriguez, R., and M. Chamberlin, eds. 1982. *Promoters: Structure and Function*. New York: Praeger.
7. Youderian, P., S. Bouvier, and M. Susskind. 1982. "Sequence Determinants of Promoter Activity." *Cell* 30:843–853.
8. Hawley, D. K., and W. R. McClure. 1983. "Compilation and Analysis of *Escherichia coli* Promoter DNA Sequences." *Nucleic Acid Res.* 11:2237–2255.
9. Melnikova, A. F., R. Beabealashvilli, and A. D. Mirzabekov. 1978. "A Study of Unwinding of DNA and Shielding of the DNA Grooves by RNA Polymerase by Using Methylation with Dimethylsulphate." *Eur. J. Biochem.* 84:301–309.
10. Gamper, H., and J. Hearst. 1982. "A Topological Model for Transcription Based on Unwinding Angle Analysis of *E. coli* RNA Polymerase Binary, Initiation, and Ternary Complexes." *Cell* 29:81–90.
11. Losick, R., and J. Pero. 1981. "Cascades of Sigma Factors." *Cell* 25:582–584.
12. Johnson, W., C. Moran, and R. Losick. 1983. "Two RNA Polymerase Sigma Factors from *Bacillus subtilis* Discriminate Between Overlapping Promoters for a Developmentally Regulated Gene." *Nature* 302:800–804.
13. Elliot, T., and E. P. Geiduschek. 1984. "Defining a Bacteriophage T4 Late Promoter: Absence of a "-35" Region." *Cell* 36:211–219.
14. Grossman, A., J. Erickson, and C. Gross. 1984. "The *htpR* Gene Product of *E. coli* Is a Sigma Factor for Heat-Shock Promoters." *Cell* 38:383–390.
15. Kassavetis, G., and E. Geiduschek. 1984. "Defining a Bacteriophage T4 Late Promoter: The T4 Gene 55 Protein Suffices for Directing Late Promoter Recognition." *Proc. Nat. Acad. Sci.* 81:5101–5105.
16. Landick, R., V. Vaughn, E. Lau, R. Van Bogelen, J. Erickson, and F. Neidhardt. 1984. "Nucleotide Sequence of the Heat Shock Regulatory Gene of *E. coli* Suggests Its Protein Product May Be a Transcription Factor." *Cell* 38:175–182.
17. Hawley, D., and W. McClure. 1980. "*In vitro* Comparison of Initiation Properties of Bacteriophage λ Wild-Type P_R and X3 Mutant Promoters." *Proc. Nat. Acad. Sci.* 77:6381–6385.
18. Lamond, A. I. and A. A. Travers. 1983. "Requirement for an Upstream Element for Optimal Transcription of a Bacterial tRNA Gene." *Nature* 305:249–250.
19. Menzel, R., and M. Gellert. 1983. "Regulation of the Genes for *E. coli* DNA Gyrase: Homeostatic Control of DNA Supercoiling." *Cell* 34:105–113.
20. Travers, A. A., and R. R. Burgess. 1969. "Cyclic Reuse of the RNA Polymerase Sigma Factor." *Nature* 222:537–540.
21. Hanna, M., and C. Meares. 1983. "Topography of Transcription: Path of the Leading End of Nascent RNA Through the *Escherichia coli* Transcription Complex." *Proc. Nat. Acad. Sci.* 80:4238–4242.
22. Masukata, H., and J. Tomizawa. 1984. "Effects of Point Mutations on Formation and Structure of the mRNA Primer for ColE1 DNA Replication." *Cell* 36:513–522.
23. Greenblatt, J., and J. Li. 1981. "The *nusA* Gene Protein of *Escherichia coli*: Its Identification and a Demonstration That It Interacts with the Gene N Transcription Anti-Termination Protein of Bacteriophage Lambda." *J. Mol. Biol.* 147:11–23.
24. Greenblatt, J., and J. Li. 1981. "Interaction of the Sigma Factor and the *nusA* Gene Protein of *E. coli* with RNA Polymerase in the Initiation-Termination Cycle of Transcription." *Cell* 24:421–428.
25. Schmidt, M., and M. Chamberlin. 1984. "Amplification and Isolation of *E. coli nusA* Protein and Studies of Its Effects on in vitro RNA Chain Elongation." *Biochemistry* 23:197–203.
26. Roberts, J. W. 1969. "Termination Factor for RNA Synthesis." *Nature* 224:1168–1174.
27. Lowery-Goldhammer, C., and J. P. Richardson. 1974. "An RNA-Dependent Nucleoside Triphosphate Phosphohydrolase (ATPase) Associated with Rho Termination Factor." *Proc. Nat. Acad. Sci.* 71:2003–2007.
28. Rosenberg, M., D. Court, H. Shimatake, C. Brady, and D. Wulff. 1978. "The Relation Between Function and DNA Sequence in an Intercistronic Regulatory Region in Phage λ," *Nature* 272:414–422.
29. Fisher, R., and C. Yanofsky. 1983. "Mutations of the β Subunit of RNA Polymerase Alter Both Transcription Pausing and Transcription Termination in the *trp* Operon Leader Region *in vitro*." *J. Biol. Chem.* 258:8146–8150.
30. Platt, T., and D. Bear. 1983. "The Role of RNA Polymerase, Rho Factor, and Ribosomes in Transcription Termination." In *Gene Function in Procaryotes*, eds. J. Beckwith, J. Davies, and J. Gallant. Cold Spring Harbor, N.Y.: Cold Spring Harbor Laboratory, pp. 123–162.
31. Ryan, T., and M. Chamberlin. 1983. "Transcription Analyses with Heteroduplex *trp* Attenuator Templates Indicate That the Transcript Stem and Loop Structure Serves as the Termination Signal." *J. Biol. Chem.* 258:4690–4693.

CHAPTER 14

Involvement of RNA in Protein Synthesis

We will now begin to look at how single-stranded RNA molecules function during the process of **translation,** or protein synthesis. When the central dogma DNA → RNA → protein was gaining validity during the mid-1950s, there was general belief that all RNA was template RNA. In addition, it was hoped that when the RNA general structure was solved, mere inspection might tell us how RNA orders amino acid sequences. Now, however, we realize that these views were naive and that protein synthesis is a much more complicated process than the synthesis of nucleic acid. Moreover, not all RNA molecules are templates. Besides a template class, there are two other classes of RNA that play vital roles in protein synthesis.

Amino Acids Have No Specific Affinity for RNA

The fundamental reason behind the complexity of translation is that there is no specific affinity between the side groups of many amino acids and the purine and pyrimidine bases found in RNA. For example, the hydrocarbon side groups of the amino acids alanine, valine, leucine, and isoleucine do not form hydrogen bonds and would be actively repelled by the amino and keto groups of the various nucleotide bases. Likewise, it is hard to imagine specific RNA surfaces with unique affinities for the aromatic amino acids phenylalanine, tyrosine, and tryptophan. It is thus impossible for these amino acids, in unmodified form, to line up passively in a specific and accurate order against an RNA template prior to peptide bond formation.

Amino Acids Are Aligned on RNA Templates by Means of Adaptors[1]

Before the amino acids line up against the RNA template, they are chemically modified to possess a specific surface capable of combining with a specific number of the bases along the template. This chemical change consists of the addition of a specific adaptor molecule to each amino acid through a single covalent bond. It is this adaptor component that interacts with the template; at no time does the amino acid side group itself make contact with the template. Adding a specific adaptor residue to an amino acid is much more economical than chemically modifying the side group itself. The latter process might require many enzymes for just a single amino acid, and a similar number might be required to change the modified side group back to its original configuration after it became part of a polypeptide chain. On the other hand, only a single enzyme is needed either to attach or to detach an amino acid from its specific adaptor.

Specific Enzymes Recognize Specific Amino Acids[2, 3]

There need not be any obvious relation between the shape of the amino acid side group and the adaptor surface. Instead, the crucial selection of an amino acid is done by a specific enzyme. The enzyme that catalyzes the attachment of the amino acid to its adaptor must be able to bind specifically to both the amino acid side group and the adaptor. Proteins are extremely suitable for this task because their active regions can be rich in either hydrophilic or hydrophobic groups. There is no difficulty in folding a suitable polypeptide chain to produce a cavity that is specific for the side group of one particular amino acid. For example, tyrosine can be distinguished from phenylalanine by an enzyme having a specific cavity containing an atom that can form a hydrogen bond to the OH group on tyrosine. The formation of one hydrogen bond yields about 4 to 5 kcal/mol of energy, which would be lost if phenylalanine were chosen instead. Thus, with the help of a physical-chemical theory, we can predict that the probability that tyrosine is found in the "tyrosine cavity" is about a thousand times greater than the probability that phenylalanine is found in the tyrosine cavity.

There is more difficulty in immediately seeing how a similar accuracy can be achieved in distinguishing between amino acids differing only by one methyl residue, a group incapable of forming either salt linkages or hydrogen bonds. For example, glycine must be distinguished from alanine, and valine from isoleucine. There is, of course, no difficulty in understanding why the larger alanine side group cannot fit into the cavity designed for the smaller amino acid glycine. The problem arises when we ask why glycine will not sometimes fit into the alanine cavity, or valine into the isoleucine hole. If this should happen, there would be loss of the van der Waals forces arising out of a snug fit around a methyl group. These forces are now thought to generate about 2 to 3 kcal/mol, by themselves too small in value to account for the accuracy with which amino acids are ordered during protein synthesis. The maximum frequency with which a wrong amino acid is inserted into a growing polypeptide chain is between 1 in 1000 and 1 in 10,000. This means that the energy gained by selecting the correct amino acid must be at least 4 to 6 kcal, about twice the energy provided by the apparent van der Waals energy. Clearly, the process that discriminates between such amino acids is more complicated than initially conceived.

This long-troublesome paradox can be resolved if all such selection processes occur in multiple steps, providing several opportunities to reject each wrongly inserted amino acid. The combination of two steps, each of which utilizes 2 to 3 kcal of binding energy, would produce the 4 to 6 kcal necessary to explain the 10^{-3} error rate. Later we will see that many supposedly one-step discriminations do in fact involve several successive steps. This is true both of the chemical reactions involved in attaching amino acids to their adaptors and of the processes that align the adaptors on the template.

The Adaptor Molecules Are Themselves RNA Molecules[1,4]

The molecules to which the amino acids attach are a group of relatively small RNA molecules called **transfer RNA (tRNA)**. It is really not surprising that the adaptors are also RNA molecules, since a prime requirement for a useful adaptor is the ability to interact specifically with the free keto and amino groups on the single-stranded template RNA molecules. This interaction is ideally accomplished if the adaptor is also a single-stranded RNA molecule, since this allows the possibility of having very specific hydrogen bonds (perhaps of the base-pair variety) hold the template and adaptor together temporarily. The tRNA adaptors for the different amino acids all have different structures, each uniquely adapted for fitting a different nucleotide sequence on the template. Thus, there are many different types of tRNA.

Transfer RNA Molecules Contain Approximately Seventy-Five Nucleotides[5,6]

By now, the sequences of several hundred tRNAs from a variety of different organisms have been determined. Each tRNA molecule con-

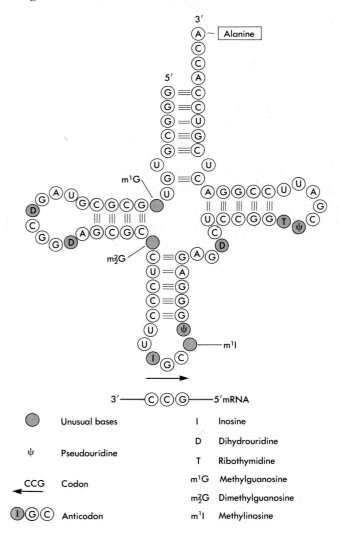

Figure 14-1
The complete nucleotide sequence of yeast alanine tRNA showing the unusual bases and anticodon position.

tains between 73 and 93 nucleotides (with a total molecular weight of 25 to 30 kdal) linked together in a single, covalently bonded chain (Figure 14-1). Although the exact sequence of residues varies, certain positions are conserved from one tRNA to another (Figure 14-2). For instance, one end of the chain, the 3' end, always terminates in a CCA sequence (cytidylic acid, cytidylic acid, adenylic acid). The other end, the 5' end, always carries a 5'-terminal monophosphate group and is often guanylic acid.

A striking aspect of all tRNA sequences is their high content of unusual bases (Figure 14-3). (By "unusual" we mean a base other than A, G, C, or U.) Many of these unusual bases differ from normal bases by the presence of one or more methyl (CH_3) groups. These methyl groups are added enzymatically after the nucleotides are linked together by 3'–5' phosphodiester bonds. Likewise, the other unusual bases arise by the enzymatic modification of a preexisting polynucleotide. Although the function of most of the unusual bases is not yet clear, some have been shown to play important regulatory roles in tRNA function.

Cloverleaf Representation of tRNA Molecules[6]

Although there is only one chain in a tRNA molecule, a majority of the bases are hydrogen-bonded to each other. Hairpin folds bring bases on the same chain into a double-helical arrangement where short stretches of nucleotides are complementary to one another. However, some nucleotides do not base-pair internally and are available to interact with other molecules during the functioning of tRNA in protein synthesis.

The exact nucleotide sequence of any one tRNA does not provide sufficient information for us to guess the way various bases hydrogen-bond to each other. A number of possible hairpin configurations can be imagined. But after the first few tRNA sequences had

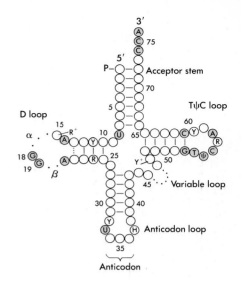

Figure 14-2
A diagram of all tRNA sequences except for initiator tRNA. The positions of invariant and semi-invariant bases are shown. The numbering system is that of yeast tRNA[Phe]. Y stands for pyrimidine, R for purine, and H for a hypermodified purine. R_{15}^+ and Y_{48}^+ are usually complementary. The dotted regions α and β in the D loop and in the variable loop contain different numbers of nucleotides in various tRNA sequences.

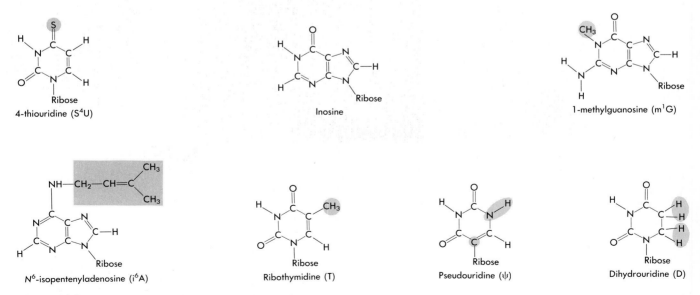

Figure 14-3
The structures of six common modified nucleotides found in yeast alanine or other tRNAs.

been determined, it was realized that there was only one way to fold the chain in two dimensions that would maximize the number of base pairs and lead to a common shape—the cloverleaf (see Figures 14-1 and 14-2).

Each cloverleaf contains four hydrogen-bonded stems and a number of non-hydrogen-bonded sections:

1. The 3' end consists of CCA$_{OH}$. This sequence plus a variable fourth nucleotide extend beyond the stem formed by base-pairing the 5' and 3' segments of the molecule. The amino acid always becomes attached to the 3'-terminal A.

2. As we move along the backbone away from the 3' end, we encounter TψC loop, the first loop of the cloverleaf. It is made up of 7 unpaired bases and nearly always contains the sequence 5'-TψCG-3'. Binding to the ribosomal surface may involve this region of the tRNA.

3. Next appears a loop of highly variable size, called the extra, or variable, loop.

4. The third loop, which consists of 7 unpaired bases, contains the **anticodon**—the 3 adjacent bases that bind (by base pairing) to the successive bases that make up a codon in messenger RNA. The anticodon is bracketed on its 3' end by a purine and on its 5' end by U. The purine is often alkylated.

5. The fourth loop contains 8 to 12 unpaired bases and characteristically contains the modified base dihydro-U (hence the name D loop).

Note that nearly all the invariant nucleotides in tRNA molecules occur in these non-hydrogen-bonded regions (see Figure 14-2).

Figure 14-4
Tertiary interactions in yeast phenylalanine tRNA. (a) The molecule is drawn in the conventional cloverleaf configuration with solid lines connecting bases that are hydrogen-bonded. (b) Sequence rearranged to show continuous stacking of the anticodon stem on the D stem and of the acceptor stem on the TψC stem. Note the close interaction between the TψC and D loops. [After A. Rich and S. H. Kim, *Ann. Rev. Biochem.* 45 (1976):805.]

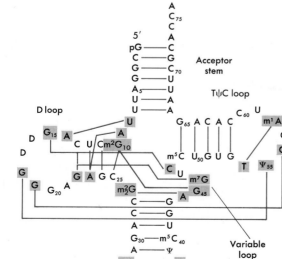

(a)

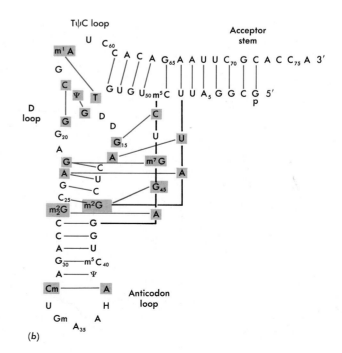

(b)

The Three-Dimensional Structure of Yeast Phenylalanine tRNA[6–9]

Discovery of the cloverleaf pattern does not by itself tell us how the various tRNA loops are arranged in space. This information can only come from X-ray diffraction studies of tRNA crystals. Fortunately, very good crystals of yeast phenylalanine tRNA (tRNA^{Phe}) obtained in several laboratories allowed elucidation of its three-dimensional (tertiary) structure to atomic resolution. Much delight came from the finding that all the double-helical stems predicted by the cloverleaf model for tRNA^{Phe} do in fact exist (Figures 14-4 and 14-5). Moreover, these stem regions conform to the A-type helix structure expected for RNA-RNA duplexes (Chapter 9). Also, an actual GU base pair was visualized in the acceptor stem of the tRNA^{Phe} structure (see Figure 14-4).

In addition to the base-pairing interactions in the stems, there is a series of additional hydrogen bonds that bends the cloverleaf into a stable tertiary structure with a roughly L-shaped appearance (Figures 14-4, 14-5, 14-6, and 14-7). Examination of this structure reveals the following:

- The amino acid acceptor CCA group is located at one end of the L, some 70 Å from the anticodon, which occurs at the opposite end.

- The dihydrouracil-rich (D) and TψC loops form the corner of the L.

- Many tertiary hydrogen-bonding interactions involve bases that are invariant in all known tRNAs (see Figure 14-4a), strongly support-

Figure 14-5
Photographs of (a) a backbone and (b) a space-filling molecular model of yeast phenylalanine tRNA. The CCA acceptor stem is at the upper right, with the anticodon loop at the bottom. [Reproduced from Kim et al., *Science* 185 (1974):435, courtesy of S. H. Kim.]

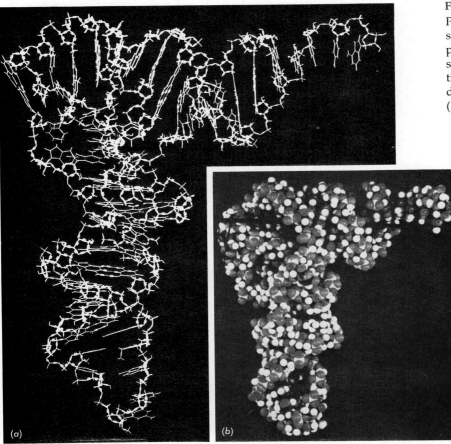

(a) (b)

Figure 14-6
Diagram illustrating the folding of the yeast tRNA^{Phe} molecule. The ribose-phosphate backbone is drawn as a continuous ribbon, and internal hydrogen bonding is indicated by crossbars. Positions of single bases are indicated by rods that are intentionally shortened. The anticodon is at the bottom of the figure, while the amino acid acceptor end is at the upper right. The anticodon and acceptor arms are shaded. The numbering of nucleotides is the same as in Figure 14-4. [After Kim et al., *Proc. Nat. Acad. Sci.* 71 (1974):4970.]

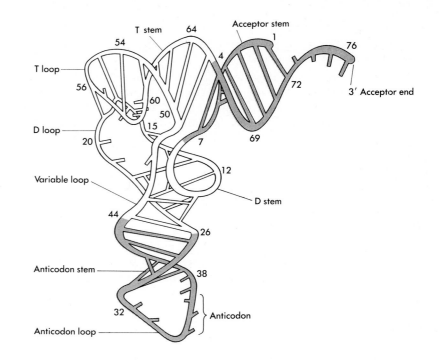

ing the belief that all tRNAs have basically the same tertiary configurations.

- Most of the hydrogen bonds that form the tertiary structure involve base pairs (or triplets) different from the conventional AT and GC pairs found in double-helical DNA (Figure 14-8). Other tertiary interactions involve groups of the ribose-phosphate backbone, including the 2' OH of the ribose sugars.

- Almost all the bases are so oriented that they can stack next to each other, thereby maximizing interaction between their hydrophobic flat faces (see Figure 14-7). Even the apparent unstabilized anti-

Figure 14-7
Diagram illustrating the hydrophobic stacking interactions between the nucleotides of yeast tRNA^{Phe}. Full stacking and partial stacking are indicated. Where adjacent stacking nucleotides are connected by a ribose-phosphate chain, the linkage is noted by a thin line. The heavy, solid lines attached to the nucleotide symbols represent in schematic fashion the purine or pyrimidine bases, while the hydrophobic interaction is indicated by the blocks between the bases. The connectivity of the molecule is indicated only by the numbering scheme. This accounts for the fact that many nucleotides appear to be unconnected. Bases not involved in stacking (bases 16, 17, 20, and 47) are omitted from the figure. [After Kim et al., *Proc. Nat. Acad. Sci.* 71 (1974):4970.]

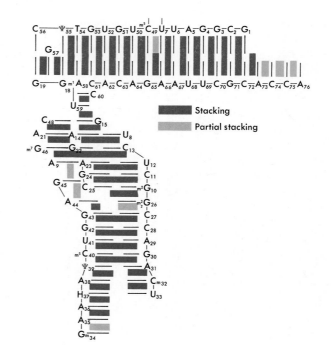

codon region is held relatively rigid by stacking interactions. Stacking is probably as major a factor in stabilizing the tRNA configuration as are the tertiary hydrogen bonds.

- Only a few tertiary hydrogen bonds hold the anticodon stem to the remainder of the molecule, raising the possibility that the relative orientation of the anticodon region may change during protein synthesis.

Other tRNAs Have Similar Three-Dimensional Structures[10, 11, 12]

An important question raised by the three-dimensional structure of tRNA^Phe is whether all other tRNAs fold into the same L-shaped configuration. Indeed, subsequent studies of crystals of several other tRNAs show the same basic shape. Only the angle between the two helical arms of the L varies slightly. There are indications that this hinge may be somewhat flexible, perhaps allowing the angle to change during different steps of tRNA function. Thus, it is not surprising that all tRNAs can fit into the same cavities on the ribosome during protein synthesis or that some tRNA molecules can be activated by enzymes from quite unrelated organisms. On the other hand, proteins that discriminate between tRNAs must detect very subtle differences in the arrangement of chemical groups on the surface of the L-shaped molecule. Among the proteins that can select among tRNAs are the enzymes that activate tRNAs (aminoacyl-tRNA synthetases), those that trim and add modifications to tRNA precursor molecules during their synthesis, and those that modify and deliver a particular tRNA to the ribosome to start protein synthesis.

Attachment to tRNA by Aminoacyl-tRNA Synthetase Activates the Amino Acid[13–16]

The link between the 3'-terminal adenosine and the amino acid is a covalent bond between the amino acid carboxyl group and the terminal ribose component of the RNA (Figure 14-9). The use of the amino acid carboxyl group to attach to the adaptor has several interesting implications. In the first place, before the carboxyl group can form a peptide bond, the adaptor must be released. Thus, peptide bond formation and adaptor removal occur in a coordinated fashion. In the second place, the bond linking the tRNA to its specific amino acid is a high-energy bond, making these complexes "activated" precursors. The energy in the amino acid–tRNA bond (an aminoacyl bond) can be used in the formation of the lower-energy peptide bond.

The energy required for forming the aminoacyl bond comes from a high-energy pyrophosphate linkage ℗~℗ in ATP. Prior to formation of the AA~tRNA compounds, the amino acids are activated by enzymes called **aminoacyl-tRNA synthetases** to form amino acid adenylates (AA~AMP), in which the amino acid carboxyl group is attached by high-energy bonding to an adenylic acid (AMP) group (see Figure 14-9):

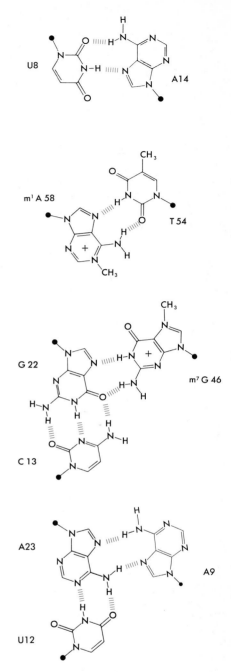

Figure 14-8
Various types of tertiary hydrogen bonding that are seen in yeast tRNA^Phe. The filled black circles represent the 1'-carbon of the ribose residues.

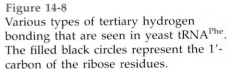

$$\text{AA} + \text{ATP} \underset{\text{synthetase}}{\overset{\text{Aminoacyl-tRNA}}{\rightleftharpoons}} \text{AA} \sim \text{AMP} + ℗ \sim ℗ \qquad (14\text{-}1)$$

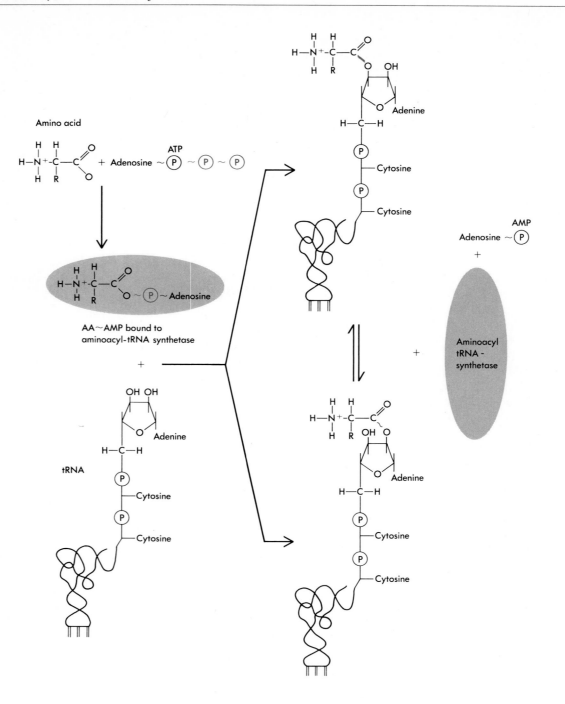

Figure 14-9
Activation of an amino acid by ATP and its transfer to the CCA end of its specific tRNA adaptor. Note that once attached, the amino acid can move rapidly between the 3'- and 2'-OH positions of the terminal adenosine.

The AA~AMP intermediate normally remains tightly bound to the activating enzyme until collision with a tRNA molecule specific for that amino acid. Then the same activating enzyme transfers the amino acid to the terminal adenylic acid residue of the tRNA:

$$\text{AA~AMP} + \text{tRNA} \underset{\text{synthetase}}{\overset{\text{Aminoacyl-tRNA}}{\rightleftharpoons}} \text{AA~tRNA} + \text{AMP} \quad (14\text{-}2)$$

A tRNA carrying its cognate amino acid is said to be "charged."

We therefore see that the activating enzymes are able to recognize and bind to both a specific amino acid and its tRNA adaptor. For this purpose, the enzymes must have two different combining sites: one that recognizes the side group of an amino acid and another that recognizes the tRNA specific for that amino acid. Similarly, the part of a tRNA molecule that is recognized by its activating enzyme can be totally different from the part that interacts with a specific group of template nucleotides. It follows that the amino acid side group itself never has to come into contact with a template molecule. It only needs to bind specifically to the correct activating enzyme.

Every cell needs at least 20 different kinds of activating enzymes and at least 20 kinds of tRNA molecules. There must be at least one of each for every amino acid. In fact, there are usually several types of tRNA for the same amino acid. Such tRNAs are called **isoacceptors.** Their existence is connected with the fact that the genetic code often uses more than one nucleotide sequence (codon) to specify a given amino acid, a characteristic called degeneracy (Chapter 15). Frequently, there is a unique tRNA molecule for each functional codon. There need not be, however, a separate activating enzyme for each of the several tRNAs corresponding to a given amino acid. In fact, isoacceptor tRNA molecules that bind to different codons are activated by the same aminoacyl-tRNA synthetase.

Many studies have indicated that aminoacyl-tRNA synthetases bind in and around the inside of the L shape assumed by tRNA molecules. The invariant uridine (or modified U) at position 8 of the tRNA chain (see Figure 14-2) is ideally positioned to interact (and may even form a transient covalent bond) with the synthetase. Recently, cocrystals of tRNA bound to an aminoacyl-tRNA synthetase have been obtained. It is hoped that these will tell us how activating enzymes can discriminate among tRNAs for different amino acids and yet bind the several isoacceptors for a particular amino acid.

AA~tRNA Formation Is Very Accurate[17-21]

How given aminoacyl-tRNA synthetases accurately select among very similar amino acids at first seemed very mysterious. So the discovery of how isoleucyl-tRNA synthetase precisely discriminates isoleucine from valine was particularly encouraging. These amino acids differ by only single methyl groups, and so the difference in their binding energy to isoleucyl-tRNA synthetase is only 2 to 3 kcal, an amount of energy that at first seems insufficient to prevent frequent mistakes. For example, when isoleucyl-tRNA synthetase is presented with a mixture of equimolar amounts of isoleucine and valine, approximately one valine~AMP activation occurs for every 100 correctly activated isoleucine~AMP complexes. If all these activated valine molecules were later transferred to isoleucine-tRNA molecules, an unacceptably high level of mistakes would result. Such transfers, however, are very infrequent, for when isoleucine tRNA binds to a valine~AMP–isoleucyl-tRNA synthetase complex, almost all valine~AMP molecules fall apart into their valine and AMP components (Figure 14-10). This form of proofreading demands continued binding of the amino acid to the active site of the synthetase through both enzymatic steps catalyzed by the synthetase. Discrimination between isoleucine and valine can thus occur twice, giving rise to an acceptable $10^{-2} \times 10^{-2} = 10^{-4}$ final error level.

It now appears that proofreading occurs only in those enzymes where it is not possible to achieve sufficient accuracy in a single recognition step. For instance, we have recently learned from the three-dimensional structure of tyrosyl-tRNA synthetase that this enzyme selects between the two very similar amino acids phenylalanine and

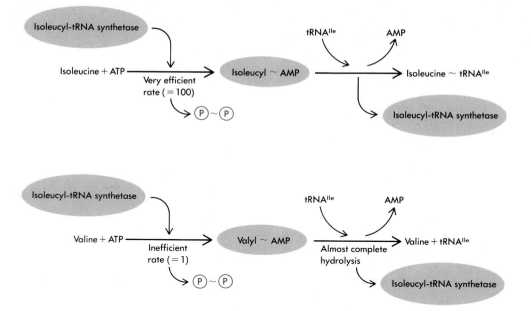

Figure 14-10
Discrimination between isoleucine and valine in the formation of isoleucyl-tRNA^{Ile}.

tyrosine by making specific hydrogen bonds to the OH group of tyrosine (Figure 14-11). Other amino acids that can be similarly discriminated without proofreading include asparagine, aspartic acid, cysteine, histidine, lysine, and tryptophan.

Once a particular amino acid is attached to its adaptor tRNA, it is not specifically recognized in any of the subsequent steps of protein synthesis. Evidence for this passive role of the amino acid came from an experiment where cysteine was first charged onto tRNA^{Cys} and then chemically converted to alanine (Figure 14-12). When this alanyl-tRNA^{Cys} was added to a hemoglobin-synthesizing reticulocyte lysate, all cysteine positions in the newly made hemoglobin were found to be occupied by alanine. Thus, the tRNA molecule is a true adaptor in that only it is selected by the template.

Aminoacyl-tRNA synthetases are an extremely diverse group of proteins with respect to both the size and number of their polypeptide subunits. Some synthetases attach the amino acid to the 3′ OH, while others attach it to the 2′ OH of the terminal ribose sugar of the tRNA. Once charged, however, the amino acid probably moves rap-

Figure 14-11
Hydrogen bonds made between tyrosyl-tRNA synthetase and tyrosyl adenylate. The three-dimensional structure, known from X-ray crystallography (ref. 18), reveals eleven possible hydrogen bonds that may be formed between the enzyme and the AA~AMP. In the synthetase, groups on the amino acid side chains as well as groups that are part of the main chain (MC) are involved. Not only does the tyrosine hydroxyl group make two specific hydrogen bonds, but there are van der Waals' interactions between the hydroxyl group and the enzyme that would not form if phenylalanine were bound instead. [After Fersht et al., *Nature* 314 (1985):235.]

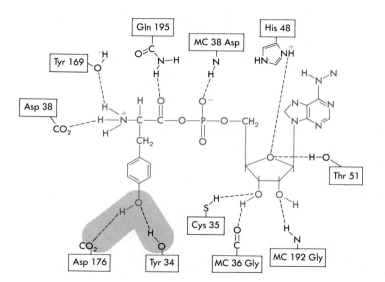

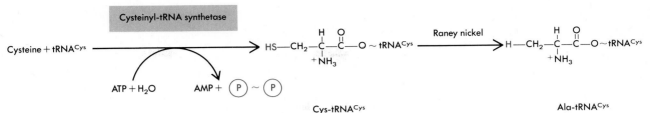

Figure 14-12
How the side chain of cysteine can be converted into alanine after attachment to tRNA^{Cys}. Alanines are subsequently incorporated into cysteine positions in hemoglobin molecules.

idly back and forth, since these two positions are chemically very similar (see Figure 14-9). Only later, when the amino acid is transferred to the growing polypeptide chain, is the position important. Then the amino acid must leave from the 3' OH group.

Peptide Bond Formation Occurs on Ribosomes[22–25]

Once the amino acids have acquired their adaptors, they diffuse to the **ribosomes,** the spherical particles on which protein synthesis occurs. Protein synthesis never takes place free in solution, but only on the surfaces of the ribosomes, which might be regarded as miniature factories for making proteins. Their chief function is to orient the incoming AA~tRNA precursors and the template RNA so that the genetic code can be read accurately. Ribosomes thus contain specific surfaces that bind the template RNA, the AA~tRNA precursors, and the growing polypeptide chain in suitable stereochemical positions.

There are approximately 15,000 ribosomes in a rapidly growing *E. coli* cell. Each ribosome has a molecular weight of slightly less than 3 million daltons. Together the ribosomes account for about one-fourth the total bacterial cell mass, and hence a very sizable fraction of the total cellular synthesis is devoted to the task of making ribosomes. Only one polypeptide chain can be formed at a time on a single ribosome. Under optimal conditions, the production of a chain of 400 amino acids (molecular weight of about 40,000) requires about 10 seconds. The finished polypeptide chain is then released, and the free ribosome can be used immediately to make another protein.

All ribosomes are constructed from two subunits, the larger subunit being approximately twice the size of the smaller one (Figure 14-13). Both subunits contain both RNA and protein. In bacterial ribosomes, the RNA:protein ratio is about 2:1; in many other organisms, it is about 1:1. Both the large and small subunits contain a large number of different proteins. Intensive work has been done with the ribosomal proteins from *E. coli*. The 21 S proteins (labeled S1 to S21) from the smaller (30S) subunit (see Figure 14-13) show a variety of sizes (Figure 14-14). Recently, all have been sequenced, and it has become clear that each is present in only one copy per ribosome. Likewise, of the 34 L proteins (L1 to L34) in the larger (50S) subunit, most are present only once in a given ribosome. An exception is the protein called L7/L12, which is present in four copies, some of which possess acetyl groups at their N termini. L7/L12 plays a central role in the energy-utilizing steps of protein synthesis that occur on the ribosome.

Ribosomal RNA Does Not Carry Genetic Information

When ribosomes first became implicated in protein synthesis in 1953, it seemed natural to suppose that their tightly bound RNA component was the template that ordered the amino acids. In fact, it was

Figure 14-13
The structure of the *E. coli* ribosome. It is usually called the 70S ribosome, since 70S is a measure of how fast this ribosome sediments in the centrifuge. (S stands for the Svedberg, a measure of sedimentation rate.) Likewise, the designation 30S and 50S are the sedimentation constants of the smaller and larger ribosomal subunits; and 5S, 16S and 23S are the sedimentation constants of the smaller and larger ribosomal RNA molecules. All bacterial ribosomes have sizes similar to those of *E. coli*, possessing 30S and 50S subunits. In higher organisms, ribosomes are somewhat larger (80S), with 40S and 60S subunits.

Figure 14-14
Two-dimensional gel electropherograms of the 21 proteins from the small (left) and the 34 proteins from the large (right) *E. coli* ribosomal subunits. S20 and L26 have turned out to be the same protein; L8 represents a complex of L7/L12 with L10. In general, the size of the protein decreases with increasing number. That is, S1 is the largest protein (MW about 60,000) and S21 the smallest protein (MW about 8000) on the 30S ribosome, and L1 is the largest protein (MW about 25,000) and L34 the smallest protein (MW about 5000) on the 50S ribosome. [Courtesy of E. Kaldschmidt and H. G. Wittmann, *Proc. Nat. Acad. Sci.* 67 (1970):1276.]

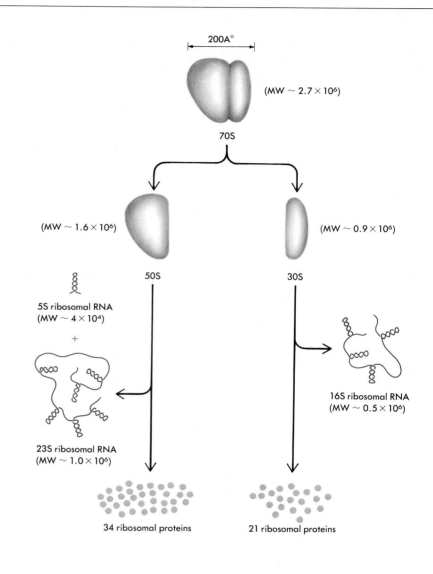

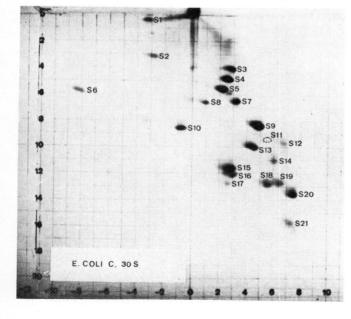

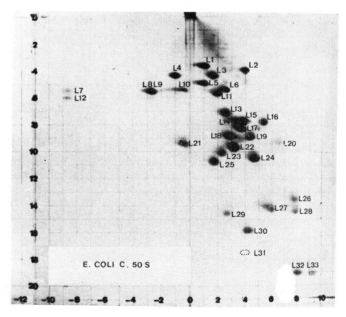

initially thought that all cellular RNA was located in the ribosomes and that the lighter, slowly sedimenting, soluble fraction (about 20 percent of total RNA) was a degradation product of the ribosomal RNA templates. The identification in 1956 of tRNA adaptor molecules in the soluble fraction corrected this latter assumption, but did not remove the belief that the remaining 80 percent of cellular RNA functioned as templates. In 1960, however, the RNA isolated from purified ribosomes (**ribosomal RNA,** or **rRNA**) was unambiguously shown not to have a template role.

Messenger RNA (mRNA) Reversibly Associates with Ribosomes[26, 27]

The active templates are an RNA fraction that account for only one to several percent of the total cellular RNA. This RNA reversibly binds to the surface of the smaller ribosomal subunit and in media of low magnesium ion concentration can be removed without affecting ribosome integrity. Because it carries the genetic message from the gene to the ribosomal factories, this particular kind of RNA is called **messenger RNA** (**mRNA**). By moving across the ribosomal site of protein synthesis, mRNA brings successive codons into position to select the appropriate AA~tRNA precursors.

The existence of mRNA was first established in experiments with T2-infected *E. coli* cells. After T2 DNA enters a host cell, it must turn out RNA templates for the many virus-specific proteins needed for viral reproduction (Chapter 17). Many biochemists were therefore surprised by the 1959 finding that no new rRNA chains and hence no new ribosomes were synthesized following T2 infection. This result could only mean that T2-specific proteins are not synthesized on rRNA templates, also suggesting that in normal cells, rRNA chains are not the templates for protein synthesis. Subsequent research quickly revealed the presence of mRNA, first in virus-infected cells and soon after in normal *E. coli* cells.

Ribosomal RNA Comes in Three Basic Sizes[28]

Two large rRNA molecules and one small rRNA molecule are found in every bacterial ribosome. They are integral components, and unlike mRNA, cannot be removed without the complete collapse of the ribosome structure. The 16S rRNA molecule, found in the smaller ribosomal subunit, has a chain length of 1542 nucleotides, whereas the 23S molecule, a component of the larger ribosomal subunit, contains 2904 nucleotides. Each larger subunit contains, in addition, one very short rRNA molecule that sediments at 5S and has 120 nucleotides. All three rRNAs are single-stranded and have unequal amounts of guanine and cytosine and of adenine and uracil. Nonetheless, there is enough equivalence of base pairs so that many rRNA bases on the same chain are hydrogen-bonded into the hairpin-type turns found in tRNA.

Sequences Reveal Conserved Folding Patterns for rRNA Molecules[28, 29, 30]

We have recently arrived at a good picture of how the three ribosomal RNAs fold, despite the fact that ribosome crystal structures have not yet been worked out (see Figure 14-15a and b). The key to this under-

E. coli 16S rRNA

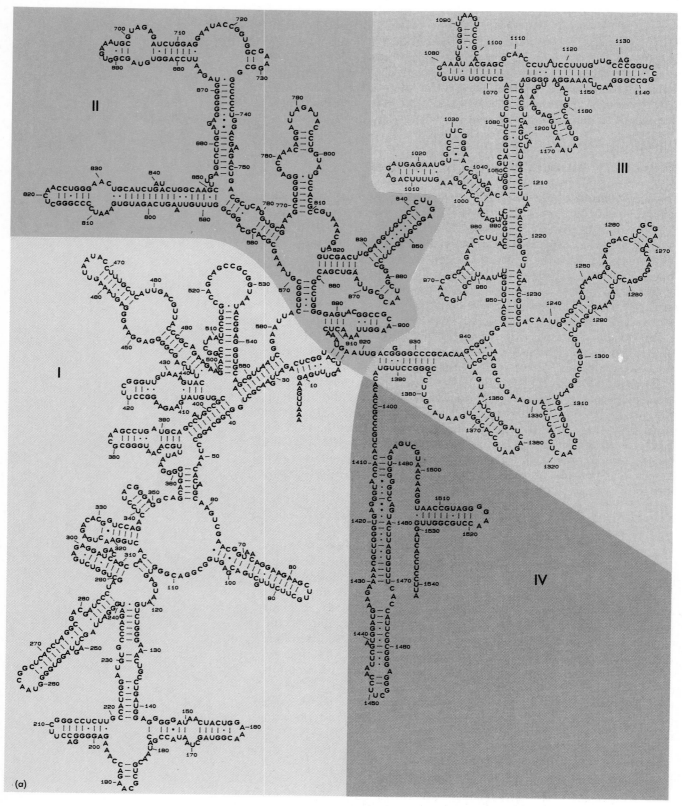

(a)

Bovine mitochondria 12S rRNA

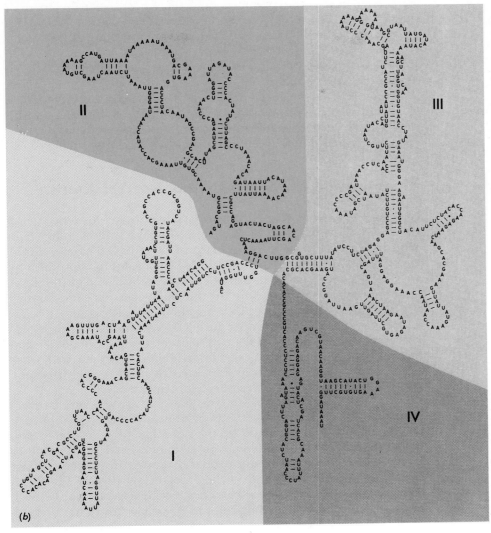

Figure 14-15
A comparison of the structures
of the small subunit rRNAs
from (a) *E. coli* (16S rRNA) and
(b) bovine mitochondria (12S
rRNA). All stems are sup-
ported by comparative se-
quence evidence from different
species. Long-range base-pairing interactions that define the
four domains of this rRNA are
indicated by I, II, III, and IV.
Note that the domain structure
is conserved despite the extreme difference in length between the two rRNAs (1542
nucleotides for 16S and 954
nucleotides for 12S). (Courtesy
of R. Gutell and H. Noller.)

standing was the elucidation of complete rRNA sequences from a
number of different organisms, some closely and some distantly related to *E. coli*. Although the chain lengths of the two large ribosomal
RNAs vary significantly among organisms, the corresponding RNAs
are often referred to as 16S-like or 23S-like.

When the 1542-nucleotide sequence of *E. coli* 16S rRNA was first
determined in 1978, it was possible to propose a folding pattern based
both on sequence complementarity and on the sensitivity of certain
parts of the RNA sequence to ribonucleases and chemical reagents.
(Single-stranded portions of the chain are more susceptible to such
probes than are tightly base-paired regions.) Sequences of rRNAs
from other bacterial species revealed that most of the proposed helical
regions are conserved even if the exact nucleotide sequence is not.
That is, if the order of residues on one side of a particular base-paired
stem differs from that in *E. coli*, then a compensating change occurs
on the other strand to restore complementarity. Now it appears that
the 16S-like RNA from the small ribosomal subunit of all organisms
folds into the same four-domain structure (Figure 14-15). These do-

mains, each containing many stem and loop regions, are closed by long-range base-pairing interactions. Even where the chain length of the rRNA differs twofold from *E. coli*, the extra or missing nucleotides are simply accommodated by extending or truncating various loop structures (see Figure 14-15).

Similar conserved folding patterns can be formulated for the 23S-like and 5S rRNAs that reside in the large ribosomal subunit. These structures are shown in Figure 14-16. The striking conservation of each of the three ribosomal RNA folding patterns confirms our suspicion that all ribosomes are constructed in essentially the same way, despite extreme differences in size and protein composition.

The Function of Most rRNA Is Not Yet Known[28]

Until recently, there was no satisfactory hypothesis for why such a large fraction of the mass of ribosomes should be rRNA rather than protein. But now we believe that many of the unpaired bases in rRNA are involved in the binding of other RNA molecules to ribosomes. As we shall see later, several unpaired bases near the 3' end of 16S rRNA form temporary base pairs with the initiation sites (ribosome binding sites) of mRNA molecules. Likewise, rRNA sequences may interact with invariant tRNA sequences during tRNA binding. Also, rRNA-rRNA bonds may help hold the two ribosomal subunits together during translation. Other rRNA residues may form base pairs with small RNA-protein complexes involved in directing newly synthesized proteins to compartments beyond the cell membrane.

Despite all these RNA-RNA interactions, we are still very far from understanding what the hundreds of unpaired nucleotides of each rRNA component are doing. Even the ribosomal proteins that help to hold the rRNA in a well-defined three-dimensional shape seem to prefer to bind to the paired nucleotides that form the conserved stem structures of the folded rRNAs. The constancy of rRNA folding patterns across all species underscores the essential, albeit not yet well understood, role of RNA molecules in ribosome function.

Self-Assembly of Ribosomes in the Test Tube[23, 31–33]

The complete reassembly of the smaller *E. coli* subunit from its RNA and protein constituents was first accomplished in 1968. Such reconstituted particles are active in protein synthesis, showing identical behavior to normal 30S subunits. The fact that no additional components except for the 16S rRNA and the 21 proteins present in the final particle are required shows that the test-tube reconstitution of 30S ribosomes is a true "self-assembly" process. These studies further revealed that certain proteins can bind directly to the rRNA, while the assembly of others requires the prior association of other S proteins (Figure 14-17). This does not necessarily mean that one protein directly contacts another in the ribosome. Rather, the first proteins that assemble may induce conformational changes in the rRNA that create or expose binding sites for the remaining proteins. For instance, protein S4, which is required early in the reconstitution of 30S ribosomes, is believed to bind and stabilize the two helical stems that close domains I and II of the 16S folded structure shown in Figure 14-15.

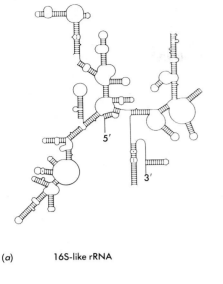

(a) 16S-like rRNA

(b) 5S rRNA

(c) 23S-like rRNA

Figure 14-16
"Minimal" conserved folding patterns for the three ribosomal RNAs. Sequences from both kingdoms of bacteria (the Archebacteria and Eubacteria) as well as from higher cells and their organelles have been included in formulating these structures. Gaps indicate regions where non-conserved sequences of various lengths are present in individual molecules. Currently over 20 complete 16S-like, about 15 23S-like, and dozens of 5S structures are known. (Parts (a) and (c) courtesy of R. Gutell, C. Woese, and H. Noller.)

Test-tube reconstitution of functional 50S *E. coli* subunits has been much harder to achieve. Not only must a larger number of proteins be assembled in an orderly fashion, but two rRNAs (the 23S and 5S) must be built into the particle. Again, there is evidence that after about half the proteins have assembled with the two RNAs, a conformational change must occur before the remaining proteins can be properly positioned (see Figure 14-17). It is worth noting that the earliest proteins to assemble bind near the 5' end of the 23S RNA, while those that are added later interact more with the 3' portion of the molecule. Thus, ribosome assembly in the cell can occur coordinately with the 5'→3' transcription of the 23S rRNA itself.

Finding the right conditions for the test-tube assembly of the two ribosomal subparticles has been crucial in understanding the structure and function of the ribosome. For instance, partial reconstitution of particles lacking one or more specific proteins has enabled us to identify proteins involved in specific ribosomal functions (e.g., peptide bond formation and GTP hydrolysis). Similarly, mixed reconstitutions using proteins from mutant and normal ribosomes or from different bacterial species can reveal which protein carries the mutation or which proteins are functionally equivalent in ribosomes from different organisms. Finally, the ability to reconstitute ribo-

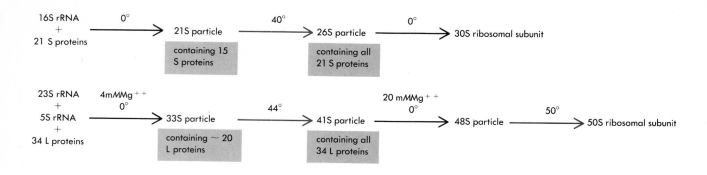

Figure 14-17
In vitro assembly pathways for the 30S and 50S *E. coli* ribosomal subparticles, illustrating the steps where energy input (conformational change) is required. In the case of 30S assembly, between 7 and 13 S proteins can be demonstrated to bind directly to the 16S rRNA, depending on the method of rRNA preparation. About 17 L proteins are thought to bind directly to the 23S rRNA and 3 to the 5S rRNA.

somes containing only a few components appropriately labeled with isotopes has been essential for the application of newer approaches to deciphering the overall structure of the ribosome.

Pre-tRNA and Pre-rRNA[34-37]

In bacteria, molecules of tRNA and rRNA are synthesized exactly as mRNA molecules are, by RNA polymerase using DNA as template. But even though rRNA and tRNA together make up over 98 percent of all RNA in the cell, less than 1 percent of the total DNA functions as their templates. This is because mature tRNAs and rRNAs are quite stable, whereas mRNAs break down rapidly.

However, neither tRNAs nor the three ribosomal RNAs are transcribed as such. Instead, they are derived from larger short-lived precursor molecules that nucleases break down into mature tRNA or rRNA molecules. The 16S, 23S, and 5S rRNAs are transcribed in that order into a single 30S pre-rRNA transcript of about 6500 nucleotides. The pre-rRNA contains additional sequences at both its 5' and 3' ends (called **leader** and **trailer regions**) as well as in the **spacer regions** between the three rRNA sections. All tRNAs are likewise transcribed in the form of larger precursors (Figure 14-18). Sometimes, only one tRNA with 5' and 3' flanking sequences is present; sometimes as many as seven different tRNAs are contained in a common pre-tRNA transcript.

In the case of pre-rRNA, it is clearly advantageous to include 16S, 23S, and 5S sequences in a single transcript to ensure that equivalent numbers of the three molecules will be available for ribosome assembly. It is not obvious why tRNAs should be made in precursor form. However, it is known that mutant pre-tRNAs that cannot be correctly converted to mature molecules often have defects that would interfere with the folding of the tRNA portion into the correct L-shaped conformation. Such mutant pre-tRNAs are rapidly degraded in the cell, providing a way of discarding defective molecules before they become mature tRNAs.

Pre-tRNAs Are Cleaved by an Enzyme That Contains a Catalytically Active RNA[34, 36, 38-41]

Since pre-tRNA molecules contain variable lengths of 5' and 3' flanking sequences, they must be acted upon by specific RNA processing enzymes (RNases) to create the exact termini found in mature tRNA. Most is known about an enzyme called **ribonuclease P (RNase P)**,

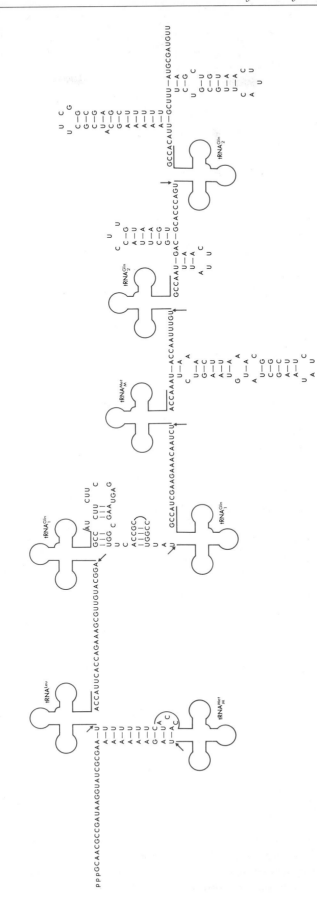

Figure 14-18

Sequence and proposed secondary structure for the pre-tRNA transcript of the *E. coli* operon called *sup*B-E. The seven tRNA sequences are indicated schematically as cloverleaf structures. The duplicated genes for tRNA$_M^{Met}$, tRNA$_1^{Gln}$, and tRNA$_2^{Gln}$ have identical sequences. Note that the sequences and structures around the sites of RNase P cleavage (indicated by arrows) are all different. In *E. coli*, the CCA-3' ends of tRNAs are always encoded. The spacer regions are of different lengths and sequences. [After N. Nakajima, H. Ozeki, and Y. Shimura, *Cell 23* (1981):239.]

which removes the extra 5′ nucleotides (see Figure 14-18). RNase P is an endonuclease that makes a single cut, simultaneously producing the 5′-P end of the tRNA and releasing the 5′ flanking sequence with a 3′-OH group. Since it acts on every *E. coli* pre-tRNA (whether containing one or several tRNAs), it must recognize some feature common to all these molecules. Yet, no particular nucleotide sequence appears either 5′ or 3′ to each cut site. Instead, RNase P probably detects certain aspects of the three-dimensional tRNA structure already correctly folded in its precursor molecule.

Purification of RNase P led to the realization that the enzyme is not a pure protein. Rather, it is a noncovalent complex of a small RNA molecule (377 residues long) and a small protein (MW about 20,000). Reconstitution studies, as well as the identification of mutants in the RNA and in the protein, have shown that both components contribute to RNase P activity under physiological conditions. A suggested secondary structure for the RNA component of *E. coli* RNase P is shown in Figure 14-19.

Most astounding is the recent discovery that the RNA component *alone* is sufficient to carry out pre-tRNA cleavage! In nonphysiological buffers containing high concentrations of magnesium ion or spermidine, the isolated RNA cuts as accurately and nearly as rapidly as the RNA-protein complex does in physiological buffers. This novel finding illustrates that proteins are not the only macromolecules that can bind substrates and catalyze reactions. Deletion and point mutation analyses of the RNase P RNA component are currently under way to pinpoint the regions that align the pre-tRNA substrate and carry out the cleavage reaction. The role of the RNase P protein also remains to be elucidated. RNA catalysis may provide an important clue to the early evolution of the gene expression apparatus.

Another tRNA processing enzyme called **RNase D** trims away the extra 3′ nucleotides from *E. coli* pre-tRNA molecules. It is an exonuclease that releases 5′ monophosphates and somehow knows to stop when it reaches the CCA end. Meanwhile, other specific enzymes introduce the many base and sugar modifications found in mature tRNA molecules. If the invariant CCA-3′ end of a mature tRNA becomes nibbled away, there also exists a specific CCA nucleotidyl transferase enzyme responsible for repairing this sequence so that an amino acid can be attached.

Pre-rRNA Molecules Also Contain tRNAs[42, 43]

Generally, only one or two specific regions on the *E. coli* chromosome encode each of the 30 to 40 different pre-tRNA molecules. In contrast, pre-rRNA chains are coded by seven different gene sets (called *rrn* operons) located at separate sites on the *E. coli* chromosome. Apparently, this number is required to provide the large amounts of ribosomal RNA needed for the rapid growth of the bacterium. The sections of the pre-rRNA that will become the mature 16S, 23S, and 5S ribosomal RNAs differ in only a very few positions (about 1 percent of the nucleotides). However the leader, trailer, and spacer regions can be strikingly different in both length and sequence from one *rrn* operon to another.

The extreme variation possible in pre-rRNA spacer and trailer regions has been underscored by the discovery of tRNA genes embedded in *rrn* operons. In the spacer between 16S and 23S sequences,

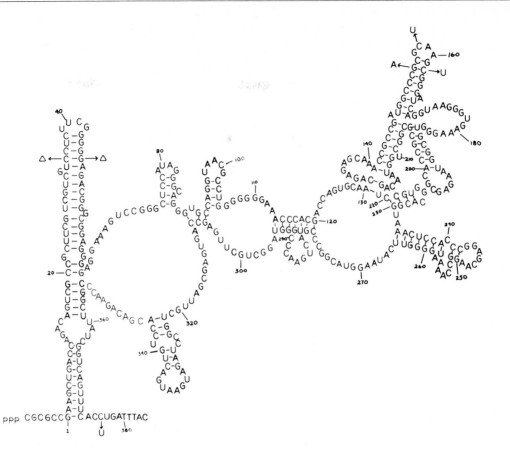

either one or two tRNA genes are found; none, one, or two tRNAs are coded in the trailer region beyond the 5S rRNA (Figure 14-20). Their presence in pre-rRNA molecules requires a coupling of tRNA and rRNA processing events. As in the processing of other pre-tRNAs, RNase P makes the cleavage at the 5′ ends of tRNAs in the pre-rRNA transcripts. Why certain tRNAs have been selected for inclusion in pre-rRNA transcripts is not clear. They are, as expected, among the more abundant tRNA species in the *E. coli* cell.

Pre-rRNA Is First Cleaved in Double-Stranded Regions[43–47]

The 30S pre-rRNA transcript containing 16S, 23S, and 5S rRNA (as well as one to three tRNA sequences) is normally very short-lived. However, in mutant *E. coli* lacking an enzyme called **ribonuclease III** (**RNase III**), unusually high levels of the 30S pre-rRNA accumulate, suggesting that RNase III plays a critical role in pre-rRNA processing. RNase III was initially characterized as an endonuclease specific for double-stranded RNA. Therefore, it was anticipated that analysis of the leader, spacer, and trailer regions of pre-rRNA would reveal adjacent complementary sequences able to fold into long hairpin loop structures that could be cut by RNase III. Instead, the complementary sequences were unexpectedly discovered to lie thousands of nucleotides apart on the pre-rRNA transcript (see Figure 14-20). Leader sequences before the 16S coding region can base-pair with sequences lying between 16S and the (first) spacer tRNA. Likewise, sequences

Figure 14-19
A possible secondary structure for M1 RNA, the RNA subunit of RNase P from *E. coli*. The structure shown was derived from studies in solution of the sensitivity of M1 RNA to digestion with nucleases and comparison with the analogous RNA from *S. typhimurium*. The sequence of mature M1 RNA ends at nucleotide 377. Δ indicates nucleotides deleted and the arrows point to base differences in the M1 RNA of *S. typhimurium*. (Courtesy of S. Altman.)

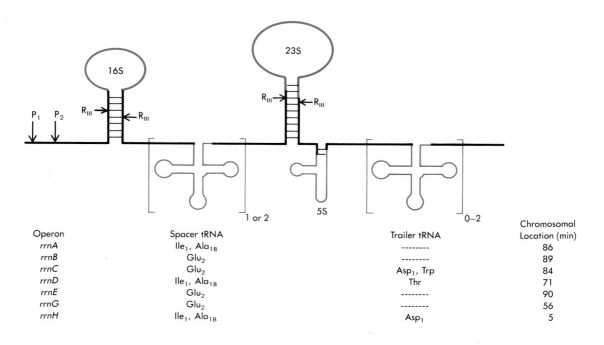

Operon	Spacer tRNA	Trailer tRNA	Chromosomal Location (min)
rrnA	Ile_1, Ala_{1B}	--------	86
rrnB	Glu_2	--------	89
rrnC	Glu_2	Asp_1, Trp	84
rrnD	Ile_1, Ala_{1B}	Thr	71
rrnE	Glu_2	--------	90
rrnG	Glu_2	--------	56
rrnH	Ile_1, Ala_{1B}	Asp_1	5

Figure 14-20

The seven ribosomal RNA (*rrn*) operons of *E. coli* K-12. Color indicates mature rRNA or tRNA sequences. R_{III} indicates sites of primary processing by RNase III. The discarded spacer regions vary in length and sequence between the different operons. P_1 and P_2 indicate the two transcription start sites located about 100 nucleotides apart. Although the *rrn* operons are distributed around the *E. coli* chromosome, the direction of their transcription is always the same as that of DNA replication. In some *E. coli* strains, *rrnH* is replaced by another operon, *rrnF*, located at 74 min; but all *E. coli* strains have a total of seven *rrn* operons. [Data from M. Ellwood and M. Nomura, *J. Bacteriol.* 149 (1982):458.]

between the (last) spacer tRNA and the beginning of 23S are complementary to sequences between the 23S and 5S rRNAs. These regions come together in the pre-rRNA transcript to form two stem structures, each of which closes a giant loop including the entire 16S or 23S rRNA molecule. RNase III then cuts at two precise sites within each stem to create separate pre-16S and pre-23S molecules (Figure 14-21).

Although the naked 30S pre-rRNA molecule contains recognition sites for RNase III cleavage, this is not the case for subsequent steps in rRNA processing. Ribosomal proteins begin to assemble on pre-rRNA soon after its transcription starts. Since complete synthesis of the 16S or the 23S sequence must precede RNase III cutting, many proteins will have already bound by the time the separate pre-16S and pre-23S rRNAs are created. Further 5' and 3' trimming of these intermediates requires the presence of ribosomal proteins. The same appears to be true of the processing steps that produce mature 5S rRNA from the 3' piece cleaved off by RNase III (see Figure 14-21). Thus, in the *E. coli* cell, the processing of the pre-rRNA and the assembly of ribosomal proteins occur concomitantly, probably via a precisely ordered series of steps, each corresponding to a slightly different ribosomal precursor particle.

Messenger RNA Molecules Come in a Large Variety of Sizes[48]

In contrast to tRNA molecules, which have molecular weights of about 2.5×10^4, and to rRNA molecules, which also have defined sizes (4×10^4, 5×10^5, and 10^6 for 5S, 16S, and 23S, respectively), mRNA molecules vary greatly in chain length and hence in molecular weight. Some of this heterogeneity reflects the large size spread in the length of polypeptide chain products. Most polypeptide chains contain 100 or more amino acids, and so most mRNA molecules must contain at least 100×3 nucleotides (because there are three nucleotides in a codon). Additional variations arise from unequal lengths of

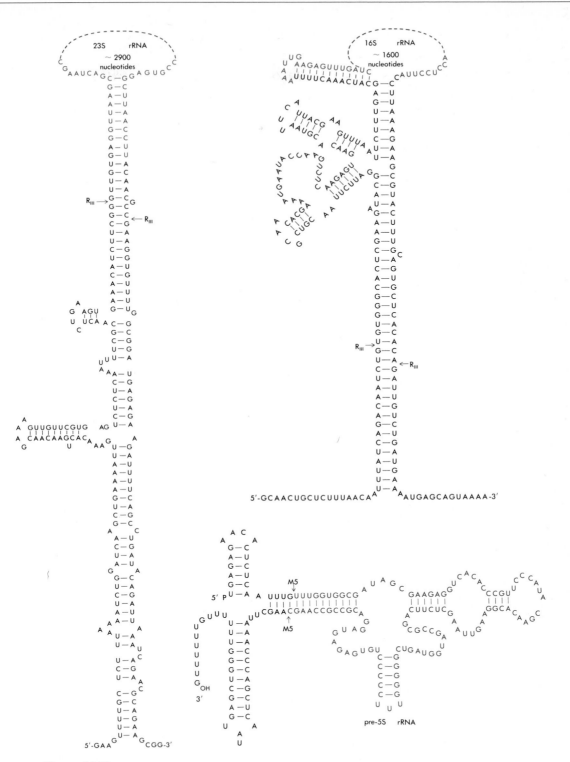

Figure 14-21
Sequences around sites where pre-rRNA is cleaved by *E. coli* RNase III (R$_{III}$) or by
B. subtilis RNase M5 (M5). The RNase III cuts are displaced by two base pairs within
a perfect or nearly perfect helical region. Subsequently, other enzymes trim the ends
down to mature 16S and 23S sequences (indicated by color type). RNase M5 makes
two cuts one base pair apart in a *B. subtilis* 5S RNA precursor to create the 5' and 3'
ends of mature 5S rRNA; its action requires the binding of other proteins. [After R. J.
Bram, R. A. Young, and J. A. Steitz, *Cell* 19 (1980):393; and M. L. Sogin, N. R. Pace,
M. Rosenberg, and S. Weissman, *J. Biol. Chem.* 251 (1976):3480.]

leader sequences at the 5′ ends and trailer sequences at the 3′ ends. These are sequences that do not code for any amino acids but which are essential for the proper synthesis and functioning of an mRNA. For example, the leader of *E. coli* β-galactosidase mRNA is 38 residues long, while that of *E. coli* tryptophan mRNA contains 162 bases. At its 3′ end, the *E. coli* tryptophan mRNA extends 35 nucleotides beyond the coding region. In *E. coli*, therefore, the mRNAs that code for average-size polypeptides of 300 to 500 amino acids usually contain between 1000 and 2000 nucleotides.

Still more heterogeneity in size comes from the fact that some mRNAs code for more than one polypeptide. Besides leader and trailer sequences, these polygenic messengers contain intergenic regions that are sometimes as long as leader sequences. In most, if not all, of these polygenic messengers, the polypeptide products have related functions. For example, a single mRNA molecule codes for the five specific enzymes needed to synthesize the amino acid tryptophan. It has recently been completely sequenced and contains about 6800 nucleotides, or an average of 1400 nucleotides coding for each enzyme and its adjacent intergenic regions.

Ribosomes Come Apart into Subunits During Protein Synthesis[22–24]

The construction of all ribosomes from easily dissociable subunits suggested a cycle during which the large and small subunits come apart at some stage in protein synthesis. This hunch was confirmed by experiments with *E. coli* and yeast that showed that most ribosomes are constantly dissociating into subunits and then reforming. Growth of cells in heavy isotopes followed by transfer to a light medium was the technique used to settle this point. Soon after transfer to the light medium, hybrid ribosomes (heavy 50S/light 30S and light 50S/heavy 30S) began to appear, the rate of their appearance suggesting that the subunits exchange once during every cycle of polypeptide synthesis. Confirmation comes from in vitro studies of protein synthesis (see Chapter 15) where the fate of heavy ribosomes can be followed in the presence of a great excess of light ribosomes. Within a minute or so (the time required to synthesize a complete polypeptide chain), almost all heavy ribosomes disappear, with the simultaneous appearance of hybrid ribosomes (Figure 14-22). The obligatory dependence of subunit exchange on protein synthesis is confirmed by in vitro experiments with the antibiotic sparsomycin. This compound, which blocks polypeptide chain elongation, also prevents any subunit exchange.

Polypeptide Chain Growth Begins at the Amino-Terminal End[49]

A completed polypeptide chain has at one end an amino acid bearing a free carboxyl group (the C terminus) and at the other end an amino acid bearing a free amino group (the N terminus). Protein synthesis by ribosomes always involves stepwise addition of single amino acids, starting with the amino terminus and ending with the carboxyl terminus (Figure 14-23). This point is clearly shown by brief exposure

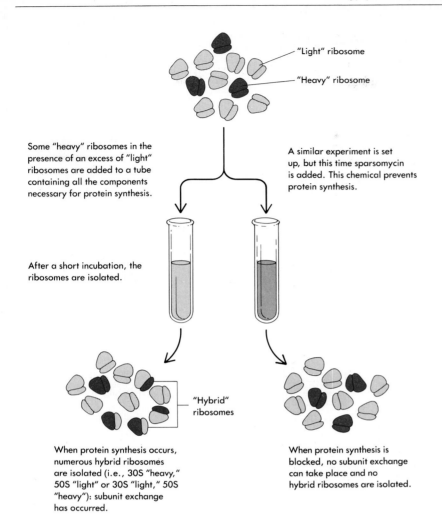

"Light" ribosome

"Heavy" ribosome

Some "heavy" ribosomes in the presence of an excess of "light" ribosomes are added to a tube containing all the components necessary for protein synthesis.

A similar experiment is set up, but this time sparsomycin is added. This chemical prevents protein synthesis.

After a short incubation, the ribosomes are isolated.

"Hybrid" ribosomes

When protein synthesis occurs, numerous hybrid ribosomes are isolated (i.e., 30S "heavy," 50S "light" or 30S "light," 50S "heavy"): subunit exchange has occurred.

When protein synthesis is blocked, no subunit exchange can take place and no hybrid ribosomes are isolated.

Figure 14-22
The obligatory dependence of subunit exchange on protein synthesis.

of hemoglobin-synthesizing reticulocytes to radioactively labeled amino acids, followed by immediate isolation of newly completed hemoglobin chains. If the time of labeling is short relative to the time required to complete a chain, very little radioactivity is found in the

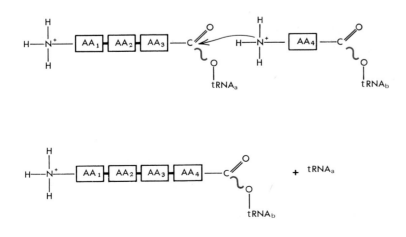

Figure 14-23
Stepwise growth of a polypeptide chain. Initiation begins at the free NH_3^+ end, with the growing point of the chain (the carboxyl end) always terminated by a tRNA molecule. (Peptide bonds between amino acids are indicated by a heavy line.)

Figure 14-24
Experimental demonstration that hemoglobin chains grow in the $NH_3^+ \rightarrow COO^-$ direction. [After H. M. Dintzis, *Proc. Nat. Acad. Sci.* 47 (1961):247.]

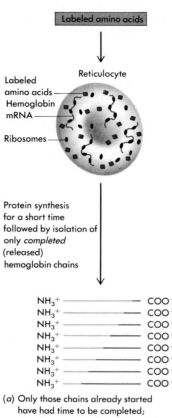

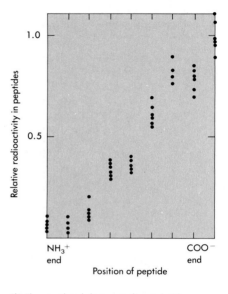

(a) Only those chains already started have had time to be completed; they show a gradient of radioactivity, with more label at the COO⁻ end; no complete chains are found labeled at the NH₃⁺ end.

(b) The completed chains are digested with protease, and the radioactivity in the resulting peptides is measured.

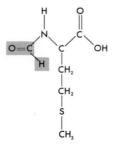

N-formylmethionine (fMet)

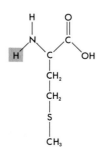

Methionine (Met)

Figure 14-25
The structure of *N*-formylmethionine compared with that of methionine.

amino acids of the amino-terminal end; most is found in the amino acids of the carboxyl end (Figure 14-24). Moreover, a clear gradient of increasing radioactivity is observed as one moves from the amino to the carboxyl end.

Bacterial Polypeptide Chains Start with *N*-Formylmethionine[22, 50]

The starting amino acid in the synthesis of virtually all bacterial polypeptides is **N-formylmethionine (fMet)**. This is a modified methionine that has a formyl group attached to its terminal amino group (Figure 14-25). A blocked amino acid like *N*-formylmethionine can be used only to start protein synthesis. The absence of a free amino group would prevent this amino acid from being inserted during chain elongation. The formyl group is enzymatically added onto methionine after the methionine has become attached to its tRNA adaptor.

Not all methionine tRNA molecules can be formylated. Instead, there are two types of tRNA^{Met}, only one of which (called tRNA$_F^{Met}$) permits the formylation reaction. Sequence analysis of tRNA$_F^{Met}$ and tRNA$_M^{Met}$ (for internal methionine) reveals that they both have the same anticodon sequence, posing the dilemma of how the same codon (AUG) can code both for the initiating amino acid (*N*-formylmethionine) and for internal methionines in the chain (Figure 14-26).

The discovery that synthesis of bacterial proteins starts with a blocked amino acid was unexpected, because isolation of pure protein from growing bacteria revealed essentially no formylated end groups. This means that a **deformylase** enzyme exists that removes the formyl group from the growing chain very soon after synthesis commences (Figure 14-27). In addition, another enzyme (an **aminopeptidase**) subsequently removes the terminal methionine from many proteins. However, it does not act on all proteins, so a large number of bacterial proteins do have methionine at their N termini.

Messenger RNA Binding by the Smaller Ribosomal Subunit Involves rRNA-mRNA Pairing[50–53]

Initiation of protein synthesis in bacteria starts with the formation of a complex between the smaller 30S ribosomal subunit, fMet-tRNA$_F^{Met}$, and an mRNA molecule. Then a 50S subunit joins to form the functional 70S ribosome. Every bacterial mRNA molecule has one **ribosome binding site** for each of its independently synthesized polypeptide products. For example, three such sites are found on the RNA genome of phage R17, which codes for three major proteins. Within each site are nucleotide sequences whose function is the correct positioning of mRNA molecules on the ribosomal surface prior to the commencement of protein synthesis.

How does the ribosome distinguish between AUG codons at the start of a gene and those coding for internal methionines? The first known initiator sequences were established by binding ribosomes to

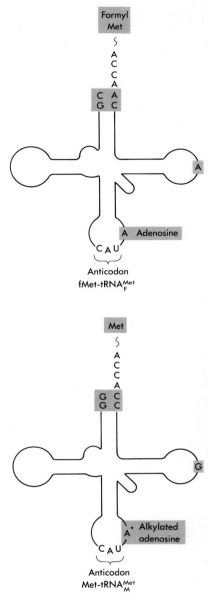

Figure 14-26
The three main points of difference between *N*-formyl-methionyl-tRNA$_F^{Met}$ and methionyl-tRNA$_M^{Met}$.

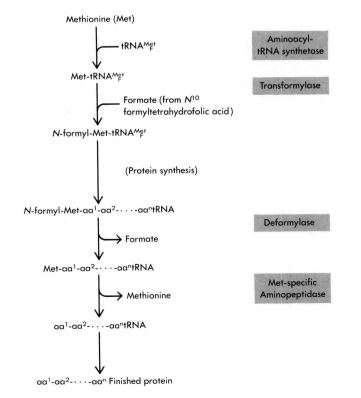

Figure 14-27
Enzymatic steps involving formylmethionine as the initiator of protein synthesis.

Phage Qβ A protein	CUG	AGU	AUA	AGA	GGA	CAU	AUG	CCU	AAA	UUA
Phage Qβ coat	CUU	UGG	GUC	AAU	UUG	AUC	AUG	GCA	AAA	UUA
Phage Qβ replicase	UUA	CUA	AGG	AUG	AAA	UGC	AUG	UCU	AAG	ACA
Phage λ Cro	AUG	UAC	UAA	GGA	GGU	UGU	AUG	GAA	CAA	CGC
Phage fl coat	UUU	AAU	GGA	AAC	UUC	CUC	AUG	AAA	AAG	UCU
Phage φX174 A	AAU	CUU	GGA	GGC	UUU	UUU	AUG	GUU	CGU	UCU
Phage φX174 A*	UUG	CUG	GAG	GCC	UCC	ACU	AUG	AAA	UCG	CGU
Phage φX174 B	AGG	UCU	AGG	AGC	UAA	AGA	AUG	GAA	CAA	CUC
Phage φX174 E	GCG	UUG	AGG	CUU	GCG	UUU	AUG	GUA	CGC	UGG
Lipoprotein	AUC	UAG	AGG	GUA	UUA	AUA	AUG	AAA	GCU	ACU
RecA	GGC	AUG	ACA	GGA	GUA	AAA	AUG	GCU	AUC	G
GalE	AGC	CUA	AUG	GAG	CGA	AUU	AUG	AGA	GUU	CUG
GalT	CCC	GAU	UAA	GGA	ACG	ACC	AUG	ACG	CAA	UUU
LacI	CAA	UUC	AGG	GUG	GUG	AAU	GUG	AAA	CCA	GUA
LacZ	UUC	ACA	CAG	GAA	ACA	GCU	AUG	ACC	AUG	AUU
Ribosomal L10	CAU	CAA	GGA	GCA	AAG	CUA	AUG	GCU	UUA	AAU
Ribosomal L7/L12	UAU	UCA	GGA	ACA	AUU	UAA	AUG	UCU	AUC	ACU
RNA polymerase β subunit	AGC	GAG	CUG	AGG	AAC	CCU	AUG	GUU	UAC	UCC

16S 3' end   HO AUUCCUCCACUAG-5'

Figure 14-28

Examples of the several hundred known ribosome-binding sites (protein synthesis initiator regions) that function in *E. coli.* The nucleotides complementary to the 3' end of 16S rRNA are in shaded boxes; lighter shading indicates GU base pairs. The initiator codon is colored. Note the variation in the length and positioning of the mRNA region that can base-pair and also the variation in exactly which 16S nucleotides interact.

specific mRNAs and then adding ribonuclease to trim away all mRNA sequences except those protected by ribosome attachment. When sequences of these 30-nucleotide-long protected fragments were analyzed, they were found to contain AUGs roughly near the middle, followed by nucleotides that coded for the first three or four amino acids of their respective polypeptide products. These are not the nucleotides specifically recognized by the 30S ribosome, for they vary from one gene to another. Instead, a group of 3 to 9 purine nucleotides appearing in the 5' half of the protected fragment is critical for ribosome binding. Almost all ribosome binding sites have a sequence such as AGGA or GAGG centered some 8 to 13 nucleotides upstream from the initiator codon (Figure 14-28). These form base pairs with complementary residues found in a pyrimidine-rich sequence at the 3' end of 16S rRNA chains (...GAUCACCUCCUUA$_{OH}$-3'). This pairing brings the initiating AUG (or GUG) codon into position so that it can bind to the anticodon of the initiator tRNA bound to the 30S subunit. Thus, the capacity to form two separate base-pairing interactions simultaneously (16S rRNA-mRNA and mRNA-tRNA) distinguishes a true initiator region from other sites containing the many internal AUGs in an mRNA.

As expected, mutations that change the purine-rich region of a ribosome binding site lower the efficiency of mRNA translation; translation can be restored by subsequent mutations that restore comple-

Figure 14-29

Sequence of the initiator region at the beginning of the mRNA of the phage T7 gene *0.3.* Its interaction with the 3' end of 16S rRNA is abolished by a G to A transition, which creates a mutant phenotype. Function is restored by a second transition (A to G) at an adjacent position. A different set of base pairs with 16S rRNA are formed by the revertant. The initiator AUG codon is shown in color. [After J. J. Dunn, E. Buzash-Pollert, and F. W. Studier, *Proc. Nat. Acad. Sci.* 75 (1978):2743.]

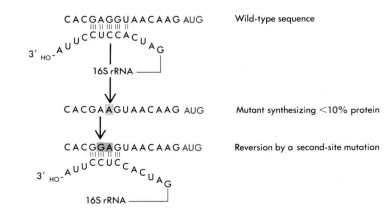

mentarity to the 3' end of 16S rRNA, even if the sequence is different from that of the original mRNA (Figure 14-29). This raises the question of whether the frequency of initiation on different mRNAs can be directly correlated with the length or stability of the mRNA-rRNA interaction. Unfortunately, no simple relationship can be seen, even though ribosome binding sites do differ by factors of 10 to 100 in their ability to start protein synthesis. Rather, the sequence of the entire ribosome-bound region, as well as the folding of the mRNA molecule, appears to contribute in complex ways to the efficiency of initiation.

Once elongation of the polypeptide chain commences, the 16S rRNA-mRNA base pairs must somehow dissociate, leaving the mRNA free to move across the ribosomal surface. Such movement brings the ribosome into contact with mRNA regions to which it would not bind unless protein synthesis had been initiated upstream. As the ribosome moves along, it must temporarily disrupt the double-helical hairpin regions that characterize so many sections of free mRNA, thus creating single-stranded regions that can correctly select the anticodons of the AA~tRNA precursors.

Initiation Factors[50, 54]

A mixture of N-formyl-Met-tRNA$_F^{Met}$, mRNA, and the 30S and 50S subunits is not by itself sufficient for initiation. At least three separate proteins that are only loosely bound by ribosomes, called IF1, IF2, and IF3, must also be present (Figure 14-30). In the first step, these three **initiation factors** attach to the 30S subunit. GTP (guanosine triphosphate) stabilizes their binding and is probably directly bound by IF2. IF3 prevents the association of 30S subunits with 50S subunits and was therefore first identified as a ribosome dissociation factor. It is now believed that IF3 causes a subtle change in the shape of the 30S subunit that both prevents its association with 50S and aids mRNA binding. IF1 assists the binding of IF2 and IF3. Table 14-1 lists the molecular weights of the three initiation factors, as well as those of factors involved in the elongation and termination steps of protein synthesis.

In the second step, fMet-tRNA$_F^{Met}$ and the mRNA attach in either order to the IF-30S-GTP aggregate. In binding, the fMet-tRNA$_F^{Met}$ makes close contact with the IF2-GTP complex. Once the complete **30S initiation complex** is formed, IF3 is released. Then the joining of a 50S subunit leads to GTP hydrolysis as well as the release of the other two initiation factors (see Figure 14-30). The final complex is called a **70S initiation complex.**

The Direction of mRNA Reading is 5'→3'[22]

After an initiator region of an mRNA molecule has been correctly bound to a ribosome, it always moves in a fixed direction during protein synthesis. It does not have the option of moving either to the right or to the left, reflecting the fact that RNA molecules have a direction defined by the relative orientations of their 5' and 3' ends. The end that is read first, the 5' end, is also synthesized first. Thus, a ribosome can attach to an incomplete mRNA molecule in the process of being synthesized on its DNA template. If, on the other hand, polypeptide synthesis went 3'→5', then a length of mRNA corre-

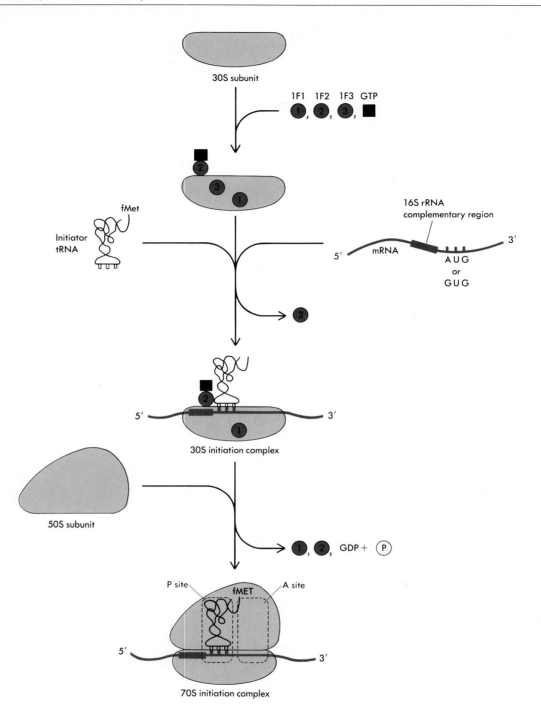

Figure 14-30
View of the initiation process in protein synthesis. Although the fMet-tRNA$_F^{Met}$ is shown as entering the P site directly, it is not known whether it initially contacts the A site. Once the 70S initiation complex is formed, the initiator is clearly bound in the P site.

sponding to a complete polypeptide chain would have to be completely synthesized before a ribosome could attach. The fact that protein synthesis goes in a 5'→3' direction means that long sections of mRNA unattached to ribosomes do not normally occur in rapidly growing bacterial cells.

Each Ribosome Has Two tRNA Binding Sites[22, 55]

Each 70S ribosome contains two cavities into which tRNA molecules can be inserted (Figure 14-31). They are the P (peptidyl) site and the

Table 14-1 Properties of Soluble Factors Involved in *E. coli* Protein Synthesis

	Approximate Molecular Weight	Ability to Bind GTP (GDP)	Approximate Abundance Relative to Number of Ribosomes
Initiation			
IF1	9,000	No	1/7
IF2	120,000	Yes	1/7
IF3	22,000	No	1/7
Elongation			
EF-Tu	45,000	Yes	10†
EF-Ts	30,000	Yes	1
EF-G	80,000	Yes	1
Termination			
RF1	36,000	No	1/20
RF2	38,000	No	1/20
RF3*	46,000	Yes	Not known

*The role of RF3 is not firmly established.
†Amounts of EF-Tu relative to ribosomes can vary from 3-fold to 14-fold, depending on the exact growth conditions. Because it is needed in such high amounts (up to 5 percent of the total cell protein), it is encoded by two different genes called *tufA* and *tufB*.

A (aminoacyl) site. Each tRNA binding site is formed partly by the 30S ribosome and partly by the 50S ribosome. The shape of the site is made specific for a given AA~tRNA by the presence of a particular mRNA codon. That is, although the ribosomal cavities can accept any of the AA~tRNAs, the mRNA-containing surface of the site makes it specific for a unique tRNA molecule.

We do not yet know whether the initiating fMet-tRNA$_F^{Met}$ enters the P site directly or whether it first goes into the A site and subsequently moves into the P site. It is clear, however, that it must end up in the P site and that IF2 must be released by GTP hydrolysis before the second AA~tRNA can bind to the ribosome. For all AA~tRNAs except fMet-tRNA$_F^{Met}$, it has been firmly established that entry to the ribosome occurs through the A site. After the second AA~tRNA is correctly placed in the A site, a peptide bond is formed to yield a dipeptide (two amino acids linked by a peptide bond) attached to the tRNA adaptor of the second amino acid. Then, in a step called **translocation,** the peptidyl-tRNA (tRNA with the growing peptide chain attached to its 3' end) and the mRNA codon to which it is bound are coordinately shifted to the P site. This process of amino acid addition repeats again and again, adding one amino acid at a time to form a complete chain (see Figure 14-31). In these events, the following steps should be emphasized:

1. The growing carboxyl end is always terminated by a tRNA molecule. The binding of this terminal tRNA to either the P or A site is the main force holding a growing polypeptide chain to the ribosome.

2. Formation of the peptide bond moves the attachment point of the growing chain from the P site to the A site. During this step, the carboxyl group of the amino acid in the P site is transferred from its

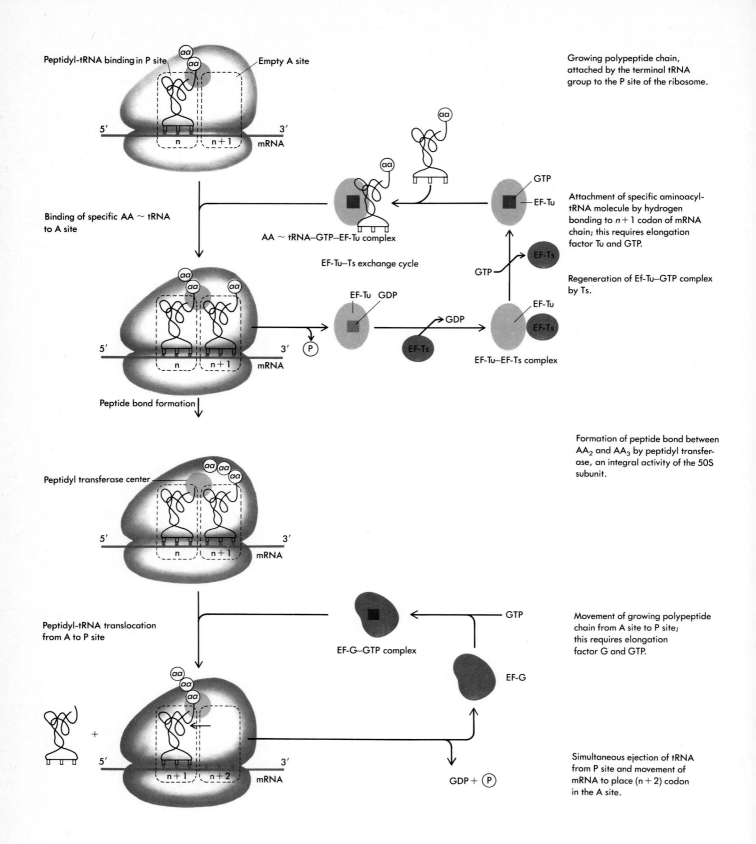

Peptidyl-tRNA binding in P site

Empty A site

5′   3′
        n   n+1   mRNA

Binding of specific AA ∼ tRNA to A site

AA ∼ tRNA–GTP–EF-Tu complex

EF-Tu–Ts exchange cycle

GTP

EF-Tu

EF-Tu   GDP

P

EF-Ts

GDP

EF-Tu

EF-Ts

GTP

EF-Ts

EF-Tu–EF-Ts complex

5′   3′
        n   n+1   mRNA

Peptide bond formation

Peptidyl transferase center

5′   3′
        n   n+1   mRNA

Peptidyl-tRNA translocation from A to P site

EF-G–GTP complex

GTP

EF-G

GDP + P

5′   3′
        n+1   n+2   mRNA

Growing polypeptide chain, attached by the terminal tRNA group to the P site of the ribosome.

Attachment of specific aminoacyl-tRNA molecule by hydrogen bonding to $n+1$ codon of mRNA chain; this requires elongation factor Tu and GTP.

Regeneration of Ef-Tu–GTP complex by Ts.

Formation of peptide bond between AA_2 and AA_3 by peptidyl transferase, an integral activity of the 50S subunit.

Movement of growing polypeptide chain from A site to P site; this requires elongation factor G and GTP.

Simultaneous ejection of tRNA from P site and movement of mRNA to place $(n+2)$ codon in the A site.

Figure 14-31
View of one cycle of peptide bond formation showing the role of the elongation factors, EF-Tu, EF-Ts, and EF-G, and the involvement of GTP hydrolysis.

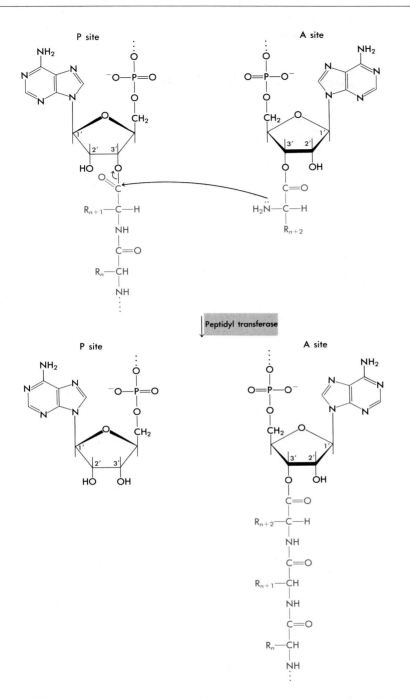

Figure 14-32
The peptidyl transferase reaction. The elongation of the polypeptide chain by one amino acid is shown. Upon chain termination, the ester bond between the 3′ adenosine of the tRNA and the amino acid undergoes nucleophilic attack by water instead of by the amino group of the next amino acid.

tRNA to the amino group of the amino acid in the A site (Figure 14-32).

3. The new terminal tRNA molecule then moves (translocates) from the A site to the P site. At the same time, the mRNA template bound to the smaller ribosome subunit moves to place codon $n + 1$ in the position previously occupied by codon n.

4. Coincident with step 3, the tRNA molecule that was released from its amino acid during peptide bond formation is ejected from the P site.

5. The now vacant A site becomes free to accept a new AA~tRNA molecule whose specificity is determined by correct base pairing between its anticodon and the relevant mRNA codon.

Chain Elongation Requires GTP[22, 56]

Because the amino acid carboxyl groups are activated by attachment to their tRNA adaptors, it was initially thought that perhaps no more energy would be needed for making the peptide bond. This hunch, however, was wrong. In addition to the one molecule of GTP required for initiation, two molecules of GTP are hydrolyzed for each amino acid subsequently added to the growing polypeptide chain. GTP utilization can be directly traced to the participation of three proteins (the elongation factors) that are not normally part of ribosomes but which are necessary for the elongation steps in protein synthesis. Surprisingly, none of these proteins participates directly in the actual formation of peptide bonds.

Binding of AA~tRNA to the A Site Requires Elongation Factors EF-Tu and EF-Ts[22, 56]

The attachment of the AA~tRNA precursor to ribosomes was initially believed to be a nonenzymatic event occurring when the correct AA~tRNA randomly bumped an A site bounded by its particular mRNA codon. Further scrutiny, however, indicated that the binding reaction is far from simple. It starts when one of the **elongation factors (EF-Tu)** reacts with GTP and AA~tRNA to form an AA~tRNA–EF-Tu–GTP complex. EF-Tu is an exceedingly abundant protein in *E. coli,* present in approximately as many copies as there are tRNA molecules (see Table 14-1). It can bind every AA~tRNA except for fMet-tRNA$_F^{Met}$; the basis of this discrimination is not understood. After binding to EF-Tu, the AA~tRNA component is transferred to the ribosomal A site with release of a free EF-Tu–GDP complex and phosphate (see Figure 14-31).

The reutilization of EF-Tu to bring another AA~tRNA to the ribosomal A site requires the participation of another factor, **EF-Ts**. EF-Ts displaces GDP from EF-Tu by itself forming a complex with EF-Tu. When this complex encounters GTP, a GTP–EF-Tu complex is formed, releasing EF-Ts. EF-Tu–GTP can then bind another AA~tRNA and repeat the cycle (see Figure 14-31).

The Peptide-Bond-Forming Activity Is an Integral Component of the 50S Ribosome[23, 57]

Enzymatic catalysis of the peptide bond itself has been assigned to the larger ribosomal subunit. This activity is called **peptidyl transferase,** but all attempts so far to dissociate it from 50S particles have failed. Rather, reconstitution experiments indicate that about six L proteins plus the 23S rRNA are essential for peptidyl transferase activity; another six proteins and 5S rRNA contribute strongly. Thus, peptidyl transferase might best be pictured as an active *center* or site on the ribosome that simply aligns the two AA~tRNAs in precisely the correct way to allow peptide bond formation (see Figure 14-32). Since the AA~tRNAs represent activated forms of the amino acids, this reaction should occur spontaneously, without the input of additional energy.

Peptidyl-tRNA Translocation Requires Elongation Factor G[22, 56]

The movement of peptidyl-tRNA from the A site to the P site is brought about by elongation factor G (**EF-G**), often called the **translocase**. In this process, an EF-G–GTP–ribosome complex first forms. Translocation then occurs coupled with ejection of the free tRNA from the P site (see Figure 14-31). Subsequent release of free reusable EF-G requires the hydrolysis of GTP to GDP and Ⓟ. The splitting of the high-energy bond is obligatory for entry of the ribosome into the next elongation cycle.

Movement of mRNA Across the Ribosomal Surface[22, 58]

Normally, at each translocation step, the mRNA template is advanced precisely three nucleotides. One can imagine the existence of a machinery within the ribosome that obligatorily advances the mRNA chain three nucleotides at a time. Alternatively, the mRNA may not be moved by the ribosome per se, but instead may be pulled by the process that moves peptidyl-tRNA from the A site to the P site. According to this hypothesis, the movement of a codon triplet is a consequence of its binding to an anticodon in tRNA. Strongly favoring this latter proposal is the discovery of frameshift suppressor tRNAs (Chapter 15) that have anticodons containing four nucleotides and so bind to groups of four adjacent mRNA nucleotides. When such unusual tRNAs are translocated, the attached mRNA template is advanced by four bases, showing that mRNA movement is directly coupled to tRNA movement.

Inhibition of Specific Steps in Protein Synthesis by Antibiotics[59, 60]

A number of antibiotics have proved very useful in delineating the steps by which proteins are built. For example, puromycin, a very powerful inhibitor of the growth of all cells, acts by interrupting chain elongation (Figure 14-33). Its structure resembles the 3' end of a charged tRNA molecule, and so it enters the ribosomal A site very efficiently, competitively inhibiting the normal entry of AA~tRNA precursors. More importantly, peptidyl transferase will use the antibiotic as a substitute, transferring nascent chains to puromycin acceptors. Since the puromycin residue is very small compared to a tRNA, it binds only weakly to the A site. Thus, the puromycin-terminated nascent chains fall off the ribosomes, producing incomplete polypeptides of varying lengths. Note that the incorporation of puromycin into the carboxy-terminal ends of growing chains proves that protein synthesis proceeds from the N terminus to the C terminus.

Other antibiotics act to inhibit other steps of protein synthesis. For instance, fusidic acid interferes with the release of EF-G after it has functioned in translocation. The stuck EF-G–GDP complex very effectively blocks the binding of the next AA~tRNA–EF-Tu–GTP to the ribosomal A site, which must precede a subsequent round of elongation. The peptidyl transferase reaction is inhibited by the antibiotics

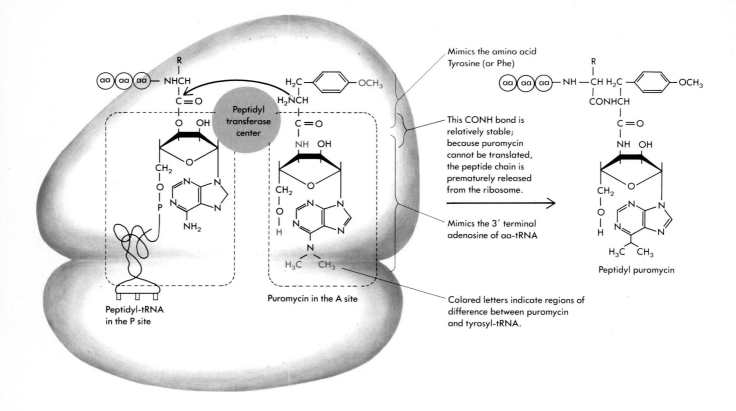

Figure 14-33
Premature polypeptide chain termination by puromycin.

chloramphenicol and sparsomycin, both of which specifically bind to the 50S subunit. In contrast, streptomycin, which binds to the 30S subunit, is a powerful inhibitor of chain initiation.

Polypeptide Chains Fold Up Simultaneously with Synthesis

Under optimal conditions, the time required for the synthesis of an *E. coli* polypeptide chain containing 300 to 400 amino acids is 10 to 20 seconds. During this time, the elongating chain does not remain a random coil, but quickly assumes much of its final three-dimensional shape through the formation of many of its secondary bonds. Thus, with many proteins, most of the final shape may be assumed before the last few amino acids are added on. As a result, trace amounts of many enzyme activities can be found on ribosomes that have not yet released their polypeptide products. Likewise, antibodies, which recognize specific three-dimensional shapes on the surfaces of proteins (Chapter 23), can often bind to growing polypeptide chains still attached to ribosomes.

Chain Release Depends on Specific Release Factors That Read Chain-Terminating Codons[61]

Two conditions are necessary for the termination of protein synthesis. One is the presence of a codon that specifically signals that polypeptide elongation should stop. The other is the presence of a protein

release factor (RF) that reads the chain-terminating signal. Behind all this complexity is the fact that after a polypeptide chain has reached its full length, its carboxyl end is still bound to its tRNA adaptor. Termination must therefore involve the splitting off of the terminal tRNA. When this happens, the nascent chain quickly dissociates, since its binding to the ribosome occurred principally through its adaptor tRNA.

Elucidation of the genetic code (Chapter 15) revealed three codons specifically signifying "stop." Their existence initially led to the expectation of chain-terminating tRNA molecules; that is, tRNAs that would specifically bind to the stop codon but which had no amino acid attached to their 3' adenosine and which in some way promoted the release of the terminal tRNA group. Experiments, however, conclusively ruled out the existence of such molecules. The stop codons are instead read by proteins called release factors. In *E. coli*, there are two well-characterized RFs (see Table 14-1), one (RF1) that recognizes the stop codons UAG and UAA and another (RF2) that acts with UGA and UAA. They function by inducing the peptidyl transferase to transfer the growing chain to water instead of to an amino acid activated by attachment to tRNA (see Figure 14-32). Whether GTP hydrolysis is required for termination in *E. coli* is not yet firmly established, although a third factor (sometimes called RF3), which does bind GTP and GDP, appears to stimulate the reaction. After chain termination, the release of the 50S ribosomal subunit from the mRNA probably precedes that of the 30S particle; and still another protein factor may be involved in ribosome release. The ribosomal subunits are then available to begin synthesis of another polypeptide chain.

GTP May Function by Causing Conformational Changes[56, 62]

The repeated use of GTP at so many stages of the translation process (initiation, elongation, and possibly termination) calls for some explanation. There is no evidence for its use in the formation of any covalent bond, and so it does not appear to function as an ATP-like energy donor. Instead, it plays a key role in the noncovalent binding of the various translational factors (IF2, EF-Tu, and EF-G) to the ribosomal surface. Such binding processes require only the presence of GTP, while the GTP → GDP cleavage reaction is the signal for the release of the bound factor. An attractive hypothesis is that GTP attachment induces the translation factors to undergo shape changes that are necessary for their binding to ribosomes. Subsequent hydrolysis to GDP leads to a return of the original free configuration, resulting in an ejection of the respective factor from the ribosomal surface.

Reconstitution experiments have shown that the four copies of protein L7/L12 on the 50S ribosome are intimately involved in GTP hydrolysis. These ribosome components are essential for the binding of the various protein synthesis factors that utilize the energy stored in GTP during their functioning. However, there are reports that protein synthesis can proceed, albeit at a very slow rate, in the absence of GTP hydrolysis and elongation factors. Thus, ribosomes themselves must possess an innate capacity to bind AA~tRNA and translocate. The participation of factors and GTP increases the rate at which they do so to a biologically meaningful level.

GTP Hydrolysis Also Contributes to the Fidelity of Translation[2, 63, 64, 65]

Another extremely important way in which GTP functions on the ribosome is the role it plays in the accuracy of translation. Evidence has now accumulated for a proofreading step that occurs after AA~tRNA binding. When AA~tRNA is first brought to the A site complexed with EF-Tu–GTP, the initial mRNA-tRNA base-pairing interaction is not sufficiently specific to account for the low error rate (about 10^{-4}) of protein synthesis. Proofreading improves the error rate by discarding an incorrectly selected AA~tRNA. This consumes an AA~tRNA–EF-Tu–GTP complex (which involves GTP $\rightarrow$ GDP conversion) but does not extend the polypeptide chain. Thus, editing steps require the input of additional energy.

To study the proofreading mechanism, it has been necessary to devise in vitro systems that translate nearly as accurately as cells do. This has been achieved by carefully choosing ionic conditions that mimic those in vivo and by ensuring that a high GTP/GDP ratio is maintained throughout the experiment. Recent results suggest that the initial tRNA selection and subsequent proofreading steps contribute approximately equally to the fidelity of translation (each about 10^{-2}). It is worth noting that some mutations in ribosomal proteins can significantly affect the proofreading ability of the ribosome (Chapter 15).

Production of ppGpp on Ribosomes by an Idling Reaction in the Absence of Charged tRNA[66, 67, 68]

If, by accident, an uncharged tRNA molecule occupies an A site, not only does polypeptide growth temporarily cease, but an idling reaction occurs on the affected ribosome, which uses ATP as a Ⓟ~Ⓟ donor to convert pp5'-G-3'OH (GDP) into the unusual guanine nucleotide pp5'-G-3'pp, often called "magic spot." Production of ppGpp is stimulated only if the uncharged tRNA can correctly pair with the mRNA codon exposed in the A site. While during normal bacterial growth only very low levels of ppGpp are present, large amounts accumulate during amino acid starvation. These ppGpp molecules do not play a passive role, but act as signals that specifically stop the synthesis of rRNA and tRNA chains. They thus act as the control molecules that keep cells from wastefully producing more ribosomes than they can employ. This specific blockage of rRNA and tRNA synthesis under conditions of amino acid starvation is called the **stringent response**. Its existence is not obligatory, however, since there are *relaxed mutants* that continue to make rRNA and tRNA during amino acid starvation. Some relaxed cells lack a 75,000-dalton enzyme, the **stringent factor**, that can bind to ribosomes and carry out the conversion of ppG to ppGpp. Other relaxed mutants are altered in one of the L proteins (L11) of the 50S ribosome.

Several Ribosomes Translate an mRNA Molecule Simultaneously

The section of an mRNA molecule that is in contact with a single ribosome is relatively short (about 30 nucleotides). This allows a given

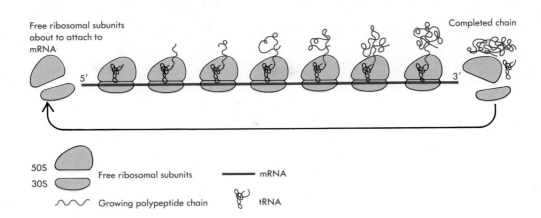

Free ribosomal subunits about to attach to mRNA

Completed chain

5'   3'

| 50S | Free ribosomal subunits |
| 30S | |

Growing polypeptide chain

mRNA

tRNA

mRNA molecule to move over the surfaces of several ribosomes at once, functioning simultaneously as a template for several identical polypeptide chains. The collection of ribosomes bound to a single mRNA chain is called a **polysome** or **polyribosome.** At a given time, the lengths of polypeptide chains attached to successive ribosomes in the polyribosome vary in direct proportion to the fraction of the messenger tape each ribosome has already read (Figure 14-34). This means that at any moment, the polypeptide chains being produced along the length of the mRNA are shortest at the 5' end and gradually lengthen toward the 3' end. There is great variation in polysome size, which depends both on the size of the mRNA chain and how closely the ribosomes are packed. The latter obviously depends on how frequently ribosomes can initiate at the beginning of a particular gene; this varies from one ribosome binding site to another. At maximum utilization of an mRNA chain, there is one ribosome for every 80 mRNA nucleotides. Thus, the polysomes that make hemoglobin molecules usually contain 4 to 6 ribosomes, while approximately 12 to 20 ribosomes are attached to the mRNA molecules concerned with the synthesis of proteins in the 30,000 to 50,000-dalton range (300 to 500 amino acids).

The ability of several ribosomes to function on a single mRNA simultaneously explains why a cell needs relatively little mRNA. Before polysomes were discovered, and when only one ribosome was thought to be attached to a given mRNA molecule, the fact that mRNA accounts for only 1 to 2 percent of the total cellular RNA seemed highly paradoxical. This followed from the calculation that given this amount of mRNA and an average chain length of about 1700 nucleotides, then at any instant, no more than 10 percent of the ribosomes in a cell could be making protein.

Proteins Designed for Export Are Made on Membrane-Bound Ribosomes and Are Cleaved After Synthesis[69-74]

Bacteria synthesize not only proteins that function inside the cell, but also proteins whose roles demand that they be properly localized in or beyond the cell membrane. In fact, the cell envelope of a Gram-negative bacterium such as *E. coli* is made up of three different compartments, each characterized by different populations of specific proteins. For instance, the inner membrane contains proteins in-

Figure 14-34
Schematic picture of a polyribosome during protein synthesis. The mRNA molecule is moving from right to left.

Table 14-2   Signal Sequences of Some Exported Bacterial Proteins*

Protein	Charged Segment	Hydrophobic Segment
Inner Membrane Proteins		
Phage fd, major coat protein	Met Lys Lys Ser Leu Val Leu Lys	Ala Ser Val Ala Val Ala Thr Leu Val Pro Met Leu Ser Phe Ala ↓ Ala Glu Gly
Phage fd, minor coat protein	Met Lys Lys	Leu Leu Phe Ala Ile Pro Leu Val Val Pro Phe Tyr Ser His Ser ↓ Ala Glu Thr
Periplasmic Proteins		
Alkaline phosphatase	Met Lys	Gln Ser Thr Ile Ala Leu Ala Leu Leu Pro Leu Leu Phe Thr Pro Val Thr Lys Ala ↓ Arg Thr Pro
Maltose binding protein	Met Lys Ile Lys Thr Gly Ala Arg	Ile Leu Ala Leu Ser Ala Leu Thr Thr Met Met Phe Ser Ala Ser Ala Leu Ala Lys ↓ Ile Glu
Leucine-specific binding protein	Met Lys Ala Asn Ala Lys	Thr Ile Ile Ala Gly Met Ile Ala Leu Ala Ile Ser His Thr Ala Met Ala ↓ Asp Asp Ile
β-lactamase of pBR322	Met Ser Ile Gln His Phe Arg	Val Ala Leu Ile Pro Phe Phe Ala Ala Phe Cys Leu Pro Val Phe Ala ↓ His Pro Glu
Outer Membrane Proteins		
Lipoprotein	Met Lys Ala Thr Lys	Leu Val Leu Gly Ala Val Ile Leu Gly Ser Thr Leu Leu Ala Gly ↓ Cys Ser Ser
LamB	Met Met Ile Thr Leu Arg Lys	Leu Pro Leu Ala Val Ala Val Ala Ala Gly Val Met Ser Ala Gln Ala Met Ala ↓ Val Asp Phe
OmpA	Met Lys Lys	Thr Ala Ile Ala Ile Ala Val Ala Leu Ala Gly Phe Ala Thr Val Ala Gln Ala ↓ Ala Pro Lys

*The hydrophobic portion (color) of the signal sequence is invariably preceded by a short charged segment. The arrows indicate the sites of cleavage by signal peptidase.
SOURCE: S. Michaelis and J. Beckwith, *Ann. Rev. Microbiol.* 36 (1982):435.

volved in energy metabolism and nutrient transport; the outer membrane has proteins that facilitate ion and nutrient entry and serve as bacteriophage receptors; the space in between (called the *periplasm*) contains hydrolytic enzymes such as nucleases and proteases, as well as nutrient-binding proteins.

How are these proteins conveyed across the inner membrane to their correct destination? Most exported proteins are synthesized in the form of longer precursors, with between 15 and 30 additional amino acids at their N termini. This N-terminal region, called the **signal sequence**, is invariably rich in uncharged, usually hydrophobic amino acids (Table 14-2). They function to initiate association of the nascent polypeptide with the membrane as soon as the signal sequence has emerged from the ribosome. It has been suggested that the signal sequence and the residues just beyond fold into two short helical segments that then bend into an antiparallel helical hairpin. The properties of such a hairpin are ideal for insertion into the hydrophobic lipid bilayer (Figure 14-35). Once the N terminus of a to-be-exported protein is anchored in the membrane, continued synthesis will extrude the remainder of the polypeptide through to the other side. Or, if another hydrophobic stretch is synthesized, this stretch may stick as it crosses the membrane and permanently fix the protein in the membrane. Meanwhile, the short N-terminal signal sequence is usually cleaved off by a specific **signal peptidase.** How proteins can be specifically targeted for the outer membrane rather than the periplasm is not yet known.

Ribosomes translating an mRNA whose product is to be exported are membrane-bound. This attachment is at least partially via the growing polypeptide chain. What other interactions are required between the ribosome and the membrane is not yet clear. However,

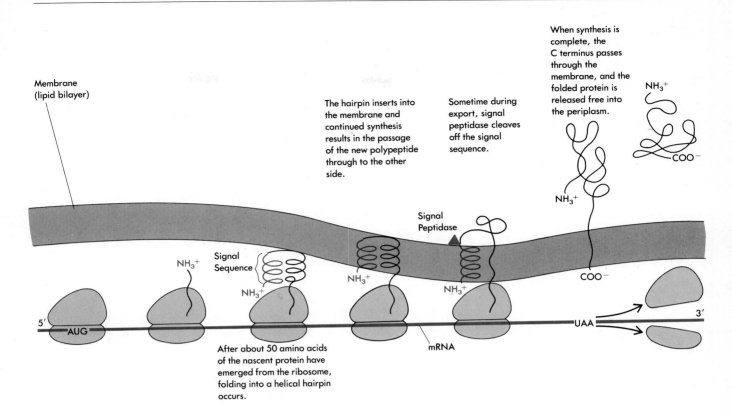

Membrane (lipid bilayer)

The hairpin inserts into the membrane and continued synthesis results in the passage of the new polypeptide through to the other side.

Sometime during export, signal peptidase cleaves off the signal sequence.

When synthesis is complete, the C terminus passes through the membrane, and the folded protein is released free into the periplasm.

Signal Peptidase

Signal Sequence

NH_3^+

NH_3^+

NH_3^+

NH_3^+

NH_3^+

NH_3^+

COO^-

COO^-

5′ AUG

mRNA

UAA 3′

After about 50 amino acids of the nascent protein have emerged from the ribosome, folding into a helical hairpin occurs.

mutations in several genes that affect the export of many proteins in *E. coli* are currently under scrutiny. Their properties hint that the situation may be quite similar to that in mammalian cells, where firm evidence points to the participation of a small RNA-protein complex both in initiating secretion and in controlling the elongation of proteins designed for export (Chapter 21).

Figure 14-35
Steps in the process of cotranslational secretion. The diagram illustrates how polysomes engaged in synthesizing a protein designed for export become membrane-bound via the nascent polypeptide chain. [After D. M. Engelman and T. A. Steitz, *Cell* 23 (1981):411.]

Potent Methods for Studying Ribosome Structure[24, 28, 75–82]

It is very likely that the general outline of protein synthesis is now known. Meanwhile, progress on relating each of the many reactions to the structure of the ribosome itself has lagged far behind. The situation is clearly not hopeless, however, for data obtained using a number of techniques have recently begun to converge. These give us a good picture of the shapes of the two *E. coli* ribosomal subunits and where many of their important functional sites are located.

Two different kinds of electron microscopy have provided us with information concerning ribosome shape. One is direct visualization of fields containing isolated 70S ribosomes, 50S subunits, and 30S subunits (Figure 14-36). The other applies mathematical reconstruction methods to interpret images generated from naturally occurring ribosome microcrystals. There is general agreement that the 50S subunit resembles a hemisphere with three protuberances emerging from its flat face. The 30S subunit is more flattened and appears to be separated into an upper one-third and a lower two-thirds, with a platform protruding on one side just below the constriction. The deduced shapes for the two subunits and how they might fit together to form the 70S ribosome are shown in Figures 14-36 and 14-37.

Figure 14-36
Direct visualization of ribosome shape using electron microscopy. (a) A field of ribosomes from *E. coli* includes some that are oriented like A or B in the model in (b). A few ribosomal subunits are also present, identified by the letters s and l, which stand for small and large. The field has been negatively stained for electron microscopy; each ribosome is defined by an outline of the salt of a heavy metal, which in this case is uranium. [Adapted with the permission of the author from J. A. Lake, "The Ribosome," *Sci. Amer.* 245 (1981):86.]

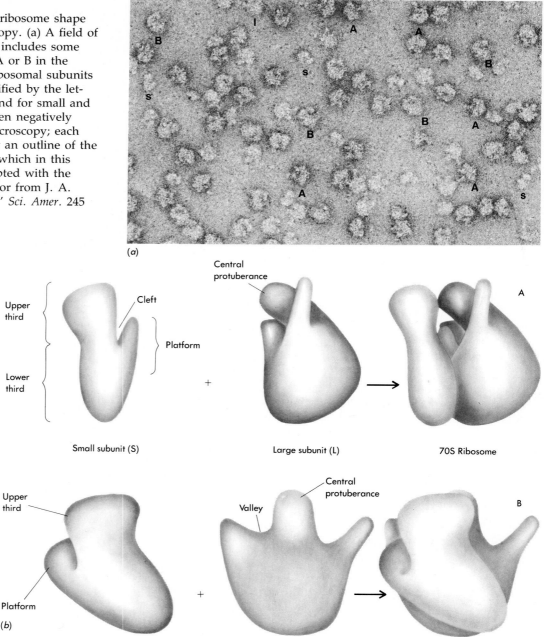

A much broader spectrum of methodologies has contributed to the assignment of functional sites to specific locations on the surfaces of these shapes. For instance, analysis of chemical crosslinks between proteins (Figure 14-38), between proteins and rRNA or tRNA sequences, and between different parts of an rRNA molecule have told us what is near what. The ability to reconstitute ribosomal subunits has allowed the labeling of certain proteins with chromophores for fluorescence energy transfer or with deuterium for neutron-scattering experiments. Such techniques yield measurements of the distances between proteins in or bound to the ribosome (Figure 14-39). Finally, antibodies have been raised against individual ribosomal proteins or

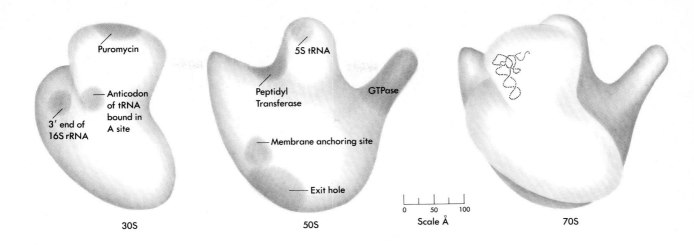

Puromycin

Anticodon
of tRNA
bound in
A site

3' end of
16S rRNA

30S

5S tRNA

Peptidyl
Transferase

GTPase

Membrane anchoring site

Exit hole

0 50 100
Scale Å

50S

70S

(the rather rare) modified nucleotides in rRNA. Using the electron microscope, researchers can visualize these antibodies bound to ribosomes and can therefore localize particular sites on the two ribosomal subunits (Figure 14-40).

The picture that emerges is quite satisfying, since components that are known to interact occupy nearby positions (see Figure 14-37).

Figure 14-37
Shapes of the two subunits of *E. coli* ribosomes. Functional sites were localized by a combination of methods. The diagram of the 70S ribosome shows how a tRNA bound in the A site would be oriented. The A site is located between the 30S and 50S subunits. [After J. A. Lake, *Sci. Amer.* 245 (1981):84; H. M. Olson, P. G. Grant, B. S. Cooperman, and D. G. Glitz, *J. Biol. Chem.* 257 (1982):2649.]

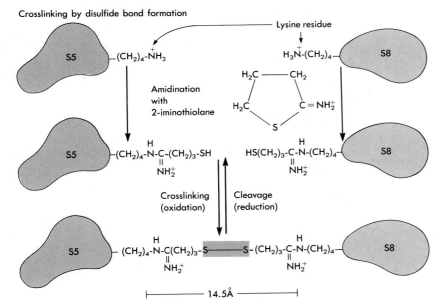

Crosslinking by disulfide bond formation

Lysine residue

S5 —(CH$_2$)$_4$-$\overset{+}{\text{NH}_3}$

H$_3$$\overset{+}{\text{N}}$-(CH$_2$)$_4$— S8

Amidination
with
2-iminothiolane

H$_2$C —— CH$_2$

H$_2$C   C = NH$_2^+$

S

S5 —(CH$_2$)$_4$-N-C-(CH$_2$)$_3$-SH
$\overset{\text{H}}{\underset{\text{NH}_2^+}{|}}$

HS(CH$_2$)$_3$-C-N-(CH$_2$)$_4$— S8
$\overset{\text{H}}{\underset{\text{NH}_2^+}{|}}$

Crosslinking
(oxidation)

Cleavage
(reduction)

S5 —(CH$_2$)$_4$-N-C(CH$_2$)$_3$-S —— S-(CH$_2$)$_3$-C-N-(CH$_2$)$_4$— S8
$\overset{\text{H}}{\underset{\text{NH}_2^+}{|}}$ $\overset{\text{H}}{\underset{\text{NH}_2^+}{|}}$

⊢——— 14.5Å ———⊣

Figure 14-38
Illustration of how chemical crosslinking can identify neighboring proteins in the ribosome. Introduction of reversible crosslinks allows analysis by a "diagonal" method employing a two-dimensional gel. Recall that mobility on a sodium dodecyl sulfate (SDS) gel is related to the molecular weight of the protein. [After Traut *et al.*, in *Ribosomes*, ed. G. Chamblis, G. R. Craven, J. Davis, L. Kahan, and M. Nomura (Baltimore, Md.: University Park Press, 1980), p. 89.]

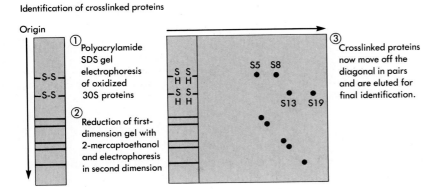

Identification of crosslinked proteins

Origin

① Polyacrylamide
SDS gel
electrophoresis
of oxidized
30S proteins

—S-S—

—S-S—

② Reduction of first-
dimension gel with
2-mercaptoethanol
and electrophoresis
in second dimension

S S
H H
S S
H H

S5 S8

S13 S19

③ Crosslinked proteins
now move off the
diagonal in pairs
and are eluted for
final identification.

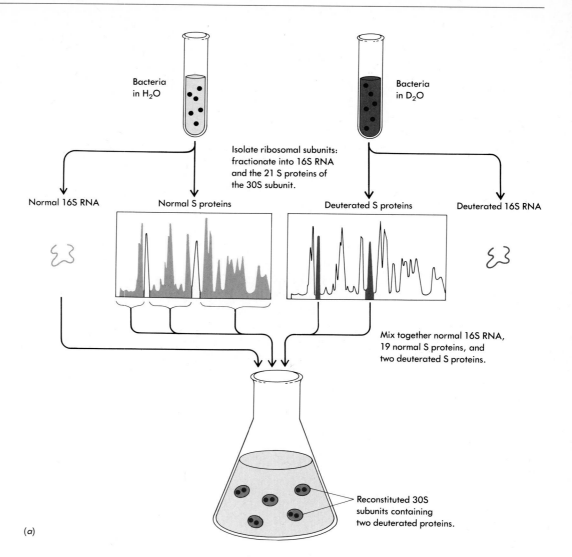

Bacteria
in H_2O

Bacteria
in D_2O

Isolate ribosomal subunits:
fractionate into 16S RNA
and the 21 S proteins of
the 30S subunit.

Normal 16S RNA   Normal S proteins   Deuterated S proteins   Deuterated 16S RNA

Mix together normal 16S RNA,
19 normal S proteins, and
two deuterated S proteins.

Reconstituted 30S
subunits containing
two deuterated proteins.

(a)

Figure 14-39
How neutron-scattering studies can determine distances between specific proteins in the ribosome. [After D. M. Engelman and P. B. Moore, *Sci. Amer.* 235 (1976):44.]

1. The mRNA binding site and the 3′ end of the 16S rRNA are located on the platform between the upper and lower portions of the 30S subunit. Initiation factors also bind in this region.

2. The two tRNA binding sites are in the cleft formed by the platform and upper third of the 30S ribosome.

3. The GTPase center (consisting of the four molecules of protein L7/L12) has been assigned to the most fingerlike of the three protuberances of the 50S subunit.

4. The peptidyl transferase center lies in the valley between the other two 50S protuberances.

5. The exit site for the growing polypeptide chain is almost on the exact opposite side of the 50S subunit from the transferase center. This suggests that the 30 amino acids of the nascent polypeptide chain passing through the interior of the ribosome must be in a nearly completely extended conformation.

6. The site where the 50S subunit interacts with membranes is located very near the polypeptide exit hole.

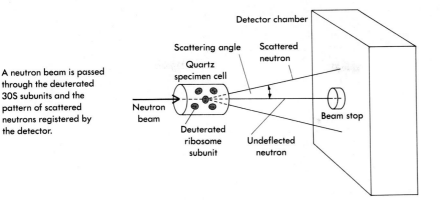

A neutron beam is passed through the deuterated 30S subunits and the pattern of scattered neutrons registered by the detector.

The interference pattern appears as a set of concentric rings in which the scattered neutron intensity is alternately higher and lower than the background level.

The neutron intensity is graphed as a function of the angular deflection of the beam; the distance between successive interference rings is inversely related to the distance, l, between the centers of mass of the two deuterated proteins in the 30S ribosome.

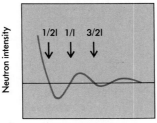

(b)

7. The two ribosomal subunits join to bring the 30S platform that binds mRNA and tRNA in close proximity to the 50S surface that contains the GTPase and peptidyl transferase activities.

   Despite these advances in understanding the topography of the ribosome, the methods mentioned here cannot tell us how the ribosome works in atomic detail. Such answers must ultimately come from X-ray diffraction analysis. Only recently have we obtained good three-dimensional crystals of *E. coli* ribosomal subunits. This, however, is just the first step in a monstrous endeavor, for X-ray methods have not yet been used to solve structures of anywhere near this complexity. Meanwhile, crystals of individual ribosomal proteins, the 5S rRNA, and a 5S rRNA–protein complex have been prepared, and their structures are being solved. They will tell us much about RNA structure and RNA-protein interactions. Nonetheless, nagging questions, such as why the ribosome is two-thirds RNA, will probably be answered only after the complete three-dimensional structure of this complicated protein synthesis machine is known.

Figure 14-40

Localization of proteins on ribosomal subunits by immune electron microscopy. Ribosome and ribosomal subunit pairs are joined by antibody molecules (which have a Y shape) against specific ribosomal proteins: (a) anti-S14, (b) anti-L7/L12, (c) anti-L27. [Courtesy of J. A. Lake, *J. Mol. Biol.* 161 (1982):95.]

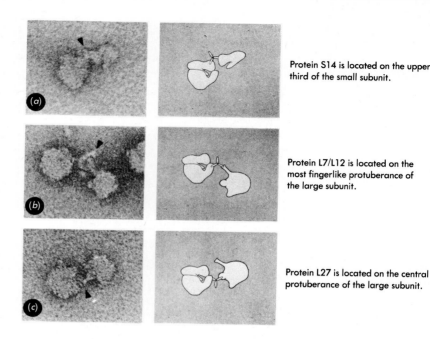

Protein S14 is located on the upper third of the small subunit.

Protein L7/L12 is located on the most fingerlike protuberance of the large subunit.

Protein L27 is located on the central protuberance of the large subunit.

Summary

Amino acids do not attach directly to RNA templates. They first combine with specific adaptor molecules to form aminoacyl (AA) adaptor complexes. The adaptor component has a strong chemical affinity for RNA nucleotides. All the adaptors are transfer RNA molecules (tRNA) with molecular weights of about 25,000. A given tRNA molecule is specific for a given amino acid.

All tRNA chains are folded into semirigid L-shaped molecules. About half the tRNA bases are found in double-helical stems, with the remainder forming loops stabilized by interloop tertiary bonds and stacking forces. One of the loops contains the anticodon, the group of three bases that base-pair to three successive template bases (the codon). Amino acids are attached through their carboxyl groups to the 3'-terminal adenosine of tRNA molecules by high-energy covalent bonds and thus become activated. There is a specific activating enzyme (aminoacyl-tRNA synthetase) for each amino acid; and so it is an enzyme, not tRNA, that recognizes the amino acid. The overall accuracy of protein synthesis can thus be no greater than the accuracy with which the activating enzymes can selectively recognize the various amino acids.

After activation, the AA~tRNA molecules diffuse to the ribosomes, which are roughly spherical particles on which the peptide bonds form. Ribosomes have molecular weights of about 3×10^6 daltons and in bacteria consist of about one-third protein and two-thirds ribosomal RNA (rRNA). They are always constructed from a large (50S) subunit and a small (30S) subunit. In a single bacterial ribosome there are 52 different proteins, in addition to three RNA chains (16S, 23S, and 5S rRNA). Ribosomal RNA contains no genetic information, and the function of most of its nucleotides remains a mystery. The template itself is a third form of RNA, messenger RNA (mRNA). Messenger RNA attaches to ribosomes and moves across them to bring successive codons into position to select the correct AA~tRNA precursors.

Protein chains grow stepwise, beginning at the amino-terminal end. This means that the growing end is always terminated by a tRNA molecule, which serves to hold the nascent chain on the ribosomal surface. Each ribosome has two tRNA-binding sites, the P (peptidyl) site and the A (aminoacyl) site. AA~tRNA molecules first enter the A site, allowing the subsequent formation of a peptide bond with the growing chain held in the P site. This transfers the nascent chain to the A site. A process called translocation then moves the tRNA carrying the nascent chain into the P site and simultaneously advances the mRNA to the next codon and ejects the free tRNA from the P site. Thus, another cycle can begin. Both the attachment of AA~tRNA to the A site and the translocation process require the splitting of GTP to GDP and the participation of protein factors. Initiation of synthesis requires separate factors from those contributing to chain elongation and also involves GTP hydrolysis. With bacterial ribosomes, initiation requires *N*-formylmethionyl-tRNA. Chain termination results when specific stop signals in the mRNA are read by protein release factors.

Completion of chain synthesis is followed by the dissociation of ribosomes into the large and small ribosomal subunits. Rejoining of subunits occurs only after the initiating aminoacyl-tRNA and the mRNA have first bound to the small ribosomal particle. A given mRNA molecule usually works simultaneously on many ribosomes (a polysome). Thus, at any given moment, many codons of the same template are "at work."

Bibliography

General References

Altman, S., ed. 1978. *Transfer RNA.* Cambridge, Mass: MIT Press. A collection of articles on tRNA structure and function.

Chambliss, G., G. R. Craven, J. Davies, K. Davis, L. Kahan, and M. Nomura, eds. 1980. *Ribosomes.* Baltimore, Md.: University Park Press. A compendium of excellent reviews and summaries, not only of ribosome structure but of all aspects of protein synthesis as of 1980.

Clark, B. and H. Petersen, eds. 1984. "Gene Expression: The Translational Step and Its Control." *Alfred Benzon Symposium,* vol. 19. Copenhagen: Munksgaard. A number of excellent reviews on aspects of translation currently under study.

"Protein Synthesis." 1969. *Cold Spring Harbor Symposium.* 34. Cold Spring Harbor, N.Y.: Cold Spring Harbor Laboratory. Summarizes the state of our understanding of protein synthesis three years after the genetic code was solved.

Nomura, M., A. Tissieres, and P. Lengyel, eds. 1974. *Ribosomes.* Cold Spring Harbor, N.Y.: Cold Spring Harbor Laboratory. An early but valuable treatise on ribosome structure and function.

P. R. Schimmel, D. Söll, and J. N. Abelson, eds. 1980. *Transfer RNA: Structure Properties and Recognition, and Biological Aspects.* Cold Spring Harbor, N.Y.: Cold Spring Harbor Laboratory. Two volumes with contributions on all aspects of tRNA.

Watson, J. D. 1963. "The Involvement of RNA in the Synthesis of Proteins." *Science* 140: 17–26. A history of work between 1953 and 1962 that established how RNA participates in protein synthesis.

Cited References

1. Crick, F. H. C. 1958. "On Protein Synthesis. Biological Replication of Macromolecules." *Symp. Exp. Biol.* 12:138–163.
2. Hopfield, J. J. 1974. "Kinetic Proofreading: A New Mechanism for Reducing Errors in Biosynthetic Processes Requiring High Specificity." *Proc. Nat. Acad. Sci.* 71:4135–4139.
3. Ninio, J. 1975. "Kinetic Amplification of Enzyme Discrimination." *Biochimie* 57:587–595.
4. Hoagland, M. B., M. L. Stephenson, J. F. Scott, L. I. Hecht, and P. C. Zamecnik. 1958. "A Soluble Ribonucleic Acid Intermediate in Protein Synthesis." *J. Biol. Chem.* 231:241–257.
5. Holley, R. W. 1968. "The Nucleotide Sequence of a Nucleic Acid." 1966. *Sci. Amer.* 214:30–39.
6. Rich, A., and U. L. RajBhandary. 1976. "Transfer RNA: Molecular Structure, Sequence, and Properties." *Ann. Rev. Biochem.* 45:805–860.
7. Kim, S. H., F. L. Suddath, F. L. Quigley, A. McPherson, J. L. Sussman, A. H. J. Wang, N. C. Seeman, and A. Rich. 1974. "Three-Dimensional Tertiary Structure of Yeast Phenylalanine Transfer RNA." *Science* 185:435–440.
8. Robertus, J. D., J. E. Ladner, J. T. Finch, D. Rhodes, R. S. Brown, B. F. C. Clark, and A. Klug. 1974. "Structure of Yeast Phenylalanine tRNA at 3 Å Resolution." *Nature* 250:546–551.
9. Rich, A., and S. H. Kim. 1978. "The Three-Dimensional Structure of Transfer RNA." *Sci. Amer.* 238:56–62.
10. Schevitz, R. W., A. D. Podjarny, N. Krishnamachari, J. J. Hughs, P. B. Sigler, and J. L. Sussman. 1979. "Crystal Structure of a Eukaryotic Initiator tRNA." *Nature* 278:188–190.
11. Woo, W., and A. Rich. 1980. "The Three Dimensional Structure of E. coli Initiator tRNA^{Met}." *Nature* 286:346–351.
12. Moras, D., M. B. Comarmond, J. Fisher, R. Weiss, J. C. Thierry, J. P. Ebel, and R. Geige. 1980. "Crystal Structure of Yeast tRNA^{Asp}." *Nature* 288:669–674.
13. Hoagland, M. B., E. B. Keller, and P. C. Zamecnik. 1956. "Enzymatic Carboxyl Activation of Amino Acids." *J. Biol. Chem.* 218:345–358.
14. Schimmel, P. R., and D. Soll. 1979. "Aminoacyl tRNA Synthetases: General Features and Recognition of Transfer RNAs." *Ann. Rev. Biochem.* 48:601–648.
15. Igloi, G. L., and F. Cramer. 1979. "Interaction of Aminoacyl tRNA Synthetases and Their Substrates with a View to Specificity." In *Transfer RNA,* ed. S. Altman. Cambridge, Mass.: MIT Press, pp. 294–349.
16. Schimmel, P., S. Putney, and R. Starzyk. 1982. "RNA and DNA Sequence Recognition and Structure-Function of Aminoacyl tRNA Synthetases." *Trends in Biochem. Sci.* 7:209–212.
17. Baldwin, A. N., and P. Berg. 1967. "Transfer Ribonucleic Acid-Induced Hydrolysis of Valyladenylate Bound to Isoleucyl Ribonucleic Acid Synthetase." *J. Biol. Chem.* 241:839–845.
18. Blow, D. M., and P. Brick. 1985. "Aminoacyl-tRNA Synthetases." In *The Structure of Biological Macromolecules and Assemblies,* vol. 2, *Nucleic Acid Binding Proteins,* ed. F. A. Jurnak and A. McPherson. New York: Wiley, pp. 442–469.
19. Fersht, A. R., J. P. Shi, J. Knill-Jones, D. M. Lowe, A. J. Wilkerson, D. M. Blow, P. Brick, P. Carter, M. M. Y. Waye, and G. Winter. 1985. "Hydrogen Bonding and Biological Specificity Analyzed by Protein Engineering." *Nature* 314:235–238.
20. Chapeville, F., F. Lipmann, G. V. Ehrenstein, B. Weisblum, W. J. Ray, Jr., and S. Benzer. 1962. "On the Role of Soluble Ribonucleic Acid in Coding for Amino Acids." *Proc. Nat. Acad. Sci.* 48:1086–1092.
21. Hecht, S. M., and A. C. Chinault. 1976. "Position of Aminoacylation of Individual *Escherichia coli* and Yeast tRNAs." *Biochem.* 73:405–409.
22. Hershey, J. W. B. 1980. "The Translational Machinery." In *Cell Biology: A Comprehensive Treatise,* vol. 4, ed. D. M. Prescott and L. Goldstein. New York: Academic Press.
23. Nierhaus, K. H. 1982. "Structure, Assembly, and Function of Ribosomes." *Current Topics in Microbiol. and Immun.* 97:82–155.
24. Wittmann, H. G. 1983. "Architecture of Prokaryotic Ribosomes." *Ann. Rev. Biochem.* 52:35–65.
25. Wittmann, H. G. 1982. "Components of Bacterial Ribosomes." *Ann. Rev. Biochem.* 51:155–183.
26. Brenner, S., F. Jacob, and M. Meselson. 1961. "An Unstable Intermediate Carrying Information from Genes to Ribosomes for Protein Synthesis." *Nature* 190:576–581.
27. Gros, F., H. Hiatt, W. Gilbert, C. G. Kurland, R. W. Risebrough, and J. D. Watson. 1961. "Unstable Ribonucleic Acid Revealed by Pulse Labelling of *Escherichia coli.*" *Nature* 190:581–585.
28. Noller, H. F. 1984. "Structure of Ribosomal RNA." *Ann. Rev. Biochem.* 53:119–162.
29. Woese, C. R., R. Gutell, R. Gupta, and H. F. Noller. 1983. "Detailed Analysis of the Higher-Order Structure of 16S-Like Ribosomal Ribonucleic Acids." *Microbiol. Rev.* 47:621–669.
30. Brimacombe, R. 1984. "Conservation of Structure in Ribosomal RNA." *Trends in Biochem. Sci.* 9:273–277.
31. Traub, P., and M. Nomura. 1969. "Studies on the Assembly of Ribosomes *in Vitro.*" *Cold Spring Harbor Symp. Quant. Biol.* 34:63–67.
32. Held, W. A., B. Ballou, S. Mizushima, and M. Nomura. 1974. "Assembly Mapping of 30S Ribosomal Proteins from *Escherichia coli.*" *J. Biol. Chem.* 249:3103–3111.
33. Rohl, R., and K. H. Nierhaus. 1982. "Assembly Map of the Large Subunit (50S) of *Escherichia coli* Ribosomes." *Proc. Nat. Acad. Sci.* 79:729–733.
34. Altman, S., C. Guerrier-Takada, H. Frankfort, and H. Robertson. 1982. "RNA-Processing Nucleases." In *Nucleases,* ed. S. Linn and R. Roberts. Cold Spring Harbor, N.Y.: Cold Spring Harbor Laboratory, pp. 243–274.
35. Gegenheimer, P., and D. Apirion. 1981. "Processing of Procaryotic Ribonucleic Acid." *Microbiol. Rev.* 45:502–541.
36. Abelson, J. 1979. "RNA Processing and the Intervening Sequence Problem." *Ann. Rev. Biochem.* 48:1035–1069.
37. Nakajima, N., H. Ozeki, and Y. Shimura. 1981. "Organization and Structure of an *E. coli* tRNA Operon Containing Seven tRNA Genes." *Cell* 23:239–249.
38. Guerrier-Takada, C., K. Gardiner, T. Marsh, N. Pace, and

S. Altman. 1983. "The RNA Moiety of Ribonuclease P is the Catalytic Subunit of the Enzyme." *Cell* 35:849–857.

39. Deutscher, M. 1984. "Processing of tRNA in Prokaryotes and Eukaryotes." *Crit. Rev. Biochem.* 17:45–71.

40. Guthrie, C. 1980. "Folding Up a Transfer RNA Molecule Is Not Simple." *Quart. Rev. Biol.* 55:335–352.

41. Deutscher, M. P. 1985. "*E. coli* RNases: Making Sense of Alphabet Soup." *Cell* 40:731–732.

42. Lund, E., J. E. Dahlberg, L. Lindahl, R. Jaskunas, P. P. Dennis, and M. Nomura. 1976. "Transfer RNA Genes Between 16S and 23S rRNA Genes in rRNA Transcription Units of *E. coli*." *Cell* 7:165–177.

43. Schlessinger, D. 1980. "Processing of Ribosomal RNA Transcripts in Bacteria." In *Ribosomes: Structure, Function and Genetics.*, ed. G. Chambliss, G. R. Craven, J. Davies, K. Davis, L. Kahan, and M. Nomura. Baltimore, Md.: University Park Press, pp. 767–780.

44. Dunn, J. J., and F. W. Studier. 1973. "T7 Early RNAs and *Escherichia coli* Ribosomal RNAs Are Cut from Large Precursor RNAs *in Vivo* by Ribonuclease III." *Proc. Nat. Acad. Sci.* 70:3296–3300.

45. Bram, R. J., R. A. Young, and J. A. Steitz. 1980. "The Ribonuclease III Site Flanking 23S Sequences in the 30S Ribosomal Precursor RNA of *E. coli*." *Cell* 19:393–401.

46. Robertson, H. D. 1982. "*Escherichia coli* Ribonuclease III Cleavage Sites." *Cell* 30:669–672.

47. Sogin, M. L., N. R. Pace, M. Rosenberg, and S. Weissman. 1976. "Nucleotide Sequence of a 5S Ribosomal RNA Precursor from *Bacillus subtilis*." *J. Biol. Chem.* 251:3480–3488.

48. Yanofsky, C., T. Platt, I. P. Crawford, B. P. Nicholas, G. E. Christie, H. Horowitz, M. VanCleemput, and A. M. Wu. 1981. "The Complete Nucleotide Sequence of the Tryptophan Operon of *Escherichia coli*." *Nucleic Acid Res.* 9:6647–6668.

49. Dintzis, H. M. 1961. "Assembly of the Peptide Chain of Hemoglobin." *Proc. Nat. Acad. Sci.* 47:247–261.

50. Kozak, M. 1983. "Comparison of Initiation of Protein Synthesis in Procaryotes, Eucaryotes, and Organelles." *Microbiol. Rev.* 47:1–45.

51. Shine, J., and L. Dalgarno. 1974. "The 3'-Terminal Sequence of *E. coli* 16S rRNA: Complementarity to Nonsense Triplets and Ribosome Binding Sites." *Proc. Nat. Acad. Sci.* 71:1342–1346.

52. Steitz, J. A. 1980. "RNA-RNA Interactions During Polypeptide Chain Initiation." In *Ribosomes: Structure, Function and Genetics*, ed. G. Chambliss, G. R. Craven, J. Davies, K. Davis, L. Kahan, and M. Nomura. Baltimore, Md.: University Park Press, pp. 479–495.

53. Gold, L., D. Pribnow, T. Schneider, S. Shinedling, B. S. Singer, and G. Stormo. 1981. "Translational Initiation in Prokaryotes." *Ann. Rev. Microbiol.* 35:365–403.

54. Grunberg-Manago, M. 1980. "Initiation of Protein Synthesis as Seen in 1979." In *Ribosomes: Structure, Function and Genetics*, ed. G. Chambliss, G. R. Craven, J. Davies, K. Davis, L. Kahan, and M. Nomura. Baltimore, Md.: University Park Press, pp. 445–478.

55. Watson, J. D. 1964. "The Synthesis of Proteins Upon Ribosomes." *Bull. Soc. Chim. Biol.* 46:1399–1425.

56. Weissbach, H. 1980. "Soluble Factors in Protein Synthesis." In *Ribosomes: Structure, Function and Genetics*, ed. G. Chambliss, G. R. Craven, J. Davies, K. Davis, L. Kahan, and M. Nomura. Baltimore, Md.: University Park Press, pp. 377–412.

57. Garrett, R. A., and P. Woolley. 1982. "Identifying the Peptidyl Transferase Centre." *Trends in Biochem. Sci.* 7:385–386.

58. Johnson, A., H. Adkins, E. Matthews, and C. Cantor. 1982. "Distance Moved by Transfer RNA During Translocation from the A Site to the P Site on the Ribosome." *J. Mol. Biol.* 156:113–140.

59. Traut, R. R., and R. E. Monro. 1964. "The Puromycin Reaction and Its Relation to Protein Synthesis." *J. Mol. Biol.* 10:63–72.

60. Cundliffe, E. 1980. "Antibiotics and Prokaryotic Ribosomes: Action, Interaction, and Resistance." In *Ribosomes: Structure, Function and Genetics*, ed. G. Chambliss, G. R. Craven, J. Davies, K. Davis, L. Kahan, and M. Nomura. Baltimore,

Md.: University Park Press, pp. 377–412.

61. Caskey, C. T., W. C. Forrester, and W. Tate. 1984. "Peptide Chain Termination." In *Alfred Benzon Symposium*, vol. 19, ed. B. Clark and H. Petersen. Copenhagen: Munksgaard, pp. 457–466.

62. Kaziro, Y. 1978. "The Role of Guanosine 5'-Triphosphate in Polypeptide Chain Elongation." *Biochim. Biophys. Acta* 505:95–127.

63. Kurland, C. G. 1982. "Translational Accuracy *in Vitro*." *Cell* 28:201–202.

64. Yarus, M., and R. Thompson. 1983. "Precision of Protein Biosynthesis." In *Gene Function in Prokaryotes*, ed. J. Beckwith, J. Davies, J. A. Gallant. Cold Spring Harbor, N.Y.: Cold Spring Harbor Laboratory, pp. 23–63.

65. Kurland, C. G., and J. Gallant. 1985. "The Secret Life of the Ribosome." In *Accuracy in Molecular Processing; Its Control and Relevance to Living Systems*, ed. D. Galas, R. Rosenburger, T. Kirkwood. London: Chapman and Hall, 1985.

66. Gallant, J. A. 1979. "Stringent Control in *E. coli*." *Ann. Rev. Genetics* 13:393–415.

67. Nomura, M., R. Gourse, and G. Baughman. 1984. "Regulation of the Synthesis of Ribosomes and Ribosomal Components." *Ann. Rev. Biochem.* 53:75–117.

68. Lamond, A. I., and A. A. Travers. 1985. "Stringent Control of Bacterial Transcription." *Cell* 41:6–8.

69. Silhavy, T. J., S. A. Benson, and S. D. Emr. 1983. "Mechanisms of Protein Localization." *Microbiol. Rev.* 47:313–344.

70. Michaelis, S., and J. Beckwith. 1982. "Mechanism of Incorporation of Cell Envelope Proteins in *Escherichia coli*." *Ann. Rev. Microbiol.* 36:435–465.

71. Wickner, W. 1983. M13 Coat Protein as a Model of Membrane Assembly." *Trends in Biochem. Sci.* 8:90–94.

72. Randall, L. L., and S. J. S. Hardy. 1984. "Export of Protein in Bacteria: Dogma and Data." In *Modern Cell Biology*, vol. 3, ed. B. Satir. New York: Liss, pp. 1–20.

73. Davis, N. G., and P. Model. 1985. "An Artificial Anchor Domain: Hydrophobicity Suffices To Stop Transfer." *Cell* 41:607–614.

74. Engelman, D. M., and T. A. Steitz. 1981. "The Spontaneous Insertion of Proteins into and Across Membranes: The Helical Hairpin Hypothesis." *Cell* 23:411–422.

75. Lake, J. A. 1981. "The Ribosome." *Sci. Amer.* 245:84–97.

76. Engelman, D. M., and P. B. Moore. 1976. "Neutron Scattering Studies of the Ribosome." *Sci. Amer.* 235:43–54.

77. Ofengand, J. 1980. "The Topography of tRNA Binding Sites on the Ribosome." In *Ribosomes: Structure, Function and Genetics*, ed. G. Chambliss, G. R. Craven, J. Davies, K. Davis, L. Kahan, and M. Nomura. Baltimore, Md.: University Park Press, pp. 497–530.

78. Cooperman, B. S. 1980. "Functional Sites on the *E. coli* Ribosome as Defined by Affinity Labeling." In *Ribosomes: Structure, Function and Genetics*, ed. G. Chambliss, G. R. Craven, J. Davies, K. Davis, L. Kahan, and M. Nomura. Baltimore, Md.: University Park Press, pp. 531–554.

79. Traut, R. R., J. M. Lambert, G. Boileau, and J. W. Kenny. 1980. "Protein Topography of *Escherichia coli* Ribosomal Subunits as Inferred from Protein Crosslinking." In *Ribosomes: Structure, Function and Genetics*, ed. G. Chambliss, G. R. Craven, J. Davies, K. Davis, L. Kahan, and M. Nomura. Baltimore, Md.: University Park Press, pp. 89–110.

80. Stoffler, G., R. Bald, B. Kastner, R. Luhrmann, M. Stoffler-Meilicke, and G. Tischendorf. 1980. In *Ribosomes: Structure, Function and Genetics*, ed. G. Chambliss, G. R. Craven, J. Davies, K. Davis, L. Kahan, and M. Nomura. Baltimore, Md.: University Park Press, pp. 171–206.

81. Cantor, C. R., and P. R. Schimmel. 1980. *Biophysical Chemistry*, Part II. San Francisco: Freeman, pp. 448–454.

82. Zimmermann, R. A. 1980. "Interactions Among Protein and RNA Components of the Ribosome." In *Ribosomes: Structure, Function and Genetics*, ed. G. Chambliss, G. R. Craven, J. Davies, K. Davis, L. Kahan, and M. Nomura. Baltimore, Md.: University Park Press, pp. 135–170.

CHAPTER 15

The Genetic Code

By 1960, the general outline of how RNA participates in protein synthesis had been established. Nevertheless, there was little optimism that we would soon have a detailed understanding of the genetic code itself. It was believed that identification of the codons for a given amino acid would require exact knowledge of both the nucleotide sequences of a gene and the corresponding amino acid order in its protein product. At that time, the elucidation of amino acid sequences, although a laborious process, was already a very practical one. On the other hand, the then-current methods for determining DNA sequences were very primitive. Fortunately, this apparent roadblock did not hold up progress. In 1961, just one year after the discovery of mRNA, the use of artificial messenger RNAs partially cracked the genetic code, with the unambiguous demonstration of a codon for the amino acid phenylalanine. To explain how this discovery was made, we must first describe how biochemists study protein synthesis in cell-free systems.

Addition of mRNA Stimulates in Vitro Protein Synthesis[1]

The need for three forms of RNA—tRNA, rRNA, and mRNA—for protein synthesis was demonstrated largely by experiments using cell-free extracts prepared from cells actively engaged in protein synthesis. In these experiments, carefully disrupted cells were fractionated to see which cell components were necessary for incorporation of amino acids into proteins. All these experiments utilized radioactively labeled (^{3}H, ^{14}C, or ^{35}S) amino acids because in vitro protein synthesis is quite inefficient; that is, there is no detectable net synthesis compared to the amount of protein present in the extract. Only by using labeled precursors can the incorporation of amino acids into new proteins be convincingly demonstrated (Figure 15-1).

A typical time course of the in vitro incorporation of radioactive amino acids into proteins is illustrated by the graph in Figure 15-2. It shows that synthesis in an *E. coli* extract proceeds rapidly for several minutes and then gradually stops. During this interval, there is a corresponding loss of mRNA, owing to the action of degradative enzymes present in the extract. This suggests that one cause of the inefficiency of cell-free protein synthesis is loss of the template component. The correctness of this supposition is shown by adding new mRNA to extracts that have stopped making protein. Such an addition causes an immediate resumption of synthesis. Extracts depleted of mRNA are very valuable in testing mRNA activity. Because they contain very little functional mRNA, they can be used to detect small amounts of template activity in externally added mRNA.

Figure 15-1
Experimental details of in vitro studies of protein synthesis.

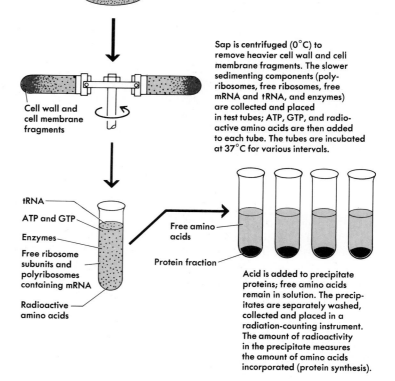

Rapidly growing *E. coli* cells are collected by centrifugation in the cold (0°C) and broken open to yield a cell sap. The enzyme deoxyribonuclease is added to break down the cellular DNA.
Sap is centrifuged (0°C) to remove heavier cell wall and cell membrane fragments. The slower sedimenting components (polyribosomes, free ribosomes, free mRNA and tRNA, and enzymes) are collected and placed in test tubes; ATP, GTP, and radioactive amino acids are then added to each tube. The tubes are incubated at 37°C for various intervals.
Cell wall and cell membrane fragments
tRNA
ATP and GTP
Enzymes
Free ribosome subunits and polyribosomes containing mRNA
Radioactive amino acids
Free amino acids
Protein fraction
Acid is added to precipitate proteins; free amino acids remain in solution. The precipitates are separately washed, collected and placed in a radiation-counting instrument. The amount of radioactivity in the precipitate measures the amount of amino acids incorporated (protein synthesis).

Amount of amino acids incorporated
Amount of mRNA
A
B
0
20
40
60
Time (mins.)

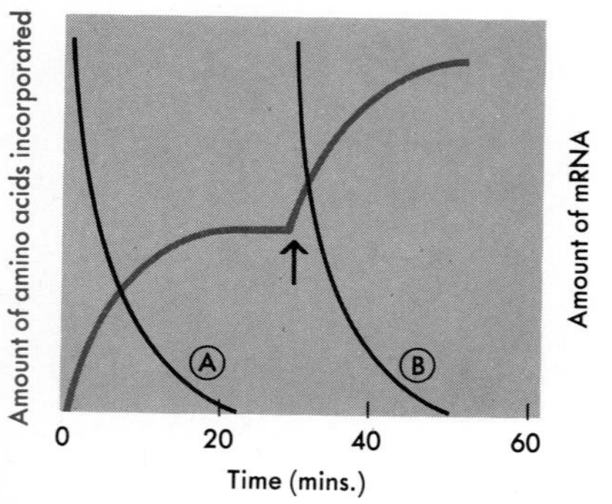

Figure 15-2
Ⓐ indicates the breakdown of endogenous mRNA, while Ⓑ shows the breakdown of exogenously added mRNA whose time of addition is indicated by the arrow.

Specific Proteins Can Be Made in Cell-Free Systems[2,3,4]

Even with the addition of excess mRNA, there is still only a small amount of amino acid incorporation into polypeptide chains in the in vitro systems studied. Therefore, doubts were initially raised about whether the newly made proteins had structures at all similar to those of natural proteins. Fortunately, the use of specific viral RNA showed that these doubts were unfounded and that the genetic code can be accurately read in cell-free systems. In the first successful experiments, which took place in 1962, RNA isolated from the bacterial virus f2 was added to preincubated *E. coli* extracts. The RNA acted as a template and promoted the incorporation of amino acids into protein. Some complete polypeptide chains were made and released from the ribosomes. These newly made protein products were then compared with the f2 coat protein (MW about 14,000) and were found to have amino acid sequences identical to that of the coat protein synthesized in vivo. Thus, under cell-free conditions, mRNA molecules on the ribosomal surface can select the appropriate AA~tRNA precursors.

Over the ensuing decade, the cell-free synthesis of a variety of other viral and bacterial proteins was achieved. One of the earliest of these successes was that of the T4 lysozyme, a 20,000-dalton enzyme that helps digest bacterial cell walls. Its cell-free synthesis in enzymatically active form further showed the great fidelity of translation possible in vitro. Although at first there was some difficulty in making very long polypeptide chains, by 1969 the synthesis of enzymatically active β-galactosidase (a protein of 1300 amino acids) was convincingly demonstrated. Subsequently, a number of important mammalian proteins (e.g., hemoglobin, light and heavy immunoglobulin chains, actin, collagen, and myosin) have been made in the test tube, proving that the conditions for in vitro synthesis of higher cell proteins are now as well understood as those of their bacterial counterparts.

Stimulation of Amino Acid Incorporation by Synthetic mRNA[5]

While the foregoing experiments with natural mRNAs told us that reading specificity is preserved in cell-free systems, by themselves they said nothing about the exact form of the genetic code. They could not tell us which three-letter words (out of the possible 64) code for any particular amino acid. To obtain such data, use was made of synthetic polyribonucleotides, whose formation involves the enzyme polynucleotide phosphorylase. This enzyme, found in all bacteria, catalyzes the reaction

$$\text{RNA} + \textcircled{P} \rightleftharpoons \text{ribonucleoside—}\textcircled{P}\sim\textcircled{P} \qquad (15\text{-}1)$$

With normal cell metabolite concentrations, the equilibrium conditions favor RNA degradation to nucleoside diphosphates, so that the main cellular function of polynucleotide phosphorylase may be to control mRNA lifetime (Chapter 16). By use of high initial nucleoside diphosphate concentrations, however, this enzyme can be made to catalyze the formation of internucleotide 3'–5' phosphodiester bonds and thus make RNA molecules (Figure 15-3). No template DNA or RNA is required for RNA synthesis with this enzyme; the base composition of the synthetic product depends entirely on the ratio of the various ribonucleoside diphosphates added to the reaction mixture. For example, when only adenosine diphosphate is used, the resulting RNA contains only adenylic acid and is thus called polyadenylic acid or poly A. It is likewise possible to make poly U, poly C, and poly G. Addition of two or more different diphosphates produces mixed copolymers such as poly AU, poly AC, poly CU, and poly AGCU. In all these mixed polymers, the base sequences are approximately random, with the nearest-neighbor frequencies determined solely by the relative concentrations of the reactants. For example, poly AU molecules with two times as much A as U have sequences like UAAUAUAAAUAAUAAAAUAUU. . . .

Figure 15-3
Synthesis (degradation) of RNA molecules using the enzyme polynucleotide phosphorylase.

Poly U Codes for Polyphenylalanine[6]

Under the right conditions in vitro, almost all synthetic polymers will attach to ribosomes and function as templates. Luckily, high concentrations of magnesium were used in the early experiments. A high magnesium concentration circumvents the need for initiation factors and the special initiator fMet-tRNA$_F^{Met}$, allowing chain initiation to take place without the proper signals in the mRNA (Chapter 14).

Poly U was the first synthetic polyribonucleotide discovered to have mRNA activity. It selects phenylalanyl tRNA molecules exclusively, thereby forming a polypeptide chain containing only phenylalanine (polyphenylalanine). Thus, we know that a codon for phenylalanine is composed of a group of three uridylic acid residues, UUU. (That a codon has three nucleotides was known from the genetic experiments described in Chapter 8.)

On the basis of analogous experiments with poly C and poly A, CCC was assigned as a proline codon and AAA as a lysine codon. Unfortunately, this type of experiment did not tell us what amino acid GGG specifies. The guanine residues in poly G firmly hydrogen-bond to each other and form multistranded triple helices that do not bind to ribosomes.

Mixed Copolymers Allowed Additional Codon Assignments[7]

Poly AC molecules can contain eight different codons, CCC, CCA, CAC, ACC, CAA, ACA, AAC, and AAA, whose proportions vary with the copolymer A/C ratio. When AC copolymers attach to ribosomes, they cause the incorporation of asparagine, glutamine, histidine, and threonine, in addition to the proline previously assigned to CCC codons and the lysine previously assigned to AAA codons. The proportions of these amino acids incorporated into polypeptide products depend on the A/C ratio. Thus, since an AC copolymer containing much more A than C promotes the incorporation of many more asparagine than histidine residues, we conclude that asparagine is coded by two As and one C and that histidine is coded by two Cs and one A (Table 15-1). Similar experiments with other copolymers allowed a number of additional assignments. Such experiments, however, did not reveal the order of the different nucleotides within a codon. There is no way of knowing from random copolymers whether the histidine codon containing two Cs and one A is ordered CCA, CAC, or ACC.

Transfer RNA Binding to Defined Trinucleotide Codons[8]

A direct way of ordering the nucleotides within some of the codons was developed in 1964. This method utilized the fact that even in the absence of all the factors required for protein synthesis, specific aminoacyl-tRNA molecules can bind to ribosome-mRNA complexes. For example, when poly U is mixed with ribosomes, only phenylalanyl-tRNA will attach. Correspondingly, poly C promotes the binding of prolyl-tRNA. Most important, this specific binding does not demand

Table 15-1 Amino Acid Incorporation into Proteins*

Amino Acid	Observed Amino Acid Incorporation	Tentative Codon Assignments	Calculated Triplet Frequency				Sum of Calculated Triplet Frequencies
			3A	2A1C	1A2C	3C	
Poly AC (5:1)							
Asparagine	24	2A1C		20			20
Glutamine	24	2A1C		20			20
Histidine	6	1A2C			4.0		4
Lysine	100	3A	100				100
Proline	7	1A2C, 3C			4.0	0.8	4.8
Threonine	26	2A1C, 1A2C		20	4.0		24
Poly AC (1:5)							
Asparagine	5	2A1C		3.3			3.3
Glutamine	5	2A1C		3.3			3.3
Histidine	23	1A2C			16.7		16.7
Lysine	1	3A	0.7				0.7
Proline	100	1A2C, 3C			16.7	83.3	100
Threonine	21	2A1C, 1A2C		3.3	16.7		20

*The amino acid incorporation into proteins was observed after adding random copolymers of A and C to a cell-free extract similar to that described in Figure 15-1. The incorporation is given as a percentage of the maximal incorporation of a single amino acid. The copolymer ratio was then used to calculate the frequency with which a given codon would appear in the polynucleotide product. The relative frequencies of the codons are a function of the probability that a particular nucleotide will occur in a given position of a codon. For example, when the A/C ratio is 5:1, the ratio of AAA/AAC = $5 \times 5 \times 5 : 5 \times 5 \times 1 = 125:25$. If we thus assign to the 3A codon a frequency of 100, then the 2A and 1C codon is assigned a frequency of 20. By correlating the relative frequencies of amino acid incorporation with the calculated frequencies with which given codons appear, tentative codon assignments can be made.

the presence of long mRNA molecules. In fact, the binding of a *trinu-cleotide* to a ribosome is sufficient. The addition of the trinucleotide UUU results in phenylalanyl-tRNA attachment, whereas if AAA is added, lysyl-tRNA specifically binds to ribosomes. The discovery of this trinucleotide effect provided a relatively easy way of determining the order of nucleotides within many codons. For example, the trinu-cleotide 5'-GUU-3' promotes valyl-tRNA binding, 5'-UGU-3' stimu-lates cysteinyl-tRNA binding, and 5'-UUG-3' causes leucyl-tRNA binding (Table 15-2). Although all 64 possible trinucleotides were syn-

Table 15-2 Binding of Aminoacyl tRNA Molecules to Trinucleotide-Ribosome Complexes

Trinucleotide						AA~tRNA Bound
5'-UUU-3'	UUC					Phenylalanine
UUA	UUG	CUU	CUC	CUA	CUG	Leucine
AAU	AUC	AUA				Isoleucine
AUG						Methionine
GUU	GUC	GUA	GUG	UCU*		Valine
UCU	UCC	UCA	UCG			Serine
CCU	CCC	CCA	CCG			Proline
AAA	AAG					Lysine
UGU	UGC					Cysteine
GAA	GAG					Glutamic acid

*Note that this codon was misassigned by this method.

Figure 15-4
Using a combination of organic synthesis and copying by DNA polymerase I, double-stranded DNA with simple repeating sequences can be generated. RNA polymerase will then synthesize long polyribonucleotides corresponding to one or the other DNA strand, depending on the choice of ribonucleoside triphosphates added to the reaction mixture.

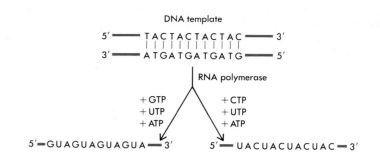

thesized with the hope of definitely assigning the order of every codon, not all codons were determined in this way. Some trinucleotides bind to ribosomes much less efficiently than UUU or GUU, making it impossible to know whether they code for a specific amino acid.

Codon Assignments from Repeating Copolymers[9]

At the same time that the trinucleotide binding technique became available, a combination of organic chemical and enzymatic techniques were being used to prepare synthetic polyribonucleotides with known repeating sequences (Figure 15-4). Ribosomes start protein synthesis at random points along these regular copolymers; yet they incorporate specific amino acids into polypeptides. For example, the repeating sequence CUCUCUCU . . . is the messenger for a regular polypeptide in which leucine and serine alternate. Similarly, UGUGUG . . . promotes the synthesis of a polypeptide containing two amino acids, cysteine and valine. And ACACAC . . . directs the synthesis of a polypeptide alternating threonine and histidine (Table 15-3). Use of the copolymer built up from repetition of the three-nucleotide sequence AAG (AAGAAGAAG) directs the synthesis of three types of polypeptides: polylysine, polyarginine, and

Table 15-3   Assignment of Codons Using Repeating Copolymers Built from Two or Three Nucleotides

Copolymer	Codons Recognized	Amino Acids Incorporated or Polypeptide Made	Codon Assignment
$(CU)_n$	CUC|UCU|CUC . . .	Leucine Serine	5'-CUC-3' UCU
$(UG)_n$	UGU|GUG|UGU . . .	Cysteine Valine	UGU GUG
$(AC)_n$	ACA|CAC|ACA . . .	Threonine Histidine	ACA CAC
$(AG)_n$	AGA|GAG|AGA . . .	Arginine Glutamine	AGA GAG
$(AUC)_n$	AUC|AUC|AUC . . . UCA|UCA|UCA . . . CAU|CAU|CAU . . .	Polyisoleucine Polyserine Polyhistidine	5'-AUC-3' UCA CAU

Table 15-4   The Genetic Code

		Second Position					
		U	C	A	G		
First Position (5′ End)	U	UUU ⎤ UUC ⎦ Phe UUA ⎤ UUG ⎦ Leu	UCU ⎤ UCC ⎥ UCA ⎥ Ser UCG ⎦	UAU ⎤ UAC ⎦ Tyr UAA* *Stop* UAG* *Stop*	UGU ⎤ UGC ⎦ Cys UGA* *Stop* UGG Trp	U C A G	Third Position (3′ End)
	C	CUU ⎤ CUC ⎥ CUA ⎥ Leu CUG ⎦	CCU ⎤ CCC ⎥ CCA ⎥ Pro CCG ⎦	CAU ⎤ CAC ⎦ His CAA ⎤ CAG ⎦ Gln	CGU ⎤ CGC ⎥ CGA ⎥ Arg CGG ⎦	U C A G	
	A	AUU ⎤ AUC ⎥ Ile AUA ⎦ AUG† Met	ACU ⎤ ACC ⎥ ACA ⎥ Thr ACG ⎦	AAU ⎤ AAC ⎦ Asn AAA ⎤ AAG ⎦ Lys	AGU ⎤ AGC ⎦ Ser AGA ⎤ AGG ⎦ Arg	U C A G	
	G	GUU ⎤ GUC ⎥ Val GUA ⎥ GUG† ⎦	GCU ⎤ GCC ⎥ GCA ⎥ Ala GCG ⎦	GAU ⎤ GAC ⎦ Asp GAA ⎤ GAG ⎦ Glu	GGU ⎤ GGC ⎥ GGA ⎥ Gly GGG ⎦	U C A G	

*Chain-terminating, or "nonsense," codons.

†Also used to specify the initiator formyl-Met-tRNA$_F^{Met}$. The Val triplet GUG is therefore "ambiguous" in that it codes both valine and methionine.

polyglutamic acid. Poly (AUC)$_n$ behaves the same way, being a template for polyisoleucine, polyserine, and polyhistidine (Table 15-3). Further codon assignments were obtained from repeating tetranucleotide sequences. The sum of all these observations permitted the definite assignments of specific amino acids to 61 out of the possible 64 codons (Table 15-4). The remaining three codons, as shown in Table 15-4, code for chain termination.

The Code Is Degenerate

Many amino acids are specified by more than one codon, a phenomenon called **degeneracy.** For example, both UUU and UUC code for phenylalanine, while serine is encoded by UCU, UCC, UCA, UCG, AGU, and AGC. In fact, when the first two nucleotides are identical, the third nucleotide can be either cytosine or uracil and the codon will still code for the same amino acid. Often, adenine and guanine are similarly interchangeable. However, not all degeneracy is based on equivalence of the first two nucleotides. Leucine, for example, is coded by UUA and UUG, as well as by CUU, CUC, CUA, and CUG (Figure 15-5). Codon degeneracy, especially the frequent third-place equivalence of cytosine and uracil or guanine and adenine, explains how there can be great variation in the AT/GC ratios in the DNA of various organisms (Chapter 9) without correspondingly large changes in the relative proportion of amino acids in their proteins.

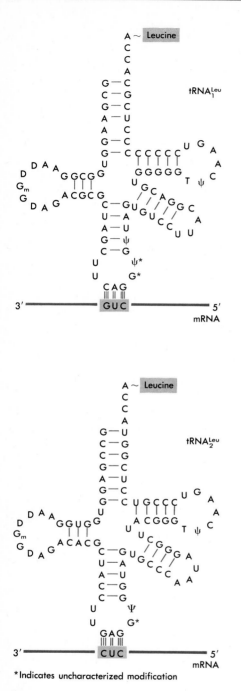

Figure 15-5
Two different *E. coli* tRNA molecules that accept leucine residues. They recognize different code words.

*Indicates uncharacterized modification

Perceiving Order in the Makeup of the Code[10]

Inspection of the distribution of codons in the genetic code suggests that the code evolved in such a way as to minimize the deleterious effects of mutations. For instance, mutations in the first position of a code word will usually give a similar (if not the same) amino acid. Furthermore, code words with pyrimidines in the second position specify mostly hydrophobic amino acids, while those with purines in the second position correspond mostly to polar amino acids (see Table 15-4). Hence, since transitions are the most common type of point mutations (Chapter 11), a change in the second position of a code word will usually replace one amino acid with a very similar one. Finally, if a code word suffers a transition mutation in the third position, almost never will a different amino acid be specified. Even a transversion mutation in this position will have no consequence about half the time.

Another consistency noticeable in the code is that whenever the first two positions of a codon are both occupied by G or C, each of the four nucleotides in the third position specifies the same amino acid (e.g., proline, alanine, arginine, or glycine). On the other hand, whenever the first two positions of the codon are both occupied by A or U, the identity of the third nucleotide does make a difference. Since GC base pairs are stronger than AU base pairs, mismatches in pairing the third codon base are often tolerated if the first two positions make strong GC base pairs. Thus, having all four nucleotides in the third position specify the same amino acid may have evolved as a safety mechanism to minimize errors in the reading of such codons.

Wobble in the Anticodon[11]

It was first proposed that a specific tRNA anticodon would exist for every codon. If that were the case, at least 61 different tRNAs, possibly with an additional three for the chain-terminating codons, would be present. Evidence began to appear, however, that highly purified tRNA species of known sequence (e.g., alanyl-tRNA^{Ala}) could recognize several different codons. Cases were also discovered in which an anticodon base was not one of the four regular ones, but a fifth base, inosine. Like all the other minor tRNA bases, inosine arises through enzymatic modification of a base present in an otherwise completed tRNA chain. The base from which it is derived is adenine, whose carbon 6 is deaminated to give the 6-keto group of inosine.

To explain these observations, the **wobble concept** was devised in 1966. It states that the base at the 5' end of the anticodon is not as spatially confined as the other two, allowing it to form hydrogen bonds with any of several bases located at the 3' end of a codon. Not all combinations are possible, with pairing restricted to those shown in Table 15-5. For example, U at the wobble position can pair with either adenine or guanine, while I can pair with U, C, or A (Figure 15-6). The pairings permitted by the wobble rules are those that give ribose-ribose distances close to that of the standard AU or GC base pairs. Purine-purine or pyrimidine-pyrimidine pairs would give ribose-ribose distances that are too long or too short, respectively.

The wobble rules do not permit any single tRNA molecule to recognize four different codons. Three codons can be recognized only when inosine occupies the first (5') position of the anticodon.

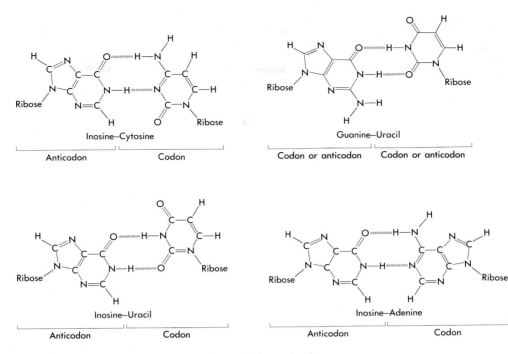

(a) "Wobble" enables the anticodon base to form hydrogen bonds with bases other than those in standard base pairs.

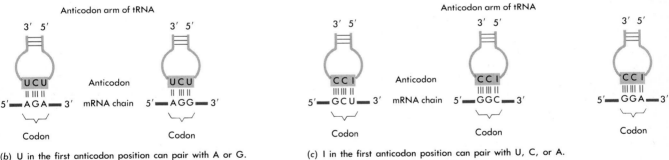

(b) U in the first anticodon position can pair with A or G.

(c) I in the first anticodon position can pair with U, C, or A.

Figure 15-6
Examples of wobble pairing. Note that the ribose-ribose distances for all the wobble pairs are close to that of the standard AU or GC base pair.

Almost all the evidence gathered since 1966 supports the wobble concept. For example, the concept correctly predicted that at least three tRNAs exist for the six serine code words (UCU, UCC, UCA, UCG, AGU, and AGC). The other two amino acids (leucine and arginine) that are encoded by six codons also have different tRNAs for the sets of codons that differ in the first or second position.

In the recently established three-dimensional tRNA structures, the three anticodon bases, as well as the two following (3′) bases in the anticodon loop, all point in roughly the same direction, with their exact conformations largely determined by stacking interactions between the flat surfaces of the bases (see Figures 14-6 and 14-7). Thus, the first (5′) anticodon base is at the end of the stack and is perhaps less restricted in its movements than the other two anticodon bases—hence, wobble in the third (3′) position of the codon. By contrast, not only does the third (3′) anticodon base appear in the middle of the

Table 15-5 Pairing Combinations with the Wobble Concept

Base in Anticodon	Base in Codon
G	U or C
C	G
A	U
U	A or G
I	A, U, or C

Table 15-6 Codon Usage Observed for Some *E. coli* Ribosomal Proteins*

Second Position

First Position (5' End)		U	C	A	G	Third Position (3' End)
U		10 UUU ⎤ Phe 23 UUC ⎦ 1 UUA ⎤ Leu 2 UUG ⎦	18 UCU ⎤ 18 UCC ⎥ Ser 1 UCA ⎥ 1 UCG ⎦	3 UAU ⎤ Tyr 13 UAC ⎦ UAA *Stop* UAG *Stop*	1 UGU ⎤ Cys 6 UGC ⎦ UGA *Stop* 3 UGG Trp	U C A G
C		4 CUU ⎤ 3 CUC ⎥ Leu 0 CUA ⎥ 79 CUG ⎦	3 CCU ⎤ 0 CCC ⎥ Pro 4 CCA ⎥ 36 CCG ⎦	3 CAU ⎤ His 15 CAC ⎦ 9 CAA ⎤ Gln 33 CAG ⎦	48 CGU ⎤ 26 CGC ⎥ Arg 0 CGA ⎥ 0 CGG ⎦	U C A G
A		13 AUU ⎤ 51 AUC ⎥ Ile 0 AUA ⎦ 30 AUG Met	36 ACU ⎤ 26 ACC ⎥ Thr 3 ACA ⎥ 0 ACG ⎦	3 AAU ⎤ Asn 42 AAC ⎦ 90 AAA ⎤ Lys 24 AAG ⎦	1 AGU ⎤ Ser 12 AGC ⎦ 1 AGA ⎤ Arg 0 AGG ⎦	U C A G
G		54 GUU ⎤ 6 GUC ⎥ Val 40 GUA ⎥ 16 GUG ⎦	93 GCU ⎤ 10 GCC ⎥ Ala 45 GCA ⎥ 28 GCG ⎦	17 GAU ⎤ Asp 45 GAC ⎦ 61 GAA ⎤ Glu 16 GAG ⎦	49 GGU ⎤ 34 GGC ⎥ Gly 0 GGA ⎥ 0 GGG ⎦	U C A G

*1209 codons are tabulated.
SOURCE: T. Ikemura, *J. Mol. Biol.* 158 (1982):573.

stack, but the adjacent base is always a bulky modified purine residue (called H in Figure 14-7). Thus, restriction of its movements may explain why wobble is not seen in the first (5') position of the code.

Transfer RNAs Are Synthesized in Varying Amounts[12, 13]

Sometimes, several tRNA molecules with distinctly different sequences have the same anticodon. Often, one of these species is present in very large amounts compared to the other species (as much as 11 times greater). In *E. coli*, for example, there are a major tyrosine tRNA and a minor tyrosine tRNA, both with the anticodon 3'-AUG-5'. Origin of the minor species by enzymatic modification of the major species following gene transcription is ruled out by the nature of the sequence differences. Rather, the two tRNAs are encoded by different genes, as is the case for the other major-minor tRNA pairs. Different promoter strengths or differences in the survival of various processing intermediates account for the different levels of mature tRNA produced. The normal function played by minor tRNA species is not yet known. Conceivably these tRNAs play regulatory roles, explaining why only small numbers of molecules need to be present. And as we will soon see, minor tRNAs are frequently involved in suppressor gene action.

Table 15-7 Codon Usage in the Genes of Animals*

		Second Position					
		U	C	A	G		
First Position (5' End)	U	13 UUU ⎤ Phe 28 UUC ⎦ 2 UUA ⎤ Leu 9 UUG ⎦	16 UCU ⎤ 18 UCC ⎥ Ser 9 UCA ⎥ 2 UCG ⎦	10 UAU ⎤ Tyr 23 UAC ⎦ UAA *Stop* UAG *Stop*	10 UGU ⎤ Cys 13 UGC ⎦ UGA *Stop* 12 UGG Trp	U C A G	Third Position (3' End)
	C	9 CUU ⎤ 27 CUC ⎥ Leu 7 CUA ⎥ 47 CUG ⎦	14 CCU ⎤ 17 CCC ⎥ Pro 10 CCA ⎥ 5 CCG ⎦	10 CAU ⎤ His 21 CAC ⎦ 10 CAA ⎤ Gln 28 CAG ⎦	8 CGU ⎤ 11 CGC ⎥ Arg 4 CGA ⎥ 5 CGG ⎦	U C A G	
	A	11 AUU ⎤ 24 AUC ⎥ Ile 4 AUA ⎦ 16 AUG Met	15 ACU ⎤ 28 ACC ⎥ Thr 11 ACA ⎥ 6 ACG ⎦	8 AAU ⎤ Asn 28 AAC ⎦ 19 AAA ⎤ Lys 49 AAG ⎦	12 AGU ⎤ Ser 21 AGC ⎦ 8 AGA ⎤ Arg 10 AGG ⎦	U C A G	
	G	9 GUU ⎤ 21 GUC ⎥ Val 5 GUA ⎥ 33 GUG ⎦	28 GCU ⎤ 38 GCC ⎥ Ala 14 GCA ⎥ 6 GCG ⎦	16 GAU ⎤ Asp 24 GAC ⎦ 21 GAA ⎤ Glu 34 GAG ⎦	22 GGU ⎤ 32 GGC ⎥ Gly 16 GGA ⎥ 11 GGG ⎦	U C A G	

*2244 codons are tabulated.
SOURCE: R. Grantham, C. Gautier, M. Gouy, M. Jacobzone, and R. Mercier, "Codon Catalog Usage Is a Genome Strategy Modulated for Gene Expressivity," *Nucleic Acid Res.* 9 (1981):r43–r74.

Correlation of tRNA Abundance with Codon Usage in Natural mRNAs[13, 14]

The fact that most amino acids are specified by more than one codon poses the question of whether the alternative codons are used with approximately equal frequencies or whether some are preferentially employed. For example, in organisms containing AT-rich DNA, we might expect that most codons have U or A at the third position. In fact, the complete sequence of the single-stranded DNA genome of bacteriophage ϕX174, which consists of 24 percent A, 22 percent C, 23 percent G, and 31 percent T, does reveal that U is the preferred base utilized in the third position of codons.

Even in organisms where there is no bias in base composition, codon usage is not random. Now that the sequence of many *E. coli* genes are known, it has become apparent that some codons are utilized repeatedly, while others almost never appear. This is illustrated in Table 15-6, which summarizes the codon frequency in a number of *E. coli* ribosomal protein genes. In the genes of higher organisms, codons are likewise used nonrandomly, but those that are favored or discriminated against are not necessarily the same as in *E. coli* (Table 15-7).

Careful analysis has revealed that the occurrence of codons in *E. coli* mRNAs is strongly correlated with the relative abundance of

Figure 15-7
Correlation between tRNA amount and the frequency of tRNA usage in genes expressed in *E. coli.* The amounts of 26 different *E. coli* tRNA species were measured relative to the amount of $tRNA_1^{Leu}$, arbitrarily set at 1.0. Based on the known anticodon sequences of these tRNAs, the frequency of appearance of the codons they decode in several sequenced genes (*recA*, β-lactamase) or sets of genes (ribosomal proteins, φX174) was calculated. Note that there is a very good correspondence between tRNA abundance and usage for the highly expressed genes-(*recA* and ribosomal protein genes) and a much more scattered distribution for the transposon gene (β-lactamase) and the phage genes (φX174). If every tRNA species were used equally, a horizontal distribution would be observed. [After T. Ikemura, *J. Mol. Biol.* 146 (1981):1.]

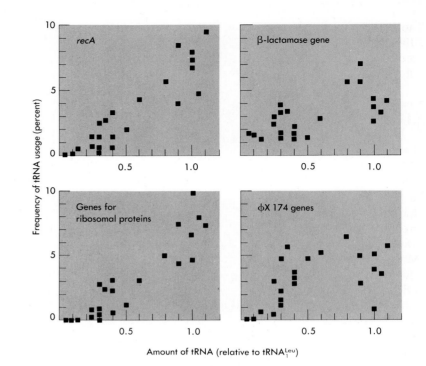

their respective tRNA species (Figure 15-7). This is especially true of the mRNAs for proteins that are synthesized in very large amounts (for instance, RecA protein and ribosomal proteins). Such mRNAs contain codons read by the most abundant isoacceptor tRNAs and lack codons corresponding to minor tRNA species. The same correlation is seen, but to a lesser degree, with other *E. coli* proteins, like those encoded by transposons or phage genomes. In the cells of higher organisms also, codons corresponding to the major tRNA isoacceptors are almost always used with the highest frequency. Two especially good examples are the abundant yeast proteins alcohol dehydrogenase isozyme I and glyceraldehyde phosphate dehydrogenase, for which more than 96 percent of the amino acids are coded by only 25 triplets, whose corresponding tRNAs are highly abundant. Hence, in strongly expressed genes, codon choice is constrained by tRNA availability and seems to be designed to optimize translational efficiency. This hints that codons corresponding to rare tRNAs may in fact slow down translation.

Even when a single tRNA can wobble to read several different codons, one codon is usually preferred. Analysis of such preferences has led to several interesting realizations regarding codon-anticodon interaction:

- Certain modifications of U in the 5' (wobble) position of the anticodon give a preference for A over G in the third position of the codon.

- I at the 5' (wobble) position of the anticodon preferentially pairs with U or C, rather than with A.

- When the first two letters of a codon form AU base pairs with the anticodon, the third base is preferentially C (giving a GC base pair) rather than U (giving a third AU pair). In other words, the more stable codon-anticodon interaction is preferred.

Knowing these rules in addition to the relative abundances of the different types of tRNA in any particular cell now permits fairly good prediction of codon usage in the mRNAs frequently translated in that cell.

AUG and GUG as Initiation Codons[15]

The discovery of a tRNA specific for *N*-formylmethionine (tRNA$_F^{Met}$) initially suggested that it would recognize a codon different from the AUG methionine codon. But as shown in Figure 14-26, elucidation of the tRNA$_F^{Met}$ sequence revealed that its anticodon 3'-UAC-5' was identical to that of tRNA$_M^{Met}$. Both fMet and Met must therefore be directly coded by AUG. Discrimination between the two forms occurs through their differential binding by protein synthesis factors. Only fMet-tRNA$_F^{Met}$ can bind to the initiation factor IF2 to form the 30S initiation complex; and only Met-tRNA$_M^{Met}$ is able to bind to the elongation factor EF-Tu during polypeptide chain elongation.

A further complication is the realization, coming first from in vitro studies and later from mRNA sequences, that fMet-tRNA$_F^{Met}$ can bind to and initiate synthesis at GUG as well as AUG codons. GUG is used as an initiator codon only about one-thirtieth as frequently as AUG initiators in *E. coli*. Normally, GUG is a valine codon and its recognition by tRNA$_F^{Met}$ would require an unusual sort of wobble in the codon-anticodon interaction. The ambiguity is at the first (5') as opposed to the third (3') position of the codon and would therefore involve the 3' base of the anticodon. A possible explanation for the unexpected ability of the initiator tRNA to engage in first-position wobble lies in the tRNA$_F^{Met}$ sequence. The nucleotide adjacent to the 3' end of the anticodon is an unmodified adenine, not the bulky alkylated derivative found in virtually all other tRNAs (H in Figure 14-2). In fact, UUG and CUG codons also specify initiation, although even more rarely than GUG. Thus, the unique wobble capability of the third (3') position of the tRNA$_F^{Met}$ anticodon seems to extend to all possible partners.

Three Different Codons Direct Chain Termination[16]

The three codons UAA, UAG, and UGA do not correspond to any amino acid. Instead, they signify chain termination. As mentioned in Chapter 14, these codons are read not by special chain-terminating tRNAs but by specific proteins, the release factors (see Chapter 14). Two release factors, RF1 and RF2, have been identified, each of which recognizes two codons. One is specific for UAA and UAG and the other for UAA and UGA. Knowledge of how two different codons can be recognized by each release factor must await detailed knowledge of the release factors' three-dimensional structures. The use of proteins to read the stop signals emphasizes the fact that not only polynucleotides can specifically interact with other polynucleotides. The specific hydrogen-bond-forming groups of the bases can also be recognized by proteins containing specific constellations of amino acids that have groups prone to hydrogen bonding.

Why three different stop codons exist is not at all obvious, and initially much speculation arose as to which codon would be prefer-

Figure 15-8
Nucleotide sequences at the 3′ ends of coding regions of mRNAs translated in *E. coli*. The stop codons are indicated by color boxes. Tandem stop codons do occur, but are rare. Likewise, when many mRNAs are compared, the average distance from the terminator to the next in-frame stop codon is about as expected on a random basis. [After J. Kohli and H. Grosjean, *Mol. Gen. Genetics* 182 (1981):430.]

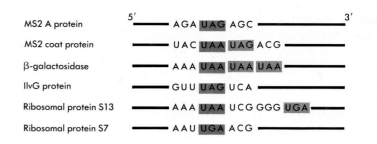

entially used to end polypeptide coding regions. Now that the nucleotide sequences at the ends of a number of *E. coli* genes have been elucidated, it is clear that UAA is the preferred, but by no means exclusive, signal. In most cases, only a single stop codon appears. But some genes end with two or even three successive stop signals (Figure 15-8). The presence of more than one stop codon may be a precaution against the rare case in which the first codon fails; but why this device is used only occasionally is unclear. For example, in the RNA phages, homologous coat protein genes terminate with either one stop codon (in phage Qβ) or two stop codons (in phages R17 and MS2).

Nonsense Versus Missense Mutations[17, 18]

An alteration that changes a codon specific for one amino acid to a codon specific for another amino acid is called a **missense mutation**. The change to a chain-termination codon is known as a **nonsense mutation**. Given the existence of only three chain-termination codons, most mutations involving single-base replacements (**point mutations**) are likely to result in missense rather than nonsense. Since a new protein arising by missense mutation contains only a single amino acid replacement, it frequently possesses some of the biological activity of the original protein. Often, missense proteins fail to function only at higher than normal temperatures and are therefore known as **temperature-sensitive mutations** (Chapter 7). Many of the abnormal hemoglobins (see Figure 3-12) are the result of missense mutations. Amino acid replacement data obtained from these changed hemoglobin molecules strongly support the idea that these mutations result from the substitution of single nucleotides.

Nonsense Mutations Produce Incomplete Polypeptide Chains[17, 18]

When a nonsense mutation occurs in the middle of a genetic message, an incomplete polypeptide is released from the ribosome owing to premature chain termination. The size of the incomplete polypeptide chain depends on the location of the nonsense mutation. Mutations occurring near the beginning of a gene result in very short fragments, while mutations near the end produce fragments of almost normal length. Most incomplete chains have no biological activity, making most nonsense mutations in vital genes easily detectable. In contrast, the majority of missense mutations have some biological activity and can be easily overlooked. Thus, after treating *E. coli* with a mutagen, a sizable fraction of the *detectable* mutations is of the nonsense variety.

Suppressor Mutations Can Reside in the Same or a Different Gene[17-21]

Often, the effects of harmful mutations can be reversed by a second genetic change. Some of these subsequent mutations are very easy to understand, being simple **reverse** (back) **mutations,** which change an altered nucleotide sequence back to its original arrangement. Much more difficult to understand are the mutations occurring at different locations on the chromosome that suppress the change due to a mutation at site A by producing an additional genetic change at site B. Such **suppressor mutations** fall into two main categories: those occurring within the same gene as the original mutation but at a different site in this gene (**intragenic suppression**) and those occurring in another gene (**intergenic suppression**). Genes that cause suppression of mutations in other genes are called **suppressor genes**.

Now we realize that both types of suppression work by causing the production of good (or partially good) copies of the protein made inactive by the original harmful mutation. For example, if the first mutation caused the production of inactive copies of one of the enzymes involved in making arginine, then the suppressor mutation allows arginine to be made by restoring the synthesis of some good copies of this same enzyme. However, the mechanisms by which intergenic and intragenic suppressor mutations cause the resumption of the synthesis of good proteins are completely different.

Those mutations that can be reversed through additional changes in the same gene often involve insertions or deletions of single nucleotides. These shift the reading frame (Chapter 8) so that all the codons following the insertion (or deletion) are completely changed, thereby generating new amino acid sequences. Often, the shifted reading frame will contain nonsense codons, and as a result, prematurely terminated polypeptides will be produced by the mutant cell. Intragenic suppression may occur when a second mutation deletes (or inserts) a new nucleotide near the original change and thus restores the original codon arrangement beyond the second change (Figure 15-9). Even though there are still scrambled codons between

Figure 15-9
Intragenic suppression of a nucleotide deletion or insertion mutation. (a) The effects of a single-nucleotide deletion mutation upon the reading of the genetic message. (b) The mechanism by which a nucleotide addition mutation can suppress the havoc caused by the previous deletion mutation. In a similar way, the effect of a nucleotide addition could be overcome by a subsequent nucleotide deletion.

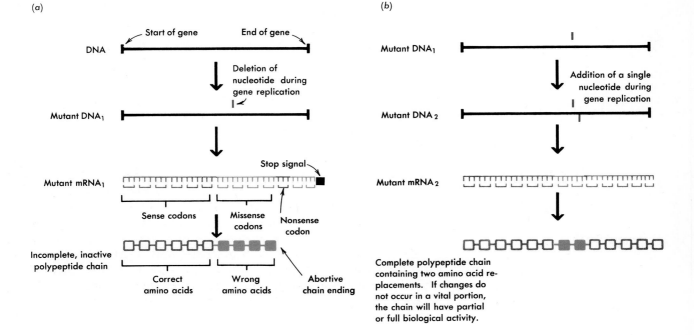

(a)

(b)

the two changes, there is a good probability, because of degeneracy, that the scrambled codons all code for some amino acid. If so, full-length, often functional proteins may be produced.

When the original mutation is a missense mutation, intragenic suppression can also result from a second missense mutation. In these cases, the original loss of enzymatic activity is due to an altered three-dimensional configuration resulting from the presence of a wrong amino acid. A second missense mutation in the same gene brings back biological activity if it somehow restores the original configuration around the functional part of the molecule. An example of this type of suppression in the tryptophan synthetase system was shown in Chapter 8 (see Figure 8-17).

Suppressor Genes Upset the Reading of the Genetic Code[17, 18, 20, 21]

Suppressor genes do not act by changing the nucleotide sequence of a mutant gene. Instead, they change the way the mRNA template is read. There are a number of different suppressor genes in *E. coli*. Since each causes the misreading of a specific nonsense or missense codon, suppressor genes can reverse the effects of only a small fraction of the point mutations that might arise within a given gene. For example, if we collect a large number of mutations blocking the synthesis of the enzyme β-galactosidase (Chapter 16), only several percent of these mutations will be suppressed by suppressor gene *a*. These few mutations will have nucleotide replacements in codons whose reading is specifically altered by gene *a*. Similarly, a completely different small fraction of β-galactosidase mutations can be suppressed by suppressor gene *b*. Thus, we see that specific codons are misread by specific suppressor genes.

On the other hand, since each suppressor gene causes the misreading of a specific codon, it is easy to understand how a given suppressor gene can suppress mutations in a number of different protein-coding genes. For example, the ability to synthesize both arginine and tryptophan in certain double mutants unable to make either amino acid can be restored by the presence of a single suppressor gene. We merely need to postulate that both these growth requirements are caused by the same specific codon change to missense or nonsense.

Nonsense Suppression Involves Mutant tRNAs[21–27]

There are suppressor genes for each of the three chain-terminating codons. They act by reading a stop signal as if it were a signal for a specific amino acid. There are, for example, three well-characterized genes that suppress the UAG codon. One suppressor gene inserts serine, another glutamine, and a third tyrosine at the nonsense position. In each of the three UAG suppressor strains, the anticodon of a tRNA species specific for one of these amino acids has been altered. For example, the tyrosine suppressor arises by a mutation within a tRNA^{Tyr} gene that changes the anticodon from 3'-AUG-5' to 3'-AUC-5', thereby enabling it to recognize UAG codons (Figure 15-10). The serine and glutamine suppressor tRNAs also arise by single-base

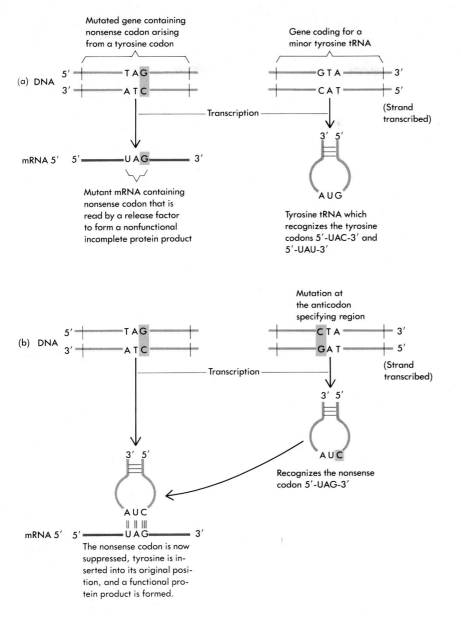

Figure 15-10
How a minor tyrosine tRNA component acts as a nonsense suppressor.

changes in their anticodons. An example of tRNA-mediated UAG suppression was first demonstrated with a nonsense mutation that blocked synthesis of the coat protein of the RNA phage R17. When mutant RNA containing this nonsense codon was used in an in vitro system, no coat protein was produced until purified tRNA from the correct suppressor strain was added.

The suppression of UAA nonsense codons is also mediated by mutant tRNAs. Tyrosine and lysine tRNAs, which normally read codons only one base different from UAA, are examples of tRNAs that can easily mutate to nonsense-suppressor tRNAs (see Table 15-4). Note that wobble in the third position of the codon enables these UAA suppressor tRNAs also to read (suppress) the UAG codon. The converse is not true.

UGA suppression likewise involves mutant tRNA molecules, with the active agent a tryptophan tRNA (tRNA^{Trp}) that inserts tryptophan at the position of UGA nonsense triplets. Normally, tRNA^{Trp} reads only the UGG codon, but as a result of the suppressor mutation, it becomes able to read UGA as well as UGG. Much to everyone's surprise, the basis for this change is not a base change at the anticodon but instead a G to A replacement at position 24 of the tRNA. Exactly how this change leads to ambiguity in the anticodon-codon interaction is not understood in molecular detail. However, since position 24 would reside in the bend of the L-shaped three-dimensional structure assumed by tRNA^{Trp}, its effect on codon-anticodon interactions some 15 Å away supports the notion that small conformational changes in tRNA molecules may be essential for accurate decoding of mRNAs on the ribosome. Perhaps not surprisingly, it turns out that even the normal tRNA^{Trp} can read the UGA terminator codon with low but detectable frequency. UGA stop signals are therefore termed "leaky."

The discovery that cells with nonsense suppressors contain mutationally altered tRNAs raised the question of how their codons corresponding to these tRNAs could continue to be read normally. In the case of the tyrosine UAG suppressor, the answer comes from the discovery that three separate genes code for tRNA^{Tyr}. One codes for the major tRNA^{Tyr} species, while the other two are duplicate genes coding for a species present in smaller amounts. One or the other of the two duplicate genes, which reside only 200 base pairs apart on the *E. coli* chromosome, is always the site of the suppressor mutation. No such dilemma exists for UGA suppression, since the suppressing tRNA^{Trp} retains its capacity to read UGG (tryptophan) codons.

Nonsense Suppressors Also Read Normal Termination Signals[16, 21, 28]

The act of nonsense suppression can be viewed as a competition between the suppressor tRNA and the release factor. When a stop signal comes into the ribosomal A site, either read-through or polypeptide chain termination will occur, depending on which arrives first. Suppression of both the UAG and UGA codons is quite efficient. In the presence of the suppressor tRNAs, over half the chain-terminating signals are read as specific amino acid codons. In contrast, suppression of the UAA codon usually averages between 1 and 5 percent. At first it was believed that the efficient suppression of UAG and UGA meant that they seldom, if ever, served as normal terminator signals. If they were frequently used, the presence of their specific suppressors might prevent much normal chain termination, leading to the production of aberrantly long polypeptides and cessation of cell growth. Now, however, the accumulation of sequence data tells us that UAG and UGA codons appear quite frequently at the ends of genes. Moreover, the use of a double stop signal is rather rare. Thus, it remains a mystery why UAA suppression is so inefficient and why UAA suppressor strains are particularly slow growers. Equally unresolved is why the UAG and UGA suppressor tRNAs do not generate an unacceptable number of abnormally long proteins.

Another aspect of nonsense suppression that is not yet understood in molecular terms is the contribution that the nucleotide sequence surrounding a stop signal makes to the efficiency of suppression (read-through). The response of a particular suppressor tRNA to a

Maturation
Coat
Replicase
AUG
UGA
UAGUAA
131 amino acids
~95% Major coat protein
+
Trp
327 amino acids
~5% A₁ protein

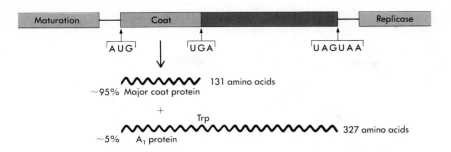

Figure 15-11
Schematic diagram of the portion of the phage Qβ genome where read-through (natural suppression) leads to production of a minor coat protein (A_1 protein). After tryptophan is inserted at the normal UGA termination signal for the major coat protein, another 195 codons are read before tandem stop codons are encountered.

particular nonsense codon can vary as much as tenfold, depending on its location in an mRNA. Moreover, such "context effects" on the reading of stop codons are not confined to cells containing suppressor genes. In fact, several cases are known where low-level read-through of a normal termination signal is biologically important. The best documented example occurs during translation of the RNA genome of bacteriophage Qβ. This phage has two coat proteins, a major component (MW about 14,000) and a minor component (MW about 38,000), with identical N-terminal amino acid sequences. The minor protein is derived by the occasional insertion of tryptophan at the UGA termination codon for the major protein (Figure 15-11). Thus, about 5 percent of the time, a much-extended polypeptide is produced that is essential for infectivity of the virus particle.

Mutations in Normal Stop Signals[29]

The frequent occurrence of single stop signals at the ends of genes means that mutations should occasionally be detected that convert stop codons to codons that specify amino acids. Since the 3′ ends of normal messages contain untranslated nucleotide sequences, translation of such mutant messages should lead to abnormally long proteins. Indeed, several such mutations have been detected, one of the first being in the human gene for the α chain of the hemoglobin molecule. Normally, the α chain is 141 amino acids long, with translation ceasing upon the reading of a single UAA stop codon. But when the stop codon is eliminated by a U to C change, glutamine becomes inserted at position 142 and translation proceeds to produce a chain of 172 amino acids (Figure 15-12). Thus, even before the sequence of the α chain gene was determined, we knew that the 3′ end of its mRNA molecule contains at least 93 nucleotides normally never translated.

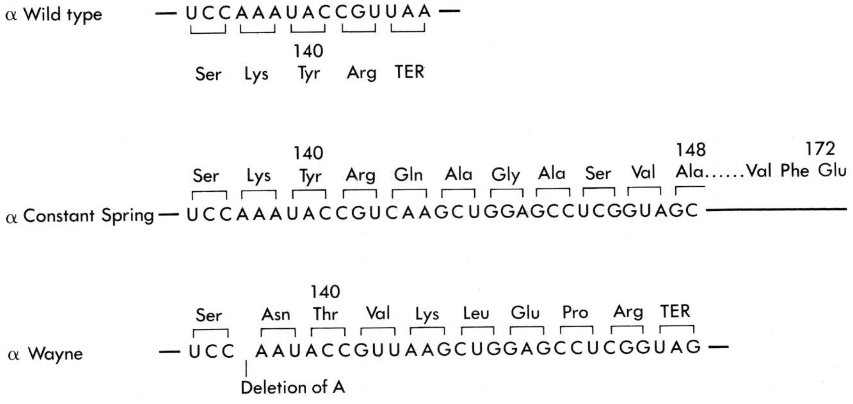

α Wild type — UCCAAAUACCGUUAA —
140
Ser Lys Tyr Arg TER
140 148 172
Ser Lys Tyr Arg Gln Ala Gly Ala Ser Val Ala......Val Phe Glu
α Constant Spring — UCCAAAUACCGUCAAGCUGGAGCCUCGGUAGC
140
Ser Asn Thr Val Lys Leu Glu Pro Arg TER
α Wayne — UCC AAUACCGUUAAGCUGGAGCCUCGGUAG —
Deletion of A

Figure 15-12
The origin of an extralong human hemoglobin α chain (hemoglobin "Constant Spring") through a U to C change in its UAA stop codon. An abnormally long α chain also characterizes hemoglobin "Wayne," which has a deletion of an A in the 139 lysine codon. The deletion shifts the reading frame so as to abolish the normal UAA stop codon. A new UAG stop codon is created at position 147, leading to a "Wayne" α chain of 146 residues.

Transfer RNA-Mediated Missense Suppression[20, 30]

Suppression of missense as well as nonsense mutations can be mediated by mutant tRNAs. For instance, there is a mutation in the tryptophan synthetase A gene in which a replacement of glycine with arginine gives rise to an inactive enzyme (Chapter 8). But there are also suppressor mutations that cause the insertion of glycine at the mutant arginine site and thus restore enzyme function. The efficiency of suppression is low, however, so even in the presence of the suppressor gene, both active and inactive forms of the enzyme are made.

The suppressor activity was traced to tRNAs that cochromatographed with glycyl-tRNA fractions, suggesting that the mutations had altered a tRNA^{Gly} anticodon. In *E. coli*, there are three different tRNA^{Gly} isoacceptors, each containing a different anticodon and thus recognizing different subsets of the four glycine codons (Table 15-8). One, called tRNA$_1^{Gly}$, recognizes only GGG. But this codon is also recognized by tRNA$_2^{Gly}$, whose anticodon can wobble to pair with either GGG or GGA. Thus, tRNA$_1^{Gly}$ is not essential for protein synthesis, and so a change in its anticodon from 3'-CCC-5' to 3'-UCC-5' allows it to decode the AGG arginine codon without simultaneously depriving the cell of the ability to read the GGG glycine codon.

In other cases, suppression arises because a mutation in a tRNA leads to its mischarging. Such changes, which can occur at various positions within the molecule, cause the tRNA to be recognized by the wrong aminoacyl-tRNA synthetase.

Frameshift Suppression[31–37]

There are also suppressor genes that mask the effects of certain frameshift mutations created by the insertion of nucleotides. Again, mutant tRNAs, such as glycine and proline tRNAs, are responsible. Complete sequence analysis of one of the frameshift glycine tRNAs (called tRNA$_{SufD}^{Gly}$, or tRNA$_{GGGG}^{Gly}$) reveals the presence of an extra base in its anticodon, with a CCC sequence being replaced by CCCC (Figure 15-13). The remainder of its sequence is identical with that of tRNA$_1^{Gly}$, indicating that it arose from an insertion mutation in *glyU*, the gene that specifies this tRNA (see Table 15-7). Possession of this extra base not only allows tRNA$_1^{Gly}$ to pair with a four-base codon, but also causes the subsequent translocation of a four-nucleotide length of mRNA from the A site to the P site of the ribosome. Thus, the reading frame is restored to the correct position.

Table 15-8   The Glycine-tRNAs of *E. coli*, Showing the Origin of Missense Suppressor tRNAs That Insert Glycine at Arginine Codons

Gene	tRNA Product	Anticodon	Glycine Codons Recognized	Derived Missense Suppressor Gene	Anticodon in tRNA Product	Arginine Codons Recognized
glyU	tRNA$_1^{Gly}$	3'-CCC-5'	GGG	*glyUsu*	3'-UCC-5'	AGG
glyT	tRNA$_2^{Gly}$	3'-CCU-5'	GGA	*glyTsu**	3'-UCU-5'	AGA
			GGG			AGG
glyV	tRNA$_3^{Gly}$	3'-CCA-5'	GGU			
			GGC			

*Strains with this mutation grow very badly (the mutation is semilethal). The origin of the residual ability to read GGA codons is unclear.

More recently, additional insertion suppressor strains of both bacteria and yeast have been characterized; some of these have tRNAs that appear to read four bases at a time, even if proper base pairing cannot occur in the fourth position. Other frameshift suppressor tRNAs act on deletions, suggesting that there is also a way of reading only two bases at a given time. Finally, frameshifting can be induced both in vivo and in vitro by a lack or overabundance of a normal charged tRNA. In other words, any condition that gives a wrong AA~tRNA a competitive advantage for binding to the ribosomal A site can potentially bring about a frameshift or misreading of the codon.

Ribosomal Mutations Also Affect the Reading Accuracy[38, 39]

Specific mutations in several of the 30S ribosomal proteins greatly influence the accuracy of reading. These effects of exact ribosome structure can be seen both in normal cells and in cells containing nonsense or missense suppressor tRNAs. A number of different amino acid replacements each must distort the ribosome so that an incorrect aminoacyl-tRNA molecule is more frequently used to elongate the polypeptide chain (Figure 15-14). One such mutation, *ram* (ribosomal ambiguity), will weakly suppress all three nonsense codons as well as missense mutations in the absence of any suppressing tRNAs. Recent evidence suggests that the proofreading step in polypeptide chain elongation, rather than the initial tRNA selection, is affected in *ram* mutants.

Streptomycin Causes Misreading[38, 39]

The finding that distorted ribosomes can misread the genetic code is strongly supported by experiments showing that the addition of the antibiotic streptomycin to either in vitro systems or living cells pro-

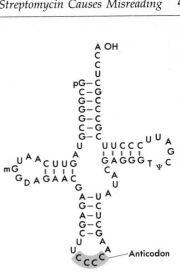

Figure 15-13
The nucleotide sequence of the frameshift suppressor tRNA$^{Gly}_{SufD}$ drawn in the cloverleaf configuration. Note the additional nucleotide in the anticodon loop.

Figure 15-14
How a missense mutation in a gene coding for one of the ribosomal proteins acts as a suppressor mutation.

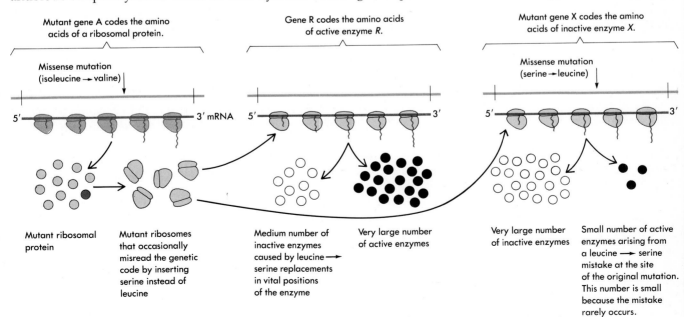

Mutant gene A codes the amino acids of a ribosomal protein.

Missense mutation (isoleucine → valine)

Mutant ribosomal protein

Mutant ribosomes that occasionally misread the genetic code by inserting serine instead of leucine

Gene R codes the amino acids of active enzyme R.

Medium number of inactive enzymes caused by leucine → serine replacements in vital positions of the enzyme

Very large number of active enzymes

Mutant gene X codes the amino acids of inactive enzyme X.

Missense mutation (serine → leucine)

Very large number of inactive enzymes

Small number of active enzymes arising from a leucine → serine mistake at the site of the original mutation. This number is small because the mistake rarely occurs.

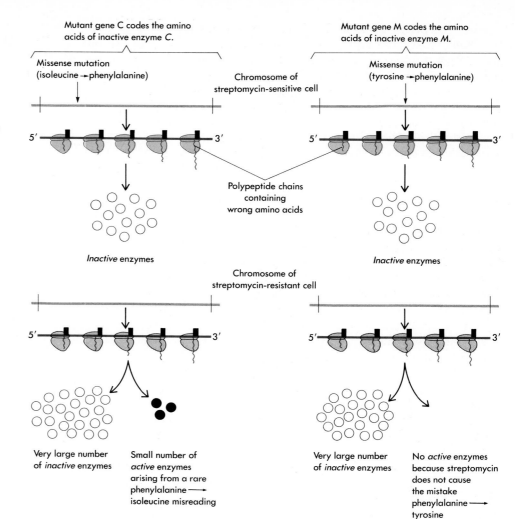

(a) Growth of streptomycin-sensitive cells in the presence of streptomycin ■ Massive misreading of the genetic code is caused by the binding of streptomycin to normal ribosomes. ⊖

Mutant gene C codes the amino acids of inactive enzyme C.

Missense mutation (isoleucine →phenylalanine)

Chromosome of streptomycin-sensitive cell

Mutant gene M codes the amino acids of inactive enzyme M.

Missense mutation (tyrosine →phenylalanine)

5'   3'

Polypeptide chains containing wrong amino acids

5'   3'

Inactive enzymes

Inactive enzymes

(b) Growth of streptomycin-resistant cells in the presence of streptomycin ■ The mutation to streptomycin resistance results in mutant ribosomes ⊖ that bind streptomycin but only occasionally misread.

Chromosome of streptomycin-resistant cell

5'   3'

5'   3'

Very large number of *inactive* enzymes

Small number of *active* enzymes arising from a rare phenylalanine ⟶ isoleucine misreading

Very large number of *inactive* enzymes

No *active* enzymes because streptomycin does not cause the mistake phenylalanine ⟶ tyrosine

Figure 15-15
The action of streptomycin on streptomycin-sensitive and streptomycin-resistant *E. coli* cells.

motes mistakes in the translation of the genetic code. Streptomycin does this by binding to the 30S ribosome and somehow disturbing the normal mRNA-tRNA-ribosome interaction. The extent of misreading depends on whether the streptomycin is added to streptomycin-sensitive or streptomycin-resistant cells. Addition of the antibiotic to sensitive cells results in large-scale misreading. Since streptomycin also acts to inhibit polypeptide chain initiation, sensitive cells become so defective in protein synthesis that they die. The mutation to streptomycin resistance alters protein S12 on the 30S ribosome in such a way that misreadings occur much less often. Nonetheless, they occur frequently enough that the addition of the drug to such resistant strains suppresses a number of mutations by allowing the synthesis of a small number of active enzyme molecules (Figure 15-15).

Suppressor Genes Also Cause Misreading of Good Genes[21, 40, 41]

The products of suppressor genes do not act selectively on those mRNA templates made on mutant genes. They also function during the decoding of wild-type mRNA templates, interfering with the synthesis of sound proteins. These changes are generally not very harmful to the growing cell, since many more good copies of each protein than bad ones are produced. There is, however, no advantage for a normal cell to harbor suppressor mutations that cause it to produce even a small fraction of bad proteins. Suppressors tend to be selected against in evolution unless a harmful mutation is present for whose effect they must compensate.

It is now possible to make a general prediction about the function of suppressor genes. A gene becomes a suppressor gene by mutation. Before this mutation occurs, the gene is normal and active. It may code for a specific tRNA or for one of the ribosomal proteins or for an aminoacyl-tRNA synthetase. Even a tRNA-modifying enzyme may be involved; there is now evidence that certain modifications of bases in or adjacent to anticodons can affect suppression (Figure 15-16). Presumably, all such genes have evolved so that their products have the optimal configuration for ensuring accurate reading of the genetic code. If a mutation yields an altered product that increases the misreading level, the gene becomes a suppressor gene. Only when an increased mistake level is necessary for cellular existence do such mutant genes have a selective advantage over their normal counterparts.

The Code Is Nearly Universal

Poly U stimulates phenylalanine incorporation in cell-free extracts from a variety of different organisms ranging from bacteria to mammals. Likewise, poly C promotes proline incorporation and poly A causes lysine incorporation in all extracts tested, regardless of their cellular source. Such indications of the universality of the code among contemporary organisms have recently been confirmed by extensive DNA sequence data on protein-coding genes from many different species. Thus, the genetic code appears to have remained constant over a long evolutionary period.

Invariability in most of the code is expected. Consider what might happen if a mutation changed the genetic code. Such a mutation might, for example, alter the sequence of the serine tRNA molecule of the class that corresponds to UCU, causing them to recognize UUU sequences instead. This would be a lethal mutation in haploid cells containing only one gene directing the production of tRNA^{Ser}, for serine would not be inserted into many of its normal positions in proteins. Even if there were more than one gene for tRNA$^{Ser}_{UCU}$ (e.g., in a diploid cell), this type of mutation would still be lethal, since it would cause the simultaneous replacement of many phenylalanine residues by serine in cell proteins.

An Altered Code in Mammalian Mitochondria[42, 43, 44]

In view of what we have just said, it was completely unexpected to find that in certain subcellular organelles, the genetic code is in fact slightly different from the "universal" code. This realization came during the elucidation of the entire DNA sequence of the 16,569-base-pair human mitochondrial genome. Mitochondria are the cellular sites of oxidative phosphorylation; they generate ATP using enzymes bound to the inner side of their double membrane exterior. Most proteins in mitochondria, including those necessary for RNA and protein

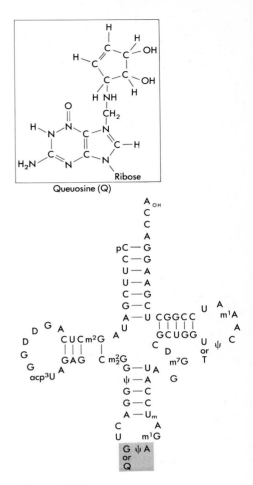

Figure 15-16
Structure of *Drosophila* tRNA^{Tyr}, in which the modification state of an anticodon base affects suppression. Whereas the tRNA containing an unmodified G in the first position of the anticodon can occasionally read (suppress) the UAG stop codon, introduction of a Q base modification at this position abolishes suppressor activity. The structure of the Q base is shown in the inset. In another system, tRNA undermethylation is responsible for the acquisition of UGA suppressor activity. (Courtesy of E. Kubli.)

Figure 15-17

Map of the 16,500-base-pair mammalian mitochondrial genome. The map is split into two parts, the outer circle indicating those genes transcribed from the H (heavy)-strand DNA and the inner circle indicating those genes transcribed from the L (light)-strand DNA. RNA genes (for 16S and 12S rRNA or for tRNAs) are indicated in black, whereas protein-coding genes are colored. Note that whenever a space exists on one strand, it is filled by a gene on the other strand. In the designation of codons recognized by certain tRNAs, Y stands for pyrimidine, R stands for purine, and N stands for any base. ATPase 6 and 8 are components of the mitochondrial ATPase complex. CO I, CO II and CO III are cytochrome oxidase subunits. Cyt *b* stands for cytochrome *b*. ND1-5 code for components of the respiratory chain NADH dehydrogenase and were previously designated URF1-5 (for *unidentified* open *reading frame*). URF6 remains an open reading frame of unknown function. O_H and O_L designate the origins of DNA replication for the H (heavy) strand (which is initiated first) and for the L (light) strand, respectively.) (Courtesy of G. Attardi.)

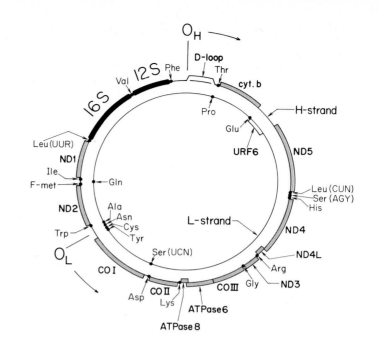

synthesis, are encoded by nuclear genes and are somehow specifically transported across the mitochondrial membrane. Mitochondria contain a double-stranded circular DNA genome, which, oddly enough, gets smaller as one proceeds from lower to higher eucaryotes. (The yeast mitochondrial DNA is five times larger than that of human mitochondria.) Mitochondrial genomes encode two rRNA molecules (which are exceptionally small in mammalian cells), a few proteins, and all the tRNAs necessary to synthesize these few proteins.

In the DNA of human (and mouse) mitochondria, the genes are extremely closely packed, with one or more tRNAs serving as spacers between the regions that code for protein or rRNA (Figure 15-17). Sequences of the regions known to specify proteins have revealed the following differences between the "universal" and the mammalian mitochondrial genetic code (Table 15-9):

- UGA is not a stop signal, but codes for tryptophan. Hence, the anticodon of mitochondrial tRNA^{Trp} recognizes both UGG and UGA, as if obeying the traditional wobble rules.

- Internal methionine is encoded by both AUG and AUA; initiating methionines are specified by AUG, AUA, AUU, and AUC.

- AGA and AGG are not arginine codons (of which there are six in the "universal" code), but specify chain termination. Thus, there are four stop codons (UAA, UAG, AGA, and AGG) in the mitochondrial code.

Perhaps not surprisingly, mitochondrial tRNAs are likewise unusual both in their structure and with respect to the rules by which they decode mitochondrial messages. Only 22 tRNAs are present in mammalian mitochondria, whereas a minimum of 32 tRNA molecules are required to decode the "universal" code according to the wobble rules. Consequently, when an amino acid is specified by four codons (with the same first and second positions), only a single mitochondrial tRNA is involved. (Recall that a minimum of two tRNAs

Table 15-9   Genetic Code of Mammalian Mitochondria*

		Second Position				
First Position (5' End)		U	C	A	G	Third Position (3' End)
	U	UUU ⎤ UUC ⎦ Phe (GAA)† UUA ⎤ UUG ⎦ Leu (UAA)	UCU ⎤ UCC ⎥ UCA ⎥ UCG ⎦ Ser (UGA)	UAU ⎤ UAC ⎦ Tyr (GUA) UAA Stop UAG Stop	UGU ⎤ UGC ⎦ Cys (GCA) UGA ⎤ UGG ⎦ Trp (UCA)	U C A G
	C	CUU ⎤ CUC ⎥ CUA ⎥ CUG ⎦ Leu (UAG)	CCU ⎤ CCC ⎥ CCA ⎥ CCG ⎦ Pro (UGG)	CAU ⎤ CAC ⎦ His (GUG) CAA ⎤ CAG ⎦ Gln (UUG)	CGU ⎤ CGC ⎥ CGA ⎥ CGG ⎦ Arg (UCG)	U C A G
	A	AUU ⎤ AUC ⎦ Ile (GAU) AUA ⎤ AUG ⎦ Met (CAU)‡	ACU ⎤ ACC ⎥ ACA ⎥ ACG ⎦ Thr (UGU)	AAU ⎤ AAC ⎦ Asn (GUU) AAA ⎤ AAG ⎦ Lys (UUU)	AGU ⎤ AGC ⎦ Ser (GCU) AGA Stop AGG Stop	U C A G
	G	GUU ⎤ GUC ⎥ GUA ⎥ GUG ⎦ Val (UAC)	GCU ⎤ GCC ⎥ GCA ⎥ GCG ⎦ Ala (UGC)	GAU ⎤ GAC ⎦ Asp (GUC) GAA ⎤ GAG ⎦ Glu (UUC)	GGU ⎤ GGC ⎥ GGA ⎥ GGG ⎦ Gly (UCC)	U C A G

*Differences between the mitochrondial and "universal" genetic code (Table 15-4) are shown by shading.

†Each bracketed group of codons is read by a single tRNA whose anticodon, written 5'→3', is in parentheses. Each four-codon group is read by a tRNA with a U in the first (5') position of the anticodon. Two-codon groups with codons ending in either U/C or A/G are read with GU wobble by tRNAs, with G or U, respectively, in the first position of the anticodon. The anticodons often contain modified bases.

‡Note that the C in the first anticodon position engages in unusual pairing.

would be required by nonmitochondrial systems.) Such mitochondrial tRNAs all have in the 5' (wobble) position of their anticodons a U residue, which somehow can engage in pairing with any of the four nucleotides in the third codon position. In cases where purines in the third position of the codon correspond to different amino acids from pyrimidines in that position, a modified U in the first position of the anticodon of the mitochondrial tRNA restricts wobble to pairing with the two purines only.

In addition to obeying unusual wobble rules, mitochondrial tRNAs often have bizarre structures. For instance, the invariant GTψCRA sequence is in most cases missing (see Figure 14-2). Other invariant nucleotides located in the D and T loops are altered. The TψC loop can have from three to nine residues, rather than the seven nucleotides found consistently in the cloverleaf structure of nonmitochondrial tRNAs. An extreme case is one of the serine tRNAs, which completely lacks the D arm of the usual cloverleaf structure (Figure 15-18). Such irregularities hint that mitochondrial tRNAs differ significantly from other tRNAs both in their three-dimensional folding and in the way they interact with the mitochondrial ribosome.

Finally, the exceedingly economic usage of the genetic material in mammalian mitochondria creates still other unexpected features of mRNA synthesis and function:

• Both strands of the mitochondrial DNA are transcribed from promoters residing near the origin of DNA replication (see Figure

Figure 15-18
Gene sequences for two human mitochondrial tRNAs folded into proposed cloverleaf forms. CCA is added post-transcriptionally to the 3' ends.

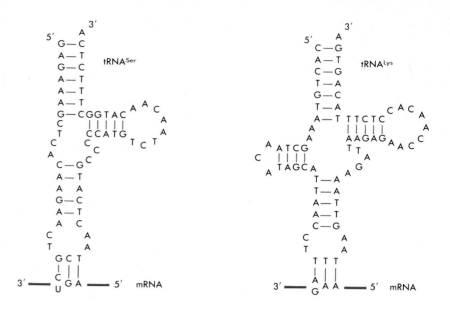

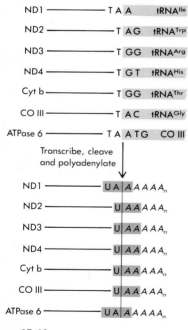

Figure 15-19
RNA processing and polyadenylation model for generating translational stop signals at the ends of the coding regions for mitochondrial proteins. The top presents the DNA sequences, showing how the coding regions abut tRNA or other protein-coding sequences. Below are the RNA sequences that would arise by transcription, site-specific cleavage at the vertical line, and addition of A residues. [After S. Anderson et al., *Nature* 290 (1981):457.]

15-17) into very long RNA molecules. Separate mRNAs are then generated by specific processing at the 5' and 3' ends of the spacer tRNA sequences.

- Initiator codons lie directly at or very close to the 5' ends of these processed mRNAs. No upstream sequences that might base-pair with the 3' end of the mitochondrial small subunit rRNA are apparent.

- The coding regions run all the way to the 3' processing sites of the mRNAs. Usually, a complete translational stop codon is not even encoded in the DNA. Rather, after the 3' end of the mRNA has been freed by cleaving off the adjacent tRNA, a string of A residues is added by a poly A–adding enzyme. This creates a UAA stop signal at the junction between the mRNA and its poly A tail (Figure 15-19).

More recently, the mitochondria of other organisms (e.g., yeast) have also been found to have slightly altered genetic codes (see Chapter 18). Moreover, they differ in detail even from the code of mammalian mitochondria (see Table 15-9). Mitochondrial DNA is known to evolve exceedingly rapidly. Hence, it is not clear whether the apparently simpler mitochondrial codes represent remnants of a primitive genetic code or whether they are more streamlined versions of the widely adopted "universal" code.

The Unexpected Existence of Overlapping Genes and Signals[45, 46, 47]

In most of the phage and bacterial genes that were sequenced earliest, the genome showed a neat sequential ordering of signals specifying the different steps in gene expression. That is, the promoter signal was followed by a 5' leader region containing the bases complementary to the 16S rRNA needed for the initiation of protein synthesis; then there appeared the polypeptide coding region; and beyond the

3' stop codon(s) were trailer nucleotides containing sequences specifying termination of RNA synthesis.

The discovery in 1977 of a gene embedded completely within another gene therefore came as a total surprise. It clearly demonstrated that a given DNA sequence can be multifunctional. Work leading to the elucidation of the complete sequence of the single-stranded DNA genome of phage φX174 first revealed that gene *E*, specifying a lysis protein, is translated in the +1 frame relative to the *D* gene, within which it resides entirely. Subsequently, it was realized that another φX174 gene, called *B* (involved in single-stranded DNA synthesis), is encoded in the −1 frame within the *A* gene. And still a third gene, *K* (also involved in lysis), overlaps both the *A* and *C* genes (Figure 15-20). Preceding the coding regions for each of the three overlapping genes there appears the expected short complementarity to 16S rRNA, always appropriately positioned relative to the initiating AUG (see Figure 14-28). Thus, the sequence of the "parent" gene not only encodes its own protein, but also provides the ribosome binding signal for initiating translation of the overlapping gene. Similarly, termination of translation is specified by an out-of-frame nonsense codon in the "parent" gene. Most remarkable is a short stretch of DNA found in both φX174 and its close relative G4 where the *K* gene is read in a third frame relative to the *A* and *C* genes, which already overlap slightly at their boundary (Figure 15-21). The N terminus of *K*, like all of *E*, is read in the +1 frame relative to its "parent" gene. Since φX174 DNA preferentially uses Ts in the third positions of its codons, both *K* and *E* polypeptides have a high proportion of hydrophobic amino acids (codons with U in the second position) and associate with the cell membrane during their function.

In addition to extensive overlapping protein-coding regions, other types of signals residing within protein-coding regions were discovered during the DNA sequence analyses of φX174 and G4 (see Figure 15-20). For example, promoters for phage mRNA transcription have been identified in the *A* and *C* protein-coding regions. The origin of single-stranded DNA synthesis is located early in the *A* gene. The *A* protein itself is translated from two different AUGs in the same reading frame, giving rise to two polypeptides (called *A* and *A**) of different size and function. Again, the ribosome binding signal for the shorter (*A**) protein is simply buried within the coding region of the larger protein.

Subsequent work has shown that the small, single-stranded DNA phages are not alone in indulging in extremely economical utilization of their DNA sequences. Completely overlapping genes have so far been identified only in viruses. But in bacterial mRNAs that encode more than one protein (polycistronic mRNAs), it is quite common to find overlap of a few nucleotides where one coding region stops and another starts (Figure 15-22). Likewise, minor promoters for RNA transcription are often buried within protein-coding regions of *E. coli* genes.

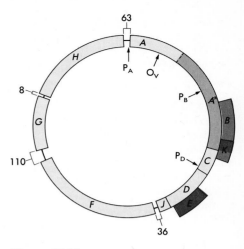

Figure 15-20
Map of the 5375-nucleotide genome of phage φX174. The reading frames of the overlapping genes relative to their "parent" genes are indicated by colored shading. Light color means translation in the −1 frame; dark color indicates translation in the +1 frame; gray means translation in the same frame. [Adapted from Sanger et al. *Nature* 265 (1977):687.]

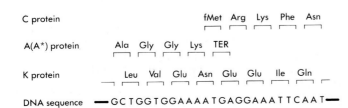

Figure 15-21
Region of the phage G4 genome where the C, A(A*), and K genes all overlap. Although this situation was first discovered in G4, an analogous region is present in φX174 DNA (see Figure 15-20). [After G. N. Godson, B. G. Barrell, R. Staden, and J. C. Fiddes, *Nature* 276 (1978):236.]

```
trpE-trpD    CAGGAGACTTTCTGATGGCT
trpD-trpC    CACGAGGGTAAATGATGCAA
trpC-trpB    TAAGGAAAGGAACAATGACA
trpB-trpA    CGAGGGGAAATCTGATGGAA
```

Figure 15-22
Gene overlap at several intercistronic junctions in the tryptophan operon of *E. coli*. Initiator and stop codons are boxed in color and gray, respectively; nucleotides complementary to 16S rRNA are shown in color. [After C. Yanofsky et al., *Nucleic Acid Res.* 9 (1981):6647–6669.]

Natural Frameshifting Allows Translation of Some Overlapping Genes[48, 49, 50]

The very existence of overlapping genes poses the dilemma of how they are translated. Does a polysome formed on a single mRNA have ribosomes decoding the same gene in two different reading frames? Or are separate mRNAs utilized to generate the alternative protein products? As of yet, we do not have the complete answer. However, studies of the translation of an overlapping gene present on the RNA bacteriophage MS2 messenger suggest that decoding errors leading to termination of synthesis of the "parent" polypeptide may be essential for initiating translation at the beginning of the overlapping gene.

The MS2 phage RNA contains three major genes separated by intercistronic spaces (Figure 15-23). A fourth protein, involved in host cell lysis, is specified by an overlapping gene whose translation begins in the second gene (the coat protein gene), extends through the intercistronic space, and ends early in the third (replicase) gene. The lysis protein is translated in the +1 frame relative to both the coat and replicase proteins. It was noticed that expression of the lysis gene is obligatorily coupled to the production of coat protein. Mutant analysis employing recombinant DNA techniques (Chapter 19) led to the conclusion that a small fraction (several percent) of the ribosomes engaged in translating the coat protein make a translocation (frameshift) error somewhere before the beginning of the lysis gene. As a consequence, they encounter out-of-phase nonsense codons preceding the lysis gene and are released. Translation can then be initiated at the AUG beginning the lysis gene subsequent to termination of the frameshifted coat polypeptide chain. Since the initiation of protein synthesis is slow compared to elongation, such a mechanism circumvents the problem that frequent readout of the parent gene is bound to interfere with initiation complex formation at the start of an overlapping gene.

It is not yet known whether the translation of all overlapping genes requires that a frameshifting error be made in translating the "parent" gene. Alternatively, ribosomes may bind de novo, as they do at the beginnings of most genes in polycistronic mRNAs.

Ribosomal frameshifting has also been shown to be utilized by other small genomes, multiplying their coding potential by allowing the synthesis of two different polypeptide chains from the same DNA sequence. For instance, bacteriophage T7, which has a linear, double-stranded DNA genome containing about 30 genes, makes a major and a minor coat protein from the same gene. The two proteins have identical sequences at their N termini but differ in both sequence and length at their C termini. The longer, minor coat protein arises because about 10 percent of the ribosomes shift into the −1 frame near the end of the coding region for the major protein. They then continue beyond the normal stop and encounter a nonsense codon in the −1 frame some 20 amino acids downstream. In this case, the position where the ribosomes make the frameshifting error has been localized

Figure 15-23
Genetic map of the RNA phage MS2. [Nucleotide numbers after W. Fiers et al., *Nature* 260 (1976):500–507.]

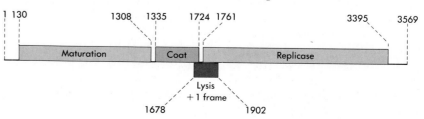

to the sequence U UUC AAA. A possible scenario is that either the tRNA^Phe or the tRNA^Lys decoding this region occasionally moves the mRNA only two residues along during translocation, resulting in a shift to the −1 reading frame for continued translation.

Evolution of the Code[51-56]

The question of how the present-day genetic code evolved has commanded much attention ever since the structure of DNA was elucidated in 1953. Exactly how each of the 20 amino acids came to be paired with its codon(s) remains a matter of speculation. However, by the time some of these pairings became fixed, a primitive coding system must have been in operation. Presumably, the system continued to evolve through many stages until the current level of sophistication was reached.

Recent analyses of DNA sequences from many genomes provide hints of how such an early coding system may have been structured. The common characteristic found in today's DNA is a constancy of the relative positions of purines and pyrimidines within protein-coding regions. Specifically, codons of the form RNY predominate (where R is a purine, Y is a pyrimidine, and N can be either), suggesting that all codons may have been of this type originally. A primitive message composed exclusively of RNY codons could have been translated in only one of the three possible frames, circumventing the need for special start signals to fix the reading frame. Interestingly, among the eight amino acids specified by RNY in today's code (glycine, isoleucine, threonine, asparagine, serine, valine, alanine, and aspartic acid) are amino acids that are most likely to have been generated by prebiotic synthesis, as well as those that often appear in meteorites.

If today's genomes are searched for RNY periodicity, the extent of coding regions and their correct reading frames can usually be identified. As shown in Figure 15-24, genes that are well expressed seem to have best preserved the RNY pattern over their entire length (e.g., yeast glyceraldehyde-3-phosphate dehydrogenase [G3PDH]). In other genes, the original RNY message can still be detected but appears badly mutated (e.g., the beginning of chicken lysozyme) or is shifted over parts of the coding region into other reading frames by appropriate deletion-insertion pairs (e.g., *E. coli* TufA). The overlapping genes of bacteriophages are most interesting in this regard. Either they correspond to a shifted region in the "parent" gene (and are themselves read in the RNY frame), or they arise by translation of the NYR frame, producing proteins with a high proportion of hydrophobic amino acids. When *E. coli* 16S and 23S rRNA sequences are subjected to a search for RNY periodicity, their purine-pyrimidine distributions appear random, suggesting either that they never carried a coding region or that any primeval message has been mutated to randomness.

Why might the RNY pattern have been favored during early evolution over other possible codon constellations? RNY is a self-complementary sequence. Moreover, a repeating RNY pattern can be perceived in a master tRNA sequence compiled from the several hundred tRNAs analyzed so far. Thus, it has been suggested that primitive tRNAs may have served dual roles as both adaptors and mRNAs. According to this scheme, before ribosomes appeared, the entire process of gene expression could have been carried out by a single class of RNA molecules.

Figure 15-24
Search for the primeval RNY message
in three gene-containing DNA sequences. The degree of mutation (m%)
away from RNY periodicity is plotted
against the base sequence number for
the three reading frames in each case.
The reading frame f that best fits the
RNY message (has the smallest m%)
will give the line shown in color.
Below, f is plotted and can be compared with the actual coding region,
designated by an arrowed line. In each
of the three cases illustrated, the frame
and extent of the coding region are well
predicted. The chicken lysozyme gene
appears badly mutated at its beginning,
whereas *E. coli* TufA (one of the two
genes for EF-Tu) has a short region in
its middle where the reading appears to
have shifted from one frame to another.
[After J. C. W. Shepherd, *Cold Spring
Harbor Symp. Quant. Biol.* 47
(1983):1099.]

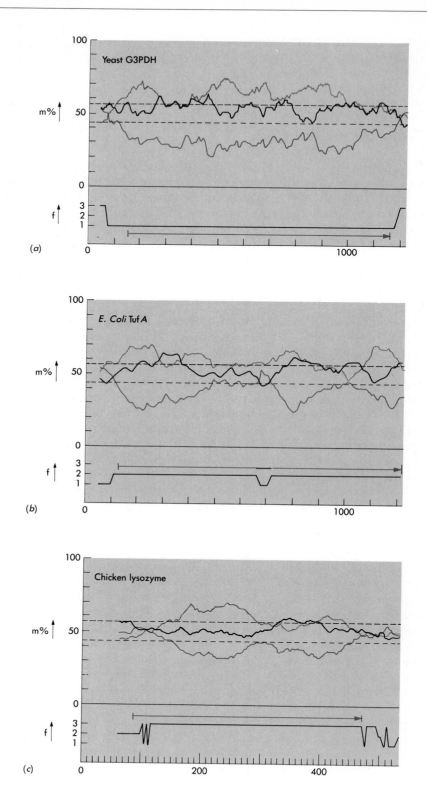

Summary

The genetic code was determined through the study of protein synthesis in cell-free extracts. Addition of new mRNA to an extract depleted of its original messenger component results in the production of new proteins whose amino acid sequences are determined by the externally added mRNA. The first (and probably most important) step in cracking the genetic code occurred when the synthetic polyribonucleotide poly U was found to code specifically for polyphenylalanine. Use of other synthetic polyribonucleotides, both homogeneous (poly C, and so on) and mixed (poly AU, and so on), then assigned the base composition of codons for the various amino acids. Determination of the exact order of nucleotides in codons subsequently came from a study of specific trinucleotide-tRNA-ribosome interactions and the use of regular copolymers as messengers.

In the "universal" genetic code used by every organism from bacteria to humans, 61 codons signify specific amino acids; the remaining three are stop signals. The code is highly degenerate, with several codons usually corresponding to a single amino acid. A given tRNA can sometimes specifically recognize several codons. This ability arises from wobble in the base at the 5' end of the anticodon. The codon for the starting amino acid, *N*-formylmethionine, is usually AUG, which also specifies internal methionine, or more rarely GUG, which normally codes for valine. The stop codons UAA, UAG, and UGA are read by specific proteins, not specialized tRNA molecules.

Certain mutations, called intergenic suppressor mutations, appear to increase the frequency of mistakes in reading the genetic code. As a result of this increase in the mistake level, a mutant gene may occasionally produce a normal product. Suppressor genes exist for both missense (amino acid replacement) and nonsense (chain-terminating) codons, as well as for frameshift mutations caused by single-nucleotide insertions. Transfer RNAs with altered-anticodons are the molecular basis of most suppressor gene actions, although mutant ribosomal proteins can also affect reading accuracy.

A slightly different genetic code is utilized in certain rapidly evolving cellular organelles such as mitochondria. Fewer tRNAs are required to read this altered code. Many mitochondrial tRNAs have bizarre structures.

Overlapping genes have been recognized in many viral genomes. Other economies in genome utilization involve occasional read-through of termination signals or ribosome frameshifting to produce more than one polypeptide from the same coding region.

Bibliography

General References

Celis, J. E., and J. D. Smith, eds. 1979. *Nonsense Mutations and tRNA Suppressors.* New York: Academic Press. A most impressive collection of reviews on termination and suppression.

Clark, B., and H. Petersen, eds. 1984. "Gene Expression: The Translational Step and Its Control." *Alfred Benzon Symposium,* vol. 19. Copenhagen: Munksgaard. A number of current reviews on frameshifting and accuracy in translation are included.

Crick, F. H. C. 1963. "The Recent Excitement in the Coding Problem." *Prog. Nucleic Acid Res.* 1:164. A superb analysis of the state of the coding problem as of late 1962.

"The Genetic Code." 1966. *Cold Spring Harbor Symp. Quant. Biol.* 31. A historic collection of papers, presented in June 1966, just after the general features of the code became clear.

Söll, D., J. N. Abelson, and P. R. Schimmel, eds. 1980. *Transfer RNA, Biological Aspects.* Cold Spring Harbor, N.Y.: Cold Spring Harbor Laboratory. This volume of a two-volume compendium on tRNA contains many valuable articles on the role of tRNA in suppression.

Ycas, M. 1969. *The Biological Code.* New York: Wiley (Interscience). The most complete monograph on the universal code, written by one of the early workers in the field.

Cited References

1. Zamecnik, P. C. 1969. "An Historical Account of Protein Synthesis, with Current Overtones—A Personalized View." *Cold Spring Harbor Symp. Quant. Biol.* 34:1–16.
2. Nathans, D., G. Notani, J. H. Schwartz, and N. D. Zinder. 1962. "Biosynthesis of the Coat Protein of Coliphage f2 by *E. coli* Extracts." *Proc. Nat. Acad. Sci.* 48:1424–1431.
3. Salser, W., R. F. Gesteland, and B. Ricard. 1969. "Characterization of Lysozyme Messenger and Lysozyme Synthesized *in Vitro.*" *Cold Spring Harbor Symp. Quant. Biol.* 34:771–780.
4. Pelham, H. R. B., and R. J. Jackson. 1976. "An Efficient mRNA-Dependent Translation System from Reticulocyte Lysates." *Eur. J. Biochem.* 67:247–256.
5. Grunberg-Manago, M., and S. Ochoa. 1955. "Enzymatic Synthesis and Breakdown of Polynucleotides: Polynucleotide Phosphorylase." *J. Amer. Chem. Soc.* 77:3165–3166.
6. Nirenberg, M. W., and J. H. Matthaei. 1961. "The Dependence of Cell-Free Protein Synthesis in *E. coli* upon Naturally Occurring or Synthetic Polyribonucleotides." *Proc. Nat. Acad. Sci.* 47:1588–1602.
7. Speyer, J. F., P. Lengyel, C. Basilio, A. J. Wahba, R. S. Gardner, and S. Ochoa. 1963. "Synthetic Polynucleotides and the Amino Acid Code." *Cold Spring Harbor Symp. Quant. Biol.* 28:559–568.
8. Nirenberg, M., and P. Leder. 1964. "The Effect of Trinucleotides upon the Binding of sRNA to Ribosomes." *Science* 145:1399–1407.
9. Nishimura, S., D. S. Jones, and H. G. Khorana. 1965. "The *in Vitro* Synthesis of a Copolypeptide Containing Two Amino Acids in Alternating Sequence Dependent upon a DNA-Like Polymer Containing Two Nucleotides in Alternating Sequence." *J. Mol. Biol.* 13:302–324.
10. Lagerkvist, U. 1981. "Unorthodox Codon Reading and the Evolution of the Genetic Code." *Cell* 23:305–306.
11. Crick, F. H. C. 1966. "Codon-Anticodon Pairing: The Wobble Hypothesis." *J. Mol. Biol.* 19:548–555.
12. Smith, J. D. 1976. "Transcription and Processing of Transfer RNA Precursors." *Prog. Nucleic Acid Res.* 16:25–73.
13. Ikemura, T., and H. Ozeki. 1982. "Codon Usage and Transfer RNA Contents: Organism-Specific Codon-Choice Patterns in Reference to the Isoacceptor Contents." *Cold Spring Harbor Symp. Quant. Biol.* 47:1087–1096.
14. Grosjean, F., and W. Fiers. 1982. "Preferential Codon Usage in Prokaryotic Genes: The Optimal Codon-Anticodon Interaction Energy and the Selective Codon Usage in Efficiently Expressed Genes." *Gene* 18:199–209.
15. Kozak, M. 1983. "Comparison of Initiation of Protein Synthesis

in Procaryotes, Eucaryotes, and Organelles." *Microbiol. Revs.* 47:1–45.

16. Kohli, J., and H. Grosjean. 1981. "Usage of the Three Termination Codons: Compilation and Analysis of the Known Eukaryotic and Prokaryotic Translation Termination Sequences." *Mol. Gen. Genetics* 182:430–439.

17. Garen, A. 1968. "Sense and Nonsense in the Genetic Code." *Science* 160:149–159.

18. Brenner, S., A. O. W. Stretton, and S. Kaplan. 1965. "Genetic Code: The Nonsense Triplets for Chain Termination and Their Suppression." *Nature* 206:994–998.

19. Crick, F. H. C., L. Barnett, S. Brenner, and R. J. Watts-Tobin. 1961. "General Nature of the Genetic Code for Proteins." *Nature* 192:1227–1232.

20. Hill, C. W. 1975. "Informational Suppression of Missense Mutations." *Cell* 6:419–427.

21. Steege, D. A., and D. G. Soll. 1979. "Suppression." In *Biological Regulation and Development I,* ed. R. F. Goldberger. New York: Plenum, pp. 433–486.

22. Capecchi, M. R., and G. Gussin. 1965. "Suppression *in Vitro*: Identification of Serine-sRNA as a 'Nonsense' Suppressor." *Science* 149: 417–422.

23. Engelhardt, D. L., R. Webster, R. Wilhelm, and N. Zinder. 1965. "*In Vitro* Studies on the Mechanism of Suppression of a Nonsense Mutation." *Proc. Nat. Acad. Sci.* 54:1791–1797.

24. Goodman, H. M., J. Abelson, A. Landy, S. Brenner, and J. D. Smith. 1968. "Amber-Suppression: A Nucleotide Change in the Anticodon of a Tyrosine Transfer RNA." *Nature* 217:1019–1024.

25. Hirsh, D. 1971. "Tryptophan Transfer RNA as the UGA Suppressor." *J. Mol. Biol.* 58:439–458.

26. Buckingham, R. H., and C. G. Kurland. 1980. "Interactions Between UGA-Suppressor tRNA^{Trp} and the Ribosome: Mechanisms of tRNA Selection." In *Transfer RNA, Biological Aspects,* ed. D. Soll, J. N. Abelson, and P. R. Schimmel. Cold Spring Harbor, N.Y.: Cold Spring Harbor Laboratory, pp. 421–426.

27. Ozeki, H., H. Inokuchi, F. Yamao, M. Kodaira, H. Sakano, T. Ikemura, and Y. Shimura. 1980. "Genetics of Nonsense Suppressor of tRNAs in *Escherichia coli.*" In *Transfer RNA, Biological Aspects,* ed. D. Soll, J. N. Abelson, and P. R. Schimmel. Cold Spring Harbor, N.Y.: Cold Spring Harbor Laboratory, pp. 341–349.

28. Weiner, A. M., and K. Weber. 1973. A single UGA Codon Functions as a Natural Termination Signal in the Coliphage Qβ Coat Protein Cistron." *J. Mol. Biol.* 80:837–855.

29. Bunn, H. F., and B. G. Forget. 1986. *Hemoglobin: Molecular, Genetic and Clinical Aspects.* Philadelphia: Saunders, pp. 407–414.

30. Carbon, J., P. Berg, and C. Yanofsky. 1966. "Missense Suppression Due to a Genetically Altered tRNA." *Cold Spring Harbor Symp. Quant. Biol.* 31:487–497.

31. Riddle, D. L., and J. Carbon. 1973. "Frameshift Suppression: A Nucleotide Addition in the Anticodon of a Glycine Transfer RNA." *Nature New Biol.* 242:230–234.

32. Roth, J. R. 1981. "Frameshift Suppression." *Cell* 24:601–602.

33. Bossi, L., and D. Smith. 1984. "Suppressor *sufj*: A Novel Type of tRNA Mutant That Induces Translational Frameshifting." *Proc. Nat. Acad. Sci.* 81:6105–6109.

34. Atkins, J., R. Gesteland, B. Reid, and C. Anderson. 1979. "Normal tRNAs Promote Ribosomal Frameshifting." *Cell* 18:1119–1131.

35. Weiss, R. 1984. "Molecular Model of Ribosome Frameshifting." *Proc. Nat. Acad. Sci.* 81:5797–5801.

36. Weiss, R., J. Murphy, G. Wagner, and J. Gallant. 1984. "The Ribosome's Frame of Mind." *Alfred Benzon Symposium,* vol. 19, ed. B. Clark and H. Petersen. Copenhagen: Munksgaard, pp. 208–220.

37. Kurland, C. G. 1979. "Reading Frame Errors on Ribosomes." In *Nonsense Mutations and tRNA Suppressors,* ed. J. E. Celis and J. D. Smith. San Francisco: Academic Press, pp. 97–108.

38. Gorini, L. 1974. "Streptomycin and Misreading of the Genetic Code." In *Ribosomes.* Cold Spring Harbor, N.Y.: Cold Spring Harbor Laboratory, pp. 791–803.

39. Yarus, M., and R. Thompson. 1983. "Precision of Protein Biosynthesis." In *Gene Function in Prokaryotes,* ed. J. Beckwith, J. Davies, J. A. Gallant. Cold Spring Harbor, N.Y.: Cold Spring Harbor Laboratory, pp. 23–63.

40. Glass, R. E., V. Nene, and M. G. Hunter. 1982. "Informational Suppression as a Tool for the Investigation of Gene Structure and Function." *Biochem. J.* 203:1–13.

41. Bienz, M., and E. Kubli. 1981. "Wild-Type tRNA$_G^{Tyr}$ Reads the TMV RNA Stop Codon, but Q Base-Modified tRNA$_Q^{Tyr}$ Does Not." *Nature* 294:188–190.

42. Anderson, S., A. T. Bankier, B. G. Barrell, M. H. L. deBruijn, A. R. Coulson, J. Drouin, I. C. Eperon, D. P. Nierlich, B. A. Roe, F. Sanger, P. H. Schreier, A. J. H. Smith, R. Staden, and I. G. Young. 1981. "Sequence and Organization of the Human Mitochondrial Genome." *Nature* 290:457–465.

43. Attardi, G. 1985. "Animal Mitochondrial DNA: An Extreme Example of Genetic Economy." *Internatl. Rev. Cytol.* 93:93–145.

44. Battey, J., and D. A. Clayton. 1980. "The Transcription Map of Human Mitochondrial DNA Implicates Transfer RNA Excision as a Major Processing Event." *J. Biol. Chem.* 255:11599–11606.

45. Sanger, F., G. M. Air, B. G. Barrell, N. L. Brown, A. R. Coulson, J. C. Fiddes, C. A. Hutchinson, P. M. Slocombe, and M. Smith. 1977. *Nature* 265:687–695.

46. Godson, G. N., B. G. Barrell, R. Staden, and J. C. Fiddes. 1978. "Nucleotide Sequence of Bacteriophage G4 DNA." *Nature* 276:236–247.

47. Yanofsky, C., T. Platt, I. P. Crawford, B. P. Nichols, G. E. Christie, H. Horowitz, M. Van Cleemput, and A. M. Wu. 1981. *Nucleic Acid Res.* 9:6647–6668.

48. Atkins, J. F., J. A. Steitz, C. W. Anderson, and P. Model. 1979. "Binding of Mammalian Ribosomes to MS2 Phage RNA Reveals an Overlapping Gene Encoding a Lysis Function." *Cell* 18:247–256.

49. Kastelein, R. A., E. Remaut, W. Fiers, and J. van Duin. 1982. "Lysis Gene Expression of RNA Phage MS2 Depends on a Frameshift During Translation of the Overlapping Coat Protein Gene." *Nature* 295:35–41.

50. Dunn, J. J., and F. W. Studier. 1983. "Complete Nucleotide Sequence of Bacteriophage T7 DNA and the Locations of T7 Genetic Elements." *J. Mol. Biol.* 166:477–535.

51. Shepherd, J. C. W. 1982. "From Primeval Message to Present-Day Gene." *Cold Spring Harbor Symp. Quant. Biol.* 47:1099–1108.

52. Crick, F. H. C., S. Brenner, A. Klug, and G. Pieczenik. 1976. "A Speculation on the Origin of Protein Synthesis." *Origins Life* 7:389–397.

53. Woese, C. 1980. "Just So Stories and Rube Goldberg Machines: Speculations on the Origin of the Protein Synthetic Machinery." In *Ribosomes: Structure, Function and Genetics,* ed. G. Chambliss et al. Baltimore, Md.: University Park Press, pp. 357–376.

54. Eigen, M., and R. Winkler-Oswatitsch. 1981. "Transfer-RNA: The Early Adaptor." *Naturwissenschaften* 68:217–228.

55. Eigen, M., and R. Winkler-Oswatitsch. 1981. "Transfer-RNA: An Early Gene?" *Naturwissenschaften* 68:282–292.

56. Crothers, D. M. 1982. "Nucleic Acid Aggregation Geometry and the Possible Evolutionary Origin of Ribosomes and the Genetic Code." *J. Mol. Biol.* 162:379–391.

VI

REGULATION OF GENE FUNCTION IN BACTERIAL CELLS

CHAPTER 16

Regulation of Protein Synthesis and Function in Bacteria

Having described the general features of DNA and RNA function in protein synthesis, we are now ready to examine the cellular mechanisms that control the rate of synthesis of proteins. Since under any given environmental condition there is great variation in the number of different protein molecules in any cell, there must be devices to ensure the selective synthesis of proteins required in large numbers. There also must be ways to change the relative numbers of protein molecules, so that cells can respond efficiently to varying nutritional and other environmental conditions. We now realize that the rate at which each step in the synthesis of a protein occurs, from the binding of RNA polymerase as it initiates mRNA synthesis to the completion of the polypeptide on a ribosome, is partially under internal genetic control and partially determined by the external chemical environment. To show how these factors can interact, we shall focus our attention on bacterial systems, since they have been the basis of most of the important concepts formulated up to now.

Different Proteins Are Produced in Different Numbers

We have already estimated from its length that the *E. coli* chromosome codes for 2000 to 4000 different polypeptide chains. Exactly how many different proteins are simultaneously present in a given cell is not yet known. Based on the probable number of enzymes needed to make the necessary metabolites, general estimates argue for the presence of at least 600 to 800 different enzymes in a cell growing with glucose as its sole carbon source. Some of these enzymes, particularly those connected with the first steps in glucose degradation and with the reactions that make the common amino acids and nucleotides, are present in relatively large amounts. Also required in large amounts are the enzymes needed to produce ATP. In contrast, other enzymes, particularly those involved in making the much smaller amounts of the necessary coenzymes, are present only in trace quantities. There must also be relatively large amounts of the various structural proteins used to construct the cell wall, the cell membrane, and the ribosomes.

It is useful to keep in mind how many copies of both scarce and abundant proteins there are in an *E. coli* cell. One well-studied protein present in moderately high amounts is the enzyme β-galactosidase (MW 4.6×10^5 daltons), which splits the sugar lactose

465

Figure 16-1
The sugar lactose can be hydrolytically cleaved to galactose and glucose by the enzyme β-galactosidase. Mutants that fail to make this protein cannot utilize lactose as a carbon source.

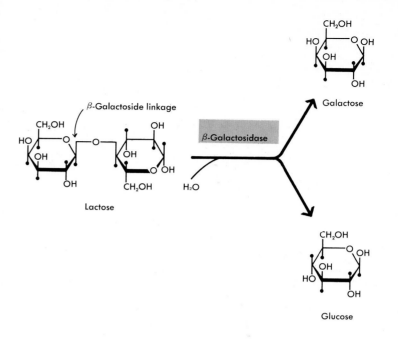

into its glucose and galactose moieties (Figure 16-1). Each active molecule has a tetrameric structure, being composed of four identical polypeptide chains of molecular weight 116,000. This is a very important enzyme, because lactose cannot be used as either a carbon or energy source unless it is first broken down to the simpler sugars glucose and galactose. An *E. coli* cell growing with lactose as its exclusive carbon source generally contains about 3000 molecules of β-galactosidase, which represents about 1.5 percent of the total protein. This is the maximum quantity that can be synthesized if just one gene coding for the β-galactosidase amino acid sequence is present on each *E. coli* chromosome. If this gene is present in two copies, 3 percent of the total protein produced by the cell can be β-galactosidase.

Now we know it is possible for a single gene that is transcribed and translated at an optimal rate to be expressed many times faster than β-galactosidase and to account for perhaps half of the protein synthesized by the cell (a situation contrived by genetic engineering). However, the cell is unhealthy with so much of its resource directed to making a single protein, and no cell does this naturally.

Other abundant proteins that have been accurately counted and cataloged include components of the protein synthesis machinery. There are 53 different ribosomal proteins (average MW 20,000) that collectively account for about 20 percent of the total protein in cells using glucose as a carbon source. Thus, the average ribosomal protein makes up 0.3 percent of the total *E. coli* protein, and there are about 20,000 copies of each. One of the most abundant *E. coli* proteins is the protein synthesis elongation factor Ef-Tu (see Chapter 14), present in around 100,000 molecules per cell.

In contrast to these prominent molecules, regulatory proteins and enzymes that make minor cellular components are present in only a few to a few hundred copies per cell. Thus, they are more difficult to detect and purify, and we often have less precise information about their amounts.

Relation Between Amount of and Need for Specific Proteins

When a protein is needed, it can be present in a vastly greater amount than when environmental conditions do not require it. For example, there are approximately 3000 β-galactosidase molecules in each normal *E. coli* cell growing in the presence of β-galactosides, such as lactose, and less than one-thousandth of this number in cells growing on other carbon sources. Substrates like lactose, whose introduction into a growth medium specifically increases the amount of an enzyme, are known as **inducers;** their enzymes are called **inducible enzymes.** The opposite response is shown by many enzymes involved in cellular biosynthesis. For example, *E. coli* cells growing in a medium without any amino acids contain all the enzymes necessary for the biosynthesis of the 20 necessary amino acids. When, on the other hand, the growth medium contains these amino acids, their biosynthetic enzymes are almost entirely missing. Biosynthetic enzymes whose amount is reduced by the presence of their *end products* (e.g., histidine is the end product of the histidine biosynthetic enzymes) are called **repressible enzymes.** Those end-product metabolites whose introduction into a growth medium specifically decreases the amount of a specific enzyme are known as **corepressors.** The inductive and repressive responses are equally useful to bacteria: When enzymes are needed to transform a specific food molecule or to synthesize a necessary cell constituent, they are present; when they are unnecessary, they are effectively absent.

Adaptation is not, however, an all-or-nothing response. Under conditions of intermediate need, there may be an intermediate enzyme level (Figure 16-2). Similar variation is found in the quantities of structural proteins, best shown by the variation in the number of ribosomes themselves. When bacteria are growing at their maximum rate, ribosomes amount to 25 to 30 percent of the cell mass. If, however, their growth rate is cut down by unfavorable nutritional conditions, the bacteria need fewer ribosomes to maintain their slower rate of protein synthesis, and the ribosome content can drop to as little as one-fifth of its maximum value.

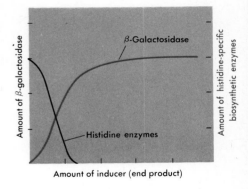

Figure 16-2
Variation in the amount of enzyme per cell as a function of the amount of inducer (end product) present in the growth medium.

Variation in Protein Amount Can Reflect the Number of Specific mRNA Molecules

In actively dividing bacteria, most individual protein molecules, once synthesized, are quite stable. Thus, variation in the amounts of proteins usually reflects rates of synthesis, not relative stability, although there are important exceptions. As the cell adapts to new conditions of growth, the change in the amount of a protein is often the result of a change in the number of available mRNA templates. The number of β-galactosidase templates in cells actively making β-galactosidase, for example, greatly exceeds the number found in cells not engaged in synthesizing this enzyme. Now our best estimate is that during maximum β-galactosidase synthesis, about fifty β-galactosidase mRNA molecules are present in each cell. In contrast, when no lactose is present, the average cell contains fewer than one mRNA molecule specific for β-galactosidase synthesis.

Constitutive Proteins Are Not Under Direct Environmental Control

There are a number of proteins within the cell whose amounts do not seem to be influenced by the external environment. Such proteins are called **constitutive proteins.** In *E. coli,* for example, the amounts of the enzymes controlling the degradation of glucose do not radically change when glucose is either removed from or added to the growth medium. Perhaps this means that *E. coli* so often finds glucose in nature that there is no advantage in evolving a means to repress the production of these enzymes. Similarly, it appears to be false economy for the cell to control proteins made in very small amounts; the very low copy number lactose repressor, for example, is made constitutively. The rate of synthesis of a constitutive protein is set genetically by: 1) the natural efficiency of the promoter from which its messenger is made; 2) the rate at which ribosomes read the messenger; and 3) the sensitivity of its messenger to degradation by nucleases.

Gene-Specific Regulatory Proteins Control Much RNA Synthesis

Special proteins called *regulatory proteins* determine when many inducible and repressible enzymes are to be made. Each regulatory protein affects the expression of one or more specific genes. They are of two fundamental types, *positive regulators* and *negative regulators*, distinguished by the contrasting effect of mutations in the genes that encode them (Figure 16-3). Thus, mutational inactivation of a negative regulator, or *repressor*, leads to synthesis of its target proteins irrespective of need. Such mutants are called *constitutive mutants*. In contrast, mutational loss of a positive regulator, sometimes called an *activator*, abolishes any substantial synthesis of the proteins it regulates, even when they are required in greater amounts by a change in cellular conditions.

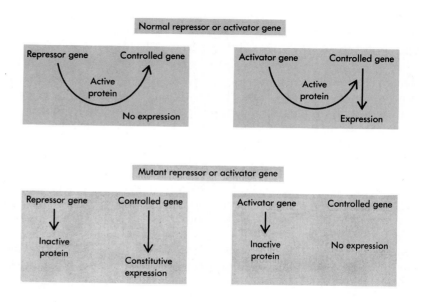

Figure 16-3
How repressors and activators are genetically distinguished.

Note that the terms *repressible* and *inducible* refer to the action of an external controlling molecule, like an amino acid or the sugar lactose, whereas the names *repressor* and *activator* reflect the action of a controlling gene in the cell. Therefore, both repressible and inducible genes could be controlled by either a repressor or an activator.

Repressors Prevent Initiation of mRNA Synthesis by Binding to DNA[1, 2]

Our first real insight into how genes can be regulated came from the isolation of two repressors: the lactose operon repressor (*lac* repressor) and the bacteriophage λ repressor (Chapter 17). Since a cell contains only 10 to 20 copies of lactose repressor and 100 to 200 copies of λ repressor, their detection and isolation in 1967 was a remarkable feat. Now about ten repressors are known, but this original pair is by far the best studied and understood.

Like most that have been studied, the *lac* and λ repressors are *transcriptional* repressors: They act by binding at specific sites on their respective DNA molecules, blocking the initiation of transcription of the corresponding mRNA molecules. The specific nucleotide sequences that bind the repressors are called *operators*. In general, an operator must be at least 10 to 12 bases long in order to interact specifically with the appropriate recognition site on a repressor. A base number this large greatly lessens the chance that a similar sequence will exist somewhere else along the same chromosome. If a smaller number of bases were used, too many false bindings would occur. The large number of specific interactions that a repressor can make with a DNA segment this long has the added benefit that the binding can be very strong. Once a *lac* repressor has bound to DNA, it effectively remains attached until subsequent interaction with its inducer.

Powerful Methods to Identify Protein Binding Sites on DNA[3, 4]

How can a protein binding site in DNA, such as an operator, be identified? Originally, operators were characterized through isolation and sequencing of DNA fragments protected from nuclease attack by specific binding of repressors. But this is a very laborious procedure. Fortunately, our ability to make and sequence specific restriction nuclease fragments has given rise to much quicker methods to outline the sites where proteins act and to learn exactly which chemical groups in DNA (methyl, amino, or phosphate) a protein contacts.

One basic principle, identical to that which the chemical method of DNA sequencing uses, underlies these methods: If a DNA fragment is labeled with a radioactive atom only at one end of one strand, the location of any break in this strand can be deduced merely from the size of the labeled fragment that results. The size, in turn, can easily be determined by high-resolution electrophoresis in a polyacrylamide gel. In the **footprinting** method (named facetiously after fingerprinting methods of chromatographic identification), the binding site is marked by internucleotide bonds that are shielded from the cutting action of a nuclease by the binding protein (Figure 16-4). The related **chemical protection** method relies on the ability of a bound protein to modify the reactivity of bases in the binding site to those base-specific

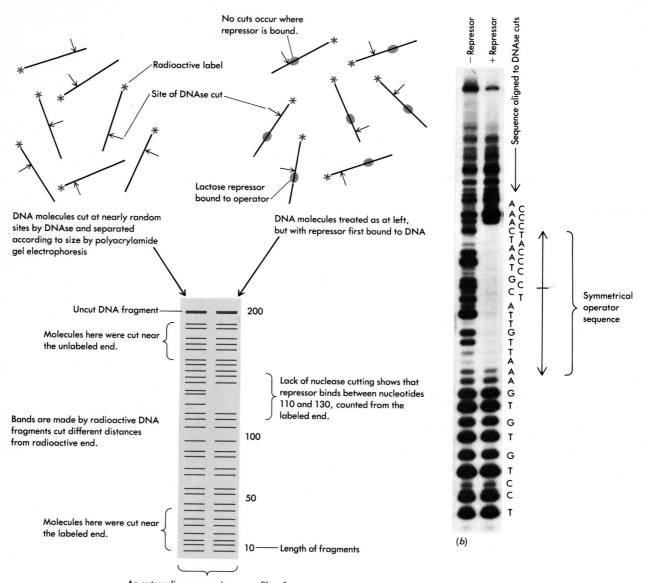

Figure 16-4
The footprinting method to map protein binding sites in DNA. (a) How footprinting reveals the DNA site to which a protein binds. Often, the same DNA fragment is subjected to chemical DNA sequencing and electrophoresed alongside the footprinted DNA samples. In this way, the exact position in the DNA sequence of a nuclease cut can be determined. (b) Photograph of an actual footprint, showing binding of lactose repressor to its operator. The DNA sequence is aligned correctly to the gel, so that radioactivity in a band indicates a nuclease cut at the nucleotide written alongside. The DNA strand labeled is the lower strand shown in Figure 16-5. In this case, the DNA fragment used was 48 nucleotides long. [Courtesy of D. Galas and A. Schmitz, *Nucleic Acid Res.* 5 (1978):3161. Used with permission.]

reagents that (after a further reaction) give rise to backbone cuts in the standard chemical DNA-sequencing method.

By changing the order of the first two steps, a third method determines which features of the DNA structure are *necessary* for the pro-

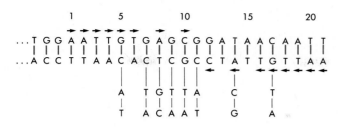

Figure 16-5
The *lac* operator sequence, showing its relation to the starting point for transcription of *lac* mRNA (nucleotide 1). Base-pair changes that lead to poor binding of the *lac* repressor are shown below their wild-type equivalents. The horizontal arrows designate bases having identical counterparts located symmetrically in the other half of the operator. The center of symmetry is at nucleotide 11.

tein to bind. An average of one chemical change per DNA is made, and then protein-DNA complexes are isolated. If a modification at a particular site does not prevent binding, DNA isolated from the complex will contain the modified chemical group, and the harmless modification allows the DNA to be broken at this site by further chemical treatment. If, on the other hand, binding is blocked, then no DNA modified at the site will be found complexed to the binding protein and the isolated fragments will not be broken at this site by subsequent chemical treatment. By all three methods, we can learn where a protein makes specific contacts both with bases and with the phosphates in the sugar-phosphate backbone of DNA.

The Structure of Operators

Many operators have now been characterized through footprinting and chemical protection experiments, as well as DNA sequence analysis of operator mutations that affect repressor binding. Their most important feature is an inverted repeat (or near repeat) that gives most operators a symmetrical shape. For example, of 24 base pairs in the lactose operator fragment isolated after nuclease digestion of DNA complexed with *lac* repressor, some 16 are related by a twofold axis of symmetry centered at the base pair in position 11 (Figure 16-5). It was realized as soon as operator sequences were determined that such symmetry could allow two subunits of a multimeric repressor to bind simultaneously to the operator. That the symmetry is not perfect was originally interpreted as meaning that not all base pairs on a given side of the operator bind to the repressor. This idea led to the speculation that only the symmetrical pairs are involved in the protein–nucleic acid interactions. However, analysis of mutant operators reveals that base-pair changes that directly reduce the strength of repressor-operator binding occur in the nonsymmetrical as well as symmetrical regions.

How Repressors Bind Operators[5–12]

Recent success in determining by X-ray crystallography the three-dimensional structure of several specific DNA binding proteins, provided in abundant amounts by genetically engineered plasmids, has given a striking view of how regulatory proteins like repressors bind specific DNA sites. The first structure determined was that of *E. coli* CAP protein, followed soon by λ Cro protein, λ repressor, and the repressor of phage 434, a close relative of λ. As protein crystallographers had expected, the active binding unit is a dimer of two identical globular polypeptide chains oriented oppositely in space to give

Figure 16-6

(a) A drawing of λ repressor bound to operator, showing both the NH₂-terminal DNA binding domains and the associated COOH-terminal domains (as they might appear). (b) How two λ repressor monomers fit onto the operator. The diagram on top shows the portion of each polypeptide chain at the operator binding surface. There are two polypeptides, placed symmetrically on the halves of the operator. Cylinders are α helices, and they are connected by less structured segments of polypeptide. The amino terminus of the polypeptide is behind the DNA and connects to α helix 1. The bottom diagram shows which bases and backbone phosphates contact the repressor, as determined by chemical protection experiments. [After C. Pabo and R. Sauer, *Ann. Rev. Biochem.* 53 (1984):300, with permission.] (c) The DNA binding faces of three DNA binding regulatory proteins. In each case, the number 3 α helices (named "F" in CAP protein) in the dimer are separated by 34 Å, the length of one turn of B-form DNA. (Courtesy of T. A. Steitz and I. T. Weber.)

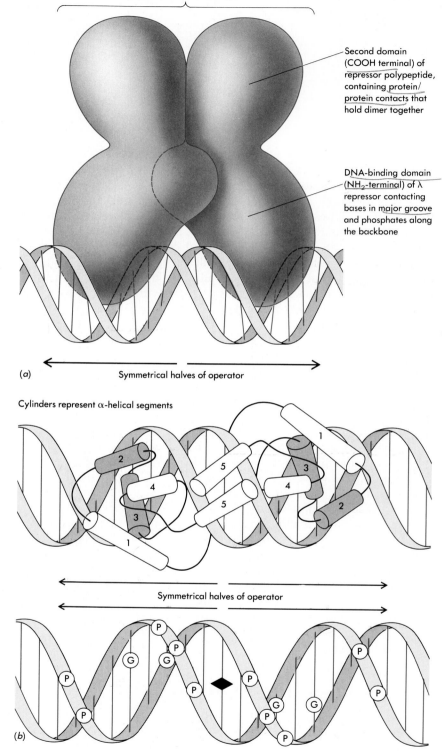

λ repressor dimer bound to operator

Second domain (COOH terminal) of repressor polypeptide, containing protein/protein contacts that hold dimer together

DNA-binding domain (NH₂-terminal) of λ repressor contacting bases in major groove and phosphates along the backbone

(a)   Symmetrical halves of operator

Cylinders represent α-helical segments

Symmetrical halves of operator

(b)

an overall shape with a twofold axis of symmetry matching that of the binding site in DNA. In the case of λ repressor, for example, the amino terminus of each chain folds into a domain that binds to one-half of the operator (Figure 16-6a).

Figure 16-6 (*Continued*)

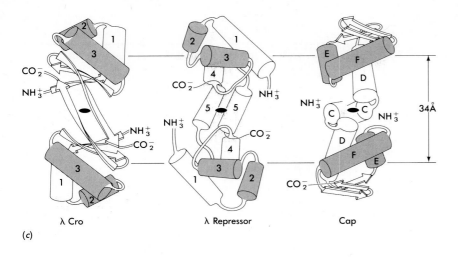

λ Cro   λ Repressor   Cap

(c)

The most critical DNA contacts are made by two short, adjacent α helices located at the binding face of each protein monomer; these are α helices numbers 2 and 3 in the λ repressor structure of Figure 16-6b and are shown for CAP, Cro, and λ repressor together in Figure 16-6c. In λ repressor, helix 3 protrudes from the protein surface, and it is separated from its counterpart in the second subunit of a λ repressor dimer by exactly the helix repeat distance of B-DNA (34 Å). Model building suggests that helix 3 fits snugly into the major groove, so that the repressor dimer binds one side of the DNA through the contacts made by the helix 3 of each monomer with half of the operator (Figure 16-6b). Most of the contacts that provide specific recognition of the λ operator result from the hydrogen-bonding complementarity between the edges of the base pairs in the major grooves and amino acid side chains of helix 3 (Figure 16-7). Strong evidence for this model is the fact that most mutations affecting λ repressor binding to the operator change amino acids in the two adjacent α helices. Very recently, the model has been confirmed directly by X-ray analysis of the repressor of phage 434 (a close relative of λ) bound to a synthetic DNA oligonucleotide containing the 434 operator sequence.

A knowledge of the actual structure and amino acid sequence of the critical pair of α helices in these several cases allows us now to predict whether this structure is likely to form simply by inspecting the amino acid sequence of a protein. Examination of more than ten DNA binding proteins indicates that a pair of α helices is a general feature

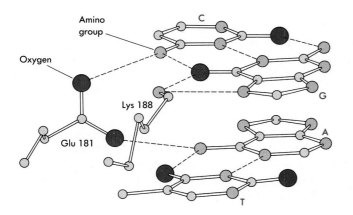

Figure 16-7
Postulated hydrogen-bond contacts between the side chains of two amino acids of CAP protein, glutamic acid at position 181 and lysine at position 188, and two base pairs in the CAP binding site. (Courtesy of T. A. Steitz and I. T. Weber.)

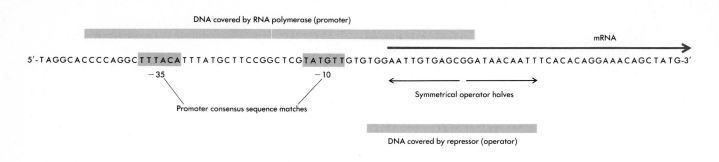

Figure 16-8
Overlap between the lactose operon promoter and operator sites. The DNA strand identical in sequence to the mRNA is shown.

of the active sites of regulatory proteins that bind specific DNA sequences. The discovery of this "helix-turn-helix" motif in DNA binding regulatory proteins is a most important advance in our understanding of the structural basis of genetic regulation.

Repressors Prevent RNA Polymerase Binding

The identification of DNA binding sites for both RNA polymerase (which bind to promoters) and repressors (which bind to operators) immediately suggested how repressors work: They physically prevent RNA polymerase from binding and initiating RNA synthesis at the promoter. The lactose operator (Figure 16-8) is made up of the same 21 base pairs of DNA that is the template for the first 21 nucleotides of lactose operon mRNA. Thus, *lac* repressor covers the template just where the catalytic center of RNA polymerase would lie, thereby denying it access to the template. Other operators are found to overlap different parts of the promoter, but all interfere with RNA polymerase binding. As we would expect, the repressor prevents only initiation, having no effect on chain growth once elongation has commenced.

Corepressors and Inducers Determine the Functional State of Repressors

Repressors are not always able to prevent specific mRNA synthesis. Otherwise, they would permanently inhibit the synthesis of their specific proteins. Instead, many repressor molecules can exist in both an active and an inactive form, depending on whether they are combined with their appropriate inducers (or corepressors). The attachment of an inducer inactivates the repressor. For example, when combined with a β-galactoside sugar like lactose or allolactose (a metabolite of lactose that is the natural inducer), the *lac* repressor cannot bind to its specific operator. Thus, the addition of β-galactosides to growing cells permits β-galactosidase synthesis by decreasing the concentration of active *lac* repressor molecules. In contrast, the binding of a corepressor changes an inactive repressor into an active form. For example, the addition of the amino acid tryptophan to cells activates the repressor that controls the synthesis of enzymes involved in tryptophan biosynthesis. This quickly shuts off synthesis of their specific mRNA molecules (Figure 16-9).

No covalent bond is formed between repressors and their specific inducers or corepressors. Instead, a portion of each repressor mole-

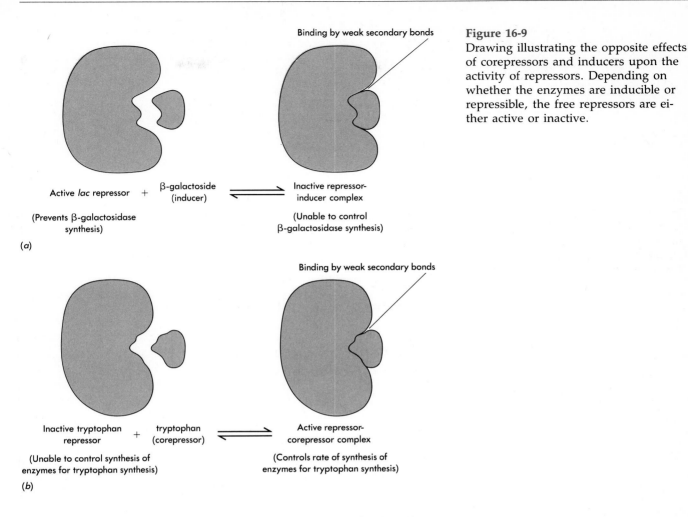

Active *lac* repressor + β-galactoside (inducer) ⇌ Inactive repressor-inducer complex

(Prevents β-galactosidase synthesis) — Binding by weak secondary bonds — (Unable to control β-galactosidase synthesis)

(a)

Inactive tryptophan repressor + tryptophan (corepressor) ⇌ Active repressor-corepressor complex

(Unable to control synthesis of enzymes for tryptophan synthesis) — Binding by weak secondary bonds — (Controls rate of synthesis of enzymes for tryptophan synthesis)

(b)

Figure 16-9
Drawing illustrating the opposite effects of corepressors and inducers upon the activity of repressors. Depending on whether the enzymes are inducible or repressible, the free repressors are either active or inactive.

cule is complementary in shape to a specific portion of its inducer or corepressor, and the two are joined by weak secondary bonds (hydrogen bonds, salt linkages, or van der Waals forces). Since these bonds are weak, they are rapidly made and broken, allowing the repressor state (active or inactive) to adjust quickly to the physiological need. For example, the synthesis of β-galactosidase mRNA ceases almost immediately after the removal of lactose. The reversible change in shape that a repressor undergoes upon binding an inducer is another example of an allosteric transformation (Chapter 4).

Repressors Can Control More Than One Protein

In some cases, repressors control the synthesis of only one protein. Often, however, a single repressor affects the synthesis of several enzymes. The *lac* repressor of *E. coli*, for example, controls at least three enzymes from its single operator: β-galactosidase itself; galactoside permease, which controls the rate of entry of β-galactosides into the bacteria; and galactoside acetylase, whose function is not yet known. When active β-galactoside permease is absent, *E. coli* cells are unable to concentrate β-galactosides within themselves. Since β-galactosidase, β-galactoside permease, and galactoside acetylase (?) are ordinarily needed to metabolize β-galactosides, their *coordinated*

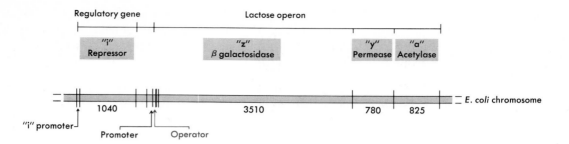

Figure 16-10
The lactose operon and its associated regulatory genes, drawn to scale based on known sizes of their gene products. The number of base pairs in each gene is shown.

synthesis is clearly desirable. Coordinated synthesis is brought about by having the enzymes coded by adjacent genes, which are transcribed into a single mRNA molecule carrying all three genetic messages (Figures 16-10 and 16-11). An even larger number of genes is coordinately repressed by the repressor of tryptophan biosynthesis. Again, this is achieved by having a single mRNA molecule carry the messages of all these genes. Such a collection of adjacent genes controlled from a single operator is called an *operon*.

Absence of an Operator Leads to Constitutive Synthesis

Like repressors, the operators to which they bind have *negative* functions: If a functional operator is absent, the corresponding repressor cannot inhibit the synthesis of the specific mRNA, and as a result, there is constitutive synthesis of its corresponding protein product(s).

The existence of operators was first revealed by genetic analysis. The structure of the operator can mutate to an inactive form, preventing the binding of its repressor. When this happens, constitutive enzyme synthesis results. These mutants are therefore called O^c (constitutive) mutants. O^c mutations can easily be distinguished from mutations in the repressor genes by measuring enzyme synthesis in special, partially diploid cells containing two copies of the relevant chromosomal regions. The lactose operons of cells containing one nonfunctional and one functional repressor gene are still repressible,

Figure 16-11
How the interaction of repressor, inducer, and operator controls the synthesis of the *E. coli* proteins β-galactosidase, β-galactoside permease, and galactoside acetylase.

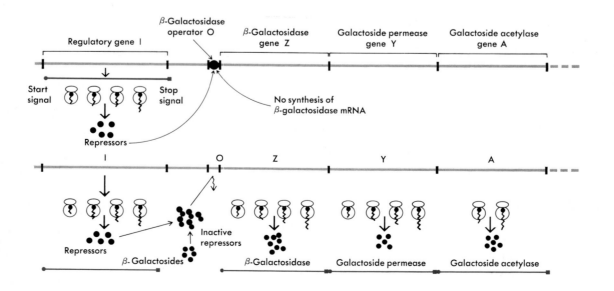

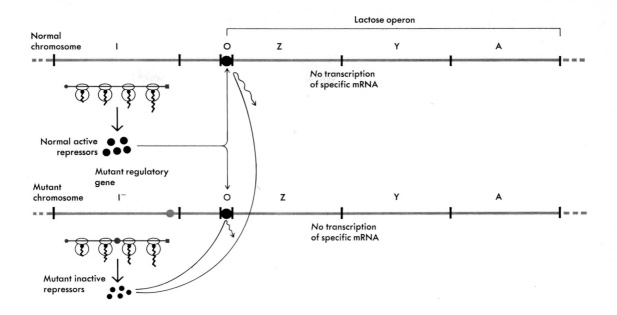

since good repressor molecules can act on both operators (Figure 16-12). In contrast, cells containing only one bad operator will express lactose operon enzymes constitutively, no matter what the condition of the repressor gene (Figure 16-13).

Figure 16-12
The use of partially diploid cells to show that the presence of functional repressors is dominant over the presence of inactive repressors. No significant amounts of β-galactosidase molecules will be produced in these cells in the absence of externally added β-galactosides.

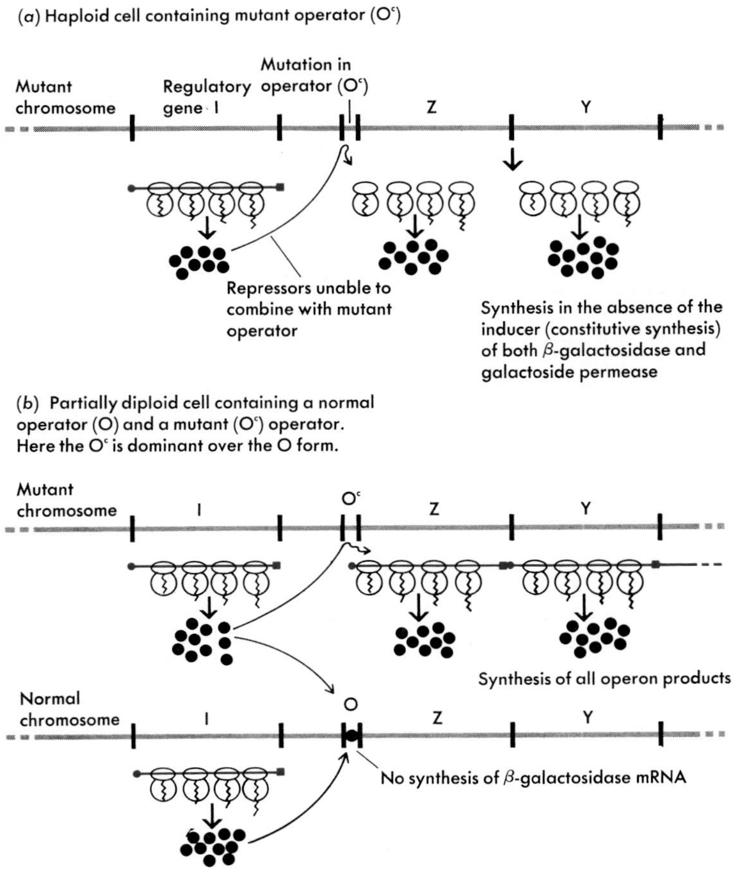

(a) Haploid cell containing mutant operator (O^c)

Mutant chromosome — Regulatory gene I — Mutation in operator (O^c) — Z — Y

Repressors unable to combine with mutant operator

Synthesis in the absence of the inducer (constitutive synthesis) of both β-galactosidase and galactoside permease

(b) Partially diploid cell containing a normal operator (O) and a mutant (O^c) operator. Here the O^c is dominant over the O form.

Mutant chromosome — I — O^c — Z — Y

Synthesis of all operon products

Normal chromosome — I — O — Z — Y

No synthesis of β-galactosidase mRNA

Figure 16-13
The control of specific mRNA synthesis by normal and mutant operators.

Positive Control of Lactose Operon Functioning

Although it was initially thought that the lactose operon was solely under negative control by the lactose repressor, it is now known that even in the presence of an inducer that neutralizes its repressor, a protein-mediated positive control signal also must be received for normal expression of the operon. The existence of this positive control protein was discovered during the study of how glucose blocks the functioning of a number of operons (**glucose-sensitive operons**), each controlling the breakdown (catabolism) of a specific sugar (e.g., lactose, galactose, arabinose, and maltose). For example, when *E. coli* is grown in the presence of both glucose and lactose, only the glucose is utilized, and the lactose operon proteins are made at a low level. Similarly, when both glucose and galactose are present, the galactose operon is inactive. The reason that glucose is preferred over other sugars can only be guessed. Perhaps, during evolution, bacteria more often found themselves in a glucose-rich environment than in one dominated by other sugars.

In any case, glucose inhibition does not operate by influencing the rate of entry of the various sugars into the cells. Instead, it acts by influencing the transcription pattern. This was first suggested by the discovery of mutations in the lactose operon control region that make the lactose operon insensitive to glucose effects. With such mutants, the lactose operon can be maximally induced, even in the presence of large amounts of glucose.

Glucose Catabolism Affects the Cyclic AMP Level

The action of glucose on transcription is not direct. Instead, one of its breakdown products (catabolites) acts by lowering the intracellular amount of *cyclic AMP* (cAMP). This key metabolite is required for the transcription of all the operons that are inhibited by glucose catabolism. The manner in which a glucose catabolite controls the cyclic AMP level within a cell is still unknown. ATP is the direct metabolic precursor of cAMP, and the enzyme responsible for this transformation, **adenylcyclase**, may be directly inhibited by a specific catabolite (Figure 16-14). On the other hand, since there is an enzyme (**phosphodiesterase**) that specifically converts cAMP to AMP, inhibition might be controlled by the rate of cAMP breakdown.

Activation of the Catabolite Activator Protein (CAP) by cAMP Binding[13]

Cyclic AMP does not directly promote *lac* mRNA synthesis, but works instead by binding to the **catabolite gene activator protein** (**CAP**), a dimeric molecule with a molecular weight of 45,000 daltons. CAP has no influence on transcription until cAMP has bound to it. It then acquires the ability to bind to very specific sites on DNA, and by so doing increases the rate of transcription of adjacent operons. CAP is the positive control element for all the glucose-sensitive operons; so cells with mutations that block its functioning are simultaneously unable to utilize a large number of different sugars.

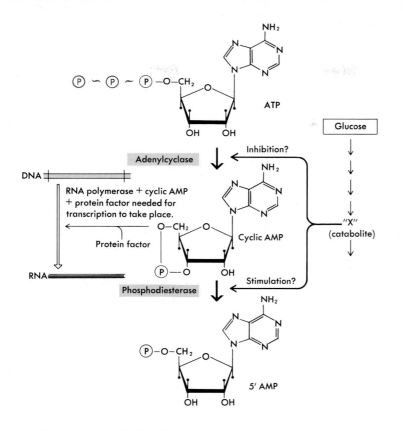

Figure 16-14
Control of catabolite-sensitive transcription through cyclic AMP.

CAP Controls RNA Polymerase Binding at the Lactose Promoter[13, 14]

The binding of neither CAP nor a specific repressor has any influence on the rate at which mRNA chains grow. Instead, both act—one positively, the other negatively—by controlling the rate at which RNA polymerase molecules attach to promoters. Whereas repressor blocks binding of RNA polymerase, CAP (when complexed with cyclic AMP) *assists* RNA polymerase in binding effectively to the lactose promoter, allowing much more frequent initiation of RNA synthesis. Like the operator, the site where CAP binds to lactose operon DNA is marked both by mutations that prevent its function (no stimulation of lactose enzymes in conditions of high cellular cyclic AMP) and by footprinting and chemical protection experiments, which identify a specific binding sequence. The important result is that CAP complexed with cyclic AMP binds next to RNA polymerase at the promoter (Figures 16-15 and 16-16).

How CAP enhances RNA polymerase binding at the promoter is a question of current interest and much importance for understanding the way regulatory proteins work. There are two main possibilities. First, RNA polymerase is close enough that it might directly complex with CAP as it binds. Such a protein-protein contact would provide extra energy to greatly strengthen RNA polymerase binding. Second, CAP binding could change the detailed structure of the DNA helix surrounding its binding site into a shape that RNA polymerase recognizes much better than unbound DNA. We know that DNA is flexible enough to allow such local distortions from the normal double-helical structure.

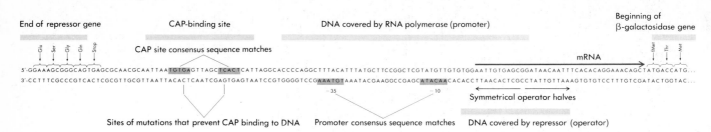

Figure 16-15
The nucleotide sequence and organization of the *E. coli* lactose operon control region.

In Vitro Analysis of Promoter Functioning

An important step in establishing our concepts about how promoters and operators function has come from in vitro transcription experiments. For example, when *lac* DNA is used as the in vitro template, the binding of RNA polymerase and the initiation of *lac* mRNA synthesis requires the simultaneous presence of cAMP and CAP. And when the DNA templates have been isolated from mutant cells whose *lac* operon is transcribed even when glucose is present, no requirement for CAP is found in the in vitro systems. Such experiments provide the clearest demonstrations that cAMP, CAP, and repressors act at the transcriptional level and not through effects on membrane permeability.

DNA Repair Genes at Many Different Sites on the Chromosome Are Regulated by a Single Repressor[15, 16, 17]

The *lac* repressor binds to only a single operator, and this operator happens to be near the repressor gene itself. But neither feature is essential to the function of the repressor. Because a repressor diffuses throughout the cell, it can regulate any number of genes, including some that are distant from its own. By inactivating such a repressor, the cell can invoke expression of a variety of cellular genes to serve a common function. A striking example is provided by the set of about fifteen *SOS genes*, which encode proteins that help the cell repair damage to its DNA, such as that caused by ultraviolet light (Figure 16-17). Each SOS gene has an operator recognized by the LexA repressor (encoded by the *lexA* gene). DNA damage is detected by the recombination enzyme RecA protein, which binds the single-stranded DNA left by replication of damaged regions (and begins to repair this damage by recombination; Chapter 11). It was a surprising discovery that the inducer of the SOS genes is not a small molecule like lactose or tryptophan, but is instead RecA protein itself. Binding to single-stranded DNA activates a site on the RecA polypeptide, distinct from those involved in strand pairing, that has a protease activity specific for LexA repressor (and phage repressors; Chapter 17). LexA molecules are cut into two fragments, so that their operator binding activity is lost and mRNA for the SOS genes can be made. Unlike the case of the lactose repressor, inactivation of LexA is irreversible: Repression can be reestablished only when more LexA is made.

The details of SOS gene induction show how flexibly and subtly a repressor control can serve the needs of the cell. Besides the SOS genes, LexA also represses the *recA* and *lexA* genes themselves. Here

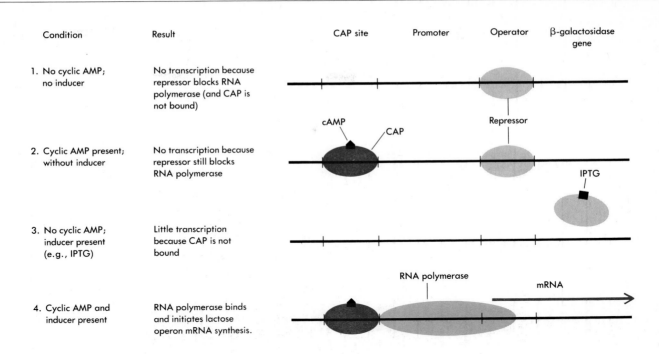

repression is not complete: A substantial basal level of both proteins is made, even when a normal level of LexA repressor is present. Destruction of LexA then greatly increases the rate of synthesis of RecA and LexA proteins, an increase that serves a different purpose in each case. More RecA protein accumulates to satisfy the need for a high rate of DNA repair by recombination. In contrast, the increased rate of LexA synthesis is not sufficient to allow it to accumulate in conditions of DNA damage, because the abundant RecA protease activity easily destroys all new LexA made at the higher rate. But this faster synthesis is important when damage is repaired and the protease activity disappears: LexA then is made rapidly so that repression can

Figure 16-16
Control of initiation of lactose operon transcription by CAP, repressor, cyclic AMP, and an inducer (like allolactose, the natural inducer, or isopropyl-thiogalactoside [IPTG], a useful synthetic inducer).

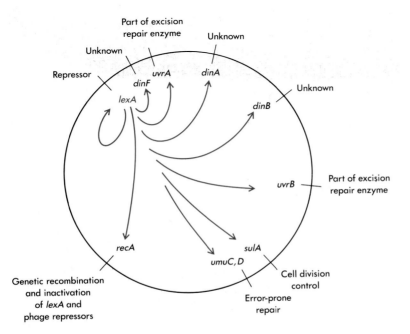

Figure 16-17
Location of SOS DNA repair genes on the circular chromosome of *E. coli*. All the SOS genes are regulated by the LexA repressor. The names of the genes are given inside the circle, and their functions (where known) outside. (After G. Walker, *Microbiological Reviews* 48 (1984):67, with permission.)

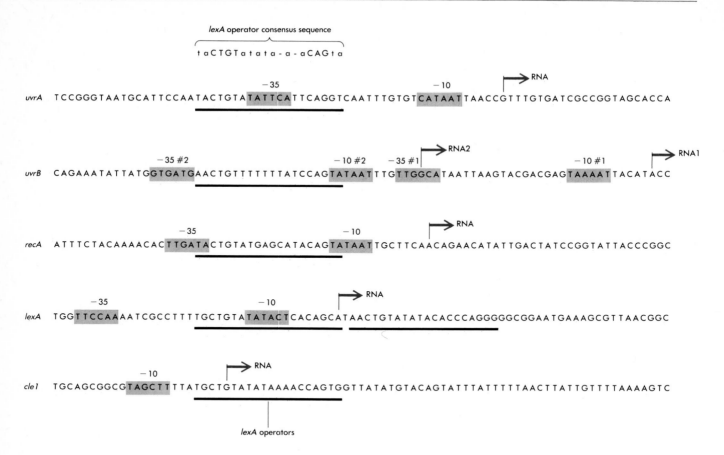

Figure 16-18
DNA sequences of five SOS gene promoter regions, showing the relative positions of promoters and operators. Promoter 1 of *uvrB* is repressed by LexA, whereas promoter 2 is not. The *lexA* gene has two adjacent operators to which its own product binds. The *cle1* gene of plasmid ColE1 encodes colicin E$_1$, an antibacterial toxin. In the *lexA* operator consensus sequence, highly conserved bases are shown in black, and less highly conserved bases are in color. (After A. Sancar, G. Sancar, W. Rupp, J. Little, and D. Mount, *Nature* 298 (1982):98, with permission.)

be reestablished efficiently. Another regulatory variation is found in the expression of *uvrB*, an SOS gene that encodes part of the major enzyme of excision repair. UvrB protein also is made at a high basal level in the presence of active LexA repressor, but for a different reason. Its mRNA is initiated from two adjacent but separate promoters, only one of which is repressed by LexA; the second is always active, providing the cell a substantial level of the excision enzyme at all times so that it can begin this type of repair as soon as the damage occurs (Figure 16-18). Still a further refinement is that operators of the various SOS genes bind LexA repressor with different affinities. Some bind it weakly, allowing their genes to be expressed after mild DNA damage that yields moderate levels of protease activity and reduces the LexA concentration only perhaps two- or three-fold. Other operators bind LexA more avidly and are induced only by extensive damage that reduces the LexA concentration much further.

A comparison of DNA sequences at the beginning of several SOS genes illustrates an important aspect of the way repressors work: The operator can be almost anywhere within the promoter, as long as repressor bound to it blocks DNA to which RNA polymerase binds (see Figure 16-18). Thus, some operators cover the −35 hexamer, some straddle the −35 and −10 hexamers, one covers the −10 hexamer, and some are mostly within the transcribed region. Since the site of RNA polymerase binding extends beyond the consensus hexamers on either side (see Chapter 13), this variety of locations makes sense. The fact that LexA represses promoter 2 but not promoter 1 of the *uvrB* gene shows that a repressor can approach within about 7 or 8 bases of the −35 hexamer without preventing RNA polymerase function.

Arabinose C Protein Is Both an Activator and a Repressor[14, 18–22]

Whereas the catabolite gene activator protein (CAP) passes the cellular signal for glucose starvation to many operons and genes, some positive regulators (activators) deliver specific signals to only one or a few genes. The arabinose *C* gene encodes one such protein, which activates transcription of the arabinose *BAD* operon, consisting of genes *araB*, *araA*, and *araD* (Figure 16-19). C protein must be complexed with arabinose to do this, so that the enzymes are made only when their substrate is present. Like the lactose operon, the arabinose operon is sensitive to glucose and thus is controlled by cyclic AMP and CAP in addition to arabinose and C protein. How can *two* activators, C protein and CAP, regulate one operon?

Perhaps not surprisingly, the mechanism is a little complicated. C protein is not only an activator but also a repressor, and in fact, it acts as a repressor in two different ways. C protein binds three separate sites in the arabinose operon control region, *araI*, *araO$_1$*, and *araO$_2$* (Figure 16-20). The site in the middle, *araO$_1$*, overlaps the promoter for gene *C* and serves only to repress synthesis of gene *C* mRNA when sufficient C protein has accumulated in the cell; thus, the *C* gene is *autoregulated*. The other two C protein binding sites, *araI* and *araO$_2$*, act along with the CAP binding site to mediate repression and activation of the *BAD* operon. It is thought that when the CAP binding site is empty because the cyclic AMP concentration is low, one C protein molecule (perhaps a dimer or tetramer) binds both *araI* and the distant site *araO$_2$*, thereby drawing the DNA into a loop (see Figure 16-20). In this *repressing* configuration, C protein cannot activate RNA polymerase at the *BAD* promoter, whether or not arabinose is bound to it. However, the binding of CAP-cAMP next to *araI* changes the activity of C protein at *araI*, perhaps by touching it directly. C protein now releases the DNA site *araO$_2$*, thus undoing the loop; moreover, it can also activate RNA polymerase to initiate synthesis at the adjacent *BAD* promoter. For C protein to activate *BAD*

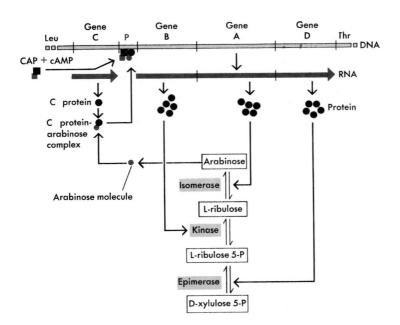

Figure 16-19

The arabinose operon of *E. coli*: an example of positive control.

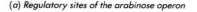

(a) *Regulatory sites of the arabinose operon*

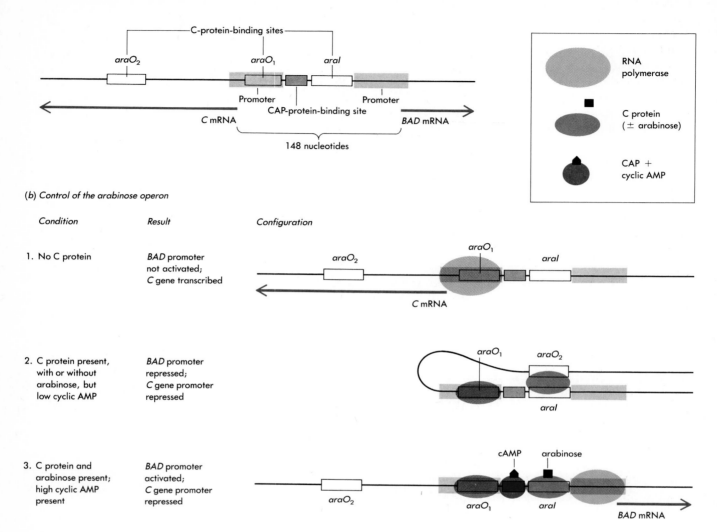

(b) *Control of the arabinose operon*

Figure 16-20
Regulation of the *E. coli* arabinose operon. (a) Regulatory sites of the arabinose operon. (b) Control of the arabinose operon.

transcription, it first must bind arabinose; therefore, two small molecules, cyclic AMP and arabinose, are required for efficient *BAD* expression. Exactly why a distant site should be required for repression is not clear, but evidence for this mechanism is quite persuasive. In particular, the function of CAP-cAMP in breaking the looped repressing configuration is shown by the fact that deletion of site *araO₂* allows the *BAD* promoter to work in the absence of CAP and cyclic AMP, although it still needs C protein and arabinose.

As with the lactose operon, the cell uses an operon-specific regulator (*lac* repressor or C protein) along with CAP protein to detect the presence of a specific sugar and the absence of glucose, although it has evolved a very different mechanism for the arabinose operon. It will be surprising if many other variations do not appear when more regulatory schemes are known in this detail. Like arabinose C protein, any DNA binding protein could be either an activator or a repressor, depending on whether it blocks or stimulates binding of RNA polymerase to a promoter. Indeed, the phage λ repressor does both (Chapter 17). Sometimes, CAP protein also acts as a repressor, preventing expression of genes not needed when glucose is limiting. For example, the complex control of the galactose operon of *E. coli* involves CAP protein acting both as activator and repressor. In addition, there is a separate galactose-regulated repressor that binds two operators and might thereby draw the DNA into a loop like that envisioned for the arabinose operon. Very recent evidence suggests that CAP may have a secondary role as a repressor even in the lactose operon. Besides the major promoter of the lactose operon, there is a second, upstream promoter where RNA polymerase binds strongly but initiates RNA synthesis very inefficiently. Because the two promoters overlap, this binding blocks productive interaction of RNA polymerase with the major promoter. The CAP binding site overlaps the second promoter, so that CAP protein prevents obstructive binding of RNA polymerase to the second promoter at the same time it directly stimulates RNA polymerase binding to the major promoter. Both functions of CAP are important for *lac* operon expression.

The CAP and arabinose C proteins represent two classes of activators found in bacteria: general activators like CAP, which each control many genes whose functions are required in certain conditions (e.g., glucose, nitrogen, or phosphate limitation), and specific activators like arabinose C, which control a few genes whose products metabolize particular compounds (e.g., arabinose, maltose, or serine). In Chapter 17, we describe another transcription activator whose mechanism is well understood: the phage λ cII protein.

Genes for Heat-Shock Proteins of *E. coli* Are Recognized by a σ Factor (σ³²) That Specifically Recognizes Their Promoters[23–29]

Repressors and operon-specific activators like the arabinose C protein that control relatively few genes usually act by changing the availability or initiation efficiency of promoters. However, they do not change the activity of RNA polymerase, which remains available to any receptive promoter. A different strategy is used when conditions require a massive redirection of gene expression. Such is the case, for example, after a sudden rise in the growth temperature, called **heat shock**. In almost all kinds of cells subjected to heat shock, certain proteins (about 17 in *E. coli*) begin to be made much faster than usual. If the new temperature is not too high (e.g., 42°C for *E. coli*), the rate of heat shock protein synthesis soon declines and the normal pattern of protein synthesis soon resumes (in about 20 minutes for *E. coli*). But at temperatures too high for growth (50°C for *E. coli*), synthesis of only heat-shock proteins continues. Remarkably, some heat-shock proteins of widely different species are closely related; in fact, there are even similarities between those of bacteria and those of eucaryotes. For example, the gene for the 66,000-dalton *E. coli* heat-shock protein DnaK is identical in almost half its nucleotides to the gene for the 70,000-dalton *Drosophila* heat-shock protein Hsp70. Although we know that some heat-shock proteins are required in *E. coli* for phage growth—the proteins GroEL and GroES are required for phage λ particle assembly, and DnaK for λ DNA synthesis—we have no idea what function they have in common that is essential to save cells from the rigors of a sudden temperature rise or certain other stressful treatments such as the addition of ethanol to the culture that also induce the heat shock genes. (Perhaps the function of heat-shock proteins in phage assembly somehow reflects the fact that phage infection is itself a shock.) What makes heat shock particularly interest-

Table 16-1   Sequences of Heat Shock Promoters*

Promoter	−35 Region		−10 Region	+1
groE	TTTCCCCCTTGAA	GGGGCGAAGCCAT	CCCCATTTCTCTGGTCAC	
dnaK promoter 1	TCTCCCCCTTGAT	GACGTGGTTTACGA	CCCCATTTAGTAG TCAA	
dnaK promoter 2	TTGGGCAGTTGAA	ACCAGACGTTTCG	CCCCTATTACAGACTCAC	
C62.5 gene	GCTCTCGCTTGAA	ATTATTCTCCCTTGT	CCCCATCTCTCCCACATC	
σ^{70}	TGCCACCCTTGAA	AAACTGTCGATGTGG	GACGATATAGCAG ATAA	
Heat-shock promoter (σ^{32}) consensus	T - - C - C - CTTGAA	13–15 bp	CCCCAT- T	
Standard promoter (σ^{70}) consensus	TTGACA		TATAAT	

*DNA sequences of the RNA-like strand of *E. coli* heat-shock promoters. Note that the "standard" σ, σ^{70}, is itself a heat-shock protein. Bases matching the σ^{32} consensus sequence are underscored.
SOURCE: Adapted from D. W. Cowing, J. C. A. Bardwell, E. A. Craig, C. Woolford, R. W. Hendrix, and C. A. Gross, "Consensus for *Escherichia coli* Heat Shock Gene Promoters," *Proc. Nat. Acad. Sci.* 82 (1985):2679–2683, with permission.

ing is the fact that the entire pattern of gene expression can be so abruptly shifted.

A critical discovery about the mechanism of this shift is that promoters for *E. coli* heat-shock proteins are recognized by a unique form of RNA polymerase holoenzyme containing a sigma (σ) factor with a molecular weight of 32,000 daltons (σ^{32}), encoded by gene *rpoH* (previously called *htpR*). Factor σ^{32} is a minor protein, much less abundant than the primary σ factor, σ^{70}; it binds tightly to core enzyme to form a holoenzyme that can be purified separately from holoenzyme containing σ^{70}. This minor holoenzyme acts exclusively on promoters of heat-shock proteins, and not on the "standard" promoters of other genes. Correspondingly, heat-shock promoters have different sequences from the standard promoter (Table 16-1), particularly in the −10 segment that σ factors may recognize (Chapter 13).

Although it is clear that σ^{32} is responsible for recognizing heat-shock promoters, the critical regulatory event that causes it to function after a temperature shift is still a mystery. Also completely unexplained is why σ^{32} stops acting on promoters of heat shock genes when cells adapt to a moderately higher temperature of growth.

Separate σ factors also control sporulation in *B. subtilis* (Chapter 22) and phage growth in both *B. subtilis* and *E. coli* (Chapter 17). *B. subtilis* contains at least four σ factors, possibly because sporulation is a radical and prolonged transformation of cell structure that implies a correspondingly large change in gene expression. There is recent evidence that the control of genes required for nitrogen metabolism in *E. coli* and other bacteria, including genes for nitrogen fixation, involves a distinct σ factor, which is encoded by gene *ntrA* in *E. coli*.

Amino Acid Biosynthetic Operons Are Regulated by Varying Termination of mRNA Synthesis Before the Structural Genes[30, 31]

Initially, it seemed likely that the synthesis of five enzymes of the tryptophan *(trp)* operon (see Figure 8-10) would also be regulated primarily by a repressor. Here the important feature is that transcription of the five contiguous *trp* genes is needed only when tryptophan is limiting (Figure 16-21). Thus, tryptophan acts not as an inducer but

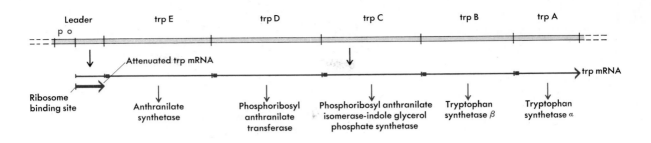

as a corepressor that activates its specific repressor so that it can bind to the *trp* operator and prevent transcription of its respective operon. When the tryptophan concentration is low, the *trp* operator is free and the synthesis of *trp* mRNA commences at the adjacent promoter. Surprisingly, however, a *trp* mRNA molecule, once so initiated, does not automatically grow to a full length. Instead, most *trp* mRNA molecules stop growing before the transcription of even the first *trp* gene (*trpE*) begins, unless a second and novel device confirms that little tryptophan is available.

The key to understanding this unexpected result came from sequence analysis of the 5′ end of tryptophan operon mRNA, which revealed that 161 nucleotides of RNA are made from the tryptophan promoter before RNA polymerase encounters the first codon of *trpE* (Figure 16-22). Near the end of the sequence and before *trpE* is a transcription terminator (Chapter 13), composed of a characteristic hairpin loop in the RNA (made from sequences in regions 3 and 4 of Figure 16-22), followed by 8 uridine residues. At this so-called **attenuator**, RNA synthesis usually stops (and, we might have thought, should always stop), yielding a leader RNA 139 nucleotides long. How, then, can mRNA for the whole operon ever be made? Three features of the leader sequence allow the attenuator to be passed by RNA polymerase when the cellular concentration of tryptophan is

Figure 16-21
The tryptophan operon of *E. coli*, showing the relation of the leader to the structural genes that code for the *trp* enzymes.

Figure 16-22
Features of the nucleotide sequence of the *trp* operon leader RNA.

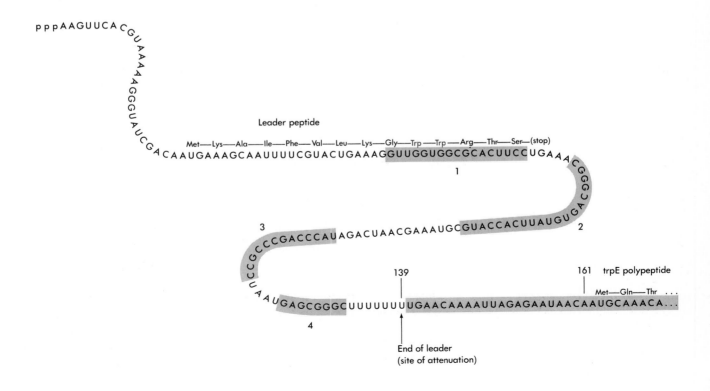

Figure 16-23
How transcription termination at the *trp* operon attenuator is controlled by the availability of tryptophan.

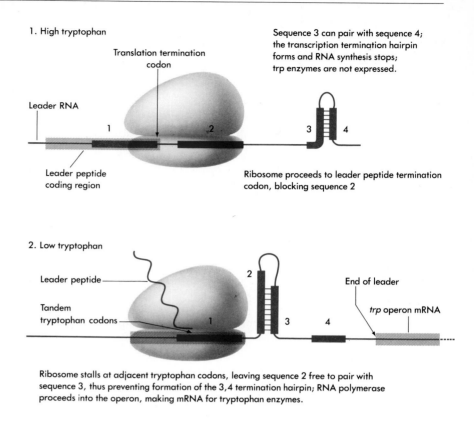

1. High tryptophan

Translation termination codon

Sequence 3 can pair with sequence 4; the transcription termination hairpin forms and RNA synthesis stops; trp enzymes are not expressed.

Leader RNA

Leader peptide coding region

Ribosome proceeds to leader peptide termination codon, blocking sequence 2

2. Low tryptophan

Leader peptide

Tandem tryptophan codons

End of leader

trp operon mRNA

Ribosome stalls at adjacent tryptophan codons, leaving sequence 2 free to pair with sequence 3, thus preventing formation of the 3,4 termination hairpin; RNA polymerase proceeds into the operon, making mRNA for tryptophan enzymes.

3. No protein synthesis

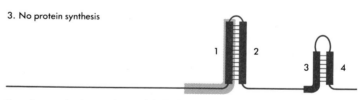

If no ribosome begins translation of the leader peptide AUG, hairpin 1,2 forms, preventing formation of 2,3 and thus allowing the terminator hairpin 3,4 to form; tryptophan enzymes are not expressed.

low. First, there is a second hairpin besides the terminator hairpin that can form, made of regions 1 and 2 of the leader (Figure 16-23). Second, region 2 also is complementary to region 3; thus, yet another hairpin, consisting of regions 2 and 3, can form, thereby preventing the terminator hairpin (3, 4) from forming. Third, the leader RNA codes for a short leader peptide of 14 amino acids that is preceded by a strong ribosome binding site. The leader peptide has a striking sequence feature: two tryptophan codons in a row. Their importance is underscored by corresponding sequences found in similar leader peptides of other operons encoding enzymes that make amino acids (Table 16-2). Thus, the leucine operon leader peptide has four adjacent leucine codons in a row, and the histidine operon leader peptide has seven histidines in a row!

The function of these codons is to stop a ribosome attempting to translate the leader peptide; thus, when tryptophan is scarce, little charged tryptophan tRNA is available, and the ribosome must stall when it reaches the tryptophan codons. RNA around the tryptophan codons stays bound to the ribosome and cannot be part of a hairpin loop. Figure 16-23 shows the consequence. A ribosome caught at a

Table 16-2 Leader Peptides of Attenuator-Controlled Operons Containing Genes for Amino Acid Biosynthesis*

Operon	Amino Acid Sequence of Leader Peptides
Tryptophan	Met Lys Ala Ile Phe Val Leu Lys Gly Trp Trp Arg Thr Ser
Threonine	Met Lys Arg Ile Ser Thr Thr Ile Thr Thr Thr Ile Thr Ile Thr Thr Gly Asn Gly Ala Gly
Histidine	Met Thr Arg Val Gln Phe Lys His His His His His His His Pro Asp
Isoleucine-valine GEDA	Met Thr Ala Leu Leu Arg Val Ile Ser Leu Val Val Ile Ser Val Val Val Ile Ile Ile . . . . . . Pro Pro Cys Gly Ala Ala Leu Gly Arg Gly Lys Ala
Leucine	Met Ser His Ile Val Arg Phe Thr Gly Leu Leu Leu Leu Asn Ala Phe Ile Val Arg Gly Arg Pro . . . . . . Val Gly Gly Ile Gln His
Phenylalanine	Met Lys His Ile Pro Phe Phe Phe Ala Phe Phe Phe Thr Phe Pro
Isoleucine-valine B	Met Thr Thr Ser Met Leu Asn Ala Lys Leu Leu Pro Thr Ala Pro Ser Ala Ala Val Val Val . . . . . . Val Arg Val Val Val Val Val Gly Asn Ala Pro

*The biosynthesis of isoleucine and valine is complex: The genes are encoded in several operons, and the pathway to leucine synthesis is a branch of the valine pathway. Thus, isoleucine, valine, and leucine are all involved in attenuation of the isoleucine-valine operons. (After C. Bauer, J. Carey, L. Kasper, S. Lynn, D. Waechter, and J. Gardner in *Gene Function in Prokaryotes*, Cold Spring Harbor, 1983, p. 68, with permission.)

tryptophan codon is placed so that region 2 is left free to pair with region 3; then the terminator hairpin (3, 4) cannot be made, and RNA polymerase passes the attenuator into the operon, allowing *trp* enzyme expression. If, on the other hand, there is enough tryptophan (and therefore enough charged *trp* tRNA) for the ribosome to proceed through the tryptophan codons, the ribosome blocks sequence 2, allowing the 3, 4 terminator hairpin to form and aborting transcription at the end of the leader RNA. A third possibility is that translation stops before the tryptophan codons (because some other amino acid is lacking) or fails to initiate altogether; then, sequence 1 is free to pair with sequence 2, and the terminator structure again forms to stop *trp* operon expression. The leader peptide is thought to be important only for positioning ribosomes during their synthesis. In fact, it is immediately destroyed by cellular proteases.

It is possible that the coexistence of repression and attenuation controls allows a finer tuning to the level of intracellular tryptophan, resulting in a two-stage response to progressively more stringent tryptophan starvation—the initial response being the cessation of repressor binding, with greater starvation leading to relaxation of attenuation. Other amino acid operons like *his* and *leu* have no repressors; instead, they rely entirely on attenuation for their control.

RNA Synthesis and Protein Synthesis Are Coupled in Bacteria[32, 33]

A polypeptide chain-terminating codon in a gene can have a surprising effect on genes transcribed later in the same operon: These genes may be expressed much less efficiently than normal, even though their own coding sequences are not interrupted by any terminating codons. This interference in downstream gene expression by an upstream translation stop is called **genetic polarity.** Its cause is termination of RNA synthesis induced by the failure of translation, and the cellular agent responsible is the rho (ρ) transcription termination factor (Chapter 13). This is shown by the fact that mutations isolated to prevent polarity, that is, to allow downstream gene expression after a normally polar mutation, map in the *rho* gene and reduce the activity of ρ protein. Rho must bind the RNA being synthesized in order to stop RNA polymerase, and usually this is prevented by ribosomes bound to the RNA. But discharge of ribosomes at a nonsense codon leaves the RNA free to bind ρ factor. Not surprisingly, the greater the distance between a nonsense codon and a new ribosome binding site, the more likely ρ is to act and the more severe is polarity. Polarity is thought to represent a cellular mechanism to prevent futile transcription, as might happen, for example, if the cell were starved and unable to support efficient protein synthesis.

Although ρ may be able to bind any substantial stretch of newly transcribed RNA that is not covered by ribosomes or enclosed in secondary structure, it cannot stop RNA polymerase just anywhere on the DNA. Instead, there are distinct sites where RNA polymerase pauses and allows ρ to release the RNA. Thus, RNA synthesis stops not where the nonsense codon that interrupts translation happens to be, but further downstream where the first RNA terminator is encountered.

The likelihood that most genes of substantial size have ρ-dependent terminators has suggested that some means exist to prevent ρ-dependent termination in genes encoding RNA that is not a messenger, such as ribosomal RNA genes. There is evidence that RNA polymerase initiating at ribosomal RNA promoters does transcribe through terminators, suggesting that the enzyme has been modified in some way. A precedent for operon-specific factors that regulate termination is provided by bacteriophage λ: It encodes two **antitermination** proteins, which modify RNA polymerase transcribing phage genes to allow it to ignore terminators (Chapter 17). Perhaps λ needs antiterminators because its genes are gathered in a few very long operons in which natural polarity caused by untranslated or sparsely translated segments would stop RNA synthesis prematurely without a specific device to defeat it.

Unequal Production of Proteins Coded by a Single mRNA Molecule

Variation in the number of molecules of different proteins arises also from the fact that the proteins coded by a single mRNA molecule need not be produced in similar numbers. This point is demonstrated by the study of the lactose operon proteins: Many more copies of β-galactosidase are synthesized than of galactoside permease or galactoside acetylase. The ratios in which these proteins appear are $1 : \frac{1}{2} : \frac{1}{5}$, respectively. This may mean that ribosomes attach to the different starting points along a given mRNA molecule at different rates, depending on the critical binding sequence preceding the AUG initiation codon (Chapter 14). Alternatively, the ribosomes may attach only at the β-galactosidase gene, with translation of subsequent genes depending on the frequency of ribosome detachment following the reading of a chain-terminating signal. This hypothesis fits nicely with the observation that the β-galactosidase gene is translated the most often and the acetylase the least often. Another factor that affects translation is the existence of codons whose corresponding tRNA species are present in very limiting amounts. Ribosomes must pause longer at such codons than at more common ones, waiting while scarce tRNAs diffuse to them.

It seems reasonable that there should be mechanisms to permit differential reading rates along single mRNA molecules. Although the coordinated appearance of related enzymes is obviously of great advantage to a cell, there is no reason why equal amounts should be produced. Equal numbers of molecules would be useful only if the specific catalytic activity rates (turnover numbers) of related proteins were equal. In general, however, there are great variations in individual turnover numbers.

Bacterial mRNA Is Often Metabolically Unstable

When corepressor or inducer molecules are added to or removed from growing bacteria, the rate of synthesis of the respective proteins is rapidly altered. This rapid adaptation to a changing environment is possible not only because growth requires continual synthesis of new mRNA molecules, but, even more significantly, because many bacterial mRNA molecules are metabolically unstable. The average lifetime of many *E. coli* mRNA molecules at 37°C is about 2 minutes, after which they are enzymatically broken down. The resulting free nucleotides are then phosphorylated to the high-energy triphosphate level and reutilized in the synthesis of new mRNA molecules.

Thus, there is almost complete replacement of the templates for many proteins every several minutes. For example, within several minutes after addition of suitable β-galactosides, *E. coli* cells synthesize β-galactosidase at the maximum rate possible for that particular inducer level. This is true because the rate of destruction soon matches the rate of synthesis of lactose mRNA, and the *amount* of mRNA therefore stays constant. If, on the other hand, all mRNA molecules were metabolically stable, the maximum rate of β-galactosidase synthesis would not be reached until a generation or so had passed. Moreover, the existence of unstable lactose mRNA also means that once β-galactosides are removed, synthesis of β-galactosidase quickly halts and does not resume until it is again needed (Figure 16-24).

The average lifetimes of mRNA molecules for different operons vary considerably; some, which perhaps encode proteins needed constantly, are very stable. This means that mRNA lifetime is itself genetically programmed by a nucleotide sequence (or sequences) that determines the chance that a nuclease attacks the molecule. Exactly what features of mRNA structure are important, and what nucleases are responsible, are only partly understood. One signal for nuclease attack is a specific type of hairpin loop that is a substrate for ribonuclease III, the enzyme responsible for processing both stable (e.g., ribosomal) and messenger RNA in *E. coli* (Chapter 14). Cleavage at these loops by RNase III opens the RNA to degradation by other nucleases, thus accelerating its destruction and decreasing its lifetime. The use of mRNA as a messenger also may affect its stability, since there are suggestions that bound ribosomes protect mRNA from degradation. Perhaps degradative enzymes must compete with ribosomes for sites where breakdown can begin. It is important to realize that the half-life of a messenger may be comparable to the time needed by RNA polymerase to transcribe a large operon. Thus, deg-

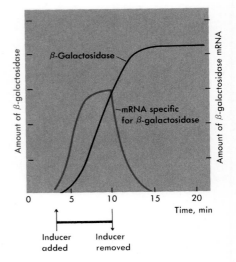

Figure 16-24
Rapid rise (or fall) of β-galactosidase mRNA upon the addition (or removal) of β-galactosidase inducers. The *E. coli* cells in this experiment were grown at 37°C under conditions where the cells divided every 40 minutes.

radation may begin before transcription finishes, in which case an entire mRNA molecule never exists.

Ribosome Attachment to mRNA Can Be Regulated[34]

From our first understanding that information is passed from DNA through RNA to proteins, it has been clear that synthesis of a particular protein could be switched on or off at two stages: when its messenger is made (transcriptional regulation), or when the messenger is read on ribosomes (translational regulation). Genetic studies led first to the discovery of transcriptional control, the mode we believe to predominate in both bacterial and higher cells. But we now know that ribosome binding to initiation sites on mRNA can be regulated, and we suspect that this will turn out to be a common mode of regulation, especially where quick and subtle changes in gene expression are required.

Messenger RNA Structure Determines the Availability of the Translation Initiation Site[35, 36]

Unlike promoters, which we believe to be uniformly accessible to RNA polymerase because they exist in the regular structure of duplex DNA, ribosome binding sites may be trapped in the secondary structure into which most single-stranded RNA molecules fold (Chapter 14). If this happens, ribosomes cannot initiate protein synthesis until the structure is broken, either by thermodynamic fluctuations or by a more directed mechanism. Sometimes, ribosome progression along the RNA from an upstream protein-coding sequence is required to open the RNA at an initiation site—a fact that explains the interdependence of translation of different genes in RNA phage (Chapter 17).

Such translation-dependent exposure of an enclosed ribosome binding site can be a specific regulatory device, linked to a controlling signal. This type of mechanism allows cells to express a gene specifying resistance to erythromycin, an antibiotic that binds the ribosomes of sensitive cells and inhibits protein synthesis. The gene for erythromycin resistance, carried on a plasmid that confers bacterial resistance to the antibiotic (Chapter 11), encodes an enzyme that methylates an adenine residue in 23S ribosomal RNA, thereby preventing the antibiotic from binding. Sequence analysis reveals a regulatory structure quite similar to that of the tryptophan operon, in that a leader polypeptide precedes the methylase gene. The controlling mechanism also involves RNA folding into alternative hairpins, but these RNA structures serve a different end. Here the "repressed" conformation prevents initiation of protein synthesis rather than causing termination of RNA synthesis. When a ribosome translates the leader polypeptide freely and is released, an RNA hairpin can form that buries the initiation sequence for methylase, preventing its translation (Figure 16-25). But when erythromycin binds and slows the progress of a ribosome making leader polypeptide, a different secondary structure forms that exposes the methylase initiation region, allowing methylase expression. Presumably, there is always a low level of methylase expression and thus a few erythromycin-

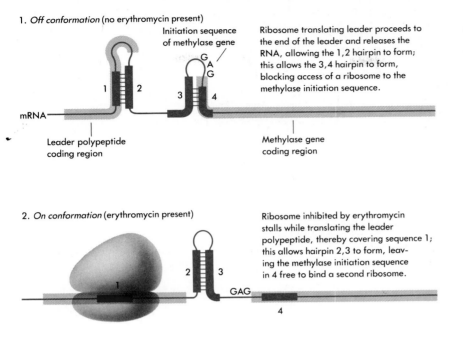

1. *Off conformation* (no erythromycin present)

Initiation sequence of methylase gene

Ribosome translating leader proceeds to the end of the leader and releases the RNA, allowing the 1,2 hairpin to form; this allows the 3,4 hairpin to form, blocking access of a ribosome to the methylase initiation sequence.

mRNA

Leader polypeptide coding region

Methylase gene coding region

2. *On conformation* (erythromycin present)

Ribosome inhibited by erythromycin stalls while translating the leader polypeptide, thereby covering sequence 1; this allows hairpin 2,3 to form, leaving the methylase initiation sequence in 4 free to bind a second ribosome.

GAG

Figure 16-25
Control of erythromycin methylase gene expression by a leader peptide and by alternate RNA structures that determine whether ribosomes can initiate translation of its gene. (After S. Horinouchi and B. Weisblum, *Mol. Gen. Genet.* 182 (1981):343, with permission.)

resistant ribosomes that can complete translation of the methylase gene when its initiation site is available. Induction is autocatalytic, since more methylase protects more ribosomes from erythromycin, thus allowing more methylase translation. Eventually, enough ribosomes are protected that none stall in the leader, and methylase synthesis is shut down.

A Small RNA Is a Translational Repressor of Tn10 Transposase[37]

Just as a ribosome binding site can be blocked if it is joined in a secondary structure with another part of its own mRNA, it also can be base-paired to a separate complementary RNA that prevents ribosome attachment by covering the critical sequences. Transposon Tn10 encodes such an RNA, which serves to inhibit the expression of the Tn10 transposase enzyme and in fact any related transposase in the cell that has the same sequences at its ribosome binding site. Two adjacent promoters near the beginning of the transposase gene of Tn10, designated pIN and pOUT, are important to this regulation (Figure 16-26a). Promoter pIN directs synthesis of transposase mRNA. RNA is made from pOUT in the opposite direction, but it overlaps the region transcribed from pIN by 35 nucleotides. Thus, the first 35 nucleotides of the pOUT transcript are complementary in sequence to the beginning of the transposase mRNA, a region that includes its translation initiation site. These overlapping sequences are believed to base-pair in the cell, covering the transposase initiation site and preventing ribosome binding (Figure 16-26b). The consequence is that expression of transposase is very low, and transposition happens only rarely (Chapter 11).

It might seem that the same end would be achieved if pIN were simply a less efficient promoter. However, one advantage of regulation by translational repression is that the cell is protected from acci-

Figure 16-26
Transposition of Tn10 is controlled by an RNA that prevents translation of transposase mRNA. (a) Map of Tn10, including the overlapping, opposed promoters pIN and pOUT. (b) How pOUT RNA is thought to cover the ribosome binding site of transposase mRNA made from pIN, thereby preventing synthesis of transposase.

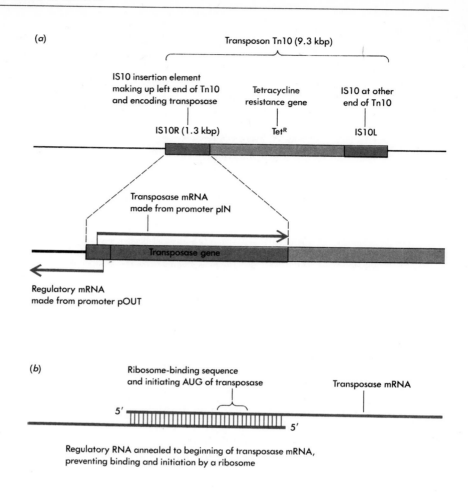

dental transcription passing through the transposase gene from outside the transposon; even if too much transposon mRNA is suddenly made, the RNA repressor will be available in some excess to inhibit its translation. Another way that Tn10 transposase is regulated is discussed in Chapter 11.

It is clear that an RNA repressor like this could be encoded anywhere in the genome and thus could be a repressor of distant as well as adjacent genes. In fact, it is possible to construct, by genetic engineering, a plasmid that makes an RNA complementary to the translation initiation region of a known gene. This strategy has been used to prevent expression of particular genes, and it may become a powerful way to artificially manipulate genetic expression in cells.

Ribosomal Proteins Are Translational Repressors of Their Own Synthesis[38-41]

Correct expression of ribosomal protein genes poses an interesting regulatory problem for the cell. Each ribosome contains some 50 distinct proteins that must be made at exactly the same rate. Furthermore, the rate at which a cell makes protein, and thus the number of ribosomes it needs, is tied closely to the cell's growth rate; a change in growth conditions quickly leads to an increase or decrease in the rate of synthesis of all ribosomal components. How is all this coordinated regulation accomplished?

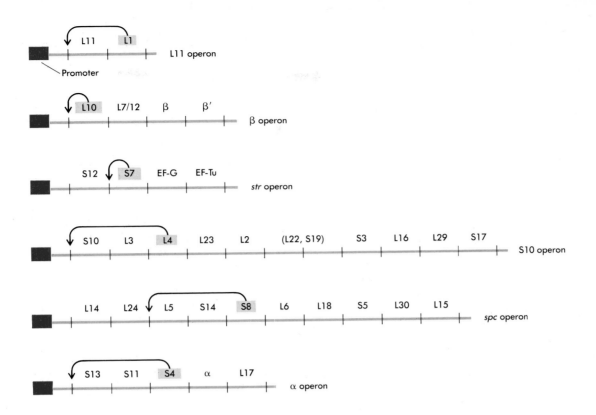

Control of ribosomal protein genes is simplified by their organization into several different operons, each containing genes for up to 11 ribosomal proteins (Figure 16-27). Some nonribosomal proteins that also are required according to growth rate are contained in these operons, including subunits α, β, and β' of RNA polymerase. Like other operons, these may sometimes be regulated at the level of RNA synthesis. However, the primary control of ribosomal protein synthesis is on *translation* of the mRNA, not on its synthesis. This is shown by a simple experiment in which extra copies of a ribosomal protein operon are put into the cell. The amount of mRNA increases correspondingly, but synthesis of the proteins stays nearly the same; thus, the cell compensates for extra mRNA by curtailing its activity as a template. This happens because ribosomal proteins are repressors of their own translation (and thus are *autogenous* regulators). For each operon, one (or a complex of two) of the proteins binds the messenger near the translation initiation sequence of one of the first genes of the operon, preventing ribosomes from binding and initiating translation (see Figure 16-27). Repression of translation of the first gene also prevents expression of some or all of the rest. We believe that this happens because translation of several successive genes in the messenger is linked, each ribosome binding site being opened by a ribosome passing from the preceding gene.

The advantage to the cell in having a protein repress its own translation is to allow quick and precise adjustments of expression. A few unused molecules of protein L4, for example, will shut down its synthesis, as well as synthesis of the other ten ribosomal proteins in its operon. In this way, these proteins are made just at the rate they are needed for assembly into ribosomes (Chapter 14).

How one protein can function both as a ribosomal component and as a regulator of its own translation is shown by comparing the sites

Figure 16-27
Ribosomal protein operons of *E. coli*. The protein that acts as repressor of the other proteins is shaded gray, and its site of action (on the mRNA) is shown by an arrow. (After M. Nomura, R. Gourse, and G. Baughman, *Ann. Rev. Biochem.* 53 (1984):82, with permission.)

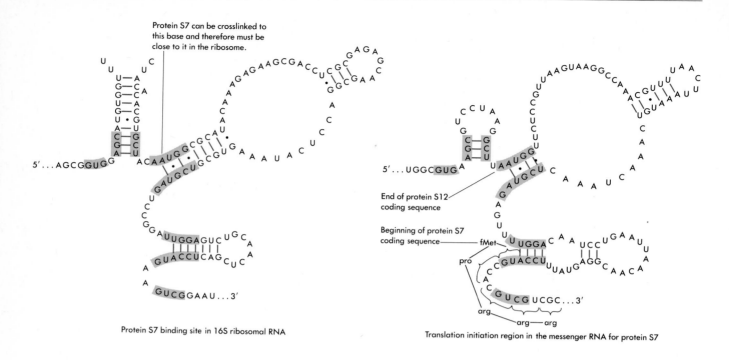

Protein S7 binding site in 16S ribosomal RNA

Translation initiation region in the messenger RNA for protein S7

Figure 16-28
A comparison of the region where ribosomal protein S7 (encoded by the *str* operon; Figure 16-27) binds 16S RNA in the ribosome, with the translation initiation site in its mRNA. Similar sequences, presumably parts of the binding site, are shaded. [After M. Nomura, J. Yates, D. Dean, and L. Post, *Proc. Nat. Acad. Sci.* 77 (1980):7086, with permission.]

where it binds in ribosomal and in messenger RNA. They are similar in sequence and in secondary structure, indicating that the same binding specificity accounts for both activities (Figure 16-28). The comparison also suggests a precise mechanism of regulation. Since the binding site in the messenger includes the initiating AUG, mRNA bound by excess protein S7 (in this example) clearly cannot attach to ribosomes to initiate translation. However, binding is stronger to ribosomal RNA than to mRNA, so that translation is repressed only when all need for the protein in ribosome assembly is satisfied.

Protein Synthesis Termination Factor RF2 Regulates Its Own Translation at the Termination Step[42]

Although most well-documented translational control mechanisms operate at the stage of polypeptide chain initiation, a remarkable exception has been recently discovered. The nucleotide sequence of the gene for RF2, one of the two *E. coli* factors necessary for polypeptide chain termination, suggests that RF2 regulates its own synthesis not at the first, but instead at the final, step of translation.

RF2 is the protein synthesis termination factor that catalyzes release of nascent chains at UGA and UAA stop codons, while the factor RF1 acts at UAG and UAA triplets (Chapter 14). Surprisingly, the carefully sequenced gene for RF2 does not contain a continuous open reading frame. Instead, the first 25 amino acids are encoded by one stretch of DNA, and the remaining 315 amino acids are specified by a sequence occurring in the +1 reading frame (Figure 16-29). Between these two coding regions lie a UGA stop codon and a C nucleotide, which somehow are read as aspartic acid whose codon is GAU or GAC, during synthesis of the full-length (38 kdal) RF2 protein.

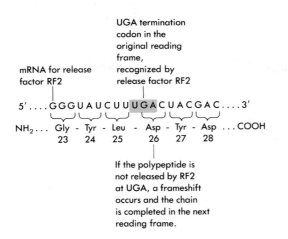

UGA termination codon in the original reading frame, recognized by release factor RF2

mRNA for release factor RF2

5′ . . . . G G G U A U C U U U G A C U A C G A C . . . . 3′

NH₂ . . . Gly - Tyr - Leu - Asp - Tyr - Asp . . . COOH
         23   24   25   26   27   28

If the polypeptide is not released by RF2 at UGA, a frameshift occurs and the chain is completed in the next reading frame.

Figure 16-29
Translation of mRNA for the protein synthesis release factor RF2 involves a shift in reading frame. The occurrence of this frameshift is thought to provide regulation of RF2 gene expression because the RF2 protein is itself required for release of the polypeptide at UGA; if RF2 protein is scarce, release is slow and there is time for the relatively inefficient frameshifting step to occur. (After W. Craigen, R. Cook, W. Tate, and C. Caskey, *Proc. Nat. Acad. Sci.*, 82 (1985):3619.)

We suppose that this translation frameshift regulates expression of the gene for RF2 in the following way. If ample RF2 is available in the cell, the 25-amino-acid polypeptide is released at the UGA codon and no functional RF2 is made. If RF2 is scarce, on the other hand, release of the polypeptide is delayed and there is time for the (presumably) slow frameshift to occur: Aspartic acid is inserted in position 26 when the ribosome shifts to GAC, and translation continues to the final UAG stop codon. Conveniently, UAG is recognized by the other release factor, RF1. Why this elegant translational self-control circuit is utilized only by the RF2 gene and not by the RF1 gene is not known; RF1 is encoded by a conventional continuous open reading frame terminated by UGA (which, as we have seen, is recognized by RF2).

ppGpp Is a Cellular Signal of Amino Acid Starvation[41, 43–46]

If ribosomal proteins are made from their mRNA only when ribosomal RNA is available for their incorporation into ribosomes, then what really determines how fast ribosomes appear is the rate of ribosomal RNA synthesis. How, then, does ribosomal RNA synthesis change according to the cell's need? Here we have no clear answers, although fairly strong circumstantial evidence says that the nucleotide ppGpp (Chapter 14) is an important agent of the regulation. Thus, starving a cell for an amino acid both causes the cellular level of ppGpp to rise abruptly and turns down rRNA synthesis and tRNA synthesis. Similarly, more subtle changes in ppGpp concentration that occur as growth conditions improve or degenerate may lead to adjustments in the rate of ribosomal RNA synthesis. Whether ppGpp acts directly on RNA polymerase to change its rate of initiation at rRNA and tRNA promoters or whether it acts through some activator or repressor is not yet known. However, it is likely that the ability of ribosomal RNA promoters to be controlled is conferred by sequences within the promoters themselves, since in vitro fusion of an rRNA promoter (those sequences between the RNA start site and −50; Chapter 13) allows any gene to be controlled like the rRNA genes.

The effects of ppGpp are not limited to ribosomal RNA synthesis. High concentrations of the nucleotide not only reduce expression of many genes besides those for ribosomal RNA, but also stimulate expression of many others. Presumably, these other genes encode proteins needed at higher levels when the cell is starved. Furthermore, ppGpp directly affects the activity of many enzymes, including those of the ribosome that catalyze peptide bond formation; here its function is thought to be in slowing the rate of protein synthesis, thus reducing translation errors by allowing more time for a correct (but scarce) aminoacyl-tRNA to find the ribosome. Therefore, ppGpp is probably a general messenger molecule that integrates a complex cellular response to amino acid starvation, just as the cellular level of cyclic AMP (acting through CAP protein) signals to the cell a sufficiency of glucose.

Alarmones: Possible Cellular Signals of Distress[45, 47, 48]

The discovery of ppGpp as a signal of amino acid starvation suggested that other unusual metabolic products made as a direct result of some specific deficiency could have similar roles. Named **alarmones** (*alar*m + hor*mones*), they would affect cellular enzymes or proteins so as to counteract the problem whose metabolic consequence they represent, moving to different sites in the cell, just as hormones move to different tissues of more complex organisms. Clearly, both ppGpp and cyclic AMP, the signal of carbon source starvation that affects expression of many genes through binding CAP protein, act in this way. Two other compounds have been considered as possible alarmones (Table 16-3). Diadenosine tetraphosphate *(ApppA)* is made by bacterial aminoacyl-tRNA synthetases in the absence of tRNA, as an aberration of the tRNA-charging reaction; it thus might accumulate when tRNA is unavailable in the cell, thereby signaling this condition. In eucaryotic cells, ApppA stimulates the activity of DNA polymerase α and has been suggested as a signal of disrupted replication that promotes new DNA synthesis. Folate starvation of bacteria increases synthesis of *ZTP*, a possible alarmone for this serious deficiency.

The Concentration of a Protein Can Be Determined by Its Sensitivity to Proteolytic Degradation[49–52]

Most bacterial proteins are stable during normal growth. That is, if a protein's synthesis stops, such as by addition of a corepressor to a growing culture, then its concentration in the cell is reduced only through dilution by other cell components that continue to be made. Many generations may be required to remove the protein effectively from the cells. For an enzyme like β-galactosidase, this is of little consequence. But some proteins must be removed more quickly if cell growth is to succeed. Among these are the LexA repressor of the SOS genes (see page 480) and the phage λ repressor (Chapter 17), which are cleaved into two segments and inactivated by RecA protein when the cell's DNA is damaged. Another such protein is an inhibitor of cell division in *E. coli* encoded by the *sulA* gene. This is itself an SOS gene whose function is to stop further cell division until the chromo-

Table 16-3 Metabolites That May Be Alarmones

Compound	Conditions that Induce Its Synthesis	Known or Suggested Function
1. cAMP (3'-5' cyclic AMP)	Absence of carbon source	To stimulate synthesis of enzymes that metabolize other sugars; to control expression of other proteins required when carbon is limiting.
2. ppGpp, pppGpp (guanosine tetraphosphate, guanosine pentaphosphate)	Amino acid starvation	To inhibit stable RNA synthesis; to control other cellular processes so as to favor adaptation to condition of limiting amino acids
3. ApppppA (diadenosine tetraphosphate)	Bacteria: shortage of tRNA (?), oxidative damage (?)	?
	Eucaryotes: replication fork arrest; normal transition to S phase of cell cycle	To stimulate DNA synthesis, cell proliferation
4. ZTP (5-amino 4-imidazole carboxamide riboside 5'-triphosphate)	Folate starvation	?

some is repaired. The SulA protein is especially sensitive to attack by the cellular *Lon protease;* only its high rate of synthesis allows SulA to accumulate when the SOS genes are induced. As soon as SOS genes are no longer needed and are once again repressed, remaining SulA protein is quickly destroyed by Lon protease and cell division and growth resume. It is therefore clear why a mutant cell lacking Lon protease is killed by very small doses of ultraviolet light: The accumulated SulA division inhibitor is not destroyed, and the cell can never divide again, even though induction of other SOS genes allows repair of the cell's DNA.

What structural feature makes some proteins very sensitive to proteases, whereas others are completely resistant? Most proteins have probably evolved a tight tertiary structure that prevents the active site of a protease molecule, which is of comparable size, from approaching its peptide bonds. In contrast, proteins like SulA may possess exposed segments that a protease can reach. A second role of Lon protease supports this idea: It is responsible for destroying polypeptide fragments made when translation stops prematurely at a nonsense codon or aborts by accident. Such pieces of protein could be loosely or incorrectly folded, allowing regions to protrude and to be attacked by the protease.

Summary

Cells have control mechanisms to ensure that proteins are synthesized in the required amounts. Most of our detailed understanding of these mechanisms at a molecular level is limited to bacterial cells, in particular *E. coli*. Bacteria contain many enzymes whose rate of synthesis depends on the availability of external food molecules. These external molecules, called inducers and corepressors, usually determine the rate of synthesis of enzymes by controlling the synthesis of their mRNA templates. Inducers and corepressors act by binding to regulatory proteins, called activators and repressors. Activators are positive regulators, because their presence is required for the regulated enzyme to be made; on the other hand, repressors act negatively, because their regulatory activity is to prevent synthesis of proteins. Thus, when lactose is absent, the lactose operon repressor prevents synthesis of enzymes that metabolize lactose; however, upon binding an inducer (a molecule related to lactose), the repressor loses this ability and allows the enzymes to be made. The arabinose operon C protein, an activator, causes arabinose enzymes to be made when it binds the inducer arabinose. In contrast to inducers, corepressors prevent enzyme synthesis; tryptophan is a corepressor that gives the tryptophan repressor an active configuration by binding to it, so that enzymes that synthesize tryptophan are no longer made.

Regulatory proteins that control mRNA synthesis usually act by binding DNA near or in the promoters of genes they regulate. Thus, operators, the binding sites of repressors, overlap promoters and directly prevent RNA polymerase binding. Many activators, on the other hand, bind beside RNA polymerase so as to stimulate its binding to the promoter. One class of activators fundamentally changes RNA polymerase itself: new σ factors, present in place of the "standard" σ factors (σ^{70} in *E. coli*) in a small fraction of the holoenzyme, recognize promoters distinct from the "standard" promoter recognized by σ^{70}. For example, σ^{32} directs RNA polymerase to bind heat-shock promoters when the cell's temperature is abruptly raised, thereby greatly increasing expression of the heat-shock proteins.

The length of DNA controlled by a specific promoter, the operon, often consists of several genes with related metabolic functions (e.g., the production of successive enzymes in the synthesis of an amino acid or nucleotide). Enzymes in an operon are not necessarily (or even usually) made at the same rate, since ribosomes may read different parts of the messenger unequally, or some segments of the messenger may be more stable than others. One repressor or activator can control more than one promoter; the LexA repressor of *E. coli*, for example, controls about 15 operons encoding DNA repair enzymes. Messenger RNA is destroyed by nuclease attack, with a half-life that is typically a few minutes, so that the template disappears quickly when a regulatory signal changes.

The rate of synthesis of many proteins is not controlled, but is instead constitutive, fixed at a rate that may be very low or very high. This rate is determined by the inherent strength of the promoter, the lifetime of its messenger, and the rate at which ribosomes read the messenger.

A distinct class of regulation acts by determining the efficiency of RNA polymerase termination rather than the rate of RNA chain initiation. Operons encoding enzymes of amino acid biosynthesis are controlled at termination sites (in this case called attenuators) that separate a leader RNA from mRNA for the enzyme structural genes; the signal of amino acid deficiency is transmitted by ribosomes that, in translating the leader RNA, prevent formation of an RNA hairpin that is essential for termination. Although they are best characterized for the *E. coli* bacteriophage λ, operon-specific antiterminator proteins also are thought to exist in *E. coli*.

Translational control is mediated by devices that determine whether the translation initiation site of an mRNA is available to ribosomes. For example, erythromycin resistance is produced by a methylase whose synthesis depends on its ribosome binding site being uncovered from RNA secondary structure when the antibiotic is present. Expression of the Tn10 transposase is repressed by a complementary RNA that covers and thus inactivates its ribosome binding site. Ribosomal proteins are their own translational repressors, acting by blocking the ribosome initiation sites of their own mRNAs when they accumulate in excess.

A small molecule can signal important cellular conditions to many different genes or operons at the same time. Carbon source deficiency causes the enzyme adenylcyclase to make cyclic AMP, which in turn stimulates CAP protein to act as a transcriptional activator of many promoters, including promoters of the arabinose and lactose operons (the *lac* operon, like the *ara* operon, is therefore regulated both negatively and positively). Amino acid starvation leads to synthesis of ppGpp, which affects both the transcription of various genes and the activity of many enzymes whose functions are required when an amino acid is scarce. Small molecules such as these have been called alarmones, because they integrate the cellular response to particular conditions of stress.

Although proteins are generally stable in bacteria, some are destroyed specifically and quickly when regulatory change subjects them to proteolytic degradation; in this way, the concentration of a protein can be decreased abruptly, whereas a stable protein disappears only through dilution during cell growth. The LexA repressor and phage λ repressor are cleaved by RecA protein activated when cellular DNA is damaged, and the SulA cell division inhibitor is attacked by the Lon protease, so that it disappears quickly when its synthesis stops.

Bibliography

General References

Beckwith, J., J. Davies, and J. Gallant, eds. 1983. *Gene Function in Prokaryotes.* Cold Spring Harbor, N.Y.: Cold Spring Harbor Laboratory. Current reviews of many subjects discussed in the chapter, including attenuation, translational repression, termination and polarity, and global control systems.

Gottesman, S. 1984. "Bacterial Regulation: Global Regulatory Networks." *Ann. Rev. Genetics* 18:415–441.

Ingraham, J., O. Maaløe, and F. Neidhardt. 1983. *Growth of the Bacterial Cell.* Sunderland, Mass.: Sinauer. Considers physiology, biochemistry, and genetics and emphasizes overall regulation of cell growth.

Jacob, F., and J. Monod. 1961. "Genetic Regulatory Mechanisms in the Synthesis of Proteins." *J. Mol. Biol.* 3:318–356. A classic review of great historical interest that ties together the concept of messenger RNA with the problem of the control of protein synthesis.

Miller, J., and W. Reznikoff, eds. 1980. *The Operon.* Cold Spring Harbor, N.Y.: Cold Spring Harbor Laboratory. Everything about the lactose operon, as well as chapters about several others (tryptophan, arabinose, and phage λ early operons).

Raibaud, O., and M. Schwartz. 1984. "Positive Control of Transcription Initiation in Bacteria." *Ann. Rev. Genetics* 18:173–206.

Cited References

1. Gilbert, W., and B. Müller-Hill. 1966. "Isolation of the Lac Repressor." *Proc. Nat. Acad. Sci.* 56:1891–1898.
2. Ptashne, M. 1967. "Specific Binding of the λ Phage Repressor to λ DNA." *Nature* 214:232–234.
3. Galas, D., and A. Schmitz. 1978. "DNase Footprinting: A Simple Method for the Detection of Protein-DNA Binding Specificity." *Nucleic Acid Res.* 5:3157–3170.
4. Siebenlist, U., R. Simpson, and W. Gilbert. 1980. "E. coli RNA Polymerase Interacts Homologously with Two Different Promoters." *Cell* 20:269–281.
5. Pabo, C., and M. Lewis. 1982. "The Operator-Binding Domain of Lambda Repressor: Structure and DNA Recognition." *Nature* 298:443–447.
6. Hecht, M., H. Nelson, and R. Sauer. 1983. "Mutations in λ Repressor's Amino-Terminal Domain: Implications for Protein Stability and DNA Binding." *Proc. Nat. Acad. Sci.* 80:2676–2680.
7. Steitz, T., D. Ohlendorf, D. McKay, W. Anderson, and B. Matthews. 1982. "Structural Similarity in the DNA-Binding Domains of Catabolite Gene Activator and *cro* Repressor Proteins." *Proc. Nat. Acad. Sci.* 79:3097–3100.
8. Anderson, W., Y. Takeda, D. Ohlendorf, and B. Matthews. 1982. "Proposed α-Helical Super-Secondary Structure Associated with Protein: DNA Recognition." *J. Mol. Biol.* 159:745–751.
9. Sauer, R., R. Yocum, R. Doolittle, M. Lewis, and C. Pabo. 1982. "Homology Among DNA-Binding Proteins Suggests Use of a Conserved Super-Secondary Structure." *Nature* 298:447–451.
10. Pabo, C., and R. Sauer. 1984. "Protein-DNA Recognition." *Ann. Rev. Biochem.* 58:293–321.
11. Wharton, R., and M. Ptashne. 1985. "Changing the Binding Specificity of a Repressor by Redesigning an σ-Helix." *Nature* 316:601–605.
12. Anderson, J., M. Ptashne, and S. Harrison. 1985. "A Phage Repressor-Operator Complex at 7Å Resolution." *Nature* 316:596–601.
13. DeCrombrugghe, B., S. Busby, and H. Buc. 1984. "Cyclic AMP Receptor Protein: Role in Transcription Activation." *Science* 224:831–838.
14. Malan, T. P., and W. R. McClure. 1984. "Dual Promoter Control of the *Escherichia coli* Lactose Operon." *Cell* 39:173–180.
15. Walker, G. 1984. "Mutagenesis and Inducible Responses to Deoxyribonucleic Acid Damage in *Escherichia coli.*" *Microbiol. Revs.* 48:60–93.
16. Little, J. W., S. H. Edmiston, L. Z. Pacelli, and D. W. Mount. 1980. "Cleavage of the *Escherichia coli lexA* Protein by the *recA* Protease." *Proc. Nat. Acad. Sci.* 77:3225–3229.
17. Sancar, G., A. Sancar, J. Little, and W. Rupp. 1982. "The *uvrB* Gene of *Escherichia coli* Has Both *lexA*-Repressed and *lexA*-Independent Promoters." *Cell* 28:523–530.
18. Dunn, T., S. Ogden, and R. Schleif. 1984. "An Operator at −280 Base Pairs That Is Required for Repression of *araBAD* Operon Promoter: Addition of DNA Helical Turns Between the Operator and Promoter Cyclically Hinders Repression." *Proc. Nat. Acad. Sci.* 81:5017–5020.
19. Lee, N., W. Gielow, and R. Wallace. 1981. "Mechanism of *araC* Autoregulation and the Domains of Two Overlapping Promoters P_C and P_{BAD}, in the L-Arabinose Regulatory Region of *Escherichia coli.*" *Proc. Nat. Acad. Sci.* 78:752–756.
20. Raibaud, O., and M. Schwartz. 1984. "Positive Control of Transcription Initiation in Bacteria," *Ann. Rev. Genetics* 18:415–441.
21. Majumdar, A., and S. Adhya. 1984. "Demonstration of Two Operator Elements in *gal*; *In Vitro* Repressor Binding Studies." *Proc. Nat. Acad. Sci.* 81:6100–6104.
22. Busby, S., H. Aiba, and B. de Crombrugghe. 1982. "Mutations in the *Escherichia coli* Galactose Operon That Define Two Promoters and the Binding Site of the Cyclic AMP Receptor Protein." *J. Mol. Biol.* 154:211–227.
23. Neidhardt, F. C., R. A. VanBogelen, and V. Vaughn. 1984. "Genetics and Regulation of Heat Shock Proteins." *Ann. Rev. Genetics* 18:295–329.
24. Grossman, A., J. Erickson, and C. Gross. 1984. "The *htpR* Gene Product of *E. coli* is a Sigma Factor for Heat-Shock Promoters." *Cell* 38:383–390.
25. Landick, R., V. Vaughn, E. T. Lau, R. A. VanBogelen, J. W. Erickson, and F. C. Neidhardt. 1984. "Nucleotide Sequence of the Heat Shock Regulatory Gene of *E. coli* Suggests Its Protein Product May Be a Transcription Factor." *Cell* 38:175–182.
26. Yura, T., T. Tobe, K. Ito, and T. Osawa. 1984. "Heat Shock Regulatory Gene (*htpR*) of *Escherichia coli* is Required for Growth at High Temperature but is Dispensable at Low Temperature." *Proc. Nat. Acad. Sci.* 81:6803–6807.
27. Cowing, D. W., J. C. A. Bardwell, E. A. Craig, C. Woolford, R. W. Hendrix, and C. A. Gross. 1985. "Consensus Sequence for *Escherichia coli* Heat Shock Gene Promoters." *Proc. Nat. Acad. Sci.* 82:2679–2683.
28. Bardwell, J. C. A., and E. A. Craig. 1984. "Major Heat Shock Gene of *Drosophila* and the *E. coli* Heat Inducible *dnaK* Gene Are Homologous." *Proc. Nat. Acad. Sci.* 81:848–852.
29. Hirschman, J., P. K. Wong, K. Sei, J. Keener, and S. Kustu. 1985. "Products of Nitrogen Regulatory Genes *ntrA* and *ntrC* of Enteric Bacteria Activate *glnA* Transcription in Vitro; Evidence that the *ntrA* Product is a σ Factor." *Proc. Nat. Acad. Sci.* 82:7525–7529.
30. Yanofsky, C., and R. Kolter. 1982. "Attenuation in Amino Acid Biosynthetic Operons." *Ann. Rev. Genetics* 16:113–134.
31. Bauer, C., J. Carey, L. Kasper, S. Lynn, D. Waechter, and J. Gardner. 1983. "Attenuation in Bacterial Operons." In *Gene Function in Prokaryotes*, ed. J. Beckwith, J. Davies, and J. Gallant. Cold Spring Harbor, N.Y.: Cold Spring Harbor Laboratory, pp. 65–89.
32. Adhya, S., and M. Gottesman. 1978. "Control of Transcription Termination." *Ann. Rev. Biochem.* 47:967–996.
33. Li, S., C. L. Squires, and C. Squires. 1984. "Antitermination of *E. coli* rRNA Transcription Is Caused by a Control Region Segment Containing Lambda *nut*-like Sequences." *Cell* 38:851–861.
34. Campbell, K., G. Stormo, and L. Gold. 1983. "Protein-Mediated Translational Repression." In *Gene Function in Prokaryotes*, ed. J. Beckwith, J. Davies, and J. Gallant. Cold Spring Harbor, N.Y.: Cold Spring Harbor Laboratory, pp. 185–210.

35. Horinouchi, S., and B. Weisblum. 1980. "Posttranscriptional Modification of mRNA Conformation: Mechanism That Regulates Erythromycin-Induced Resistance." *Proc. Nat. Acad. Sci.* 77:7079–7083.

36. Horinouchi, S., and B. Weisblum. 1981. "The Control Region for Erythromycin Resistance: Free Energy Changes Related to Induction and Mutation to Constitutive Expression." *Mol. Gen. Genet.* 182:341–348.

37. Simons, R., and N. Kleckner. 1983. "Translational Control of IS10 Transposition." *Cell* 34:683–691.

38. Nomura, M., J. Yates, D. Dean, and L. Post. 1980. "Feedback Regulation of Ribosomal Protein Gene Expression in *Escherichia coli*: Structural Homology of Ribosomal RNA and Ribosomal Protein mRNA." *Proc. Nat. Acad. Sci.* 77:7084–7088.

39. Baughman, G., and M. Nomura. 1983. "Localization of the Target Site for Translational Regulation of the L11 Operon and Direct Evidence for Translational Coupling in *Escherichia coli*." *Cell* 34:979–988.

40. Baughman, G., and M. Nomura. 1984. "Translational Regulation of the Lll Ribosomal Protein Operon of *Escherichia coli*: Analysis of the mRNA Target Site Using Oligonucleotide-Directed Mutagenesis." *Proc. Nat. Acad. Sci.* 81:5389–5393.

41. Nomura, M., R. Gourse, and G. Baughman. 1984. "Regulation of the Synthesis of Ribosomes and Ribosomal Components." *Ann. Rev. Biochem.* 53:75–117.

42. Craigen, W. J., R. G. Cook, W. P. Tate, and C. T. Caskey. 1985. "Bacterial Peptide Chain Release Factors: Conserved Primary Structure and Possible Frameshift Regulation of Release Factor 2." *Proc. Nat. Acad. Sci.* 82:3616–3620.

43. Sarmientos, P., and M. Cashel. 1983. "Carbon Starvation and Growth Rate-Dependent Regulation of the *Escherichia coli* Ribosomal RNA Promoters: Differential Control of Dual Promoters." *Proc. Nat. Acad. Sci.* 80:7010–7013.

44. Ryals, J., R. Little, and H. Bremer. 1982. "Control of rRNA and tRNA Synthesis in *Escherichia coli* by Guanosine Tetraphosphate." *J. Bacteriol.* 151:1261–1268.

45. Stephens, J., S. Artz, and B. Ames. 1975. "Guanosine 5'-Diphosphate 3'-Diphosphate (ppGpp): Positive Effector for Histidine Operon-Transcription and General Signal for Amino-Acid Deficiency." *Proc. Nat. Acad. Sci.* 72:4389–4393.

46. Lamond, A. I., and A. A. Travers. 1984. "Genetically Separable Functional Elements Mediate the Optimal Expression and Stringent Regulation of a Bacterial tRNA Gene." *Cell* 40:319–326.

47. Lee, P., B. Bochner, and B. Ames. 1983. "Diadenosine 5',5'''-P^1,P^4-Tetraphosphate and Related Adenylylated Nucleotides in *Salmonella typhimurium*." *J. Biol. Chem.* 258:6827–6834.

48. Bochner, B., and B. Ames. 1982. "ZTP (5-Amino 4-Imidozole Carboxamide Riboside 5'-Triphosphate): A Proposed Alarmone for 10-Formyl-Tetrahydrofolate Deficiency." *Cell* 29:929–937.

49. Mizusawa, S., and S. Gottesman. 1983. "Protein Degradation in *Escherichia coli*: The *lon* Gene Controls the Stability of sulA Protein." *Proc. Nat. Acad. Sci.* 80:358–362.

50. Charette, M. R., G. W. Henderson, and A. Markovitz. 1981. "ATP Hydrolysis-Dependent Protease Activity of the *lon(capR)* Protein of *Escherichia coli* K-12." *Proc. Nat. Acad. Sci.* 78:4728–4732.

51. Chung, C. H., and A. L. Goldberg. 1981. "The Product of the *lon(capR)* Gene in *E. coli* Is the ATP-Dependent Protease La." *Proc. Nat. Acad. Sci.* 78:4931–4935.

52. Baker, T. A., A. D. Grossman, and C. A. Gross. 1984. "A Gene Regulating the Heat Shock Response in *Escherichia coli* Also Affects Proteolysis." *Proc. Nat. Acad. Sci.* 81:6779–6783.

CHAPTER 17

The Replication
of Bacterial Viruses

Molecular geneticists have often focused their study on viruses because of the apparent simplicity of the virus. In the beginning, viruses were thought of as naked genes; but gradually it became clear that the correct analogy is the naked chromosome, since many of the viruses first thought to be so uncomplicated are now known to contain several hundred genes. And when we began to discover the various ways bacteria can control gene expression, it seemed probable that the essence of a virus's existence was the lack of any such regulatory devices. Being so constituted, they would be able to multiply rapidly at the expense of their host cells' metabolism. Again the first hunches proved wrong. The replication of even the smallest of viruses is a very complicated affair, achieved only with the aid of highly evolved regulatory systems designed to ensure that the right molecules are synthesized at just the right time in the life cycle of a virus. We now understand the molecular interactions that underlie many of the basic regulatory mechanisms that direct growth of bacterial viruses (bacteriophages), and comparable descriptions of eucaryotic virus development are in sight (Chapter 24). In this chapter, we first consider some general principles of virus structure and multiplication and then describe in detail the growth of a few important bacteriophages that have provided our most advanced knowledge of molecular genetics.

The Core and Coating of Viruses[1-4]

Both the size and structural complexity of viruses show great variation. Some viruses have molecular weights as small as several million, whereas others approach the size of very small bacteria. However, all viruses differ fundamentally from cells, which have both DNA and RNA, in that viruses contain only one type of nucleic acid, which may be either DNA or RNA. The genetic nucleic acid component is always present in the center of the virus particle, surrounded by a protective coat, or shell, called a capsid. Some of the shells are quite complex; they contain many proteins arranged in several layers and have complex appendages like phage λ tails (Figure 17-1). Other viruses have lipid or carbohydrate molecules as well as protein (Figure 17-2). In still other cases, such as the small RNA bacterial viruses MS2, f2, R17, and Qβ, the shell contains only one type of protein and no lipid or carbohydrate (Figure 17-3).

All capsids contain many copies of the protein components, often arranged either in an elongated particle with helical symmetry or in a

Figure 17-1
The structure of the λ phage particle. (a) An electron micrograph of several negatively stained λ particles. (b) A schematic drawing showing detailed features revealed by electron microscopy and biochemical analysis. [Courtesy R. W. Hendrix, University of Pittsburgh.]

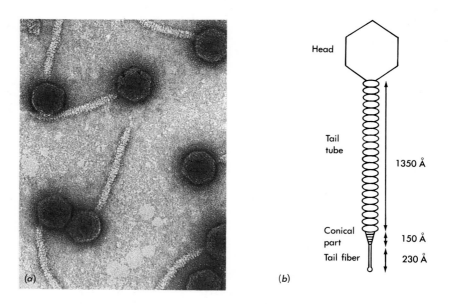

(a)   (b)

nearly spherical particle, which is actually an icosahedron (a polyhedron with 20 faces) and has cubical (or quasi-cubical) symmetry. Thus, tobacco mosaic virus, TMV (Chapter 5) has about 2150 identical protein molecules (MW about 17,000 daltons) helically arranged around a central RNA molecule containing approximately 6000 nucleotides, whereas the icosahedral bacteriophage MS2 (as well as f2, R17, and Qβ) has 180 identical proteins (MW about 14,000 daltons) cubically arranged about an inner RNA molecule with 3569 nucleotides (Figure 17-3).

The use of a large number of identical protein molecules in the construction of the capsid is an obligatory feature of the structures of all viruses. This follows from their limited nucleic acid content, which in turn places a restriction on the maximum number of amino acids in the proteins coded by the viral chromosome. For example, the 3569 nucleotides in an MS2 RNA chain could code for about 1200 amino acids, corresponding to a protein molecular weight of about 1.3×10^5 daltons. This is much smaller than the molecular weight of the MS2 protein shell (2.5×10^6). Thus, even if the entire MS2 RNA chain coded for its coat protein (which we know it does not), approximately 20 identical protein molecules would be needed. This use of a large number of identical protein subunits is why the simpler viruses, which often contain only one major type of protein molecule in their coat, have either helical or cubical (or quasi-cubical) symmetry. Only these two types of symmetry permit the identical protein subunits to be packed together in a regular (or quasi-regular) fashion and thus to have virtually identical chemical environments, except for their contacts with the nucleic acid core.

Nucleic Acid: The Genetic Component of All Viruses

Viruses originally provided the clearest demonstration that genetic specificity is carried by nucleic acid molecules. Many viral nucleic acids, both DNA and RNA, are easily isolated from their protein shell and prepared in highly purified form. When they are added to host cells, new infective virus particles are produced, identical to those from which the nucleic acid was isolated. Such experiments showed definitively that the viral nucleic acid carries the genetic specificity to

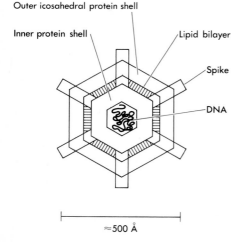

Figure 17-2
The structure of PM2, a lipid-containing bacteriophage that infects the bacterium *Pseudomonas*. [After R. Schaefer and R. M. Franklin, *J. Mol. Biol.* 97 (1975):32, with permission].

code both for its own replication and for the amino acid sequences in its specific coat proteins.

Viral Nucleic Acid May Be Either Single- or Double-Stranded

The nucleic acid of most viruses has the form of its cellular counterparts. Thus, all the best-known DNA viruses, such as smallpox (or its harmless relative *Vaccinia*), SV40, bacteriophage λ, and the T2, T4, and T6 group of bacteriophage, have DNA in the double-stranded helical form. Correspondingly, the RNA of TMV, influenza virus, poliomyelitis virus, and the bacteriophages MS2 and R17 is single-stranded. There are, however, several groups of bacterial viruses in which the DNA is single-stranded, as well as at least two groups of RNA viruses (e.g., the reoviruses) in which the RNA is double-stranded.

It does not really matter whether the genetic message is initially present as a single strand or as the double helix, for the single strand can quickly be used to form a complementary DNA or RNA chain soon after it enters a suitable host cell. The really important fact is that the genetic information is present as a sequence of nucleotide bases.

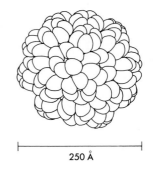

| 250 Å |

Figure 17-3
Capsid structure of a small icosahedral virus (e.g., R17, f2, MS2, or Qβ) that contains 180 subunits of a single type of coat protein.

Viral Nucleic Acid and Protein Synthesis Occur Independently

Exactly what happens after a viral nucleic acid molecule enters a susceptible host cell depends on the specific viral system. Particularly important is whether the virus contains DNA or RNA. If the genetic component is DNA, then the DNA serves as a template both for its own replication and for the virus-specific RNA necessary for the synthesis of its specific proteins. Similarly, if RNA is the genetic component, the RNA molecules have two template roles, the first to make more RNA molecules and the second to make the virus-specific proteins. In both cases, the end result of virus infection is the same: the production of many new copies of both the viral nucleic acids and the coat proteins.

The new progeny molecules then spontaneously aggregate to form mature virus particles. For most viruses, enzymes play no role in these final aggregation events, since the formation of new covalent bonds is not needed either to build a stable protein coat or to affix it firmly to the nucleic acid core. Only weak secondary bonds (salt linkages, van der Waals forces, and hydrogen bonds) need be involved. This last point is shown clearly with TMV. Here the rod-shaped particles can be gently broken down and their free RNA and coat protein components separated. When the components are again mixed together, new infectious particles, identical to the original rods, quickly form. We thus see that the essential aspects of viral multiplication often are known once we understand the principles by which viral nucleic acid and protein components are individually synthesized.

Viral Nucleic Acids Code for Both Enzymes and Coat Protein

The viral nucleic acid genetic component must code the amino acid sequences in the protein(s) that make(s) up the protective coat. These coat proteins are never found in normal uninfected cells, and are

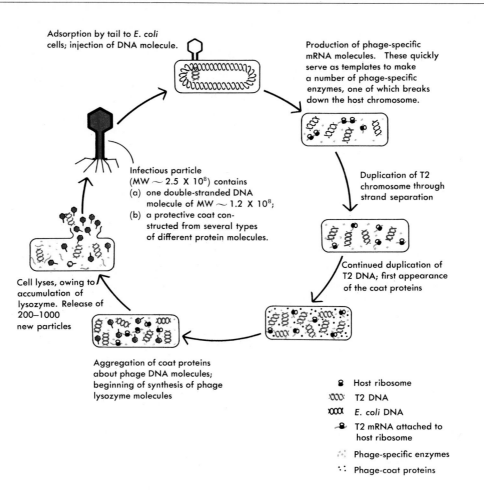

Adsorption by tail to *E. coli* cells; injection of DNA molecule.

Production of phage-specific mRNA molecules. These quickly serve as templates to make a number of phage-specific enzymes, one of which breaks down the host chromosome.

Duplication of T2 chromosome through strand separation

Continued duplication of T2 DNA; first appearance of the coat proteins

Infectious particle (MW ∼ 2.5 X 10⁸) contains (a) one double-stranded DNA molecule of MW ∼ 1.2 X 10⁸; (b) a protective coat constructed from several types of different protein molecules.

Cell lyses, owing to accumulation of lysozyme. Release of 200–1000 new particles

Aggregation of coat proteins about phage DNA molecules; beginning of synthesis of phage lysozyme molecules

Host ribosome
T2 DNA
E. coli DNA
T2 mRNA attached to host ribosome
Phage-specific enzymes
Phage-coat proteins

Figure 17-4
Chemical details in the life cycle of the double-stranded DNA virus T2 (T4).

completely specific to a given virus. In addition, one or more new enzymes are usually synthesized to permit successful viral multiplication.

For example, some DNA viruses carry information for the amino acid sequences of enzymes that synthesize precursor nucleotides for viral DNA. One of the most striking cases involves bacteriophage T4 multiplication (Figure 17-4). No cytosine is present in its DNA, a fact that initially suggested that T4 DNA might be very different from normal DNA. Instead, there is always present the closely related base 5-OH-methylcytosine, which, like cytosine, forms base pairs with guanine (Figure 17-5). The three-dimensional structure of T4 DNA is thus basically the same as that of normal double-helical DNA.

No 5-OH-methylcytosine is found in uninfected *E. coli* cells, and so the several new enzymes required for its biosynthesis must be coded by the T4 DNA. In addition, the rate of DNA synthesis in T4-infected cells is several times faster than in normal cells. This faster rate is achieved by having other T4 genes code for many of the enzymes involved in normal nucleotide metabolism, as well as for a new DNA polymerase (Figure 17-6).

New virus-specific enzymes are also frequently needed to ensure the release of progeny virus particles from the host cell. This is a vital need for those viruses multiplying in bacteria with rigid cell walls. Since these cell walls do not spontaneously disintegrate, they could effectively inactivate progeny particles by preventing their release

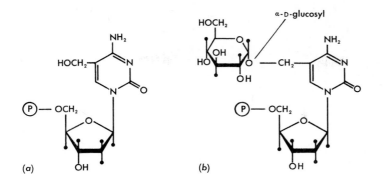

Figure 17-5
Phages of the T-even group do not contain cytosine in their DNA. Instead they contain the related base 5-OH-methylcytosine (a), which base-pairs exactly like cytosine. One or more glucose residues are attached to some of their 5-CH₂-OH groups. (b) The base 5-OH-methylcytosine with one glucose molecule attached. The biological significance of these unusual bases has not yet been clearly established. One hypothesis asserts that their function is to protect T-even DNA from a phage-specific enzyme that only breaks down unmodified DNA. This hypothesis would explain how the *E. coli* DNA is selectively broken down during phage growth.

and transfer to new host cells. To solve this problem, many phages have a gene that codes for the amino acid sequence of lysozyme, a cell wall–destroying enzyme. This enzyme begins to be synthesized when the coat proteins appear and causes the rupture of the cell wall at about the time virus maturation is complete.

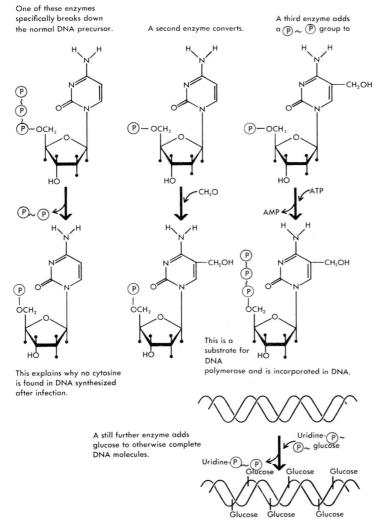

Figure 17-6
The biochemical mechanism that brings about the synthesis of DNA lacking cytosine and containing instead 5-OH-methylcytosine and its glucose derivatives. Immediately after infection, a number of specific enzymes are synthesized. These are coded for by the viral DNA, and each has a specific role in ensuring the successful multiplication of the phage.

Figure 17-7

The morphogenetic pathway leading to the formation of the T4 particle. The numbers refer to the various genes involved in each step. [After W. B. Wood and R. A. Crowther, in *Bacteriophage T4*, ed. C. K. Mathews, E. M. Kutter, G. Mosig, and P. B. Berget (Washington, D.C.: American Society for Microbiology, 1983), p. 263, with permission.]

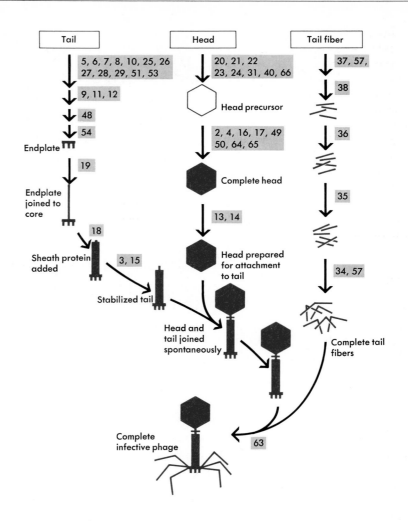

Virus Assembly Follows Morphogenetic Pathways

The assembly of structurally complex viruses like λ and T4 is much more involved than the simple aggregation process needed for simple viruses like TMV. Some 40 different T4 gene products, each coded by a specific T4 gene, interact to produce the mature virus particle. Many of the genes code for the various structural proteins, while at least one other codes for an enzyme that probably converts a precursor protein into the form found in mature T4 particles. In the assembly process, the various components do not associate with one another at random. Instead, the assembly occurs in a definite sequence (a **morphogenetic pattern**), possibly because each step provides the essential substrate for the next. Three different branches, the first concerned with the head, the second with the tail, and the third with the tail fibers, come together as shown in Figure 17-7 to produce the final infectious particles.

Viral Infection Often Radically Changes Host Cell Metabolism

Sometimes, the synthesis of virus-specific nucleic acids and proteins coincides with normal cell synthesis, and the host genome functions more or less normally during the viral life cycle. This is true for the filamentous single-stranded DNA phages fd and M13, cellular parasites that are continuously replicated and extruded through the membrane of the bacterium as it grows. In many cases, however, soon after infection, most cellular metabolism is redirected to the exclusive synthesis of viral components. In the most extreme cases, all DNA and RNA synthesis on the host chromosome ceases, the preexisting RNA templates are degraded, and all subsequent protein synthesis occurs on new RNA templates coded by the viral nucleic acid.

The extent to which a virus is able to control its host's synthetic facilities varies greatly, depending both on the nature of the infecting virus and on the type of host cell. For bacteriophage in particular, the larger the viral nucleic acid content, the larger the number of viral genes directed toward stopping host cell functions unnecessary for the production of new virus particles and replacing bacterial enzymes with phage-encoded enzymes more suited to the needs of the virus. For example, during T4 multiplication, the host *E. coli* chromosome is enzymatically broken down by enzymes coded for by T4 DNA. However, the T4 chromosome is not attacked enzymatically, probably because it contains the unusual base 5-OH-methylcytosine. Furthermore, T4 provides its own enzymes for DNA replication, repair, and recombination and encodes proteins that change the specificity of *E. coli* RNA polymerase to favor T4 promoters over those of *E. coli*.

Synthesis of Virus-Specific Proteins

Although many viruses radically redirect replication and transcription to their own DNA in infected cells, virus-specific proteins are synthesized in the same basic way as normal cellular proteins. Viral messenger RNA molecules attach to host ribosomes, forming polysomes to which the AA ~ tRNA precursors are attached. Viral proteins, like host proteins, start with fMet-tRNA, (bacteria) or Met-tRNA (higher cells), and the termination codons for viral and host protein synthesis are the same. For bacteriophage, the only virus-encoded change that we know in the cellular machinery of protein synthesis is in tRNAs. Some new tRNA species arise by enzymatic modification of preexisting tRNAs, and other completely new tRNAs are encoded by phage DNA. Their significance is still unclear. Perhaps they are connected with differences in base composition between the phage chromosome and host chromosome. For example, T4 DNA has twice as many AT base pairs as GC base pairs, while *E. coli* DNA has roughly similar amounts of both types. As not all potential wobble pairs are equally strong, growth of an AT-rich virus using tRNAs adapted for a chromosome with equal amounts of the four bases might slow down the translation of many virus-specific proteins. Hence, there may be a selective advantage of new tRNA species with a different set of favored anticodons.

The Distinction Between Early and Late Proteins

After a viral chromosome enters a cell, its genes do not usually begin working all at the same time. Instead there is a time schedule by which they function. Some viral proteins appear immediately after infection, whereas the synthesis of others may not begin until more than halfway through a viral growth cycle. Also, some genes start working early and continue to do so throughout viral growth, while others function for only several minutes and then shut off (Figure

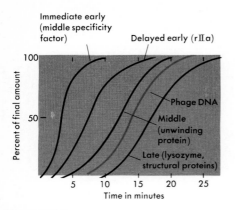

Figure 17-8

Time of synthesis of various T4-specific proteins and DNA at 37°C.

17-8). There are obvious advantages to such sequential appearance. Some of the early gene products are enzymes necessary to redirect cellular synthesis from host to viral genes (they clearly must be turned on early), while others, such as the viral coat proteins or the lysozyme for bacterial cell lysis, must only appear later. Hence, even the most cursory knowledge of a viral life cycle suggests that quite intricate control systems must exist in order to guarantee that the right proteins appear just when they are needed. As far as we know, these control systems usually operate at the transcriptional level. Thus, some specific mRNA molecules appear early, while others are found only late in the growth cycle. At first it was convenient to subdivide the various mRNAs into just two categories—early and late mRNAs—and likewise to classify a particular protein as either early or late. But often the situation is more complex, and, particularly for bacteriophage, the literature is filled with terms like "immediate early," "delayed early," "middle," "quasi-late," and so on.

Controlling the Time of Bacteriophage Gene Expression Through Gene Order

Under optimal growth conditions, the time needed to synthesize a bacterial mRNA molecule with a molecular weight of about 2 million daltons is of the order of 3 or 4 minutes. Molecules of this size (6000 nucleotides) can code for four or five average-size proteins. But synthesis of these proteins will not begin at the same time; the proteins coded at the 5' end can be completely translated before the transcription of the gene at the 3' end has even started. Genes that need to function first thus tend to occur at the start of operons. Likewise, when the optimal time for the beginning of a gene's function is 3 to 4 minutes after phage infection, it is likely to be placed toward the 3' end of a fairly large operon.

The delayed appearance of the "late" bacteriophage proteins, however, is not completely explained by the existence of very long operons. In T4, for example, most operons are relatively short, and few have more than several thousand base pairs. Yet, perhaps one-third of its 200 or so genes do not begin to function until about 10 minutes after its chromosome has entered the host cell. Part of the delay must be due to the time required for the accumulation and function of phage-specific proteins that direct the expression of particular sets of phage genes.

New σ Factors Direct RNA Polymerase to New Promoters During Phage Development[5–11]

The central strategy that most complex bacteriophages (e.g., T4, λ, T7, Mu and the *B. Subtilis* phage SPO1) use to ensure efficient and correctly timed expression of their genes is to modify the host RNA polymerase with phage proteins that direct it to specific phage genes, or in some cases to make an entirely new phage-specific RNA polymerase. Since the σ subunit endows RNA polymerase with the specificity to recognize a particular promoter sequence (Chapter 13), a phage can program the enzyme by providing new and distinct σ factors that recognize only phage promoters of a particular sequence. As an example, the proteins encoded by genes *33* and *55* of phage T4 bind core RNA polymerase in place of the predominant cellular σ^{70}, causing it to transcribe only phage late genes. Correspondingly, the T4 late promoters have a consensus sequence different from that recognized by σ^{70}.

Orderly development of the *B. subtilis* phage SPO1 is provided by two or three σ factors that are made in sequence. Several genes of SPO1, identified by suppressible nonsense mutations, encode positive regulatory factors required for the expression of other SPO1 genes. The 26 kdal gene *28* product is a phage-specific σ factor (σ^{gP28}) required for synthesis of mRNA for SPO1 middle genes (those expressed beginning about midway into the infection cycle), whereas the gene *33* and *34* proteins act (possibly together) as a σ factor for SPO1 late genes. Gene *28* is itself an early gene, and its promoter is recognized by the standard *B. subtilis* σ factor σ^{55} (the counterpart of the *E. coli* standard σ factor σ^{70}). When the gene *28* protein appears early in infection, it displaces σ^{55} from the RNA polymerase core, possibly by binding the core more tightly; this new holoenzyme then recognizes SPO1 middle promoters. After σ^{gP28} is made, genes *33* and *34* are expressed to provide yet another holoenzyme, which is specific for late gene promoters. The timing of their expression suggests that genes *33* and *34* are controlled by middle promoters, although this has not yet been shown directly. Thus the program of gene expression in phage SPO1 infection is determined by a cascade of σ factors, each controlled by the σ factor active in the previous stage of phage growth.

DNA sequence analysis of SPO1 middle and late promoters shows that they differ not only from the "standard" consensus promoter recognized by the predominant holoenzyme containing σ^{55}, but also from each other (Table 17-1). Distinct consensus sequences can be derived for middle and for late promoters, and it is thought that each sequence defines those bases that the corresponding σ factor (σ^{gP28} or $\sigma^{gP33-34}$) recognizes to direct specific promoter recognition. In fact, phage SPO1 provided the first evidence that σ factors confer DNA sequence specificity to RNA polymerase in recognizing promoters (Chapter 13).

Table 17-1 Consensus Sequences for Phage SPO1 Middle and Late Promoters, Compared with the "Standard" Bacterial Consensus Sequence*

Promoter Class	−35 Consensus Sequence	−10 Consensus Sequence	σ Factor Recognizing Promoter
Major bacterial; SPO1 early	TTGACA	TATAAT	σ^{55} (bacterial σ factor)
SPO1 middle	AGGAGA	TTTNTTT	σ^{gP28} (SPO1 encoded middle gene σ factor)
SPO1 late	CGTTAGA	GATATT	$\sigma^{gP33-34}$ (SPO1 encoded [putative] late gene σ factor)

*In the phage promoters, T stands for hydroxymethyl uracil, which replaces thymine in phage DNA. Adapted from M. Costanzo, "Bacteriophage SP01 Regulatory Genes." 1983. Ph.D. thesis, Harvard University.

The Complete 39,936-Nucleotide Sequence of Phage T7 DNA Is Known[12]

Despite their importance in the history of molecular genetics, the *E. coli* phages T2 and T4 have lost some prominence to smaller but still complex phages that are easier to study. Arguably the best-understood complex phage is T7, which also infects *E. coli*. Its entire 39,936-nucleotide DNA sequence is known, and the functions of enough of its genes are understood that we can give a satisfying description of its life cycle at the molecular level (Figure 17-9). The T7 DNA sequence reveals 55 possible genes; the proteins encoded by 44 of these have been identified, although a function is known for only about 30 of them. Like other relatively small genomes that have been sequenced, T7 has no waste, virtually all of its DNA occurring either as genes or regulatory sites such as promoters and ribosome binding sites. As an example of this economy, the DNA sequence in Figure 17-10, about 0.5 percent of T7, contains the longest segment between two genes in the entire DNA. These 193 nucleotides include a promoter for the T7 RNA polymerase, sequences encoding mRNA hairpins recognized by the RNA processing enzyme RNase III (Chapter

Figure 17-9

An accurately scaled map of bacteriophage T7, with a detailed view of the first third.

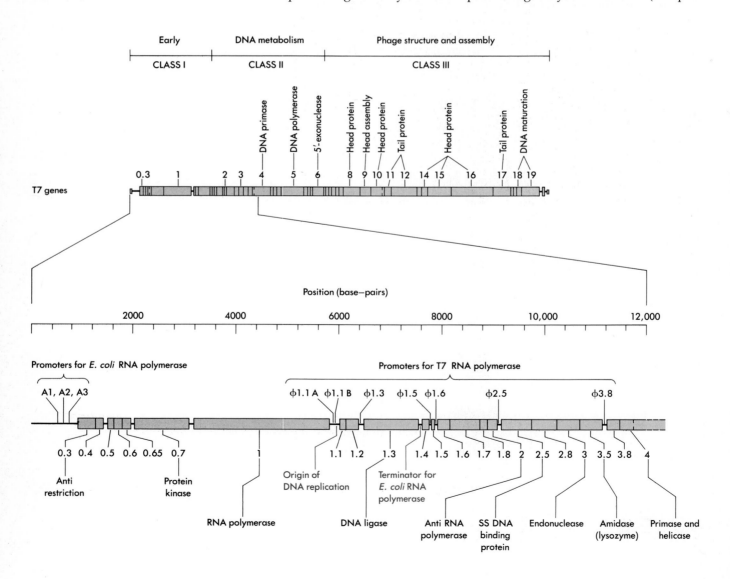

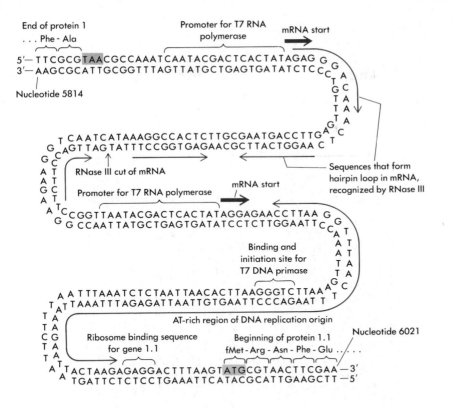

Figure 17-10
Regulatory signals in the DNA sequence between T7 genes *1* and *1.1*.

14), a second T7 RNA polymerase promoter, an essential AT-rich section of the origin of DNA replication (Chapter 10), an initiation site for T7 DNA primase, and sequences encoding the ribosome binding site in mRNA for gene *1.1*. At most, five or six nucleotides separate these distinct sites.

T7 Encodes Most Enzymes Required for Its Replication and Recombination[12]

T7 provides its own replication proteins to speed and direct the multiplication of its DNA; 200 copies of the infecting DNA are made within 15 minutes after T7 enters the bacterial cell. These proteins have all the essential activities to replicate double-stranded DNA, but they are a condensed, minimal set, as compared to the complex group of 30 or so polypeptides apparently required to replicate *E. coli* DNA (Chapter 10). The DNA polymerase polypeptide encoded by gene *5* is active when it is complexed with only one other protein, the bacterial protein thioredoxin, whose function in this role is unknown. A primase and a helicase (DNA-unwinding enzyme) combined in one protein encoded by gene *4*, a DNA ligase (encoded by gene *1.3*), a single-stranded DNA binding protein (encoded by gene *2.5*), and a nuclease to remove RNA primers (encoded by gene *6*) complete the set.

Crossing over between different T7 DNA molecules that infect one cell also depends on proteins made by T7. In fact, the gene *6* nuclease again is required, along with a second nuclease encoded by gene *3*. These probably act by partially degrading replicating molecules, exposing single-stranded regions in different molecules that can anneal to initiate heteroduplex formation (Chapter 11).

A New RNA Polymerase Coded by T7 Recognizes Phage-Specific Promoter Sequences[13,14,15]

Like SPO1, T7 makes proteins that redirect cellular transcription to viral DNA. Its principal strategy is different, however. Instead of encoding new σ factors for the host core RNA polymerase, T7 makes an entirely new RNA polymerase (encoded by T7 gene *1*) that is highly specific for a set of distinct phage promoter sequences, the T7 late promoters. Of course, *E. coli* RNA polymerase must make the messenger RNA for gene *1* (as well as a few other early genes), but the new enzyme can then take over transcription of phage late genes. Compared to the *E. coli* enzyme which has a molecular weight of about 450,000 daltons, the T7 RNA polymerase is small, containing a single polypeptide chain with a molecular weight of 98,000 daltons. The T7 RNA polymerase is a very active enzyme and is highly specific for the T7 late promoters. Corresponding to the distinct structure of the enzyme, the consensus sequence derived from comparison of the 17 T7 late promoters is quite different from the *E. coli* RNA polymerase consensus; instead of separate −35 and −10 regions, the T7 late promoters consist of a single highly conserved sequence from −17 to +6 relative to the RNA start site (Figure 17-11). Only one T7 promoter differs from the 23-nucleotide consensus by as many as 7 nucleotides, and the five strongest promoters (class III, to be discussed shortly) have exactly the consensus sequence. Its great specificity and activity make the T7 RNA polymerase (and a related enzyme from the *Salmonella typhimurium* phage SP6) very useful for in vitro synthesis of RNA specific to any DNA segment that can be cloned next to a T7 or SP6 promoter.

Figure 17-11
Nucleotide sequence of promoters for T7 RNA polymerase. The replication promoters are found in the origin of DNA replication and are thought to be required for transcriptional activation of replication by T7 RNA polymerase. Bases that differ from the consensus sequence are shaded.

		−10	+1 (start of RNA)
Conserved sequence		TAATACGACTCACTATAGGGAGA	
Class II promoters		+1	
φ1.1A	AACGCCAAAT	CAATACGACTCACTATAGAGGGA	CA
φ1.1B	TTCTTCCGGT	TAATACGACTCACTATAGGAGAA	CC
φ1.3	GGACTGGAAG	TAATACGACTCAGTATAGGGACA	AT
φ1.5	AGTTAACTGG	TAATACGACTCACTAAAGGAGGT	AC
φ1.6	TGGTCACGCT	TAATACGACTCACTAAAGGAGAC	AC
φ2.5	AGCACCGAAG	TAATACGACTCACTATTAGGGAA	GA
φ3.8	CGTGGATAAT	TAATTGAACTCACTAAAGGGAGA	CC
φ4c	CCGACTGAGA	CAATCCGACTCACTAAAGAGAGA	GA
φ4.3	AGTCCCATTC	TAATACGACTCACTAAAGGAGAC	AC
φ4.7	TTCATGAATA	CTATTCGACTCACTATAGGAGAT	AT
Class III promoters			
φ6.5	GTCCCTAAAT	TAATACGACTCACTATAGGGAGA	TA
φ9	GCCGGGAATT	TAATACGACTCACTATAGGGAGA	CC
φ10	ACTTCGAAAT	TAATACGACTCACTATAGGGAGA	CC
φ13	GGCTCGAAAT	TAATACGACTCACTATAGGGAGA	AC
φ17	GCGTAGGAAA	TAATACGACTCACTATAGGGAGA	GG
Replication promoters			
φOL	TTGTCTTTAT	TAATACAACTCACTATAAGGAGA	GA
φOR	CACGATAAAT	TAATACGACTCACTATAGGGAGA	GG

T7 Genes Are Expressed in Sequence[12]

T7 illustrates nicely the way that timing of phage gene expression can be controlled through ordered and sequential transcription. T7 genes are all transcribed from one strand; RNA polymerase travels rightward on the DNA, according to the conventional genetic map (Figure 17-12). When T7 infects the cell, *E. coli* RNA polymerase initiates synthesis of T7 early gene mRNA from the extreme left end; here are situated three closely spaced promoters, which have the standard *E. coli* consensus sequence at −35 and −10. A terminator between genes 1.3 and 1.4 stops the enzyme to yield an mRNA about 7000 nucleotides long, which is cut between gene-coding segments by the processing nuclease RNaseIII to form several smaller RNAs. This early mRNA encodes the class I proteins: T7 RNA polymerase, some T7 proteins that inactivate host enzymes, T7 DNA ligase, and other proteins of unknown function. Synthesis of class I proteins begins and ends in the first 6 minutes, as can be seen easily by examining radioactively labeled proteins displayed on a gel (Figure 17-13).

The late mRNA of T7 is made by the class I gene product T7 RNA polymerase. Two groups of late mRNAs and their corresponding proteins are made: those for class II genes, encoding mostly enzymes required for DNA metabolism, and those for class III genes, which encode mostly structural proteins of the phage particle. Class II proteins are made between about 6 and 12 minutes after infection, whereas synthesis of class III mRNA and proteins begins at 8 minutes and then continues vigorously until the cell lyses. Since both are transcribed from promoters recognized by T7 RNA polymerase, how can class II and class III mRNAs be made at different times? It is believed

Figure 17-12
The pattern of mRNA synthesis from T7 DNA by *E. coli* RNA polymerase and T7 RNA polymerase. Upward vertical lines designate promoters, and downward lines, terminators. Vertical bars through the RNAs are sites of processing by RNase III. The readthrough RNA, made by T7 RNA polymerase that initiates at a class II promoter and proceeds to the end of the DNA, is minor but important, because it is the only source of mRNA for genes *11* and *12*. [From J. J. Dunn and F. W. Studier, *J. Mol. Biol.* 166 (1983):477, with permission.]

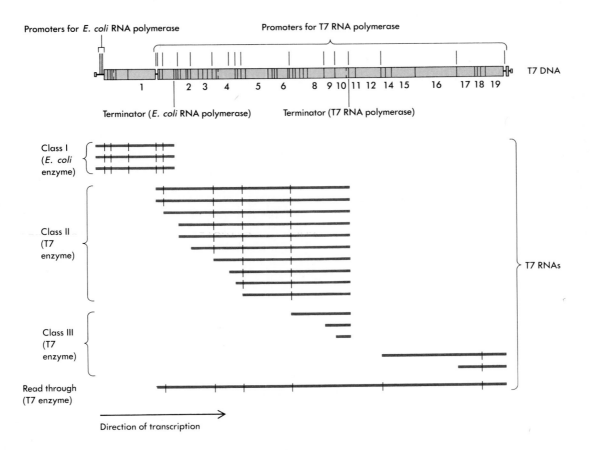

Figure 17-13
The appearance of T7-specific proteins during infection. Proteins being synthesized at different times during infection were made radioactive with a short pulse of radioactive amino acid and were separated in a denaturing polyacrylamide gel and detected by autoradiography. Numbers at the side designate the T7 genes encoding the indicated proteins. Many bacterial proteins are made at the beginning of infection, but after a few minutes, host synthesis is shut down and only T7 proteins are labeled. Inspection shows that proteins of different classes are made at the indicated times (see Figure 17-9); however, not all proteins are visible here, because some are made in very small amounts. [From J. J. Dunn and F. W. Studier, *J. Mol. Biol.* 166 (1983):477, with permission.]

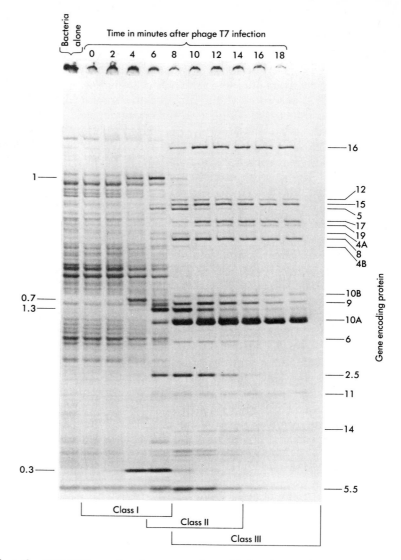

that the T7 DNA molecule enters the cell slowly, and always from its left end, so that for a few minutes only the weaker class II promoters are inside; when more DNA is injected, the stronger class III promoters are available and they begin to dominate late transcription. The nearly exclusive synthesis of class III proteins at later times may result simply from the high concentration of class III mRNA, which saturates the available ribosomes.

It is clear that the major T7 proteins are made when they are needed: RNA polymerase early, to begin expression of the remainder of T7 genes; replication proteins next, to begin the production of new DNA; and structural proteins at the end when they are needed for the final step of phage growth, the assembly of new particles.

Viral Proteins Attack Specific Cellular Components and Help T7 Take Over the Cell[16–19]

T7 encodes proteins that disable both host RNA and host DNA synthesis. The small (7-kilodalton) gene *2* protein binds and inactivates *E. coli* RNA polymerase, although this does not happen until enough early phage RNA has accumulated to ensure successful phage growth. A protein kinase (encoded by gene *0.7*) may contribute to this inactivation by phosphorylating the β and β' subunits of RNA polymerase; the kinase also phosphorylates many other cellular proteins, but it is not known what effect this has.

A drastic attack is reserved for components of cellular DNA synthesis. Here, T7 destroys the bacterial DNA by the combined effects of an endonuclease (encoded by gene 3) and an exonuclease (encoded by gene 6). In fact, the bacterial DNA is degraded to its nucleotide components, which provide the essential precursors for synthesis of T7 DNA. It is not clear why T7 DNA is not also destroyed by the nucleases, but one possibility is that phage DNA is repaired more efficiently after nuclease attack.

Bacteria defend themselves against phages and other foreign DNA with DNA restriction enzymes, against which cellular DNA is protected by methylation (Chapter 9); T7 evades the cellular restriction nucleases *Eco* B and *Eco* K with its gene *0.3* protein, a specific inhibitor that binds and inactivates the nucleases. Curiously, the bacteriophage T3, which is closely related in structure and gene function to T7, has a gene at the site corresponding to that of T7 *0.3* that achieves the same goal by a totally different enzymatic mechanism: It hydrolyzes the compound *S*-adenosylmethionine (SAM), an essential cofactor of the restriction enzymes.

Finally, T7 lysozyme destroys the bacterial cell wall and allows newly formed phage particles to escape and renew their growth cycle in other bacteria.

Bacteriophage N4 Virion RNA Polymerase Recognizes Promoters in Single-Stranded DNA[20,21]

The relatively unknown *E. coli* phage N4 makes an unusual and interesting RNA polymerase that selects initiation sites differently from other known RNA polymerases. N4 virion RNA polymerase is a single large polypeptide chain of 300,000 daltons that is packaged in the phage particle at the end of one growth cycle and then injected along with phage DNA to act in the next generation. Its function is to transcribe the earliest phage genes from three promoters at one end of the phage DNA, and it recognizes only these promoters. Most surprisingly, N4 virion RNA polymerase is inactive on native double-helical DNA, but it binds and initiates RNA synthesis at specific promoter sites on single-stranded N4 DNA in vitro. The promoters include the 18 nucleotides before each RNA start site, a region that is nearly identical in sequence in all three. Early N4 transcription requires the cellular activities of DNA gyrase and SSB (single-stranded DNA binding protein), suggesting that the promoter region denatures when gyrase introduces negative supercoils in the DNA and that a melted region is stabilized by binding to SSB. Since the promoter includes one of two hairpin loops that could form in single-stranded DNA around the RNA start sites, the DNA structure recognized by virion RNA polymerase might be at least partly helical.

Phage λ: A Simple Model of Developmental Regulation[22]

Phage λ is rarely regarded as simple, but only because so many facts are known about it, including its complete DNA sequence. However, the important regulatory events in its life cycle can be summarized

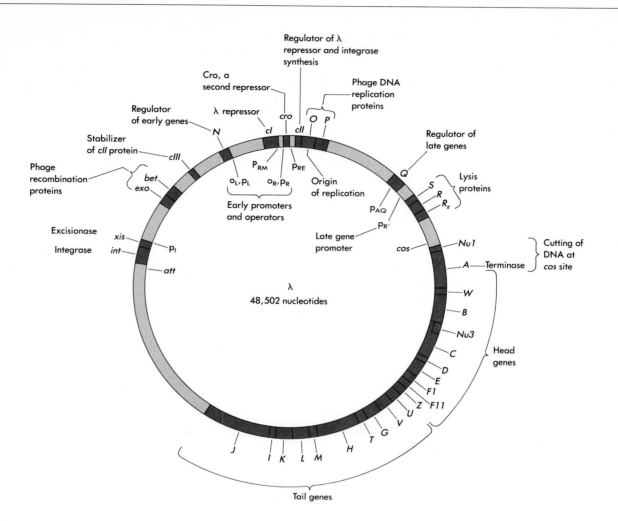

Figure 17-14
A map of phage λ in the intracellular circular form, showing the major λ genes. Before it is packaged into the phage coat, the DNA is cut at the site named *cos;* this means that the genetic map is linear, with one end near gene *A* and the other end near gene *R*.

quite easily, as we do in the following section. Besides providing the best-understood example of repressors and operators, λ is important because it is a **temperate phage** that follows either of two distinct developmental pathways. First, λ can undergo **lytic growth** like T7, reproducing itself manyfold while it kills the cell. Second, it can enter **lysogenic growth,** in which its DNA is combined with the bacterial chromosome and carried passively by the growing bacterium until it is roused by a special signal. We can give an accurate description of the basic regulatory mechanics that drive each developmental pathway, although understanding exactly how the choice between them is made is still an important challenge.

An Outline of λ Growth[23]

To understand how each mode of growth is achieved, we first should know the endpoints of each. Lytic growth culminates with release of newly made phage from the dead cell. Lysogenic growth is established when λ repressor binds the operators that control the two early promoters p_L and p_R (Figure 17-14), and when phage DNA is inserted into the host DNA. Because expression of all λ genes (except that for repressor itself and another minor gene) ultimately requires that RNA polymerase start at p_L and p_R, phage genes are inactive during lysogenic growth. Now the questions are: What happens in each pathway of λ growth and why does a particular phage DNA molecule take one pathway or the other?

As we might expect, the two directions of development reflect the expression of different genes. However, the first stages of growth when λ infects a cell are the same, regardless of the eventual outcome. The phage DNA ends meet and become connected through the annealing of complementary single-stranded tails 12 nucleotides long ("sticky ends"), much like the joining of restriction fragments; the remaining nicks are sealed by DNA ligase, forming a covalent circle. Because there is no repressor in the beginning, RNA synthesis starts immediately from p_L and p_R (Figure 17-15). Gene *N*, the first in the leftward operon, encodes a transcription antiterminator that allows RNA synthesis to extend leftward of gene *N* and rightward of gene *cro*, to include the remainder of the early genes. Early genes on the right include *O* and *P*, whose products direct phage DNA replication, and *Q*, whose product turns on the late genes encoding lysis and virion proteins. Only if the lytic pathway predominates and RNA synthesis continues from p_R for some time does enough Q protein accumulate to act effectively.

Two early λ genes, *cro* and *cII*, are critical to determining the direction of phage development. Both Cro and cII proteins accumulate during early infection, but they have opposite effects. The **cII protein** activates transcription from a special promoter to provide a burst of phage repressor (encoded by the *cI* gene), thereby favoring lysogenic growth. If enough repressor is made, it shuts off the early promoters, thus repressing all phage genes required for lytic growth; in particular, the late gene regulator Q is unavailable to provide lysis proteins and phage coats. **Cro protein,** on the other hand, favors lytic growth, although its mechanism is a little more subtle. First, it turns down (but not off) RNA synthesis from p_L and p_R, reducing synthesis of cII protein and disfavoring repressor synthesis. Cro also turns down synthesis of the O, P, and Q proteins, but enough of them is made to complete lytic growth. The second important effect of Cro is to inhibit repressor synthesis directly by blocking the promoter p_{RM} from which repressor is made during lysogenic growth (Figures 17-15 and 17-16).

Thus, Cro acts in opposition to repressor (and to cII, which stimulates repressor synthesis). In any given infected cell, only one will prevail; in normal conditions of cell growth, the odds are about equal for each. We will consider how the outcome is decided.

Repressor and Cro Both Act by Binding the Early Operators O_L and O_R[24–30]

As described in Chapter 16, λ and other repressors are dimers that bind a symmetrical operator structure in DNA. Cro is also a dimeric repressor, although for historical reasons only the *cI* gene product is called the λ repressor. Both proteins control RNA synthesis from promoters p_L and p_R by binding the operators o_L and o_R. How can they bind the same sites on DNA, yet have such different biological effects?

The answer lies in the complex structure of the operators o_L and o_R. We will describe how repressor and Cro work at o_R, keeping in mind that the structure and function of o_L is quite similar. Operators o_L and o_R are each made up of three adjacent suboperators; in o_R, these are

Both pathways of phage growth begin when RNA polymerase binds the early promoters p_L and p_R, and makes mRNA for the *N* and *cro* genes.

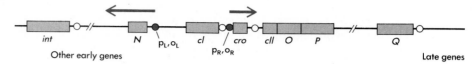

N protein then acts at three sites to extend RNA synthesis to other early genes; O and P proteins allow DNA synthesis to begin; cII protein is made and begins to act as described below.

Know this

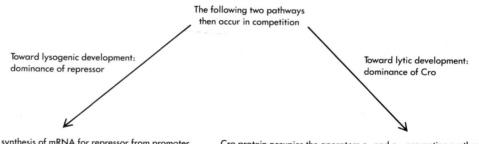

The following two pathways then occur in competition

Toward lysogenic development: dominance of repressor

Toward lytic development: dominance of Cro

cII protein stimulates synthesis of mRNA for repressor from promoter p_{RE} and mRNA for integrase from promoter p_I.

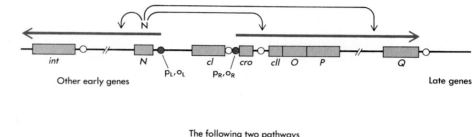

If enough repressor is made, it binds o_R and o_L, blocking RNA synthesis from the early promoters; furthermore, repressor bound to o_R stimulates synthesis of more repressor mRNA from p_{RM}, so that its concentration is maintained in the cell; integrase promotes integration of phage DNA, and lysogeny is established.

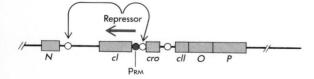

Cro protein occupies the operators o_R and o_L, preventing synthesis of repressor mRNA from promoter p_{RM}, but allowing enough rightward transcription for Q protein to accumulate.

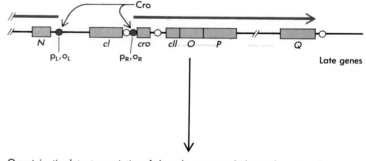

Q protein stimulates transcription of phage late genes, which encode structural proteins that package DNA into new phage particles.

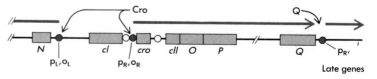

Figure 17-15
Gene expression during phage λ development along two distinct pathways.

named o_{R1}, o_{R2}, and o_{R3}. Both repressor and Cro bind to all three suboperators (Figures 17-16 and 17-17) and thereby control the overlapping promoters p_R and p_{RM}. Since each suboperator has two symmetrical elements, there are a total of six sequences related, *but not identical*, to a consensus binding sequence (Figure 17-17). It is this nonidentity of the binding sites that allows repressor and Cro to work

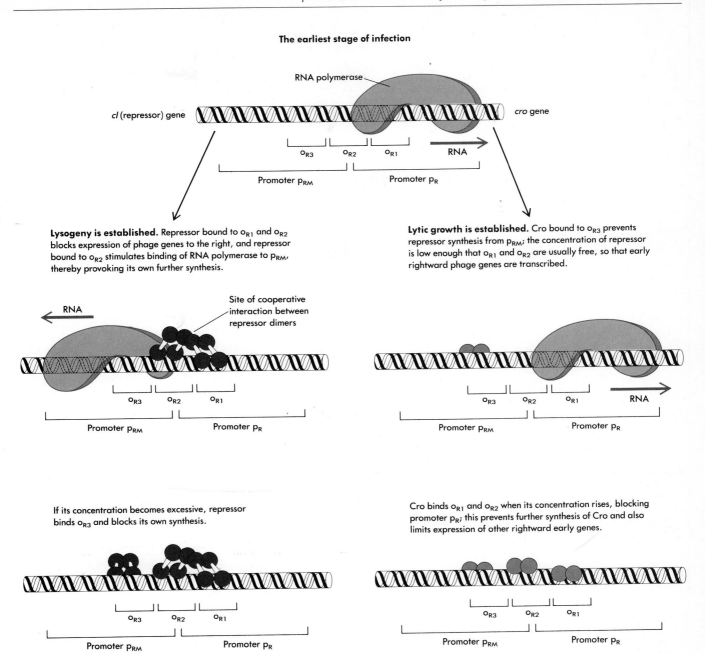

The earliest stage of infection

RNA polymerase

cl (repressor) gene

cro gene

o_{R3} o_{R2} o_{R1} RNA

Promoter p_{RM} Promoter p_R

Lysogeny is established. Repressor bound to o_{R1} and o_{R2} blocks expression of phage genes to the right, and repressor bound to o_{R2} stimulates binding of RNA polymerase to p_{RM}, thereby provoking its own further synthesis.

Lytic growth is established. Cro bound to o_{R3} prevents repressor synthesis from p_{RM}; the concentration of repressor is low enough that o_{R1} and o_{R2} are usually free, so that early rightward phage genes are transcribed.

RNA

Site of cooperative interaction between repressor dimers

o_{R3} o_{R2} o_{R1}

Promoter p_{RM} Promoter p_R

o_{R3} o_{R2} o_{R1} RNA

Promoter p_{RM} Promoter p_R

If its concentration becomes excessive, repressor binds o_{R3} and blocks its own synthesis.

Cro binds o_{R1} and o_{R2} when its concentration rises, blocking promoter p_R; this prevents further synthesis of Cro and also limits expression of other rightward early genes.

o_{R3} o_{R2} o_{R1}

Promoter p_{RM} Promoter p_R

o_{R3} o_{R2} o_{R1}

Promoter p_{RM} Promoter p_R

differently, because the sites do not bind the proteins with equal strength. Therefore, if the concentration of repressor or Cro is low, the tightest binding sites are filled but the others are empty. Furthermore, the nonidentity of binding sequences has opposite effects on the binding of repressor and Cro to o_R. The consequence of these different binding strengths for the control of promoters p_R and p_{RM} is shown in Figure 17-16, and the corresponding biological effects are described in Figure 17-15.

When neither repressor nor Cro is present, RNA polymerase binds p_R and makes mRNA for genes to the right. In this condition, RNA polymerase does *not* bind p_{RM}, for the reason discussed shortly. When repressor is made in abundance, it binds o_{R1} and o_{R2}, thereby blocking p_R (Figure 17-16). The repressor dimer at o_{R2} contacts not

Figure 17-16
The occupancy of the λ rightward operator and promoters p_R and p_{RM} by repressor, Cro, and RNA polymerase during successive stages of λ development along two pathways. [Adapted from M. Ptashne, A. Jeffrey, A. D. Johnson, R. Maurer, B. J. Meyer, C. O. Pabo, T. M. Roberts, and R. T. Sauer, *Cell* 19 (1980):1, with permission.]

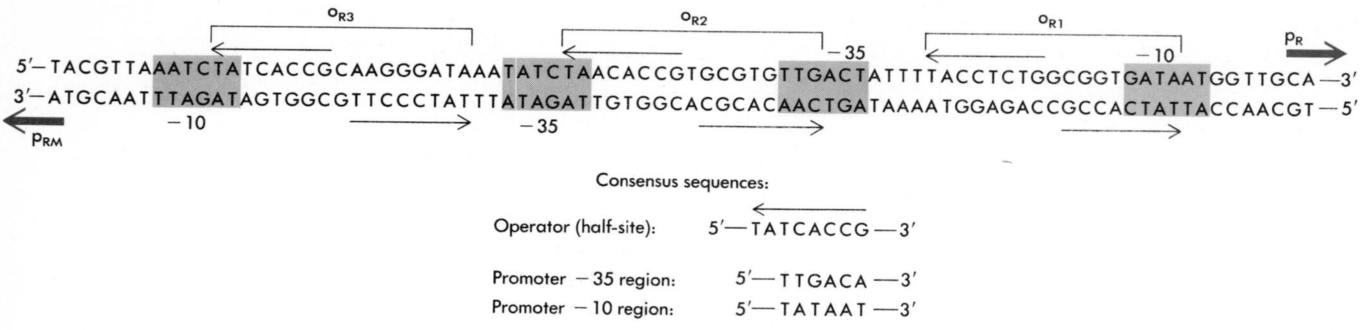

o_{R3} o_{R2} o_{R1} P_R

5'—TACGTTAAATCTATCACCGCAAGGGATAAATATCTAACACCGTGCGTGTTGACTATTTTACCTCTGGCGGTGATAATGGTTGCA—3'
3'—ATGCAATTTAGATAGTGGCGTTCCCTATTTATAGATTGTGGCACGCACAACTGATAAAATGGAGACCGCCACTATTACCAACGT—5'

P_{RM} −10   −35   −35   −10

Consensus sequences:

Operator (half-site):   5'—TATCACCG—3'

Promoter −35 region: 5'—TTGACA—3'
Promoter −10 region: 5'—TATAAT—3'

Figure 17-17
DNA sequence of the λ rightward operator and its associated promoters, and the consensus nucleotide sequences of each.

only DNA, but also the repressor dimer at o_{R1}; it therefore binds more tightly than it would if o_{R1} were not occupied. This interaction between repressor dimers is called cooperativity, and it is important for effective repression.

Repressor binding to o_{R2} has a second important effect besides ensuring repression of the rightward promoter. Repressor also acts as an *activator* for RNA polymerase binding to the promoter p_{RM}, which is the promoter from which mRNA for repressor itself is made in a lysogen; thus, repressor stimulates its own synthesis. This requirement for self-activation explains why only p_R, and not p_{RM}, is active during early infection when no repressor is present.

Finally, at very high concentrations, repressor binds o_{R3} and blocks P_{RM}, thereby preventing its own overproduction. Synthesis of the λ repressor is therefore self-regulated both positively and negatively.

If Cro also binds o_{R1} o_{R2}, and o_{R3}, then why does it act differently from repressor? The reason is that Cro binds most tightly to o_{R3}, not o_{R1} and o_{R2}. Therefore, its first effect is to block p_{RM}, preventing the synthesis of repressor that is necessary for lysogeny. At higher concentrations, Cro also fills o_{R1} and o_{R2}, but this only happens part of the time, so that the rightward promoter stays active enough for the essential genes on the right to be expressed. There is no cooperativity between Cro dimers in binding to the operators, nor does Cro act like λ repressor as an activator of either promoter at o_R.

The λ cII Protein Directs Phage Development According to Cellular Growth Conditions[31–35]

Our description of the competition between repressor and Cro left unanswered why one or the other prevails in a particular infected cell. Clearly, the amount of each protein must be critical to the balance. As the system has evolved, it is the extra burst of repressor synthesis provided by cII protein that gives the lysogenic pathway a good chance to succeed. The cII protein is an activator that causes RNA polymerase to initiate synthesis at the promoter p_{RE}, located between genes *cro* and *cII* itself (see Figure 17-15). Transcription from P_{RE} goes leftward (and through the *cro* gene backward) and makes mRNA for repressor. Once repressor is made from this mRNA, it then can bind o_{R1} and o_{R2}, thus activating the promoter p_{RM} which functions thereafter to provide repressor during lysogenic growth. But the initial function of cII protein is necessary for lysogeny to be established: A *cII* mutant phage almost always grows lytically.

The activity of cII protein is not firmly programmed, but instead changes with certain conditions of cell growth. For example, when

5'—ATCTAAGGAAATACTTACATATGGTTCGTGCAAACAAACGCAACGAGGCT—3'
3'—TAGATTCCTT TATGAAT GTATACCAAGCA CGTTTGTTTGCGTT GCTCCGA—5'

(TAATAT)   (ACAGTT)   Standard promoter
−10   −35   consensus sequences

Beginning of   cII protein binding site
PRE mRNA

glucose is plentiful and the cyclic AMP level is low (Chapter 16), there is less cII activity, so that lytic growth is favored over lysogenic growth. The cII protein is very unstable in the cell; its concentration, and thus the total cII activity, is determined by the rate at which it is degraded by cellular proteases. The effect of cyclic AMP is thought to be mediated by these proteases, although the details of this regulation are still unknown. Because its level is controlled by such environmental factors as the presence of glucose, it is possible that cII protein acts primarily as a sensor of cellular growth conditions for the phage, measuring how favorable it would be to join the cell in lysogenic growth rather than to multiply lytically.

The reason why promoter p_{RE} is essentially inactive unless cII is present is apparent in the nucleotide sequence of the promoter (Figure 17-18). Although its −10 region matches the promoter consensus sequence in three out of six positions, including two of the most strongly conserved, there is no match to the −35 consensus sequence at all. However, evidence from mutations preventing cII function, as well as DNA protection experiments, shows that cII protein binds DNA at two 5'-TTGC-3' sequences that flank the site where RNA polymerase contacts DNA at −35. Thus, cII protein might bind both DNA and RNA polymerase here, substituting for the attraction that a good −35 sequence otherwise provides. The cII binding regions are separated by 10 nucleotides (one turn of DNA helix), and each is about 5 nucleotides (one-half turn of DNA helix) from the center of the contact region at −35. Thus, cII protein binds the other side of the DNA from RNA polymerase, so that it could reach around to touch the RNA polymerase, but still allow the enzyme to contact the DNA backbone in the −35 region.

A second cII protein-dependent promoter, p_I, has a nucleotide sequence similar to that of p_{RE} and is located before gene *int* (see Figure 17-16); it provides a high concentration of the integrase enzyme that catalyzes site-specific recombination of λ DNA into the bacterial chromosome to form the prophage (Chapter 11). A third cII-dependent promoter, p_{AQ}, located in the middle of gene *Q*, acts in an unknown way to retard lytic development and thus to promote lysogenic development.

Positive Control by the λ Gene *N* and Gene *Q* Antiterminator Proteins[36-43]

Phage λ provided the first example of a type of positive transcriptional regulation entirely distinct from promoter activation by regulators like cII protein, CAP, and the arabinose operon C protein (Chapter 16). The transcripts that λ N and Q proteins affect are initiated perfectly well in their absence, but then terminate a few hundred to a thousand nucleotides downstream unless RNA polymerase has been modified by the regulator; λ N and Q proteins are therefore called antiterminators.

Figure 17-18
DNA sequence of λ promoter p_{RE}, including the binding site for the activator cII. The promoter is directed leftward so that the consensus sequences are on the bottom strand in order that the direction agree with the conventional genetic map of λ.

N protein recognition site in the λ early rightward operon, and the first terminator at which it acts

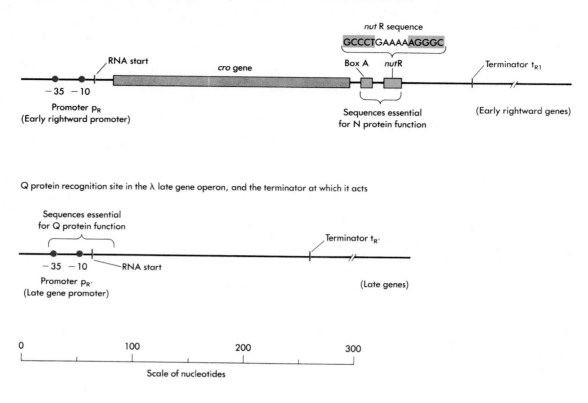

Q protein recognition site in the λ late gene operon, and the terminator at which it acts

Figure 17-19
Recognition sites and sites of action of the λ N and Q protein transcription antiterminators. Box A is a seven nucleotide sequence associated with *nut* sites that is probably required for efficient N protein function. The sequence of the RNA-like strand of *nutR* is shown above *nutR*. Complementary regions of *nutR* that might pair to form a hairpin loop are indicated in color.

N protein regulates early gene expression by acting at three terminators: one to the left of gene *N* itself, one to the right of gene *cro*, and one between genes *P* and *Q* (Figures 17-15 and 17-19). Q protein has one known target, a terminator that is 200 nucleotides downstream from the single late gene promoter, p$_{R'}$, located between genes *Q* and *S* (see Figures 17-14, 17-15, and 17-19). This late gene operon of λ is remarkably large for a procaryotic transcription unit: about 26 kilobases, a distance that takes about 10 minutes for RNA polymerase to traverse. Possibly in this great expanse of DNA there are other, unidentified terminators at which Q protein activity is also required.

Although Q protein–modified RNA polymerase transcribes the whole late region evenly, and all late mRNA has about the same stability, the λ late proteins are made at rates that differ vastly by up to a thousandfold. This illustrates how important the efficiency of mRNA translation by ribosomes can be in determining the rate of gene expression.

We are just beginning to understand how these antiterminators act. Like other regulatory proteins, N and Q proteins recognize specific nucleotide sequences, and they affect only those operons that contain these sequences. Thus, N protein prevents termination in the early operons of λ, but not in other bacterial or phage operons. The λ N protein does not even work in the early operons of phages closely related to λ, like φ80 or the *Salmonella typhimurium* phage P22, which have their own specific "N" proteins.

Genetic experiments first showed that the specific recognition sequences for antiterminators do not occur in the terminators where they act, but instead occur in the operons well before the terminators. N protein recognizes sites named *nut* (for *N utilization*) that are 60 and 200 nucleotides downstream from the early leftward and rightward RNA start sites (Figure 17-19), but it then prevents transcription termination at terminators hundreds to thousands of nucleotides further downstream. This means that N protein must modify RNA polymerase as it passes the *nut* site, perhaps becoming a firmly attached subunit of the enzyme and thereby changing its ability to recognize terminators. Genetic experiments have shown that the products of the bacterial genes *nusA*, *nusB*, and *nusE* act along with λ N protein. The NusA protein is an important cellular transcription factor (Chapter 13). NusE is the small ribosomal subunit protein S10, but its role in N protein function is unknown. No cellular function of NusB protein is known.

Unlike N protein, the λ Q protein recognizes sequences adjacent to and partly within the late gene promoter (see Figure 17-19). Fortunately, Q protein is active in vitro during RNA synthesis by purified RNA polymerase, so that progress has been made in understanding how Q protein prevents termination: It changes the ability of the enzyme to be stopped temporarily (to "pause") at certain nucleotide sequences in DNA that are believed to be essential components of terminators (Chapter 13).

Retroregulation: An Interplay of Controls on RNA Synthesis and Stability Determines *int* Gene Expression[44]

The cII protein activates the promoter p_I that serves the *int* gene, as well as the promoter p_{RE} responsible for repressor synthesis (see Figure 17-15). Therefore, conditions favoring cII protein activity give rise to a burst of both repressor and integrase enzyme. However, the *int* gene is transcribed from p_L as well as from p_I, so that we would expect integrase to be made even in the absence of cII protein. This does not happen. The reason is that *int* messenger RNA initiated at p_L is degraded by cellular nucleases, whereas messenger RNA initiated at p_I is stable and can be translated into integrase protein. This occurs because the two messengers have different structures at their 3' ends. RNA initiated at p_I stops at a terminator about 300 nucleotides after the end of the *int* gene; it has a typical stem-and-loop structure followed by six uridine nucleotides at the end (Figure 17-20). When RNA synthesis is initiated at p_L, on the other hand, RNA polymerase is modified by the N protein antiterminator and thus goes through and beyond the terminator. Then a more extended RNA stem can form that is a substrate for nucleases (Figure 17-20). It is attacked first by the RNA processing nuclease RNaseIII, which leaves an end from which exonucleases can degrade the mRNA and thus destroy it. Because the site responsible for this negative regulation is downstream of the gene it affects, and because degradation proceeds backward through the gene, this process is called **retroregulation.**

The biological function of retroregulation is clear. When cII activity is low and lytic development is favored, there is no need for integrase enzyme; thus, its template is destroyed. However, when cII activity is high and lysogeny is favored, the *int* gene is expressed to promote recombination of the repressed phage DNA into the bacterial chromosome.

There is yet a further subtlety in this regulatory device. When a prophage is induced, it needs to make integrase (as well as excisionase; Chapter 11) to catalyze reformation of free phage DNA by recombination out of the bacterial DNA, and it must do this whether or not

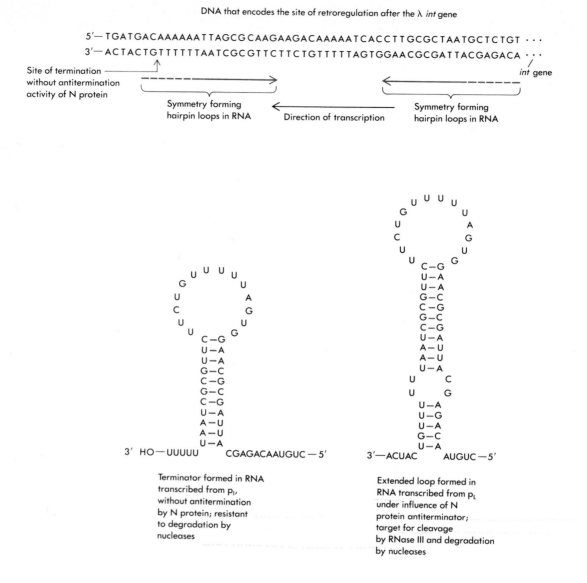

DNA that encodes the site of retroregulation after the λ *int* gene

5'— TGATGACAAAAAATTAGCGCAAGAAGACAAAAATCACCTTGCGCTAATGCTCTGT ···
3'— ACTACTGTTTTTTAATCGCGTTCTTCTGTTTTTAGTGGAACGCGATTACGAGACA ···

Site of termination without antitermination activity of N protein

Symmetry forming hairpin loops in RNA

Direction of transcription

Symmetry forming hairpin loops in RNA

int gene

3' HO—UUUUU   CGAGACAAUGUC — 5'

Terminator formed in RNA transcribed from p_L, without antitermination by N protein; resistant to degradation by nucleases

3'—ACUAC   AUGUC — 5'

Extended loop formed in RNA transcribed from p_L under influence of N protein antiterminator; target for cleavage by RNase III and degradation by nucleases

Figure 17-20
DNA coding site and transcribed RNA structures active in retroregulation of Int protein synthesis.

cII activity is high. In this case, the phage must make stable integrase mRNA from p_L despite the antitermination activity of N protein. Stable integrase mRNA is made, because the phage attachment site at which recombination occurs is *between* the end of the *int* gene and those sequences encoding the extended stem from which mRNA degradation is begun. Thus, the site causing degradation is removed from the end of the *int* gene, and *int* mRNA made from p_L in the prophage is stable; ample integrase is made to help catalyze excision.

Specific Initiation Factors for Viral DNA Replication: The λ O and P Proteins[45, 46]

Like φX174 (Chapter 10), phage λ encodes initiation factors, the products of its *O* and *P* genes, that direct the bacterial replication complex to a specific origin of replication on λ DNA. O protein recognizes the λ origin and is therefore the counterpart of the *E. coli* dnaA protein.

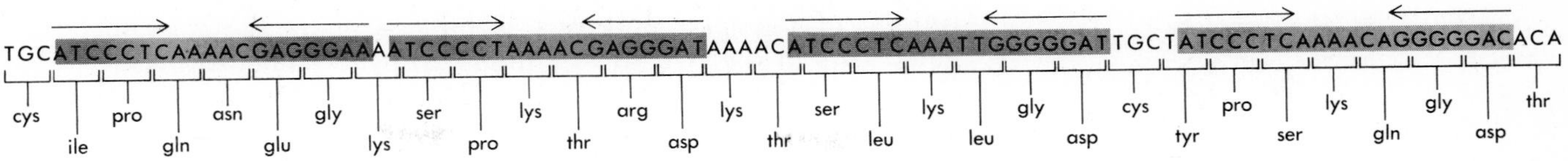

```
      →           ←             →         ←              →             ←               →             ←
TGCATCCCTCAAAACGAGGGAAAATCCCCTAAAACGAGGGATAAAACATCCCTCAAATTGGGGGATTGCTATCCCTCAAAACAGGGGGACACA
```

cys		pro		asn		gly		ser		lys		arg		lys		ser		lys		gly		cys		pro		lys		gly		thr
	ile		gln		glu		lys		pro		thr		asp		thr		leu		leu		asp		tyr		ser		gln		asp	

Figure 17-21
Binding sites for λ O protein in the origin of replication, showing their overlap with the coding sequence for the O protein. Each half-site in the four symmetrical binding sequences is indicated by an arrow.

For this reason, λ does not require dnaA protein for its replication. O protein also binds P protein, which in turn interacts with the E. *coli* replication protein dnaB; thus, P protein links λ to the bacterial replication enzymes. Except for dnaA and dnaC (possibly the counterpart of the λ P protein), λ is thought to utilize most of the normal E. *coli* replication proteins.

Remarkably, the λ origin of replication is *within* the O gene, woven into the coding sequence of the O protein (Figure 17-21). The flexibility of the genetic code, due largely to the multiplicity of codons available for most amino acids, allows this DNA to be at the same time a complex protein binding sequence and a continuous protein coding segment. The origin includes four adjacent and nearly identical 20 nucleotide repeated sequences each containing two symmetrically related half-sites, like the λ operators (Figure 17-21). Four O protein dimers bind to the origin, as shown by the fact that purified O protein protects the four sites from DNase digestion in vitro. Electron microscopy reveals the O protein–origin complex to be a large globular structure similar to that made by integrase protein binding to the attachment site (Chapter 11).

How O protein binding causes DNA replication to initiate at the λ origin is now beginning to be understood. Even when O and P proteins are already present, replication starts only if RNA polymerase is actively transcribing DNA near or through the origin. One guess is that transcription melts the DNA, exposing a template to which the initiation complex can direct DNA primase, so that primers for DNA polymerase are made. An AT-rich region in the origin, to the right of the O protein binding sites, might be involved in this melting. Recent success in showing replication of λ DNA in vitro, dependent on O and P proteins, RNA polymerase, and other cellular replication proteins, should lead to rapid progress in understanding the biochemical details.

Sticky Ends Are Made as Unit-Length λ DNA Molecules Are Cut from Concatemers During Packaging into Phage Heads[47]

The nature of λ DNA replication changes during the growth cycle. Early in infection, the circular infecting DNA is replicated bidirectionally from its origin in gene *O*, like the DNAs of SV40 or E. *coli* itself (see Figure 10-4), yielding more circular molecules. Later, there is a switch to rolling circle replication, similar to that of φX174 (see Figure 10-22) but with two differences. First, there is no enzyme like φX174 A protein to circularize each DNA as it is made, so that a continuous long tail of tandemly duplicated λ DNA molecules (a concatemer) forms. Second, DNA primase acts continually to initiate lagging strand synthesis as the circle rolls, so that the product is double-stranded. As they are packaged into phage heads, unit-length DNA molecules are cut from the concatemer by the enzyme **terminase,** the

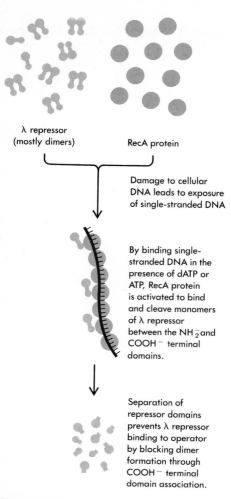

λ repressor
(mostly dimers)

RecA protein

Damage to cellular DNA leads to exposure of single-stranded DNA

By binding single-stranded DNA in the presence of dATP or ATP, RecA protein is activated to bind and cleave monomers of λ repressor between the NH_2^- and $COOH^-$ terminal domains.

Separation of repressor domains prevents λ repressor binding to operator by blocking dimer formation through $COOH^-$ terminal domain association.

Figure 17-22
Activation of *E. coli* RecA protein by DNA to bind and destroy λ repressor by proteolytic cleavage.

product of λ genes *A* and *Nu1*. Terminase makes 12-nucleotide staggered cuts in the two DNA strands at a site named *cos* (see Figure 17-14), thereby generating the "sticky ends" that allow an infecting molecule to circularize in the cell. The concerted reactions of cutting and packaging occur only on concatemers, in which each unit length of DNA is bounded by a *cos* site on either side. It is not clear how or why two *cos* sites separated by this considerable length of DNA might be required for DNA packaging. However, one consequence is that the monomer circles made early in infection cannot be packaged.

Destruction of the λ Repressor by Proteolytic Cleavage Leads to Prophage Induction[48,49]

Although the state of lysogeny is stable, it is not irreversible. Abrupt inactivation of λ repressor in a lysogen leads to induction of lytic growth of the prophage, just as though it were a newly infecting phage. Prophage DNA still must be freed from host DNA through reversal of the integration event, but this is taken care of by the enzymes integrase and excisionase (Chapter 11). Although a λ prophage can be induced to grow in trivial ways—heat inactivation of a temperature-sensitive repressor, for example—prophage induction is also part of an important physiological process: induction of the SOS DNA repair genes (Chapters 12 and 16). Single-stranded DNA exposed by damage to cellular DNA binds the *E. coli* RecA protein, activating in it a protease function that destroys LexA protein, the repressor of the SOS genes. A likely source of single-stranded DNA is the gap left in cellular DNA when replication is stopped by damaged bases (Chapter 11). The λ repressor is also cleaved by this activated RecA protein (Figure 17-22); in fact, it was the study of λ repressor that led to the discovery of the mechanism of SOS gene induction. Treatment of a lysogen with agents that damage DNA, such as ultraviolet irradiation or the crosslinking chemical mitomycin C, causes induction of the prophage and eventual lysis of the cell. It is thought that this gives the prophage a means of escaping a cell subjected to damage so severe that its survival is threatened.

Phage P22 Antirepressor Inactivates Phage Repressors by Noncovalent Binding[50,51]

The *Salmonella* phage P22, one of the related lambdoid phages (including λ, φ80, 82, and 434), makes a unique protein that inactivates repressors in a way entirely different from *recA*-dependent cleavage. P22 **antirepressor** binds specifically to repressors of lambdoid phages, including its own, thereby preventing them from complexing with phage operators. Thus, P22 induces the prophage in a lysogen it infects; furthermore, P22 can infect a lysogen of P22 itself, because antirepressor inactivates the P22 repressor made by the prophage. In contrast, lysogens of other lambdoid phages, which have no antirepressor genes, are immune to infection by the same phage, because repressor made by the prophage shuts down the infecting phage just as it does the prophage. It may seem paradoxical that P22 itself can form a lysogen if it makes antirepressor; but in conditions favoring lysogenic growth, enough P22 repressor is made from the entering phage to overwhelm the antirepressor, and a separate mechanism prevents further synthesis of antirepressor from the prophage.

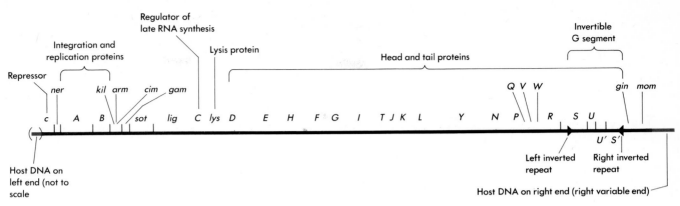

Figure 17-23
Genetic and physical map of bacteriophage Mu. [Adapted from F. J. Grundy and M. M. Howe, *Virology* 134 (1984):296, with permission.]

Phage Mu Is a Transposon[52]

Phage Mu is a temperate bacteriophage, like λ in many ways, but distinct in one important respect: In the major transactions of its DNA—integration into bacterial DNA and replication—Mu acts as a transposon and not as a typical phage. In effect, Mu is a transposon that happens to have not only genes that regulate its transposition, but also genes that encode structural proteins that package its DNA. It is unique also in transposing much more often than other transposons, providing about a hundred new copies of Mu during its hour-long life cycle.

Since Mu is a transposon, the ends of its DNA are never free but instead are embedded in other DNA, namely, that of its bacterial host. This is true not only of the integrated Mu prophage, but also of packaged phage DNA, which has short (50 to 2000 base pairs) segments of bacterial DNA at either end (Figure 17-23). Like other transposons, Mu has inverted repeats at and near the ends of its DNA, and it makes a short (5-base-pair) duplication of the bacterial DNA into which it is inserted.

Mutagenesis by Nearly Random Insertion of Mu Prophage into Cellular Genes[52–57]

As is true for many other transposons (and completely different from phage λ), the transpositional integration of Mu into bacterial DNA to form a prophage is almost random. Thus, each lysogen made when Mu infects a bacterial cell is likely to have the prophage inserted at a different site. Since a bacterial gene interrupted by Mu DNA is usually inactive, Mu makes mutations in the bacterial host as it integrates. This is particularly useful, since once a lysogen has been isolated in which the prophage has interrupted some gene of interest, it is a straightforward process to clone either of the separated pieces of this gene; one simply screens among randomly cloned segments of lysogen DNA for those containing one or the other end of the Mu phage DNA.

Specially constructed variants of Mu named *Mud* (shortened from Mu d*lac*, meaning Mu defective *lac*tose operon) allow genes that can be induced by some treatment of interest to be isolated. At one end, *Mud* phages have part of the lactose operon containing the β-galactosidase gene but lacking the lactose operon promoter. Integra-

Figure 17-24
Prophage integration and replication of phage Mu by transposition.

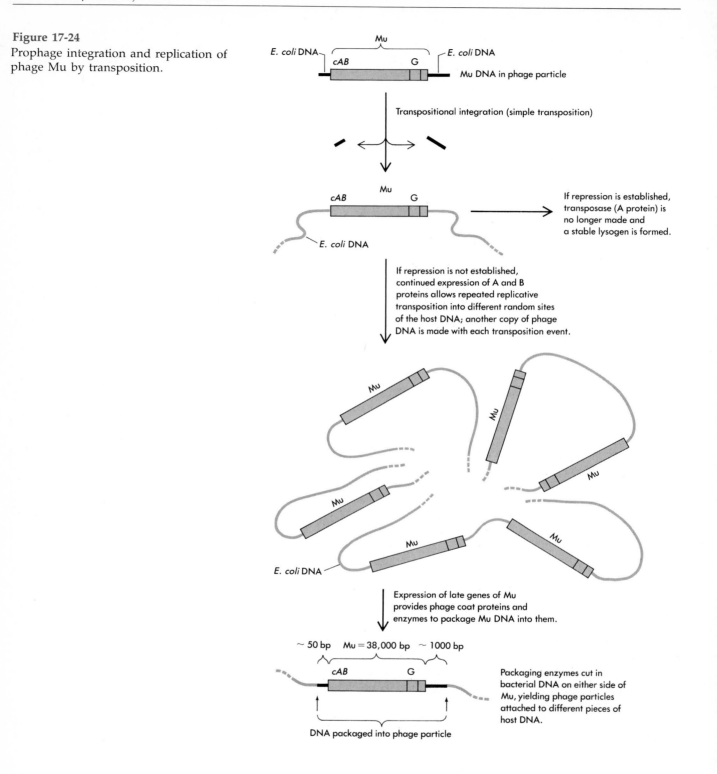

tion of *Mud* into a bacterial gene in the correct direction puts β-galactosidase expression under control of the promoter of the bacterial gene in place of the unknown gene product, thereby revealing and marking that gene. As an example, the characterization of *Mud* lysogens in which β-galactosidase is induced by DNA-damaging agents like ultraviolet light led to identification of many of the SOS genes (Chapters 12 and 16).

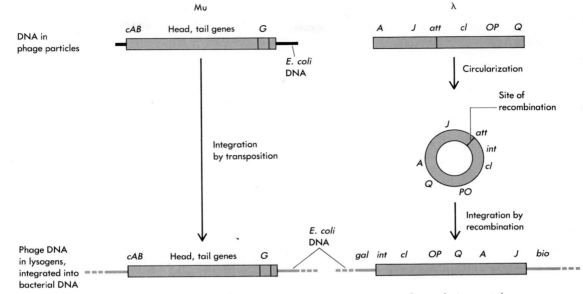

Gene order remains the same   Gene order is permuted

Figure 17-25
A comparison of prophage integration by phages Mu and λ. Phage Mu integrates by moving from its ends, so that its gene order always remains the same. The linear λ DNA molecule circularizes when it enters the cell and then integrates by reciprocal recombination between specific sites on phage and bacterial DNA; the λ gene order is thereby permuted.

Phage Mu Replicates by Repeated Transposition into Cellular DNA[58-62]

We can understand the life cycle of Mu by remembering how transposons move and replicate (Chapter 11). Integration and replication of Mu are examples, respectively, of *simple transposition* and *replicative transposition*. As for most other transposons, the essential elements are a transposase, here the Mu gene *A* protein, and the ends of the transposon, which are at the junction between Mu and bacterial DNA and contain binding sites for the gene *A* protein. When Mu infects a cell, the gene *A* protein directs transposition of Mu from the fragments of bacterial DNA that bounded it into a new site in the bacterial host's DNA (Figure 17-24). The old end pieces of bacterial DNA are probably degraded by nucleases in the cell. Since Mu moves precisely from its ends, its gene order is the same as an infecting particle or prophage; in contrast, λ integrates as a circle by crossing over from a site near the middle of the virion DNA molecule, so that its gene order becomes permuted (Figure 17-25).

If enough Mu repressor is made, Mu genes are repressed and this newly integrated copy remains as a prophage. If, on the other hand, Mu repressor is not made efficiently, then the lytic cycle continues.

Phage Mu replicates during its lytic cycle by continual replicative transposition into different (nearly) random sites of the bacterial DNA. Since one copy remains at the old site and one appears at the new, each transposition event yields another copy of Mu. In addition to transposase (A protein), the protein encoded by gene *B* is required for Mu replication, although its function is not known. No resolvase (Chapter 11) is made by Mu to divide the cointegrates that are formed. Instead, each progeny Mu is packaged into a phage coat along with some of the bacterial DNA that happens to be adjacent— 50 to 100 base pairs on the left, and 1000 to 2000 base pairs on the right. Since Mu inserts at random sites during replication, almost all phage particles are bounded by different bacterial DNA sequences.

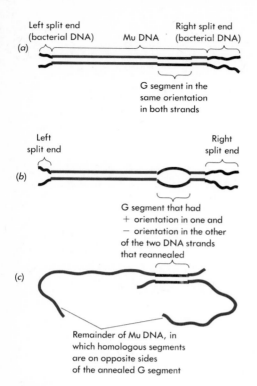

Figure 17-26
DNA structures resulting from denaturing and reannealing Mu virion DNA molecules. Straight double lines indicate double-stranded (i.e., homologous) DNA, and wavy lines indicate single-stranded DNA.

The presence of bacterial DNA at the ends of Mu virion DNA was shown originally by electron microscopy of denatured and reannealed Mu DNA (Chapter 9). For phage like λ and T7, whose virion DNAs consist of discrete and invariable sequences, reannealing of denatured virion DNA produces only homoduplexes—that is, continuous duplex DNA molecules. However, since Mu DNA molecules have different bacterial sequences at their ends, and since the two strands that meet to reanneal almost certainly were not partners originally, reannealed molecules of Mu phage DNA are partial heteroduplexes bearing single-stranded segments of bacterial DNA at their ends—the so-called split ends (Figure 17-26a). (The shorter single-stranded pieces at the left end are not visible in the electron microscope.)

A Surprise in the Structure of Mu DNA: An Invertible DNA Segment That Controls Mu Host Range[63, 64]

Some reannealed Mu DNA molecules have a second region of nonhomology: a bubble formed by two 3000-base-pair segments of single-stranded DNA near the right end of the DNA (Figure 17-27b). Furthermore, in a few molecules, only this segment of DNA has reannealed, and all the rest is left as long, single-stranded tails (Figure 17-26c). These structures arise because the 3000-base-pair **G segment** of Mu has been inverted in about half of the molecules; thus, about half of the reannealing structures cannot pair across the G segment (Figure 17-26b), or instead pair only in the G segment and then cannot pair in the rest (Figure 17-26c). Inversion occurs by a mechanism like that of phase variation in *Salmonella* (p. 331), by recombination across inverted repeats (of 34 base pairs in this case), and it is catalyzed by a similar enzyme, the product of the Mu *gin* gene.

The consequence of G segment inversion is to substitute one pair of phage proteins involved in tail fiber formation (S and U) for a second pair (S' and U') that is equivalent in forming tails, but different in detailed structure. The same segment of DNA, situated to the left of the G segment, encodes the beginning of tail fiber genes S and S' (Figure 17-27). However, the ends of genes S and S', and all of genes U and U', are different and are on the G segment oriented oppositely. When genes S and U are pointed rightward [the G(+) orientation] and are thus expressed along with the other late genes, tail fibers contain the S (and maybe U) protein; then Mu can bind surface receptors of *E. coli* strain K12 and infect it, but cannot bind certain other bacteria. On the other hand, when S' and U' proteins are made [the G(−) orientation], Mu tail fibers contain S' (and maybe U'), and Mu can adsorb to and infect *E. coli* strain C, as well as the bacteria *Shigella sonnei* and *Serratia marcescens*. Thus, inverting its G segment allows Mu to expand or control its host range. G segment inversion does not occur frequently during lytic growth of the phage, but it does occur during growth of lysogens. Thus, a population of lysogens grown for many generations contains about half G(+) and half G(−) prophage.

Expression of the Phage Mu *mom* Gene Is Determined by the State of DNA Methylation[65–69]

Like T7, phage Mu makes an enzyme that protects its DNA against attack by cellular restriction endonucleases. After a Mu lysogen is

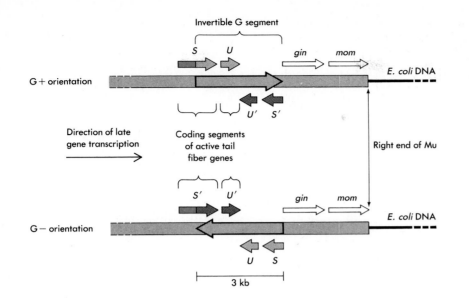

G + orientation

G − orientation

Direction of late gene transcription

Coding segments of active tail fiber genes

Invertible G segment

S *U*

gin *mom*

E. coli DNA

U' *S'*

Right end of Mu

S' *U'*

gin *mom*

E. coli DNA

U *S*

3 kb

Figure 17-27
Structure and coding segments of the invertible G segment of phage Mu in its two orientations. [Adapted from F. J. Grundy and M. M. Howe, *Virology* 134 (1984):296, with permission.]

induced, the enzyme encoded by the Mu *mom* gene modifies about 15 percent of the adenine in phage and bacterial DNA to N^6-(1-acetamido)-adenine; this happens wherever adenine exists in the sequence 5'-C/G-A-C/G-N-Py-3' (where N is any nucleotide and Py is a pyrimidine). The recognition sites of many restriction endonucleases include this sequence, and the modification spoils them as substrates. As is true of other late genes, *mom* expression requires the Mu gene C protein, the activator of late gene transcription. But the *mom* gene is also controlled by the bacterial Dam methylase, the enzyme that methylates adenine in DNA wherever it occurs in the sequence 5'-GATC-3' (Chapter 12). Three GATC sequences in the hundred nucleotides before the *mom* gene promoter must be methylated by the Dam methylase for the *mom* gene to be transcribed. So far, it is not clear how the regulation works or what its function may be, but it will be important to understand this kind of transcriptional regulation by methylation of DNA.

The Single-Stranded DNA Phages: The Smallest DNA Viruses[70,71,72]

Single-stranded DNA (ssDNA) phages like φX174 and M13 have been important for studies of both DNA replication (Chapter 10) and genome structure at the DNA sequence level (Chapter 15). Phage φX174 was the first DNA phage to be sequenced completely, followed soon by its relatives G4, f1, fd, and M13. All are relatively simple, containing 10 or 11 genes that mediate DNA replication and assembly of progeny virions (Figure 17-28). The level of gene expression is determined mainly by the efficiencies of several promoters and terminators on the double-stranded replicative form DNA (Chapter 10), which is the template for synthesis of phage mRNA by the host RNA polymerase.

There are two classes of ssDNA phages, distinct in virion structure and in their effect on the cell in which they grow. The **spherical** (actually polyhedral) **ssDNA phages**, including φX174, G4, and S13, are packaged in an icosahedral shell made of one major and several minor proteins (Figure 17-29a). Like typical phages, they infect by

Figure 17-28
Genetic maps of typical spherical
(ϕX174) and filamentous (fd) ssDNA
phages. [Map of ϕX174 after
A. Kornberg, *1982 Supplement to DNA
Replication*, W. H. Freeman, San Fran-
cisco, p. 162; map of fd after N. D.
Zinder and K. Horiuchi, *Microbiol. Revs.*
49 (1985):102.]

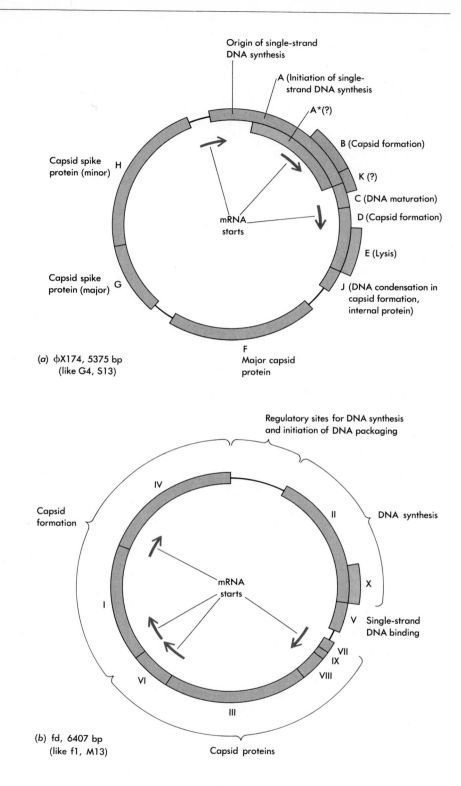

(a) ϕX174, 5375 bp
(like G4, S13)

(b) fd, 6407 bp
(like f1, M13)

adsorbing to receptors on the outside of the cell, replicate within the
cell, and destroy the cell by lysis to escape. The **filamentous ssDNA
phages**, including M13, f1, and fd, are thin particles as long as the cell
(1 μm), within which phage DNA forms an extended loop (Figure
17-29b). They also infect by adsorbing an external receptor, in this
case the F pilus carried by male bacteria (Chapter 7). Although their
DNA is replicated and transcribed like that of spherical phages, fila-

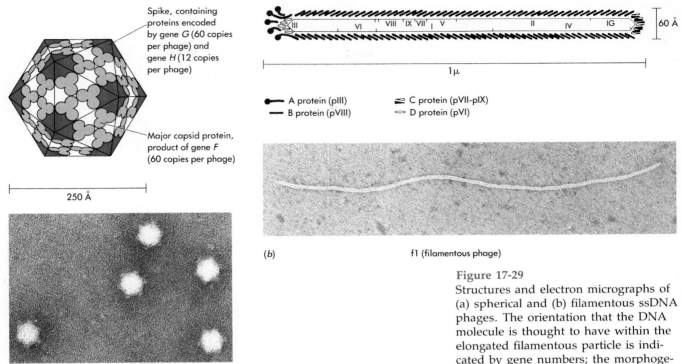

(a) φX174 (spherical phage)

(b) f1 (filamentous phage)

Figure 17-29
Structures and electron micrographs of (a) spherical and (b) filamentous ssDNA phages. The orientation that the DNA molecule is thought to have within the elongated filamentous particle is indicated by gene numbers; the morphogenetic signal sequence is believed to be in the intergenic (IG) region at the right, possibly bound to gene *VII* and *IX* proteins. The diagram also shows how the major coat protein is polymerized in helical array around the DNA. [Electron micrograph in (a) courtesy of Professor R. C. Williams; electron micrograph in (b) courtesy of Professor R. Webster; structures in (a) after A. Kornberg, *DNA Replication,* W. H. Freeman, San Francisco, 1980, p. 499; structure in (b) courtesy of Professor R. Webster.

mentous ssDNA phages do not kill the cell or even make a lysis protein. Instead, phage particles are extruded continuously as the cell continues to grow (albeit more slowly than normal), and the synthesis of phage proteins and DNA continues at a moderate rate that allows cellular metabolism to continue.

Filamentous Single-Stranded DNA Phages Are Secreted Through the Cell Membrane[73–78]

Most phages kill cells because infection destroys the vital cell membrane, equilibrating the cell with its surroundings. Filamentous ssDNA phages, on the other hand, preserve the cell's integrity by adapting its natural protein secretion mechanism for their escape. The major coat protein of phage fd, encoded by gene *VIII*, does not complex with phage DNA as it is made, but instead inserts into the cell membrane—the first step of protein secretion (Chapter 14). As is true of many proteins that go into or through membranes, insertion of coat protein is guided by a signal sequence (of 23 amino acids in this case), which is removed by a protease as insertion occurs. The mature coat protein of 50 amino acids has a basic portion within the cell, a hydrophobic segment that spans the membrane, and an acidic portion on the outside.

Newly synthesized phage DNA is covered first not by the gene *VIII* coat protein in the membrane, but by the abundant, cytoplasmic phage gene *V* protein; about 1500 molecules of gene *V* protein bind to each newly made phage DNA strand. This complex of DNA and gene *V* protein is then drawn to the inner surface of the membrane, where phage assembly begins. The phage particle is formed in the mem-

Figure 17-30

Pathway of filamentous phage particle morphogenesis. The particle is secreted from the cell as it is formed through replacement of gene *V* protein bound to the newly synthesized single-stranded DNA by the membrane-bound gene *VIII* major capsid protein. [After R. E. Webster and J. S. Cashman, in *The Single-Stranded DNA Phages*, ed. D. T. Denhardt, D. Dressler, and D. S. Ray (Cold Spring Harbor; N.Y.: Cold Spring Harbor Laboratory, 1978), p. 558.]

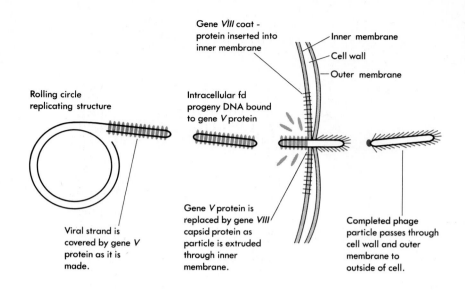

brane through replacement of gene *V* protein by the gene *VIII* coat protein, a process that occurs as the growing filament emerges from the cell (Figure 17-30). Attraction between the basic end of the coat protein within the cell and the acidic DNA probably helps drive the assembly process. Although no phage-specific DNA sequences are required for gene *V* protein to bind, only phage DNA containing a **morphogenetic signal sequence** located between genes *IV* and *II* is assembled into particles at the membrane. It is thought that the minor phage proteins encoded by genes *VII* and *IX* bind to the morphogenetic sequence and are responsible for initiating the assembly process.

Unlike polyhedral phages, filamentous phages have no definite length and thus can accommodate variable amounts of DNA. This flexibility, as well as the fact that phage DNA can be prepared either double-stranded (as the circular replicative form isolated from the cell) or single-stranded (as DNA from phage particles), makes them very useful as cloning vectors. Isolated M13 or fd replicative form DNA can be used like plasmid DNA to make recombinants in vitro; the recombinant DNAs are put into the cell by transformation, again like plasmid DNA, whereupon they give rise to phage particles whose DNA carries the inserted DNA sequence. Even phage λ DNA can be cloned into M13 DNA, although it is eight times as long as M13 itself!

Gene *V* protein of fd has a regulatory function in addition to its role in phage assembly: It represses translation of mRNA for phage genes *II* and *X*. The fd gene *II* protein, like the φX174 A protein, is responsible for initiating phage DNA synthesis by nicking the replicative form DNA, and the control of its expression is believed to limit phage DNA synthesis to a level that the cell can live with.

Viral RNA Self-Replication: Requirements for a New Virus-Specific Enzyme

Cellular RNA molecules never serve as templates for the formation of new RNA strands. Therefore, the replication of most RNA viruses demands the participation of a completely new enzyme capable of

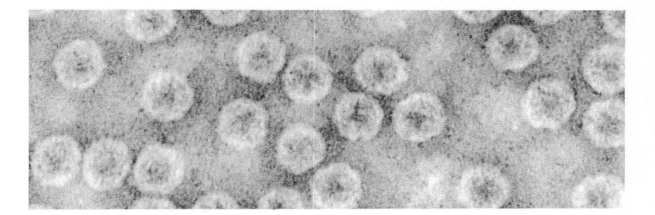

forming new RNA strands upon parental RNA templates. This enzyme, called **RNA replicase** (or **synthetase**), is usually formed just after the viral RNA enters the cell and attaches to host ribosomes. Like both DNA polymerase and RNA polymerase, RNA replicase catalyzes the formation of a complementary strand upon a single-stranded template. The fundamental mechanism for the copying of all nucleic acid base sequences is thus the same. Pairing of complementary bases is always used to achieve accurate replication of specific nucleotide sequences.

After their formation, the complementary RNA strands in turn serve as templates for new rounds of RNA synthesis. In this process, some of the progeny strands become part of new virus particles, while others are used as templates for specific protein formation. Exactly how this happens depends on whether the mature virus contains single-stranded or double-stranded RNA. In both cases, however, only one of the two complementary strands serves as mRNA for a given gene. Thus, there is a similarity in the formation of mRNA in DNA and RNA viral systems.

During the reproduction of most single-stranded RNA viruses (e.g., polio and the various RNA phages), the strand with the template function is the same strand found in the mature particles (i.e., +). This allows the infectious strand to code for the RNA replicase molecules necessary to initiate viral RNA replication. The single-stranded RNA animal virus vesicular stomatitis virus (VSV) behaves quite differently, however. The VSV particle contains the − strand, so that the infecting RNA cannot encode the replicase. Instead, each VSV particle has packaged within it a replicase molecule, made in the previous cycle of viral infection, that enters the cell along with the viral RNA molecule and immediately begins to copy it into a + strand that acts as the mRNA.

RNA Phages Are Very Simple[79–86]

The well-characterized RNA phages of *E. coli*—MS2, R17, f2, and Qβ—are very small (about 250 Å in diameter) and are the simplest viruses known, containing only four genes in their 3600- to 4200-nucleotide-long chromosomes (Figures 17-31 and 17-32). They are prolific, each infecting particle giving 5000 to 10,000 progeny after 30 to 60 minutes' growth at 37°C (Figure 17-33). For the nearly identical phages of the MS2 family (which also includes R17 and f2), two of the

Figure 17-31
Electron micrograph of a collection of f2 RNA phage particles negatively stained so that they appear light against a dark background. (Courtesy of Dr. Norton Zinder of Rockefeller University.)

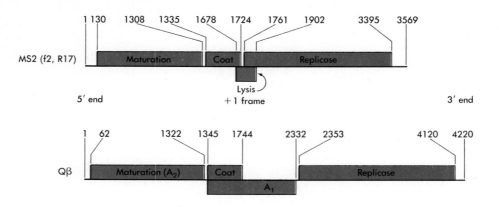

Figure 17-32
Genetic maps of RNA phages MS2 (f2, R17) and Qβ. Gene boundaries are given in nucleotides from the 5′ end of the RNA. [After R. B. Winter and L. Gold, *Cell* 33 (1983):878, with permission.]

four genes code for their two types of structural polypeptide chains. One, the 129-amino-acid coat protein, has a molecular weight of 13,700 daltons and is present 180 times in each virus particle. The other structural protein is the A (attachment) protein needed for adsorption of these phages to their host bacteria, as well as for the subsequent penetration of the viral RNA through the bacterial cell wall and membrane. One copy is found in each virus particle. It has a molecular weight of 44,000 and is made from 393 amino acids. Thus, almost half of RNA phage genetic material is used to code for its structural proteins. Most of the remainder encodes the virally specified component of the replicase, a 544-amino-acid polypeptide of molecular weight 61,000. Overlapping the coat protein and replicase-coding segments is the 75-amino-acid lysis protein, which is encoded in a different reading frame from both and is made by a small fraction of the ribosomes that shift reading frames while they are translating the coat protein–coding segment (Chapter 15). The slightly larger genome of Qβ (see Figure 17-32) is different in two ways: It has no separate lysis gene, the maturation protein A_2 having a second role as a lysis protein, and it encodes a second minor virion protein (A_1) made by ribosomes that occasionally read through the UGA termination codon of the coat protein.

The Nucleotide Sequence of MS2 Suggests Extensive Secondary Structure[87, 88]

Our detailed understanding of RNA phage structure required learning the 3569-nucleotide-long sequence of MS2 RNA, an impressive achievement requiring methods much more difficult than those required for modern DNA sequencing (Figure 17-34). The sequence confirms earlier evidence from physical experiments that phage RNA has much internal structure. Although the base-paired structure shown in Figure 17-34 is hypothetical, the existence of some of the base-paired segments is shown by their resistance to digestion by the nuclease T1. Besides the predominantly local pairing shown in the figure, it is likely that distant segments also join to give the RNA an overall tight structure, accounting for its distinctive behavior as a messenger.

The sequence shows that translation does not start at the 5′ end; instead, 129 nucleotides precede the GUG codon specifying the initiating formyl-methionyl group of A protein. Furthermore, there is no termination codon at the 3′ end: The UAG termination signal for the

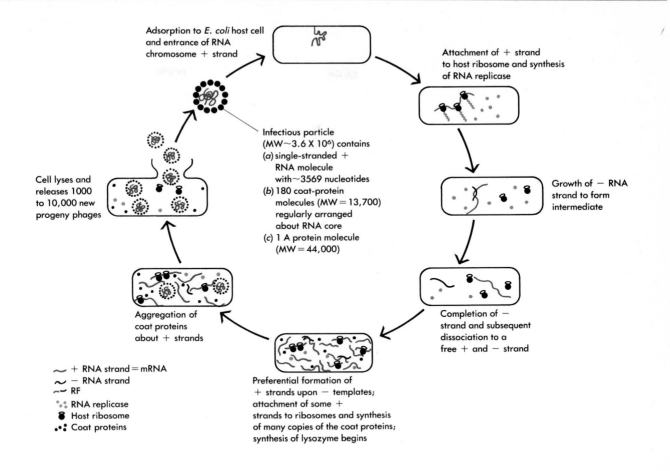

Figure 17-33
The life cycle of a single-stranded RNA phage of the MS2 (R17, f2) family. Growth of phage Qβ is essentially the same, except that the particle contains both A₁ and A₂ proteins, and the A₁ protein is also the phage lysis protein.

replicase gene is followed by an untranslated sequence of 174 nucleotides. It is common for mRNA to have untranslated leading and trailing sequences, which may act in part to determine its resistance to nucleases and thus its stability; in this case, they may also be important for replicase function or in forming the structure of the phage particle.

Initial Binding of Ribosomes to One Site on the Phage RNA

After entry of an RNA phage genome into a host cell, ribosomes attach only to the ribosome binding site at the beginning of the coat protein gene (Figure 17-35a). No attachment occurs at the beginning of either the A protein gene or the replicase gene, because their ribosome binding sites are blocked by the secondary structure of the viral RNA. In contrast, the initiating AUG for coat protein appears to be accessible to ribosomes because it is exposed at the end of a stem (see Figure 17-34).

The first of the two blocked sites to become available is the replicase site, which opens up when reading of the coat protein gene temporarily disrupts the inhibitory hydrogen-bonded hairpin loops (Figure 17-35b). Thus, functioning of the replicase gene always depends on prior ribosome attachment at the coat protein binding site. This dependence was first suggested by the discovery that a nonsense mutation early in the coat protein gene, at the sixth amino acid, is "polar"—

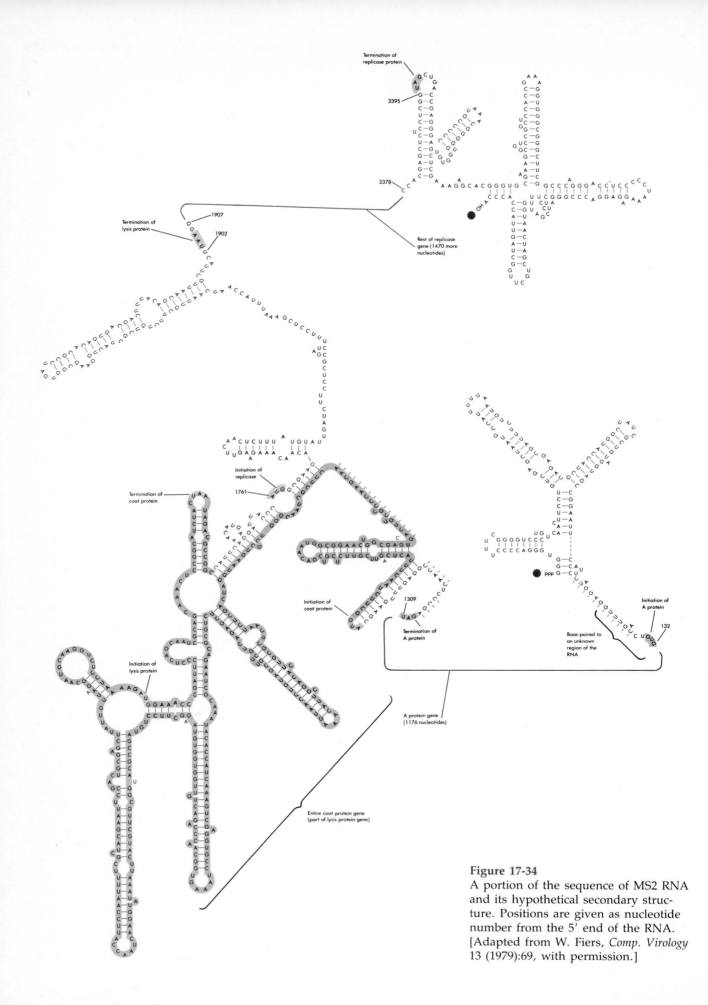

Figure 17-34
A portion of the sequence of MS2 RNA and its hypothetical secondary structure. Positions are given as nucleotide number from the 5′ end of the RNA. [Adapted from W. Fiers, *Comp. Virology* 13 (1979):69, with permission.]

meaning that it prevents replicase expression even though the replicase-coding sequence is intact. It is clear that according to the structural model of Figure 17-34, a ribosome stopping after six amino acids would not open enough RNA to expose the replicase initiation site. The fact that nonsense mutations at positions 50, 54, or 70 are *not* polar also is consistent with the model, since the replicase initiation sequence would already be open when a ribosome had gone this far.

Coat Protein Chains Can Repress the Translation of the Replicase Gene[89, 90]

The RNA phage coat protein polypeptide chains are made in much greater numbers than the replicase polypeptide chains, even though the replicase ribosome binding site is opened each time the coat cistron is translated. This reflects an in vivo control device: Newly made coat protein subunits specifically attach to the replicase gene ribosome binding site and prevent ribosome attachment (Figure 17-36). Thus, the coat protein is a specific **translational repressor** of the replicase gene. Within 10 minutes after infection, enough coat protein has accumulated to block all further replicase synthesis.

Stopping replicase synthesis less than halfway through the life cycle makes obvious sense. By this time, enough enzyme has accumulated to support viral RNA replication, and the primary process needed in the latter phases of the life cycle is the production of the structural proteins of the virus.

The RNA segment that coat protein binds to in order to repress replicase initiation has a structure different from that believed to exist in the virion RNA molecule (see Figure 17-34). It is a 20-nucleotide hairpin loop, a structure about the size of one coat protein molecule, which may be able to form in vivo only when the original base pairing is opened by ribosomes reading the coat protein gene (Figure 17-37). Although this hairpin apparently is not tight enough by itself to prevent ribosomes from initiating translation of replicase, binding of coat protein is thought to stabilize it enough to block ribosome entry and thereby repress replicase synthesis.

This RNA hairpin has been an important tool for understanding the specific interaction of a protein with a polynucleotide different from double-stranded DNA, which is the target of most known regulatory proteins. Measurements of coat protein binding to variants of the hairpin, synthesized in vitro with the enzymes polynucleotide phosphorylase and T4 RNA ligase, show that both the single-stranded loop and the helical stem are recognized by coat protein. Only four of the particular bases present in the hairpin are necessary for coat protein binding, although appropriate bases must exist in other positions so that the structure can form (see Figure 17-37). Unlike many other DNA or RNA binding proteins, coat protein contacts mostly the bases and not the phosphates in the backbone chain.

Functional Complexes of Virally Coded Replicase Chains and Host Proteins[91, 92, 93]

The 61,000-dalton polypeptide product of the replicase gene does not function by itself but only after it has combined with three different host polypeptide chains (Figure 17-38). Two of the host polypeptide chains (with molecular weights of 30,000 and 45,000) are the Tu and

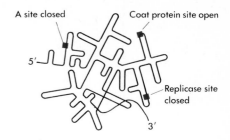

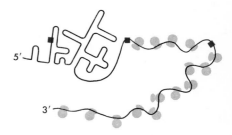

(*a*) In the absence of protein synthesis, the A and replicase sites are blocked.

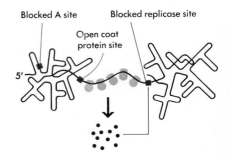

(*b*) Passage of ribosomes over the coat protein gene opens up the secondary structure normally blocking the replicase site, thereby making the replicase gene available for translation.

Figure 17-35
Schematic diagram showing how secondary structure affects the accessibility of ribosome binding sites of RNA phage RNA.

Figure 17-36
Repression of replicase synthesis by binding of a coat protein molecule to the initiation site of the replicase gene.

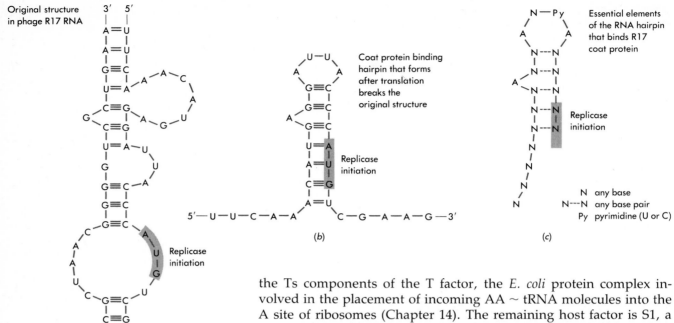

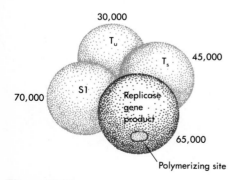

Figure 17-37
Postulated secondary structure of MS2 RNA containing the initiation site for replicase (a) before ribosomes bind and (b) after the original structure is opened by translation of the coat protein gene, allowing a hairpin to form. Essential elements of the coat protein binding RNA hairpin are shown in (c). [After O. C. Uhlenbeck, J. Carey, P. J. Romaniuk, P. T. Lowary, and D. Beckett, *J. Biomolecular Structure and Dynamics* 1 (1983):539, with permission.]

Figure 17-38
Schematic representation of Qβ replicase, showing its construction from one phage coded polypeptide chain and three host components.

the Ts components of the T factor, the *E. coli* protein complex involved in the placement of incoming AA ~ tRNA molecules into the A site of ribosomes (Chapter 14). The remaining host factor is S1, a 70,000-dalton protein that is usually found tightly bound to the 30S ribosomal subunit. We are certain only of the role of the virally coded component. It contains the catalytic site responsible for making the internucleotide 3′–5′ phosphodiester bonds between a growing chain and ribonucleoside triphosphate precursors. The T factor probably functions in chain initiation, helping to recognize the 3′ ends of the RNA templates. Favoring this hypothesis is the presence of tRNA-like sequences (e.g., CCA–OH) at the 3′ ends of both the viral + and − strands. As T factors are adapted specifically to recognize the tRNA molecules, their presence within the complete replicase complex may have evolved to help replicase molecules preferentially attach to 3′ ends of viral as opposed to cellular RNA molecules. The S1 component of replicase is required only for the initiation stage of replication, and only on + strand templates; thus, S1 may help replicase to recognize a specific binding site on the viral RNA molecule.

RNA Phage RNA Self-Replication Does Not Involve Fully Double-Helical Intermediates[94, 95, 96]

Replication of viral RNA starts with the attachment of replicase to the infecting + strand. Surprisingly, the enzyme binds not only to the 3′ end, where 5′ → 3′ synthesis of the product − strand must begin, but also to two internal sites, one that includes the ribosome binding site of the coat gene and one that is within the replicase gene. It is likely that these sites are close to one another in the complex three-dimensional structure of the viral RNA molecule. Binding of replicase to the coat gene has regulatory function: It prevents ribosomes from attaching and translating coat protein, a process that might interfere with replication.

Although base pairing determines the − strand sequences, no double-helical **replicative intermediate** is generated, and a free − strand is the product of the first round of synthesis (Figure 17-39). Only at the growing end of the − strand is a short double-helical stretch present within the replicative intermediate. However, if the replicase is removed by treatments with protein-denaturing agents, all the nascent strands base-pair with their parental complements to form double-helical sections. Therefore, replicase itself is responsible for keeping the strands apart, despite their potential for full annealing if the protein is removed.

Both the complete progeny − strand and the original + strand in turn serve as templates to make more + and − strands. In this process, many more + strands are made than − strands, indicating that replicase molecules preferentially bind to the 3′ end of − strands, as opposed to 3′ ends of + strands.

Laboratory-made + strands have the same length as those made in vivo and have equal infectivity when added to growing bacteria. Moreover, the frequency of mutants is similar to that in nature. This means that the copying achieved in vitro is just as accurate as that occurring within cells.

Only Nascent Stretches of + RNA Serve as Templates for the A Protein

Reading of the A gene is limited to the periods of viral RNA replication that are characterized by incomplete stretches of nascent RNA. Just after a viral + strand has started to grow, the newly formed A binding site is accessible for ribosome attachment (Figure 17-40). But it quickly becomes closed as extensive patches of secondary structure form upon further chain elongation. Open A sites bind ribosomes even better than open coat protein sites, so that their inherent affinity for ribosomes by itself does not determine how often they will be translated; the degree to which they are not blocked by secondary structure is more important.

Self-Assembly of Progeny Particles and the Formation of Intracellular Viral Crystals

Assembly of progeny phages spontaneously starts as soon as the A protein, coat protein, and progeny + strands begin to accumulate. Only progeny + strands are ever encapsulated; thus the newly made − strands are left free to produce still more progeny + strands. In the aggregation process, the A protein is thought to bind early to the phage RNA, forming a tight complex that persists through the next cycle of phage growth when the A protein enters the bacterial cell along with its respective RNA molecule. Conceivably, binding of the A protein converts progeny + strands into forms that hasten the aggregation of coat protein around them.

As infection progresses, the progeny phage often form such a substantial fraction of the internal bacterial contents that they begin to form tiny crystals that are easily visualized by the electron microscope (Figure 17-41). Since RNA phage crystals have also been made in vitro, there is promise that the molecular structure of RNA phage will soon be known from x-ray diffraction analysis.

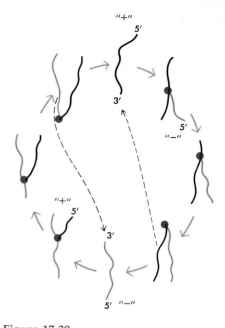

Figure 17-39
Schematic model for the replication of RNA phage (Qβ, MS2, R17, f2) RNA directed by the phage replicase. Here, for simplicity of visualization, the single-stranded regions are shown in an extended rather than tightly collapsed form.

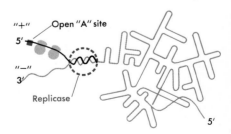

Figure 17-40
Translation of the RNA phage A protein gene from nascent + strands.

Figure 17-41
Electron micrograph of a thin section of *E. coli* cells infected with the RNA phage f2. Most of the progeny particles lie next to each other in crystalline array. (Courtesy of Dr. Norton Zinder of Rockefeller University.)

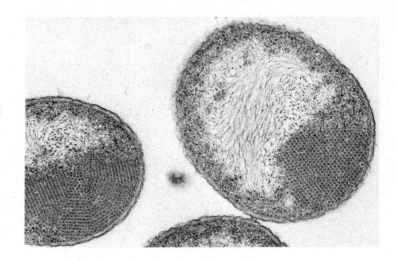

The Smallest Known Viruses Are Almost at the Lower Limits for a Virus

The need ordinarily to code for both virus-specific RNA replicase and viral coat protein effectively places a lower limit on how small an RNA-containing virus can be. First, it would be surprising if a replicase much smaller than, say, 30,000 daltons could carry out all the tasks involved in RNA self-replication. Second, it is hard to imagine stable coat protein subunits made up of less than 50 to 75 amino acids. Thus, the smallest complete RNA virus likely to exist probably contains at least 1500 nucleotides (half the MS2 number). The expectation of anything much smaller seems highly unrealistic.

Analogous reasoning for the minimal size of a DNA virus is not yet possible, since it is possible that a circular viral DNA molecule might be replicated using only host enzymes. If so, then someday we might find a DNA virus whose only gene codes for its coat protein. Conceivably, such a DNA chain might contain as few as 300 to 500 base pairs.

Making intelligent guesses of the upper size for a virus is much more difficult. The largest viruses now known code for some 200 to 300 genes, but there seems to be no reason why a much larger number could not be coded for. Perhaps, above a certain size, there is no substantial need for new genes to manipulate the host cell in favor of viral products.

Always very clear, however, is the absolute difference between a virus and a cell. As little as 20 years ago, there was much confusion. At that time, a key criterion for a virus was still its ability to pass through membrane filters that hold back bacteria. Under this definition, the very small disease-causing bodies known as *Rickettsiae* were generally treated as viruses. But they since have been found to contain DNA and RNA, as well as a protein-synthesizing system, and so are realized to be very small bacteria. The essence of viruses is thus the absence of protein-synthesizing systems. Hence, their multiplication must always involve the breakdown of their surrounding protective coats, thereby letting their chromosomes come into contact with cellular enzymes.

Summary

All known viruses contain a core of either DNA or RNA, surrounded by a protective capsid of protein. More complicated viruses also contain lipids or carbohydrates, or both. The bulk of the viral coat is invariably constructed of one or a few types of polypeptides that are present in many copies. Viral DNA or RNA encodes both viral coat proteins and enzymes that are required to replicate the genome. After coat proteins and copies of the viral nucleic acid are made in an infected cell, these components specifically aggregate to form new infectious viruses identical to the infecting parental particle. When RNA is the genetic component, it is always replicated by a virus-specific enzyme, whereas DNA viruses often are replicated by cellular DNA polymerases. The viral genome may be either single-stranded or double-stranded. Single-stranded viral DNA or RNA first is copied into a complementary strand, which in turn serves as a template for synthesis of more viral nucleic acid; thus, the basic rules by which nucleic acids are duplicated through complementary base pairing are always the same.

Viruses often specify proteins that subvert cellular metabolism to promote selective expression of the viral genome. The more complex DNA bacteriophages (bacterial viruses) either alter the host RNA polymerase so that it initiates only at phage promoters or provide their own virus-specific RNA polymerase. Furthermore, they may encode nucleases that destroy the bacterial DNA and lysozyme to disrupt the bacterial cell wall and liberate progeny phage particles. Virus-encoded regulatory proteins act in a defined sequence to establish a correctly timed program of gene expression during viral growth.

Bacteriophages have provided our best understanding of the molecular basis of gene regulation, particularly the regulation of transcription. Phages SPO1 and T4 encode new σ factors that direct the host RNA polymerase to distinct phage-specific promoter sequences. In contrast, phages T7 and N4 encode entirely new RNA polymerases specific for phage promoter sequences.

The temperate phage λ can either grow lytically upon infection, expressing all genes required for its replication, or instead join the host chromosome by recombination and grow as a repressed prophage that is replicated passively by the host cell. The opposing activities of λ Cro protein and λ repressor direct these alternate developmental pathways; the outcome of their competition is influenced by the separate regulator cII, which senses the environmental condition of the cell and activates transcription of λ repressor if conditions are favorable for lysogeny. The positive regulators of early and late gene transcription of λ are antiterminators that modify RNA polymerase and thereby allow it to pass transcription terminators that precede regulated genes in specific operons.

Phages encode proteins that direct their own replication. The λ O and P proteins, like the ϕX174 A protein, initiate replication at a specific phage origin. Phage Mu replicates by transposition, so that its specific replication protein is a transposase. In lytic growth, Mu transposes repeatedly into different sites of the host genome, and it is cut from the chromosome for packaging into phage coats with random bits of bacterial DNA still attached to its ends.

Phage Mu is a temperate phage that also integrates into host DNA by transposition, entering nearly random sites of the bacterial DNA and making mutations in the host as it does. The invertible G segment of Mu specifies different tail fiber proteins in its two orientations, thereby allowing the host adsorption specificity of Mu to change.

The small, single-stranded DNA phages like ϕX174 and fd encode mostly structural proteins, except for the initiator of single-stranded DNA synthesis. DNA of filamentous single-stranded DNA phages is packaged through a process of secretion, in which coat protein waiting in the cell membrane coalesces with newly replicated phage DNA to form a particle that is extruded from the cell as it is made.

RNA phages encode a replicase that first synthesizes a − strand copy of the infecting + strand and then more viral + copies from the − strand template. The efficiency with which RNA phage genes are translated is determined by the complex secondary structure of the phage RNA; at first only its coat protein initiation codon is accessible to ribosomes, until the act of translating the coat gene opens the RNA for translation of the replicase. RNA phage coat protein is a specific translational repressor of replicase synthesis, acting by binding a hairpin loop in phage RNA that includes the replicase translation initiation site.

Bibliography

General References

Denhardt, D. T., D. Dressler, and D. S. Ray, eds. 1978. *The Single-Stranded DNA Phages.* Cold Spring Harbor, N.Y.: Cold Spring Harbor Laboratory.

Hendrix, R. W., J. W. Roberts, F. W. Stahl, and R. A. Weisberg, eds. 1983. *Lambda II.* Cold Spring Harbor, N.Y.: Cold Spring Harbor Laboratory.

Luria, S. E., J. E. Darnell, Jr., D. Baltimore, and A. Campbell. 1978. *General Virology.* 3rd ed. New York: Wiley.

Mathews, C. K., E. M. Kutter, G. Mosig, and P. B. Berget, eds. 1983. *Bacteriophage T4.* Washington, D.C.: American Society for Microbiology.

Stent, G. S. and R. Calendar. 1978. *Molecular Genetics: An Introductory Narrative.* 2nd ed. San Francisco: W. H. Freeman.

Zinder, N. D., ed. 1975. *RNA Phages.* Cold Spring Harbor, N.Y.: Cold Spring Harbor Laboratory.

Cited References

1. Crick, F. H. C., and J. D. Watson, 1957. "Virus Structure: General Principles." *CIBA Found. Symp. Nature Viruses.* 5–13.
2. Caspar, D. L. D., and A. Klug. 1962. "Physical Principles in the Construction of Regular Viruses." *Cold Spring Harbor Symp. Quant. Biol.* 27:1–241.
3. Wood, W. B., and R. S. Edgar. 1973. "Building a Bacterial Virus." *Scientific American* 1967. In *The Chemical Basis of Life*, ed. P. C. Hanawalt and R. H. Haynes. San Francisco: Freeman.
4. Harrison, S. C. 1983. "Virus Structure: High-Resolution Perspectives." *Adv. Virus Res.* 28:175–240.
5. Losick, R. and J. Pero. 1981. "Cascades of Sigma Factors." *Cell* 25:582–584.
6. Chelm, B. K., J. R. Greene, C. Beard, and E. P. Geiduschek. 1982. "The Transition of Early and Middle Gene Expression in the Development of Phage SPO1: Physiological and Biochemical Aspects." In *Molecular Cloning and Gene Regulation in* Bacilli, ed. A. T. Ganesan, S. Chang, and J. A. Hoch. San Diego: Academic Press, pp. 345–359.
7. Fox, T. D., R. Losick, and J. Pero. 1976. "Regulatory Gene *28* of Bacteriophage SPO1 Codes for a Phage-Induced Subunit of RNA Polymerase." *J. Mol. Biol.* 101:427–433.
8. Duffy, J. J., and E. P. Geiduschek. 1977. "Purification of a Positive Regulatory Subunit from Phage SPO1-Modified RNA Polymerase." *Nature* 270:28–32.
9. Costanzo, M., L. Brzustowicz, N. Hannett, and J. Pero. 1984. "Bacteriophage SPO1 Genes *33* and *34*: Location and Primary Structure of Genes Encoding Regulatory Subunits of *Bacillus Subtilis* RNA Polymerase." *J. Mol. Biol.* 180:533–547.
10. Costanzo, M. 1983. "Bacteriophage SPO1 Regulatory Genes." Ph.D thesis, Harvard University.
11. Lee, G., and J. Pero. 1981. "Conserved Nucleotide Sequences in Temporally Controlled Bacteriophage Promoters." *J. Mol. Biol.* 152:247–265.
12. Dunn, J. J., and F. W. Studier. 1983. "Complete Nucleotide Sequence of Bacteriophage T7 DNA and the Locations of T7 Genetic Elements." *J. Mol. Biol.* 166:477–535.
13. Chamberlin, M., J. McGrath, and L. Waskell. 1970. "New RNA Polymerase from *E. coli* Infected with Bacteriophage T7." *Nature* 228:227–231.
14. Bailey, J. N., J. R. Klement, and W. T. McAllister. 1983. "Relationship Between Promoter Structure and Template Specificities Exhibited by the Bacteriophage T3 and T7 RNA Polymerases." *Proc. Nat. Acad. Sci.* 80:2814–2818.
15. Davanloo, P., A. H. Rosenberg, J. J. Dunn, and F. W. Studier. 1984. "Cloning and Expression of the Gene for Bacteriophage T7 RNA Polymerase." *Proc. Nat. Acad. Sci.* 81:2035–2039.
16. Mark, K., and F. W. Studier. 1981. "Purification of the Gene *0.3* Protein of Bacteriophage T7, an Inhibitor of the DNA Restriction System of *Escherichia coli*." *J. Biol. Chem.* 256:2573–2578.
17. Hesselbach, B. and D. Nakada. 1977. "'Host Shutoff' Function by Bacteriophage T7: Involvement of T7 Gene 2 and Gene *0.3* in the Inactivation of *Escherichia coli* RNA Polymerase." *J. Virology* 24:736–745.
18. Zillig, W., H. Fujiki, W. Blum, D. Janeković, M. Schweiger, H. Rahmsdorf, H. Ponta, and M. Hirsch-Kauffman. 1975. "*In Vivo* and *in Vitro* Phosphorylation of DNA-Dependent RNA Polymerase of *Escherichia coli* by Bacteriophage-T7-Induced Protein Kinase." *Proc. Nat. Acad. Sci.* 72:2506–2510.
19. Center, M. S., F. W. Studier, and C. C. Richardson. 1970. "The Structural Gene for a T7 Endonuclease Essential for Phage DNA Synthesis." *Proc. Nat. Acad. Sci.* 65:242–248.
20. Haynes, L. L., and L. B. Rothman-Denes. 1985. "N4 Virion RNA Polymerase Sites of Transcription Initiation." *Cell* 41:597–605.
21. Falco, S. C., W. Zehring, and L. B. Rothman-Denes. 1980. "DNA-Dependent RNA Polymerase from Bacteriophage N4 Virions, Purification and Characterization." *J. Biol. Chem.* 255:4339–4347.
22. Hendrix, R. W., J. W. Roberts, F. W. Stahl, and R. A. Weisberg, eds. 1983. *Lambda II.* Cold Spring Harbor, N.Y.: Cold Spring Harbor Laboratory.
23. Friedman, D., and M. Gottesman. 1983. "Lytic Mode of Lambda Development." In *Lambda II*, ed. R. W. Hendrix, J. W. Roberts, F. W. Stahl, and R. A. Weisberg. Cold Spring Harbor, N.Y.: Cold Spring Harbor Laboratory, pp. 21–51.
24. Gussin, G., A. Johnson, C. Pabo, and R. Sauer. 1983. "Repressor and Cro Protein: Structure, Function, and Role in Lysogenization." In *Lambda II*, ed. R. W. Hendrix, J. W. Roberts, F. W. Stahl, and R. A. Weisberg. Cold Spring Harbor, N.Y.: Cold Spring Harbor Laboratory, pp. 93–121.
25. Ptashne, M. 1986. *A Genetic Switch.* Palo Alto: Blackwell Scientific Publications and Cell Press.
26. Johnson, A. D., A. R. Poteete, G. Lauer, R. T. Sauer, G. K. Ackers, and M. Ptashne. 1981. "λ Repressor and Cro-Components of an Efficient Molecular Switch." *Nature* 294:217–223.
27. Meyer, B. J., R. Maurer, and M. Ptashne. 1980. "Gene Regulation at the Right Operator (O_R) of Bacteriophage λ. II. O_R1, O_R2, and O_R3: Their Roles in Mediating the Effects of Repressor and cro." *J. Mol. Biol.* 139:163–194.
28. Hochschild, A., N. Irwin, and M. Ptashne. 1983. "Repressor Structure and the Mechanism of Positive Control." *Cell* 32:319–325.
29. Eisen, H., P. Brachet, L. Pereira da Silva, and F. Jacob. 1970. "Regulation of Repressor Expression in λ." *Proc. Nat. Acad. Sci.* 66:855–862.
30. Reichardt, L., and A. D. Kaiser. 1971. "Control of λ Repressor Synthesis." *Proc. Nat. Acad. Sci.* 68:2185–2189.
31. Wulff, D. L., and M. Rosenberg. 1983. "Establishment of Repressor Synthesis." In *Lambda II*, ed. R. W. Hendrix, J. W. Roberts, F. W. Stahl, and R. A. Weisberg. Cold Spring Harbor, N.Y.: Cold Spring Harbor Laboratory, pp. 53–73.
32. Herskowitz, I., and D. Hagen. 1980. "The Lysis-Lysogeny Decision of Phage λ; Explicit Programming and Responsiveness." *Ann. Rev. Genetics* 14:399–445.
33. Shimatake, H., and M. Rosenberg. 1981. "Purified λ Regulatory Protein cII Positively Activates Promoters for Lysogenic Development." *Nature* 292:128–132.
34. Hoyt, M. A., D. M. Knight, A. Das, H. I. Miller, and H. Echols. 1982. "Control of Phage λ Development by Stability and Synthesis of cII Protein: Role of the Viral *cIII* and Host *hflA*, *himA* and *himD* Genes." *Cell* 31:565–573.
35. Hoopes, B. C., and W. R. McClure. 1985. "A cII-Dependent Promoter Is Located Within the *Q* Gene of Bacteriophage λ." *Proc. Nat. Acad. Sci.* 82:3134–3138.
36. Roberts, J. W. 1969. "Termination Factor for RNA Synthesis." *Nature* 224:1168–1174.
37. Adyha, S., M. Gottesman, and B. de Crombrugghe. 1974. "Release of Polarity in *Escherichia coli* by Gene N of Phage λ: Termination and Antitermination of Transcription." *Proc. Nat. Acad. Sci.* 71:2534–2538.
38. Franklin, N. 1974. "Altered Reading of Genetic Signals Fused to the N Operon of Bacteriophage λ: Genetic Evidence for Modification of Polymerase by the Protein Product of the N Gene." *J. Mol. Biol.* 89:33–48.
39. Friedman, D. I., G. S. Wilgus, and R. J. Mural. 1973. "Gene N Regulator Function of Phage λimm21: Evidence That a Site of N Action Differs from a Site of N Recognition." *J. Mol. Biol.* 81:505–516.
40. Salstrom. J. S., and W. Szybalski. 1978. "Coliphage λ *nut* L⁻: A Unique Class of Mutants Defective in the Site of Gene N Product Utilization for Antitermination of Leftward Transcription." *J. Mol. Biol.* 124:195–221.
41. Herskowitz, I., and E. R. Signer. 1970. "A Site Essential for Expression of All Late Genes in Bacteriophage λ." *J. Mol. Biol.* 47:545–556.
42. Grayhack, E. J., X. Yang, L. F. Lau, and J. W. Roberts. 1985. "Phage Lambda Gene Q Antiterminator Recognizes RNA Polymerase Near the Promoter and Accelerates It Through a Pause Site." *Cell* 42:259–269.
43. Ray, P. N. and M. L. Pearson. 1974. "Evidence for Posttranscriptional Control of the Morphogenetic Genes of Bacteriophage Lambda." *J. Mol. Biol.* 85:163–175.
44. Gottesman, M., A. Oppenheim, and D. Court. 1982. "Retroregulation: Control of Gene Expression from Sites Distal to the Gene." *Cell* 29:727–728.

45. Tsurimoto, T., and K. Matsubara. 1981. "Purified Bacteriophage λ O Protein Binds to Four Repeating Sequences at the λ Replication Origin." *Nucleic Acid Res.* 9:1789–1799.

46. LeBowitz, J. H., M. Zylicz, C. Georgopoulos, and R. McMacken. 1985. "Initiation of DNA Replication on Single-Stranded DNA Templates Catalyzed by Purified Replication Proteins of Bacteriophage λ and *Escherichia coli*." *Proc. Nat. Acad. Sci.* 82:3988–3992.

47. Feiss, M. and A. Becker. 1983. "DNA Packaging and Cutting." In *Lambda II*, ed. R. W. Hendrix, J. W. Roberts, F. W. Stahl, and R. A. Weisberg. Cold Spring Harbor, N.Y.: Cold Spring Harbor Laboratory, pp. 305–330.

48. Roberts, J. W., and C. W. Roberts. 1975. "Proteolytic Cleavage of Bacteriophage Lambda Repressor in Induction." *Proc. Nat. Acad. Sci.* 72:147–151.

49. Craig, N. L., and J. W. Roberts. 1980. "E. coli recA Protein-Directed Cleavage of Phage λ Repressor Requires Polynucleotide." *Nature* 283:26–30.

50. Susskind, M. M., and P. Youderian. 1983. "P22 Antirepressor and Its Control." In *Lambda II*, ed. R. W. Hendrix, J. W. Roberts, F. W. Stahl, and R. A. Weisberg. Cold Spring Harbor, N.Y.: Cold Spring Harbor Laboratory, pp. 347–363.

51. Susskind, M. M., and D. Botstein. 1975. "Mechanism of Action of *Salmonella* Phage P22 Antirepressor." *J. Mol. Biol.* 98:413–424.

52. Shapiro, J. A., ed. 1983. *Mobile Genetic Elements.* New York: Academic Press.

53. Taylor, A. L. 1963. "Bacteriophage-Induced Mutation in *Escherichia coli*." *Proc. Nat. Acad. Sci.* 50:1043–1051.

54. Bukhari, A., and D. Zipser. 1972. "Random Insertion of Mu-1 DNA Within a Single Gene." *Nature New Biol.* 236:240–243.

55. Daniell, E., R. Roberts, and J. Abelson. 1972. "Mutations in the Lactose Operon Caused by Bacteriophage Mu." *J. Mol. Biol.* 69:1–8.

56. Daniell, E., D. E. Kohne, and J. Abelson. 1975. "Characterization of the Inhomogeneous DNA in Virions of Bacteriophage Mu by DNA Reannealing Kinetics." *J. Virology* 15:739–743.

57. Castilho, B. A., P. Olfson, and M. J. Casadaban. 1984. "Plasmid Insertion Mutagenesis and *lac* Gene Fusion with Mini-Mu Bacteriophage Transposons." *J. Bacteriol.* 158:488–495.

58. Faelen, M., A. Toussaint, and M. Couturier. 1971. "Mu-1 Promoted Integration of a λ-gal Phage in the Chromosome of *E. coli* K12." *Mol. Gen. Genetics* 113:367–370.

59. Van de Putte, P., and M. Gruijthuijsen. 1972. "Chromosomal Mobilization and Integration of F Factors in the Chromosome of a *recA* Strain of *E. coli*." *Mol. Gen. Genetics* 118:173–183.

60. Ljungquist, E., and A. I. Bukhari. 1977. "State of Prophage Mu Upon Induction." *Proc. Nat. Acad. Sci.* 74:3143–3147.

61. Harshey, R. M. 1984. "Transposition Without Duplication of Infecting Bacteriophage Mu DNA." *Nature* 311:580–581.

62. Mizuuchi, K. 1983. "*In vitro* Transposition of Bacteriophage Mu: A Biochemical Approach to a Novel Replication Reaction." *Cell* 35:785–794.

63. Van de Putte, P., S. Cramer, and M. Giphart-Gassler. 1980. "Invertible DNA Determines Host Specificity of Bacteriophage Mu." *Nature* 286:218–222.

64. Grundy, F. J., and M. M. Howe. 1984. "Involvement of the Invertible G Segment in Bacteriophage Mu Tail Fiber Biosynthesis." *Virology* 134:296–317.

65. Toussaint, A. 1976. "The DNA Modification Function of Temperate Phage Mu-1." *Virology* 70:17–27.

66. Hattman, S. 1979. "Unusual Modification of Bacteriophage Mu DNA." *J. Virology* 34:277–279.

67. Hattman, S., M. Goradia, C. Monaghan, and A. I. Bukhari. 1983. "Regulation of the DNA Modification Function of Bacteriophage Mu." *Cold Spring Harbor Symp. Quant. Biol.* 47:647–654.

68. Plasterk, R. H. A., M. Vollering, A. Brinkman, and P. Van de Putte. 1984. "Analysis of the Methylation-Regulated Mu *mom* Transcript." *Cell* 36:189–196.

69. Swinton, D., S. Hattman, P. F. Crain, C.-S. Cheng, D. L. Smith, and J. A. McCloskey. 1983. "Purification and Characterization of the Unusual Deoxynucleoside, α-N-(9-β-D-2'-Deoxyribofuranosylpurin-6-yl) Glycinamide, Specified by the Phage Mu Modification Function." *Proc. Nat. Acad. Sci.* 80:7400–7404.

70. Denhardt, D. T., D. Dressler, and D. S. Ray, eds. 1978. *The Single-Stranded DNA Phages.* Cold Spring Harbor, N.Y.: Cold Spring Harbor Laboratory.

71. Fiers, W., and R. L. Sinsheimer. 1962. "The Structure of the DNA of Bacteriophage φX174. I. The Action of Exopolynucleotidases." *J. Mol. Biol.* 5:408–419.

72. Sanger, F., G. M. Air, B. G. Barrell, N. L. Brown, A. R. Coulson, J. C. Fiddes, C. A. Hutchison III, P. M. Slocombe, and B. Smith. 1977. "Nucleotide Sequence of Bacteriophage φX174 DNA." *Nature* 265:687–695.

73. Webster, R. E., and J. S. Cashman. 1978. "Morphogenesis of the Filamentous Single-Stranded DNA Phages." In *The Single-Stranded DNA Phages*, ed. D. T. Denhardt, D. Dressler, and D. S. Ray. Cold Spring Harbor, N.Y.: Cold Spring Harbor Laboratory, pp. 557–569.

74. Zinder, N. D., and K. Horiuchi. 1985. "Multiregulatory Element of Filamentous Bacteriophages." *Microbiol. Revs.* 49:101–106.

75. Webster, R. E., R. A. Grant, and L. W. Hamilton. 1983. "Orientation of the DNA in the Filamentous Bacteriophage f1." *J. Mol. Biol.* 152:357–374.

76. Messing, J., B. Gronenborn, B. Müller-Hill, and P. H. Hofschneider. 1977. "Filamentous Coliphage M13 as a Cloning Vehicle: Insertion of a *Hin*dII Fragment of the *lac* Regulatory Region in M13 Replicative Form *in Vitro*." *Proc. Nat. Acad. Sci.* 74:3642–3646.

77. Model, P., C. McGill, B. Mazur, and W. D. Fulford. 1982. "The Replication of Bacteriophage f1: Gene V Protein Regulates the Synthesis of Gene II Protein." *Cell* 29:329–335.

78. Benedict Yen, T. S., and R. E. Webster. 1982. "Translational Control of Bacteriophage f1 Gene II and Gene X Proteins by Gene V Protein." *Cell* 29:337–345.

79. Zinder, N. D., ed. 1975. *RNA Phages.* Cold Spring Harbor, N.Y.: Cold Spring Harbor Laboratory.

80. Fiers, W. 1979. "Structure and Function of RNA Bacteriophages." *Comp. Virology* 13:69–204.

81. Atkins, J. G., J. A. Steitz, C. W. Anderson, and P. Model. 1979. "Binding of Mammalian Ribosomes to MS2 Phage RNA Reveals an Overlapping Gene Encoding a Lysis Function." *Cell* 18:247–256.

82. Beremand, M. N., and T. Blumenthal. 1979. "Overlapping Genes in RNA Phage; a New Protein Implicated in Lysis." *Cell* 18:257–266.

83. Model, P., R. E. Webster, and N. D. Zinder. 1979. "Characterization of Op3, a Lysis-Deficient Mutant of Bacteriophage f2." *Cell* 18:235–246.

84. Blumenthal, T., and G. C. Carmichael. 1979. "RNA Replication: Function and Structure of Qβ-Replicase." *Ann. Rev. Biochem.* 48:525–548.

85. Winter, R. B., and L. Gold. 1983. "Overproduction of Bacteriophage Qβ Maturation (A₂) Protein Leads to Cell Lysis." *Cell* 33:877–885.

86. Kastelein, R. A., E. Remaut, W. Fiers, and J. van Duin. 1982. "Lysis Gene Expression of RNA Phage MS2 Depends on a Frameshift During Translation of the Overlapping Coat Protein Gene." *Nature* 295:35–41.

87. W. Fiers, R. Contreras, F. Duerinck, G. Haegemeean, D. Iserentant, J. Merregaert, W. Min Jou, F. Molemans, A. Raeymaekers, A. Van den Berghe, G. Volckaert, Y. M. Ysebaert. 1976. "Complete Nucleotide Sequence of Bacteriophage MS2 RNA: Primary and Secondary Structure of the Replicase Gene." *Nature* 260:500–507.

88. Gussin, G. N., M. R. Capecchi, J. M. Adams, J. E. Argetsinger, J. Tooze, K. Weber, and J. D. Watson. 1966. "Protein Synthesis Directed by RNA Phage Messengers." *Cold Spring Harbor Symp. Quant. Biol.* 31:257–272.

89. Campbell, K. M., G. D. Stormo, and L. Gold. 1983. "Protein-Mediated Translational Repression." In *Gene Function in Prokaryotes*, ed. J. Beckwith, J. Davies, and J. A. Gallant. Cold Spring Harbor, N.Y.: Cold Spring Harbor Laboratory, pp. 185–187.

90. Uhlenbeck, O. C., J. Carey, P. J. Romaniuk, P. T. Lowary, and D. Beckett. 1983. "Interaction of R17 Coat Protein with Its RNA

Binding Site for Translational Repression." *J. Biomol. Structure and Dynamics* 1:539–552.

91. Spiegelman, S., I. Haruna, I. B. Holland, G. Beaudreau, and D. Mills. 1965. "The Synthesis of a Self-Propagating and Infectious Nucleic Acid with a Purified Enzyme." *Proc. Nat. Acad. Sci.* 54:919–927.

92. Blumenthal, T., T. A. Landers, and K. Weber. 1972. "Bacteriophage Qβ Replicase Contains the Protein Biosynthesis Elongation Factors EFTu and EFTs." *Proc. Nat. Acad. Sci.* 69:1313–1317.

93. Kamen, R. I. 1975. "Structure and Function of the Qβ RNA Replicase." In *RNA Phages*, ed. N. D. Zinder. Cold Spring Harbor, N.Y.: Cold Spring Harbor Laboratory, pp. 203–234.

94. Kolakofsky, D., and C. Weissmann. 1971. "Possible Mechanism for Transition of Viral RNA from Polysome to Replication Complex." *Nature New Biol.* 231:42–46.

95. Vollenweider, H. G., T. H. Koller, H. Weber, and C. Weissmann. 1976. "Physical Mapping of Qβ RNA." *J. Mol. Biol.* 101:367–377.

96. Meyers, F., H. Weber, and C. Weissmann. 1981. "Interactions of Qβ Replicase with Qβ RNA." *J. Mol. Biol.* 153:631–660.

VII

FACING UP TO EUCARYOTIC CELLS

CHAPTER 18

Yeasts as the *E. coli* of Eucaryotic Cells

The very concentrated effort that has gone into the study of all aspects of *E. coli* is one of the major reasons why molecular biology has advanced so rapidly over the past 40 years. As a procaryotic organism, however, *E. coli* possesses a structural organization that is fundamentally different from the more complex eucaryotic organisms. In possessing chromosome-containing nuclei, eucaryotic cells are so arranged that the fundamental processes of transcription and translation occur in physically separate compartments. Thus, by the time the central dogma was an established fact, serious opinions were already voiced that the molecular devices used to control the patterns of gene expression would be radically different between procaryotes and eucaryotes. By now, these suspicions have matured into hard facts. Therefore, no matter how well we understand *E. coli*, it cannot serve as the intellectual scaffold from which to approach the organization and functioning of higher cells. Instead, we need to understand one or more of the simplest eucaryotic organisms at the same level of detail that we presently comprehend *E. coli*.

The organisms that stand out as the obvious subjects for study are the eucaryotic microorganisms that superficially grow and divide like bacteria. Of these, the best understood by far are several of the yeasts. Their well-dissected genetic systems provide a virtually autocatalytic stimulus for their further study by the community of molecular biologists. Most importantly, these yeasts have only 3½ times more DNA in their haploid genetic complements than *E. coli* cells, while *Drosophila* contains 25 times more DNA and higher vertebrate cells contain at least 1000 times more DNA. Still another factor favoring the emphasis on yeasts is economic. Work with higher vertebrate cells is far more expensive because of the high cost of the media required to grow them in culture. If we can solve the same problem equally well with yeasts or with human cells, common sense tells us to stick with the simpler and less expensive system.

For these reasons, an ever-increasing number of molecular biologists have taken up yeasts as their primary research system over the last decade. At this rate, the interest expressed in yeasts will quickly match the maximal interest ever generated by *E. coli*.

New Cells Are Produced Through Either Budding or Fission

The best-characterized yeast, genetically speaking, is bakers' yeast, *Saccharomyces cerevisiae*. This species has oblately spheroid or ovoid shaped cells some 3 μm in diameter, which under optimal nutritional

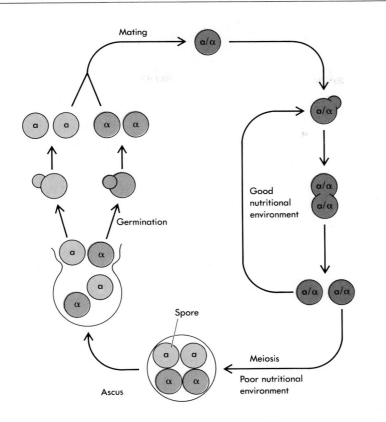

Figure 18-1
Life cycle of *Saccharomyces cerevisiae*. The two mating types are **a** and α. In a rich nutritional environment, both haploid and diploid cells will reproduce by mitosis; in a poor environment, diploid **a**/α cells (formed by mating between *a* and α haploid cells) will undergo meiosis. The life cycle of *Schizosaccharomyces pombe* is very similar, the main difference being that these cells divide by fission rather than budding.

conditions double their mass every 90 minutes. Each cell does this by forming a single budlike appendage into which a full complement of daughter chromosomes moves during a mitotic division (Figures 18-1 and 18-2). The bud is then pinched off to yield a smaller daughter cell, which increases in size until it becomes as large as its parent cell. Then it, in turn, begins to grow a bud. Like all other yeasts, *S. cerevisiae* can exist in either the haploid or diploid state, with the haploid cells being either of two sexes (mating types), designated **a** and α. Fusion of **a** and α cells yields **a**/α diploid cells, which under satisfactory nutritional conditions will grow and divide, maintaining the diploid state. Under conditions of nutritional deficiency, however, the diploid cells undergo meiosis to yield four progeny haploid cells, which become encapsulated as spores within a thick, cell-walled, saclike body called an **ascus** (plural **asci**). Upon rupture of the ascus, the spores are released and subsequently germinate to commence new rounds of haploid existence (see Figure 18-1).

Also very well studied is the fission yeast *Schizosaccharomyces pombe*, whose two haploid sexes are called h^+ and h^-. This particular species does not form buds, but grows in size until an equatorial cell division process produces two equally sized daughter cells. Otherwise, its life cycle is similar to *S. cerevisiae*.

The Primary Products of Meiosis Are Constrained Within Tetrads[1-4]

Yeasts were initially only of applied interest to most biologists because of their use in the baking and brewing industries. However, when their genetic analysis began in the 1930s, yeasts became valued as tools for studying recombination, in particular, the details of crossing over. In meiosis, the diploid number of chromosomes is reduced

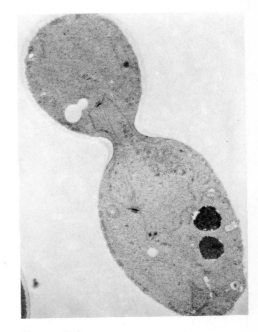

Figure 18-2
Budding yeast cell. (Courtesy of Barbara J. Stevens.)

to a haploid number. Thus, a diploid cell will yield meiotic products that have only a single representative of each of its chromosomes. Even though meiosis reduces the chromosome number, the process begins with a doubling of chromosomes. It is at this time that crossing over in yeast occurs, just as it does in all other eucaryotic organisms. No further DNA replication occurs during the two meiotic divisions that produce the four haploid spores, each of which therefore carries one copy of each yeast chromosome.

In most organisms, the haploid products from the various meiotic divisions become intermixed (e.g., a human sperm sample randomly arises from many meiotic events). However, in many fungi, including yeasts, the four primary spore products of a given meiosis are contained as a **tetrad** within a single ascus that separates them from the products of other meiotic events. Each spore within an ascus can be individually dissected away with the help of a micromanipulator, and its genetic constitution can then be analyzed. Yeasts thus provide a system for examining the two primary DNA products of individual crossing over events, something that cannot be accomplished with most other genetically studied organisms. For example, they made it possible to examine whether crossing over always yields daughter chromosomes whose genetic constitutions are reciprocally related. In the early 1940s, analysis of such yeast tetrads revealed that in a certain percentage of asci (approximately 4%), a given locus did not show the anticipated 2:2 segregation pattern but instead displayed occasional 3:1 patterns (see Figure 11-18). This phenomenon became known as **gene conversion** because one of the two original genes converted to its partner allele. An even rarer event—occurring in approximately 0.2% of asci—was also observed. In this case, spores—which are expected to contain only one form of the gene or the other—were not genetically pure; when they divided, they produced some progeny with one type of gene and some progeny with the other type of gene.

Today, these once baffling results are seen as valuable clues as to how crossing over occurs at the DNA level. First of all, studies of yeast tetrads show that although the number of chromosomes that are present at the time of crossing over is four, only two of these chromosomes engage in a single crossing over event. It is further important to remember that each of the participating chromosomes is a DNA duplex. Hence a single crossing over involves interactions among four strands of DNA. Those loci that segregate aberrantly are located at or very near to the physical sites of crossing over. Analysis of these aberrant types of tetrads has shown that a crossing over event often involves formation of joint DNA molecules containing DNA heteroduplexes (Chapter 11). Failure to repair the heteroduplex leads to formation of a spore that contains this heteroduplex and that will yield a mixed colony upon germination and subsequent mitotic divisions. In contrast, if the heteroduplex is removed by DNA repair a gene conversion event may occur.

Establishing Chromosomal Assignments and Locations Through Tetrad Analysis[5, 6, 7]

Tetrad analysis is now routinely employed to assign newly found mutant genes to linkage groups (chromosomes) and then to map each gene in relation both to the chromosomal centromere and to the other genes on the same chromosome. This is accomplished by analyzing the segregation patterns of different heterozygous markers among the four haploid spores. Initially, such mapping was much more time-consuming than mapping *E. coli* genes. The intensively studied yeast *S. cerevisiae* has 16 established chromosomes, while *S. pombe* has 3 chromosomes. Until they were all found and each characterized by suitable markers, the mapping of a new mutant gene could consume many months. Now, however, there are several classes of genetically identified "tester" strains that greatly speed up mapping. Especially helpful are tester strains containing extra copies of a given chromosome, which lead to aberrant segregation patterns when tested against strains carrying mutations on this chromosome. Also very useful are strains in which specific, marked chromosomes are unstable and whose loss after an appropriate genetic cross can immediately tell us that a given gene is on that chromosome.

Through the use of such "tricks," the number of mapped genes in *cerevisiae* is now 568, roughly half the number of mapped *E. coli* genes, and is increasing rapidly (Figure 18-3). However, since yeasts contain some four times more DNA than *E. coli*, the mapping of the yeast genome still remains in relative infancy.

Separation of Daughter Chromosomes on Microtubule-Containing Spindles[8, 9, 10]

Cell division in yeasts, as in all other eukaryotes, involves the participation of spindles upon which the chromosomes line up prior to the regular segregation processes that pull daughter sets of chromosomes in opposite directions. A major constituent of those spindles are **microtubules,** hollow 240 Å diameter filaments composed of the monomeric building blocks α and β tubulins (molecular weights about 55,000 daltons). Unlike that of most eukaryotic cells, the nuclear membrane of yeasts does not break down prior to formation of the spindle. Instead, the **spindle pole bodies (SPBs),** microtubule-organizing centers analogous to the centrosomes of animal cells, are integrated constituents of the nuclear membrane; microtubules emanate from the SPBs into the nucleoplasm. Each yeast cell contains one SPB that duplicates prior to the initiation of mitosis to generate a daughter SPB immediately adjacent to its parent on the nuclear membrane. As these become positioned to the opposite side of the nuclear membrane, the microtubules elongate, extending into the nucleus and eventually interdigitating to form a complete spindle (Figures 18-4 and 18-5).

Two types of spindle microtubules span the nucleus. One type runs continuously across the entire spindle, while the other, more common form consists of shorter microtubules attached to only one SPB. It is this second group of tubules that attaches the chromosomes to the spindle and whose subsequent shortening pulls the two sets of homologous chromosomes to opposite poles. At the same time, the longer pole-to-pole microtubules grow even further, thereby elongating the spindle greatly and positioning the daughter sets of chromosomes into positions that favor the subsequent cleavage of the nucleus into similar-size daughter nuclei.

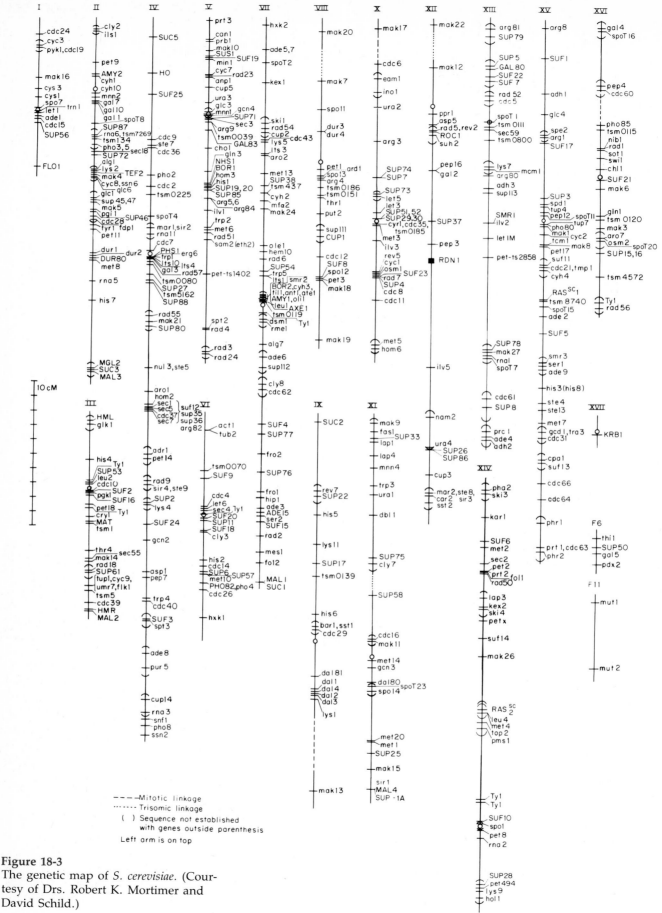

Figure 18-3
The genetic map of *S. cerevisiae.* (Courtesy of Drs. Robert K. Mortimer and David Schild.)

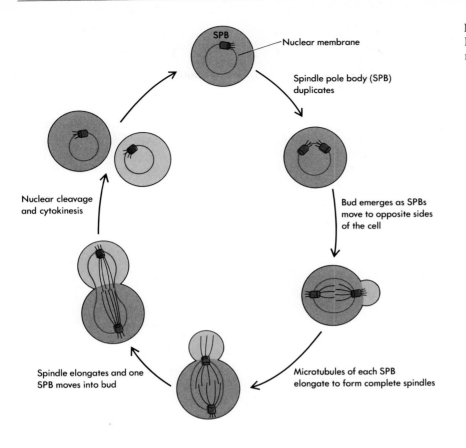

Figure 18-4
Mitotic spindle formation and movement in yeast.

SPB

Nuclear membrane

Spindle pole body (SPB)
duplicates

Nuclear cleavage
and cytokinesis

Bud emerges as SPBs
move to opposite sides
of the cell

Spindle elongates and one
SPB moves into bud

Microtubules of each SPB
elongate to form complete spindles

While in higher eucaryotic cells the microtubule attachment sites at the centromeric regions on chromosomes are morphologically conspicuous and are called **kinetochores,** no recognizable equivalents occur on yeast chromosomes. This difference probably reflects the much smaller size of individual yeast chromosomes, each of whose attachment to the spindle at the centromere is mediated by just single microtubules rather than the microtubule bundles that emanate from the kinetochores of higher cells.

The cell division cycle terminates with the division of the cytoplasm into two daughter cells (cytokinesis) by the formation of an ever-constricting ring of circumferential filaments containing the proteins *actin* and *myosin,* the same proteins involved in muscle contraction. The relative motion of these two molecules along each other during cytokinesis may underlie the contractive and constrictive movements of the filaments.

Mutations That Help Define the Cell Cycle[11–14]

During a cell cycle (Figure 18-6), the continuous processes that duplicate most cellular molecules (e.g., glycolytic enzymes) are punctuated by stage-specific events or landmarks. Mutations causing stage-specific blockage of the cell cycle are an increasingly important tool in elucidating these landmarks (Table 18-1). The most useful cell division cycle (*cdc*) mutations are temperature-sensitive mutations in genes essential for completing the cell cycle. Many gene products of

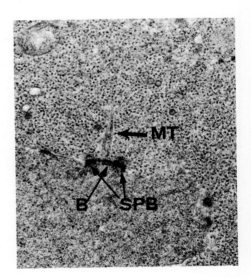

Figure 18-5
Formation of spindles in yeast. Here, the bridge (B) of the duplicated SPBs bears cytoplasmic microtubules (MT). (Courtesy of Breck Byers.)

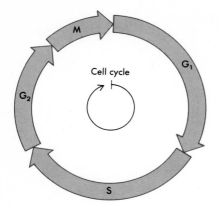

Figure 18-6
Phases in the life cycle of a yeast cell.
M = mitosis, G₁ = pre-DNA synthesis,
S = DNA synthesis, G₂ = period be-
tween DNA synthesis and mitosis.

such mutated genes are inactive at high temperatures but active at low temperatures. When such a *cdc* mutant is subjected to restrictive hot conditions, it becomes blocked from moving through the cell cycle at that stage where its normal equivalent product begins to function. Thus, if we raise the temperature of an asynchronously growing yeast population containing a heat-sensitive *cdc* gene, the population becomes converted into a synchronous population of cells all blocked at the same stage of the cell cycle (Figure 18-7). There are also the opposite type of temperature-sensitive *cdc* mutant genes that are inactive at low temperatures while being active at high temperatures.

By now, some 50 different *cdc* genes have been defined in terms of the landmark events beyond which they do not pass in the cell cycle. They all function during the growth of **a** and α haploid cells, as well as during **a**/α diploid cell growth, with many also playing vital roles during meiosis and spore formation. A particularly important problem is to determine the roles of these proteins. As of now, only a few are known. One such gene codes for the enzyme thymidylate synthetase, which provides thymine residues needed for DNA synthesis. Another gene codes for one of the two forms of β tubulin. Still another codes for DNA ligase, an enzyme that, among other functions, joins together the fragments forming the lagging strand during DNA synthesis. How all these *cdc* genes function harmoniously and in the right order is only beginning to be understood.

The beginning event in the cell cycle, called *start*, takes place in the GI stage and has a unique function. The decision or commitment to undergo another cell cycle is made at start. Once a cell has passed

Table 18-1   Various *cdc* Genes

Gene	Function and Remarks
cdc4	Required for initiation of DNA synthesis in the mitotic cell division cycle and for premeiotic DNA synthesis in meiosis; required for replication of nuclear chromosomes and the 2 micron plasmid but not mitochondrial DNA.
cdc7	Similar to *cdc4* but not required for premeiotic DNA synthesis.
cdc8	Structural gene for thymidylate kinase; required for DNA precursor synthesis and hence replication of all DNA.
cdc9	Structural gene for DNA ligase.
cdc16	Required for completion of chromosome distribution; possibly codes for a microtubule-associated protein.
cdc21	Structural gene for thymidylate synthetase; required for DNA precursor synthesis and hence replication of all DNA.
cdc25	Necessary for completion of synthesis of the spindle pole body and exit from G1; involved in cyclic AMP metabolism.
cdc28	Necessary for duplication of the spindle pole body and exit from G1; homologous to mammalian protein kinases and has protein kinase activity.
cdc31	Necessary for duplication of the spindle pole body; Ca⁺⁺ binding protein.
tub2	Structural gene for β-tubulin; necessary for nuclear migration and chromosome distribution.

SOURCES: Sclafani, R. A. and W. L. Fangman. 1984. "Yeast Gene *cdc8* Encodes Thymidylate Kinase and Is Complemented by Herpes Thymidine Kinase Gene *TK*." *Proc. Nat. Acad. Sci.* 81:5821–5825. S. I. Reed, J. A. Hadwiger, and A. T. Lorincz. 1985. "Protein Kinase Activity Associated with the Product of the Yeast Cell Division Cycle Gene *cdc28*." *Proc. Nat. Acad. Sci.* 82:4055–4059.

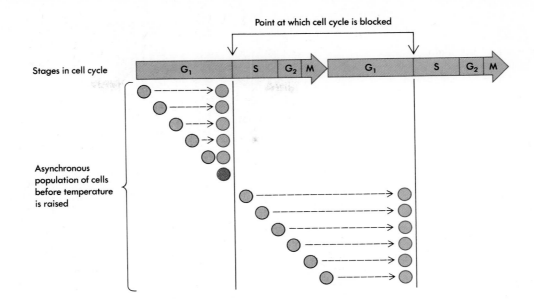

Point at which cell cycle is blocked

Stages in cell cycle

Asynchronous population of cells before temperature is raised

through start, it will commence DNA synthesis and pass through all the successive stages in the cell cycle until its daughters reach start. These cells will then initiate another cell cycle (passing through start), unless they become arrested in the cell division cycle by unfavorable nutritional conditions or by an appropriate sex hormone. The completion of start is marked by the duplication of the spindle pole body. Once that event occurs in *S. cerevisiae*, DNA synthesis commences, the bud begins to grow, and the normal sequential passage of cell cycle events is ensured.

Yeasts Display All the Characteristic Internal Compartments of Eucaryotic Cells[15-19]

Yeast cells are organized into the same major membrane-bounded compartments as all other nonphotosynthetic eucaryotic cells (Figure 18-8). In addition to the *nucleus*, yeast cells contain:

- The *endoplasmic reticulum (ER)* with its attached ribosomes that synthesize the integral membrane proteins, as well as proteins destined for secretion, largely into the outer cell wall.

- The *Golgi apparatus*, to which many proteins made by the endoplasmic reticulum are first transported and where the initial stages of protein modification (e.g., addition of sugar groups) occur; somehow in the Golgi sacs, newly synthesized proteins are tagged with markers that direct them to their correct cellular destinations.

- The energy-generating *mitochondria*.

- The *cytosol*, which in rapidly growing cells is tightly packed with essential enzymes as well as with very large numbers of free polyribosomes engaged in protein synthesis.

- The *vacuole* (analogous in several respects to the *lysosome* of mammalian cells), which contains a variety of degradative enzymes and is a storage area for certain cellular nutrients.

- The *peroxisomes*, which sequester enzymes (e.g., catalase) that destroy the free radicals derived from oxidative reactions.

Figure 18-7
Heat-sensitive *CDC* mutants become blocked at a specific stage in the cell cycle as the temperature is raised. Therefore, an asynchronous population of these cells becomes converted to a culture all blocked at the same stage. When the temperature is lowered, the cells in the culture will grow synchronously for a short time. (Redrawn from B. Alberts et al, *Molecular Biology of the Cell*, p. 641.)

Figure 18-8
Electron micrograph of a typical yeast cell, showing some of the major membrane-bounded organelles. (Courtesy of Drs. R. Schekman, P. Novick, and P. C. Esmon.)

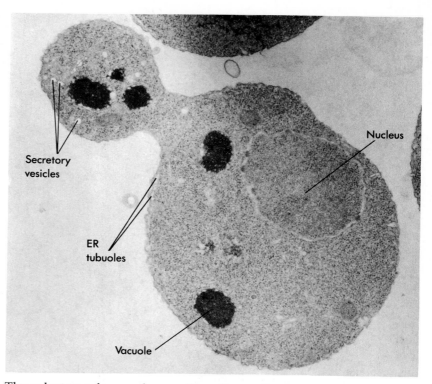

These last two forms of organelles allow cells to segregate potentially harmful hydrolytic enzymes or metabolic intermediates from most of the cellular cytoplasm. For example, cellular function would rapidly cease if the proteolytic enzymes found within vacuoles came into contact with most cellular proteins. Likewise, free peroxide (H_2O_2) would oxidatively destroy many cellular components if it had access to them. Because the generation of H_2O_2 is limited to the interior of peroxisomes, where catalase almost instantly breaks it down, the cellular level of H_2O_2 never reaches dangerous levels.

In yeast, most of the proteins made in the endoplasmic reticulum are rapidly passed through the Golgi apparatus to other intracellular membranes, to the periplasmic space, and to the outermost layer of the cell wall. Consequently, the endoplasmic reticulum and Golgi compartments are not swollen with protein products and thus are not morphologically conspicuous, as they are in liver cells, for example. In fact, only after yeast cells containing mutations in genes involved in protein transport were examined, did the Golgi compartments of yeast become unambiguously identified (Figure 18-9). It is hard to believe, however, that the Golgi would have remained unidentified for long. They allow the highly complex glycoproteins of the surface layers to be assembled in several discrete steps at internal locations, transferred to secretory granules, and finally delivered to their cellular destinations.

Like all other eucaryotic cells, yeasts contain the key protein *clathrin,* a major outer component of the *coated vesicles* that transport proteins from one membrane site to another (e.g., from the plasma membrane to lysosomes). Through possession of clathrin, yeasts have the capability to selectively endocytose those proteins of their external membranes to which extracellular molecules have attached. Which extracellular molecules are taken up in such a manner remains to be discovered. However, now that we know that clathrin is present in yeasts, it is difficult to imagine that it does not sometimes function as

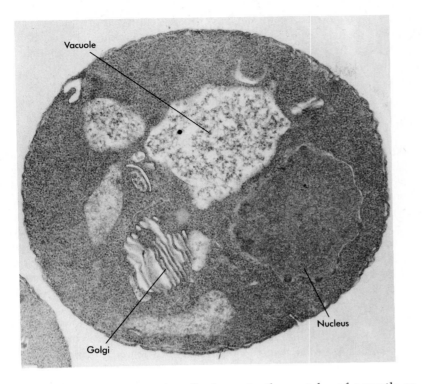

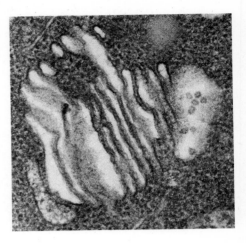

Figure 18-9
Electron micrograph showing an accumulation of Golgi in a yeast cell with a mutation affecting protein transport. (Courtesy of Drs. R. Schekman, P. Novick, and P. C. Edmon.)

it does in higher eucaryotic cells (e.g., in the uptake of growth or antigrowth factors).

Signals on Proteins Determine Proper Cellular Location

Yeast proteins contain signals that are responsible for their proper localization within the cell. These signals determine whether the protein should go to the vacuole or the mitochondria, be secreted from the cell, or return to the nucleus. One way in which these localization determinants have been identified is by producing hybrid proteins, through *in vitro* methods, which consist of joining one part of the protein that is localized to one cellular compartment to a test protein that is ordinarily found in a different cellular compartment. Test proteins are, for example, *E. coli* β-galactosidase or dihydrofolate reductase. A segment is considered to have localization information if it determines who wins this molecular tug-of-war. Proteins destined to the vacuole or to be secreted from the cell both proceed first to the endoplasmic reticulum, under direction of a hydrophobic signal at their amino terminus. These proteins then must contain further routing information, since their intracellular paths diverge after the Golgi. Mitochondrial proteins do not pass through the ER, but rather go directly to the mitochondria. In all of these cases, the signal is likely to be a binding site for guiding the protein species to its proper place in the cell. Finally, proteins that concentrate in the nucleus (regulatory proteins such as α2 and *GAL4*, and ribosomal proteins) all contain short amino-terminal segments that can drag β-galactosidase into the nucleus. In this case, it is not known whether the localization determinant on these nuclear proteins is responsible for entry of the protein into the nucleus, or for retaining the protein after it has diffused into the nucleus. In the one case in which this distinction has been made (nucleoplasmin in *Xenopus*), the signal is responsible for uptake rather than retention. Whatever the molecular mechanism, it is remarkable that a protein the size of β-galactosidase is able to enter the nucleus, presumably through the nuclear pores.

Cytoskeletal Elements Within Yeast Cells[20, 21]

Because yeast cells are small and appear to be morphologically simple, there initially was no reason to presume that the cytosol of yeast would contain the multiplicity of filamentous elements contained in higher eucaryotic cells. Recall that the **cytoskeleton** of higher eucaryotic cells is composed of actin-containing microfilaments, intermediate (100 Å) filaments, and tubulin-containing microtubules. However, the finding that some microtubules are present in the yeast cytosol, as well as within the spindles of nuclei, prompted the question of whether they singly or in conjunction with other filamentous elements help direct the intracellular flow of molecules and membrane-bounded organelles. The question has become even more significant with the recent finding that actin- and myosin-containing filaments, as well as intermediate (100 Å) filaments, are components of the circumferential rings whose formation leads to the final separation of the daughter cytoplasm to conclude a cell division cycle. Such filaments may also act earlier in the cell cycle, when they are directed into growing *S. cerevisiae* buds. There they may serve to direct the secretory granules bringing newly made molecules for insertion into the expanding outer components of the buds.

The Thick and Tough Cell Walls of All Yeasts[22]

Surrounding the plasma membrane of yeast cells is the yeast cell wall, a thick, complex mixture of secreted proteins and elongated, intertwined polysaccharides. On its outer surface is a tight aggregation of highly immunogenic *mannoproteins,* which consist of large numbers of mannose groups covalently bound to *N*-acetylglucosamine groups that in turn are attached to specific threonine and serine side groups of a number of cell wall–specific polypeptide chains. Between this mannoprotein layer and the plasma membrane is the more open periplasmic region. In this region, a number of degradative enzymes (e.g., the sucrose-hydrolyzing agent invertase and the phosphate-releasing agent alkaline phosphatase) lie among the long, thin polysaccharides glucan (a polymer of D-glycopyranose groups) and chitin (a polymer of *N*-acetylglucosylamine), which in some way are anchored to both the plasma membrane and the mannoprotein layer (Figure 18-10).

For years, their thick and largely impenetrable outer surfaces prevented much research on yeasts. Cytological work, for example, was difficult because embedding substances could not penetrate. Nor was there any way to introduce externally added DNA into intact cells or to isolate from ruptured cells even semi-intact organelles. Many of these problems can now be circumvented by simultaneously employing several cell wall–degrading enzymes that sever the restraining glucan chains many times and make many holes in the external man-

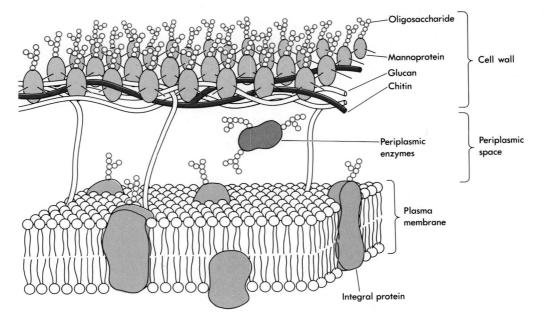

Oligosaccharide
Mannoprotein — Cell wall
Glucan
Chitin

Periplasmic enzymes — Periplasmic space

Plasma membrane

Integral protein

Figure 18-10
The yeast cell surface.

noprotein layer to produce much less rigid **spheroplasts** (protoplasts).

Partition of the Yeast Genome into Linear Chromosomes[23–28]

S. cerevisiae nuclear DNA (14×10^3 kilobase pairs) is distributed among 16 discrete chromosomes, each of which contains a single DNA molecule (Table 18-2). Until recently, even the smallest of these DNA molecules was too large to be effectively isolated in pure form. Now, however, the development of pulse field electrophoretic

Table 18-2 Genetic and Physical Sizes of Yeast Chromosomes

Band	Chromosome	Recombinant length (cM)	DNA (kb)*
1	1	98	198
2	6	137	225
3	3	137	311
4	9	198	495
5a	8	183	635
5b	5	233	709
6	11	242	763
7	10	200	835
8	14	283	864
9	2	244	948
10	13	229	991
10b	16	176	1047
11a	15	361	1146
11b	7	391	1285
12	4	485	1571
13	12	294	2194

*DNA content/chromosome, assuming 14,000 kb per haploid genome (96).
SOURCE: Data from R. K. Mortimer and D. Schild, *Microbiol. Revs.* (1985) 49:206.

Figure 18-11
Identification of the bands corresponding to chromosomes II, XIII, XV, and
XVI by DNA·DNA hybridization (strain
AB972). DNA from the gel whose
ethidium bromide staining pattern is
shown on the left was transferred to a
single sheet of nitrocellulose. This sheet
was then cut into four strips that were
separately hybridized to four chromo-
some-specific probes, and the filter
strips were then positioned in their
original alignment before the
autoradiogram on the right was exposed. The ethidium bromide staining
pattern and the autoradiogram are
printed to the same scale and are
aligned appropriately with one another.
[Courtesy of G. F. Carle and M. V.
Olson. Printed from *Proc. Nat. Acad.
Sci.* (1985) 82:3758, with permission.]

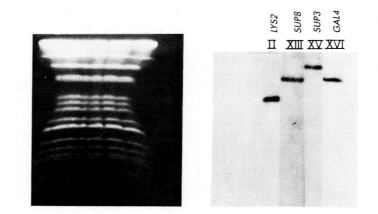

methods allows the almost routine separation on agarose gels of 15 of
their chromosomes. Only chromosome 12, over half of which contains an average of 125 tandemly arranged ribosomal genes (rDNA),
does not cleanly separate. This is most likely due to the wide variation
in the average number of the 8.6 kbp rDNA genes found per chromosome. Variation of this magnitude is to be expected because of unequal crossing over events (Chapter 7) occurring within the rDNA
region. Given the existence of a DNA probe for a given yeast gene, it
is now a simple matter to map this gene to a precise chromosome
using DNA-DNA hybridization techniques (Chapter 19) (Figure
18-11). Interestingly, there is significant variation in the lengths of
individual chromosomes from one strain of *S. cerevisiae* to another
(chromosomal length polymorphisms). The molecular basis for these
polymorphisms remains to be established. One possibility is variation
in the length of the telomeric sequences found at the ends of all chromosomes (see below).

Only three chromosomes carry the genetic information of *S. pombe*;
all its chromosomes are also linear. Many yeast cells also possess
multiple copies of autonomously replicating small, circular plasmid
DNAs, 2μ circles (see Chapter 19), that normally do not carry any
essential genes. Because they lack centromeres, they are not regularly
partitioned on the mitotic spindle during cell division. Thus, still to be
explained is why these yeast plasmids tend to be maintained rather
than lost in a random fashion from a small fraction of all daughter
cells. The function of these plasmids is not known.

Yeast DNA, like that of all other eucaryotic cells, is complexed with
the four main histones (H2A, H2B, H3, and H4) used to construct the
146-base-pair nucleosomal cores of chromatin (a nucleosome is a globular subunit of chromosomal DNA complexed with histones; see
Chapter 21). Apparently absent in *S. cerevisiae* is the H1 histone, which
in higher eucaryotes plays a key role in the compaction of chromatin
to the highly supercoiled state that exists during mitosis (Chapter 21).
The lack of any yeast equivalent to H1 may explain why, with the
exception of *S. pombe*, individual yeast chromosomes cannot be visualized during mitosis. Yeast chromatin, however, does contain major
amounts of an acidic, nonhistonelike group of proteins whose amino
acid composition and charge closely resemble those of similar proteins found in the chromosomes of higher eucaryotes. What their
function is remains a mystery despite the several decades in which
their existence has been known.

Also still to be established is why, despite the linear nature of their underlying DNA molecules, sections of yeast chromatin appear to be rotationally restrained into superhelical regions. Conceivably, structures similar to those that maintain the circular *E. coli* DNA in some 40 to 50 independently negatively supercoiled domains are also essential ingredients in the functioning of yeast chromatin.

Absence of Conventional Nucleosomes at the Centromeric Regions[29]

The centromeres of mitotic chromosomes display such morphological discontinuities with the chromosome arms that it was originally suspected that they represented regions separating two (or more) discrete DNA molecules. However, the discovery that a single DNA molecule runs through an entire yeast chromosome made it clear that chromosomal DNA was continuous through the centromere and that there were specialized DNA sequences to promote interaction with the mitotic spindle. The first of such *CEN* (centromeric) **sequences** was isolated from a shortened aberrant circular form of chromosome III of *S. cerevisiae*, formed by accidental crossing over between two of the mating-type genes that reside on this chromosome. Because of its smaller circular form, this DNA can easily be separated from other yeast DNA and used as an enriched source of restriction fragments for the "chromosome walks" (discussed later in the next chapter) used to find the *CEN3* sequence. When this DNA is inserted into yeast plasmid DNAs, it promotes their regular segregation during mitosis and meiosis in the manner expected for centromere-linked genes. Subsequently, *CEN* sequences have been isolated from several other yeast chromosomes, with each example showing sequences homologous to *CEN3*. All the essential *CEN* sequences are contained within a 130-base-pairs long, very AT-rich region encompassing two shorter blocks of high homology (Figure 18-12).

The *CEN* sequences reside within a region 220 to 250 base pairs long that is more resistant to nuclease digestion than the vast bulk of yeast chromatin. On both sides of this region, the nucleosomes are singly placed along the DNA. The resistance to nuclease digestion suggests that at the *CEN* sequences there is either a modified nucleosome or a complex of proteins, other than the normal histone components of nucleosomes, replacing the *CEN* sequences. In any case, the protein-bound region of the centromere (the kinetochore) mediates attachment to the microtubules of the mitotic spindle (Figure 18-13). Several yeast proteins that bind specifically to *CEN* sequences have already been isolated, and it should soon become clear whether their presence prevents the simultaneous association with any of the four main histone classes. Eventually, we will succeed with the in vitro reconstitution of the centromere-microtubule complexes and gain a more detailed understanding of how chromosomes attach to the mitotic spindle.

Figure 18-12
Sequence elements common to three yeast centromeres. Elements I and III are regions of very high homology flanking a longer central element of high AT content.

	Element I		Element II		Element III
CEN3	ATAAGTCACATGAT	⟵	88 bp (93% AT)	⟶	TGATTTCCGAA
CEN11	ATAAGTCACATGAT	⟵	89 bp (94% AT)	⟶	TGATTTCCGAA
CEN4	AAAGGTCACATGCT	⟵	82 bp (93% AT)	⟶	TGATTACCGAA

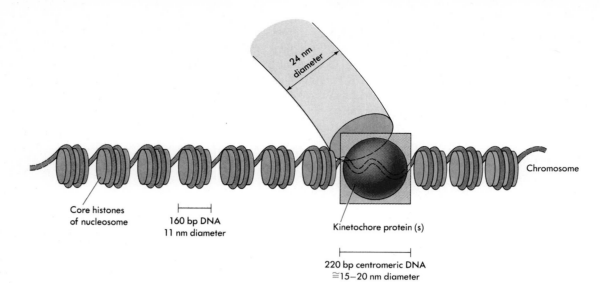

Figure 18-13
Diagram of *S. cerevisiae* chromosome, showing the nuclease-resistant centromere core (shaded box) and the nucleosomes that are placed along the DNA on either side of the centromere core. At the *CEN* region, the nucleosomes are replaced with other proteins that attach to microtubules of the mitotic spindle. This proteinaceous region of the centromere is called the *kinetochore*. [After J. S. Bloom and J. Carbon, *Cell* 29 (1982):305.]

Common Telomeric Sequences Are at the Ends of All Yeast Chromosomes[30-34]

The ends of all yeast chromosomes, like those of all other linear eucaryotic chromosomes, have unique structures that are called **telomeres.** Each yeast telomere ends with approximately 100 base pairs of irregularly repeated sequences, of the form

$$5'\text{-}C_{1-3}A \ldots$$
$$3'\text{-}G_{1-3}T, \ldots$$

attached to specific X and Y sequences.

Within these specific terminal X and Y sequences are additional internal sections of the repetitive elements (Figure 18-14). All known telomeres contain similar tandemly repeated sequences. As we pointed out when we discussed DNA replication (Chapter 10), these repetitive elements have evolved as a device to preserve the integrity of the ends of DNA molecules, which often cannot be finished by the conventional mechanisms of DNA replication. The ends of yeast chromosomes are maintained by enzymatically adding $3'\text{-}G_{1-3}T\text{-}5'$ groups to the 3' ends of DNA chains as well as by one or more primaselike enzymes that make the complementary $5'\text{-}C_{1-3}A\text{-}3'$ polyribonucleotide primers.

What determines the length of such terminal additions is not known. But the process must be genetically controlled, since the average number of $C_{1-3}A$ repeats is fixed for a given yeast strain. This number varies between different yeast strains, and more than one yeast gene is involved in determining the number.

Broken chromosome ends that have completely lost their telomeres can often be repaired. This long-puzzling observation can now be explained by the internal copies of the repetitive sequences. When a yeast chromosome becomes broken in or near a poly $C_{1-3}A$ tract, it can acquire new telomeric sequences through a repair process (Chapter 11) in which a broken end invades an intact double helix at a

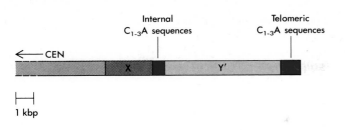

Figure 18-14
Structure of yeast telomeres and adjacent sequences. Not all yeast telomeres contain a Y' element.

region of sequence homology. The genetic information of the intact strand is then used to replace that missing from the broken end (Figure 18-15). Such events lead to constantly changing telomeric structures, giving the sequences at the end the potential for continued dispersal and amplification.

Artificial Yeast Chromosomes Can Be Made by Ligation of Telomeres, Origins of Replication, and Centromeres[35–38]

The isolation and characterization of the three elements essential for eucaryotic chromosome replication—origins of DNA replication, centromeres, and telomeres (see Chapter 19 for more details)—opened up the possibility of linking together single copies of each of these units with appropriate specific yeast DNA segments to form completely new, artificial chromosomes. Such chromosomes can be tested for their functioning by inserting them into appropriate yeast strains.

Very short, linear, artificial chromosomes segregate incorrectly much more frequently than larger (300 kilobase-pair) artificial chromosomes, which segregate with the same degree of accuracy as their similarly sized normal counterparts. Increasing their lengths may lead to more accurate chromosome partitioning by increasing the time the newly duplicated chromosomes remain attached to each other. Conceivably, if daughter chromosomes separate from each other too soon, they cannot line up correctly on the spindle. Favoring this possibility is the fact that very small circular chromosomes (e.g., plasmids that replicate in yeasts) segregate more accurately than their linear counterparts. Replication of circular DNA molecules initially leads to intertwined daughter double helices whose eventual separation depends on topoisomerase I action, allowing them to segregate accurately (only 1 in 100 mistakes).

DNA Synthesis Does Not Occur Throughout the Cell Cycle[39]

In contrast to bacterial DNA synthesis, DNA synthesis in yeasts does not normally occur over the entire mitotic cycle. Upon separation of a daughter cell (bud) from its parent, an interval (designated the G1 phase) of some 10 to 30 minutes passes before DNA synthesis commences. The **S phase** (the period of DNA synthesis) in turn occupies some 20 to 30 minutes. The final 30 minutes occupied by the G2 and

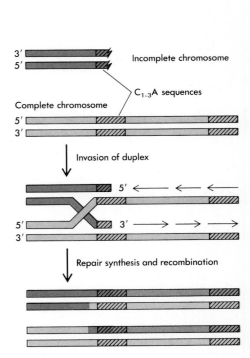

Figure 18-15
Chromosomes broken in or near internal $C_{1-3}A$ repetitive sequences (///) can complete their ends by a repair process in which they recombine with an intact chromosome at homologous sequences.

M phases is the time needed for growth of the spindle, chromosome segregation, and the terminal cytokinesis events that split off the daughter cell (see Figure 18-6).

DNA Replication Commences at Many Sites Along Each Yeast Chromosome[40]

Yeast DNA molecules, like DNA molecules of all eucaryotic organisms, contain multiple origins of replication. Electron microscopic examination reveals that origins are generally spaced about 35 μm apart, suggesting the existence of some 400 origins in the entire yeast genome. Replication is normally bidirectional, proceeding 2 kilobase pairs per minute (at 30°C) or faster at each replicating fork. It has been suggested that synthesis starts at approximately the same time at all origins. Assuming that this were true, then a fork movement of 2 kilobase pairs per minute might mean that the interval for DNA replication would occupy no more than 10 minutes—an interval considerably less than 25 to 30 minutes usually taken for DNA replication within cells doubling their masses every 90 to 100 minutes. It is possible that some replication origins are much more widely spaced from each other, with a correspondingly longer time needed to complete replication between them; or it is possible that the times at which DNA replication starts are staggered.

In *S. cerevisiae*, DNA replication often begins just as budding becomes observable; but there is no strict correlation between these two events, as proved by cell cycle mutants that dissociate them. The enzymes involved in yeast DNA synthesis are still largely unknown. It is hoped that this situation will soon be corrected, as yeast should provide ideal mutants for mastering eucaryotic DNA synthesis.

Yeast Genomes Are Transcribed by Three Different Forms of RNA Polymerase[41]

Yeast DNA, like the DNA of all eucaryotic cells, is transcribed by three different and easily separated forms of RNA polymerase, each of which transcribes different sets of genes. The function of RNA polymerase I is to transcribe the ribosomal genes (rDNA) into a precursor rRNA that eventually yields the two large rRNAs (25S and 17S) and one of the two small RNAs (5.8S) in ribosomes. RNA polymerase II makes all the RNA that is to serve as mRNA and some smaller-sized RNAs whose functions are just recently being elucidated (Chapter 20). RNA polymerase III transcribes the genes coding for tRNA, 5S ribosomal RNA, and a number of other small RNAs. The three polymerases can conveniently be distinguished from one another by their different sensitivities to the fungal toxin α-amanitin. Each of the yeast RNA polymerases, like the more multipurpose bacterial RNA polymerase (which transcribes all the genes of a given bacterial genome), is large (MW 4 to 6 $\times$ 10^5) and is composed of a number of different polypeptide chains. The exact roles of most of these polypeptides are still unclear, but some subunits may be shared between the three polymerases.

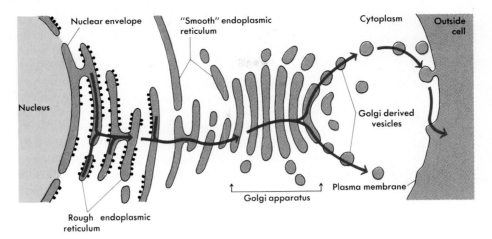

Newly Made RNA Molecules Move Through the Pores of the Nuclear Membrane[42]

The chromosomes of yeast function within a nuclear envelope identical in appearance to that of higher eucaryotic cells. The nuclear envelope can be regarded as a specialized outpocketing of the endoplasmic reticulum, since its external surface is covered with ribosomes that secrete proteins into the cavity separating the two unit membranes (Figure 18-16). Moreover, electron micrographs often show this cavity to be connected to the main cavities of the endoplasmic reticulum. Distributed over the surface of the nuclear envelope of most eucaryotic cells are very large numbers of uniformly sized (100 nm) nuclear pores. These pores have eight-sided polygonal shapes and usually appear to be filled with plugs made up of an amorphous material (Figure 18-17).

Through these pores must pass all the RNA that moves from the nucleus to the cytosol to function during protein synthesis. The forces that direct this unidirectional RNA transport have until recently been completely mysterious. An important new clue has arisen from experiments showing that tRNA molecules are exported not by random diffusion but by some carrier-mediated process. It is conceivable that ribosome-size granules that line the nuclear pores are involved in the transport of both tRNAs and mRNAs across the nuclear membrane (Figure 18-17c).

Molecular Specialization for Eucaryotic Protein Synthesis[43–49]

The protein-synthesizing apparatus in all eucaryotic cells, including yeast cells, shows striking differences when compared with the protein-synthesizing apparatus of bacterial cells. In the first place, the ribosomes of all eucaryotic cells are larger, sedimenting roughly at 80S in comparison to the 70S value found for procaryotes. Correspondingly, eucaryotic ribosomes have bigger subunits that sediment at 60S and 40S. The 60S subunit contains a 28S (25S in yeast) rRNA chain, while the 40S particle has an 18S (17S in yeast) rRNA chain. Two small RNAs are present in the large ribosomal subunit: one 5S rRNA, analogous to 5S RNA in *E. coli* ribosomes, and an additional 5.8S molecule. Eucaryotic ribosomes are also somewhat more protein-

Figure 18-16
Schematic representation of the various membrane-bounded cavities in the eucaryotic cell and their possible role in synthesis and excretion. The rough endoplasmic reticulum bears ribosomes on its outer surfaces. The Golgi apparatus, which secretes vesicles from one face, is formed on the other face by elements of the smooth endoplasmic reticulum. The solid-colored arrow indicates both the general membrane flow within the cell and also the probable route of proteins that are to be secreted. Areas cut off from the cytoplasm by membranes are shaded in color.

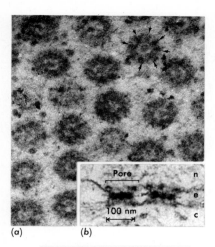

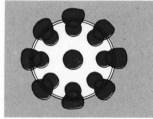

Figure 18-17
Nuclear pores. (a) Electron micrograph of part of a nuclear envelope with a number of pores. (b) Cross section through two pores (n = nucleus, e = envelope, c = cytoplasm). (c) Diagram of a pore, showing the eight pairs of granules that line it and a central granule. Although yeast does have nuclear pores, detailed electron microscopic studies have not yet been carried out. (Courtesy of K. Roberts, John Innis Institute, Norwich, England.)

rich; proteins account for about one-half their mass (as compared to one-third in bacterial ribosomes) and represent more individual polypeptide species. Why these differences exist is not known. The thing to remember is that on a structural level, there is a larger gap in ribosome size between procaryotic and eucaryotic microorganisms (e.g., between bacteria and the yeasts, *Neurospora*, or *Aspergillus*) than between the most evolutionarily divergent eucaryotic cells.

Each mature yeast rRNA chain, with the exception of that of 5S rRNA, is initially synthesized as part of a much longer rRNA precursor (37S) that is processed to give rise to the two large (25S and 17S) rRNA species and one of the small (5.8S) rRNA species. In yeast, but not in higher eucaryotes, the 5S gene is located close to the rDNA units and is part of the same repeating unit, even though it is transcribed by a different RNA polymerase and in the opposite direction. Because the rDNA units are tandemly located next to each other, intramolecular crossing over between them occasionally generates yeast cells containing small circles of rDNA, which are rapidly lost since they do not possess centromeres. Tandem repetition, however, is important for maintaining the sequence homogeneity of the individual repeats within the larger unit by mechanisms that will be discussed in Chapter 20.

Within all eucaryotic cells, the actively transcribed rDNA units are somehow compacted into dense-appearing nucleolar bodies. In yeasts, these nucleoli are crescent shaped and closely associated with the nuclear membrane.

Caps at the 5' Ends of Eucaryotic mRNAs[50–53]

Eucaryotic mRNAs, including those of yeast, have several important features that differentiate them from procaryotic mRNAs. First, they are modified at their 5' ends by the addition of a guanine nucleotide. This alone would not be considered unusual if the linkage were a conventional 3'–5' phosphodiester bond. Instead, GTP reacts with the triphosphate at the 5' end of an mRNA chain in a 5'–5' condensation to form the structure 3'-G-5'ppp5'-N-3'p . . . , generally known as a **cap**. Thus, the 5' and 3' ends of most eucaryotic mRNAs are terminated by ribose moieties with free 2'- and 3'-OH groups. Subsequent to cap formation, a methyl group is added to the backward guanine residue (at the 7 position of the purine ring) and often also to the 2'-OH groups of the first and/or second adjacent nucleotides (Figure 18-18).

Why do the 5' ends of eucaryotic mRNAs need to be so blocked? One possible reason is that these caps (and apparently specific proteins that bind to them) help ribosomes attach to mRNA chains so that they start translation at the correct AUG codon. Favoring this hypothesis is the absence of sequences complementary to 18S rRNA preceding coding regions in eucaryotic mRNA molecules. Specific ribosome binding sequences analogous to those of procaryotic mRNAs thus do not exist on eucaryotic mRNA molecules. Instead, eucaryotic ribosomes search out AUG initiator codons by binding to the caps and then migrating to the closest (5'-most proximal) AUG codon to start translation. Apparent exceptions to this rule are several viral mRNAs (e.g., polio RNA) that function perfectly normally in eucaryotic cells but lack the 5' cap structure. Their ends are blocked instead by specific proteins that perhaps substitute as positioning agents.

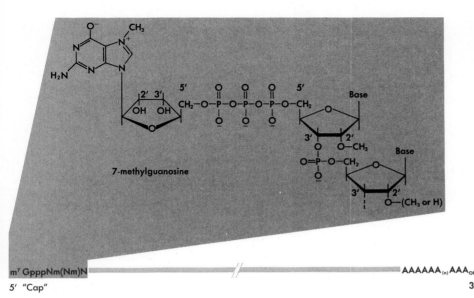

Yeast mRNA Molecules Code for Single (Never Multiple) Polypeptide Chains[54, 55]

A given yeast mRNA molecule, like most other eucaryotic mRNAs, carries the genetic message for only a single polypeptide chain. Unlike many bacterial mRNAs, where translation can begin and end at several sites to give rise to several independent polypeptide chains, eucaryotic mRNAs are constructed so that translation usually commences at the first AUG codon following the capped 5' end. (Exceptions will be discussed in Chapter 24.) Thus, eucaryotic mRNAs are generally **monocistronic,** as opposed to the frequently **polycistronic** mRNAs of bacteria. Like bacterial mRNAs, however, eucaryotic mRNA molecules usually have extensive untranslated sequences before and after their protein-coding regions (leader and trailer sequences).

The inability of eucaryotic mRNAs to encode multiple proteins and thereby ensure their coordinate regulation probably explains why many eucaryotic polypeptides consist of several independent domains having related enzymatic functions. The yeast *HIS4* mRNA, for example, codes for a single protein that carries out three different enzymatic steps in histidine biosynthesis. Yet, it is rare that a single polypeptide carries out all the steps involved in a given metabolic pathway. Thus, regulatory molecules that coordinate the transcription of several functionally related mRNAs must act at multiple chromosomal sites. It is important to note that this latter feature is not unique to eucaryotes. For example, the enzymes that carry out the biosynthesis of arginine in *E. coli* are encoded by several different mRNA molecules whose expression is coordinately regulated, rather than being regulated by a single polycistronic mRNA.

Poly A at the 3' Ends of Eucaryotic mRNAs[56–60]

Still very mysterious is the observation that most yeast mRNA molecules, like those of all other eucaryotic cells, contain relatively long stretches (about 200 residues) of poly A at their 3' ends. These poly A

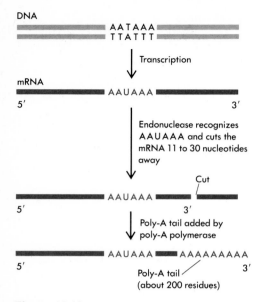

Figure 18-19
An AATAAA sequence appears toward the 3' end of many eucaryotic genes, which encodes an AAUAAA sequence that instructs an endonuclease to cleave the strand at a site 11 to 30 bases downstream; a poly A tail is then added.

Table 18-3   Wobble Rules for Yeast tRNAs

	Third Position of Anticodon	
Third Position of Codon	*E. coli*	*Yeast*
U	A, G, or I	G or I
C	G or I	G or I
A	U or I	U*
G	C or U	C

*U indicates a modified uridine residue.
SOURCE: Data from C. Guthrie and J. Abelson in J. N. Strathern, E. W. Jones, and J. R. Broach, eds., *The Molecular Biology of the Yeast Saccharomyces, Metabolism and Gene Expression* (Cold Spring Harbor, N.Y.: Cold Spring Harbor Laboratory, 1982), p. 492.

tracts (also called *tails*) are not specified by the genomic DNA, but are added after transcription to the newly made mRNA molecules released from the DNA into the nucleus. The enzyme responsible, poly A polymerase, specifically recognizes mRNA, as evidenced by the fact that poly A is not found at the 3' ends of rRNA or tRNA. However, the poly A does not add onto the 3' ends of completed mRNA chains.

Instead the poly A is added upstream from actual transcription termination sites onto 3' ends created by a very specific endonuclease that, after recognizing certain specific AAUAAA sequences, cleaves the bound mRNA chains some 11 to 30 bases downstream (3') to the AAUAAA recognition sequence (Figure 18-19). Still to be worked out are the signals (DNA sequences) used to specify actual termination of yeast mRNA synthesis.

For reasons that are still unknown, higher-cell histone mRNAs (not, however, those in *S. cerevisiae*) lack poly A, as do a few other mRNAs. Thus, whatever the function of poly A, it is not essential for mRNA synthesis or export from the nucleus. Poly A remains attached to the 3' end after an mRNA binds to ribosomes and initiates translation. There is evidence, that as polypeptides are repeatedly made off the same mRNA molecule, the length of its poly A tract may shorten.

S. cerevisiae mRNA Molecules Are Translated by Only Forty-Six Different tRNA Molecules[61, 62]

Given the existence of 61 different sense codons in the genetic code, it was initially thought that an equivalent number of tRNA molecules would be required for their reading. But with the realization that *wobble* (a degree of allowed nonspecificity) occurred in the pairing between codons and anticodons at the third position, the possibility arose that many fewer than 61 distinct tRNAs might suffice (Chapter 15). The yeast *S. cerevisiae* now provides the cleanest system for understanding wobble, because (1) a majority of its tRNAs have been isolated, sequenced, and their anticodons determined; (2) the cloning of many of its tRNA genes has already occurred; and (3) the cloning of many protein-coding genes has revealed the precise codons used to signify their various amino acids.

There are some 20 species of *S. cerevisiae* tRNAs present in relatively large amounts (one for every amino acid) and another 26 species present in lesser amounts. Collectively, they are coded by approximately 360 separate tRNA genes. So on the average, each tRNA is coded by eight separate genes of virtually identical sequence. However, each amino acid is not represented by equal numbers of tRNA genes. Those amino acids that are relatively more abundant in proteins tend to be represented by more tRNA genes. There is almost no tandem clustering of the tRNA genes; members of a given set are usually located on different chromosomes. Thus, the number of copies of a given tRNA gene is effectively fixed, unlike the rRNA genes, which constantly fluctuate in number because of unequal crossing over within the rDNA tandem array.

Examination of established tRNA anticodons shows that wobble in yeast is more restricted than in *E. coli* (Table 18-3), and so the minimum number of tRNA species needed to translate the yeast mRNAs must be correspondingly larger. Single tRNAs recognize the pairs of related codons ending in either U or C, while two different tRNAs are needed to recognize the related codons ending in A or G. Thus, three different tRNAs are required for those amino acids like proline or threonine that are specified by blocks of four codons. Since two different tRNAs are needed for the initiating methionines and the internally located methionines, 46 different yeast tRNAs (Table 18-3) can be predicted to exist, given the caveat that still to be discovered modified bases within certain tRNA anticodons might upset the simple rules outlined here.

As the exact codons used to specify yeast proteins began to be known, it became clear that they do not represent a random collection. Instead, only 26 different codons typically specify the amino acids of yeast proteins, these being the codons recognized by the 20 more abundant tRNAs (see Table 18-4). The choice of these common codons follows three general rules:

1. When an amino acid is specified by a block of three or four related codons, C and U appear in the wobble position with similar frequencies, with A or G almost never used (with the exception of proline).

2. In cases of pyrimidine degeneracy (UUU and UUC specifying the same amino acid), either U or C can occupy the wobble site, with C being the preferred base when the other codon members are U and/or A.

3. In twofold purine degeneracy, either A or G in the wobble position is used for any given amino acid.

In each situation, the preferred codon-anticodon pairings are those which use conventional base pairing and which maximize the hydrogen-bonding strengths (e.g., UAC is preferred over UAA to specify tyrosine, since it avoids the use of the weaker AU hydrogen bond).

When less favored codons are used in yeast mRNAs, they must be recognized by one of the less abundant tRNA species. The more than occasional use of these less favored codons within a given coding sequence might thus be expected to slow down the rate of translation and thus be a characteristic of a protein made in small amounts. Conceivably, the most abundant proteins make almost exclusive use of the favored codons. Also important is the apparent inability of a cell to mutate away one of its minor tRNA species and still remain viable. Loss of the tRNA species would lead to untranslatability of the corresponding codon, and so many vital cell proteins would not be made.

Control Elements at the Ends of Yeast Genes[63, 64]

Because of the absence of operons in yeasts, there is a one-to-one correspondence between the transcriptional units along chromosomes and the genes that yeast geneticists have defined in terms of cellular function. A yeast gene can thus be defined as a DNA segment that is transcribed into a single mRNA product together with the adjacent *control elements* at its two ends (Figure 18-20). These control elements are specific groups of base pairs that by binding to specific cellular proteins, direct the starting and stopping of RNA chains at correct positions along the chromosomes (see also Chapter 21). Gene boundaries are thus effectively set by the location of these control elements.

Those control elements used to start yeast RNA chains are *promoters* (using the terminology first used to describe bacterial genes in Chap-

Table 18-4   The 46 Different Predicted Yeast tRNAs and Their Corresponding Amino Acids

Amino acid	Codon	Codon Usage	Anticodon Predicted	Found
Ala	GCU	208	IGC	IGC
	GCC	76		
	GCA	3	U*GC	
	GCG	2	CGC	
Arg	CGU	7	ICG	ICG
	CGC	0		
	CGA	0	U*CG	
	CGG	0	CCG	
	AGA	85	U*CU	U*CU
	AGG	2	CCU	
Asn	AAU	5	GUU	
	AAC	101		
Asp	GAU	51	GUC	GUC
	GAC	106		
Cys	UGU	22	GCA	GCA
	UGC	1		
Gln	CAA	61	U*UG	
	CAG	4	CUG	
Glu	GAA	150	U*UC	U*UC
	GAG	7	CUC	
Gly	GGU	214	ICC	GCC
	GGC	6		
	GGA	3	U*CC	
	GGG	2	CCC	
His	CAU	11	GUG	GUG
	CAC	57		
Ile	AUU	75	IAU	
	AUC	71		
	AUA	1	U*AU	
	UUA	14	U*AA	ZAA
	UUG	170	CAA	CAA

Amino acid	Codon	Codon Usage	Anticodon Predicted	Found
Leu	CUU	3	IAG	
	CUC	1		
	CUA	7	U*AG	UAG
	CUG	0	CAG	
Lys	AAA	15	U*UU	U*UU
	AAG	184	CUU	CUU
Met$_m$	AUG		CAU	CAU
		56		
Met$_i$	AUG		CAU	CAU
Phe	UUU	12	GAA	G$_m$AA
	UUC	74		
Pro	CCU	16	IGG	
	CCC	1		
	CCA	85	U*GG	
	CCG	0	CGG	
Ser	UCU	103	IGA	IGA
	UCC	94		
	UCA	6	U*GA	U*GA
	UCG	1	CGA	CGA
	AGU	5	GCU	
	AGC	1		
Thr	ACU	75	IGU	IGU
	ACC	75		
	ACA	8	U*GU	
	ACG	4	CGU	
Trp	UGG	27	CCA	C$_m$CA
Tyr	UAU	10	GUG	GψA
	UAC	76		
Val	GUU	121		
	GUC	97	IAC	IAC
	GUA	1	U*AC	N*AC
	GUG	2	CAC	CAC

m = methylated; *indicates modified; underscoring denotes the more abundant tRNAs.
SOURCE: Data from C. Guthrie and J. Abelson in J. N. Strathern. E. W. Jones, and J. R. Broach, eds., *The Molecular Biology of the Yeast Saccharomyces, Metabolism and Gene Expression* (Cold Spring Harbor, N.Y.: Cold Spring Harbor Laboratory, 1982), p. 490.

ter 8). But yeast promoters differ in a fundamental way from bacterial promoters through their possession of **upstream activating sequences (UASs)** and **upstream repressing sequences (URSs),** which can be several hundred base pairs 5' to the start of transcription. Binding of positive control proteins to UAS sequences turns up the rate of transcription of their respective genes, while binding of negative control proteins to URS sequences turns down the transcription rate of those genes that need to be negatively regulated. Given the very subtle devices that we know are utilized in the regulation of bacterial DNA transcription, we must anticipate that many yeast

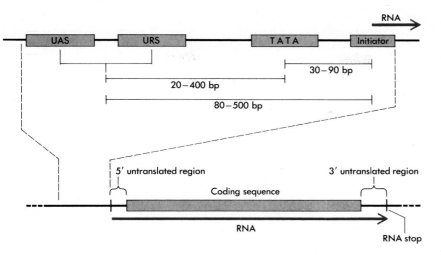

genes will contain both UAS and URS elements, thereby allowing their expression to be complexly tuned to reflect a multiplicity of external environmental conditions.

In addition to this obligatorily upstream element, yeast promoters have two other essential components: (1) TATAAA-like blocks (called **TATA** sequences) usually placed some 20 to 40 base pairs 5' to the transcription start site (TATA sequences may play similar roles to the block at −35 base pairs of the *lac* promoter) and (2) an initiator element encompassing the start site, which precisely determines where transcription begins.

Many yeast genes are transcribed at constant rates throughout the cell cycle. The speed of such **constitutive** synthesis is a function of the specific base pairs at the yeast promoters. Depending on the promoter sequence, the rate of constitutive transcription can vary from very slow to very fast. Other promoters work at variable rates that depend on the extent of their binding to specific control proteins, many of whose amounts are environmentally controlled. Such promoters control the speed at which "regulated" genes function.

Regulated Expression of the Galactose-Utilizing Genes[65-68]

Gene regulation in yeasts, as in bacteria, was first discovered when mutations were found in genetic elements that block the functioning of still other genes. The easiest such systems to analyze are those whose functioning is increased or decreased by externally added nutrients (e.g., sugars, amino acids, and phosphate ion). Already very well investigated are several regulated genes of *S. cerevisiae* that code for three very well characterized enzymes involved in the utilization of the sugar galactose (yeasts do not contain the lactose-hydrolyzing enzyme β-galactosidase). They are galactokinase, a product of the *GAL1* gene; galactose transferase, the product of the *GAL7* gene; and galactose epimerase, encoded by the *GAL10* gene (Figure 18-21). Together they transform galactose into glucose-1-phosphate, which becomes converted to glucose-6-phosphate and enters the glycolytic pathway using enzymes that are always present. Yeast cells produce these three enzymes only when they are grown in the presence of galactose; these genes are not transcribed when yeast is grown in other carbon sources, such as glycerol or glucose.

Although these three *GAL* genes are located close together, they do not comprise an operon; they are each transcribed into separate

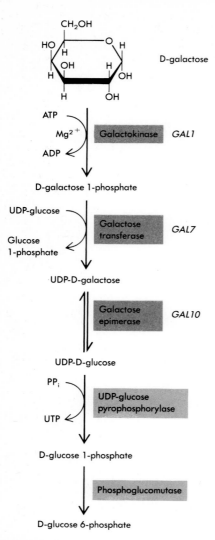

Figure 18-21
Galactose metabolism in yeast.

Figure 18-22
Sequence organization of the *GAL1–GAL10* promoter region (color) and the *GAL7* promoter region (gray boxes). *GAL1* and *GAL10* are transcribed in opposite directions away from a common promoter element, while a separate promoter functions for the adjacent *GAL7* gene.

mRNA molecules. The *GAL1* and *GAL10* mRNAs are divergently transcribed away from a common promoter element, while a separate promoter functions for the immediately adjacent *GAL7* gene (Figure 18-22). Regulating these three genes are two other genes, *GAL4* and *GAL80*. Genetic analysis first suggested that *GAL4* is a positive regulator, which activates the transcription of *GAL1*, *GAL7*, and *GAL10* genes. In contrast, *GAL80* is thought to act as a negative regulator.

The function of *GAL4* as a positive regulator has been directly confirmed by isolation of the *GAL4* product; it is a 99 kdal protein that binds to specific "upstream" sequences within the *GAL1–GAL10* and *GAL7* promoters. Each promoter contains two 30-base-pair blocks that are protected from nuclease digestion when bound to the *GAL4* protein. *GAL4* binding is mediated by a stretch of 74 amino acids at its amino-terminal end. The sequences within the 30-base-pair blocks are closely homologous, showing 18 to 22 identical base pairs within a consensus sequence of 23 residues. As expected from its DNA binding function, the *GAL4* protein is normally found within the yeast nucleus.

How the regulatory protein encoded by *GAL80* works is not known. It is argued to be a negative regulator of transcription of the *GAL* genes because yeast mutants that lack *GAL80* function express the *GAL* genes constitutively, that is, even in the absence of galactose. Somehow, in the absence of galactose, the product of *GAL80* prevents *GAL4* from functioning. It might code for a repressorlike protein that binds to the *GAL1-GAL10* promoter region. The best guess now is that the *GAL80* protein is not a DNA binding protein but somehow directly inactivates the *GAL4* product when galactose is absent (but not when it is present!). In this way, the presence of galactose would turn on galactose-utilizing genes. If this scheme is correct, then we should expect the *GAL80* protein to inhibit the binding of *GAL4* to DNA. This proposal will be experimentally tested in the near future.

Different Promoter Elements for Constitutive and Regulated Expression[69–73]

Many highly regulated genes that are turned on by positive regulatory proteins also maintain finite basal levels of constitutive expression that is not under environmental control. Examination of the promoter elements of one of these genes, the well-studied *HIS3* gene, suggests that the two types of expression involve quite distinct promoter elements. Constitutive *HIS3* expression uses upstream, TATA, and initiation elements that are different from those involved in its regulated expression (Figure 18-23). Most interesting is the upstream element for constitutive expression, 17AT base pairs, which can act bidirectionally. Such poly dA:dT blocks are also found upstream of several other constitutive start sites, suggesting that they may be a general signal for much, if not all, constitutive expression. Favoring this idea is the fact that the level of constitutive expression rises with the length of the poly dA:dT tract. Since poly dA:dT tracts cannot

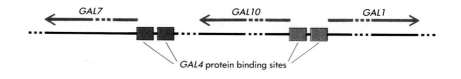

GAL4 protein binding sites

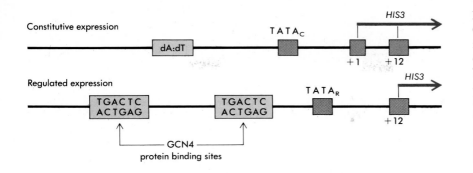

Figure 18-23
Constitutive expression of *HIS3* differs
from regulated expression in the control
elements involved: poly dA:dT versus
TGACTC sequences, TATA elements (T_c
and T_R), and initiation sites (1 and 12).
In regulated expression, the protein
coded by the gene *GCN4* binds to pairs
of TGACTC control blocks to turn on
the synthesis of amino acids in conditions of amino acid starvation.

form nucleosomes, the obvious hypothesis is that they provide a region of DNA free of nucleosomes for interacting with the proteins
involved in initiating RNA chains.

High-level regulated expression of many genes coding for enzymes
involved in amino acid biosynthesis requires the protein product of
the *GCN4* gene. The regulation of this large set of genes is termed
"general amino acid control." Activation of these genes occurs when
cells are starved for one of the amino acids, for example, histidine in
the case of *HIS3*. The upstream elements for regulated expression of
HIS3 are pairs of TGACTC blocks that bind to the *GCN4* protein.
Similar TGACTC blocks are found upstream of all the promoters
under *GCN4* control. It is interesting to note that the transcription of
the *GCN4* gene itself is not regulated by amino acid starvation; its
mRNA product is made at a constant rate. Instead, the expression of
the *GCN4* gene is regulated at the translational level, with the efficiency of translation increasing with amino acid starvation. The *GCN4*
mRNA has an unusually long 5' untranslated region that contains
four AUG start codons. If these codons are removed, the efficiency of
translation dramatically increases. It is possible that these untranslated AUG sites bind ribosomes to prevent other ribosomes from initiating at the correct *GCN4* start sites. How amino acid starvation might
effectively cover up these inhibitory AUG codons remains a mystery.

Only a Minority of the S. cerevisiae Genes Are Split[74, 75]

While the vast majority of the genes of higher eucaryotes are split into
exons and introns (see Chapter 20), only a small minority of the genes
of *S. cerevisiae* contain introns. Among those that do are the genes
coding for actin and for many tRNA molecules. However, the *S.
pombe* genome has a much larger number of split genes. This strongly
hints that the two forms of yeast may not be as closely related as first
believed. That *S. cerevisiae* and *S. pombe* may be evolutionarily very
distant is also suggested by comparisons of their rDNAs. Conceivably, these two examples of unicellular fungi are as distant from each
other as from the genomes of higher vertebrates.

Insertion of Transposable Genetic Elements Affects Gene Expression[76–81]

Mutated control elements frequently arise through the insertion of
large blocks of base pairs that are present in virtually identical form at

Figure 18-24
A map of a Ty1-like genome. ENH = an enhancer element promoting gene expression; ORF = open reading frame. LTR = long terminal repeat, found in retroviruses and related elements such as Tyl. The LTRs of Tyl are called "delta." (Courtesy of Dr. Gerry Fink.)

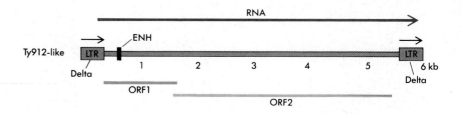

a number of other sites on the yeast genome. These *repetitive sequences* are usually movable genetic elements (Chapters 7 and 11) that carry internal genes, some of which code for the enzymes that transpose (move) them to other chromosomal locations.

Why the placement of a movable element within a promoter might destroy promoter activity is easy to comprehend. Initially, however, it was hard to understand how such an insertion event radically increased the expression of previously dormant genes. The explanation came with the finding that the flanking sequences of movable elements often contain strong promoter elements. They increase gene expression by providing much more effective (and sometimes less strongly regulated) start signals for transcription.

The most common movable element in yeast genomes is called Ty1 (Figure 18-24); structurally, it is closely related to the copia movable element of *Drosophila* and to the genomes of retroviruses that multiply in vertebrate cells (Chapter 24). Some 35 copies of complete 6-kilobase-pair Ty1 elements are present within the genomes of most yeast strains. Each contains a large central region with two large, open reading frames that code for two major protein products, one of which is a reverse transcriptase. It is flanked by two identical terminal 334-base-pair repeats called delta (δ) elements, each of which contains a promoter element as well as sequences recognized by the transposing enzymes. Transcribed from each Ty1 element is a single, 6000-nucleotide mRNA molecule that begins at the promoter located at the 5' end of Ty1. Usually, these Ty1 elements are frequently transcribed within host cells; the total Ty1 RNA comprises 5 to 10 percent of the total yeast mRNA.

The close similarity in genome structure and function between the Ty1 element and the retrovirus genome strongly suggests not only that these two elements have a common evolutionary origin, but also that they move to new chromosomal locations by essentially the same molecular mechanism. It seemed very likely, from the moment its genome structure was worked out, that the Ty1 element would not move like bacterial transposons (by a transfer of information from DNA to DNA). Instead, it seemed highly probable that Ty1 would follow the retroviral pattern and transpose by a reverse transcription of its RNA chain to yield a complementary DNA chain. After conversion of this DNA chain to a double helix, it would then insert into a new chromosomal location.

Proof that such reverse transcription does in fact occur comes from the discovery that one of the two Ty1 genes codes for a reverse transcriptase, as well as from the finding that when a Ty1 is genetically modified to contain an intron (see Chapter 20) and subsequently moves to a new site, the intron is lost. This is to be expected if transposition involves an RNA intermediate, since RNA molecules with an intron will have the intron spliced out following transcription. Thus, Ty1 elements can appropriately be referred to as **retrotransposons**.

Rapid Reversals of Sex That Do Not Obey Conventional Genetic Rules[82]

Very important information about the expression of yeast genes also comes from the study of the mating types (sexes) of *S. cerevisiae*. An allele present in the mating type (*MAT*) locus on chromosome 3 determines whether a haploid cell is of the **a** or α mating type (sex). The *MAT***a** allele specifies the **a** mating type, while the α mating type is manifested when the *MAT*α allele occupies the *MAT* locus. These two alternative alleles set into motion two different pathways of gene expression that are beginning to help us understand how differentiated cells choose to produce one of several mutually incompatible sets of proteins. Such studies by themselves are bound to attract much future attention to the *MAT* locus.

Already, the *MAT* locus has been the focus of a great deal of interest through studies on why many strains of yeast do not have stable sexes but instead rapidly alternate between the **a** and α sexes. In **homothallic yeasts,** the mating-type genes of haploid cells interconvert much more rapidly (almost every other cell division) than could be anticipated by any mechanism involving conventional spontaneous mutations. Such rapid switches, however, are absent in **heterothallic strains,** where the *MAT* locus behaves like a conventional genetic element. An allele present at a second locus, *HO,* located on chromosome 4, determines whether a yeast strain is homothallic or heterothallic.

Most *S. cerevisiae* strains found in nature are homothallic, with heterothallic strains usually restricted to laboratory variants that have been selected for this trait. The prevalence of the homothallic state is not limited to *S. cerevisiae,* but is characteristic of virtually all other yeasts, as well as many other classes of fungi whose life cycles also alternate between the haploid and diploid states. Why homothallism is so widespread is still unclear. Perhaps it is favored because it promotes the early diploidization of the descendants of all spores, thereby sheltering the yeasts' own progeny from the burden of harmful mutations that would almost immediately be expressed in the haploid state. Moreover, since diploidy is a prerequisite for the meiotic cyclic generation of spores, sex reversals that lead to mating and diploidization have the further advantage of leading to a more rapid sporulation response to unfavorable environmental circumstances.

The Cassette Model for Mating-Type Switches[83–87]

Initially, it was guessed that the rapid reversals of sex might be due to flip-flop inversions of the *MAT* locus, like those that lead to the alternative expression of the two different proteins found within the flagella of *Salmonella* bacteria (see Figure 11-25). Genetic analysis, however, ruled out such explanations and instead led to an unorthodox proposal for which there was no precedence. The key came from genetic crosses that showed that chromosome 3 possesses both an active *MAT* gene and two unexpressed mating-type loci. One unexpressed locus, *HML,* is situated 200 kilobase pairs to the left of the *MAT* locus and contains a silent copy of the α information; the other silent gene, to the right of *MAT,* is *HMR,* and it contains **a** specificity. The switching of these genes involves the donation of genetic information (called a **cassette**) from one of the unexpressed genes to the

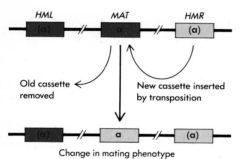

Figure 18-25
The cassette model of mating-type switching. Here the **a** cassette replaces the α cassette in the mating-type locus that is expressed (*MAT*). Parenthesis shading indicates that a gene is silent. (Redrawn from I. Herskowitz and Y. Oshima, in *The Molecular Biology of the Yeast Saccheromyces, Life Cycle and Inheritance*, p. 194.)

Figure 18-26
Sequence homology in the *MAT*, *HMLα*, and *HMRa* loci as determined by heteroduplex analysis. Note that the X and Z segments are common to all, while *HML* has the same Y segment as *MATα*, and *HMRa* has the same Y segment as *MATa*. Thus, the silent loci contain the exact information needed to generate the postulated cassettes. (Redrawn from I. Herskowitz and Y. Oshima, in *The Molecular Biology of the Yeast Saccheromyces, Life Cycle and Inheritance*, p. 196.)

MAT locus (Figure 18-25). In this process, the original *MAT* gene is somehow displaced. However, transposition of a cassette from a silent gene does not lead to the loss of any genetic information from that site. Thus, the genetic information needed to promote subsequent mating-type switches is preserved.

The cassette model, bizarre as it first seemed, is supported by the cloning of the two forms (α and **a**) of active *MAT* genes. *MATα* and *MATa* have easily distinguishable structures; there is no sequence homology within their long (about 600 base pairs) internal Ya and Yα segments, which are bounded by common X and Z segments (Figure 18-26). These differences in the Y segments of the alternative forms of *MAT* are faithfully reproduced in the two silent genes; *HMLα* has a Y segment identical to that of *MAT*, and *HMRa* has the Y sequence found in MATa. The silent loci thus possess exactly the information needed to generate the postulated cassettes.

SIR Genes Turn Off the *HMLα* and *HMRa* Genes[88–91]

The *HMLα* and *HMRa* genes were originally thought to be silent because they lacked essential promoter sites required for their expression. But this hypothesis became untenable when the transcription of the specific mRNAs of *MATα* and *HMRa* was shown to start within the internally located Y segments that have identical structures both in *MAT* loci and in the silent *HM* loci. The silent genes must therefore possess promoters identical to those found within their active equivalents. At around the same time, geneticists discovered four unlinked genes—*SIR1, 2, 3,* and *4,*—whose gene products act together in *trans* (they do not have to be on the same chromosome) to prevent the expression of the silent mating-type genes. If any one of these *SIR* genes fails to act, the *HMLα* and *HMRa* genes are transcribed at rates equal to that of the gene at the *MAT* locus. To find the sites where the *SIR* products act, deletions were first created within *HMLα* and *HMRa* by in vitro mutagenesis. Then the genetically altered DNAs were inserted into appropriate yeast strains using DNA transformation procedures (Chapter 19).

Such studies reveal essential (E) regions that are adjacent to both the *HML* and *HMR* genes and that control the expression of those genes, making them silent (Figure 18-27). E regions act in *cis*, only affecting genes on the same chromosome. Within E is a block of base pairs (common to both *HML* and *HMR* E regions) that is the likely site where the *SIR* products act. However, these blocks are located some 1500 base pairs away from the promoters they control, raising the question of how the *SIR* proteins binding to the E region can inhibit (silence) a promoter as far as 2 kilobase pairs away. Equally puzzling is the fact that an E region can work in either orientation and that its

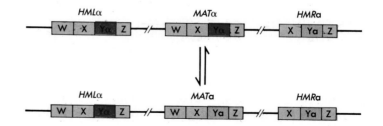

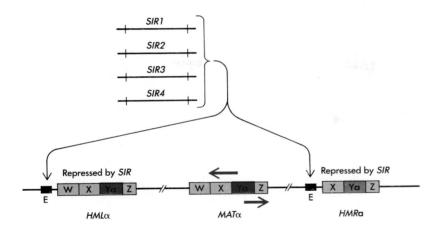

Figure 18-27
SIR proteins act at essential E regions to turn off the expression of the *HML* and *HMR* genes. The E regions are located some 1500 base pairs away from the promoters they control to silence the *HML* and *HMR* genes. Transcription from *MAT* starts at a point within Y and proceeds in both directions. (Redrawn from A. J. S. Klar, J. N. Strathern, and J. B. Hicks, R. Losick and L. Shapiro, eds., *Microbial Development*, p. 161.)

relative location (3′ or 5′) with regard to the promoters that it silences does not matter. Perhaps the *SIR* proteins work by influencing the structure of chromatin within the *HML* and *HMR* genes. Favoring this proposal are experiments showing that when *SIR* control is absent, the chromatin within *HML* and *HMR* is more susceptible to nuclease attack, perhaps because of a failure of nucleosomes to form within these regions.

Double-Stranded Breaks at the *MAT* Locus Initiate Cassette Movements[92–95]

It was initially unclear how a copy of a silent *HM* locus replaces the resident allele of the *MAT* locus. Then the seminal observation was made of double-stranded cuts at the junction between the Y and Z segments, events that have now been proved to initiate cassette switching. Such breaks are made by a site-specific endonuclease that recognizes bases within the sequence CAGCTTTCCGCAACAG-TAAA that is present at the Y/Z junction. This endonuclease is coded by the *HO* gene and is termed the Y/Z endonuclease. Such breaks are made only in HO^+ (homothallic) strains, a finding that quickly led to experiments showing that HO^- (heterothallic) strains do not switch mating types because they fail to make this enzyme. The discovery that a double-stranded break initiated switching events came soon after the discovery that double-stranded DNA scissions, and not just single-chain cuts, can stimulate crossing over in yeasts. Many of the fundamental steps involved in cassette movement may thus be identical to those that carry out the crossing events that exchange blocks of genes between homologous chromosomes. The *RAD52* gene is one such example; it is needed not only for mating-type interconversion, but also for repair of X-ray induced damages and for recombination.

Following an initiating double-stranded cut, the *MAT* Y region becomes degraded, which explains the loss of the parental *MAT* allele. To replace the missing Y segment, the two DNA free ends at the Z end of the cut together with the free ends generated by the removal of the Y segment invade the homologous regions of the appropriate *HM* gene. DNA replication onto the 3′ ends of the invading double helices then repair the missing Y segment; in the process, the *MAT* locus acquires the genetic specificity of the invaded *HM* locus (Figure 18-28). By this stage, the two interlocked mating-type loci are held

Figure 18-28
Detailed mechanism of cassette switching. A double-stranded cut is made, and the *MAT* Y segment is degraded. The free ends of the Z and X segments invade homologous regions of the appropriate *HM* gene, creating short heteroduplexes. The 3′ extension during DNA repair replaces the missing Y segment with that of the invading *HM* gene. As shown in other figures, the *HM* and *MAT* genes involved in cassette switching are located on the same chromosome. [After A. J. S. Klar, J. N. Strathern, and J. B. Hicks, R. Losick and L. Shapiro, eds., *Microbiological Development* (Cold Spring Harbor, N.Y.: Cold Spring Harbor Laboratory), p. 175.]

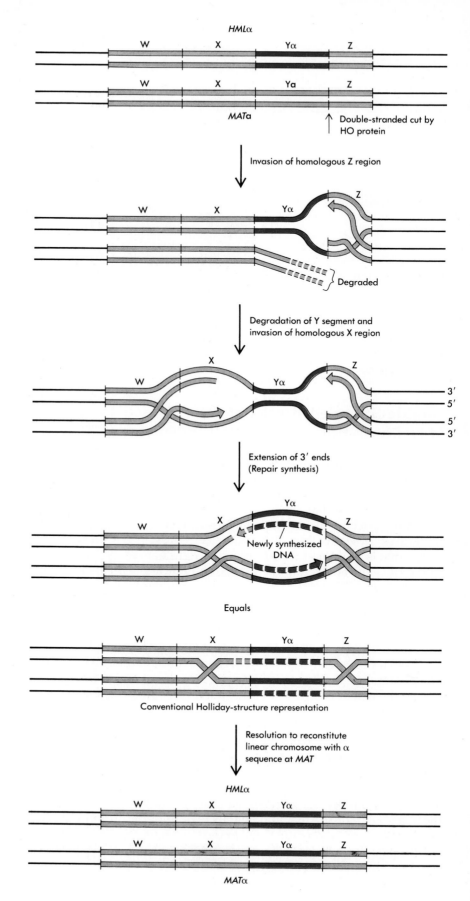

together by two regions of strand exchanges that must be resolved by subsequent single-stranded breaks similar to those that separate the joined intermediates of conventional crossing over (Chapter 11). These final resolving breaks must occur in a way that does not allow parts of the *MAT* locus to become exchanged with segments adjacent to the *HML* or *HMR* loci. Thus, isomerization processes that would lead to the exchange of flanking markers (Chapter 11) and that would produce a defective chromosome, do not occur.

The site that is recognized and cut by the *HO*-coded endonuclease is present not only at *MAT* but also at *HML* and *HMR*. However, only the site at *MAT* is cut *in vivo*. The *SIR* gene products are responsible for distinguishing between the cutting sites at these different locations: in mutants defective in *SIR*, cutting can occur at *HML* and *HMR*. Thus *SIR*, which was suggested above to be responsible for condensing the DNA in the *HML* and *HMR* regions to turn off transcription from these loci, also has the effect of preventing these loci from being cut by an endonuclease.

Programming the Direction of Cassette Movements[96, 97]

Switching is a highly programmed phenomenon that is limited to haploid cells and that only occurs in the parent cells, as opposed to the daughter, or bud, cells. Bud cells must divide once before they become capable of switching. Any "experienced" cells (those that have divided once) have a very high probability (80 percent) of switching. A switch is never observed in only one of a pair, suggesting that the cassettes move before the time of *MAT* DNA synthesis, with the switched cassette being subsequently replicated and distributed to both daughter cells (Figure 18-29). That switching is restricted to a predetermined point in the cell cycle is confirmed by recent experiments demonstrating that Y/Z endonuclease is made in the G1 phase and acts only during the next S phase of the cell cycle. The activity of *HO* Y/Z endonuclease must thus be inherently unstable, perhaps being inactivated while executing its function.

The switching of almost all experienced cells indicates that the direction of switching has been programmed to move from **a** to α and from α to **a** mating types. If the choice existed for the gapped *MAT* gene to randomly invade either the *HML* or *HMR* loci with equal probabilities, then the maximum frequency of observable switches would be 50 percent. This point was revealed through observing the switching behavior of strains in which *a* sequences occur at the *HML* loci and α sequences at the *HMR* locus. In such strains, observable switches occur much less often (about 5 percent). The lower values do not represent a less frequent switching, but instead indicate that a given cassette has been replaced by one of identical specificity (Figure 18-30). Thus, the two silent *HM* loci compete with each other as donors of the genetic information that will fill in the gapped *MAT* locus. Apparently, the *MAT* α locus usually preferentially associates with and then invades the *HMR* locus, while the *MAT***a** locus preferentially associates with and invades the *HML* locus. Such pairings are brought about by DNA binding proteins directly or indirectly controlled by the mating-type loci.

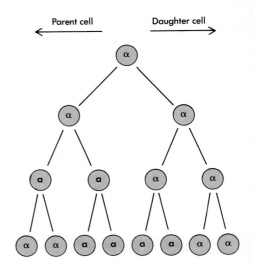

Figure 18-29
Division of a cell containing the *HO* gene results in either two cells with the same mating-type phenotype as the parental cell or two cells of switched phenotype. The switches always occur in pairs.

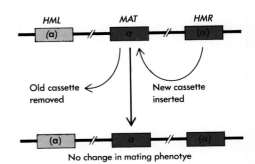

No change in mating phenotye

Figure 18-30
In yeast strains in which *HML* has **a** cassette and *HMR* has α cassette, the frequency of phenotypic switches from α to **a** is much lower than when HML is α and HMR is **a**. Although the actual switches occur at the normal frequency, the *MAT*α gene is most often replaced with one of identical specificity.

MATα and *MATa* Genes
Code for Control Proteins[98–103]

α cells contain a number of proteins, "α-specific proteins," that are not produced by **a** cells. Correspondingly, **a** cells contain a set of **a**-specific proteins that are not produced by α cells. The genes for the α-specific proteins are transcribed only in α cells, and the genes for the **a**-specific proteins are transcribed only in **a** cells. The mating-type locus is responsible for the expression of the appropriate set of cell-type-specific genes: *MATa* and *MATα* code for proteins that either activate or inhibit expression of genes that are scattered throughout the yeast genome. The mating-type locus is thus the master control locus for cell specialization. The precise manner in which the products of the mating-type locus govern expression of the specialized gene sets is shown in Figure 18-31. The α-specific genes are turned on in α cells by a protein coded by *MATα*, the α1 protein. In contrast, the **a**-specific genes are synthesized constitutively in **a** cells—there is no activator protein encoded by *MATa* that is needed. Therefore, the set of **a**-specific genes needs to be turned off in α cells. This task is performed by a protein coded by the second gene within *MATα*, the α2 protein. α2 protein is repressor, a site-specific DNA binding protein that recognizes a target site (an **a**-specific gene operator) located upstream of each of the members of the **a**-specific gene set.

This scheme leaves no obvious function for the *MATa* gene in haploid cells, and, in fact, it has no known function in these cells. In diploids, however, the α-specific genes must be turned off and the pathway of meiosis and sporulation must be turned on. This is accomplished by a third regulatory activity coded by *MAT*, which requires both the α2 protein and a product coded by *MATa*, the **a**1 protein. This **a**1-α2 activity is responsible for many of the properties of **a**/α cells and turns off expression of a wide variety of genes. For example, it turns off expression of the α-specific genes by blocking transcription of the α1 protein. Thus **a**/α cells do not express the α-specific gene set. Furthermore, **a**/α cells do not undergo mating-type interconversion because **a**1-α2 turns off transcription of the *HO* gene. In addition, **a**1-α2 turns on meiosis and sporulation by repressing transcription of yet another gene (*RME1*). The *RME1* product is an inhibitor of meiosis and sporulation and is produced in **a** and in α cells; hence, they are unable to enter the pathway of meiosis and sporulation. In **a**/α cells, however, *RME1* synthesis does not occur, and cells can undergo meiosis and sporulation. Yet another example of repression by **a**1-α2 is the Ty1 element, whose transcript is repressed (for unknown physiological reasons) by **a**1-α2. The molecular nature of **a**1-α2 is not presently known. A plausible hypothesis is that the **a**1 polypeptide chain associates with the α2 polypeptide chain to form a repressor with novel recognition properties. Possible binding sites for **a**1-α2 are related sequences that are found in the upstream regions of genes repressed by **a**1-α2.

The three regulatory activities encoded by the yeast mating-type locus—α1, α2, and **a**1-α2—provide the first opportunity to study a class of regulatory proteins that do not respond to the environment (for example, to nutritional signals such as the presence of sugars or amino acids) but rather confer on a cell the proper expression of sets of genes responsible for differentiation. It is of course of interest to know whether higher eucaryotes have master regulatory loci and master regulatory proteins similar to those of yeast that are responsible for maintaining cellular differentiation.

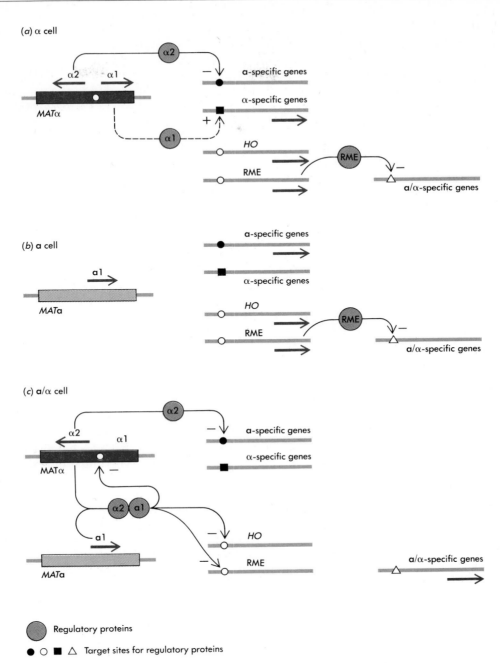

Regulatory proteins

● ○ ■ △ Target sites for regulatory proteins

Pheromones Make Haploid Cells Competent to Mate[104–107]

When mixed, α and **a** cells do not automatically bind strongly to each other and fuse to form α/**a** diploid cells. They must first convert each other into cells that are competent for mating by releasing specific hormonelike molecules called **pheromones,** which bind to receptors on cells of the opposite mating type. α cells secrete into the medium α-factor, which is 13 amino acids long. α-factor binds to a specific receptor on the **a** cell and induces the **a** cell to produce **a**-factor, which is 11 amino acids long. **a**-factor binds to a specific receptor on the α cell. A series of biochemical reactions follows that promotes the clumping of the cells to each other in α-**a** pairs and stops their movement through the cell cycle at the G1 stage. In this stage, the cells

Figure 18-31
The function of *MAT* genes as master control loci in (a) α haploid, (b) **a** haploid, and (c) **a**/α diploid cells. Positive control functions are denoted by dotted arrows, and negative control functions are denoted by solid arrows. **a**/α specific genes include those necessary for meosis and spore formation.

Figure 18-32
Processing of initial *MFα*1 protein product into biologically active pheromones. S1, S2, S3, and S4 are spacers, which are cleaved as the polypeptide is processed. αF1, αF2, αF3, and αF4 are four identical mature α-factor pheromone molecules. Several different proteolytic enzymes catalyze the processing. [Courtesy of J. Kurjan and I. Herskowitz from *Cell* 30 (1982):940.]

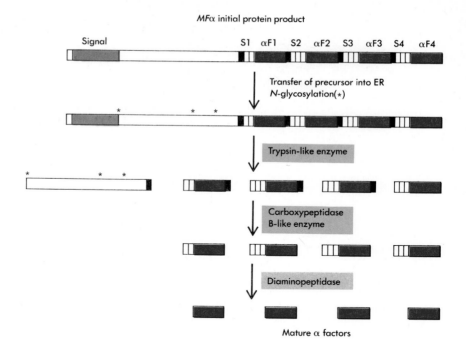

cannot begin to replicate their DNAs, but they retain the capacity to synthesize the other major classes of cellular molecules. Finally, these biochemical reactions lead to specific changes in the cell walls and membranes that permit fusion between the paired cells.

The α pheromone is coded by the *MFα* gene, while the **a** pheromone is coded by the *MF***a** gene. The initial protein products of the *MF* genes are not yet active, but become so upon their degradation by specific proteolytic enzymes to produce the relatively short and biologically active oligopeptide pheromones (Figure 18-32). Conceivably, the relatively small size of both these pheromones permits their penetration through the cell wall to reach the outer surface of the plasma membrane of their target cells. If they function like other known polypeptide hormones, they bind to specific receptor molecules inserted into the plasma membrane of the target cell.

Meiosis Involves Proteins from the Mitotic Cell Cycle and Others That Are Unique to Meiosis[108–112]

Yeast cells undergo the process of meiosis in a manner very much like the meiosis of higher eucaryotes: diploid cells give rise to haploid meiotic products in a characteristic pathway in which the chromosome number is duplicated and then reduced in two successive meiotic segregations. In yeast, the meiotic products are then encased in spore coats to form spores. The entire process of meiosis and ascospore formation in yeast is termed "sporulation." Studies of yeast are providing considerable information about how cells make a decision to enter the pathway of meiosis, and how the numerous events of meiosis are orchestrated.

For **a**/α diploid yeast cells to enter the meiotic pathway, the cells first must receive a nutritional signal (nitrogen starvation and a poor carbon source), which causes cyclic AMP levels to drop. In subse-

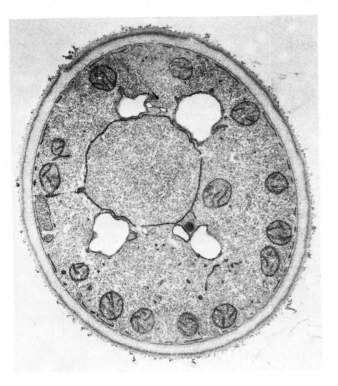

Figure 18-33
Electron micrograph of a yeast cell, showing multiple mitochondria distributed throughout its cytoplasm. (Courtesy of Dr. Barbara J. Stevens.)

quently initiating meiosis, they exit the mitotic cycle at the GI phase carrying out a round of DNA replication. This is followed by recombination, which is especially high in meiotic cells (compared to mitotic cells), and then the two meiotic chromosome distributions. Many of the same proteins that are used for the mitotic cycle are also needed in meiosis. For example, there are several *CDC* genes needed for both meiotic and mitotic DNA replication and for nuclear division (chromosome distribution), which presumably make up the DNA replication apparatus and the nuclear spindle. Likewise, there are at least two genes needed for both mitotic and meiotic recombination. There are, however, also many genes whose roles are specific to the meiotic process. Notable examples are genes needed for initiation of meiotic DNA replication, for meiotic recombination, and for meiotic chromosome distribution. There are estimated to be at least 50 genes whose expression is specifically triggered during meiosis.

Mitochondria Have Their Own Chromosomes[113, 114]

Each haploid yeast cell contains about 10 to 50 mitochondria distributed throughout its cytoplasm (Figure 18-33). This variation in number does not usually represent a variation in the amount of mitochondrial matter per cell, which ordinarily accounts for some 10 percent of the cellular mass. Instead, it reflects a variation in the size of individual mitochondria, which can both divide to form smaller mitochondria or fuse together to create the very long and often branched mitochondria found in many rapidly growing cells.

Their tiny, sometimes bacterium-like shapes and the fact that they grow in length and then divide suggested to early-twentieth-century cytologists that mitochondria might have their own genetic elements and might have evolved during the early days of eucaryotic life from a procaryotic-like progenitor. Then, around 1950, evidence appeared for a special class of yeast mutation that did not show Mendelian segregation during mitosis that is characteristic of nuclear genes. These unusual mutants instead segregated as if they were genes located in the cytoplasm. These so-called *"petite" mutations* create mitochondria that are missing several vital enzymes, rendering them unable to generate ATP. As a result, the mutant cells grow slowly and form smaller ("petite") colonies on solid agar surfaces. The simplest way to explain these abnormal mitochondria was to postulate the existence of genetic molecules within normal mitochondria that code for the missing proteins. This conjecture became fact with the discovery that all eucaryotic mitochondria possess relatively tiny DNA molecules (mitochondrial DNA, or mtDNA) that in almost all organisms have circular shapes. Even though the individual molecules are quite small, the amount of these mtDNA molecules per cell is quite appreciable. Mitochondrial DNA is present in 50 to 100 copies per diploid cell and amounts to about 20 percent of the total DNA in a yeast cell. At around the same time, it was shown that chloroplasts also contain DNA molecules that code for many of their constituent proteins.

Synthesis of mtDNA Occurs Throughout the Yeast Life Cycle[115, 116]

Replication of circular yeast mtDNA is not limited to the S phase of the cell cycle. This may reflect the fact that mtDNA synthesis is carried out, at least in part, by different proteins than those that replicate nuclear DNA. Once started, mitochondrial DNA elongation occurs at a slower rate than that of its chromosomal equivalents. Single molecules of normal mtDNA require replication times equal to that of the entire cell cycle. Conceivably, the minimal length of the yeast cell cycle is dictated by the size of its mtDNA. Alternatively, there may have been no strong selective advantage for yeasts to reduce the mtDNA genome below the minimum size that can be replicated during the cell cycle.

Another interesting difference between nuclear and mitochondrial DNA replication is the absence of any proofreading (exonuclease) activity of the mitochondrial DNA polymerase. This leads to a much higher mutation rate within the mitochondrial genome than within chromosomal genes, and so mtDNA can evolve extremely rapidly. Perhaps this lack of an error repair mechanism is partly compensated by the large number of mitochondria within a single cell.

Much less is known about the replication of the DNA from yeast mitochondria than about the much smaller vertebrate mitochondrial DNA. The much larger size of yeast mtDNA has so far prevented the isolation of intact replicating molecules. Therefore, it is not known whether yeast mtDNA employ displacement-type replication like that employed by human mtDNA (Chapter 10).

The mitochondrial genome can undergo many different types of gross chromosomal rearrangements. A particularly common one is intramolecular recombination between repeated sequences. This generates deletion mutations of mtDNA and thus leads to mutants with a petite growth behavior. In some cases, the defective mtDNA can outgrow normal mitochondria in the same cell and create a cellular defect in mitochondrial function. A remarkable example of this type of deletion is one deletion mutant that lacks virtually all the normal mitochondrial genome, containing only many repetitions of those DNA sequences that make up the origin of DNA replication. A second type of gross chromosomal rearrangement is a specially programmed transposition of a segment of DNA throughout the population of mtDNA molecules. Some mtDNAs contain a 1.1 kbp insert ("omega") in the ribosomal RNA genes. When mtDNAs carrying omega are introduced into a cell that contains mtDNAs lacking omega, the omega element is rapidly transferred as if it were a movable genetic element to the omega-deficient mtDNAs. This occurs because the omega element codes for a site-specific endonuclease that produces a double-stranded break at a site within the ribosomal RNA gene of omega-deficient strains. Just as in mating-type interconversion, an efficient repair process ensues after a double-stranded break is produced. Repair of this break results in a copying in of the omega element as part of the repair process.

Noncoding Segments
Along the Mitochondrial Genome[117, 118, 119]

When mitochondria from two different haploid cells are placed together in a diploid cell as a result of mating, they become physically intermixed, allowing genetic recombination to occur between them. Because such crossing-over events are very frequent, genetic crosses can be used only to show linkage between very closely spaced genetic units. The only way to assign the location of the genes on an mtDNA molecule is to isolate and sequence the relevant DNA segments. Such sequence analyses reveal a most unexpected structure for yeast mtDNA. Not only are the segments corresponding to well-defined genes frequently separated by long segments of apparently useless DNA (Figure 18-34), but some genes themselves are interrupted by

Figure 18-34
Physical map of a long mitochondrial genome unit of wild-type *S. cerevisiae.* Black areas correspond to mitochondrial genes or their exons, colored areas to intervening sequences and intergenic open reading frames (ORF1–ORF5). Radial lines indicate tRNA genes (the *thr*1 gene is the only one to have an anticlockwise orientation). Among mitochondrial genes, OXI1, 2 and 3 encode sub-units II, III, and I, respectively, of cytochrome c oxidase; cob, cytochrome b; *aap*1, *oli*2 and *oli*1, sub-units 8, 6, and 9 of ATPase; *var*1, a protein associated with the small mitochondrial ribosome sub-unit; 9S corresponds to the central part of tRNA synthesis locus; 15S and 21S are the genes for the small and large ribosomal RNAs, respectively. Triangles indicate the location of ori sequences 1–8. Mitochondrial transcripts are represented on an inner circle. They have a clockwise orientation except for two short transcripts in the ORF1-*thr*1 region. Dots correspond to initiation sites, arrows to termination sites, crosslines to processing sites. For some transcripts, initiation and/or termination sites are not known so far. [Adapted from M. de Zamaroczy and G. Bernardi, *Gene* 37 (1985):2, with permission.]

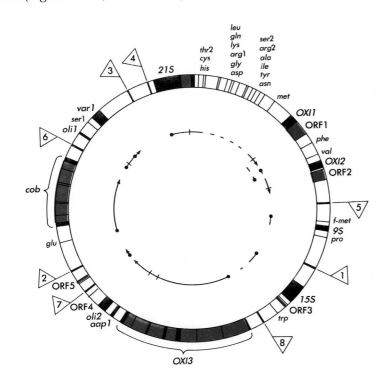

Table 18-5   Genetic Code for Yeast Mitochondrial DNA

		Second Position				
		U	C	A	G	
First Position	U	UUU ⎤ Phe UUC ⎦ UUA ⎤ Leu UUG ⎦	UCU ⎤ UCC ⎦ Ser UCA ⎤ UCG ⎦	UAU ⎤ Tyr UAC ⎦ UAA ⎤ Ter UAG ⎦	UGU ⎤ Cys UGC ⎦ UGA ⎤ Trp UGG ⎦	U C A G
	C	CUU ⎤ CUC CUA ⎦ Leu CUG (Thr)	CCU ⎤ CCC CCA Pro CCG ⎦	CAU ⎤ His CAC ⎦ CAA ⎤ Gln CAG ⎦	CGU ⎤ CGC CGA Arg CGG ⎦	U C A G
	A	AUU ⎤ AUC Ile AUA ⎦ AUG Met	ACU ⎤ ACC ACA Thr ACG ⎦	AAU ⎤ Asn AAC ⎦ AAA ⎤ Lys AAG ⎦	AGU ⎤ Ser AGC ⎦ AGA ⎤ Arg AGG ⎦	U C A G
	G	GUU ⎤ GUC GUA Val GUG ⎦	GCU ⎤ GCC GCA Ala GCG ⎦	GAU ⎤ Asp GAC ⎦ GAA ⎤ Glu GAG ⎦	GGU ⎤ GGC GGA Gly GGG ⎦	U C A G

Third Position

Grey shading denotes those codons that are read differently from the universal genetic code.

long, noncoding regions (introns) more reminiscent of the organization of eucaryotic DNA than of procaryotic DNA (Chapter 20). This observation suggests that the original procaryotic DNA that gave rise to mtDNA may have been more like today's eucaryotic DNA than the DNA in currently existing procaryotes.

Mitochondria Contain Their Own Specific Protein-Synthesizing Systems[120–124]

The proteins encoded by mitochondrial DNA are synthesized within the mitochondria themselves on a special class of small ribosomes called **mitoribosomes,** whose rRNA components are transcribed from two separate stretches of mtDNA. The tRNAs involved have unique structures that are also specified by mtDNA. Only 24 or 25 different tRNA molecules are used to translate yeast mitochondrial mRNA. Most of the 64 different codons specify the same amino acids and stop signals as specified by the genetic code for chromosomal DNA (Table 18-5). Only one codon is always read differently by the yeast tRNAs; UGA specifies tryptophan rather than a stop signal.

Also, the codon CUN, normally read as leucine, can sometimes be read as threonine. Moreover, as yet, all yeast mitochondrial genes have been found to terminate with UAA. The fact that only 25 tRNAs exist means that more *wobble* is possible than with the nuclearly coded tRNAs. When a single amino acid is specified by four codons with the same first and second positions (e.g., the four proline codons CCU, CCC, CCA, CCG), a single tRNA is sufficient; wobble permits pairing of any third codon base with the U that invariably resides in the first anticodon position of the tRNA. Furthermore, pairs of codons with pyrimidines in their third position are read by G in the tRNA anticodon, while pairs with purines in the third position are read by U in the tRNA.

Yeast mitoribosomes, like all other ribosomes, have two separable subunits (37S and 50S), each of which contains a single rRNA chain (15S and 21S, respectively) and a large number of different ribosomal proteins, each specific for the mitochondrial ribosomes.

The Genomes of Mitochondria Code Only for a Small Fraction of Their Proteins[125]

Given the almost bacterial size of mitochondria, it would not be surprising if mtDNAs were as large as the genomes of small bacteria. However, even the largest of mitochondrial DNAs can encode only a small percentage of mitochondrial proteins. Yeasts have among the largest mtDNAs (about 50 kilobase pairs) of any organism, and yet they code for only a small percentage of the proteins present in their mitochondria. Not even the majority of the enzymes involved in the generation of ATP are coded by mtDNA. Only seven well-defined proteins, in addition to the 15S and 21S rRNA molecules and the 25 different tRNAs, are known to be encoded by the yeast mtDNA. In contrast, all of the 60 or so proteins found in mitoribosomes, 20 different aminoacyl synthetases, and all the enzymes needed to replicate and transcribe the mtDNA are coded by chromosomal genes located in the nucleus. The mRNA molecules transcribed off these nuclear genes are translated by cytoplasmic ribosomes, and their protein products are later transported to their sites of function within the mitochondria.

The functioning of its protein-synthesizing system is not even necessary for mitochondrial growth and division. There are mutants that fail to make functional mitoribosomes and so cannot translate any of the genetic messages transcribed off mtDNA. Yet, the resulting mitochondria contain all of their usual chromosomally coded components and have essentially normal appearance. So it is still unknown why mitochondria need to contain any genetically active DNA at all.

We have seen that the yeast mitochondrial genome presents somewhat of a puzzle. It does some things, but not all that many. And it carries out these activities with a genome that is vastly larger than necessary. Why has the large mitochondrial genome persisted? One way in which yeast differs from higher eucaryotes is that the nuclear genome in higher eucaryotes contains vast amounts of noncoding DNA—satellites and introns. Yeast contains only modest amounts of such sequences. In contrast, the yeast mtDNA is ten times larger than the mammalian mtDNA. Perhaps the mitochondrial genome plays the additional role in yeast as a reservoir of genetic diversity, capable of serving the nuclear genome by contributing evolved sequences.

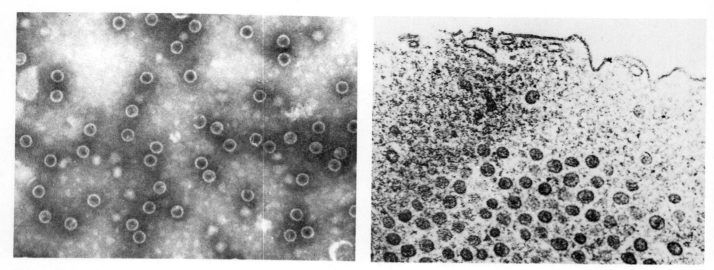

Figure 18-35

A cluster of viruslike killer particles (a) and retroviruslike particles (b) in the cytoplasm of yeast cells. [(a) Courtesy of J. D. Welsh. (b) Courtesy of D. J. Garfinkel.]

No Infectious Viruses Have Been Found for Yeasts[126, 127]

No infectious viruses are known to exist for any form of yeast. However, two different viruslike particles that contain RNA are found within many yeasts and may represent descendants of true viruses of past evolutionary epochs that today have lost their ability to effectively deliver their RNA genomes through the cell walls of yeast. The most commonly observed particles contain double-stranded RNA, are surrounded by a protein-containing capsid, and exist in multiple copies in the cytoplasm of most yeast strains. This viruslike RNA encodes a "killer" protein toxin that is secreted through the cell wall and is lethal to yeast cells that lack these viruslike particles. The killer toxin apparently acts by binding to the cell membrane of non-killer cells, inducing the release of ATP and K^+ into the surrounding medium.

Also in many yeast strains are particles morphologically similar to the retroviruses of higher cells (Figure 18-35). These particles have recently been shown to contain the RNA transcripts of the *Ty*-type movable DNA elements (transposons) that exist within virtually all yeast strains. The number of copies of these retrovirus-like particles depends on the rate of transcription of the yeast genome and varies with its surrounding environment. Why yeast cells produce such retroviruslike particles if they are never used as infectious agents remains a mystery. The same dilemma exists for the encapsulation of the "killer" RNA genomes.

Summary

Saccharomyces cerevisiae and *Schizosaccharomyces pombe* are now being extensively studied at the molecular level for the fundamental insights they can reveal about the structure and function of eucaryotic genes. Both these yeasts have well-studied, simple life cycles that involve both haploid and diploid phases, with the diploid cells undergoing conventional meiotic divisions to yield the haploid spores. All four products of a given meiotic division are encased within a thick-walled ascus. This allows the examination of the reciprocal chromosome partners of individual crossing-over events. Such tetrad analysis reveals the phenomenon of gene conversion, making yeasts among the preferred organisms for studying crossing over at the DNA level. *S. cerevisiae* cells divide through a budding process that separates parental cells from their daughter (bud) cells. In contrast, *S. pombe* is a fission yeast in which parental cells grow and then divide to form two morphologically identical daughter cells.

The average haploid yeast cell contains 14×10^6 base pairs; about 3½ times more DNA than is found in the *E. coli* chromosome. In *S. cerevisiae*, the DNA is distributed among 16 different chromosomes, while *S. pombe* has three different chromosomes. Already, more than 560 different *S. cerevisiae* genes have been mapped. During mitosis and meiosis, the yeast chromosomes are separated on microtubule-containing spindles, which remain completely encased by intact nuclear membranes and which, unlike those of higher eucaryotic cells, do not break down at the time of spindle formation. The microtubules of the spindle emanate from spindle pole bodies that are inserted into the nuclear membranes and which themselves divide at the time the decision for cell division is made. How the yeast cell cycle operates is beginning to be revealed through the analysis of *CDC* (cell division cycle) mutants blocked at specific stages of the cell cycle.

Like all other eucaryotic cells, yeasts contain internally located structural proteins that come together to form organized cytoskeletal elements. The contractile proteins actin and myosin already have been identified, as well as the α and β tubulins that aggregate to form the microtubules. Because of their small size, it was only recently discovered that yeasts contain all the major membrane-bounded internal compartments found in higher eucaryotic cells. Characterization of several of these organelles (e.g., the endoplasmic reticulum) was not straightforward until the discovery of mutants that block their functioning. Outside of all yeast cells is a thick cell wall impermeable to large molecules. Enzymes that specifically digest away much of the cell wall enable the generation of osmotically unstable protoplasts, now particularly useful because of their ability to take up externally added DNA. The DNA of the yeast chromosome is largely organized into nucleosomes through the coiling of DNA around histone cores made of H2A, H2B, H3, and H4 units. So far, there is no evidence for H1-like histones. Each chromosome contains a single centromeric DNA sequence that does not bind histones, but instead binds specialized proteins that in turn have the potential to attach to spindle microtubules. At the ends (telomeres) of all yeast chromosomes are some 100 base pairs of tandemly repeated $C_{1-3}A$ sequences that serve to maintain the ends of DNA molecules during DNA replication. DNA synthesis is initiated at many separate sites (origins of replication) along each chromosome. Synthesis occurs at a limited phase (S) during the cell cycle and proceeds bidirectionally at some 2 to 4 μm per minute. Nothing is known about the precise signals used to initiate DNA synthesis.

Three different forms of RNA polymerase are used to transcribe the yeast genome. RNA polymerase I transcribes the ribosomal DNA; RNA polymerase II transcribes the genes used to make the mRNA chains that code for proteins; and RNA polymerase III makes tRNA, 5S RNA, and several still functionally obscure forms of small RNA chains. Some primary RNA yeast transcripts have introns that become spliced out within the nucleus to yield smaller products. All mRNA molecules acquire caps at their 5' ends and poly A tails at their 3' ends within the nucleus before movement of their final functional forms through the nuclear pores into the cytoplasm. Each yeast mRNA molecule is monocistronic, coding for only a single polypeptide chain. A number of yeast polypeptides become folded into several enzymatic sites, each of which catalyzes a different chemical reaction.

Translation of mRNA chains on ribosomes requires the participation of 46 different tRNA molecules. Those tRNA molecules used to translate the more abundantly used codons are present in many more copies than those that translate infrequently utilized codons.

The rate at which a given gene is expressed is controlled at the level of both transcription and translation. Transcription levels are determined in part by the promoter sequences located 5' to the initiation sites for transcription. Each promoter contains sequences near the initiation site (initiation sequences and TATA blocks) as well as upstream activating sequences (UASs) or URSs (upstream repressing sequences). Some upstream elements act independently of the particular intracellular environment (constitutive expression), while others are regulated by specific control proteins whose activity is metabolically controlled (regulated expression). A given gene can contain both constitutive and regulatable upstream activating elements. The proteins that regulate these upstream elements can function either positively (turn on gene expression) or negatively (turn it down). Promoters also can have their activity either increased or decreased through the insertion of movable genetic elements. The main movable element in *S. cerevisiae* is Ty1, an element that moves by reverse transcription of its RNA product into DNA chains that can reinsert themselves back into many other chromosomal sites. Thus, Ty1 is an example of a retrotransposon.

The two alternative forms (*MAT*α and *MAT*a) of the genes determining mating type in *S. cerevisiae* code for control proteins, some acting positively and others acting negatively. The two forms of *MAT* are inherently unstable, being rapidly interchanged through the insertion of genetic cassettes emanating from normally unexpressed loci. These silent genes are turned off by products of the *SIR* genes, which act on sequences far removed from the promoter elements that they control. How this occurs remains unknown. The instability of *MAT* cassettes results from the action of the *HO* gene, which codes for the specific endonuclease that makes the double-stranded break at the *MAT*α (or *MAT*a) allele that initiates cassette movement.

Yeast mitochondria, like all mitochondria, contain genetically active DNA that replicates within the mitochondria, using enzymes imported into the mitochondria from the cytoplasm. Mitochondrial DNA (mtDNA) codes for the rRNA of its ribosomes, the tRNAs it needs to translate its own mRNA, and a small number of the enzymes involved in the mitochondrial generation of ATP. Mutations occur very frequently during mtDNA replication because the DNA polymerase involved has no proofreading capability. Many mtDNA mutants have lost most of their functional base sequences and in extreme cases contain only hundreds of tandemly repeated copies of their origin of replication. Such mitochondria have no protein-synthesizing capability, yet divide much more quickly and efficiently than normal mitochondria under conditions where oxidative ATP generation is not necessary. Why mitochondria have maintained genetically active DNA remains a puzzle.

Bibliography

General References

Egel, R., J. Kohli, P. Thuriaux, and K. Wolf. 1980. "Genetics of the Fission Yeast *Schizosaccharomyces pombe*." *Ann. Rev. Genetics* 14:77–108.

Fincham, J. R. S., and P. R. Day. 1971. *Fungal Genetics*. Philadelphia: Davis.

Klar, A. J. S., J. N. Strathern, and J. B. Hicks. 1984. "Developmental Pathways in Yeast." In *Microbial Development*, ed. R. Losick and L. Shapiro. Cold Spring Harbor, N.Y.: Cold Spring Harbor Laboratory, pp. 151–196.

Strathern, J. N., E. W. Jones, and J. R. Broach. 1981. *The Molecular Biology of the Yeast Saccharomyces, Life Cycle and Inheritance*. Cold Spring Harbor, N.Y.: Cold Spring Harbor Laboratory.

Strathern, J. N., E. W. Jones, and J. R. Broach, eds. 1982. *The Molecular Biology of the Yeast Saccharomyces. Metabolism and Gene Expression*. Cold Spring Harbor, N.Y.: Cold Spring Harbor Laboratory.

Cited References

1. Fogel, S., and D. D. Hurst. 1967. "Meiotic Gene Conversion in Yeast Tetrads and the Theory of Recombination." *Genetics* 57:455–481.
2. Esposito, M. S. 1971. "Post Meiotic Segregation in *Saccharomyces*." *Mol. Gen. Genetics* 111:297–299.
3. Fogel, S., R. K. Mortimer, and K. Lusnak. 1983. "Meiotic Gene Conversion in Yeast: Molecular and Experimental Perspectives." In *Yeast Genetics: Fundamental and Applied Aspects*, ed. J. F. T. Spencer, D. M. Spencer, and A. R. W. Smith. New York: Springer-Verlag, pp. 65–107.
4. Davidow, L. S., and B. Byers. 1984. "Enhanced Gene Conversion and Post Meiotic Segregation in Pachytene-Arrested *Saccharomyces cerevisiae*." *Genetics* 106:165–183.
5. Mortimer, R. K., and D. Schild. 1981. "Genetic Mapping in *Saccharomyces cerevisiae*." In *The Molecular Biology of the Yeast Saccharomyces, Life Cycle and Inheritance*, ed. J. N. Strathern, E. W. Jones, and J. R. Broach. Cold Spring Harbor, N.Y.: Cold Spring Harbor, Laboratory, pp. 11–26. Contains an excellent explanation of tetrad analysis and other mapping techniques.
6. Kuroiwa, T., H. Kojima, I. Miyakawa, and N. Sando. 1984. "Meiotic Karyotypes of the Yeast *Saccharomyces cerevisiae*." *Exp. Cell Res.* 153:259–265.
7. Mortimer, R. K., and D. Schild. 1985. "Genetic Map of *Saccharomyces cerevisiae*, Edition 9." *Microbial Rev.* 49:181–213.
8. Moens, P. B., and E. Rapport. 1971. "Spindles, Spindle Plaques and Meiosis in the Yeast *Saccharomyces cerevisiae*." *J. Cell Biol.* 50:344–361.
9. Byers, B. 1975. "The Behavior of Spindles and Spindle Plaques in the Cell Cycle and Conjugation of *Saccharomyces cerevisiae*." *J. Bacteriol.* 124:511–523.
10. Byers, B., K. Shriver, and L. Goetsch. 1978. "The Role of Spindle Pole Bodies and Modified Microtubule Ends in the Initiation of Microtubule Assembly in *Saccharomyces cerevisiae*." *J. Cell Sci.* 30:331–352.
11. Hartwell, L. H. 1974. "*Saccharomyces cerevisiae* Cell Cycle." *Bacteriol. Revs.* 38:164–198.
12. Game, J. C. 1976. "Yeast Cell Cycle Mutant cdc21 Is a Temperature Sensitive Thymidylate Auxotroph." *Mol. Gen. Genetics* 146:313–315.
13. Johnston, L. H., and K. A. Nasmyth. 1978. "*Saccharomyces cerevisiae* Cell Cycle Mutant cdc9 Is Defective in DNA Ligase." *Nature* 274:891–893.
14. Hiroaka, Y., T. Toda, and M. Yanagida. 1984. "The *NDA3* Gene of Fission Yeast Encodes β-Tubulin: A Cold-Sensitive *nda3* Mutation Reversibly Blocks Spindle Formation and Chromosome Movement in Mitosis." *Cell* 39:349–358.
15. Schekman, R., and P. Novick. 1982. "The Secretory Process and Yeast Cell Surface Assembly." In *The Molecular Biology of the Yeast Saccharomyces, Metabolism and Gene Expression*, ed. J. N. Strathern, E. W. Jones, and J. R. Broach. Cold Spring Harbor, N.Y.: Cold Spring Harbor Laboratory, pp. 361–398.

16. Bitter, G. A., K. K. Chen, A. R. Banks, and P.-H. Lai. 1984. "Secretion of Foreign Proteins from *Saccharomyces cerevisiae* Directed by α-Factor Gene Fusions." *Proc. Nat. Acad. Sci.* 81:5330–5334.
17. Brake, A. J., J. P. Merryweather, D. G. Coit, U. A. Heberlein, F. R. Masiarz, G. T. Mullenbach, M. S. Urdea, P. Valenzuela, and P. J. Barr. 1984. "α-Factor-Directed Synthesis and Secretion of Mature Foreign Proteins in *Saccharomyces cerevisiae*." *Proc. Nat. Acad. Sci.* 81:4642–4646.
18. Hall, M. N., L. Hereford, and I. Herskowitz. 1984. "Targeting of *E. coli* Beta-Galactosidase to the Nucleus in Yeast." *Cell* 36:1057–1065.
19. Hurt, E. C., B. Pesold-Hurt, K. Suda, W. Oppliger, and G. Schatz. 1985. "The First Twelve Amino Acids (Less Than Half of the Presequence) of An Imported Mitochondrial Protein Can Direct Cytosolic Dihydrofolate Reductase into the Yeast Mitochondrial Matrix." *EMBO J.* 4:2061–2068.
20. Adams, A. E. M., and J. R. Pringle. 1984. "Relationship of Actin and Tubulin Distribution to Bud Growth in Wild-Type and Morphogenic-Mutant *Saccharomyces cerevisiae*." *J. Cell Biol.* 98:934–945.
21. Kilmartin, J. V., and A. E. M. Adams. 1984. "Structural Rearrangements of Tubulin and Actin During the Cell Cycle of the Yeast *Saccharomyces*." *J. Cell Biol.* 98:922–933.
22. Ballou, C. E. 1982. "The Yeast Cell Wall and Cell Surface." In *The Molecular Biology of the Yeast Saccharomyces, Metabolism and Gene Expression*, ed. J. N. Strathern, E. W. Jones, and J. R. Broach. Cold Spring Harbor, N.Y.: Cold Spring Harbor Laboratory.
23. Lohr, D., and K. E. Van Holde. 1975. "Yeast Chromatin Subunit Structure." *Science* 188:165–166.
24. Moll, R., and E. Wintersberger. 1976. "Synthesis of Yeast Histones in the Cell Cycle." *Proc. Nat. Acad. Sci.* 73:1863–1867.
25. Carle, G. F., and M. V. Olson. 1984. "Separation of Chromosomal DNA Molecules by Orthogonal-Field-Alteration Electrophoresis." *Nucleic Acid Res.* 12:5647–5664.
26. Carle, G. F., and M. V. Olson. 1985. "An Electrophoretic Karyotype for Yeast." *Proc. Nat. Acad. Sci.* 82:3756–3760.
27. Schwartz, D. C., and C. R. Cantor. 1984. "Separation of Yeast Chromosome-Sized DNAs by Pulsed Field Gradient Gel Electrophoresis." *Cell* 37:67–75.
28. Strathern, J. S., C. S. Newlon, I. Herskowitz, and J. B. Hicks. 1979. "Isolation of a Circular Derivative of Yeast Chromosome III: Implications for the Mechanism of Mating Type Interconversion." *Cell* 18:309–319.
29. Carbon, J. 1984. "Yeast Centromeres: Structure and Function." *Cell* 37:351–353.
30. Blackburn, E. 1984. "Telomeres: Do the Ends Justify the Means?" *Cell* 37: 7–8.
31. Dunn, B., P. Szauter, M. L. Pardue, and J. W. Szostak. 1984. "Transfer of Yeast Telomeres to Linear Plasmids by Recombination." *Cell* 39:191–201.
32. Shampay, J., J. W. Szostak, and E. H. Blackburn. 1984. "DNA Sequences of Telomeres Maintained in Yeast." *Nature* 310:154–157.
33. Walmsley, R. M., C. S. M. Chan, B.-K. Tye, and T. D. Petes. 1984. "Unusual DNA Sequences Associated with the Ends of Yeast Chromosomes." *Nature* 310:157–160.
34. Walmsley, R. M., and T. D. Petes. 1985. "Genetic Control of Chromosome Length in Yeast." *Proc. Nat. Acad. Sci.* 82:506–510.
35. Murray, A. W., and J. W. Szostak. 1983. "Pedigree Analysis of Plasmid Segregation in Yeast." *Cell* 34:961–970.
36. Murray, A., and J. W. Szostak. 1983. "Construction of Artificial Chromosomes in Yeast." *Nature* 305:189–193.
37. Hieter, P., C. Mann, M. Snyder, and R. W. Davis. 1985. "Mitotic Stability of Yeast Chromosomes: A Colony Color Assay That Measures Nondisjunction and Chromosome Loss." *Cell* 40:381–392.
38. Koshland, D., J. C. Kent, and L. H. Hartwell. 1985. "Genetic Analysis of the Mitotic Transmission of Minichromosomes." *Cell* 40:393–403.

39. Williamson, D. H. 1965. "The Timing of Deoxyribonucleic Acid Synthesis in the Cell Cycle of *Saccharomyces cerevisiae*." *J. Cell Biol.* 25:517–510.

40. Burke, W., and W. L. Fangman. 1975. "Temporal Order in Yeast Chromosome Replication." *Cell* 5:263–269.

41. Sentenac, A., and B. Hall. 1982. "Yeast Nuclear RNA Polymerases and Their Role in Transcription." In *The Molecular Biology of the Yeast* Saccharomyces, *Metabolism and Gene Expression*, ed. J. N. Strathern, E. W. Jones, and J. R. Broach. Cold Spring Harbor, N.Y.: Cold Spring Harbor Laboratory, pp. 561–606.

42. Zazloff, Michael. 1983. "tRNA Transport from the Nucleus in a Eucaryotic Cell: Carrier-Mediated Translocation Process." *Proc. Nat. Acad. Sci.* 80:6436–6440.

43. Bell, G. I., L. J. DeGennaro, D. H. Gelfand, R. J. Bishop, P. Valenzeula, and W. J. Rutter. 1977. "Ribosomal RNA Genes of *Saccharomyces cerevisiae*. I. Physical Map of the Repeating Unit and Location of the Regions Coding for 5S, 5.8S, 18S Ribosomal RNAs." *J. Biol. Chem.* 252:8118–8125.

44. Petes, T. D., and D. B. Botstein. 1977. "Simple Mendelian Inheritance of the Reiterated Ribosomal DNA of Yeast." *Proc. Nat. Acad. Sci.* 74:5091–5095.

45. Philippsen, P., M. J. Thomas, R. A. Kramer, and R. W. Davis. 1978. "Unique Arrangement of Coding Sequences for 5S, 5.8S, 18S and 25S Ribosomal RNA in *Saccharomyces cerevisiae* as Determined by R-Loop and Hybridization Analysis." *J. Mol. Biol.* 123:387–404.

46. Petes, T. D. 1979. "Yeast Ribosomal DNA Genes Are Located on Chromosome XII." *Proc. Nat. Acad. Sci.* 76:410–414.

47. Egel, R. 1981. "Intergenic Conversions and Reiterated Genes." *Nature* 290: 191–192.

48. Jackson, J. A., and G. R. Fink. 1981. "Gene Conversion Between Duplicated Genetic Elements in Yeast." *Nature* 292:306–311.

49. Pringle, J. R., and L. H. Hartwell. 1981. "The *Saccharomyces cerevisiae* Cell Cycle." In *The Molecular Biology of the Yeast* Saccharomyces, *Life Cycle and Inheritance*, ed. J. N. Strathern, E. W. Jones, and J. R. Broach. Cold Spring Harbor, N.Y.: Cold Spring Harbor Laboratory.

50. Perry, R. P., and D. E. Kelly. 1976. "Kinetics of the Formation of 5′ Terminal Caps in mRNA." *Cell* 8:433–442.

51. Adams, J. M., and S. Cory. 1975. "Modified Nucleosides and Bizarre 5′ Termini in Mouse Melanoma Messenger RNA." *Nature* 255:28–31.

52. Furuichi, Y., M. Morgan, S. Muthukrishman, and A. J. Shatkin. 1975. "Reovirus Messenger RNA Contains a Methylated Blocked 5′-Termini Structure: m⁷G(5′)ppp(5′)-GᵐpCp." *Proc. Nat. Acad. Sci.* 72:362–356.

53. Kozak, M. 1978. "How Do Eucaryotic Ribosomes Select Initiation Regions in Messenger RNA?" *Cell* 15:1109–1123.

54. Keesey, J. K., Jr., R. Bigelis, and G. R. Fink. 1979. "The Product of the *his4* Gene Cluster in *Saccharomyces cerevisiae*; A Trifunctional Polypeptide." *J. Biol. Chem.* 254:7427–7433.

55. Sherman, F., and J. W. Stewart. 1982. "Mutations Altering Initiation of Translation of Yeast Iso-1-Cytochrome c; Contrasts Between the Eucaryotic and Procaryotic Initiation Process." In *The Molecular Biology of the Yeast* Saccharomyces, *Metabolism and Gene Expression*, ed. J. N. Strathern, E. W. Jones, and J. R. Broach. Cold Spring Harbor, N.Y.: Cold Spring Harbor Laboratory, pp. 301–333.

56. McLaughlin, C. S., J. R. Warner, M. Edmonds, H. Nakazato, and M. H. Vaughn. 1973. "Polyadenylic Acid Sequences in Yeast Messenger Ribonucleic Acid." *J. Biol. Chem.* 248:1466–1471.

57. Hereford, L. M., and M. Roshbash. 1977. "Number and Distribution of Polyadenylated RNA Sequences in Yeast." *Cell* 10:453–462.

58. Fahrner, K., J. Yarger, and L. Hereford. 1980. "Yeast Histone mRNA Is Polyadenylated." *Nucleic Acid Res.* 8:5725–5737.

59. Bennetzen, J. L., and B. D. Hall. 1982. "The Primary Structure of the *Saccharomyces cerevisiae* Gene for Alcohol Dehydrogenase I." *J. Biol. Chem.* 257:3018–3025.

60. Zaret, K. S., and F. Sherman. 1982. "DNA Sequence Required for Efficient Transcription Termination in Yeast." *Cell* 28:563–573.

61. Crick, F. H. C. 1966. "Codon-Anticodon Pairing: The Wobble Hypothesis." *J. Mol. Biol.* 19:548–555.

62. Guthrie, C., and J. Abelson. 1982. "Organization and Expression of tRNA Genes in *Saccharomyces cerevisiae*." In *The Molecular Biology of the Yeast* Saccharomyces, *Metabolism and Gene Expression*, ed. J. N. Strathern, E. W. Jones, and J. R. Broach. Cold Spring Harbor, N.Y.: Cold Spring Harbor Laboratory, pp. 487–528.

63. Guarente, L. 1984. "Yeast Promoters: Positive and Negative Elements." *Cell* 36:799–800.

64. Brent, R. 1985. "Repression of Transcription in Yeast." *Cell* 42:3–4.

65. St. John, T. P., and R. W. Davis. 1981. "The Organization and Transcription of the Galactose Gene Cluster of *Saccharomyces*." *J. Mol. Biol.* 152:285–315.

66. Oshima, Y. 1982. "Regulatory Circuits for Gene Expression: The Metabolism of Galactose and Phosphate." In *The Molecular Biology of the Yeast* Saccharomyces, *Metabolism and Gene Expression*, ed. J. N. Strathern, E. W. Jones, and J. R. Broach. Cold Spring Harbor, N.Y.: Cold Spring Harbor Laboratory, pp. 159–180.

67. Silver, P. A., L. P. Keegan, and M. Ptashne. 1984. "Amino Terminus of the Yeast *GAL4* Gene Product Is Sufficient for Nuclear Localization." *Proc. Nat. Acad. Sci.* 81:5951–5955.

68. Bram, R. J., and R. D. Kornberg. 1985. "Specific Protein Binding to Far Upstream Activating Sequences in Polymerase II Promoters." *Proc. Nat. Acad. Sci.* 82:43–47.

69. Donahue, T. F., R. S. Daves, G. Lucchini, and G. R. Fink. 1983. "A Short Nucleotide Sequence Required for Regulation of *his4* by the General Control Systems of Yeast." *Cell* 32.89–98.

70. Thireos, G., M. D. Penn, and H. Greer. 1984. "5′ Untranslated Sequences Are Required for the Translational Control of a Yeast Regulatory Gene." *Proc. Nat. Acad. Sci.* 81:5096–5100.

71. Struhl, K. 1984. "Genetic Properties and Chromatin Structure of the Yeast *gal* Regulatory Element: An Enhancer-Like Sequence." *Proc. Nat. Acad. Sci.* 81:7865–7869.

72. Hinnebusch, A. G., G. Lucchini, and G. R. Fink. 1985. "A Synthetic *HIS4* Regulatory Element Confers General Amino Acid Control on the Cytochrome c Gene (*CYC1*) of Yeast." *Proc. Nat. Acad. Sci.* 82:498–502.

73. Struhl, K., W. Chen, D. E. Hill, I. A. Hope, and M. A. Dettinger. 1985. "Constitutive and Coordinately Regulated Transcription of Yeast Genes: Promoter Elements, Positive and Negative Regulatory Sites, and DNA Binding Proteins." *Cold Spring Harbor Symp. Quant. Biol.* In press.

74. Mao, J., B. Appel, J. Schaack, S. Sharp, H. Yamada, and D. Soll. 1982. "The 5S RNA Genes of *Schizosaccharomyces pombe*." *Nucleic Acids Res.* 10:487–498.

75. Hindley, I., and G. A. Phear. 1984. "Sequence of the Cell Division Gene *CDC2* from *Schizosaccharomyces pombe*; Patterns of Splicing and Homology to Protein Kinases." *Gene* 31:129–134.

76. Roeder, G. S., P. J. Farabaugh, D. T. Chaleff, and G. R. Fink. 1980. "The Origins of Gene Instability in Yeast." *Science* 209:1375–1380.

77. Scherer, S., C. Mann, and R. W. Davis. 1982. "Reversion of a Promoter Deletion in Yeast." *Nature* 298:815–819.

78. Boeke, J. D., D. J. Garfinkel, C. A. Styles, and G. R. Fink. 1985. "Ty Elements Transpose Through an RNA Intermediate." *Cell* 40:491–500.

79. Clare, J., and P. Farabaugh. 1985. "Nucleotide Sequence of a Yeast Ty Element: Evidence for an Unusual Mechanism of Gene Expression." *Proc. Nat. Acad. Sci.* 82:2829–2833.

80. Kostriken, R., J. N. Strathern, A. J. S. Klar, J. B. Hicks, and F. Heffron. 1983. "A Site-Specific Endonuclease Essential for Mating-Type Switching in *Saccharomyces cerevisiae*." *Cell* 35:167–174.

81. Mellor, J., S. M. Fulton, M. J. Dobson, W. Wilson, S. M. Kingsman, and A. J. Kingsman. 1985. "A Retrovirus-Like Strategy for Expression of a Fusion Protein Encoded by Yeast Transposon Ty1." *Nature* 313:243–246.

82. Herskowitz, I., and Y. Oshima. 1981. "Control of Cell Type in *Saccharomyces cerevisiae*: Mating Type and Mating-Type Interconversion." In *The Molecular Biology of the Yeast* Saccharomyces, *Life Cycle and Inheritance*, ed. J. N. Strathern, E. W. Jones, and J. R. Broach. Cold Spring Harbor, N.Y.: Cold Spring Harbor Laboratory, pp. 181–210.

83. Hicks, J. B., J. N. Strathern, and I. Herskowitz. 1977. "The

Cassette Model of Mating Type Interconversion." In *DNA Insertion Elements, Plasmids and Episomes,* ed. A. I. Bukhari, J. A. Shapiro, and S. L. Adhya. Cold Spring Harbor Laboratory Cold Spring Harbor, N.Y.

84. Strathern, J. N., E. Spatola, C. McGill, and J. B. Hicks. 1980. "The Structure and Organization of the Transposable Mating Type Cassettes in *Saccharomyces.*" *Proc. Nat. Acad. Sci.* 77:2839–2843.

85. Astell, C., L. Ahlstrom-Jonasson, M. Smith, K. Tatchell, K. Nasmyth, and B. D. Hall. 1981. "The Sequence of the DNAs Coding for the Mating-Type Loci of *Saccharomyces cerevisiae.*" *Cell* 27:15–23.

86. Beach, D. H. 1983. "Cell Type-Switching by DNA Transposition in Fission Yeast." *Nature* 305:682–688.

87. Beach, D. H., and A. J. S. Klar. 1984. "Rearrangements of the Transposable Mating-Type Cassettes of Fission Yeast." *EMBO J.* 3:603–610.

88. Nasmyth, K. 1982. "Regulation of Yeast Mating-Type Chromatin Structure by *SIR:* An Action at a Distance Affecting Both Transcription and Transposition." *Cell* 30:567–578.

89. Abraham, J., K. A. Nasmyth, J. N. Strathern, A. J. S. Klar, and J. B. Hicks. 1984. "Regulation of Mating Type Information in Yeast: Negative Control Requiring Sequences Both 5′ and 3′ to the Regulated Region." *J. Mol. Biol.* 176:307–331.

90. Miller, A. M., and K. A. Nasmyth. 1984. "Role of DNA Replication in Repression of Silent Mating Type Loci in Yeast." *Nature* 312:247–250.

91. Brand, A. H., L. Breeden, J. Abraham, R. Sternglanz, and K. Nasmyth. 1985. "Characterization of a "Silencer" in Yeast: A DNA Sequence with Properties Opposite to Those of a Transcriptional Enhancer." *Cell* 41:41–48.

92. Strathern, J. N., A. J. S. Klar, J. B. Hicks, J. A. Abraham, J. M. Ivy, K. A. Nasmyth, and C. McGill. 1982. "Homothallic Switching of Yeast Mating Type Cassettes Is Initiated by a Double-Stranded Cut in the MAT Locus." *Cell* 31:183–192.

93. Kostriken, R., J. N. Strathern, A. J. S. Klar, J. B. Hicks, and F. Heffron. 1983. "A Site-Specific Endonuclease Essential for Mating-Type Switching in *Saccharomyces cerevisiae.*" *Cell* 35:167–174.

94. Klar, A. J. S., and J. N. Strathern. 1984. "Resolution of Recombination Intermediates Generated During Yeast Mating Type Switching." *Nature* 310:744–747.

95. Klar, A. J. S., J. N. Strathern, and J. A. Abraham. 1984. "The Involvement of Double-Strand Chromosomal Breaks for Mating-Type Switching in *Saccharomyces cerevisiae.*" *Cold Spring Harbor Symp. Quant. Biol.* 49:77–88.

96. Rine, J., R. Jensen, D. Hagen, L. Blair, and I. Herskowitz. 1981. "Pattern of Switching and Fate of the Replaced Cassette in Yeast Mating-Type Interconversion." *Cold Spring Harbor Symp. Quant. Biol.* 45:951–960.

97. Klar, A. J. S., J. B. Hicks, and J. N. Strathern. 1982. "Directionality of Yeast Mating-Type Interconversion." *Cell* 28:551–561.

98. Strathern, J. N., J. B. Hicks, and I. Herskowitz. 1980. "Control of Cell Type in Yeast by the Mating Type Locus: The α1–α2 Hypothesis." *J. Mol. Biol.* 147:357–372.

99. Klar, A. J. S., J. N. Strathern, J. R. Broach, and J. B. Hicks. 1981. "Regulation of Transcription in Expressed and Unexpressed Mating Type Cassettes of Yeast." *Nature* 289:239–244.

100. Nasmyth, K. A. 1982. "Molecular Genetics of Yeast Mating Type." *Ann. Rev. Genetics* 16:439–500.

101. Sprague, Jr., G. F., L. C. Blair, and J. Thorner. 1983. "Cell Interactions and Regulation of Cell Type in the Yeast *Saccharomyces cerevisiae.*" *Ann. Rev. Microbiol.* 37:623–660.

102. Johnson, A. D., and I. Herskowitz. 1985. "A Repressor (MATα2 Product) and its Operator Control Expression of a Set of Cell-Type Specific Genes in Yeast." *Cell* 42:237.

103. Miller, A. M., V. L. MacKay, and K. A. Nasmyth. 1985. "Identification and Comparison of Two Sequence Elements That Confer Cell-Type Specific Transcription in Yeast." *Nature* 314:598–602.

104. Thorner, J. 1981. "Pheromonal Regulation of Development in *Saccharomyces cerevisiae.*" In *The Molecular Biology of the Yeast* Saccharomyces, *Life Cycle and Inheritance,* ed. J. N. Strathern, E. W. Jones, and J. R. Broach. Cold Spring Harbor, N.Y.: Cold Spring Harbor Laboratory, p. 143.

105. Kurjan, J., and I. Herskowitz. 1982. "Structure of a Yeast Pheromone Gene (MFα): A Putative α-Factor Precursor Contains Four Tandem Copies of Mature α-Factor." *Cell* 30:933–943.

106. Jenness, D. D., A. C. Burkholder, and L. Hartwell. 1983. "Binding of α-Factor Pheromone to Yeast **a** Cells: Chemical and Genetic Evidence for an α-Factor Receptor." *Cell* 35:521–529.

107. Hagen, D. C., and G. F. Sprague, Jr. 1984. "Induction of the Yeast α-Specific *STE3* Gene by the Peptide Pheromone **a**-Factor." *J. Mol. Biol.* 178:835–852.

108. Wagstaff, J. E., S. Klapholz, and R. E. Esposito. 1982. "Meiosis in Haploid Yeast." *Proc. Nat. Acad. Sci.* 79:2986–2990.

109. Clancy, M. J., B. Buten-Magee, D. J. Straight, A. L. Kennedy, R. M. Partridge, and P. T. Magee. 1983. "Isolation of Genes Expressed Preferentially During Sporulation in the Yeast *Saccharomyces cerevisiae.*" *Proc. Nat. Acad. Sci.* 80:3000–3004.

110. Matsumoto, K., I. Uno, and T. Ishikawa. 1983. "Initiation of Meiosis in Yeast Mutants Defective in Adenylate Cyclase and Cyclic AMP-Dependent Protein Kinase." *Cell* 32:417–423.

111. Percival-Smith, A., and J. Segall. 1984. "Isolation of DNA Sequences Preferentially Expressed During Sporulation in *Saccharomyces cerevisiae.*" *Mol. Cell. Biol.* 4:142–150.

112. Mitchell, A. P., and I. Herskowitz. 1986. "Meiosis and Sporulation Are Activated by Repression of the *RME1* Product in Yeast." *Nature,* 319:738-742.

113. Stevens, B. 1981. "Mitochondrial Structure." In *The Molecular Biology of the Yeast* Saccharomyces, *Life Cycle and Inheritance,* ed. J. N. Strathern, E. W. Jones, and J. R. Broach. Cold Spring Harbor, N.Y.: Cold Spring Harbor Laboratory, pp. 471–504.

114. Slonimski, P., P. Borst, and G. Attardi, eds. 1982. *Mitochondrial Genetics.* Cold Spring Harbor, N.Y.: Cold Spring Harbor Laboratory.

115. Williamson, D. H. 1965. "The Timing of Deoxyribonucleic Acid Synthesis in the Cell Cycle of *Saccharomyces cerevisiae.*" *J. Cell Biol.* 25:517–528.

116. Zinn, A. R., and R. A. Butow. 1985. "Nonreciprocal Exchange Between Alleles of the Yeast Mitochondrial 21S rRNA Gene: Kinetics and the Involvement of a Double-Strand Break." *Cell* 40:887–895.

117. Dujon, B., P. P. Slonimski, and L. Weill. 1974. "Mitochondrial Genetics. IX. A Model for Recombination and Segregation of Mitochondrial Genomes in *Saccharomyces cerevisiae.*" *Genetics* 78:415–437.

118. Bernardi, G. 1976. "Organization and Evolution of the Mitochondrial Genome of Yeast." *J. Mol. Evol.* 9:25–35.

119. Borst, P., and L. A. Grivell. 1978. "The Mitochondrial Genome of Yeast." *Cell* 15:705–723.

120. O'Brien, T. W., and D. E. Matthews. 1976. "Mitochondrial Ribosomes." In *Handbook of Genetics,* vol. 15, ed. R. C. King. New York: Plenum, pp. 535–550.

121. Martin, N., and M. Rabinowitz. 1978. "Mitochondrial tRNAs in Yeast: Identity of Isoaccepting tRNAs." *Biochemistry* 17:1628–1634.

122. Faye, G., N. Dennebouy, C. Kujawa, and C. Jacq. 1979. "Inserted Sequences in the Mitochondrial 23S Ribosomal RNA Gene of the Yeast *Saccharomyces cerevisiae.*" *Mol. Gen. Genetics* 168:101–109.

123. Tabak, H. F., N. B. Hecht, H. Menke, and C. P. Hollenberg. 1979. "The Gene for the Small Ribosomal RNA on Yeast Mitochondrial DNA: Physical Map, Direction of Transcription and Absence of an Intervening Sequence." *Curr. Genetics* 1:33–43.

124. Bonitz, S. G., R. Berlani, G. Coruzzi, M. Li, G. Macino, F. G. Nobrega, M. P. Nobrega, B. E. Thalenfeld, and A. Tzagoloff. 1980. "Codon Recognition Rules in Yeast Mitochondria." *Proc. Nat. Acad. Sci.* 77:3167–3170.

125. Schatz, G. 1979. "How Mitochondria Import Proteins from the Cytoplasm." *FEBS Lett.* 103:203–211.

126. Garfinkel, D. J., J. D. Boeke, and G. R. Fink. 1985. "Ty Element Transposition: Reverse Transcriptase and Virus-like Particles." *Cell* 42:507–517.

127. Wickner, R. B. 1981. "Killer Systems in *Saccharomyces cerevisiae.*" In: J. N. Strathern, E. W. Jones, and J. R. Broach, eds. *The Molecular Biology of the Yeast* Saccharomyces, *Life Cycle and Inheritance,* pp. 415–444.

CHAPTER 19

Recombinant DNA at Work

Only through the employment of recombinant DNA procedures have the molecular features of eucaryotic genomes become effectively open to experimental analysis. The more than 14,000 kilobase pairs of DNA found within even the simplest of eucaryotic genomes intimidated the most motivated of molecular geneticists as late as the early 1970s. And as we mentioned in earlier chapters, the 4,000 kilobase-pair genome of *E. coli* would have frightened away even the most determined adventurer had it not been for its viruses, which were powerful tools for genetic analysis.

From the very start, the key attraction of *E. coli*, as opposed to any small eucaryotic microorganism, was the great abundance of phages that multiplied within it and which happily turned out to include those with genomes of RNA as well as of DNA. Moreover, it turned out not to be too hard to devise methods for isolating many of these viral chromosomes as intact molecules whose sizes could be exactly determined. Thus, no radical surprises were expected, and as we have related earlier, none were found when the chromosome of several *E. coli* phages and then the chromosome of *E. coli* itself were exposed to the ultimate examination of DNA sequencing.

Because of their lack of phagelike genetic parasites, yeasts were relatively ignored until about 10 years ago; until then, only a few molecular biologists, upon deciding to move up from *E. coli* to eucaryotic cells, chose to work with either *S. cerevisiae* or *S. pombe*. Most molecular biologists settled upon the cells of higher vertebrates, within which a large variety of different classes of viruses were known to multiply. By concentrating on these vertebrate viruses, they hoped to reveal the essential tricks that govern the expression of all eucaryotic genomes. We now realize that such a restriction to viral genomes would have provided only limited knowledge had not the vast powers of recombinant DNA appeared on the scene.

When in 1973 the first practical procedure became available for making recombinant DNA molecules using *E. coli* vectors, the then only slowly expanding world of yeast genetics immediately saw the multiple advantages that would accrue if similar systems could be developed for cloning molecules within yeast cells. By 1977, a practical method for introducing genetically active DNA into yeast cells had been worked out, and the extraordinary effort to understand the structure and functioning of the yeast genome had begun. The procedures now available for genetically engineering yeast cells are the most powerful so far developed for any eucaryotic organism. In discussing these recombinant DNA methods, we shall see that some remain restricted to work with yeasts. Others, often those developed

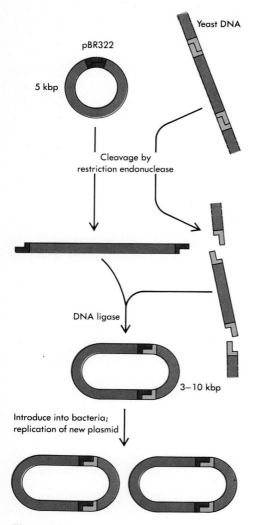

Figure 19-1
Cloning a yeast restriction fragment into plasmid pBR322. Fragments of at least 5 kilobase pairs can be stably introduced and are retained throughout plasmid replication. The boxes indicate the restriction endonuclease recognition sites.

for use with other life forms, are broadly applicable to all known organisms.

Libraries of Yeast Genes Within *E. coli* Vectors[1,2]

Recombinant DNA research often starts with the preparation of a **gene library** containing all the DNA of a given organism inserted as discrete restriction enzyme–generated fragments into many thousands of different **plasmid**, λ, or **cosmid vectors**. The most useful libraries are those containing genome insertions of the sizes needed to encompass complete genes. The construction of a gene library depends on the use of restriction enzymes like *Eco*RI that do not make too frequent cuts within DNA (since they recognize only groups of six consecutive base pairs). Even then, it is usually necessary to use incompletely digested DNA when the desired genes are suspected of being extremely large (the situation with many of the genes of vertebrates; Chapter 20).

Most gene libraries are made using *E. coli* plasmids or phage components as vectors. Not only were they the first vectors to be developed, but they are constantly being improved. Moreover, they are particularly useful when the genome inserts come from other organisms. Once a yeast DNA fragment is cloned and grown as part of an *E. coli* vector, it becomes isolated from all other yeast sequences, and much less care is needed to purify it than if it were present on a yeast plasmid in a yeast cell.

Genomic fragments of at least 5 kilobase pairs are very stably reproduced when inserted into plasmids like pBR322 (Figure 19-1). In contrast, larger inserts, which slow down the multiplication rates of their host plasmids, tend to be replaced by smaller derivative plasmids that have lost sections of their DNA inserts. Larger fragments, however, can be stably grown within specially tailored phage λ systems that require DNA inserts of some 22 kilobase pairs in order to be packaged within mature particles. Even larger fragments can be stably inserted into **cosmid vectors** (Chapter 8) that can hold some 40 to 50 kilobase pairs of foreign DNA.

Because the yeast genome is approximately 14,000 base pairs, fragments representing the entire yeast genome can thus be contained in either 3000 pBR322 plasmids, 1000 λ particles, or 300 cosmids. The very much larger sizes of higher plant and animal genomes means that the complete libraries of their genomes are also larger. A λ library requires some 250,000 particles to contain all the human genome. Constructing a genomic library of even this size, however, is now essentially a trivial problem.

Ingenuity, while not required in the making of a gene library, is still often needed for screening libraries for desired genes. Such searches are usually routine only for those genes in which mutations have been found that can be complemented by the introduction of the appropriate library member (DNA sequence) into the mutant cell (see Chapter 8's discussion of the complementation test). In the absence of such mutants, the usual way to pinpoint the appropriate library member is by means of its DNA sequence. A number of genes, for example, have been detected by their ability to hybridize to specific cellular mRNA molecules, known to promote the synthesis of a specific protein product, such as a histone (Figure 19-2). Such techniques are generally very laborious, and on many occasions success in employing recombinant DNA requires an elegant trick (rather than brute force) to pick out the desired gene.

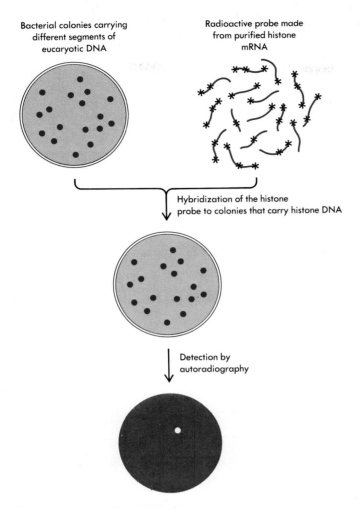

Bacterial colonies carrying
different segments of
eucaryotic DNA

Radioactive probe made
from purified histone
mRNA

Hybridization of the histone
probe to colonies that carry histone DNA

Detection by
autoradiography

Figure 19-2
Identification of a specific DNA library
member by its ability to hybridize to
specific mRNA molecules whose protein product can be identified.

The Unexpected Expression
of Many Yeast Genes in *E. coli*[3, 4]

It initially seemed obvious that intact eucaryotic genes would not
function properly, if at all, in procaryotic *(E. coli)* cells. There was no
reason why the promoters recognized by the three types of eucaryotic
RNA polymerases would be recognized by the single bacterial *(E. coli)*
RNA polymerase. Nor did any reason exist for the eucaryotic RNA
termination sites to be recognized by the proteins that bring about
RNA chain termination in *E. coli.* Even if synthesized, eucaryotic
mRNA molecules may not possess the correct ribosome binding site
sequences to allow them to associate with bacterial ribosomes.

For these reasons, much surprise was generated when the *E. coli*
mutant strain *leuB*, which has lost the ability to make the enzyme
β-isopropylate dehydrogenase and so cannot make leucine, was observed to lose its leucine requirement after the introduction of a plasmid carrying the yeast gene *(LEU2)* that codes for the homologous
enzyme in yeast. Apparently, the yeast genome contains many sequences that mimic functional *E. coli* promoters, with one of them
properly situated upstream of the *LEU2* gene. These sequences also
occur in front of many other yeast genes, and it is possible that as
many as 30 percent of the yeast genes are transcribed into RNA when
they are inserted into the *E. coli* chromosome. For such RNAs to function as messengers, they must also possess ribosome binding-site like
sequences immediately adjacent to their AUG translational initiation

Figure 19-3
Integrative transformation of yeast by homologous recombination between a recombinant plasmid carrying a yeast gene (in this case *LEU2⁺*) and a yeast chromosome that has a mutation in the *LEU2* gene (*LEU2⁻*) (arrow indicates position of crossover). Single crossover events are most common, although double crossovers occasionally create transformants. [After D. Botstein and R. W. Davies.]

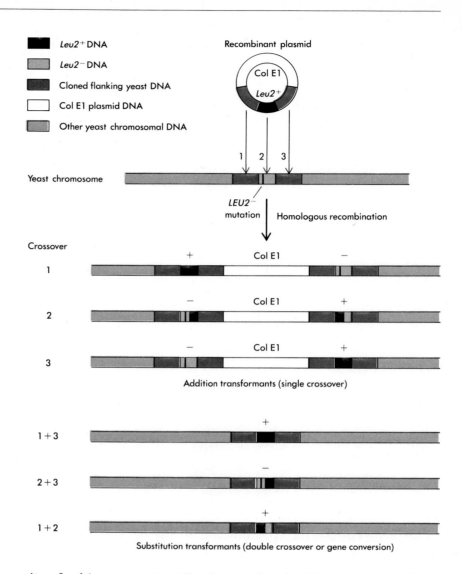

sites. In this way, mutant *E. coli* genes involved in tryptophan, histidine, arginine, and uracil synthesis have been complemented by yeast genes encoding equivalent proteins. The functioning of such a complementary yeast gene in *E. coli* can quickly lead to its isolation (cloning) and amplification. By so purifying the newly resident *E. coli* plasmid, complete copies of the yeast genes are automatically generated.

The Return of Cloned Genes to Yeast Using DNA Transformation[5, 6]

Only limited information can be obtained by cloning a eucaryotic gene within *E. coli*. For more complete answers as to how their functioning is controlled, eucaryotic genes must be reintroduced into the cells where they normally function. Free DNA molecules, however, are not taken up by normal yeast cells; their entry requires exposure of potential host cells to a set of enzymes that collectively degrade away much of the cell wall to produce the more permeable **spheroplasts**. DNA is then added in the presence of calcium ions (Ca^{2+}) and polyethylene glycol, which somehow makes the plasma membrane permeable and further encourages the passage of DNA through it.

The first demonstration that externally added DNA could be introduced into yeast cells in a genetically active form utilized the yeast gene $LEU2^+$. Because it had been cloned away from the remainder of the yeast genome on a bacterial plasmid (ColE1), very high copy numbers of the gene could be added to mutant $LEU2^-$ haploid cells. All the resulting transformed $LEU2^+$ cells that subsequently grew represented the descendants of primary transformants in which the added DNA became incorporated into the host cell chromosome carrying the $LEU2^-$ gene by recombination enzymes that mediate homologous crossing-over events. On no occasion was the added $LEU2^+$ gene inserted at a nonhomologous region.

The most common event that inserts the $LEU2^+$ gene is a single reciprocal crossover event involving homologous yeast sequences on the plasmid and respective chromosomes (Figure 19-3). Such events lead to the placement of the entire plasmid with its $LEU2^+$ gene insert into the homologous chromosome (**integrative transformation**). No genetic information is lost from either the chromosome or the plasmid, and the resulting chromosome contains two copies of the yeast gene: one usually the original unmodified $LEU2^-$ gene, the other the wild-type $LEU2^+$ gene that came along with the plasmid DNA. By so acquiring an intact copy of the wild-type gene, the cells acquire the ability to make leucine and can easily be selected because they no longer have a leucine growth factor requirement. Less frequently, two crossovers occur in the homologous regions leading to substitution events in which the mutant $LEU2^-$ gene is replaced by DNA sequences originating from the added plasmid. When the substituted sequences overlap those sequences causing loss of function, the ability to make leucine is restored.

After these original positive results, a number of other genes were successfully reintroduced into yeast using genes available in cloned form on bacterial plasmids. In all cases, the transformation frequencies are low, with the most common of these rare events being the insertion of the entire donor plasmid sequence into the appropriate homologous chromosome.

Increasing Transformation Frequencies Using the 2μ DNA Plasmid[6, 7, 8, 9]

Much higher transformation frequencies (one hundred to a thousand times higher) result when genes like $LEU2$ are introduced into yeast cells on plasmids containing replication origins that let them multiply autonomously in yeast. The added genes (e.g., $LEU2$) do not have to be inserted into host chromosomes by homologous crossing-over events in order for their replication to be ensured once every cell generation. Thus, the very low transformation rates observed when genes are added on nonreplicating plasmids does not reflect only inefficient uptake of added DNA. Instead, another limiting factor is the naturally very low frequency of homologous crossing over in mitotically growing yeast cells.

The first sequences used to make a functional yeast vector were those of the so-called **2μ plasmid,** the only plasmid that has so far been found to occur in yeast (Figure 19-4). While most strains of *Saccharomyces cerevisiae* contain this circular, 6300-base-pair plasmid, its function has not yet been discovered. No consistent difference in properties has been observed in those relatively rare strains that lack the 2μ plasmid. When present, the plasmids number some 30 copies

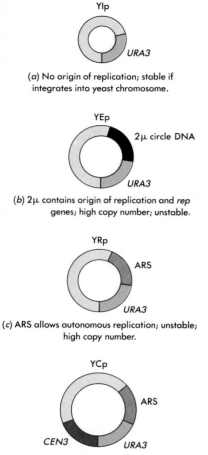

(a) No origin of replication; stable if integrates into yeast chromosome.

(b) 2μ contains origin of replication and *rep* genes; high copy number; unstable.

(c) ARS allows autonomous replication; unstable; high copy number.

(d) Contains centromere; stable; single copy.

Figure 19-4
The four different plasmids found in yeast. All contain a bacterial origin of replication and selectable marker such as Amp^R for selection of bacteria.

per haploid cell and are located in the nucleus, where their replication is largely accomplished by the same cellular enzymes used to replicate the linear yeast chromosomes. For this reason, they duplicate only during the S phase, the time of replication for the cellular chromosomes. When a single 2μ circle is introduced into a plasmid-free yeast, however, its replication, as well as that of its immediate descendants, occurs more than once every cell cycle until the normal complement of some 30 copies per cell is reached. Allowing this override of the normal cell cycle control of replication are two *Rep* proteins coded by two specific genes found on the 2μ DNA. In the absence of functional *Rep* genes, newly acquired 2μ circles can be present at a maximum of 1 to 2 copies per cell. And because they are not regularly segregated on mitotic (or meiotic) spindles, they are randomly lost from many daughter cells. Only when they carry a gene (e.g., *LEU2$^+$*) whose presence is needed for the growth of their host *(LEU2$^-$)* cell is their stability within a given strain ensured. This is done by growing the transformed yeast strain in minimal medium lacking leucine.

Plasmids carrying 2μ DNA sequences are called **YEp** or **yeast episomal plasmids,** while those which transform by integration (crossing over) into a yeast chromosome are called **YIp** or **yeast integrative plasmids** (see Figure 19-4).

The Construction of New Yeast Plasmids Using ARS Sequences[10]

Another source of yeast sequences for making new plasmids that multiply in yeast cells are the many genomic segments at which DNA replication is ordinarily initiated along the yeast chromosome. Each of the 200 to 400 such regions have the potential to act as the origin of replication for a circular plasmid. Because these regions are present in so many copies per genome, they frequently lie very close to genes we wish to study. In fact, the first **autonomously replicating sequence (ARS)** was discovered when the bacterial plasmids into which the *TRP1* gene was inserted displayed the capacity to replicate in yeast (see Figure 19-4). These yeast replicating plasmids **(YRp),** however, are not stably maintained in transformed cells. For unknown reasons, YRp plasmids tend to remain associated with the mother cell and are not efficiently distributed to the daughter cell. The replication of YRp plasmids, like that of any chromosomal DNA segment, occurs but once every cell generation. Rarely, stably transformed cells emerge through insertion of the intact YRp into one of the yeast chromosomes through crossing over within autonomously replicating sequences.

Further Stabilizing Plasmids Using *CEN* Sequences[11, 12, 13]

Highly stable yeast plasmid vectors can be made by including, in addition to autonomously replicating sequences, certain yeast DNA sequences that function as centromeres. These sequences are necessary for the attachment of the chromosomes to the microtubules of the mitotic spindle, presumably through some specific connector proteins that join the microtubules to the centromere (see Figure 19-4). Such *CEN*-containing plasmids (YCp) function as true chromosomes and segregate accurately during both mitosis and meiosis. *CEN* sequences can also be used to construct linear chromosomes if circularization is prevented by the addition of the telomeric sequences that terminate chromosomes.

Shuttling DNA Between Yeasts and Bacteria

All versatile vectors for use in yeast cells contain origins of replication active in both yeast and *E. coli* as well as one or more markers (e.g., for antibiotic resistance) that can be used to select those *E. coli* cells into which the vectors have entered. The vectors can thus be amplified (purified) in *E. coli* before being used to transform yeast cells. Furthermore, such **shuttle vectors** allow genetic sequences to be routinely transferred back and forth between yeast and *E. coli* (Figure 19-5).

YEp, YRp, and YCp vectors now provide the ideal tools to isolate yeast genes by virtue of their ability to complement mutations in yeast. Ordinarily, there is no longer any reason to clone a yeast gene through its ability to function in *E. coli*. Instead, fragments of the yeast genome are used to make libraries in yeast vectors that carry a marker (such as *LEU2*); the marker can then be used to select for the vector's transfer into a corresponding yeast mutant. The resulting library can then be amplified in *E. coli* or directly used to transform yeast strains that have defective genes of interest. Once the appropriate transformed yeast cell has been found, its descendants can be used directly as a source of the appropriate plasmid or as a source of DNA for amplification and subsequent isolation from *E. coli* cells.

Double-Stranded Cuts Promote High-Level Directed Integration into the Yeast Genome[14]

Initially, it seemed that the best way to optimize the insertion of yeast genes into their homologous chromosomes was to add them to appropriate cells in the form of intact, circular, integrative plasmids. In this way, the possibility of unwanted digestion of the free ends of linear DNA molecules by cellular nucleases would be avoided. Then, most unexpectedly, it was discovered that the frequency of crossing over–mediated integrative transformation into a specific chromosomal location was enhanced by a single double-stranded cut in the plasmid DNA. In such cases, integration occurs at frequencies some hundredfold higher and always into the chromosomal site homologous to the cut site (Figure 19-6). The vectors of choice for many transformation experiments have thus become linearized, yeast integrative plasmids, which only transform by homologous crossing over.

The observation that a double-stranded cut in DNA promotes homologous crossing over initiated a profound rethinking of the way crossing over occurs in eucaryotic cells. The free ends so made are highly recombinogenic, invading homologous genetic segments to form the heteroduplex intermediates. These intermediates are then resolved when appropriate nucleases cut single-stranded DNA to yield recombined DNA molecules.

Gene Retrievals Brought About by Gapped Shuttle Vectors[15]

Specific genes can be retrieved from their normal chromosomal locations through homologous crossing-over events that transfer them to specially constructed classes of **retriever vectors.** The most versatile

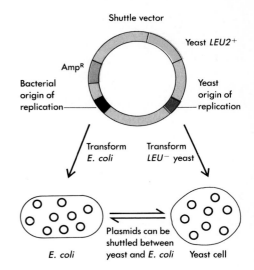

Figure 19-5
The possession of both bacterial and yeast origins of replication allows shuttle vectors to transform both bacteria and yeast and be transferred back and forth between them.

Figure 19-6
When double-stranded cuts are made in yeast integrating plasmids (YIp), integration into the yeast chromosomes always occurs at the homologous chromosomal site and at frequencies 100 times greater than with circular YIp.

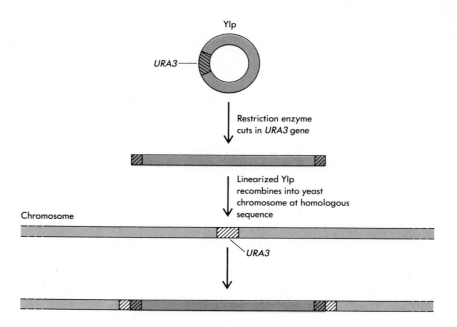

retriever vectors contain origins of replication that let them multiply in *E. coli* as well as yeast, two selectable markers that favor their selective growth in either yeast or *E. coli*, and a homolog to the gene to be recovered from the yeast chromosome. Once such a retriever vector has been constructed, the gene in question is removed by appropriate restriction enzyme cuts, leaving behind the flanking yeast sequences on its two sides. The now "gapped vector" is added to the cell possessing the mutant gene to be recovered. Once it enters the nucleus, its two highly recombinogenic ends invade the gene's flanking segments and so position their 3' ends to begin copying homologous sequences and thereby replace the lost (gapped) sequences with those of the chromosomal gene (Figure 19-7). These repair events generate two points of strand exchange (crossover) between the plasmid and the chromosome, which subsequently will be resolved by nuclease cuts to yield an intact plasmid as well as a complete chromosome. Only those plasmids so repaired are able to multiply and to complement their host yeast cell's deficiency with their selectable markers (e.g., $LEU2^+$). The stage is thus set for the mutant chromosomal gene to be shuttled to *E. coli* for amplification and sequencing.

Retriever vectors are a powerful weapon in the armory of the yeast geneticist because they greatly simplify the recovery of mutations produced *in vivo* onto plasmids so that they can be subjected to molecular scrutiny—e.g., sequencing. In particular, they provide a means to establish the precise structures of the mating-type genes, whose unusual behavior implies the existence of novel control elements (Chapter 18).

The Test-Tube Production of Mutated Control Elements[16–24]

Once a gene has been cloned and can be returned in a functional form to its normal cellular environment, the DNA sequences that control its expression can be directly worked out using a broad array of in vitro mutagenesis techniques. Until several years ago, the only way

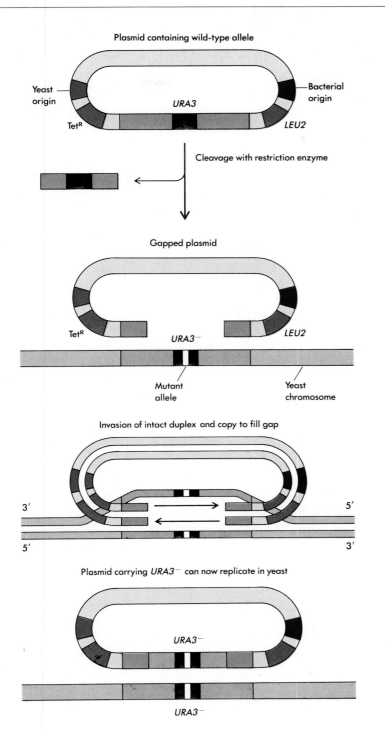

Plasmid containing wild-type allele

Yeast origin

Bacterial origin

URA3

Tet^R

LEU2

Cleavage with restriction enzyme

Gapped plasmid

Tet^R

URA3^-

LEU2

Mutant allele

Yeast chromosome

Invasion of intact duplex and copy to fill gap

Plasmid carrying *URA3^-* can now replicate in yeast

URA3^-

URA3^-

Figure 19-7
A retriever vector. A specific restriction enzyme is used to splice out the desired gene. This "gapped vector" is added to the cell possessing the mutant gene desired. Once inside the yeast cell, it invades the gene's flanking segments and copies the sequence of the gene homologous to the one it lost.

to identify control sequences was to find mutations that affected their functioning. This approach, as indicated earlier (Chapters 16 and 17), has been most successful with *E. coli* and its phages. Naturally occurring mutations have also led to most of our current understanding of gene expression in yeast, and the discovery of additional key mutants will undoubtedly yield much new information. On the other hand, it is now becoming so easy to do *in vitro* mutagenesis procedures that we must anticipate a growing use of such procedures, even with organisms whose genetic analysis is relatively quick and inexpensive. A precise definition of the blocks of base pairs that make up control

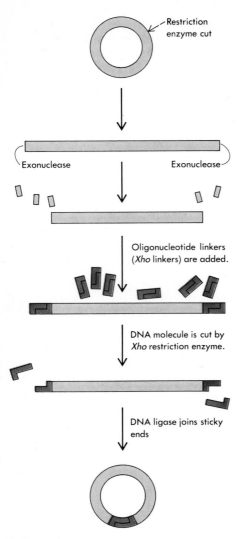

Figure 19-8

The creation of short deletions by the use of oligonucleotide linkers. A plasmid is first cut with a restriction enzyme at one site and then treated with a nuclease to create a slightly shorter linear DNA. Oligonucleotide linkers are added to the ends, and the plasmid is cut by the appropriate restriction enzyme to yield sticky ends that are then joined by DNA ligase.

elements will come from an almost routine creation of deletions, insertions, and base-pair replacements at predetermined locations or within a defined region in a given DNA.

Of all the changes, long deletions between two restriction enzyme sites are the easiest to make. By light digestion of a plasmid with a restriction enzyme, large numbers of partially digested molecules lacking just one fragment can be isolated. Their subsequent circularization yields smaller plasmids that frequently have biological activity. Much shorter deletions can be made by first cutting a plasmid at one restriction site to generate linear DNAs that can then be treated with nucleases to remove small groups of bases at each end. Ligating these slightly shorter linear fragments back into circular forms is often best accomplished by adding specific oligonucleotide linkers to the ends (see Figure 9-44). When subsequently cut by the appropriate restriction enzyme, these yield sticky ends that routinely can be joined together by DNA ligase (Figure 19-8). The making of precise insertion mutations also benefits from the use of such oligonucleotide linkers. These linkers can be added at cut sites without using nuclease to yield insertions of known sizes.

Single-base-pair changes at cytosine residues can be created using chemicals such as bisulfite, which deaminates cytosine to uracil, which base-pairs like thymine. Such chemical reactions thereby convert GC base pairs into AT base pairs. These reagents, however, only work on single-stranded DNA, where the base-pairing groups on the bases are freely accessible to the chemical. Thus, the ends of a DNA molecule cut by a restriction enzyme must first be exposed to exonucleases, which degrade from either the 3' end or the 5' end. Desired single-base-pair replacements can also be made by allowing DNA repair or DNA synthesis to occur either in the presence of base analogs that frequently mispair or in the total absence of one of the four nucleotides, which leads to incorporation of wrong bases at predetermined sites (Figure 19-9).

Now, however, these essentially hit-and-miss procedures for making single-base-pair substitutions are rapidly being replaced by much more powerful procedures that use 15- to 20-base-long oligonucleotide chains to make all possible classes of base-pair changes at any determined site along a DNA molecule. Oligonucleotides of this length form very stable DNA-DNA double helices with complementary segments of DNA. Less stable but nonetheless firm unions can also be made with DNA sequences that contain one to two mismatches. In the presence of DNA polymerase and the four nucleotide precursors of DNA, these slightly mismatched short DNA segments can function as primers for DNA extension when added to the circular, single-stranded chains of phages like M13. The resulting complete double helices will be perfectly complementary except for mismatched bases at the primer site. After their entry into *E. coli*, DNA replication will generate two types of daughter DNA strands: wild-type strands and mutant strands, the latter containing the changes imposed by the mismatched primer (Figure 19-10). The strands can easily be distinguished by the way they bind the corresponding oligonucleotides. The wild-type daughter DNAs will bind more strongly to wild-type oligonucleotide, while the short DNA segment that primed the mutant sequences will bind best to the mutated progeny chains.

Until recently, such oligonucleotide-directed mutagenesis was restricted to the laboratories of organic chemists capable of synthesizing the required probes. Now, however, such oligonucleotides are easily

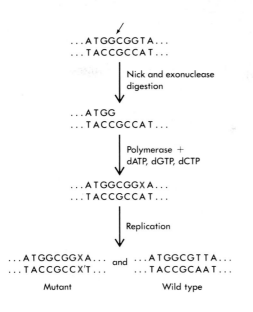

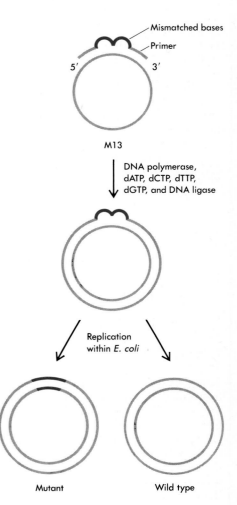

Figure 19-9
Production of single-base-pair replacements by creating a single-stranded section in the DNA and allowing repair to occur in the absence of one of the nucleotide bases. An incorrectly inserted base (in this case either A, G, or C) is represented by X, and X' represents its complement.

synthesized by the nonprofessional chemist using commercially available "DNA synthesizers" to polymerize the blocked nucleotide precursors that are also widely available to all who can afford their increasingly modest prices.

Using Integrative Transformation to Replace Chromosomal Genes with Mutated Derivatives[25, 26, 27]

Not only can inactive mutant genes be replaced by active wild-type genes, but integrative transformation provides a powerful way to replace active genes with inactive mutant genes. This genetic tool was first employed using a YIp containing both a deleted, nonfunctional *HIS3⁻* gene and an active *URA3* gene together with their immediately flanking sequences. Transformation of yeast with this plasmid results in 50 percent of the plasmids being inserted at the *HIS3* locus, while the other 50 percent of the plasmids insert into the *URA3* gene, which is located on a different chromosome. Insertions adjacent to the *HIS3* locus result in both the wild-type and nonfunctional forms of the *HIS3* locus in tandem on the yeast chromosome (Figure 19-11). Subsequent growth of these transformants leads to the occasional crossover between mutant and wild-type genes. Such crossovers remove the intervening vector and *URA3⁺* sequences, events that can be detected in the cells by the resumption of the *URA3⁻* phenotype. Of these *URA3⁻* revertants, some will retain the *HIS3⁺* gene, while others will possess the mutant form (see Figure 19-11). Using this general procedure, we can insert any mutant allele into yeast cells, as long as the gene either is not essential or (like *HIS3*) has the possibility of its product being supplied externally.

Integrative transformation can be used in an even simpler procedure to totally inactivate any chromosomally located gene whose active form has already been cloned on a plasmid. The trick is to transform using a segment of the gene that lacks both ends of the gene; in other words, an internal fragment. When crossing over occurs be-

Figure 19-10
Creation of base-pair mutants by using oligonucleotide DNA primers containing mismatched base pairs. These primers are added together with the four nucleotide precursors, DNA polymerase, and DNA ligase to a single-stranded phage such as M13 to create double-stranded M13 molecules. These are then introduced into *E. coli*, where replication of the double-stranded DNA creates one mutant and one wild-type daughter molecule.

Figure 19-11
Replacement of the active *HIS3* gene
with the mutant *HIS3⁻* gene by integrative transformation. A YIp containing
mutant *HIS* and a functional *URA* gene
may transform by separating at either
the *HIS* or *URA* locus on the yeast
chromosome. (a) Transformation near
the *HIS* locus results in either transformed or wild-type yeast chromosomes, depending on where the plasmid cuts out. (b) Integration of the
plasmid near the chromosomal *HIS3⁺*
locus leads to production of two *HIS3*
genes in tandem. Excision of the plasmid (with loss of the plasmid *URA3*
marker) can occur by the reverse
process—recombination between the
two copies of the *HIS3* gene. If recombination occurs in interval 1, then the
HIS3⁻ mutation originally on the plasmid is left behind in the genome. If
recombination occurs in interval 2, then
the functional *HIS3* gene is left in the
genome.

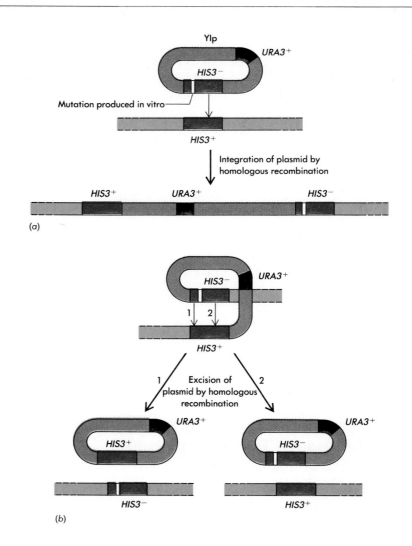

(a)

(b)

Figure 19-12
Using integrative transformation to determine whether a given chromosomal
gene (in this case the yeast *actin* gene)
is essential. A YIp was created in which
the two ends of the yeast *actin* gene
were removed and which contained the
URA3⁺ gene. Integration of the plasmid
by homologous recombination between
the *actin* fragment of the plasmid and
the chromosomal *actin* gene leads to a
URA3⁻ transformant. The integration
event leads to production of two versions of the *actin* gene, both of them
defective. Yeast lacking a functional
actin gene do not survive. Consequently, the type of insertional inactivation described here was done into a
diploid strain, so that one active *actin*
gene was present. Half of the haploid
meiotic products—those carrying the
inactivated *actin* gene—were inviable.
Thus, the *actin* gene is shown to be essential in yeast.

tween the fragment and the chromosome, both of the resulting duplicated gene copies will be defective, one because it lacks the beginning
of the gene, the other because it lacks the end. This procedure was
first used to show that the gene coding for yeast actin is essential. In

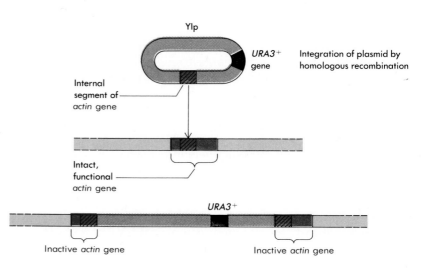

these experiments, a diploid *URA3*⁻ cell was transformed to the *URA3*⁺ phenotype by a plasmid carrying both a *URA3*⁺ gene and the incomplete internal fragment of the actin gene. When the 50 percent of the *URA3*⁺ cells in which integration had occurred within the actin gene (as opposed to within the *URA3* gene) sporulated, only *URA3*⁻ spores were produced. No *URA3*⁺-containing chromosomes survived in the haploid state, since they lacked any functioning actin genes (Figure 19-12). This procedure provides a powerful way to test whether genes of still unknown function carry out essential roles.

Measuring the Activity of Fused Genes[28, 29]

Gene expression is best studied when the genes concerned code for products that can be assayed quickly and accurately with minimal expense. That the lactose operon of *E. coli* is still the best understood of all genetic segments is no accident. There is a simple colorimetric test for its principal product, the enzyme β-galactosidase, that unambiguously detects the enzyme's presence either in permeabilized cells or in cell-free extracts. Were β-galactosidase with its 1021 amino acids an unstable, hard-to-measure protein, we would never have concentrated on its underlying gene.

Yeasts, however, do not metabolize lactose and so normally do not produce any β-galactosidase-like enzyme. However, a close derivative of β-galactosidase can be made in yeasts using genetic engineering tricks to fuse the 5' upstream control elements and short amino-terminal sequences of selected yeast genes to the very long DNA fragments coding the carboxyl-terminal 990 amino acids of β-galactosidase (Figure 19-13). Since the first 30 amino acids at the amino-terminal end of β-galactosidase are not required for enzyme activity, the resulting fusion proteins have full β-galactosidase activity and effective stability. Such fused genomes allow the activity of mutated yeast 5' control elements to be monitored through the amount of β-galactosidase activity that is synthesized, rather than by the much more laborious methods needed, say, to quantitate the amount of the uracil biosynthetic enzyme coded by the *URA3* gene or the amount of cytochrome c1, the product of the *CYC1* gene.

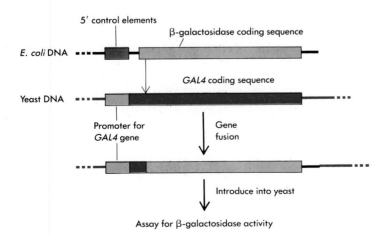

Figure 19-13
Assaying for unstable, hard-to-measure proteins using gene fusion techniques. Here the *GAL4* promoter is assayed by fusing the promoter to segments of the β-galactosidase gene that code for the carboxy-terminal 990 base pairs of enzyme. A simple colorimetric test is available to measure the amount of β-galactosidase activity, and since the genes are fused, this is also a measure of the expression of the *GAL4* gene.

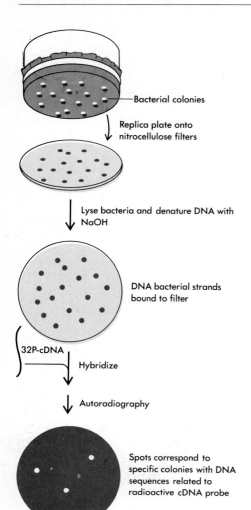

Figure 19-14
Detection of related DNA sequences by hybridization of single-stranded DNA on nitrocellulose filters. Bacterial colonies are replica plated onto nitrocellulose filters and treated with sodium hydroxide (NaOH) to lyse the bacteria and denature the DNA, which remains attached to the filter after baking. The filter is then exposed to specific radioactively labeled cDNA probes. Colonies with DNA sequences related to the cDNA probe hybridize to the probe and can be detected by their radioactivity.

Detecting Related DNA Sequences Through Hybridization Procedures[30, 31, 32]

Once a gene has been cloned, its DNA can be used to make a restriction map and to determine whether similar DNA sequences exist elsewhere. The degree of relatedness between two DNA segments is best approached by measuring the ability of their component single strands to hybridize under appropriate conditions to form complementary double helices. A given cloned gene (restriction fragment) actually becomes a "probe" to search out other DNA segments of identical or nearly identical sequences. Now there are simple procedures in which such probes are used for the mass screening of many thousands of plaques or bacterial colonies. Such procedures always involve the production of replicas of the plaques or bacterial colonies on nitrocellulose paper. The DNA within these replicas is first fixed on the paper and then exposed to radioactively labeled DNA probes to pinpoint those plaques or colonies that contain related DNA sequences (Figure 19-14).

Probes can also be made of RNA. The addition of labeled rRNA or tRNA chains to appropriate library replicas quickly reveals those members carrying the respective rDNA or tDNA genes. Messenger RNA molecules also can be used as probes when they are available in partially purified form. But since most cells contain more than a thousand different mRNA molecules, it is seldom possible to obtain worthwhile probes from the ordinary cell. Only when a single cell devotes much of its protein-synthesizing capacity to making single proteins (e.g., hemoglobin and immunoglobin) can the respective mRNA probes pick out their corresponding genes from a genomic library. To obtain pure mRNA probes, the mRNA molecules themselves must be cloned through their reverse transcription into complementary DNA (cDNA) and subsequent insertion into appropriate vectors.

Analyzing Related Genes Through "Southern" Blotting[33]

Given the availability of a suitable mRNA, cDNA, or genomic probe, important preliminary data on the structure of related genes are generated through a technique developed in Edinburgh by the molecular biologist E. M. Southern and which is now called **Southern blotting.** It starts with the digestion of a DNA population by one or several restriction enzymes. The resulting fragments, after separation on gels according to their size, are transferred to nitrocellulose sheets so that an exact replica of the DNA fragments in the gel is present on the nitrocellulose filter. A specific radioactive probe is then applied to the filter under hybridizing conditions. Subsequent autoradiography of the filter results in a specific pattern of bands corresponding to the discrete restriction fragments that are complementary to the DNA or RNA probe (Figure 19-15).

In conjunction with recombinant DNA technology, Southern blots can easily provide a physical map of restriction sites within a gene in its normal chromosomal location, as well as provide information on the number of copies of the gene in the genome, and the degree of similarity of the gene under study to other known homologs.

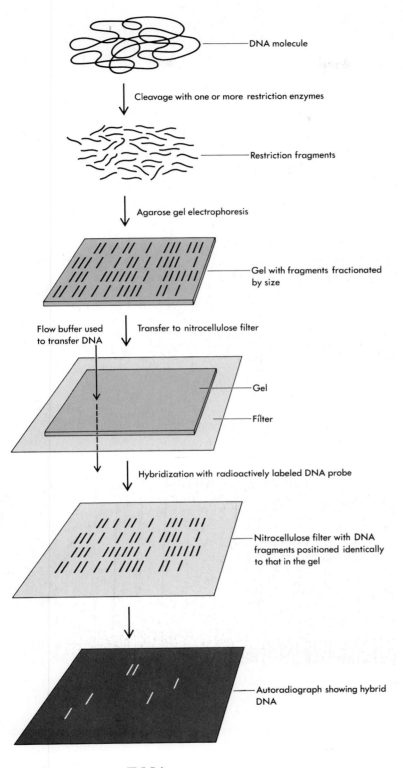

DNA molecule

Cleavage with one or more restriction enzymes

Restriction fragments

Agarose gel electrophoresis

Gel with fragments fractionated by size

Flow buffer used to transfer DNA

Transfer to nitrocellulose filter

Gel

Filter

Hybridization with radioactively labeled DNA probe

Nitrocellulose filter with DNA fragments positioned identically to that in the gel

Autoradiograph showing hybrid DNA

Figure 19-15
Southern blotting method of analyzing DNA segments that share homology with a nucleic acid probe.

Complementary DNA
Cloning Using Poly A Tails[34]

There are many important uses for the gene libraries made by cloning the population of mRNA molecules present within specialized cells. In particular, they afford a way to isolate those gene segments that

Figure 19-16
Poly A tails on mRNA may be used to form oligo dT primers for reverse transcription to create double-stranded DNA. This DNA can then be inserted into a plasmid, which is then inserted into *E. coli*. In this way, a desired gene or gene segment can be isolated and amplified.

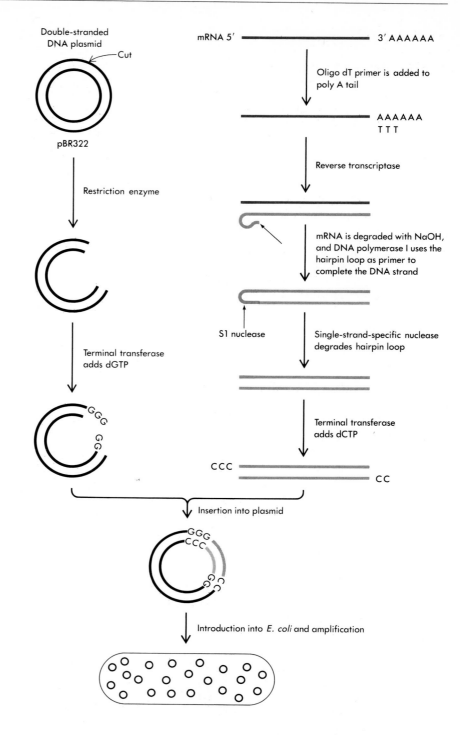

code for exons away from their noncoding, intron components. In this procedure, the information in mRNA is copied into cDNA (complementary DNA) replicas using the retrovirus enzyme *reverse transcriptase*, which normally functions to convert infecting retroviral RNA chains into DNA copies (Chapter 24). Reverse transcriptase, like all other DNA polymerases, can initiate DNA synthesis only by adding onto primers. Fortunately, it is easy to prime cDNA synthesis by utilizing the stretches of poly A (called a **poly A tail**) at the 3' ends of most mRNA molecules. Added oligo dT binds to these poly A tails and serves as the necessary primer. Synthesis of a single-stranded complementary DNA chain can run to the 5' end of its template, where it often makes a hairpin loop turn, which serves as a primer for the synthesis of a complementary DNA strand. The loop of the hairpin can subsequently be dissected away by a nuclease specific for single-stranded DNA.

The resulting double-stranded cDNA can then be inserted into a plasmid like pBR322 either by adding complementary tails (e.g., poly G and poly C) using the enzyme terminal transferase or by attaching chemically synthesized restriction enzyme sites (linkers) using the enzyme DNA ligase (Figure 19-16). The cDNA-containing plasmids so constructed can then be introduced into *E. coli* and amplified.

Complementary DNA libraries will of necessity be heavily biased toward members representing the more abundant classes of mRNA found in given cells. It should not be taken for granted, however, that they must always contain the member we expect. The reverse transcription of long mRNA molecules into complete cDNA copies frequently aborts, producing short fragments representing only part of the gene. The making of a highly representative large-copy-number **cDNA library** is thus a much harder task than the construction of its equivalent **genomic library**.

Identifying the Products of cDNA Clones[35, 36, 37]

One of the most direct ways to screen a cDNA library is to use individual members as traps for the collection of specific mRNA molecules that in turn are added to *in vitro* systems for protein synthesis. With luck, the mRNA class selected out by its hybridization to the denatured strands of a given cDNA library member will direct the synthesis of a protein that can be identified by its position on a two-dimensional protein gel (Figure 19-17). Such procedures work best when the proteins concerned have been previously well characterized and are major cellular constituents (e.g., the hemoglobin molecules synthesized by red blood cells, or one of four major histones). But with organisms like yeast, whose proteins have not yet been well characterized, this form of screening is not a practical way to find most desired cDNA clones. Once a cDNA clone has been so identified, however, it is very straightforward to pick out the corresponding genomic clone using DNA-DNA hybridization procedures.

Those cDNAs made off less abundant mRNAs will obviously be much harder to identify. Directly looking for the protein products necessarily involves the testing of many hundreds or thousands of clones, often a costly task necessarily limited to industries. If, however, the proteins of interest have been characterized by partial or full amino acid sequencing, then an important shortcut may be used. Given an amino acid sequence, intelligent guesses can be made as to its corresponding mRNA (DNA) sequence. Because all amino acids but one are specified by more than one codon (Chapter 15), it is not possible to go from an amino acid sequence to a DNA sequence unambiguously. By focusing, however, on sequences that mainly contain the less common amino acids, it is usually possible to define a small collection of oligonucleotides, one of which should be exactly complementary to the segment of interest (Figure 19-18). Such a restricted collection can then be used as probes to identify the complementary cDNA clones by hybridization. Already this approach has

Figure 19-17
Identification of a cDNA-containing plasmid by translation of complementary mRNA selected from cellular mRNA by hybridization to the cDNA insert. The protein product is detected by gel chromatography.

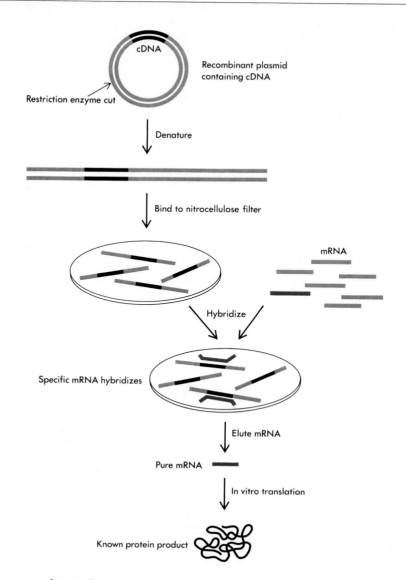

been used to isolate a number of important cDNA clones, and while it never is as simple as described here, it nevertheless is a practical technique that is bound to be increasingly employed.

Identifying cDNA Clones Through Their Expression in *E. coli*[38-43]

It is increasingly possible to find desired eucaryotic cDNA clones through techniques that allow their coding regions to function in *E. coli* strains that have been especially tailored to express foreign genes. Such strains are in fact called **expression strains.** In these techniques, the eucaryotic sequences are usually attached just downstream from a strong bacterial (or phage) promoter-ribosome binding site complex like those that govern the tryptophan and lactose operons or the expression of the phage λ genome. These gene fusions can easily be made to yield fusion proteins containing short bacterial amino-terminal segments linked to virtually all the amino acids of the eucaryotic protein (Figure 19-19). Often these hybrid proteins can be detected because they possess almost the total functional potential

Known amino acid sequence	Phe	Met	Glu	Trp	His	Lys	Asn
Possible mRNA sequence	UUU UUC	AUG	GAA GAG	UGG	CAU CAC	AAG AAA	AAU AAC

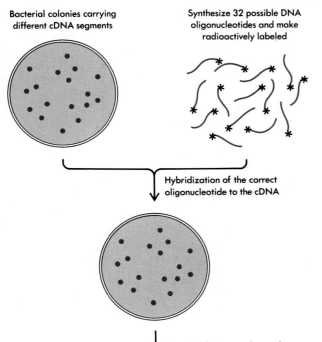

Bacterial colonies carrying different cDNA segments

Synthesize 32 possible DNA oligonucleotides and make radioactively labeled

Hybridization of the correct oligonucleotide to the cDNA

Detection by autoradiography

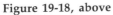

Figure 19-18, above

Cloning genes by use of a probe designed from a known amino acid sequence. Given the amino acid sequence, it is possible to synthesize DNA oligonucleotides for the possible mRNA sequences. The correct oligonucleotide will hybridize to the appropriate cDNA that we wish to identify that is carried in the plasmid. Since the oligonucleotide probe is radioactively labelled, the hybrid DNA can be detected by autoradiography.

Figure 19-19, right

Expression of eucaryotic DNA in *E. coli* allows high-level expression and identification of its protein product. Eucaryotic DNA can be introduced into a plasmid just after a strong bacterial promoter. This allows its subsequent translation into mRNA when introduced into *E. coli*. The mRNA with eucaryotic sequences will be translated into functional proteins, since *E. coli* ribosome binding sites are also present on the plasmid. An amino-terminal bacterial segment is usually attached to the protein product, but this often can be enzymatically cleaved away.

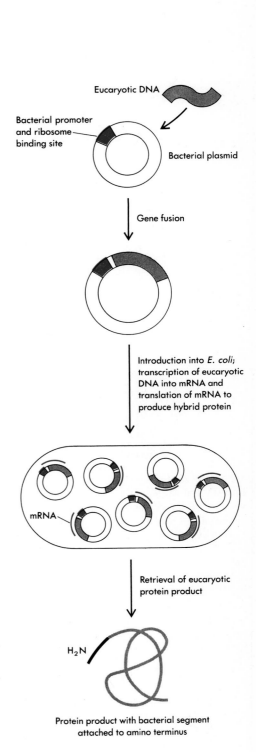

Eucaryotic DNA

Bacterial promoter and ribosome binding site

Bacterial plasmid

Gene fusion

Introduction into *E. coli;* transcription of eucaryotic DNA into mRNA and translation of mRNA to produce hybrid protein

mRNA

Retrieval of eucaryotic protein product

H₂N

Protein product with bacterial segment attached to amino terminus

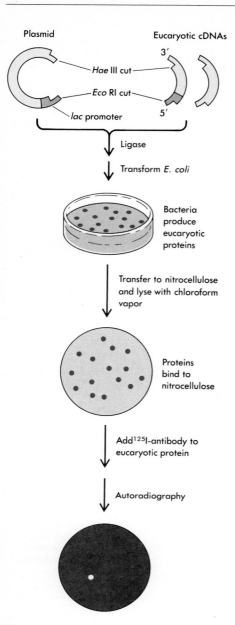

Figure 19-20
Immunological screening of expression vectors to detect desired eucaryotic-bacterial hybrid proteins. Proteins produced by bacteria are transferred to nitrocellulose filters and treated with chloroform to make them permeable to antibodies. An antibody specific for the hybrid protein we want to isolate is radioactively labeled and added to the filter, specifically binding to the correct protein. The protein can then be detected by autoradiography.

(e.g., enzymatic activity) of their pure eucaryotic equivalent. However, the hybrid proteins can usually be found more quickly by screening very large numbers of **expression vectors** for their possession of fusion products that specifically bind to appropriate antibodies. In such procedures, bacterial colonies that have been treated with chloroform to make them permeable to antibodies are exposed to radioactively labeled antibodies of the correct specificity. The usually rare colonies that are making immunologically cross-reacting proteins are easily picked out by autoradiography (Figure 19-20). To replace cleanly all of the *E. coli* coding sequences with those coding for an eucaryotic protein, appropriate restriction enzyme sites must occur within a very short and specific sequence of base pairs. It is now possible to produce such sites by in vitro synthesis. Consequently, new expression vectors have been constructed that greatly simplify the task of producing eucaryotic proteins without the bacterial peptides, and thus often make possible the synthesis in *E. coli* of perfect replicas of desired eucaryotic proteins. These vectors are often required when the primary purpose is not the detection of a given eucaryotic gene but the synthesis of its effectively pure product. Expression in *E. coli* is proving especially valuable for the purpose of making large amounts of normally rare eucaryotic proteins (e.g. the various interferons and growth factors) for their subsequent molecular (functional) characterization.

The bacterial promoters used in *E. coli* expression vectors usually have their expression controlled either like the lactose promoter, by an externally added molecule (e.g., an inducer), or like mutant λ, by a temperature-sensitive repressor that functions at 31°C but not at 37°C. In this way, an expression vector can be instructed at appropriate moments to make very large amounts of a protein whose continuous presence in large amounts during the growth of a bacterial population would be deleterious.

The routine use of **cDNA inserts** as opposed to genomic inserts of eucaryotic DNA into bacterial expression vectors is the consequence of the split nature of so many eucaryotic genes. As we explained in the previous chapter, while only a small fraction of the genes of *S. cerevisiae* are split, in higher eucaryotic cells only the exceptional gene is not split. Thus, many (if not most) of the genes of a higher eucaryotic organism will produce abnormally large, usually nonfunctional protein products when placed in an *E. coli* expression vector. Of course, the hassle of cDNA cloning can be avoided when the overexpression of an *E. coli* gene (none of which are split) itself is desired. Then the much more routinely cloned genomic segments can be inserted into the appropriate expression vector.

Eucaryotic Expression Vectors[44]

The development of multiple transforming systems (e.g., YEp, YRp, and YCp) for yeasts automatically creates the potential for using yeasts as hosts for expression vectors. The more we learn about how gene expression is controlled in eucaryotes, the more intelligently we can develop expression vector systems that can rival those already in use in *E. coli*. Several factors already encourage the development of eucaryotic systems for the expression of eucaryotic genes. First, such systems may more readily allow the passage through their external

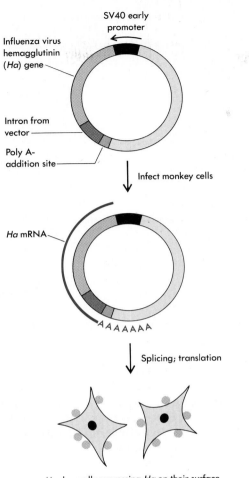

SV40 early promoter

Influenza virus hemagglutinin (*Ha*) gene

Intron from vector

Poly A-addition site

Infect monkey cells

Ha mRNA

AAAAAAA

Splicing; translation

Monkey cells expressing *Ha* on their surface

Figure 19-21
Many eucaryotic proteins cannot pass through procaryotic membranes. Therefore, expression vectors have been created that can be expressed in eucaryotic cells. Here, the SV40 early promoter is used to express flu virus hemagglutinin in monkey cells.

plasma membrane of proteins that normally are exported outside the cells in which they are translated. Such secreted proteins are frequently held together by large numbers of internal disulfide bonds (S–S) that are formed only after the passage of their polypeptide chains through cell membranes. Many of these exported eucaryotic proteins, for reasons that are still unclear, cannot pass through the *E. coli* membrane and, as a consequence, never form the correct sets of S–S bonds. The functional expression of many eucaryotic proteins thus demands their synthesis within the proper eucaryotic (as opposed to procaryotic) cells (Figure 19-21).

Also favoring the use of eucaryotic cells for the expression of eucaryotic genes is the ability of eucaryotic cells to process (splice) the newly synthesized pre-mRNA products of split genes. There is no reason why a yeast gene on a yeast vector ever needs to be converted into cDNA. In contrast, when such genomic insertions are placed in *E. coli*, only those genes lacking noncoding inserts (called introns) are likely to produce functional products. Thus, it is hoped that expression vectors for use in all genetically well-characterized organisms will soon be developed. When it comes to choosing which vector to use for a stated purpose, the fact that eucaryotic microorganisms like yeast can be grown both faster and cheaper will favor their use whenever possible.

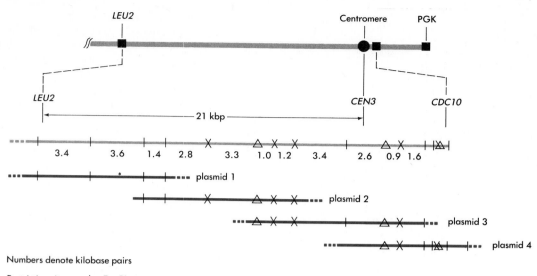

Figure 19-22
The centromere region of *S. cerevisiae* chromosome 3 was cloned by the method of chromosome walking. Starting with plasmids that contain genes flanking the centromere (plasmid 1 and plasmid 4), the region in between was cloned by overlap hybridization. The four yeast DNA inserts obtained in this chromosome walk and the regions of overlap are indicated as colored lines. [Redrawn from L. Clarke and J. Carbon, *Nature* 287 (1980):505, with permission.]

Analyzing Long Stretches of Eucaryotic DNA Using Chromosome Walks[45, 46, 47]

Eventually, the structure of very long stretches of eucaryotic DNA, if not complete eucaryotic chromosomes, will be worked out. In accomplishing this task, DNA hybridization techniques will provide the easiest way to pick out clones from gene libraries that have overlapping sequences. Thanks to such "chromosome walking," (also called "overlap hybridization") the first centromeric sequences on a yeast chromosome has already been isolated. Two cloned genes *CDC*10 and *LEU2*, that had been mapped to opposite sides of the centromere were used as starting points for two walks that finally reached each other in a total of 4 steps (Figure 19-22). Each of the clones so identified were then tested to see which one(s) ensured regular plasmid segregation during mitosis and meiosis. Walks have also been used to establish the structure of some very large *Drosophila* genes that are much too long even to be encompassed on cosmid clones.

Unfortunately, chromosome walks are almost impossible to carry out in higher plants and animals, since their chromosomes contain large numbers of nearly identical repetitive sequences (Chapter 20). Their presence adjacent to so many genes leads to large numbers of cross-hybridizing clones, whose related sequences reflect the presence of common repetitive elements rather than overlapping unique chromosomal DNA. Hence, an easy and effective method to link up restriction enzyme–generated fragments along the chromosomes of higher organisms is not yet available.

Summary

Effective experimentation on the structure and functioning of the eucaryotic genome is only possible using recombinant DNA techniques. To understand how eucaryotic genes work, we must first isolate them in pure form through gene-cloning procedures and then reintroduce them into cells where their expression can be studied. The first step in the cloning of a gene is usually its insertion into a gene library made with an appropriate plasmid, phage λ, or cosmid vector. The members of these libraries are then individually propagated in cells where the function of the respective library member can be tested.

A simple way to screen a library is to see whether it contains members that can complement the functions of genetic defects present in mutant cells. In this way, it is relatively easy to isolate intact copies of any *E. coli* gene for which a functionally deficient mutant already exists. It was surprising to find that many yeast genes complement their functionally equivalent mutated *E. coli* genes. This shows not only that some transcriptional start signals on the yeast genome can be read by *E. coli* RNA polymerase, but also that the ribosomes and tRNAs of *E. coli* can translate, albeit sometimes poorly, yeast mRNAs.

The identification of the first cloned yeast gene allowed transformation procedures to be developed for the introduction of yeast DNA into yeast protoplasts. When such yeast DNA is present on an *E. coli* plasmid, the reintroduced DNA becomes stabilized within the yeast genome by crossing over with a homologous section of yeast DNA (integrative recombination). Yeast DNA can also be functionally introduced into cells using any of three different plasmids: a derivative of the resident 2μ yeast episomal plasmid (YEp); the yeast replicating plasmid (YRp), which contains a yeast origin of DNA replication called an autonomously replicating sequence (ARS); and a plasmid (YCp) containing both a centromere and a yeast origin of replication. Vectors containing two different origins of replication that allow them to replicate in two different organisms (e.g., in both *E. coli* and yeast) are called shuttle vectors.

The frequency of integrative recombination is greatly increased by making a double-stranded cut in the DNA of the gene we wish to integrate. The resulting free ends promote recombination by invading homologous DNA segments to form heteroduplex intermediates. These subsequently become resolved through appropriate nuclease cuts to yield genetically recombined DNA. By using shuttle vectors containing gaps in the gene being studied, any desired normal or mutant gene can be retrieved and its structure subsequently worked out. Integrative recombination now allows any yeast gene to be replaced with its mutated derivative and so provides a powerful way to probe the physiological role of genes whose function is not known.

After a gene is cloned, the manner in which its expression is controlled is studied using in vitro mutagenesis procedures. The most powerful of these procedures allows specified base changes to be made at precise locations along already sequenced DNA molecules. Also very useful are gene fusion procedures in which the promoter for one gene is attached to the amino acid coding element of another gene. In this way, the activity of mutated promoters can be measured through the presence of an easily assayable gene product (e.g., the color test for *E. coli* galactosidase activity) instead of by laborious measurements of enzymatic activity.

Many of the elegant recombinant DNA techniques now available for work with yeast genes remain restricted to this system. In most higher eucaryotic organisms, the recombinant DNA procedures used both for initial gene cloning and for later understanding of the rules by which eucaryotic gene expression is controlled are much more laborious. When practical, hybridization procedures (DNA-DNA or DNA-RNA) are used not only to search for genes with homologous sequence, but also to work out the sequences of long DNA stretches of a given chromosome (chromosome walks). When a gene to be cloned is suspected of being split (containing introns), then the most efficient way to isolate it usually involves reverse transcription of its mRNA product into cDNA molecules. These cDNAs are then placed into appropriate expression vectors, which allow expression of the DNA. Although the first good expression vectors involved *E. coli* systems, there are now available equally useful expression vectors for a variety of eucaryotic organisms.

Bibliography

General References

Botstein, D., and R. W. Davis. 1982. "Principles and Practice of Recombinant DNA Research with Yeast." In *The Molecular Biology of the Yeast* Saccharomyces, *Metabolism and Gene Expression*, ed. J. N. Strathern, E. W. Jones, and J. R. Broach. Cold Spring Harbor, N.Y.: Cold Spring Harbor Laboratory, pp. 607–638.

Davis, R. W., D. Botstein, and J. R. Roth. 1980. *Advanced Bacterial Genetics.* Cold Spring Harbor, N.Y.: Cold Spring Harbor Laboratory.

Maniatis, T., E. F. Fritsch, and J. Sambrook. 1982. *Molecular Cloning.* Cold Spring Harbor, N.Y.: Cold Spring Harbor Laboratory.

Silhavy, T. J., M. L. Berman, and L. W. Enquist. 1984. *Experiments with Gene Fusions.* Cold Spring Harbor, N.Y.: Cold Spring Harbor Laboratory.

Watson, J. D., J. Tooze, and D. T. Kurtz. 1983. *Recombinant DNA: A Short Course.* San Francisco: Freeman.

Cited References

1. Collins, J., and B. Hohn. 1978. "Cosmids: A Type of Plasmid Gene Cloning Vector That Is Packageable *in Vitro* in Bacteriophage Heads." *Proc. Nat. Acad. Sci.* 75:4242–4246.

2. Maniatis, T., R. C. Hardison, E. Lacy, J. Lauer, C. O'Connell, D. Quon, G. K. Sim, and A. Efstratiadis. 1978. "The Isolation of Structural Genes from Libraries of Eucaryotic DNA." *Cell* 15:687–701.

3. Struhl, K., J. R. Cameron, and R. W. Davies. 1976. "Functional Genetic Expression of Eucaryotic DNA in *E. coli.*" *Proc. Nat. Acad. Sci.* 73:1471–1475.

4. Ratzkin, B., and J. Carbon. 1977. "Functional Expression of Cloned Yeast DNA in *E. coli.*" *Proc. Nat. Acad. Sci.* 74:487–491.

5. Hinnen, A., J. B. Hicks, and G. R. Fink. 1978. "Transformation of Yeast." *Proc. Nat. Acad. Sci.* 75:1929–1933.

6. Beggs, J. D. 1978. "Transformation of Yeast by a Relicating Hybrid Plasmid." *Nature* 275:104–109.

7. Gerbaud, C., P. Fournier, H. Blanc, M. Aigle, H. Heslot, and M. Guerineau. 1979. "High Frequency Yeast Transformation by Plasmids Carrying Part or Entire 2μm Yeast Plasmid." *Gene* 5:233–237.

8. Broach, J. R. 1981. "The Yeast Plasmid 2μ Circle." In *The Molecular Biology of the Yeast* Saccharomyces, *Life Cycle and Inheritance*, eds. J. N. Strathern, E. W. Jones, and J. R. Broach. Cold Spring Harbor, N.Y.: Cold Spring Harbor Laboratory, pp. 445–470.

9. Struhl, K., D. T. Stinchcomb, S. Scherer, and R. W. Davies. 1979. "High-Frequency Transformation of Yeast: Autonomous Replication of Hybrid DNA Molecules." *Proc. Nat. Acad. Sci.* 76:1035–1039.

10. Stinchcomb, D. T., K. Struhl, and R. W. Davies. 1979. "Isolation and Characterization of a Yeast Chromosomal Replicator." *Nature* 282:39–43.

11. Hsiao, C.-L., and J. Carbon. 1981. "Characterization of a Yeast Replication Origin *(ARS2)* and Construction of Stable Minichromosomes Containing Cloned Yeast Centromere DNA *(CEN3)*." *Gene* 15:157–166.

12. Clarke, L., and J. Carbon. 1980. "Isolation of a Yeast Centromere and Construction of Functional Small Circular Chromosomes." *Nature* 287:504–509.

13. Clarke, L., and J. Carbon. 1985. "The Structure and Function of Yeast Centromeres." *Annual Review of Genetics* 19:29–57.

14. Orr-Weaver, T. L., J. W. Szostak, and R. L. Rothstein. 1981. "Yeast Transformation: A Model System for the Study of Recombination." *Proc. Nat. Acad. Sci.* 78:6354–6358.

15. Hicks, J. B., J. N. Strathern. A. J. S. Klar, and S. L. Dellaporta. 1982. "Cloning by Complementation in Yeast: The Mating Type Genes." In *Genetic Engineering—Principles and Methods,* vol. 4, ed. J. K. Setlow and A. Hollaender. New York: Plenum, pp. 219–248.

16. Shortle, D., and D. Nathans. 1979. "Regulatory Mutants of Simian Virus 40: Constructed Mutants with Base Substitutions at the Origin of Viral DNA Replication." *J. Mol. Biol.* 131:801–817.

17. Shortle, D., D. DiMaio, and D. Nathans. 1981. "Directed Mutagenesis." *Ann. Rev. Genetics* 15:265–294.

18. McKnight, S. L., and R. Kingsbury. 1982. "Transcriptional Control Signals of a Eukaryotic Protein-Coding Gene." *Science* 217:316–324.

19. Winter, G., A. R. Fersht, A. J. Wilkinson, M. Zoller, and M. Smith. 1982. "Redesigning Enzyme Structure by Site-Directed Mutagenesis." *Nature* 296:756–758.

20. Zakour, R. A., and L. A. Loeb. 1982. "Site-Specific Mutagenesis by Error-Directed DNA Synthesis." *Nature* 195:708–710.

21. Smith, M. (1985). "In vitro Mutagenesis." *Annual Review of Genetics* 19:423–463.

22. Heffron, F., M. So, and B. J. McCarthy. 1978. "*In vitro* mutagenesis of a circular DNA molecule by using synthetic restriction sites." *Proc. Nat. Acad. Sci.* 75:6012–6016.

23. Totchell, K., K. A. Nasmyth, B. D. Hall, C. Astell, and M. Smith. 1981. "*In vitro* mutation analysis of the mating-type lows in yeast." *Cell* 27:25–35.

24. McKnight, S. L., and R. Kingsbury. 1982. "Transcriptional Control Signals of a Eukaryotic Protein-Coding Gene." *Science* 217:316.

25. Scherer, S., and R. W. Davies. 1979. "Replacement of Chromosome Segments with Altered DNA Sequences Constructed *in Vitro.*" *Proc. Nat. Acad. Sci.* 76:4951–4955.

26. Shortle, D., J. Haber, and D. Botstein. 1982. "Lethal Disruption of the Yeast Actin Gene by Integrative DNA Transformation." *Science* 217:371–373.

27. Rothstein, R. J. 1983. "One-step Gene Disruption in Yeast." *Methods in Enzymology* 101:202–211.

28. Guarente, L., and M. Ptashne. 1981. "Fusion of *E. coli lacZ* to the Cytochrome *c* Gene of *Saccharomyces cerevisae.*" *Proc. Nat. Acad. Sci.* 78:2199–2203.

29. Rose, M., M. J. Casadaban, and D. Botstein. 1981. "Yeast Genes Fused to Beta-Galactosidase in *E. coli* Can Be Expressed Normally in Yeast." *Proc. Nat. Acad. Sci.* 78:2460–2464.

30. Grunstein, M., and D. S. Hogness. 1975. "Colony Hybridization: A Method for the Isolation of Cloned DNAs That Contain a Specific Gene." *Proc. Nat. Acad. Sci.* 72:3961–3965.

31. Efstratiadis, A., F. C. Kafatos, A. M. Maxam, and T. Maniatis. 1976. "Enzymatic *in Vitro* Synthesis of Globin Genes." *Cell* 7:279–288.

32. Benton, W. D., and R. W. Davis. 1977. "Screening λgt Recombinant Clones by Hybridization to Single Plaques *in Situ.*" *Science* 196:180–182.

33. Southern, E. M. 1975. "Detection of Specific Sequences Among DNA Fragments Separated by Gel Electrophoresis." *J. Mol. Biol.* 98:503–517.

34. Efstratiadis, A., and L. Villa-Komaroff. 1979. "Cloning of Double-Stranded DNA." In *Genetic Engineering,* vol. 1, ed. J. K. Setlow and A. Hollaender. New York: Plenum, pp. 15–27.

35. Paterson, B. M., B. E. Roberts, and E. L. Kuff. 1977. "Structural Gene Identification and Mapping by DNA-mRNA Hybrid-Arrested Cell-Free Translation." *Proc. Nat. Acad. Sci.* 74:4370–4374.

36. Noyes, B. E., M. Mevarech, R. Stein, and K. L. Agarwal. 1979. "Detection and Partial Sequence Analysis of Gastrin mRNA by Using an Oligodeoxynucleotide Probe." *Proc. Nat. Acad. Sci.* 76:1770–1774.

37. Wallace, R. B., J. Schaffer, R. F. Murphy, J. Bonner, T. Hirose, and K. Itakura. 1979. "Hybridization of Synthetic Oligodeoxynucleotides to 0X174 DNA: The Effect of Single Base Pair Mismatch." *Nucleic Acid Res.* 6:3543–3557.

38. Backman, K., and M. Ptashne. 1978. "Maximizing Gene Expression on a Plasmid Using Recombinant DNA *in Vitro.*" *Cell* 13:65–71.

39. Goeddel, D. V., H. L. Heyneker, T. Hozumi, R. Arentzen, K. Itakura, D. G. Yansura, M. J. Ross, G. Miozzari, R. Crea, and P. H. Seeburg. 1979. "Direct Expression in *E. coli* of a DNA Sequence Coding for Human Growth Hormone." *Nature* 281:544–548.

40. Guarente, L., G. Lauer, T. M. Roberts, and M. Ptashne. 1980. "Improved Methods for Maximizing Expression of a Cloned Gene: A Bacterium That Synthesizes Rabbit β-Globin." *Cell* 20:543–553.

41. Remaut, E., P. Stanssens, and W. Fiers. 1981. "Plasmid Vectors for High-Efficiency Expression Controlled by the p_L Promoter of Coliphage Lambda." *Gene* 15:81–93.

42. Silhavy, T. J., M. L. Berman, and L. W. Enquist. 1984. *Experiments with Gene Fusions.* Cold Spring Harbor, N.Y.: Cold Spring Harbor Laboratory.

43. Young, R. A., and R. W. Davies. 1983. "Efficient Isolation of Genes Using Antibody Probes." *Proc. Nat. Acad. Sci.* 80:1194–1198.

44. Gluzman, Y. 1982. *Eukaryotic Viral Vectors.* Cold Spring Harbor, N.Y.: Cold Spring Harbor Laboratory.

45. Bender, W., P. Spierer, and D. Hogness. 1979. "Gene Isolation by Chromosomal Walking." *J. Supra. Mol. Struc.* 10 (suppl.):32.

46. Chinault, A. C., and J. Carbon. 1979. "Overlap Hybridization Screening: Isolation and Characterization of Overlapping DNA Fragments Surrounding the *LEU2* Gene on Yeast Chromosome III." *Gene* 5:111–126.

47. Clarke, L., and J. Carbon. 1980. "Isolation of a Yeast Centromere and Construction of Small Circular Chromosomes." *Nature* 287:504–509.

VIII

THE FUNCTIONING OF EUCARYOTIC CHROMOSOMES

CHAPTER 20

The Unexpected Structures of Eucaryotic Genomes

After many years of work on bacteria and their viruses, even the most skeptical scientists were certain that our understanding of how proteins are encoded in DNA would be automatically transferable to higher plants and animals. Early studies had established that the central dogma (DNA makes RNA and RNA makes protein) holds for mammalian cells as it does for bacteria. Moreover, the genetic code appeared to be universal. Hence, no one would have expected that another fundamental tenet of gene expression might not apply to eucaryotic cells. This was the concept of colinearity (Chapter 8), first deduced from genetic studies and later made absolutely firm when DNA, RNA, and protein sequences demonstrated a direct correspondence between the sequence of a gene, its mRNA, and its protein product.

The discovery of interruptions in eucaryotic genes therefore came as a complete shock to the scientific world in 1977. It was a unique sort of discovery in that experimental data accumulated in several laboratories forced a conclusion that no one could have foreseen. During the early years of the same decade, intensified studies had correctly indicated that every aspect of gene expression and its regulation in higher cells would be more complex than in bacteria. But only with the advent of cloning and rapid DNA-sequencing technology did the revolution in our understanding of the structure and functioning of eucaryotic genes begin.

The Amount of DNA per Cell Increases About Eight Hundredfold from *E. coli* to Mammals[1, 2]

For years, biologists had been puzzled about the very large DNA content found in most eucaryotic cells. For instance, the amount of DNA present per mammalian cell is approximately 800 times that in *E. coli*. Since there was no reason to believe that gene size increased from the lower to the higher forms of life, this number was used to calculate an upper limit of the number of different genes. Using the assumption that each gene is present in only one copy per haploid genome, a mammalian cell was predicted to be capable of synthesizing over two million different proteins. Thus, the task of relating a given protein to its gene was viewed as much, much more formidable than the corresponding job with bacteria.

Figure 20-1

Haploid amounts of DNA in various cells, expressed as multiples of the amounts found in *E. coli* (4×10^{-12} mg = 2.4×10^9 MW). Many of these values should be taken as approximate. [After R. Holliday, *Symp. Soc. Gen. Microbiol.* 20 (1970):362.]

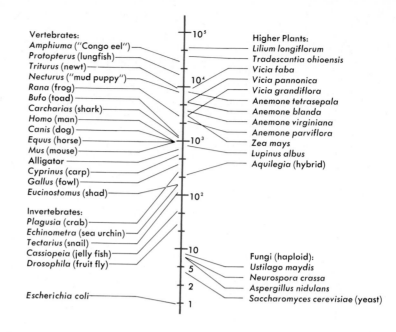

In some cases, however, the amount of DNA appeared misleading. There are groups of amphibians that contain 25 times more DNA in their cells than is present in mammalian cells (Figure 20-1). Here, there is no reason to believe that greater biological complexity is involved. Rather, this situation hints that we should be very cautious about relating DNA content directly to the number of different proteins that may be synthesized by a given cell. Morphological examination with the electron microscope tells us that a much larger variety of subcellular structures do exist in the vertebrate cell than in *E. coli* (compare Figure 20-2 with Figure 4-7). Correspondingly, a larger number of structural proteins and enzymes should be necessary for their construction and function. This expectation is borne out by two-dimensional gel displays of proteins from mammalian cells, which do reveal many times more polypeptide spots than those of *E. coli* proteins (see Figure 4-6). Nonetheless, it would be surprising if the vertebrate cell were more than 50 to 100 times more complex genetically than *E. coli*.

The Surprisingly Long Length of Pre-mRNA Molecules[3, 4]

Other mystifying observations were made when scientists began to examine mammalian RNA synthesis in the late 1960s. When labeled RNA precursors were used and nuclear RNA was analyzed in sucrose gradients, the nonribosomal, non-tRNA molecules sedimented as if they were very large and very heterogeneous in size. This gave rise to the term **heterogeneous nuclear RNA (hnRNA)**. Heterogeneous nuclear RNA possesses many properties expected of lengthy messenger precursors. Its average size is about ten times the length of cytoplasmic mRNAs. It has a very short lifetime; about half the labeled nucleotides incorporated during a short pulse into hnRNA disappear within a 3-minute chase period ($t_{1/2} \cong 3$ minutes). And somewhat less than 10 percent of the label originally in hnRNA eventually appears in the cytoplasm in the form of mRNA-like molecules.

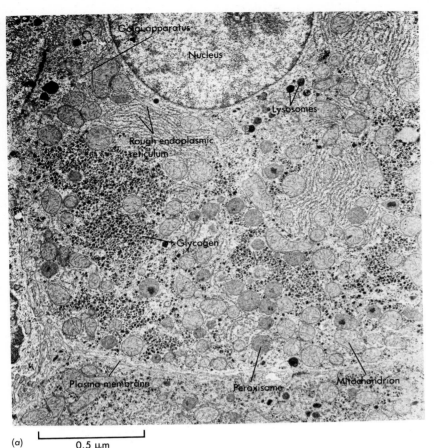

(a)   0.5 μm

Figure 20-2
The complexity of a vertebrate cell.
(a) An electron micrograph of part of a
mammalian liver cell is seen in cross
section. The major cellular compartments and structures are indicated.
(Courtesy of Daniel Friend.) (b) A
schematized version. [After B. Alberts
et al., *Molecular Biology of the Cell* (New
York: Garland, 1983), 322.] Also see
Figures 1-1 and 1-2, demonstrating similar structures in a higher plant cell.

Extracellular matrix

Plasma membrane

Endoplasmic reticulum

Smooth endoplasmic
reticulum

Mitochondrion

Golgi apparatus

Centrioles

Nucleus

Filamentous cytoskeleton

Lysosomes, peroxisomes

Cytosol

Nuclear envelope

(b)   10–30 μm

Once poly A tails were identified on eucaryotic mRNAs in the early 1970s, hnRNA was also immediately checked. It, too, possessed 100- to 200-nucleotide-long stretches of A residues at its 3′ termini. This served to strengthen further the belief that hnRNA is indeed a nuclear precursor form of cytoplasmic message. Thus, it seemed that extraordinarily long 5′ leaders must be trimmed away by RNA processing events, sending 3′ coding sequences with their poly A tails already attached out to the cytoplasm to be translated by the ribosomes.

This neat picture, however, did not survive long. The 5′ cap structures on eucaryotic mRNAs were discovered shortly thereafter. Careful analyses revealed that the number of 5′ caps on hnRNA, like the number of poly A tails, was approximately equal to the number of molecules present. How could both the 5′ and 3′ ends of hnRNA be preserved during processing into smaller mRNA molecules? The facts were thus unreconcilable with any simple model for the generation of mRNA from hnRNA. There the matter rested until 1977, when the discovery of interruptions in genes finally provided a rationale for the existence of extraordinarily long pre-mRNA molecules.

Amazing Sequence Arrangement at the 5′ Ends of Adenovirus mRNAs[5-9]

Studies of the abundant messenger RNAs produced late in infection of human cells by adenovirus 2 were among the first to provide incontrovertible evidence that parts of the same mRNA molecule can be specified by noncontiguous segments of eucaryotic DNA. The genome of adenovirus is a double-stranded linear molecule of about 36 kilobase pairs (Chapter 24). Coding regions for the late (structural) proteins are located in the rightmost two-thirds of the viral DNA. RNA sequencing had revealed that several different late mRNAs possess not only the expected cap structures, but also an identical 11-nucleotide sequence at their 5′ ends. Surprisingly, these 5′ regions were not resistant to ribonuclease digestion in hybrids formed between the mRNAs and restriction fragments from the late region of the adenovirus genome, as would be expected if the DNA and RNA sequences were complementary along their entire length.

These hints that the 5′ ends might not be encoded adjacent to the main bodies of the adenovirus late mRNAs were followed up in a series of elegant **R-looping experiments.** If an RNA molecule is incubated with a complementary double-stranded DNA under conditions where RNA-DNA hybrids form in preference to DNA duplexes, the DNA strand that possesses the same sequence as the RNA will be displaced. Such displacement loops (R loops) can be easily visualized using the electron microscope and their ends mapped relative to restriction endonuclease cut sites in the DNA (Figure 20-3a). When the late adenovirus mRNAs were hybridized under R-loop conditions with DNA fragments known to contain their protein-coding regions, projections were indeed formed by the 5′ ends (as well as by the 3′ poly A tails) of the mRNAs (Figure 20-3b). By preparing single-stranded restriction fragments from other parts of the adenovirus genome and adding them each in turn during the preparation of R loops, researchers were able to locate the DNA regions specifying the 5′ projections (Figure 20-3c). Amazingly, several different late mRNAs had 5′ ends derived from the same three separate locations in

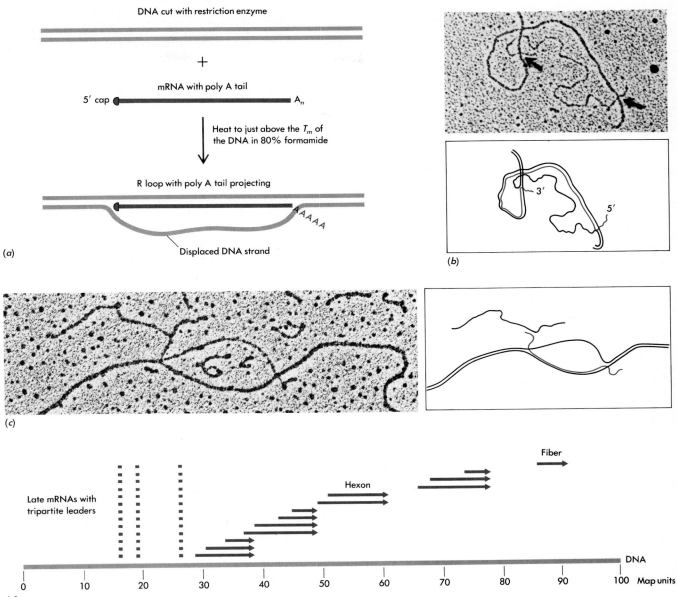

Figure 20-3
How R-loop mapping showed the origins of 5' leader sequences on adenovirus-2 late mRNAs. (a) The formation of R-loop structures in which the DNA strand with the sequence corresponding to the mRNA is displaced. (b) Electron micrograph and schematic diagram of an R loop observed after incubating hexon mRNA with a complementary DNA restriction fragment from the late region of the adenovirus-2 genome. (The black lines are DNA; the color line is mRNA.) Note that both 5' and 3' extensions are seen. [Electron micrograph courtesy of S. M. Berget, C. Moore, and P. A. Sharp, *Proc. Nat. Acad. Sci.* 74 (1977):3171–3175.] (c) Electron micrograph and schematic diagram of an R loop observed after incubating fiber mRNA with two DNAs: the adenovirus genome and a restriction fragment from the early region. (The Ad2 DNA and the restriction fragment are shown in black; the color line is mRNA.) The structure formed shows that part of the 5' extension is specified by the DNA between positions 11.3 and 18.1 on the viral genome. [Electron micrograph courtesy of L. T. Chow, R. E. Gelinas, T. R. Broker, and R. J. Roberts, *Cell* 12 (1977):1–8.] (d) Map of the adenovirus-2 genome showing the location of DNA specifying the late transcripts and the tripartite leader (arising from positions 16.6, 19.6, and 26.6) found on each of the late mRNAs.

the leftmost one-third of the adenovirus genome. Short sequences occurring at map positions 16.6, 19.6, and 26.6 seemed to be joined to form a leader of 150 to 200 nucleotides common to many of the adenovirus late mRNAs! (Figure 20-3d.) Clearly, an unprecedented mechanism had to underlie the biosynthesis of these particular mRNAs. Was the phenomenon peculiar to viruses or a manifestation of some general mechanism for the expression of genes in mammalian cells?

Discovery of Interruptions in Ovalbumin and β-Globin Genes[10–17]

At the same time that the preceding data were being collected on adenovirus mRNAs, comparably bizarre observations were being made concerning several cellular genes, including the chicken gene for ovalbumin. The major component of egg white, this protein is synthesized only by specialized cells of the oviduct. The ovalbumin gene was being investigated because its expression is regulated by female sex hormones. To understand this control process at the molecular level, several groups had set out to clone the ovalbumin gene using the newly available recombinant DNA techniques. The first step was isolation of the mRNA, only a moderately formidable task because of its abundance in the oviduct cells of laying hens. From the mRNA, a complementary DNA (cDNA) was produced (Chapter 19), cloned, and sequenced. The structure deduced for the 1872-nucleotide mRNA provided no real surprises: 1158 coding nucleotides (corresponding to the 386 amino acids of the protein) were preceded by a 64-nucleotide 5' leader sequence and followed by a nontranslated stretch of 650 nucleotides at the 3' end. The next step was to use the cloned cDNA to investigate possible differences in the structure of the ovalbumin gene between the cells of the laying-hen oviduct and the cells of tissues not engaged in synthesizing ovalbumin. Here the blot hybridizations (see Chapter 19) revealed a totally unanticipated situation. It was expected that analysis of genomic restriction fragments generated by enzymes that did not cut the ovalbumin cDNA would show only a single radioactive band containing the entire ovalbumin gene. Instead, several bands appeared in the autoradiograph of the hybridized blot! (Figure 20-4.) Moreover, the multiple-band pattern was the same whether the chromosomal DNA was derived from oviduct cells or erythrocytes (red blood cells, which do not make ovalbumin). Initially, such puzzling results were ascribed to experimental artifacts.

Meanwhile, similar experiments mapping restriction sites in the β-globin genes of the rabbit and mouse were also generating unanticipated results. The data could be explained only by assuming the existence of an interruption of about 600 base pairs in the middle of the protein-coding sequence of the gene from both animals. Again, the insertion did not appear to be involved in inactivating the β-globin gene; it was present in DNA from both erythroid (globin-producing) and nonerythroid tissue.

Visual evidence for split gene organization soon came from the examination of R loops formed between ovalbumin or globin mRNAs and cloned DNA segments derived from these genes. In the ovalbumin case, seven unhybridized DNA regions looped out of the DNA-RNA hybrid molecule, having failed to find complementary RNA sequences with which to base-pair (Figure 20-5). Thus, at least

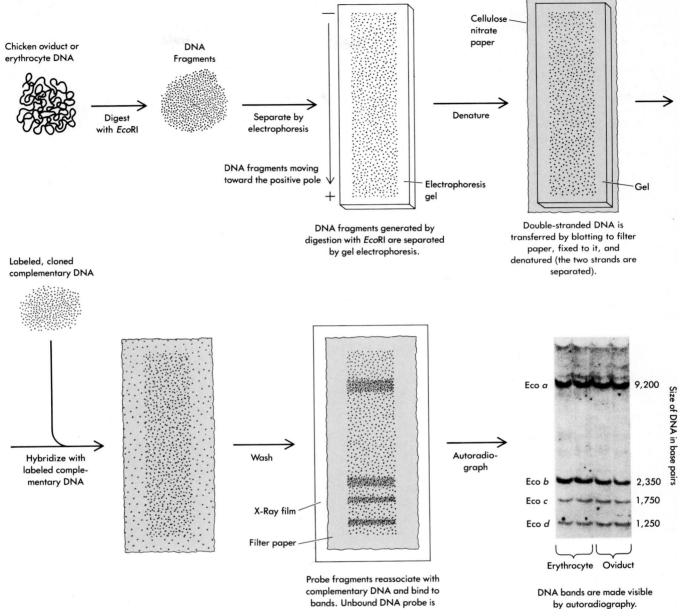

DNA fragments generated by digestion with *Eco*RI are separated by gel electrophoresis.

Double-stranded DNA is transferred by blotting to filter paper, fixed to it, and denatured (the two strands are separated).

Probe fragments reassociate with complementary DNA and bind to bands. Unbound DNA probe is washed away.

DNA bands are made visible by autoradiography.

eight discontinuous segments of genomic DNA specify the ovalbumin mRNA. Electron micrographs of R loops between β-globin mRNA and its cloned gene revealed a single, large, looped-out segment (corresponding to the 600-base-pair insert deduced from mapping restriction sites). Careful measurements suggested the additional presence of a second much smaller interruption closer to the beginning of the β-globin gene.

Many Genes Are More Intron Than Exon[18, 19, 20]

Sequence analysis subsequently defined the exact extent and nature of the discontinuities in the ovalbumin and β-globin genes. Regions not represented in the mRNA are called **introns** or **intervening**

Figure 20-4
Demonstration of the interrupted structure of the ovalbumin gene by a blotting experiment in which chromosomal DNA from chicken erythrocytes or oviduct cells was cleaved with *Eco* RI. This restriction enzyme does not cleave the ovalbumin cDNA. Thus, the four bands seen in the autoradiograph reveal a split gene structure that appears to be the same in the two tissues. [After P. Chambon, *Sci. Amer.* 244 (1981):60–71.]

Figure 20-5
The organization of the chicken ovalbumin gene. DNA containing the gene was allowed to hybridize with ovalbumin mRNA under R-loop conditions. The eight exons (L, 1–7) of the gene pair with the complementary regions of RNA, and the seven introns (A–G) loop out from the DNA-RNA hybrid. The 5' and 3' ends of the mRNA are indicated, as is the poly A tail. [After P. Chambon, *Sci. Amer.* 244 (1981):60–71.]

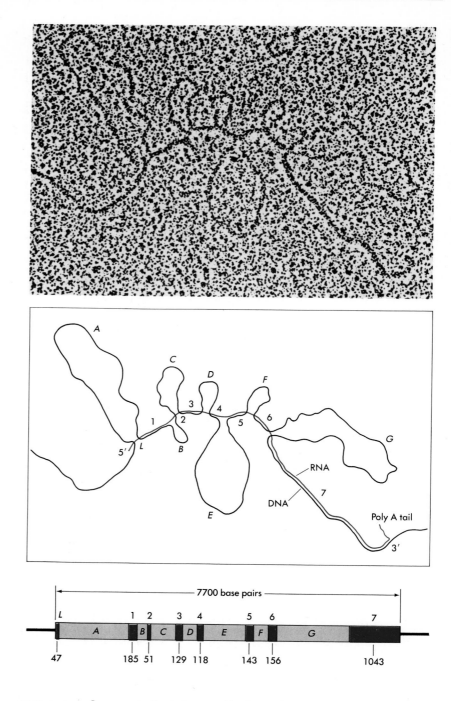

sequences. Segments that do specify the mRNA are called **exons** (because they are expressed). Exons include not only the protein-coding sequences of a gene, but also the untranslated 5' and 3' residues (leader and trailer sequences) present in the mRNA. Recall that the poly A tail of mRNA is not encoded in the genome, but added post-transcriptionally (Chapter 18).

Over three-quarters of the 7700 base pairs of the chicken ovalbumin gene are contained in introns (see Figure 20-5). As indicated by the R-loop patterns, the introns are seven in number. They range in size from about 250 to 1600 nucleotides. Whereas one intron resides in the untranslated 5' leader sequence, the other six interrupt the protein-coding portion of the gene. Like the introns, the eight ovalbumin exon segments vary considerably in length. The shortest is 47 base pairs and the longest is 1043 base pairs.

A simpler distribution of introns and exons was revealed by sequencing several mammalian β-globin genes (Figure 20-6). They all contain one large intron (about 600 base pairs) and one small intron (about 100 base pairs), as had been predicted by the examination of R loops. This gene is therefore approximately equally divided between introns and exons. If the nucleotide sequences of the mouse, rabbit, and human β-globin genes are compared in detail, little divergence is seen in their protein-coding regions, reflecting the evolutionary conservation of the β-globin polypeptide. The introns, by contrast, differ between species in both length and exact sequence. Yet, they occur at precisely the same positions within the β-globin-coding region of each species, with the small intron between the codons specifying amino acids 30 and 31, and the large intron between the codons for residues 104 and 105.

Now that many more genes have been analyzed, it is clear that the split gene patterns exemplified by chicken ovalbumin and the mammalian β-globins are typical of the majority of vertebrate protein-coding genes. In a few cases, the distribution of residues between introns and exons is quite extraordinary. For instance, the chromosomal gene for mouse dihydrofolate reductase extends over about 31,000 base pairs, but gives rise to an mRNA of only about 1600 nucleotides, which in turn possesses only 558 coding residues. The chicken collagen gene contains at least 50 exons, which divide its coding region into regular repeating units that reflect comparable repeats of the amino acid sequence along the polypeptide chain (Figure 20-7).

Introns can be of many different lengths and sequences. They appear in what will become the 3' trailer region as well as the protein-coding and 5' leader regions of various mRNAs. They need not reside neatly between codons, as they do in the β-globin or collagen genes. Often they split a triplet, so that the information needed to specify a particular amino acid lies thousands of base pairs apart on the chromosomal DNA.

Not All Genes Are Split

Even as the roster of split genes grows daily, firm evidence accumulates that some protein-coding genes are not interrupted. For example, the multiple genes encoding the histone proteins that package eucaryotic DNA into nucleosomes (Chapter 21) only rarely have introns. The interferons, proteins produced by vertebrates in response to viral infections (Chapter 24), are also often specified by uninterrupted genes. Many mRNAs synthesized by viruses that infect mammalian cells are colinear with their respective DNA genomes. Thus, the presence of introns cannot be essential for gene expression, nor does there appear to be any obvious correlation with how the expression of a gene is regulated.

Although the best initial guess for any uncharacterized mammalian protein-coding gene is that it contains introns, this is not true of genes from all eucaryotic organisms. Generally, as one proceeds down the evolutionary ladder, fewer and fewer protein-coding genes are interrupted, and the introns that do exist tend to be shorter. For instance, whereas the human and rat cytochrome *c* genes are interrupted, the yeast cytochrome *c* gene lacks introns. The *Drosophila* alcohol

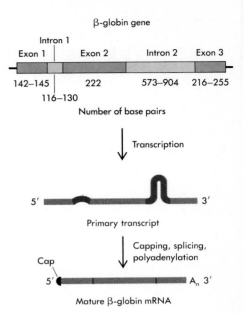

Figure 20-6
Structure of the mammalian β-globin gene, showing how introns are removed and exons preserved during the maturation of the mRNA. Exon 1 specifies the 5' nontranslated region and amino acids 1–30; exon 2, amino acids 31–104; and exon 3, amino acids 105 to the terminus plus the 3' nontranslated region.

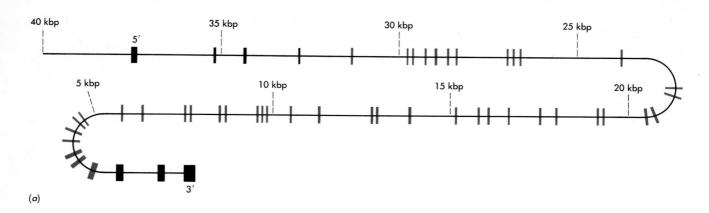

(a)

Figure 20-7
The chick type 1α2 collagen gene has over 50 introns. (a) In the schematic representation, exons are indicated by bars and introns by the connecting line. The 5′ and 3′ ends of the 38-kilobase-pair gene are located. Much of the collagen polypeptide is arranged in a triple helical structure with glycine appearing every third residue. The exons encoding this portion of the chain (colored) always contain a multiple of 9 base pairs; most are 54 base pairs, but others are as short as 45 or as long as 108 base pairs. (b) Electron micrograph and schematic diagram of R-loop structures formed between α2 collagen mRNA and cloned DNA derived from the central portion of the gene. For historical reasons, the exons are numbered beginning at the 3′ end of the gene. When two numbers appear, the segment contains two exons separated by a small intron often seen as a pimple. [Courtesy of H. Ohkubo et al., *Proc. Nat. Acad. Sci.* 77 (1980):7059–7063. Used with permission.]

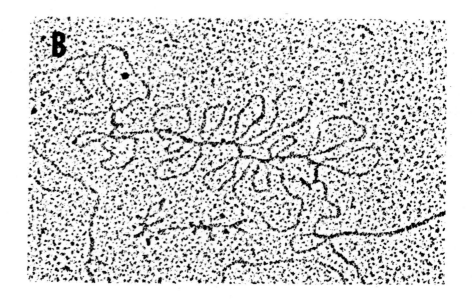

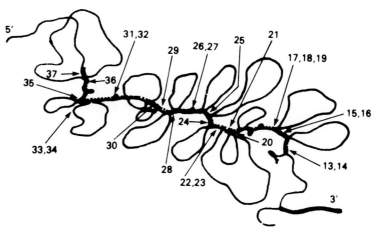

(b)

dehydrogenase gene has introns, while the comparable yeast gene does not. Thus, in the yeast *Saccharomyces cerevisiae*, a gene containing introns is the exception rather than the rule. It now appears that this is the case for many other lower eucaryotes as well.

tRNA and rRNA Genes Also Contain Introns[21, 22, 23]

At the same time that interruptions in protein-coding genes were being analyzed in 1977, the first tRNA genes to have introns were discovered. The seemingly dull task of sequencing the gene for a yeast suppressor tRNA^{Tyr}, whose RNA sequence was already known, turned out to be quite interesting when 14 extra base pairs (not present in the RNA) appeared in the DNA sequence specifying the anticodon loop. Since the suppressor strain had arisen by spontaneous mutation and the anticodon sequence showed the expected single-base change, it seemed highly likely that this must be an active gene. Equally surprising was the finding that three other distinct genes for tRNA^{Tyr} all had the identical sequence (except in the anticodon), including the 14-base-pair insertion. Soon after, 19-base-pair insertions were identified in yeast tRNA^{Phe} genes, also located just 3' to the anticodon sequence.

We now know that (nuclear) tRNA genes from many eucaryotic organisms contain introns. Transfer RNA introns come in lengths of 14 to 60 base pairs. They almost always reside between the first and second nucleotides 3' to the anticodon. Curiously, they appear confined to a particular subset of tRNAs: those accepting tyrosine, phenylalanine, serine, leucine, isoleucine, tryptophan, proline, and lysine. The introns present in any family of tRNA genes from one organism (e.g., the eight yeast tRNA^{Tyr} genes) are identical or nearly identical. But intron sequences bear no resemblance between genes specifying different tRNAs (Figure 20-8).

Figure 20-8
Three yeast tRNA precursors containing introns. The anticodons (boxed in gray) often (but not always) can be drawn to pair with intron sequences (shown in color). Arrows indicate the sites where the tRNA-splicing machinery must cut and rejoin the pre-tRNA to form the mature tRNA molecule. Positions of sequence variation among the several genes specifying each tRNA species are shown. Such tRNA precursors can be isolated from yeast strain ts36 and have been used to decipher the biochemistry of yeast tRNA splicing.

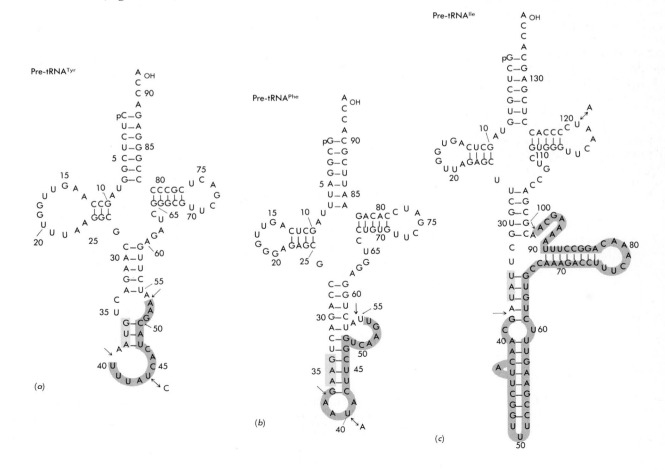

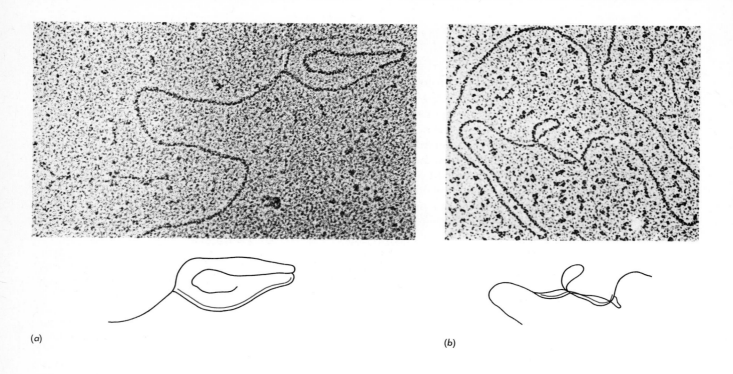

(a)

(b)

Figure 20-9
Demonstration that the large intron of mouse β-globin gene is transcribed. Electron micrographs and schematic diagrams of the R-loop structures formed by incubating the cloned gene with either (a) 15S (precursor) RNA or (b) 10S (messenger) RNA are shown. (Black lines indicate DNA; colored lines indicate RNA.) The 15S RNA had previously been characterized as having many of the features expected of a precursor to the mature mRNA, including a 5′ cap and a 3′ poly A tail. In (b), only the larger of the two β-globin introns is big enough that the nonhybridized DNA can be seen buckling out from the R loop. [Electron micrographs courtesy of S. M. Tilghman et al., *Nat. Acad. Sci.* 75 (1978):1309–1313. Used with permission.]

Some ribosomal RNA genes from eucaryotic organisms also contain introns. So far, known active (nuclear) rRNA genes with introns all specify the large (23S-like) ribosomal RNA and are from lower eucaryotic species.

Thus, we see that introns are not restricted to any particular type of eucaryotic gene. Such interruptions can appear in genes whose terminal products are RNA as well as in protein-coding genes. Even mitochondrial and chloroplast genomes sometimes contain introns in regions producing mRNA, rRNA, or tRNA.

Introns Are Removed at the RNA Level[23–27]

Once it became clear that many chromosomal genes contain interruptions, much speculation arose as to how the corresponding intronless RNAs might be generated. An initial notion that the creation of such interruptions might be a mechanism for gene inactivation was quickly dispelled when genes from tissues active in the production of particular proteins were shown to have exactly the same discontinuities as DNA from inactive tissues. The possibility of "hopping polymerases" that somehow knew to transcribe only the exon portions of genes was also considered. A third possibility was RNA splicing, whereby the introns would be removed by RNA processing after the entire gene had been transcribed into RNA. Whatever the mechanism, it clearly had to be exquisitely precise. If intron removal were off by even one nucleotide, then missense or frameshifted proteins would result.

Many lines of evidence quickly converged to establish that **RNA splicing** must be the mechanism of intron removal for both protein-coding and RNA-specifying genes. For example, when R loops were prepared using nuclear RNA (hnRNA) and an interrupted gene like β-globin or ovalbumin, occasional hybrid molecules appeared where the DNA regions corresponding to introns were not looped out (Fig-

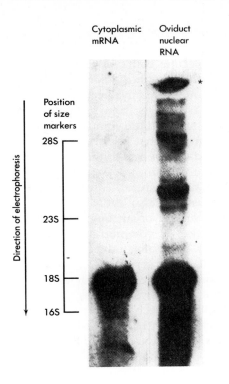

Cytoplasmic mRNA   Oviduct nuclear RNA

Direction of electrophoresis

Position of size markers

28S
23S
18S
16S

Figure 20-10
Detection of high-molecular-weight ovalbumin mRNA precursors by **Northern blotting.** In this procedure an RNA sample is first fractionated on a gel and then transferred to a specially modified paper with which it makes covalent crosslinks. [See J. C. Alwine, D. J. Kemps, and G. R. Stark, *Proc. Nat. Acad. Sci.* 74 (1977):5350–5354.] Then, as in Southern blotting (Figure 19-15), bands of complementary RNA are detected by hybridization with labeled DNA probes. Here, note the presence of many discrete nuclear RNA species up to four times longer than the mature ovalbumin mRNA. These have been detected with a cDNA probe (containing only mRNA sequences). Hybridization with probes containing only intron sequences from the cloned gene can reveal the sizes of nuclear RNAs that still retain various introns. (The asterisk indicates an autoradiographic artifact, not a reproducibly observed RNA.) [Courtesy of D. R. Roop, J. L. Nordstrom, S. Y. Tsai, M. J. Tsai, and B. W. O'Malley, *Cell* 15 (1978):671–685. Used with permission.]

ure 20-9). Rather, a complete R loop was formed, indicating that a continuous transcript of the entire gene was being detected. Likewise, hybridization experiments revealed the existence of tiny amounts of nuclear poly A–containing RNA that was not only significantly longer than the corresponding cytoplasmic mRNA but also contained both intron and exon sequences (Figure 20-10). Usually, such longer nuclear molecules are detected in several different sizes. These apparent processing intermediates occur in varying amounts, suggesting that some introns in a particular pre-mRNA are removed more quickly than others. Yet, the deduced order of intron removal for several closely scrutinized genes correlates neither with intron size nor with intron position in the transcript.

At long last, the giant hnRNA molecules of mammalian cell nuclei had been explained. At least many of them must be mRNA precursors undergoing progressive RNA splicing before export from the nucleus. Their short lifetimes, large sizes, and equivalent contents of 5' caps and 3' poly A tails are totally consistent with a scenario whereby long internal stretches of nucleotides are excised from pre-mRNAs as they are fashioned into mature messenger molecules.

Once it was discovered that some tRNA genes are interrupted, tRNA precursors containing introns were also immediately detected. A yeast strain that had long been known to be defective in the production of functional RNAs was found to contain unspliced tRNAs. These pre-tRNA molecules, which are matured at their 5' and 3' ends but still retain intron sequences, proved to be ideal substrates for characterizing tRNA-splicing enzymes.

Similarly, after analysis of rRNA gene structure had revealed extra sequences in the region specifying the large rRNA, rRNA precursors containing introns were detected, most notably in the ciliated protozoan *Tetrahymena* (Figure 20-11). This organism provided experimental material that allowed the discovery of self-splicing RNA, which will be discussed shortly.

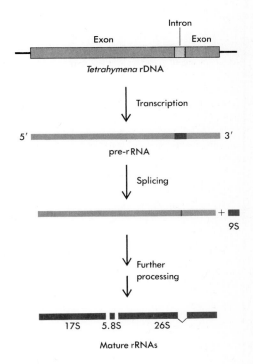

Exon   Intron   Exon

Tetrahymena rDNA

Transcription

5'   3'
pre-r RNA

Splicing

+ ▪
9S

Further processing

17S   5.8S   26S

Mature rRNAs

Figure 20-11
Synthesis of *Tetrahymena* rRNAs. A 9S fragment (about 0.4 kilobase) is excised from the 35S pre-rRNA as the first step in rRNA processing. The caret in the 26S rRNA represents the region from which the intron has been removed. [After A. J. Zaug and T. R. Cech, *Cell* 19 (1980):331–338.]

Deciphering Splicing Systems[28]

The availability of RNA precursor molecules containing introns challenged biochemists to ask whether accurate splicing could be carried out in the test tube. Indeed, systems now exist in which introns can be excised from nuclear pre-mRNA as well as from pre-tRNA and pre-rRNA molecules in vitro. From such studies, much has been learned about the biochemical mechanisms that achieve precise cutting at the **splice junctions** (the boundaries between intron and exon segments) and the subsequent ligation (joining) of exon sequences.

Meanwhile, an understanding of the structural features required for splicing RNA precursor molecules has also evolved. DNA sequences of split genes from many eucaryotic organisms have rapidly accumulated, allowing the identification of common sequence elements. Both naturally occurring and artificially induced mutations that affect splicing have been systematically scrutinized. These mutant analyses have confirmed our original guesses about which sequences are necessary, have uncovered other sequences, and have raised additional questions about splicing mechanisms.

We now believe that splicing reactions can be grouped into three distinct classes. One mechanism applies to rRNA molecules, another to mRNA molecules, and a third to tRNA molecules. This division corresponds exactly to the synthesis of these three kinds of transcripts by the three different eucaryotic RNA polymerases I, II, and III, respectively. Although these three types of splicing reactions all occur in eucaryotic cell nuclei, splicing also sometimes takes place in the other DNA-containing cellular compartments—the mitochondria and chloroplasts. The splicing of some mitochondrial mRNA and rRNA precursors closely resembles that of nuclear pre-rRNA.

The three splicing systems can be readily distinguished by examining the nature of the signals recognized by the splicing apparatus and their location in the precursor RNA. In pre-mRNAs, short sequences located directly at (or near) the splice junctions seem to be the principal determinants of where and how splicing occurs. In pre-tRNAs, splice junction sequences are not conserved. Rather, the common feature of tRNA introns is their location 3' to the anticodon. This placement would not disrupt the highly conserved three-dimensional folding of tRNA molecules (see Figures 14-5 and 14-6), arguing strongly that exon sequences and structures are major recognition elements for tRNA splicing. In sharp contrast, certain sequences internal to the intron are essential for pre-rRNA splicing (and for the excision of introns from at least some mitochondrial transcripts). These segments apparently interact to stabilize a precise folding of the RNA that juxtaposes the two splice junctions in such a way that a defined series of cleavage-ligation reactions occurs. The contrasting features of the three nuclear splicing systems are summarized in Table 20-1.

A Separate Nuclease and Ligase Act in tRNA Splicing[29, 30, 31]

The biochemistry of tRNA splicing has been best described for yeast, where the enzymes involved have been purified (Figure 20-12). In the first stage, two cuts precisely at the ends of the intron are executed by a splicing endonuclease associated with the nuclear membrane. Using a mechanism common to many ribonucleases, the endonucle-

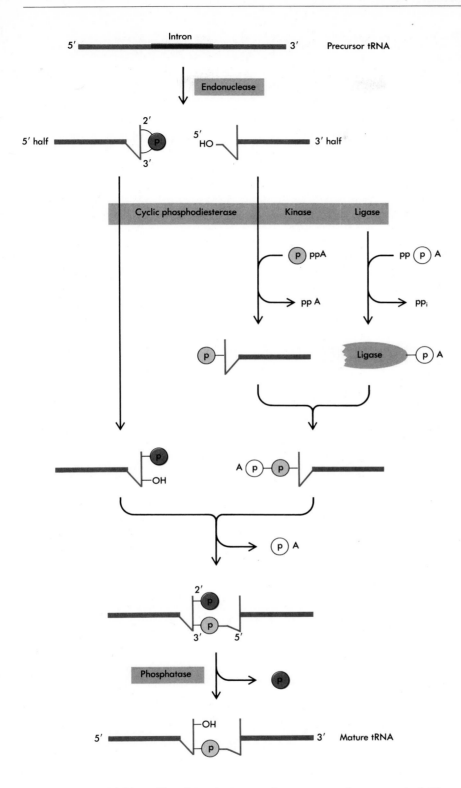

Figure 20-12
The splicing of yeast tRNA. Although the splicing of mammalian pre-tRNAs proceeds via a distinctly different pathway, adenylylated enzyme intermediates also appear to be involved. [After C. L. Greer, C. L. Peebles, P. Gegenheimer, and J. Abelson, *Cell* 32 (1983):537–546.]

ase leaves a 2'-3' cyclic phosphate on the two newly generated 3' ends and a hydroxyl group at the new 5' ends. In the second stage of splicing, the tRNA halves are joined by a splicing ligase in a series of reactions requiring ATP. The process begins when a phosphate group is added to the 5'-OH group of the 3' tRNA exon with the concomitant conversion of ATP to ADP. Next, the 5' phosphate end is activated by the transfer of an AMP group from the splicing ligase itself.

Table 20-1   Major Features of RNA Splicing in Eucaryotes[†]

Final Product	RNA Polymerase Synthesizing Pre-RNA	Location of Signals Important for Splicing	Distinguishing Characteristics
rRNA	I	Intron and splice sites	RNA catalysis, guanosine cofactor
mRNA	II	Splice sites, intron branch point	Lariat formation, involvement of small nuclear RNPs
tRNA	III	Exons	ATP-activated enzyme intermediates
Organelle RNA (mitochondrial, chloroplast)	Organellar		
Group I Introns* (rRNA, tRNA, mRNA)		Intron and splice sites (structure of intron same as rRNA above)	RNA catalysis, guanosine cofactor, intron-encoded splicing proteins
Group II Introns* (mRNA)		Intron and splice sites (sequences at splice sites similar to mRNA above)	RNA catalysis, lariat formation, intron-encoded splicing proteins

[†]See Chapter 28 for a discussion of splicing in procaryotic organisms.
*For a discussion of the generality of the Group I and Group II designations, see T. R. Cech, *Cell,* 44 (1986):207.

This AMP was originally derived from a second ATP, but first appears in the form of the adenylylated enzyme. Meanwhile, a cyclic phosphodiesterase activity has opened the 2'-3' cyclic phosphate on the 5' exon to form a 2' phosphate. The final reaction that joins the two tRNA exons is remarkable in that it creates a 2' phosphate 3'-5' phosphodiester linkage in the newly spliced tRNA. A hallmark of this unusual mechanism is that the phosphate group at the spliced position is not one that was present in the pre-tRNA. Rather, it was donated during the kinase reaction by ATP. Both the kinase and ligase activities, as well as the cyclic phosphodiesterase activity, appear to be carried by the single polypeptide chain of the splicing ligase (MW about 90,000). After the two halves of the tRNA are joined, a separate phosphatase enzyme removes the 2' phosphate, for it is not found in mature yeast tRNA molecules.

The separate nuclease and ligase enzymes isolated from yeast cells are specific for pre-tRNA molecules and do not act on other RNAs. Furthermore, the two tRNA halves generated by excision of the intron must be properly hydrogen-bonded together (in the tRNA cloverleaf form) for ligation to occur; in other words, the 5' exon of one tRNA molecule cannot be joined to the 3' exon of another tRNA molecule. Even nucleotide changes at one of the splice junctions or insertion of sizable sequences into the intron do not interfere with correct tRNA splicing. Together, these observations reinforce our notion that the similar three-dimensional structure into which all tRNA exons fold is critical for their recognition by the tRNA-splicing machinery.

Not all tRNA splicing proceeds via the bizarre 2' phosphate 3'-5' phosphodiester intermediate identified in yeast. Although this intermediate probably also occurs in other lower eucaryotes and plants, in vertebrates the phosphate group that joins the two exons in the final spliced tRNA is not donated by ATP. Instead, the phosphate that originally linked the 5' exon and the intron in the pre-tRNA is retained at the spliced junction. The details of the biochemical mechanism must therefore be different from that diagrammed in Figure 20-12. Yet, the three-dimensional structural features of tRNAs that allow pre-tRNA recognition by splicing enzymes appear to be conserved in all organisms, as yeast pre-tRNAs containing introns are readily spliced by the vertebrate machinery.

The Self-Splicing of rRNA: G Nucleoside Involvement and the Creation of a Circular Intron[32, 33, 34]

Ribosomal RNA processing in several lower eucaryotes involves excision of intron sequences from a site near the 3' end of the region that becomes the longest (23S-like) mature rRNA. Dissection of this splicing reaction in *Tetrahymena thermophila* has revealed that proteins are not essential. Instead, the splicing activity is intrinsic to the precursor rRNA molecule itself.

In this fascinating mechanism, a 413-nucleotide intron is excised and circularized by a series of **phosphoester transfer** reactions (Figure 20-13). The train of events begins when the 3'-OH group of a G nucleoside attacks the phosphodiester bond at the 5' end of the intron and adds itself to the intron via a normal 3'-5' phosphodiester bond. Next, the newly generated 3'-OH group on the 5' exon similarly attacks the phosphodiester bond at the other end of the intron, creating a spliced joint and releasing the intron (terminated again by a 3'-OH). Finally, the linear intron undergoes autocyclization: Its 3'-OH group attacks a bond near its 5' end, releasing a 15-nucleotide-long fragment that starts with the G residue added in the initial step.

Elegant proof that the pre-rRNA can splice itself has been obtained by cloning DNA sequences specifying the intron and its flanking regions downstream from the *E. coli lac* promoter, amplifying this recombinant DNA plasmid in *E. coli*, and then transcribing with *E. coli* RNA polymerase in vitro. The RNA synthesized readily undergoes self-splicing in the test tube, despite the fact that it never contacted any *Tetrahymena* proteins.

The requirements for this amazing RNA-catalyzed reaction are minimal. Because the 5' end of the G nucleoside cofactor does not participate directly in any of the phosphoester transfer reactions, it can be modified in a number of ways; for example, 5'-GMP, GDP, or GTP work as well as a simple G nucleoside. By contrast, nucleosides modified at the 2'- or 3'-OH groups are inactive as cofactors. G analogs in which those groups on the purine ring that normally hydrogen-bond to C have been altered do not work as well, suggesting that standard GC base-pairing interactions may be important for positioning the G cofactor properly on the RNA surface. Both monovalent and divalent (usually Mg^{2+}) cations are required, but no GTP or ATP hydrolysis is involved. The lack of an energy requirement may be explained by the linking of each ligation step to a cleavage step: In the overall splicing reaction, phosphodiester bonds are neither lost nor gained. Since every one of the phosphoester transfer reactions is in theory reversible, some explanation is needed for why splicing proceeds in the forward direction. In vivo, rapid degradation of the intron (observed for both its linear and circular form) or export of the spliced rRNA to the cytoplasm may be responsible.

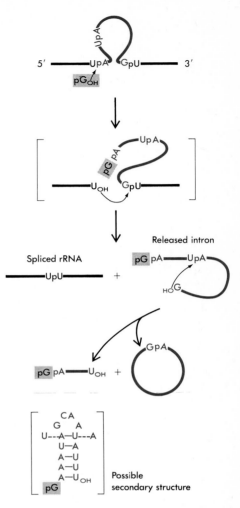

Figure 20-13
The self-splicing of *Tetrahymena thermophila* rRNA. Here the initiating G nucleoside is GMP (boxed). Intron sequences are in color. Square brackets indicate a postulated intermediate that has not been isolated. A possible secondary structure for the 15-mer released by the final reaction is shown below, with sequence differences in the intron of *Tetrahymena pigmentosum* indicated. [After A. J. Zaug, P. J. Grabowski, and T. R. Cech, *Nature* 30 (1983):578–583.]

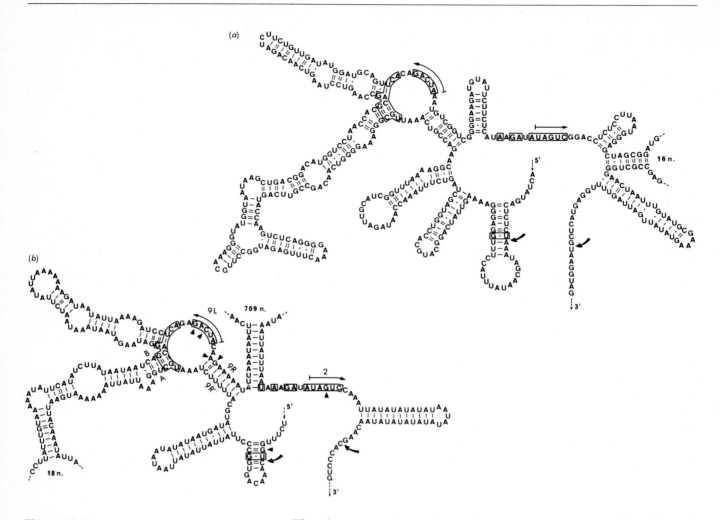

Figure 20-14
Similar sequences and secondary structures found in the *Tetrahymena thermophila* rRNA intron (a) and a yeast mitochondrial protein intron (the fourth in the pre-mRNA for cytochrome oxidase) (b) Large arrows designate the splice sites. Boxed sequences are invariant between these and some other mitochondrial introns. Triangles point to the sites of splicing defective mutants. Sequence elements A and B, 9R and 9R', and 9L and 2 are believed to base pair, establishing a defined intron tertiary structure. [Courtesy of F. Michel and B. Dujon, *EMBO J.* 2 (1983):33–38. Used with permission as modified by T. R. Cech and B. L. Bass.]

The observation that the excised intron is capable of undergoing an internal splicing reaction to produce a covalent circle tells us that the cleavage-ligation activity of the pre-rRNA must reside within the intron itself. Indeed, certain small deletions of the intron far from the splice junctions totally prevent splicing. A secondary structure model for the *Tetrahymena thermophila* intron that brings its 5' and 3' ends into close proximity is illustrated in Figure 20-14. Also indicated are certain short sequences found in common with some mitochondrial introns. In the mitochondrial introns, these sequences have been pinpointed by mutational analyses as essential elements in splicing.

Mitochondrial Introns Encode Splicing Proteins[35–40]

Whereas the mitochondrial DNA of mammalian cells does not contain introns, many mitochondrial genomes in lower eucaryotes do. Interruptions have been detected in both protein-coding and rRNA-specifying genes. Analyses of numerous mutations affecting expression of such mitochondrial genes suggest that the splicing of their transcripts may be carried out in various ways; even intron removal from two different pre-mRNAs can have different requirements. Among the components identified as essential for splicing are nuclear-encoded proteins, specific sequences within introns, and polypeptides encoded by the intron sequence itself (or by the intron of another gene).

Participation of intron-encoded proteins is a unique aspect of the mitochondrial splicing process. In some interrupted genes in yeast mitochondria, an open reading frame that begins in the first exon simply continues through the splice junction into intron sequences (Figure 20-15). Nonsense mutations introduced into these lengthy intron-coding regions not only lead to the accumulation of prematurely terminated polypeptide chains, but also block the splicing of that intron. Hence, such intron-encoded polypeptides are agents in the destruction of their own mRNA, which unavoidably occurs upon intron excision. So far, this scenario seems limited to situations where transcription and translation occur in the same cellular compartment, such as the mitochondrion. Recall that the other splicing systems that act on nuclear pre-mRNAs, tRNA, or rRNA precursors operate in the ribosome-free environment of the nucleus.

Yet, it is not intron-encoded proteins but another intron feature that ties together nuclear rRNA splicing in *Tetrahymena* and the removal of some mitochondrial mRNA and rRNA introns in yeast and other fungi. Four short regions (each 10 to 12 nucleotides long) of sequence homology have been recognized in all these introns, despite their diverse origins (see Figure 20-12). The four conserved sequences can be base-paired to form two stems that appear crucial in defining a particular three-dimensional structure for the intron; splicing mutations occurring in any one of these regions can be suppressed by changing the complementary region to restore complementarity. Two additional intron regions that are not conserved in exact sequence but are always complementary to each other are also sites of splicing mutations. The remarkable phylogenetic conservation of these three pairs of short complementary sequences argues strongly that precise intron folding is essential for splicing. Not surprisingly, the excision of (at least some) mitochondrial introns also appears to involve RNA catalysis. Additional requirements for intron- and nuclear-encoded proteins may be explained by their role in stabilizing the proper intron conformation for self-splicing.

Conserved Sequences at Exon-Intron Boundaries in Pre-mRNA[16, 41–44]

In contrast to the importance of exon sequences for tRNA splicing or intron sequences for rRNA and mitochondrial splicing, signals crucial for the splicing of mammalian pre-mRNAs reside directly at the splice junctions themselves. (The exception to this rule is a short, conserved intron sequence in yeast.) Even when only a few sequences of split protein-coding genes had accumulated, it became obvious that the nucleotides at exon-intron boundaries are not random. Introns always seemed to begin with GT and end with AG.

Now that the sequences of several hundred interrupted genes have been compiled, it is clear that the **GT-AG rule** is nearly always obeyed and that somewhat longer consensus sequences can be written for both the 5' and 3' splice junctions. The nine-nucleotide consensus for 5' junctions extends three residues upstream and six residues downstream from the splice point (Figure 20-16). The average 5' junction matches this consensus in at least seven positions; only rarely does a 5' splice junction with as little as a five-residue match

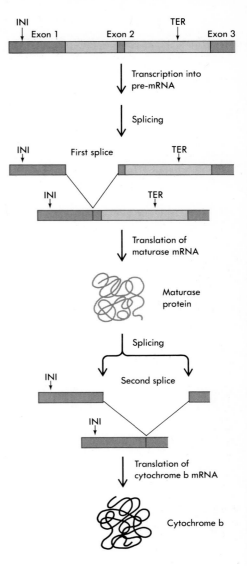

Figure 20-15
The action of an intron-encoded protein in yeast mitochondrial splicing. Only the first third of the cytochrome *b* gene is shown. After the first intron is excised, the RNA can be translated into maturase. Action of the maturase in excising the second intron creates the cytochrome *b* mRNA, but concomitantly destroys the maturase mRNA. Thus, no more synthesis of maturase takes place. INI and TER indicate translational start and stop codons. [After J. Lazowska, C. Jacq, and P. P. Slonimski, *Cell* 22 (1980):333–348.]

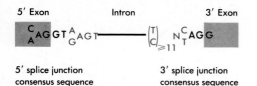

5′ splice junction consensus sequence

3′ splice junction consensus sequence

Figure 20-16
Consensus sequences for 5′ and 3′ splice sites in nuclear protein-coding genes from the whole range of eucaryotic organisms. The nearly invariant GT and AG dinucleotides are indicated. [For details, see S. M. Mount, *Nucleic Acid Res.* 10 (1982):459–472.]

occur. The consensus sequence for 3′ splice junctions is different from the 5′ consensus (Figure 20-16). It is composed of a pyrimidine-rich region of variable length (but always greater than ten nucleotides) followed by a short consensus sequence extending only three residues upstream and one residue downstream from the splice point. The pyrimidine-rich region is characteristically devoid of the sequence AG, the invariant dinucleotide that occurs just preceding the 3′ intron-exon boundary. Note that at both splice junctions, the conserved sequences extend farther into the intron than the exon, thereby placing only minimal constraints on the sequence of the mature mRNA. Quite remarkably, splice junction sequences from genes representing the whole range of eucaryotic species (yeast to human) conform well to the same 5′ and 3′ consensus sequences. This suggests that whatever recognizes these splicing signals must likewise have been highly conserved over evolutionary time.

Some of the most potent evidence supporting the importance of conserved junction sequences for pre-mRNA splicing has come from analyses of mutant human genes. **Thalassemias** are genetic diseases in which the synthesis of hemoglobin polypeptides is defective. Recently, a number of thalassemias have been traced to mutations that interfere with the correct removal of introns from globin pre-mRNAs (Figure 20-17). Use of a 5′ splice site is abolished by changes in the invariant GT sequence and is decreased by mutations elsewhere in the nine-nucleotide consensus. The surprising result, observed with both the thalassemias and other 5′ splice site mutants, is that splicing is not simply prevented. Rather, new sites called **cryptic splice sites,** which match the 5′ consensus but are normally not used, become activated. Also leading to aberrant splicing of the pre-mRNA molecule in other thalassemias are mutations that create a new 5′ or 3′ splice junction consensus sequence. While the bizarre results of splice site mutations do not tell us how the splicing machinery discriminates between true splice junctions and other regions that fit the consensus sequences, they do emphasize that junction recognition is basic to pre-mRNA splicing.

It is therefore not surprising that huge blocks of sequence can be deleted from the interior of most introns with little noticeable effect on splicing. Only when deletions begin to impinge on the conserved splice junction sequences do aberrations in splicing occur. Moreover, chimeric introns—created by cutting the DNA within two different introns and joining the 5′ and 3′ halves one to the other—function quite nicely, showing that splice junction signals are interchangeable. All these observations underscore the versatility of the pre-mRNA splicing machinery and are entirely compatible with the huge variations known to occur in intron size and sequence.

Small Nuclear RNA-Protein Complexes in Search of Functions[45, 46, 47]

A distinctive feature of mammalian cells, as compared to bacteria, is their great abundance of small RNA species that are neither transfer RNAs nor ribosomal RNAs. Many of these species reside in the nucleus rather than the cytoplasm. Cells of all vertebrate species contain six different **small nuclear RNAs (snRNAs)** called U1, U2, U3, U4, U5, and U6, which range in size from about 100 (U6) to 215 (U3) nucleotides (Figure 20-18). These RNAs are relatively stable (having lifetimes at least as long as a cell generation), possess unusual

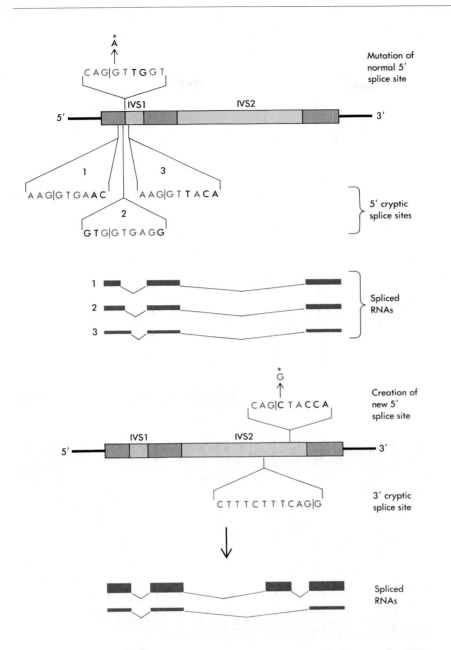

Figure 20-17
Human β-thalassaemia genes that produce wrongly spliced β-globin RNAs. Mutation of the first nucleotide (G to A) of the first intron (IVS 1) leads to no correct splicing but instead utilization of three cryptic 5′ splice sites. Mutation of a single base pair within the second intron (IVS 2) creates a new 5′ splice site and uncovers a cryptic 3′ splice site, leading to a predominance of misspliced over correctly spliced RNA. The more abundant products are indicated by thicker bars. Nucleotides conforming to the 5′ or 3′ splice junction consensus sequences (Figure 20-16) are in color. Vertical lines indicate the splice point. None of the misspliced RNAs can be translated into functional β-globin chains. [After R. Treisman, S. H. Orkin, and T. Maniatis, *Nature* 302 (1983):591–596.]

trimethylated ($m_3^{2,2,7}G$) cap structures at their 5′ ends (except for U6), and contain some other modified nucleotides (as do tRNAs and rRNAs). They all exist not as naked RNA molecules but stably complexed with certain abundant nuclear proteins. The resulting particles are referred to as **small nuclear ribonucleoproteins (snRNPs)**. Although snRNPs are only about one-fifth the mass of small ribosomal subunits (10 to 12S) and have higher protein/RNA ratios, like ribosomes, each snRNP contains several different protein species. Some of these polypeptides are shared by those snRNPs containing U1, U2, U5, and U4/U6 RNAs (only in the last case do two RNAs coexist in the same snRNP particle), whereas other polypeptides are unique to particular snRNPs. The existence of common and unique protein components was first recognized with the discovery that some patients with the disease systemic lupus erythematosus make **autoantibodies** (normally forbidden antibodies that bind one's own cellular components) directed against snRNP proteins. For instance, some autoantibodies precipitate U1, U2, U5, and U4/U6 snRNPs and therefore recognize

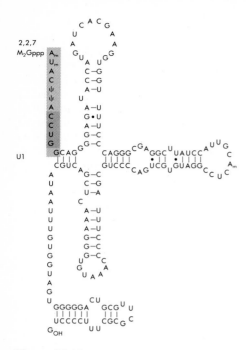

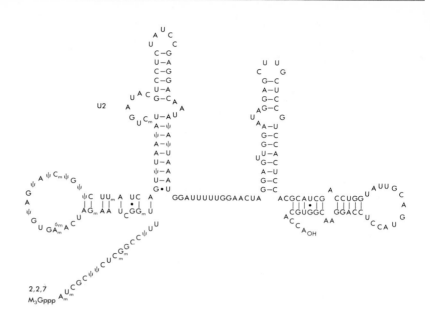

Figure 20-18

Sequences and probable secondary structures for human U1 and U2 snRNAs. U1 nucleotides complementary to the 5' splice junction consensus sequence (Figure 20-16) are boxed in color; those complementary to the 3' splice junction consensus sequence are boxed in gray. This region of U1 snRNA is not only available for interaction with other RNAs but is completely conserved in sequence from mammals to sea urchins. U1 and U2 are the most abundant of all snRNAs in mammalian cells. Both appear to be involved in pre-mRNA splicing. Note the presence of many 2'-O-methyl modifications (m) as well as pseudouridines (ψ), 6-methyladenosine (6m), and the trimethyl cap structure at the 5' ends. [See H. Busch, R. Reddy, and L. Rothblum, *Ann. Rev. Biochem.* 5 (1982):617–654.]

shared polypeptides; other autoantibodies precipitate U1, U2, or U3 snRNPs singly because they bind polypeptides unique to each of these particles.

Several properties of snRNPs argue that they play important roles in nuclear function. First, both their protein components and the sequences of their RNAs are highly conserved across all higher eucaryotic species (e.g., mammals, amphibians, and insects). Even in lower eucaryotes like yeast or the slime mold *Dictyostelium*, small nuclear RNAs with homologous sequences have been characterized. Second, snRNPs are very plentiful, particularly in mammalian cells, where the number of each type is equal to 1 to 10 percent the number of ribosomes. Since the U3 snRNP has been localized to the nucleolus, it is expected to participate somehow in ribosome biogenesis. The remaining snRNPs reside in the nucleoplasm (the remainder of the nucleus) and are often found associated with hnRNA. Hence, it is suspected that they contribute to various aspects of pre-mRNA processing.

Involvement of the U1 snRNP in Pre-mRNA Splicing[48–51]

U1 RNA is the most abundant of all the small nuclear RNAs in vertebrate cells. It has at its 5' end a sequence that is precisely complementary to the nine-nucleotide consensus for 5' splice junctions (see Figure 20-18). This generated the idea that the U1-containing small RNA–protein complex (U1 snRNP) might use base pairing to interact with splice sites and thus be an essential component of the pre-mRNA splicing apparatus. That this is the case has been verified by the use of patient autoantibodies directed against protein components of the U1 snRNP. When added to in vitro splicing systems, such antibodies effectively block intron removal. Moreover, digesting away only the 5' end of U1 RNA (just the region complementary to the 5' splice junction consensus sequence) prevents in vitro splicing. Studies performed in a more purified system demonstrate that U1 snRNPs specifically bind and protect from nuclease digestion a stretch of pre-mRNA sequences just at the 5' end of the intron. Both

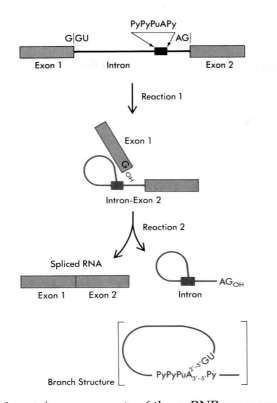

Figure 20-19
Formation of a branched RNA intermediate during pre-mRNA splicing. The reaction sequence applies to both mammalian and yeast in vitro splicing, but the sequence at the branch point is more rigorously defined in yeast. The phosphate group at the spliced junction in the newly matured mRNA is the one present at the 5′ end of exon 2 in the pre-mRNA. [After E. Brody and J. Abelson, *Science* 228 (1985):963–967.]

the RNA and protein components of the snRNP are essential for this selective binding, reminiscent of the requirements for 30S ribosomes to interact with start sites on procaryotic mRNAs during polypeptide chain initiation (Chapter 14).

What recognizes the 3′ ends of introns during pre-mRNA splicing is not yet clear. Another most intriguing unsolved problem is how the splicing machinery faithfully joins one end of an intron to the other and avoids mistakenly excising exons from a transcript that contains multiple introns. One possibility is that some splicing component tracks along the intron, ensuring that neighboring splice sites are paired. Yet, even if such a mechanism exists, it is difficult to explain why cryptic sites in nonmutant pre-mRNAs are ignored and how the occasional intron that utilizes multiple 5′ or 3′ splice sites functions (Chapter 21).

Lariat Structures Are Formed During Pre-mRNA Splicing[52–58]

As was the case for tRNA and rRNA splicing, probing the biochemical mechanism of pre-mRNA splicing has uncovered the existence of unusual intermediates leading to intron removal. An early step, which requires prior U1 snRNP binding and ATP hydrolysis, is a cut at the 5′ splice junction accompanied by formation of an unusual 2′-5′ phosphodiester bond between the 5′-terminal G residue of the intron and a site about 30 nucleotides upstream from the 3′ end of the intron. This creates an RNA lariat structure in which an A residue invariably forms the branch point (Figure 20-19). The branched A nucleotide engages in three phosphodiester bonds: the two ordinary 3′-5′ bonds plus a novel 2′-5′ bond. In mammalian pre-mRNAs, the sequence surrounding this A residue is only very loosely defined (see Figure 20-19), and intron sequences can be deleted nearly at random.

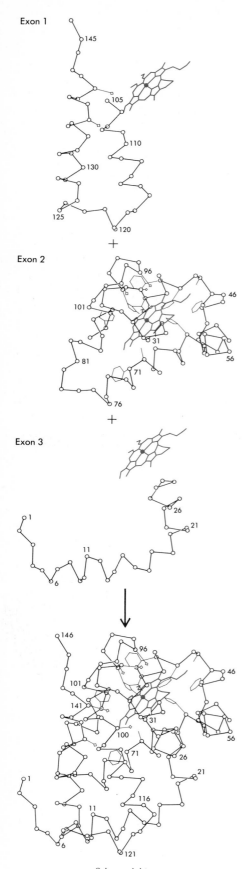

Exon 1

Exon 2

Exon 3

β-hemoglobin

Figure 20-20
A backbone protein model showing how the three exons of the β-hemoglobin gene specify different domains of the protein. The heme group is shown in color. The amino acid side chains contacting the heme are drawn symbolically. [After M. Go, *Nature* 291 (1981):90–92.]

In yeast, on the other hand, the branch always occurs at the last A within a particular sequence (TACTAAC), which appears in every intron 20 to 55 nucleotides upstream of the 3' splice site. Deletion of this yeast sequence or certain mutations within it abolish excision of that intron. Apparently, the yeast splicing machinery is more demanding than the mammalian apparatus in requiring that a defined sequence occur at the branch point.

Many components must be required to assemble an active splicing complex, for lariat intermediates formed in in vitro splicing reactions sediment very rapidly—in the range of 40S (for yeast) to 60S (for mammalian systems). U2 as well as U1 snRNPs appear to be essential components of this complex (**spliceosome**). During the splicing reaction, U2 snRNPs associate with the intron branch point while U1 snRNPs interact with the 5' splice site and some other component (another snRNP?) binds the 3' splice junction in the pre-mRNA. All these participants seem to assemble along an ordered pathway, just as ribosomal subunits do during the initiation of protein synthesis (Chapters 14 and 21). Very soon, a complete description of the many components required for the pre-mRNA splicing process should become available.

Exons Sometimes Correspond to Protein-Folding Domains[59, 60]

In many instances where we know both the three-dimensional structure of a protein and the distribution of introns and exons in its gene, an intriguing correlation can be perceived; namely, exons often correspond to distinct **protein-folding** (or functional) **domains** of the encoded polypeptide (Chapter 5). A vivid example is provided by the immunoglobulins (Chapter 23), where exon boundaries precisely coincide with junctions between the separate folded domains of the protein molecule.

Another interesting case is the β-globin gene, whose three exons can be identified with particular functional as well as structural regions of the protein (Figure 20-20). First, it was shown that the amino acid residues specified by the central exon alone are capable of binding heme tightly and specifically. Second, a detailed analysis of the course of the polypeptide chain in the β-globin three-dimensional structure suggested that the molecule possesses four distinct subdomains or compact structures, each composed of continuous stretches of amino acid residues (Chapter 5). One subdomain corresponds to the first exon of the gene and one to the third exon, but two subdomains are specified by the central exon. Quite remarkably, when the sequence of the gene for soybean leghemoglobin (a related oxygen-binding protein in leguminous plants) was determined, a third intron was found precisely at the division between the two subdomains of what is the central exon of animal globins.

Likewise in the enzyme alcohol dehydrogenase, introns neatly divide structural and functional domains of the protein. Alcohol dehy-

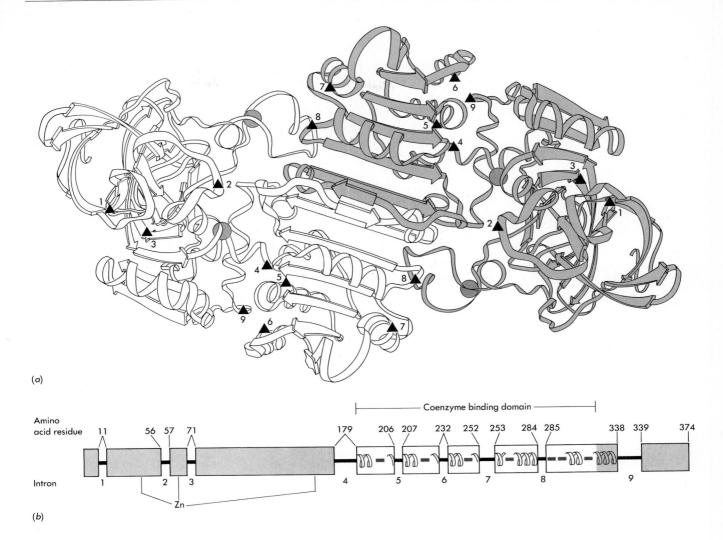

(a)

(b)

Figure 20-21

Division of the alcohol dehydrogenase gene into structural domains. (a) The three-dimensional structure of horse liver alcohol dehydrogenase. One of the two identical subunits is shaded, and the zinc ligands are shown as balls. (b) The amino acid sequence of the liver enzyme is aligned with intron positions (gaps) in the maize alcohol dehydrogenase gene. The catalytic domain is shaded gray and its contacts with one of the zinc atoms are shown. The α-helical regions are coiled and strands of β-pleated sheet are indicated by colored lines. The introns are numbered and their positions in the three-dimensional structure are indicated by ▲ in (a). [After C.-I. Branden, H. Eklund, C. Cambillau, and A. J. Pryor, *EMBO J.* 3 (1984):1307–1310.]

drogenase utilizes the coenzyme NAD^+, which binds to a domain of the folded polypeptide chain similar in its three-dimensional structure to the NAD^+ binding domains of two other dehydrogenases (lactate dehydrogenase and glyceraldehyde-3-phosphate dehydrogenase). Exon-intron junctions in the gene mark the edges of the coenzyme binding domain: this portion of the protein is encoded by five of the ten exons (numbers 5 to 9), whereas amino acid residues involved in catalysis are specified by exons 1 to 4 and 10 (Figure 20-21). Moreover, within the part of the gene encoding the coenzyme binding domain itself, introns are distributed in an interesting way: Intron 7 separates two structurally similar mononucleotide binding units (recall from Chapter 6 that NAD^+ is a dinucleotide), and introns 5 and 6 in turn subdivide the first of these units into three repeating sections, each composed of a short stretch (about 20 residues) of α helix followed by β strand.

Split Genes May Enhance the Rate of Evolution[61–66]

The existence of introns immediately raises evolutionary questions. Were these interruptions in genes acquired by eucaryotic cells because they fulfilled some special purpose? Or did they exist in primordial messages, to be later discarded by bacteria under pressure to streamline their genomes for maximal reproduction rates? As yet, the evidence in favor of one or the other of these alternatives is not conclusive. However, several aspects of the structures of interrupted genes hint that the presence of introns (coupled with the ability to excise them at the RNA level) could have been used to advantage during the evolution of higher eucaryotic genomes.

Since exons can correspond to separate structural or functional domains of polypeptide chains, mixing strategies may have been employed during evolution to create new proteins. Having individual protein-coding segments separated by lengthy intron sequences spreads a eucaryotic gene out over a much longer stretch of DNA than the comparable bacterial gene. This will simultaneously increase the rate of recombination between one gene and another and also lower the probability that the recombinant joint will fall within an exon and create some unacceptable aberrant structure in the new protein. Instead, recombination will most likely take place between intron sequences, thereby generating new combinations of independently folded protein domains. The transcript of such a new gene would have at least one hybrid intron, but this should pose no problem for the splicing machinery, since internal intron sequences are largely dispensable and splice junction signals interchangeable.

As yet, there are no clear-cut examples where different exons of one gene can be traced to exons present in two other genes. However, a number of protein-coding genes are known to contain two or more exons that are obviously closely related to one another. These presumably were generated by unequal crossing over between intron sequences followed by divergence of the duplicated exons to provide new functions. Both the immunoglobulins (Chapter 23) and collagen (see Figure 20-7) are conspicuous examples of genes that contain multiple copies of related exons. The chicken ovomucoid gene appears to have arisen by triplication of an ancestral protein-coding DNA segment split by one intron. Homologies between the three domains (which are themselves separated by introns) are quite striking (Figure 20-22).

Perhaps even more important for understanding the evolution of higher organisms is the realization that introns allow a new protein to be tested while enabling the organism to retain expression of the original (perhaps essential) product. We know that even a point mutation can occasionally create a new 5' or 3' splice junction signal where none previously existed. Because the pre-mRNA splicing machinery has the capacity to join a single 3' splice site to either of two (or more) 5' splice junctions (or vice versa), it is conceivable that a mutated gene might then be spliced to yield two different mRNAs. Thus, both the old and a new protein could be synthesized simultaneously. Depending on the position of the mutation, coding regions could be either expanded or contracted in this manner. Eucaryotic cells thus enjoy an evolutionary advantage, for in procaryotes a mutation necessarily destroys the old gene product when creating a new product, unless a duplication of that particular gene has already taken place (Chapter 7).

Examples of genes that have added coding sequences at splice junctions abound if one compares homologous proteins from different organisms. Interestingly, where three-dimensional protein structures are known, the added sequences often correspond to loops on the surface of the folded polypeptide chain and thereby do not perturb

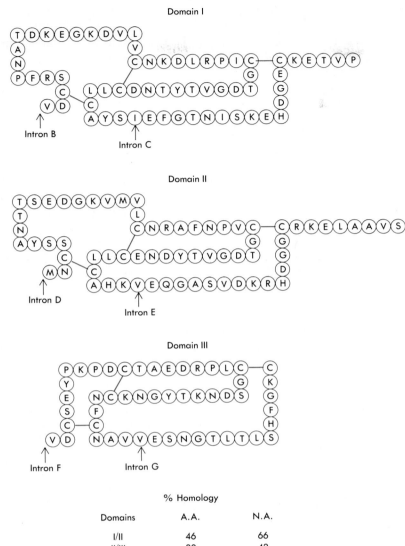

Figure 20-22
Alignment of the three domains of chicken ovomucoid, showing the similar placement of disulfide bonds (indicated by colored lines) and positions of introns in its gene. The amino acid (A.A.) and nucleic acid (N.A.) sequence homologies are listed below. [After J. P. Stein, J. F. Catterall, P. Kristo, A. R. Means, and B. W. O'Malley, *Cell* 21 (1980):681–687.]

| | % Homology | |
Domains	A.A.	N.A.
I/II	46	66
II/III	30	42
I/III	33	50

the internal structure of the molecule (Figure 20-23). Recently, the α-crystallin gene of the mouse has been shown to specify two polypeptide products that differ by a 23-amino-acid insertion; two mRNAs, one with an additional internal exon, are responsible. This is clearly an alternative use of splice junctions (Chapter 21) and could represent a gene caught in the act of testing a new product.

Single Copies of Most Protein-Coding Genes per Haploid Complement[67]

The presence of lengthy intron sequences within most protein-coding genes multiplies our estimate of the amount of DNA required to encode a vertebrate protein by about tenfold. Yet, this factor can account for only a fraction of the excess DNA known to reside in higher eucaryotic genomes. Another way of expanding the DNA content would be to have multiple copies of some or all genes.

Figure 20-23

How the creation of new splice junctions can account for length differences in related proteins. On the left, a hypothetical gene containing two introns labeled A and B is transcribed; the introns are excised precisely, and the RNA is spliced to form the functional mRNA. The mRNA is then translated into protein. The letters A and B on the protein mark the splice junctions. On the right, a mutation alters the point of excision of intron B. This results in an extension of the exon and hence an added segment in the functional mRNA. On translation, this produces a protein that has an additional surface loop at the splice junction B. Such a situation is found in the family of serine proteases. [After C. S. Craik; W. J. Rutter, and R. Fletterick, *Science* 220 (1983):1125–1129.]

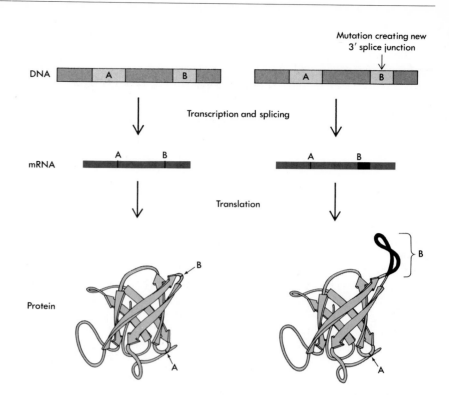

Figure 20-24

Arrangement of human α- and β-globin gene clusters and temporal expression of the individual genes. The colored boxes are functional genes, while those designated ψ are pseudogene regions. At all times in development, two α-like and two β-like polypeptide chains assemble to form functional hemoglobin tetramers. [After T. Maniatis, P. Melon, V. Parker, N. Proudfoot, and B. Seed, in *Molecular Genetics of Neuroscience*, ed. S. J. Bird and F. E. Bloom (New York: Raven Press, 1982), pp. 87–101.]

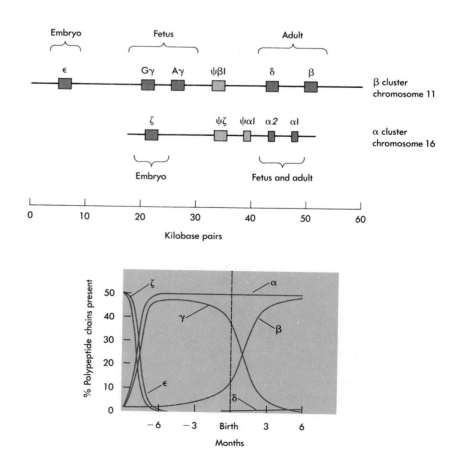

The number of DNA-RNA hybrid molecules that are formed when purified mRNA molecules are mixed with cellular DNA provides a method for measuring the number of times a given gene appears within a haploid genetic complement. Such measurements became possible when the first specific mRNA molecules were isolated from cells engaged in the selective synthesis of specific proteins. For example, studies using the 10S mRNA molecules that code for the α or β polypeptide chains of hemoglobin revealed that only one copy of each of these genes is present per haploid genetic complement. This conclusion holds not only for the egg and sperm but also for the cells (erythroblasts) differentiated for the almost exclusive synthesis of hemoglobin. Likewise, the gene encoding the silkworm protein fibroin is present in only one copy per haploid complement, even in the silk gland cells actively making silk. These examples are not exceptional; now that cDNAs corresponding to many mRNAs have been isolated and used to probe genomic DNA, we repeatedly find that only one hybridizing region is detected. These solitary numbers reflect the fact that the final level of protein synthesis is usually the result of two levels of amplification: the first occurring when a gene is transcribed many times to produce a large number of messenger RNA molecules, and the second occurring when these mRNA molecules in turn act as templates for many, many rounds of protein synthesis. The one fibroin gene of the silkworm, for example, can give rise in four days to some 10^5 fibroin mRNA molecules, which in turn serve as the templates for translation of some 10^{10} fibroin polypeptides—enough to spin an entire cocoon!

Developmentally Expressed Globin Genes Are Arranged in Sequence Along the Human Genome[68]

Although there is only one gene specifying the α or β polypeptide chain of hemoglobin molecules found in the adult mouse or human, this does not mean that a single α- or β-globin gene suffices during development of a complex mammalian organism. In fact, the requirement for slightly different hemoglobin proteins at various stages of development is fulfilled by having a whole family of related α-globin genes and another family of related β-globin genes. The expression of these genes is turned on and off coordinately and sequentially in response to changing oxygen tension, pH, and intracellular milieu as the embryo becomes a fetus and the fetus is born.

In humans, the α-like globin genes are clustered together on chromosome 16, while the β-like globin genes reside on chromosome 11 (Figure 20-24). Between the genes are long stretches of nontranscribed DNA devoid of other known genes. Also included are **pseudogene** sequences ($\psi\beta1$, $\psi\zeta$, and $\psi\alpha1$), regions that are highly homologous to the globin genes but do not give rise to detectable mRNA. Whether these sequences are former genes that have been mutated to nonfunctionality or whether they represent regions undergoing evolution to some new function is not known. Overall, both the α- and β-globin gene families encompass many kilobases of DNA, most of which cannot be considered to be genetic in the strictest sense of the word. This situation contrasts sharply with the much more economical usage of DNA in the genomes of procaryotes (like E. *coli*) and even lower eucaryotes (like yeast). There, the spaces between genes are generally much more modest in length.

In both the α-globin and β-globin gene clusters in humans, the genes themselves are arranged in order of their expression during development, with the embryonic genes followed by genes that function during fetal and adult life (see Figure 20-22). A similar order of genes is found in most other mammals, although the total size of the α- and β-globin gene clusters vary considerably. In contrast, chicken β-globin genes expressed in the adult are flanked by the embryonic genes, showing that there is nothing obligatory about this particular gene order.

Sequences of the several α-globin genes or β-globin genes contained in each cluster are very similar within their protein-coding regions. For instance, the two fetal β-like polypeptide chains (G_γ and A_γ) differ in only one amino acid out of 141. In contrast, their noncoding regions (both introns and leader and trailer sequences) hardly appear related, except that the second intervening sequence is always larger than the first. These findings suggest that the two globin gene clusters arose by repeated duplication and divergence of original α- and β-globin genes, which themselves were apparently derived from a common ancestral gene. We expect the protein-coding sequences to be less divergent than the noncoding sequences because protein function (including the ability of each kind of α-globin chain to form heterologous dimers with each kind of β-globin) is a major constraint during evolution.

Other Families of Related Genes Also Show Common Evolutionary Origin[65, 69, 70]

Intensive work on both protein and DNA sequences has revealed that many proteins in higher organisms are encoded by families of highly homologous but nonidentical genes. The different members of a gene family may be switched on and off at various stages of development, as in the case of globins. Or, the needs of different tissues within an organism may call for the synthesis of slightly different variants of a protein at the same time but in different cells.

Both these demands are met by the actin gene family in mammals. At least six closely related actin polypeptides are synthesized in various cells at various times. Actin is a major component of the contractile apparatus in muscle cells; but one type of α actin is specific for skeletal muscle, another α actin is found only in cardiac muscle, and both types are found in the smooth muscle of the gut. Still other actins (called β and γ) are constituents of the cytoskeleton of nonmuscle cells. Furthermore, slightly different α actins are present in fetal and adult skeletal muscle. As many as 20 to 30 different actin-related sequences can be detected in the human genome by hybridization. These probably lie on different chromosomes. This situation contrasts with that in yeast, where only a single actin gene suffices in fulfilling the needs of this single-celled organism.

Other well-studied examples of proteins encoded by gene families in mammals are albumin and α fetoprotein (major components of blood plasma), the serine proteases, the interferons (which are instrumental in our defense against viral infection; Chapter 24), and the immunoglobulins (Chapter 23). In insects, the chorion proteins—which constitute major components of the egg shell—are encoded by

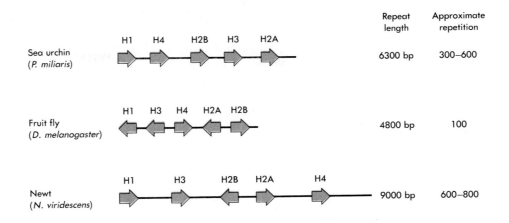

	Repeat length	Approximate repetition
Sea urchin (P. miliaris) H1 H4 H2B H3 H2A	6300 bp	300–600
Fruit fly (D. melanogaster) H1 H3 H4 H2A H2B	4800 bp	100
Newt (N. viridescens) H1 H3 H2B H2A H4	9000 bp	600–800

Figure 20-25
Organization of highly repeated histone gene clusters in several organisms. The genes (colored) are separated by nontranscribed spacer regions. The direction of transcription is indicated by arrows. The clusters are repeated tandemly in the sea urchin and fruit fly; in the newt, they are interspersed with 10 to 50 kilobase pairs of satellite DNA. [Data from C. C. Hentschel and M. L. Birnstiel, *Cell* 25 (1981):301–313.]

families of genes with many divergent members specialized for different roles (Chapter 22). Sometimes, the members of a gene family all reside in a single cluster on one chromosome. In other cases, they have been dispersed through evolution onto two or more chromosomes. Examination of their coding sequences, however, makes it clear that all genes in a family, regardless of how prodigious their numbers, have evolved from a single ancestral gene.

Multiple Copies of the Histone Genes[71, 72]

In some gene families, not only is there a need to encode related proteins, but many copies of identical genes are required. The genes for the histone proteins that package DNA into nucleosomes (Chapter 21) are a prime example. In most eucaryotes, there are five major histones called H1, H2A, H2B, H3, and H4. Since higher eucaryotes have relatively large amounts of DNA per cell, single-copy genes cannot produce sufficient numbers of histone proteins to package all the newly synthesized DNA that is made during the S phase of the cell cycle. Instead, higher eucaryotic genomes contain multiple copies of their histone genes—ranging from 10 to 20 in birds and mammals, and to 600 to 800 in sea urchins and newts. Yeast, in contrast, possesses only two copies of each of its four histone genes (for H2A, H2B, H3, and H4).

Organisms such as sea urchins and amphibians have the most highly repetitive histone genes because of their need for rapid cell division (and accompanying rapid DNA synthesis) during early development. Histone gene copy number must be high to supply the enormous amounts of histone mRNAs both stored in the egg and synthesized soon after fertilization. In these organisms, the five histone genes are grouped together into a basic unit that is tandemly repeated in a large head-to-tail array (Figure 20-25). However, the organization of the five individual histone genes within the basic repeat differs from one organism to another (and even between species). Sometimes, all five genes are transcribed off the same strand of DNA; in other cases, the genes have different orientations (see Figure 20-25). Because the sequence of the five histone proteins has been very highly conserved through evolution, the DNA sequences of the histone-coding regions in the basic repeat are quite conserved (despite the degeneracy of the genetic code). However, the spacer regions that separate the genes within the basic repeat unit are very different in different organisms and are at least as long as the protein-coding regions themselves.

In mammals and most other higher animals, cell division during early embryogenesis is not so rapid, and fewer copies of the histone genes are needed. The organization of these genes is quite variable. Although vertebrate histone genes are often clustered, the gene order is not fixed, nor are there tandem arrays. In addition, some individual histone genes are located far away from any other histonelike sequences. Nonetheless, in all eucaryotic organisms, histone genes share two features not shared by most other genes: Their transcripts almost always lack both introns and poly A tails.

Tandem Repeats of rDNA[73-77]

Even in *E. coli*, seven copies of the rRNA genes are required to produce the 15,000 ribosomes needed by a rapidly growing bacterial cell (Chapter 14). Not surprisingly, higher eucaryotic cells have many more identical copies (50 to 5000) of the genes that specify the 18S, 5.8S, and 28S rRNA components in their 10 million ribosomes. These rRNA genes are tandemly arranged in that order into huge blocks situated on one or more chromosomes. The regions that are transcribed into large pre-rRNAs range in size from 8 kilobases (in yeast, *Drosophila*, and *Xenopus*) to 13 kilobases (in mammals) and are separated by **nontranscribed spacers** (Figure 20-26). In some organisms like the toad *Xenopus*, repetitious shorter sequences within these spacers appear to be essential for controlling or activating rDNA transcription (Chapter 21). Like the transcribed regions, the nontranscribed spacers are highly conserved from one tandem repeat unit to the next within an organism. The rDNA repeat units are usually looped off the main chromosomal fiber masses as highly extended threads. These loops then coalesce with specific proteins to form the **nucleoli,** where rRNA synthesis and processing take place and where assembly of the processed rRNA with ribosomal proteins begins. Some organisms have only one nucleolus organizer (rDNA repeat) region per haploid complement, while others have several.

The genes for 5S rRNA (120 nucleotides) also exist in tandemly repeated units separated by nontranscribed spacer regions. However, in higher eucaryotes, they are not closely linked to the large blocks of rDNA. In *Drosophila*, for instance, there is only one cluster of about 160 5S rRNA genes with a basic repeat unit of 375 base pairs that varies only slightly in length and sequence from one unit to another (see Figure 20-26). The 5S genes in *Xenopus* are distributed into many smaller clusters located near the ends (telomeres) of perhaps 18 different toad chromosomes. Here, there are two different sets of 5S genes. One encodes the 5S rRNA found in oocyte ribosomes, while the other specifies the 5S rRNA in somatic cell ribosomes, with only eight nucleotide differences distinguishing their two 120-nucleotide-long sequences. There are several hundred copies of the somatic-type 5S genes, but over 10,000 copies of the oocyte variety. The difference in gene copy number reflects the need to synthesize the large amounts of 5S rRNA that are stored in the egg during oogenesis (Chapters 21 and 22). Depending on which species of *Xenopus* and which 5S gene is being examined, the repeating units can contain internally repetitious subrepeats of a short sequence or 5S pseudogenes, sequences that are highly homologous to real genes but do not give rise to detectable RNA products (see Figure 20-26).

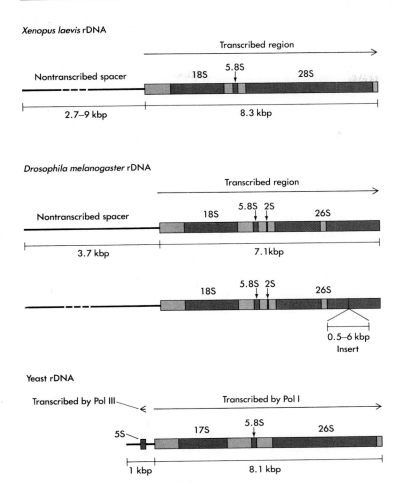

Xenopus laevis rDNA

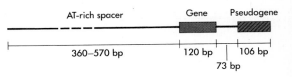

Xenopus borealis, somatic 5S DNA

Xenopus laevis, major oocyte 5S DNA

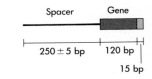

Drosophila melanogaster 5S DNA

Figure 20-26

Arrangements of rRNA gene repeating units in various eucaryotic organisms. Note that in yeast, the repeating unit also contains the 5S gene, even though it is transcribed from the opposite strand by an RNA polymerase (III) different from the one (I) that synthesizes the large rRNA precursor. *Drosophila* rDNA is unusual in several ways. *Drosophila* 60S ribosomes contain an additional 2S rRNA species, and the 26S molecule is interrupted about in the middle. Moreover, insertions are present in a large fraction of *Drosophila* rDNA repeats, but are not introns, since these genes are not transcribed. The arrangement of the large rDNA repeat in mammals is similar to that in *Xenopus* except that both the transcribed and nontranscribed spacers are longer. [After N. V. Fedoroff, *Cell* 16 (1979):697–710.]

Clusters of Different tRNA Genes[78–82]

The eucaryotic cell's demand for tRNA molecules is so high that single genes do not suffice. Generally, ten to several hundred genes for each tRNA species are present in the haploid genome, depending on the organism examined. In organisms with lower numbers of tRNA gene copies, identical genes are usually widely dispersed and have flanking sequences that are totally unrelated—even in those immediately adjacent 5' and 3' regions that are transcribed as part of their respective tRNA precursor molecules. In organisms (like *Xenopus*) with more highly reiterated tRNA genes, tandem repeats of relatively long sequences containing the genes for several different tRNAs can be detected. For instance, two tRNA$_i^{Met}$ genes and one gene each for six other tRNAs lie on a 3.2-kilobase-pair repeated fragment (Figure 20-27). In other organisms, tRNA genes are often grouped into heteroclusters that contain the genes for several completely different tRNA species within regions of several thousand base pairs of DNA. Sometimes, tRNA pseudogene sequences are also included in these regions. Only in rare instances, however, does it appear that neighboring tRNA genes are cotranscribed into multimeric tRNA precursor molecules. (The well-documented examples are in yeast.) Hence, it is not at all clear why different tRNA genes are so clustered.

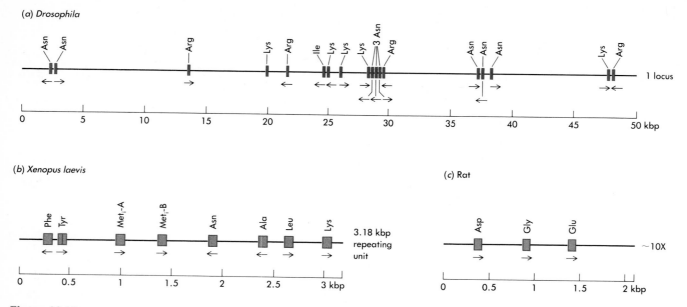

Figure 20-27
Arrangement of tRNA genes in higher eucaryotic organisms. (a) One of the 50 to 60 sites on *Drosophila* chromosomes where tRNA genes are clustered. [After P. H. Yen and N. Davidson, *Cell* 22 (1980):137–148.] (b) A 3.18-kilobase-pair unit from *Xenopus* DNA that is repeated about 150 times. The tRNA^{Tyr} gene contains an intron. One of the tRNA$_i^{Met}$ genes differs from the other by only one base pair in the coding region, but it is believed to be a pseudogene. [After J. Fostel et al., *Chromosoma* 90 (1984):254–260.] (c) This region of rat DNA is present about ten times in the haploid genome. However, the exact sequences are nonidentical, and it is likely that one or more of the genes in some of the repeats are pseudogenes. [After T. Sekiya, Y. Kuchino, and S. Nishimura, *Nucleic Acid Res.* 9 (1981): 2239–2250.]

Gene Amplification in Response to Environmental Pressure[83, 84]

The existence in eucaryotic DNA of gene families containing many identical members can be explained by the repeated duplication of specific DNA segments over evolutionary time. Subsequent divergence (i.e., the accumulation of mutations in such duplicated regions) could similarly explain the creation of gene families in which the members are related to one another but not identical. Whether the members of a particular gene family are identical or simply related, we have seen that they tend to be clustered together in the genome and arranged in a head-to-tail fashion separated by stretches of spacer DNA (which can be anywhere from short to very long).

These observations raise the possibility that tandem gene duplication is a relatively frequent event in all cells, a consequence of occasional mistakes made during the normal processes of DNA replication and recombination. Usually, such DNA duplications will confer no selective advantage on the organism, and the extra gene copy will be rapidly lost. (A simple crossover event between adjacent homologous sequences would excise the extra gene copy as a small DNA circle; the circle would then be diluted out of the cell population because it cannot replicate by itself.)

Duplication (amplification) of a specific gene can, however, be selected and stabilized by applying appropriate external pressure. If, for instance, mammalian cells growing in culture are treated with the drug methotrexate (an inhibitor of the essential enzyme dihydrofolate reductase, DHFR), most of the cells will die. Among the rare survivors are cells that produce significantly larger amounts of DHFR and therefore have enough uninhibited enzyme to generate tetrahydrofolate for survival and growth. By stepwise selection, cell variants can eventually be obtained that have multiplied the normally single copy (per haploid genome) of the DHFR gene 100- to 1000-fold! The amplified genes (and concomitant drug-resistant state) can be either stable or unstable. In the unstably amplified state, the DHFR genes reside

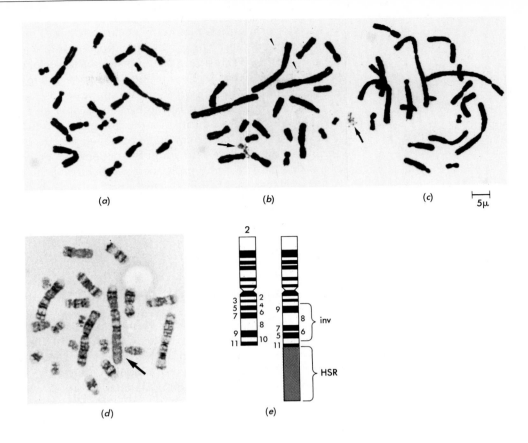

on small, paired extrachromosomal elements known as **double minute chromosomes** (Figure 20-28). Because the double minutes lack centromeres, they are rapidly lost when the selective drug (methotrexate) is removed. In contrast, stably amplified DHFR genes have been incorporated directly into one of the cell's chromosomes, where they often appear as elongated chromosomal bands called **homogeneously staining regions** (see Figure 20-28). These can occur either at the natural site of the unamplified DHFR gene (resulting from amplification without excision) or on other chromosomes (resulting from translocation, which could occur either before or after amplification).

Whether the amplified genes are located on normal chromosomes or on double minutes, the DNA sequences surrounding the DHFR gene are usually unchanged from the original unamplified gene, so that regulation and transcription are normal. However, the size of the amplified unit is highly variable, being sometimes as small as 25 kilobase pairs (just enough to encompass the DHFR gene itself) and sometimes as large as 2000 kilobase pairs! We do not know whether specific DNA sequences are involved in amplification, nor do we understand the mechanism. However, we do know that even in the absence of drug treatment, the rate of spontaneous duplication of the DHFR gene is about 1×10^{-3}, so that one in a thousand cells carries at least two DHFR gene copies. This observation could explain the tendency of eucaryotic organisms to meet the demand for increased gene product during the course of evolution by gene amplification (forward mutation rate 10^{-3}) rather than by devising a more powerful promoter or by stabilizing the gene product (forward mutation rate of less than 10^{-5}).

Figure 20-28
Double minute chromosomes and a homogeneously staining region (HSR) in Chinese hamster cells that have amplified the dihydrofolate reductase (DHFR) gene. Metaphase chromosome spreads from (a), a methotrexate-sensitive cell, are compared to (b) and (c), two resistant cells that contain extrachromosomal elements (indicated by arrow). [Courtesy of R. T. Schimke *et al.* (1981) *Cold Spring Harbor Symp. Quant. Biol.* 45, 785–797. Used with permission.] In (d) we see trypsin-Giemsa banded chromosomes from a (stably) methotrexate-resistant cell with a large HSR located on one end of chromosome 2 (diagrammed in e). [Courtesy of J. L. Biedler and B. A. Spengler, *Science* 191 (1976):185–187. Used with permission.]

Selective amplification of many eucaroytic genes has been observed after treatment with drugs or other toxic substances. For example, exposure of tissue culture cells to poisonous doses of cadmium (Cd^{2+}) selects for cells that have amplified the gene for metallothionein, a protective metal-binding protein. Programmed gene amplification is also occasionally utilized by higher organisms as a mechanism for synthesizing large amounts of a specific gene product in response to special needs at a particular stage of development. Examples are the amplification of rDNA in amphibian oocytes, of chorion genes in *Drosophila,* and of many protozoan genes during the formation of the macronucleus that directs vegetative growth (Chapter 22). The polytenization of dipteran chromosomes in the larval salivary glands can also be considered a case of gene amplification, although in that case, almost the entire chromosomal DNA is uniformly amplified (Chapter 21).

Gene Sequences Drift Through Time

Despite the existence of sophisticated cellular DNA repair mechanisms, all DNA sequences in the genome, and thus all genes, are constantly subject to mutation. Mutations are caused by errors made during DNA replication and recombination, by chemical DNA-damaging agents, by the chemical instability of certain groups within DNA, and even by errors made by DNA repair mechanisms themselves. Many of these mutations represent single-base substitutions (transitions and transversions; Chapter 12). Other mutations arise from insertions or deletions of a few base pairs. However, some mutations represent gross changes in the DNA, resulting from large deletions and insertions, from inversion, or from translocations (Chapter 12). Given a relatively high rate of mutation, how are the sequences of genes conserved through evolutionary time?

We have no trouble understanding how natural selection can maintain a functional single-copy gene like globin or insulin. If the gene product is defective in any serious way, the organism producing it will be immediately subjected to a selective disadvantage; it will either die prematurely or produce fewer progeny than its unmutated siblings. Thus, natural selection does not preserve the DNA sequence of the gene per se but rather its functional product. In the case of protein-coding genes, even significant changes in the DNA sequence may not affect the functional product because of the degeneracy of the genetic code (Chapter 15). In the case of structural RNAs like ribosomal RNA or transfer RNA, the RNA transcript *is* the final gene product, and the rate of sequence change tolerated is often much lower than that for protein-coding genes. Natural selection even acts on the parts of genes that are noncoding because certain sequences are critical for function. For example, splice junction signals (lying partly within introns) must be maintained, and spurious initiation codons for protein synthesis must not be introduced into the 5' untranslated leaders of mRNAs.

But we must also remember that gene sequences do ultimately change through time. New genes may be assembled from fragments of preexisting genes, rather than created de novo. In some cases, poorly understood recombination events result in tandem duplication of DNA sequences. The vertebrate α- and β-globin clusters (see Figure 20-24) clearly arose by some such mechanism. Once a gene has been duplicated, one copy is sufficient for survival of the organism, while the other gene copy is free to change. These changes will generally be useless or even harmful, and the extra gene copy will probably be lost. However, occasionally one of the gene copies will mutate in an advantageous way, and natural selection will preserve both the new gene and the old. In other cases, several mutations may be paired or grouped in such a way that the gene product (whether protein or RNA) is significantly changed without affecting its function. For example, the folding (and function) of structural RNA species depends critically on the existence of specific internally base-paired regions, but the actual RNA sequence within the stems may be relatively unimportant (Chapter 14). In such cases, *compensatory* mutations occurring on both strands of a duplex region can preserve the base pairing. Similar compensatory changes can occur in the amino acid sequence of a protein as well (intragenic suppressor mutations), so that overall structure and function are preserved (Chapter 15). Thus, the molecular evolution of a gene like globin or insulin resembles the predicament of the Red Queen in *Alice in Wonderland,* who had to run fast in order to stay in place. Mutations are constantly changing gene sequences, while natural selection keeps pace by preserving functional gene products.

The Apparent Paradox of Multigene Families[85]

Even more baffling than how a gene duplication originates is how virtual identity can be maintained among the many gene copies that constitute a multigene family. We have seen how natural selection can maintain the sequence of an essential single-copy gene like α- or β-globin; any deleterious mutation in the gene immediately places the organism at a selective disadvantage relative to its normal (wild-type) competitors. However, in the case of a large multigene family like ribosomal RNA (500 gene copies in the human genome) or amphibian oocyte 5S ribosomal RNA (20,000 gene copies in the *Xenopus* genome), it is very difficult to imagine how a mutation in any single gene copy would endanger the survival of the organism. In fact, our biological intuition rejects the notion that a defect in one out of 500 gene copies (0.2 percent of the genes) could have any significant effect on the fitness of the organism. Transcription of the remaining 499 unmutated copies would still produce 99.8 percent of the normal level of rRNA required to assemble functional ribosomes. Moreover, if a mutation in one out of 500 genes is harmless, why should mutations in 2 or even 20 out of 500 gene copies (96 percent normal genes) be harmful? Thus, we might have expected that large multigene families would quickly accumulate mutations because the many wild-type genes would shelter the organism from deleterious changes in any particular gene copy. In addition, because many (and perhaps most) mutations are selectively "neutral" rather than harmful (i.e., they do not affect the function of the gene product), we might also have expected the many gene copies in a multigene family to accumulate many neutral as well as occasional deleterious mutations. But in sharp contrast to such predictions, most stable cellular RNA species encoded by multigene families (including 18S, 5.8S, 28S, and 5S rRNAs, and U1 and U2 small nuclear RNAs) have been found to be essentially homogeneous at the RNA sequence level. This discrepancy between our theoretical expectation (that genes within a multigene family will diverge from each other) and experimental observations (that genes within a multigene family are usually homogeneous) is the apparent paradox of multigene families.

Keeping Multigene Families Homogeneous[85–88]

We can resolve the paradox of multigene families by postulating that eucaryotic organisms have one or more mechanisms for maintaining homogeneity among the members of repeated DNA sequence families. In fact, there is both circumstantial and experimental evidence for the existence of such homogenization mechanisms.

One convincing reason for believing that eucaryotic cells have mechanisms for homogenizing tandemly repeated sequences is that the length of the basic repeat unit in most tandemly repeated multigene families is much larger than the protein- or RNA-coding sequences within it. For example, the 10 to 20 human genes for U2 small nuclear RNA are tandemly repeated in a single cluster on chromosome 17, but the 188-base-pair U2-coding region constitutes only 3 percent of the 6000-base-pair repeat unit! How could natural selection maintain the essentially perfect homogeneity of the remaining 97 percent of the repeat unit? One might have argued that this 5880-nucleotide DNA sequence is conserved because it contains important signals controlling the start and stop of transcription as well as the RNA processing of the primary U2 transcript to the size of mature U2 snRNA. However, we shall see in Chapter 21 that such regulatory signals in the DNA and RNA of eucaryotes occupy remarkably few nucleotides, just as do comparable regulatory signals in bacteria (Chapters 13, 14, and 16). Thus, essential control regions could conceivably account for conservation of another tiny fraction of the 6000-nucleotide U2 repeat unit, but there is no obvious way that selective pressure could bring about homogenization of the remaining 5700 nucleotides. The inescapable conclusion is that any DNA that is organized in a tandem array will automatically be homogenized, regardless of the genetic information contained within it. Thus, spacer sequences that cannot be subject to direct natural selection (because mutations within them would not harm the organism) are automatically cohomogenized along with the selectable sequences (coding and regulatory regions) to which they are attached. Although the mechanism of homogenization is a matter of great controversy among geneticists, it is clear that such a mechanism must exist to account for the homogeneity of tandemly repeated gene families, as well as of tandemly repeated "satellite" sequences having no obvious function.

More compelling evidence for a gene homogenization mechanism comes from recent experimental work on tandemly repeated genes encoding ribosomal rRNA in yeast. Located between each transcription unit encoding the 35S precursor of yeast 17S, 5.8S, and 25S ribosomal RNAs is a sequence that serves as a localized stimulator of recombination. Although the mechanism by which this sequence works is still unknown, it is not difficult to imagine how such recombinational hot spots could ensure homogenization of the tandemly repeated rDNA array by promoting gene conversion, unequal sister chromatid exchange (homologous but out-of-phase recombination between repeat units), or even gene amplification (Figure 20-29).

But even multigene families that are not tandemly repeated are kept homogeneous. For example, the eight identical genes encoding tyrosine tRNA in yeast are located on eight different chromosomes.

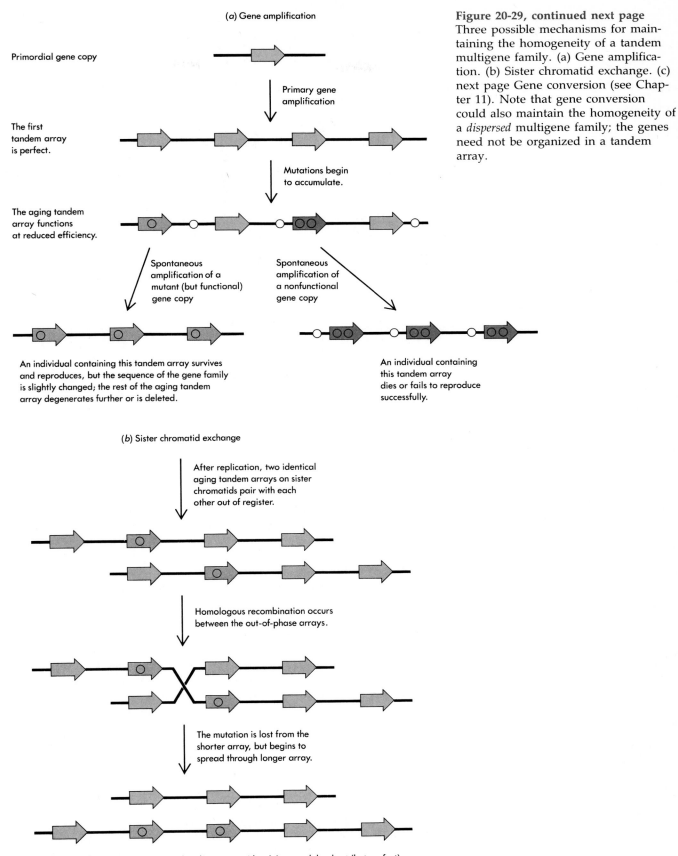

(a) Gene amplification

Primordial gene copy

Primary gene amplification

The first tandem array is perfect.

Mutations begin to accumulate.

The aging tandem array functions at reduced efficiency.

Spontaneous amplification of a mutant (but functional) gene copy

Spontaneous amplification of a nonfunctional gene copy

An individual containing this tandem array survives and reproduces, but the sequence of the gene family is slightly changed; the rest of the aging tandem array degenerates further or is deleted.

An individual containing this tandem array dies or fails to reproduce successfully.

(b) Sister chromatid exchange

After replication, two identical aging tandem arrays on sister chromatids pair with each other out of register.

Homologous recombination occurs between the out-of-phase arrays.

The mutation is lost from the shorter array, but begins to spread through longer array.

Further cycles of sister chromatid exchange can either (a) expand the short (but perfect) tandem array to its original length or (b) spread the mutation through the longer array, eventually producing a new tandem array with a slightly different sequence.

Figure 20-29, continued next page
Three possible mechanisms for maintaining the homogeneity of a tandem multigene family. (a) Gene amplification. (b) Sister chromatid exchange. (c) next page Gene conversion (see Chapter 11). Note that gene conversion could also maintain the homogeneity of a *dispersed* multigene family; the genes need not be organized in a tandem array.

(c) Gene conversion

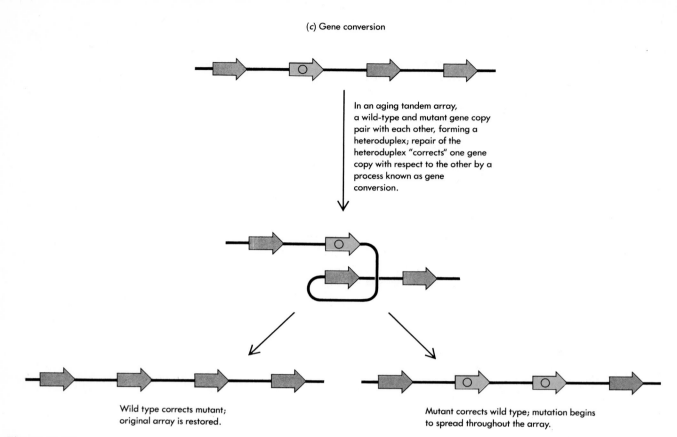

In an aging tandem array, a wild-type and mutant gene copy pair with each other, forming a heteroduplex; repair of the heteroduplex "corrects" one gene copy with respect to the other by a process known as gene conversion.

Wild type corrects mutant; original array is restored.

Mutant corrects wild type; mutation begins to spread throughout the array.

Figure 20-29
(Continued)

Although the tRNA-coding sequence and the 14-nucleotide intervening sequence are identical except at one position at each locus (see Figure 20-8), the 5' and 3' flanking sequences display no obvious similarities. How can we account for such localized homogeneity when our biological intuition would anticipate that the eight individual tRNA genes in this small multigene family would have diverged slightly from each other through the accumulation of various minor or neutral mutations? The absence of such divergence indicates that these eight genes are **coevolving;** that is, their sequences all change together. But how might eight dispersed genes communicate with each other? The alert reader will have noticed that gene conversion, as detailed in Figure 20-29c for a tandemly repeated multigene family, could also homogenize a dispersed multigene family; tandem repetition may simply accelerate the rate of gene conversion by increasing the local concentration of related genes. Repeated rounds of pairwise gene conversions may also help to explain why the 5' and 3' flanking sequences of the eight tyrosine tRNA genes are divergent. Although gene conversion can homogenize two closely related DNA sequences, gene conversion is unlikely to proceed through unrelated sequences.

Tandemly Repeated DNA Sequences Are Naturally Unstable

We might have guessed that tandem repeats of an identical DNA sequence would be an ideal substrate for *intra*chromosomal recombination (i.e., recombination within a single chromosome) and that tandemly repeated multigene families would therefore be rapidly lost from a eucaryotic genome (in much the same way that bacteriophage λ excises from the bacterial chromosome by homologous recombination between the two *att* sites; Chapter 17). In fact, certain mutants of the fruit fly *Drosophila* are known to have deleted a part of their tandemly repeated array of rDNA. Such flies grow slowly, because they cannot synthesize rRNA at the normal rate; however, over the course of many generations, the mutant flies gradually expand the number of rDNA repeat units and regain a normal growth rate. We can understand this type of expansion by noting that during unequal sister chromosome exchange (i.e., *inter*chromosomal exchange) one daughter chromosome acquires the repeat units lost by the other (Figure 20-29b). This process constitutes one mechanism for regaining the original number of repeated units within a tandem gene cluster that has undergone accidental contraction. Another such expansion mechanism is provided by gene amplification (Figure 20-29a). Thus, natural selection can maintain the overall gene copy number within a tandem gene cluster at an approximately constant level to meet the needs of the organism.

Abundance of Pseudogenes for Small Nuclear RNAs[89, 90]

The small RNAs like U1, U2, and U3, which are such numerous constituents of the nuclei of higher eucaryotic cells, are (not surprisingly) encoded by multiple copies of identical genes. Usually, the genes for these RNAs are organized into tandem arrays that contain only one type of small nuclear RNA (snRNA) gene surrounded by relatively conserved spacer DNA varying in length from 800 base pairs to greater than 45 kilobase pairs, depending on the organism.

In humans and other mammals, a most unexpected feature of the genomic regions complementary to the snRNAs is that pseudogenes outnumber real genes by at least ten to one. There are some 30 true U1 snRNA genes located on human chromosome 1; yet 500 to 1000 other regions hybridize to U1 RNA but cannot encode it. Some pseudogenes differ from real genes by as little as one nucleotide (out of 164 in the case of U1 snRNA) and have homologous flanking sequences. Other pseudogenes possess only a portion of the snRNA sequence and are flanked by DNA that is totally unrelated to the sequences adjacent to real genes. The 500 to 1000 U1 pseudogenes are dispersed at many locations in the human genome. The abundance of pseudogenes for U6 snRNA may be even greater.

The most intriguing snRNA pseudogenes are those that lack the region specifying the 3' end of their respective RNA and are surrounded by DNA sequences unrelated to those next to real genes. Such pseudogenes are usually flanked by short (6 to 20 base pairs), nearly perfect direct repeats. This suggests that these pseudogenes may have arisen by insertion of a partial copy of the snRNA sequence between staggered breaks in genomic DNA (as occurs during the movement of transposable elements; Chapter 11). For instance, a frequently occurring variety of U3 pseudogene exhibits homology to only the first 70 nucleotides at the 5' end of the RNA molecule. This is just the size predicted if the 3' terminus of U3 RNA had been utilized as a self-primer for a reverse transcriptase-like activity, producing a partial cDNA copy of U3 sequences that was then inserted into the human genome (Figure 20-30).

Figure 20-30

Schemes for the generation of a frequently occurring type of U3 snRNA pseudogene. The resulting pseudogene exhibits homology only to the first 70 nucleotides of the RNA and is flanked by short direct repeats. [After L. B. Bernstein, S. M. Mount, and A. M. Weiner, *Cell* 32 (1983):461–472.]

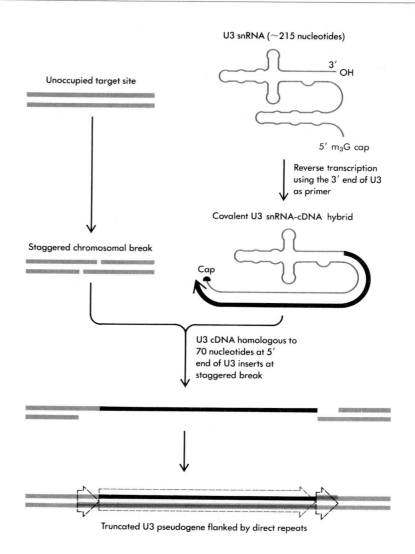

U3 snRNA (~215 nucleotides)

3′ OH

5′ m₃G cap

Unoccupied target site

Reverse transcription using the 3′ end of U3 as primer

Covalent U3 snRNA-cDNA hybrid

Staggered chromosomal break

Cap

U3 cDNA homologous to 70 nucleotides at 5′ end of U3 inserts at staggered break

Truncated U3 pseudogene flanked by direct repeats

Processed Genes: Intronless Replicas of Protein-Coding Genes[91–94]

As a mechanism for producing pseudogenes, reverse transcription does not appear limited to snRNAs. An increasing number of pseudogenes identified in the course of analyzing mammalian gene families bear distinctive hallmarks of RNA processing. These pseudogene sequences have lost their introns precisely and also possess 3′-terminal poly A tracts. The mRNA-like sequence is flanked by short direct repeats and lacks the upstream promoter sequences required for transcription of the corresponding real gene. In most cases, **processed pseudogenes** have also accumulated point mutations and small deletions or insertions. For example, the human β tubulin gene family consists of about 15 to 20 members. Seven of the 10 of these that have been studied are pseudogenes, of which 5 are intronless, contain poly A, and are flanked by short direct repeats. In contrast to the ability of U3 snRNA to prime its own reverse transcription (see Figure 20-30), it is not obvious how synthesis of a full-length cDNA copy of a polyadenylated mRNA could have been primed beginning on the poly A tract. For this reason, the proposal has been made that reverse transcription might be primed by the 3′ end of the DNA at the chromosomal break into which the cDNA will be inserted. However, un-

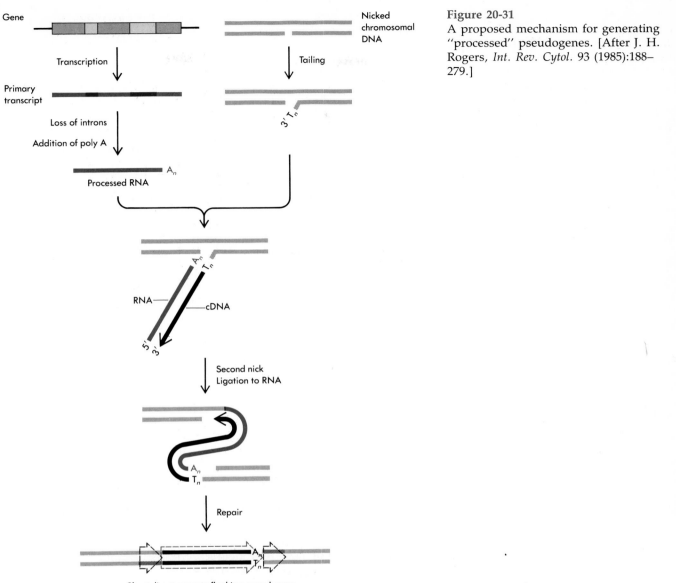

Gene

Primary

Loss of introns

Addition of poly A

Processed RNA

Nicked
chromosomal
DNA

Tailing

RNA

cDNA

Second nick
Ligation to RNA

Repair

Short direct repeats flanking pseudogene

Figure 20-31
A proposed mechanism for generating "processed" pseudogenes. [After J. H. Rogers, *Int. Rev. Cytol.* 93 (1985):188–279.]

less we postulate further that such chromosomal breaks almost always occur within tracts of T (thymine) residues or that there is a mechanism for tailing the DNA with Ts (Figure 20-31), it is difficult to understand why priming occurs on the poly A tract of the mRNA rather than internally. Nonetheless, reinsertion of copies of processed RNAs provides an appealing explanation for the existence of these curious sequences in mammalian DNA. Interestingly, processed pseudogenes appear less frequently in other eucaryotic genomes, even those of lower vertebrates.

Several Classes of Transposable Elements Cause Mutations in *Drosophila*[95-98]

Bacterial and lower eucaryotic (yeast) genomes are not the only genomes inhabited by transposons. In *Drosophila*, for example, at least three distinct types of mobile elements have been found to be respon-

Figure 20-32

Structures of three classes of *Drosophila* transposable elements. The repeated sequences that make up the ends of each of these elements are shown in color. They all give rise to short direct repeats of target DNA upon insertion. Open reading frames (ORF) are shown for a representative copia element (other copias have different gene orders) and the complete P factor; gag (a nucleic acid binding protein), int (integrase), and RT (reverse transcriptase) refer to sequences homologous to retroviral genes; LTR stands for long terminal repeat. [After G. M. Rubin, in *Mobile Genetic Elements*, ed. J. A. Shapiro (New York: Academic Press, 1983), p. 329; D. Finnegan, *Int. Rev. Cytol.* 93 (1985):281–326; S. M. Mount and G. M. Rubin, *Mol. Cell Biol.* 5 (1985):1630–1638.]

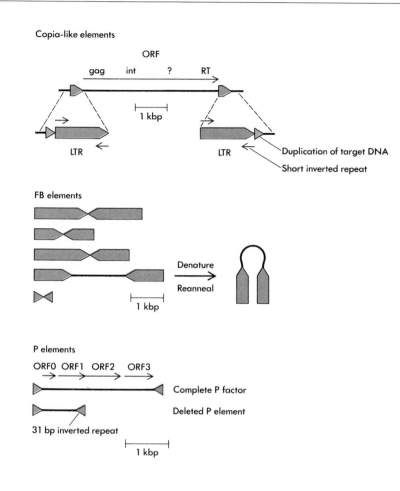

sible for mutations and other genetic rearrangements. These DNA sequences move repeatedly and can distribute themselves, apparently at random, around the genome. As the DNAs of more and more higher organisms are closely scrutinized, sequences similar to one or another of these categories continue to be recognized, arguing that all genomes are subject to remodeling by transposition.

Copia elements in *Drosophila melanogaster* possess long terminal direct repeats and are in many ways analogous to the Ty1 transposable elements of yeast (Chapter 18) and the integrated proviral forms of RNA tumor viruses (Chapter 24). Actually, there are several families (more than 11) of similar "copia-like" elements; the members of each family are well conserved and are located at 5 to 100 different sites in the *Drosophila* genome. Evidence that copia-like sequences are mobile comes not only from observations that the number and location of any particular element varies between *Drosophila* strains but also from the finding that stable mutations in a number of genes are due to their insertion. Copia-like elements are about 5000 base pairs long, with long terminal repeats (LTRs) of several hundred base pairs that vary in length and sequence between families (Figure 20-32). At the extreme ends of each element are short imperfect inverted repeats (about 10 base pairs). As with all other movable elements, insertion of copia-like elements into a new site is accompanied by duplication of a short stretch (3 to 6 base pairs) of target DNA; the length, but not the sequence, of the direct repeats that consequently appear immediately before and after the element is the same for all members of the same family. A distinctive property of copia-like elements is that their entire length is transcribed at a very high rate to produce polyadenylated RNAs; curiously, many of these transcripts are confined to the cell nucleus. Recent sequencing of a complete copia element has revealed that it possesses one long open reading frame that encodes proteins homologous to those of RNA tumor viruses. Homologies to reverse transcriptase, integrase, and nucleic acid-binding proteins are seen (Chapter 24), constituting strong evidence that copia moves via RNA, as does the yeast Ty1 element (Chapter 18). A transposable element with properties very similar to copia also occurs in the mouse genome; called the IAP element, it clearly encodes a protein that encapsulates its own RNA into noninfectious viruslike structures called **intracisternal A-type particles.**

A second distinct class of *Drosophila* transposable elements is known as the **fold-back (FB) elements.** These possess long inverted repeats either directly adjacent to one another or separated by up to several thousand base pairs of DNA (see Figure 20-32). When such sequences are denatured, they readily fold back to form hairpin structures. In contrast to copia, FB elements are very heterogeneous in both length and sequence and cause unstable mutations, resulting in a high frequency of deletion and chromosome rearrangement. The variably sized inverted repeats of FB elements are made up of internally repetitious sequences; yet, their extreme ends are highly conserved for about 30 base pairs and routinely produce 9-base-pair duplications of target-site DNA. There is evidence that any segment of DNA flanked by FB elements can transpose, underscoring the potential importance of this class of transposons to genome evolution. Although several percent of the genome of *Drosophila* and other higher eucaryotes is made up of inverted repeat sequences that rapidly fold back, probably only a small fraction of such sequences represents genuine FB transposable elements. The remaining sequences may be more similar to the *Alu* repeats of primate genomes.

P elements are a third class of *Drosophila* transposable elements. They are responsible for a genetic phenomenon known as **hybrid dysgenesis** and have recently been developed as potent vectors for introducing genes into new locations in the *Drosophila* genome (Chapter 22). P elements reside at multiple sites in most *Drosophila* DNA and are 0.5 to 1.4 kilobase pairs in size, bounded by perfect inverted repeats of 31 base pairs (see Figure 20-32). They represent internally deleted versions of a larger element of about 3 kilobase pairs called a P factor, which occurs in one or a few copies only in so-called P strains of *Drosophila*. Sequencing has revealed that the large P factors possess four major open reading frames, all of which are required for transposition. They presumably specify functions (e.g., transposase and repressor) similar to those required for the movement of bacterial transposons (Chapter 11). When P strain males (whose DNA contains complete P factors as well as P elements) are mated with M strain females (which contain only deleted P elements, if any), massive mobilization of the many copies of P elements results, presumably because of the sudden production of transposase encoded by the entering male DNA (Figure 20-33). In addition to frequent mutations caused by new P element insertions, the progeny of such dysgenic crosses exhibit lowered fertility and chromosome aberrations. Crosses of M strain males with P strain females (or P × P crosses) are not dysgenic, since transposition of the many short P elements is repressed when complete P factor(s) resides on the female genome: Presumably, the egg cytoplasm is already filled with repressor-type molecules. Upon insertion into a new site in *Drosophila* DNA, both P factors and P elements create 8-base-pair duplications of the target

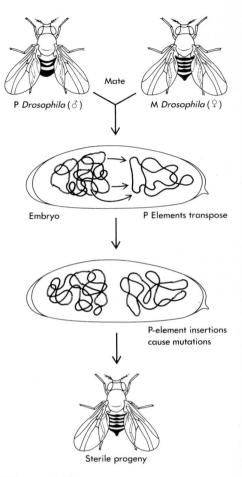

Figure 20-33
The phenomenon known as hybrid dysgenesis results from the mobilization of DNA sequences called P elements in *Drosophila* embryos. When a sperm from a P-carrying strain fertilizes an egg from a non-P (M) strain, the P elements transpose throughout the genome, usually disrupting vital genes. [After J. D. Watson, J. Tooze, D. T. Kurtz, *Recombinant DNA: A Short Course* (New York: W. H. Freeman, 1983).]

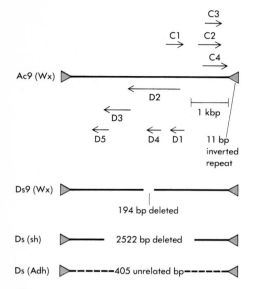

Figure 20-34
Comparison of a complete Ac element with several Ds elements. Note that one Ds element contains a completely unrelated sequence between its terminal 11-base-pair inverted repeats. All open reading frames (C1–C4 and D1–D5) encoding proteins greater than 10 kdal are indicated. [After R. F. Pohlmann, N. V. Fedoroff, and J. Messing, *Cell 37* (184):635–643.]

sequence. P factors and other transposon systems that manifest hybrid dysgenesis appear to play an important role in the process of speciation because they effectively serve to isolate breeding populations and thus allow rapid incorporation of new mutations into the germ line.

Use of P factors as vectors is accomplished by inserting a gene into the middle of a deleted P factor and injecting this DNA into an M strain embryo along with a helper P element made defective for integration by deleting one terminal inverted repeat. A remarkably high fraction of cells in the germ line of the resulting fly will have acquired at some location(s) in their genomes the new gene, always flanked by P factor sequences (see Chapter 22 for details).

Controlling Elements in Maize Are Also Transposons[99, 100, 101]

As early as the first decades of this century, corn geneticists recognized the existence of unstable mutations that affect kernel color in maize. Today we realize that all the characteristics manifested by the "controlling elements" responsible for these mutations are precisely those exhibited by transposable elements in other systems. Insertion of a controlling element into or adjacent to a gene inhibits its expression and produces a colorless kernel. Excision of the element reverses the effect and results in colored spots on a colorless background. Even when their insertion does not affect gene expression, the controlling elements can be detected as sites of frequent chromosome breakage.

One of the best-studied maize controlling elements is the **Ac-Ds system** (Figure 20-34). Just as P elements in *Drosophila* are internally deleted verions of P factors, Ds elements come in several different sizes (0.4 to 4 kilobase pairs) and usually represent deleted forms of a larger complete element called Ac (4.5 kilobase pairs). Whereas Ac elements can move independently, Ds elements remain stationary unless an Ac element is also present. Sequencing a complete Ac element and several (shorter) Ds elements has revealed perfect inverted repeats of 11 base pairs at the termini flanked by 6- to 8-base-pair direct repeats of the target-site DNA. Interestingly, the open reading frames in the Ac element diverge from a central intergenic region, similar to the bacterial transposon Tn3 (Chapter 11). When an Ac or Ds element excises, it leaves behind imperfect but recognizable duplications of the 6- to 8-base-pair target sequence—a clear footprint of its former presence at a particular site in maize DNA.

Satellite DNA Sequences Near the Centromere[102, 103]

The chromosomal region adjacent to the centromere in higher eucaryotes is composed of very long blocks of **highly repetitive DNA,** in which simple sequences are repeated thousands of times or more. In many cases, these repeating simple sequences have compositions unlike that of most of the organism's DNA and so are easily separated by centrifuging slightly fragmented DNA in a cesium chloride (CsCl) density gradient (Figure 20-35). DNA separated by this process is referred to as **satellite DNA.** Often, several different satellites of

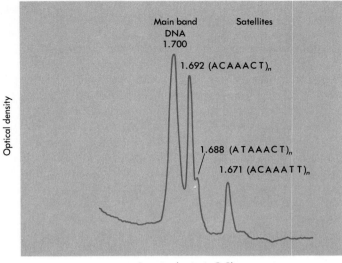

Figure 20-35
The satellite DNAs of *Drosophila virilus*. Fragmented DNA was centrifuged to equilibrium in a cesium chloride gradient. [After J. G. Gall and D. D. Atherton, *J. Mol. Biol.* 85 (1974):633–634.]

centromeric origin occur in a given species. The three satellites from *Drosophila virilis* have highly related sequences. One has a 5'-ACAAACT-3' basic repeat; a second contains 5'-ATAAACT-3'; and the sequence repeated in the third is 5'-ACAAATT-3'. Why several sequences have so evolved is not clear, and none of these satellites is a completely pure repeat of the basic sequence. Often, satellites present in one species appear to be absent even in a closely related species, but this can be explained, since a change in only one base pair of the short repeating unit can significantly alter its base composition and therefore its density in cesium chloride.

Blocks of satellite sequences are readily localized to regions around the centromeres of metaphase chromosomes by a technique called **in situ hybridization** (Figure 20-36). Here, radioactively labeled RNA copies of the satellite sequence made in vitro are hybridized to squashed metaphase cells after they have been treated with alkali to

Figure 20-36
How satellite sequences can be localized to centromeres by in situ hybridization. (a) The experimental procedure. (b) An actual autoradiograph of a mouse chromosome preparation after hybridization. Note that mouse chromosomes have terminal centromeres and thus do not form conventional x-shaped metaphase chromosomes. [Courtesy of M. L. Pardue and J. G. Gall, *Chromosomes Today* 3 (1972):47–52. Used with permission.]

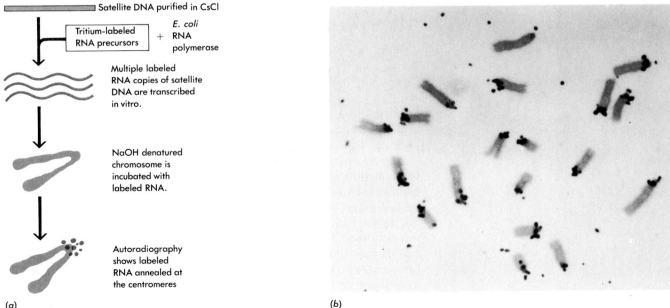

(a)

(b)

denature the chromosomal DNA. Subsequently, the position of the satellite sequences can be visualized by autoradiography.

In *Drosophila*, some 25 percent of the total DNA is of the simple sequence variety, with most higher eucaryotes having 10 to 20 percent of their genomes as such DNA. Satellite DNAs are usually not transcribed, for their restricted nucleotide sequences lack promoter sites on which RNA polymerase can initiate RNA chains. In contrast, there must be nucleotide sequences that are recognized by DNA polymerase, since centromeric DNA is replicated as rapidly as "unique sequence" DNA. Presumably, satellite sequences near the centromere are involved in binding particular proteins essential for centromere function. This includes holding the two daughter chromatids together during metaphase, assembling the kinetochores that allow subsequent attachment to the spindle, and guiding topoisomerases to the site(s) where they must act to separate the intertwined DNA duplexes of sister chromatids at the conclusion of replication. Perhaps the highly repetitive nature of satellite sequences at the centromeres of higher eucaryotes is related to the fact that many spindle fibers attach to each chromosome. In yeast, only one spindle fiber attaches to a centromere, each of which has a unique, nonrepetitive sequence (Chapter 18). Satellite sequences are thus clear examples of what might be called structural rather than genetic regions of the DNA contained in eucaryotic genomes. The nicked, not so highly repetitive sequences that appear at the ends of chromosomes (telomeres) are another example (Chapters 18 and 22).

Other Highly Repetitive Sequences Are Dispersed Throughout the Genome[94, 104–110]

The genomes of almost all higher eucaryotes also contain highly repetitive sequences that are not clustered together as are satellite sequences at centromeres. Rather, they are more evenly distributed throughout the genome, interspersed with longer stretches of unique (or moderately repetitive) DNA.

In the human genome, the majority of such sequences belong to a single family called the **Alu family.** Its members are about 300 base pairs long and are recognizably related but not precisely conserved in sequence. Their name derives from the fact that most contain a single site of cleavage for the restriction enzyme *Alu* I near their middle. Almost a million *Alu* sequences are present in the human genome, accounting for some 3 to 6 percent of the total DNA. Thus, any particular DNA segment of 5000 base pairs or longer has a high probability of containing an *Alu* sequence. Accordingly, many introns and sequences adjacent to genes (in both transcribed and nontranscribed regions) harbor *Alu* sequences.

The widespread dispersion of *Alu* family repeats throughout the human genome is believed to have occurred in an RNA-mediated fashion, similar to the mechanism envisioned for the creation of processed genes and of many small nuclear RNA pseudogenes. *Alu* sequences terminate at their 3' ends with repetitive A-rich tails and are often flanked by direct repeats of 8 to 19 base pairs. Although they are not transcribed in vivo in proportion to their abundance in the genome, isolated *Alu* sequences are usually excellent templates for in vitro transcription by RNA polymerase III. This suggests a scenario in which a rare *Alu* sequence transcript could be reverse transcribed by self-priming. Its U-rich 3' end (U-rich sequences are found at the ex-

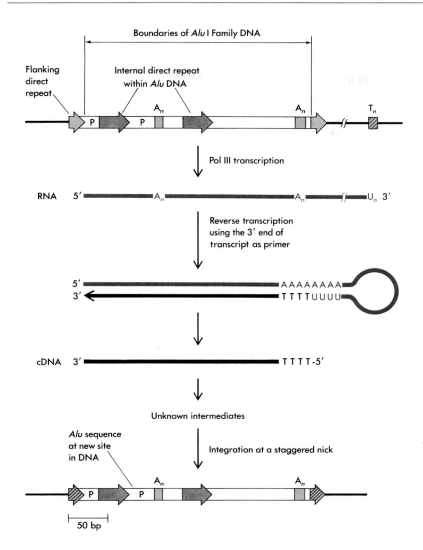

Figure 20-37
How *Alu* sequences may have multiplied and dispersed themselves throughout the human genome. A salient point is that sequences important for RNA synthesis (P) are carried within the sequence itself (mobile promoter). [After S. W. Van Arsdell, R. A. Denison, L. B. Bernstein, and A. W. Weiner, *Cell* 26 (1981):11–17; P. Jagdeeswaran, B. G. Forget, and S. M. Weissman, *Cell* (1981):141–142.]

treme 3′ ends of all RNA polymerase III products; Chapter 21) would fold back on its lengthier A-rich tail (Figure 20-37). The resulting cDNA copy would then insert between staggered breaks at a new site in the genome. Since the DNA sequences important for initiation of transcription by RNA polymerase III reside within the transcribed region rather than in upstream DNA (Chapter 21), an *Alu* cDNA would carry its transcription signals along to its new location. Thus, it would be capable of repeated rounds of transcription, cDNA synthesis, and insertion. The structure of many *Alu* family members can be explained if insertion occurs preferentially into the A-rich tails of preexisting *Alu* sequences.

A most curious property of *Alu* sequences in the human genome (and of related sequences in other mammalian genomes) is that they are in part homologous to an abundant small RNA known as 7SL. This 300-nucleotide-long RNA associates with six proteins to form a small cytoplasmic RNA—protein complex (called SRP) that is essential for the translocation of newly synthesized proteins across membranes of the rough endoplasmic reticulum in higher eucaryotic cells (Chapter 21). 7SL RNA is transcribed by RNA polymerase III from three or four identical copies of its gene in the human genome. It exhibits homology to *Alu* sequences at both its 5′ and 3′ ends, but is interrupted by 155 nucleotides of unrelated sequence in its middle (Figure 20-38). This suggests that mammalian *Alu* sequences may

Figure 20-38

The relationship between human 7SL RNA and human *Alu* DNA. The different shadings indicate areas of sequence homology between the RNA and the DNA. The right half of human *Alu* DNA contains an insertion (uncolored) not present in the left half. The dark gray areas correspond to the most perfectly repeated regions, which are indicated by arrows in Figure 20-37. [After E. Ullu and C. Tschudi, *Nature* 312 (1984):171–172.]

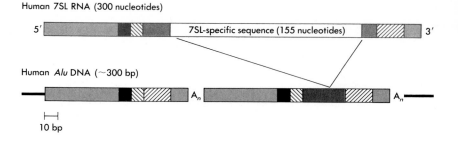

have been derived from 7SL RNA (or DNA) by deletion of the central 7SL-specific sequence. Recently, the 7SL RNA of frogs and flies has been shown strongly to resemble mammalian 7SL RNA, although these lower eucaryotes completely lack *Alu* sequences. Hence, human *Alu* sequences have all the earmarks of a very large family of "processed 7SL pseudogenes" that arose subsequent to the divergence of mammals.

In addition to members of the *Alu* family, there are other highly repetitive sequences (present in more than 10^4 copies) widely dispersed in mammalian genomes and other higher eucaryotic genomes. Some are similar to *Alu* in that they are relatively short (less than 500 base pairs) and are frequently flanked by direct repeats. Included in this group are sequences recently discovered to be homologous to tRNA genes, suggesting that they may have arisen via tRNA intermediates. Other repetitive sequences are much longer (greater than 5 kilobase pairs), have A-rich 3' ends, and sometimes contain open reading frames. Thus, the distinction between bona fide transposons, pseudogenes, and repetitive DNA sequences in eucaryotic genomes has become progressively blurred as DNA sequence analysis has revealed the details of their structure. Determining whether and how the several types of repetitive sequences move is a current challenge for molecular biologists.

Variation in DNA Amounts Between Closely Related Species[111]

Before much was known about the actual sequences that make up higher eucaryotic genomes, it was expected that the amount of DNA per haploid genome would increase in proportion to the biological complexity of the organism. Higher plants and animals do have much more DNA than lower forms (see Figure 20-1), and we have already seen that this is in part due to the much more frequent occurrence of introns in the genes of higher as compared to lower eucaryotes. Yet, no one anticipated the finding that certain fish and amphibia would have 25 times more DNA than any mammalian species. And as more and more animals and plants are examined, closely related species are sometimes found to vary in their DNA content by a factor of five to ten (Figure 20-39).

These variations usually do not arise because of increases in chromosome number but rather because of increases in the amount of DNA per chromosome. The amounts of simple sequence satellite DNA, mildly repetitious DNA, and unique sequence DNA all increase, sometimes in roughly proportional amounts. Very large genomes need to be studied in more detail before we can understand

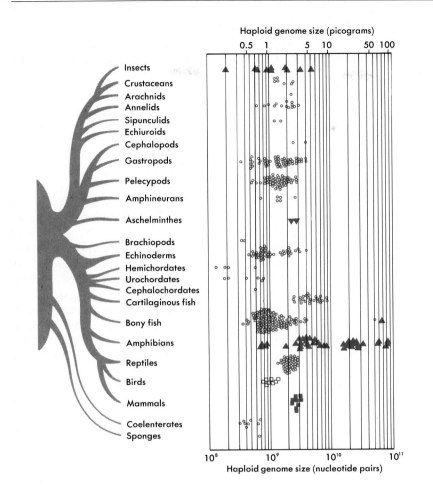

Haploid genome size (picograms)

Haploid genome size (nucleotide pairs)

Figure 20-39
Variation in the genome size among related members of evolutionarily related animal groups. [After R. J. Britten and E. H. Davidson, *Quant. Rev. Biol.* 46 (1971):111–138.]

the true meaning of such variations in genome size. However, we do know that the salamander *Necturus*, which has some 25 times more DNA than the toad, contains only single copies of its various hemoglobin genes. Hence, our best current guess is that the extra DNA simply represents longer spacer regions between true genes; these spacers are subject to little selective pressure and provide an ample home for all types of DNA sequences.

Selfish DNA?[112, 113, 114]

We have seen that the unexpected structures that contribute to the large size of higher eucaryotic genomes take many forms. The existence of introns multiplies the number of base pairs required to encode a protein by as much as tenfold. In addition, many mRNAs possess mysteriously long untranslated regions (particularly at their 3′ ends) of as yet unknown function. Many protein-coding genes have given rise to gene families producing several closely related products, while other gene families consist of tens or thousands of identical gene copies. Within the lengthy sequences that separate genes are found recognizable repeated sequence elements: transposons of various types, other mobile repetitive sequences like the *Alu* family, pseudogenes, and complete or partial copies of certain viral genomes (Chapter 24). Furthermore, some DNA is devoted to specialized structural roles such as satellite sequences at centromeres

Figure 20-40

Increase in occurrence of P elements in *Drosophila melanogaster* collected worldwide. Plotted are the fractions of flies exhibiting P or M phenotypes upon mating (see Figure 20-33). The number of strains actually tested is given above each point on the graph. [After M. G. Kidwell, *Proc. Nat. Acad. Sci.* 80 (1983):1655–1659.]

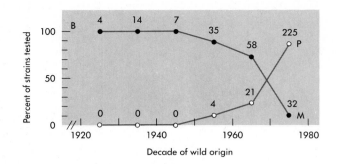

or the shorter nicked repeated sequences at the ends of chromosomes (telomeres).

Still, we cannot account for all the sequences that make up a eucaryotic genome nor do we understand the enormous variations in DNA amount that occur between closely related species. This has led some scientists to speculate that DNA is inherently "selfish," constantly endeavoring to expand itself to the limit tolerated by each organism. Surely, transposons and *Alu* family sequences have properties most closely fitting what might be expected of **selfish DNA**: built-in mechanisms for reproduction and movement. Yet, even their proliferation seems to be limited in some way. For instance, the spread of P elements through wild-type strains of *Drosophila* has been a quite recent and remarkably rapid occurrence (Figure 20-40). Nevertheless, examination of the genomes of individual P strain flies reveals only one to several complete P factors and less than 50 shorter P elements, rather than the hundreds or thousands of copies that might be expected if their multiplication were unchecked. As with all parasites, advantages gained by replication must be balanced by ensuring survival of the host organism. Therefore, to what extent purely selfish DNA sequences have molded eucaryotic genomes remains to be determined.

Summary

The genomes of higher eucaryotes are much bigger than those of bacteria. Far more DNA is present than needed to encode even a hundredfold more proteins. We now realize that this excess DNA represents a number of different kinds of sequences. Most of them are unique and unanticipated features of higher eucaryotic genomes.

Cloning and sequencing higher eucaryotic genes led to the discovery of introns (intervening sequences), interruptions that expand the sizes of genes as much as 20-fold. Introns are highly variable in length and sequence. They can occur in any part of a protein-coding gene: 5' leader, 3' trailer, or the body. Although most protein-coding genes in vertebrates are split by one or more introns, the presence of introns is not required. Some eucaryotic tRNA and rRNA genes also contain introns.

In all cases, split genes are transcribed into precursor RNA molecules that contain intron as well as exon sequences. Exons are those parts of the gene that are represented in its final RNA product (mRNA, tRNA, or rRNA). Introns are removed by a process called RNA splicing. At least three different splicing systems can be distinguished

on the basis of the location of the signals recognized in the precursor and the biochemistry of splicing reaction. Ribosomal RNA splicing is an RNA-catalyzed reaction for which certain intron sequences (and their precise folding into a three-dimensional structure) are critical. Transfer RNA introns (which always reside in the anticodon loop) are removed by two separate (protein) enzymes, an endonuclease and a ligase; they recognize some aspect of the folded tRNA (exon) structure. Pre-mRNAs are spliced by a complex machinery that includes small nuclear RNA-protein complexes, which bind to conserved sequences at the junctions between introns and exons; an important splicing intermediate is a lariat structure in which the 5' end of the intron forms a branch with a sequence near the 3' end. Some introns in mitochondrial DNA encode proteins essential for splicing, but RNA catalysis (as with rRNA) appears to be fundamental to intron removal.

Introns in protein-coding genes often occur at boundaries between separately folded domains or functionally distinct regions of the polypeptide chain. Evolution of new proteins may therefore be accelerated by mixing strategies or by adding (or deleting) material at intron-exon junctions as a consequence of mutations that create (or abolish) splice sites.

Many genes in higher eucaryotic genomes belong to multigene families. Either different members encode related, but nonidentical gene products that fulfill slightly different functions in a multicellular or developing organism, or multiple copies of identical genes are necessary simply to synthesize massive amounts of a particular gene product. Multigene families can be either dispersed or clustered into a single or tandemly repeating array. The nontranscribed spacer regions that separate genes are lengthier than those in bacterial genomes.

Amplification of segments of higher eucaryotic DNA containing particular genes readily occurs when cells are subjected to selective growth conditions. Amplification was probably instrumental in the original creation of multigene families and greatly contributes to their maintenance. Tandem arrays of identical DNA sequences are particularly susceptible to contraction (resulting from intrachromosomal crossing over). Lost copies could be regained either by amplification or by unequal sister chromosome exchange. Amplification and unequal sister chromosome exchange, along with gene conversion, may play further roles in ensuring the homogeneity of tandemly repeated DNA sequence families.

Pseudogenes are another unexpected feature of higher eucaryotic genomes. These are sequences that exhibit high homology to real genes, but lack signals necessary for expression (e.g., promoters). Often their presence can be rationalized by assuming that a copy of a processed RNA transcript of the original gene became inserted into the genome.

Transposons also inhabit higher eucaryotic genomes. They are present in multiple copies (but not very large numbers). They fall into two main classes: those that move as DNA and encode transposase- and repressor-like proteins and those that move via RNA and encode reverse transcriptase and sometimes viral capsidlike proteins.

Some highly repetitive sequences in higher eucaryotic DNA reside at the centromeres, where they serve structural roles in chromosome replication and sorting. Other repetitive sequences are dispersed randomly throughout the genome.

Together, all these different kinds of nongenetic DNA account for the large and variable sizes of higher eucaryotic genomes. It may be that DNA seeks to expand itself to the maximum tolerated by each organism.

Bibliography

General References

Apirion, D., ed. 1984. *Processing of RNA.* Boca Raton, Fla.: C.R.C. Press. Comprehensive reviews on all types of RNA processing.

Busch, H. ed. *The Cell Nucleus.* New York: Academic Press. A series of volumes containing review articles on nuclear structure and function. The most valuable here is volume X (1982) on rDNA.

MacLean, N., S. P. Gregory, and R. A. Flavell, eds. 1983. *Eukaryotic Genes: Their Structure, Activity and Regulation.* London: Butterworths. A compendium of excellent reviews on the structure and function of eucaryotic genes.

Schimke, R. T., ed. 1982. *Gene Amplification.* Cold Spring Harbor, N.Y.: Cold Spring Harbor Laboratory. Gives details on all aspects of amplification phenomena.

Shapiro, J. A., ed. 1983. *Mobile Genetic Elements.* New York: Academic Press. Includes several valuable chapters on eucaryotic transposons, including those in plants.

Structures of DNA. 1982. *Cold Spring Harbor Symp. Quant. Biol.* 47. Includes valuable articles on gene families, repetitive DNA sequences, and pseudogenes.

Cited References

1. Britten, R. J., and E. H. Davidson. 1969. "Gene Regulation for Higher Cells: A Theory." *Science* 165:349–357.
2. Thomas, C. A., Jr. 1971. "The Genetic Organization of Chromosomes." *Ann. Rev. Genetics* 5:237–256.
3. Darnell, J. E. 1976. "mRNA Structure and Function." *Prog. Nucl. Acid Res. Mol. Biol.* 19:493–511.
4. Darnell, J. E., Jr. 1983. "The Processing of RNA." *Sci. Amer.* 249:89–99.
5. Berget, S. M., C. Moore, and P. A. Sharp. 1977. "Spliced Segments at the 5′ Terminus of Adenovirus 2 Late mRNA." *Proc. Nat. Acad. Sci.* 74:3171–3175.
6. Chow, L. T., R. E. Gelinas, T. R. Broker, and R. J. Roberts. 1977. "An Amazing Sequence Arrangement at the 5′ Ends of Adenovirus 2 Messenger RNA." *Cell* 12:1–8.
7. Berget, S. M., A. J. Berk, T. Harrison, and P. A. Sharp. 1977. "Spliced Segments at the 5′ Termini of Adenovirus-2 Late mRNA: A Role for Heterogeneous Nuclear RNA in Mammalian Cells." *Cold Spring Harbor Symp. Quant. Biol.* 42:523–529.
8. Broker, T. R., L. T. Chow, A. R. Dunn, R. E. Gelinas, J. A. Hassell, D. F. Klessig, J. B. Lewis, R. J. Roberts, and B. S. Zain. 1977. "Adenovirus-2 Messengers—An Example of Baroque Molecular Architecture." *Cold Spring Harbor Symp. Quant. Biol.* 42:531–553.
9. Thomas, M., R. L. White, and R. W. Davis. 1976. "Hybridization of RNA to Double-Stranded DNA: Formation of R-loops." *Proc. Nat. Acad. Sci.* 73:2294–2298.
10. Breathnach, R., J. L. Mandel, and P. Chambon. 1977. "Ovalbumin Gene Is Split in Chicken DNA." *Nature* 270:314–319.
11. Doel, M. T., M. Houghton, E. A. Cook, and N. H. Carey. 1977. "The Presence of Ovalbumin mRNA Coding Sequences in Multiple Restriction Fragments of Chicken DNA." *Nucleic Acid Res.* 4:3701–3713.
12. Weinstock, R., R. Sweet, M. Weiss, H. Cedar, and R. Axel. 1978. "Intragenic DNA Spacers Interrupt the Ovalbumin Gene." *Proc. Nat. Acad. Sci.* 75:1299–1303.
13. Lai, E. C., S. L. Woo, A. Dugaiczyk, J. F. Catterall, and B. W. O'Malley. 1978. "The Ovalbumin Gene: Structural Sequences in Native Chicken DNA Are Not Contiguous." *Proc. Nat. Acad. Sci.* 75:2205–2209.
14. Jeffreys, A. J., and R. A. Flavell. 1977. "The Rabbit Beta-Globin Gene Contains a Large Large Insert in the Coding Sequence." *Cell* 12:1097–1108.
15. Tilghman, S. M., D. C. Tiemeier, J. G. Seidman, B. M. Peterlin, M. Sullivan, J. V. Maizel, and P. Leder. 1978. "Intervening Sequence of DNA Identified in the Structural Portion of a Mouse Beta-Globin Gene." *Proc. Nat. Acad. Sci.* 75:725–729.
16. Breathnach, R., and P. Chambon. 1981. "Organization and Expression of Eucaryotic Split Genes Coding for Proteins." *Ann. Rev. Biochem.* 50:349–383.
17. Chambon, P. 1981. "Split Genes." *Sci. Amer.* 244:60–71.
18. Crouse, G. F., C. C. Simonsen, R. N. McEwan, and R. T. Schimke. 1982. "Structure of Amplified Normal and Variant Dihydrofolate Reductase Genes in Mouse Sarcoma S180 Cells." *J. Biol. Chem.* 257:7887–7897.
19. Yamada, Y., V. E. Avvedimento, M. Mudryj, H. Ohkubo, G. Vogeli, M. Irani, I. Pastan, and B. de Crombrugghe. 1980. "The Collagen Gene: Evidence for Its Evolutionary Assembly by Amplification of a DNA Segment Containing an Exon of 54 bp." *Cell* 22:887–892.

20. Wozney, J., D. Hanahan, V. Tate, H. Boedtker, and P. Doty. 1981. "Structure of the Pro Alpha (2) (I) Collagen Gene." *Nature* 294:129–135.

21. Goodman, H. M., M. V. Olson, and B. D. Hall. 1977. "Nucleotide Sequence of a Mutant Eukaryotic Gene: The Yeast Tyrosine-Inserting Ochre Suppressor SUP4-o." *Proc. Nat. Acad. Sci.* 74:5453–5475.

22. Guthrie, C., and J. Abelson. 1982. "Organization and Expression of tRNA Genes in *Saccharomyces cerevisiae*." In *The Molecular Biology of the Yeast Saccharomyces: Metabolism and Gene Expression*, ed. J. N. Strathern, E. W. Jones, and J. R. Broach. Cold Spring Harbor, N.Y.: Cold Spring Harbor Laboratory, p. 487–528.

23. Abelson, J. 1979. "RNA Processing and the Intervening Sequence Problem." *Ann. Rev. Biochem.* 48:1035–1069.

24. Tilghman, S. M., P. J. Curtis, D. C. Tiemeier, P. Leder, and C. Weissmann. 1978. "The Intervening Sequence of a Mouse Beta-Globin Gene Is Transcribed Within the 15S Beta-Globin mRNA Precursor." *Proc. Nat. Acad. Sci.* 75:1309–1313.

25. Roop, D. R., J. L. Nordstrom, S. Y. Tsai, M. J. Tsai, and B. W. O'Malley. 1978. "Transcription of Structural and Intervening Sequences in the Ovalbumin Gene and Identification of Potential Ovalbumin mRNA Precursors." *Cell* 15:671–685.

26. Alwine, J. C., D. J. Kemps, and G. R. Stark. 1977. "Method for Detection of Specific RNAs in Agarose Gels by Transfer to Diazobenzyloxymethyl-Paper and Hybridization with DNA Probes." *Proc. Nat. Acad. Sci.* 74:5350–5354.

27. Hopper, A. K., F. Banks, and V. Evangelidis. 1978. "A Yeast Mutant Which Accumulates Precursor tRNAs." *Cell* 14:211–219.

28. Cech, T. R. 1983. "RNA Splicing: Three Themes with Variations." *Cell* 34:713–716.

29. Greer, C. L., C. L. Peebles, P. Gegenheimer, and J. Abelson. 1983. "Mechanism of Action of a Yeast RNA Ligase in tRNA Splicing." *Cell* 32:537–546.

30. Filipowicz, W., and A. J. Shatkin. 1983. "Origin of Splice Junction Phosphate in tRNAs Processed by HeLa Cell Extract." *Cell* 32:547–557.

31. Laski, F. A., A. Z. Fire, U. L. RajBhandary, and P. A. Sharp. 1983. "Characterization of tRNA Precursor Splicing in Mammalian Extracts." *J. Biol. Chem.* 258:11974–11980.

32. Cech, T. R. 1985. "Self-Splicing RNA: Implications for Evolution." *Int. Rev. Cytol.* 93:3–22.

33. Kruger, K., P. J. Grabowski, A. J. Zaug, J. Sands, D. E. Gottschling, and T. R. Cech. 1982. "Self-Splicing RNA: Autoexcision and Autocyclization of the Ribosomal RNA Intervening Sequence of *Tetrahymena*." *Cell* 31:147–157.

34. Cech, T. R., N. K. Tanner, I. Tinoco, Jr., B. R. Weir, M. Zuker, and P. S. Perlman. 1983. "Secondary Structure of the *Tetrahymena* Ribosomal RNA Intervening Sequence: Structural Homology with Fungal Mitochondrial Intervening Sequences." *Proc. Nat. Acad. Sci.* 80:3903–3907.

35. Lazowska, J., C. Jacq, and P. P. Slonimski. 1980. "Sequence of Introns and Flanking Exons in Wild-Type and BOX-3 Mutants of Cytochrome *b* Reveals an Interlaced Splicing Protein Coded by an Intron." *Cell* 22:333–348.

36. Borst, P., and L. A. Grivell. 1981. "One Gene's Intron Is Another Gene's Exon." *Nature* 289:439–440.

37. Lamb, M. R., P. Q. Anziano, K. R. Glaus, D. K. Hanson, H. J. Klapper, P. S. Perlman, and H. R. Mahler. 1983. "RNA Processing Intermediates in *cis*- and *trans*-Acting Mutants in the Penultimate Intron of the Mitochondrial Gene for Cytochrome *b*." *J. Biol. Chem.* 258:1991–1999.

38. Davies, R. W., R. B. Waring, J. A. Ray, T. A. Brown, and C. Scazzocchio. 1982. "Making Ends Meet: A Model for RNA Splicing in Fungal Mitochondria." *Nature* 300:719–724.

39. Michel, F., and B. Dujon. 1983. "Conservation of RNA Secondary Structures in Two Intron Families Including Mitochondrial-, Chloroplast-, and Nuclear-Encoded Members." *EMBO J.* 2:33–38.

40. Garriga, G., and A. L. Lambowitz. 1984. "RNA Splicing in Neurospora Mitochondria: Self-Splicing of a Mitochondrial Intron *in Vitro*." *Cell* 39:631–641.

41. Mount, S. M. 1982. "A Catalogue of Splice Junction Sequences." *Nucleic Acid Res.* 10:459–472.

42. Treisman, R., S. H. Orkin, and T. Maniatis. 1983. "Specific Transcription and RNA Splicing Defects in Five Cloned β-Thalassaemia Genes." *Nature* 302:591–596.

43. Wieringa, B., F. Meyer, J. Reiser, and C. Weissmann. 1983. "Unusual Splice Sites Revealed by Mutagenic Inactivation of an Authentic Splice Site of the Rabbit Beta-Globin Gene." *Nature* 301:38–43.

44. Chu, G., and P. A. Sharp. 1981. "A Gene Chimaera of SV40 and Mouse Beta-Globin Is Transcribed and Properly Spliced." *Nature* 289:378–382.

45. Lerner, M. R., and J. A. Steitz. 1979. "Antibodies to Small Nuclear RNAs Complexed with Proteins Are Produced by Patients with Systemic Lupus Erythematosus." *Proc. Nat. Acad. Sci.* 76:5495–5499.

46. Busch, H., R. Reddy, and L. Rothblum. 1982. "SnRNAs, SnRNPs, and RNA Processing." *Ann. Rev. Biochem.* 51:617–654.

47. Mattaj, I. W. 1984. "snRNA: From Gene Architecture to RNA Processing." *Trends in Biochem. Sci.* 9:435–437.

48. Lerner, M. R., J. A. Boyle, S. M. Mount, S. L. Wolin, and J. A. Steitz. 1980. "Are snRNPs Involved in Splicing?" *Nature* 283:220–224.

49. Rogers, J., and R. Wall. 1980. "A Mechanism for RNA Splicing." *Proc. Nat. Acad. Sci.* 77:1877–1879.

50. Krämer, A., W. Keller, B. Appel, and R. Lührmann. 1984. "The 5' Terminus of the RNA Moiety of U1 Small Nuclear Ribonucleoprotein Particles Is Required for the Splicing of Messenger RNA Precursors." *Cell* 38:299–307.

51. Mount, S. M., I. Pettersson, M. Hinterberger, A. Karmas, and J. A. Steitz. 1983. "The U1 Small Nuclear RNA-Protein Complex Selectively Binds a 5' Splice Site *in Vitro*." *Cell* 33:509–518.

52. Keller, W. 1984. "The RNA Lariat: A New Ring to the Splicing of mRNA Precursors." *Cell* 39:423–425.

53. Ruskin, B., A. R. Krainer, T. Maniatis, and M. R. Green. 1984. "Excision of an Intact Intron as a Novel Lariat Structure During Pre-mRNA Splicing *in Vitro*." *Cell* 38:317–331.

54. Padgett, R. A., M. M. Konarska, P. J. Grabowski, S. F. Hardy, and P. A. Sharp. 1984. "Lariat RNAs as Intermediates and Products in the Splicing of Messenger RNA Precursors." *Science* 225:898–903.

55. Brody, E., and J. Abelson. 1985. "The 'Spliceosome'; Yeast Premessenger RNA Associates with a 40S Complex in a Splicing Dependent Reaction." *Science* 228:963–967.

56. Grabowski, P. J., S. R. Seiler, and P. A. Sharp. 1985. "A Multicomponent Complex Is Involved in the Splicing of Messenger RNA Precursors." *Cell* 42:345–353.

57. Frendewey, D., and W. Keller. 1985. "The Stepwise Assembly of a Pre-mRNA Splicing Complex Requires U-snRNPs and Specific Intron Sequences." *Cell* 42:355–367.

58. Black, D. L., B. Chabot, and J. A Steitz. 1985. "U2 as Well as U1 Small Nuclear Ribonucleoproteins Are Involved in Premessenger RNA Splicing." *Cell* 42:737–750.

59. Go, M. 1981. "Correlation of DNA Exonic Regions with Protein Structural Units in Haemoglobin." *Nature* 291:90–92.

60. Branden, C.-I., H. Eklund, C. Cambillau, and A. J. Pryor. 1984. "Correlation of Exons with Structural Domains in Alcohol Dehydrogenase." *EMBO J.* 3:1307–1310.

61. Gilbert, W. 1978. "Why Genes in Pieces?" *Nature* 271:501.

62. Doolittle, W. F. 1978. "Genes in Pieces: Were They Ever Together?" *Nature* 272:581–582.

63. Crick, F. 1979. "Split Genes and RNA Splicing." *Science* 204:264–271.

64. Craik, C. S., W. J. Rutter, and R. Fletterick. 1983. "Splice Junctions: Association with Variation in Protein Structure." *Science* 220:1125–1129.

65. Gilbert, W. 1985. "Genes-in-Pieces Revisited." *Science* 228:823–824.

66. Stein, J. P., J. F. Catterall, P. Kristo, A. R. Means, and B. W. O'Malley. 1980. "Ovomucoid Intervening Sequences Specify Functional Domains and Generate Protein Polymorphism." *Cell* 21:681–687.

67. Suzuki, Y., J. P. Gage, and D. D. Brown. 1972. "The Genes for Silk Fibroin in *Bombyx mori*." *J. Mol. Biol.* 70:637–649.

68. Maniatis, T., P. Melon, V. Parker, N. Proudfoot, and B. Seed.

1982. "Molecular Genetics of Human Globin Gene Expression." In *Molecular Genetics of Neuroscience*, ed. F. O. Schmitt, S. J. Bird, and F. E. Bloom. New York: Raven Press, pp. 87–101.

69. Vandekerckhove, J., and K. Weber. 1979. "The Complete Amino Acid Sequence of Actins from Bovine Aorta, Bovine Heart, Bovine Fast Skeletal Muscle, and Rabbit Slow Skeletal Muscle. A Protein-Chemical Analysis of Muscle Actin Differentiation." *Differentiation* 14:123–133.

70. Engel, J. N., P. W. Gunning, and L. Kedes. 1981. "Isolation and Characterization of Human Actin Genes." *Proc. Nat. Acad. Sci.* 78:4674–4678.

71. Maxson, R., R. Cohn, and L. Kedes. 1983. "Expression and Organization of Histone Genes." *Ann. Rev. Genetics* 17:239–277.

72. Hentschel, C. C., and M. L. Birnstiel. 1981. "The Organization and Expression of Histone Gene Families." *Cell* 25:301–313.

73. Birnstiel, M. L., M. Chipchase, and J. Speirs. 1971. "The Ribosomal RNA Cistrons." *Prog. Nucleic Acid Res. Mol. Biol.* 11:351–389.

74. Sinclair, J. H., and D. D. Brown. 1971. "Retention of Common Nucleotide Sequences in the Ribosomal Deoxyribonucleic Acid of Eukaryotes and Some of Their Physical Characteristics." *Biochemistry* 10:2761–2769.

75. Long, E. O., and I. B. Dawid. 1980. "Repeated Genes in Eukaryotes." *Ann. Rev. Biochem.* 49:727–764.

76. Fedoroff, N. V. 1979. "On Spacers." *Cell* 16:697–710.

77. Planta, R. J., and J. H Meyerink. 1980. "Organization of the Ribosomal RNA Genes in Eukaryotes." In *Ribosomes: Structure, Function and Genes*, ed. G. Chambliss, G. R. Craven, J. Davies, K. Davis, L. Kahan, and M. Nomura. Baltimore, Md.: University Park Press, pp. 871–887.

78. Fostel, J., S. Narayanswami, B. Hamkalo, S. G. Clarkson, and M. L. Pardue. 1984. "Chromosomal location of a major tRNA gene cluster of *Xenopu laevis.*" *Chromosoma* 90:254–260.

79. Clarkson, S. G. 1983. "Transfer RNA genes." In *Eukaryotic Genes: Their Structure, Activity and Regulation*, ed. N. Maclean, S. P. Gregory, R. A. Flavell. London: Butterworths, pp. 239–261.

80. Yen, P. H., and N. Davidson. 1980. "The Gross Anatomy of a tRNA Gene Cluster at Region 42A of the *D. melanogaster* Chromosome." *Cell* 22:137–148.

81. Hovemann, B., S. Sharp, H. Yamada, and D. Soll. 1980. "Analysis of a *Drosophila* tRNA Gene Cluster." *Cell* 19:889–895.

82. Sekiya, T., Y. Kuchino, and S. Nishimura. 1981. "Mammalian tRNA Genes: Nucleotide Sequence of Rat Genes for tRNA^{Asp}, tRNA^{Gly} and tRNA^{Glu}." *Nucleic Acid Res.* 9:2239–2250.

83. Schimke, R. 1984. "Gene Amplification in Cultured Animal Cells." *Cell* 37:705–713.

84. Stark, G. R., and G. M. Wahl. 1984. "Gene Amplification." *Ann. Rev. Biochem.* 53:447–491.

85. Weiner, A. M., and R. A. Denison. 1982. "Either Gene Amplification or Gene Conversion May Maintain the Homogeneity of the Multigene Family Encoding Human U1 Small Nuclear RNA." *Cold Spring Harbor Symp. Quant. Biol.* 47:1141–1149.

86. Keil, R. L., and G. S. Roeder. 1984. "*Cis*-Acting, Recombination-Stimulating Activity in a Fragment of the Ribosomal DNA of *S. cerivisiae.*" *Cell* 39:377–386.

87. Jeffreys, A. J., V. Wilson, and S. L. Thein. 1985. "Hypervariable Minisatellite Regions in Human DNA." *Nature* 314:67–73.

88. Van Arsdell, S. W., and A. M. Weiner. 1985. "Human Genes for U2 Small Nuclear RNA Are Tandemly Repeated." *Mol. Cell. Biol.* 4:492–499.

89. Bernstein, L. B., T. Manser, and A. M. Weiner. 1985. "Human U1 Small Nuclear RNA Genes: Extensive Conservation of Flanking Sequences Suggests Cycles of Gene Amplification and Transposition." *Mol. Cell. Biol.* 5:2159–2171.

90. Bernstein, L. B., S. M. Mount, and A. M. Weiner. 1983. "Pseudogenes for Human Small Nuclear RNA U3 Appear To Arise by Integration of Self-Primed Reverse Transcripts of the RNA into New Chromosomal Sites." *Cell* 32:461–472.

91. Hollis, G. F., P. A. Hieter, O. W. McBride, D. Swan, and P. Leder. 1982. "Processed Genes: A Dispersed Human Immunoglobulin Gene Bearing Evidence of RNA-Type Processing." *Nature* 296:321–325.

92. Lee, M. G.-S., S. A. Lewis, C. D. Wilde, and N. J. Cowan. 1983. "Evolutionary History of a Multigene Family: An Expressed Human Beta-Tubulin Gene and Three Processed Pseudogenes." *Cell* 33:477–487.

93. Soares, M. B., E. Schon, A. Henderson, S. K. Karathanasis, R. Cate, S. Zeitlin, J. Chirgwin, and A. Efstratiadis. 1985. "RNA-Mediated Gene Duplication: The Rat Preproinsulin I Gene Is a Functional Retroposon." *Mol. Cell Biol.* 5:2090–2103.

94. Rogers, J. H. 1985. "The Origin and Evolution of Retroposons." *Int. Rev. Cytol.* 93:188–279.

95. Rubin, G. M. 1983. "Dispersed Repetitive DNAs in *Drosophila.*" In *Mobile Genetic Elements*, ed. J. A. Shapiro. New York: Academic Press, pp. 329–361.

96. Finnegan, D. J. 1985. "Transposable Elements in Eukaryotes." *Int. Rev. Cytol.* 93:281–326.

97. Mount, S. M., and G. M. Rubin. 1985. "Complete Nucleotide Sequence of the *Drosophila* Transposable Element Copia: Homology Between Copia and Retroviral Proteins." *Mol. Cell Biol.* 5:1630–1638.

98. Karess, R. E., and G. M. Rubin. 1984. "Analysis of P Transposable Element Functions in *Drosophila.*" *Cell* 38:135–146.

99. Fedoroff, N. 1983. "Controlling Elements in Maize." In *Mobile Genetic Elements*, ed. J. A. Shapiro. New York: Academic Press, pp. 1–63.

100. Pohlmann, R. F., N. V. Fedoroff, and J. Messing. 1984. "The Nucleotide Sequence of the Maize Controlling Element *Activator.*" *Cell* 37:635–643.

101. Sutton, W. D., W. L. Gerlach, D. Schwartz, and W. J. Peacock. 1984. "Molecular Analysis of Ds Controlling Mutations at the Adhl Locus of Maize." *Science* 223:1265–1268.

102. Gall, J. G., and D. D. Atherton. 1974. "Satellite DNA Sequences in *Drosophila virilis.*" *J. Mol. Biol.* 85:633–634.

103. Pardue, M. L., and J. G. Gall. 1970. "Chromosomal Localization of Mouse Satellite DNA." *Science* 168:1356–1358.

104. Jelinek, W. R., and C. W. Schmid. 1982. "Repetitive Sequences in Eukaryotic DNA and Their Expression." *Ann. Rev. Biochem.* 51:813–844.

105. Van Arsdell, S. W., R. A. Denison, L. B. Bernstein, and A. W. Weiner. 1981. "Direct Repeats Flank Three Small Nuclear RNA Pseudogenes in the Human Genome." *Cell* 26:11–17.

106. Jagdeeswaran, P., B. G. Forget, and S. M. Weissman. 1981. "Short Interspersed Repetitive DNA Elements in Eucaryotes: Transposable DNA Elements Generated by Reverse Transcription of RNA Pol III Transcripts." *Cell* 26:141–142.

107. Ullu, E., and C. Tschudi. 1984. "Alu Sequences Are Processed 7SL RNA Pseudogenes." *Nature* 312:171–172.

108. Weiner, A. M., P. L. Deininger, and A. Efstratiadis. 1986. "Nonviral Retroposons: Genes, Pseudogenes, and Transposable Elements Generated by the Reverse Flow of Genetic Information." *Ann. Rev. Biochem.* 55: 631–661.

109. Singer, M. F., and J. Skowronski. 1985. "Making Sense Out of LINES: Long Interspersed Repeat Sequences in Mammalian Genomes." *Trends in Biochem. Sci.* 10:119–122.

110. Sakamoto, K., and N. Okada. 1985. "Rodent Type 2 *Alu* Family, Rat Identifier Sequence, Rabbit C Family, and Bovine or Goat 73-bp May Have Evolved From tRNA Genes." *J. Mol. Evolution* 22:134–140.

111. Britten, R. J., and E. H. Davidson. 1971. "Repetitive and Non-Repetitive DNA Sequences and a Speculation on the Origins of Evolutionary Novelty." *Quant. Rev. Biol.* 46:111–138.

112. Orgel, L. E., and F. H. C. Crick. 1980. "Selfish DNA: The Ultimate Parasite." *Nature* 284:604–607.

113. Doolittle, W. F., and C. Sapienza. 1980. "Selfish Genes, the Phenotype Paradigm and Genome Evolution." *Nature* 284:601–603.

114. Kidwell, M. G. 1983. "Evolution of Hybrid Dysgenesis Determinants in *Drosophila melanogaster.*" *Proc. Nat. Acad. Sci.* 80:1655–1659.

CHAPTER 21

The Functioning of Higher Eucaryotic Genes

The careful reader will have noticed that in our discussion of the functioning of bacterial and bacteriophage genes, many statements were qualified; our understanding of how bacterial cells control the synthesis of proteins is less complete than our knowledge of the mechanisms of transcription and of protein synthesis. Yet, concentration on bacterial systems has hardened many previous speculations into facts. The more we learn, the more we appreciate the ingenious multiplicity of mechanisms that bacteria use to fine-tune the expression of each of their genes precisely to meet the cell's momentary needs.

To gain an understanding of the functioning of higher eucaryotic genes at the same level of detail is a much more formidable task. Studies with yeast (Chapter 18) have informed us of some of the fundamental aspects of gene expression, which are different in procaryotes and eucaryotes, since procaryotes lack internal membranes, whereas eucaryotes have a nuclear membrane that segregates the transcription of the genetic material from the translational machinery in the cytoplasm. But in another sense, yeast cells are not so different from bacteria. Both are unicellular organisms that grow by adjusting their metabolism (gene expression) to the availability of nutrients in a changing environment. Moreover, although there is a clear leap in size between bacterial cells and yeast cells (20- to 50-fold), as well as between the bacterial genome and the yeast genome, the differences can largely be accounted for by the increased complexity of the eucaryotic cell.

The cells of higher eucaryotic organisms differ from yeast cells in ways that place many further demands on the functioning of their genes. We have already seen that their amounts of DNA are enormous, varying between species in ways that are not yet completely understood. Their genomes are composed of different classes of sequences: true gene regions interrupted by numerous and lengthy introns, structural sequences necessary for chromosome sorting during cell division, and various kinds of repetitive sequences that seem to have a life of their own. Moreover, higher eucaryotic cells are significantly larger than yeast cells (up to about 50 times as large), requiring that the end products and intermediates in gene expression be transported much greater distances. Perhaps most important, different cells in different tissues of the same organism serve different functions, demanding that they make quite different protein products. Thus, cells in a vertebrate organism are programmed not so much to cope with a quickly changing environment as to stably express only a subset of the cell's genes, for example, those needed to make a muscle cell, or a bone cell, or a skin cell. We calculate that only about 1 percent of the genome is being expressed in any given higher eucaryotic cell, at a given time. But that 1 percent is made up of different—though overlapping—sets of genes in different cells of the same organism.

Eucaryotic DNA Is Normally Packaged in Chromatin[1, 2, 3]

Each individual eucaryotic chromosome contains a single DNA molecule, and the huge size of these molecules demands that the DNA be folded in some orderly fashion. This is especially true during mitosis and meiosis, when the DNA must be compacted into separate chromosomes that can be tidily segregated pairwise into the newly forming daughter cells. But even during interphase (the G1, S, and G2 phases of the cell cycle), some folding is necessary simply to pack such long molecules into the cell's nucleus. Moreover, slight differences in the folding in different regions of the genome appear to play an important role in determining which genes are active in any particular cell. DNA folding is the task of specific nuclear proteins. The tight complexes formed between eucaryotic DNA and these proteins are referred to as **chromatin.**

The key proteins in chromatin that serve to fold eucaryotic DNA are the **histones.** These are relatively small proteins that contain a high proportion of positively charged amino acids (lysine and arginine), helping them to bind tightly to the negatively charged DNA double helix. Histones are among the most highly conserved proteins known. They are present in enormous amounts in virtually all eucaryotic cells, their combined mass being approximately equal to that of the cell's DNA. There are five basic types of histone molecules, called **H1, H2A, H2B, H3,** and **H4** (Table 21-1). They occur in equal numbers, except for H1, which is present at one-half the level of each of the others.

When chromatin is viewed in the electron microscope, it often has a beaded appearance built up of spherical, 100 Å particles connected by thin fibers (Figure 21-1a). The amount of DNA associated with each of the so-called **nucleosome** beads is approximately 200 base pairs. (The number varies from about 150 to 250 base pairs, depending on the organism and tissue being examined.) This periodicity can be easily demonstrated by lightly treating chromatin with DNase (micrococcal nuclease). Since the **linker DNA** that connects adjacent nucleosomes in the chromatin fiber is most available to the nuclease, a ladder of DNA fragments is generated, with each fragment differing in length

Table 21-1 Properties of Histones from Calf Thymus

Histone	Composition	MW	Relative Molar Abundance
H1	Lys rich	21,000	1
H2A	Slightly Lys rich	14,500	2
H2B	Slightly Lys rich	13,700	2
H3	Arg rich	15,300	2
H4	Arg rich	11,300	2

Figure 21-1
Demonstration that each nucleosome is associated with approximately 200 base pairs of DNA. (a) An electron micrograph of chromatin in the extended beads-on-a-string configuration. (Courtesy of Victoria Foe.) (b) The products of progressive digestion of chromatin with micrococcal nuclease. First, nucleosomes containing H1 (whose position is shown by a dotted line) and an average length of 200 base pairs of DNA are produced. Further digestion releases H1 and yields core particles containing 146 base pairs of DNA. (c) A ladder of DNA fragments differing by 200 base pairs in length is produced upon light digestion with nuclease. [Photograph courtesy of R. D. Kornberg.]

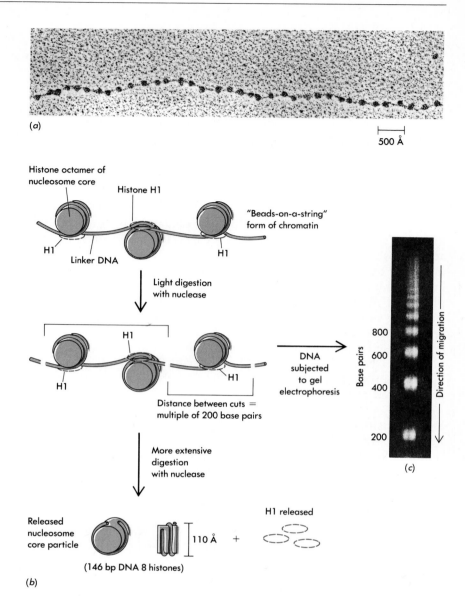

from the next by 200 base pairs (Figure 21-1b and c). More extensive digestion then chews away the remaining tails of linker DNA, yielding highly protected DNA segments of 146 base pairs, regardless of which organism or tissue is examined. This DNA fragment and its bound histones are called the **nucleosome core particle.** It contains two each of histones H2A, H2B, H3, and H4 (the **histone octamer**), as revealed by protein crosslinking analyses. The single molecule of histone H1 associated with each intact nucleosome is lost when the linker DNA is degraded (see Figure 21-1b).

In addition to the histones, chromatin contains an almost equal mass of other proteins (many acidic), more or less firmly attached to the repeating nucleosome array. These nonhistone chromosomal proteins are of many different types and must include various DNA and RNA polymerases as well as regulatory proteins. Since they do not participate in the basic chromatin structure, both the abundance and the identity of nonhistone chromosomal proteins can vary greatly from one cell type to another.

Irregular Bending of the DNA Double Helix Around the Nucleosome Core Particle[4, 5, 6]

In the nucleosome, the DNA winds around the outside of the octamer of histone molecules. Hence, it is quite accessible for interaction with other molecules, such as regulatory proteins. One side of the DNA helix, however, is protected by close association with the histones. We know this because treatment of nucleosomes with high concentrations of DNase I nicks each of the two DNA strands preferentially on the outer, unprotected side of the helix, producing a ladder of single-stranded fragments differing in length by 10 nucleotides (one turn of the helix) (Figure 21-2).

In wrapping about the outside of the disk-shaped histone octamer, the DNA makes nearly two full turns. This compacts it by a factor of about 5: 146 base pairs of extended DNA double helix would be about 500 Å long compared to the 100 Å diameter of the nucleosome particle. Wrapping also introduces one left-handed (negative) superhelical turn into the DNA associated with each nucleosome (Chapter 9). The histone molecules contact the 146 base pairs of DNA in the following order: H2A, H2B, H4, H3, H3, H4, H2B, and H2A. Thus, the nucleosome core particle clearly has a bipartite structure (Figure 21-3).

One of the most fascinating realizations to emerge from the recent three-dimensional analysis of nucleosome core particles is that the DNA does not take a perfectly regular path in winding about the histone octamer. Instead, the double helix is bent fairly sharply at several specific locations (Figure 21-4). The histones interact with the phosphodiester backbone at nearly every turn of the DNA on the inner face of the superhelix. The sharp bends occur adjacent to points of substantial contact between the DNA and histones H3 and H4. At these sites, the regular B helix conformation is distorted over several neighboring base pairs rather than being kinked abruptly.

Sharp bending of the DNA at certain points in its path around the nucleosome may explain the observation that nucleosomes are "phased" on certain DNA sequences. **Phasing** means that along any particular stretch of the genome, nucleosomes in all cells of a tissue or organism are positioned in exactly the same way relative to the nucleotide sequence. Although nucleosomes in many cases appear to be randomly placed, nucleosome phasing has been detected, not only in regions of repeating sequence, such as satellite DNA or tandemly repeated gene families, but also on nonrepetitive DNA. Perhaps the relative flexibility of some nucleotide sequences allows them to be preferentially accommodated at the points where the helix bends in its path around the nucleosome, while other sequences tend to be excluded. Particularly if the DNA sequence repeats every 200 base pairs (or a multiple thereof), such preferences could serve to locate nucleosomes at precise repeating positions. Phasing can also arise adjacent to sites where assembly of the DNA into a nucleosome is prevented by the tight binding of some nonhistone protein(s) to a specific sequence. We currently do not know whether phasing is a fortuitous consequence of the properties of nucleosomes, or whether phasing preferentially exposes certain DNA sequences for interaction with regulatory factors.

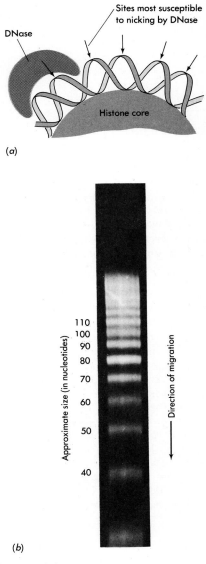

Figure 21-2
How nuclease digestion studies first suggested that DNA is wrapped around the outside of the nucleosome. (a) The sites most susceptible (exposed) to nuclease (DNase I) are spaced an average of ten nucleotides apart on each strand of the DNA. (b) Digested DNA shows a ladder of bands differing by about ten nucleotides in length after fractionation on a polyacrylamide gel under denaturing conditions. (Courtesy of R. D. Kornberg.)

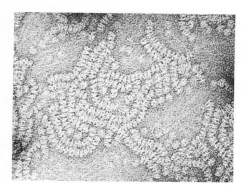

(a)

Figure 21-3
The nucleosome core particle has a bipartite structure. (a) An electron micrograph of a nucleosome core particle preparation used for crystallographic studies. Nucleosome cores are flat, disk-like objects (about 57 × 110 Å) that stack in arcs because they are wedge shaped. (Courtesy of J. Finch.) (b) A model for the nucleosome core particle derived from low-resolution (22 Å) crystallographic studies and later confirmed by high-resolution (7 Å) studies. The DNA makes two turns about the histone octamer, giving the particle its bipartite structure. (The dyad axis, which intersects the 146 base pairs of DNA in its middle, is shown.) The histone positions were assigned on the basis of histone-histone and histone-DNA crosslinking results. [After R. D. Kornberg and A. Klug, *Sci. Amer.* 244 (1981): 52–64.]

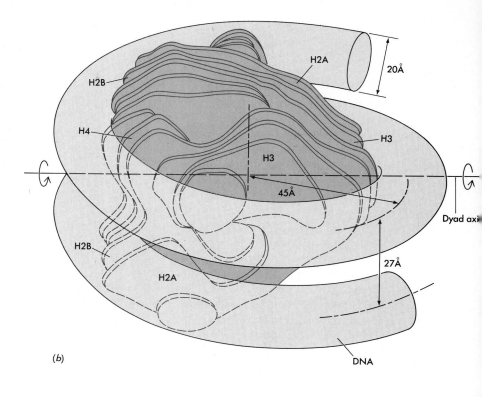

(b)

Higher-Order Chromatin Structure and the Role of Histone H1[1, 2, 7]

The beads-on-a-string structure, created when the DNA is wrapped around individual nucleosomes, is only the first order of packing experienced by eucaryotic DNA. When interphase nuclei are gently broken open, most chromatin is seen as a 300 Å filament in which the nucleosomes are further assembled into a structure of higher order (Figure 21-5a). Local decondensation of the 300 Å filament occurs during transcription of the DNA. Accordingly, the amount of chromatin existing as 300 Å filament provides a rough measure of the relative level of gene *in*activity in cells from various tissues.

Critical for the assembly of nucleosomes into the compact 300 Å filament is the presence of histone H1. This histone molecule is very elongated, with both N- and C-terminal arms extending from a more globular central core. In chromatin, these arms may extend along the linker DNA between nucleosomes. Relative to the nucleosome core particle, H1 is positioned such that it serves to seal the two turns of DNA about the histone octamer. Histone H1 is the most variable of all histone molecules and is found in several different subtypes. (One of these is a histone called H5, which condenses the inactive chromatin in nucleated avian erythrocytes.) The existence of histone H1 variants may underlie the differences in linker lengths seen in chromatin from different tissues or organisms.

Although the exact structure of the 300 Å filament has not yet been determined, most evidence strongly favors the model shown in Figure 21-5b. Here the chromatin is compacted by winding into a solenoid containing six nucleosomes per turn. In this model, the linker DNA is on the inside, as are the molecules of histone H1, which are known to contact one another closely in chromatin. By assembling

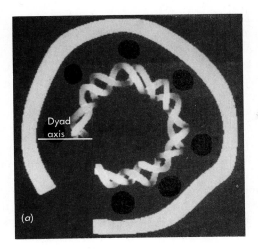

(a)

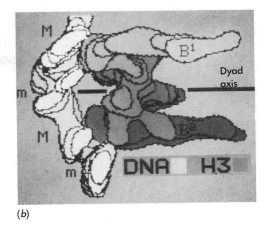

(b)

nucleosomes into the 300 Å filament, the DNA is compacted overall by a factor of about 40 (significantly greater than in the extended beads-on-a-string conformation).

Condensation of Inactive DNA into Heterochromatin[8]

Heterochromatin is the term frequently used to designate highly compacted chromatin that remains visible in the light microscope during interphase. Early genetic experiments suggested that such

Figure 21-4
Irregular bending of the DNA around the nucleosome core. (a) The structure of the right-handed DNA double helix deduced from high-resolution (7 Å) crystallographic studies of nucleosome core particles. Only one-half of the DNA is shown, but the remainder is related to the portion shown by dyad (twofold) symmetry. The DNA is essentially in the B-form helix but is distorted at several points. (b) The region at the center of the 146 base pairs of DNA where the two H3 histones contact the DNA. M stands for the major and m for the minor groove of the DNA helix. Long rods of density labelled B_1 and B_2 in the protein (presumably α helices) make extensive contacts with the minor groove of the DNA one-half turn of the helix away on either side of the dyad axis. [Courtesy of T. J. Richmond, J. T. Finch, B. Rushton, D. Rhodes, and A. Klug, *Nature* 311 (1984):532–537. Used with permission.]

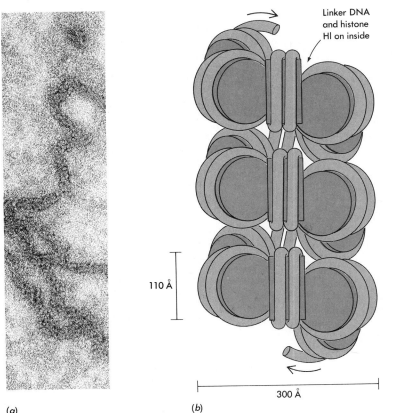

Linker DNA and histone HI on inside

110 Å

300 Å

(a)

(b)

Figure 21-5
Structure of the 300 Å filament. (a) An electron micrograph of chromatin filaments. (Courtesy of B. A. Hamkalo and J. B. Rattner.) (b) The model most consistent with all available data for how nucleosomes pack into the 300 Å filament. [After Alberts et al., *Molecular Biology of the Cell* (New York: Garland, 1983), Fig. 8-5.]

regions contain very few genes, giving rise to the idea that hetero-chromatic DNA does not have a genetic role. Indeed, visibly compacted DNA is not transcribed into RNA, whereas the much more open form of chromatin called **euchromatin,** which is difficult to visualize during interphase, is vigorously copied into RNA chains. Heterochromatin and euchromatin contain approximately the same DNA/histone ratio, confirming that the DNA is packaged into the basic nucleosome structure in both cases. The biochemical basis for the difference between heterochromatin and euchromatin thus remains unknown; but during mitosis, euchromatin is temporarily converted into a form that is almost indistinguishable from heterochromatin.

Not only is heterochromatin relatively inactive, but genes can be inactivated by chromosomal rearrangements that insert them into a heterochromatic region. This suggests that there are control devices that can turn on or off the functioning of a very long section of a given chromosome. This point was first demonstrated for *Drosophila*, but holds equally well for mammalian systems.

The most striking findings concern the sex chromosomes. Initially, the unexpected discovery was made that in female mammals, the two homologous *X* chromosomes look quite different. One always appears highly condensed (hinting that it does not function), whereas the other is extended. This suggestion was confirmed by biochemical analysis, which showed that in each female cell, only the genes on one of the two *X* chromosomes are active. Surprisingly, the inactive chromosome varies from one cell to another; so female tissue is in reality a mosaic containing mixtures of two different cell types. Again, the molecular basis for this bizarre phenomenon remains unknown, although it may be related to methylation of the DNA (discussed later). Nevertheless, *X* chromosome inactivation is important for demonstrating that there are devices that can specifically block the expression of an entire chromosome.

While the other eucaryotic chromosomes do not show all-or-none effects, they also can have extended regions of heterochromatin, whose length depends on when in the lifetime of the organism they are examined. For example, many regions that are completely heterochromatic in later life appear as euchromatin early in development. Such regions probably contain genes programmed to function in one of the early embryonic stages. DNA adjacent to the centromere, which contains highly repetitive satellite sequences (Chapter 20), is an example of heterochromatin that is never transcribed but seems instead to play a role in chromosome structure and maintenance.

Organizing Chromatin onto the Metaphase Scaffold[9, 10]

As in yeast, each chromosome of a higher eucaryotic cell is believed to contain just one enormously long DNA molecule (approximately 5 cm in length for an average mammalian DNA). Yet, the metaphase chromosomes that become visible during mitosis are only about 5 μm long, the diameter of a typical mammalian cell nucleus. This means that there must be several additional higher orders of DNA folding beyond those provided by packaging first into nucleosomes and then into 300 Å filaments. Again, histone H1 may play a pivotal role, for it is found to be extensively phosphorylated (at up to five or six serine residues per molecule) in mitotic chromosomes. Following comple-

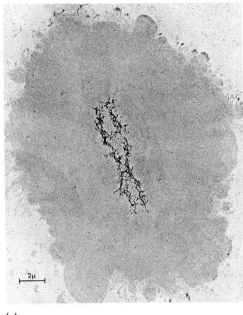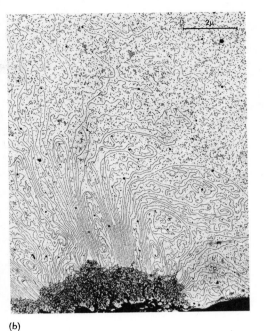

(a)   (b)

tion of mitosis, H1 phosphorylation drops back to less than 20 percent of its peak value and again reaches a maximum level only late in interphase, just before the next cycle of chromosome condensation.

It has become clear that the incredible compaction observed in metaphase chromosomes is accomplished by the organization of the chromatin into large, looped domains attached at their bases to a protein scaffold called the **metaphase scaffold.** This has been demonstrated in experiments in which histones and most of the nonhistone proteins have been removed from a metaphase chromosome by treatment with polyanions (such as heparin or dextran sulfate), which compete with DNA for protein binding (Figure 21-6a). Such depleted chromosomes still retain much of their familiar metaphase morphology, but are surrounded by a halo of unfolded DNA. Upon closer examination (Figure 21-6b), the halo can be seen to consist of many discrete DNA loops, each anchored to the scaffold at its base. The lengths of most of the DNA loops are 10 to 30 μm (35 to 100 kilobase pairs). Examination of the protein content of the histone-depleted scaffold structure reveals a predominance of two proteins of high molecular weight (170 and 135 kdal). These proteins presumably form a network that constitutes the central axis linking together the bases of the DNA loops. How this elegant metaphase scaffold assembles is not known, but the recent realization that the 170 kdal protein is topoisomerase II (Chapter 9) argues that a change in the superhelicity of the DNA may be critical to the process.

Figure 21-6
Electron micrographs of histone-depleted metaphase chromosomes from human HeLa cells showing (a) the familiar morphology of the residual scaffold and (b) that the DNA is attached to the scaffold in loops. [Courtesy of J. R. Paulson and U. K. Laemmli, *Cell* 12 (1977):817–828. Used with permission.]

Protamines Replace Histones to Compact DNA into Sperm Heads[11]

Although higher eucaryotic genomes remain packaged as nucleosomes throughout all stages of the cell cycle in most cells of an organism, there is one situation in which this is not true. The compaction required to condense an entire genome's worth of DNA down to the restricted volume of a sperm head is so great that even histones are

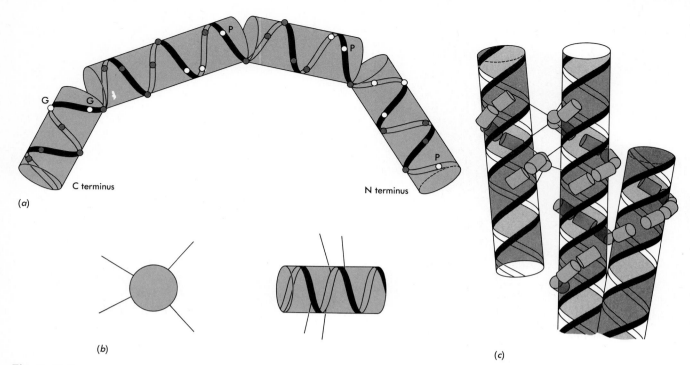

(a)

C terminus

N terminus

(b)

(c)

Figure 21-7

Model showing how protamine molecules can crosslink DNA double helices. (a) A representative protamine molecule (Salmine A1) folding into four α-helical segments connected by partially flexible joints. Each segment is about 8 amino acids long; the positions of glycine (G) and proline (P), which are known helix-breakers, are shown. Black dots represent arginines. (b) Two orientations showing how the side chains of four consecutive arginine residues would radiate out from an α-helical segment. (c) A model showing how protamine molecules could wrap around DNA helices in the major groove. Each α-helical segment could use its arginine side chains to make two contacts across the major groove (gray) of the DNA molecule it has wrapped around, while making two others across the minor groove of an adjacent DNA helix. Alternatively, different α-helical segments of a single protamine molecule could interact with two adjacent DNA helices using only major groove contacts. [After R. W. Warrant and S.-H. Kim, *Nature* 271 (1978):130–135.]

too bulky and must be displaced. Instead, **protamines,** another special class of small, basic, DNA binding proteins, serve to bring the double helices into the closest possible proximity.

Nearly two-thirds of the amino acid residues in protamines are basic (mostly arginine). These are usually found clustered, four or five together along the polypeptide chain. After their synthesis in the cytoplasm, protamines are phosphorylated on their serine residues by a special protamine kinase. They then migrate to the cell nucleus, where they replace histones and become dephosphorylated, concomitantly condensing the DNA. In the final protamine-DNA complexes, the number of positive charges on the protamine amino acid side chains just balances the number of negative charges on the phosphate backbone of the DNA.

Most satisfying is the finding that protamine molecules undergo a dramatic conformational change as they complex with nucleic acid. (The crystal structure of a protamine-tRNA complex is what has been analyzed.) Whereas protamine polypeptide chains occur as random coils free in solution, they become largely α-helical upon binding. Specifically, they form three or four short α-helical segments (depending on the particular protamine) connected by two or three flexible joints (Figure 21-7a). Each helical domain contains four or more consecutive arginine residues and can lie along one of the grooves of double-helical DNA (the major groove is more likely). As a result, the two arginines facing the groove form hydrogen bonds to and neutralize one negatively charged phosphate group from each of the two strands of the double helix. The two remaining arginines in the same α-helical domain of the protamine molecule are then ideally situated to interact in the same way with the backbone phosphates of a neighboring DNA helix (Figure 21-7c). Hence, the protamine molecules serve to crosslink the DNA helices, effecting the high degree of condensation accomplished in sperm heads.

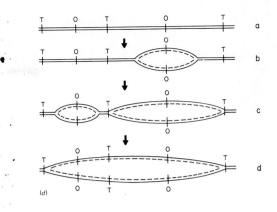

Different DNA Sequences Are Scheduled to Replicate at Different Times During S Phase[12, 13]

As in yeast, the DNA of higher eucaryotic genomes is (usually) replicated bidirectionally starting from multiple origins (Figure 21-8). This means that as many as 100 replication forks are operating on each mammalian chromosome (DNA molecule) at any time during S phase of the cell cycle. Adjacent origins are 50 to 300 kilobase pairs apart, approximately the same length as the DNA loops that extend from histone-depleted metaphase chromosomes. Thus, individual replication units or **replicons** may correspond directly to the independently supercoiled domains of DNA that lie between the attachment points to the metaphase scaffold.

As the DNA is duplicated, histones are quickly deposited and ensure that each new daughter strand becomes properly packaged into nucleosomes. Although little is known about the assembly of nucleosomes in vivo, the one fact that seems clear is that the old nucleosomes segregate conservatively at the fork. Despite the existence of an axis of twofold symmetry in the histone octamer (H2A, H2B, H3, H4)$_2$, nucleosomes do not split down the middle (semiconservatively like DNA), with half of each nucleosome going to each daughter duplex. Instead, one daughter duplex receives the old nucleosome intact, while the other daughter duplex acquires an entirely new nucleosome.

Not all replication origins are activated at the same time during S phase. Rather, clusters of adjacent replicons (between 25 and 100 replicons per cluster in mammalian cells) initiate and terminate replication at a specific time that is characteristic of that particular cluster. Again, a striking relationship between the organization of interphase and metaphase chromatin is evident, for each cluster of coordinately timed replicons appears to correspond to a distinct metaphase chromosome band. After staining with Giemsa (a DNA binding dye), some 2000 light and dark bands can be visualized alternating along mammalian metaphase chromosomes. Adjacent areas stain differently only in part because of differences in base composition; instead,

Figure 21-8
Bidirectional growth of a mammalian chromosome. (a) An autoradiograph of a Chinese hamster DNA molecule from a cell that has briefly been labeled with tritiated thymidine. The tandem arrays of exposed grains indicate the existence of several replication points in the portion of the DNA fiber under view. (b) Tandem arrays seen after a pulse exposure to label followed by a chase period in nonradioactive medium. Here the grain density declines from the center to the ends, suggesting that the growing points move in opposite directions. (c) An example of sister replicating molecules held together by a replicating fork. (d) A schematic model interpreting pictures (a) through (c). [Photographs (a), (b), and (c) are courtesy of J. Huberman; they originally appeared in J. A. Huberman and A. D. Riggs, *J. Mol. Biol.* 32 (1968):327.]

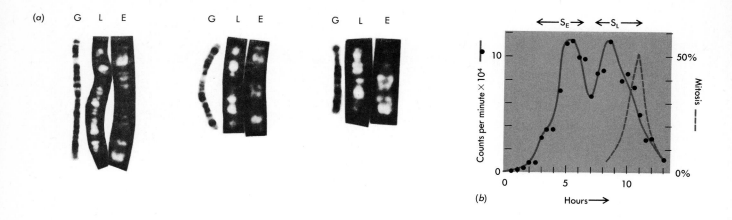

Figure 21-9
Different chromosome bands replicate at different times during S phase.
(a) Fluorescence micrographs of three representative hamster metaphase chromosomes, showing correspondence between Giemsa staining and late-replicating bands (L). The chromosomes labeled G were treated with trypsin and Giemsa. Late-replicating bands were visualized by staining with acridine orange after incorporating BUdR (bromodeoxyuridine) early in S phase; BUdR quenches the green acridine orange fluorescence of the labeled (early-replicating) DNA. Conversely, early-replicating bands (E) were visualized by incorporating BUdR late. (b) The bimodality of DNA synthesis during S phase. Synchronized cells were pulsed with ³H-thymidine and the incorporation plotted against time. [Photographs courtesy of G. P. Holmquist et al., *Cell* 31 (1982):121–129. Used with permission.] To ascertain when a specific gene is replicated, BUdR-containing (heavy) DNA can be fractionated from nonisotopically labeled DNA in a density gradient and the fractions hybridized with the appropriate cloned probe to establish whether the gene incorporated BUdR early or late during S phase. [For more information, see M. A. Goldman, G. P. Holmquist, M. C. Gray, L. A. Caston, and A. Nag, *Science* 224 (1984):686–692.]

the bands clearly give an indication of slightly different modes of DNA compaction. When cells are exposed to isotopic DNA precursors (BUdR) during the first half of S phase, the light-staining chromosome bands become labeled; during the second half of S phase, the dark bands label (Figure 21-9a). Between the two halves of S phase, the rate of DNA synthesis dips significantly (Figure 21-9b). It has been known for some time that blocks of heterochromatin, including regions near the centromere (satellite sequences) and the inactive X chromosome in female mammals, correspond to dark-staining chromosome regions and replicate late during S phase. Conversely, genes that are transcriptionally active in all cells (often called "housekeeping genes") replicate early. Tissue-specific genes usually replicate late in cells in which they are not expressed, but early in cells in which they are potentially expressed. Hence, gene activity seems to be correlated with early replication and light-staining bands. However, how the timing is controlled and what ensures that every origin of replication is utilized only once during each S phase remain to be deciphered.

Specific Attachment Sites for Interphase DNA to the Nuclear Scaffold[14]

In contrast to the networks of fibrous structures (microtubules, microfilaments, etc.) that are readily visualized crisscrossing the eucaryotic cytoplasm, direct examination of nuclei reveals no such filamentous components (Figure 21-10). This, however, does not mean that the interphase nucleus is totally devoid of defined structures. There are one or more nucleoli (singular, nucleolus), discrete bodies that reform after each mitosis at chromosome regions containing repeated rDNA sequences. The nucleoli are responsible for the transcription, processing, and protein packaging of ribosomal RNAs. Also present is the nuclear lamina, a fibrous protein meshwork lying on the inner surface of the nuclear membrane. It appears to hold certain regions of the interphase chromatin (predominantly highly condensed heterochromatin) tightly against the membrane. Yet, chromatin is specifically excluded from the regions around the nuclear pores (Chapter 18), clearing a path for the passage of molecules to and from the cytoplasm. Beyond such examples, no clear organizing principles are evident in the remainder of the interphase nucleus (**nucleoplasm**). Nevertheless, there has long been a strong belief that some as-yet-undefined underlying structure must impose order on nuclear contents and processes.

Thus, it is very satisfying that solid evidence has recently emerged for stable attachment of specific DNA sequences to some sort of scaffolding structure throughout the various stages of the cell cycle. When nuclei from *Drosophila* tissue culture cells are exposed to lithium diiodosalicylate, the histones and many nonhistone proteins are extracted. Subsequent treatment of the remaining **nuclear scaffold** preparation with restriction endonucleases reveals that the DNA is attached at specific sites along the genome. For instance, if we examine the tandemly repeated histone gene family by asking which restriction fragments remain with the residual scaffold and which are released, tight binding can be localized to the AT-rich nontranscribed spacer region between the H1 and H3 genes (Figure 21-11a). Only one attachment point per histone gene repeat (made up of the H1, H3, H4, H2A, and H2B genes, in that order) is detected. This site must be the same in all 100 histone repeat units, since the relevant restriction fragments are either completely retained or completely released from the scaffold. Moreover, since the cells examined have not been synchronized, this attachment site must be stable throughout most stages of the cell cycle.

With several other *Drosophila* genes examined so far, attachment to the nuclear scaffold likewise occurs within five kilobase pairs upstream of (and sometimes also 3′ to) the transcribed region. One or more genes may reside between adjacent attachment sites. In general, active or potentially active genes possess an attachment site in close proximity, whereas certain unexpressed genes are not closely associated with the scaffold. Current topics of investigation are: defining more exactly the attachment sites (do they correspond to recognition sequences for topoisomerase II?); asking whether the interphase attachment points correspond to the bases of the DNA loops associated with the metaphase scaffold; determining if the attachment points around specific genes change during the development of an organism; and defining the protein composition of the preparations to which the DNA remains specifically attached.

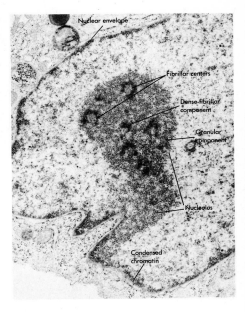

Figure 21-10
Electron micrograph of a thin section of the nucleus from a human fibroblast cell. Within the nucleolus, the fibrillar centers contain the rDNA, the dense fibrillar component represents rRNA transcripts, whereas the granular component is composed of maturing ribosomal subunits. [Courtesy of E. G. Jordan and J. McGovern.]

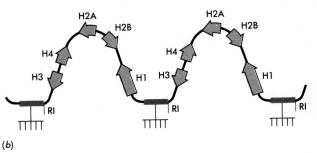

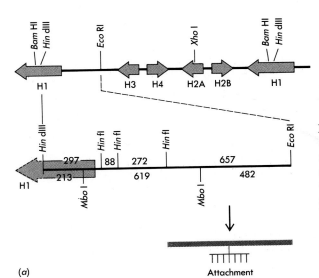

Figure 21-11
One specific attachment site to the nuclear scaffold exists in each repeat of the cluster of *Drosophila* histone genes. (a) A restriction endonuclease map of the major histone repeating unit is shown. Only the 657-base-pair *Hin* fI/*Eco* RI fragment remains attached to scaffold preparations, while all other DNA fragments generated by the enzymes indicated are released into the supernatant. (b) A schematic representation showing how the *Drosophila* histone repeat (present in about 100 tandem copies on chromosome 2) might be attached to the nuclear scaffold. [After J. Mirkovitch, M.-E. Mirault, and U. K. Laemmli, *Cell* 39 (1984):223–232.]

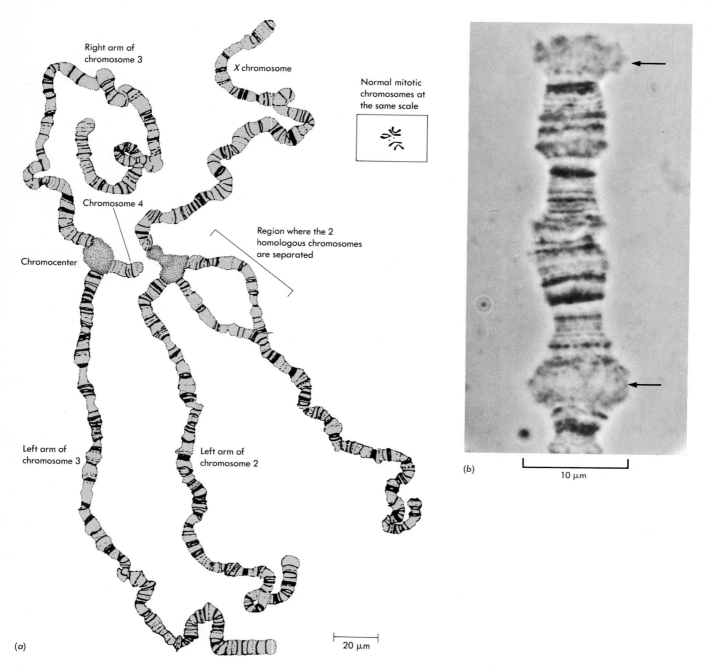

Right arm of
chromosome 3

X chromosome

Normal mitotic
chromosomes at
the same scale

Chromosome 4

Chromocenter

Region where the 2
homologous chromosomes
are separated

Left arm of
chromosome 3

Left arm of
chromosome 2

(a)

20 μm

(b)

10 μm

Figure 21-12
Drosophila polytene chromosomes.
(a) Sketch of the entire set of polytene chromosomes in
one *Drosophila* salivary cell. These chromosomes have
been spread out for viewing by squashing them against
a microscope slide. Note that there are four different
chromosome pairs present. Each chromosome is tightly
paired with its homolog, and the four chromosome pairs
are linked together by regions near their centromeres
that have aggregated to create a single large "chromocenter." Here, this chromocenter has been split into two
halves by the squashing procedure used. [After T. S.

Painter, *J. Hered.* 25 (1934):465–476.] (b) Light micrograph
of polytene chromosome 3L showing the distinct patterns recognizable in different chromosome bands. These
bands occur in interphase chromosomes and are a special property of the giant polytene chromosome. They
should not be confused with the much coarser bands
seen in Figure 21-9, which are revealed by special staining techniques used on normal mitotic chromosomes.
Bands that are puffed are indicated by arrows. (Courtesy
of Joseph G. Gall.)

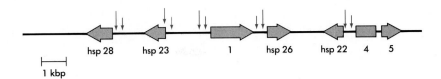

Figure 21-13
The clustering of seven genes within 15 kilobase pairs at cytological locus 67B of the *Drosophila* genome. Four of these genes are vigorously transcribed after heat shock and give rise to heat-shock proteins hsp 28, 23, 26, and 22. Other subsets of this gene cluster are known to be developmentally regulated: hsp 26 and hsp 28 are expressed during oogenesis, while hsp 22, hsp 26, and transcripts 1, 4, and 5 are expressed maximally in the early pupa and late third instar larva. The directions of transcription are indicated by arrows when known. Vertical arrows above the line indicate DNase hypersensitive sites mapped by Keene et al., *Proc. Nat. Acad. Sci.* 78 (1981):143–146. [Data from K. Sirotkin and N. Davidson, *Dev. Biol.* 89 (1982):196–210, and V. Corces, R. Holmgren, R. Freund, R. Morimoto, and M. Meselson, *Proc. Nat. Acad. Sci.* 77 (1980):5390–5393.]

Most attractive is the idea that this newly described nuclear scaffold functions to lure and retain proteins important for transcription. Since the genome is attached near promoters, but not within transcribed regions (Figure 21-11b), we can envision RNA polymerases proceeding along the free DNA loop regions, being recaptured by the scaffold after terminating RNA synthesis, and then—with the help of appropriate transcription factors—reinitiating at sites near the scaffold.

Polytene Chromosomes Allow Visualization of Gene Expression[15–19]

The salivary glands of *Dipteran* flies like *Drosophila* and *Chironomus* contain giant chromosomes that are visible during the interphase stage of the cell cycle. When these chromosomes are stained and examined in the light microscope, an alternating pattern of dark and light bands (bands and interbands) is revealed (Figure 21-12). Band sizes vary tremendously. Some are so thin that they are hardly visible, whereas others occupy several percent of the length of a given chromosome. Each such giant chromosome, called a **polytene chromosome**, consists of a large number of partially replicated chromosomes neatly stuck together in lateral array. Polytene chromosomes arise by repeated cycles of DNA duplication, uninterrupted by separation of the daughter chromosomes or cell division, so that most chromosomal regions are replicated some ten times to yield 2^{10} (or 1024) copies. Remaining unreplicated is the centromeric DNA, while the genes specifying ribosomal RNA are duplicated less often (probably because they are already present in multiple tandemly repeated copies).

In salivary gland cells, some 5000 bands (and interbands) can be detected in the four polytene chromosomes that make up the *Drosophila* genome. The different staining properties of the bands and interbands are highly suggestive of different states of compaction or organization of the chromatin in these regions. Thus, it was initially guessed that each band would encode a single protein. Indeed, early exhaustive genetic mapping of selected regions of several *Drosophila* chromosomes revealed that the number of complementation groups corresponds very closely to the number of visible bands. However, the average band contains some 30,000 base pairs of DNA, much more than enough to encode a single polypeptide chain (even if the gene's transcript contains several lengthy introns). Recent analyses using cloning and sequencing have in fact revealed that many bands do contain several (up to at least seven) genes, which give rise to independent transcripts and are separated by the usual nontranscribed regions (Figure 21-13). Such genes are frequently related, being members of the same gene family. But this is not always the case, nor are the genes in a single band always coordinately expressed, leaving the basis of their clustering a mystery. Whatever the reason, we can now confidently state that the number of genes in the

Figure 21-14
An autoradiogram that illustrates the location of ³H-uridine incorporation sites on polytene chromosomes. The highest grain densities are found over expanded regions on the chromosomes (chromosome puffs), indicating that these are the most active sites of RNA synthesis. [Courtesy of J. J. Bonner and M. L. Pardue, *Cell* 12 (1977):227–234. Used with permission.]

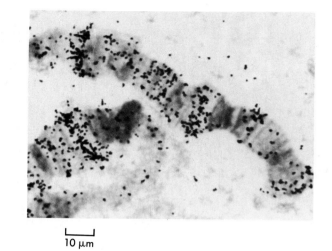

10 μm

Drosophila genome probably does not exceed the number of polytene chromosome bands by more than a factor of 5 or 10.

A particularly useful property of the giant polytene chromosomes is that obvious changes in morphology can be seen when the genes in a particular band become activated. The DNA decondenses into a much more open state, forming a distinctive **puff**. The larger and more diffuse a puff appears, usually the higher the rate of its transcription. This can be shown by briefly labeling cells with a radioactive precursor of RNA (³H-uridine) and localizing the newly made transcripts by autoradiography. Whereas a compacted DNA band synthesizes very little RNA, intense labeling occurs over the largest puffs (Figure 21-14). As the organism progresses through development, puffs arise and regress (recondense) in a defined order as the synthesis of different mRNAs is turned on and off. Experimentally, puffing of different sets of bands can be induced by heat shock or by the addition of the insect hormone ecdysone, which stimulates the synthesis of proteins required for molting and pupation (also see Chapter 22). Electron microscopic examination of serial sections through large puffs can reveal the ribonucleoprotein structure of the nascent RNA transcripts (Figure 21-15). Use of antibodies directed against various nuclear proteins can show that molecules such as RNA polymerases, hnRNP proteins, snRNPs (Chapter 20), and topoisomerases (Figure 21-16) become specifically concentrated in puff regions.

DNase I Sensitivity Characterizes Active Chromatin Regions[20–27]

The changes seen in *Drosophila* polytene chromosomes make it obvious that active (or potentially active) genes undergo some dramatic change in their chromatin structure relative to quiescent regions of the genome. The decondensation of the DNA contained in a particular band to form a puff is often a necessary, but not a sufficient, condition for transcription. (For instance, the multigene band shown in Figure 21-13 puffs both developmentally and upon heat shock, but different subsets of its genes are expressed under the two conditions.) Likewise in vertebrate cells, the two distinguishable forms of chromatin do not correlate directly with gene activity. Heterochromatin does

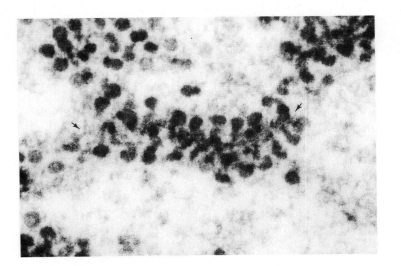

Figure 21-15

Electron micrograph of a portion of a loop within the Balbiani ring 2 of chromosome 2 of a salivary gland cell of *Chironomus tentans*. It shows spherical ribonucleoprotein granules attached by "stalks" to the thin chromatin thread (arrow). Almost simultaneous with RNA synthesis, specific proteins attach to the nascent RNA chains, leading to the formation of ribonucleoprotein granules whose shapes reflect the secondary structure of their underlying RNA chains. [Reproduced with permission from B. Daneholt, *Cell* 4 (1975):1.]

remain both visibly condensed and inactive in RNA synthesis during interphase. Yet, the remainder of the DNA (the euchromatin) cannot all be transcriptionally active, since on average only about 7 percent of the sequences in the genome of a typical higher eucaryotic cell are ever copied into RNA molecules. Thus, we must ask whether other methods can detect changes in the structure of euchromatin in those regions expressing specific genes.

One important indication of altered chromatin structure within active gene regions is a heightened sensitivity to DNase I. Recall that when bulk chromatin is subjected to digestion using low concentrations of nuclease, a ladder of fragments differing by about 200 base pairs in length is generated (see Figure 21-1c). This pattern reflects the rather regular placement of successive nucleosomes along the chromatin fiber. One way of determining whether any particular gene region is organized into such a repeating nucleosome array is to hybridize the ladder of DNA fragments with radioactive DNA derived from the cloned gene. In tissues where the gene of interest is inactive, a defined ladder appears (Figure 21-17). In contrast, hybridizing fragments are much smaller on the average and irregularly sized in tissues where the gene is actively expressed. This greater sensitivity to

Figure 21-16

Specific association of topoisomerase I with transcriptionally active regions of *Drosophila* polytene chromosomes. Antibodies prepared against *Drosophila* topoisomerase I were incubated with polytene chromosome preparations and their binding visualized by applying a fluorescently tagged second antibody directed against the anti-topoisomerase antibodies. A phase-contrast image (c) of part of chromosome 3 is compared to fluorescent images of the same chromosome region from (a) a larva grown at 25°C and (b) a larva heat shocked at 37°C for 20 minutes prior to dissection. 87A, 87C, 93D, and 95D are known locations of heat-shock puffs; their fluorescence is clearly enhanced in (b) relative to (a). [Courtesy of G. Fleishmann et al., *Proc. Nat. Acad. Sci.* 81 (1984):6958–6962. Used with permission.]

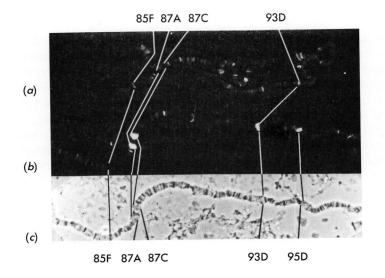

85F 87A 87C   93D

(a)

(b)

(c)

85F 87A 87C   93D 95D

Figure 21-17

Demonstration of heightened DNase sensitivity in actively expressed versus unexpressed ribosomal DNA genes of *Tetrahymena*. Nuclei from vegetative cells (V) synthesizing ribosomal RNA or from slug cells (S), which synthesize little ribosomal RNA, were digested for various times with micrococcal nuclease. The stained gels reveal regularly spaced ladders of fragments, indicating that most DNA is organized into nucleosomes in both types of cells. Examining rRNA genes with a specific probe, however, shows that rDNA fragments are much smaller and more irregularly sized in vegetative cells than in slug cells. (Especially compare the 60-second patterns.) [Courtesy of J. P. Ness, P. Labhart, E. Banz, T. Koller, and R. W. Parish, *J. Mol. Biol.* 166 (1983):361–381. Used with permission.]

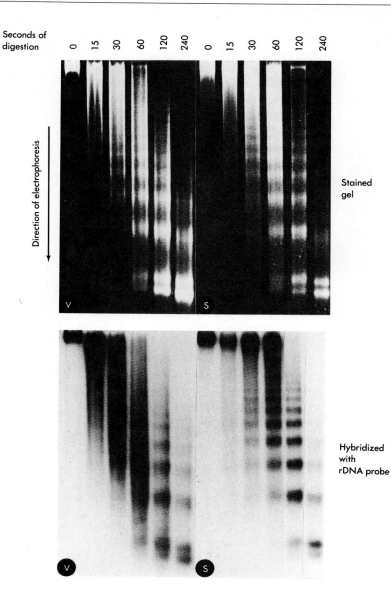

nuclease strongly suggests an unfolding or disorganization of the chromatin upon gene activation. The sensitive region always extends some distance upstream and downstream from the sequences being copied into RNA. If the nuclease used is DNase I, sensitivity is observed for all active genes, regardless of their extent of expression, and it is detected not only when a gene is being transcribed, but also in tissues where a gene is potentially activatable. When micrococcal nuclease is used, the results are more variable. Thus, as with puffing, DNase I sensitivity appears to be a necessary but not sufficient condition for gene expression.

Superimposed on the generally heightened DNase I sensitivity of active chromatin are so-called **hypersensitive sites**. These are localized regions (limited to less than 100 to 200 base pairs) where the nuclease has an extremely high probability of cutting either strand of

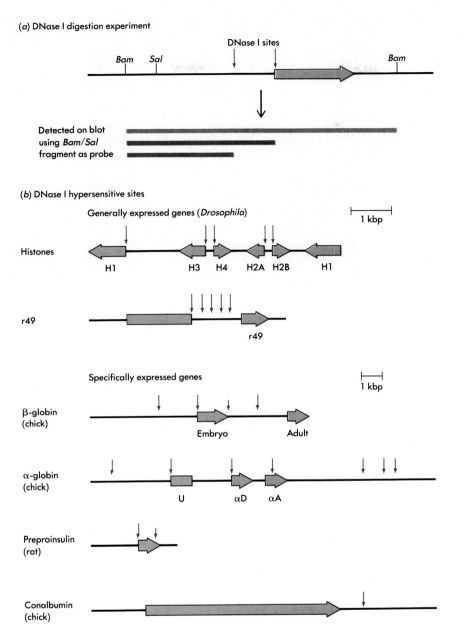

(a) DNase I digestion experiment

(b) DNase I hypersensitive sites

Figure 21-18
The location of DNase I–hypersensitive sites in the vicinity of various genes. (a) How the positions of sensitive sites are mapped. Nuclei are first treated with DNase I, and then the isolated DNA is digested with a suitable restriction enzyme (here *Bam*). After fractionating and transferring the digest to nitrocellulose, the Southern blot is probed with a shorter subfragment (here *Bam/Sal*) located near the gene of interest. In addition to the full-length *Bam/Bam* fragment, shorter bands with one end generated by DNase I will appear in the blot. (b) Some hypersensitive sites are within or 3′, as well as 5′, to the genes examined. See also Figure 21-3 for hypersensitive sites appearing near some *Drosophila* heat-shock genes. [After S. C. R. Elgin, *Cell* 27 (1981):413–415.]

the DNA molecule (Figure 21-18). Hypersensitive sites are generally found within 1000 base pairs flanking the 5′ end of genes that are active or can be activated. Usually, more than one such site is present per gene. The first clues as to how DNase hypersensitivity might arise came from examination of the minichromosome of SV40 (Figure 21-19). Here, the hypersensitive sites are all clustered within a region of the DNA that is free of nucleosomes. The absence of nucleosomes could reflect either an altered DNA conformation, which cannot bind nucleosomes (such as the formation of a Z helix), or the tight binding of other proteins to the DNA. Thus, it is intriguing that hypersensitive sites correspond very closely to the locations of several types of upstream control signals that are known to bind activator proteins required for RNA polymerase II transcription.

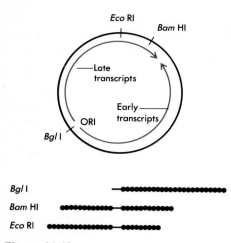

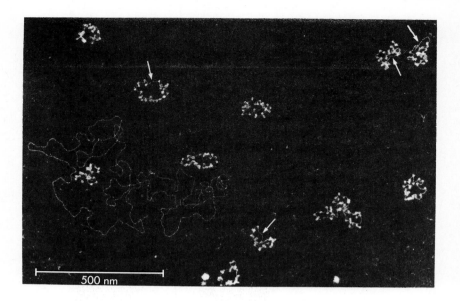

Figure 21-19

A nucleosome-free gap in SV40 chromatin is localized at the origin of DNA replication (ORI). In the electron micrograph, several "minichromosomes" display a gap (indicated by arrows). The gap was localized by cleaving the circular minichromosomes with each of several restriction enzymes to produce the expected patterns of linearized complexes shown below. A variety of DNases preferentially cleave SV40 chromatin within several hundred base pairs of ORI, showing that the nucleosome-free and DNase-hypersensitive sites correspond. [Photograph courtesy of S. Saragosti, G. Moyne, and M. Yaniv, *Cell* 20 (1980):65–73. Used with permission.]

Gene Activity Also Correlates with Undermethylation of the DNA[20, 28, 29, 30]

In most higher eucaryotic genomes, including those of mammalian cells, approximately 5 percent of the C residues are modified by methylation at the 5 position of the cytosine ring. Such methylations occur almost exclusively at CpG sequences to yield mCpG (mC = 5-methylcytosine) and appear symmetrically on both strands of the DNA (since the CpG dinucleotide is always base-paired with another CpG sequence in the antiparallel double helix). Curiously, CpG sequences are themselves underrepresented in genomes that contain mC, being present at only about one-third the expected frequency.

What has become apparent in recent years is that certain CpG sites in the vicinity of many genes are undermethylated in tissues where the gene is expressed, compared to tissues where the gene is inactive. The use of particular restriction enzyme pairs has provided a convenient and simple assay for assessing the state of methylation of some subsets of CpG sequences. For example, both *Hpa* II and *Msp* I cut the sequence CCGG, but methylation renders the site resistant to the *Hpa* II nuclease only. Thus, by hybridizing *Hpa* II and *Msp* I digests of genomic DNA with probes corresponding to a particular gene region, the degree of methylation at nearby CCGG sequences can be easily measured. Demethylation of active genes is not complete. Up to 30 percent of the CpG sequences in transcribed regions are methylated, compared to about 70 percent methylation of all CpG sequences in animal cell DNA. As with DNase sensitivity, undermethylation is usually detected over a significantly larger region than the transcribed portion of a particular gene and is often found in tissues where the gene is not (yet) active but potentially activatable.

Although demethylation and expression are not coupled for every gene that has been examined, in most cases the correspondence is quite striking. This prompts the question of whether changes in DNA methylation are the cause or effect of gene activation. Methylation could regulate gene expression in two main ways. First, addition of a

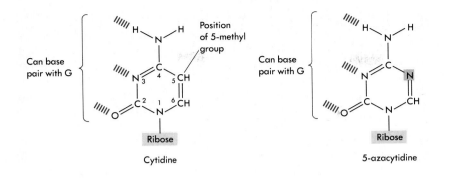

Figure 21-20
The difference between C and 5-azaC. Although 5-azaC lacks a methylatable atom in the 5 position, the extent of demethylation it causes in DNA far exceeds the amount of 5-azaC incorporation. This may reflect the fact that DNA methylases are processive enzymes, so that the presence of a single 5-azaC interferes with methylation at nearby C residues.

5-methyl group to cytosine could increase (or decrease) the interaction between the DNA and proteins such as repressors and activators, for the 5-methyl group protrudes into the major groove of the double helix, where specific DNA-protein recognition often takes place (Chapters 9 and 16). Second, because the addition of a 5-methyl group to C leads to molecular crowding in the major groove, methylation tends to shift the conformational equilibrium of the DNA away from the standard B-form toward other forms (notably the Z form, in which a larger major groove can relieve some of the steric constraints; Chapter 9). Since DNA binding proteins are generally sensitive to the configuration of the sugar-phosphate backbone as well as to the base sequence at their recognition sites, such conformational changes could dramatically alter repressor (or activator) binding.

Some evidence that demethylation may indeed induce gene activation comes from studies using 5-azacytidine (5-azaC), an analog of C whose incorporation into DNA inhibits methylation in vivo (Figure 21-20). Strikingly, 5-azaC induces differentiation (turning on of genes) in cultured mouse embryo cells. Similarly in the case of a number of cloned genes that have been reintroduced into mammalian cells, the unmethylated form proves to be more active than the deliberately methylated gene. However, there are also examples where demethylation lags behind gene expression. Thus, the exact relationship between methylation and gene expression remains unclear, perhaps because only a few of the potential methylation sites in DNA are critical for gene control. Also unclear is how selective demethylation of a critical site might be accomplished when gene activation is desired. A final puzzle arises from the realization that some eucaryotic organisms (notably insects) regulate their genes perfectly well without any DNA methylation whatsoever.

Active Genes Have Altered Nucleosomes or None at All[20, 31–37]

Underlying the heightened DNase sensitivity of active gene regions must be some dramatic change in the wrapping of successive 200-base-pair stretches of DNA around nucleosome beads. This disorganization cannot be ascribed solely to the passage of RNA polymerase molecules, since neither the size of the affected regions nor the time of appearance of DNase sensitivity is strictly coupled to transcription. Instead, either an unfolding of the histone octamers or a disruption of the regular spacing between nucleosomes must occur. Experimentally, both biochemical fractionation and examination in the electron microscope have detected differences in the nucleosome structure of "active" and "inactive" chromatin.

Differentiating active from inactive chromatin are changes in the protein composition of nucleosomes, as well as certain covalent modifications of the histone proteins. The critical role played by histone H1 in condensing the basic beads-on-a-string configuration of chromatin into an inactive 300 Å filament has already been discussed. Consistent is the finding that most (about 80 percent) of the mC in mammalian cell DNA is localized in nucleosomes that contain histone H1. On the other hand, nucleosomes recovered from transcriptionally active regions are deficient in histone H1 and often contain instead the highly abundant nonhistone chromosomal proteins HMG 14 or 17. (HMG stands for **high mobility group**; these proteins are both small and unusually highly charged, therefore moving rapidly upon gel electrophoresis.) Some of the known covalent modifications of the core histone proteins (H2A, H2B, H3, H4) occur selectively in "active" nucleosomes. Included are acetylation, which may decrease the tendency of neighboring nucleosomes to aggregate, and the attachment of a polypeptide called ubiquitin (76 amino acids) to a fraction of H2A molecules, perhaps targeting them for specific protein degradation.

Examination by electron microscopy suggests that even more dramatic changes occur in the packaging of some active genes. For instance, the genes specifying ribosomal RNA—which are among the most transcriptionally active known—appear to be completely devoid of nucleosomes (Figure 21-21). Similarly, in the case of at least one intensely transcribed protein-coding gene region—the Balbiani ring 2 of *Chironomus* polytene chromosomes—the DNA is studded with RNA polymerase molecules, not nucleosomes. These observations raise the question of whether, in biochemical experiments, the "active" genes examined might have been truly active in only a relatively small fraction of the cells. Or, it may be that with anything less than the very highest rates of transcription, the DNA can occasionally be (partially) assembled into nucleosomes between successive passages of RNA polymerase molecules. Of course, just because nucleosomes cannot be visualized does not mean that the DNA is entirely free of bound protein(s). Much remains to be clarified concerning the molecular changes that accompany the transitions of chromatin from the inactive to the potentially active to the transcribing state.

Positive Versus Negative Control of Gene Expression in Higher Eucaryotes[38]

So far, we have presumed that most gene regulation in higher eucaryotes takes place at the level of transcription, as it does in bacteria (Chapter 16). Evidence that this is true has come from examining the functioning of many genes. Yet, also as in bacteria, control of gene expression at other levels cannot be overlooked. Later we will consider how regulatory mechanisms involving RNA processing, mRNA stability, translation, and even protein degradation serve to fine-tune the levels of gene products available for the functioning of higher eucaryotic cells.

Nevertheless, since RNA synthesis is the first step in gene expression, its control overrides regulation at all other levels. Most relevant,

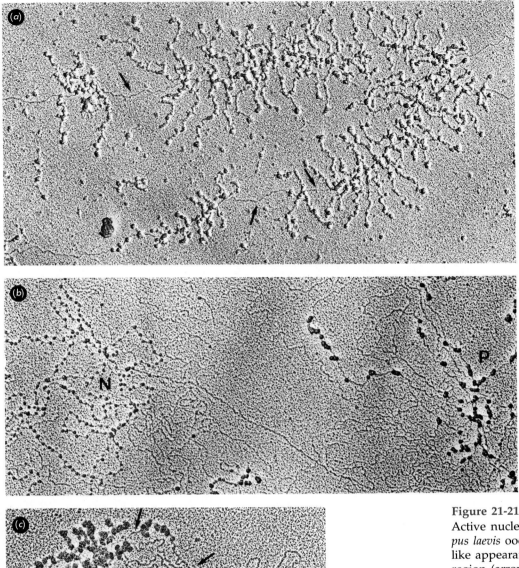

Figure 21-21
Active nucleolar chromatin from *Xenopus laevis* oocytes has a smooth, DNA-like appearance within the transcribed region (arrows in (a) and (c)).
(a) Oocyte nuclei were lysed and the samples centrifuged through a sucrose step gradient onto an electron microscope grid revealing Christmas-tree–like structures. (b) Comparison of presumptive RNA polymerases (P) to nucleosomes (N) in an RNase treated sample. (c) Comparison of the nucleolar chromatin fiber to purified DNA. Note the similar appearance between the chromatin axis (arrow) and the coprepared circular viral DNA on the right side of the figure. [Courtesy of T. Koller. See P. Labhart and T. Koller, *Cell* 28 (1982):279–292, for details.]

therefore, is the question of whether positive or negative mechanisms determine which RNAs are transcribed from higher eucaryotic genomes. Bacteria utilize both activators (e.g., cAMP binding protein) and repressors (e.g., *lac* repressor) to control their RNA synthesis. These regulatory proteins recognize specific DNA sequences upstream from genes, thereby either preventing or facilitating the binding of RNA polymerases (Chapters 13 and 16). In addition, they have a general but much lower affinity for nonspecific (random) DNA sequences and spend part of their time binding to irrelevant stretches of the genome. To circumvent this problem, bacteria simply adjust the numbers of regulatory proteins they synthesize so that although some repressor (or activator) molecules are bound to random DNA, others are available to repress (or activate) transcription completely by binding to the correct specific DNA sequence.

The molecular problems posed by positive versus negative regulatory strategies differ for higher eucaryotic cells with their huge DNA contents. First, much greater numbers of repressor (activator) molecules are required to compensate for regulatory proteins being sopped up by more nonspecific DNA. Here, the correspondingly larger sizes of higher eucaryotic cells can probably accommodate the necessary increase in numbers of molecules. And second, the tremendously large size of the genome makes it much more probable that DNA sequences identical (or very similar) to a specific binding site will appear more than once (and perhaps in front of a gene that should not be controlled by that particular regulatory molecule).

There are two obvious ways out of this dilemma. One is to invoke the ability of histones and other chromosomal proteins to conceal DNA sequences. (Recall that only 7 percent of the DNA in the genome of a typical higher eucaryotic cell is ever copied into RNA.) But in this hall of mirrors, how would the chromosomal proteins know what sequences to cover up? The other way out is to invoke *positive* control of gene expression executed by the simultaneous binding of several regulatory factors to nearby specific sites on the DNA. Only if all sites are appropriately occupied would an RNA chain be initiated. And the probability of several regulatory molecules being bound in the correct juxtaposition by random DNA would be low. Negative control, on the other hand, would not be made more specific by multiple sites, because repressor binding to just one sequence would probably be enough to block RNA polymerase access. As we shall see, most (but not all) well-characterized transcriptional control mechanisms in higher eucaryotes do turn out to be positive and to involve multiple DNA binding factors. This is true whether the gene is transcribed by RNA polymerase I, II or III.

The Nucleolar Location of Ribosome Biogenesis[39, 40, 41]

Just as ribosomes are factories for making proteins in all cells, nucleoli are factories for making ribosomes in eucaryotic cells. These distinctive nuclear substructures form in somatic cells at the one or more chromosomal sites where multiple copies of rRNA genes are present in tandem array (Chapter 20). Although no membrane separates a nucleolus from the remainder of the nuclear contents (see Figure 21-10), nucleoli have distinctive protein compositions representing high concentrations of all the enzymes required for synthesizing and processing precursor ribosomal RNA into the 28S, 18S, and 5.8S molecules found in mature ribosomes. 5S rRNA molecules synthesized elsewhere in the nucleus are somehow specifically transported to the nucleolus for assembly into nascent ribosomes, as are almost all of the 70 to 80 different ribosomal proteins. They, like all other proteins, are synthesized on previous generations of ribosomes present in the cell cytoplasm. Assembly of these many components begins while the ribosomal RNA is still incompletely synthesized and processed, and apparently proceeds along a defined pathway similar to (but more

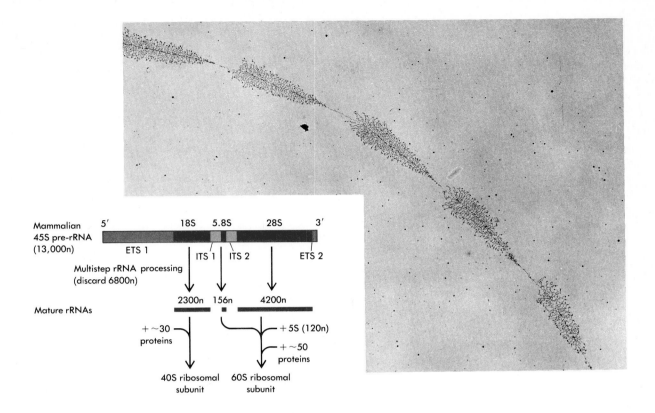

complicated than) that deduced for *E. coli* ribosomes (Chapter 14). One important difference may be the participation of proteins that are not present in the final structures but remain in the nucleolus to catalyze further rounds of ribosome assembly. Nearly completed 40S and 60S subunits are then exported by unknown mechanisms from the nucleolus to the cytoplasm. There they acquire a few remaining proteins to become functional ribosomes.

Although we are still largely ignorant of the details of eucaryotic ribosome assembly, many insights into the mechanisms of rRNA synthesis and processing have been gained in recent years. Examination of spread nucleolar chromatin in the electron microscope (Figure 21-22) reveals tandem arrays of rRNA genes studded with closely packed RNA polymerase I molecules from which emanate growing rRNA precursor chains (about 100 per gene). At the point of each Christmas tree–like formation is the promoter region for that unit, while the base of the tree corresponds to the 3′ ends of the rRNA transcripts. The active units are separated by the so-called **nontranscribed spacer regions** of the rDNA repeat (Chapter 20).

The 18S, 5.8S, and 28S sequences are transcribed into a single rRNA precursor molecule (in that order), ensuring that these three rRNAs will be available in equimolar amounts for assembly into ribosomes. The length of the pre-rRNA chain in mammalian cells is about 13,000 nucleotides (45S), of which about half (denoted **transcribed spacer**) is discarded during rRNA processing (see Figure 21-22). Recently, it has been discovered that the 5.8S RNA molecule (156 nucleotides) present in all eucaryotic ribosomes is homologous in both sequence and structure to the 5′ end of bacterial 23S rRNA (Figure 21-23). Accordingly, the region specifying 5.8S RNA is located just upstream from 28S rRNA sequences in the rDNA, and 5.8S RNA very likely base-pairs with 28S RNA in the ribosome, using sequences at

Figure 21-22
A tandemly arranged series of nucleolar rRNA genes from *Triturus viridescens*, showing growing 45S rRNA precursor chains. The inset at the lower left shows the steps in the transformation of the mammalian 45S pre-rRNA molecule into mature rRNA molecules. ETS stands for external transcribed spacer and ITS for internal transcribed spacer. (Micrograph courtesy of O. L. Miller and B. Beatty.)

Figure 21-23

Proposed universal secondary structure model for 5.8S rRNA and its correspondence to the 5' end of *E. coli* 23S rRNA. Both secondary structures were based on extensive phylogenetic data. The numbering in the 23S sequence corresponds to residues from the 5' end. The sizes of the rRNAs in the inset are not drawn to scale. [After J. C. Vaughn, S. J. Sperbeck, W. J. Ramsey, and C. B. Lawrence, *Nucleic Acid Res.* 12 (1984):7479–7502.]

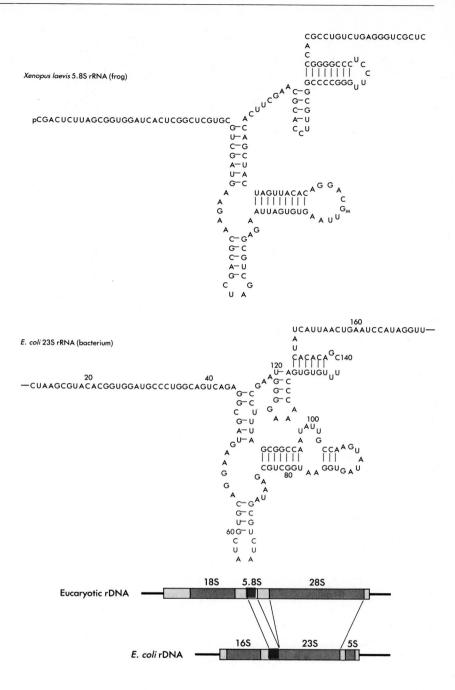

both its 5' and 3' ends. Why eucaryotes clip off this section of the molecule to form a separate RNA chain is unknown.

Whereas mature rRNA molecules show remarkable similarity between lower eucaryotes (yeast) and higher eucaryotes (mammals) and even between procaryotes and eucaryotes (Chapter 14), both the sizes and the sequences of the transcribed spacer segments are highly diverged. The number of nucleotides discarded during processing of the pre-rRNA increases from about 1400 in *E. coli* to 6800 in humans. Lacking in eucaryotic spacer regions are the unusual double-stranded stems that bring together distant sequences of *E. coli* pre-rRNA for recognition and cleavage by the enzyme RNase III (Chapter 14). Instead, even the first cleavages experienced by eucaryotic pre-rRNAs seem to occur in single-stranded regions and often directly generate

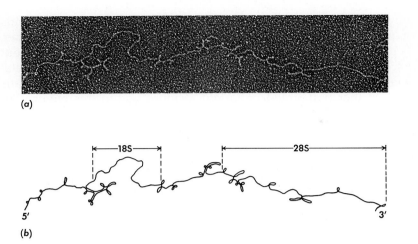

(a)

(b)

Figure 21-24
(a) Electron micrograph of a 45S pre-rRNA molecule spread in 80 percent formamide and 4 *M* urea to preserve the stronger hairpin loops. (b) A tracing of (a), which shows the regions destined to become 28S and 18S rRNA. [Reproduced with permission from P. Wellauer and I. Dawid, *Cold Spring Harbor Symp. Quant. Biol.* 38 (1974):325.]

the 5′ and 3′ termini of the mature rRNA species. The highly conserved, stable RNA secondary structures observed in vertebrate 45S pre-rRNAs are confined to interior portions of the 28S and spacer sequences (Figure 21-24). It has been proposed that a small RNA-protein complex (Chapter 20) containing U3 RNA, which is specifically confined to the nucleolus, participates in rRNA processing; but how it might act has not been firmly established.

Species-Specific Signals for rDNA Transcription by RNA Polymerase I[42-50]

In contrast to RNA polymerases II and III, each of which is responsible for the transcription of many different genes with many different sequences, RNA polymerase I synthesizes only a single product, the pre-rRNA transcript. Thus, it need recognize only one promoter and one terminator signal, but must do so very efficiently to achieve the high rate of transcription typical of active rRNA genes. For these reasons, it was initially assumed that it would be easy to decipher the sequences required for initiation and termination of synthesis by RNA polymerase I simply by comparing rDNA sequences from a number of eucaryotic species.

This strategy, however, quickly failed when completely different sequences were discovered around rRNA start sites in the DNA from different organisms (Figure 21-25). Strong homologies are detected

Figure 21-25
Limited sequence homologies around start sites for RNA polymerase I in rDNA of various eucaryotic organisms. No recognizable homologies appear where the sequence is not given. [After J. Sommerville, *Nature* 310 (1984):189–190.]

	−40	−30	−20	−10	+1	+10
Mouse		ATCTTT	TATTG		TACTGACACGC	
Rat		ATCTTT	TATTG		TACTGACACGC	
Human	ATCTTT		TTTGG		TGCTGACACGC	
Xenopus laevis		CTACGCTTTT	GTGCGGGCAGGAAGGTAGGG			
Xenopus clivii		CTACGCGTTT	GTGCCGACAGGAAGGTAGGG			
Xenopus borealis		CTACGCTTTT	GTGCGGACAGGAAGGTAGGG			
Drosophila melanogaster		TTT	G		TAGGT	
Tetrahymena pyriformis			G		T	
Dictyostelium discoideum			T	ATACATATA		
Physarum polycephalum			A	ATACATATA		
Saccharomyces carlsbergensis			A	AGGTACTTCATGCGAAAGC		
Saccharomyces rosei			G	AGGAACTTCATGCGAAAGC		

only for very closely related species (e.g., two mammals or two kinds of toads). Yet, an essential role for sequences located in the immediate vicinity of the point where rRNA synthesis initiates (between about −45 and +10) has been established by examining the effects of deletions coming closer and closer to the start site. Single-base mutations have further pinpointed important residues within this region. For instance, in mammalian rDNA, a TTT sequence around position −30, a conserved G at position −16, and the first few transcribed nucleotides may all act together to achieve accurate and efficient initiation by RNA polymerase I. These sequences surrounding the rRNA start site are sufficient for the initiation of transcription by RNA polymerase I and can therefore be said to constitute a promoter for RNA polymerase I.

The rate of rRNA transcription is also dramatically influenced by more distant sequences. This aspect of rRNA gene expression has been most thoroughly analyzed in several species of the toad *Xenopus*, but similar situations probably exist in other eucaryotes, including *Drosophila* and even mammals. In *Xenopus laevis*, the nontranscribed spacers separating rRNA genes are of variable length and are composed mainly of several types of repeated DNA sequences. One type of repeated sequence (about 200 base pairs long) is a 90 percent perfect duplication of the rRNA promoter itself (here defined to lie between positions −142 and +6). This repeat appears as many as eight times within a single nontranscribed spacer. Between the reduplicated promoters and preceding the true rRNA promoter are even more copies of either a 60- or 81-base-pair repeated unit, both of which contain a 42-base-pair sequence that appears between positions −73 and −114 of the rRNA promoter (Figure 21-26). Not only have the 60/81-base-pair repeats been shown to enhance initiation at a downstream rRNA promoter, but the promoter duplications can (at least under certain conditions) initiate the synthesis of short transcripts. That the 60/81-base-pair repeats may bind some limiting factor essential for rRNA expression is suggested by the observation that rRNA promoters preceded by more of these elements outcompete templates with fewer repeats in in vitro transcription experiments. Such competition may in fact be the molecular basis for the phenomenon of **nucleolar dominance** observed when *Xenopus laevis* is mated with *Xenopus borealis*. The hybrid embryos express only *laevis* rRNA genes, whose spacer regions contain many more of the repeating 60/81-base-pair elements than *borealis* rDNA spacers. Whether the putative binding protein participates directly in transcription initiation or (more likely) prevents the binding of interfering molecules (e.g., histones) to DNA upstream from the promoter is not yet known.

On the other hand, differences in actual transcription initiation factors (and in the DNA sequences they recognize) are probably the basis of the species specificity observed in rRNA synthesis when components from two more distantly related species are mixed. Mouse cell extracts transcribe only mouse rDNA, and human cell extracts transcribe only human rDNA. But when the extracts are fractionated, it is not the RNA polymerase I itself that is species-specific, but rather some additional required factor(s). At least one required factor (perhaps in conjunction with RNA polymerase I) binds to the DNA in the vicinity of the start site, forming a complex so stable that a subsequently added rDNA promoter fragment cannot be utilized for transcription. It remains to be worked out exactly how many factors there are, which are species-specific, and what rDNA sequences they rec-

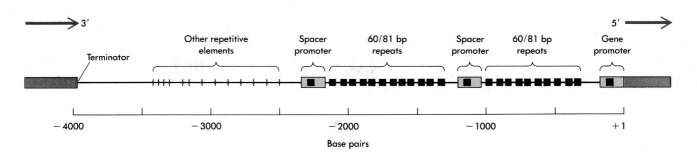

ognize. Even without this knowledge, it seems safe to ascribe to such transcription initiation factors a second type of nucleolar dominance that is observed when cells from different species (such as humans and mouse) are fused in the test tube. The resulting **somatic cell hybrids** (which have been widely used for gene-mapping experiments) specifically lose chromosomes of one parental species; in hybrids between human and rodent cells, human chromosomes are gradually lost. Although both types of rRNA genes are often retained, most hybrid cells express only the rRNA of the parent whose chromosomes are not lost. Loss of the chromosome(s) containing the gene(s) coding for the species-specific rRNA transcription factor(s) may be responsible.

How has the species specificity of RNA polymerase I transcription factors evolved? We have seen in Chapter 20 that tandemly repeated multigene families are generally very homogeneous because eucaryotic organisms possess mechanisms for homogenizing closely related DNA sequences. Thus, a random mutation in any individual rDNA repeat unit will have a finite chance of spreading to all the other repeat units. If such a mutation occurs in a region of the rDNA that binds an essential transcription factor, only those organisms that have sustained a compensatory mutation in the factor survive. In this fashion, mutations in the template select for mutations in the binding factors (and vice versa), so that the DNA and protein factors *coevolve*. Increasing (but coupled) divergence of both promoter sequences and interacting transcription factors ultimately leads to species specificity.

There is not enough information about rRNA termination signals to ascertain whether these, too, are species specific or can be generally recognized by RNA polymerase I. The terminators in vertebrate rDNA reside several hundred base pairs beyond the end of 28S sequences and are only now being characterized.

A most tantalizing question that has not yet begun to be tackled is how the production of all the components of eucaryotic ribosomes is coordinated. Ribosome synthesis responds to many conditions, such as the growth state of the cell, nutrient starvation, and viral infection. Surely there must be some ingenious mechanism to couple rDNA transcription by RNA polymerase I not only to 5S RNA production by RNA polymerase III but also to the synthesis of 70 to 80 ribosomal protein mRNAs by RNA polymerase II.

Figure 21-26
A typical nontranscribed spacer region from *Xenopus laevis* rDNA, showing that the majority of its length is composed of sequences related to the true promoter. The small black boxes are imperfect copies of a 42-base-pair sequence that appears between positions -72 and -114 relative to the transcription start site. The white boxes (which include a black box) are imperfect reduplications of the entire promoter region consisting of about 200 base pairs. At the end of the spacer that abuts the site of transcription termination are other unrelated repetitive sequences. [After R. H. Reeder, *Cell* 38 (1984):349–351.]

Promoter Sequences for RNA Polymerase II Reside Upstream from the Start Site[51–57]

In sharp contrast to RNA polymerase I, which synthesizes only a single type of RNA (pre-rRNA), RNA polymerase II is responsible for

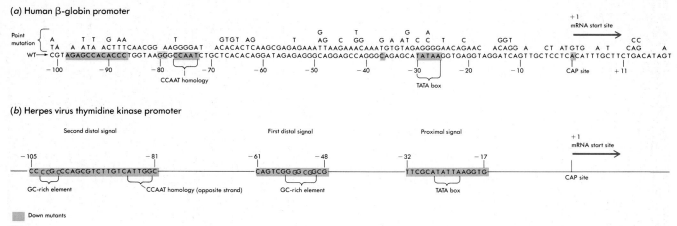

Figure 21-27
Essential sequences in two of the best-characterized mammalian RNA polymerase II promoters. (a) How saturation mutagenesis of the human β-globin promoter revealed that sequences homologous to those in other promoters are required for transcription. Point mutations were introduced into cloned DNA using nitrous acid, formic acid, or hydrazine, and the level of RNA synthesis measured after introduction of the mutant promoters into *HeLa* cells. Positions where mutations caused lower promoter activity are shown in color; positions where mutations enhanced RNA synthesis are shown in gray. All other substitutions gave wild-type transcription levels. Note that the RNA start site, the TATA box, and the CCAAT homology are all promoter regions sensitive to mutation. [Data courtesy of R. M. Myers, K. Tilly, and T. Maniatis; see also R. M. Myers, L. S. Lerman, and T. Maniatis, *Science* 229 (1985):242–247.] (b) How clustered and single-base substitution mutagenesis revealed the location of essential sequences in the herpes virus thymidine kinase promoter. Clustered base substitutions were made by the linker scanning technique [see S. L. McKnight and R. Kingsbury, *Science* 217 (1982):316–324], and single-base substitutions (bold letters) by oligonucleotide mutagenesis [see S. L. McKnight and R. C. Kingsbury, *Cell* 37 (1984): 253–262]. Mutant DNAs were assayed for transcriptional level by injection into *Xenopus* oocytes (see Figure 21-36a). Mutations within the three regions whose sequences are shown resulted in lower promoter activity; mutation of regions indicated by straight lines caused no reduction in transcription efficiency. In this promoter, regions sensitive to mutation include two GC-rich regions, the TATA box and the CCAAT homology. Binding of the GC-rich sequences by transcription factor SP1 (see Figure 21-29) and the CCAAT homology by some other cellular protein has recently been demonstrated in vitro [See K. R. Jones, K. R. Yamanoto, and R. Tjian, *Cell* 42 (1985):559–572.]

the regulated synthesis of a multitude of different transcripts in all eucaryotic cells. RNA polymerase II acts on genes specifying certain small nuclear RNAs (U1, U2, U3, etc.) as well as on every known protein-coding gene, whether the gene is always expressed (a constitutive or **housekeeping gene**) or is expressed only at a particular stage of development in one tissue of a multicellular organism (a regulated gene). Perhaps not surprising, therefore, are two basic features of RNA polymerase II promoters. First, essential sequences lie almost entirely outside of (5' to) the region copied into RNA. Hence, the structure (and function) of RNA polymerase II transcripts can vary almost infinitely. Second, certain sequences within promoter regions are strikingly conserved across species boundaries. Since many thousands of different genes are transcribed by RNA polymerase II, many mutations would have to occur coordinately for species specificity to evolve; the probability of such complex coevolution is very low.

The sequences of hundreds of protein-coding genes from many organisms reveal certain homologies lying within several hundred nucleotides 5' to the initiation site for RNA polymerase II. A frequently occurring homology, the so-called **TATA box**, is centered around position −25 (Figure 21-27). It appears to be essential for accu-

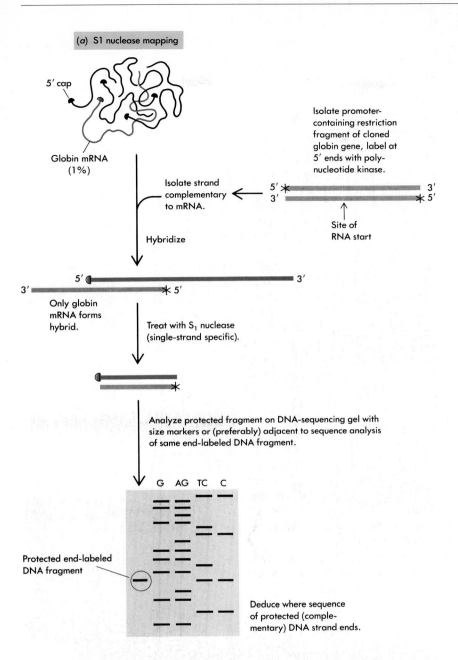

(a) S1 nuclease mapping

5' cap

Globin mRNA (1%)

Isolate strand complementary to mRNA.

Isolate promoter-containing restriction fragment of cloned globin gene, label at 5' ends with poly-nucleotide kinase.

Site of RNA start

Hybridize

Only globin mRNA forms hybrid.

Treat with S₁ nuclease (single-strand specific).

Analyze protected fragment on DNA-sequencing gel with size markers or (preferably) adjacent to sequence analysis of same end-labeled DNA fragment.

G   AG   TC   C

Protected end-labeled DNA fragment

Deduce where sequence of protected (comple-mentary) DNA strand ends.

Figure 21-28, continued next page
Two indirect methods for locating the 5' end of a particular transcript in a mixture of RNA molecules (a) **S₁ nuclease mapping,** (b) **primer extension.** Note that S1 nuclease mapping can also be used to locate the 3' end of an RNA by labeling the DNA probe at its 3' (instead of 5') end. This method is also widely used to locate the position of splice sites in an mRNA. (See Figure 9-47 for DNA sequencing by the Maxam-Gilbert method.)

rately positioning the start of transcription (Figure 21-28) and is in this sense analogous to the −10 region of bacterial promoters (Chapter 13) but different from the TATA sequences in yeast promoters (Chapter 18). A purine is usually (but not always) the first nucleotide incorporated into an RNA chain by RNA polymerase II (again similar to bacterial RNA polymerases; Chapter 13). The 5' end of the transcript then invariably becomes modified by the rapid addition of a cap moiety (Chapter 18).

Other sequence homologies found for RNA polymerase II promoters reside in more variable positions farther upstream from the start site. This is quite different from the stringent spacing requirement between the −10 and −35 regions displayed by bacterial promoters (Chapter 13). For instance, a CCAAT homology occurs in both promoters of Figure 21-27, but it resides on different strands of the DNA.

Figure 21-28
(*Continued*)

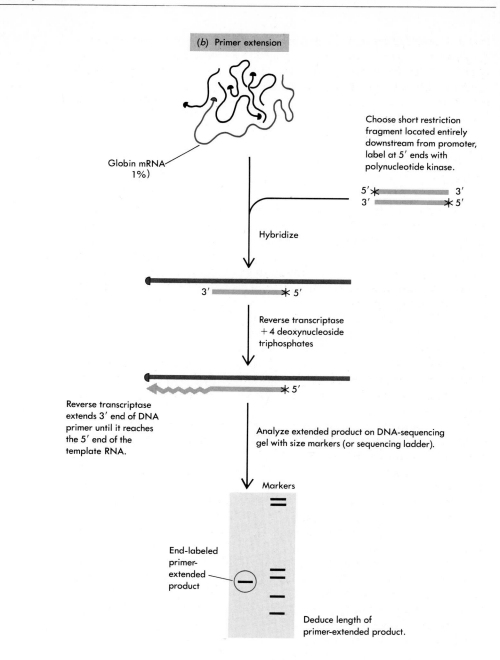

Also included among the upstream promoter elements are specialized signals that seem to characterize certain subsets of genes. Evidence is accumulating that all "housekeeping" genes are preceded by multiple copies of the GC-rich element shown in Figure 21-27b.

That these conserved sequences are essential for binding and initiation by RNA polymerase II when they do appear has been demonstrated by examination of the effects of deleting them or altering them by single or clustered point mutations (see Figure 21-27). An emerging rule is that at least two upstream sequence domains, in addition to the −25 (TATA) region, are required for efficient transcription of every gene. Sometimes, the position and orientation of these elements relative to one another (and to the start site) are critically important; sometimes not. Only now are we beginning to comprehend how these disparate promoter sequences and the proteins they bind contribute to the initiation of transcription by RNA polymerase II. That we still have much to learn is amply demonstrated by the all too frequent identification of yet another gene that seems to lack all conserved sequences previously identified.

Sorting Out Factors for Initiation by RNA Polymerase II[58-64]

The existence of multiple essential sequence elements whose positions are not stringently fixed within RNA polymerase II promoters suggests that the binding of more than one component is necessary for initiating transcription. Only through in vitro fractionation studies can we hope to arrive at a complete picture of the protein-DNA and protein-protein interactions that allow expression of the vast majority of eucaryotic genes. For example, we would like to know which (if any) of the conserved sequences shown in Figure 21-27a and b are recognized directly by RNA polymerase II (which itself contains multiple polypeptide chains; Chapter 18) and which bind auxiliary factors. How many factors are there? Which are generalized transcription factors required by all RNA polymerase II promoters, and which are specific for certain subsets of genes? Is there a defined order of addition of components required for the assembly of an active transcription complex? In addition to positively acting factors, are there negative regulatory proteins? Are cofactors, as well as nucleotide substrates, required for transcription?

The earliest fractionation studies showed that RNA polymerase II cannot by itself recognize a promoter and selectively initiate transcription. At least several separable factor fractions are also required. One fraction that has been partially purified from both mammalian and *Drosophila* cell extracts contains DNA binding protein(s) that interact with sequences from about -40 to $+30$ relative to the RNA start site. Whether this factor is needed by all RNA polymerase II promoters or is one that recognizes the TATA box (included in this region) is not yet clear. Another highly purified mammalian factor called SP1 is required for the transcription of genes whose promoters include multiple copies of the GC-rich regions shown in Figure 21-27b and binds tightly to these sequences (Figure 21-29). Thus, it is tempting to speculate that the observed undermethylation of C residues within such clustered GC-rich sequences in mammalian DNA might be due to this factor's binding to the promoters of constitutively active "housekeeping" genes. Finally, it has recently become clear that some factor binds the CCAAT sequence that resides between the CG-rich elements in the thymidine kinase promoter (see Figure 21-27b); isolation and characterization of this DNA binding component are under way.

In vitro experiments using fractionated systems have also shown that stable complexes including RNA polymerase II and factors form on the DNA prior to initiation of the RNA chain. This means that after a certain combination of factors have bound, they cannot be easily displaced by the addition of a second competing promoter. Interestingly, the hydrolysis of ATP to ADP is required for some very early step in transcription; this reaction is distinct from the incorporation of ATP into the nascent RNA chain. Though we are still far from a

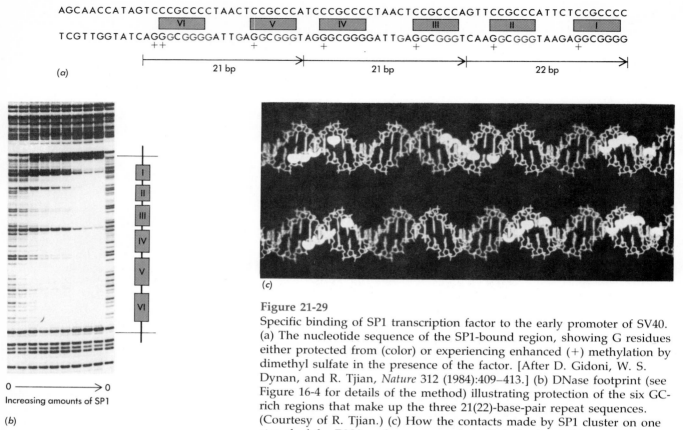

AGCAACCATAGTCCCGCCCCTAACTCCGCCCATCCCGCCCCTAACTCCGCCCAGTTCCGCCCATTCTCCGCCCC

| | VI | | V | | IV | | III | | II | | I |

TCGTTGGTATCAGGGCGGGGATTGAGGCGGGTAGGGCGGGGATTGAGGCGGGTCAAGGCGGGTAAGAGGCGGGG
  ++   +   +   +   +   +

|← 21 bp →|← 21 bp →|← 22 bp →|

(a)

(b)

0 ——————→ 0
Increasing amounts of SP1

(c)

Figure 21-29
Specific binding of SP1 transcription factor to the early promoter of SV40.
(a) The nucleotide sequence of the SP1-bound region, showing G residues
either protected from (color) or experiencing enhanced (+) methylation by
dimethyl sulfate in the presence of the factor. [After D. Gidoni, W. S.
Dynan, and R. Tjian, *Nature* 312 (1984):409–413.] (b) DNase footprint (see
Figure 16-4 for details of the method) illustrating protection of the six GC-
rich regions that make up the three 21(22)-base-pair repeat sequences.
(Courtesy of R. Tjian.) (c) How the contacts made by SP1 cluster on one
strand of the DNA and are arranged similarly in the major groove of both
the SV40 promoter (above) and an SP1-bound promoter from monkey
DNA (below). (Courtesy of R. Tjian.)

complete description of the initiation of transcription by RNA
polymerase II and its requirement for various factors, we can be confi-
dent that a clearer picture will soon emerge.

Transcriptional Enhancers Are Tissue-Specific and Can Act at a Distance[65-70]

In addition to the sequences that constitute a promoter itself, there
are outside elements that drastically alter the efficiency of transcription by RNA polymerase II. The most intriguing of these are the activating regions known as **enhancers**. Enhancer sequences were first
discovered in DNA tumor viruses such as SV40 and polyoma, where
they are located near the origin of DNA replication, which is also the
region that initiates RNA synthesis both early and late after infection
(Figure 21–30). Deletion of the enhancer region (which is composed
primarily of a tandem repeat of 72 base pairs) from SV40 DNA reduces
early gene expression by a factor of over 100. Subsequently, it was
found that the SV40 enhancer region augments transcription not only
from its own promoters but also from almost any other promoter to
which it is linked. Included are a variety of mammalian and avian

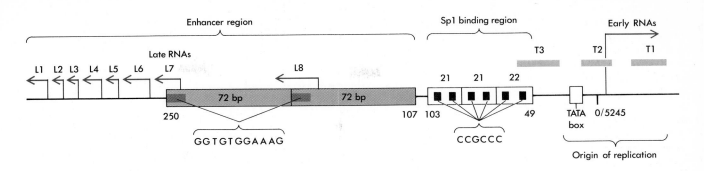

Figure 21-30
Schematic diagram of the elements essential for SV40 early gene expression, including the well-studied SV40 enhancer. The region of SV40 DNA shown corresponds to the nucleosome-free gap in SV40 chromatin (Figure 21-19) and includes the origin of DNA replication, the TATA box, three 21(22)-base-pair repeats, and the 72-base-pair repeat (which represents most of the enhancer region). The start points for early and late transcripts are shown. The GC-rich sequences duplicated within each 21-base-pair repeat are conserved essential promoter elements analogous to those present in the herpes virus thymidine kinase gene promoter (Figure 21-27). The sequence indicated at the 5' end of each 72-base-pair repeat is essential for enhancer function and is homologous to sequences present in other viral enhancer elements (e.g., polyoma, murine sarcoma virus). Also included are the binding sites for T antigen (Chapter 24), one of the rare transcriptional repressors known in a higher eucaryotic system. [After W. S. Dynan and R. Tjian, *Nature* 316 (1985):774–777; and W. Schaffner, E. Serfling, and M. Jasin, *Trends in Genetics* 1 (1985):224–230.]

genes as well as genes from other viruses. Thus, enhancer sequences have been exploited as experimental tools; by linking on an enhancer, genes with intrinsically weak promoters can be readily investigated.

Several remarkable features distinguish enhancer elements from the sequences essential for initiation by RNA polymerase II:

- They appear to be relatively large elements, often including several hundred base pairs and sometimes containing repeated sequences that are independently functional (see Figure 21-30).

- They can act over considerable distances, up to several thousand base pairs.

- They function in either orientation.

- They are position-independent, and they need not reside 5' to a gene, but can be located within or even downstream from the transcribed region. Yet, the enhancer element and the gene must always be present on the same DNA molecule. If several promoters are nearby, the enhancer acts preferentially on the closest.

- A particular enhancer element functions preferentially or exclusively in certain cell types. This tissue specificity (which also manifests itself as species specificity) may in part explain the host range of some animal viruses (Chapter 24). Likewise, tissue-specific enhancers could provide the basis for differential expression of genes during development or in the various cell types of a mature organism.

One stunning example of enhancer action is the turning on of immunoglobulin gene expression after DNA rearrangement brings a promoter into the proximity of a B-cell-specific enhancer element (Chapter 23). Another is the action of steroid hormones, which is discussed later.

Accounting for the properties of enhancers must be the presence of specific cellular proteins whose interaction with an enhancer element somehow renders the DNA in the vicinity of a nearby promoter region more accessible for the binding of RNA polymerase II or its associated factors. Indeed, the presence of an enhancer element is usually correlated with the location of a DNase I hypersensitive site (e.g., the origin region of SV40 DNA; see Figure 21-19). In one attractive model, the specific interaction of enhancer sequences with protein(s) creates a three-dimensional assemblage that acts as a high-affinity target site for polymerase, somewhat like a nest. Alternatively, a gyraselike enzyme might be bound and introduce supercoils, making a localized region of DNA easier for the polymerase to melt. Either model would

Figure 21-31

A control region in the center of the 5S gene directs specific initiation of transcription. The *Xenopus* somatic 5S rRNA gene is diagrammed with the extent of deletions impinging on the gene from its 5' or 3' end indicated by horizontal bars. Those deletion mutations that support initiation of transcription in vitro are designated +. Note that when upstream (5') deletions extend into the gene (but not into the internal control region), RNA synthesis is initiated on vector sequences. When the T-rich sequence at the 3' end of the gene (see inset) is deleted, transcription proceeds beyond the point where termination normally occurs (around position 120; 5S RNAs ending with from one to four U residues have been detected). [After D. F. Bogenhagen, S. Sakonju, and D. D. Brown, *Cell* 19 (1980):27–35.]

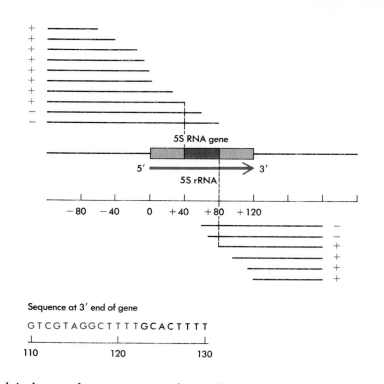

explain how enhancers can work at a distance, in either orientation, and even downstream of a promoter; protein assemblies could bring together components bound to disparate sites on the DNA, and the strain of supercoiling could be transmitted in either direction from the point of gyrase action along the DNA. Experimental tests of these models have so far been hampered by the fact that enhancer elements exert only marginal effects on in vitro transcription. Very recently, however, proteins that bind selectively to portions of enhancer regions have been detected. Thus, research aimed at deciphering how these important controlling elements activate genes should now proceed rapidly.

Internal Control Regions Direct Initiation by RNA Polymerase III[57, 71–79]

Like RNA polymerase II, RNA polymerase III acts on a number of different cellular genes. It synthesizes ribosomal 5S RNA, all tRNAs, and certain other small RNAs (like the 7SL RNA present in a small RNA-protein complex called SRP, described later). In addition, some genes carried by viral DNAs are transcribed by RNA polymerase III: Adenovirus specifies the two VA RNAs, and Epstein-Barr virus specifies two EBER RNAs (Chapter 24). Yet, RNA polymerase III products are not nearly so diverse as RNA polymerase II transcripts:

- They are almost always short, usually containing fewer than 300 nucleotides.

- In no known case do they encode proteins. Rather, they fulfill structural roles, in most cases as components of the cellular protein-synthesizing machinery.

- They are usually (but not always) transcribed from multiple copies of identical genes (Chapter 20). Thus, they are quite abundant in cells (10^4 to 10^7 copies).

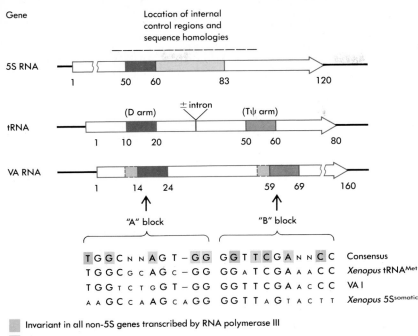

Gene

Location of internal
control regions and
sequence homologies

5S RNA

1   50 60   83   120

± intron

(D arm)   (Tψ arm)

tRNA

1   10 20   50 60   80

VA RNA

1   14   24   59   69   160

"A" block   "B" block

T	G	G	C	N	N	A	G	T	–	G	G	G	G	T	T	C	G	A	N	N	C	C	Consensus
T	G	G	C	G	c	A	G	c	–	G	G	G	G	A	T	C	G	A	A	A	C	C	*Xenopus* tRNA^Met
T	G	G	T	c	T	G	G	T	–	G	G	G	G	T	T	C	G	A	A	c	C	C	VA I
A	A	G	C	c	A	A	G	c	A	G	G	G	G	T	T	A	G	T	A	c	T	T	*Xenopus* 5S^somatic

(Consensus / *Xenopus* tRNA^Met / VA I / *Xenopus* 5S^somatic labels at right)

■ Invariant in all non-5S genes transcribed by RNA polymerase III

□ Invariant in all tRNA

Figure 21-32
Summary of the internal control regions required for genes transcribed by RNA polymerase III. Numbering is from the 5' end of the mature RNAs. In the consensus sequences for "A" block and the "B" block, the nucleotides in color are invariant in all non-5S genes, whereas those in gray are invariant in all tRNA genes. Regions defined as essential for transcription by deletion analyses are designated by rectangles. The broken line over the 5S gene shows the region protected from DNase by factor A (TFIIIA) binding. [After A. B. Lassar, P. L. Martin, and R. G. Roeder, *Science* 222 (1983):740–748.]

- They exhibit a common structural feature: It is nearly always possible to propose an RNA secondary structure that brings together the 5' and 3' ends of the molecule to form a stable base-paired stem [e.g., tRNAs (Figure 14-2) and 5S rRNA (Figure 14-15)]. Yet, some of these RNAs (like tRNAs) have undergone extensive processing, including trimming at both their 5' and 3' termini, splicing to remove intervening sequences, and nucleotide modification. Other RNAs function as primary transcripts (like VA and EBER RNAs) and retain the 5' triphosphate and 3' U-rich termini, hallmarks of initiation and termination by RNA polymerase III.

Since bacterial promoters, as well as promoters for eucaryotic polymerases I and II, all reside upstream from the start site, it was quite unexpected to learn that the synthesis of RNA polymerase III transcripts requires sequences located inside of the genes themselves. This was first discovered upon analysis of deletions that come closer and closer to the 5' end of the *Xenopus* 5S rRNA gene (Figure 21-31). No effect on the level of transcription was observed until the deletions extended more than 50 nucleotides into the 5S coding region itself! When the deletion boundary lay between positions +1 and +50 of the 5S gene, RNAs were simply initiated on the newly introduced DNA to yield a transcript of approximately normal length. Likewise, deletions impinging on the 5S gene from its 3' side lowered the transcription level only upon removal of more than 40 nucleotides of the 5S sequence itself. In this way, an **internal control region** lying between positions +50 and +84 of the 5S gene was defined as essential for initiating synthesis of the 5S RNA molecule.

Other genes transcribed by RNA polymerase III also possess internal control elements. Two stretches of conserved sequences (called the A and B blocks) have been demonstrated to be important by the introduction of deletions and point mutations into them. Their sequences and positions within various genes are shown in Figure 21-32. The distance between A and B blocks can vary considerably (for instance, when an intervening sequence is present in the anticodon loop region of a tRNA gene). It is most intriguing that the A and B

Figure 21-33
Scheme for interaction of the 5S rRNA gene with transcription factors based on in vitro analyses. [After J. J. Bieker, P. L. Martin, and R. G. Roeder, *Cell* 40 (1985):119–127.]

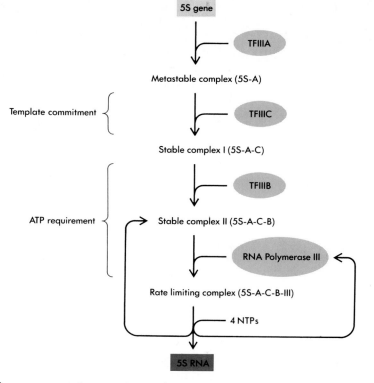

blocks correspond to sections of tRNA molecules that are highly conserved not only in eucaryotes but also in procaryotes, where tRNAs are transcribed from upstream promoters. Whichever arrangement evolved first, this situation in eucaryotic tRNA genes underscores the versatility of nucleic acid sequences. The very same region can function structurally (in the RNA product) and biosynthetically (in the DNA template).

The signals that direct termination by RNA polymerase III also reside within the transcribed region. A group of three or more T residues on the nontemplate strand, often preceded (or surrounded) by GC-rich sequences, specifies the 3' end of the RNA chain (see Figure 21-31). As for bacterial RNA polymerases, termination at such a sequence is believed to be facilitated by the intrinsic instability of a DNA-RNA hybrid made up of polyribo U and polydeoxy A (Chapter 13).

In addition to internal control elements, specific upstream and downstream flanking sequences are essential for the transcription of some genes by RNA polymerase III. Both these external regions and the better-studied internal signals appear to serve as recognition sites for proteins. Which of these DNA binding proteins are factors that cooperate directly with RNA polymerase III to achieve accurate transcription and which might act simply to render the gene more accessible (for instance, by preventing nucleosome formation) are now under intense scrutiny in in vitro systems.

Oocyte 5S rRNA Synthesis Is Autoregulated by a Transcription Factor That Binds to the Gene Product[78, 79]

A most important realization to emerge from in vitro studies with RNA polymerase III is that stable complexes formed between factors

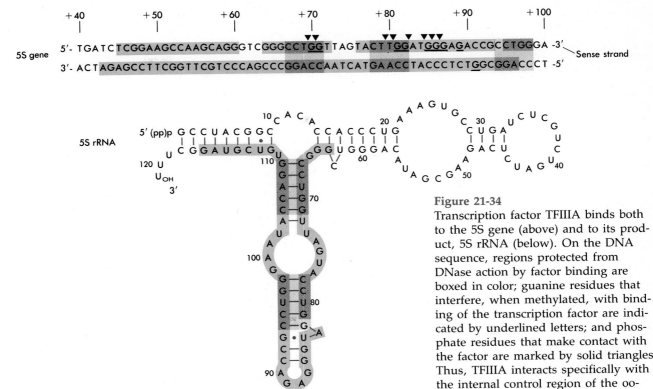

Figure 21-34
Transcription factor TFIIIA binds both to the 5S gene (above) and to its product, 5S rRNA (below). On the DNA sequence, regions protected from DNase action by factor binding are boxed in color; guanine residues that interfere, when methylated, with binding of the transcription factor are indicated by underlined letters; and phosphate residues that make contact with the factor are marked by solid triangles. Thus, TFIIIA interacts specifically with the internal control region of the oocyte-specific 5S gene [see D. R. Engelke, S.-Y. Ng, B. S. Shastry, and R. G. Roeder, *Cell* 19 (1980):717–728; S. Sakonju and D. D. Brown, *Cell* 31 (1982):395–405] making closest contacts with the sense (RNA-like) strand. On the 5S RNA structure, the regions protected from the nuclease α-sarcin [see I. Wool, *Trends in Biochem. Sci.* 9 (1984):14–17] by TFIIIA binding are shown. Dark shading indicates the striking repetition of the sequence CCUGG in double-stranded regions of both the RNA and DNA binding sites; in fact, another such pentamer appears at positions 104 to 108 of the 5S gene. Note that although the sequence of the somatic 5S gene (on which the DNA studies were done) differs in a few positions from that of the oocyte 5S gene (the oocyte 5S RNA-protein complex was studied), the contacts made by the factor with the oocyte 5S gene are very likely to be the same as those shown. [After S. Sakonju and D. D. Brown, *Cell* 31 (1982):395–405, and courtesy of I. Wool.]

and the DNA allow many rounds of RNA synthesis to occur before dissociation. Such stable transcription complexes have been best characterized for the oocyte-specific 5S rRNA genes of the African clawed toad *Xenopus* (Figure 21-33). Recall that the *Xenopus* genome has two different sets of genes encoding 5S ribosomal RNA (Chapter 20). The oocyte-specific 5S RNA genes are expressed exclusively during the long period of oogenesis, when the maturing oocyte stores various macromolecular components for use after fertilization. The somatic 5S genes are expressed primarily after fertilization and throughout subsequent development of the frog. (Because of its aquatic existence, molecular biologists have fallen into the habit of calling this toad a frog.) How is the expression of oocyte-specific 5S RNA turned on during oogenesis and turned off in the mature egg?

Transcription of oocyte-specific 5S genes requires three factors in addition to RNA polymerase III. Two of these (called TFIIIB and TFIIIC) are general factors that also function with other genes such as tRNA or VA genes (see Figure 21-32). In contrast, the third transcription factor (called TFIIIA) is an oocyte-specific protein (40 kdal) that binds tightly to the internal control region, the middle third of the 5S gene (see Figures 21-31 and 21-32). Once the immature oocyte has synthesized a sufficient quantity of TFIIIA, its binding to oocyte-specific 5S genes activates them for transcription by RNA polymerase III (see Figure 21-33). But how are these same genes turned off in the mature oocyte?

For many years, it was known that oocyte-specific 5S RNA was stored in the cytoplasm of the egg as a ribonucleoprotein (RNP) particle consisting of 5S RNA complexed with a specific protein molecule. Remarkably, the 5S storage protein and TFIIIA turn out to be one and the same. This suggests an extremely elegant mechanism for turning the oocyte-specific genes off as soon as a sufficient quantity of 5S RNA has been synthesized (Figure 21-34). Early in oogenesis, when there is an excess of TFIIIA over 5S RNA, most of the factor will bind to 5S genes and stimulate transcription. Later in oogenesis, as 5S RNA accumulates, TFIIIA will be sequestered in the form of a 5S storage particle, and 5S RNA synthesis will be proportionately reduced. Eventually, almost all of the TFIIIA will be present in storage particles and 5S RNA synthesis will cease. Ultimately, after fertilization, the 5S RNA is used to build ribosomes, and the TFIIIA is somehow programmed for destruction.

Thus, we see that the transcription of oocyte-specific 5S rRNA genes is a direct consequence of the amount of TFIIIA available for DNA binding. But once a controlling factor such as TFIIIA has been identified, we are then obliged to ask how *its* synthesis is regulated. In other words, finding the "cause" for one "effect" leads us to wonder about the "cause" of the "cause," and this can only be found if we delve deeper into the mechanisms of eucaryotic gene expression.

Transfer RNA Populations Are Optimized for the Synthesis of Particular Protein Products in Specialized Tissues[80, 81, 82]

That transcription by RNA polymerase III can be controlled in a tissue-specific manner is also evident from examining tRNA populations in different cells of the same organism. In Chapter 15, we saw that the relative abundance of various tRNA species in microorganisms (bacteria and yeast) correlates very well with codon usage in highly expressed genes. Thus, it appears that the tRNA and mRNA promoters in these genomes have coevolved to yield levels of products that maximize the rates of synthesis of abundant proteins. But in a multicellular organism, different tissues are programmed to synthesize different abundant proteins, and a tRNA population tailored to the needs of one cell type might be quite unsuitable for another. Hence, a mechanism for differentially adjusting tRNA populations has evolved in higher eucaryotes. It exists in every tissue examined that is specialized for the production of protein(s) with unusual amino acid compositions.

The most dramatic example of a skewed tRNA population is found in the silk gland of *Bombyx mori* (the silkworm). This tissue is designed for the massive synthesis of two classes of silk polypeptides (called fibroin and sericin), both of which are extremely rich in certain amino acids (Gly, Ala, and Ser in fibroin; Ser, Gly, Asp, and Gly in sericin). At the time that fibroin and sericin mRNAs are being produced, the silk gland radically alters its relative transcription of tRNA genes. Over 30-fold increases in the amounts of tRNA^{Gly} and tRNA^{Ala} result, including the appearance of a new silk-gland-specific species of tRNA^{Ala}. Although the mechanism by which these tRNA genes are turned on has not been worked out to the same extent as the control of oocyte-specific 5S RNA synthesis, some sort of tissue-specific transcription factor must be involved. Again we must ask, What turns on the synthesis of this factor? Could it be that a single silk-gland-specific factor activates both the fibroin and sericin genes for RNA polymerase II transcription and the relevant tRNA genes for RNA polymerase III transcription?

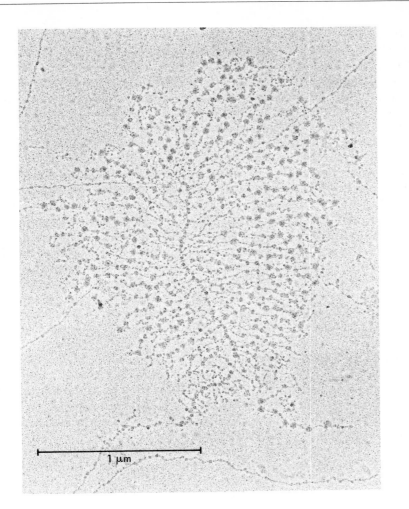

Figure 21-35
The packaging of nascent RNA transcripts into hnRNP produces a beads-on-a-string appearance. An active gene with emanating transcripts from a *Drosophila melanogaster* follicle cell is shown. (Courtesy of Y. Osheim and A. Beyer.)

Messenger RNA Precursors Are Packaged by Specific Nuclear Proteins[83, 84, 85]

Even while pre-mRNAs are nascent and remain attached to their DNA templates via RNA polymerase II molecules, they quickly become (noncovalently) associated with certain RNA binding proteins. This ubiquitous protein-bound state of nuclear pre-mRNA is referred to as **heterogeneous nuclear ribonucleoprotein (hnRNP),** derived from hnRNA (*h*eterogeneous *n*uclear RNA). It is in this form that pre-mRNAs undergo splicing and the other processing reactions that convert them into mature mRNA molecules. When viewed in the electron microscope, hnRNP has the appearance of beads on a string, much like the nucleosome arrays observed in extended chromatin (Figure 21-35). But the hnRNP particles are larger than nucleosomes (about 200 Å across) and more heterogeneous in size. They are connected by RNase-sensitive links. Mild ribonuclease digestion converts hnRNP into individual particles that sediment as a surprisingly homogeneous peak at 30S to 40S, much like small ribosomal subunits. Yet, in contrast to ribosomes, hnRNP particles are very protein rich, containing about three-fourths protein and only one-fourth RNA (corresponding to an approximately 600-nucleotide-long fragment of pre-mRNA per 30S particle).

In higher eucaryotes, the major proteins present in hnRNP are a set of highly conserved polypeptides with molecular weights of 30 to 40 kdal. They contain unusual modified amino acids (e.g., dimethylarginine) and are very closely related to each other, since monoclonal antibodies directed against one protein often cross-react with others. Not surprisingly, the hnRNP polypeptides are among the most abundant nonhistone proteins of the cell nucleus. Somehow they know to associate only with pre-mRNA and not with rRNA or RNA polymerase III transcripts.

How and why are pre-mRNA molecules packaged into hnRNP? Because hnRNP complexes are relatively unstable during isolation and analysis (compared to nucleosomes, for example), much less is known about their structural organization. What seems certain is that the bound RNA fragment wraps several times about the exterior of an essentially proteinacious particle. In the few cases where a specific pre-mRNA has been examined, both intron and exon segments were found in hnRNP. There are recent indications that assembly into hnRNP is essential for splicing and the other processing events that fashion mature mRNA molecules. Alternatively, hnRNP proteins may act primarily to prevent tangling of the RNA; this could make the pre-mRNA more accessible to the processing machinery, or could enable the cell to more easily degrade introns once they are excised. Whatever the case, the job of hnRNP proteins must be strictly nuclear, for they are replaced upon mRNA export from the nucleus. In the cytoplasm, a completely different set of RNA binding proteins associates with mRNA molecules to form RNA-protein complexes called **mRNP** (*messenger ribonucleoprotein*).

Cleavage and Polyadenylation, Rather Than Termination, Generate the 3' Ends of Most mRNAs[52, 86–89]

The tracts of A residues found at the 3' ends of the majority of eucaryotic mRNAs (and their nuclear precursors) are not encoded in the DNA. Rather, they are added by a template-independent enzyme called poly A polymerase after the gene has been copied into pre-mRNA by RNA polymerase II. Some 10 to 30 nucleotides upstream from the resulting poly A tail, within the genome-specified part of the RNA, the sequence AAUAAA (or a slight variant of it) always appears. However, this hexanucleotide cannot be the complete signal for polyadenylation, since mRNAs and pre-mRNAs contain many other AAUAAA sequences that are not followed by poly A tails.

Originally, it was assumed that RNA polymerase II responds to a transcription termination signal somewhere in the vicinity of the correct polyadenylation site, releasing the pre-mRNA transcript with a 3' end that was then acted on by poly A polymerase. That this is not the case has recently become apparent. Instead, RNA polymerase II continues elongating the RNA chain beyond the polyadenylation site (sometimes for thousands of nucleotides); exactly what DNA sequences do signal transcription termination are not yet known. The nascent pre-mRNA is then cut by a specific endonuclease, which creates the correct 3' terminus for polyadenylation and releases downstream sequences for rapid degradation. In this process, the 5'-AAUAAA hexanucleotide upstream from the cut site is recognized as part of the signal essential for cleavage; other relatively conserved sequences downstream from (3' to) the cut site are probably also involved. What recognizes these signals is only now being investigated with the aid of in vitro systems that accurately cleave and polyadenylate pre-mRNAs. Perhaps an essential role is played by one or more of the small nuclear RNA-protein complexes of the class to which U1 snRNPs belong (Chapter 20).

The 3' Ends of Histone mRNAs Are Produced by RNA Processing Involving a Small RNA-Protein Complex[87]

Unlike other eucaryotic mRNAs, histone gene transcripts in most organisms do not acquire poly A tails, nor do they have near their 3' termini the hexanucleotide AAUAAA that signals polyadenylation. Instead, histone mRNAs from a wide variety of species terminate with a highly conserved palindromic sequence that presumably forms a hairpin loop near the 3' end of RNA (Figure 21-36). In addition, an ACCA sequence often occurs just beyond the hairpin loop preceding the 3' end. Hints that these conserved features might be signals for RNA processing rather than for transcription termination by RNA polymerase II were first obtained by progressively deleting the 3' flanking sequence of the H2A histone gene. These studies showed that sequences as far as 80 nucleotides downstream beyond the mRNA terminus were required to generate the 3' end correctly and efficiently. Included in this portion of the spacer DNA between histone genes is another sequence that is highly conserved (CAAGAAAGA), particularly among the well-studied sea urchin histone genes.

Conclusive proof that the 3' ends of histone mRNA are generated by RNA processing came from experiments in which cloned H2A histone genes were used to synthesize long, radioactively labeled transcripts that possessed the correct 5' end of histone mRNA but proceeded far beyond the mature 3' end of the message. When these elongated histone mRNA precursors with extra 3' nucleotides were injected into *Xenopus* oocytes, the oocytes were able to convert them to mature mRNA (see Figure 21-36).

Correct 3'-end processing of histone mRNAs requires the participation of a small nuclear RNA-protein complex (snRNP) containing a 60-long RNA called U7. This has been established in experiments where sea urchin histone genes were injected into *Xenopus* oocytes (see Figure 21-36). Paradoxically, it was most fortunate that one of the sea urchin transcripts (the histone H3 precursor) *fails* to be processed correctly in this mixed system. This provided a convenient assay for isolating from sea urchin extracts the essential factor missing from frog oocytes. Activity has been traced to the U7 snRNP, which shares common antigenic proteins with the U1 and U2 snRNPs implicated in pre-mRNA splicing (Chapter 20). The sequence of U7 RNA suggests that it may interact with the pre-mRNA by forming base pairs with the conserved sequences just preceding and following the 3' end of the mature histone mRNA (see Figure 21-36). Since the mRNA precursors for all five histone species share almost identical 3'-terminal stem-and-loop structures and 3' flanking CAAGAAAGA sequences, it remains a mystery why *Xenopus* oocytes can successfully process four of the sea urchin histone mRNA precursors (H2A, H2B, H4, and

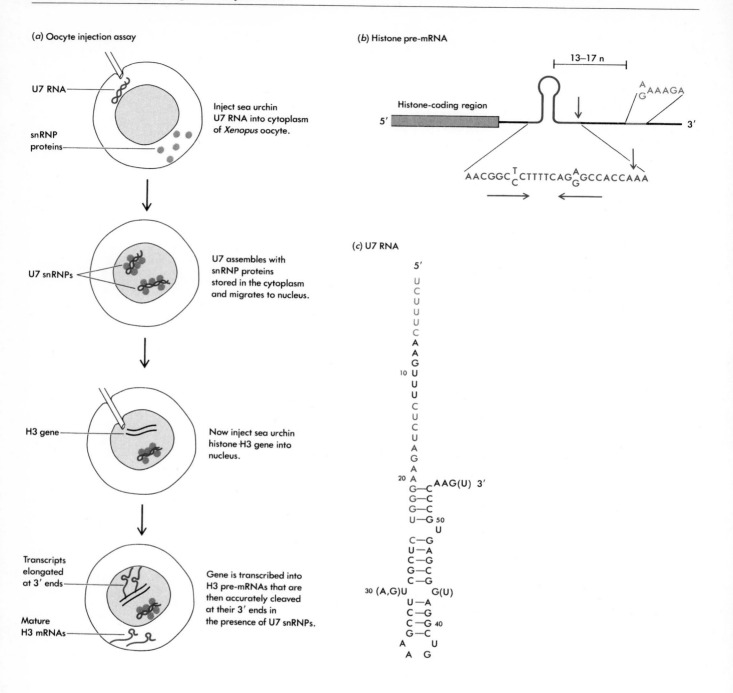

(a) Oocyte injection assay

U7 RNA

snRNP proteins

Inject sea urchin U7 RNA into cytoplasm of *Xenopus* oocyte.

U7 snRNPs

U7 assembles with snRNP proteins stored in the cytoplasm and migrates to nucleus.

H3 gene

Now inject sea urchin histone H3 gene into nucleus.

Transcripts elongated at 3′ ends

Mature H3 mRNAs

Gene is transcribed into H3 pre-mRNAs that are then accurately cleaved at their 3′ ends in the presence of U7 snRNPs.

(b) Histone pre-mRNA

13–17 n

Histone-coding region

5′   3′

A
G AAAGA

AACGGC ^T_C CTTTTCAG ^A_G GCCACCAAA

(c) U7 RNA

Figure 21-36

Processing the 3′ ends of histone pre-mRNAs requires U7 snRNA. (a) How it was established that U7 is involved in histone mRNA processing. (b) A schematic diagram of a histone pre-mRNA, showing the two consensus sequences that are conserved from sea urchins to mammals. The protein-coding region is not drawn to scale. The arrow indicates the site cleaved to generate the 3′ termini of histone mRNAs. The distance between the palindrome (base of the hairpin) and the downstream conserved sequence is also highly conserved. (c) The sequence of sea urchin U7 RNA with the regions complementary to the two histone pre-mRNA consensus sequences indicated by matching colors. [After M. L. Birnstiel, M. Busslinger, and K. Strub, *Cell* 41 (1985):349–359; C. Birchmeier, W. Folk, and M. L. Birnstiel, *Cell* 35 (1983):433–440; and K. Strub, G. Galli, M. Busslinger, and M. L. Birnstiel, *EMBO J.* 3 (1984):2801–2807.]

H1) but not the fifth (H3). Whether the RNA processing nuclease is contained in the snRNP itself and whether its action is endo- or exonucleolytic should soon be determined.

Different mRNAs from the Same Gene in Different Tissues: Exploiting Differences in Both Transcription and RNA Processing[90, 91, 92]

The existence of enhancer elements and putative tissue-specific factors that recognize them can easily explain how the various members of a gene family might be differentially expressed in different tissues or at different stages of development. Thus, the discovery that eucaryotic organisms have also evolved the capacity to use single genes to encode different protein products in different tissues was unanticipated. As with the related proteins specified by different members of a multigene family (e.g., the α- or β-globins), the two or more polypeptides encoded by a single gene are homologous but distinct—presumably tailoring them to the individual needs of the cells in which they are made. We now realize that this additional level of versatility exploits tissue-specific differences in both the transcriptional apparatus and the RNA processing machinery of eucaryotic cells.

If a single gene is preceded by two promoter regions, each with its own tissue-specific regulatory element(s), then two transcripts that differ in length at their 5' ends will be produced in a tissue-dependent fashion. If, further, the extra 5' nucleotides include splicing signals (5' and 3' junction sequences) and coding regions, two polypeptides differing in their N-terminal amino acid sequences may be produced in the relevant tissues. This is precisely the strategy adopted by the chicken genome to generate two different myosin light chains (important muscle proteins) from a single gene. Cardiac muscle contains one type of light chain (called LC_1) and gizzard muscle another (called LC_3), whereas skeletal muscle contains both LC_1 and LC_3. As shown in Figure 21-37, two promoters about 10,000 base pairs apart account for the synthesis of two pre-mRNAs; these undergo differential splicing, yielding the mRNAs for LC_1 and LC_3, which have identical 3' but distinct 5' coding regions. The finding that different myosin chains in rats are likewise encoded by single genes whose expression is identically regulated (by differential transcription and splicing) argues that this ingenious scenario evolved even before the divergence of birds and mammals.

In the case of the myosin LC_1/LC_3 gene, it is unclear whether structural differences between the two pre-mRNA transcripts dictate differential splice site utilization or whether tissue-specific changes in the splicing machinery are also involved. A growing list of other genes, however, strongly hints that the same transcript can be processed differently in different tissues to yield mRNAs with different coding potentials. One example is the rat calcitonin gene transcript, which is polyadenylated at one site in cells of the thyroid gland and at another site in the brain (see Figure 21-37). The two transcripts are spliced in distinct ways. We do not yet know whether calcitonin gene expression is controlled primarily by the choice of splice sites or by the choice of polyadenylation sites. If polyadenylation precedes splicing (as seems likely from the study of many other genes), then the cell might control polyadenylation in a tissue-specific manner, and the

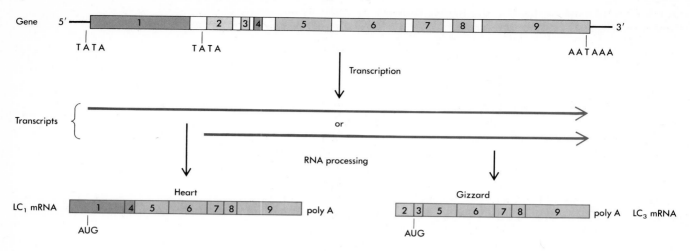

(*a*) Two promoters and different splicing patterns: the chicken myosin alkali light-chain gene.

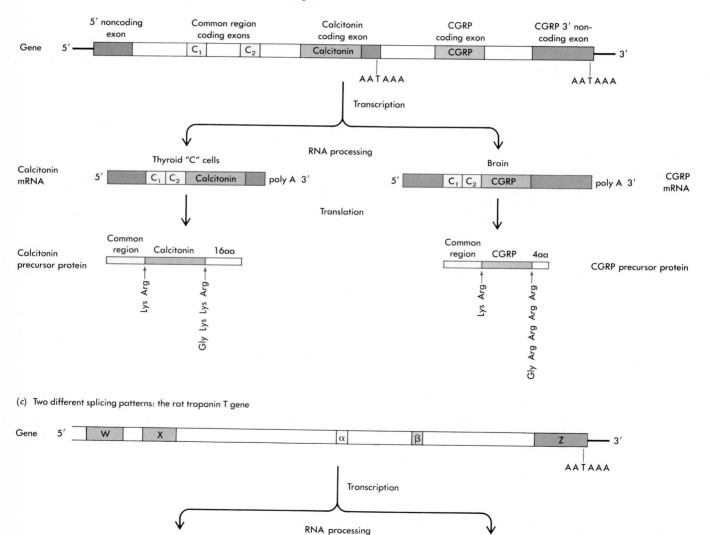

(*b*) Two poly A sites and different splicing patterns: the rat calcitonin gene.

(*c*) Two different splicing patterns: the rat troponin T gene

Figure 21-37
Three different strategies for producing more than one protein product from the same gene. In the case of the calcitonin gene, the protein product is further processed at the basic residues indicated to yield either calcitonin or a neuropeptide called calcitonin gene-related peptide (CGRP).

structure of the resulting mRNA precursor would dictate the subsequent choice of splice sites. Alternatively, if splicing can sometimes occur on the nascent mRNA transcript, then the cell might control the choice of splice sites, and the structure of the nascent transcript would influence the subsequent choice of polyadenylation site. In any case, the end result is two calcitonin mRNAs with three common 5' exons but different 3' exons, thus specifying different polypeptides (and eventually different hormones) in the two cell types.

Finally, there are situations where both the 5' end (promoter site) and 3' end (polyadenylation site) of a pre-mRNA appear to be identical in different tissues, but alternative splicing produces two different mRNAs. For example, mammalian troponin T, a regulatory protein of the contractile apparatus, occurs in several forms whose appearance depends on the developmental stage and the type of striated muscle examined. Part of this heterogeneity can be traced to a divergent troponin T gene family, including at least three separate genes. But one of these genes, in rats, has been shown to encode two different polypeptides whose amino acid sequences differ only in a small internal section (positions 229 to 242 in the 259-residue chain). This difference results from the mutually exclusive incorporation of one or another internal exon into the mRNA (see Figure 21-37). Tissue-specific mRNA processing could reflect fundamental (albeit subtle) changes in the splicing machinery of different cell types. However, a simpler model would be that the splicing machinery itself is universal, but different cell types contain different factors (protein or RNA or both) that can bind to specific sequences (or structures) within mRNA precursors, thereby facilitating or blocking the activity of the splicing machinery on particular splice junctions.

We now suspect that similar manipulation of RNA processing pathways may generate multiple RNAs from the transcripts of genes that play pivotal roles in the developmental program of *Drosophila* (the homeotic loci; Chapter 22). Likewise, immunoglobulin-producing cells utilize differential RNA processing to convert antibody molecules from their membrane-bound to secreted forms (Chapter 23). Introns and RNA splicing therefore confer on eucaryotic organisms a unique potential for the functional diversity of single genes, as well as an advantage in evolving new genes (Chapter 20).

Nearly All RNA Processing Is Nuclear

So far, all the processes we have considered in higher eucaryotic gene expression take place in the nucleus. The synthesis of RNA chains must occur there, since the cell's genome is confined to this compartment. RNA processing proceeds both during and directly after completion of chain growth. Thus, it is within the nucleolus that rRNA molecules are extensively trimmed and that they receive certain nucleotide modifications concomitant with ribosomal subunit assembly. Likewise, the nucleoplasm is where pre-mRNA molecules undergo

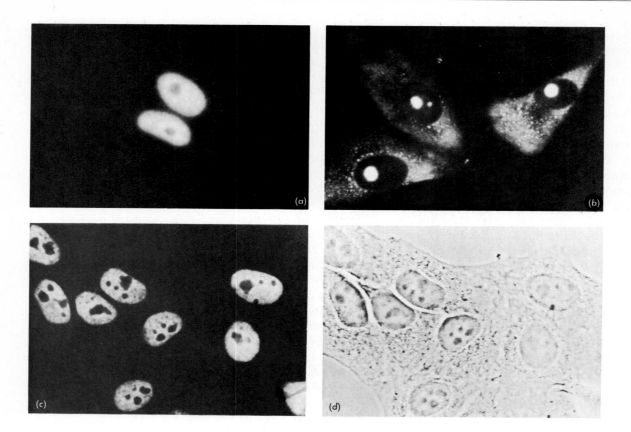

Figure 21-38

The nuclear, nonnucleolar location of both U1 snRNPs and hnRNP, as revealed by indirect immunofluorescence. Cells have been incubated with either (a) a patient serum containing autoantibodies directed against U1 snRNP protein components, (b) a monoclonal antibody directed against ribosomal RNA, or (c,d) a monoclonal antibody reactive with several major proteins (C proteins) of the hnRNP. [Note that (d) is a phase-contrast micrograph of the same field as in (c).] Then, a fluorescently tagged antibody directed against the first antibody is applied, and the cells are visualized in a fluorescent microscope. (a) and (b) show African green monkey kidney cells, whereas (c) and (d) show human *HeLa* cells. Note that the nucleolar and cytoplasmic location of rRNA is the exact inverse of the location of snRNP and hnRNP proteins. [Parts (a) and (b) courtesy of E. Lerner; (c) and (d) courtesy of G. Dreyfuss.]

splicing and acquire their 5′ caps and some internal m^6A residues (of unknown function); indirect immunofluorescence studies demonstrate that both hnRNP proteins and snRNPs (some of which act in splicing) are localized there (Figure 21-38). Polyadenylation could be linked to export (and therefore take place on the nuclear envelope), but this seems unlikely for several reasons. First, some mRNAs (notably histone messengers) lack poly A, so the two processes cannot be obligatorily coupled. Second, poly A tails are always complexed with specific proteins, but these are distinct in the nucleus and the cytoplasm; why should there be a special nuclear poly A binding protein if poly A is acquired only during export? What is clear is that the export apparatus somehow surveys mRNA molecules to ensure that every intron that is meant to be removed has been removed before passage to the cytoplasm. How this is accomplished remains a mystery. Finally, in the case of tRNA splicing, there are hints that introns may be removed only as tRNA molecules pass into the cytoplasm; both the splicing endonuclease and ligase appear to be concentrated on the nuclear membrane. Whether other processing enzymes that act on RNA polymerase III transcripts are also so located is not yet known. There are only a few exceptions to the general rule of nuclear RNA processing; for example, certain tRNA base modification reactions are known to occur in the cytoplasm.

Messenger RNA Lifetimes in Higher Cells[52, 86]

Compared to what we know about bacterial messengers, our knowledge of the lifetimes of functioning mRNA in higher cells is relatively sparse. It seems clear that many mRNA species have a limited lifetime, and average half-life estimates of 3 hours are frequently given for the mRNA in rapidly growing vertebrate cells, which divide about once a day. If these figures are correct, then their *relative* lifetimes are not unlike those of many bacterial mRNA molecules, which turn over some ten times during a cell cycle.

Although many cellular mRNAs encode "housekeeping" proteins that must be present at all times, the synthesis of a significant number of proteins (and RNAs?) must be turned on and off in a strict order to enable the cell to proceed with regularity through the cell cycle. This is particularly true for yeast, where over 50 genes are known to control the cell division cycle (CDC). Temperature-sensitive lesions in any of these CDC genes cause the cells to arrest in a defined stage of the cell cycle at elevated temperatures (Chapter 18). Comparable CDC genes must occur in higher eucaryotes, and it is likely that at least some of them encode unstable mRNAs whose brief existence advances the cell from one stage in the cell cycle to the next.

In contrast, much of the mRNA of the terminally differentiated cells of higher animals is relatively stable. Reticulocytes (immature red blood cells) provide a good example. These cells produce virtually no RNA while they are synthesizing their principal protein, hemoglobin. If their mRNA molecules were rapidly made and broken down, it would be possible to measure incorporation of RNA precursors into RNA; none is detected. This stability has an obvious advantage, particularly since the very constant environment of reticulocytes (as contrasted with the highly fluctuating growth conditions of bacteria) makes rapid changes in the mRNA population unnecessary. Reticulocytes synthesize mostly (more than 90 percent) hemoglobin. There is no reason to break down hemoglobin mRNA chains only to resynthesize them. Similarly, most of the mRNAs found in the cytoplasm of the adult liver encode plasma proteins that are released into the circulatory system. As expected, liver cell mRNA also appears to be relatively stable.

We understand almost nothing about what determines mRNA half-life. As a general rule, polyadenylated mRNA species live longer than nonpolyadenylated mRNAs (such as histone messengers). But we do not know whether the extraordinary stability of certain mRNAs (e.g., hemoglobin mRNA in a reticulocyte) is an intrinsic property of the mRNA itself, or whether it is acquired by complexing with specific stabilizing proteins (e.g., mRNP proteins). Most recent evidence favors the latter alternative and raises the intriguing possibility that mRNA (and even pre-mRNA) stability may be yet another level on which gene expression is actively controlled. For example, we will soon see that the mechanism of steroid hormone action may include the stabilization of specific mRNAs.

Myriad Initiation Factors for Eucaryotic Protein Synthesis[93, 94]

In contrast to the three well-defined factors required to start translation in *E. coli* are a multitude of **eucaryotic initiation factors** (designated **eIFs**). How they contribute to the initiation of protein chains is diagrammed in Figure 21-39. The process is in some ways similar to that in procaryotes. The small ribosomal subunit (here 40S) is the first to form an initiation complex with the mRNA and initiator tRNA. Also, GTP is cleaved only upon joining of the large ribosomal subunit

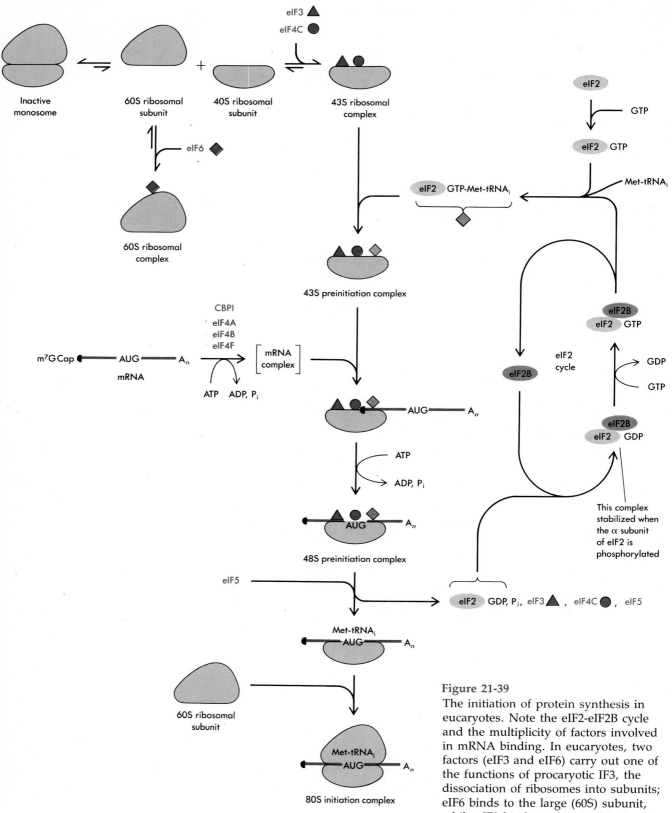

Figure 21-39

The initiation of protein synthesis in eucaryotes. Note the eIF2-eIF2B cycle and the multiplicity of factors involved in mRNA binding. In eucaryotes, two factors (eIF3 and eIF6) carry out one of the functions of procaryotic IF3, the dissociation of ribosomes into subunits; eIF6 binds to the large (60S) subunit, while eIF3 binds to the small (40S) subunit. CPB stands for cap binding protein; eIF4F is also called CPBII. (Scheme courtesy of W. Merrick.)

(60S) to form a complex (80S) that possesses an open tRNA binding site (A site) and can thus enter the elongation phase of protein synthesis.

However, there are many essential ways in which eucaryotic differs from procaryotic initiation (see Figure 14-30):

- Again, a special tRNA^{Met} (here called Met-tRNA$_i^{Met}$) is utilized, but it does not acquire an N-terminal formyl group.
- The initiator Met-tRNA$_i^{Met}$ and GTP form an isolatable complex with eIF2, independent of the small ribosomal subunit. (Recall that in procaryotes, IF2 binds initiator tRNA on the 30S ribosome.)
- The initiator tRNA is always bound to the 40S ribosomal subunit prior to the mRNA (rather than in either order).
- Hydrolysis of ATP to ADP is required for mRNA binding.
- Not only are there many more initiation factors, but many eIFs are themselves multisubunit proteins. Most striking is eIF3 (involved in mRNA binding, as is IF3 in bacteria); it has about ten subunits and weighs 10^6 kdal (one-third as much as the 40S subunit itself)!
- The presence of a capped 5' end on the mRNA is required for initiation, except for certain viral mRNAs, some of which substitute a covalently bound protein (Chapter 24).

Pivotal to the regulation of eucaryotic protein synthesis is a final difference in the initiation process: The factor eIF2 undergoes GTP-GDP exchange via a cycle involving another protein called **eIF2B,** also called **GEF,** for **GTP-GDP exchange factor** (Figure 21-39). Enzymatically, the steps are virtually identical to the EF-Tu/Ts cycle in procaryotic chain elongation (see Figure 14-31). After eIF2 has deposited the initiator tRNA on the ribosome, it is released in an inactive form containing bound GDP. GDP is displaced only when eIF2 interacts with eIF2B to form an eIF2-eIF2B complex. In this complex, eIF2 has a heightened affinity for GTP and can thereby reenter the active pool and again bind Met-tRNA$_i^{Met}$. The progress of eIF2 through this cycle is sensitive to the phosphorylation of one of its three polypeptide subunits (called eIF2α): Phosphorylated eIF2 interacts more stably with eIF2B, and therefore both factors become sequestered into an eIF2-eIF2B complex. The consequent lowered availability of eIF2 for participation in initiation forms the common basis of several mechanisms in which seemingly disparate agents control mammalian protein synthesis.

40S Ribosomal Subunits Almost Always Select the First AUG[93, 94, 95]

In only one respect does the initiation of protein synthesis appear simpler in eucaryotes than in procaryotes, and that is in choosing where to start on the mRNA. First, AUG alone serves as an initiator codon (whereas GUG, UUG, and even AUU are also utilized in *E. coli*). Second, the AUG located nearest the 5' end of the mRNA (usually within 10 to 100 nucleotides) is almost always selected. The simplest model consistent with these facts proposes that 40S ribosomes attach to the capped 5' terminus and scan along the mRNA, searching for the first AUG and consuming ATP in the process. Stopping would be the consequence of correct codon-anticodon interaction between AUG and the Met-tRNA$_i^{Met}$, which is invariably prebound to the 40S subunit (see Figure 21-39). That AUG is exclusively recognized can be explained by the presence of a bulky modified A residue just 3' to the anticodon of eucaryotic tRNA$_i^{Met}$. (This modification apparently disallows first-position wobble in codon-anticodon interactions; recall that it is lacking in *E. coli* tRNA$_F^{Met}$, allowing more flexibility in its interaction with initiator codons. See Figure 14-26.) Moreover, an additional upstream signal (such as the procaryotic mRNA region that base-pairs with the 3' terminus of 16S rRNA; see Figure 14-28) would not be needed if ribosomes only very rarely start at internal initiation sites.

Yet, closer scrutiny of the mechanisms underlying initiation site selection by eucaryotic ribosomes uncovers certain complexities. There are, in fact, some eucaryotic mRNAs where it is not the first AUG but a downstream AUG codon that begins the open reading frame. Such exceptions to the first-AUG rule can be explained by the effect of sequence context, which was revealed by comparing hundreds of mRNA sequences. Namely, ^A_GNNAUGG is optimal for initiation by 40S ribosomes; if pyrimidines appear instead at position -3 or $+1$ relative to the AUG, the site has a high probability of being bypassed. Another unresolved question is how 40S ribosomes load onto the mRNA. One possibility is that they literally thread on, for it has been shown that a free end on the mRNA (whether capped or not) is absolutely required; eucaryotic ribosomes are completely incapable of initiating on circular mRNAs, whereas procaryotic ribosomes do so quite nicely. Finally, do 40S ribosomes really move physically from the cap to the first AUG, propelled by the energy of ATP hydrolysis? Such a scanning mechanism is consistent with all available data. An alternative possibility implicates the specific cap binding protein (CBPI), which is known to be present among the initiation factors and to be required for initiation (see Figure 21-39). Upon binding to the 5' end of the mRNA, CBPI may facilitate the assembly of other initiation factors (probably including the huge eIF3) onto the adjacent leader sequence. There is some evidence that this assembly process itself is ATP-dependent and serves to create a pathway for easy access of the 40S ribosome to the first AUG.

Elongation and Termination by 80S Ribosomes[93, 96–99]

Compared to initiation factors, the factors required for elongating and terminating polypeptide chains in eucaryotes are disarmingly simple. Their contributions are summarized in Table 21-2.

EF1α (50 kdal) replaces EF-Tu of procaryotes. This factor forms a complex with GTP and aminoacyl-tRNA and then donates the latter to ribosomes. Less well characterized is an EF-Ts-like factor (called **EF1βγ**) that catalyzes GDP-GTP exchange and readies EF1α for another cycle of aminoacyl-tRNA binding. The eucaryotic factor **EF2** (about 100 kdal) corresponds to procaryotic EF-G. It hydrolyzes GTP and catalyzes translocation of the aminoacyl-tRNA from the A site to the P site on the ribosome with concomitant movement of the message.

The termination of polypeptide chains by 80S ribosomes involves only a single factor called **RF** (115 kdal). It replaces both RF1 and RF2 of procaryotes in that it can recognize all three stop codons (UAA, UGA, and UAG). Eucaryotic termination involves GTP hydrolysis,

Table 21-2   Factors Involved in the Elongation and Termination of Polypeptide Chains in Eucaryotes

Factor	Function	Comparable Procaryotic Factor*
EF1α	Binds animoacyl-tRNA to ribosomes	EF-Tu
EF1βγ	Assists the recycling of EF1α	EF-Ts
EF2	Translocation	EF-G
RF	Chain release	RF1
		RF2

*See Figures 14-31 and 14-32.

which is required for RF dissociation from the ribosome after the factor has activated peptidyl transferase to release the polypeptide chain.

An unusual property of the aminoacyl-tRNA synthetases from higher eucaryotic cells is that they are generally isolated as aggregates. These high-molecular-weight complexes contain multiple synthetase activities, as well as other components whose identity suggests that synthetases may be membrane-bound in vivo. An association with the cell membrane for at least some fraction of the synthetase population makes sense, since higher eucaryotic cells often depend on amino acids supplied externally (by the circulatory system). This location could also explain the remarkable observation that mammalian cells incorporate exogenously added amino acids directly into proteins without mixing with the large amino acid pools already present.

Specific Inhibitors of Eucaryotic Translation[100, 101, 102]

Not surprisingly, some antibiotics that inhibit translation in procaryotes are equally effective against eucaryotic protein synthesis. Puromycin, for example, also fools 80S ribosomes into attaching it to the growing end of polypeptide chains, thereby prematurely terminating translation (see Figure 14-33). Another inhibitor of both systems is fusidic acid, which prevents the release of procaryotic EF-G from the ribosome after it has functioned in translocation; it likewise inhibits EF2 release, suggesting a high degree of evolutionary conservation between these homologous GTP binding factors.

Other inhibitors are quite specific for eucaryotic protein synthesis. In their modes of action, they mainly exploit differences between the procaryotic and eucaryotic translational apparatus. For instance, m7Gp inhibits initiation only in eucaryotic in vitro systems, because it competes with the 5' ends of mRNAs for cap binding protein. A multitude of other agents bind specifically to some essential component of the eucaryotic machinery. They therefore block progression to the next step in the translation pathway, as summarized in Figure 21-40. Whereas most of these inhibitors are small molecules, some are polypeptides (e.g., ricin, a specific nuclease that halts elongation by catalytic inactivation of the 60S ribosomal subunit).

The deadly toxin produced by the diphtheria bacterium *Corynebacterium diphtheriae* is a well-studied polypeptide inhibitor of eucaryotic protein synthesis. Diphtheria toxin is a 65 kdal protein encoded not

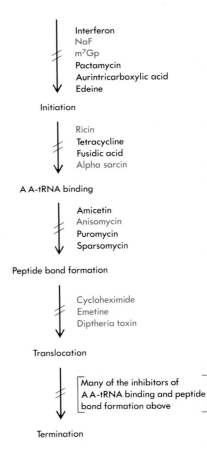

Figure 21-40
Sites of action of some inhibitors of eucaryotic protein synthesis. Inhibitors that target the eucaryotic translation system exclusively are shown in color.

Figure 21-41
The ADP-ribosylation reaction that inactivates EF2. Cells probably contain endogenous ADP-ribosyltransferases as well as inhibitors of this reaction that prevent catastrophic inactivation of EF2.

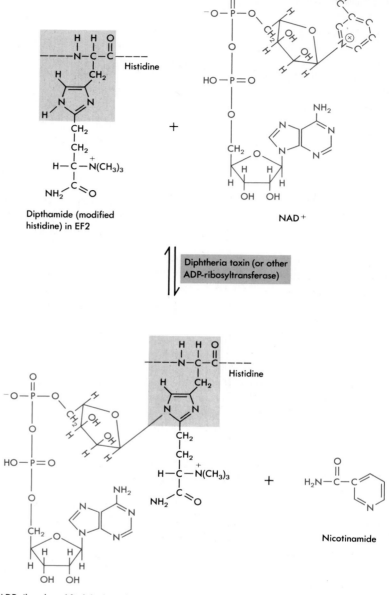

Dipthamide (modified histidine) in EF2

NAD$^+$

Diphtheria toxin (or other ADP-ribosyltransferase)

Histidine

Nicotinamide

ADP-ribosyl modified dipthamide

by the genome of the bacterium, but by a phage that lysogenizes it (called phage β). The toxin is excreted by the bacteria, becomes transported throughout the body of the infected victim, and gains entry into the cells of virtually all tissues. Once inside, it catalyzes the attachment of ADP-ribose to elongation factor EF2. The reaction is shown in Figure 21-41. ADP-ribose is donated by NAD$^+$ and becomes attached to an unusually modified histidine residue in EF2. In vitro, the reaction can be readily reversed by the addition of nicotinamide. Once ADP-ribosylated, EF2 is completely inactive in elongating polypeptide chains. Since the toxin acts catalytically, only a tiny amount (perhaps as little as one molecule) is effective in blocking a cell's entire protein synthesis capability and producing cell death. Presumably, the special modified histidine residue in EF2 is not present simply so that diphtheria toxin can kill mammalian cells; it may also be the target of cellular mechanisms designed to regulate protein synthesis in meaningful ways.

Finally, interferon must be mentioned as an important inhibitor of the initiation of translation. Because interferon operates on virus- infected cells, its synthesis and mode of action are described in some detail in Chapter 24. At least two components required for polypeptide chain initiation are targets in interferon-treated cells. Viral mRNAs are cleaved by a special endonuclease; and initiation factor eIF2 is inactivated by phosphorylation (see Figure 21-39).

Translational Controls: Fact, Not Fantasy[103]

The importance of translational controls to the lives of eucaryotic cells has been long debated. The obvious question is, Why waste energy making a messenger you may never use? Regulation of translation makes some sense for bacteria with their characteristically polycistronic mRNAs; if one protein product is needed and another not, then expression can only be controlled at the translational level (Chapters 16 and 17). By contrast, mRNAs are monocistronic in nucleated cells. Moreover, their production requires many more steps, several of which are already known to be regulated (promoter utilization, RNA splicing, polyadenylation) and others that may be (pre-mRNA stability, export from the nucleus). Thus, regulating eucaryotic translation might seem unnecessary. Yet, it provides a very rapid way to control gene expression and is used widely by higher eucaryotic organisms.

Sometimes, translational levels are manipulated by changing essential components of the cell's translational machinery. We would expect such alterations to affect the activity of every mRNA in the cell. For instance, it appears that phosphorylation of ribosomal proteins (particularly S6, in the 40S subunit) is correlated with higher polysome levels after mammalian cells are stimulated by a variety of different growth factors. Likewise, upon fertilization, the sudden appearance of a previously missing initiation factor activity mobilizes all the mRNAs stored in the brine shrimp egg for incorporation into polysomes (see also Chapter 22). Or, as we shall see, modifying initiation factor eIF2 is a device used by the reticulocyte to link its protein synthesis to the availability of heme (the prosthetic group of hemoglobin).

Other strategies are employed to alter the translation of only one mRNA (or some subset of cellular messengers). Many such mechanisms act at or before the initiation step. For instance, there is the inherent variability in eucaryotic mRNA lifetimes, as well as instances where mRNA stability is selectively altered in response to certain agents. Alternatively, mRNAs can exist stably in the cytoplasm but be somehow sequestered away from the translational apparatus (perhaps by mRNA binding proteins). Polio virus uses the device of shutting off host cell protein synthesis by destroying the cap binding protein essential for initiation on capped mRNAs, while its own mRNAs are covalently bound to protein rather than being 5'-capped (Chapter 24). That the elongation phase of mammalian protein synthesis can also be controlled in a messenger-specific way is neatly demonstrated by the example of SRP action, which we will discuss shortly.

Thus, there can no longer be any doubt that translational controls do contribute importantly to the functioning of eucaryotic cells. Already it is apparent that such mechanisms operate on a variety of components of the translational apparatus. Presumably, it is only a matter of time until some regulatory strategy affecting every conceivable stage of eucaryotic protein synthesis is uncovered.

Translational Control of Hemoglobin Synthesis by Heme[104, 105]

One of the earliest documented examples of regulation at the level of translation occurs in mammalian reticulocytes (immature red blood cells). These cells have lost their nuclei, but retain high levels of stable mRNAs encoding mostly hemoglobin chains. Reticulocytes engage in exceedingly high rates of protein synthesis. Hence, reticulocyte lysates have long been the favorite system of biochemists interested in eucaryotic translation. Indeed, much of what we know about the mechanisms involved in protein synthesis is based on studies using reticulocyte lysates.

In reticulocyte lysates, protein synthesis starts at a high rate but declines sharply unless heme is added. (In cells, heme is synthesized by mitochondria, which are lacking in the in vitro system.) The decline occurs because heme deficiency activates an inhibitor of protein synthesis initiation called **heme-controlled inhibitor** (**HCI**). In the presence of heme, HCI remains inactive; in its absence, HCI becomes activated. Although its mechanism of action was mysterious for many years, we now know that HCI is a specific kinase that phosphorylates the α subunit of eIF2. Because of the eIF2 GTP-GDP exchange cycle depicted in Figure 21-39, phosphorylation of only a fraction (20 to 30 percent) of eIF2 is sufficient to stop initiation of protein chains. Apparently, all of the eIF2B (which is present in lower molar amounts than eIF2) becomes sequestered into eIF2-eIF2B complexes and is no longer available to recycle the remaining (unphosphorylated) eIF2. Moreover, as already explained, the phosphorylated eIF2 contained in the complex with eIF2B has a lowered affinity for GTP.

Thus, we see that hemoglobin translation is regulated at the level of initiation by phosphorylation of eIF2. The intermediary is a heme-sensitive kinase. There also exists in mammalian cells another kinase that acts similarly but is distinct. It is activated by double-stranded RNA and is present in elevated amounts in interferon-treated cells (Chapter 24). What role it plays in normal cells has not been established. Yet, the characterization of two such kinases strongly hints that there may be others, activated by different agents but all exerting their regulatory effect by phosphorylating the same factor (eIF2) crucial for the initiation of polypeptide chains.

Signal Recognition Particle (SRP): A Small RNA-Protein Complex That Initiates Export of Secreted Proteins[106, 107]

All cells must not only synthesize proteins in a controlled fashion but also deliver them to the appropriate site inside (or outside) the cell. In so doing, eucaryotic cells face many more difficulties than bacteria. First, the cells themselves are much larger. Hence, specific transport mechanisms may be required where bacteria can rely on simple diffusion. Second, there are numerous membrane-bound compartments, each with its own distinctive protein composition. Thus, for many newly synthesized eucaryotic proteins, a decision must be made about *which* membrane to cross, rather than just whether or not to cross a membrane.

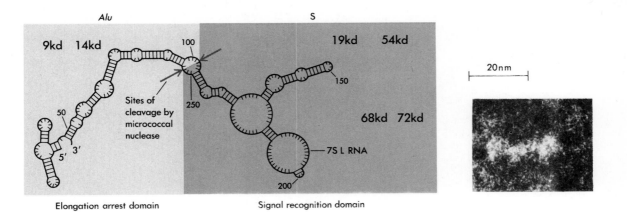

Elongation arrest domain   Signal recognition domain

The eucaryotic proteins whose delivery to a particular cellular compartment is best understood are those that are synthesized by polysomes bound to the extensive intracellular membrane system called the **endoplasmic reticulum (ER)**. This class includes secreted proteins, lysosomal proteins, and certain membrane proteins. Translocation across the membrane of the endoplasmic reticulum is initiated while the polypeptide is still incompletely synthesized (**cotranslational secretion**). A common feature of virtually all proteins that enter the endoplasmic reticulum is a special N-terminal extension called the **signal sequence** (or **signal peptide**). Eucaryotic signal sequences, like those in procaryotes (see Table 14-2), do not possess a particular sequence, but are all about 15 to 30 amino acids long and are rich in hydrophobic amino acids. They presumably fold into a helical hairpin conformation that readily inserts itself into the membrane bilayer to initiate transport (see Figure 14-35). Sometime during the subsequent passage of the growing polypeptide chain through the membrane, the signal sequence is clipped off by a special protease called **signal peptidase.** Once inside the lumen of the endoplasmic reticulum, the protein is further modified by the addition of carbohydrate (and other) groups that facilitate its ultimate delivery to the proper final destination.

That a small RNA-protein complex plays a pivotal role in this translocation process has been appreciated only recently. The **signal recognition particle (SRP)** is so named because it interacts specifically with the signal sequences of nascent secretory proteins. SRP is made up of 7SL RNA (about 300 nucleotides) (see Figure 20-38) and six different tightly bound protein components, ranging in size from 9 to 72 kdal (Figure 21-42). A fascinating aspect of its mode of action is SRP's ability to bind to ribosomes and halt protein synthesis (Figure 21-43). Translational arrest occurs not at initiation but when the nascent polypeptide chain is about 70 amino acids long. This is just the length at which the signal sequence will have emerged completely from the ribosome. (Recall that 30 to 40 amino acids adjacent to the growing carboxyl terminus are sequestered within the ribosome.) SRP does not arrest translation of all mRNAs, but specifically acts on those encoding to-be-secreted proteins. The translational block is reversed only when the SRP makes contact with the **docking protein** (72 kdal), an integral component of the ER membrane. Then, SRP diffuses away to initiate transport of another polypeptide chain. During translocation, the attachment of polysomes to the membrane of the endoplasmic reticulum is primarily mediated by the nascent polypeptide chains being translocated, aided by ribosome-receptor proteins

Figure 21-42
Schematic diagram of the mammalian signal recognition particle (SRP), showing which proteins associate with the two domains of the folded RNA. *Alu* refers to sequences in 7SL RNA that exhibit homology to the *Alu* repetitive sequence family in primate genomes, while S indicates the unique central region (see Figure 20-38). The inset shows a dark-field electron micrograph of unstained SRP; the long dimension of the rod-shaped particle is nearly equivalent to the diameter of a ribosome. [SRP model courtesy of P. Walter; micrograph courtesy of D. Andrews, P. Walter, and P. Ottensmeyer, *Proc. Nat. Acad. Sci.* 82 (1985):785–789. Used with permission.]

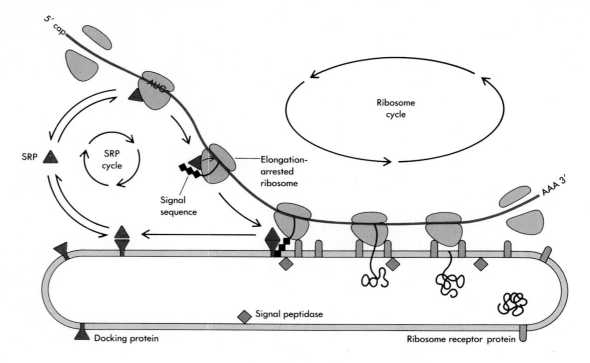

Figure 21-43
The functioning of SRP and the docking protein (also called the SRP receptor) in the translocation of proteins across the endoplasmic reticulum. [After P. Walter, R. Gilmore, and G. Blobel, *Cell* 38 (1984):5–8.]

(sometimes called ribophorins) in the endoplasmic reticulum that interact specifically with the large ribosomal subunit (see Figure 21-43). SRP is present in very high amounts in the cytoplasm, despite its only transitory role in the translocation process (about 10^6 copies, or one-tenth the number of ribosomes in mammalian cells).

SRP's activity as a negative regulator of translation has an important biological consequence. Many proteins exported by mammalian cells are degradative enzymes (e.g., nucleases and proteases). Even their occasional presence within the cytoplasm might be expected to wreak havoc within the cell that made them. By stopping the translation of such proteins midway, SRP ensures that they will not be completed until the correct membrane (across which they can be translocated) is at hand. Most recently, it has been found that the two activities of the SRP can be assigned to separate domains of the oblong-shaped particle (see Figure 21-42). Clipping the RNA with nuclease produces a large subcomplex that retains SRP's ability to bind to ribosomes (perhaps via RNA-RNA interactions) and to facilitate secretion; but this subcomplex cannot carry out translational arrest. It is fascinating that the two SRP domains correspond almost precisely to the two types of sequences present in 7SL RNA (see Figure 20-38).

Delivery of Proteins to Other Cellular Compartments[108–115]

Analyses of protein transport into other compartments of eucaryotic cells are only now beginning to catch up with our understanding of translocation into the endoplasmic reticulum. Best understood is the nature of import into mitochondria and plant cell chloroplasts; nuclei, because of their nuclear pores, present a somewhat different problem. Yet each of these compartments is surrounded by a double-membrane system (making protein translocation more complex than with the single membrane of the endoplasmic reticulum). Moreover, their outer membranes can be observed to be studded with ribosomes. Does this mean that protein translocation into these three compartments is also cotranslational? Is a special small RNA-protein complex likewise involved? (There do exist a number of small cytoplasmic RNA-protein complexes distinct from SRP; their structures have been defined to varying extents, but their functions remain unknown.) So far, it appears that a different mechanism of delivery is employed in each case.

Import of most proteins into mitochondria involves the following steps: synthesis as a precursor polypeptide (with an N-terminal extension that is usually positively charged); binding of this precursor to a receptor on the mitochondrial surface; and translocation into or across the mitochondrial membranes. There are differences, however, depending on whether the protein is destined for the outer membrane, the intermembrane space, or the inner membrane/matrix. Import of both intermembrane and inner membrane/matrix proteins requires an electrochemical potential across the inner membrane. The initial "targeting" interaction with a protein receptor in the outer mitochondrial membrane appears to be mediated by the precursor's N-terminal extension, which is cleaved off after import. The cleavages experienced by intermembrane proteins, however, are often more complex than those of proteins bound for the inner membrane/matrix. By contrast, delivery of proteins to the outer mitochondrial membrane neither requires an energized inner membrane nor is accompanied by proteolytic processing. All three mechanisms require much additional scrutiny. What is clear in each case is that import can occur after translation is complete. Moreover, mitochondrial proteins are synthesized principally on free (rather than membrane-bound) polysomes. Only the most indirect evidence hints at the involvement of a small RNA-protein complex in protein translocation into mammalian mitochondria.

Specific transport of plant proteins into chloroplasts is superficially similar to mitochondrial import. Chloroplast proteins are likewise synthesized as longer precursors, and insertion takes place after translation is complete. Once the additional sequences in the precursor have functioned—presumably in the recognition of specific receptors on the chloroplast surface—they are removed by chloroplast protease(s) to generate mature proteins. Interestingly, transport into chloroplasts is stimulated by light. This apparently reflects a requirement for energy in the form of ATP, which is produced in chloroplasts by photosynthetic phosphorylation. Much still remains to be learned about the specificity of precursor-receptor interactions and the passage of polypeptide chains into the chloroplast's interior.

The recurring motif that characterizes proteins delivered to the endoplasmic reticulum, to mitochondria, and to chloroplasts—a polypeptide extension that is critical for transport but is later discarded to form the mature protein—does not hold for proteins destined for the nucleus. Instead, nuclear proteins are synthesized in their mature form. Many nuclear proteins (e.g., histones, hnRNP proteins, and polymerases and their associated factors) bind strongly to either DNA or RNA. Hence, one theory is that they arrive in the nucleus by free diffusion through the pores and are selectively retained because of their high affinity for nuclear components. That this is probably not the case is now emerging from the results of deletion analyses of nuclear proteins. Instead, part of the amino acid sequence—which can in some cases be quite short—appears to target the protein for the nucleus. Whether there is a defined sequence common to all nuclear proteins is not yet clear. One intriguing possibility

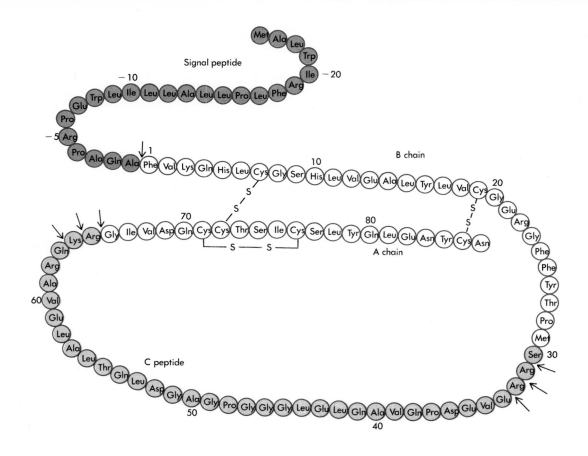

Figure 21-44
The amino acid sequence of rat preproinsulin II, showing the sequences lost during its processing to the mature hormone, which contains only the A chain (30 amino acids) and the B chain (21 amino acids). Why the sequence of the released C peptide (31 amino acids) has been conserved over evolution is not known. Arrows show the sites of proteolytic cleavage during processing.

is that some proteins may be transported into the nucleus by specifically associating with another protein that does possess the nuclear targeting sequence. For instance, nucleoplasmin (29 kdal), a highly abundant protein of the *Xenopus* oocyte that forms specific complexes with histones, may serve to carry the histones into the nucleus as well as to assemble them onto the DNA. RNAs may also be transported into the nucleus by protein carriers. Small nuclear RNAs, which travel to the cytoplasm during their maturation, are an example. They do not become properly relocalized in the nucleus if their binding sites for certain associated proteins (snRNP proteins) have been mutated.

Polyprotein Precursors of Peptide Hormones[116-119]

Once synthesized, a eucaryotic protein is much more likely than its procaryotic counterpart to undergo extensive covalent modification. In both cases, the initiating methionine residue is normally removed by a special exopeptidase. Likewise, N-terminal signal sequences are cut off concomitant with translocation across membranes in all cell types. But additionally in eucaryotic cells, membrane and exported proteins acquire numerous carbohydrate moieties during their transit through the endoplasmic reticulum. Further, endopeptidases frequently cut specific internal peptide bonds to generate active polypeptides from larger precursors.

Cleavage of precursor proteins is a recurring theme in the synthesis of peptide hormones. A classic example is that of insulin (Figure 21-44), which has been known for several decades to be constructed of

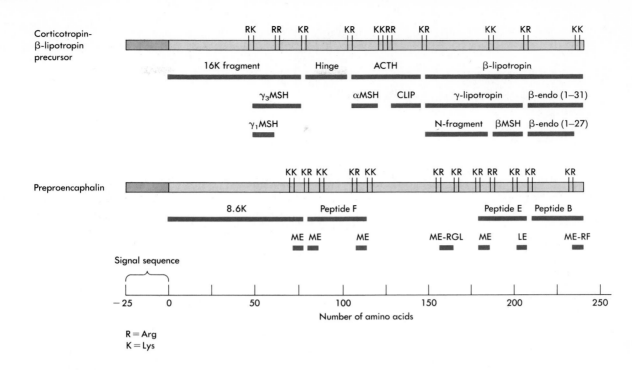

Figure 21-45
The structures of two bovine multihormone precursors: The corticotropin-β-lipotropin precursor and preproencephalin. In both cases, the data are based on cloned cDNA sequences. The active peptide hormones are always bounded by paired basic amino acid residues, which are the sites of protein processing. The abbreviations for the various component peptides of the corticotropin-β-lipotropin precursor are: ACTH = corticotropin, endo = endorphin, MSH = melanotropin, and CLIP = corticotropin-like intermediate lobe peptide. Preproencephalin contains four copies of Met-encephalin (Tyr-Gly-Gly-Phe-Met) (ME in the figure) and one copy each of Leu-encephalin (which substitutes Leu for Met at the C terminus) (LE in the figure) and of two extended Met-encephalins (one a heptapeptide and one an octapeptide). [After S. Nakanishi et al., *Nature* 278 (1979):423–427; and R. E. Mains, B. A. Eipper, C. C. Glembotski, and R. M. Dores, *Trends in Neurosci.* 4 (1983):229–235.]

two polypeptide chains. Only much later was it discovered that insulin is initially synthesized as an 86-amino-acid single chain (preproinsulin). After removal of its signal sequence to create proinsulin (62 amino acids) and folding into a specific three-dimensional shape stabilized by disulfide (S–S) bonds, internal cuts excise 31 amino acids from the middle of the chain to create the active insulin hormone.

Even more dramatic is the protein processing that occurs in pituitary and adrenal gland cells to generate the opiate peptides (natural hormones whose activity is partially mimicked by certain plant metabolites). Here, it has been shown that lengthy polypeptides (e.g., the corticotropin-β-lipotropin precursor and preproencephalin) are cleaved to yield a number of different hormones. As shown in Figure 21-45 (see also Figure 21-37), cleavage always occurs adjacent to pairs of basic amino acids (Lys-Arg or Lys-Lys). The majority of amino acid residues present in the original multihormone precursor are then discarded. Although the cotranslation of so many active molecules into one precursor would appear to be a device for producing equal amounts of hormones, this is not necessarily so. The pattern of peptide production in the fetus is different from the pattern in the adult, and there are even differences in different lobes of the pituitary gland. Disturbance of the normal cleavage pathways in the adult could also conceivably lead to altered neuronal activity and mental states. Hence, proteolytic processing is yet another control point for gene expression in higher eucaryotic cells.

Targeted Protein Degradation Is Essential for Cell Survival and Growth[120–123]

The amount of a gene product available for a cell's functioning depends not only on the many control mechanisms that operate during its synthesis. Protein lifetimes, like RNA lifetimes, are highly variable and subject to manipulation when a cell's environment changes. For instance, upon severe nutritional deprivation, significant proteolysis

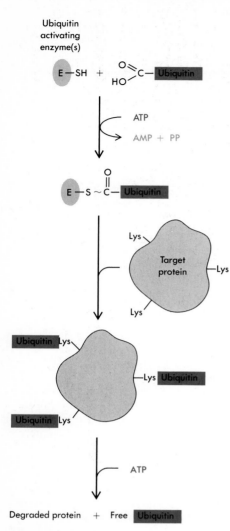

Figure 21-46
Sequence of events in ATP- and ubiquitin-dependent degradation of proteins. Recent work suggests that the first step of the process (the conjugation of ubiquitin to certain substrates) requires tRNA. [After A. Hershko and A. Ciechanover, *Ann. Rev. Biochem.* 51 (1982):335–364.]

of *intra*cellular proteins occurs in the lysosomes, cytoplasmic organelles normally devoted to the degradation of proteins and other molecules taken up from the exterior. But even during normal cell growth and maintenance, protein turnover is high. It is not unreasonable to expect that certain proteins needed at one stage of the cell cycle might have to be eliminated at another, or that a protein's rapid disappearance might be as crucial for survival as its rapid appearance when conditions change. The systems involved in such selective protein turnover are nonlysosomal. Generally, they require ATP, whose energy of hydrolysis is presumably channeled into making degradation specific (since peptide bond cleavage is not an energy-consuming reaction).

One important ATP-requiring proteolytic system in mammalian cells utilizes an unusual protein called **ubiquitin** to target other proteins for destruction. Ubiquitin is a 76-amino-acid polypeptide, whose sequence is nearly invariant from yeast to human. (It is perhaps the most highly conserved eucaryotic protein known.) It becomes activated for attachment to a targeted protein in a series of reactions similar to those catalyzed by the aminoacyl-tRNA synthetases (Figure 21-46). Once activated, usually several ubiquitin molecules are transferred to the protein substrate, forming isopeptide bonds between the ϵ-amino group of lysine residues and the carboxy-terminal glycine of ubiquitin. The ubiquitinated protein is subsequently degraded, releasing the ubiquitin molecules intact for use in further rounds of protein degradation. Both ubiquitin attachment and the less well characterized degradation steps require ATP (see Figure 21-46). The possible involvement of this proteolytic system in turning on gene expression is suggested by the finding that 10 to 20 percent of histone H2A molecules are ubiquitinated. This modification occurs preferentially in nucleosomes associated with transcribed regions, suggesting that selective histone degradation may contribute to the transition from inactive to active chromatin. Further, a mouse cell cycle mutant that arrests at the S/G2 boundary at high temperatures has been shown to be temperature-sensitive for ubiquitin activation (and hence for protein degradation). Most intriguing is the recent demonstration that ubiquitin-dependent proteolysis requires tRNA(s), conceivably linking this protein degradation system to amino acid levels in mammalian cells.

Histone Gene Expression Is Cell Cycle–Dependent[124,125]

Histones provide a conspicuous example of proteins that are needed in bulk during only one stage of the cell cycle. This is S phase, when an additional full complement of histone molecules must be supplied to package the newly replicated DNA into nucleosomes. Rather than storing histones for later use, eucaryotic cells synthesize histone molecules almost exclusively during S phase. The coupling with DNA replication appears to be direct, for inhibition of DNA synthesis abruptly terminates histone protein synthesis as well. Although the coupling mechanism is not understood in detail, growing evidence indicates that control of histone production is exerted at both the transcriptional and posttranscriptional levels.

There is little if any histone mRNA present in the cytoplasm just after mitosis. The dramatic increase observed when the cell enters S phase is due first to an increase (about tenfold) in the rate of synthesis of histone mRNAs. This has been traced (at least in part) to the ap-

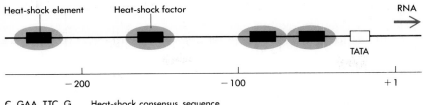

Heat-shock element   Heat-shock factor   RNA

C..GAA..TTC..G   Heat-shock consensus sequence

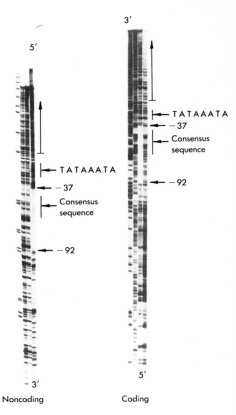

Noncoding   Coding

Figure 21-47
Binding of the heat-shock factor to sites upstream from the *Drosophila* hsp 70 gene. [After W. S. Dynan and R. Tjian, *Nature* 316 (1985):774–777; footprint courtesy of C. Parker.]

pearance of S phase–specific transcription factor(s) that selectively stimulate initiation by RNA polymerase II at histone gene promoters. (Whether there are different or common factors for the various histone genes is not yet known.) Second, once synthesized, histone mRNAs experience slightly different fates, depending on the stage of the cell cycle. Because they are not polyadenylated, histone mRNAs have unusually short half-lives, which contribute to their rapid disappearance when S phase is complete. But during S phase, their half-lives are lengthened (from 8 minutes to 40 minutes in cultured human HeLa cells) by as yet unknown mechanisms. Contrasting with the behavior of most histone mRNAs are mRNAs for minor, variant histones, whose synthesis and stability are not linked to DNA replication. They may provide a low level of new histone synthesis required during interphase because of DNA repair or nucleosome turnover accompanying gene expression.

Turning On of Heat-Shock Genes: Control at Both the Transcriptional and Translational Levels[126–131]

The cells of all organisms, from bacteria to human, respond to a sudden elevation of temperature (**heat shock**) by changing their pattern of gene expression. Usually, the synthesis of a few specific proteins (**hsp,** or **heat-shock proteins**) is dramatically turned up, while that of the majority of cellular polypeptides is sharply curtailed. In bacteria, a new σ factor accounts for an altered transcriptional pattern (Chapter 16). In eucaryotes, the heat-shock phenomenon has been most intensively scrutinized in *Drosophila*. Within one minute after temperature increase, puffs are induced at a few distinct sites along the polytene chromosomes (the heat-shock loci). Simultaneously, other puffs active at the time of heat shock regress. Indeed, we now know that the puffing reflects a conversion of the transcriptional apparatus to a state in which the seven heat-shock genes are almost exclusively transcribed. An accompanying change in the cell's translational machinery ensures that the heat-shock mRNAs are also preferentially translated.

The activation of heat-shock gene transcription is mediated by specific sequence elements present in the heat-shock promoters. Deletion analyses initially revealed that a 20-base-pair region located about 20 nucleotides upstream of (5′ to) the TATA box was essential for induction of one *Drosophila* heat-shock gene *(hsp70)*. A consensus sequence for this regulatory element was derived by comparing sequences upstream of all seven heat-shock genes (Figure 21-47). It shows dyad symmetry, characteristic of sites recognized by multimeric DNA binding proteins. The one heat-shock gene that does not conform well to this consensus is only weakly inducible. That a positively acting transcription factor binds to the DNA region containing this consensus sequence has been suggested by recent DNase footprinting studies (see Figure 21-47). In fact, upstream of the *hsp70* gene there are four regions, each with its own consensus sequence, to which the factor(s) bind(s); binding to the two contiguous sites (sites 1 and 2) is cooperative and appears to be closely linked to transcriptional activation. How factor activity itself is regulated upon heat shock remains to be clarified.

The translational discrimination that occurs in heat-shocked cells is not yet as well understood as the transcriptional regulation. What is clear is that preexisting mRNAs are stable, but are less efficiently utilized probably because of lower rates of initiation. In fact, some signal contained near the 5′ end of the heat-shock mRNA seems to confer efficient translation on the induced heat-shock mRNAs. How the heat-shock proteins themselves help cells combat the deleterious effects of high temperature remains unknown. Most remarkable is the evolutionary conservation of the components of the heat-shock regulatory system: *Drosophila* genes introduced into mammalian cells are faithfully induced upon heat shock, but respond at a temperature characteristic of mammalian cells. Complete elucidation of the mechanisms underlying the heat-shock response therefore promises to provide general insights into gene regulation in all cells.

Positive Control of Gene Expression by Steroid Hormones: Enhancer Action and Messenger Stabilization[132, 133, 134]

Steroid hormones are important to the development and physiological regulation of animals ranging from water molds to humans. When a sensitive target cell is treated with a steroid, the specific synthesis of new RNAs is stimulated, usually within minutes. This initial response appears to be direct, for inhibitors of protein synthesis have no effect. Steroids, however, do not act by themselves. Instead, they bind tightly to highly specific receptor proteins (called **steroid hormone receptors**) that are present in several thousand copies per cell. The hormone-receptor complexes then interact with particular genomic loci to turn on transcription of the appropriate gene(s).

Best understood is the action of the glucocorticoid receptor, a large protein made up of four identical 94 kdal subunits. When complexed with dexamethasone, a synthetic glucocorticoid hormone, it binds to specific DNA sequences that can be considered hormone-dependent transcriptional enhancers. In the well-studied mouse mammary tumor virus genome, there are five such binding regions (one upstream and four downstream from the start point of transcription; Figure 21-48). In vitro, binding of the hormone-receptor complex occurs independently to these sites. In vivo, these receptor-binding sequences confer stimulation by glucocorticoids when fused in either orientation to another promoter whose expression is not normally hormone regulated. Several alterations in the chromatin structure of the hormone-activated gene(s) can be detected. First, general DNase I sensitivity increases upon dexamethasone addition and persists long after hormone withdrawal. Second, discrete DNase I–hypersensitive regions develop near the sites of hormone-receptor binding; these disappear when hormone is removed. Near genes whose expression is regulated by more than one steroid hormone (e.g., the chicken lysozyme gene responds to both glucocorticoids and progesterone), there are distinct (but sometimes overlapping) sites for the binding of

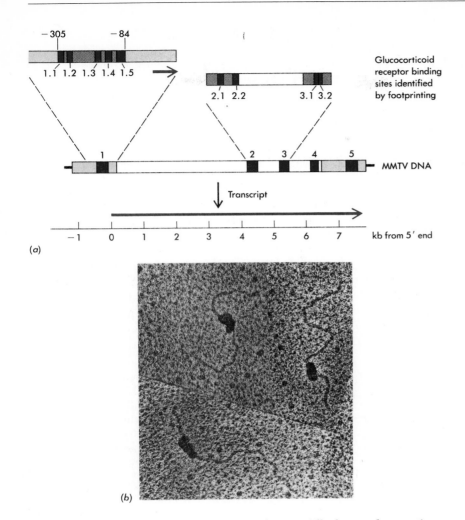

Glucocorticoid receptor binding sites identified by footprinting

MMTV DNA

Transcript

kb from 5' end

(a)

(b)

Figure 21-48
Each of the five glucocorticoid receptor binding regions in mouse mammary tumor virus (MMTV) DNA contains multiple interaction sites. (a) Binding sites for the purified receptor protein were mapped by DNase footprinting. Each site (colored boxes) contains sequences closely related to a consensus octanucleotide $AGA^A_TCAG^A_T$. Certain mutations introduced into the protected sequences have been shown to alter both receptor binding and enhancer activity. Gray boxes at the ends of the DNA denote the right and left LTR (long terminal repeat) of the MMTV genome. [After K. R. Yamamoto, *Molecular Developmental Biology: Expressing Foreign Genes*, ed. L. Bogorad and G. Adelman (New York: Alan Liss, in press); and F. Payvar et al., *Cell* 35 (1983):381–392.] (b) Electron micrograph of receptor bound to region 1 DNA. Both the size and shape of the bound protein cluster suggest that many subunits are involved. (Courtesy of K. Yamamoto.)

the different hormone-receptor complexes. All these observations once again raise the question of how enhancer elements operate. Yet, they strongly argue that the glucocorticoid and other steroid hormone receptor proteins can be considered prototypes of the class of tissue-specific proteins that activate transcription through their interaction with enhancer-like elements.

Also contributing to steroid hormone regulation are mechanisms acting at other levels of gene expression. Most striking of these is the selective stabilization of hormone-stimulated transcripts against cytoplasmic degradation. For instance, in *Xenopus* liver, the half-life of vitellogenin mRNA (vitellogenin is an egg yolk precursor protein) is about 3 weeks in the presence of the inducing hormone estrogen; it drops to 16 hours after estrogen withdrawal. Again, it seems that the hormone acts through some sort of receptor protein, which appears to be different from the receptor that binds to the vitellogenin gene region to activate transcription. Selective mRNA stabilization has also been observed for a number of other genes that are stimulated by steroid hormones. In addition, after hormone treatment of sensitive cells, RNA processing has been reported to be altered, and increased rates of overall RNA synthesis are sometimes observed, leading to proliferative growth. Thus, we see that steroid hormones achieve massive accumulation of specific mRNAs by concerted manipulation of both transcriptional and posttranscriptional processes. Deciphering the precise underlying mechanisms is a current challenge for molecular biologists.

Summary

The huge genomes of higher eucaryotic cells are associated with specific proteins that serve to pack and organize the DNA within the nucleus. The first level of packing involves the four core histone proteins that wrap approximately 150 base pairs of DNA around the exterior of the nucleosome core particle. The next higher order of packing requires in addition histone H1, which associates preferentially with linker DNA between nucleosomes and yields a 300 Å filament. Even more highly condensed forms of eucaryotic DNA are found in interphase heterochromatin, in the metaphase scaffold, and in sperm heads, where the histones are replaced by another class of DNA binding proteins.

During interphase, different segments of the genome are replicated at different times. Replication early in S phase is correlated with transcriptional activity (or potential activity) and light-staining Giemsa bands on metaphase chromosomes. During all stages of the cell cycle, the DNA remains attached to an underlying structure called the nuclear scaffold. Points of attachment are located at specific sites in repeating gene clusters, close to but not within sequences that are copied into RNA.

Active chromatin is distinguished from inactive regions of higher eucaryotic genomes by a number of morphological and biochemical criteria. In the giant polytene chromosomes of *Drosophila* salivary glands, bands puff visibly as they become transcriptionally active. However, since a single band may contain several genes that are not cotranscribed, puffing and transcription are not strictly linked. Examination in the electron microscope has suggested that the regular packing of DNA into nucleosomes is altered or absent in very actively transcribed genes. Biochemically, active chromatin exhibits a heightened general sensitivity to DNase, hypersensitivity to DNase at a few more localized sites in or near the gene, and (usually) undermethylation of C residues relative to other parts of the genome.

The synthesis, processing, and assembly of ribosomal rRNA occur in nucleoli. Promoter regions and other activating sequences for RNA polymerase I lie upstream from the start site of transcription. Both in vivo and in vitro, the initiation of rRNA synthesis is highly species-specific because one or more required transcription factors have coevolved to recognize promoters only in the rDNA of the same species.

Several types of conserved sequence elements are present in promoter regions for RNA polymerase II, which synthesizes all cellular messengers. They are located upstream from the start site and are tightly bound by specific factors required for initiating transcription. More distant activating sequences known as enhancers can function in either orientation, can reside even within or downstream from the transcribed region, and are bound by tissue-specific proteins that somehow facilitate transcription from a nearby promoter.

Essential sequences within the genes themselves are a hallmark of promoters for RNA polymerase III. These internal control regions are bound by factors to form stable transcription complexes that synthesize multiple copies of product before dissociation from the DNA. The existence of gene-specific transcription factors that selectively activate certain 5S or tRNA genes in particular tissues is exemplified by a *Xenopus* factor that binds both to the oocyte-specific 5S gene and to its 5S rRNA product.

Transcripts synthesized by RNA polymerase II undergo a number of covalent modifications before export to the cytoplasm as mature messenger RNAs. Their 3' ends do not arise by transcription termination but from specific cleavage and (in most cases) the addition of poly A tracts. Small RNA-protein complexes participate in these processing events as well as in splicing. Meanwhile, the pre-mRNA is packaged by a special set of nuclear RNA binding proteins. Differences in promoter utilization, in splice site selection, and in 3'-end placement can enable a single gene to yield a number of different mRNAs (and protein products) in a tissue-specific or developmentally specific fashion.

Protein synthesis by 80S ribosomes proceeds along the same general pathway as with bacterial ribosomes. But the number of factors involved (particularly for initiation) is much greater. Usually, only one AUG per message serves as an initiator codon. Polypeptide chain initiation is regulated by a number of agents using a common mechanism: Phosphorylation of one of the initiation factors prevents its reutilization in bringing the initiator tRNA to the ribosome.

Newly synthesized proteins are specifically delivered to a variety of membrane-bound compartments within the higher eucaryotic cell. A small RNA-protein complex called SRP recognizes an N-terminal signal sequence to initiate the transfer of secreted proteins into the endoplasmic reticulum; SRP blocks completion of these polypeptide chains if the correct membrane is not present. For delivery to chloroplasts, to mitochondria or to the nucleus, other specific signals within the protein are essential, but their role and the targeting mechanisms involved are only now being worked out.

Even after their translation is complete, the proteins of higher eucaryotic cells are often modified further. Some are extensively cleaved to yield smaller functional products, such as peptide hormones. Also operating are energy-dependent systems that selectively degrade particular proteins; changes in cell or tissue metabolism requires not only the appearance of new proteins but also the disappearance of preexisting ones.

Gene function can be regulated by manipulating any of the many steps required for gene expression in higher eucaryotic cells. Clear evidence has emerged for control at both transcriptional and posttranscriptional levels for several well-studied gene systems. Histone mRNAs are both transcribed at higher rates and selectively stabilized during the S phase of the cell cycle. The *Drosophila* genes that are selectively turned on after heat shock require a specific transcription factor that binds to multiple sites on the DNA upstream from each RNA start point; the translational machinery also discriminates in favor of the mRNAs encoding the heat-shock proteins. Steroid hormones interact with special receptor proteins, which then bind DNA regions that function as enhancers to activate transcription from a nearby promoter. Striking stabilization of hormone-induced mRNAs as well as changes in RNA processing also contribute to the hormone response.

Bibliography

General References

Alberts, B., D. Bray, J. Lewis, M. Raff, K. Roberts, and J. D. Watson. *Molecular Biology of the Cell*. 2nd edition, in press. New York: Garland. This valuable text presents all aspects of the functioning of higher eucaryotic cells, as well as a more detailed consideration of the structures and functional states of chromosomes.

Busch, H., ed. *The Cell Nucleus*. New York Academic Press. A whole series of volumes (the first published in 1970) containing review articles on nuclear structure and function. Especially relevant here are Vol. IV–VIII on chromatin, Vol. VIII and IX on nuclear particles, and Vol. XI and XII on nucleolar and other rDNA-containing chromatin.

Cold Spring Harbor Symp. Quant. Biol. XLVII. 1982. *Structures of DNA*. Cold Spring Harbor, N.Y.: Cold Spring Harbor Laboratory. A two volume publication that includes many articles on chromatin structure as well as its organization into transcribing genes.

Gluzman, Y., ed. 1985. *Eukaryotic Transcription: The Role of cis- and trans-acting Elements in Initiation*. Cold Spring Harbor, N.Y.: Cold Spring Harbor Laboratory. An even more current version of the next reference (*Enhancers and Eucaryotic Gene Expression*) that focuses not only on enhancers, but also on other transcription factors and the signals that they recognize.

Gluzman, Y., and T. Shenk. 1983. *Enhancers and Eucaryotic Gene Expression*. Cold Spring Harbor, N.Y.: Cold Spring Harbor Laboratory. Current research articles on all aspects of enhancer structure and function.

Hadjiolav, A. A. 1985. *The Nucleolus and Ribosome Biogenesis*. New York, N.Y.: Springer-Verlag. A complete and informative treatise on the nucleolar structure and function, but lacking the very most recent references.

Perez-Bercoff, R., ed. 1982. *Protein Biosynthesis in Eucaryotes*. New York: Plenum. A compendium of valuable contributions on all aspects of eucaryotic translation.

Razin, A., H. Cedar, and A. D. Riggs, eds. (1984) *DNA Methylation: Biochemistry and Biological Significance*. New York, N.Y.: Springer-Verlag. An authoritative and balanced treatment of the molecular biology of DNA methylation.

Schlesinger, M. J., M. Ashburner, and A. Tissières, eds. 1982. *Heat Shock, from Bacteria to Man*. Cold Spring Harbor, N.Y.: Cold Spring Harbor Laboratory. Everything about the heat-shock response as of 1982.

Cited References

1. Kornberg, R. D., and A. Klug. 1981. "The Nucleosome." *Sci. Amer.* 244:52–64.
2. McGhee, J. D., and G. Felsenfeld. 1980. "Nucleosome Structure." *Ann. Rev. Biochem.* 49:1115–1156.
3. Igo-Kemenes, T., W. Horz, and H. G. Zachau. 1982. "Chromatin." *Ann. Rev. Biochem.* 51:89–121.
4. Richmond, T. J., J. T. Finch, B. Rushton, D. Rhodes, and A. Klug. 1984. "Structure of the Nucleosome Core Particle at 7 Å Resolution." *Nature* 311:532–537.
5. Wang, J. C. 1982. "The Path of DNA in the Nucleosome." *Cell* 724:724–726.
6. Simpson, R. T., and D. W. Stafford. 1983. "Structural Features of a Phased Nucleosome Core Particle." *Proc. Natl. Acad. Sci. USA* 80:51–55.
7. Thoma, F., T. Koller, and A. Klug. 1979. "Involvement of Histone H1 in the Organization of the Nucleosome and of the Salt-Dependent Superstructures of Chromatin." *J. Cell Biol.* 83:403–427.
8. Brown, S. W. 1966. "Heterochromatin." *Science* 151:417–425.
9. Paulson, J. R. and U. K. Laemmli. 1977. "The Structure of Histone-Depleted Metaphase Chromosomes." *Cell* 12:817–828.
10. Lewis, C. D., and U. K. Laemmli. 1982. "Higher-Order Metaphase Chromosome Structure: Evidence for Metalloprotein Interactions." *Cell* 29:171–181.
11. Warrant, R. W., and S.-H. Kim. 1978. "α-Helix-Double Helix Interaction Shown in the Structure of a Protamine-Transfer RNA Complex and a Nucleoprotamine Model." *Nature* 271:130–135.
12. Hand, R. 1978. "Eucaryotic DNA: Organization of the Genome for Replication." *Cell* 15:317–325.
13. Goldman, M. A., G. P. Holmquist, M. C. Gray, L. A. Caston, and A. Nag. 1984. "Replication Timing of Genes and Middle Repetitive Sequences." *Science* 224:686–692.
14. Mirkovitch, J., M.-E. Mirault, and U. K. Laemmli. 1984. "Organization of the Higher-Order Chromatin Loop: Specific DNA Attachment Sites on Nuclear Scaffold." *Cell* 39:223–232.
15. Ashburner, M., and G. Richards. 1976. "The Role of Ecdysone in the Control of Gene Activity in the Polytene Chromosomes of *Drosophila*." In *Insect Development*, ed. P. A. Lawrence. New York: Wiley, pp. 203–225.
16. Bonner, J. J. and M. L. Pardue. 1977. "Polytene Chromosome Puffing and *in Situ* Hybridization Measure Different Aspects of RNA Metabolism." *Cell* 12:227–234.
17. Lamb, M. M., and B. Daneholt. 1979. "Characterization of Active Transcription Units in Balbiani Rings of *Chironomus tentans*." *Cell* 17:835–848.
18. Sirotkin, K., and N. Davidson. 1982. "Developmentally Regulated Transcription from *Drosophila melanogaster* Chromosomal Site 67B." *Dev. Biol.* 89:196–210.
19. Corces, V., R. Holmgren, R. Freund, R. Morimoto, and M. Meselson. 1980. "Four Heat Shock Proteins of *Drosophila melanogaster* Coded Within a 12-Kilobase Region in Chromosome Subdivision 67B." *Proc. Nat. Acad. Sci.* 77:5390–5393.
20. Weisbrod, S. 1982. "Active Chromatin." *Nature* 297:289–294.
21. Elgin, S. C. R. 1981. "DNAase I-Hypersensitive Sites of Chromatin." *Cell* 27:413–415.
22. Weintraub, H., and M. Groudine. 1976. "Chromosomal Subunits in Active Genes Have an Altered Conformation." *Science* 193:848–856.
23. Garel, A., and R. Axel. 1976. "Selective Digestion of Transcriptionally Active Ovalbumin Genes from Oviduct Nuclei." *Proc. Nat. Acad. Sci.* 73:3966–3970.
24. Wu, C., 1980. "The 5' Ends of *Drosophila* Heat Shock Genes in Chromatin Are Hypersensitive to DNase I." *Nature* 286:854–860.
25. Saragosti, S., G. Moyne, and M. Yaniv. 1980. "Absence of Nucleosomes in a Fraction of SV40 Chromatin Between the Origin of Replication and the Region Coding for the Late Leader RNA." *Cell* 20:65–73.
26. Elgin, S. C. R. 1984. "Anatomy of Hypersensitive Sites." *Nature* 309:213–214.
27. Emerson, B. M., C. D. Lewis, and G. Felsenfeld. 1985. "Interaction of Specific Nuclear Factors with the Nuclease-Hypersensitive Region of the Chicken Adult β-Globin Gene: Nature of the Binding Domain." *Cell* 41:21–30.
28. Felsenfeld, G., and J. McGhee. 1982. "Methylation and Gene Control." *Nature* 296:602–603.
29. Doerfler, W. 1983. "DNA Methylation and Gene Activity." *Ann. Rev. Biochem.* 52:93–124.
30. Bird, A. P. 1984. "DNA Methylation—How Important in Gene Control?" *Nature* 307:503–504.
31. Allfrey, V. G. 1977. "Post-Synthetic Modifications of Histone Structure: A Mechanism for the Control of Chromosome Structure by the Modulation of Histone-DNA Interactions." In *Chromatin and Chromosome Structure*, ed. H. J. Li and R. Eckhardt. New York: Academic Press, pp. 167–191.
32. Ball, D. J., D. S. Gross, and W. T. Garrard. 1983. "5-Methylcytosine Is Localized in Nucleosomes That Contain Histone H1." *Proc. Nat. Acad. Sci.* 80:5490–5494.
33. Weintraub, H. 1985. "Assembly and Propagation of Re-

pressed and Derepressed Chromosomal States." *Cell* 42:705–711.

34. Varshavsky, A., L. Levinger, O. Sundin, J. Barsoum, E. Oz-kaynak, P. Swerdlow, and D. Finley. 1983. "Cellular and SV40 Chromatin: Replication, Segregation, Ubiquitination, Nuclease-Hypersensitive Sites, HMG-Containing Nucleosomes, and Heterochromatin-Specific Protein." *Cold Spring Harbor Symp. Quant. Biol.* 47:511–520.

35. Labhart, P., and T. Koller. 1982. "Structure of the Active Nucleolar Chromatin of *Xenopus laevis* Oocytes." *Cell* 28:279–292.

36. Widmer, R. M., R. Lucchini, M. Lezzi, B. Meyer, J. M. Sogo, J.-E. Edstrom, and T. Koller. 1984. "Chromatin Structure of a Hyperactive Secretory Protein Gene (in Balbiani Ring 2) of *Chironomus*." *EMBO J.* 3:1635–1641.

37. Foe, V. E., L. E. Wilkinson, and C. D. Laird. 1976. "Comparative Organization of Active Transcription Units in *Oncopeltus fasciatus*." *Cell* 9:131–146.

38. Lin, S., and A. D. Riggs. 1975. "The General Affinity of lac Repressor for *E. coli* DNA: Implications for Gene Regulation in Prokaryotes and Eukaryotes." *Cell* 4:107–111.

39. Hadjiolav, A. A. and N. Nikolaev. 1976. "Maturation of Ribosomal Ribonucleic Acids and the Biogenesis of Ribosomes." *Prog. Biophys. Mol. Biol.* 31:95–144.

40. Vaughn, J. C., S. J. Sperbeck, W. J. Ramsey, and C. B. Lawrence. 1984. "A Universal Model for the Secondary Structure of 5.8S Ribosomal RNA Molecules, Their Contact Sites with 28S Ribosomal RNAs, and Their Prokaryotic Equivalent." *Nucleic Acid Res.* 12:7479–7502.

41. Bowman, L. H., W. E. Goldman, G. I. Goldberg, M. B. Hebert, and D. Schlessinger. 1983. "Location of the Initial Cleavage Sites in Mouse Pre-mRNA." *Mol. Cell. Biol.* 3:1501–1510.

42. Sommerville, J. 1984. "RNA Polymerase I Promoters and Transcription Factors." *Nature* 310:189–190.

43. Reeder, R. H. 1984. "Enhancers and Ribosomal Gene Spacers." *Cell* 38:349–351.

44. Moss, T. 1982. "Transcription of Cloned *Xenopus laevis* Ribosomal DNA Microinjected into *Xenopus* Oocytes, and the Identification of an RNA Polymerase I Promoter." *Cell* 30:835–842.

45. Grummt, I. 1982. "Nucleotide Sequence Requirements for Specific Initiation of Transcription by RNA Polymerase I." *Proc. Nat. Acad. Sci.* 79:6908–6911.

46. Yamamoto, O., N. Takakusa, Y. Mishima, R. Kominami, and M. Muramatsu. 1984. "Determination of the Promoter Region of Mouse Ribosomal RNA Gene by an *in Vitro* Transcription System." *Proc. Nat. Acad. Sci.* 81:299–303.

47. Boseley, P., T. Moss, M. Machler, R. Portmann, and M. Birnstiel. 1979. "Sequence Organization of the Spacer DNA in a Ribosomal Gene Unit of *Xenopus laevis*." *Cell* 17:19–31.

48. Reeder, R. H., and J. C. Roan. 1984. "The Mechanism of Nucleolar Dominance in *Xenopus* Hybrids." *Cell* 38:39–44.

49. Miesfeld, R., and N. Arnheim. 1984. "Species-Specific rDNA Transcription Is Due to Promoter-Specific Binding Factors." *Mol. Cell. Biol.* 4:221–227.

50. Learned, R. M., S. Cordes, and R. Tjian. 1985. "Purification and Characterization of a Transcription Factor That Confers Promoter Specificity to Human RNA Polymerase I." *Mol. Cell Biol.* 5:1358–1369.

51. Breathnach, R., and P. Chambon. 1981. "Organization and Expression of Eucaryotic Split Genes Coding for Proteins." *Ann. Rev. Biochem.* 50:349–383.

52. Nevins, J. R., 1983. "The Pathway of Eukaryotic mRNA Formation." *Ann. Rev. Biochem.* 52:441–466.

53. McKnight, S. L. and R. Kingsbury. 1982. "Transcriptional Control Signals of a Eukaryotic Protein-Coding Gene." *Science* 217:316–324.

54. Myers, R. M., L. S. Lerman, and T. Maniatis. 1985. "A General Method for Saturation Mutagenesis of Cloned DNA Fragments." *Science* 229:242–247.

55. Cochran, M. D. and C. Weissmann. 1984. "Modular Structure of the β-Globin and the TK Promoters." *EMBO J.* 3:2453–2459.

56. McKnight, S. L., R. C. Kingsbury, A. Spence, and M. Smith. 1984. "The Distal Transcription Signals of the Herpes Virus *tk* Gene Share a Common Hexanucleotide Control Sequence." *Cell* 37:253–262.

57. Shenk, T. 1981. "Transcriptional Control Regions: Nucleotide Sequence Requirements for Initiation by RNA Polymerase II and III." In *Current Topics in Microbiology and Immunology*, vol. 93. New York: Springer-Verlag, pp. 25–46.

58. Dynan, W. S., and R. Tjian. 1985. "Control of Eukaryotic Messenger RNA Synthesis by Sequence-Specific DNA Binding Proteins." *Nature* 316:774–777.

59. Parker, C. S., and J. Topol. 1984. "A *Drosophila* RNA Polymerase II Transcription Factor Contains a Promoter-Region-Specific DNA-Binding Activity." *Cell* 36:357–369.

60. Davison, B. L., J.-M. Egly, E. R. Mulvihill, and P. Chambon. 1983. "Formation of Stable Preinitiation Complexes Between Eukaryotic Class B Transcription Factors and Promoter Sequences." *Nature* 301:680–686.

61. Gidoni, D., W. S. Dynan, and R. Tjian. 1984. "Multiple Specific Contacts Between a Mammalian Transcription Factor and Its Cognate Promoters." *Nature* 312:409–413.

62. Bird, A., M. Taggart, M. Frommer, O. J. Miller, and D. Macleod. 1985. "A Fraction of the Mouse Genome That Is Derived from Islands of Nonmethylated, CpG-Rich DNA." *Cell* 40:91–99.

63. Fire, A., M. Samuels, and P. A. Sharp. 1984. "Interactions Between RNA Polymerase II, Factors, and Template Leading to Accurate Transcription." *J. Biol. Chem.* 259:2509–2516.

64. Sawadogo, M., and R. C. Roeder. 1984. "Energy Requirement for Specific Transcription Initiation by the Human RNA Polymerase II System." *J. Biol. Chem.* 259:5321–5326.

65. Khoury, G., and P. Gruss. 1983. "Enhancer Elements." *Cell* 33:313–314.

66. Weiher, H., M. Konig, and P. Gruss. 1983. "Multiple Point Mutations Affecting the Simian Virus 40 Enhancer." *Science* 219:626–631.

67. Schaffner, W., E. Serfling, and M. Jasin. 1985. "Enhancers and Eukaryotic Gene Transcription." *Trends in Genetics* 1:224–230.

68. Sergeant, A., D. Bohmann, H. Zentgraf, H. Weiher, and W. Keller. 1984. "A Transcription Enhancer Acts *in Vitro* Over Distances of Hundreds of Base-Pairs on Both Circular and Linear Templates but Not on Chromatin-Reconstituted DNA." *J. Mol. Biol.* 180:577–600.

69. Sassone-Corsi, P., A. Wideman, and P. Chambon. 1985. "A *trans*-Acting Factor Is Responsible for the Simian Virus 40 Enhancer Activity *in Vitro*." *Nature* 312:458–463.

70. Scholer, H. R., and P. Gruss. 1984. "Specific Interaction Between Enhancer-Containing Molecules and Cellular Components." *Cell* 36:403–411.

71. Ciliberto, G., L. Castagnoli, and R. Cortese. 1983. "Transcription by RNA Polymerase III." *Curr. Top. Dev. Biol.* 18:59–88.

72. Sakonju, S., D. F. Bogenhagen, and D. D. Brown. 1980. "A Control Region in the Center of the 5S RNA Gene Directs Specific Initiation of Transcription I: The 5' Border of the Region." *Cell* 19:13–25.

73. Bogenhagen, D. F., S. Sakonju, and D. D. Brown. 1980. "A Control Region in the Center of the 5S RNA Gene Directs Specific Initiation of Transcription II: The 3' Border of the Region." *Cell* 19:27–35.

74. Hall, B., S. G. Clarkson, and G. Tocchini-Valenti. 1982. "Transcription Initiation of Eucaryotic Transfer RNA Genes." *Cell* 29:3–5.

75. Stillman, D. J., and E. P. Geiduschek. 1984. "Differential Binding of a *S. cerevisiae* RNA Polymerase III Transcription Factor to the Promoter Segments of a tRNA Gene." *EMBO J.* 3:847–853.

76. Wilson, E. T., D. Larson, L. S. Young, and K. U. Sprague. 1985. "A Large Region Controls tRNA Gene Transcription." *J. Mol. Biol.* 183:153–163.

77. Schaack, J., S. Sharp, T. Dingermann, D. J. Burke, L. Cooley, and D. Soll. 1984. "The Extent of a Eucaryotic tRNA Gene." *J. Biol. Chem.* 259:1461–1467.

78. Brown, D. D. 1984. "The Role of Stable Complexes That Repress and Activate Eukaryotic Genes." *Cell* 37:359–365.
79. Lassar, A. B., P. L. Martin, and R. G. Roeder. 1983. "Transcription of Class III Genes: Formation of Preinitiation Complexes." *Science* 222:740–748.
80. Garel, J. P., R. L. Garber, and M. A. Siddiqui. 1977. "Transfer RNA in Posterior Silk Gland of *Bombyx mori*: Polyacrylamide Gel Mapping of Mature Transfer RNA, Identification and Partial Structural Characterization of Major Isoacceptor Species." *Biochemistry* 16:3618–3624.
81. Sprague, K. U., O. Hagenbuchle, and M. C. Zuniga. 1977. "The Nucleotide Sequence of Two Silk Gland Alanine tRNAs: Implications for Fibroin Synthesis and for Initiator tRNA Structure." *Cell* 11:561–570.
82. Goldsmith, M. R., and F. C. Kafatos. 1984. "Developmentally Regulated Genes in Silkmoths." *Ann. Rev. Genetics* 18:443–487.
83. Beyer, A. L., M. E. Christensen, B. W. Walker, and W. M. LeStourgeon. 1977. "Identification and Characterization of the Packaging Proteins of Core 40S hnRNP Particles." *Cell* 11:127–138.
84. Martin, T. E., J. M. Pullman, and M. D. McMullen. 1980. "Structure and Function of Nuclear and Cytoplasmic Ribonucleoprotein Complexes." In *Cell Biology, a Comprehensive Treatise*, vol. 4., ed. L. Goldstein and D. N. Prescott. New York: Academic Press, pp. 137–174.
85. Dreyfuss, G., Y. D. Choi, and S. A. Adam. 1984. "Characterization of Heterogeneous Nuclear RNA-Protein Complexes *in Vivo* with Monoclonal Antibodies." *Mol. Cell. Biol.* 4:1104–1114.
86. Darnell, J. E., Jr. 1982. "Variety in the Level of Gene Control in Eukaryotic Cells." *Nature* 297:365–370.
87. Birnstiel, M. L., M. Busslinger, and K. Strub. 1985. "Transcription Termination and 3' Processing: The End Is in Site!" *Cell* 41:349–359.
88. Falck-Pedersen, E., J. Logan, T. Shenk, and J. E. Darnell, Jr. 1985. "Transcription Termination Within the E1A Gene of Adenovirus Induced by Insertion of the Mouse β-Major Globin Terminator Element." *Cell* 40:897–905.
89. Moore, C. L. and P. A. Sharp. 1985. "Accurate Cleavage and Polyadenylation of Exogenous RNA Substrate." *Cell* 41:845–855.
90. Amara, S. G., V. Jonas, M. G. Rosenfeld, E. S. Ong, and R. M. Evans. 1982. "Alternative RNA Processing in Calcitonin Gene Expression Generates mRNAs Encoding Different Polypeptide Products." *Nature* 298:240–244.
91. Nabeshima, Y., Y. Fujii-Kuriyama, M. Muramatsu, and K. Ogata. 1984. "Alternative Transcription and Two Modes of Splicing Result in Two Myosin Light Chains from One Gene." *Nature* 308:333–338.
92. Breitbart, R. E., H. T. Nguyen, R. M. Medford, A. T. Destree, V. Mahdavi, and B. Nadal-Ginard. 1985. "Intricate Combinational Patterns of Exon Splicing Generate Multiple Regulated Troponin T Isoforms from a Single Gene." *Cell* 41:67–82.
93. Moldave, K. 1985. "Eukaryotic Protein Synthesis." *Ann. Rev. Biochem.* 54:1109–1119.
94. Kozak, M. 1983. "Comparison of Initiation of Protein Synthesis in Procaryotes, Eucaryotes, and Organelles." *Microbiol. Rev.* 47:1–45.
95. Shatkin, A. J. 1985. "mRNA Cap Binding Proteins: Essential Factors for Initiating Translation." *Cell* 40:223–224.
96. Weissbach, H. 1980. "Soluble Factors in Protein Synthesis." In *Ribosomes: Structure, Function and Genetics*, ed. G. Chambliss, G. R. Craven, J. Davies, K. Davis, L. Kahan, and M. Nomura. Baltimore: University Park Press, pp. 377–411.
97. Hershey, J. W. B. 1980. "The Translational Machinery: Components and Mechanism." In *Cell Biology, a Comprehensive Treatise*, vol. 4, ed. L. Goldstein and D. N. Prescott. New York: Academic Press, pp. 1–68.
98. Caskey, C. T. H. 1980. "Peptide Chain Termination." *Trends in Biochem. Sci.* 5:234–237.
99. Deutscher, M. P. 1984. "The Eukaryotic Aminoacyl-tRNA Synthetase Complex: Suggestions for Its Structure and Function." *J. Cell. Biol.* 99:373–377.
100. Vazquez, D., E. Zaera, H. Dolz, and A. Jimenz. 1982. "Action of the Inhibitors of Protein Biosynthesis." In *Protein Biosynthesis in Eucaryotes*, ed. R. Perez-Bercoff. New York: Plenum, pp. 311–337.
101. Pappenheimer, A. M., Jr., and D. M. Gill. 1973. "Diphtheria." *Science* 182:353–358.
102. Lee, H., and W. J. Iglewski. 1984. "Cellular ADP-Ribosyltransferase with the Same Mechanism of Action as Diphtheria Toxin and *Pseudomonas* Toxin A." *Proc. Nat. Acad. Sci.* 81:2703–2707.
103. Hunt, T. 1985. "False Starts in Translational Control of Gene Expression." *Nature* 316:580–581.
104. Ochoa, S. 1983. "Regulation of Protein Synthesis Initiation in Eucaryotes." *Archives of Biochem. and Biophysics* 223:325–349.
105. Safer, B. 1983. "2B or Not 2B: Regulation of the Catalytic Utilization of eIF-2." *Cell* 33:7–8.
106. Walter, P., R. Gilmore, and G. Blobel. 1984. "Protein Translocation Across the Endoplasmic Reticulum." *Cell* 38:5–8.
107. Meyer, D. I., E. Krause, and B. Dobberstein. 1982. "Secretory Protein Translocation Across Membranes—the Role of the 'Docking Protein.'" *Nature* 297:647–650.
108. Schatz, G., and R. A. Butow. 1983. "How Are Proteins Imported into Mitochondria?" *Cell* 32:316–318.
109. Grossman, A., S. Bartlett, and N. H. Chua. 1980. "Energy-Dependent Uptake of Cytoplasmically Synthesized Polypeptides by Chloroplasts." *Nature* 285:625–628.
110. Dingwall, C., S. V. Sharnick, and R. A. Laskey. 1982. "A Polypeptide Domain That Specifies Migration of Nucleoplasmin into the Nucleus." *Cell* 30:449–458.
111. Kalderon, D., B. L. Roberts, W. D. Richardson, and A. E. Smith. 1984. "A Short Amino Acid Sequence Able To Specify Nuclear Location." *Cell* 39:499–509.
112. Hall, M. N., L. Hereford, and I. Herskowitz. 1984. "Targeting of *E. coli* β-Galactosidase to the Nucleus in Yeast." *Cell* 36:1057–1065.
113. Silver, P. A., L. P. Keegan, and M. Ptashne. 1984. "Amino Terminus of the Yeast GAL4 Gene Product Is Sufficient for Nuclear Localization." *Proc. Nat. Acad. Sci.* 81:5951–5955.
114. Mattaj, I. W., and E. M. DeRobertis. 1985. "Nuclear Segregation of U2 snRNA Requires Binding of Specific snRNP Proteins." *Cell* 40:111–118.
115. Wolin, S. 1985. "Small Cytoplasmic Ribonucleoproteins." *Trends in Genetics* 1:201–204.
116. Steiner, D. F. 1984. "The Biosynthesis of Insulin: Genetic, Evolutionary, and Pathophysiologic Aspects." Harvey Lecture, Series 78. New York: Academic Press, pp. 191–228.
117. Noda, M., Y. Furutani, H. Takahashi, M. Toyosato, T. Hirose, S. Inayama, S. Nakanishi, and S. Numa. 1982. "Cloning and Sequence Analysis of cDNA for Bovine Adrenal Preproenkephalin." *Nature* 295:202–206.
118. Mains, R. E., B. A. Eipper, C. C. Glembotski, and R. M. Dores. 1983. "Strategies for the Biosynthesis of Bioactive Peptides." *Trends in Neurosci.* 4:229–235.
119. Douglass, J., O. Civelli, and E. Herbert. 1984. "Polyprotein Gene Expression: Generation of Diversity of Neuroendocrine Peptides." *Ann. Rev. Biochem.* 53:665–715.
120. Hershko, A., and A. Ciechanover. 1982. "Mechanisms of Intracellular Protein Breakdown." *Ann. Rev. Biochem.* 51:335–364.
121. Busch, H. 1984. "Ubiquitination of Proteins." *Meth. Enzymol.* 106:238–262.
122. Ciechanover, A., D. Finley, and A. Varshavsky. 1984. "The Ubiquitin-Mediated Proteolytic Pathway and Mechanisms of Energy-Dependent Intracellular Protein Degradation." *J. Cell. Biochem.* 24:27–53.
123. Ciechanover, A., S. L. Wolin, J. A. Steitz, and H. F. Lodish. 1985. "Transfer RNA Is an Essential Component of the Ubiquitin- and ATP-Dependent Proteolytic System." *Proc. Nat. Acad. Sci.* 82:1341–1345.
124. Heintz, N., and R. G. Roeder. 1984. "Transcription of Human Histone Genes in Extracts from Synchronized HeLa Cells." *Proc. Nat. Acad. Sci.* 81:2713–2717.

125. Sittman, D. B., R. A. Graves, and W. F. Marzluff. 1983. "Histone mRNA Concentrations Are Regulated at the Level of Transcription and mRNA Degradation." *Proc. Nat. Acad. Sci.* 80:1849–1853.

126. Ashburner, M., and J. J. Bonner. 1979. "The Induction of Gene Activity in *Drosophila* by Heat Shock." *Cell* 17:241–254.

127. Pelham, H. R. B. 1982. "A Regulatory Upstream Promoter Element in the *Drosophila* Hsp 70 Heat-Shock Gene." *Cell* 30:517–528.

128. Parker, C. S., and J. Topol. 1984. "A *Drosophila* RNA Polymerase II Transcription Factor Binds to the Regulatory Site of an Hsp 70 Gene." *Cell* 37:273–283.

129. Ballinger, D. G., and M. L. Pardue. 1983. "The Control of Protein Synthesis During Heat Shock in *Drosophila* Cells Involves Altered Polypeptide Elongation Rates." *Cell* 33:103–114.

130. Klemenz, R., D. Hultmark, and W. Gehring. 1985. "Selective Translation of Heat Shock mRNA in *Drosophila Melanogaster* Depends on Sequence Information in the Leader." *EMBO J.* 4:2053–2060.

131. McGarry, T. J., and S. Lindquist. 1985. "The Preferential Translation of *Drosophila hsp70* mRNA Requires Sequences in the Untranslated Leader." *Cell* 42:903–911.

132. Yamamoto, K. R. 1985. "Hormone-Dependent Transcriptional Enhancement and Its Implications for Mechanisms of Multifactor Gene Regulation." In *Molecular Developmental Biology: Expressing Foreign Genes*, 43rd Symposium of the Society for Developmental Biology, ed. L. Bogorad and G. Adelman. New York: Alan Liss, pp. 131–148.

133. Zaret, K. S., and K. R. Yamamoto. 1984. "Reversible and Persistent Changes in Chromatin Structure Accompany Activation of a Glucocorticoid-Dependent Enhancer Element." *Cell* 38:29–38.

134. Brock, M. L., and D. J. Shapiro. 1983. "Estrogen Stabilizes Vitellogenin mRNA Against Cytoplasmic Degradation." *Cell* 34:207–214.

Index

Page numbers in bold face refer to a definition and major discussion of the entry. F after a page number indicates a figure; T after a page number indicates a table.

protein synthesis, mammalian, 453–456, 454F, 456F; yeast, 588–589
signals on proteins, 559
topoisomerases in, 307
transcription in, 455–456, 456F
translation in, 456, 456F
in yeast, 557, 558F, 585, 585F
Mitochondrial DNA, evolution, 456
genetic code, mammalian, 453–454, 455T, 456F; yeast, 588T
high mutation rate, 342
lack of perceptible function, 589
mammalian, genetic map, 454F; genome, unusual features, 454–456, 454F
replication, enzymes, mechanism, 306–307, 307F; mutation and rearrangements in yeast, 586–587
yeast, amount per cell, 586; introns, 587–588, 587F; "omega" inserts, 587
Mitochondrial RNA, pre-mRNA and pre-rRNA, splicing, 638–639, 639F
pre-mRNA sequences for some, 638, 638F
self-splicing of some, 639
sequence for 12S rRNA of bovine, 397F
tRNA, 454–455, 456F
Mitomycin C, as carcinogen, 352
Mitoribosomes, in yeast, 588–589
Mitosis, 5–6, 6F
accuracy of chromosome separation, 565
centromere structure in, 563, 563F, 564F
chromosome condensation in, 682–683, 683F
in yeast, 551, 551F, 553, 555–557, 557F
Molecular models, limitations for building, 138
Monocistronic, mRNA, eucaryotic, **569**
Morphogenetic pathways, virus assembly by, 508, 508F
Morphogenetic pattern, of virus assembly, **508,** 508F
Morphogenetic signal sequence, of filamentous ssDNA phages, **536**
Mouse, differentiation and 5-azaC, 695
in situ hybridization, centromeres, 667F
Mouse mammary tumor virus, activation by glucocorticoid hormone, 738–739, 739F
Movable genetic elements. *See* Transposable genetic elements
mRNP. *See* Messenger ribonucleoprotein
Mud lysogens, use in gene isolation and identification, 529–530
Multigene families, mechanism for homogenization, 658–660, 659F, 660F
paradox of gene stability in, 657–658
Multimeric proteins, palindromic DNA sequences and, **276**
Murine sarcoma virus, enhancer region, 708, 709F
Mutable sites, 223–226, 224F, 225F
adjacent, for given amino acid, 225–226, 226F, 227F
Mutagenesis, base-pair changes, 604–605, 605F
in vitro, techniques, 602–605, 604F, 605F
oligonucleotide-directed, 604–605, 605F
Mutagens, **180**
adaptive response to, 346–347
Ames test, 355–356, 355F
as carcinogens, 355–356
kinds, 180
oxidants as, 356
Mutagens, chemical, base analogs, 344F, 345
DNA-reactive chemicals, 343, 344F, 346F
frameshift mutagens, 343F, 345
importance of studying, 343
Mutant genes, nonfunctional protein products, 221
retriever vectors in obtaining, 602, 603F
Mutant proteins, gel electrophoresis of, 221, 221F
Mutants, detection with dyes, 183, 184F
of *E. coli*, 178–184, 178F, 179F, 181F
genes in *Drosophila*, 18T
isolation by screening colonies, 183–184
phenotypes, diversity of, 181–182; timing of expression, 182

techniques, for in vitro genesis, 602–605, 604F, 605F; for increasing yield, 182–183, 182F, 183F
use in gene isolation, 208–209
Mutation, **17**
to antibiotic resistance, 179F, 180, 182
average frequency of, 340–342, 351
blocking DNA synthesis, 296, 297T, 297F
cancer and, 355–356
cdc genes and, 555–557, 556T, 557F
conditional lethal in phages, 188
in consensus sequences of promoters, 370, 370F
in control elements, 22
deletion, 223, 225F, 228–229, 230F, 231F
deletions in *rII* region of phage T4, 216, 217F
demonstration of spontaneous, 178–179, 178F
error-prone repair and, 353–354
frameshift, 228–229, 230F
genetic code and, 438
hot spots, 216F
importance of limiting, 339, 355
induction, by chemical mutagens, 343–345, 344F; in vitro methods, 210, 704F; by X-rays, 20
insertion, 223, 225F, 228–229, 230F, 231F
as key to genetic research, 176–177
large DNA rearrangements, 342–343
5-methylcytosine hot spots, 349, 350F
mismatch correction enzyme and, 349, 351–352, 351F
missense, 444
molecular evolution and, 656–657
multigene families and, 657–658
nature of, 339
nonsense, 444
by phage Mu insertion, 529–530
plaque-type in phages, 188, 188F
prevalence in recessive alleles, 182
requiring growth factors, 180–181, 181F, 184, 185F
reverse, 216, 225–226, 226F, 228, 229F
reversibility of single-base switches, 340
in *rII* genes of phage T4, 215
single-base changes, 223–224, 225F, 339–342, 341F, 343F
single-base changes, assay for, 340, 341F
splice site, 640, 641F
suppressor, 228, 229F, 445–453, 450T, 445F–453F
temperature sensitive, 444; *cdc*, 555–556, 556T, 557F
transposable elements in *Drosophila* and, 664–666, 664F, 665F
in yeast mtDNA, 586
Mutators, **342**
Mycoplasmas, distinction from viruses, 187
Mycoplasma myocoides, 123, 123F
Myoglobin, 3-D structure, 48F
O₂ binding by, 50
Myosin, cell division and, 555
Myosin LC1/LC3 gene, differential transcription and splicing, 719, 720F
Myosin light chains, types and synthesis, 719, 720F

N protein, as λ antiterminator protein, 523–525, 524F
mechanism of action, 524–525, 524F
N-formylmethionine, translation in bacteria and, **408,** 408F, 409F
Nalidixic acid, gyrase inhibition by, 374
Natural selection, maintenance of gene sequences and, 656–657
multigene families and, 657–658
Negative control, procaryotes, operators and, 476–477, 477F
Negative control, procaryotes, repressors and, 469–477, 470F–477F, 480–485, 480F–484F
Negative cooperativity, **159**
Neurospora crassa, growth factor mutants, 177
Neutron-scattering, to study ribosome structure, 424, 426F, 427F

Nick translation, **350**
Nicking, of DNA, **259,** 259F
Nicotinamide, diphtheria toxin effects and, 728, 728F
Nicotinamide adenine dinucleotide (NAD+), 29F
as coenzyme, 30–31, 32F
dehydrogenase domains and binding by, 645
dinucleotide fold structure and, 144–145, 148F
glucose fermentation and, 40F
Nitrocellulose paper, recombinant DNA procedures and, 608, 608F, 609F
Nitrogen metabolism, σ factor and control of, 486
Nitrosamines, as mutagens, 346–347
structural formula, 346F
Nitrosoguanidine, as mutagen, 180–181, 181F, 346–347
structural formula, 346F
Nitrous acid (HNO₂), in vitro mutagenesis using, 704F
as mutagen, 343, 344F
Nonhistone chromosomal proteins, amounts per chromosome, 678–679
examples, 151, 687
hnRNP polypeptides as, 716
mode of interaction with DNA, 151, 152F
removal with polyanions, 683
Nonpolar molecules, 130–**131**
Nonpolar side groups, protein stabilization and, 140–141
Nonsense codons, 87
Nonsense mutations, **444**
results of, 444
suppression of, 446–448, 447F
Nontranscribed spacer regions, in rDNA repeats, **699,** 699F
repeated promoter-like DNA in *Xenopus*, 702, 703F
variations among organisms, **652,** 653F
Northern blotting, **633,** 633F
Nuclear dominance, in *Xenopus* hybrid embryos, **702**
Nuclear lamina, interaction with chromatin, 686
Nuclear membrane, 4, 5F
during mitosis, yeast, 553
electron micrograph of LIS extracted, 687F
lack in procaryotes, 96
nuclear scaffold and, 686, 687F
structure and function, yeast, 567, 567F, 568F
Nuclear pores, β-galactosidase entry through, 559–560
electron micrograph and diagram, 568F
transport through, 567
Nuclear scaffold, locations of histone genes and, 687, 687F
preparation with LIS, 687
transcription and attachment sites on, **687,** 689, 687F
Nucleases, **266**
breakdown of host DNA by phage, 517
in recombination, 318, 323–325, 324F
synthesis on endoplasmic reticulum, 732
Nucleic acid-binding proteins, homology to copia element ORF protein, 665
Nucleic acids, denaturation, 44
as genetic molecules, 68
nucleotides as building blocks, 68, 69F
presence in all viruses, 71
Nucleoid region, in *E. coli*, 104F
"Nucleoid" bodies, 96
Nucleoids, **110**
Nucleoli, **6**
composition, **652**
electron micrograph, 4F, 687F
function, **652,** 686, 687F
ribosome biogenesis in, 698–703, 699F
structure, 686, 687F
yeast, function and location, 568
Nucleolus organizer region, definition, 652
Nucleoplasm, **686**
Nucleoplasmin, as protein carrier, 734